Sustainable Engineering Forum

Core Programming Topic at the 2012 AIChE Annual Meeting

Pittsburgh, Pennsylvania, USA
28 October - 2 November 2012

ISBN: 978-1-62276-741-0

Printed from e-media with permission by:

Curran Associates, Inc.
57 Morehouse Lane
Red Hook, NY 12571

CURRAN ASSOCIATES INC.
proceedings
.com

Some format issues inherent in the e-media version may also appear in this print version.

Printed by Curran Associates, Inc. (2013)

AIChE
3 Park Avenue
New York, NY 10016-5991

Phone: (203) 702-7660
Fax: (203) 775-5177

www.aiche.org

Additional copies of this publication are available from:

Curran Associates, Inc.
57 Morehouse Lane
Red Hook, NY 12571 USA
Phone: 845-758-0400
Fax: 845-758-2634
Email: curran@proceedings.com
Web: www.proceedings.com

TABLE OF CONTENTS

(13a) Biomasspyrolysisrefinery: A Two Step Approach for Liquid Energy Carrier Production From Lignocellulose

Nikolaus Schwaiger[1], Hannes Pucher[1], Roland Feiner[1], Kerstin Zahel[2], Peter Pucher[2], Hartmuth Schröttner[3] and Matthäus Siebenhofer[4], (1)Institute of Chemical Engineering and Environmental Technology, Graz University of Technology, Graz, Austria, (2)R&D, BDI BioEnergy International AG, Grambach, Austria, (3)Institute of Electron Microscopy, Graz University of Technology, Graz, Austria, (4)Institute of Chemical Engineering and Environmental Technology, Graz University of Technology, 8010 Graz, Austria

BiomassPyrolysisRefinery:

A two step approach for liquid energy carrier production from lignocellulose

Schwaiger, N. Feiner, R. Pucher, H. Pucher, P. Zahel*, K. E. Wilhelm[+], P. Schroettner[+], H. Siebenhofer, M. Graz University of Technology Institute of Chemical Engineering and Environmental Technology, *BDI-BioEnergy International AG, [+]Graz University of Technology Institute for Electron Microscopy*

Lignocellulose is expected to contribute to production of liquid and solid combustibles with high potential for CO_2 neutral processing. The "BiomassPyrolysisRefinery" is a two step concept for biomass liquefaction and hydrogenation, as shown in figure 1. First step of this concept is liquid phase pyrolysis: In this step a significant amount of oxygen is removed under thermal load and the calorific value of the lignocellulosic feed is partially condensed. Liquid phase pyrolysis is an exothermic process. During dehydration at T=350°C about 25-28% liquid CHO-products (biocrude) and about 38-42% BioChar is formed. [1]

Biocrude oil from liquid phase pyrolysis passes then a dewatering step where almost 90% of water from dehydration is separated. Afterwards the biocrude oil undergoes hydrodeoxygenation.

After liquid phase pyrolysis biochar is separated from the heat carrier and undergoes hydrogenation as well gasification and water gas shift reaction. Through hydrogenation of Biochar Hexane, Toluene (Asphaltenes) and THF (Preasphaltenes) solubles are formed.

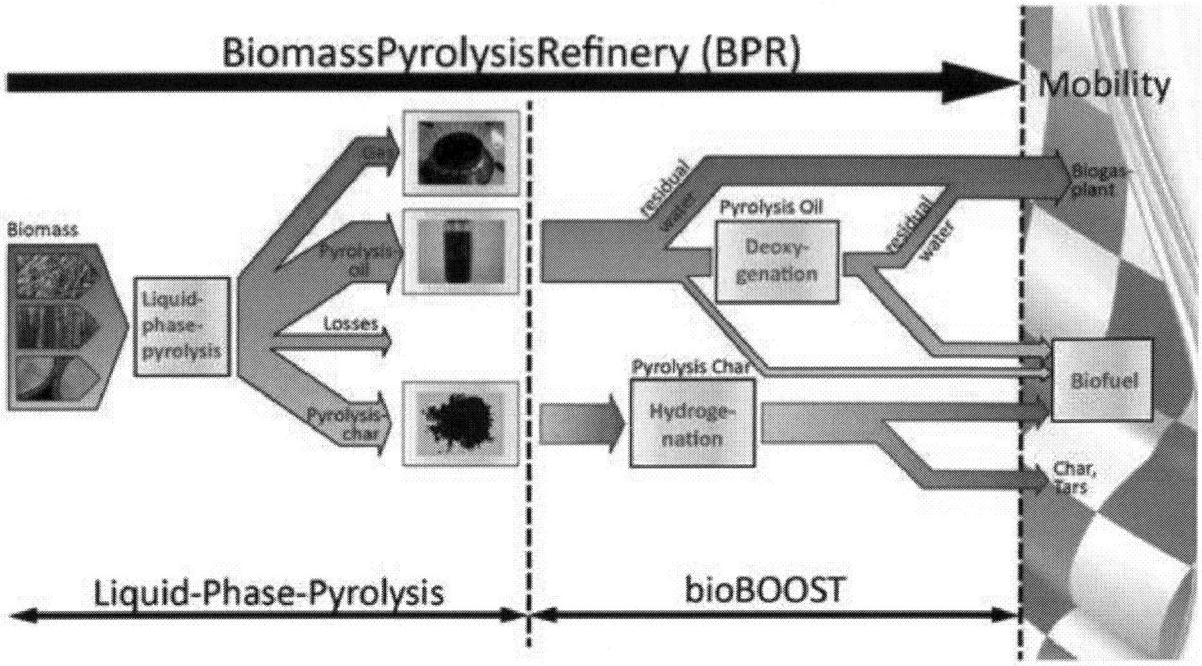

Figure 1: BiomassPyrolysisRefinery concept

1. Schwaiger, N., et al., *Liquid and Solid Products from Liquid-Phase Pyrolysis of Softwood.* Bioenergy Research, 2011. **4**(4): p. 294-302.

CONVERSION OF ORGANIC WASTE TO LIQUID FUEL BY PYROLYSIS OVER ALUMINA CATALYST

Alessandro A. Casazza, Bahar Aliakbarian, Patrizia Perego and Guido Busca

Department of Chemical Engineering "G.B. Bonino", Genoa University, via Opera Pia 15, 16145 Genoa, Italy

Introduction

In recent years strong efforts were being devoted to the substitution of fossil raw materials with renewable sources in the production of energy. Bioethanol produced by amidaceous raw materials and biodiesel, i.e. fatty acid methyl esters produced by transesterification of vegetable oils, are usually denoted as first generation biofuels (Sannita et al., 2012). Different kinds of treatments were considered to produce second generation biofuels in order to overcome these drawbacks (Naik et al., 2010).

Most interest is concerned to the development of gasification technologies followed by syngas purification and Fischer-Tropsch synthesis (Damartzis and Zabaniotou, 2011): however efficiency of this processes are estimated to be quite low (Manganaro et al., 2011).

Simple thermal treatments, with or without catalysts, were also being considered to produce either fuel gases, gasoline or bio-oils applicable as fuels.

Both vegetable oils and animal fats are mostly constituted by triglycerides and are considered to be a promising source of raw materials for the production of renewable fuels by thermal processes (Sannita et al. 2012).

Vegetable oils are mainly from coconut, corn, cottonseed, canola, olive, peanut, safflower, sesame, soybean, and sunflower. For instance, soybean and sunflower oil have a particularly high content of linoleic acid and are mainly made of C_{18} fatty acids, whereas in palm oil palmitic acid and oleic acid are especially present. Animal fats are normally constituted by triglycerides with a saturated acid content higher than that of vegetable oils (Demirbas, 2009).

Maher and Bressler (2007) summarized literature results on the thermal conversion of vegetable oils. Similar thermal processes could also be applied to other residues, including free fatty acids and plastic wastes, in order to recover energy and reduce waste material. High temperature processes may allow the production of gasoline and gases potentially applicable as fuels (Buekens and Huang, 1998; Al-Salem et al., 2009; Al-Salem et al., 2010). The focus of this work was on the utility of low-temperature pyrolysis to maximize the production of liquids from organic waste like vegetable and animal fats and auto shredder residues. The samples were pyrolyzed using a batch reactor process working at 350 °C for 3 hours.

Material and methods

Materials

Two vegetable fats (VFa and VFb), chicken fat (CF), beef tallow (BT) and Auto Shredder Residue (ASR) were used in this study. Diathermic oil (Thermex 32) was purchased from Essex. Siralox 1.5/40 was used as catalyst.

Pyrolysis reaction

The samples were pyrolyzed in a stirred reactor (250 ml) (Figure 1) at a temperature of 350 °C for 3 hours. The liquid product (LP) was distilled during the reaction using an integrated distiller and was characterized by gas chromatography mass spectrometry (GC-MS). A heating mantle (in stainless steel) was used to achieve the reaction temperature. The top of the reactor vessel was equipped with a water cooling jacket (condenser) aimed to condensate the gaseous products (GP) formed during the reaction. At the top of the condenser a special pure latex balloon was positioned in order to collect the incondensable gaseous products. The bulk temperature of the vessel was monitored by a thermometer positioned in the backside of the reactor. For VFa, VFb, CF, BT and VFa/CF/BT (in the same amount) the vessel was loaded with approximately 107 g of the samples, while for ASR, 21 g of sample were dissolved in 86 g of diathermic oil. In all experiments 3% (wt/wt) of catalyst was used. After replacing the reaction atmosphere with nitrogen the heating jacket was turned on. At the end of the reaction period, the residue product (RP) was distilled separately at 350 °C. The liquid distillate (LD) was analyzed by GC-MS.

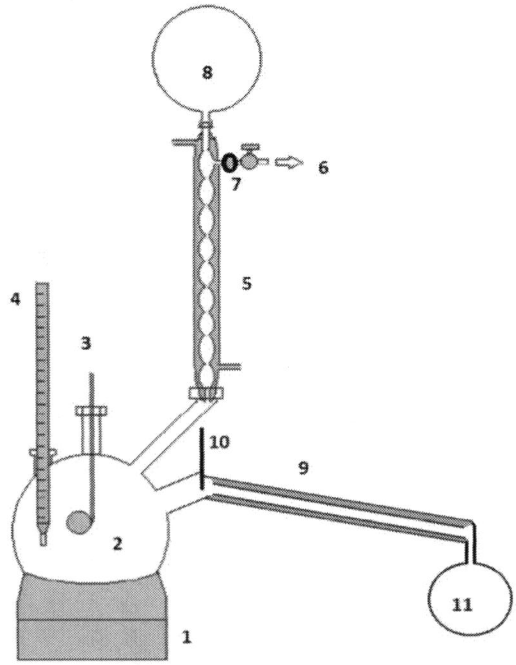

Figure 1. Scheme of bench-scale catalytic pyrolysis reactor. 1) heater; 2) Pyrex glass tank; 3) stirrer; 4) thermometer; 5) bubble condenser; 6) Nitrogen and gas outlet; 7) silicon membrane for gas analysis; 8) latex balloon; 9) condenser; 10) thermocouple; 11) Pyrex container for distilled sample.

Analysis

For the GC-MS analysis an Agilent 4890D gas chromatograph with Electron Ionization Detector (250 °C) and heated injection equipped with a Hewlett Packard G1800B-GCD system with a quadrupole mass detector was used. The gas chromatograph utilizes a HP-VOC (stationary phase volatiles) 30 m × 0.32 mm, film thickness 1.8 μm, capillary column. Helium (>99.9%) as carrier gas at a flow rate of 1.6 mL min^{-1} was used and the maximum column temperature was 250 °C. The initial temperature of the column was 70 °C the following temperature ramp: one level of ten minutes at 130 °C and a second level at 230 °C with a rate of temperature increase of 10 °C/min.

A Nicolet 380 FT-IR Spectrometer was used to analyse the reaction gases and the residues from distillation. The gas fractions were analysed immediately at the end of the reaction using a glass cell which collected them directly into the instrument, while the analysis of the residues were performed by deposing the distillation residues on a pressed disk of potassium bromide.

Results and discussions

Table 1 represents the yield of gas, liquid and residue products from pyrolysis reaction and distillation of residues. GP yields shows similar values in all experiments while LP varies notably using different kinds of biomass. The highest LP yield was achieved using CF (36.0 g/100g) as biomass followed by PO/CF/BT (20.6 g/100g) and ASR (20.3 g/100g).

After distillation of the reaction residue (RP), VFb resulted to produce highest amount of liquid (38.9 g/100g). Considerable amount of gas was produced using BT (26.1 g/100g).

Table 1. Yields (g/100g) in gas, liquid and residue products from catalytic pyrolysis reaction and from the distillation of the residue products

	GP[1] (g/100g)	LP[2] (g/100g)	RP[3] (g/100g)	Distillation of RP[3]		
				GD[4] (g/100g)	LD[5] (g/100g)	RD[6] (g/100g)
VFa	7.2	5.7	87.1	4.5	21.5	61.1
VFb	4.3	3.1	92.6	2.9	38.9	50.8
CF	9.6	36.0	54.4	3.9	7.0	43.4
BT	10.0	5.6	84.4	26.1	10.5	47.7
PO/CF/BT	9.2	20.6	70.2	5.9	5.8	58.5
ASR	11.8	20.3	67.9	0.4	0.0	67.5

[1] gas product; [2] liquid product; [3] residue product; [4] gas from distillation; [5] liquid from distillation [6] residue from distillation

Figure 2, reported the FT-IR spectrum related to the VFb gas product analysis. In general, for the gas fractions of vegetable oil and animal fats, the FT-IR analysis revealed the presence of carbon dioxide, carbon monoxide, methane, acrolein, hydrocarbons, olefin and some oxygenated compounds. While for the ASR gas fraction the FT-IR analysis showed the presence of carbon monoxide, hydrocarbons, carbon dioxide, oxygenated compouds, HCl, HCN and NH_3.

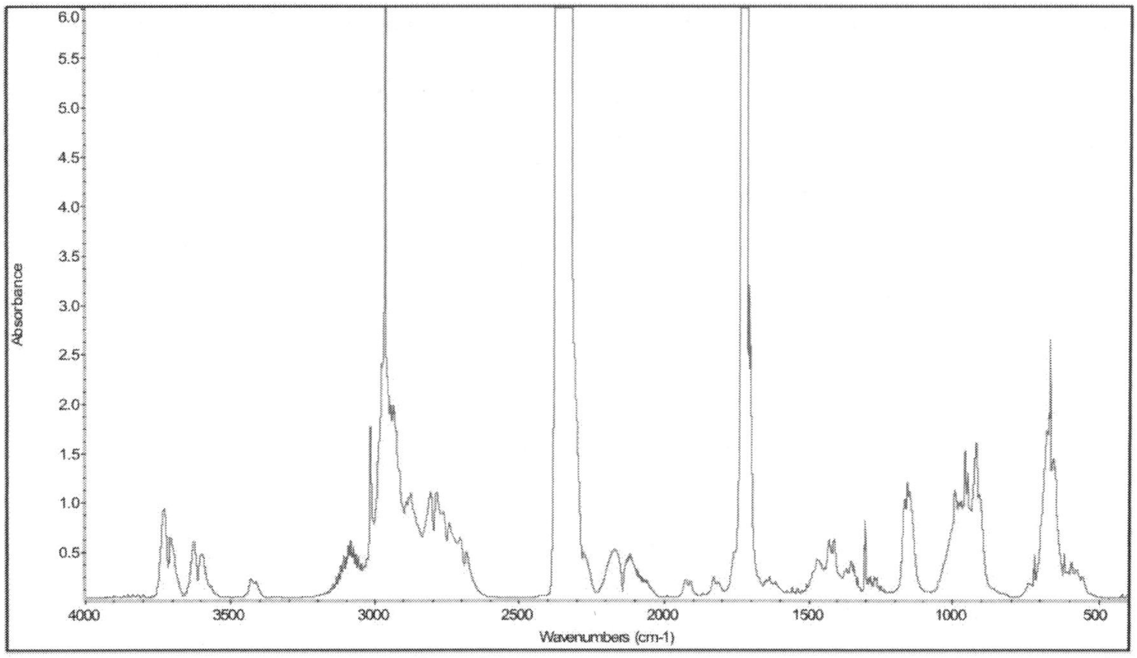

Figure 2. FT-IR spectrum of gas produced from the catalytic pyrolysis reaction of vegetable oil (b) using Siralox 1.5/40 as the catalyst.

Table 2 summarizes the main classes of substances of liquids coming from the reaction and distillation identified by GC-MS.

Table 2. Total peak area of liquid products (LP) and liquid distillated (LD) identified by GC-MS analysis.

	Hydrocarbons (%)	Organic acids (%)	Other oxygenated compounds (%)	Nitrogenated compounds (%)	Chlorinated compounds (%)	Not identified (%)
VFa (LP)	76.4	7.7	15.9	-	-	-
VFa (LD)	66.1	15.3	18.6	-	-	-
VFb (LP)	65.8	8.3	22.4	-	-	3.5
VFb (LD)	31.9	32.8	21.1	-	-	14.2
CF (LP)	29.6	18.8	38.8	12.7	-	-
CF (LD)	68.3	3.6	28.1	-	-	-
BT (LP)	60.7	35.4	2.6	1.3	-	-
BT (LD)	74.7	6.9	18.4	-	-	-
PO/CF/BT (LP)	43.6	22.4	19.6	14.3	-	-
PO/CF/BT (LD)	74.0	10.9	15.1	-	-	-
ASR (LP)	82.4	2.0	11.8	3.3	0.2	-
ASR (LD)	-	-	-	-	-	-

GC-MS analysis of distillates revealed that they mainly contain hydrocarbons followed by oxygenated and nitrogenated compounds. Chlorinated compounds were identified only in ASR (0.2%)

Generally, pyrolysis of vegetable fats did not result in a considerable amounts of gas and liquid yields while distillation of residue produced high quantity of LD.

Chicken fat produced great amount of liquid during the reaction. Distillation of RP and BT was concluded with considerable quantity of LD and GD. In the case of chicken and beef fats, the overall gas and liquid yield were almost similar (56.6 and 52.3 g/100g, respectively) and were higher than the other biomasses. ASR resulted to the less amount of overall yield which was 32.5 g/100g.

In the residues of vegetable oils (Figure 3) and animal fats the FT-IR analysis showed the presence of triacylglycerides, carboxylic acids, alcohols, ketones and hydrocarbons.

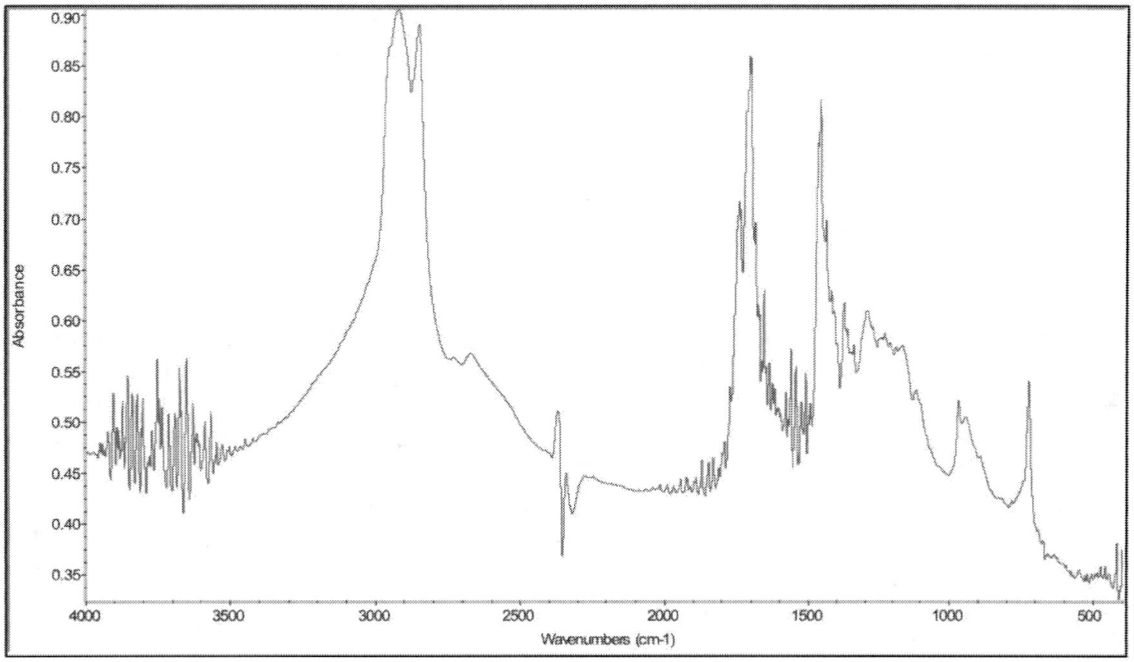

Figure 3. FT-IR spectrum of residue product for the reaction with vegetable oil (a) and Siralox 1.5/40

Conclusions

From the results of this study it can be concluded that catalytic pyrolysis of different organic wastes at relatively high temperature (350 °C) resulted to produce acceptable amounts of biofuels (gas-liquid) higher than 50 % (wt/wt). In order to optimize these catalytic conversions, considerable attention must be taken into account about the interaction between different compounds from organic waste mixtures

References

Al-Salem, S.M., Lettieri, P., Baeyens, J. (2012). The valorization of plastic solid waste (PSW) by primary to quaternary routes: from re-use to energy and chemicals. Prog. Energ. Combust. 36, 103–129.

Al-Salem, S.M., Lettieri, P., Baeyens, J. (2009). Recycling and recovery routes of plastic solid waste (PSW): A review. Waste Manage. 29, 2625-2643.

Buekens, A.G., Huang, H. (1998). Catalytic plastics cracking for recovery of gasoline-range hydrocarbons from municipal plastic wastes. Resour. Conserv. Recy. 23, 163–181.

Damartzis, T., Zabaniotou, A. (2011). Thermochemical conversion of biomass to second generation biofuels through integrated process design - A review. Renew. Sust. Ene. Rev. 15, 366-378.

Demirbas, A., (2009). Progress and recent trends in biodiesel fuels. Energ. Convers. Manage. 50, 14–34.

Maher, K.D., Bressler, D.C., (2007). Pyrolysis of triglyceride materials for the production of renewable fuels and chemicals. Bioresource Technol. 98, 2351–2368.

Manganaro, J., Chen, B., Adeosun, J., Lakhapatri, S., Favetta, D., Lawal, A., Farrauto, R., Doraio, L., Rosse, D.J. (2011). Conversion of residual biomass into liquid transportation fuel: an energy analysis. Energy Fuels 25, 2711-2720.

Naik, S.N., Goud, V.V., Rout, P.K., Dalai, A.K. (2010). Production of first and second generation biofuels: A comprehensive review. Renew. Sust. Energ. Rev. 14, 578-597.

Sannita, E., Aliakbarian, B., Casazza, A.A., Perego, P., Busca, G. (2012). Medium-temperature conversion of biomass and wastes into liquid products. A review. Renew. Sust. Energ. Rev. 16, 6455-6475.

(13c) Hydropyrolysis of Loblolly Pine to Produce a Low-Oxygen Pyrolysis Oil

John R. Carpenter, Justin Farmer, Hua Song and David Dayton, Center for Energy Technology, RTI International, Research Triangle Park, NC

Pyrolysis oil prepared by conventional (non-catalytic) fast pyrolysis is not suitable for integrating into the current fuel infrastructure or into a petroleum refinery because it is : a) thermally unstable with a high fouling tendency, b) corrosive due to high carboxylic acid content (pH 2.2 – 2.4 typically), and c) not miscible with refinery feedstocks. Fast pyrolysis oil tends to have high oxygen content similar to the solid biomass feedstock used as the starting material. The thermal instability and high water content require specialized hydroprocessing to lower the oxygen content of the oil before use. Hydropyrolysis is a biomass thermal liquefaction process option that combines biomass, hydrogen, and a catalyst to integrate pyrolysis and a degree of hydroprocessing into a single unit operation.

The National Advanced Biofuels Consortium (NABC) is working to develop sustainable, cost-effective processes to produce advanced "drop-in" biofuels that are compatible with today's transportation infrastructure. As part of the NABC, the hydropyrolysis of loblolly pine has been explored to produce low-oxygen content bio-oil. An entrained flow reactor is utilized to test catalysts and reaction conditions (T, P, and pH_2). Testing was performed with a nominal loblolly pine feed rate of 100 g/hr at temperatures between 350°C to 500°C and pressure up to 350 psig. Full mass closures are obtained by measuring all liquid, solid, and gas products. Bio-oils are characterized by ultimate analyses, Karl-Fischer titration, total acid number (TAN), pH, and GC/MS.

A selected catalyst was tested over an extended trial of 21h on stream with over 2kg of biomass. Initial high activity led to lower liquid yields early in the trial, however liquid yield increased with time on stream. The product distribution and composition was consistent over the latter half of the trial. The liquid product separated into an aqueous layer that was 97wt% water and a bio-oil that was less than 1wt% water and had an oxygen content of 4wt%. Hydropyrolysis has the potential to produce bio-oils with improved properties and suitability for integration into refinery processes.

(13d) Technoeconomic Analysis of Pyrolysis of Forest Residues

Jose L. Carrasco[1], Nicole L. Hammer[2], Andrew McAloon[3], Akwasi A. Boateng[4], William J. DeSisto[1] and M. Clayton Wheeler[5], (1)Chemical and Biological Engineering, University of Maine, Orono, ME, (2)Eastern Regional Research Center, Agricultural Research Service, USDA, Wyndmoor, PA, (3)USDA-ARS-ERRC, Wyndmoor, PA, (4)U.S. Department of Agriculture, Crop Conversion Science & Engineering, Agricultural Research Service, Wyndmoor, PA, (5)Chemical and Biological Engineering and Forest Bioproducts Research Institute, University of Maine, Orono, ME

Because of its simplicity, fast pyrolysis is an attractive biomass conversion process which is scalable from mobile units to thousands of tons per day. However, the utility of the oils produced by fast pyrolysis is limited because of the well-known chemical and physical properties of these highly oxygenated oils. It is possible to improve the quality of the oils by employing catalysts in the pyrolysis reactor and by performing downstream hydrodeoxygenation with additional catalysts and hydrogen. It is important to understand the effects these additional processes have on the technoeconomics and ability to do distributed processing. Also, biomass feedstock composition significantly affects the product distributions and, therefore, the thermal integration within the process.

As the most heavily forested state in the country, Maine has a large pulp and paper industry. Along with the 12 million green tons of pulpwood which are harvested annually,[1] there is likely the potential to collect an equal quantity of associated biomass (residues) if a suitable conversion technology and product market were to exist. These residues, which consist of tree tops, limbs, stumps and bark, have a higher percentage of lignin and ash than round-wood chips. This talk will present results of a technoeconmic analysis based on Aspen simulations and data from pyrolysis of forest residues. The pyrolysis oil is simulated using a set of model compounds which can be optimized for different oil products based on weighted parameters such as higher heating value, CHO ratios and chemical functionality.

[1]2010 Wood Processor Report, Department of Conservation, Maine Forest Service, Forest Policy and Management Division www.maine.gov/doc/mfs/pubs/pdf/wdproc/10wdproc.pdf .

(13e) Pilot Scale Production of Biocrude Oils From Biomass: Effect of Feedstock

Ofei Mante and Foster Agblevor, Biological Engineering, Utah State University, Logan, UT

The production of high quality biocrude oils was performed in a direct step via fractional catalytic pyrolysis with recycling of the non-condensable gases. For the purpose of investigating the effect of biomass type, various feedstocks including hybrid polar, pine, corn stover, switchgrass and pine bark were pyrolyzed with HZSM-5 zeolite in a 10 cm bubbling fluidized bed reactor at 475 °C. TGA-FTIR studies were also conducted in support of the work. The product distribution and the quality of the biocrude oils were feedstock dependent. The overall liquid yields were 30 - 44.5 wt.% and char/coke ranged between 15 wt% and 25 wt%. Generally, the oils had relatively higher pH (3.25-5.05), lower viscosity (23-95 cSt) and lesser density (1.06-1.14 g/cm^3) compared to the non-catalytic experiments. The oils were also analyzed for elemental composition (CHNSO) and higher heating value. The gas analysis showed that the rejection of oxygen in the form of CO was predominant in the woody biomass compared to the herbaceous feedstocks. The average CO/CO_2 ratio for hybrid poplar was 2.0 and that of switchgrass was 1.1. ^{13}C-NMR and FT-IR analysis of the biocrude oils showed cracking of carbohydrate decomposition products such as levoglucosan and acetic acids. The concentration of phenols and aromatic hydrocarbons in the biocrude oils were dependent on the feedstock used.

(13g) Wet Torrefaction As a Pretreatment for Pyrolytic Bio-Oil Production From Microalgae

Zhenyi Du, Bioproducts and Biosystems Engineering, University of Minnesota, St Paul, MN, Michael Mohr, University of Minnesota, St Paul, MN, Paul Chen, Dept. of Biosystems & Agricultural Engineering, University of Minnesota, Saint Paul, MN and Roger Ruan, Department of Biosystems & Agricultural Engineering, University of Minnesota, St. Paul, MN

Pyrolytic bio-oil derived from microalgae has been explored in recent studies. The resultant bio-oil contains significant amounts of aliphatic and aromatic hydrocarbons; however, pyrolytic conversion requires energetically unfavorable drying to remove excess water before treatment. Additionally, high protein feedstocks, like those from microalgae, yield in bio-oils with large N content which lowers the heating value and leads to potential NOx emission during combustion. Therefore, wet torrefaction, a pretreatment for pyrolysis, was performed on microalgae *Nannochloropsis oculata*, which aimed to reduce the N content in the algal feedstock by hydrolyzing the proteins. The wet torrefaction process treats the algal biomass in pressurized water at elevated temperature (ca. 200 °C), thus eliminating the need for drying and reducing overall process energy. The effects of various reaction conditions on the yield and composition of torrefied algae were investigated by varying temperature from 150 to 225 °C and reaction time from 10 to 60 min. Compared with untreated algae, torrefied samples had higher carbon content and enhanced heating values under all reaction conditions. Also, decreased N content was observed under relatively severe conditions. It was found that 73%-99% of the fatty acids in the starting algae were retained in the torrefied samples, which is important because fatty acids can be mainly converted to hydrocarbons during pyrolysis. Gas chromatography- Mass spectrometry analysis confirmed that pyrolytic bio-oil from torrefied algae contained less N-containing compounds and was dominated with long-chain fatty acids which could be readily converted into hydrocarbon fuels in the presence of simple catalysts. Energy consumption ratio analysis showed that wet torrefaction has a more favorable energy balance than directly drying of algae. The aqueous phase products, obtained after the separation of torrefied algae, contained significant amount of nutrients and can be recycled for algae cultivation.

(44b) Conceptual Net Energy Output for Biofuel Production from Lignocellulosic Biomass Through Biorefining: A Review

J.Y. Zhu, USDA Forest Service, Forest Products Laboratory, Madison, WI

Conceptual net energy output for biofuel production from lignocellulosic biomass through biorefining: A review

J.Y. Zhu[1,2]

[1] USDA Forest Service, Forest Products Laboratory, Madison, WI, USA

[2] Dept. Biological Systems Engineering, University of Wisconsin-Madison, WI, USA

Abstract

The concept of producing lignocellulosic biofuel, bioproducts, and chemicals through a biorefinery using lignocelluloses has been around for 70 years or more. The recent interest in this old concept arises from the promises of mitigating climate change by substituting some biomass energy for petroleum or other fossil-fuel energy; achieving national energy independence and sustainable economic development; and advances in biotechnology in the past 2-3 decades. Geopolitical and national security reasons have greatly contributed to the need for seeking alternative energy, especially using domestic renewable and sustainable sources.

However, there is a lack of comprehensive information in the retrievable literature on pilot scale process and energy data using promising process technologies and commercially scalable and available capital equipment for lignocellulosic biomass biorefining, despite several pilot scale and demonstration plants are under construction. Unfortunately, bioenergy production is not sustainable without net energy production. Recognizing the heated debate on energy production from lignocellulosic biomass in the scientific community and most arguments were not based on actual laboratory data for biorefinery operation, this presentation is intended to provide an objective view on energy production from lignocellulosic biomass. The process data from approximately a dozen studies that represent state-of-the-art in cellulosic biofuel production concepts, along with literature energy input data for agriculture operations, were analyzed to provide estimates of net energy production. Several scenarios were created for proper allocation energy input to biomass and major agriculture co-product, such as corn or lumber.

For biorefinery operations, increasing solids loading in pretreatment to at least 25% is critical to reducing energy input in a biorefinery. Post thermo-chemical pretreatment size reduction approach should be adopted for energy efficient woody biomass processing. When appropriate pretreatment technologies are used, woody biomass can be processed as efficiently as herbaceous biomass and agricultural residues. Net energy output for cellulosic ethanol was estimated to range approximately from -500 to 2000 MJ/ton biomass (HHV base); indicating that the energy input/output ratio is approximately 1:1 for cellulosic ethanol. However, net energy can reach approximately 4000-7000 MJ/ton of biomass when energy from lignin is included.

(44c) The R&D of Cellulosic Ethanol in China

Xinshu Zhuang, Bio-chem conversion lab, Guangzhou institute of energy conversion,Chinese Academy of Sciences, Guangzhou, WI, China

As the biggest developing country, China needs fuel ethanol to meet its demands of its social and economy development. Basing on the "Mid-long term plan for renewable energy" established in 2007, the annual consumption of fuel ethanol will reach 10 million tons in 2020. Cellulosic ethanol, one of the most important second generation bio-fuel, gains more and more attentions for its substitution of the transportation liquid, lessen the air pollution by the vehicles and promote the rural area economy. In China, more than five demonstration projects have been set up since 2008. In this paper, the main cellulosic ethanol processes applied in China are illustrated, simultaneously the R&D of a green and high efficient process pretreated with hot liquid water is presented.

(70a) Study On Stability of Pyrolysis Bio-Oil From Torrefied Loblolly Pine

Jiajia Meng[1], David Tilotta[1], Sushil Adhikari[2] and Sunkyu Park[1], (1)Forest Biomaterials, North Carolina State University, Raleigh, NC, (2)Biosystems Engineering, Auburn University, Auburn, AL

The unstable nature of pyrolysis oil produced from fast pyrolysis of biomass is one of the undesirable properties that limit its application. Mild thermal pretreatment, torrefaction, has been proven to be an effective approach to reduce oxygen content in bio-oil; however, its effect on bio-oil stability has not yet been investigated. In this study, bio-oils generated from torrefied loblolly pine were aged at both accelerated (80 °C for 24 hour) and natural (25 °C for 12 months) aging conditions and the resulting bio-oil was characterized by several wet-chemistry and analytical methods. In order to monitor the property change of bio-oil during natural aging, oil samples were taken every three months for characterization. These characterizations measured bio-oil properties including water content, viscosity, acidity, molecular weight distribution, chemical composition from solvent fractionation of bio-oil, and GC/MS quantification of pyrolysis marker compounds. The preliminary results showed different aging behaviors among bio-oils under investigation; for instance less viscosity increase was observed for aged bio-oil from thermally pretreated loblolly pine when compared to that of non-treated one. By comparing these aging behaviors of bio-oils prepared from torrefied wood to those of non-treated wood, the effect of torrefaction on bio-oil stability can be elucidated as well as possible aging mechanisms.

(70b) Intrinsic Chemistry of Biomass Pyrolysis for Biofuels: Primary and Secondary Reactions of Cellulose Decomposition

Paul Dauenhauer, Chemical Engineering, University of Massachusetts Amherst, Amherst, MA

The next generation of biomass pyrolysis models must be capable of predicting the molecular-level chemistry occurring during the fast pyrolysis process. The first step in developing molecular prediction requires elucidating the primary and secondary reactions of biopolymers such as cellulose during pyrolysis. Experimental evidence has indicated that the reactive intermediates for carbohydrates, lignin, and lignocellulose at temperatures exceeding 500 °C exist within a short-lived intermediate liquid which subsequently evaporates to vapors and gases. We utilize a novel 'thin-film pyrolysis' technique to elucidate the primary and secondary reaction pathways of cellulose to volatile products including anhydrosugars, pyrans, furans, and light oxygenates. The distribution of products from carbohydrate polymers is shown to depend upon the chain length, with longer chains producing more anhydrosugars such as levoglucosan. Finally, the secondary, condensed-phase pathways of levoglucosan are revealed using isotopically-labeled samples to depend upon the residence time within molten cellulose. Combination of these results provides the first insight into the chemical pathways which dictate the composition of bio-oil.

(70c) The Pyrolysis of Lignin

Raffaella Ocone[1], Mustafa Adam[1], Franco Berruti[2] and Cedric Briens[3], (1)Chemical Engineering, Heriot-Watt University, Edinburgh, United Kingdom, (2)Chemical and Biochemical Engineering, The University of Western Ontario, London, ON, Canada, (3)University of Western Ontario, London, Canada

Lignin is an organic polymer and it is the second most abundant renewable carbon source on Earth, after cellulose. Lignin is not one compound but many complex polymers; the commonality between all of them is their phenylpropane structure, that is, a benzene ring with a tail of three carbons. In their natural unprocessed form, their molecular weights may reach 15,000 or more.

In this work the kinetics of lignin pyrolysis is described by using a lumping model (G. Astarita, R.Ocone, "Lumping Nonlinear Kinetics", AIChE Journal, **34**, 129, 1988), where the lignin structure is simplified and assumed to be formed by repeating units (monomer) which can be identified with the basic benzene ring with the tail of three carbons. The bio-oil, produced by the pyrolysis of lignin, is then fractionated to obtain lighter components. Two different modelling approaches are attempted and comparisons are made; the bio-oil is considered as the feedstock for the fractionation and as one of the products from the primary thermal cracking. To model the products of the primary cracking we consider a discrete 3-lump approach and assume that the oil is one of the lumps (the other being the gas and the char). Successively, we model the fractionation in a continuum fashion by extending the approach of previous works on catalytic and thermal hydrocracking and we propose two different kinetic models. Prediction of the product distribution after pyrolysis is reported and the benefit of the lumping approach is assessed. Results are validated against experimental data taken at ICFAR, University of Western Ontario.

(70d) Fast Pyrolysis of Biomass with Microwave Heating: A Preliminary Analysis with CFD Modeling

Fernanda C. Borges, Juliana da S. Espindola, Lucas M. Raymundo, Luis Antônio R. Muniz, Marcelo Farenzena and Jorge Otávio Trierweiler, Chemical Engineering, Federal University of Rio Grande do Sul, Porto Alegre, Brazil

The search for sustainability instigates the investigation for alternative sources of energy, in order to mitigate economic, social and environmental problems. Thus, biomass appears as one important renewable source for the production of food, materials, chemicals, fuels and energy, being essential to develop processes and equipment to convert its resources efficiently.

Several biochemical and thermochemical processes have been investigated for biomass processing, but thermochemical methods have been shown as the easiest to adapt to the existing energy infrastructure.

In this context, fast pyrolysis has emerged as a promising alternative, consisting in the conversion of biomass into a complex mixture of organic compounds and fractions of char and gas [1; 2]. The main product of this process is the bio-oil, which can be used in engines and turbines, or fed into refineries to obtain products with higher added value [2].

Fluid beds are the most popular configuration used in fast pyrolysis, due to their ease of operation and ready scale-up, allowing to obtain high yields of bio-oil [3]. At laboratory scale, this reactor is conventionally heated by electric resistances, and at larger scales, by burning by-products of pyrolysis.

Recently, an alternative that is being proposed is the microwave heating, which has been proven to be energy efficient and has been widely accepted as an easy to control technology, resulting in a bio-oil with interesting characteristics for a sustainable manufacturing of chemicals and biofuels, it has also shown favorable energy balances for the process [4; 5; 6; 7; 8; 9; 10; 11].

Aiming to understand and develop this process, the software COMSOL Multiphysics® was used for the simulation of thermal behavior in the fluid bed with microwave heating. The results permit to make a preliminary analysis of the thermal behavior of biomass, the distribution of electric field and resistive losses in the oven, and also its influence on the homogeneity of heating and on the bio-oil yield obtained.

References

[1] BAHNG, M. K. *et al.* Current technologies for analysis of biomass thermochemical processing: A review. Analytica Chimica Acta, v. 651, n. 2, p. 117-138, **2009**.

[2] BRIDGWATER, A. V.; PEACOCKE, G. V. C. Fast pyrolysis processes for biomass. Renewable and Sustainable Energy Reviews, v. 4, n. 1, p. 1-73, **2000**.

[3] BRIDGWATER, A. V. An Introduction to Fast Payrolysis of Biomass for Fuel and Chemicals. In: PRESS, C. (Ed.). Fast Pyrolysis of Biomass: A Handbook. Second. Newbury, UK, **2008**. cap. 1, p.1-13.

[4] BUDARIN, V. L. *et al.* The preparation of high-grade bio-oils through the controlled, low temperature microwave activation of wheat straw. Bioresource Technology, v. 100, n. 23, p. 6064-6068, **2009**.

[5] CHEN, M. *et al.* The effects of additives on yields and distribution of gasand liquid from pyrolysis of biomass by microwave heating. Hefei, **2008**. 513-516 p.

[6] CLARK, D. E.; SUTTON, W. H. Microwave processing of materials. Annual Review of Materials Science, v. 26,

n. 1, p. 299-331, **1996**.

[7] DE LA HOZ, A.; DÍAZ-ORTIZ, Á.; MORENO, A. Microwaves in organic synthesis. Thermal and non-thermal microwave effects. Chemical Society Reviews, v. 34, n. 2, p. 164-178, **2005**.

[8] DOMÍNGUEZ, A. *et al.* Production of bio-fuels by high temperature pyrolysis of sewage sludge using conventional and microwave heating. Bioresource Technology, v. 97, n. 10, p. 1185-1193, **2006**.

[9] DU, Z. *et al.* Microwave-assisted pyrolysis of microalgae for biofuel production. Bioresource Technology, v. 102, n. 7, p. 4890-4896, **2011**.

[10] HUANG, Y. F. *et al.* Total recovery of resources and energy from rice straw using microwave-induced pyrolysis. Bioresource Technology, v. 99, n. 17, p. 8252-8258, **2008**.

[11] LUQUE, R. *et al.* Microwave-assisted pyrolysis of biomass feedstocks: The way forward? Energy and Environmental Science, v. 5, n. 2, p. 5481-5488, **2012**.

Acknowledgements

The authors thank to CNPq and CAPES for encouraging the development of this work and for financial support.

(70e) Modeling of Pyrolysis and Boudouard Reactions of Various Coals, Biomasses and Coal-Biomass Blends Using Thermogravimetric Analysis, Experimental Moving Bed Gasification System and Stable Carbon Isotope Ratio Mass Spectroscopy

Venkata Abhijit Bhagavatula[1], Naresh Shah[1], Christopher Romanek[2] and Gerald Huffman[1], (1)Consortium for Fossil Fuel Science, Department of Chemical and Materials Engineering, University of Kentucky, Lexington, KY, (2)3. Department of Earth and Environmental Sciences, University of Kentucky, Lexington, KY

Modeling of pyrolysis and Boudouard reactions of various coals, biomasses and coal-biomass blends using thermogravimetric analysis, experimental moving bed gasification system and stable carbon isotope ratio mass spectroscopy

Abhijit Bhagavatula[1], Naresh Shah[2], Gerald Huffman[2] Christopher Romanek[3]

1. **Corresponding author:** Department of Chemical and Materials Engineering, University of Kentucky, Lexington, KY, 40506-0043, USA **email:** abh222@uky.edu, **abhijitbv@gmail.com**

2. Department of Chemical and Materials Engineering, University of Kentucky, 533 S.Limestone St, Suite 107, Whalen Building, Lexington, KY, 40506-0043, USA

3. Department of Earth and Environmental Sciences, University of Kentucky, 216 Slone Building, Lexington, KY, 40506-0053, USA

Abstract

Non-isothermal thermogravimetric analysis has been performed at different heating rates of 5, 10, 20 and 40 °C/min to investigate the thermal decomposition kinetics of two coals: lignite and sub-bituminous; four biomass materials: pinewood, poplar, corn stover and switchgrass; and blends of each biomass type with both coals at 10 % and 30 % by weight. The pyrolytic behavior of the samples was analyzed under these conditions using an inert nitrogen atmosphere while a carbon dioxide atmosphere was employed for studying the gasification kinetics, or the Boudouard reaction kinetics. A distributed activation energy model was utilized to estimate the kinetic parameters (order, activation energy and pre-exponential factor) for both sets of experimental data. Predicted results from the optimum kinetic parameters have been compared with the experimental data. The DAEM equation predicts the experimental data very well for different heating rates.

Apart from this, all the feedstock materials have been gasified in a moving bed reactor running in a completely auto-thermal batch mode using several air/oxygen/steam ratios. A laboratory-scale gasification system has been designed and constructed. The main components are schematically represented in Figure 1. The core of the system is an updraft gasifier, where pressure and temperature profiles are measured by a pressure transducer and a set of thermocouples, respectively. Coal/biomass is fed at the top of the gasifier by means of a quick-open flange. Air/oxygen and steam is fed at the bottom of the gasifier and its rate is measured by a rotameter. A safety valve is located on the gas exit tube. Here, the gas stream enters two condensers, where steam and tars are condensed and collected at the bottom before sampling for gas chromatographic (GC) analysis. The product gas from the gasifier contains mainly hydrogen, carbon monoxide, carbon dioxide and small amounts of methane. Finally, the exit gas flow rate is again measured by a wet test meter.

The gasifier is a cylindrical stainless steel modular flange assembly having an internal diameter of 1.37 inches with quartz/stainless steel tubing of 0.075 inches on the inside and a height of about 10 inches from the grate. The internal tubing is fitted with a stainless steel grate with holes large enough to let the ash pass through but small enough to hold the feed material. The grate is connected to a mechanical rotary linear feedthrough to periodically remove ash. The bottom zone, under the grate, has another cylindrical stainless steel flange with a height of about 5 inches to collect and then discharge the ash produced in the process. The grate at the bottom of the gasifier is used not only for holding the solid particles together but also as an oxidant distributor. The oxidant, fed at the bottom of the reactor, flows along the channels and exits through the small holes along the grate, so it is distributed across the whole section of the gasifier. At the bottom, a small tube allows the use of pre-heated steam to enter the bed. Temperature profiles along the gasifier axis are measured by a set of K-type thermocouples placed within a steel protective tube.

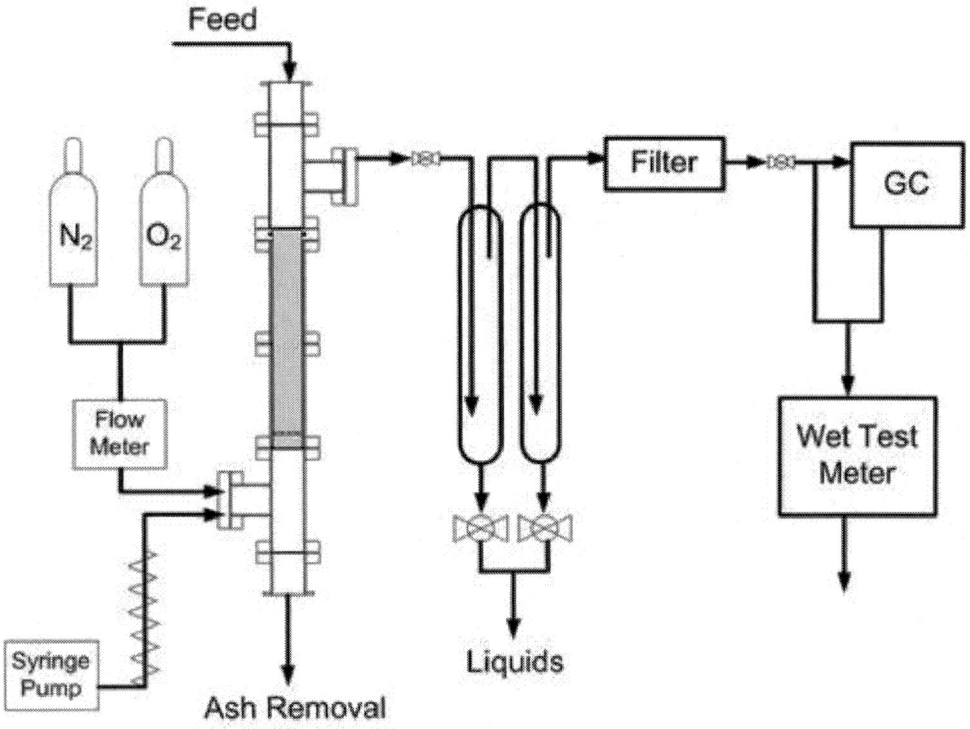

Figure 1: Schematic of the laboratory scale moving bed gasification system.

Also, a stable carbon isotope analysis was performed on both the feedstocks as well as the effluent gases from the moving bed gasifier in order to determine their $^{13}C/^{12}C$ ratios. The difference in the $^{13}C/^{12}C$ ratios of the feedstocks acts as a natural isotopic tracer for determining the amount of coal and biomass converted to each component in the product gases. This novel technique can be utilized for quantitatively determining the individual contribution or source apportionment of coal and biomass to the product gases by calculating the isotopic carbon mass balance for each product gas.

(70f) Thermal Conversion of Lignin to Transportation Fuels and Chemicals Through Formate Assisted Fast Pyrolysis

Saikrishna Mukkamala, William J. DeSisto, M. Clayton Wheeler and Adriaan van Heiningen, Chemical engineering, University of maine, orono, ME

Thermal conversion of lignin to transportation fuels and chemicals through formate assisted fast pyrolysis

Biomass currently supplies 14% of the world's energy need. Forest biomass is comprised of cellulose, hemicellulose and lignin. Cellulose and hemicellulose are used in pulping industries, however lignin is essentially considered as a waste product, which is burned for energy. We are introducing a novel green process, Formate Assisted Fast Pyrolysis (FAFP) that is very effective in producing liquid fuels and value-added chemicals from lignin. In this method, lignin is mixed with formate salts and pyrolyzed at 500°C in a fluidized bed reactor. The addition of calcium formate to lignin prior to fast pyrolysis results in deoxyhydrogenation of the lignin during pyrolysis. The calcium formate thermally decomposes in the inert atmosphere to hydrogen and carbon monoxide that facilitates the chemical transformation in-situ. The process occurs at atmospheric pressure and without catalysts. There are several immediate benefits to this method for fast pyrolysis of lignins including improvements in lignin feeding to the pyrolysis reactor, an increase in liquid yield, and an increase in the carbon yield in the oil phase of the liquid product. The primary products of this method are alkylated phenols. GC-MS analysis indicated that methoxy groups were removed from the lignin with the addition of calcium formate. Bio- oils, which were collected, have very low oxygen content, and high heat values of 41MJ/Kg approaching crude oil. Results from this method will be presented and evaluated with respect to the advantages relative to competing technologies.

(70g) Investigation of the Pyrolytic Decomposition Pathways of Lignin Model Compounds

Michelle K Kidder and A. C. Buchanan III, Chemical Sciences Division, Oak Ridge National Laboratory, Oak Ridge, TN

Lignin is an abundant biopolymer that constitutes ca. 25-30% of dry biomass, that is rich in aromatic oxygen functional groups (phenols, ethers, alcohols, etc) making it a potential renewable energy and chemical resource that would replace or supplement fossil fuel-based resources. However it is a very complex cross-linked biopolymer material which makes it challenging to deconvolute and control the ability to form desirable products formed from thermochemically decomposition without energy intensive upgrading processes. Much of the fundamental pyrolysis behavior that occurs from the thermochemical decay pathways involving oxygen substituted molecules is not understood. To approach this complicated problem we investigate the simplified model compound that represents 46-60% of the major arylglycerol-beta-aryl ether, β-O-4 interunit linkage, phenethyl phenyl ether (PPE; $Ph^{\alpha}CH_2{}^{\beta}CH_2OPh$). We then systematically rebuild the structure with its lignin-like functionalities, i.e., alcohols, ethers, etc., and investigate the pyrolysis decomposition and substituent influences on product selectivity and reaction kinetics. It has been seen that under liquid phase pyrolysis conditions at ca. 345 °C, decomposition occurs through a free radical chain pathway which involves homolysis of the weak C-O bond, followed by competitive hydrogen abstraction at either the α or β carbon which can be used to describe the product selectivity as α/β, and a rearrangement step that follows the β-hydrogen step. We have recently investigated the pyrolysis of oxygen substituents on the aromatic positions of PPE, and discovered that substitution on the aromatic positions has an effect on the rate and selectivity. We now seek to understand the influence on the thermodynamic reaction rates and selectivities of substitutuents at the aliphatic alpha and beta carbons.

Climate Change and Sustainable Development Legal Update

Monday, October 29, 2012

ENVIRONMENTAL & CARBON REPORTING & DISCLOSURE:

NOT BUSINESS AS USUAL

By

John. P. Fillo, Ph.D. CPEA

EXTENDED ABSTRACT

Climate change and carbon risk management are complex environmental and societal issues, as well as being fundamental to business sustainability. Increased regulation and reporting of environmental and carbon exposures, liabilities and risks are driven by national, regional, state/provincial and local requirements in the U.S. and abroad. These obligations have evolved from participation in voluntary initiatives to legislative and regulatory-driven mandatory actions. While reporting of greenhouse gas (GHG) emissions is only one component of these requirements, statutory financial disclosure requirements for publicly held companies and even country-specific requirements for environmental/ sustainability reporting extend these obligations to address strategy, operations, business risk and opportunity management, legal compliance and supply chain issues, as well as disclosures in financial statements and reports.

This presentation will provide an overview of and insight into the complexity of these requirements and their relevance to the chemical engineering profession. While it will provide a brief overview of the mandatory 'environmental' reporting programs and requirements as a baseline, the focus will be on statutory, industry and voluntary environmental, carbon and sustainability reporting and disclosure in the U.S. and internationally. These requirements and voluntary venues represent complex carbon reporting and disclosure requirements, and can extend beyond just simply reporting a GHG inventory. Such disclosures often require that corporate strategy, risk and opportunity management, and programs and initiatives be disclosed as a basis for assessing the veracity of such disclosures.

Mandatory Reporting & Disclosure

The U.S. is faced with a patchwork quilt of GHG reporting and disclosure requirements at the Federal, regional and state levels:

- The U.S. EPA's GHG Mandatory Reporting Rule, for which annual reporting of Scope 1 GHG emissions is required for numerous industry sectors.

- Regional GHG programs are a work in progress, with the most active being the Regional GHG Initiative (RGGI), the first market-based regulatory program in the U.S. to reduce GHG emissions.
- Several states have or are developing GHG reporting and regulatory programs, with California's Global Warming Solutions Act (AB-32) enacted, with GHG inventory reporting and verification requirements in place and cap and trade program requirements recently issued in 2012.

Additionally, U.S. Federal sustainability and carbon planning and reporting requirements have been in place since late 2009 under Executive Order 13514, *Federal Leadership in Environmental, Energy and Economic Performance*. The Executive Order requires annual reporting of Scope 1, 2 and 3 GHG emissions, as well as setting and achieving goals for energy use and GHG emissions reductions, and improvement of broad-based sustainability performance.

And our neighbors to the North have GHG reporting requirements, as well. Canada has Federal annual GHG emissions reporting requirements. Several provinces require industry-wide reporting of GHG emissions (e.g., Alberta, British Columbia, Quebec and Ontario), with some requiring formal verification of these emissions above stipulated thresholds.

Environmental, social and carbon reporting and disclosure requirements are typically embedded in securities reporting regimes throughout the world:

- Environmental and carbon risk reporting and disclosure are mandated by the U.S. Securities and Exchange Commission under Regulation S-K. Similar statutory requirements exist in Canada through Provincial securities commission requirements.
- European companies are required to include non-financial information in their annual and consolidated reports per the European Accounts Modernization Directive, and country-specific enabling legislation that can include requirements for reporting and disclosure of environmental, social and governance information.
- Country-specific requirements can mandate environmental and/or sustainability reporting (e.g., Japan, Brazil).

And while in the early stages of development, additional climate change and sustainability related reporting and disclosure mechanisms are being developed, which have the potential to become mandatory in the future:

- Global carbon disclosure and attestation standards, including those from the Climate Disclosure Standards Board, provide guidance on carbon risk disclosure for companies that are expected to follow International Financial Reporting Standards.

- The International Integrated Reporting Council is leading the development of a global framework for corporate reporting that demonstrates the linkages between an organization's strategy, governance and financial performance, and the social, environmental and economic context within which it operates.
- The Sustainability Accounting Standards Board, a US based, non-profit organization, intends to establish industry-based sustainability standards for the recognition and disclosure of 'material' environmental, social and governance impacts by companies traded on U.S. stock exchanges.

Voluntary/Industry Programs

While mandatory programs developed over the past few years in the U.S., GHG reporting and disclosure has been driven by a variety of voluntary programs and initiatives. These include, but are not limited to:

- The Carbon Disclosure Project, an independent not-for-profit organization, headquartered in the UK that seeks disclosure on climate change business risks and opportunities and GHG data from the world's largest companies since 2000. Reporting programs have extended into the areas of supply chain, water, and a cities project, as well.
- The Climate Registry, established in North America to measure, report and verify GHG emissions consistently across industry sectors.
- The Carbon Principles is a collaboration between investor-owned utilities, banks and NGOs to provide a consistent approach for banks and their U.S. power clients to evaluate and address carbon risks in financing electric power projects.
- The National Association of Insurance Commissioners has developed a reporting and disclosure framework that applies to insurers with aggregate annual premiums over $300 million, who are required to submit annual climate risk reports.

And there are additional voluntary venues, including GHG mitigation project development through carbon offset registries (e.g., Climate Action Reserve, Verified Carbon Standard, and others) and industry trade organization programs (e.g., Global Environmental Management Initiative, Asia-Pacific Partnership on Clean Development and Energy, and others).

Sustainability Indices

Sustainability indices can be an important venue through which to achieve reporting, disclosure and recognition. While these are frequently voluntary, representativeness, completeness, consistency, accuracy, and balance of disclosures speaks to corporate credibility and a company may be evaluated even if you choose not to formally participate. Corporate sustainability reporting is typically one of many criteria utilized to rank the included companies. Such indices include but are not limited to the Dow Jones Sustainability Index, FTSE4Good Index, Calvert Social Index, Morgan Stanley KLD 400 Social Index, the Newsweek Green Rankings and others.

Implications

And are these relevant to the chemical engineering profession? They most certainly are. For a chemical engineer whose career has evolved down a path of EHS/sustainability management or into the legal profession, they will be directly involved in the reporting and disclosure process. For the chemical engineer who is focused to product/process R&D or design and engineering, understanding these issues is essential to assure they are addressed progressively through the process and product development lifecycle. Engineering efficiencies and pollution prevention into processes and products, and assuring continuous improvement, is essential so as to improve a corporation's top and bottom line, and to ensure that the necessary information is available to support these required disclosures.

(77b) Epa's Clean Air Act Authority to Regulate Greenhouse Gases

Mary Ellen Ternes, McAfee& Taft, Oklahoma City, OK

Ever since the passage of the 1990 Clean Air Act Amendments, regulation of greenhouse gases has been controversial. Throughout the George W. Bush administration, EPA creatively avoided adopting Clean Air Act regulations governing greenhouse gases. Then, with the 2007 landmark United States Supreme Court case, Massachusetts v. EPA, the EPA was required to directly address greenhouse gas emission sources pursuant to the Clean Air Act. In 2009 and 2010, EPA initiated several rulemakings governing greenhouse gas emissions, including the Greenhouse Gas Reporting Rule, the Endangerment and Cause and Contribute Finding, the mobile source emission standards (the "Tailpipe" rule), the Tailoring and Timing Rules. Throughout this period, the U.S. House of Representatives entertained several proposals to eliminate EPA, combine EPA with the Department of Energy or simply modify the Clean Air Act to remove any authority to regulate greenhouse gas emissions. Then, in 2012, the D.C. Circuit reviewed all the legal challenges to the suite of greenhouse gas rulemakings and issued its decision. This presentation will review the D.C. Circuit's decision regarding the validity of EPA's greenhouse gas regulations, impacts from the decision, and implications for future policy.

(77c) Synthesis of CaO-Scaffold Adsorbent with Superior Stability for High-Temperature CO_2 Capture

Ming Zhao, School of Chemical & Biomolecular Engineering, The University of Sydney, Sydney, Australia, Nicholas H. Florin, Chemical Engineering, Imperial College, London, London, United Kingdom, Paul S Fennell, Chemical Engineering and Chemical Technology, Imperial College London, London, United Kingdom and Andrew T. Harris, Chemical Engineering, University of Sydney, Sydney, Australia

Introduction

Reducing greenhouse gas emissions to the atmosphere is a key environmental issue facing the world in general. Carbon dioxide (CO_2), primarily from the combustion of fossil fuels for energy production, is the most common greenhouse gas emitted by human activities and the one with the most significant impact on our climate. Calcium looping is one of the most promising technologies for both pre- and post-combustion CO_2 captures at a wide temperature range (400-900 °C) due to the high CO_2 carrying capacity and economical availability [1]. However, the natural CaO-based sorbents (limestone, calcite, dolomite, etc.) and even the synthetic species with tuned structures are suffering from a rapid decay in activity because of a serious sintering and fraction deduction, both of which happen to a pure CaO sorbent [2].

To solve the problem, we developed a sol-gel method to graft CaO onto an inert scaffold (silicon oxide) in order to mitigate sintering and enhance the mechanical properties. This method gives a new approach to a porous and stable CaO-based sorbent that would not only improve CO_2 capture but also enable the introduction of extra metallic sites to form a sorbent/catalyst composite.

Experimental and results

A saturated ethanol solution of tetraethoxysilane (TEOS) and $Ca(NO_3)_2$ was added dropwise into the mixture of 2.5 M ammonium hydroxide, hexadecyltrimethyl ammonium bromide (CTAB) and excess Na_2CO_3. The mixture was gelled and aged at 60 °C over a total of 24 h. The precipitate was washed, dried and calcined at 550 °C in air for 4 h. This CaO-scaffold sorbent was prepared with $n_{Si}:n_{Ca}$ = 1:1. A commercial sorbent that is stored as micrometer-sized $CaCO_3$ (Sigma-Aldrich) was used for comparison.

$$Si(OC_2H_5)_4(TEOS) + 2\,H_2O \xrightarrow{OH^{-1},\,\Delta} SiO_2\downarrow + 4\,C_2H_5OH \qquad (1)$$

$$Na_2CO_3 + Ca(NO_3)_2 \longrightarrow CaCO_3\downarrow + 2NaNO_3 \qquad (2)$$

CTAB was used as a template agent due to its positively charged molecules and formed growing micelles by combining negatively charged hydrated silica that derived from hydrolysis of TEOS [3]. At a higher temperature (60 °C) than conventional MCM-41 synthesis condition (25 °C), the aggregation and precipitation of SiO_2 (Eq. 1) was observed much accelerated so that a co-precipitation together with the formation of $CaCO_3$ (Eq. 2) was initiated. The resultant solid product was found with a one-fold framework instead of separately grown mixture. The morphology imaged by SEM is shown in Fig. 1 where numerous spherical nano granules (20-50 nm) can be observed with regularly arrangement simultaneously forming porosity in nano scale. Therefore, this CaO-scaffold sorbent gave a much higher surface area (111.1 m^2/g by BET model) than a pure-CaO-based sorbent (mostly <10 m^2/g). The local ESD spectra indicate an even distribution of Si and Ca that also reflects the one-fold framework. Free $CaCO_3$ phases are confirmed by powder XRD technique, which predicts a reactivity of capturing CO_2 molecules.

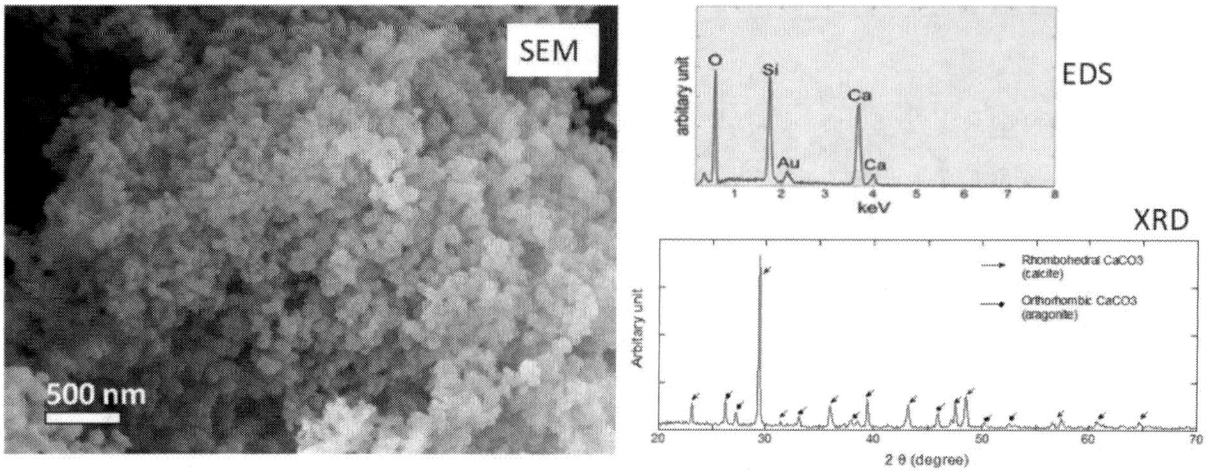

Fig. 1 Characterisation of the CaO-scaffold sorbent

The stability in multiple CO_2-capture-and-release runs was tested using a thermogravimetric analyser (TA Q5000IR). The sorbents were decarbonated in N_2 at 700 °C for 10 min followed by a cooling-down step to 600 °C. A 60-min carbonation in 60 vol.% CO_2 in N_2 then commenced. The decarbonation-carbonation process was repeated up to 13 cycles. The solid curves in Fig. 2 show the CO_2 carrying capacities (per unit sorbent) of the CaO-scaffold and commercial CaO sorbent, respectively, while the red dash line indicates the CO_2 uptake per unit CaO of CaO-scaffold, the value of which is different from the CO_2 uptake per unit sorbent due to the inert component. The new CaO-scaffold sorbent showed an excellent stability over multi-cycle tests despite the lower CO_2 uptake per unite sorbent compared to the commercial one. This stability reflected a superior ability to mitigate sintering deactivation that significantly decreases the practical potential of CaO-based CO_2 capture and is still investigated by current researchers. As shown by the red dash curve, the CO_2 binding capacity of free CaO on the new sorbent has surpassed the commercial one since the 2nd cycle and remained a nearly flat dropping track. This is likely attributed to the strong interaction between free Ca species with the scaffold [4].

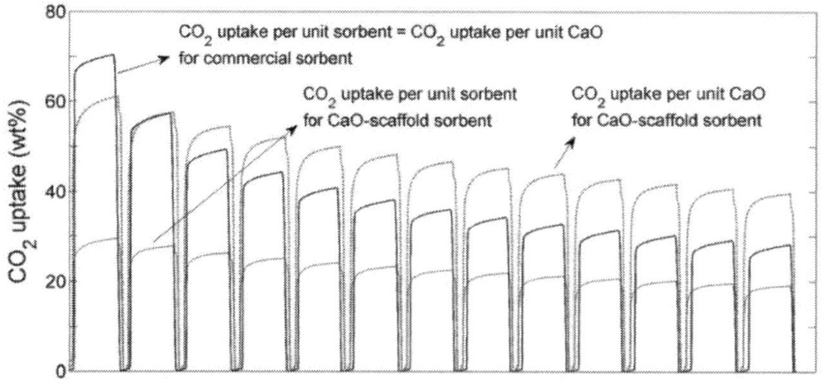

Fig. 2 CO_2-capture-and-release multiple tests using TGA

Conclusions

The demonstrated sol-gel method formed interaction between CaO and scaffold (SiO_2) and consequently a stable, porous, and ordered structure, which favored the sintering-resistant stability over multiple usages. Further investigations will be conducted for the deep understanding of the structure and optimization of the synthesis. The porous framework also provides space to house extra metallic catalyst, which would give a dual enhancement for hydrogen production from coal or biomass.

Acknowledgement. Authors are grateful to the Australian Research Council under DP0666488 and University of Sydney for International Program Development Fund.

References

[1]. Yang, Z.; Zhao, M.; Florin, N. H.; Harris, A. T., Synthesis and Characterization of CaO Nanopods for High Temperature CO_2 Capture. *Industrial & Engineering Chemistry Research* **2009,** *48*, (24), 10765-10770.

[2]. Florin, N. H.; Harris, A. T., Screening CaO-Based Sorbents for CO_2 Capture in Biomass Gasifiers. *Energy & Fuels* **2008,** *22*, (4), 2734-2742.

[3]. Cheng, C.-F.; Ho Park, D.; Klinowski, J., Optimal parameters for the synthesis of the mesoporous molecular sieve [Si]-MCM-41. *Journal of the Chemical Society, Faraday Transactions* **1997,** *93*, (1), 193-197.

[4]. Zhao, M.; Yang, X.; Church, T. L.; Harris, A. T., Novel CaO¨CSiO_2 Sorbent and Bifunctional Ni/Co¨CCaO/SiO_2 Complex for Selective H_2 Synthesis from Cellulose. *Environmental Science & Technology* **2012,** *46*, (5), 2976-2983.

In-Situ Synthesis of Useful Polyamines for CO_2 Capture from Piperazine

Omkar Namjoshi[1], *Yang Du*[1], *Han Li*[1,2], *Mark Goldman*[1], *Gary Rochelle*[1]

[1]The University of Texas at Austin, Austin, Texas, USA; [2]Tsinghua University, Beijing, PRC

Abstract

Concentrated piperazine (PZ) is considered to be a promising solvent for use in post-combustion CO_2 capture from coal-derived flue gas due to its fast rate of CO_2 absorption, high CO_2 capacity, and low degradation rates. However, this solvent has a limited solubility window at conditions corresponding to the operating temperature of a CO2 capture plant, limiting its use in commercial CO_2 capture systems. Blending PZ with a triamine analogue of PZ, such as aminoethylpiperazine (AEP), can widen the solubility window while maintaining many of the properties that make concentrated PZ attractive. With the exception of AEP, many of the triamine PZ analogues are expensive and are not commercially available. They can, however, be made in-situ in the regeneration section of the CO_2 capture plant by the selective reaction of PZ with an amine that forms an oxazolidinone (either a primary or secondary alkanolamine) as its initial thermal degradation byproduct.

The data suggests that the structure of the oxazolidinone is a strong function of reactivity with PZ. Increasing steric hindrance – such as increasing the chain length between the amino group or hydroxyl group or by adding alkyl functional groups to the carbon chain between the hydroxyl and amino groups – can reduce reaction rates by a factor of three to five. Secondary amines that form oxazolidinones react four times as fast as primary oxazolidinone-forming amines. Piperazine and solvent alkalinity losses are reduced significantly (~0.2%/week) once the oxazolidinone-forming amine is completely consumed, indicating that the synthesized amine blend is thermally stable.

Keywords: Carbon Capture, Amine Synthesis, Piperazine, Oxazolidinones

1. Introduction and Theory

A novel method for synthesizing triamines and tetramines in the regeneration section (stripping column) of an amine CO_2 capture plant was investigated. This process uses piperazine (PZ) or a PZ derivative and an amine whose initial degradation product is an oxazolidinone.

Oxazolidinones are the initial thermal degradation byproducts of primary or secondary alkanolamine with two or three carbon atoms between the oxygen and nitrogen atoms. They form when the amine carbamate (which, along with a protonated amine, is formed when CO_2 reacts with a primary or secondary amine) first undergoes electron abstraction followed by ring closing (Davis, 2009; Gouedard 2012). This scheme is shown in figure 1.1.

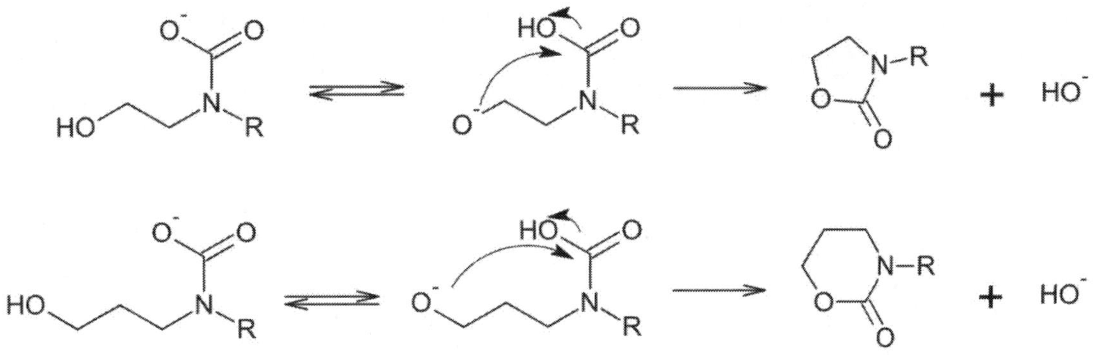

Figure 1.1. General scheme of oxazolidinone formation.

Primary and secondary amines can react rapidly with oxazolidinones to form substituted amine synthesis products. This general scheme is shown in figure 1.2.

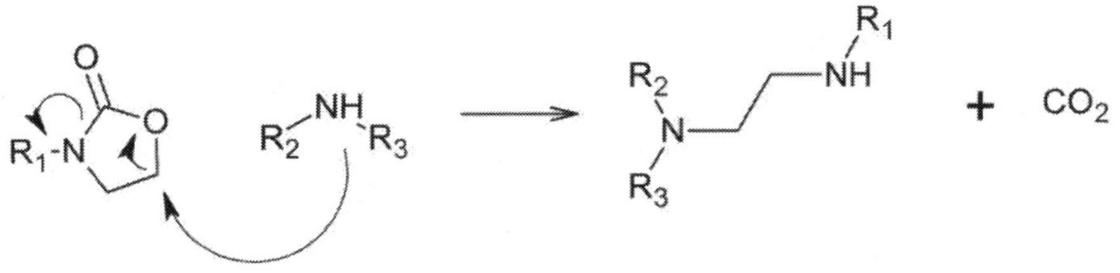

Figure 1.2. General scheme of reaction a primary or secondary amine with an oxazolidinone to form a synthesis product and CO2.

The oxazolidinone-forming amines studied in this paper include monoethanolamine (MEA), propanolamine (MPA), methylisopropanolamine (MIPA), 2-amino-2-methyl-1-propanol (AMP), and methylaminoethanol (MAE). Structures of these amines are shown in table 1.1. An example reaction scheme between PZ and MAE to form the synthesis product methylaminoethylpiperazine is shown in figure 1.3. The structures of the synthesis products formed by reaction with PZ are shown in table 1.2.

Table 1.1. List of the oxazolidinone forming amines used in this study

Amine Name	Abbreviation	Amine Structure	Structure of the Oxazolidinone
Monoethanolamine	MEA	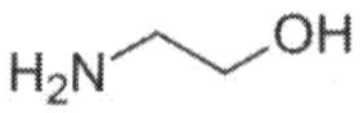	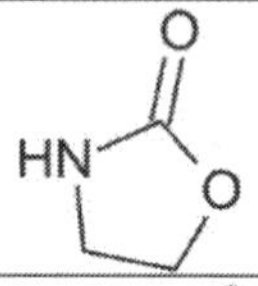
Methylaminoethanol	MAE	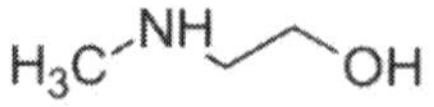	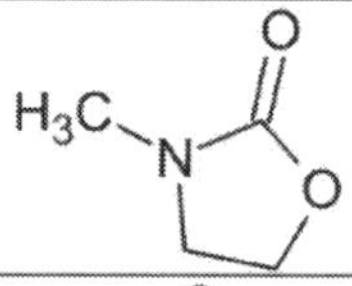
Monoisopropanolamine	MIPA	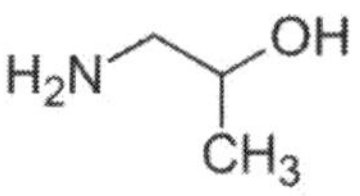	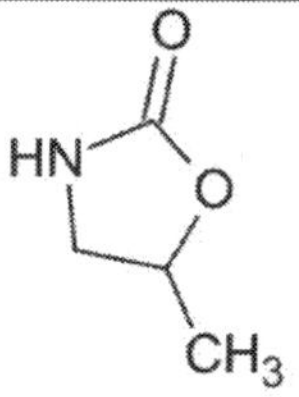
2-Amino-2-methyl-1-propanol	AMP	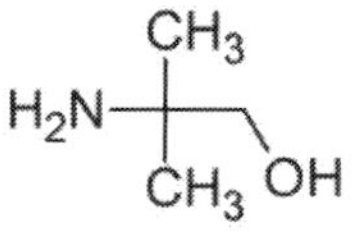	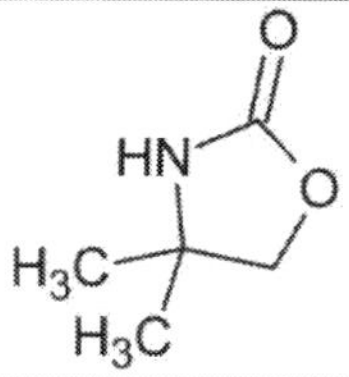
Monopropanolamine	MPA		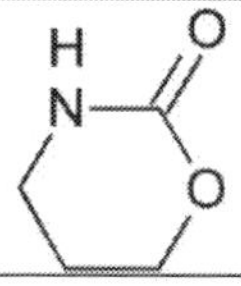

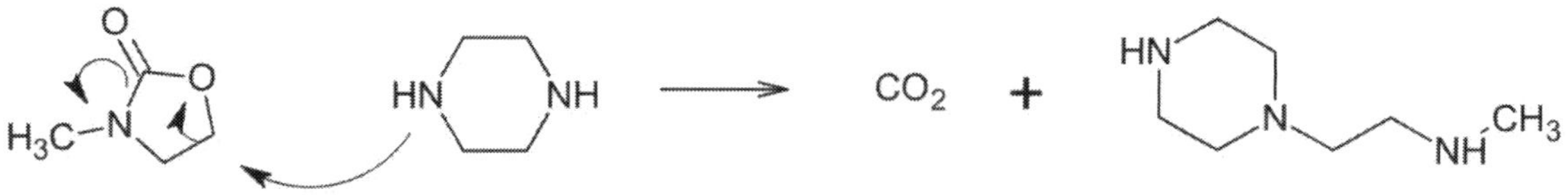

Figure 1.3. Formation of methylaminoethylpiperazine by reaction of PZ with MAE.

Table 1.2. Structures of the principal synthesis products seen in reaction between PZ and the oxazolidinone-forming amine

Amine	Triamine Synthesis Product	Tetramine Synthesis Product
Monoethanolamine	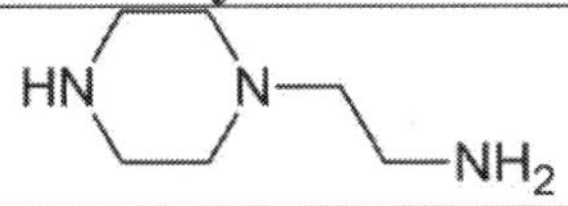	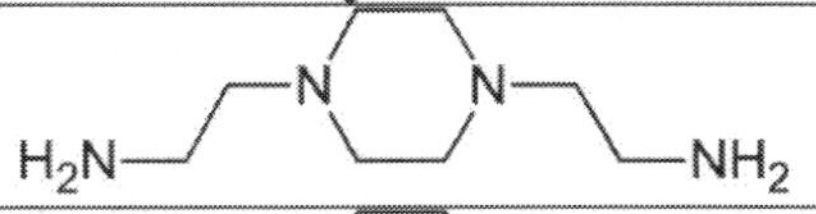
Methylaminoethanol	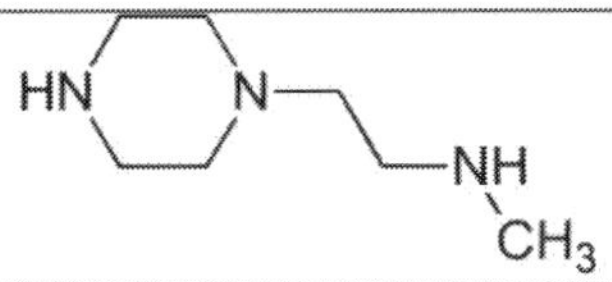	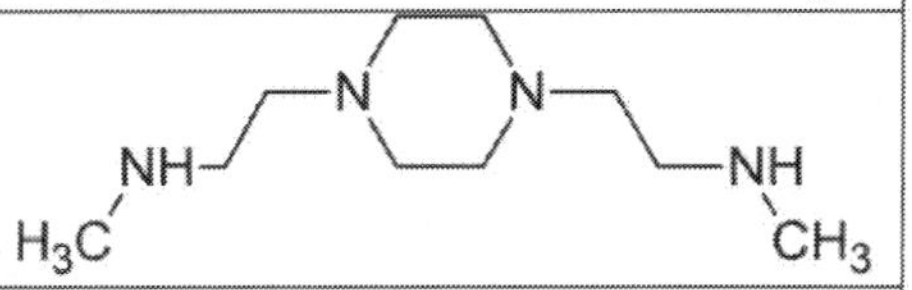
Monoisopropanolamine	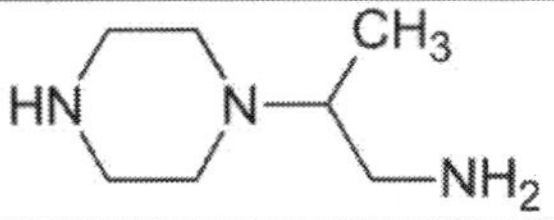	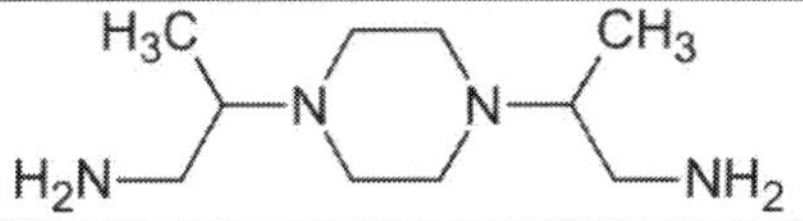
2-Amino-2-methyl-1-propanol	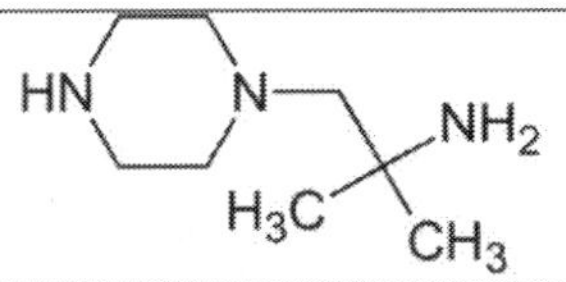	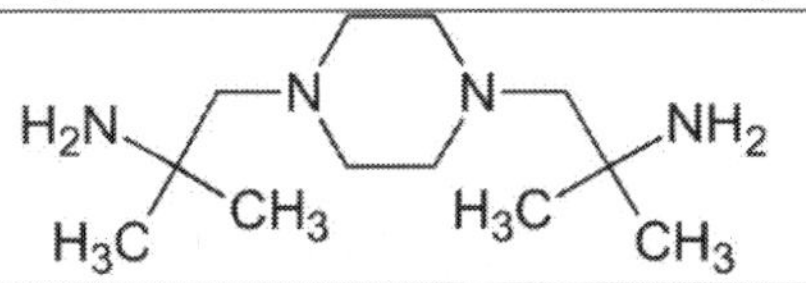
Monopropanolamine	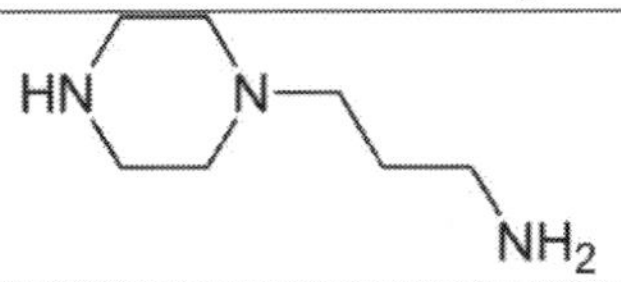	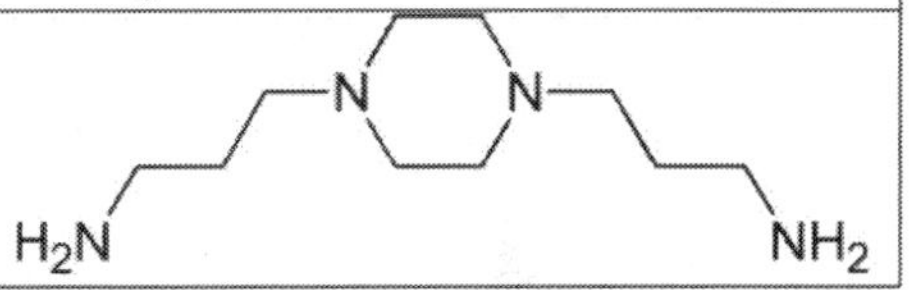

AEP has been studied as a potential capture solvent by Chen (Chen, 2011) and Du (Rochelle, 2012a; Rochelle, 2012b). 6 m AEP and 5 m PZ/2.3 m AEP both have lower cyclic CO_2 capacities than 8 m PZ but have competitive CO_2 absorption rates, slightly higher heats of absorption and, in the case of 5 m PZ/2.3 m AEP, have comparable thermal degradation characteristics as 8 m PZ. AEP, when blended with PZ, has better solid solubility characteristics than 8 m PZ, a key limitation of the concentrated PZ solvent. AEP (and other triamines and tetramines) are less volatile than PZ and thus will have reduced amine emissions from the CO_2 absorption column and/or the water wash column. Although AEP is commercially available, the other synthesis products, which are shown in in table 1.2, are not readily available. These synthesis products, however, can be produced in the regeneration section of the CO_2 capture plant.

2. Experimental Methods

Samples were prepared gravimetrically. CO_2 was sparged into the solution and measured gravimetrically. An initial CO_2 loading of 0.3 mol CO_2/mol alkalinity was used for each solution that was prepared. 7 ml of solution was loaded into 10 ml stainless-steel Swagelok® cylinders, which were then sealed and placed in convection ovens at set temperatures of 135 and 150 degrees C. These temperatures correspond to those encountered in the regeneration sections of proposed CO_2 capture plants using amine solvents. Cylinders were removed periodically throughout the experimental run.

A Dionex ICS-2100 chromatography unit was used to analyze for parent amine concentrations and synthesis product concentrations. Samples were diluted 10000X by mass. A Dionex CG17 50 mm guard column and a CS17 250 mm analytical column connected in series were used to carry out the separation 18.2 umho deionized water spiked with a gradient of methylsulfonicacid was used as the mobile phase. A Metrohm potentiometric autotitrator was used to estimate the loss of alkalinity in each sample. Samples were diluted 300X by mass and 0.2N H_2SO_4 was added until the pH reading reached approximately 3.9, corresponding to total protonation of all amine compounds present in the sample. The experimental methods are essentially similar to those of Freeman (Freeman, 2011).

A list of chemical suppliers and product purity is presented in table 2.1.

Table 2.1. List of chemicals suppliers and product purity

Chemical name	Purity	Supplier
Piperazine (PZ)	99	Sigma-Aldrich
Monoethanolamine (MEA)	99	Acros Organics
Monopropanolamine (MPA)	99	Acros Organics
Monoisopropanolamine (MIPA)	93	Sigma-Aldrich
2-Aminomethylpropanol (AMP)	99	Acros Organics
Methylaminoethanol (MAE)	99	Acros Organics

The supplied MIPA likely contains an isomer as its principal impurity (Bedell, personal communication, May 2012). This isomer is shown in figure 2.1 and can also participate in the amine synthesis like MIPA.

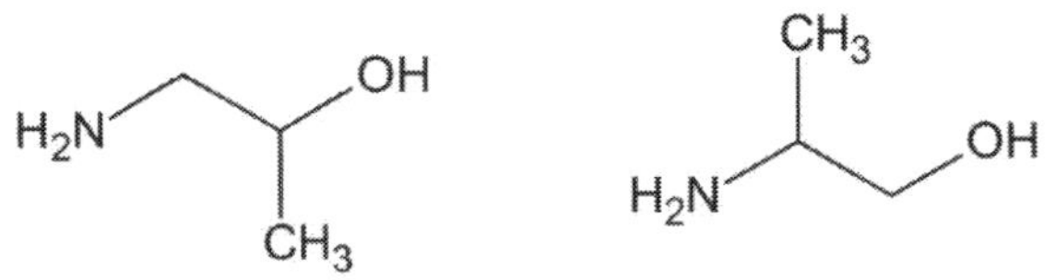

Figure 2.1. Isomers of MIPA; the principal product is on the left

3. Results and Analysis

Unified Calibration Curve

Standards were unavailable for several of the synthesis compounds made. The calibration data, however, for AEP and PZ can be modeled well using a single calibration curve. These calibration data were used to estimate the quantities of the synthesis products present in each sample; the curve is presented in figure 3.1.

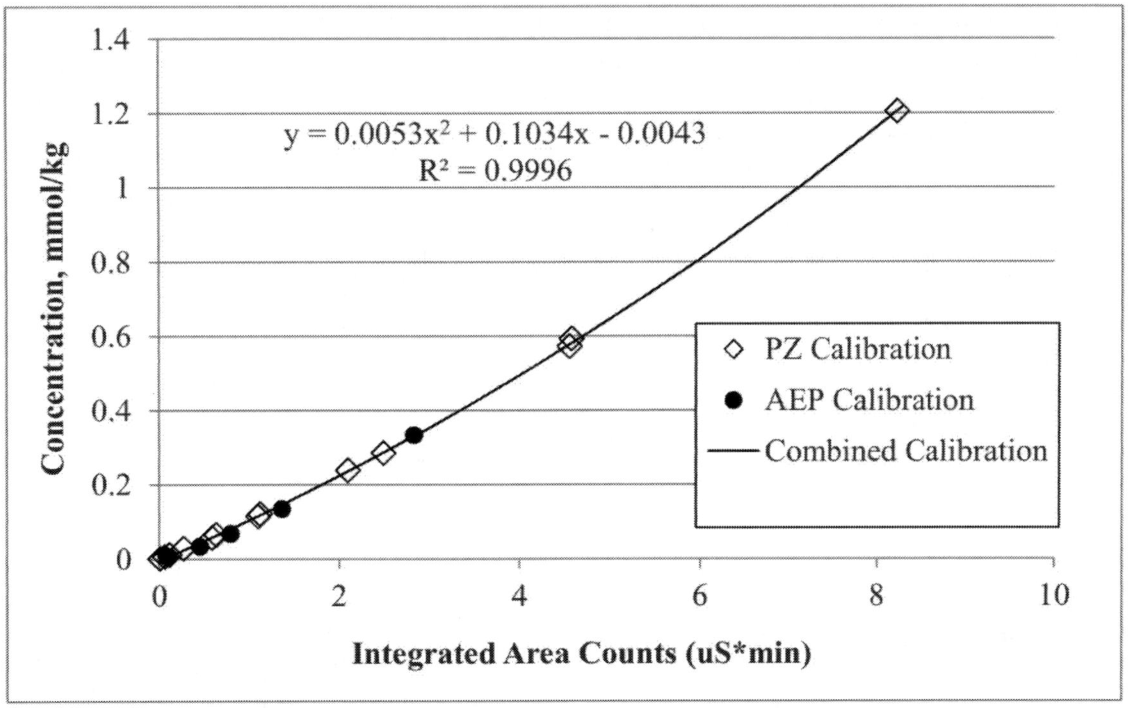

Figure 3.1. Unified Calibration Curve for PZ and AEP

Representative Chromatograms

A sample chromatogram of the synthesis reaction between 7 m PZ/2 m AMP initially at 0.3 mol CO_2/mol alkalinity after 1030 hours and 150°C is shown in figure 3.2. The parent amines and the synthesis product have been highlighted in the chromatogram.

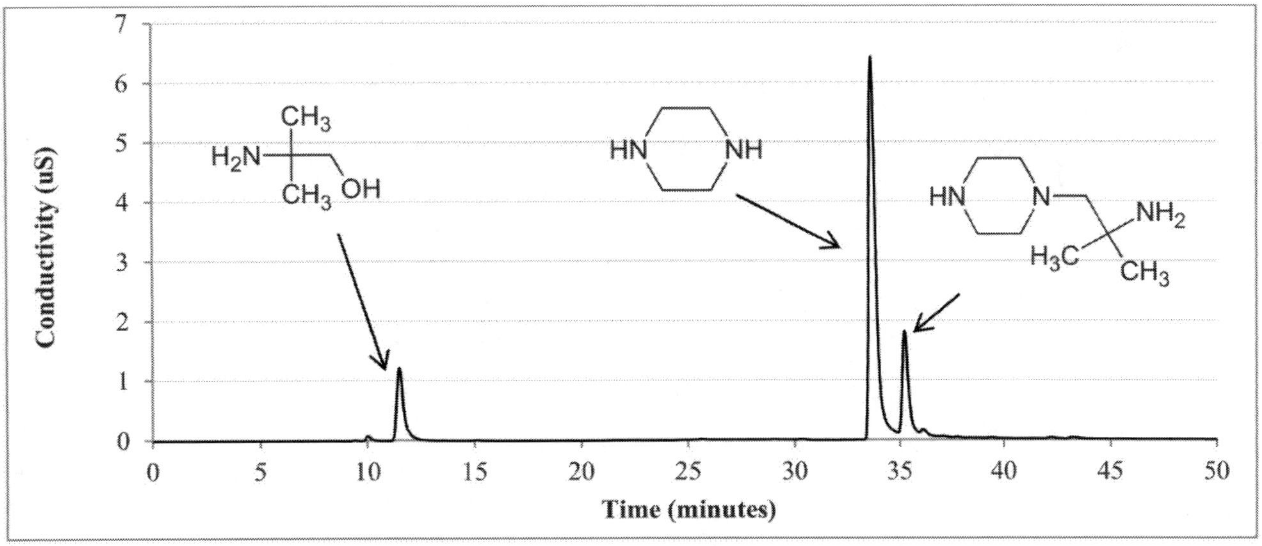

Figure 3.2. Sample chromatogram of AMP-PZ synthesis

Representative Reaction Behavior – First Order Modeling & Arrhenius Behavior

A first order rate model was used to estimate the loss of the oxazolidinone-forming amine throughout the reaction run. Figures 3.3 and 3.4 suggest that this model approximates amine loss reasonably well.

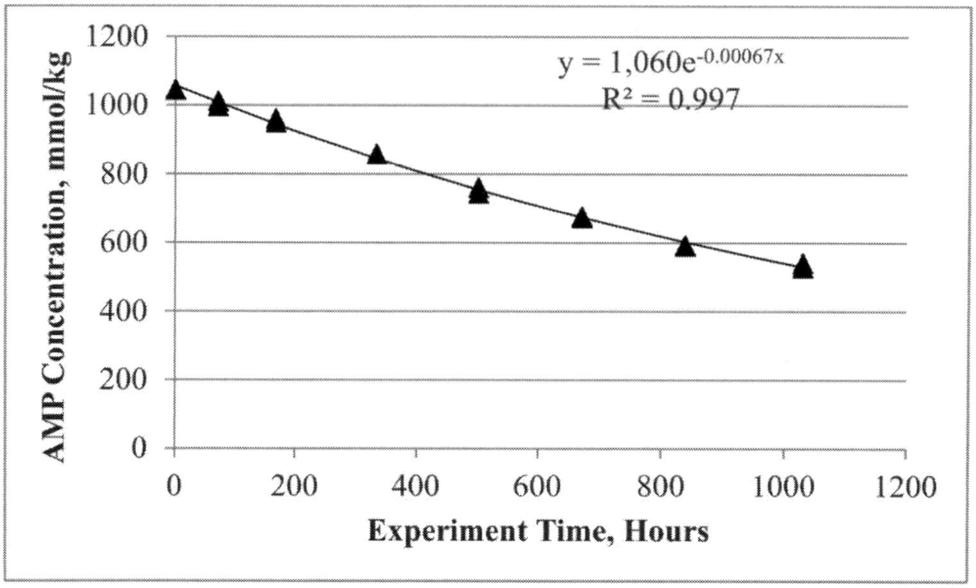

Figure 3.3. Concentration vs time plot of AMP in 7 m PZ/2 m AMP at 150°C, initially at 0.3 mol CO_2/mol alkalinity loading

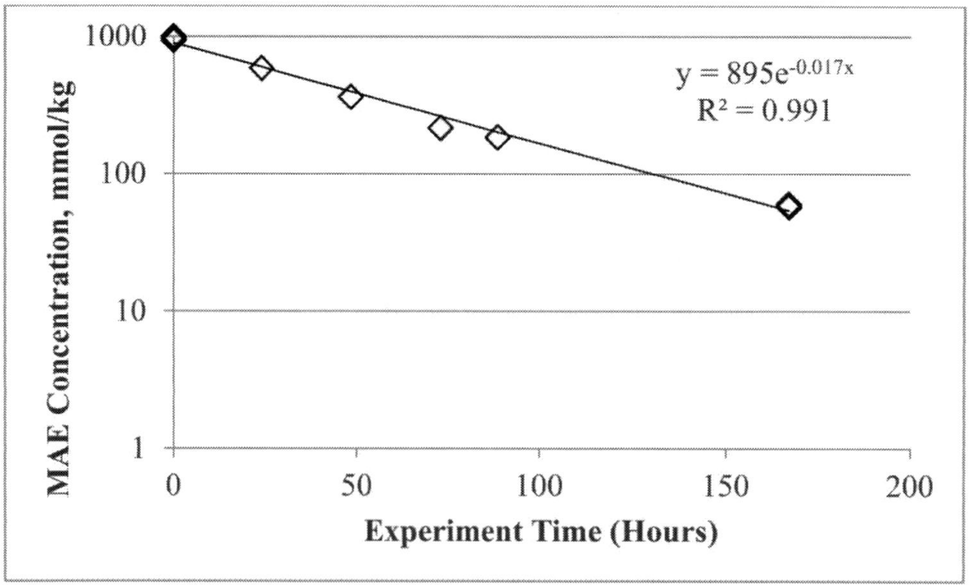

Figure 3.4. Concentration vs time plot of MAE in 7 m PZ/2 m MAE at 150°C, initially at 0.3 mol CO_2/mol alkalinity loading

The activation energy was estimated by using the Arrhenius relationship. Although this work only incorporates data from two different temperatures, data from Freeman suggests that thermal degradation of amines (among them oxazolidinone-formers) follow this behavior (Freeman, 2011).

Selectivity to the synthesis products is defined as the following:

$$\text{Selectivity}_t = \frac{\sum C_{\text{Synthesis Product},t}}{C_{\text{Oxazoidinone Former},i} - C_{\text{Oxazolidinone Former},t}} \quad (3.1)$$

Synthesis products can constitute both the triamine synthesis product and the tetramine dimer synthesis product shown in table 1.1.

Results Summary

Table 3.1 shows the summary of the results for the various amines tested.

Table 3.1 – Results Summary
Note: Degradation data apply to 150°C

Oxazolidinone-Forming Amine	**MEA**	**MPA**	**MIPA**	**AMP**	**MAE**
Amine Conversion (at 500 hours)	85%	45%	44%	28%	100% *(occurred between 167 and 340 hours)*
Synthesis Product Selectivity (at 500 hours)	69%	>90%	87%^	93%	>90%
Alkalinity Loss, %/week	2.0	1.7	1.5	1.0	5.0 (at 167 hours)*
First-order Amine Conversion Rate, 1/hr	0.0042	0.0012	0.00088	0.00067	0.017
Activation Energy, kJ/mol	160	150	130	140	100

* alkalinity of the solution was essentially unchanged after 167 hours (decreased by 0.2 %/wk).
^ also includes synthesis products from the secondary isomer

Analysis

Amine structure appears to have a strong effect on the rate of reaction with PZ. Increasing the steric hindrance of the oxazolidinone forming amine – either by lengthening the carbon chain from two to three carbons or by adding functional groups to the carbons in the alpha position relative to the amino or hydroxyl groups – reduces the rate of reaction with PZ by a factor of three to six. On the other hand, it appears as if oxazolidinone-forming secondary amines have a rate of reaction with PZ that is about four times as fast than the primary amines.

Alkalinity loss can be correlated with the oxazolidinone-forming amine conversion rate. Increased rates of alkalinity loss are indicative of the formation of ureas. Ureas can be formed from the synthesis product that is produced when an oxazolidinone-forming amine reacts with its own oxazolidinone. This is commonly encountered in MEA thermal degradation when the MEA-oxazolidinone synthesis product, hydroxyethylethylenediamine (HEEDA), can rapidly form a urea.

A key finding is that it that the alkalinity loss rate is reduced significantly after consumption of the oxazolidinone-forming amine – especially in the case of the PZ-MAE synthesis reaction. Once the MAE was consumed, the PZ loss was reduced to approximately 30 mmol/week and alkalinity loss was reduced by a factor of 25. This implies that the amine synthesis products and piperazine form a thermally stable mixture and therefore can be used at elevated temperatures in the regeneration section without significant amine loss. A similar phenomenon was noted in PZ-MEA synthesis, with PZ loss averaging about 10 mmol/week once the MEA was consumed.

4. Conclusions

a. First-order kinetics can accurately model the consumption (loss) of the oxazolidinone-forming amine when reacted with piperazine to form substituted triamine and tetramine piperazine analogues.
b. At 150°C, an initial amine concentration of 7 m piperazine and 2 m oxazolidinone forming amine, and at an initial loading of 0.3 mol CO_2/mol alkalinity, the oxazolidinone-forming amine consumption rate is the following:
    - Monoethanolamine: 0.0042 1/hr
    - Monopropanolamine: 0.0012 1/hr
    - Monoisopropanolamine: 0.00087 1/hr
    - 2-amino-2-methylpropanol: 0.00067 1/hr
    - Methylaminoethanol: 0.017 1/hr
c. The structure of the oxazolidinone-forming amine plays a significant role in its loss rate:
    - Increasing steric hindrance – such as increasing the chain length between the amino group or hydroxyl group or by adding alkyl functional groups to the carbon chain between the hydroxyl and amino groups – can reduce reaction rates by a factor of three to five.
    - Secondary amines that form oxazolidinones react four times as fast as primary oxazolidinone-forming amines.
d. Piperazine and solvent alkalinity losses are significantly reduced (~0.2 wt%/wk) once the oxazolidinone-forming amine is completely consumed, indicating thermal stability of the synthesized amine blend.

5. Acknowledgements

The authors wish to acknowledge the Luminant Carbon Management Program for funding.

6. References

Chen X. *Carbon Dioxide Thermodynamics, Kinetics, and Mass Transfer in Piperazine Derivatives and Other Amines.* The University of Texas at Austin. Ph.D. Dissertation. 2011.

Davis JD. *Thermal Degradation of Aqueous Amines Used for Carbon Dioxide Capture.* The University of Texas at Austin. Ph.D. Dissertation. 2009.

Freeman SA. *Thermal Degradation and Oxidation of Aqueous Piperazine for Carbon Dioxide Capture.* The University of Texas at Austin. Ph.D. Dissertation. 2011.

Gouedard C, Picq D, Launay F, Carrette PL. "Amine Degradation in CO2 Capture. I. A Review." *International Journal of Greenhouse Gas Control.* 2012;10:244-270.

Rochelle GT et al. "CO_2 Capture by Aqueous Absorption, Fourth Quarterly Progress Report 2011." Luminant Carbon Management Program. The University of Texas at Austin. 2012.

Rochelle GT et al. "CO_2 Capture by Aqueous Absorption, First Quarterly Progress Report 2012." Luminant Carbon Management Program. The University of Texas at Austin. 2012.

(108a) Optimal Synthesis and Planning of Sustainable Chemical Processes

Ignacio Grossmann, Department of Chemical Engineering, Carnegie Mellon University, Pittsburgh, PA

Sustainability has recently emerged as a key issue in process systems engineering (PSE). Mathematical programming techniques such as mixed-integer programming and global optimization offer a general modeling framework for including environmental concerns in the synthesis and planning of chemical processes. Including environmental issues in the synthesis and planning of chemical processes poses significant challenges that have not yet been fully solved, and hence merit further attention. One major critical issue is how to systematize the search for alternatives leading to reductions in environmental impact. Furthermore, aside from anticipating the effect of uncertainties, which are quite pronounced in this area, there is the issue on how to cope with competing economic and environmental objectives. In the area of process synthesis we address the energy and water optimization of biofuel plants. We present an optimization framework based on superstructure optimization that effectively accounts for energy and water integration, and illustrate its application to the optimization of corn-based ethanol and lignocellulosic bioethanol plants. In the area of supply chain management we address the environmentally conscious design and planning of hydrogen supply chains for vehicle use using a multiobjective optimization approach coupled with the Eco-indicator 99 to assess the effect on human health. We also address the optimal design of chemical supply chain with uncertain emissions using a chance constrained optimization approach to account for the uncertainties associated with the parameters of LCA models (e.g. emissions, potential harm, etc.).

(108b) Sustainable Manufacturing Advances in Research and Technology: Challenges and Opportunities

Yinlun Huang, Dept. of Chemical Engineering & Material Science, Wayne State University, Detroit, MI

The quest for engineering sustainability reflects a crucial paradigm shift in the 21st century, i.e., from the traditional techno-economic paradigm to triple-bottom-lines-based sustainable system design, manufacturing, and management at various length-time scales. However, sustainability is an extremely complex area of research and practice in terms of scope, contents, and spatial/temporal aspects, and sustainability science is far from exact today. Manufacturing Sustainability needs not only evaluation of the status quo, but also prediction and strategic decision-making for the future, all under the situation of information uncertainty and insufficient knowledge. Thus, how to characterize, analyze, design, and restructure manufacturing systems, from the material-product-process level to the large-scale industrial system level, in the context of sustainability, is a new research frontier in engineering that needs much exploration.

In this presentation, the challenges, opportunities, and research frontiers in manufacturing sustainability research and practice will be discussed first. Then, a Multiscale Design-Manufacturing for Sustainability (MDMS) methodology will be introduced, which is developed by resorting to basic sustainability science, multiscale complex systems theory, uncertainty theory, and control theory. The methodology is capable of performing hierarchical system modeling, comprehensive sustainability analysis, and multiscale decision making under uncertainty. The methodological efficacy will be demonstrated by a number of applied studies, ranging from sustainable design of nanomaterials and manufacturing, to collaborative profitable pollution prevention in surface coating systems, and to industrial-zone-based technology network sustainability. Finally, an NSF funded academic-industrial collaboration involving eight countries on sustainable manufacturing research, education, and technology development will be highlighted and future directions will be discussed.

(115a) Comparison of Environmental Releases in Ethanol and Gasoline Production

Debalina Sengupta[1], Raymond L. Smith[2] and Troy R. Hawkins[2], (1)National Risk Management Research Laboratory, ORISE Post Doctoral Fellow at US Environmental Protection Agency, Cincinnati, OH, (2)National Risk Management Research Laboratory, U.S. Environmental Protection Agency, Cincinnati, OH

As the use of biofuels for transportation grows, it is important to scientifically assess the impacts associated with their production. Currently corn ethanol provides the largest volume of transportation biofuel in the United States with more than 13 billion gallons produced in 2010 and 15 billion gallons projected by 2015.

Most prior studies of fuel ethanol plants include greenhouse gas emissions and in some cases criteria air pollutants. In this contribution the focus is on assessing all reported pollutants released from ethanol plants and petroleum refineries. Ethanol is viewed as an alternative to gasoline, and a model is provided to describe the environmental releases at a facility level for ethanol and gasoline production.

Data is collected from the National Emissions Inventory (NEI) database and compiled for emissions for the species in the database. This includes criteria air pollutants and hazardous air pollutants. This research also shows a relative comparison of toxic releases in the United States with data compiled from the Toxics Release Inventory (TRI). Water and land releases are available from the TRI database.

In this work, the focus is on the dry grind ethanol production process rather than the wet grind alternative because the dry grind process is more directly associated with ethanol as the main product, is responsible for over 85% of US ethanol production, and is the primary technology anticipated for future ethanol production. This study addresses suitable methods and criteria for allocating the co-products of ethanol and gasoline production.

The present work attempts to capture these results and present them for further evaluation.

(115b) Effect of Siloxane Present in Renewable Natural Gas On the Operation of Various Natural Gas Equipment

Nitin Narayanan Nair[1], Jorge Gutierrez[2], Jack Chen[2], Alireza Divsalar[3], Arjun Vas[4], Tongyu Zhao[4], May Lew[2], Fokion N. Egolfopoulos[5] and Theodore Tsotsis[1], (1)Mork Family Department of Chemical Engineering and Materials Science, University of Southern California, Los Angeles, CA, (2)Southern California Gas Company, Los Angeles, CA, (3)Chemical Engineering, University of Southern California, CA, (4)Chemical Engineering, University of Southern California, Los Angeles, CA, (5)Department of Aerospace and Mechanical Engineering, University of Southern California, Los Angeles, CA

Effect of Siloxane Present in Renewable Natural Gas on the Operation of Various Natural Gas Equipment

N. N. Nair[a], J. Gutierrez[c], J. Chen[c], A. Divsalar[a], A. Vas[a], T. Zhao[a], M. Lew[c], F. N. Egolfopoulos[b], T. T. Tsotsis[a]

[a]Department of Chemical Engineering and Materials Science, University of Southern California, CA 90089, USA

[b]Department of Aerospace and Mechanical Engineering, University of Southern California, CA 90089, USA

[c] *Engineering Analysis Center, ML SC723B, Southern California Gas Company*

Abstract

The goal of this project is to systematically investigate, under well-controlled laboratory conditions, the operation of different appliances when using simulated renewable natural gas (RNG), particularly as it relates to the impact of siloxanes a key impurity found in such fuels. Two different devices were investigated, namely a small electric generator operating on RNG, and a typical household furnace. Both appliances have been operated for extended periods of time when using siloxane-containing RNG and their operation has been compared with similar appliances operating on conventional natural gas. The impact of the siloxane combustion (which results in solid components that tend to deposit and coat various surfaces) has been investigated in greater detail. It has been shown that the presence of siloxanes has a very detrimental impact on the life of the various sensors that are utilized to control the operation of such devices. The same is true with the impact on the operating life of catalyst beds which are connected to the devices in order to reduce the pollutants emitted. In addition to coating the various internal surfaces, silica microparticulates resulting from siloxane combustion find their way into the condensate (in the case of the furnace) and the spent oil (in the case of the engine). Studies are ongoing to identify the complete failure mode of these devices and the results will be presented at the meeting.

Keywords: Siloxanes, biomethane, biogas, renewable natural gas, SiO_2 particles

ACKNOWLEDGMENT:

The support of the support of the Southern California Gas Company is gratefully acknowledged, as are many helpful discussions with Dr. Siari Sosa, Mr. Scott Kysar, and Mr. Gilbert Ching

(115c) Population Dynamics of Microalgae

Alim Dewan, Siva A. Vanapalli and M. Nazmul Karim, Chemical Engineering, Texas Tech University, Lubbock, TX

One of the distinctive features of cell-division of microalgae is that various single-cells can divide into different numbers although they are from the same species. This has led us to study the population dynamics of the microalgae starting from single and multiple cells. The goal was to see how the population of algae cells changes over time, or if the daughter cells divide into same numbers as do the mother cells, and if it is possible to artificially select a high-growing algae cell colony. We have investigated the population dynamics of *Chlorella vulgaris* in nano-liter scale static droplet array in PDMS microfluidic chip. Each of the droplets in the chip mimicked a batch bioreactor. The change of number of cells was observed until the growth of the cells reached to a stationary phase. We found that a single *Chlorella vulgaris* cell can divide into 3, 4, 5, 6, and 16 cells. There is a significant heterogeneity in the population dynamics even if the initial cell numbers are the same. We also have observed that the daughter cells do not divide into same numbers as do the mother cells. Currently, using a dynamic droplet generation method, we are developing a method of isolating algae cells that divide into higher number of cells.

(115d) Kinetic and Environmental Impact Modeling of Reaction Products From Supercritical Water Gasification of Phenol

Chad M. Huelsman and Phillip E. Savage, Chemical Engineering, University of Michigan, Ann Arbor, MI

Biomass gasification is a potential source of sustainable fuel gas that can help mitigate growing global energy needs and concerns over fossil fuel depletion and greenhouse gas emissions. Hydrothermal gasification in supercritical water offers improved thermal efficiencies for wet biomass, which otherwise requires energy intensive dewatering. This technology has been demonstrated on many starting feedstocks, including seaweed, microalgae, wood, straw, switchgrass, corn stover, cellulose, and lignin, but, in each case, phenol and/or other aromatic compounds persisted in the liquid phase as undesired byproducts. Phenolic compounds are notoriously difficult to gasify and represent one of the main obstacles to complete gasification of biomass, but little is known about the chemistry that leads to such liquid phase byproducts. In order to better understand this underlying chemistry, kinetic and temperature studies were conducted using phenol as a model compound. Reactions were carried out in quartz batch reactors to eliminate catalytic effects from metal walls. These experiments led to the production of little gas, but instead a swath of heavier molecular weight compounds formed. A reaction network was constructed based on additional experiments and careful analysis of chemical pathways. Many of the non-gaseous byproducts identified in these reactions are polycyclic aromatic hydrocarbons (PAHs), EPA priority pollutants, and/or have known human health and environmental effects. The UNEP/SETAC environmental toxicity model "USEtox" was employed to characterize the human and ecotoxic impacts due to a hypothetical emission of this byproduct stream into freshwater. Coupling environmental impact modeling with kinetic modeling of the formation chemistry of these byproduct molecules can provide the insight needed to minimize their production, decrease the risk associated with an unintended environmental release, and simultaneously increase gas yields.

(115e) Alternatives for Anhydrous Ethanol Production an Economic and Environmental Approach

Michel Kahwaji Janho[1], Jorge E. Gatica[2], Fernando Daniel Mele[3], María Rosa Hernández[4] and Mauricio Colombo[3], (1)Chemical And Biomedical Engineering, Cleveland State University, Cleveland, OH, (2)Chemical and Biomedical Engineering, Cleveland State University, Cleveland, OH, (3)Ingenieria de Procesos y Gestion Industrial, Universidad Nacional de Tucuman, Tucuman, Argentina, (4)Chemistry and Chemical Engineering, Universidad Nacional de Tucuman, Tucuman, Argentina

The production of sugar is one of the oldest agro-industrial activities. It includes the production of sugar cane, sugar and dairy products: paper, alcohol, energy, etc. In addition to the production of sugar for domestic and international markets, one of the activities that is growing in importance is the production of alcohol from molasses. Energetic integration has also resulted in bringing power generation from bagasse as a secondary activity in modern sugar-cane processing industries.

A new chapter for bio-energy products associated with the manufacture of sugar has gained relevance as blending gasoline with ethanol is becoming part of energetic policies in most industrialized countries. Indeed, energy products from renewable sources are receiving increasing attention, and this has generated a renewed stimulus to examine alternative processes for bio-ethanol production. It has become common for sugar mills to have complementary distilleries with production capacities ranging from 50 to 400 liters per day of non-anhydrous alcohol.

At sugar plants, steam is generated in boilers using as fuel bagasse-cellulosic fiber waste obtained crushing the sugar cane- and it is used to give useful heat and shaft work to the plant, but it can also be used to generate electricity with export opportunities to the electrical network. The great number of process alternatives outlines a serious decision making problem in order to take advantage of the resources.

At steady state conditions the amount of bagasse obtained is more than enough to supply the demands of useful heat and energy. Thus these industries can advantage of the emergence of a new potential market, as they have an increased potential for generating a surplus of electricity for sale to the grid.

In this paper, a classic bio-ethanol distillery is adopted as case-study and simulated using AspenPlus® and UniSim ® commercial simulators. Alternative process flow sheets of varying complexity are analyzed aiming to reduce power consumption and effluent generation.

The process simulation allows a systematic analysis that leads to a good combination of all the values of process variables. In particular, since they demand significant energy influx, an important subsystem for analyzing is the separation of alcohol-water mixtures, There are also different technologies for this dehydration. In this paper we consider azeotropic distillation and hybrid schemes with membrane separation. The development of a simulation module for calculating the membranes performance, typically not included in commercial process simulators,

The consumed electric energy and steam depends on the adopted scheme and they have, depending to their source or generation policy, an associated cost and environmental impact. Among the energy systems that deserve special attention, cogeneration in the sugar industry must be pointed out. This part of the process makes efficient use of a common fuel for the generation of heat and power. As a result, the environmental impact and costs associated with the use of electricity and useful heat change appreciably, such as anhydrous alcohol production schemes can be achieved with a lesser environmental at the expense of decreasing the economic benefit which would result from selling electricity to the grid.

Cogeneration is not new in the sugar-cane processing industry. Indeed, typically co-generation refers to the generation of heat as the main objective and the generation of electrical and mechanical power as secondary objectives. To date, however, no versatile tool that would allow analyzing economical feasible alternatives of anhydrous alcohol production, bearing in mind environmental issues, is available.

In addition, the explicit consideration of environmental issues offers an alternative route to explore trade-offs. Life Cycle Analysis (LCA) is used in this work. Unlike most environmental approaches to design that focus on reducing the generation of residuals, but without taking into account the impact associated to related processes, LCA allows considering environmental impact as an integral part of the problem. The software used to evaluate the environmental impact is SimaPro®.

Alternative schemes that meet the same specifications of purity and recovery in effluent streams, are examined in this paper. Of particular relevance is the assessment of costs and environmental impact of these processes. A baseline case is defined in our case-study, and structural and parametric changes are made. Different scenarios with respect to economic and environmental issues taking into account the specifications of the final product. Near-optimal scenarios are formulated by defining an operating window for some of environmentally friendly technologies available. The impact of different operating policies for the production of anhydrous alcohol on the environment is examined in detail.

The membrane area, the number of plates in the distillation units, the heat exchange area, and the pressure manipulators are considered as structural variables. Among the most relevant operational variables considered one can mention reflux ratio, pressure ratio, as well as compositions and flow rates of streams.

(115f) Comprehensive Utilization of Rice Husks

Haoran Chen[1], Weixing Wang[2], Jarett C. Martin[1], Adam Oliphant[1], Katelyn Deborn[1], Lishan Liang[2], Wenxi Huang[2], Hong Wang[2] and **Luyi Sun**[1], (1)Department of Chemistry and Biochemistry, Texas State University-San Marcos, San Marcos, TX, (2)Department of Chemical Engineering, South China University of Technology, Guangzhou, China

The annual global production of rice husks (RHs) is ca. 170 million tons. The current application of RHs has been very limited, mainly owing to their tough, woody, abrasive nature, low nutritive properties, great bulk, and high ash content. Therefore, RHs are widely considered as a bio-waste. RHs are mainly composed of lignocellulose (ca. 85 wt%) and hydrated silica (ca. 15 wt%). An approach for comprehensive utilization of RHs has been developed to prepare lignocellulose and silica nanoparticles from this biowaste. Majority of the lignocellulose was firstly extracted from RHs by dissolving in ionic liquids (ILs). The dissolved lignocellulose was subsequently separated and can be used as a starting material for fabric products, biofuel, etc. The remaining RH residues after extraction that contain a high concentration of hydrated silica were thermally treated to synthesize silica nanoparticles with a high purity and surface area. The implication of the present findings is discussed.

(115g) Biomasspyrolyisrefinery: Pyrolysis Oil Hydrodeoxygenation

Hannes Pucher[1], Roland Feiner[1], Nikolaus Schwaiger[1], Peter Pucher[2] and Matthaeus Siebenhofer[1], (1)Institute of Chemical Engineering and Environmental Technology, Graz University of Technology, Graz, Austria, (2)R&D, BDI - BioEnergy International AG, Grambach, Austria

BiomassPyrolyisRefinery:

Pyrolysis Oil Hydrodeoxygenation

H. Pucher, R. Feiner, N. Schwaiger, P. Pucher*, M. Siebenhofer

Graz University of Technology, Institute of Chemical Engineering and Environmental Technology, *BDI-BioEnergy International AG

Non-renewable energy sources such as oil and coal will not suffice the increasing demand of energy in the future. It is assumed that renewable resources such as Biomass are able to fill this dawning gap because they have been considered as the most promising energy sources in the next century [1]. In addition the European Commission has released a directive to increase the volume of renewable biofuels within the field of automotive fuels to 10% by 2020 [2].

To fill the upcoming gap and meet the requirements of the directive a biomass liquefaction concept from renewable resources is presented in this work. The BiomassPyrolysisRefinery concept consists of two main process steps. In the first step lignocellulosic biomass, like wood and straw, is converted in pyrolysis oil and pyrolysis char through liquid phase pyrolysis. In the second step these two products are upgraded. Two promising upgrading technologies are investigated in this concept: hydrodeoxygenation [3] [4] of pyrolysis oil and hydrogenation of pyrolysis char. Through hydrodeoxygenation the polar, highly acidic, water and oxygen rich pyrolysis oil is converted in an appropriate, carbon rich, oxygen and water poor product phase. Then the upgraded product is ready for further processing in a conventional refining process.

The main focus of this project is to investigate and optimize the upgrading of pyrolysis oil to maximize the output of the carbon rich fraction and raise stability. For that purpose, the pyrolysis oil is dehydrated prior to the hydrodeoxygenation step. The water content can be reduced by 90%, (Table 1). Then the operation parameters for hydrodeoxygenation, like hydrogen pressure, temperature, reaction time and catalysts, are determined. Investigations have been carried out in lab size and batch mode.

Table 1: Comparison of Pyrolysis Oil, Pyrolysis Oil dehydrated and Crude Oil [5].

		Pyrolysis Oil	Pyrolysis Oil dehydrated	Crude Oil
Water Content	[%]	60	5	0,1
Density	[kg/m^3]	1130	1200	860
Polarity		polar	polar	apolar

Carbon Content	[%]	22,3	51,6	83-86
Hydrogen Content	[%]	9,6	7,0	11-14
Oxygen Content	[%]	67,7	41,0	<1
Nitrogen Content	[%]	0,4	0,4	<1

A Conversion of pyrolysis oil to carbon rich product phase of about 45% has already been obtained. This results indicate that it is possible to control and supress the unwanted reactions like repolymerisation. The composition and characterisation of the products were determined by GC-MS, elemental analysis, NDIR continuous gas analysis and by a solvent-solvent extraction procedure defined by Oasmaa [6].

[1] Wang, G. et al., "The direct liquefaction of sawdust in tetralin.," *Energy Sources, Part A,* pp. 1221-1231, 2007.

[2] Richtlinie_2009/28/EG, "EU Richtlinie zur Förderung der Nutzung von Energie aus erneuerbaren Quellen und zur Änderung und anschließenden Aufhebung der Richtlinien.".

[3] Mercader, F.M. et al., "Hydrodeoxygenation of pyrolysis oil fractions: Process understanding and quality assessment through co-processing in refinery units," *Energy & Environmental Science* , pp. 985-997, 2011.

[4] Wildschut, J. et al., "Insights in the hydrotreatment of fast pyrolysis oil using a ruthenium on carbon catalyst," *Energy Environ. Sci.,* pp. 962-970, 2010.

[5] Mortensen, P.M. et al., "A Review of catalytic upgrading of bio-oil to engine fuels," *Applied Catalysis A: General,* pp. 1-19, 2011.

[6] Anja, O., Fuel Oil Quality Properties of wood-based Pyrolysis Liquids, 2003.

(115h) Cultivation of a New Butanol-Acetone Producing Clostridium Species Strain G33

Teck Khiang Chua, National University of Singapore, Singapore, Singapore and **Jianzhong He**, CEE, National Univeristy of Singapore, Singapore, Singapore

ABSTRACT

The global energy crisis has renewed interest in producing alternative liquid fuels from clean, renewable sources, e.g., generation of biofules via biological processes. In this study, a *Clostridium* sp. strain G33 was obtained with soil inoculum from grassland in Singapore, which is capable of producing dominant butanol from monosugars (e.g., glucose and xylose). The 16S rRNA gene of strain G33 shows 99% identity with *Clostridium* sp. strain 8052, while both strains possess the ability to produce biosolvents such as acetone and butanol. In the presence of 30 g/L glucose, culture G33 could produce 8.60 g/L butanol in batch cultures with a butanol-to-acetone ratio of 2.7:1 and a butanol yield of 29%, the yield of which is the highest comparing with previous reported strain at similar conditions. No ethanol was detected during the fermentation, implying a simplified purification process of the products (butanol and acetone as end-products) in contrast to that of typical *Clostridium strains* (acetone, butanol and ethanol as end-products). Butanol dehydrogenase (*bdh*) was found to be the functional gene responsible for butanol production. Interestingly, strain G33 could utilize xylan to produce hydrogen and fatty acids under mesophilic conditions, while most clostridium could only feed on simple sugars. When supplemented with 10 g/L xylan, culture G33 generated ~1.4 mmol hydrogen and ~10 mM butyric acid. Direct utilization of hemicellulosic feedstock by culture G33 provides advantages over the traditional physicochemical pretreatment and enzymatic hydrolysis. Therefore, culture G33 shows potential to convert the renewable lignocellulosic biomass into biofuels and serves as a candidate for industrial butanol production.

(119a) Effective Application of Established Treatment Technologies for Metals, Metalloids, and Oxyanions

James Patterson, Patterson Environmental Consultants, Chicago, IL

The underlying theories and fundamentals of application of treatment technologies for control of metals, metalloids and oxyanions are well established. However, the complexity of the wastewater matrix to be treated can introduce unanticipated responses to the applications of such technologies, including when discharge limits are unusually restrictive due for example to water quality-based effluent limitations, or in scale-up from treatability investigations to full-scale application. This Presentation will provide an overview of such theories and applications, and address pollutant speciation, wastewater matrix, and treatment technology aspects which can enhance or interfere with the effectiveness of treatability performance. Such aspects include target pollutant oxidation state and speciation, interfering (or enhancing) interactive species, wastewaters containing multiple (potentially non-compatible) target species, and other issues. Effective pretreatment can be requisite to the efficient application of the appropriate treatment technology. Pretreatment options may include extraction or chemical modification of complexing or other interfering species, electrochemical pre-oxidation or reduction of a target speci, and management of adsorptive surface area and electrical charge in applications of inherent primary, or applied secondary co-precipitation. Improved awareness and understanding of such issues support cost- and performance-effective application of treatment technology.

(119b) Impact of Water Quality and Assessment of Adsorption Technologies for Arsenic Removal

Kashinath Banerjee, Veolia Water Solution & Tech, Moon Township, PA, Gary L. Amy, Water Desalination and Reuse Research Center, 4700 King Abdullah University of Science and Technology, Thuwal, Saudi Arabia and Tapas K. Das, School of Engineering, Saint Martin's University, Lacey, WA

Abstract

The sources of arsenic in drinking water are both natural and anthropogenic. In order to protect public health, the U.S. EPA, World Health Organization (WHO), and the European Commission have decided to lower the MCL of arsenic in drinking water to < 10 μg/L. This stringent arsenic standard will inevitably require many utilities to upgrade their present systems or consider new treatment options. Extensive literature and vendor surveys revealed that a large number of commercial and experimental adsorbents are available for arsenic removal. Some of the commercially available materials are matured products (e.g., activated alumina) that have been widely tested, revealing both their attributes and limitations, while others are more recent products. Results from several investigators (DeMarco *et al.*, 2003; Greenleaf and SenGupta, 2005; Jekel et al., 1999; Pena *et al.*, 2005) indicate that some of these adsorbents are very promising.

Bench-scale studies were conducted to develop a comparative assessment of several commercially available adsorbents for arsenic removal, and to assess their performance in the presence of phosphate, vanadium, and silica. The primary objective of this work was to identify the impacts of water quality, and to define adsorbent properties that will minimize the effects of water quality and/or maximize arsenic adsorption capacity. Additionally, the effect of temperature on the adsorption kinetics and capacities were investigated. Iron modified activated alumina (AA-FS50), granular ferric hydroxide (GFH), granular ferric oxide (GFO), and a titanium-based media (MetSorb G) were used as the adsorbent. In order to better understand the adsorption mechanics of the selected adsorbents, each was characterized according to its important physical/chemical properties, including particle size, bulk density, pH of zero-point charge (pH_{zpc}), surface area, and mineralogy. The concentrations of dissolved arsenic and vanadium were analyzed using Inductively Coupled Plasma – Mass Spectroscopy (ICP-MS); Ion Chromatography (IC) was used to analyze phosphorus. The concentrations of molybdate-reactive (monomeric) silica in the samples were measured, and polymeric silica was estimated by the difference between the total and monomeric silica.

Results revealed that pH has a significant impact on the adsorption capacities. AA-FS50 showed a moderate decrease in adsorption capacity as pH increased, and phosphate had a small impact. GFO media was only slightly affected by increased pH, but was significantly impacted by phosphate, and monomeric silica had some minor effects. GFH performance decreased as pH increased, with phosphate and monomeric silica exhibiting some impact. The adsorption performance of MetSorb G also decreased as pH increased, and phosphate and monomeric silica had some impact, although the interfering effects were less at a lower pH for this media. The two iron oxide media and MetSorb G generally exhibited comparable capacities. For iron oxide based media, rate data are best described by a pseudo-first order kinetic model at each temperature and pH condition studied. At lower pH values, arsenic (V) exhibits greater removal rates than arsenic (III). An increase in temperature increases the overall adsorption reaction rate constant values for both arsenic (V) and arsenic (III). An examination of thermodynamic parameters reveals that adsorption of arsenic (III) as well as arsenic (V) by iron oxide media is an endothermic process and is spontaneous at the specific temperatures investigated.

References

DeMarco, M.J., SenGupta A.K., Greenleaf J.E., 2003. Arsenic removal using a polymeric/inorganic hybrid sorbent, *Water Res.* 37, 164-176.

Greenleaf, J. E. and SenGupta, A. K., 2005, *Investigating New Arsenic Removal Technologies*, Chapter 9.3.2, p. 524-

544; In Das, T. K., (ed.), *Toward Zero Discharge: Innovative Methodology and Technologies for Process Pollution Prevention,* John Wiley and Sons, Hoboken, NJ.

Jekel, M, and Seith, R., 1999. Comparison of conventional and new techniques for the removal of arsenic in a full-scale water treatment plant. Proceedings World Water Conference, Buenos Aires, Argentina.

Pena, M.E., Korfiatis, G.P., Patel, M., Lippincott, L., Meng, X., 2005. Adsorption of As (V) and As (III) by nanocrystalline titanium dioxide. *Water Res.* 39, 2327 – 2337.

(119c) Removal of Arsenic by a Nano-Particle Blend Hollow Fiber Membrane

Jinsong He, Civil and Env Engineering, National University of Singapore, Singapore, Singapore and J. Paul Chen, Department of Civil & Environmental Engineering, National University of Singapore, Singapore, Singapore

The objective of this research is to prepare a based nano-particle blend hollow fiber membrane to adsorb arsenate from groundwater. A new nano-sized metal oxide was first prepared, which was then used in fabrication of the membrane. The hydrophobicity, porosity, mechanical strength, water flux and arsenate adsorption capacity of the membrane were investigated. The additional zirconium based nano-particle increased the hydrophilicity, which resulted in a significant enhancement on the water flux of the blend membrane. The batch adsorption experiments indicated that the membrane could effectively remove arsenate in a wide pH ranging from 2 to 9. The adsorption equilibrium could be reached in 24 h, and the maximum adsorption capacity was above 120mg/g. Ionic strength did not obviously affect the adsorption. The co-anions such as F^-, HCO_3^-, $SiO_3^{2-}$and PO_3^{2-} had different effects on the adsorption. In addition, the membrane could remove arsenate efficiently though the continuous filtration. The mechanism for the removal of arsenate was proposed based on the FTIR and XPS studies.

COMPETITIVE ADSORPTION OF NICKEL, MANGANESE, CHROMIUM AND CADMIUM FROM AQUEOUS SOLUTIONS BY DIFFERENT ADSORBENTS

Malyuba Abu-Daabes [a,*], *Hani Abu Qdais* [b], *Hatem Alsyouri* [c]

[a] *Pharmaceutical-Chemical Engineering Department, School of Applied Medical Sciences, German-Jordanian University, Amman, Jordan*

[b] *Department of Civil Engineering, Faculty of Engineering, Jordan University of Science and Technology, Irbid, Jordan*

[c] *Department of Chemical Engineering, Faculty of Engineering and Technology, University of Jordan, Amman, Jordan*

**corresponding author:*

Telephone: +962-6-530-0666, Fax: +962-6-5300-672, Email address: malyuba.abudaabes@gju.edu.jo

Abstract

Landfilling is the primary solid waste disposal method in Jordan. Since Jordan is considered one of the poorest countries with water resources, it is crucial to protect its groundwater resources from any possible risk of contamination from landfill leachates. A survey on leachates from landfills of different ages and control systems in Jordan, performed by our group during the years 2009 and 2010, showed that heavy metals concentrations in leachates were high exceeding the allowable limits of both Jordan Industrial wastewater and US Environmental Protection Agency. The main heavy metals with high risk are identified to be: Cr (0.66-20.5 mg/L), Mn (0.41-10.37 mg/L), Cd (0.48-4.5 mg/L), and Ni (0.74-6.69 mg/L). In the present study, competitive adsorption of these four metals on different adsorbents including: natural untreated shrimp shells (SH), granular activated carbon (GC), acid-treated granular activated carbon (AGC) and acid-treated powder activated carbon (APC) were investigated under simulated conditions. All adsorption isotherms were generated using the bottle point method, following the ASTM standard procedure (ASTM D3860-89a). Untreated granular carbon (GC) was the best adsorbent for Ni removal from a simulated solution with 90% removal efficiency. AGC and APC both showed very high selectivity for Cr over other present metals with removal efficiency of 99%. For Mn, shrimp shells (SH) was the best adsorbent with 79% removal efficiency, while both GC and SH adsorbents have comparable efficiency for Cd removal from the simulated solution (92%). Order of metals selectivity for SH adsorbent was Cd > Mn > Ni > Cr, for GC the selectivity was Cd > Ni > Cr > Mn, for both AGC and APC the order was Cr > Cd > Ni > Mn.

Key words: competitive metal ions, Leachate, Nickel, Manganese, Chromium, Cadmium, activated carbon, shrimp shells, chitin.

Introduction

Landfilling remains the predominant management method for the disposal of Municipal Solid Waste which leads to the formation of leachates. Leachates may act as a vehicle for the introduction of heavy metals to groundwater resources. Contamination of water resources by heavy metals is a great environmental problem, mainly due to their high toxicity even at very low concentrations (Jaramillo et al., 2009). Heavy metals are not biodegradable and cannot be removed by natural processes from the ecosystems. A previous study by our group on leachates collected from three landfills with different ages and control systems showed that leachates from landfills possess high contamination potential to the limited water sources in Jordan. Heavy metals, mainly Cr, Mn, Cd, and Ni, were found to exceed the allowable limits set by Jordan Industrial Estates Corporation and US Environmental Protection Agency for industrial wastewater discharge. Pollution of heavy metals may arise from different sources. Cd, Cr, Mn and Ni are mainly introduced to the environment from spent rechargeable and household batteries, electroplating, (Xara et al. 2009; Karnchanawong and Limpiteeprakan 2009) and tannery industry (Gotvajn et al., 2009).

One of the promising methods for heavy metals removal from wastewater is the use of natural and surface-modified adsorbents for selective removal of metal ions. Activated carbon is widely used in many applications. In this study the effect of modifying activated carbon surface with acidic groups on competitive metal ions removal from aquatic solution is evaluated. Shrimp shells are also used as a low cost natural adsorbent without any surface modification.

Materials and Methods

1) Adsorbents

Activated carbon was used both in granular (GC) and powder form. All carbon was boiled in deionised water, dried in an oven at 110 °C for 24 hours. Further surface modification was performed by boiling part of them with 4M HNO_3 solution for 1 hour with total condensation. After cooling, the solution of the acid-treated granular activated carbon (AGC) and acid-treated powder activated carbon (APC) was decanted and the solids were washed with deionised water several times until a neutral, constant pH was achieved. Shrimp shells (SH) were washed thoroughly, grounded and sieved without further modification.

2) Characterization of Adsorbents

Surface area and pore size distribution of different adsorbents were measured following the Brunauer, Emmett, and Teller (BET) method (Quantachrome Instruments, Model Autosorb-1, USA).

Table 1. Adsorbents characterization

Adsorbent	BET	HK cumulative pore volume	HK pore wedth
	m^2/g	cc/g	nm
GC	635	2.99E-01	1.398
AGC	591	2.79E-01	1.388
APC	892	4.22E-01	1.388
SH	25.6	9.95E-03	1.058

3) Adsorption isotherms

Stock solutions of heavy metals of 100 mg/L concentration (Ni(II), Mn(II), Cr(VI)) and 50 mg/L Cd(II) were prepared separately by dissolving appropriate quantities of nickel (II) nitrate hexahydrate, cadmium nitrate in nitric acid 0.5 mol/L (Fluka, Switzerland), manganese (II) sulfate monohydrate and chromium (VI) oxide (Merck, Germany). Adsorption isotherms were generated using the bottle point method, following the ASTM standard procedure (ASTM D3860-89a). Adsorption isotherms were generated at 25 °C, using simulated solution of the main heavy metals found in the landfill leachates. The composition of the simulated solution was 100 mg Ni/L, 100 mg Cr/L, 100 mg Mn/L, and 50 mg Cd/L. The pH of the solution was adjusted to 6. Ten ml of the simulated solution was added to 15-ml bottles that contained preweighed amounts (25, 50, 100, 500 and 1000 mg) of the adsorbents. All samples were agitated for 3 days in a temperature controlled water bath to ensure that equilibrium was achieved. The samples were then centrifuged and filtered using 0.45 μm syringe filters before being analysed. The samples were analyzed for their heavy metals content using Inductive Coupled Plasma Technique (ICP, Perkin-Elmer, Optima DV2000, USA). Blank samples were used with every isotherm in order to account for any loss and adsorption on the walls of the tube.

The equilibrium adsorptive quantities of the metal ions on the adsorbents surfaces were calculated using the following relation:

$$q_e = \frac{C_0 - C_e}{M} \times V \qquad (1)$$

Where q_e the amount of metal ion adsorbed per unit mass of adsorbent (mg/g), C_0 and C_e are the original ion concentration and the equilibrium solution concentration (mg/L), respectively, M is the mass of adsorbent, V is the volume of solution (L).

The percent removal of the metal ion by the adsorbent was calculated using the following mass balance relation:

$$\%\ Removal = \frac{C_0 - C_e}{C_0} \times 100\% \qquad (2)$$

Results and Discussion

1) Properties of the adsorbents

The highest surface area is reported for the powdered activated carbon treated with acid (APC) (892 m^2/g) and the lowest one for shrimp shells (25.6 m^2/g), as shown in Table 1. All measured samples have microporous structure with pore sizes less than 2 nm based on Horvath-Kawazoe (HK) model for the determination of pore size distribution in microporous sorbents such as activated carbon and zeolites. Acid modification of granular activated carbon lowered its surface area from 635 m^2/g to 591 m^2/g for GC and AGC, respectively. Previous work showed that oxygenation of activated carbon with 4M HNO_3 significantly increased the concentration of acidic surface oxygen complexes on the surface (carboxylic and carbonyl groups), reduced the total number of basic groups on the surface and caused the surface to be more hydrophilic (Franz et al., 2000; Abu-Daabes and Pinto, 2004; Jaramillo et al., 2009). These acidic sites are bonded to the edges of the graphitic basal plane.

2) *Competitive adsorption of heavy metals*

Adsorption isotherms for all four metal ions, Ni^{+2}, Cr^{+6}, Mn^{+2}, and Cd^{+2}, are shown in Figure 1 (a-d). All isotherms don't show uniform curve which could be due to the competition between the ions for the adsorption sites as well as the repulsion and synergy between them. Similar behaviour for competitive adsorption of metal ions was observed by Covelo et al.,(2004). In general, the mechanism of heavy metal ions adsorption on the surface of activated carbon is by electrostatic attraction to the carbon surface, ion-exchange and surface complexation (Jaramillo et al., 2009; Radovic et al., 2001). Adsorption of heavy metals on the unmodified carbon surface was suggested to be through interactions of the metal ion with the π electrons of the carbon basal planes according to the reactions (Jaramillo et al., 2009):

$$-C\pi + H_2O \leftrightarrow -C\pi\text{-}H^+ + OH^- \quad (3)$$

$$2(-C\pi\text{-}H^+) + M^{2+} \leftrightarrow (-C\pi)_2 - M^{2+} + 2H^+ \quad (4)$$

In the presence of acidic functional groups on the surface of activated carbon the mechanism become surface complexation or ion-exchange depending on the pH. If the solution has low pH, the weakly acidic groups are more likely to be protonated and will undergo ion-exchange reactions with the metal ions, as described in reaction (5).

$$2(-COOH) + M^{2+} \leftrightarrow (-COO)_2Cu + 2H^+ \quad (5)$$

At higher pH, the acidic groups will be nonprotonated and the metal ions will bind to them by surface complexation, releasing less than one hydrogen ion by each metal ion adsorb (Jaramillo et al., 2009).

The selectivity sequences for heavy metal ions depends on the characteristics of the ions such as ionic radius ($Cd^{2+} > Ni^{+2} = Mn^{+2} > Cr^{+6}$), ionic potential or charge-to-radius ratio ($Cr^{+6} > Ni^{+2} = Mn^{+2} > Cd^{+2}$) and electronegativity (Ni > Cd > Cr > Mn) (Sposito, 1989; McBride, 1994; Covelo et al., 2004). The selectivity of each adsorbent for heavy metals under simulated conditions is shown in Figure 2 for the 500 mg adsorbents isotherms. The results show that acid treatment of both granular carbon and powder carbon (AGC and APC, respectively) changed the order of metals selectivity compared to the basic granular carbon (GC). Furthermore, acid treatment of granular carbon (AGC) increased Cr removal efficiency by 30% while it decreased for Ni, Mn and Cd by 23%, 36% and 37%, respectively. This can be explained by the much higher ionic potential of Cr^{+6} compared to other ions which will lead to strong attraction to the acidic surface groups of activated carbon and less accessibility for the ions with lower ionic potential.

Shrimp shells compose of 20% chitin by mass of dry weight, 37% protein and 43% minerals (Einbu, 2007). Chitin is a natural aminopolysaccharide. The results of this study showed that untreated shrimp shells have very high selectivity for Cd^{2+} and Mn^{2+}. This could be due to the presence of amine groups on the chitin of SH. The amine functional groups will form metal complexation. The order of metals affinity for different adsorbents under simulated conditions is summarized in Table 2.

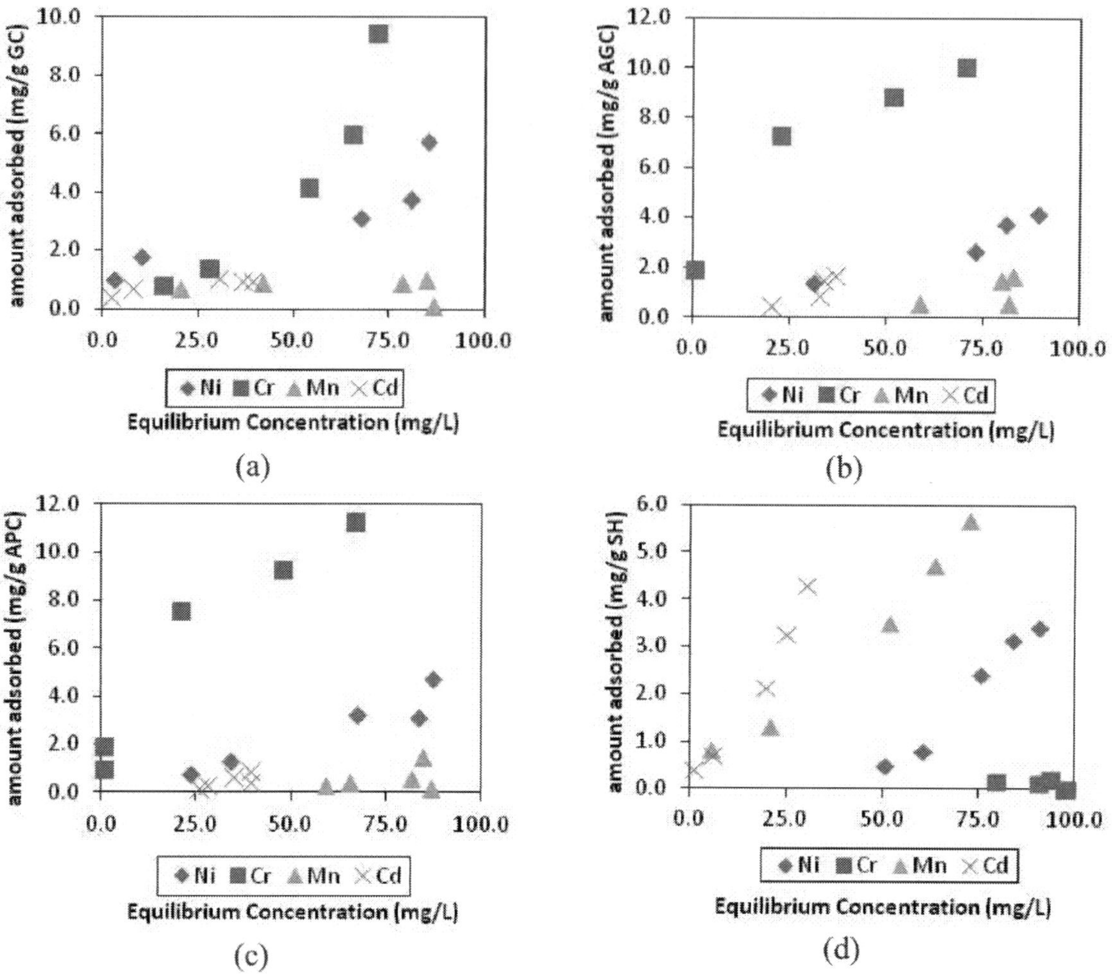

Figure 1. Adsorption Isotherms of Metal Ions on Different Adsorbents

Table 2. Order of metals selectivity for different adsorbents

Adsorbent	**Metals order**
GC	Cd > Ni > Cr > Mn
AGC	Cr > Cd > Ni > Mn
APC	Cr > Cd > Ni > Mn
SH	Cd > Mn > Ni > Cr

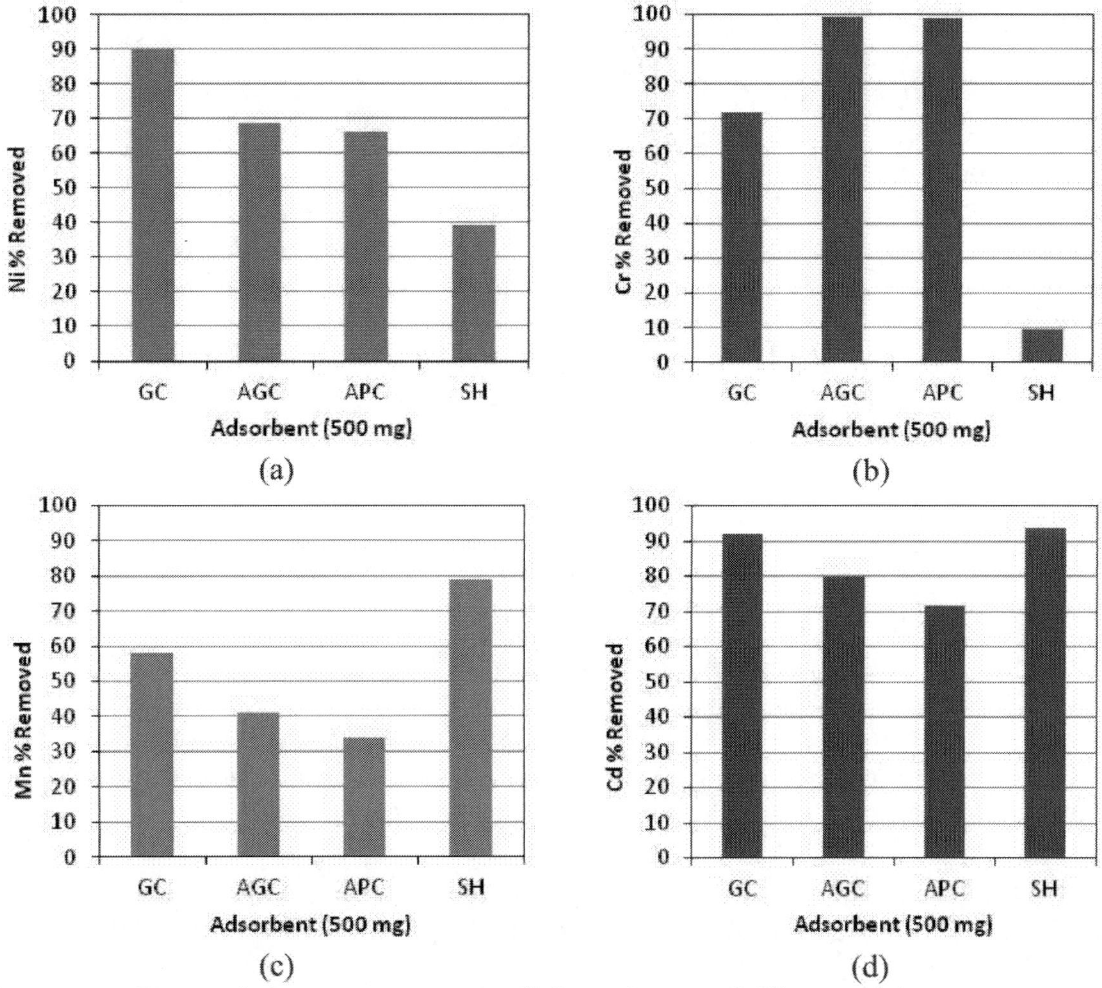

Figure 2. Percent removal of Metal Ions on Different Adsorbents

Conclusion

In this study activated carbon, surface modified activated carbon and unmodified natural shrimp shells were evaluated as potential adsorbents for competitive removal of Ni^{+2}, Cr^{+6}, Mn^{+2}, and Cd^{+2} from aqueous solutions. Untreated granular carbon was the best adsorbent for Ni^{+2} removal from a simulated solution with 90% removal. AGC and APC both showed very high selectivity for Cr^{+6} over other present metals with removal efficiency of 99%. This was due the high ionic potential of Cr^{+6} compared to other ions. For Mn^{+2}, shrimp shells (SH) was the best adsorbent with 79% removal efficiency, while both GC and SH adsorbents have comparable efficiency for Cd^{+2} removal from the simulated solution.

Acknowledgments

The authors wish to thank the German-Jordanian University, the Higher Council for Science and Technology in Jordan, and the European Union presented by Support to Research and Technological Development & Innovation Initiative & Strategies in Jordan (SRTD) Project for their financial support to this project.

References

Abu-Daabes, M., Pinto, N.G. (2004) "Effect of Surface Oxygen Complexes of Activated Carbon on Phenol Adsorption from Single and Mixed Non-Aqueous Solvents", *Separation Science and Technology*, 39, 2997-3009.

Covelo, E. F., Andrade Couce, M. L., Vega, F. A. (2004) "Competitive Adsorption and Desorption of Cadmium, Chromium, Copper, Nickel, Lead, and Zinc by Humic Umbrisols", *Communications in Soil Science and Plant Analysis*, 35, 2709-2729

Einbu, A. (2007), "Characterisation of Chitin and a Study of its Acid-Catalysed Hydrolysis", Ph.D. Thesis, Norwegian University ofScience and Technology.

Franz, M., Arafat, H.A., Pinto, N.G. (2000), "Effect of chemical surface heterogeneity on the adsorption mechanism of dissolved aromatics on activated carbon", *Carbon*, 38, 1807–1819

Gotvajn, Z., Tisler, T., Zagorc-Koncan, J. (2009), "Comparison of different treatment strategies for industrial landfill leachate". *Journal of Hazardous Materials*, 162, 1446-1456.

Jaramillo, J., Go´mez-Serrano, V., A´ lvarez P.M. (2009), "Enhanced adsorption of metal ions onto functionalized granular activated carbons prepared from cherry stones". *Journal of Hazardous Materials*, 161, 670–676

Karnchanawong, S., Limpiteeprakan, P. (2009), "Evaluation of heavy metal leaching from spent household batteries disposed in municipal solid waste". *Waste Management*, 29, 550–558.

McBride, M.B. Environmental Chemistry of Soils; Oxford University Press: New York, 1994.

Radovic, L.R., Moreno-Castilla, C., Rivera-Utrilla, J. (2001), "Carbon materials as adsorbents in aqueous solutions", in: L.R. Radovic (Ed.), Chemistry and Physics of Carbon, 27, Dekker, New York, pp. 227–405.

Sposito, G. (1989), "The Chemistry of Soils", Oxford University Press: New York,.

Strelko, V., Malik, D.J. (2002) "Characterization and metal sorptive properties of oxidized active carbon", *J. Colloid Interf. Sci.* 250 213–220

Xará, S., Delgado, J., Almeida, M., Costa, C. (2009), "Laboratory study on the leaching potential of spent alkaline batteries", *Waste Management*, 29, 2121–2131.

(119e) Arsenic Mobilization In Alluvial Soils of Punjab, North-West India Under Flooded Conditions

Kuldip Singh[1], Dhanwinder Singh[2], H. S. Hundal[1] and Raj Kumar[2], (1)Soils, Punjab Agricultural University, Ludhiana, India, (2)Soil Science, Punjab Agricultural University, Ludhiana, India

Elevated arsenic contaminant in drinking water above the WHO (1996) safe limits of 10 μgL^{-1} was reported at many locations in underground water of Punjab, North-West India. In rural areas, flooded irrigations and maintenance of standing water in paddy fields likely to create anoxic conditions in the soils over-lying above the shallow aquifers. The objective of the present investigation was to study the mechanism of arsenic mobilization in surface alluvial soils. Fifty gram of each of 18 soil samples were transferred to 200 ml polythene bottles, and 100 ml of deionized water was added to maintained flooded conditions. These soil-water systems were kept in BOD incubator at 25^0C for equilibration under anaerobic conditions of 24 and 240 hours period, respectively. After equilibration periods, pH, EC, soluble HCO_3 and Cl^- were determined for each soil-water equilibrated systems. Supernatant aliquots were withdrawn using a syringe, centrifuged, filtered and acidified with a drop of nitric acid prior to analysis for Fe, Mn, SO_4^{2-} and PO_4^{3-} on ICAP-AES. Arsenic in the filtrates solution was determined using hydride generation assembly on ICAP-AES.

Total dissolved As in 24 hour soil-water equilibrated systems varied from 3 to 16 μgL^{-1} with mean value of 9 μgL^{-1} (Figure 1), whereas its concentration range in different soil-water suspensions after 240 hour of equilibration period enhanced to 33 to 1761 μgL^{-1} with mean value of 392 μgL^{-1}. The median, 75th and 90th percentiles concentrations of As were 8, 12 and 14 μgL^{-1} after 24 hour and eventually elevated to 190, 490 and 1013 μgL^{-1}, respectively after 240 hours of anaerobic conditions. Arsenic concentration was enhanced with decreasing redox potential. The results showed antagonistic relationship of arsenic concentration with the magnitude of redox potential status of alluvial soil-water suspensions (Figure 2a). Dissolved Fe concentration versus redox status of soil-water systems determined after two equilibration periods are illustrated in Figure 2b. As the pe decreased, the Fe concentration declined in different alluvial soil water systems.

The effect of redox potential on soluble Mn concentrations determined at different flooding periods in alluvial soil-water suspensions is shown in Figure 2c. As the pe decreased, the soluble Mn concentration enhanced in soil water systems. The soluble concentrations of Mn was ranged between 0.034 to 0.407 mgL^{-1} with a mean value of 0.132 mgL^{-1} after 24 hour equilibration period, but were considerably enhanced to 0.449-44.44 with mean value of 22.398 mgL^{-1} after 240 hour of equilibration periods. The behavior of SO_4^{2-} in different alluvial soil-water systems under different redox status is shown in Figure 2d. With decrease in pe, SO_4^{2-} concentration declined in all soil water systems. Total dissolved concentration of SO_4^{2-} anion in soil-water suspensions also considerably lowered to range of 0.003-25.30 mgL^{-1} with mean value of 1.95 mgL^{-1} after 240 hour of equilibration periods from 4.56 to 81.6 mgL^{-1} with a mean value of 25 mgL^{-1} after 24 hours of equilibration periods. Under reducing conditions, sulfate concentrations decreased significantly.

Pourbaix (pe-pH) diagrams for arsenic and ferromagnesian oxides (Figure 3 and 4) were constructed by to identify the chemical species for each of these elements in different soil-water suspensions. Aqueous arsenic species, $HAsO_4^{2-}$ and $H_3AsO_3^0$ along with the possible solid phase, orpiment (amorphous As_2S_3) as were identified in soil-water systems (Figure 3). The mono valent arsenate oxy-anion ($HAsO_4$) specie was predominant in all the soil-water suspensions except soils 2, 4 and 8, where arsenite ($H_3AsO_3^0$) specie was identified after 24 hour of anaerobic condition. The predominance of neutral arsenite ($H_3AsO_3^0$) specie was observed after 240 hours of reduced incubated period in all soil water systems except soils 12, 15 and 16 in which orpiment (As_2S_3) as solid phase was identified. The neutral arsenite ($H_3AsO_3^0$) and arsenate oxy-anions ($H_2AsO_4^-$) has been found to be predominantly occurred in the alluvial aquifers of Punjab between the pH of 7.05 to 8.25 and under reduced pe value from 0.77 to -4.49, respectively. Thus, it is conclusively evident that two chemical aqueous species controlling As chemistry and its toxicity in soils are arsenate and arsenite. Arsenate is the thermodynamically stable form under aerobic conditions. Arsenite is the predominant species under anaerobic conditions and it is a neutral species at natural pH and is more soluble, mobile and phototoxic than arsenate. The orpiment (As_2S_3) solid phase domain predominates near the bottom of pe-pH diagram and lying over the H_2O / H_2g stability line in the Pourbaix diagram (Figure 3). After 240 hours of equilibrium period, arsenite ($H_3AsO_3^0$) specie was predominant in all alluvial soil-water suspensions. The increase in arsenic concentration was also primarily due to an increase in As (III) species in all the soil-water suspensions under more reduced conditions. In some soil water systems (soil number 12, 15 and 16) apparent precipitation of orpiment was identified on pe-pH diagram. It is evident that total arsenic concentration continuously enhanced in these soil-water systems with decreased magnitude of redox potential. The simultaneously drastic decline in concentration of both Fe and SO_4^{2-} in contrast to accumulation As in reduced soil-water suspensions of alluvial Holocene validated the formation of iron sulfide mineral in reduced alluvial soil-water systems of Punjab. Because of the differences in solubility of iron (DG= -117.11 Kcal mole^{-1}) versus arsenic (DG = -382.64 Kcal mole^{-1}) sulfides, precipitation of iron sulfide may remove sulfide from solution but not arsenic since the precipitation rates of iron sulfide was very rapid compared to amorphous arsenic sulfide.

The aqueous species of Fe^{2+}, Mn^{2+} and $MnHPO_4$ and solid phase of hematite (Fe_2O_3) were predominant in all the alluvial soil-water equilibrated solutions after 24 hours within the pe and pH ranges of -1.75 to 0.77 and 7.25-8.25, respectively (Figure 4). After 240 hours of incubation period, reduction potential status range in soil-water suspensions declined between -4.49 to -2.74 within pH range of 7.05 to 7.94. Aqueous species of Fe^{2+}, Mn^{2+} and $MnHPO_4$ and solid phase of hematite [Fe_2O_3] were predominant in soil-water suspensions of soils 1, 5, 6, 11 and 14. Pourbaix diagram (Figure 4) for iron illustrate that declined in iron was due to the predominantly precipitation of iron sulfide [FeS_2] and or as vivianite [$Fe_3(PO_4)_2 8H_2O$] minerals. The reduced solid phase vivianite [$Fe_3(PO_4)_2 8H_2O$] was found in soil 2, 3, 7 and 13, respectively. Pyrite [FeS_2] was found to be predominant in soil 8, 9, 10, 12, 16, 17 and 18, respectively. All the water composition falls in the stability field for aqueous Mn^{2+} ion (Figure 4) and its concentrations varied from 0.45 to 44.44 mgL^{-1}. Stability diagram for manganese showed that Mn^{2+} ion is thermodynamically more stable and predominant across a wide pe-pH range than for Fe^{2+} ion. Consequently, the Mn^{2+} ion remain in solution in aqueous environment in which Fe^{2+} ions is oxidized to Fe^{3+} and precipitated as Fe $(OH)_2.7Cl_{0.30}$ below pH 6.8 at pe > -1.8 and Fe_2O_3 above pH 6.8 at pe < -1.8 and thus eventually contributed to higher dissolved Mn^{2+} concentration under reduced environments in soils. Iron and manganese concentration in the anoxic soil-water systems were as high as 1.634 and 44.44 mgL^{-1}, respectively. The largest contributor of ferrous to the equilibrated soil-water interaction appears to be an iron oxide phase (probably maghemite, Gamma-Fe_3O_3, an alteration of magnetite). These results showed that dissolve Fe concentration were probably controlled by hematite (Fe_2O_3) under alkaline oxic condition and by iron sulfide/pyrite (FeS) or in some soils by vivianite [$Fe_3(PO_4)_2 8H_2O$] under reduced environment, where soluble phosphate concentration was high. Under reduced condition (developed after 240 hours of flooding), soil solution Fe and SO_4^{2-} are governed by newly formed precipitates of pyrite (FeS_2).

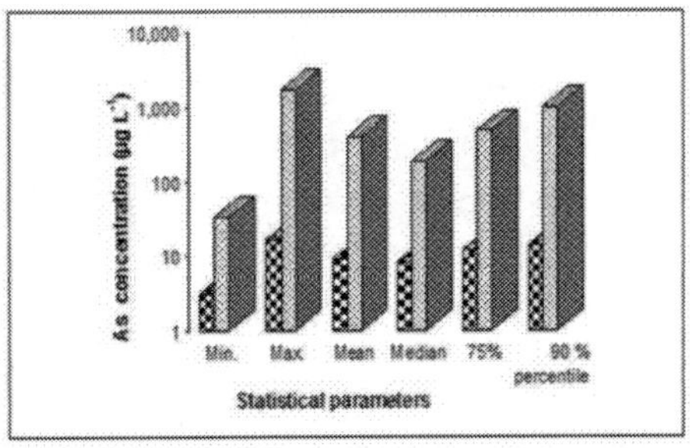

Fig.1 As concentrations measured in 18 alluvial soil-water suspensions of Punjab after equilibration of 24 (▨) and 240(▤) hours

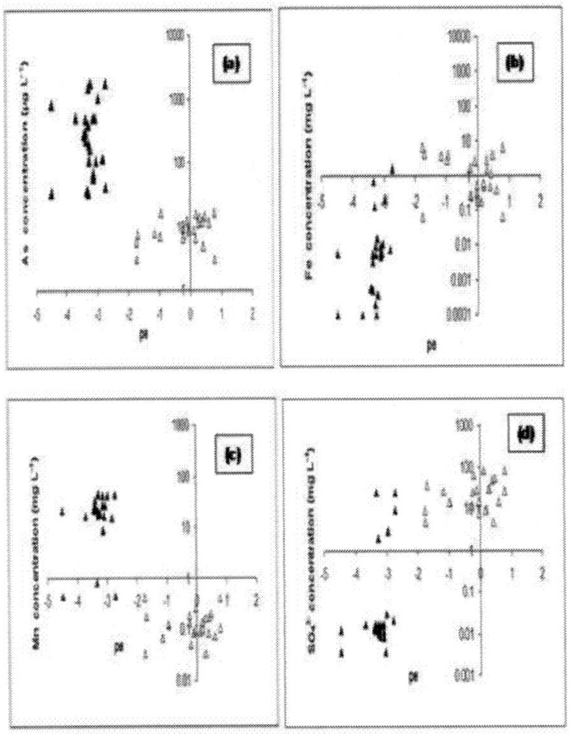

Fig. 2 Effect of redox potential on As (a), Fe (b), Mn (c) and SO_4^{2-} (d) concentrations measured in 18 alluvial soil-water suspensions after 24 (?) and 240 (?) hours of equilibration periods

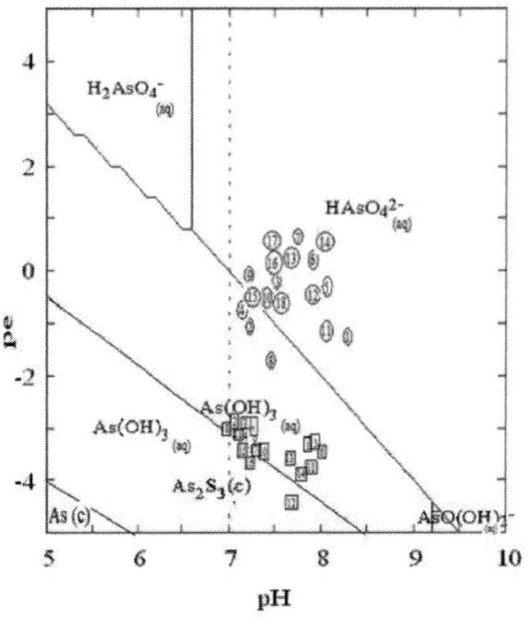

Fig. 3 Arsenic species identified in alluvial soil-water suspensions after equilibration of 24 (O) and 240 (?) hours at 25 °C. The concentration of $[AsO_4^{3-}]$ =6.70μM, $[PO_4^{3-}]$=0.37mM, $[SO_4^{2-}]$= 0.25mM$[CO_3^{2-}]$=0.20mM, $[Cl^-]$=0.13mMand ionic strength=0.020M were fixed to construct the Pourbaix diagram

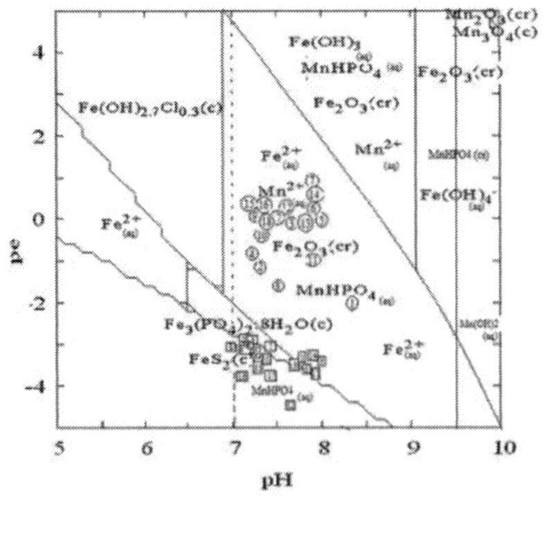

Fig.4 Iron and manganese species identified in alluvial soil-water suspensions after equilibration of 24 (O) and 240 (?) hours at 25 °C. The concentration of [Fe=15μM], [Mn=0.27mM], =6.70uM, [PO43-] =0.37mM, $[SO_4^{2-}]$= 0.25mM$[CO_3^{2-}]$=0.20mM, $[Cl^-]$=0.13mM and ionic strength=0.020M were fixed to construct the Pourbaix diagram

ADSORPTION OF LEAD FROM AQUEOUS SOLUTIONS BY COTTON FABRICS MODIFIED WITH CITRIC ACID

Refugio Bernardo García Reyes, Saúl H. Martínez Treviño, Alcione García González and Ricardo Gómez González*

Universidad Autónoma de Nuevo León, Facultad de Ciencias Químicas, Av. Universidad s/n, Cd. Universitaria, C.P. 66400, San Nicolás de los Garza, N. L., México. FAX: +52(81)83294000 Ext. 6282, TEL: +52(81)83294000 Ext. 6288
*E-mail: bernardogarciareyes@yahoo.com.mx

ABSTRACT

The objective of this research was to modify cotton (i.e. cellulose) fabrics for the removal and recovery of lead (Pb(II)) from aqueous solutions. Cotton fabrics were modified with 0.6 M citric acid solutions and characterized by infrared spectroscopy (ATR-FTIR) and potentiometric titrations. Also, adsorption, desorption, and kinetic experiments in batch systems were conducted.

When modifying cotton fabrics with citric acid, density of functional groups increased up to 2.47 mmol/g in comparison with the non-modified fabric that had 0.61 mmol/g. In addition, estimated equilibrium constants for the modified fabrics can be associated in part to carboxylic groups of the citric acid. The appearance of carboxylic groups on modified fabrics was also confirmed by ATR-FTIR analyses. Maximum adsorption capacity of Pb(II) on modified fabrics, estimated with Langmuir isotherm, was 73.3 y 42.3 mg/g at pH 4 y 3, respectively, whereas non-modified cotton fabrics did not adsorb Pb(II) from water. Decrement in adsorption capacity suggests that Pb(II) and H^+ ions compete for the same adsorption sites. On the other hand, it was possible to desorb around 92% of Pb(II) previously adsorbed on the modified cotton with using 0.1 M HCl solutions since the chemical equilibrium is shifted. Finally, adsorption kinetics can be predicted with a pseudo-second order model.

In summary, cotton fabric treated with 0.6 M citric acid has an adsorption capacity of Pb(II) up to 3.5 times greater than the activated carbon, and the adsorbed metal can be recovered almost completely with 0.1 M HCl. These results suggest that modified cotton fabric is a highly competitive adsorbent for the treatment of metal-containing wastewater.

1. INTRODUCTION

Consumption of water polluted with lead (Pb(II)) is associated to human diseases such as anaemia, kidney damage, sterility, and it is suspected to be carcinogenic. For these reasons, several treatment technologies have been applied for the removal of this pollutant from water, for example: chemical precipitation, coagulation-flocculation, ion exchange, and adsorption with activated carbon, among others (Stumm y Morgan, 1996; Metcalf and Eddy, 2004). Adsorption process is one of the most commonly used technologies because it is useful as a polishing step in a water treatment plant. Nevertheless, it is important to explore the possible application of new adsorbents for removing lead from aqueous solutions. For instance, lignocelullosic materials can be used as new adsorbents because these are worldwide available and low-cost in comparison with polymeric resins and activated carbons. Moreover, it was reported that modification of lignocelullosic materials with citric acid enhanced the adsorption capacity of copper in comparison with non-treated adsorbents (Marshall *et al.*, 2007; Zhu *et al.*, 2008). In this way, the adsorption of metal cations in chemically modified lignocelullosic materials can be enhanced significantly. Nonetheless, to the best of our knowledge, at this time is not reported the density of functional groups added to the adsorbent and their equilibrium constants before and after chemical modification.

In this work, the effect of citric acid modification of cotton fabrics on lead adsorption from aqueous solutions is explored by means of ATR-FTIR analyses, potentiometric titrations, adsorption/desorption and kinetic experiments.

2. METHODOLOGY

2.1 Chemical modification. Thirty grams of cotton fabrics were added in a precipitated flask with 210 mL of 0.6 M citric acid solution. Suspension was heated at 60°C for 1 hour and, afterwards, water was evaporated at about 100°C until dryness. The remaining residues were dried at 120°C in an oven for 24 hours to carry out the esterification on cotton fabrics. Next, modified cotton fabrics were rinsed out with deionized water at 50°C and were dried at 100°C and stored in desiccators.

2.2 Adsorbent characterization. Non-modified and modified cotton fabrics were characterized by means of attenuated total reflection Fourier transform infrared spectroscopy (ATR-FTIR) and potentiometric titrations (Yun *et al.*, 2001; Garcia-Reyes *et al.*, 2009) to identify quantitative and qualitatively the functional groups incorporated to the adsorbent materials, respectively.

2.3 Adsorption experiments. Adsorption tests were conducted in batch systems at 30°C, 150 rpm and the selected pH (3 or 4) to keep lead in solution. A mass of 40 mg of cotton fabrics (non-modified and modified) were added to 40 mL of Pb(II) solutions with different concentration (5-100 mg/L). Solution pH was daily adjusted (with HNO_3 or NaOH) along the Pb(II) adsorption tests at the chosen value (pH 3 or 4) until the equilibrium was reached. Aliquots were taken at the beginning and the ending of adsorption experiments to determine the metal concentration by atomic absorption spectroscopy (ASS).

2.4 Desorption trials. Cotton fabrics, previously used in lead adsorption experiments, were regenerated with 40 mL of 0.1 N HCl solutions after being separated by filtration and rinsed with 30 mL of deionized water at pH 4. These suspensions were continuously stirred at 150 rpm during 24 hours. After that, an aliquot was taken to determine the Pb(II) concentration in solution by AAS.

2.6 Adsorption kinetics. To determine the time required to achieve the equilibrium and to estimate the parameters of kinetic models, adsorption rate experiments of Pb(II) on cotton fabrics were carried out in a packed column with recirculation at initial Pb(II) concentration of 0.256 mM, 30°C, initial pH of 4 and a feeding flow of 10 mL/min. Previous to the adsorption kinetic experiments, modified cotton fabrics were conditioned by stirring the adsorbent in a solution with pH 4 for 24 hours.

3. RESULTS AND DISCUSSION

Figure 1 shows the infrared spectra of the adsorbents used in this research. It is observed that chemical modification of cotton fabrics shown a peak about 1740 cm^{-1} (absent in non-modified cotton fabrics) that can be associated to carboxylic groups of the citric acid (see Figure 1) covalently bound to hydroxyl groups of cellulose structure. It was reported that carboxylic functional groups play a major in adsorption of metal cations from aqueous solutions (Tiemann *et al.*, 1999; Garcia-Reyes *et al.*, 2009).

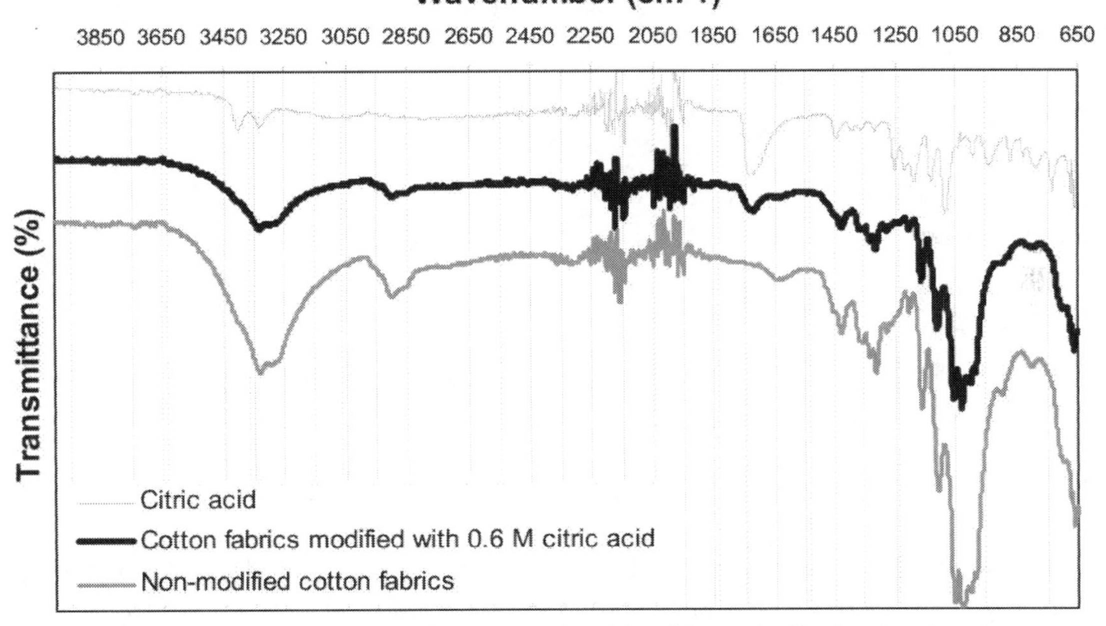

Figure 1. Infrared spectra (ATR-FTIR) of the studied adsorbents.

Table 1 shows the quantity of functional groups per unit mass of adsorbent (b_j) and their equilibrium constants (pk_{aj}) of non-modified and modified cotton fabrics. Non-modified cotton fabrics presented one functional group with an estimated quantity of 0.61 mmol/g and a pk_a value of 8.71 probably due to the hydroxyl groups of cellulose (peaks at 3350 and 1030 cm^{-1} in Figure 1) (Silverstein and Webster, 1998). In contrast, cotton fabrics modified with citric acid showed two functional groups (b_1=0.67 y b_2=1.80 mmol/g) with different pk_a values (4.06 and 8.08). These equilibrium constants are close to two of three pk_a values reported for citric acid (3.13, 4.8 y 6.40). It is important to mention that the functional group of modified cotton fabrics with pk_a=4.06 may correspond to a carboxylic group of citric acid and, therefore, these functional groups can be used as adsorption sites of positively-charged species in water such as Pb(II) species (Pb^{2+}, $PbNO_3^+$, $PbOH^+$, and Pb_2OH^{3+}) (Tiemann *et al.*, 1999; García-Reyes *et al.*, 2009) meanwhile pH of solution remains below 5.6 to avoid Pb(II) precipitation.

Table 1. Density of functional groups (b_j) and their equilibrium constants (pk_{ai}) of the studied adsorbents.

Adsorbent	b_1 (mmol/g)	pk_{a1}	b_2 (mmol/g)	pk_{a2}
Non-modified cotton fabrics	n.d.	n.d.	0.61	8.71
Modified cotton fabrics	0.67	4.06	1.80	8.08

n.d.: Not detected.

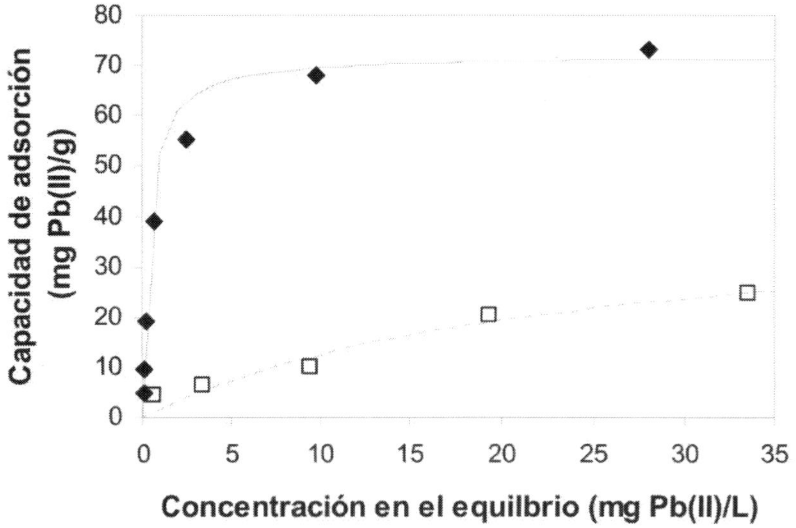

Figure 2. Adsorption isotherms of Pb(II) on cotton fabrics modified with 0.6 M citric acid at 30°C, pH 3 (□) y 4 (♦). Symbols represent experimental data and lines were predicted with the Langmuir isotherm.

Figure 3 shows the adsorption rate of Pb(II) on modified cotton fabrics without pH control. Kinetic experiments were conducted at initial pH 5 but final (at equilibrium) pH decreased to 3.36. This result seems to be reasonable because pH was not corrected along the kinetic experiments and Pb(II) species displaced H^+ ions from protonated carboxylic groups (covalently bound on the cotton fabrics). Adsorption kinetics was well predicted with the pseudo-second order model (Ho y

McKay, 1999). This result suggests that lead species interact with adsorption sites by physical forces such as electrostatic charges.

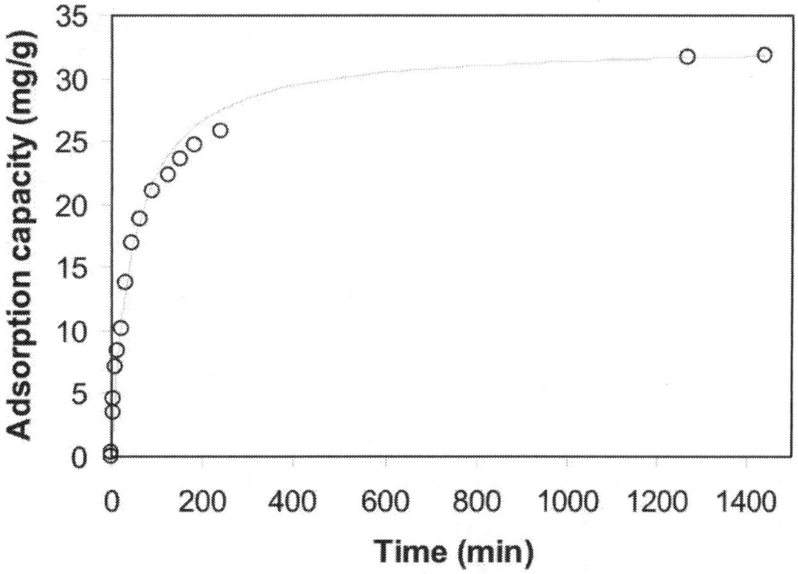

Figure 3. Adsorption kinetics of Pb(II) on modified cotton fabrics at 30°C, initial pH 5, initial concentration 53.18 mg/L, volumetric flow of 10 mL/min, and adsorbent dosage of 1 g/L. Experimental data are represented by symbols and line was predicted with the pseudo second order model.

4. CONCLUSIONS

- ✓ Chemical modification of cotton fabrics by citric acid incorporate functional groups (*i.e.* carboxyl groups) capable of binding metal cations such as Pb(II) from aqueous solutions.
- ✓ Adsorption capacity of Pb(II) on modified cotton fabrics at pH 4 was around 3.5 times greater than activated carbon.
- ✓ It is possible to recover up to 92% of the lead previously adsorbed on the modified cotton with using 0.1 M HCl solutions.
- ✓ Adsorption kinetics can be predicted by a pseudo-second order model and it suggests that lead species interact with adsorption sites by physical forces such as electrostatic charges.

5. ACKNOWLEDGMENTS

This research was financially supported in part by the Secretaría de Educación Pública (PROMEP/103.5/11/6627) y la Universidad Autónoma de Nuevo León (PAICYT IT577-10).

6. REFERENCES

Acharya, J., Sahu, J.N., Mohanty, C.R. and Meikap, B.C. (2009) Removal of lead(II) from wastewater by activated carbon developed from Tamarind wood by zinc chloride activation. Chemical Engineering Journal 149(1-3), 249-262.

Garcia-Reyes Refugio Bernardo, Rangel-Mendez Jose Rene, Alfaro-De la Torre Ma. Catalina (2009), Chromium (III) uptake by agro-waste biosorbents: chemical characterization, sorption-desorption studies, and mechanism, Journal of Hazardous Materials, 170(2-3) 845-854.

Ho, Y.S. and McKay, G. (1999) Pseudo-second order model for sorption processes. Process Biochemistry 34(5), 451-465.

Marshall, W.E., Akin, D.E., Wartelle, L.H. and Annis, P.A. (2007) Citric acid treatment of flax, cotton and blended nonwoven mats for copper ion absorption. Industrial Crops and Products 26(1), 8-13.

Metcalf and Eddy (2004), Waste Water Engineering Treatment and Reuse, 4th Ed., Ed. Mc Graw Hill, USA, 1819 + XXVIII.

Silverstein, R. M., and Webster, F. X. (1998), Spectrometric Identification of Organic Compounds. Wiley & Sons, New York.

Stumm, W. & J. J. Morgan, Aquatic Chemistry: Chemical Equilibria and Rates in Natural Waters, 3rd. ed., John Wiley & Sons, New York, 1996.

Tiemann, K. J., J. L. Gardea-Torresdey, G. Gamez, K. Dokken, S. Sias (1999), Use of X-ray absorption spectroscopy and esterification to investigate Cr(III) and Ni(II) ligands in alfalfa biomass. Environmetal Science and Technology, 33(1) 150-154.

Yun, Y. S., D. Park, J. M. Park, B. Volesky (2001), Biosorption of trivalent chromium on the brown seaweed biomass, Environmetal Science and Technology, 35(21) 4353-4358.

Zhu, B., T. X. Fan, D. Zhang (2008), Adsorption of copper ions from aqueous solution by citric acid modified soybean straw. Journal of Hazardous Materials, 153(1-2) 300-308.

ID: 275766

INVESTIGATION OF BROWN COAL CHAR-CO_2 GASIFICATION IN A LABORATORY SCALE DROP TUBE FURNACE

Md Saiful Alam, Kyushu University, Fukuoka, Japan
Kazi Bayzid Kabir, Monash University, VIC, Australia
Agung Tri Wijayanta, Kyushu University, Fukuoka, Japan
Koichi Nakaso, Kyushu University, Fukuoka, Japan
Jun Fukai, Kyushu University, Fukuoka, Japan
Sankar Bhattacharya, Monash University, VIC, Australia

Abstract

This paper provides detailed investigation of Victorian brown coal char gasification in laboratory scale drop tube furnace (DTF) under CO_2/N_2 mixtures. The DTF is a 2.0 m long quartz reactor consisting of two concentric tubes (diameters 50 mm and 80 mm). Morwell coal char with diameter of <106μm (mean value = 50μm) is used for this investigation. This char sample was prepared in a tubular furnace under pure N_2 at 823K. The char is gasified under CO_2/N_2 atmosphere at temperatures of 1173 and 1273K. The gas compositions from char gasification in the DTF are measured on-line using micro gas chromatography (GC). The results show that char conversion and CO concentration increases with increasing the inlet gas temperature. At 1273K, CO concentration increases due to increased rate of the char-CO_2 reaction. A 3D comprehensive numerical modeling is employed to investigate the char-CO_2 gasification process under various operating conditions. The calculated results are compared with experiments at two different temperatures of 1173 and 1273K using at 20 vol. % CO_2 /N_2 mixtures. The comparisons show a good agreement confirming the validity of the current model. We predict carbon conversion from char-CO_2 gasification for various char particle sizes at elevated gas temperature (>1273K). The temperature distribution in the furnace is found to be uniform under current operating conditions. It is also found that the char conversion significantly depend on the inlet gas temperature and particle size.

Keywords: Char-CO_2 gasification, char conversion, CO_2 reduction

Introduction

Coal is an important global resource due to its relative abundance. Presently about 75% of electric power generation in China, over 50% in the US and nearly 40% in the world, depend on coal [1]. Victoria in Australia has the second largest reserves of brown coal in the world, with economically recoverable reserves projected to last over 500 years at the current rate of consumption [2]. Victorian brown coals have very high moisture content of up to 70% [3]. Therefore, it becomes highly reactive and spontaneously combustible upon drying. At present, the majority of commercial utilization of this resource is for mine-mouth power generation, which suffers from very low efficiency [4]. Alternative modes of utilization are currently being investigated, including gasification and subsequent liquid fuels production. The high reactivity of dried Victorian brown coal makes it eminently suitable for conversion by gasification in which there has now been considerable interest on entrained flow gasification. However, to date, there has been no published work investigating the entrained flow gasification of Victorian brown coals. This work is first-of-its-kind and part of a wider project investigating entrained flow gasification behavior of Victorian brown coal chars.

Char-CO_2 reaction is considered to be a dominant reaction in oxygen-blown entrained flow gasification. E.S. Hecht et al. [5] explained that the presence of the char-CO_2 gasification reaction increases the char conversion rate for combustion at low O_2 concentrations, but decreases char conversion for combustion at high O_2 concentrations. Wijayanta et al., [6] reported that the formation of soot from coal volatiles under various operating conditions. The paper explained that the CO_2 input had a large effect on PAHs/soot reduction at high temperatures.

The study on gasification of Victorian brown coal char is very limited. Considering the environmental need for CO_2 reduction and higher carbon conversion by utilizing CO_2, the present paper reports some of the results on CO_2 gasification of Victorian brown coal char under entrained flow conditions. To scale-up gasification process and improve carbon conversion from the char gasification, it is necessary to conduct the numerical simulation. Therefore, the main objective of this study is to predict carbon conversion under various operating conditions and make preliminary comparison with experimental data.

Experimental

Char preparation

Coal samples were oven dried at 60 C for over 1 hour. The feed was entrained with 0.5 L/min nitrogen as the primary gas and 4.5 L/min nitrogen as the secondary gas. Pyrolysis experiments were conducted at furnace temperatures of 1000 C to prepare char samples. The resultant char samples were stored at less than 4 C to retard surface oxidation prior to further analysis and gasification experiments. The prepared char was then sieved into three size fractions: < 106 μm, 150-355 μm and 355-500 μm. The elemental composition of chars was determined using a Perkin-Elmer CHNS/O analyzer (Model 2400).

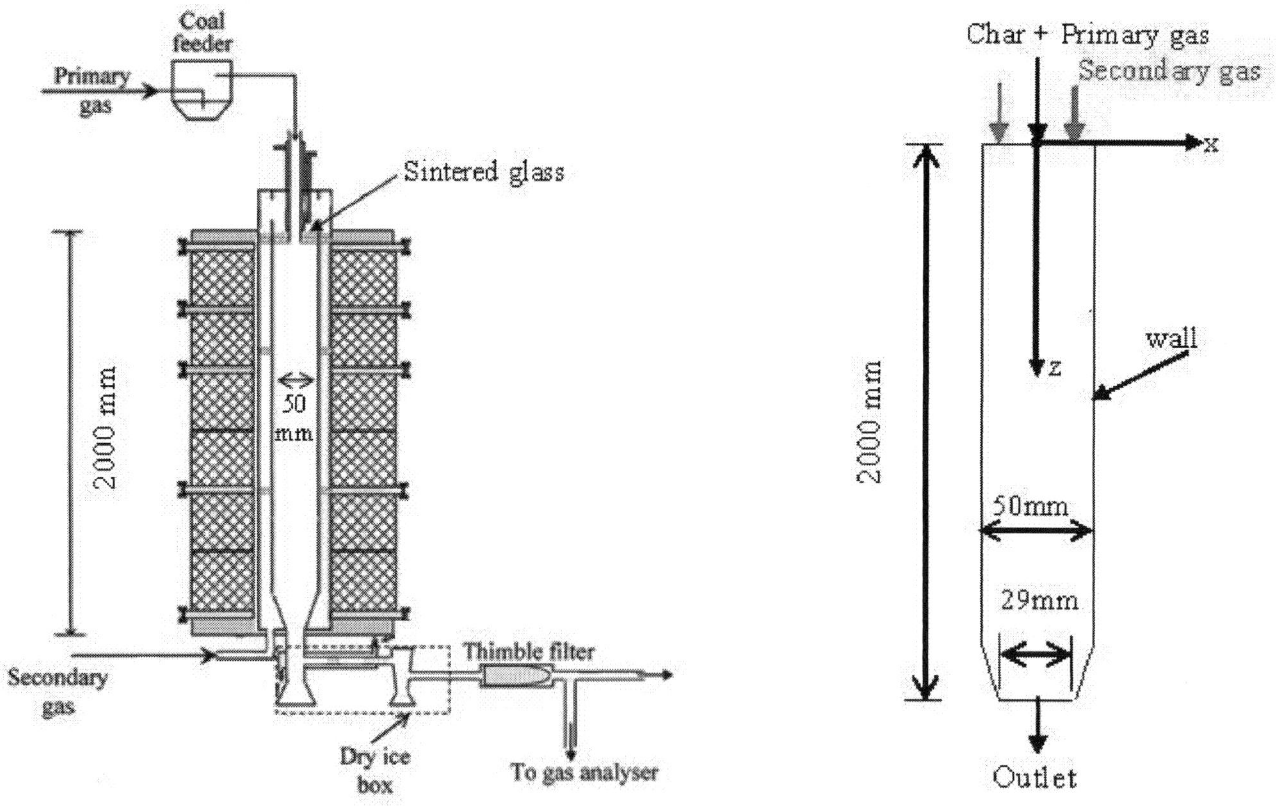

Figure 1(a): Schematic of Drop Tube Furnace (DTF) (b): Computational domain

Char gasification

An electrically heated Drop tube furnace (DTF) coupled with a transparent quartz reactor [7] shown in Figure 1a is employed for char (95.25 wt. % carbon and 4.75 wt. % ash) gasification. The primary gas (10 vol. % of the total inlet gas) at 300K flows through the feeder and entrains the char particles at a feeding rate of ~0.03 g/min before entering the top of the inner chamber (50mm in diameter) of the quartz reactor. The secondary gas (balance of total) is introduced at the bottom of the reactor through the outer annulus (80mm in diameter) of the quartz reactor. This gas is pre-heated to the reactor temperature before entering through a sintered plate and mixing with char and primary gas at the top of the inner tube. Outlet gas from the reactor flows through a char collection system where it is rapidly cooled in a dry-ice trap. Cold gas is then passed through a filter to remove particulates and any potential tars, and sent to a Varian micro-GC (GC-4900) for gas species analysis.

Numerical Study

Mathematical Model

For the fluid phase, the time-averaged steady-state Navier-Stokes equations as well as the mass and energy conservation equations are solved. The numerical domain is shown in Figure 1b. The governing equations for the conservation of mass, momentum, energy and species in 3D Cartesian coordinates are given as:

Mass: $$\nabla.\left(\rho\vec{\upsilon}\right)=S_m \qquad (1)$$

Momentum: $$\nabla.\left(\rho\vec{\upsilon}\vec{\upsilon}\right)=-\nabla p+\nabla.\left(\bar{\bar{\tau}}\right)+\rho\vec{g}+\vec{F} \qquad (2)$$

Energy: $$\nabla.\left(\vec{\upsilon}\left(\rho E+p\right)\right)=-\nabla.\left(\sum_j h_j J_j\right)+S_h \qquad (3)$$

Species: $$\nabla.\left(\rho\vec{\upsilon}Y_i\right)=-\nabla.\vec{J_i}+R_i+S_i \qquad (4)$$

where S is source term to the continuous phase from the dispersed char particle, $\rho\vec{g}$ is the gravitational body force, $\vec{F}$ is external body forces that arise from interaction with the dispersed phase, h_j is the sensible enthalpy of gas species j *and* J_j is mass flux of species j. S_h represents energy source term due to heat of chemical reaction and radiation. R_i is the net rate of production of species i by chemical reaction. ∇ and $\bar{\bar{\tau}}$ are define as:

$$\nabla=\frac{\partial}{\partial x}\vec{i}+\frac{\partial}{\partial x}\vec{j}+\frac{\partial}{\partial x}\vec{k} \qquad (5)$$

$$\bar{\bar{\tau}}=\mu\left[\left(\nabla\vec{\upsilon}+\nabla\vec{\upsilon}^{T}\right)-\frac{2}{3}\nabla.\vec{\upsilon}I\right] \qquad (6)$$

Laminar finite rates model is used for the gas phase reactions (Table 1). Laminar finite rate computes only Arrhenius rate. The net rate of production of species i is computed as :

$$R_i=M_{w,i}\sum_{r=1}^{N_R}\hat{R}_{i,r} \qquad (7)$$

where $M_{w,i}$ is the molecular weight of species i and $\hat{R}_{i,r}$ is the Arrhenius molar creation/destruction of species i in reaction r. The molar rate of creation/destruction of species i in reaction ($\hat{R}_{i,r}$) is written as follows:

$$\hat{R}_{i,r}=\left(v_{i,r}^{//}-v_{i,r}^{/}\right)\left(k_r\prod_{j=1}^{N}[C_{j,r}]^{(\eta_{j,r}^{/}+\eta_{j,r}^{//})}\right) \qquad (8)$$

where $C_{j,r}$ = molar concentration of species j in reaction r [kg mol/m^3

$\eta_{j,r}^{/}$ = rate exponent for reactant species j in reaction r

$\eta_{j,r}^{//}$ = rate exponent for product species j in reaction r

Table 1: Reactions considered in calculation

Gas phase reactions		
$CO + ½O_2$	$\rightarrow CO_2$	(R1)
$CO + H_2O$	$\rightarrow CO_2 + H_2$	(R2)
$CO_2 + H_2$	$\rightarrow CO + H_2O$	(R3)
$CO + 3H_2$	$\rightarrow CH_4 + H_2O$	(R4)
$CH_4 + H_2O$	$\rightarrow CO + 3H_2$	(R5)
$CH_4 + ½O_2$	$\rightarrow CO + 2H_2$	(R6)
$H_2 + ½ O_2$	$\rightarrow H_2O$	(R7)
Surface reactions		
$C + ½O_2$	$\rightarrow CO$	(R8)
$C + CO_2$	$\rightarrow 2\ CO$	(R9)
$C + H_2O$	$\rightarrow CO + H_2$	(R10)

In discrete phase modeling, char particles of known properties are injected into the furnace and tracked in a Lagrangian fashion throughout the computational domain. The dispersed phase is solved by tracking a large number of char particles through the calculated flow field.

The rate of depletion of carbon from coal char due to a surface reaction in Table 1 is calculated by using Smith model which is expressed as [8, 9]:

$$\overline{R} = A_p \eta\, Y_j p_j \frac{kD}{D+k} \tag{9}$$

where $\overline{R}$ represents rate of particle surface species depletion (kg/s).

The diffusion coefficient D is computed as:

$$D = C_{1,r} \frac{\left[(T_p + T_\infty)/2\right]^{0.75}}{d_p} \tag{10}$$

The kinetic rate of the reaction, k is defined as:

$$k = Ae^{-(E/RT_p)} \tag{11}$$

The char surface reaction consumes the oxidant species in the gas phase; i.e., it supplies a (negative) source term during the computation of the transport equation for this species. The surface reaction consumes or produces energy, in an amount determined by the heat of reaction and the heat balance equation is used as follows:

$$m_p c_p (dT_p / dt) = hA_p (T_\infty - T_p) - f_h (dm_p / dt) H_{reac} \tag{12}$$

where H_{reac} is the heat released by the corresponding surface reaction.

The heat transfer coefficient, h, is evaluated using the correlation of Ranz and Marshall [10-12] as follows:

$$hd_p / k_\infty = 2.0 + 0.6 \mathrm{Re}_d^{1/2} \mathrm{Pr}^{1/3} \tag{13}$$

where

k_∞ = thermal conductivity of the continuous phase [W/m-K]

Pr = Prandtl number of the continuous phase ($c_p \mu / k_\infty$)

The Discrete Ordinates (DO) radiation model is used to solve the radiative heat transfer equation. The DO radiation model solves the radiative transfer equation (RTE) for eight numbers of discrete solid angles, each associated with a vector direction fixed in the global Cartesian system (x, y, z). The DO model considers the RTE in the direction as a field equation which is written as:

$$\frac{dI(\vec{r},\vec{s})}{ds} = -(a+\sigma_s)I(\vec{r},\vec{s}) + an^2\frac{T^4}{\pi} + \frac{\sigma_s}{4\pi}\int_0^{4\pi} I\left(\vec{r},\vec{s}'\right)\Phi\left(\vec{s}.\vec{s}'\right)d\Omega' \quad (14)$$

where I is the radiation intensity, $(\vec{r},\vec{s})$ are the unit vectors in the directions of propagation, a is the local absorption coefficient, σ_s is the scattering coefficient, Φ represents a phase function used to characterize the nature of the scattering media

Boundary Conditions

Uniform distributions of inlet mass flow rate and constant temperature are given for inlet primary gas and secondary gas boundary surfaces. The wall is assumed as stationary and smooth with no slip condition. A constant wall temperature is assigned for wall boundary surface. A surface injection model is used for char inlet. The boundary condition of the discrete phase at walls is assigned as "reflect", that means the discrete phase elastically rebound off once reaching the wall. At the outlet, the discrete phase simply escapes/exits the computational domain.

Numerical Modeling

The computation is carried out using the finite-volume-based commercial CFD software FLUENT (version 12.1) under steady state condition. The simulation uses the pressure-based solver, which employs an implicit pressure correction scheme and decouples the momentum and energy equations. SIMPLEC algorithm is used to couple the pressure and the velocity. First order upwind scheme is selected for spatial discretization of the convective terms.

Results and Discussion

Experiment and calculation

To investigate the char gasification under CO_2 atmosphere, char particles diameter of 50 μm are used in the experimental work. The char is gasified with 20 vol. %CO_2 /N_2 mixtures at three different secondary gas temperatures of 1073, 1173 and 1273K. It is found that at lower temperature of 1073K, the char-CO_2 gasification doesn't proceed. As a result, carbon conversion is very low at this temperature. However, a significant increase in carbon conversion from 24% to 68% for 50μm particle has been obtained if the secondary gas temperature was ranged between 1173 and 1273K. 3D numerical simulation is conducted to investigate the char-CO_2 gasification process. The inlet conditions for calculation are shown in Table 2. The calculated outlet carbon conversion and gas species concentration are compared with experimental result in Figure 2. During experiment, it is impossible to maintain a perfectly constant char flow rate. A minute quantity of char remaining in the char hopper is difficult to collect resulting in discrepancy between experimental data and predictions.

Table 2: Experimental conditions

Primary gas temperature [K]	300	300
Secondary gas temperature [K]	1173	1273
Diameter of char particle [μm]	50	50
Flow rate of char [kg/s]	3.675×10^{-7}	4.72×10^{-7}
Flow rate of primary gas [kg/s]	2.65×10^{-6}	2.65×10^{-6}
Flow rate of secondary gas [kg/s]	2.39×10^{-5}	2.39×10^{-5}
Concentration of CO_2 [vol%]	22.52	21.2

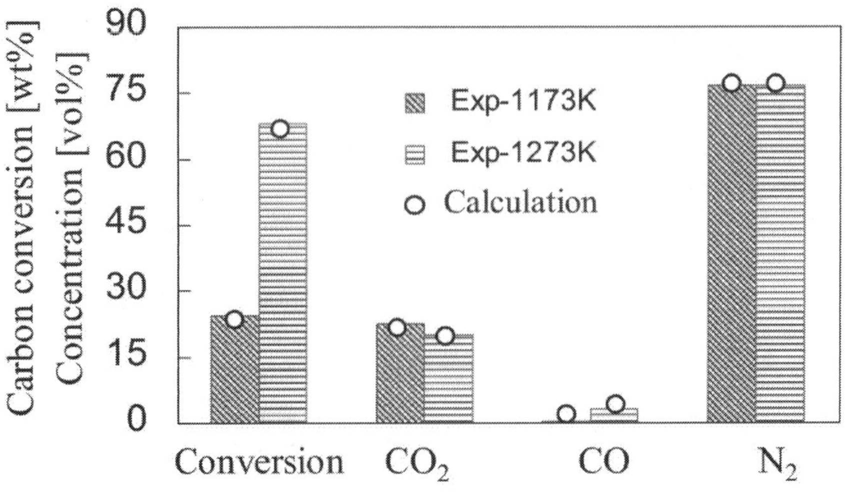

Figure 2: Comparison between experiment and calculation

Contours of gas temperature and species concentration

The contours of gas temperature and gas species concentrations produced from two different temperatures are shown in Figure 3. The temperature contour shows that there is no significant change in the furnace gas temperature. This is because of small furnace diameter and constant temperature of the furnace wall. The char is fed with primary gas at 300K and then mixed with high temperature secondary gas. The conversion of carbon by char-CO_2 reaction (Boudouard reaction) started after the char particle enters into the high temperature gas region. As a result, CO_2 concentration decreases while the CO concentration increases. It is found that the concentration of CO increases at center region because of char injection at center of the furnace where carbon reacts with CO_2.

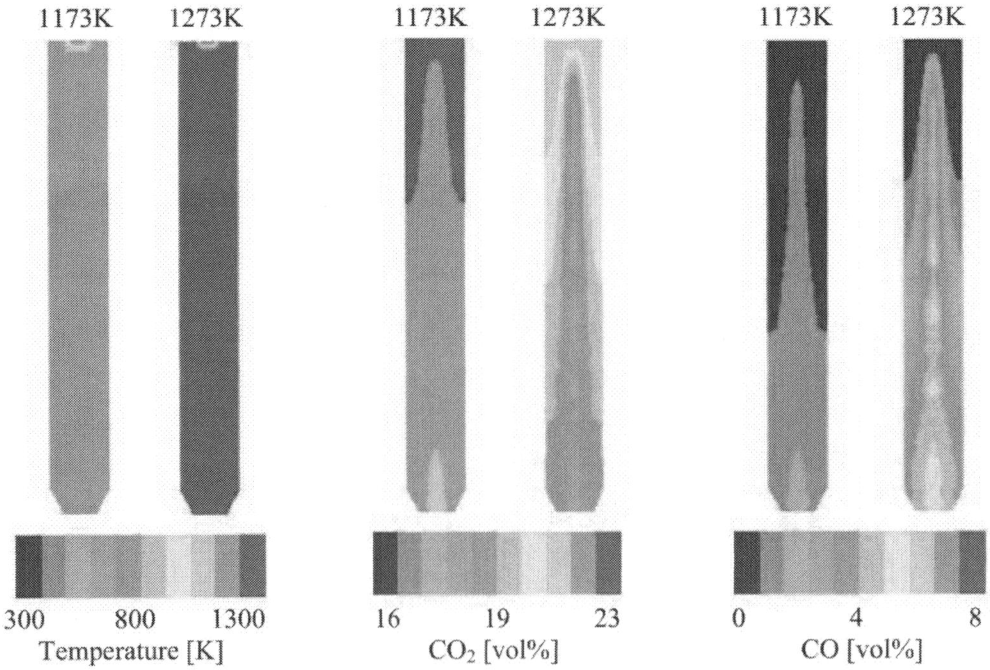

Figure 3: Contours of gas temperature and species concentration for diameter of 50 μm char at secondary gas temperature of 1173 and 1273K

Prediction of carbon conversion

A number of numerical simulations are carried out to predict the carbon conversion of char. The range of particles diameter and secondary gas temperatures considered in this study are 50-425 μm and 1073-1573K, respectively. Figure 4 shows the change of carbon conversion with secondary gas temperature for particle diameters of 50, 250 and 425 μm. At 1073K, the conversion of carbon for 50 μm particle is only 5% whereas the larger particles show no gasification at this temperature. The conversion for 50μm particle increases linearly with increasing the secondary gas temperature. But, the conversion of larger particles significantly changes at higher temperature. However, a complete conversion of carbon can be achieved with increasing the gas temperature. The calculation shows that the conversion of 50, 250 and 425 μm particles is 100% if the secondary gas temperatures > 1373K. The changes in carbon conversion with particle diameter at different temperatures are shown in Figure 5. It is found that at lower temperature (1073K), the conversion of carbon decreases gradually with increasing the particles diameter. But at higher temperature, the carbon conversion changes rapidly.

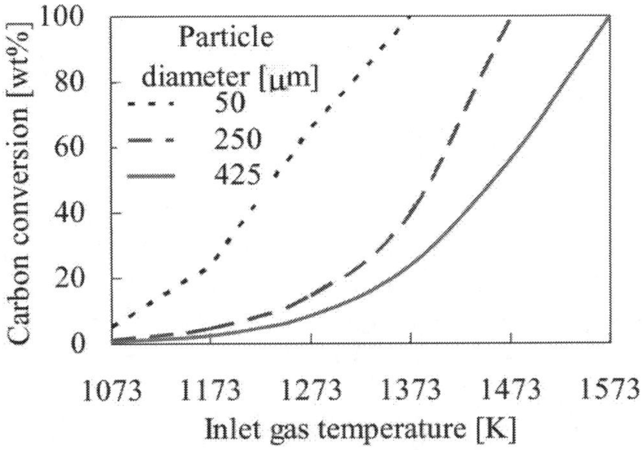

Figure 4: Effect of secondary gas temperature on carbon conversion for various sizes of char particle

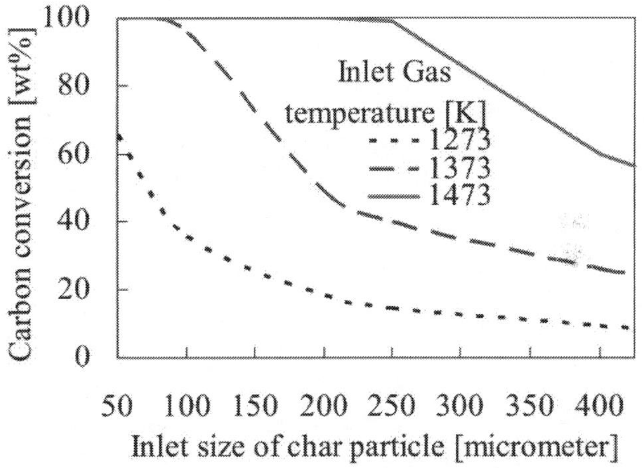

Figure 5: The changes of carbon conversion with inlet diameter of char particle at various secondary gas temperatures

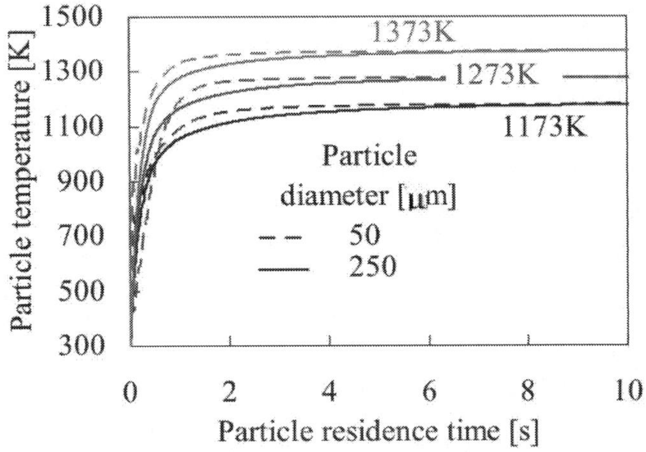

Figure 6: The change of particle temperature with particle residence time

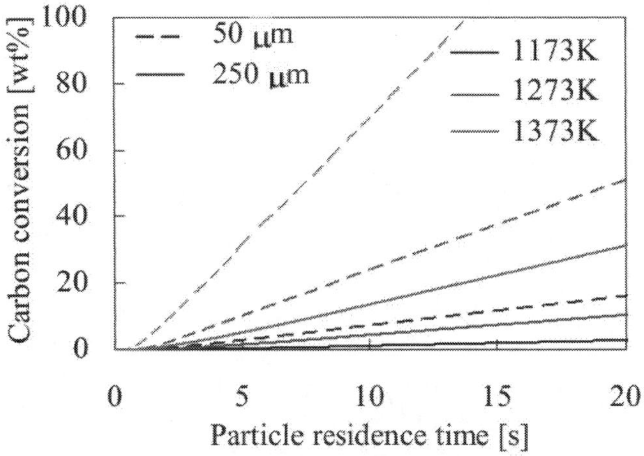

Figure 7: The change of carbon conversion with particle residence time

Change of particle temperature and conversion

The changes of char particle temperature with residence time for various sizes at various secondary gas temperatures are shown in Figure 6. It is found that the particle temperature also increases with increasing the gas temperature. However, at the same temperature, initially the rise in temperature for smaller particle is higher than the larger one. After certain time, the temperatures for both smaller and larger particle become same. The rapid increase in temperature for smaller particles makes the conversion of smaller particle faster than the larger one. As a result, at 1373K 100% carbon conversion of 50μm particle can be achieved after 15s (Figure 7). On the other hand, only 20% conversion is observed for the 250μm particle at that time. The trend is the same for all calculated conditions carried out in this study.

Concluding comments

The char-CO_2 gasification of Morwell coal char is studied at 1173 and 1273K under CO_2/N_2 atmosphere. Char conversion increases with increasing gas temperature and decreasing particle size. The comparison between experiment and calculation shows that the numerical modeling employed in this calculation is able to predict the char-CO_2 gasification accurately. Therefore, the predicted CO_2 gasification results can be used to design a new furnace for the desired gasification. Further work is in progress to refine the model with wider range of data from CO_2 and steam gasification.

Acknowledgement

This research is supported by Japan Society for the Promotion of Science (JSPS) doctoral fellowship for young researchers and GCOE, Novel carbon Resource Sciences, Kyushu University, and the Brown Coal Innovation Australia (BCIA) under the Victoria-Kyushu collaboration.

References

[1] Chui E.H. et al., Simulation of Entrained Flow Coal Gasification / Energy Procedia 1 (2009) 503–509.
[2] ABARE. Australian Energy Resource Assessment. Canberra: Geoscience Australia and ABARE; 2010.
[3] Allardice D.J. et al., Chapter 3 - Water in Brown Coal and Its Removal. Advances in the Science of Victorian Brown Coal, Amsterdam, Elsevier Science, 2004, 85-133.
[4] ABARE. Australian Energy Resource Assessment. Canberra: Geoscience Australia and ABARE; 2010.
[5] Ethan S. et al., Effect of CO_2 gasification reaction on oxy-combustion of pulverized coal char, Proceedings of the Combustion Institute 33 (2011) 1699–1706.
[6] Wijayanta A.T. et al., Numerical investigation on combustion of coal volatiles under various O_2/CO_2 mixtures using a detailed mechanism with soot formation. Fuel 93(2012), 670–676.
[7] Lian Z. et al., Experimental Investigation of the Combustion of Bituminous Coal in Air and O_2/CO_2 Mixtures: 1. Particle Imaging of the Combustion of Coal and Char, Energy Fuels, 24(2010), 4803–4811.
[8] A. Silaen and T. Wang. Int. J. Heat Mass Transfer, 53(2010)2074–2091.
[9] Smith I.W., The combustion rates of coal chars: a review, Proc. Combust. Inst. 19 (1982) 1045-1065.
[10] ANSYS FLUENT 12.0 User's Guide, ANSYS Inc. 2009-01-29
[11] Ranz W.E. and Marshall W.R., Chem. Eng. Prog., 48(3), 141-146, 1952.
[12] Ranz W.E. and Marshall W.R., Chem. Eng. Prog., 48(4), 173-180, 1952.

(128b) Kinetic and Mechanistic Studies of Microbial and Chemical Enhancement of Carbon Mineralization

Huangjing Zhao, Earth and Environmental Engineering and Chemical Engineering, Columbia University, New York, NY, Edris Taher Sr., Earth and Environmental Engineering, Columbia University, New York, NY, Kartik Chandran, Earth and Environmental Engineering, Columbia University and Ah-Hyung Alissa Park, Earth and Environmental Engineering & Chemical Engineering, Columbia University, New York, NY

Carbon mineralization is one of the safest methods of sequestering anthropogenic carbon dioxide. It is based on the exothermic reaction between carbon dioxide and the metal ions present in silicate minerals, such as antigorite, to form geologically and thermodynamically stable mineral carbonates. However, several challenges need to be overcome to successfully deploy this technology. In particular, the acceleration of the extremely slow mineral weathering step along with process optimization is essential to ensure economic feasibility as well as overall sustainability. In this study, the effect of various organic chelating agents on the dissolution rate of antigorite was explored to accelerate the weathering rate. First, the chelating agents were produced via the anaerobic digestion of food waste in a microbial reactor. As these catalysts which are enhancing the carbon storage are derived from a waste stream, their employment in the proposed carbon mineralization technology enhances the overall sustainability of the process. Magnesium extracted from antigorite was then reacted with CO_2 to form precipitated magnesium carbonates to mimic commercially available $CaCO_3$-based filler materials. The effects of various chelating agents, pH, reaction time, and reaction temperature on the mean particle size, particle size distribution, composition, and particle morphological structures of precipitated magnesium carbonates were also investigated.

(128c) CO2 Utilization: Combined Dry and Steam Reforming As an Alternative to Steam Methane Reforming

Preeti Gangadharan, Krishna Chaitanya Kanchi and Helen H. Lou, Dan F. Smith Department of Chemical Engineering, Lamar University, Beaumont, TX

Syngas platform for the production of fuels and chemicals, including methanol, dimethyl ether, ethylene, propylene and Fischer-Tropsch fuels, has been a topic of great interest in industry and academia in recent years. Steam methane reforming (SMR) which is the most widely applied method of producing syngas from natural gas, reacts natural gas with steam. Dry reforming of methane (DRM) uses waste carbon dioxide to produce syngas from natural gas. Dry reforming alone has not yet been implemented commercially; however, a combination of steam methane reforming and dry reforming of methane (SMR+DRM) has been used in certain plants for several years.

This paper deals with the simulation of both the SMR and SMR+DRM processes, and an evaluation based on economic and environmental criteria to determine whether the SMR+DRM process is competitive with the more popular SMR process. From the results, it is observed that the SMR+DRM process has a lower carbon footprint. And, with the development of more efficient DRM catalysts, this process could be economically competitive with steam methane reforming.

(128d) Liquid Fuels From Alternative Carbon Sources Minimizing Carbon Dioxide Emissions

Matthew J. Metzger[1], Benjamin J. Glasser[2], Bilal Patel[3], James A. Fox[4], Baraka Celestin Sempuga[5], Diane Hildebrandt[6] and David Glasser[6], (1)Chemical and Biochemical Engineering, Rutgers University, Piscataway, NJ, (2)Chemical and Biochemical Engineering, Rutgers, the State University of New Jersey, Piscataway, NJ, (3)Centre of Material and Process Synthesis, University of the Witwatersrand, Johannesburg, South Africa, (4)University of the Witwatersrand, Center of Material and Process Synthesis, Johannesburg, South Africa, (5)School of Chemical and Metallurgical Engineering, Centre of Material and Process Synthesis, University of the Witwatersrand, Johannesburg, South Africa, (6)Centre of Material and Process Synthesis (COMPS), University of the Witwatersrand, Johannesburg, South Africa

The energy needs of the world continue to grow, as does the resulting environmental impact. Policy makers continue to call for alternative energies to replace today's petroleum based liquid fuels. However, liquid fuels have significant advantages and it is probably unwise to abandon the existing infrastructure without appropriately exploring alternatives to lessen the environmental burden of producing liquid fuels. Biomass and coal are often proposed as alternatives to petroleum based carbon sources, but we demonstrate that those processes, by themselves, lose a significant amount of their potential product to unwanted carbon dioxide emissions. However, combining biomass and coal with cleaner natural gas yield processes with less environmental impact to produce liquid fuels with small, zero, or even negative, carbon dioxide emissions. Our process synthesis approach is applied to commonly encountered liquid fuel production methods in order to identify promising routes and to establish feasibility limits on those less promising alternatives.

(128e) Co-Pyrolysis/Gasification of Coal and Biomass

Oisik Das and Foster Agblevor, Biological Engineering, Utah State University, Logan, UT

Coal is a major primary energy source for USA, however, coal is also a major source of carbon dioxide and other associated pollutants. On the other hand, biomass is nearly carbon neutral and contain less pollutants than coal. Thus, by combining coal and biomass, potential both carbon dioxide and pollutant emissions may be reduced considerably. We investigated the pyrolysis/gasification of various biomass feedstocks (corn stover, hybrid poplar, and switchgrass) and sub-bituminous coal in both fluidized bed reactor, thermogravimetric analysis (TGA) and pyroprobe gas chromatographic analysis. Various ratios of coal to biomass were investigated. The TGA analysis showed that whereas the corn stover and and switchgrass appeared to influence the maximum weight loss at 400°C,the polar wood had minimal influence on this peak. In the case of the poplar, the coal and the wood appeared to behave independently. Data will also be presented on the co-pyrolysis/gasification in a fluidized bed reactor.

(129a) Role of Metal Halides in Enhancing the Dehydration of Xylose to Furfural: A 1H and 13C NMR Spectroscopy Study

Kristopher Enslow, Chemical and Biomolecular Engineering, University of California, Berkeley, Berkeley, CA and Alexis T. Bell, Department of Chemical and Biomolecular Engineering, University of California, Berkeley, Berkeley, CA

Role of Metal Halides in Enhancing the Dehydration of Xylose to Furfural: a ^{1}H and ^{13}C NMR Spectroscopy Study

Kristopher R. Enslow[1,2] and Alexis T. Bell[1,2*]

[1]University of California, Berkeley, California 94720-1462

[2]Energy Biosciences Institute, Calvin Laboratory,

Berkeley, California, USA 94720-1462

*bell@cchem.berkeley.edu

Introduction

High value fuel molecules (aromatics, alcohols, and alkanes) can be formed in a renewable and carbon-neutral manner from biomass [1]. These chemicals represent a wide range of fuel molecules that can be blended with gasoline and diesel or combined to create stand-alone fuels. A major component of biomass, hemicellulose (20 – 40 wt%) [1], is a complex amorphous polymer consisting of a backbone of b-1,4 glycosidically linked xylose sugars branched with arabinose, glucose, galactose, and mannose sugars. Previous work has shown that hemicellulose can be hydrolyzed to its monosaccharides [2] and those sugars can be either dehydrated to furanic chemicals [3] or hydrogenated to their respective alcohols [4]. The only large volume chemical produced industrially from biomass-derived carbohydrates is furfural, which is typically obtained through the concentrated acid-hydrolysis of agricultural waste, where selectivities never exceed 70% [5]. Other dehydration reaction systems have shown improvements over this conventional process, increasing furfural selectivity to 74% using β-zeolites (Si/Al = 12) [6], to 85% when employing a biphasic reactor (water/MIBK) under Brznsted-acidic conditions [7], and to 88% using metal halides with sulfuric acid in a biphasic reactor (water/toluene) [8]. While there have been proposed xylose dehydration mechanisms in the literature [9] suggesting the formation of 1,2-enediol (in the presence of metal halides), there has been little empirical evidence to support these mechanisms.

The objective of this study is to develop an understanding of the role metal halides play in the dehydration of xylose to furfural under acidic conditions and to elucidate a reaction mechanism supported by empirical NMR data. The kinetics of xylose dehydration in the presence of metal halides were also investigated.

Materials and Methods

Xylose (99%, from Sigma-Aldrich) was used as a reagant and standard (for quantification), and 2-furaldehyde (furfural, 99%, from Sigma-Aldrich) was used as a standard (for quantification). All salts were used as purchased from Fisher-Scientific. Typical reaction: 5 wt% of xylose was dissolved in water in an HEL high pressure Chem-SCAN autoclave at 180 °C while stirred at 1000 rpm. Toluene was added in 4:1 v/v ratio with water, and 0.5 M of salt and 50 mM H_2SO_4 were added to start the reaction.

Discussion

While adding metal halides to aqueous/organic biphasic systems are known and have been shown (see Table 1 below) to enhance the dehydration of xylose to furfural under acidic conditions, the role that the salt plays is not entirely known.

Time, min	No Salt	NaCl	KCl	NaBr	KBr	$CaCl_2$
15	25%	61%	69%	68%	67%	77%
30	32%	72%	76%	75%	80%	87%
60	25%	68%	73%	70%	78%	85%

Table 1. Yields of furural from xylose in a biphasic water/toluene system (1:4 v/v) at 180¼C with 50 mM H_2SO_4 and 500 mM salt.

The presence of these metal halides in solution may lead to an interaction, or complexation, with xylose that promotes the stablization of xylose transition states that exist in the dehydration pathway to furfural. These salts could also be salting out furfural from the aqueous phase by making that phase less hydrophillic and altering the partition coefficient between the aqueous and organic phases in favor of toluene (keeping the xylose separate from the furfural has been shown to increase furfural yields by barring the degradative coupling reactions that lead to humins formation [10]). With these two fields of thought on the role of metal halides, NMR experiments have been performed looking specifically at how changes in the concentration of metal halides effect the chemical shift of the protons of xylose and the water proton. Comparing these changes in chemical shift provides a qualitative picture of how and to what degree these salts interact with xylose and how this compares to the interaction of these salts with water. NMR labelling has been used to determine whether proton transfer occurs to initiate xylose dehydration to furfural. With this information, a reaction mechanism has been proposed, the role of metal halides has been discovered, and the reaction kinetics associated with xylose dehydration have been determined.

References

1. Huber, G.W., Iborra, S., and Corma. *A. Chem. Rev.* 106, 4044 (2006).

2. Wyman, C.E., et al. in "Hydrolysis of Cellulose and Hemicellulose (Polysaccharides: Structural Diversity and Functional Versatility)" (S. Dumitriu, Ed.) Vol. 2 p. 995. Marcel Dekker, New York, 2005.

3. Chheda, J.N, Roman-Leshkov, Y., and Dumesic, J.A. *Green Chemistry* 9, 342 (2007).

4. Wisniak, J., et al. *Ind. End. Chem. Prod. Rd.* 13, 75 (1974).

5. Moreau, C., et al. *Industrial Crops and Products*. 7 (2-3), 95 (1998).

6. Lima, S., et al. *Appl. Catal., A.* 388, 141 (2010).

7. Weingarten, R., et al. *Green Chem.* 12, 1423 (2010).

8. Marcotullio, G., and Jong, W. D. *Green Chem.* 12, 1739 (2010).

9. Marcotullio, G., and Jong, W. D. *Carb. Res.* 346, 1291 (2011).

10. Sievers, C., et al. *ChemSusChem* 2 (7): 665-671 (2009).

(129b) Design of Solid Acid Catalysts for Aqueous Phase Dehydration of Carbohydrates: The Role of Lewis and Brønsted Acid Sites

Ronen Weingarten, Joungmo Cho, Geoffrey A. Tompsett, Wm. Curtis Conner Jr. and George W. Huber, Department of Chemical Engineering, University of Massachusetts Amherst, Amherst, MA

Aqueous phase acid-catalyzed reactions offer a viable route for the production of liquid fuels and fuel precursors from lignocellulosic biomass. In particular, dehydration of carbohydrates makes up a principal class of these reactions for the production of renewable chemicals, including furfural and 5-Hydroxymethylfurfural. These furan based chemicals can be used to make a wide variety of chemical feedstocks[1] and intermediates for production of liquid alkanes, including jet and diesel fuel range alkanes, and other furan-based chemicals.[2-5] We have studied the kinetics of the aqueous phase dehydration of xylose in a biphasic reaction using a homogeneous catalyst.[6] Xylose undergoes a triple dehydration reaction to form furfural. The furfural can react further with itself to form solid humins, which are highly polymerized insoluble carbonaceous species. Furfural can also react with xylose to form humin products, which is the primary pathway to humin formation. Humin formation is a low activation energy process, whereas xylose dehydration is a high activation energy step. A biphasic system is able to quickly extract the furfural from the aqueous phase and prevent further decomposition reactions. Thus high yields of furfural (i.e. 85%) are obtained at high temperatures in biphasic systems.

It would be desirable to obtain solid acid catalysts that exhibit activities and selectivities comparable to homogeneous catalysts for aqueous phase dehydration. In this respect we have prepared a series of well characterized solid acid catalysts and tested them for aqueous phase dehydration. These catalysts include Zr-P, SiO_2-Al_2O_3, WO_x/ZrO_2, Al_2O_3, and HY zeolite. We have characterized the concentration of both Brønsted and Lewis acid sites in these catalysts and compared the catalytic activity and selectivity with that of homogeneous catalysts. This in turn enables the design and optimization of solid acid catalysts for the aqueous phase dehydration of carbohydrates in general and xylose in particular. The catalyst selectivity is a function of the Brønsted to Lewis acid ratio for both the heterogeneous and homogeneous reactions. Lewis acid sites decrease furfural selectivity by catalyzing a side reaction between xylose and furfural to form humins. Catalysts with high Brønsted to Lewis acid ratios, such as Zr-P and ion-exchange polymer resins, have comparable results with HCl. Using HY zeolite revealed a low furfural selectivity due to strong irreversible adsorption of the furfural in the pores, causing an increase in the rate of humin formation. Thus, to design more efficient aqueous phase dehydration catalysts, it is desirable to have a high ratio of Brønsted to Lewis acid sites. Furthermore, we have shown that gas phase characterization of acid sites can be used to predict catalytic activity in the aqueous phase.

References

1. R. H. Kottke, in *Kirk-Othmer Encyclopedia of Chemical Technology*, John Wiley & Sons, Editon edn., 2000, vol. 12, pp. 259-286.

2. S. Bayan and E. Beati, *La Chimica e L'Industria*, 1941, **23**, 432-434.

3. G. W. Huber, J. N. Chheda, C. J. Barrett and J. A. Dumesic, *Science*, 2005, **308**, 1446-1450.

4. R. West, Z. Liu, M. Peter and J. Dumesic, *ChemSusChem*, 2008, **1**, 417-424.

5. R. Xing, A. V. Subrahmanyam, H. Olcay, W. Qi, G. P. van Walsum, H. Pendse and G. W. Huber, *Green Chemistry*, 2010, **12**, 1933-1946.

6. R. Weingarten, J. Cho, W. C. Conner Jr and G. W. Huber, *Green Chemistry*, 2010, **12**, 1423-1429.

(129c) One-Pot Homogeneous Synthesis of Furfural From High Purity Xylulose in Aqueous Media

Bin Li, Bioengineering, The University of Toledo, Toledo, OH, Sasidhar Varanasi, Department of Chemical and Environmental Engineering, The University of Toledo, Toledo, OH and Patricia Relue, Department of Bioengineering, The University of Toledo, Toledo, OH

As prices of and environmental concerns over fossil energy increase, the utilization of renewable biomass as a source of energy and chemicals has received considerable attention. Furfural is an important chemical that can be obtained from biomass. Furfural can be used itself as a drop-in fuel or as a precursor to many important chemicals and fuels, such as furfuryl alcohol and jet and diesel fuel.

Much research in furfural production has focused on the dehydration of xylose through homogeneous or heterogeneous catalysis in either water or ionic liquid solvents. Furfural production from xylose is typically catalyzed under high temperature and pressure to minimize side-product formation and increase reaction kinetics and furfural yield.

Some pathways to furfural suggest xylulose as an intermediate in the conversion of xylose to furfural. The isomerization of xylose to xylulose can be enzymatically-catalyzed under temperature and pH conditions incompatible with dehydration, but suffers from an unfavorable xylulose yield. To date, no one has reported high-yield production of furfural directly from xylulose due to the excessive cost of high purity xylulose. Using our previously developed method for producing concentrated xylulose at high purity, we report here high-yield, high-selectivity conversion of xylulose to furfural through homogeneous synthesis under relatively low temperature conditions with minimal side-product formation.

(129d) Catalytic Conversion of Xylan to Value-Added Carboxylic Acids

Hongfei Lin, Chemical and Materials Engineering, University of Nevada, Reno, Reno, NV

The success of biofuel production relies on the effective utilization of the C5 and C6 sugars presented in hemicellulose and cellulose, respectively. Both hemicellulose and cellulose are carbohydrates with different structures and sugar monomer units. They share some similarities but significant differences exist. There are only few studies on hemicellulose partly because C5 sugars, the monomers of hemicellulose, cannot be utilized by fermentation to produce bio-ethanol. Hemicellulose and its derivatives have long been considered as the low-value by-products which are not well utilized in biorefinery industry. The aqueous phase partial oxidation (APPO) method has been demonstrated to produce levulinic acid in high yields from cellulose in our group. Since cellulose and hemicellulose are both composed of sugar monomer units through glycosidic bonds, the APPO technology herein was applied to convert hemicellulose to carboxylic acids. However, unlike cellulose with a rigid crystalline structure and strong hydrogen bonding, hemicellulose is amorphous and can be easier to break down. The APPO of hemicellulose was therefore operated at a lower temperature. And the reaction mechanism of the APPO of hemicellulose was different from the cellulose reaction mechanism. The APPO of xylan in hot temperature water is a multi-step process: firstly, xylan is hydrolyzed to xylose, and xylose is converted to furfural by dehydration, and then is further oxidized to form carboxylic acids over the heterogeneous catalyst. We found that the product distribution is highly depended on the catalysts.

(129e) Production of 5-Hydroxymethylfurfural From Glucose Using Water Compatible Lewis Acid Catalysts in a Biphasic Reactor with an Alkylphenol Solvent

Tianfu Wang, Iowa State University, Ames, IA, Jean Marcel Gallo, University of Wisconsin,Madison, Madison, WI, James A. Dumesic, Chemical and Biological Engineering, University of Wisconsin-Madison, Madison, WI and Brent H. Shanks, Chemical and Biological Engineering, Iowa State University, Ames, IA

The synthesis of 5-Hydroxymethylfurfural (HMF), a potential key biomass-derived platform chemical for the production of plastics, solvents, and transportation fuels, represents an important challenge in the economical utilization of lignocellulosic biomass. We have demonstrated that in the presence of both homogeneous Lewis (e.g., $AlCl_3$) and Bronsted acids (e.g., HCl), glucose could first isomerize to fructose catalyzed by a Lewis acid salt and then the fructose could subsequently dehydrate to HMF catalyzed by a Bronsted acid. A biphasic system with an alkylphenol solvent, sec-butyl phenol (SBP) was used to continuously extract the HMF produced from the aqueous phase into the organic phase, thereby minimizing HMF degradation. With this multifunctional reaction systems design, high HMF yields (~60%) were achieved. While a low solution pH is required for most conventional Lewis acid salts such as aluminum to maintain their Lewis acidity in the presence of water, we will also discuss a serious of lanthanide metal salts could function as water-compatible Lewis acids under a wide range of pH values. Comparison of different lanthanide metal salts showed that decreasing glucose reactivity to HMF occurred with increasing ionic radii from Yb to Dy to La likely due to the larger cations having weaker electrostatic interaction with the glucose molecule.

(129f) Bi-Functional Catalysts: A Path to Improved Catalytic Performance for the Conversion of Biomass Feedstocks

Brandon J. O'Neill, Mei Chia, David Jackson, Thomas F. Kuech and James A. Dumesic, Chemical and Biological Engineering, University of Wisconsin-Madison, Madison, WI

The conversion of sustainably derived bio-feedstocks to fuels and chemicals generally involves myriad individual reactions. To effectively realize the goal of an integrated bio-refinery for the conversion of biomass, a strategy for intensifying the number of catalysts, conditions, and separations involved in moving from plant to product without sacrificing overall yield is required. One strategy involves the development of bifunctional catalysts capable of performing two or more distinct conversions, ideally more selectively than if the two catalytic functionalities were kept discreet.

One type of bifunctional catalyst is the combination of a traditional noble metal catalyst with a more oxophilic metal. As opposed to traditional bimetallic catalysts, where the addition of another metal merely augments the rate or selectivity of the reaction, for example by reducing the binding of catalyst poisons, a bifunctional catalyst introduces an additional reactive site to the catalyst. A RhRe/C catalyst employs Rh as a traditional hydrogenation catalyst, and utilizes Re in a partially oxidized state (as $Re(OH)_x$) to introduce an acid active site. This catalyst has been used successfully for selective ring opening of compounds like 2-methylpyran and has shown acidity by dehydrating fructose to hydroxymethylfurfural. The inherent proximity of the active sites in a bifunctional catalyst allows for the potential to increase selectivity by decreasing side reactions of intermediates or allowing the quick conversion of otherwise unstable intermediates. For example, it will be shown that the use of a RhRe/C catalyst can improve the selectivity of the reaction of furfuryl alcohol to 2-methylfuran compared to a Rh/C catatlyst paired with either a heterogeneous or homogeneous acid catalyst. Additionally, the $Re(OH)_x$ functionality provides a way to augment the acidity of the catalyst by providing an anchor site for atomic layer deposition (ALD) of different functionalities, such as Nb_2O_5-xH2O or $-SO_3H$. The augmentation of the acid site in the bifunctional catalyst provides another handle with which rate and selectivity and be manipulated.

Utilizing ALD provides another route to bifunctional catalysts. The atomic layer control of ALD allows for the careful addition of the extra functionality. For example, ALD overcoating of Cu catalysts with alumina prevents the leaching of Cu in the liquid phase hydrogenation of furfural to furfuryl alcohol. However, it will be shown that by utilizing ALD Bronsted acidity can be added to the overcoating by sandwiching a layer of Nb_2O_5 among the alumina overcoat layers. This Bronsted acidity introduces a bifunctional site that allows for the hydrogenation of furfural to furfuryl alcohol on the Cu and the alcoholysis of furfuryl alcohol to levulinate esters. The bifunctionality of this catalysts allows for this conversion over a single catalyst and prevents the accumulation of the very reactive furfuryl alcohol intermediate, thus improving overall yield compared to pairing Cu with a homogeneous acid or a discrete heterogeneous solid acid in either the same or separate reactors.

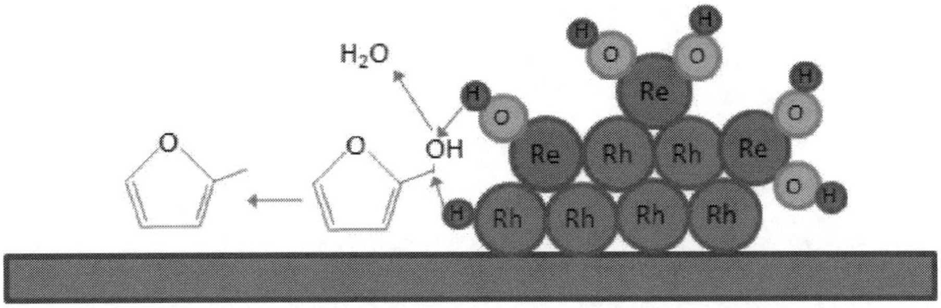

Figure 1: A potential graphical representation of how Rh and $Re(OH)_x$ work in tandem to produce 2-methylfuran from furfuryl alcohol.

(139a) Engineering Metabolic Modules for Electrofuels Production in Proteobacteria

Curt R. Fischer, JT Sauls, John Sexton, Reshma Shetty and Jason Kelly, Ginkgo Bioworks, Boston, MA

Electrofuels as a concept seeks to produce liquid transportation fuels from carbon dioxide and renewable electricity. Ginkgo Bioworks's approach to electrofuels is to use formic acid salts as a vector of electricial energy for biology-based fuels production. Unlike other vectors such as hydrogen, formate is water-soluble and easy to handle safely, which simplifies process scale up. We have designed and are building a set of metabolic modules aimed at (i) converting the electrons carried by formate into central metabolic intermediates such as NAD(P)H, (ii) converting the carbon carried by formate into central metabolic intermediates such as pyruvate, and (iii) producing liquid, water-immiscible hydrocarbons from pyruvate. Together these modules enable the biosynthesis of liquid fuels from formate salts.

Recently we have also begun to implement these modules in alternate host organisms (alphaproteobacteria instead of *E. coli*). These hosts offer a number of potential advantages over *E. coli* but require the fabrication and characterization of synthetic biology "parts" including promoters, plasmid replication and selection elements, etc. that will also be discussed.

(139b) Metabolic Engineering of Clostridium Ljungdahlii for Production of Fuels and Other Chemicals From Carbon Dioxide Via Microbial Electrosynthesis or with Carbon Monoxide or Hydrogen As the Electron Donor

Toshiyuki Ueki, Ching Leang, Kelly P. Nevin and Derek R. Lovley, Department of Microbiology, University of Massachusetts, Amherst, MA

Clostridium ljungdahlii is an acetogenic microorganism that has the natural ability to use hydrogen as the electron donor to reduce carbon dioxide as its terminal electron acceptor with the production of acetate that is excreted from the cell. Carbon monoxide can also serve as an electron donor for acetate production. Under some conditions ethanol and butanediol are produced in addition to acetate, but usually in lesser quantities. Recent studies in our laboratory have also demonstrated that *C. ljungdahlii* can directly accept electrons from a negatively poised electrode in order to reduce carbon dioxide to acetate. The ability of microorganisms like *C. ljungdahlii* to convert carbon dioxide to organic products with electrons derived from an electrode as the driving force is known as microbial electrosynthesis. Carbon dioxide is reduced to organic products in *C. ljungdahlii* via the Wood-Ljungdahl pathway. Acetyl-CoA is a key intermediate in the Wood-Ljungdahl pathway, which offers the possibility of genetically modifying *C. ljungdahlii* to make a diversity of products more valuable than acetate. Techniques for genetic manipulation of *C. ljungdahlii* were developed which permitted deleting genes from, or inserting genes into, the chromosome. Initial gene manipulation studies focused on the mechanisms for hydrogen uptake and identified a single hydrogenase gene that was required for growth on hydrogen. This hydrogenase-deficient strain continued to reduce carbon dioxide on negatively poised electrodes, demonstrating that hydrogen was not an intermediate in electron transfer from electrodes to *C. ljungdahlii.* Introducing the appropriate genes and/or deleting genes for unwanted pathways made it possible to divert metabolism in *C. ljungdahlii* away from acetate and ethanol production and toward the production of more desirable products. A strain for the production of the transportation fuel butanol was developed, as were strains that can produce other multi-carbon organic commodities from carbon dioxide with an electrode or hydrogen serving as the energy source. Furthermore, these strains could ferment sugars to desired products. These results demonstrate that *C. ljungdahlii* is a genetically tractable chassis organism for the production of fuels and other commodities from a diversity of renewable energy sources.

(139c) Bioelectrosynthetic Reduction of Carbon Dioxide to Biofuel Precursors Using a Mixed Microbial Consortium Isolated From Bovine Rumen

Ryan Hammonds[1], Kristi Whitehead[1], Stephen Creager[2], Jung Oh[3] and J. Michael Henson[1], (1)Biological Sciences, Clemson University, Clemson, SC, (2)Clemson University, Anderson, SC, (3)Chemistry, Clemson University, Clemson, SC

Microbially-catalyzed reduction of carbon dioxide to higher-molecular weight compounds through consumption of electrical current offers prospects for electrochemical production of renewable liquid transportation fuels. Recently, pure cultures of chemoautotrophic microorganisms were reported to produce electrofuel precursors, such as acetate, from carbon dioxide and electrical current. A consortium consisting of several different microbial populations acting as a stable community might offer an energetically favorable approach for production of these electrofuels. In order to evaluate this concept, bovine rumen contents were enriched under strict anaerobic conditions for autotrophic, acetogenic microbes for several generations using hydrogen and carbon dioxide. A stable acetogenic consortium was obtained and transferred to bioelectrosynthesis cells supplied with electric current from a potentiostatic controller. Current consumption and biofuel precursor production were measured. Initial electron recovery as biofuel precursors exceeded 80% efficiency. To further improve rates and efficiency of biofuel precursor production, a combinatorial approach to evaluate growth conditions, active microbial community structure, and reactor geometry is in progress.

(139d) Electroautotrophic Synthesis of Acetate and Methane

Christopher W. Marshall[1], Edward V. LaBelle[1], Erin J. Fichot[2], Daniel E. Ross[2], R. Sean Norman[2] and Harold D. May[1], (1)Marine Biomedicine & Environmental Science Center, Medical University of South Carolina, Charleston, SC, (2)Environmental Health Sciences, University of South Carolina, Columbia, SC

Electroautotrophic synthesis of chemicals from CO_2 offers an attractive way of producing fuels that would be sustainable, domestic and environmentally benign, particularly in relation to climate change. Autotrophic rates of acetate and methane production by microbes with hydrogen gas as an electron donor have been demonstrated in the multiple kg m^{-3} h^{-1} range. However, this approach requires a source of hydrogen. Generation of hydrogen by electrolysis, or direct delivery of electrons through an electrode, could serve as the source of hydrogen if sustainable energy, such as solar or wind, is used to drive the production of electrical energy. The overall method becomes even more attractive if the use of rare earth elements is avoided and only microbial catalysts are employed. This approach is especially advantageous if the microbes operate within the electrolysis/electrochemical cell. Using a community of microorganisms selected from a brewery waste we have produced high amounts of methane and acetate from reductive electrical current in electrochemical cells with only microbial catalysts. The data thus far indicate that the generation of both products is dependent upon the biological catalysts associated with the electrodes, and the production rates observed for both products exceed any published rates as of the time of this submission. The electromethane could readily be dropped into existing natural gas infrastructure and used as a transportation fuel. Acetate is a commodity chemical but is also a building block that may be used to produce other fuels. We are attempting to convert that acetate into alcohols (ethanol and butanol) with communities of microbes selected within the electrochemical cells. The rates of production, electrochemical data associated with the catalysis, reactor design, microbial catalyst identification and community characteristics (examined metagenomically) will be discussed.

Integrated Electromicrobial Conversion of CO_2 to Higher Alcohols

Han Li,[1,2] Paul H. Opgenorth,[3] David G. Wernick,[1] Steve Rogers,[4] Tung-Yun Wu,[1] Wendy Higashide,[4] Peter Malati,[5] Yi-Xin Huo,[4] Kwang Myung Cho,[4] James C. Liao[1,2,3,6]*

A man-made photovoltaic device is relatively efficient in converting sunlight to electricity, but the electrical energy generated is difficult to store. The biological photosystems, on the other hand, are limited by the intrinsic design and biomaterials available, for which no near-term improvements are in sight (*1*). One way to circumvent both problems is to link man-made solar cells to biological CO_2 fixation and fuel production. Theoretically, H_2 generated by solar electricity can drive CO_2 fixation in lithoautotrophic microorganisms engineered to synthesize high energy-density liquid fuels. However, the low solubility, low mass transfer rate, and safety issues of H_2 in microbial cultures limit the efficiency and scalability of such processes.

Compared with H_2, formic acid is a favorable energy carrier. Electrochemical production of formic acid from CO_2 and H_2O can achieve relatively high efficiency (*2*, *3*). Formate is highly soluble and is readily converted to CO_2 and NADH in the cells, providing a safe replacement for H_2. However, the high solubility of formate increases the cost of separation. Accumulated formate will decompose at the anode, decreasing the yield of the process (*2*). Therefore, simultaneous electrochemical formate production and biological formate conversion to higher alcohols is desirable (Fig. 1A). Unfortunately, introduction of electricity to microbial cultures may impede cell growth.

As such, an integrated process for reduction of CO_2 to liquid fuel powered by electricity requires (i) metabolic engineering of a lithoautotrophic organism to produce liquid fuels, (ii) electrochemical generation of formate from CO_2 in fermentation medium, and (iii) enabling microbes to withstand electricity. In this work, we chose *Ralstonia eutropha* H16 as the production host and isobutanol and 3-methyl-1-butanol (3MB) as the target fuels, which can be used in the internal combustion engines. We introduced the set of genes previously reported (*4*, *5*) for isobutanol and 3MB production into *R. eutropha* H16 (fig. S1, A to E). We also disrupted polyhydroxybutyrate synthesis (fig. S1, A and B) and used the heterologous isobutanol and 3MB production pathway as the new metabolic sink for carbon and reducing equivalents. In a pH-coupled formic acid feeding fermentor, the engineered strain LH74D produced fuels with the final titer of over 1.4 g/l (~846 mg/l isobutanol and ~570 mg/l 3MB) (fig S2B).

Next, we used an In foil cathode and a Pt anode in the culture medium bubbled with air containing 15% CO_2 to produce formate electrochemically (Fig. 1A). However, growth of *Ralstonia* was inhibited when electric current was introduced. Studies using *Escherichia coli* showed transient inhibition of electric current on cell growth (fig. S3). When the current stopped, cell growth resumed, suggesting that unstable compounds such as reactive oxygen or nitrogen species might be responsible for the growth inhibition.

To test this hypothesis, we constructed three *Ralstonia* strains with different reporter plasmids to detect the presence of reactive oxygen and nitrogen species. The plasmids each contained a *lacZ* gene driven by a promoter of *Ralstonia katG*, *sodC*, or *norA* gene, genes which have been shown to be induced by hydrogen peroxide (H_2O_2), superoxide free radicals (O_2^-), and nitric oxide (NO), respectively (*6*–*8*). When the plasmid-bearing strains were exposed to electrolysis, expression of β-galactosidase from *sodC* and *norA* promoters were induced but not from the *katG* promoter (Fig. 1B). These results suggested that O_2^- and NO trigger a stress response in *Ralstonia* cells and inhibit growth. To circumvent this toxicity problem, a porous ceramic cup was used to shield the anode (Fig. 1A). This inexpensive shield provides a tortuous diffusion path for chemicals. Therefore, the reactive compounds produced by the anode may be quenched before reaching the cells growing outside the cup. Using this approach, healthy growth of *Ralstonia* strain LH74D and production of over 140 mg/l biofuels were achieved with the electricity and CO_2 as the sole source of energy and carbon, respectively (Fig. 1C).

This integrated process to convert CO_2 to liquid fuels does not depend on biological "light reactions." Electricity generated from photovoltaic cells, wind turbines, or off-peak grid power sources can be used to drive CO_2 fixation and fuel production. Thus, this process provides a way to increase photosynthetic efficiency by coupling man-made photoelectric generation device with biological CO_2 fixation and fuel production capability. The approach demonstrated here can also be applied to produce other compounds, thus opening the possibility of electricity-driven bioconversion of CO_2 to a variety of chemicals. Furthermore, because formate is a by-product of chemical dehydration and conversion of biomass (*9*), transformation of formate into liquid fuel will also play an important role in the biomass refinery process.

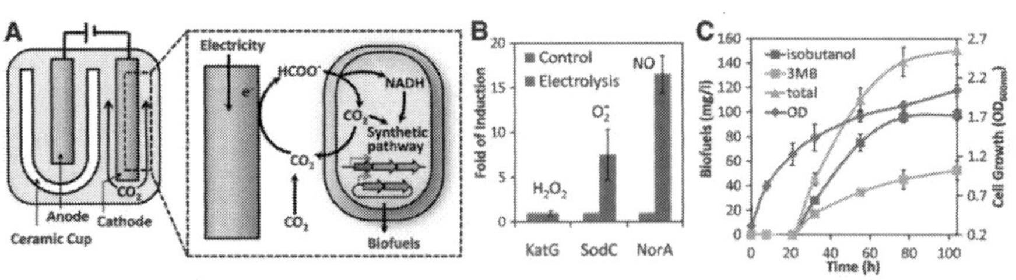

Fig. 1. An integrated electromicrobial process to convert CO_2 to higher alcohols. (**A**) Electricity powered the electrochemical CO_2 reduction on the cathode to produce formate, which is converted to isobutanol and 3MB by the engineered *R. eutropha.* (**B**) *R. eutropha* strains harboring a reporter gene driven by promoters that respond to reactive oxygen and nitrogen species showed increased reporter activity upon electrolytic exposure. (**C**) Engineered strain *R. eutropha* LH74D (supporting online material) showed healthy growth and produced over 140 mg/l biofuels in the integrated electromicrobial reactor with electricity and CO_2 as the sole energy and carbon sources, respectively. Error bars indicate SD (n = 3).

References and Notes

1. R. E. Blankenship *et al.*, *Science* **332**, 805 (2011).
2. K. S. Udupa, G. S. Subramanian, H. V. K. Udupa, *Electrochim. Acta* **16**, 1593 (1971).
3. S. Ikeda, T. Takagi, K. Ito, *Bull. Chem. Soc. Jpn.* **60**, 2517 (1987).
4. S. Atsumi, T. Hanai, J. C. Liao, *Nature* **451**, 86 (2008).
5. S. Atsumi *et al.*, *Appl. Microbiol. Biotechnol.* **85**, 651 (2010).
6. D. Touati, *J. Bacteriol.* **170**, 2511 (1988).
7. C. Michán, M. Manchado, G. Dorado, C. Pueyo, *J. Bacteriol.* **181**, 2759 (1999).
8. A. Pohlmann, R. Cramm, K. Schmelz, B. Friedrich, *Mol. Microbiol.* **38**, 626 (2000).
9. H. Mehdi *et al.*, *Top. Catal.* **48**, 49 (2008).

Acknowledgments: This work is supported by the Electrofuel program of Advanced Research Projects Agency–Energy (ARPA-E).

Supporting Online Material

www.sciencemag.org/cgi/content/full/335/6076/1596/DC1
Materials and Methods
Figs. S1 to S4
Tables S1 and S2
References (*10*–*22*)

8 December 2011; accepted 14 February 2012
10.1126/science.1217643

[1]Department of Chemical and Biomolecular Engineering, University of California, Los Angeles, CA 90095, USA. [2]The Molecular Biology Institute, UCLA, Los Angeles, CA 90095, USA. [3]Department of Chemistry and Biochemistry, UCLA, Los Angeles, CA 90095, USA. [4]Easel Biotechnologies, Los Angeles, CA 90024, USA. [5]Department of Materials Science and Engineering, UCLA, Los Angeles, CA 90095, USA. [6]Institute of Genomics and Proteomics, UCLA, Los Angeles, CA 90095, USA.

*To whom correspondence should be addressed. E-mail: liaoj@seas.ucla.edu

Rhodobacter as a platform for autotrophic fuel production

Nymul Khan[1], Alex Rajangam[1], Eric Nybo[2], Joe Chappell[2], Wayne Curtis[1]

[1] *Department of Chemical Engineering, Pennsylvania State University, University Park, PA*

[2] *Department of Plant and Soil Sciences, University of Kentucky, Lexington, KY*

Overview

Our overall ARPA-E project goal seeks to produce a renewable hydrocarbon feedstock by the genetic engineering of a C_{34} branched hyrdrocarbon of Botrococcus braunii into *Rhodobacter capsulatus*. We have shown the ability of this genetically engineered autotrophic organism to produce C_{30+} isoprenes from H_2, O_2 and CO_2. My specific component of this project has been to examine fundamantal kinetics of autotrophic growth towards the production of bioreactactor systems and multiplexed autotrophic growth capabilities for screening the genetically engineered bacteria.

Introduction

Liquid fuels are an essential part of today's economy. The transportation sector is the largest consumer of liquid fossil fuels (13.6 million bbl/day in 2011, www.eia.gov/forecasts/aeo/MT_liquidfuels.cfm) and the second largest source of CO_2 emission (1.9 billion metric tons in 2008, http://www.eia.gov/oiaf/1605/ggrpt/carbon.html) after power plants. Coal and natural gas account for 67% of US electricity generation and together generate the major portion of the CO_2 (about 2.3 billion tons in 2008, http://www.eia.gov/oiaf/1605/ggrpt/carbon.html). So, besides being an unsustainable source of energy, consumption of fossil fuels has high environmental impact due to the generation of CO_2. Furthermore, it is estimated that the demand for liquid fuels for transportation sector is likely to increase during the next 20-30 years (Figure 1).

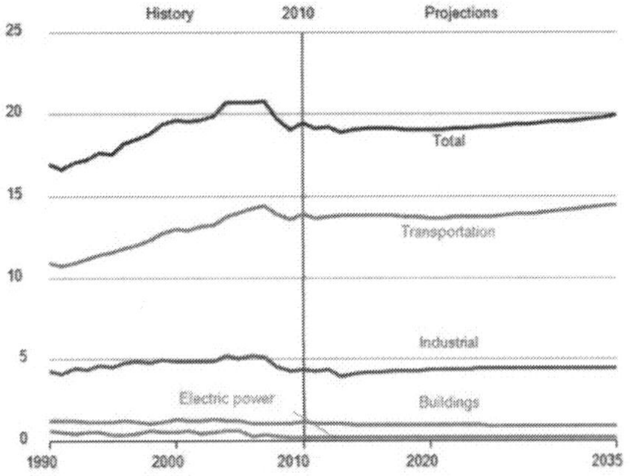

Figure 1: Consumption of petroleum and other liquids by sector, 1990-2005 (million bbl/day). http://www.eia.gov/forecasts/aeo/MT_liquidfuels.cfm#uses

In comparing liquid fuel products to other energy storage forms, it is apparent that a liquid hydrocarbon fuel is a much more efficient energy carrier by weight and volume than thermal, mechanical, battery and hydrogen energy storage. It is probably not possible to replace liquid fuels for many current end uses. Therefore, there is an urgent need to develop a renewable source of liquid fuels. Our proposed process is to take the distributed energy of solar radiation and store it in a liquid fuel molecule that can be easily used in the existing infrastructure of refining, storage and distribution. Figure 2 shows a schematic of the process. The electricity generated by photovoltaics will be used to split water to produce H_2 and O_2. These gases plus the CO_2 from power plants or other point sources will be used as the major substrates to grow an autotrophic bacterium (*Rhodobacter capsulatus*), genetically engineered to produce a liquid fuel, botryococcene.

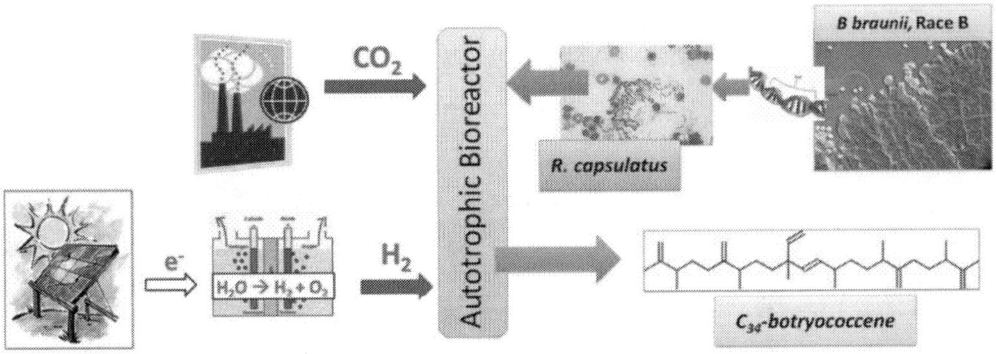

Figure 2: Schematic of the proposed biofuel production process in the context of supporting technologies of photovoltaics (P.V.) and carbon capture.

Botryococcenes are a group of C_{34}-branched hydrocarbons naturally produced by the micro-algae *Botryococcus braunii* Race B. These oxygen-free molecules are extremely energy dense (43.8 KJ/g as compared to ethanol at 29.7 KJ/g). As a hydrophobic long chain hydrocarbon, it easily separates as a distinct layer from water and has characteristics very similar to crude oil (Hillen, Pollard, Wake, & White, 1982). Hillen and co-workers also showed that botryococcene can be hydrocracked with a commercial catalyst to produce 67% gasoline, 15% aviation, 15% diesel and 3% residual oil. The octane rating of the gasoline fraction was much higher than natural crude and very low nitrogen and sulfur levels to affect the catalyst operation. As such, it can be a drop-in feedstock to the current petroleum industry. In fact, *B. braunii* oils appear to have contributed as much as 1.4% to the total hydrocarbons in oil shales (McKirdy & Cox, 1986).

Non-photosynthetic growth of organisms on CO_2 has been observed for a long time (Kelly, 1992) and has been proposed as a platform for the production of single cell protein, including an interface to H_2 production by electrolysis (Aragno & Schlegel, 1992). *Rhodobacter capsulatus* is a purple non-sulfur bacterium which has an extremely versatile metabolic capacity for growth under variety of growth conditions including aerobic heterotrophic, anaerobic photoheterotrophic and dark autotrophic using H_2 as the electron donor, O_2 as the electron source and CO_2 as the carbon source (Madigan & Gest, 1979). It has been chosen as the organism for this work for both its diverse metabolic capacity and our previous experience of working with this organism for membrane protein expression.

An autotrophic process is quite unique in its requirement for multiple gas phase substrates. While traditional aerobic fermentation processes are already limited by gas-liquid mass transfer, it is not difficult to imagine that that will be the limiting factor in this process as well. To make a rough comparison, while it takes ~0.7 L O_2 to grow 1 g of *E. coli*, it would take ~12.5 L of gas mix (in stoichiometric composition) to grow 1 g of *R. capsulatus* (based on yield of *R. capsulatus* on the different gases, Madigan & Gest 1979). This poses some unique restrictions on the bioreactor and genetic engineering of the organism. For example, from a preliminary calculation using the yield of the bacteria on the different gas species, the average cost of electricity and gasoline it was immediately clear that the growth-associated fuel production is not a feasible scenario for this process. Figure 3 shows a comparison of just the electricity cost per gallon of fuel corresponding to the fraction of fuel accumulated per unit biomass. Even at a very high fraction of fuel accumulation the cost is exceedingly high (and this does not include the operating or other downstream processing costs). So, the

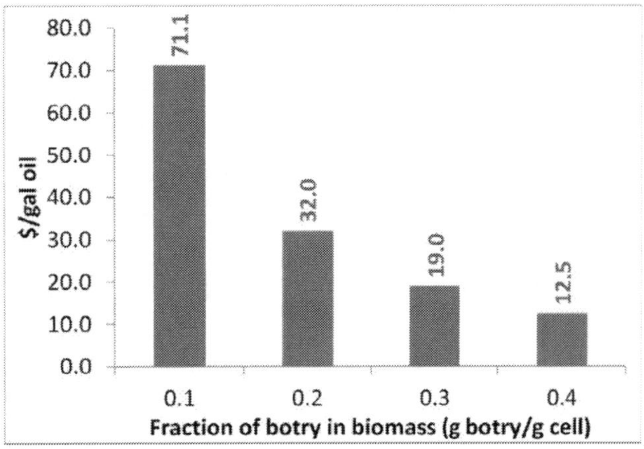

Figure 3: Cost of electricity per gallon of fuel at different fuel to biomass ratio.

fuel production must be growth-dissociated. The parameter that determines the quantity of gas that would be consumed under very low or no-growth condition is the **maintenance coefficient** (and the true yield) of the bacteria. While the maintenance coefficients of bacteria growing heterotrophically are quite abundant in the literature, those for autotrophic bacteria (like *R capsulatus*) are not very common. In this paper, we present our recent work related to the measurement of this parameter for *R capsulatus*.

True yield and maintenance coefficient

Maintenance coefficient of an organism is the amount of energy it needs to stay alive and maintain itself without cell division. On the other hand, true yield of an organism is the amount of substrates it would consume for growth only and not for maintenance. Both of these parameters are very important not only for accurately assessing the baseline economic feasibility of the proposed process but also indicate energetic efficiency of the organism. The yield and maintenance coefficient of *R capsulatus* on H_2, the most expensive substrate, is not available in the literature. In fact, values of these parameters are generally very scarce for any organism under autotrophic conditions.

In principle, maintenance coefficient and true yield can be calculated if the yield on the substrate is known as a function of the growth rate. (Pirt, 1965) derived the following equation as the inter-relationship among specific growth rate (μ) and yield (Y_{XS}) and true yield (Y_{XS}^T) and maintenance coefficient (m_S).

$$\frac{1}{Y_{XS}} = m_S \frac{1}{\mu} + \frac{1}{Y_{XS}^T} \qquad (1)$$

Y_{XS} and μ are both experimentally measurable parameters. Therefore plotting $\frac{1}{Y_{XS}}$ against $\frac{1}{\mu}$ a straight line can be obtained whose slope and intercept would be equal to the m_S and $\frac{1}{Y_{XS}^T}$ respectively. This has been traditionally used in the determination of true yield and maintenance coefficient. However, in a batch system the growth rate varies continuously, therefore an accurate measure of instantaneous growth rate and yield are not possible. At the same time, the culture goes through physiological changes as it progresses through the different growth stages and therefore the yields obtained this way may not refer to the same condition. So, true yield and maintenance are measured in continuous systems, in which the growth rate of an organism can be fixed at a particular value by fixing the dilution rate (inverse of residence time in a reactor).

Cell balance on a continuous reactor:

$$V\frac{dX}{dt} = FX_0 - FX \quad (2)$$

where V = volume of reator, $\frac{dX}{dt}$ = growth rate, F = flow rate, X_0 and X = inlet and outlet cell concentration respectively. With 0 inlet cell concentration at steady state,

$$\frac{1}{X}\frac{dX}{dt} = \frac{F}{V} \quad (3)$$

From the definitions of specific growth rate, $\mu = \frac{1}{X}\frac{dX}{dt}$ and dilution rate, $D = \frac{F}{V}$, we have $\mu = D$ in a steady state continuous flow reactor.

Multiple steady states are established by varying the dilution rate and yield values are measured at these steady states. The corresponding yield and dilution rates are plotted on a $1/Y$ vs. $1/D$ graph to get the true yield and maintenance coefficient. However, this method is very time consuming and produces a limited number of data points. These parameters can alternatively be measured in deceleration-stat or acceleration-stat (D-stat or A-stat) (Barbosa et al., 2005) rather than a true chemostat. Instead of going from one dilution rate to the next in a step and waiting for a steady state to be established, the dilution rate is varied smoothly from one to the next. If the rate is slow enough for the cells to adapt physiologically, the culture goes through a succession of pseudo steady-states and the yield and dilution rate in this period is used to calculate the maintenance and true yield. The D-stat method is followed in our determination of the yield and maintenance coefficient of *R capsulatus*.

Materials and methods

The experimental setup and schematic of the autotrophic bioreactor is shown below (Fig. 4). H_2, O_2 and CO_2 gases are mixed via three mass flow controllers (Brooks Instruments, CA) and delivered to the reactor after going through a humidifying column. The outlet gas passes through a drying column, then through a mass flow controller (Omega Engineering) that measures the flow rate and a GC that measures the composition. The composition data is collected once every hour. For safety concerns about using explosive gas mix (H_2+O_2), the reactor used in this work is a 250 ml media bottle. The liquid flow rate into and out of the reactor is controlled via a highly precise peristaltic pump (Watson-Marlow), whose rotation can be controlled to within 0.01 rpm by a 0-5 V analog input. This is very critical in maintaining an accurate and consistent flow rate and smoothly varying it for D-stat. The actual flow rate of the inlet media is calculated by periodically recording the weight of the media bottle which sits on a balance and fitting appropriate regression lines through them dynamics. The culture is recirculated via a second peristaltic pump for online optical density measurement. Samples are taken periodically from the outlet for offline OD and dry weight measurements. The media used for growth is based on a previous work in the literature (Madigan & Gest, 1979) where the autotrophic growth of *R capsulatus* was first demonstrated. A pH probe continuously measures the pH of the culture. As discussed before, the culture

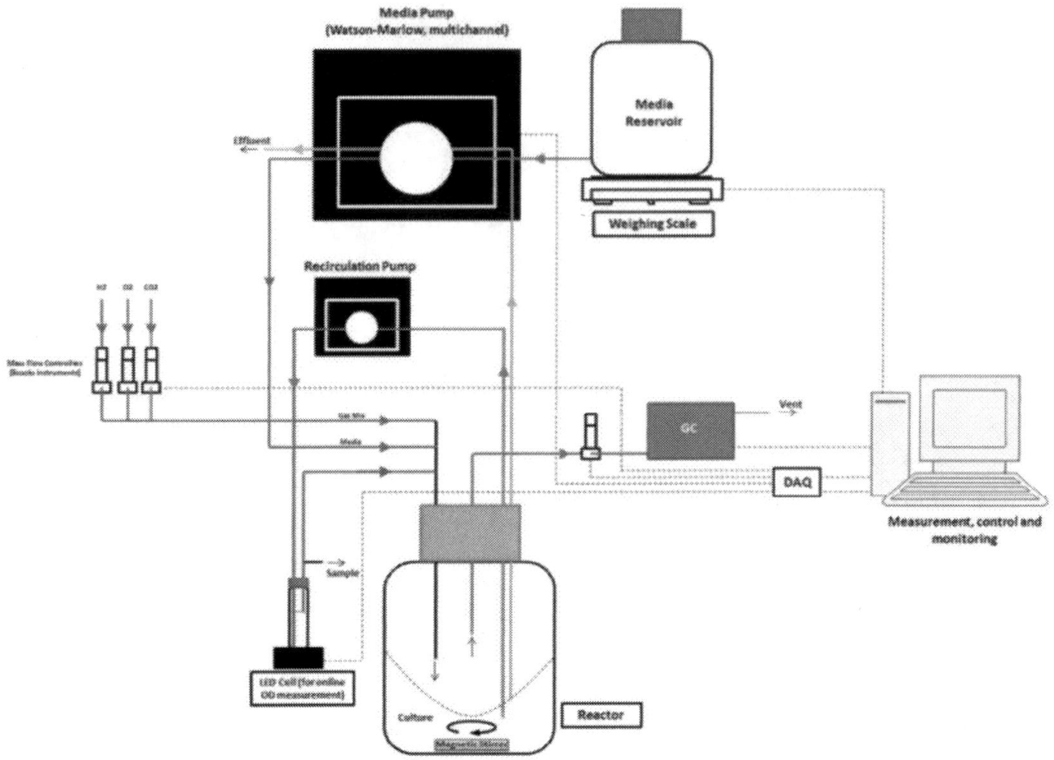

Figure 4: Schematic of the experimental setup.

is allowed to come to a steady state at a particular dilution rate and then the dilution rate is decreased slowly. The deceleration rate is based on previous literature work on *R capsulatus* (Hoekema, Douma, Janssen, Tramper, & Wijffels, 2006) and experience with running the experiment.

Another critical aspect of the experiment is the measurement of gas flow rate and composition, which is much more difficult to do accurately than liquid phase because of strong temperature and pressure dependence. Mass flow meters, although capable of measuring flow rate of a single gas very accurately, are not very accurate for a mixture of gases and at low flow rates. A soap-bubble meter was therefore placed at the end of the line to calibrate the outlet flow based on manual measurements. This was checked against the flow rate based on composition correction of the MFM and they compared very well. Furthermore, a state-of-art GC was used to analyze these gases continuously in a single pass for composition measurement.

Results and discussion

The experimental data is shown from Fig. 5 where the dilution rate was decreased from about 0.08/hr to about 0.06/hr at a rate of 0.094/day^2. This deceleration rate was based conservatively on a previous work of (Hoekema et al., 2006) where they measured the yield and maintenance of *R capsulatus* in the photoheterotrophic growth mode, producing H_2. Optical density plot (Fig 5A) shows that the expected pattern of variation is followed as the dilution rate is varied. Fig 5C shows the precision of media flow variation. The final calculated results are plotted in Fig 5D. The true yield and maintenance coefficient values are shown below (Table 1).

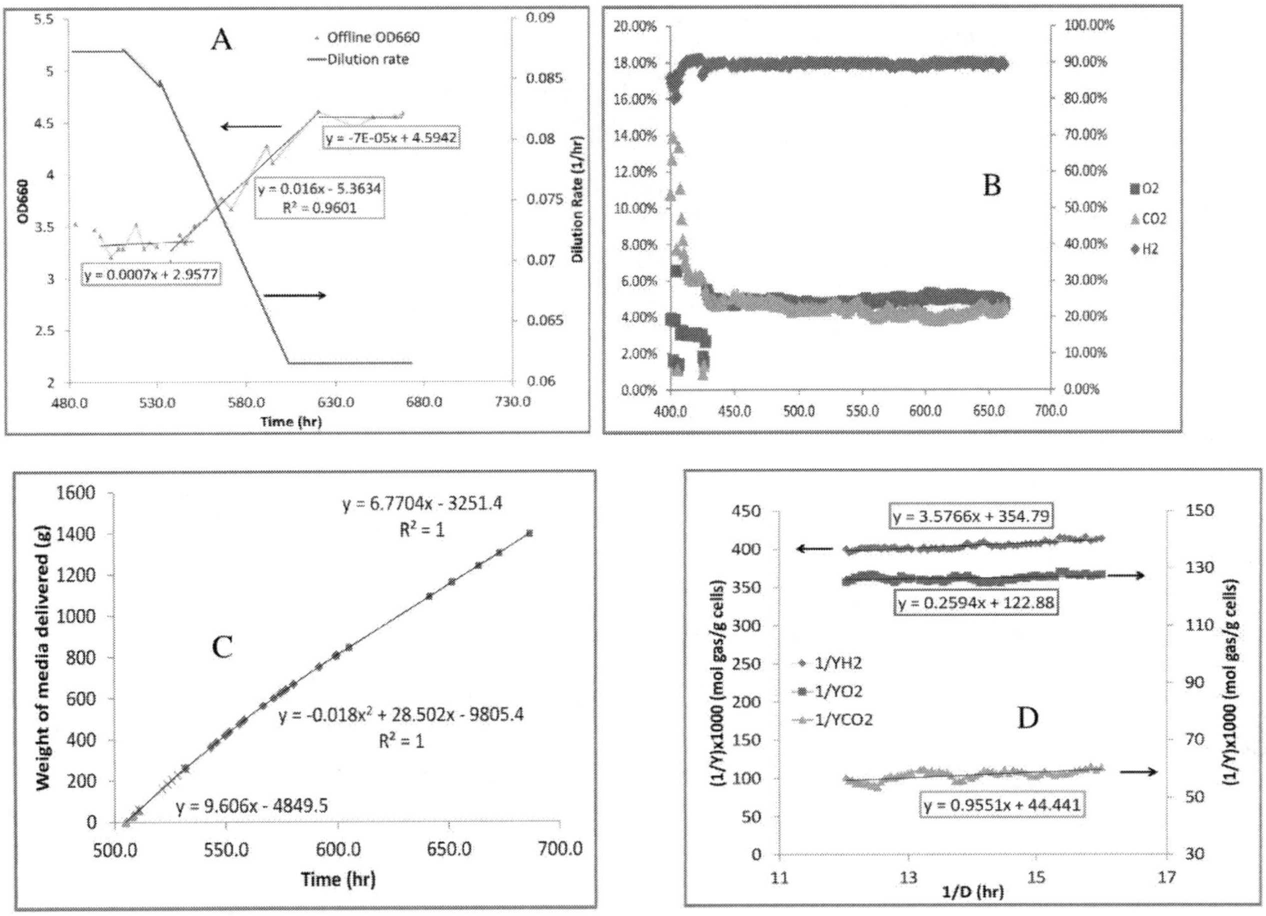

Figure 5: Experimental data for *R. capsulatus* true yield and maintenance coefficient determination. (A) Ramp down of dilution rate (red) and the corresponding OD increase (green), (B) gas phase composition, (C) media bottle weight to measure flow rate, (D) 1/Y vs. 1/D.

These values compare quite well with the values for *R eutropha* reported in the literature. Table 2 lists the true yields and maintenance coefficients of *R eutropha* and a couple of methanogens. Siegel & Ollis (1984) noted that the maintenance coefficient of *R eutropha* is a strong function of dissolved O_2 concentration. As observed by other researchers (Howells, 1982), O_2 can be both inhibitory and limiting for the growth of a bacteria at different concentrations and is strongly strain dependent. Under continuous culture conditions, the steady state in between inhibitory and limiting concentrations is unstable and tends to move either toward washout at higher concentration or towards O_2 limitation (Siegel & Ollis, 1984). We can see from the values that the maintenance coefficient increases significantly under H_2-limited conditions. The explanation offered by the authors is that the NAD-reducing hydrogenase in *R eutropha* is deactivated by O_2 and becomes an additional metabolic burden to replace the deactivated hydrogenases at high O_2 concentration, hence the increase in maintenance cost. High O_2 concentration is also found to be inhibitory to *R capsulatus* (Madigan & Gest, 1979; Siefert & Pfennig, 1979). These facts lead us to believe that this is a general phenomenon in lithotrophic CO_2 fixation and give us an idea of the possible operating regime for such a process. Future studies will examine the effect of these parameters for *R eutrophoa* as well.

Table 1: Experimental values of true yield and maintenance coefficient.

	True yield, Y_{XS} (g cells/mol gas)	Maintenance coefficient, m_S (mol gas/g cells.hr)
H_2	2.82	3.58×10^{-3}
O_2	8.13	2.60×10^{-4}
CO_2	22.52	9.55×10^{-4}

The yield and maintenance coefficient of *M. bryantii* were determined in batch growth experiments. The yield was calculated from q_{max} (maximum specific substrate utilization rate) and μ_{max} (maximum specific growth rate), which were determined under non-limiting H_2 concentration. Also, they treated the maintenance coefficient as "endogenous metabolism" i.e. oxidation of biomass to support cell maintenance, and determined the maintenance rate, b, from a biomass decay experiment where there was no H_2 present. This is clearly different from a true maintenance coefficient where the biomass is not really oxidized to provide for maintenance but the actual substrate is consumed by active cell machinery to maintain itself. Since they have not exactly provided the maintenance coefficient, we have calculated the value from yield (Y) and maintenance rate (b). Again, it is not really compatible since they were calculated under two different physiological conditions. However, it gives us something for comparison. The values for *M arboriphilus* were obtained under H_2-limited continuous culture conditions (Morii, Koga, & Nagai, 1987). The true yield and maintenance values were not given directly in the paper but the dilution rates and the substrate utilization rates were used to calculate the values using method outlined above.

Table 2: True yield and maintenance coefficient of some organisms on H_2.

Organism	True yield (g cells/mol H_2)	Maintenance coefficient (mol H_2/g cells.hr)	Experimental conditions	Reference
Ralstonia eutropha	4.1	0.001	O_2 limited continuous culture	(Siegel & Ollis, 1984)
	4.5	0.025	H_2 limited continuous culture	
Methanobacterium bryantii M.o.H	0.325	0.0115	Non-limited batch culture	(Karadagli & Rittmann, 2005)
Methanobrevibacter arboriphilus A2	0.40	0.0106	H_2 limited continuous culture	(Morii et al., 1987)

The values for these methanogens are included for comparison. It is readily apparent that the yield and maintenance are both much worse for these organisms. It is quite likely because the methanogenic process is anaerobic and the yields are expected to be lower than an aerobic process. It needs to be borne in mind that these organisms are actually producing a product that is much more reduced than biomass and that is also a reason for low yield and high maintenance. The energetic analysis of a fuel producing process therefore must include the metabolic effect of producing the fuel molecule.

Economic analysis

The preliminary economic assessment of the electricity cost per gallon of fuel has been made for the non-growth scenario for a comparison against the growth based scenario. For calculations, literature values for *R eutropha* and the experimental values shown in the previous section for *R capsulatus* were used. Cost of H_2 and CO_2 are taken to be \$3.82/kg and \$87/ton respectively. The H_2 cost is based on the current solar electricity prices but as solar energy becomes more widespread this is expected go down and affect the economics correspondingly.

Baseline scenario. Growth based. Shown in Figure 3.

Proposed scenario. At very low growth rates maintenance energy would be the main form of energy consumption. At this point, we lack the important parameter of fuel production rate by the bacteria to make an accurate estimate. A work-around to provide a quantitative perspective is to use a similar value for another organism from literature. *Chlorella vulgaris* which produces lipid under severely growth-limited conditions was chosen as a reasonable basis. From (Tuerk, 2011) this value is estimated to be 0.001256 g lipid/g cells.hr. This value was used as a baseline fuel production value and varied until 0.1 g borty/g cells.hr to observe the effect on cost of fuel (Fig 6). (Residence time in the reactor was a parameter in this calculation which was set to 10 hr. The cell retention was set to 0.9). The high fuel cost calculated for low production rates demonstrates that even in a non-growth associated production scenario maintenance energy requirements play a significant role in determining the cost of fuel production.

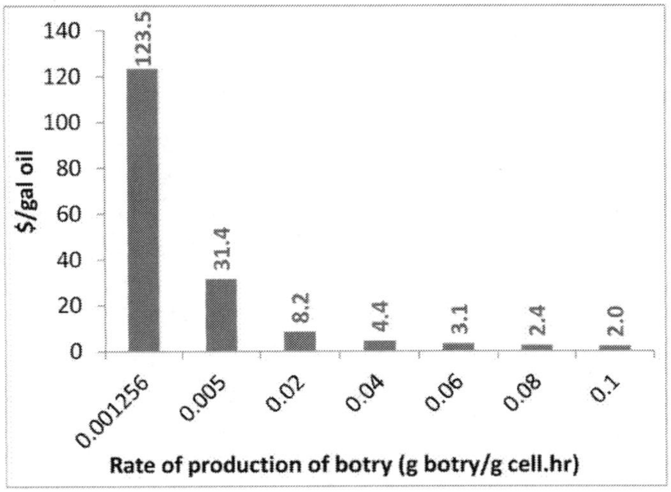

Figure 6: Cost of electricity per gallon of fuel under non-growth associated fuel production.

Future work on this will include studying the effect of various parameters (like O2 tension, nitrogen source etc.) on the maintenance coefficient of *R capsulatus*. Similar studies will be done on *R eutropha* for an unbiased comparison of the two organisms.

References

Aragno, M., & Schlegel, H. G. (1992). The chemolithotrophic prokaryotes. In A. Balows, H. G. Truper, M. Dworkin, W. Harder, & K. Schleifer (Eds.), *The Prokaryotes, 2nd Edition. A Handbook on the Biology of Bacteria: Ecophysiology, Isolation, Identification, Applications.* (pp. 332–343). New York: Springer-Verlag.

Barbosa, M. J., Zijffers, J. W., Nisworo, A., Vaes, W., van Schoonhoven, J., & Wijffels, R. H. (2005). Optimization of biomass, vitamins, and carotenoid yield on light energy in a flat-panel reactor using the A-stat technique. *Biotechnology and bioengineering*, *89*(2), 233–42. doi:10.1002/bit.20346

Hillen, L. W., Pollard, G., Wake, L. V., & White, N. (1982). Hydrocracking of the oils of Botryococcus braunii to transport fuels. *Biotechnology and bioengineering*, *24*(1), 193–205. doi:10.1002/bit.260240116

Hoekema, S., Douma, R. D., Janssen, M., Tramper, J., & Wijffels, R. H. (2006). Controlling light-use by Rhodobacter capsulatus continuous cultures in a flat-panel photobioreactor. *Biotechnology and bioengineering*, *95*(4), 613–626. doi:10.1002/bit

Howells, E. R. (1982). Single-cell protein and related technology. *Chemistry Industry*, (15), 508–511.

Karadagli, F., & Rittmann, B. E. (2005). Kinetic characterization of Methanobacterium bryantii MoH. *Environmental science & technology*, *39*(13), 4900–4905.

Kelly, D. P. (1992). The chemolithotrophic prokaryotes. In A. Balows, H. G. Truper, M. Dworkin, W. Harder, & K. Schleifer (Eds.), *The Prokaryotes, 2nd Edition. A Handbook on the Biology of Bacteria: Ecophysiology, Isolation, Identification, Applications.* (pp. 332–343). New York: Springer-Verlag.

Madigan, M. T., & Gest, H. (1979). Growth of the photosynthetic bacterium Rhodopseudomonas capsulata chemoautotrophically in darkness with H2 as the energy source. *Journal of bacteriology*, *137*(1), 524–30. Retrieved from http://www.pubmedcentral.nih.gov/articlerender.fcgi?artid=218479&tool=pmcentrez&rendertype=abstract

McKirdy, D., & Cox, R. (1986). Botryococcane in a new class of Australian non-marine crude oils. *Nature*, *320*. Retrieved from http://www.nature.com/nature/journal/v320/n6057/abs/320057a0.html

Morii, H., Koga, Y., & Nagai, S. (1987). Energetic analysis of the growth of Methanobrevibacter arboriphilus A2 in hydrogen-limited continuous cultures. *Biotechnology and bioengineering*, *29*(3), 310–5. doi:10.1002/bit.260290304

Pirt, S. J. (1965). The maintenance energy of bacteria in growing cultures, *163*(991), 224–231.

Siefert, E., & Pfennig, N. (1979). Chemoautotrophic growth of Rhodopseudomonas species with hydrogen and chemotrophic utilization of methanol and formate. *Archives of Microbiology*, *122*(2), 177–182. Retrieved from http://www.springerlink.com/index/V323851512727030.pdf

Siegel, R. S., & Ollis, D. F. (1984). Kinetics of growth of the hydrogen-oxidizing bacterium Alcaligenes eutrophus (ATCC 17707) in chemostat culture. *Biotechnology and bioengineering*, *26*(7), 764–70. doi:10.1002/bit.260260721

Tuerk, A. L. (2011). *An assessment of photosynthetic biofuels and electrofuels technologies under rate-limited conditions*. Pennsylvania State University.

(140b) "Green" Solvents: Multi-Criteria Knowledge-Based and Statistical Approaches for Their Evaluation

Stavros Papadokonstantakis, Tianhe Liu, Alireza Banimostafa and Konrad Hungerbühler, Institute for Chemical and Bioengineering, Swiss Federal Institute of Technology, Zurich (ETHZ), Zurich, Switzerland

"Green" solvents: Multi-criteria knowledge-based and statistical approaches for their evaluation

Stavros Papadokonstantakis, Tianhe Liu, Alireza Banimostafa, Konrad Hungerbühler

Institute of Chemical- and Bioengineering, Swiss Federal Institute of Technology (ETH) Zurich, Switzerland

Sustainable process design has been recognized as one of the key research challenges for process systems engineering. In fact, the experience gained so far has pointed out that the applicability of sustainability principles can be more advantageous in earlier phases of process design characterized by more degrees of freedom for decision making. Typical decisions made in these early process design phases involve the selection of chemical synthesis path, chemical auxiliaries, unit operation conditions up to a basic flowsheet design.

Solvents belong to chemical auxiliaries of particular interest for the chemical industry due to their multi-functional use (e.g., to achieve desired reaction rates while protecting from the reaction exothermicity, to ensure solid product quality requirements through effective crystallization, to perform liquid-liquid separations or azeotropic distillations etc.) and their large amounts (e.g., in fine-chemical and pharmaceutical production large amounts are used per mass of final products). Historically, solvents have been designed to maximize technical performance and minimize cost. But nowadays, an effort to find 'green' solvents gains interest, for replacing conventional solvents with environmentally benign substitutes while, of course, maintaining their effectiveness for the intended use. In this context, the concept of ''green" solvents expresses the goal to minimize the environmental impact resulting from their use in chemical production, both from a life cycle impact assessment (LCIA) and a safety, health and environmental (SHE) hazard assessment point of view.

Several shortcut, index-based methodologies have been proposed for evaluating the relevant LCIA and SHE effects, both for independent substances and for chemical production processes. These index-based frameworks encapsulate prior knowledge regarding main SHE hazards, cradle-to-gate resources consumption, and emission related effects. The indices are typically defined on the basis of substance properties, derived from available databases or estimated using available techniques (thermodynamic property estimation methods, QSAR models, etc.), and process features derived either from a preliminary process flowsheet or even on the basis of the chemical synthesis path. On the one hand, the simplicity of the index-based approaches makes them attractive, especially for early stages of process design lacking detailed process information, but on the other hand they face criticism regarding limited coverage of considered effects, subjectivities for the calculation of aggregated end-point indices, and unknown resolution of the scaled final scores according to which the ranking of process alternatives is inferred.

Especially for the problem of subjective aggregation to end-point scores principal component analysis (PCA) has been recently proposed as a potential remedy. In this approach the rows of the PCA input matrix refer to the evaluation objects (e.g., solvents, chemical process steps, chemical process sections or overall chemical processes) and the columns refer to metrics for their assessment (e.g., LCIA and/or SHE hazard assessment indicators). The evaluation objects are ranked according to the obtained principal component (PC) scores, which in the simplest PCA version can be viewed as linear combinations of the assessment metrics. The weights of these linear combinations express the importance of the original assessment metric for a specific PC and are derived in such a way that the first PC covers most of the variability of the PCA input matrix, the second PC most of the remaining variability under the condition that it is orthogonal to the first one, and so on. The concept behind this PCA-based method, e.g., when it is used for the assessment of substances, is that dependencies must exist between the assessment metrics, since the molecular structure of a substance should, in theory, provide (although most of the times via unknown functional relationships) all necessary information for the assessment metrics. The PCA-based method will, therefore, detect these dependencies in a theoretically sound way and this information will be used instead of *a priori* weighting schemes for the diverse metrics. In this way, two objectives are achieved simultaneously, i.e., the inter-relationships between the diverse metrics reveal the "true" dimensionality of the multi-criteria assessment and the level of multi-criteria similarity among individual solvents can be derived. Moreover, the same PCA-based concept can be used in an expectation-maximization iterative procedure for filling data gaps, typically occurring in the assessment metrics.

The proposed framework is demonstrated (see Figure 1) on the basis of 138 organic solvents used in industrial applications. The performance of the PCA-based method for filling data gaps is compared with other approaches based on priority settings between substance properties and reveals in which extent this should be preferred (i.e., relative amount and patterns of data gaps). Moreover, two case studies are presented in which the results of the PCA-based framework are compared to the results of other studies (from the Safety and Environmental Technology group in ETHZ and GlaxoSmithKline, respectively) using different index-based methods with *a priori* postulated weighting schemes. The results indicate that the presented framework is useful for covering most of the information contained in the other index-based methods for the evaluation of "green" solvents, while being inherently flexible with respect to the number and type of assessment metrics, as well as to the existence of data gaps. Finally, the present study discusses the advantages and limitations of the PCA-based approach in terms of simplicity, interpretability and statistical robustness.

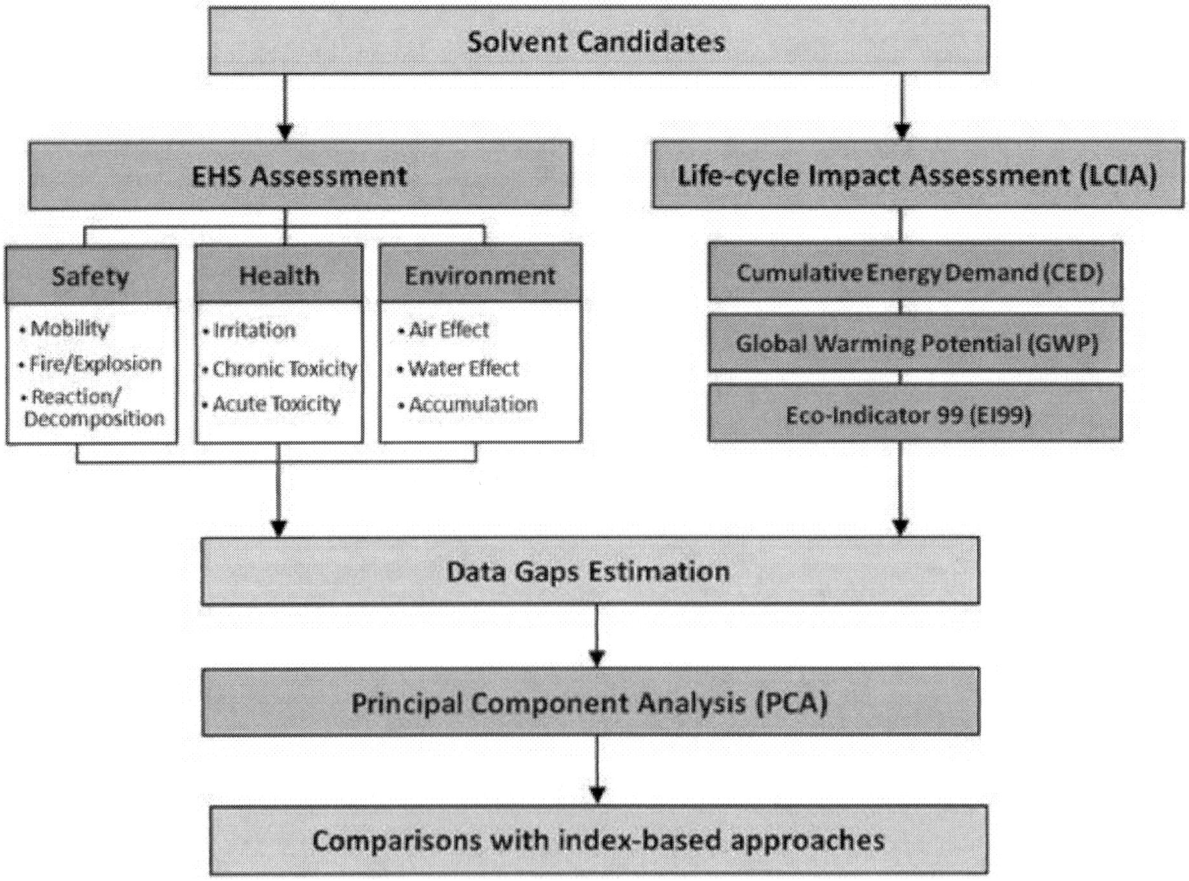

Figure 1: Integrating a PCA-based approach for the evaluation of "green solvents" and comparison with existing index-based approaches.

(140c) Resilience in Complex Economic Networks: Insights From Graph Theory and Input-Output Models

Shauhrat Chopra, Civil and Environmental Engineering, University of Pittsburgh, Pittsburgh, PA and Vikas Khanna, Department of Civil and Environmental Engineering, University of Pittsburgh, Pittsburgh, PA

Comprehensive analysis of complex systems has exposed the demand for bolstering resilience in such systems with respect to sustainable development. Economies are self-organizing, adaptive, complex systems where the constituent components or sectors are connected with each other in myriad ways. Such networks can be modeled as a directed-weighted graph where the nodes represent diverse sectors and the edges represent the monetary transactions between sectors. For example, Input-Output tables that describe the monetary exchanges between different pairs of industries in an economy are available for different countries. Determining the resilience of complex networks, like the economy, enables us to assess the capacity of the network to retain its functionality and structure while adapting to stress, a property vital for a sustainable network.

Employing a systems approach from a resilience perspective is useful for understanding the trade-offs between efficiency and the ability to cope with variations in the environment because of reorganization. Solely, observing simple cause-effect type of computation among the actors may neglect information regarding indirect effects and system-wide properties arising due to interaction. These interactions and their cascading effects can be better assessed through a systems approach. In addition, research on network robustness has elucidated that complex networks could be resilient to random perturbations but vulnerable to explicit attacks on critical nodes. Identifying vulnerable sectors of the economy, and subsequent, reinforcement of the network topology through policy measures can increase resilience of complex systems like the economy.

The goal of this study is to compare the network structure of the United States (U.S.) and the Chinese economies to identify vulnerabilities in the system and their response to disruptions. We also discern the evolution of resilience from the period of late 1980's to the mid 2000's for both economies by utilizing network theory. The Chinese and the U.S. economies represent distinctive categories, with China emerging as one of the biggest manufacturing economies, and the U.S. economy evolving into a service economy. The proposed research integrates concepts of graph theory and Economic Input-Output (EIO) model to elucidate network properties concerning resilience of economic systems. Structural and functional analysis based on graph theory sheds light on vulnerabilities in the network using various centrality and vitality measures. Moreover, using IO analysis and network efficiency performance, we evaluate the direct and indirect impact of disruptions on the economic system. We apply network analysis tools on IO tables for the Chinese and the US economy from 1980's to 2000's with an aspiration to observe the evolution of the basic backbone of these economies. It is noted that on introduction of various stresses, resilient systems exhibit low modularity and percolation. Moreover, on increasing the flexibility in the system, both service and manufacturing economies exhibit higher resilience to disruptions. The implications of the results for developing resilient and sustainable economic systems will be presented and discussed.

(140d) Role of Controllability for LONG TERM Sustainability

Pahola T. Benavides, VRI- Custom, Chicago, IL, Urmila Diwekar, Vishwamitra Research Institute, Center for Uncertain Systems: Tools for Optimization and Management, Clarendon Hills, IL; Center for Uncertain Systems: Tools for Optimization and Management (CUSTOM), Vishwamitra Research Institute, Clarendon Hills, IL; Bioengineering, University of Illinois at Chicago, Chicago, IL and Heriberto Cabezas, Office of Research and Development, U.S. Environmental Protection Agency, Cincinnati, OH

Successful implementation of sustainability ideas in ecosystem management requires a basic understanding of the often nonlinear and nonintuitive relationships among different dimensions of sustainability, particularly the system-wide implications of human actions. This basic understanding further includes a sense of the time scale of possible future events and the limits of what is and is not likely to be possible. With this understanding, systematic approaches based on control theory can then be used to develop policy guidelines for the system. Therefore, controllability of the system is very important to determining long term sustainability of system. In a recent article in nature by Liu et al. [1] presents a new analytical tool to study controllability of complex system. We apply these tools to a complex system consisting of an integrated ecological-economic social model, which comprises various ecological (natural) and domesticated compartments representing species along with a macroeconomic price setting model [2,3]. This model has been used to study sustainability of the planet under different scenarios like population explosion and per capita consumption. We argue that controllability of this system is linked with the long term sustainability and present our arguments in the light of various scenarios studied earlier.

Bibliography

1. Liu Y., J. Slotine and B. Albert-La´szlo, "Controllability of complex networks", 12 may 2011, nature, 473, 167.
2. Shastri Y. , U. Diwekar, H. Cabezas, and J. Williamson " Is Sustainability Achievable? Exploring the Limits of Sustainability with Model Systems", (2008), Environmental Science and Technology, 42, 6710.
3. Shastri Y. ,U. Diwekar, and H. Cabezas, "Optimal Control Theory for Sustainable Environmental Management", (2008),Environmental Science and Technology, 42, 5322.

(140e) Life-Cycle-Based Sustainability Assessment of Nanocoating Material

Rohan Uttarwar and Yinlun Huang, Department of Chemical Engineering and Materials Science, Wayne State University, Detroit, MI

Over the last decade, nanopaint has become a dominant coating material with a wide range of applications in the automotive, aerospace, ship-making, construction, architectural and pharmaceutical industries. Despite the obvious benefits of nanocoatings, there exist unintended health and environmental risks associated with the use of nanoproducts which are not yet fully understood. In order to develop nanocoating systems with improved product performance and reduced impact on environment and health care, a complete assessment over each stage of the life-cycle of nanocoating material is needed. Development of sustainability assessment techniques for the comprehensive evaluation of performance of nanocoating products over their life-cycle becomes urgent.

The life-cycle of nanocoating materials consists of the stages ranging from raw material selection and processing, to nanopaint/nanocoating manufacturing, product use and disposal. In this work, a model-based sustainability assessment methodology is introduced that can be used to assess the economic, environmental, and social aspects in every life-cycle stage. To perform a comprehensive assessment, different sets of sustainability metrics have been identified for use in different life-cycle stages. These metrics are analyzed to ensure the consistency of the assessment. Due to the lack of sufficient data from open resources, a number of computational models are developed that can generate data that is sound from engineering perspective. The methodology has been used to study the sustainability performance of nanopaint and its application to automotive coatings. A comprehensive case study will highlight critical issues concerning the material's development and nanoparticles emission to the environment and health impact, economic incentive and social satisfaction.

(140f) The Effect of CO2 Adsorbent to the H2 Production in Biomass Gassifier

KuZilati KuShaari, Chemical Engineering Department, Universiti Teknologi PETRONAS, Seri Iskandar, Perak, Malaysia

Biomass gasification has become one of the most popular methods to produce biohydrogen. Hydrogen yield has been proven to further increase by adding CO_2 adsorbent into the biomass gasifier. The proposed CO_2 adsorbent for biomass gasification system is Calcium Oxide (CaO) as it is converted to solid Calcium Carbonate ($CaCO_3$) as it reacts with CO_2 in the gasifier. Due to its high density, the solid $CaCO_3$ produced can be easily removed from the system and it will accumulate at the bottom of the gasifier. This paper highlights the study on the effect of CO_2 adsorbent to the H_2 production from the gasifier using CFD approach. The results show that the addition of CO_2 adsorbent helps to increase the H_2 production. The simulation was done for two different cases, which are gasification process with in-situ CO_2 adsorption and without adsorbent. The reaction temperature for both cases ranging between 500-900 ^{0}C. From the results obtained, it is observed that the addition of CO_2 adsorbent does improve the H_2 production by 12%.

(148a) Novel Materials for Carbon Capture, Utilization and Storage (CCUS)

Ah-Hyung Alissa Park, Earth and Environmental Engineering & Chemical Engineering, Columbia University, New York, NY

Historically, the atmospheric concentration of CO_2 fluctuated naturally on the timescales of ice ages. Concerns, however, stem from the recent dramatic increase in CO_2 concentration, which coincides with global industrial development. This rise is mainly due to the high use of fossil fuels. In order to meet the ever-increasing global energy demands while stabilizing the CO_2 level in the atmosphere, it is widely believed that current carbon emissions must be reduced by at least a factor of three. The containment of CO_2 involves three operations: separation, transportation, and storage. Until now, these technologies have been developed independent of one another, which has resulted in complex and economically challenging large-scale designs. The future direction of carbon management technologies now focuses on the integration of CO_2 capture and storage schemes as well as CO_2 utilization. In this seminar, two novel carbon capture, utilization and storage (CCUS) technologies will be introduced. CO_2 capture fluids based on the Nanoparticle Organic Hybrid Materials (NOHMs) are currently developed and their absorption isotherms are characterized as a function of CO_2 partial pressure and temperature (i.e., combustion and gasification conditions). NOHMs are a new class of organic-inorganic hybrids that consist of a hard nanoparticle core functionalized with a molecular organic (sometimes polymeric) corona that possesses high degree of tunability. NOHMs are non-volatile and stable over a very wide temperature range, which make them interesting materials for various energy and environmental applications. Once captured, CO_2 needs to be stored for permanent disposal. The geological storage of carbon dioxide has been considered to be the most economical method of carbon sequestration, while mineral carbonation is a relatively new and less explored method of sequestering CO_2. The advantage of carbon mineral sequestration is that it is the most permanent and safe method of carbon storage, since the gaseous CO_2 is fixed into a solid matrix of Mg-bearing minerals (e.g., serpentine) forming a thermodynamically stable solid product. These carbon sequestration technologies can be integrated into the existing or new energy conversion systems in order to achieve their overall sustainability.

(148b) On the Selection of Ionic Liquids for the Conversion of Cellulosic Biomass Feedstock

Sameer Parvathikar, Andrew R. Tadd and Johannes W. Schwank, Chemical Engineering, University of Michigan, Ann Arbor, MI

The use of ionic liquids (ILs) as solvents represents a promising route for the conversion of cellulosic biomass feedstock. 1,3-dialkylimidazolium ILs in particular, have become a popular and thoroughly-studied reaction media for this. Reactions such as the dissolution and hydrolysis of cellulose, the dehydration of glucose, etc. have been demonstrated. While these ILs have proven to be quite capable of processing these large biomolecules, their widespread adoption has been limited due to their toxicity and inhibitive costs.

It is clear that there is a need for less expensive, and environmentally benign alternatives. The objective of this work is to develop a deeper understanding of the interaction between the solute (biomolecules) and the solvent, which is necessary to effectively choose these alternative solvents. A number of solvent properties like hydrogen bond formation, acidity, etc., will be studied for both imidazolium-based ILs and lower-cost alternatives. Using probe reactions like the isomerization and dehydration of sugars, property-performance relationships will be established. The effect of using different catalysts to affect the reaction environment and thence product yields will also be examined. This work will provide insights into rationally selecting ILs to perform targeted chemistries.

(148c) Production of Hydroxymethylfurfural From Sugars or Renewable Biomass Feedstock As a Potential Platform Intermediate for Bio-Plastics

Wei Liu, Pacific Northwest National Laboratory, Richland, WA

There has been world-wide interest in making plastics out of renewable biomass feedstock for recent years. Hydroxymethylfurfural (HMF) is viewed as an attractive alternate to terephthalic acid (TPA) for production of polyesters (PET) and polyamides. An economic production process needs to be developed to make HMF at a cost competitive with current petroleum-derived feedstock. Conversion of sugars into HMF has been studied in numerous publications. In this presentation, a brief overview of those literature reports pertinent to a practical process development will be presented. A complete ionic liquid-based catalytic process recently developed at PNNL will be discussed. This new process comprises integration of ionic liquid catalytic conversion with adsorption separation. The ionic liquid catalytic reaction is conducted under moderate reaction conditions and provides both high reactivity and selectivity, which is practical for scale-up. The main challenge lied in separation of reaction mixtures. Thus, novel membrane-coated zeolite adsorbent particles are invented, and the adsorption process is demonstrated for isolation of 99% pure HMF product and complete recovery of the ionic liquid and un-converted species. The new process enables nearly 100% conversion of fresh fructose feed into HMF product in the recycled ionic liquid. A simplified process flow diagram is conceived based on these innovations, and the key equipment such as reactor and adsorbent bed is sized for a plant of 200,000 ton/year of fructose processing capacity. The proposed HMF production process is much simpler than the current paraxylene (PX) manufacturing process from petroleum oil, which suggests substantial reduction to the capital cost and energy consumption is possible. At the equivalent value to PX on the molar basis, there can be a large gross margin for HMF production from fructose and/or sugars.

(148d) Large-Scale Energy Storage: New Generation Vanadium Mixed Acid Redox Flow Battery

Liyu Li, UniEnergy Technologies, LLC, Mukilteo, WA

Electrical energy storage is considered as a key enabler to significant penetration of intermittent renewable power from solar and wind, as well as implementation of the Smart Grid. Among the most promising technologies is the all-vanadium redox flow battery (VRFB), which is capable of storing electricity on a MW/MWh scale in the form of two pairs of reduced and oxidized vanadium species (V^{2+}/V^{3+} and V^{4+}/V^{5+}) dissolved in two separate liquid electrolytes, for a discharge duration of hours or even longer. VRFBs offer the flexibility of separate design of power (kW) and energy (kWh): power depends on stack design, while energy depends on the volume and vanadium concentrations in electrolytes. With electrodes free of repetitive mechanical or structural stresses, VFRBs potentially deliver virtually unlimited cycle life. In addition, VRFBs are inherently safe as flowing electrolytes carry away the heat generated during operation. Other advantages include quick response, deep discharge ability, capability to withstand a fluctuating power supply, tolerance to over-charge and over-discharge, and fast charge and discharge. Substantial advancement has been made in VRFBs using the traditional vanadium sulfate electrolytes in the past years. However this technology cannot meet the economic and performance requirements for broad market penetration. While there is much room to improve the economy and performance by engineering and system integration, the current technologies are inherently limited by the current electrolyte chemistries. To prevent irreversible vanadium oxide precipitation, active control of operating temperatures in the range 10-35°C is essential, as is vanadium concentrations (i.e. energy capacity) $\leq$ 1.7M (1.5M in practical systems). Many impurities have to be controlled down to ppm levels to maintain the stability of the electrolytes.

Recently, new vanadium redox chemistry with SO_4^{2-}/Cl^- mixed supporting electrolytes was invented.[1] To stabilize the vanadium electrolytes, chloride was added into the aqueous sulfate electrolyte solution to form new complexes of vanadium species. The new complexes were proved to be much more stable over the current vanadium sulfate chemistries, leading to: i) an extension of operation temperatures to a range of -5 to 60°C, potentially eliminating the need for external heat management; ii) an increase in concentrations of vanadium species (V^{5+}, V^{4+}, V^{3+}, V^{2+}) to 2.5M, representing an energy capacity increase of 70% when compared to the all-vanadium sulfate chemistry; iii) achievement of nearly 90% energy efficiency at $50 mA.cm^{-2}$, with negligible degradation over a few hundred cycles. In this presentation, the structure of the vanadium species in the new electrolyte solutions and the performance and large-scale implementation potential of this advanced VRFB system will be discussed.

Acknowledgements

Part of the work was performed at Pacific Northwest National Laboratory, and supported by US Department of Energy.

References

1. Li*, Kim, Wang, Vijayakumar, Nie, Chen, Zhang, Xia, Hu, Graff, Liu, Yang*, *Advanced Energy Materials*, 1, 394~400, 2011.8th International Vanadium Symposium: Chemistry, Biological Chemistry, & Toxicology (vanadium8.org)

(148e) High Surface Area Transparent Conducting Oxide Electrodes and Their Application to Energy Technologies Based On Electrochemical Processes

Arnold J. Forman, Zhebo Chen and Thomas F. Jaramillo, Chemical Engineering, Stanford University, Stanford, CA

The ability to control the interfacial area and light trapping characteristics of structured electrodes enables new levels of performance in modern optoelectronic devices. Textured, high surface area electrodes (HSEs) combine large interfacial areas and increased utilization of active material with a light trapping morphology – key parameters for boosting device performance. Transparent conductive oxide (TCO) HSEs with tunable pore sizes on the nanometer to micron scale facilitate diffusion through the structure and serve as a broadly applicable platform for functionalization in many fields including electrochemical sensors, catalysis, (super)capacitors, batteries, photovoltaics and photoelectrochemistry. In this talk we describe the development of optically transparent, electrochemically stable, physically robust HSEs with tunable roughness factors (RF) from 1 through >100. We present a bottom-up approach to future device design and a low-cost, scalable synthetic route for production of transparent HSEs of several TCO compositions. Lastly, we demonstrate the utility of these HSEs by incorporating thin films and particles of electrochemically active materials for enhanced performance. Examples will include adopting the HSE platform for improved supercapacitors and solar photoelectrochemical water-splitting, among others.

(148f) Novel Wave-Shaped MnO2 Cathode Supported by a Partially Porous Nickel Layer for Fuel Cell/Battery

Mike Musil, Institute of Industrial Science, The University of Tokyo, Tokyo, 153-8505, Japan and Atsushi Tsutsumi, Institute of Industrial Science, The University of Tokyo, Tokyo, Japan

Increasing global demand for green energy has boosted the development of various means of energy generation and storage. Both fuel cells and batteries show big potential for an increasing amount of applications, however they still suffer from various drawbacks. Among others, fuel cells possess a low power density, which can mainly be ascribed to the reaction that occurs at a three-phase interface between electrode, electrolyte and cathodic and anodic reactants. Batteries, on the other, hand lack high energy density, whose reason lies in the limited amount of active material in the electrodes.

To eliminate each one's drawback, a so called Fuel Cell/Battery (FCB) has been suggested in previous works. As the name implies, it can work both as a fuel cell and battery. When working as latter, the best results were obtained when coating MnO_2 via electrochemical deposition on carbon fibers. However, oxygen charging – crucial when running in fuel cell mode – was fairly poor. Therefore, a novel cathode structure is presented in this work.

On the one hand, a high surface area to enhance the oxygen charging rate is obtained via a wave-shaped structure in the micrometer scale. Thanks to the obtained surface increase, a reaction rate improvement by up to one order of magnitude is expected compared to a flat electrode. To tackle the problem of having to charge electrically insulating MnO_2 via oxygen while enabling electron collection, a novel, tunnel-looking structure has been developed. Via template method, a several micrometer thick Nickel electrode with one micrometer long pillars has been obtained through electrochemical deposition. On top of the pillars, MnO_2 is deposited, creating a triple layer consisting of MnO_2, porous Nickel, and bulk Nickel.

(167a) Design and Operation of Bioethanol Production Process for Small-Scale Biorefinery

Kazuhiro Mochidzuki[1], Shin-ichi Kobayashi[1], Akiyoshi Sakoda[2], Tran Phuoc Nhat Uyen[3], Vu Le Van Khanh[3], Nguyen Dinh Quan[3], Le Thi Kim Phung[4] and Phan Dinh Tuan[5], (1)IIS, Univ. of Tokyo, Tokyo, Japan, (2)Institute of Industrial Science, University of Tokyo, Tokyo, Japan, (3)HCMUT, HCM City, Vietnam, (4)CHMUT, HCM City, Vietnam, (5)Hochiminh City Univ. of Tech., Hochiminh, Vietnam

To establish a model of sustainable integration of local agriculture and biomass industries in rural areas in South-East Asia, e.g., Vietnam, we are developing the biomass utilization technologies which meet with the satiation of actual communities. This presentation deals with a pilot-scale work of lingocellulosic bioethanol production, focusing on small-scale local biorefinery systems. We built a pilot plant which consists of bioethanol production process (with an 800 L fermentation tank) and biomass boiler. In this pilot plant, we developed the design of small-scale bioethanol production processes which utilized locally available biomass feedstock, including pretreatment system, small fermentor operation and use of biomass boiler, through the operation of the pilot process. For example, rice straw was converted into ethanol and rice husk was used for biomass boiler. Also, some byproducts from local industries were tested as nitrogen source for yeast. As wall as the technological development, the material and energy balances in the total system were analyzed to evaluate its features and benefit as a core facility in a small-scale biorefinery, focusing on the community-level availability biomass feedstock and demands of bioenergy/biofules.

(167b) Effects of Pellet Feed and Air Supply Rates On the Combustion Behavior of Spruce Pellets in a Grid-Type Stove

Bawadi Abdullah[1], Yoshimitsu Uemura[2], Noridah Osman[1], Takami Kai[3], Hirokazu Takanashi[4] and Yukio Teraoka[5], (1)Centre for Biofuel and Biochemical Research, Universiti Teknologi PETRONAS, Tronoh, Malaysia, (2)Centre for Biofuel and Biochemical Research, Universiti Teknologi PETRONAS, Tronoh, Malaysia, (3)Chemical Engineering, Kagoshima University, Kagoshima, Japan, (4)Chemistry and Biotechnology, Kagoshima University, Kagoshima, Japan, (5)Environmental Sciences and Technology, Kagoshima University, Kagoshima, Japan

The growing attentions to the utilization of biomass as a heat and power productions via conventional burning has initiated this study of using spruce materials as a feedstock. Unlike other types of biomass i.e. olive cake, apricot and peach stones, spruce is abundantly available and thus, suitable to use as a source of feedstock. In this study, the pellet of spruce was combusted in a grid type combustion chamber (5 kW stove). The stove consists of electric heater igniter, a screw feeder (2-18 g/min) and an air blower (0.6-1.0 NTP-m^3/min). To observe the combustion behavior, the concentration of O_2, CO, NO and NO_2 were measured continuously using a combustion gas composition sensor (Hodaka HT-2000). For each run, approximately 20-25 min of combustion under constant pellet feed and air supply was conducted. It is interesting to observe that combustion showed two phases cyclically with one of the cycles is flaming while other char is glowing. It is expected that the behavior of gas composition in the form of cyclic period and amplitude due to the combustion condition. The details would be discussed in the full paper.

(167c) Is Elevated Pressure Required to Achieve a High Fixed-Carbon Yield of Charcoal From Biomass?

Liang Wang[1], Oyvind Skreiberg[1], Morten Gronli[2] and **Michael J. Antal**[3], (1)SINTEF Energy Research, Trondheim, Norway, (2)Energy and Process Engineeringt, Norwegian University of Science and Technology, Trondheim, Norway, (3)Hawaii Natural Energy Institute, University of Hawaii, Honolulu, HI

Elevated pressure secures the highest fixed-carbon yields of charcoal from corncob. Operating at a pressure of 0.8 MPa a Flash-Carbonization reactor realizes fixed-carbon yields that range from 70 to 85% of the theoretical thermochemical equilibrium value from Waimanalo corncob. The fixed-carbon yield is reduced to a range from 68 to 75% of the theoretical value when whole Waimanalo corncobs are carbonized under nitrogen at atmospheric pressure in an electrically heated muffle furnace. The lowest fixed-carbon yields are obtained by the standard proximate analysis procedure for biomass feedstocks: this yield falls in a range from 49 to 54% of the theoretical value.

A round-robin study of corncob charcoal and fixed-carbon yields involving three different thermogravimetric analyzers (TGAs) revealed the impact of vapor-phase reactions on the formation of charcoal. Deep crucibles that limit the egress of volatiles from the pyrolyzing solid greatly enhance charcoal and fixed-carbon yields. Likewise, capped crucibles with pinholes increase the charcoal and fixed-carbon yields compared with values obtained from open crucibles. Large corncob particles offer much higher yields than small particles. These findings show that secondary reactions involving vapor-phase species (or nascent vapor-phase species) are at least as influential as primary reactions in the formation of charcoal.

Our results offer considerable guidance to industry for its development of efficient biomass carbonization technologies. Size reduction handling of biomass (e.g. tub grinders and chippers), which can be a necessity in the field, significantly reduces the fixed-carbon yield of charcoal. Fluidized bed and transport reactors, which require small particles and minimize the interaction of pyrolytic volatiles with solid charcoal, cannot realize high yields of charcoal from biomass. When a high yield of corncob charcoal is desired, whole corncobs should be carbonized at elevated pressure. Under these circumstances, carbonization is both efficient and quick.

(167d) Understanding the Reaction Network and Kinetics of the Hydrothermal Liquefaction of Microalgae

Peter J. Valdez, Chemical Engineering, University of Michigan, Ann Arbor, MI and Phillip E. Savage, Chemical Engineering, University of Michigan, Ann Arbor, Ann Arbor, MI

Hydrothermal liquefaction can convert wet algal biomass, at elevated temperatures and pressures, into an energy-dense biocrude oil without the need for dewatering the algae or extracting the lipids. Liquefaction occurs as the biomacromolecules react in and with water at elevated temperatures (e.g., 350 °C) and pressures to form smaller molecules that are closer to being in the range required for liquid fuels. Recent studies of microalgal liquefaction, examining some process variables, provide some clues about the reaction pathways, but not enough information to develop an entire reaction network. As the field progresses, quantitative reaction models based on the governing reaction network will be needed for process design and optimization. Developing such a model requires data from a thorough systematic study of the influence of all the relevant process variables on liquefaction of a single alga.

We present results from the liquefaction of *Nannochloropsis* sp. and the product intermediates to determine reaction pathways. From the experimental results, we determined the reaction network and constructed a model to predict the yields of the product fractions. We used first-order kinetics to describe the formation and consumption of the product fractions from liquefaction. The abundance of different compounds and the complexity of the numerous reactions necessitated the use of a global kinetic analysis to make the problem tractable. We applied this model to the liquefaction of other alga species. We will also show the fate of phosphorus in the product fractions.

Inhibition of Catalyst Poisoning in Supercritical Water Gasification of Sewage Sludge: Technical Study on Reactor Design

Osamu Sawai, Teppei Nunoura and Kazuo Yamamoto
Environmental Science Center, The University of Tokyo, Bunkyo-ku, Tokyo, Japan

Abstract

A lab-scale batch reactor with two compartments, so called two-bed reactor, was introduced to the supercritical water gasification of sewage sludge. With Ni catalyst presence, the reactor was capable of suppressing deposition of byproducts and residuals on catalyst surface attaining higher gasification efficiencies. Moreover, acceleration of hydrogen gas production was achieved when K_2CO_3 was loaded with the presence of Ni catalyst. Further studies have revealed that K_2CO_3 loading location influenced the gasification behavior dramatically. Besides Ni catalyst, RuO_2 and Ru catalyst were also employed using the reactor. In prospect of future application, discussion on heavy metal concentration in the effluent is made as well.

Introduction

Supercritical water gasification is a thermal method which enables us to transform biomass to hydrogen rich gas with reduced tar and char formation [1,2]. Easy gas separation and fast reaction rate (compared to anaerobic digestion) leading to compact equipment as well as elimination of post-drying step unlike conventional thermo-chemical treatment processes [3] are also likely to be featured as significant advantages of this technology. Although the process seems to be promising, supercritical water gasification for real biomass is still under development. The main challenges are problems concerning with salts originating from biomass. Alkaline salts tend to increase hydrogen yield, but at the same time it is associated with the risk of plugging [4] and catalyst poisoning [5]. The issue still remains as a challenge being the major obstacle to technical application.

This paper focuses on the geometry of the gasification reactor that will lead to the inhibition of catalyst poisoning in supercritical water. Experimental investigations were conducted under lab-scale batch reactors in order to exclude the plugging issues, thus focusing on the catalyst poisoning inhibition. Corresponding to the new reactor design, combination of metal supported catalyst and alkaline additives was introduced.

Methods

Materials

In this work, sewage sludge was used as real biomass feedstock. Used sewage sludge classified as excessive sludge, was supplied from the conventional activated sludge processes operated by a waste water treatment plant in Japan. Sewage sludge was dried completely in an oven kept at 105°C for at least 6 hours, and subsequently sieved into fine particles (<180 μm). All the experiments and results shown in this work were conducted using this post-treated sewage sludge (cited as "initial sewage sludge" further on). Precise profile of the initial sewage sludge has been described in our previous work [6].

Nickel (65 wt.%) on silica/alumina (35 wt.%) catalyst (cited as "Ni/SiO_2-Al_2O_3" catalyst further on) and ruthenium (5 wt.%) on alumina (95 wt.%) catalyst (cited as "Ru/Al_2O_3 catalyst" further on) used were purchased from Sigma-Aldrich, USA. Ruthenium oxide and potassium carbonate used were purchased from Wako Pure Chemical Industries Ltd., Japan. Deionized water was prepared in our laboratory by Milli-Q A10; Millipore and all materials were used without further treatment.

Experimental Set Up

All the experiments were performed by two-bed reactor, a lab-scale batch reactor manufactured from stainless steel (SUS316) tube and Swagelok® SUS316 fittings/valves. As shown in figure 1, the reactor consists of two compartments, A and B, with volume ratio of approximately 4.2:1. This reactor configuration aims to reduce poisoning caused by the solid residue. The compartment is expected to physically separate or limit direct contact of biomass/water feedstock and catalyst. SUS316 frit filter with pore size of 2 μm, purchased from Valco Instruments Co. Inc., USA is used to separate the two compartments. An identical frit filter is inserted on the other side of compartment B to limit the access of solids to equipped pressure transducer and the needle valve.

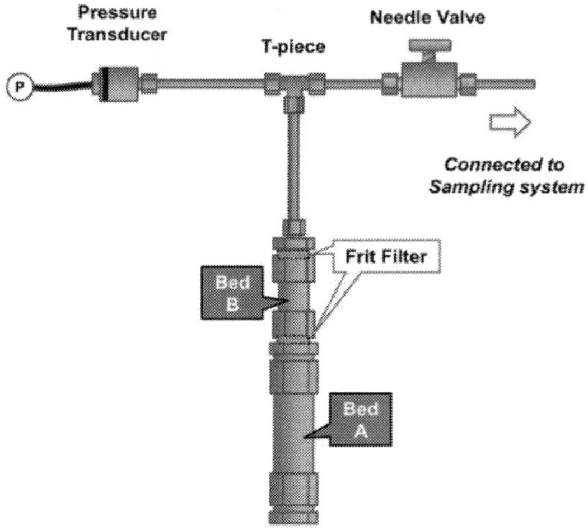

Figure 1. Schematic diagram of two-bed reactor

Char and tar which are likely to pass the filter causing catalyst poisoning, can be resolved by using an additive inhibiting their formation. In our experiments, K_2CO_3 which is reported to function as a char and tar formation inhibitor [7-9], was added to either of the compartment. Deionized water, initial sewage sludge (5 wt.% of deionized water), catalyst and K_2CO_3 additives (10 wt.% of sewage sludge respectively) were loaded in either of the bed. The gasification was conducted under the condition of 400°C and 27 MPa in standard (cited as "standard SCWG condition" further on).

The gasification reaction was initiated by immersing the two-bed reactor into temperature-controlled molten salt bath (mass ratio of salt was adjusted to K_2CO_3:$NaNO_2$:$NaNO_3$ = 6:5:1). After lapse of predefined time, the reactor was taken out of the bath and subsequently quenched by a water bath to stop the reaction. Reaction time, defined as time from which the reactor was immersed into molten salt bath until it was taken out, was set to 30 minutes. During the reaction, pressure was monitored by the transducer to make sure that the reaction media was kept at supercritical conditions.

Reaction products were collected after the reactor was cooled to ambient condition. Gas products were first transferred to the initially evacuated sampling loop with known volume. Amount of gas product was determined from pressure monitored by pressure transducers connected to the sampling loop. Produced gas was sampled from one of the sampling loop ports using a gas-tight syringe for gas analysis injection. Liquid and solid residue collected as mixtures were subsequently separated by centrifugation run at 2500 rpm for 5 minutes. Liquid was filtered by 0.45 μm pore size syringe filter (Millex LH, Millipore) and diluted by deionized water prior to the analysis. Separated solid residues were dried in an oven kept at 105°C for at least 6 hours.

Analytical Equipments

Gas analysis was carried out with gas chromatograph (GC) GC-2014; SHMADZU equipped with Shincarbon ST 50/80 column and thermal conductivity detector (TCD) to separate hydrogen, methane, carbon monoxide, and carbon dioxide. Total organic carbon (TOC) content in the liquid phase was determined by total organic carbon analyzer TOC-V; SHMADZU. Concentrations of heavy metals were analyzed by inductively coupled plasma mass spectrometry (ICP-MS) ICP-8500; SHIMADZU. Microscopic observation of solid residues was carried out by scanning electron microscope (SEM) JSM-5600; JEOL ltd. equipped with energy dispersive X-ray spectroscopy (EDX) Oxford Link ISIS; Oxford Instruments. Prior to SEM analysis, Au was spattered on the sample as a pretreatment.

Results and Discussion

Validity of Two-bed Reactor

Loading location of Ni catalyst was varied in order to demonstrate the validity of the reactor. Figure 2 shows the effect of Ni catalyst loading location to gas yield and gas efficiency. Citation such as "(SS+Ni Cat.)|None" indicates that sewage sludge and Ni catalyst is loaded in compartment A, and that compartment B is vacant. In this paper, hydrogen gas efficiency (HGE) and carbon gas efficiency (CGE) are defined as shown below.

HGE = (mol of H atom in gas products) / (mol of H atom in sewage sludge loaded) (1)
CGE = (mol of C atom in gas products) / (mol of C atom in sewage sludge loaded) (2)

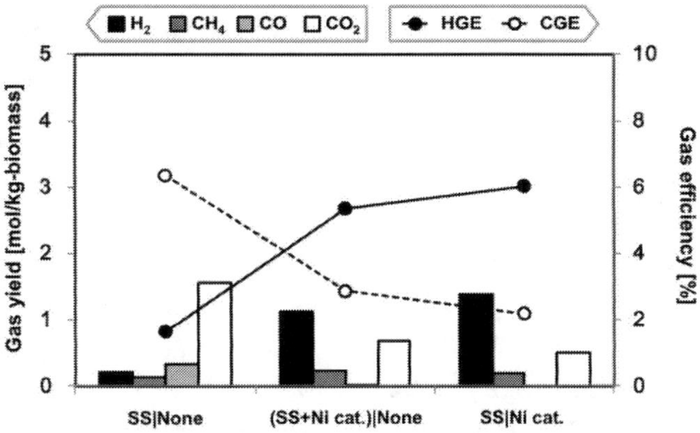

Figure 2. Effect of two-bed reactor to gas yield and efficiency

Addition of Ni catalyst increased hydrogen gas yield and decreased carbon monoxide and carbon dioxide gas yield significantly as shown in figure 2. Hydrogen gas yield increased and carbon dioxide gas yield decreased furthermore when the sewage sludge and Ni catalyst was loaded in different compartments. On the other hand, methane gas yield was not affected by the loading location of Ni catalyst. Gasification efficiency curves behaved corresponding to the component gas behavior. From this trend, it is clear that the reactor geometry adopted in this work did not limit catalyst activation, but contributed to attain higher hydrogen gas efficiency.

The two-bed reactor not only exhibited higher hydrogen gas efficiency but also inhibited both inorganic salt and byproduct deposition. SEM image of Ni catalyst after run "SS|Ni cat." is shown in figure 3. As shown in the SEM images, salts and byproducts deposited on neither of the support surfaces. Although SiO_2 support (figure 3 (a)) avoided the deposition of byproducts, its porous structure was destroyed during the supercritical gasification.

Figure 3. SEM image of Ni catalyst after reaction using two-bed reactor; (a) SiO_2 (b) Al_2O_3 support.

Effect of K_2CO_3 Addition on Gas Yield and Metal Elution

To attain higher gasification efficiency, addition of K_2CO_3, well known for accelerating water-gas shift reaction [10-13] and suppressing formation of char [7,8] and tar [9] was taken into consideration. Figure 4 shows the effect of K_2CO_3 addition or/and Ni catalyst to gas yield and gas efficiency. From the figure, addition of K_2CO_3 shows a significant increase of hydrogen gas yield in the case of Ni catalyst absence. Methane and carbon dioxide gas yield increased slightly whereas carbon monoxide gas yield decreased. HGE and CGE curve behaved accordingly to the corresponding gas component yield.

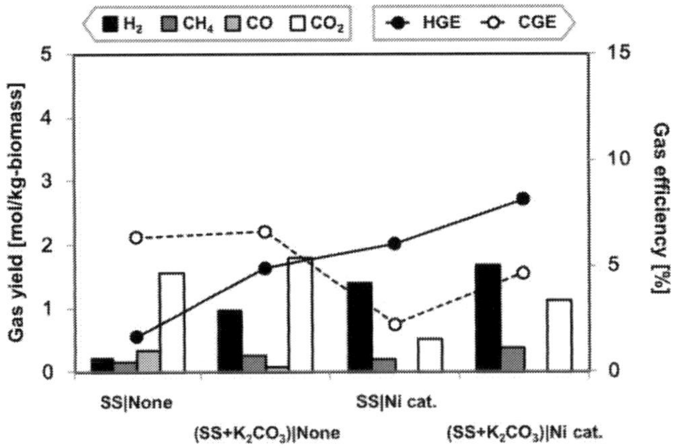

Figure 4. Effect of K_2CO_3 addition to gas yield and efficiency

Effect of K_2CO_3 addition under the presence of Ni catalyst is also shown in the figure 4. Combination of Ni catalyst and K_2CO_3 [(SS+K_2CO_3)|Ni Cat.] presented a positive effect on the hydrogen gas yield exhibiting higher yield than when K_2CO_3 or Ni catalyst was loaded alone [(SS+K_2CO_3)|None, SS|Ni Cat.]. Similar behavior was observed for methane exhibiting highest gas yield when Ni catalyst was used with K_2CO_3. HGE increased rapidly corresponding to the behavior of hydrogen and methane gas yield. The figure also indicates that Ni catalyst suppresses the formation of carbon monoxide production. K_2CO_3 addition presented a positive effect on carbon dioxide production under Ni catalyst presence, resulting in a higher CGE. Compared to the works of D'Jesus *et al.* or Schmieder *et al.* in which they reported that potassium addition did not influence the gasification yield in case of real biomass [12,14], sewage sludge presented both positive and negative effect on the yield of component gas.

In parallel, effect of K_2CO_3 addition to metal elution was investigated. From figure 5 comparing the effect of K_2CO_3 addition to metal elution, Mo exhibited higher degree of elution when K_2CO_3 was present. Elution of Ni and Mo was induced when the gasification is conducted under the presence of Ni catalyst with K_2CO_3. Zn was eluted when the gasification is conducted under catalytic conditions. On the contrary, species such as Fe and Se tended to increase when Ni catalyst was used regardless of K_2CO_3 addition.

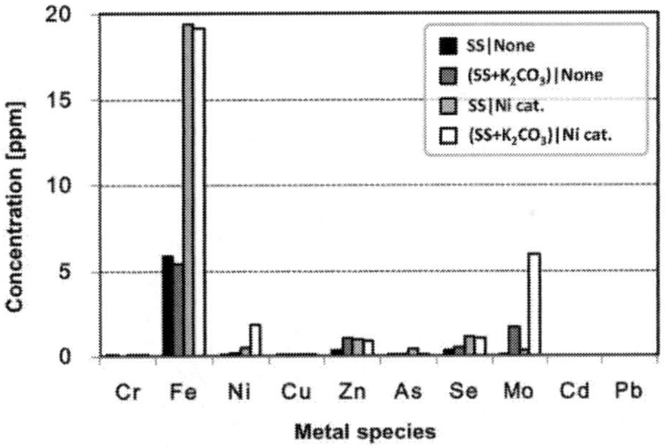

Figure 5. Effect of K_2CO_3 addition to metal concentration in liquid phase

Result shown here indicates that the enhancement of gas yield by Ni catalyst and/or K_2CO_3 addition is likely to be associated with elution of heavy metal species originating from reactor material and metals deposited on the catalyst.

Effect of K_2CO_3 Loading Location

Effect of K_2CO_3 loading location on gas yield and efficiency was explored. K_2CO_3 was loaded with either sewage sludge or Ni catalyst. As shown in figure 6 comparing the effect of K_2CO_3 loading location to gas yield and efficiency, the highest gas yield was achieved when Ni catalyst was separated from K_2CO_3 and sewage sludge loaded compartment, i.e., [(SS+K_2CO_3)|Ni Cat.] condition. On the other hand, gasification was suppressed when sewage sludge was separated with Ni catalyst and K_2CO_3 loaded compartment. Hydrogen, methane and carbon dioxide gas yield was lower than the case in which sewage sludge, Ni catalyst, and K_2CO_3 was loaded together in the same compartment. Furthermore, carbon monoxide yield showed a slight increase which implies that the catalytic activity was not efficiently exerted at the condition where Ni catalyst and K_2CO_3 were loaded in the same compartment.

In addition, elution of metal species was investigated. Effect of K_2CO_3 loading location to metal elution is shown in figure 7. As shown in the figure, metal elution can be inhibited by loading Ni catalyst and sewage sludge in separate compartments. Especially Ni concentration decreased drastically when the Ni catalyst and sewage sludge were loaded separately regardless of K_2CO_3 loading location. Elution of Mo was inhibited drastically when Ni catalyst was loaded separately from both K_2CO_3 and sewage sludge. Zn and Se showed similar trend to Mo but with moderate degree. Interaction between Ni catalyst and K_2CO_3 can assumedly provoke metal elution via behaviors such as drastic alteration of pH or dissolution of produced gas. Therefore, limitation of interaction by separating Ni catalyst and K_2CO_3 has possibly relieved the metal elution. The result obtained here implies that the two-bed reactor can also contribute to the inhibition of metal elution as well as gasification enhancement and inhibition of catalyst poisoning.

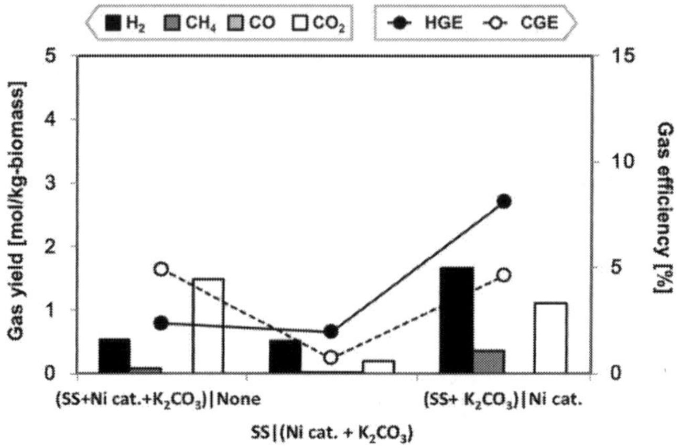

Figure 6. Effect of K_2CO_3 loading location to gas yield and efficiency

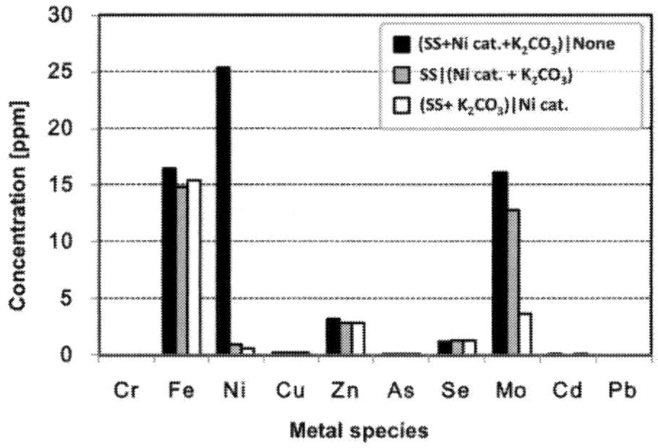

Figure 7. Effect of K_2CO_3 loading location to metal concentration in liquid phase

Variation of catalyst types

Metal supported catalysts reported to perform high catalytic activity were employed. Figure 8 compares the gas yield and efficiency of sewage sludge SCWG using RuO_2, Ru and Ni catalyst. Nickel apparently presented the highest catalytic activity in this work unlike in previous reports [15-17], since the catalyst amount (including support weight) added was fixed in this work. RuO_2, also reported to enhance biomass gasification under SCW condition, presented lower hydrogen gas yield than Ru catalyst. However, HGE was higher due to the production of methane. Comparative high methane yield is also reported by Yamamura *et al.* in which they applied RuO_2 to sewage sludge SCWG [17]. They have concluded that RuO_2 not only enhanced the gasification reaction but at the same time accelerated the methanation reaction. Meanwhile, the addition of RuO_2 in our work also enhanced the production of carbon dioxide and carbon monoxide gas yield, which as a result increased the CGE significantly.

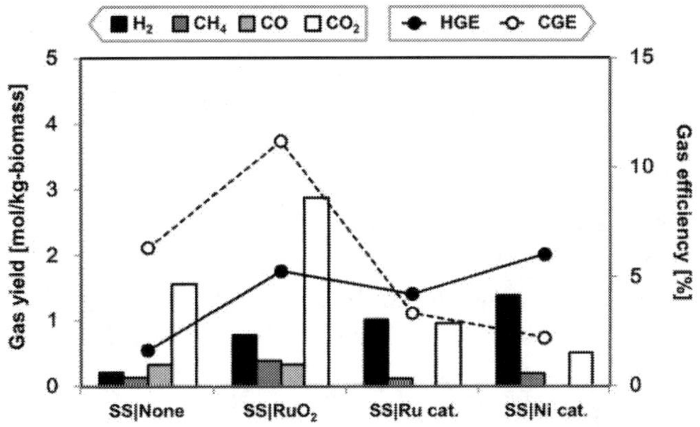

Figure 8. Comparison of gas yield using various catalysts

Conclusions

In this work, a lab-scale batch reactor with two compartments was designed to inhibit catalyst poisoning. Under the condition of 400°C, 27 MPa and reaction time of 30 minutes with Ni catalyst presence, the reactor was capable to suppress the deposition of byproducts and residuals with higher gasification efficiencies. Furthermore, acceleration of hydrogen gas production was achieved when K_2CO_3 was added to sewage sludge compartment, loaded separately from Ni catalyst. Further studies have revealed that the loading location of K_2CO_3 influenced the gasification behavior dramatically.

RuO_2 and Ru catalysts reported to promote SCWG of various biomass were also employed. Due to low metal loading of Ru, Ni catalyst tended to exhibit slightly higher hydrogen yield than Ru catalyst. RuO_2 exhibited lower catalyst activity for hydrogen production but promoted the production of carbon dioxide compared to Ni catalyst and Ru catalyst. Metal elution was also investigated throughout this work. The results have revealed that enhancement of gas yield by catalyst and/or K_2CO_3 addition is likely to be associated with elution of heavy metal species (trade-off relationship between gasification yield/efficiency and metal elution), originating from reactor material, sewage sludge and supported catalyst particles. From the view point of effluent management, elution of metal should also be taken into account along with the catalytic activity of gasification in case of selecting a catalyst in a real process.

Acknowledgments

The authors acknowledge the support of New Energy and Industrial Technology Development Organization (NEDO) and Japan Society of the Promotion of Science (JSPS) through the Funding Program for World-Leading Innovative R&D on Science and Technology (FIRST). The authors wish to gratefully appreciate the Materials Design and Characterization Laboratory, Institute for Solid State Physics (ISSP), University of Tokyo for providing SEM-EDX used in this work.

References

1. L.J. Guo, Y.J. Lu, X.M. Zhang, C.M. Ji, Y. Guan, A.X. Pei, Hydrogen production by biomass gasification in supercritical water: A systematic experimental and analytical study, Catalysis Today 129 (2007) 275–286.
2. A. Demirbas, Hydrogen-rich gas from fruit shells via supercritical water extraction, Int. J. Hydrogen Energy 29 (2004) 1237–1243.

3. D.T. Furness, L.A. Hoggett, S.J. Judd, Thermochemical treatment of sewage sludge, Water Environ. J. 14 (2000) 57–65.
4. M.J. Antal, S.G. Allen, D. Schulman, X.D. Xu, R.J. Divilio, Biomass gasification in supercritical water, Ind. Eng. Chem. Res. 39 (2000) 4040–4053.
5. M.H. Waldner, F. Krumeich, F. Vogel, Synthetic natural gas by hydrothermal gasification of biomass: Selection procedure towards a stable catalyst and its sodium sulfate tolerance, J. Supercritical Fluids 43 (2007) 91–105.
6. O. Sawai, Y. Tanaka, T. Nishihara, T. Nunoura, K. Yamamoto, Application of Supercritical Water Gasification to Municipal Wastewater Treatment Plant: Experimental Studies and Operational Challenges, 10th International Symposium on Supercritical Fluids, San Francisco, California, USA (2012) L-139.
7. A. Sinag, A. Kruse, V. Schwarzkopf, Key compounds of the hydropyrolysis of glucose in supercritical water in the presence of K_2CO_3, Ind. Eng. Chem. Res. 42 (2003) 3516-3521.
8. T. Minowa, F. Zhen, T. Ogi, Cellulose decomposition in hot-compressed water with alkali or nickel catalyst, J. Supercritical Fluids 13 (1998) 253–259.
9. A. Kruse, T. Henningsen, A. Sinag, J. Pfeiffer, Biomass gasification in supercritical water: influence of the dry matter content and the formation of phenols, Ind. Eng. Chem. Res. 43 (2004) 502-508.
10. J. Yanik, S. Ebale, A. Kruse, M. Saglam, M. Yüksel, Biomass gasification in supercritical water: II. Effect of catalyst, Int. J. Hydrogen Energy 33 (2008) 4520–4526.
11. J.A. Onwudili, P.T. Williams, Role of sodium hydroxide in the production of hydrogen gas from the hydrothermal gasification of biomass, Int. J. Hydrogen Energy 34 (2009) 5645-5656.
12. H. Schmieder, J. Abeln, N. Boukis, E. Dinjus, A. Kruse, M. Kluth, G. Petrich, E. Sadri, M. Schacht, Hydrothermal gasification of biomass and organic wastes, J. Supercritical Fluids 17 (2000) 145–153.
13. X.H. Hao, L.J. Guo, X. Mao, X.M. Zhang, X.J. Chen, Hydrogen production from glucose used as model compound of biomass gasified in supercritical water, Int. J. Hydrogen Energy 28 (2003) 55–64.
14. P. D'Jesus, N. Boukis, B. Kraushaar-Czarnetzki, E. Dinjus, Gasification of corn and clover grass in supercritical water, Fuel 85 (2006) 1032-1038.
15. P. Azadi, K.M. Syed, R. Farnood, Catalytic gasification of biomass model compound in near-critical water, Applied Catalysis A: General 358 (2009) 65–72.
16. A. Yamaguchi, N. Hiyoshi, O. Sato, K. Bando, M. Osada, M. Shirai, Hydrogen production from woody biomass over supported metal catalysts in supercritical water, Catalysis Today 146 (2009) 192–195.
17. T. Yamamura, T. Mori, K.C. Park, Y. Fujii, H. Tomiyasu, Ruthenium (IV) dioxide-catalyzed reductive gasification of intractable biomass including cellulose, heterocyclic compounds, and sludge in supercritical water, J. Supercritical Fluids 51 (2009) 43-49.

(167f) Effect of Acid and Alkali On Supercritical Water Gasification of Glucose

Yukihiko Matsumura, Hokuto Takai and Chutinan Promdej, Hiroshima University, Higashi-Hiroshima, Japan

Supercritical water gasification, which is the gasification of biomass and biomass component compounds in hot compressed water above critical temperature (647 K) and critical pressure (22.1 MPa) of water, is expected to be an effective gasification technology for wet biomass. However, its reaction kinetics is still under investigation. Especially, char and tarry material formation is a big problem, and its reaction kinetics elucidation is needed. During the study of glucose decomposition in supercritical water, the authors' group found that some of the reactions behave as Arrhenius reaction rate, and others do not, and dsitinguished ionic and radical reactions. It was then found that the char formation is ionic reaction. Char formation was expected to be enhanced by the organic acids produced by glucose decomposition. If it is the case, addition of acid or alkali should affect the char formation. In this study, the authors present the results of HCl and NaOH addition on the supercritical water gasification of glucose, clarifying the effect of acid and alkali on the reaction kinetics.

(174a) Reaction Route Selection Based On Sustainability Criteria for Glutaraldehyde Production

Preeti Gangadharan and Helen H. Lou, Dan F. Smith Department of Chemical Engineering, Lamar University, Beaumont, TX

Challenges in the specialty chemicals industry are enormous. Companies need to deliver the most profitable product mix to keep up with global competition, while also keeping in mind long term environmental goals. In most cases, problems such as waste disposal or treatment may be avoided, or their intensity reduced, by considering their effects prior to plant design. Hence, sustainability criteria should form an integral part of process design, starting from the early design stage. In this stage, the sustainability performance of each potential reaction pathway is evaluated in terms of possible profit, driving force of the pathway (Gibbs energy), inherent safety index, environmental analysis, and atom economy. For specialty chemicals, particular emphasis has to be given on safety and environmental criteria.

A structured methodology incorporating all the major criteria is applied to reaction pathway selection for producing Glutaraldehyde, a specialty chemical primarily used as a bactericide, tanning agent, and fixative. Three alternative routes are considered. The fast and comprehensive assessment is used to quantify the desirability of each route, so that inferior routes can be screened out systematically, while also identifying areas for improvement, prior to initiating detailed design. In the detailed design stage, the sustainability performance of the selected process is improved through sequential sustainability assessment covering economic, safety, environmental, and efficiency criteria.

(174b) Testing Room Temperature Ionic Liquid Solutions

Paul Brezovec, Concurrent Technologies Corp., Johnstown, PA

To avoid legacy problems associated with cadmium's carcinogenicity and high toxicity, some new Air Force weapon system steel component designs by original equipment manufacturers (OEMs) specify an aluminum plating instead, to satisfy weapon system corrosion protection requirements. When these weapon systems come to the Air Logistics Centers (ALCs) for maintenance, repair and overhaul (MRO), repair-scale processes need to be in place to refurbish the coating that is on the exterior and recesses of the components. The new technologies used by the OEMs to plate these components with aluminum require extensive infrastructure and capital expense.

Technologies that electroplate aluminum from non-aqueous chemistries such as room temperature ionic liquid solutions are being marketed to replace cadmium electroplating for MRO needs, with fewer infrastructure and capital expenses than the OEM's outlays. In general, room temperature ionic liquids are salts that remain liquid near room temperatures and have a very low vapor pressure unlike organic solvents. Room temperature ionic liquids can dissolve metal salts from which the metals such as aluminum may be electrodeposited, and have attracted significant laboratory research efforts, because they can be energy efficient and some are considered "green."

The applicability of this technology to DoD plating shops, Air Force ALC plating shops in particular, is of interest. Despite their promise, there are caveats associated with room temperature ionic liquids, including concerns with the toxicity of a variety of them, and processing difficulties. In particular, processing must consider the effects of additives on the physical properties of the ionic liquid-based electrolyte. A review of room temperature ionic liquid additives and their effects on the electrodeposition process will be presented, and screening testing results to date for potential applications will be described.

(174c) Looped Carbon Capturing and Environmental Remediation: Case Study of Magnetic Polypropylene Nanocomposites

Jiahua Zhu[1], Hongbo Gu[1], Sowjanya B. Rapole[1], Minjiao Chen[1], Suying Wei[2] and Zhanhu Guo[1], (1)Integrated Composites Laboratory (ICL), Dan F Smith Department of Chemical Engineering, Lamar University, Beaumont, TX, (2)Chemistry and Biochemistry, Lamar University, Beaumont, TX

A waste-free process to recycle Fe@Fe_2O_3/polypropylene (PP) polymer nanocomposites (PNCs) is introduced to synthesize magnetic carbon nanocomposites (MCNCs) and simultaneously produce useful chemical species which can be utilized as feedstock in petrochemical industry. The magnetic nanoparticles (NPs) are found to have an effective catalytic activity on the pyrolysis of PP. The PNCs (with a NP loading of 20.0 wt%) undergo a complete degradatisxcsxaxazxasxzasason with 2-hour pyrolysis at 500 °C in the H_2/Ar atmosphere and the degradation components exhibit a distribution of different carbon number species, while only 40% of pure PP is decomposed after applying the same pyrolytic conditions. The coked solid waste in the conventional process has been utilized as carbon source to form a protective carbon shell surrounding the magnetic NPs. The magnetic carbon nanocomposites (MCNCs) pyrolyzed from PNCs containing 20.0 wt% NPs demonstrate an extremely fast Cr(VI) removal from the wastewater with an almost complete removal of Cr(VI) within 10 min. The pH effect on the Cr(VI) removal efficiency is investigated with a preferable value of 1-3. The adsorbent exhibits much higher adsorption capacity in the acidic solutions than that in the alkali solutions. The large saturation magnetization (32.5 emu g^{-1}) of these novel magnetic carbon nanocomposites allows fast recycling of both the adsorbents and the adsorbed Cr(VI) from liquid suspension in a more energetically and economically sustainable way by simply applying a permanent magnet. The significantly reduced treatment time required to remove the Cr(VI) makes these MCNCs promising for efficient removal of the heavy metals from the wastewater. Kinetic investigation reveals a pseudo-second-order adsorption of Cr(VI) on these novel magnetic carbon nanocomposite adsorbents.

(174d) Transformation of Carbon Dioxide Into Boron-Graphene Oxide Nanocomposites Using a Reducing Agent

Jae W. Lee[1], Junshe Zhang[1], Yu Zhao[2], Daniel L. Akins[2], Ruth Stark[3], Xudong Guan[4] and Ah-young Byun[5], (1)Department of Chemical Engineering, The City College of New York, New York, NY, (2)Department of Chemistry, The City College of New York, New York, NY, (3)Chemistry, CCNY, New York, NY, (4)Chemistry Department, CCNY, New York, NY, (5)Chemical Biomolecular Eng, KAIST, Daejeon, South Korea

Carbon dioxide (CO_2), a readily available, low-cost and sustainable carbon source, could be converted to liquid fuels, useful chemicals, and carbon materials. Among the several alternatives of reducing CO_2, attractive options include transforming CO_2 to diamonds and nanotubes, either through direct CO_2 splitting or via reaction with metals subject to pressures higher than 70 MPa. Nonetheless, the possibility of CO_2 reduction to other carbon nanomaterials like graphene oxide without using metal-containing compounds and without employing extreme conditions has not been demonstrated. Graphene oxide is not only a suitable and scalable precursor of graphene, but also a flexible platform for biological, chemical, electronic, medical, and optical applications. This talk reports the conversion of CO_2 to graphene oxide nanocomposites using ammonia borane (NH_3BH_3), which is a well-known candidate material for hydrogen storage. The conversion involves two consecutive steps: carbon fixation and then graphenization. The carbon fixation is the reaction of CO_2 with NH_3BH_3 under mild temperature and pressure conditions ($T < 100\ ^oC$ and $P < 3$ MPa) to produce a solid compound that contains methoxy (OCH_3), formate (COOH) and aliphatic groups, whereas the graphenization is the pyrolysis of this solid compound at high temperatures ($T > 600\ ^oC$) and atmospheric pressure of nitrogen to generate graphene oxide-boron oxide nanocomposites. The generation of the nanocomposites is confirmed by micro-Raman spectroscopy, solid state ^{13}C and ^{11}B magic angle spinning-nuclear magnetic resonance (MAS-NMR), transmission electron microscopy (TEM), and atomic force microscopy (AFM).

(174e) Environmental Life Cycle Assessment of New Energetic Materials

Amirhossein Mehrkesh and Arunprakash T. Karunanithi, Department of Civil Engineering, University of Colorado Denver, Denver, CO

There is growing interest in the development of new energetic ionic liquids for use as aerospace propellants and explosives. Due to their negligible vapor pressure ionic liquids are usually considered as "green" alternatives to volatile organic compounds. In addition, ionic liquids and salts have tunable physical and chemical properties. Energetic materials release large amount of energy when they decompose. In the case of explosives all energy is released rapidly while in the case of fuels the energy is released in a controlled manner. These materials derive their energy content from oxidation of the carbon backbone or from their high positive heats of formation. The general requirements for energetic materials are high energy and density, thermal stability, low sensitivity to impact and low toxicity. Traditional energetic materials that are commonly used in explosive formulations are HMX (1,3,5,7-tetranitro-1,3,5,7-tetraazacyclooctane), RDX (1,3,5-trinitro-1,3,5-triazacyclohexane) and TNT (2,4,6, trinitrotoluene). Hydrazine derivatives are widely used as energetic fuel in rocket propulsion systems. Most recently several energetic ionic liquids have been proposed as high-energy materials, propellants and explosives. A large number of ionic salts and ionic liquids that are based on a triazole derivative have been proposed as energetic materials. 1,2,4 triazole and 1,2,3 triazole have heats of formation values of 109 KJ/mol and 272 KJ/mol respectively.

In this work we use environmental life cycle assessment method to study the environmental impact associated with the production of ionic liquids and ionic salts to compare with the environmental impact of common energetic material such as TNT. This will provide insights into whether the newer energetic materials that are currently being designed are indeed environmentally friendly as being assumed.

(174f) Economic Comparison of Processes for the Recovery of Lactic Acid by Reactive Distillation with Different Alcohols

Chien-Yuan Su, Cheng-Ching Yu and Jeffrey D. Ward, Department of Chemical Engineering, National Taiwan University, Taipei, Taiwan

Poly lactic acid (PLA) is a biodegradable, environmentally-friendly polymer that can be made from renewable resources and has many potential applications. Production of lactic acid (LAC) from renewable resources is usually accomplished by fermentation which has the additional advantage that an enantiomerically pure product is produced. Therefore, separation and purification of lactic acid from fermentation broth is important. However, it is also challenging because LAC is relatively nonvolatile and tends to oligomerize at high temperature. The most attractive method for purification of LAC is a two-step reaction process in which the acid is first reacted with an alcohol to form a more volatile ester which can be more easily purified by distillation. Then pure LAC is recovered by hydrolysis of the ester (Kumar et al., 2006; Kumar and Mahajani, 2007).

In this work we determine the optimal designs of esterification/hydrolysis lactic acid recovery processes using different alcohols and compare the cost of each process. We consider the presence of a heavy organic impurity and LAC oligomerization reactions in order to make the design problem more realistic. Processes are simulated using the commercial simulation package ASPEN PLUS. The processes are optimized to minimize the Total Annual Cost (TAC), with an overall lactic acid with lactic acid esterification and hydrolysis conversions of 99 mol%.

Different alcohols lead to different process configurations due to different component characteristics (especially boiling point ranking). The methanol system includes one pretreatment column, two reactive distillation columns and one separation column. The advantage of the methanol system is that methyl-lactate is the lightest compound among all lactic esters, so it is easier to separate methyl-lactate from other species. The butanol system includes one pretreatment column, two reactive distillation columns, two decanters and two separation column. The advantage of the butanol system is that there is a two-liquid phase zone, which facilitates separation of the species. It also may reduce the energy consumption of the separation system if the system is well designed. Designs using ethanol and isopropanol are also briefly discussed.

The butanol system has the lowest TAC among all alcohols due to low separation cost in the esterification and separation column because of the two-liquid-phase region. Furthermore, the methanol system has lower TAC than the ethanol and isopropanol systems due to the absence of an azeotrope between the alcohol and water.

References

Kumar R., Nanavati H., Noronha S. B. and Mahajani S. M. A continuous process for recovery of lactic acid by reactive distillation, *Journal of Chemical Technology and Biotechnology*, **81**, 1767-1777, 2006.

Kumar R. and Mahajani S. M. Esterification of Lactic Acid with n-Butanol by Reactive Distillation, *Ind. Eng. Chem. Res.*, **46**, 6873-6882, 2007.

(174g) Synthesis of Organosilicon Complexes From Rice Husk Derived Silica Nanoparticles

Weixing Wang[1], Jarett C. Martin[2], Rongcai Huang[2], Wensi Huang[1], Anhua Liu[3], Aijie Han[4] and **Luyi Sun**[2], (1)Department of Chemical Engineering, South China University of Technology, Guangzhou, China, (2)Department of Chemistry and Biochemistry, Texas State University-San Marcos, San Marcos, TX, (3)Materials Science and Engineering, South China University of Technology, Guangzhou, China, (4)Department of Chemistry, University of Texas-Pan American, Edinburg, TX

Organosilicon compounds are a group of important materials with widespread applications. However, their synthesis has been relying mainly on the carbothermal reduction of silica to silicon and subsequent reaction between silicon and alkyl or aryl chloride. This multi-step approach involves high temperature, high pressure, and strong acidity, which is energy-intensive, expensive, and eco-hazardous. An alternative approach was explored to synthesize organosilicon complexes via a green chemical process using silica nanoparticles derived from rice husk (RH) biomass as the starting material. By controlling the pre-treatment of RHs and pyrolysis conditions, silica samples with various surface areas and levels of crystallinity were synthesized. Such silica can be converted to silicon complexes via a low temperature approach. The synthesized silicon complexes were characterized by ^{1}H, ^{13}C, ^{29}Si nuclear magnetic resonance and elemental analysis. Overall, the biogenic silica nanoparticles with high surface area and low crystallinity exhibit high reactivity. Thanks to their high reactivity and low cost, such biogenic silica nanoparticles from RHs can serve as an ideal starting material for producing organosilicon complexes.

(174h) A Novel Synthesis of Odd-Numbered Fatty Acids From Corresponding Natural Fatty Acids and the Potential Application in Soap-Based Detergents

Xiaoxiao Huang[1,2] and Yun Fang[1], (1)School of Chemical and material engineering, Jiangnan University, Wuxi, China, (2)The Key Laboratory of Food Colloids and Biotechnology, Ministry of Education, Jiangnan University, Wuxi, China

The odd-numbered fatty acids (OFAs), almost negligible in nature, are potential precursors, such as lubricant agents, medicines for their special properties. OFAs were occasionally prepared in small amounts by oxidation of aldehyde, olefin, or by fermentation route. Those traditional routes have many drawbacks, such as lower conversion, more difficult separation, higher energy consumption, environmental pollution, which hinders the process in large scale. A novel synthesis route of OFAs is developed in this paper from natural fatty acids of one more carbon atom through chlorination, hydrolysis and followed by decarboxylation. The structure of OFAs was confirmed by gas chromatography-mass spectrum (GC-MS) and fourier transform infrared spectrum (FTIR). The corresponding sodium soaps were prepared from both the newly launched odd-numbered fatty acids and the parent even-numbered fatty acids. The experimental results showed that the Krafft points of the odd-numbered soaps were lower or even far lower than that of the parent fatty acids, which means the former has better solubility in cold water. This discovery shed light on formulations of soap-based detergents which were bothered longtime by poor low-temperature solubility of sodium soap derived from natural fatty acids.

(191a) Systems Biology Analysis of the Effect of Acetic Acid On S. Cerevisiae 424A(LNH-ST) and Resistant Strains During the Co-Fermentation of Glucose and Xylose

Nathan S. Mosier, Agricultural and Biological Engineering & Laboratory of Renewable Resources Engineering (LORRE), Purdue University, West Lafayette, IN, Miroslav Sedlak, Agricultural & Biological Engineering & Laboratory of Renewable Resources Engineering (LORRE), Purdue University, West Lafayette, IN and Nancy W. Y. Ho, ChemE/LORRE, Purdue University, West Lafayette, IN

Efficient conversion of hemicellulose-derived sugars to ethanol at high yields and titers are goals toward commercializing cellulosic ethanol production. *S. cerevisiae* 424A (LNH-ST) developed at Purdue University can efficiently ferment glucose and xylose. However, inhibitors present in cellulosic feedstocks (acetic acid) and the desired fermentation product (ethanol) reduce yeast growth rate and fermentation rates, especially during xylose fermentation.

Through selective culturing, we have developed new, evolved strains with improved xylose fermentation compared to the original strain. Under control fermentations (no acetic acid), the new strain consumes xylose (g/g cells/hr) 2.2 times faster than the original strain. When challenged with acetic acid (10 g/L), the rate of xylose consumption drops almost 90% in the control strain compared to a drop of only 47% in the evolved strain. We present the results from a system biology approach to analyzing differences between our original strain and newly developed strains. We report differences between the strains and the effect of acetic acid on both. Data from the concentrations of intracellular metabolites as well as gene expression for multiple points during batch fermentations are analyzed to give insights into changes in the evolved strain.

(191b) Enhanced Ethanol Tolerance in Yeast by Micronutrients

Felix Lam, Chemical engineering, Massachusetts Institute of Technology, Cambridge, MA, Gerald Fink, Whitehead Institute, Cambridge, MA and Greg Stephanopoulos, Chemical Engineering, Massachusetts Institute of Technology, Cambridge, MA

Ethanol will continue to be the dominant mass-market biofuel in the near to intermediate term. The yeast *Saccharomyces cerevisiae* is the catalyst of choice in the conversion of plant sugars to ethanol. One of the primary bottlenecks to higher ethanol production is the toxicity of ethanol itself as well as compounds present in the fermentation medium. Cell death occurs at ethanol concentrations greater than 15-20% (v/v), and hydrolysates of cellulosic biomass (widely considered to be the next generation feedstock for renewable fuels) contain a variety of organic molecules inhibitory to growth. Conditions that increase ethanol tolerance could improve both product yield and the viability of yeast for reuse in subsequent fermentations. However, the biological basis for ethanol tolerance remains elusive: whole genome studies have shown that tolerance is influenced by many genes, and thus far, none have elicited significant improvement when manipulated alone. An alternative approach is to alter extracellular conditions such that cells become less sensitized to high ethanol concentrations. To this end, we have screened and identified several micronutrients which, when added to a fermentation culture, can increase ethanol production in *S. cerevisiae* by ~50%. These modifications to the medium do not require any genetic manipulation of strains and, in fact, lead to increased ethanol output in both laboratory and industrial strains. Furthermore, we have found that the boost in ethanol production is attributable directly to an enhancement of population viability (i.e., tolerance), and occurs in both glucose- and xylose-fermenting strains alike. Our findings simultaneously hint at a novel, univariate determinant of ethanol tolerance and are a step forward in unlocking the economic potential of cellulosic biomass.

(191c) The Impacts of Glucose Repression on Invertase Expression in Various Yeast Strains During the Ethanol Fermentation of Sucrose-Containing Substrates

Lei Zhang, Institute of New Energy Technology, Tsinghua University, Beijing, China and Shi-zhong Li, Institute of Nuclear and New Energy Technology, Tsinghua University, Beijing, China

Recently, the sweet sorghum ethanol using advanced solid-state fermentation (SSF) developed by Tsinghua University in China has been applied in the largest solid-state fermenter (550 m^3) of the world, which makes sugar-based fuel ethanol production more cost-competitive, first demonstrating the SSF for fuel ethanol production is feasible in industrial scale. Although the relatively high ethanol productivity have been achieved, seeking the yeast strains with high fermentation rate is still an important means to further strengthen the cost competitiveness of the process, due to the inherently insufficient mass and heat transfer of SSF. Due to the high contents of sucrose in sweet sorghum and sugarcane, the hydrolysis rate of sucrose by yeast into fermentable sugars, including glucose and fructose, significantly determines the ethanol fermentation rate of sucrose-containing substrates. Invertase is the key enzyme for yeast involved in the cleavage of sucrose, which is subjected to the glucose repression that poses significantly adverse effects on the yeasts' fermentation of mix-sugar substrates by extension of fermentation time. The dynamic changes of glucose levels during the utilization of sucrose-containing substrates make the regulation of invertase expression in yeast more complicated. Thereby, in this study we investigated the impacts of invertase expression on the final fermentation performances of the various *Saccharomyces cerevisiae* strains through monitoring the dynamic variation of specific invertase acivities of yeasts during the fermentation of sucrose (120 g l^{-1} of sucrose) or mixed sugars (70 g l^{-1} of sucrose, 30 g l^{-1} of glucose, and 23 g l^{-1} of fructose). During the cell growth in YPD media with glucose as sole carbon source, the both commercial strains CA and AF showed much higher induction levels of invertase than both CL and AL did when the glucose levels in the media decreased to the threshold value of triggering the invertase induction. In both sucrose and mix sugars media, surprisingly, the both originally invertase-repressed CA and AF showed an obviously increasing trends of invertase activities after 7 h of fermentations even under the glucose repression conditions at the time. Moreover, the both CA and AF showed highest ethanol productivities in sucrose and mix sugars media. Meanwhile, AL showed a stable and relatively high background invertase activity during the whole process of ethanol fermentation even under the impact of glucose repression, which led the high sucrose hydrolysis rate of AL at the early stage of fermentation. On the other hand, ethanol inhibits the induction of invertase to a great extent, and the invertase induction of both CA and AF showed much stronger tolerance to ethanol. All these results suggested that the glucose repression might be not the only mechanism for the regulation of the invertase involved in the process of ethanol fermentation in *S. cerevisiae* but a net effect of multi-factors, such as glucose repression, stress response. Meanwhile, these results can provide us some useful indicators for isolating or developing the yeast strains with high ethanol productivities with sucrose-containing feedstocks as sustrates.

(191d) Effects of Dissolved Inorganic Carbon and Mixing On Growth and Lipid Formation of Chlorella Vulgaris

Jinsoo Kim, School of Energy, Environmental, Biological and Medical Engineering, University of Cincinnati, Cincinnati, OH, Zhouyang Liu, School of Energy, Environmental, Biological, and Medical Engineering , University of Cincinnati, Cincinnati, OH, Joo-Youp Lee, School of Energy, Environmental, Biological, and Medical Engineering, University of Cincinnati, Cincinnati, OH and Ting Lu, Wastewater Treatment Division, Metropolitan Sewer District of Greater Cincinnati, Cincinnati, OH

Microalgae have recently received great interests as a potential alternative biofuel source because they can generate a large amount of triacylglycerols (e.g. 20-50 % dry cell weight) for biodiesel production. *Chlorella vulgaris* has been extensively studied due to its relatively rapid growing rate and reasonable quantity of triacylglycerols (e.g. 14-22 % dry cell weight). Dissolved inorganic carbon including carbon dioxide (CO_2(aq)) and bicarbonate ion (HCO_3^-) is an important carbon source for the autotrophic growth of microalgae. However, the dissolved inorganic carbon concentration in culture mediums is not usually enough for the growth of microalgae in typical open ponds when CO_2(g) is added to the medium due to a narrow pH window suitable for the growth. In this study, sodium bicarbonate ($NaHCO_3$) was used to increase the inorganic carbon concentration. Then, different mixing speeds were employed to investigate the effect of dissolved inorganic carbon concentrations on the algal cell surface. During the entire growth phase, the lipid content in algal cell was also monitored using fluorescence spectroscopy *in vivo* to trace neutral lipid formation.

(191e) High Level Malic Acid Production From Polymalic Acid Fermentation by Aureobasidium Pullulans and Acid Hydrolysis

Xiang Zou Sr., College of Pharmaceutical Sciences, Southwest University, Chongqing, China, **Yipin Zhou**, William G. Lowrie Department of Chemical and Biomolecular Engineering, The Ohio State University, columbus, OH and S.T. Yang, Chemical Engineering, The Ohio State University, Columbus, OH

High level malic acid production from polymalic acid fermentation by *Aureobasidium pullulans* and acid hydrolysis

Xiang Zou[a], Yipin Zhou[b], Shang tian Yang[b*]

(a. *College of Pharmaceutical Sciences, Southwest University, Chongqing 400715, P.R China;*

(b. *William G. Lowrie Department of Chemical and Biomolecular Engineering, The Ohio State University, Columbus, OH 43210, USA*)

Abstract

Malic acid, a C4-dicarboxylic acid, is widely used in the food industry and a potential C4 platform chemical in biorefinery. Malic acid is commercially produced either through hydration of fumaric or maleic acids under high temperature and pressure yielding a racemic mixture or via the enzymatic transformation of fumarate producing L-malic acid by immobilized bacterial cells. However, these processes require expensive raw materials such as petroleum-derivative maleic anhydride or costly enzymes, which have limited their economical competitiveness. Compared with these methods, microbial fermentation using relatively inexpensive substrates is a promising way for economical production of malic acid. However, the low product yield from conventional malic acid fermentation limits further industrialization efforts due to product inhibition. In this work, we developed a novel process for malic acid production from fermentation produced polymalic acid (PMA) followed with acid hydrolysis. First, a high-yield strain *A. pullulans* ZX10 was isolated, which produced PMA at a high titer equivalent to 87.6 g/l malic acid and high productivity of 0.61 g/l h in free-cell fed-batch fermentation. Repeated fed-batch fermentations with cells immobilized in a fibrous-bed bioreactor achieved the highest product titer of 144.2 g/l and productivity of 0.74 g/l h. The fermentation produced PMA was separated by ethanol precipitation and then purified by adsorption with IRA-900 resin, achieving a ~100% purity. Pure malic acid was then produced from PMA by acid hydrolysis. This process provides an efficient and economical way for PMA and malic acid production, and is promising for industrial application.

(191f) Economical Production of Propionic Acid from Renewable Feedstocks by Propionibacteria in a Fibrous-bed Bioreactor

Zhongqiang Wang, Chemical & Biomolecular Engineering, The Ohio State University, Columbus, OH and Shang-Tian Yang, Department of Chemical and Biomolecular Engineering, Ohio State University, Columbus, OH

Propionic acid is an important chemical with various industrial applications. *Propionibacterium freudenreichii* subsp. *shermanii*, typically used as ripening culture for Swiss-type cheese, has been studied for its ability to produce propionic acid and vitamin B_{12}. A *P. shermanii* strain with high propionic acid yield was used for propionic acid production by cells immobilized in a fibrous-bed bioreactor (FBB). Batch and fed-batch fermentations with glycerol and glucose as carbon sources were investigated first. The results showed that the strain was able to convert both glucose and glycerol simultaneously to propionic acid at a high rate and with a high propionic acid yield. Then, the fermentation was studied using cassava bagasse hydrolysate and crude glycerol from biodiesel wastes to demonstrate the feasibility of using low-cost feedstocks to economically produce propionic acid.

(195a) Catalytic Valorization of Isosorbide As Versatile Renewable Platform Chemical

Marcus Rose, Rebecca Pfützenreuter, Katharina Thenert and Regina Palkovits, RWTH Aachen University, Aachen, Germany

Catalytic Valorization of Isosorbide as Versatile Renewable Platform Chemical

Marcus Rose, Rebecca Pfützenreuther, Katharina Thenert, Regina Palkovits
RWTH Aachen University, Aachen, Germany

The substitution of fossil resources in the production of crude oil-based products such as polymers or fuels by renewable lignocellulosic biomass is essential in the future. Cellulosic feedstock such as straw, wood residues or other bio-waste can be converted by efficient catalytic processes to platform chemicals of great importance such as isosorbide.

Isosorbide poses great potential as an intermediate in the production of biofuels, fuel additives, high boiling solvents, surfactants and polymers. Thus, the development of catalytic processes for a highly efficient isosorbide conversion is crucial. Main target is the functionalization or substitution of the two hydroxyl groups maintaining the *cis*-fused ditetrahydrofurane scaffold. Numerous investigations on feasible isosorbide transformation routes have been reported in the middle of the 20th century. Nowadays, two isosorbide derivatives are already commercially available. Isosorbide nitrates are used in medicine as vasodilating pharmaceuticals, whereas isosorbide dimethyl ether gained interest as a versatile non-toxic pharmaceutical solvent. Several publications deal with the isosorbide functionalization via the *Williamson* etherification using alkyl halides. A disadvantage of this process is the formation of huge amounts of halogen containing side product. Alternatively, the etherification by telomerization with 1,3-butadiene has been shown recently, resulting in several isomeric products.

One focus of our work is the conversion of isosorbide to defined *tert*-butyl ether derivatives heterogeneously catalyzed by acidic ion exchange resins such as Amberlyst-15. Instead of the *Williamson* method, two different approaches, the condensation reaction of isosorbide with *tert*-butanol avoiding side products other than water as well as the addition of isosorbide hydroxyl groups to isobutene, have been investigated. In the reaction with *tert*-butanol an isosorbide conversion of 27 % was achieved due to the water formation shifting the equilibrium reaction to the side of the substrates. With isobutene close to full conversion of isosorbide was observed with up to 60 % yield of the diether and two different monoethers as by-products. In our current work we develop an energy efficient process for the continuous etherification of the biogenic platform chemical isosorbide with isobutene.

In this contribution we want to give a critical overview on the different aspects of isosorbide chemistry and feasible catalytic conversions as well as present recent results on the development of efficient processes for the etherification of isosorbide to produce novel value-added products such as green solvents and fuel additives.

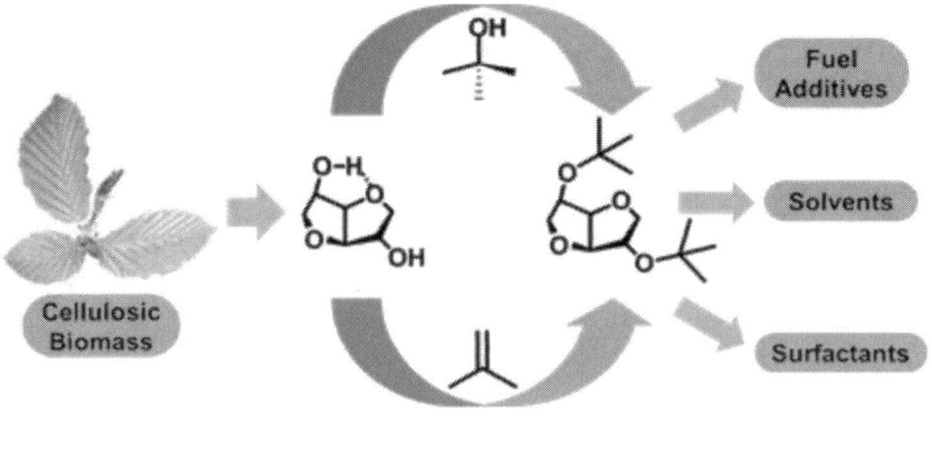

(195b) Reaction of Sugar Derivatives On Metal Surfaces

Vassili Vorotnikov, Chemical Engineering, University of Delaware, NEWARK, DE and Dionisios G. Vlachos, Catalysis Center for Energy Innovation, Chemical and Biomolecular Engineering, University of Delaware, Newark, DE

Over the past decade, the interest in theoretical methods to aid catalyst design has grown with the advances in computational resources. In this research, we apply such state-of-the-art theory to describe the adsorption and the reaction kinetics of biomass derivatives on transition metal surfaces to gain mechanistic insights into this complex chemistry. Specifically, we looked at furfural hydrogenation to methylfuran, which can be used as an additive in transportation fuel. In this work, we performed a density functional theory (DFT) study on the adsorption of furan, furfuryl alcohol, furfural, and methylfuran, which are the main products in furfural reaction on Pd. For the first time in modeling reactions on transition metal surfaces, we applied the recently-developed DFT-D3 method, which combines the standard DFT functionals with a correction for weak long-range interactions. This correction is essential in modeling multi-metallic catalysts, because these weak interactions have shown to be the dominating forces on some catalysis-relevant metals. As a part of this work, we have shown that DFT-D3 performs well in predicting geometric adsorption parameters regardless of the type of interactions involved.

Following the adsorption study, we explored the furfural interconversion on Pd surface. We obtained the heats of reaction and performed transition state search to obtain the activation energies for each reaction. The results showed that despite decarbonylation being thermodynamically favored, the hydrogenation activation energy is lower, thus allowing the chemical upgrade of the biomass derivative. This explains the hydrogenation products that were experimentally observed in Pd-catalyzed furfural reaction. Further mechanistic analysis showed full consistency with the recently reported ultra-high vacuum (UHV) experimental findings. As such, this work constitutes a great starting point in the development of predictive models for biomass derivatives' reactions on transition metal surfaces.

(195c) Acid-Catalyzed Conversion of Dimethylfuran to p-Xylene in Catalytic Zeolite-Y. A Combined Computational and Experimental Study

Nima Nikbin, Phuong T. Do, Stavros Caratzoulas, Raul F. Lobo and Dionisios G. Vlachos, Catalysis Center for Energy Innovation, Chemical and Biomolecular Engineering, University of Delaware, Newark, DE

p-Xylene, conventionally produced by catalytic reforming of petroleum naptha, is used on a large scale for the manufacture of terephthalic acid for polyethylene terephthalate, which is used to make plastic bottles and clothing; several million tonnes are produced annually. We have recently proposed a renewable pathway to p-xylene from biomass-derived carbohydrates by reacting dimethylfuran (DMF) with ethylene in zeolite-Y catalyst.[1] The reaction proceeds through Diels-Alder cyclo-addition and subsequent dehydration of the resulting oxa-norbonene derivative cyclo-adduct.

In this paper, we use electronic structure calculations to investigate the mechanism and energetics of the reaction under different types of acid catalysis – Lewis vs Brønsted – map out the network of the most important side reactions and, based on the calculated activation and reaction enthalpies, explore how the selectivity to p-xylene may be optimized via rational design of bifunctional catalysts. Some of the key findings that we shall present include:

The Diels-Alder (DA) reaction proceeds through a concerted mechanism and its rate is accelerated in the case of Lewis acid catalysis. In contrast, we find that Brønsted catalysis of the DA is not feasible and, in fact, adsorption of the dienophile (DMF) on a Brønsted site activates side reactions. The dehydration of the oxa-norbonene derivative cyclo-adduct to p-xylene requires Brønsted catalysis.

Two major side reactions that affect the selectivity to p-xylene are: (1) DMF hydrolysis to 2,5-hexanedione, a reaction that is catalyzed by Brønsted acids; and (2) Further DA cyclo-addition between DMF and the oxa-norbonene intermediate and its isomers.

Finally, by comparing the calculated activation barriers of the pathway from DMF to p-xylene with experimentally observed turnover frequencies, we conclude that the H-Y zeolite is more active than the zeolite-Y with fully exchanged protons with alkali cations (i.e., zeolite-Y with Lewis-acid activity).

[1] C. Luke Williams, Chun-Chih Chang, Phuong Do, Nima Nikbin, Stavros Caratzoulas, Dionisios G. Vlachos, Raul F. Lobo, Wei Fan, Paul J. Dauenhauer. *ACS Catalysis*, 2012 , *2*, 935-939

(195d) Economic Analysis of the Production of p-Xylene From 5-Hydroxymethyfurfural

Zhaojia Lin, Chemical and Biochemical Engineering, Rutgers, The State University of New Jersey, Piscataway, NJ, Vladimiros Nikolakis, Catalysis Center for Energy Innovation, Department of Chemical and Biomolecular Engineering, University of Delaware, Newark, DE and Marianthi G. Ierapetritou, Department of Chemical and Biochemical Engineering, Rutgers, The State University of New Jersey, Piscataway, NJ

Economic analysis of the production of p-xylene from 5-hydroxymethyfurfural

Zhaojia Lin[1], Vladimiros Nikolakis[2], Marianthi Ierapetritou[1]

[1] Department of Chemical and Biochemical Engineering, Rutgers - The State University of New Jersey

[2.] Catalysis Center for Energy Innovation & Department of Chemical & Biomolecular Engineering, University of Delaware

This work focuses on the techno-economic analysis of alternative routes for the production of p-xylene from 5-hydroxymethyfurfural (HMF), which is a biomass derived platform chemical. P-xylene is a key intermediate in the production of terepthalic acid, which can be polymerized to polyester polyethylene terephthalate (PET). PET is a polymer resin widely used in the synthesis of fibers and beverage containers. Currently, p-xylene is primarily made from petroleum-based feedstocks. However, depleting oil resources and rising prices motivates the development of routes for the efficient synthesis of fuels and chemicals from renewable sources [1]. 5-hydroxymethyfurfural (HMF), which is considered as one of the ten most promising biomass-derived chemicals [2], can be used as a raw material for p-xylene synthesis [3-6]. The conversion of HMF to p-xylene can be carried out in two steps: from HMF to 2,5-dimethylfuran (DMF) and from DMF to p-xylene. First, HMF hydrodeoxygenation can form DMF using either hydrogen [5] or formic acid as an alternative hydrogen source [6]. Then DMF can react with ethylene [3, 7] or acrolein [4], followed by a dehydration reaction, to convert p-xylene. Each route has advantages and disadvantages in terms of use of renewable feedstocks, raw materials and utility costs. The aims of this work are to propose alternative flowsheets, to evaluate the economics of the production, to determine the major contributors of the total cost, and to explore and identify potential approaches to reduce these costs.

The analysis was carried out in two stages. In the first stage, an approximate evaluation of the various alternatives is performed considering different separation methods and rough estimates of the various parameters needed, whereas at the second stage a detailed economic estimation of the most promising path is performed. It is found that conversion and selectivity, the values of which were taken from the literature [5, 7], are the most important parameters that affect recycling streams and separations; and consequently the process economics. Finally sensitivity analysis was used to examine the impact of different factors on economics and to identify the most significant catalyst, material or reactor properties, and the improvement of which will have the maximum impact on process economics.

The basic models and different flowsheets were studied using ASPEN Plus [8]. ASPEN Economic Analyzer [9] was also utilized to determine the production cost of p-xylene, considering the raw material costs of HMF, H_2, and ethylene [10, 11]. The main findings of this work contain the minimum selling cost of bio-based p-xylene, the impacts of different factors on the total cost and potential development to improve the economics of the conversion process.

References:

1. Bozell, J.J., *Feedstocks for the Future – Biorefinery Production of Chemicals from Renewable Carbon.* CLEAN – Soil, Air, Water, 2008. **36**(8): p. 641-647.

2. Bozell, J.J. and G.R. Petersen, *Technology development for the production of biobased products from biorefinery carbohydrates-the US Department of Energy's "Top 10" revisited.* Green Chemistry, 2010. **12**(4): p. 539-554.

3. Brandvold, T.A., *Carbohydrate route to para-xylene and terephthalic acid*, 2010, US Patent 2010/0331568 A1.

4. Shiramizu, M. and F.D. Toste, *On the Diels–Alder Approach to Solely Biomass-Derived Polyethylene Terephthalate (PET): Conversion of 2,5-Dimethylfuran and Acrolein into p-Xylene.* Chemistry – A European Journal, 2011. **17**(44): p. 12452-12457.

5. Roman-Leshkov, Y., et al., *Production of dimethylfuran for liquid fuels from biomass-derived carbohydrates.* Nature, 2007. **447**(7147): p. 982-985.

6. Thomas B. Rauchfuss, T.T., *Efficient method for preparing 2,5-dimethylfuran*, 2011, US 2011/0263880 A1.

7. C. Luke Williams, C.-C.C., Phuong Do, Raul F. Lobo, Wei Fan, Paul J. Dauenhauer, *Cycloaddition of biomass-derived furans for catalytic production of renewable p-xylene.* **Submitted**.

8. *Aspen Plus User Guider*, 2000, Aspen Technology Inc.

9. *Aspen Process Economic Analyzer*, 2009, Aspen Technology Inc.

10. *ICIS pricing*. [cited 2011, 28th October]; Available from: www.icispricing.com.

11. Kazi, F.K., et al., *Techno-economic analysis of dimethylfuran (DMF) and hydroxymethylfurfural (HMF) production from pure fructose in catalytic processes.* Chemical Engineering Journal, 2011. **169**(1–3): p. 329-338.

(195e) Ethanol Conversion On ZrO2: The Role of Brönsted and Lewis Acidic Sites

Changjun Liu[1], Junming Sun[1], Colin Smith[1] and Yong Wang[1,2], (1)The Gene & Linda Voiland School of Chemical Engineering and Bioengineering, Washington State University, Pullman, WA, (2)Pacific Northwest National Laboratory, Richland, WA

Ethanol Conversion On ZrO_2: The role of Brönsted and Lewis acidic sites

Changjun Liu[†], Junming Sun[†], Colin Smith[†], Yong Wang[*†‡]

[†] The Gene & Linda Voiland School of Chemical Engineering and Bioengineering, Washington State University, Pullman Washington 99164, United States

[‡] Institute for Interfacial Catalysis, Pacific Northwest National Laboratory, Richland, Washington 99352, United States

Abstract

As one of the sustainable natural resources, biomass has attracted increased attentions for its conversion into valuable chemicals in recent decades[1]. The increased availability and reduced cost of bioethanol[2] provides the potential to make value-added chemicals from ethanol in large scale. Direct conversion of bioethanol into ethylene, propene has been studied over wide range of acidic catalyst like zeolites. More recently the direct conversion of bioethanol into isobutene over nanosized $Zn_xZr_yO_z$ catalyst has also been reported[3]. Ethylene is considered as the intermediate in conversion bioethanol to propene[4] while it is considered as one of the major byproduct in isobutene production. Generally either the production of propene or isobutene from bioethanol are conducted at the temperature above 400 °C over various acidic catalysts. Although, ethanol conversion to ethylene has been studied extensively at a temperature range of 150 °C to 300 °C, the investigation on the intrinsic catalytic behavior of acidic sites of different nature and varied strength under different temperature ranges are still rare. In this work, the ethanol dehydration at a wide temperature ranges (200-500 °C) have been studied on ZrO_2 catalysts with controlled Brönsted and Lewis acidic sites. Specifically, the effect of Brönsted and Lewis acidic sites on the reaction pathway in ethanol dehydration has been investigated (Figure 1.).

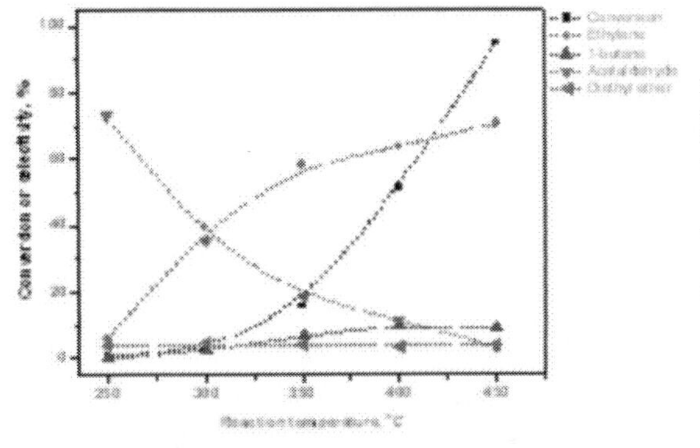

a) ZrO_2 with Lewis acidic site

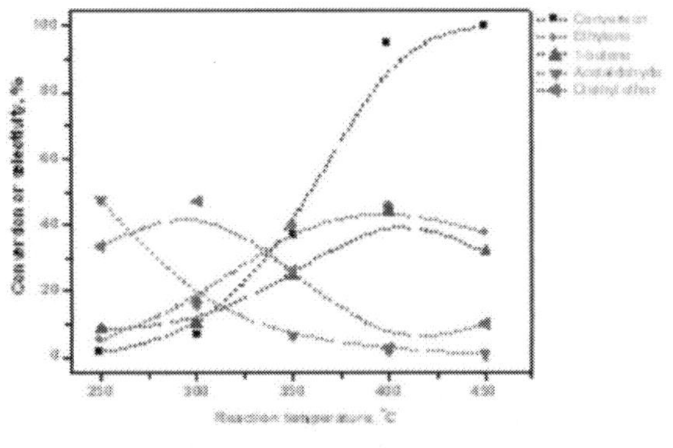

b) ZrO_2 with Brönsted and Lewis acidic sites

Figure 1. Ethanol conversion and main products distribution at various reaction temperatures

At low temperature range (T<300 ^oC), Brönsted acid site is mainly responsible for the intermolecular dehydration of ethanol to diethyl ether. While only intramolecular dehydration of ethanol to ethylene was observed on Lewis acidic site at the whole temperature range studied albeit it's low activity. Dehydrogenation of ethanol was significant at lower temperature with Lewis acidic sites being more selective. At higher temperature (T>300 ^oC), the intramolecular dehydration started taking over with intermolecular dehydration of ethanol being largely suppressed on the Brönsted acidic sites. Interestingly, ethylene dimerization to 1-butene was also observed, and its selectivity increased with reaction temperature; whereas, no dimerization was observed on Lewis acidic sites. The possible reaction mechanisms regarding Brönsted and Lewis acidic sites a catalyzed different reaction was discussed in details.

* Corresponding author

Reference

1. Serrano-Ruiz, J. C.; Luque, R.; Sepulveda-Escribano, A., Transformations of biomass-derived platform molecules: from high added-value chemicals to fuels via aqueous-phase processing. *Chem. Soc. Rev.* **2011,** *40* (11), 5266-5281.

2. Deluga, G. A.; Salge, J. R.; Schmidt, L. D.; Verykios, X. E., Renewable Hydrogen from Ethanol by Autothermal Reforming. *Science* **2004,** *303* (5660), 993-997.

3. Sun, J.; Zhu, K.; Gao, F.; Wang, C.; Liu, J.; Peden, C. H. F.; Wang, Y., Direct Conversion of Bio-ethanol to Isobutene on Nanosized ZnxZryOz Mixed Oxides with Balanced Acid¨CBase Sites. *J. Am. Chem. Soc.* **2011,** *133* (29), 11096-11099.

4. (a) Oikawa, H.; Shibata, Y.; Inazu, K.; Iwase, Y.; Murai, K.; Hyodo, S.; Kobayashi, G.; Baba, T., Highly selective conversion of ethene to propene over SAPO-34 as a solid acid catalyst. *Applied Catalysis A: General* **2006,** *312* (0), 181-185; (b) Xia, W.; Takahashi, A.; Nakamura, I.; Shimada, H.; Fujitani, T., Study of active sites on the MFI zeolite catalysts for the transformation of ethanol into propylene. *J. Mol. Catal. A: Chem.* **2010,** *328* (1¨C2), 114-118.

Enhancement of Aromatic Production from Bioethanol Dehydration Using Dual-Bed Catalytic Systems

Sak Saewong [a], Chatapong Wungtanagorn [b], Sirirat Jitkarnka [a, c]

[a] The Petroleum and Petrochemical College, Chulalongkorn University,Thailand
[b] Thaioil Public Company Limited, Chonburi,Thailand
[c] Center of Excellence on Petrochemical and Materials Technology, Thailand

Abstract

Aromatics are not only produced from conventional petroleum-based feedstocks, but they can be also converted from bioethanol via the catalytic dehydration, which is an alternative process that has gained high attention from petrochemical industries. Modified HZSM-5 catalysts were proven to be effective catalysts for the direct conversion of ethanol to aromatics. Since ethylene is an intermediate for aromatic formation via ethanol dehydration, the aromatic production was expected to be improved by using a dual-bed catalytic system, which ethylene is produced in the first bed and aromatics are formed from ethylene aromatization in the second consecutive bed. The objective of this work was to compare the activity of single-bed catalytic systems (using Ga_2O_3/HZSM-5 or ZnO/HZSM-5) with that of dual-bed systems (using $MgHPO_4/Al_2O_3$ as the ethylene-producing catalyst coupled with one of the modified HZSM-5 as aromatization catalysts). The HZSM-5 and Ga_2O_3- and ZnO-doped HZSM-5 were individually tested in the single-bed system for their activity on aromatic production under the temperature of 500 °C. For the dual-bed systems, the first catalytic bed contained 0.5 wt% $MgHPO_4/Al_2O_3$, which was fixed for all experiments, whereas the second bed contained either Ga_2O_3/HZSM-5 or ZnO/HZSM-5. The temperatures of the 1st and 2nd beds were separately controlled at 370 °C and 500 °C, respectively. The results revealed that the ethanol conversion range of 83 – 85 % was observed from all cases for both systems. The aromatic yields gained from the single-bed systems were found increasing in the order: HZSM-5 (9.1 %) < ZnO/HZSM-5 (12.7 %) < Ga_2O_3/HZSM-5 (16.8 %). Using the dual-bed systems, the aromatic yield was increased by 70.3 % and 31.5% with using the HZSM-5 and Ga_2O_3/HZSM-5 as the second-bed catalysts, respectively. In contrast, the aromatic yield obtained from using ZnO/HZSM-5 in the second bed was slightly decreased by 3.1 %. In conclusion, the aromatic yield can be significantly enhanced by using the dual-bed system of $MgHPO_4/Al_2O_3$ coupled with a catalyst that has high ethylene aromatization activity.

Introduction

Nowadays, aromatic hydrocarbons are manufactured from conventional petroleum-based feedstocks. Nevertheless, the petroleum resource shortage is soon being a big problem for the petroleum and petrochemical industries because of the unlimited energy demand. Fortunately, in the recent year aromatic production from renewable sources, like bioethanol, via a catalytic process, has been proven possible. Therefore, the alternative way has become a greatly-interesting approach for the petrochemical entrepreneurs. Earlier, the ethanol conversion to ethylene was studied by Kojima *et al.* (1981), and their result revealed that the 0.5 %wt loading of $MgHPO_4$ over an alumina catalyst can improve the catalytic activity of bioethanol dehydration as compared with the unloaded case. Both ethanol conversion and ethylene selectivity were nearly 100 %. In addition, aromatic hydrocarbon is a target group of products that have gained great attention. An effective catalyst, like HZSM-5, is frequently used for the conversion of light alcohols and light hydrocarbons to aromatics (Inaba *et al.,* 2005). A few early studies have been reported on the aromatization of ethylene over unmodified HZSM-5 and modified HZSM-5. The results showed the metals such as gallium and zinc resulted in a significant increase in aromatization activity (Lukyanov *et al.,* 1994, Qiu *et al.*, 1998, and Saha and Sivasanker., 1992). Moreover, in 2006 Barthos and co-workers showed that the aromatic selectivity can be raised in a consecutive bed system of two catalysts connected in series. However, the direct dehydration of ethanol to hydrocarbons over several catalysts earlier was commonly studied in a single-bed system; both in cases of a bed of catalyst and a consecutive bed of two catalysts. In this work, it is thus our intention to create a novel system in order to enhance the efficiency of aromatic production from bioethanol and/or to use as an alternative process for future. The new system, which is combination of the idea between the use of catalysts with various catalytic activities and a new pattern of reaction process, is differently designed under the theoretical feasibility. Ethanol transformation to aromatics was investigated in the two consecutive beds of catalysts packed in a series of two reactors controlled at two different temperatures. The dehydration of bioethanol to olefins such as ethylene was expected to occur at the first bed of $MgHPO_4$-added alumina catalyst, and the conversion of olefins to aromatics was expected to take place at the second bed of HZSM-5 based catalysts. Ga_2O_3/HZSM-5 and ZnO/HZSM-5 were specially investigated in the second bed.

EXPERIMENTAL

A. Catalyst Preparation

For the catalytic Bed #1, the physical mixing method was used to prepare the $MgHPO_4$-added alumina catalyst. The commercial alumina from NanoScale Corporation was firstly dried at 110 °C overnight. Then, 0.5 wt% $MgHPO_4$ was added into the dried alumina. The resulting mixture was kneaded together for 2 hours at room temperature prior to drying again for a few hours, and finally calcined at 600 °C for 3 hours with the heating rate of 5 °C/min. After that, the catalyst was pelletized and sieved into the particle sizes of between 425 and 400 μm (20/40 mesh).
For the catalytic Bed #2, as a zeolite support, the HZSM-5(Si/Al = 20) purchased from Tosoh was used in this experiment. The nitrate solution of gallium and zinc was used as the source of Ga and Zn, respectively. 2.0 wt% Ga was loaded on the HZSM-5 support by the incipient wetness impregnation technique. Then, the wet catalyst was dried at 110 °C for a few hours, and then calcined at 600 °C for 5 hours. Finally, the calcined catalyst was pelletized, crushed and sieved to 20/40 mesh granules. Also, the same impregnation technique was performed for the preparation of 2.0 wt % Zn supported on the HZSM-5.

B. Catalytic activity testing

The dehydration of bioethanol to aromatics was carried out under a system of dual catalytic bed at the atmospheric pressure. The U-tube fixed-bed continuous flow reactor (10 mm i.d. and 45.8 cm length) was charged with the two beds of catalysts separately located on the top-left side (Catalytic Bed #1 of $MgHPO_4$-doped alumina) and the bottom-right side (Catalytic Bed #2 of

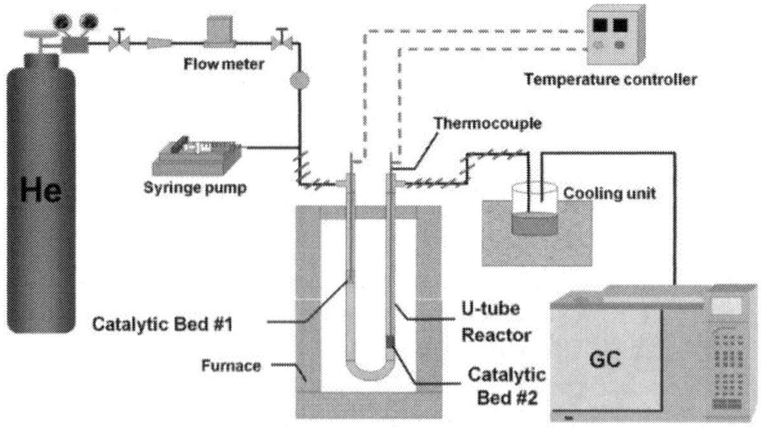

Figure 1 Schematic of experimental set-up.

HZSM-5-based catalysts) as shown in Figure 1. A 3.7 ml catalyst volume was packed in each bed. The reaction temperature was independently controlled at 370 °C and 500 °C in Bed #1 and Bed #2, respectively. The bioethanol (99.5% purity) was preheated to vaporize and then diluted with helium at the flow rate of 27 ml/min or the LHSV of 1.0 h^{-1}. The gaseous product was analyzed by an online gas chromatography equipped with a TCD detector with a Porapak Q Column for their composition, and equipped with FID detector with a DB-WAX column for oxygenate compositions. The oil products were analyzed for the simulated true boiling point curves using a Varian GC-3800 gas chromatograph (SIMDIST GC) to determine petroleum fractions. Also, a high-performance liquid chromatography (HPLC) conformed to IP 391 standard method was employed to examine mono-, di-, and tri-aromatic contents in oils.

RESULTS AND DISCUSSION

A. Aromatic Production from Single-Bed System

The performance of catalysts in the HZSM-5 family on aromatic production in the single-bed system was first investigated for comparison with the dual-bed system. Ethanol feed is converted by only one of catalytic beds, which belong to the HZSM-5 family, at 500 °C. The product distribution obtained from the catalysis over pure and modified HZSM-5 is shown in Table 1 in the Columns A – C. It is found that the range of ethanol conversion is approximately 83 - 85 %, which is for all cases not significantly different. On the other hand, the oil yield is increased when the modified HZSM-5 catalysts are used. Also, the total aromatic yield in the oil products increases slightly in the order: HZSM-5 (9.1 %) < ZnO/HZSM-5 (12.7 %) < Ga_2O_3/HZSM-5 (16.8 %). It means that the total aromatic production is enhanced significantly when HZSM-5 is promoted with ZnO and Ga_2O_3 by 39.6 % and 84.6 %, respectively. Moreover, it is noticed that in all cases the produced oils are entirely composed of aromatic hydrocarbons, which more than 87 % is classified to mono-aromatics and the rest is non-aromatics combined with di-aromatics and tri-aromatics.

A significant increase of aromatic yield over both ZnO/HZSM-5 and Ga_2O_3/HZSM-5 is advocated by the increase in dehydrogenation activity of the catalysts (Brathos *et al.*, 2006, and Qiu *et al.*, 1998). In addition, it can be explained that ethanol is initially activated on the acid sites of HZSM-5 catalyst via dehydration reaction, which the primary products are ethylene and water. Then, ethylene is further catalyzed by the acid sites of HZSM-5 again through carbenium ion mechanism. Subsequently, oligomerization is rapidly occurred, followed by isomerization, cyclization and/or cracking, resulting in aromatic formation. Moreover, the metal oxide loading can accelerate the aromatization by improving the dehydrogenation reaction of several interediates (Brathos *et al.*, 2006). Ultimately, the dehydrogenated products are possibly re-oligomerized, isomerized, cyclized, and aromatized on the acid sites of zeolite to form heavy products and aromatic hydrocarbons.

B. Aromatic Production from Dual-Bed Systems

In a dual-bed system, bioethanol was fed into the first reactor packed with an ethylene-producing catalyst, $MgHPO_4$-doped Al_2O_3, and then the ethylene-rich effluent from the first reactor was continuously fed to the second reactor packed with an ethylene aromatization catalyst (HZSM-5, Ga_2O_3/HZSM-5, or ZnO/HZSM-5). Previously, the catalytic activity of the ethylene-producing catalyst was investigated. It was found that 83 % ethanol feedstock was converted to various intermediate hydrocarbons, which are composed of 80 % ethylene, 15 % C5 HCs, and 5 % others (ethane, propane, propylene, mixed C4). In other words, the effluent having from the first reactor mainly contains ethylene, and can be also called as an ethylene-rich feed, which is used as the feed for the second reactor. Furthermore, it was confident that the catalytic activity of $MgHPO_4$/ Al_2O_3 can practically maintain during in least 5-hours of reaction (Saewong *et al.*, 2012).

(a). Product distribution

In the two-consecutive bed system, the conversion of ethanol is not much significantly changed as compared with that of the corresponding single bed system, which is about 83-85 % as results shown in Table 1; Columns D – F. The yields of aromatic in oil products are enhanced by 70.3 % and 31.5 % in the case of using HZSM-5 and Ga_2O_3/HZSM-5, respectively, as the catalysts in the second bed of the dual-bed system. Unfortunately, in the case of ZnO/HZSM-5, the aromatic yield slightly decreases from 12.7 % to 12.3% (or, decreases by 3.1 %).

Table 1. Product distribution from using single-bed and dual-bed catalytic systems

	Single-Bed System			Dual-Bed System		
	A	B	C	D	E	F
	HZ5	**GHZ5**	**ZHZ5**	**EPC:HZ5**	**EPC:GHZ5**	**EPC:ZHZ5**
Ethanol Conversion (%)	85	83.4	82.8	83.7	85.6	85.3
Product yield (%)						
Total gas	56.2	48.5	53.7	50.6	43.5	53.8
Total oil	10.3	18.3	13.2	16.1	23	12.7
Total aromatics in oil	9.1	16.8	12.7	15.5	22.1	12.3
Gas Composition (%)						
Methane	10.1	10.9	8.9	7.6	9.4	12.9
Ethylene	7.6	6.4	5.8	6.2	5.4	2.3
Ethane	20.8	17.5	21.5	21.1	21.1	26.5
C3	40.9	36.6	12.5	52	50.7	9.7
C4	4.5	0	3	6.4	3	1.6
C5+	2.4	5.7	3.6	3	4.2	2.5
CO_2	13.6	22.9	44.8	3.7	6.2	44.5
	100	100	100	100	100	100
Oil Composition (%)						
Monoaromatics	87	90.1	88.5	85.1	84	87.8
Diaromatics	1.3	1.7	7.9	11.4	12.3	8.4
Triaromatics	< 0.1	< 0.1	< 0.1	< 0.1	< 0.1	< 0.1
Non-aromatics	11.7	8.2	3.63	3.5	3.7	3.8
	100	100	100	100	100	100

Data were taken at the fifth hour of time-on-stream

Note: Catalyst packed at 1st bed

- EPC = Ethylene producing catalyst ($MgHPO_4/Al_2O_3$)

Catalysts packed at 2nd bed

- HZ5 = HZSM-5
- GHZ5 = Ga_2O_3/HZSM-5
- ZHZ5 = ZnO/HZSM-5

$$Yield\ (\%) = \frac{Total\ weight\ of\ any\ products\ (g)}{Total\ weight\ of\ converted\ bioethanol\ (g)} \times 100$$

For the explanation of the increase of aromatic production, the catalyst selection and reactor design are key issues in this study. Aromatic hydrocarbons can be produced from both direct ethanol feed and direct ethylene feed over HZSM-5 and its family. The previous study revealed that the ability of aromatic production is improved when the ethylene is directly fed through the effective catalysts instead of feeding ethanol (Brathos *et al.*, 2006). As observed, the production of aromatics from the dual-bed system gives better outcomes than that obtained from the single-bed system in terms of both oil yield and total aromatic ratio. According to Figure 2, even though there are a few pathways involving bioethanol transformation to aromatic hydrocarbons, the direct dehydration of ethanol to ethylene followed by ethylene aromatization is the shortest route. In the dual-bed system, the ethylene-rich effluent is first built up from the first reactor. Afterward, the ethylene is next converted to heavy hydrocarbons by active catalysts, which are greatly effective to ethylene aromatization, like HZSM-5 and its family packed on the second bed. The separation of catalyst beds according to their function is likely to be more beneficial because the reaction in each bed is more effective.

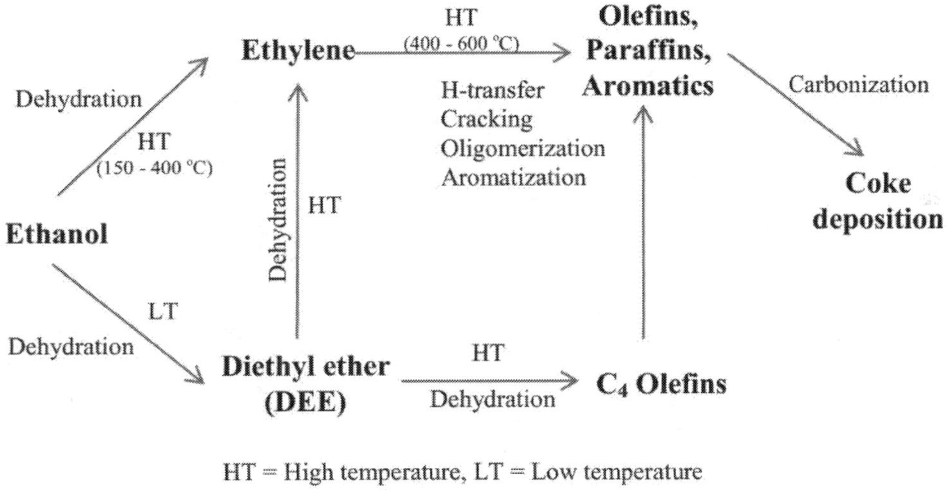

Figure 2 Reaction pathways of ethanol transformation to hydrocarbons (Inaba *et al.*, 2005).

(b). Petroleum Fractions in Oils

In addition, the true boiling point curves of oils obtained from the sing-bed and dual-bed catalytic systems were cut into petroleum fractions as illustrated in Figure 3, according to the following boiling point ranges: gasoline (< 149 °C), kerosene (149-232 °C), gas oil (232-343 °C), light vacuum gas oil (LVGO) (343-371 °C), and heavy vacuum gas oil (HVGO) (> 371 °C).

The derived oils from all catalysts are mainly distributed in the hydrocarbon range of gasoline. More than 80 % of gasoline can be produced as shown in Figure 3. When the single-bed cases in Figure 3(a) are compared with the corresponding dual-bed cases in Figure 3(b), the slight increases in gasoline fraction are observed. This suggests that the oils obtained from the dual-bed system are slightly lighter than those from the single-bed systems. Additionally, the dopants (Ga_2O_3 and ZnO) on the HZSM-5 bring about a slight increase in gasoline with a subsequent decrease in kerosene and gas oil fractions. These indicate that the addition of some dopants into the HZSM-5 catalyst can slightly improve the catalytic cracking activity on which heavier hydrocarbons are cracked into lighter hydrocarbons.

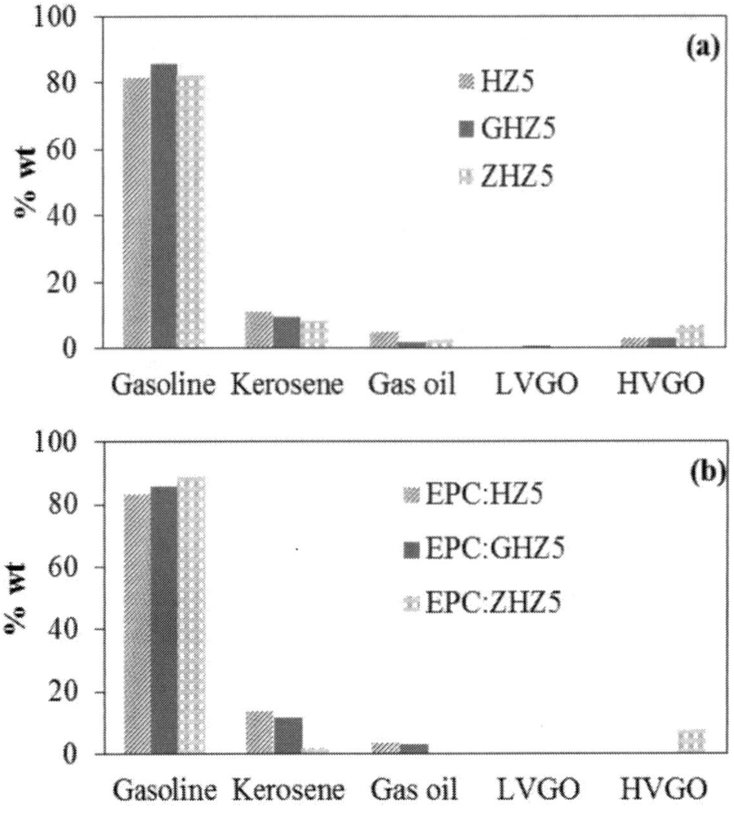

Figure 3 Petroleum fractions in oils obtained from using (a) the single-bed catalytic systems, and (b) the dual-bed catalytic systems.

CONCLUSIONS

The catalyst selection and design are crucial factors that can affect to the enhancement of aromatic production from bioethanol. The Ga_2O_3-containing HZSM-5 was found to be the best catalyst among those used in the single-bed catalytic system. Moreover, the use of dual-bed catalytic system can improve the efficiency of aromatic production from bioethanol. Also, the consecutive-bed case of $MgHPO_4/Al_2O_3$ and Ga_2O_3/HZSM-5 was the best dual-bed system for aromatic production because of the highest aromatic yield.

ACKNOWLEDGEMENTS

The authors gratefully acknowledge Thaioil Public Company Limited, the Petroleum and Petrochemical College, Chulalongkorn University, and the Center of Excellence on Petrochemical and Materials Technology, Thailand, for supporting this research work. The authors also thank Sapthip Company Limited for providing bioethanol used as the feed.

REFERENCES

Barthos, R., Sze´chenyi, A., and Solymosi, F. (2006) J. Phys. Chem. B. 10 (43), 21816-21825.

Inaba, M., Murata, K., Saito, M., and Takahara, I. (2005) React. Kinet. Catal. Lett. 88 (1), 135-142.

Kojima, M., Aida, T., and Asami, Y. (1981) US. Patent 4,302,357

Lukyanov, D., B., Gnep, S., and Guisnet, M.R. (1994) Industrial & Engineering Chemistry Research, 33, 223-234.

Qiu, P., Jack, H., Lunsford and Michael, P., and Rosynek, P. (1998) Catalysis Letters 52, 37-42.

Saha, S.K., and Sivasanker, S. (1992) Catalysis Letters, 15, 413-418.

Saewong S., Wungtanagorn C., and Jitkarnka S. (2012) Dual-bed ($MgHPO_4/Al_2O_3:Ga_2O_3$/HZSM-5) Catalytic System for Liquid Hydrocarbon Production from Bioethanol. Proceedings of the 3rd Research Symposium on Petroleum, Petrochemicals, and Advanced Materials and the 18th PPC Symposium on Petroleum, Petrochemicals, and Polymers, Queen Sirikit National Convention Centre, Bangkok, Thailand.

(196a) Catalytic Filtration of Pyrolysis Vapor Via Moving Bed Granule Filter

Lysle Whitmer, Center for Sustainable Environmental Technologies, Iowa State University, Boone, IA and Robert C. Brown, Mechanical Engineering, Iowa State University, Ames, IA

Catalytic Filtration of Pyrolysis Vapor via Moving Bed Granule Filter

Lysle Whitmer and Robert C. Brown

Center for Sustainable Environmental Technologies

Department of Mechanical Engineering

Iowa State University

Ames Iowa – 50010 U.S.A.

Abstract

The objective of this project is to simultaneously reduce the amount of particulate matter inherent in bio-oil vapors through hot gas filtration while catalytically deoxygenating the vapors to form stable bio-oil intermediates. Current filtration techniques employed in the production of bio-oil only filter particulate down to the 5-10 micron level, leaving the very fine particles in the vapor stream. Particulate matter is thought to catalyze the polymerization of bio-oil; a process that increases the viscosity and water content of the oil. Additionally, the oxygen content of bio-oil contributes to a number of secondary reactions that also result in the polymerization of bio-oil.

This work focuses on the utilization of a moving bed granule filter to remove fine particulate from a pyrolysis vapor stream using catalytically active filtration media at process temperatures reaching 500°C. The moving bed design features adjustable catalyst residence time allowing for continuous vapor processing under stable operating conditions. This research evaluates the effectiveness of the filter at reducing particulate content in the pyrolysis vapor stream and its ability to deoxygenate the resulting bio-oil.

(196b) A Response Surface Methodology Study On Catalytic Pyrolysis of Biomass with FCC Catalyst

Ofei Mante[1], Foster Agblevor[1] and Ron McClung[2], (1)Biological Engineering, Utah State University, Logan, UT, (2)BASF Catalysts LLC, Iselin, NJ

A response surface methodology was used to investigate the influence of temperature, weight hourly space velocity (WHSV) and vapor residence time on the catalytic pyrolysis of hybrid poplar wood using Y-zeolite based FCC catalyst in a 50 mm bubbling fluidized bed reactor. The levels of the independent variables were: temperature (400 - 600 °C), WHSV (1 - 3 h^{-1}), and vapor residence time (3 - 6 s). The Box-Behnken experimental design was used and a total number of 15 experimental runs including 3 center runs were generated. The analysis of variance (ANOVA) at 95 % confidence interval was performed with Minitab 16 software package and reduced models were generated for the prediction of the responses. It was found that higher levels of WHSV (> 1 h^{-1}) with shorter vapor residence time and moderate temperatures (500 ≤ Temperature < 600) favor higher yield of organic liquid fraction. Also, the employment of higher temperature (> 500 °C) and WHSV (> 2 h^{-1}) reduces the formation of coke/char. The catalytic pyrolysis with the Y-zeolite based FCC catalysts at high temperatures (> 500 °C) and long residence time enhanced the cracking of the primary vapors and increased the formation of CO, H_2 and CH_4. The operation at higher temperatures (> 500 °C) and lower WHSV (≤ 2 h^{-1}) decreased the oxygen content and increased the viscosity of the bio-oil. The reaction temperature was the most influential significant independent variable on char/coke yield, concentration of non-condensable gases, carbon content, oxygen content, pH and viscosity of the bio-oils. The WHSV was found to be the most important significant independent variable that affected the yield of organic liquid and the formation of water. The models for the prediction of the responses with the exception of viscosity were adequate and statistically significant.

(196c) Properties of Olivine Catalysts for Indirect Gasification of Biomass

Remco J. Lancee, Hans O.A. Fredriksson, Peter C. Thüne, Hubert J. Veringa and J.W. (Hans) Niemantsverdriet, Eindhoven University of Technology, Eindhoven, Netherlands

To meet the increasing demand for renewable, CO_2-neutral energy, biomass (in particular waste products) can be used for the production of fuels such as hydrogen, synthetic natural gas or Fisher-Tropsch diesel [1]. Conversion of biomass to such fuels can be performed using an indirect gasification process, where gasification and partial combustion (necessary to sustain the endothermic gasification process) take place in separate chambers. Heat generated by combustion of chars and tars derived from the biomass is transferred to the gasification chamber by a bed material, conventionally sand, that is circulated between the two chambers. By exchanging the inert sand with a catalytically active bed material it is possible to improve the efficiency of the gasification process and to reduce the content of polluting tars in the product gas. A widely used and studied catalytic bed material is the naturally occurring mineral olivine ($(Mg,Fe)_2SiO_4$). Although the exact catalytic function of the olivine is not yet fully understood, it has been observed that pretreatments can improve the performance of this catalyst[2] and that the phase composition of the mineral as well as its activity change over time, under typical process conditions[3].

We have used a flow reactor to mimic the conditions that olivine catalysts are exposed to during gasification of biomass. Oxidation (20% O_2 in Ar) and reduction (CO, H_2 and H_2O mixtures) treatments at 750°C were applied. The resulting changes in the olivine catalyst's bulk composition were characterized using X-ray Absorption Spectroscopy (XAS) and X-Ray Diffraction (XRD). Surface chemistry and morphology changes were characterized using X-ray Photoelectron Spectroscopy (XPS) and electron microscopy. The main changes in the catalyst after exposure to high temperatures, necessary for biomass gasification, concern the iron present in the olivine. After exposure to the oxidizing environment, the olivine catalysts exhibit a significant increase of surface Fe and a simultaneous decrease of Si. Furthermore, Fe_2O_3 and Fe_3O_4 phases are observed to form from Fe segregated out of the bulk mineral. On the catalyst surface, formation of crystallites is observed and the amount of Fe, the most dynamic element, at the surface is more than doubled. The catalyst entering the gasification zone is thus best described as a largely Fe-depleted olivine with the inclusion of FeOx phases at the surface.

After exposure to reducing conditions (H_2+CO+H_2O) the FeOx phases are reduced to FeO and Fe^0, both in the bulk and at the surface. The surface content of Fe is reduced to roughly half the amount present after oxidation and the surface crystallites decrease in size. In CO containing gases significant carbon deposition is observed, some of this in the form of nanofibers.

These results suggest that Fe segregates reversibly to and from the surface under oxidizing and reducing conditions and on time scales which are relevant in commercial plants.

In addition, characterization of the olivine samples using XPS depth profiling and Thermo Gravimetric Analysis (TGA) provided further insight in the dynamic behavior of iron in olivine and its chemical looping abilities under relevant process conditions.

[1] A. V. Bridgwater, A. J. Toft and J. G. Brammer, Renewable & Sustainable Energy Reviews 6, 181-248 (2002)
[2] L. Devi, K. J. Ptasinski and F. Janssen, Industrial & Engineering Chemistry Research 44, 9096-9104 (2005)
[3] J. N. Kuhn, Z. K. Zhao, L. G. Felix, R. B. Slimane, C. W. Choi and U. S. Ozkan, Applied Catalysis B-Environmental 81, 14-26 (2008)

THERMOCHEMICAL BIOMASS PROCESSING FOR H_2 AND BIODIESEL PRODUCTION

Richa Tungal, Rajesh V. Shende
South Dakota School of Mines & Technology, Rapid City, SD, USA

Abstract

This paper reports subcritical aqueous phase reforming of pine wood in presence of homogeneous $Ni(NO_3)_2$ for the production of H_2 and oxygenated hydrocarbons. Slurry of 10-20 wt% pine wood was prepared in distilled water containing 7 wt% $Ni(NO_3)_2$ and loaded in high-temperature high-pressure PARR reactor and heated to 250°C. The reaction progress was monitored for 60 min at 250°C and while the reaction was in progress, gas samples were withdrawn at regular time intervals and analyzed using Chrompak capillary column on the GC equipped with thermal conductivity detector. The analysis of gas samples indicated the presence of H_2, CO_2, CO, and CH_4. Specifically, after 60 min of reaction 13.24 vol% of H_2 was observed. The liquefied biomass (biocrude) obtained was analyzed using HPLC and found to contain sugars (glucose and xylose), HMF, furfurals and oxygenated hydrocarbons such as aldehydes, ketones and organic acids. Among organic acids produced, lactic acid was produced in higher amounts as compared to propionic, formic and acetic acids. The bio-oil was recovered and further reacted with methanol in presence mesoporous NiO material prepared by sol-gel.

Introduction

H_2 and biocrude can be produced from thermochemical treatment of biomass using several technologies such as gasification [1], pyrolysis [2], and aqueous phase reforming using supercritical [3] and subcritical [4] water. Among all the above mentioned processes for converting biomass to valuable fuels, use of subcritical or supercritical water for the aqueous phase reforming of biomass is very advantageous as it eliminates the need for the drying of the biomass. Supercritical water operates at high temperatures of > 374°C and high pressures of > 22.1 MPa. Therefore this process is less attractive for the conversion of biomass into fuels on industrial scale. Contrasting to this, subcritical water processing of biomass is getting more attention as it operates at low to moderate temperatures of 125-350°C and pressures of 0.5-21 MPa [5]. Subcritical water is a non-toxic, inexpensive and environmentally friendly solvent and has been used as the reaction medium for the liquefaction of biomass. Subcritical water has higher ionization constant ($Kw = 10^{-11}$) near the critical point and behaves as acid/base catalyst. Ionization constant of water increases with temperature and hence helps in the solubilization of lignocellulosic biomass components [6].

Recently, Cheng et al., [7] investigated rapid conversion of switchgrass in subcritical water under pyrolysis conditions at 250- 350°C and 20 MPa and reported significant H_2 (0.17 mol %) in the gaseous products. In another study, paper waste was processed under subcritical water conditions in presence of charcoal supported ruthenium catalyst at 250-300°C. This study reported production of CH_4, H_2 and other gases. Many other investigations reported production of different fuels under various hydrothermal conditions. Minowa et al., [8] investigated gasification of cellulose at 350°C in presence of a Ni catalyst and observed product gas containing H_2. Antal et al., [9] used glucose as potential feedstock for H_2 production via steam reforming in presence of different metal oxide supported metal catalyst (Ni/SiO_2) and reported H_2 rich syn gas. Hydrothermal liquefaction of empty palm fruit bunch was performed in alkaline hot compressed water and found to contain oxygenated hydrocarbons such as benzoic acid, methyl ester and phenols in the biocrude [10]. Pourali et al., [11] reported production of organic acids from rice bran at 360°C.

Most of these studies were focused on H_2 generation from model compounds, however, very little effort is noticed in the literature on the production of high energy biocrude from real biomass at low temperatures. In this paper subcritical aqueous phase reforming of pine wood was performed using homogeneous Ni-catalysts at 250°C. The reaction was monitored for 60 min and while the reaction was in progress, the gas and liquid samples were withdrawn and analyzed using GC and HPLC, respectively. The results obtained on H_2 volume generation and biocrude analysis are presented and discussed.

Experimental Methods and Analysis

Materials

$Ni(NO_3)_2.6H_2O$ was purchased from Alfa Aesar and and used without purification. N_2 from a cylinder with the minimum stated purity of 99% was purchased from Linweld Inc., SD. Pine wood was made into slurry in distilled water and used for subcritical hydrothermal processing in N_2 environment.

Experimental Set-up and Procedure

High-temperature high-pressure PARR SS316 reactor of 300 ml capacity equipped with heating and cooling assembly, gas inlet and outlet valves, sampling valves, pressure gauge, safety rupture disc, thermocouple, magnetic stirrer and temperature indicator controller was used to perform subcritical hydrothermal processing of pine wood. This reactor is rated for maximum working temperature and pressure of 350°C and 3000 psi, respectively. An additional SS gas sampling vessel was mounted on the reactor. The entire reactor set-up is shown in Figure 1.

Accurately weighed 15.0 g of pine wood was made into slurry in 150 ml distilled water containing 7wt% $Ni(NO_3)_2$ (taken on a biomass weight basis) and loaded in the high-temperature high-pressure PARR reactor. The contents in the reactor were flushed continuously with N_2 for 10-15 min to replace air and pressurized to 40 psi. The reactor was heated to 250°C under continuous magnetic stirring of biomass slurry at 1300 rpm. It took approx. 45 min to attain the set point of 250°C where the total reactor pressure was about 700 psi. Once the reactor temperature was attained, a gas sample was withdrawn. This time was taken as 'zero' time for the reaction. The reaction was continued for 60 minutes and during the course of the reaction, gas samples were withdrawn at regular time intervals and analyzed using GC.

Product Analysis

Gas samples were withdrawn with a gas tight syringe during the subcritical aqueous phase reforming of pine wood and analyzed using Agilent 6890A GC equipped with Chrompak capillary column and thermal conductivity detector. A sample injection volume of 100 µl was used for the analysis. The GC analysis was performed under the isothermal conditions where column and detector temperatures were maintained at 105°C and 200°C, respectively.

At the end of the reaction, the reactor contents were cooled down to room temperature and centrifuged to collect biocrude and biochar. The filtrate was analyzed with HPLC to analyze sugar and oxygenated hydrocarbons.

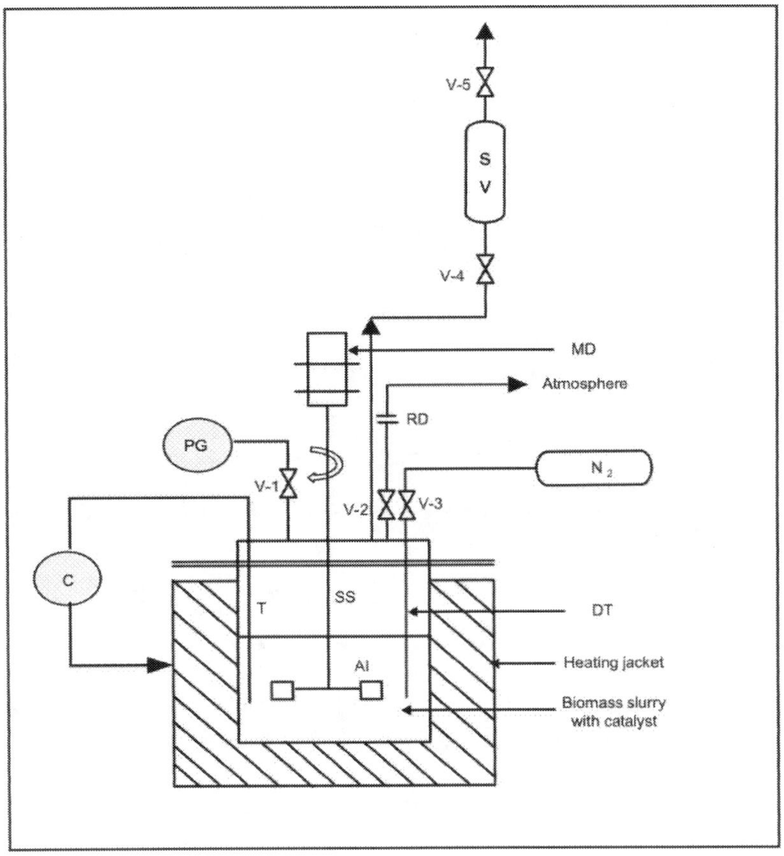

Fig. 1: Reactor set-up for subcritical hydropyrolysis of a biomass for H_2 generation. (RD-rupture disc, T-thermocouple, AI-adjustable impeller, SS-stirring shaft, PG-pressure gauge, MD-magnetic drive, V-valve, DT-dip tube, C-controller, SV-gas sampling vessel).

Results and Discussion

The product gas obtained at 250°C indicated primarily H_2, CO, CH_4 and CO_2 gases. In addition to these gases, H_2O vapor should also be present in the gas samples, however, as the gas sampling vessel was held at ambient conditions, the water vapors might have been already condensed. Moreover the Chrompak capillary column that was used for the analysis of product gases was not capable of detecting water vapors. Therefore, the concentrations were determined without accounting for H_2O vapor in terms of vol% of total headspace volume, which was 150 ml. It was observed that H_2 production from pine wood was negligible at 225°C whereas as temperature increased to 250°C, the H_2 production was observed at 'zero' reaction time. The H_2 vol% obtained is shown in Figure 2, which indicated 13.24 vol% after 60 min reaction time. The product gas contained significant amount of H_2 and CO_2 gases, lower amount of CO and negligible amount of CH_4. Along with H_2 in product gas, CO_2 was also increased with the increase in reaction time up to 60 min during aqueous phase reforming of pine wood in presence of 7 wt % Ni^{2+} catalyst at 250°C. This trend of H_2 production from pine wood suggested that H_2 levels will further increase linearly if the reaction would have been continued beyond 60 min.

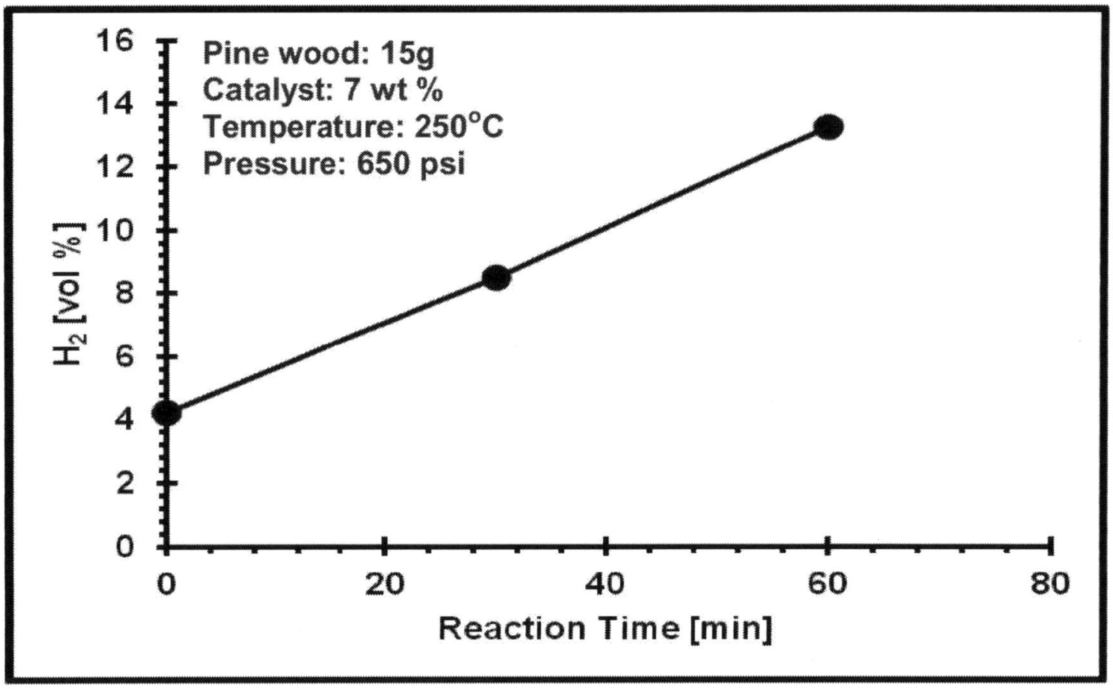

Fig. 2: Production of H_2 during subcritical aqueous phase reforming of pine wood at 250°C in presence of 7 wt % $Ni(NO_3)_2$.

After the completion of reaction, the reactor was cooled down to room temperature. The liquid slurry left after the reaction was centrifuged and the supernatant was separated from biochar. The char formation, however, was not studied in detail as it was not the goal of this investigation. The supernatant is called as biocrude (Figure 3); these biocrude samples were collected at an interval of 30 min. It can be seen from Figure 3 that at 250°C and 'zero' reaction time, pine wood was solubilized in presence of 7 wt % $Ni(NO_3)_2$ catalyst during aqueous phase reforming of pine wood. This shows that $Ni(NO_3)_2$ was not only capable of producing H_2 but was also effective in breaking down the complex wood structure. The biocrude samples collected at different time intervals were filtered and the supernatant was extracted in different solvents such as ethyl acetate, THF and ethyl ether. These extracts were evaporated on rota evaporator and the residue obtained was analyzed using FT-IR. The FT-IR results showed the presence of oxygenated hydrocarbons such as alcohols, aldehydes, ketones, phenols, esters and carboxylic acids in all the extracts.

Fig. 3: Biocrude obtained during subcritical aqueous phase reforming of pine wood at 250°C in presence of 7 wt % $Ni(NO_3)_2$ at different time intervals.

The biocrude obtained during aqueous phase reforming of pine wood at 250°C in presence of $Ni(NO_3)_2$ was analyzed using HPLC and the results obtained are shown in Table 1. The HPLC analysis of the biocrude showed the production of sugars such as glucose and xylose, HMF, furfural and oxygenated compounds such as lactic acid, formic acid and propionic acids. Among various organic acids produced, the production of lactic acid was higher as compared to formic, acetic and propionic acids in the biocrude. This shows that $Ni(NO_3)_2$ catalyst was effective for the production of H_2 as well as valuable chemicals such as oxygenated hydrocarbons in the biocrude obtained during catalytic liquefaction of pine wood.

Table 1. HPLC analysis of the biocrude obtained after aqueous phase reforming of pine wood at 250°C in presence of 7 wt % $Ni(NO_3)_2$

Compounds	Concentration (mM)
Glucose	0.146
Xylose	0.130
Lactic acid	84.626
Formic acid	6.446
Propionic acid	54.452

These organic acids are usually obtained from the catalytic hydrolysis of carbohydrates or sugars under subcritical water reactions. Glucose and fructose have been known to produce organic acids such as acetic, formic at 300-400°C [12]. Bicker et al., [13] reported 42 to 86 % conversion of aqueous sugar solution into lactic acid. The reaction for the production of lactic acid proceeds through a series of reactions where sugars (glucose and fructose) first get converted into glyceraldehydes, later to pyruvaldehyde and finally to lactic acid. Lactic acid, due the presence of bifunctionality, can be converted into other compounds such as propionic acid and H_2 through a series of reactions such as acetaldehyde can be produced via decarboxylation/decarbonylation, propionic acid via reduction, 2,3-pentanedione via condensation etc. [12]. This infers that the production of H_2 might have been routed from the degradation reactions of sugars as well as from organic acids present in the biocrude obtained during the aqueous phase reforming of pine wood under subcritical water in presence of 7 wt% $Ni(NO_3)_2$ catalyst. Based on these findings, the reaction pathways can be proposed as shown by scheme 1.

Biocrude obtained from the catalytic aquoeus phase reforming of biomass under subcritical water conditions can be converted to liquid fuels such as biodiesel. Many researchers have performed the preliminary studies on the hyrothermal treatment of biomass in hot compressed water and indicated that it is more beneficial for the production of liquid fuels as compared to conventional pyrolysis. The liquid obtained after aqueous phase reforming of biomass consists of a complex mixture of more than 300 oxygenated compounds, carbohydrates (sugars) and lignin fragments. During aqueous phase reforming, a large portion of the oxygen from biomass is removed in the form of CO_2 and the biocrude left after the reaction contains only 10-15 % of oxygen as compared to woody biomass where it contains almost 40 % of the oxygen content. However, due to the high oxygen content of biocrude, it is required to be upgraded further and the deoxygenation of biocrude can be performed over zeolites or mesoporous metal oxide catalyts to produce liquid fuels. In one of the studies by Brahmkhatri et al., [14] free fatty acids were used for the production of biodiesel in presence of heterogeneous MCM-41 catalyst. The biocrude obtained during aqueous phase liquefaction of pine wood in presence of homogeneous $Ni(NO_3)_2$ that contained fatty acids can be converted into biodiesel range hydrocarbons using heterogeneous catalyst such as NiO.

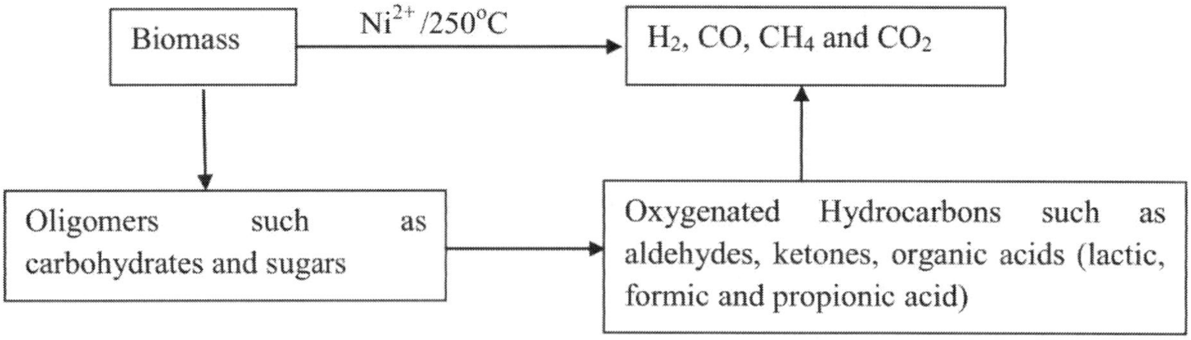

Scheme 1. Possible reaction pathways during catalytic aqueous phase reforming of pine wood at 250°C.

NiO synthesis involved dispersion of metal salt ($NiCl_2$) in ethanol containing brij-76 surfactant. To this visually clear solution, water and propylene oxide were added to achieve the gel formation. The gel was aged, dried, and calcined at 600°C. The calcined powder was characterized using SEM and TEM and found to contain mesopores in the range of about 6-50 nm. This mesoporous NiO catalyst will be used for the conversion of acids present in the biocrude obtained from the subcritical aqueous phase reforming of biomass for the production of liquid fuels such as biodiesel range hydrocarbons. The detailed study of the production of biodiesel range hydrocarbons from biocrude in presence of heterogeneous mesoporous NiO catalyst is in progress. The SEM of mesoporous NiO is shown in Figure 4.

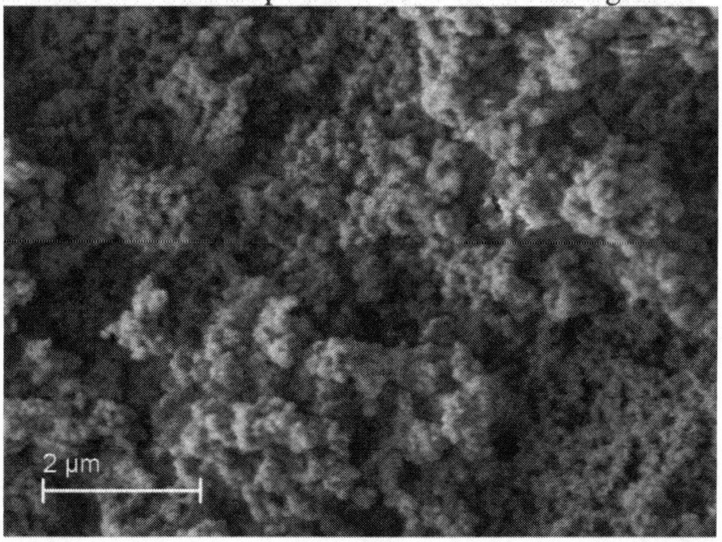

Fig. 4: SEM of NiO made with sol-gel using brij-76 surfactant and calcined at 600°C.

Conclusion

The subcritical aqueous phase reforming of pine wood was successfully performed in high-temperature and high-pressure PARR reactor at 250°C in the presence of homogeneous Ni-catalyst $Ni(NO_3)_2$. Their product gas profiles indicate H_2 rich gas containing CO_2 and CO and CH_4 gases. The concentration of CH_4 was found to be negligible. The maximum H_2 of 13.24 vol% was produced from pine wood using $Ni(NO_3)_2$ catalyst after 60 min of reaction at 250°C. The liquid phase obtained after the completion of reaction was centrifuged and the supernatant was analyzed using HPLC and FT-IR. The liquid phase was found to contain sugars, HMF, furfural and oxygenated hydrocarbons such as lactic, formic and propionic acid where lactic acid was found to be in higher amounts as compared to other acids. It is believed that the production of H_2 has been routed from the production of sugars as well as organic acids during aqueous phase reforming of pine wood under subcritical water reaction at 250°C in presence of $Ni(NO_3)_2$ catalyst. It can be concluded that $Ni(NO_3)_2$ is effective not only in the decomposition of complex lignocellulosic biomass but also the generation of H_2 as well as organic acids in the biocrude which could be converted into liquid fuels such as biodiesel range of hydrocarbons using heterogeneous catalyst such as NiO.

Acknowledgement

The authors wish to acknowledge the financial supports by the Chemical and Biological Engineering Ph.D. support grant and Tyndall AFRL.

References

1. Minowa, T., Inoue, S., (1999), "Hydrogen production from biomass by catalytic gasification in hot compressed water," *Renewable Energy* 16, pp. 114-117.
2. Dalai, A.K., Bakhshi, N.N, Valliyappan, T., (2008), "Pyrolysis of glycerol for the production of hydrogen or syn gas," *Biresource Technology*, 99, pp. 4476-4483.
3. Davda, R.R., Dumesic, J.A., (2002), "Hydrogen from catalytic reforming of biomass-derived hydrocarbons in liquid water," *Nature*, 481, pp. 964-967.
4. Onwudili, J.A., Williams, T.P, (2007), "Hydrothermal catalytic gasification of municipal solid waste, '' *Energy & Fuels,* 21, pp. 3676-3683.
5. Khajavi, S.H., Kimura, Y., Oomaori, T., Matsuno, R., Adachi, S., (2005), ''Degradation kinetics of monosaccharides in subcritical water,'' *Journal of Food Engineering,* 68, pp. 309-313.
6. Kumar, S., Gupta R.B., (2009), "Biocrude Production from Switchgrass Using Subcritical Water," *Energy Fuels*, 23, pp. 5151–5159.
7. Cheng, L., Ye, P. X., He, R. and Liu, S. (2009), "Investigation of rapid conversion of switchgrass in subcritical water," *Fuel Processing Technology,* 90, pp. 301-311.
8. Minowa, T. and Inoue , S. (1999), "Hydrogen production from biomass by catalytic gasification in hot compressed water," *Renewable Energy*, 16, pp. 1114 1117.
9. Antal, M.J, Jr., Manarungson, S, Mok, WS-L. (1994) "Hydrogen production by steam reforming glucose in supercritical water," *Adv. Thermochemical Biomass Conversion.,* pp. 367-77.
10. Akhtar, J., Kuang, S.K., Amin, A.S., (2010) *Renewable Energy*, 35, pp. 1220-1227.
11. Pourali, O., Asghari, F.S., Yoshida, H., (2009), "Sub-critical water treatment of rice bran to produce valuable materials," *Food Chemistry*, 115, pp. 1-7.
12. Serrano-Ruiz, J.C., West, R.M., Dumesic, J.A., (2010), "Catalytic conversion of renewable biomass resources to fuels and chemicals," Annu. Rev. Chem Biomol. Eng., 1, pp. 79-100.
13. Bicker, M., Endres, S., Ott. L., Vogel, H., (2005), "Catalytic conversion of carbohydrates in subcritical water: A new chemical process for lactic acid production," *Journal of Molecular Catalysis A: Chemical*, 239, pp. 151-157.
14. Brahmkhatri, V., Patel, A., (2011), "Biodiesel Production by Esterification of Free Fatty Acids over 12-Tungstophosphoric Acid Anchored to MCM-41," *Ind. Eng. Chem. Res.,* 50, pp. 6620–6628.

(196e) In-Situ Treatment of Pinewood Pyrolysis Vapors Using Zeolites

Vamshi Krishna Guda[1], Hossein Toghiani[2], Rebecca K. Toghiani[2] and Charles U. Pittman Jr.[3], (1)Mississippi State University, Mississippi State, MS, (2)Dave C. Swalm School of Chemical Engineering, Mississippi State University, Mississippi State, MS, (3)Department of Chemistry, Mississippi State University, Mississippi State, MS

Pine wood pyrolysis vapors were treated in-situ using zeolite catalysts. An auger fed reactor was used for the pinewood pyrolysis, while a packed bed reactor mounted on the top of auger reactor housed the catalyst was used to treat the pyrolytic vapors. The pyrolytic vapors produced at 450 °C were passed through zeolite catalysts maintained at 450 °C at a weight hourly space velocity (WHSV) of 12 h^{-1}. Seven zeolites, including three HZSM-5 catalysts with differing SiO_2/Al_2O_3ratio (23, 55, and 280), as well as mordenite, ferrierite, HY, and beta, were used as catalysts. The effect of catalyst type on deoxygenation of pyrolysis vapors was studied. Chemical characterization of the upgraded bio-oils was performed using GC-MS. Water content, density, viscosity, acid value, pH, and elemental composition of each upgraded bio-oil was also measured. Catalyst characterization was performed using XRD, TEM, and TGA.

Catalytic treatment of pyrolysis vapors decreased the liquid yield at the expense of gaseous product. While all the zeolites deoxygenated the pyrolysis vapors, HZSM-5 catalysts were most effective. The HZSM-5 upgraded bio-oils were less viscous, had low acid number, high pH, and an oxygen content half that of untreated pyrolysis oil. HZSM-5 catalysts also produced the highest amount of phenolics and aromatic hydrocarbons. Acidity, pore size, and the shape selective catalysis of HZSM-5 catalysts were responsible for their high activity. TG analysis revealed that the pore size of the catalysts highly influenced the coking reactions. Catalysts were successfully regenerated.

(196f) Mixed Rare Earth Oxides for Desulfurization and Tar Removal From Gasifier Effluents

Rui Li[1], Michael J. Janik[2], Matthew D. Krcha[2] and Kerry M. Dooley[1], (1)Chemical Engineering, Louisiana State University, Baton Rouge, LA, (2)Chemical Engineering, Pennsylvania State University, University Park, PA

The removal of H_2S and tars from syngas generated in biomass or coal/biomass gasification plays an important role in syngas cleanup. Rare earth oxides (REOs, e.g., Ce/LaO_x) mixed with transition metals (e.g., Mn, Fe, Ni) are promising materials for both sulfur removal and for tar reforming in the presence of some sulfur. We have synthesized such materials by a variety of methods, including templated sol-gel, hydrothermal, incipient wetness coimpregnation, and evaporation-induced self assembly (EISA). These materials have been investigated for either H_2S adsorption or tar reforming at high temperatures (~630-750°C), using model syngas mixtures, a goal being to improve heat integration throughout the syngas process. Successive desulfurization-regeneration cycles using Mn/Ce/LaO_x demonstrated good sulfur capacities can be obtained even in the presence of water and CO_2, using just air for the regenerations. The limits of Mn incorporation into the REOs without bulk phase separation have been studied, and the sulfur adsorption has been modeled using DFT. Naphthalene was used as a model compound for tar reforming/cracking, and the effects of H_2S, CO, water and CO_2 on its reactions have been examined. While H_2S, water and CO_2 are inhibitors, CO is not. Transition metal-doped REOs are more effective than the pure REOs, suggesting an important role of the transition metal in increasing the number of oxygen vacancies. We have also examined propane reforming in model syngas, because its surface behavior is easier to model by DFT. Fresh, used and regenerated sorbents and catalysts have been characterized by BET, TPR, XRD, XANES and EXAFS.

(198a) One Pot Synthesis of Lactic Acid From Sugars and Polyols

Xin Jin, Chemical and Petroleum Engineering, Center for Environmentally Beneficial Catalysis, Lawrence, KS, Debdut S. Roy, SABIC Research & Technology Pvt. Ltd, Vadodara, India, Bala Subramaniam, Department of Chemical and Petroleum Engineering, Center for Environmentally Beneficial Catalysis, University of Kansas, Lawrence, KS and Raghunath V. Chaudhari, Center for Environmentally Beneficial Catalysis, Chemical & Petroleum Engineering Department, University of Kansas, Lawrence, KS

With depleting reserves of fossil fuels (including petroleum, coal resources) worldwide, the chemical industry is actively exploring the possibility of replacing fossil-based raw materials with renewable feedstocks such as plant-based biomass. Sugars and sugar-derived polyols, which can be obtained from cellulosic biomass, are regarded as one of the most promising feedstocks for value-added commodity chemicals. Lactic acid is one such example with promising applications in food and pharmaceutical industries as well as being an intermediate to industrial chemicals. Conventional production of lactic acid employs fermentation or hydrothermal treatment of glucose at elevated temperatures (>573K). The major disadvantage of those processes is lower productivity, and several side reactions leading to lower overall yield of lactic acid[1]. In another process[2], 43% lactic acid yield from glucose is reported over Pt/C catalyst in the presence of air and excess alkali (molar OH^-/sugars = 20).

Here we report a detailed study on one-pot catalytic synthesis of lactic acid from sugars and polyols in significant yields at low temperatures (as low as 338 K). It is shown that using supported noble metal catalysts and alkali promoters, sugars (glucose, fructose) and polyols (glycerol, sorbitol) are converted to lactic acid in high yields (62~81%). The catalytic activity of Cu, Ni, Pt, Ru, Rh and Pd based catalysts for the one pot synthesis has been investigated for glycerol, xylitol and sorbitol as substrates. Catalytic conversion of sugars to lactic acid is more efficient compared to polyols as substrates at significantly low temperatures (338 K). A possible reaction pathway is discussed based on observed experimental results. Detailed kinetic study on conversion of glycerol and sorbitol to lactic acid will also be presented along with rate equations and kinetic parameters.

1. (a) Yan, X. Y.; Jin, F. M.; Tohji, K.; Kishita, A.; Enomoto, H., Hydrothermal Conversion of Carbohydrate Biomass to Lactic Acid. *AIChE J* **2010,** *56* (10), 2727-2733; (b) Ramirez-Lopez, C. A.; Ochoa-Gomez, J. R.; Fernandez-Santos, M.; Gomez-Jimenez-Aberasturi, O.; Aonso-Vicario, A.; Torrecilla-Soria, J., Synthesis of Lactic Acid by Alkaline Hydrothermal Conversion of Glycerol at High Glycerol Concentration. *Ind Eng Chem Res* **2010,** *49*(14), 6270-6278.

2. Onda, A.; Ochi, T.; Kajiyoshi, K.; Yanagisawa, K., A new chemical process for catalytic conversion Of D-glucose into lactic acid and gluconic acid. *Appl Catal a-Gen* **2008,** *343* (1-2), 49-54.

(198b) Selective Oxidation of Diols Over Supported Pt Catalysts in Base-Free Aqueous Solution

Matthew Ide and Robert J. Davis, Chemical Engineering, University of Virginia, Charlottesville, VA

Interest in producing commodity chemicals from biomass instead of fossil fuel continues to grow as the biological and chemical transformations of carbohydrate feedstocks become more economical. The selective oxidation of sugar-derived multi-functional molecules, such as glycerol and hydroxymethylfurfural, is one route to add value to biorenewable chemical feedstocks. Recent studies have demonstrated high activity of gold catalysts for alcohol oxidation reactions in the presence of added base, but that process suffers from the need to neutralize the product stream with acid to release the organic acid product. The fundamental barriers limiting the rate and selectivity during the oxidation of alcohols to acids in the absence of added base are unclear but are critical to the development of industrially-relevant biorenewable chemical oxidation processes.

This work explores the overall conversion, product selectivity, oxidation rate, and stability of supported metal catalysts as a function of pH and alcohol structure. In particular, supported Pt catalysts were evaluated in the acidic aqueous-phase oxidation of diols to mono- and di-acids at 343 K in a semibatch reactor. The influence of solution pH (acidic versus basic) on the identifiable intermediate reaction products was also determined. The effect of diol chain length on the initial turnover frequency and the final selectivity to diacids was investigated at 10 bar O_2 and 0.1 M diol. Results from the kinetic studies of diols were compared to those from mono-alcohols and polyols at similar conditions. The initial rate of reaction was significantly higher for diols with four to six carbons in the chain compared to diols with only two or three carbons. In addition, the longer chain diols were very selective to the diacid at high conversions, while shorter chain diols were unable to selectively convert to the diacid, even at high catalyst loadings.

Deactivation of the Pt catalyst was observed at low pH conditions. A typical reaction profile at moderate catalyst loading showed considerable deactivation compared to that observed with supported gold catalysts in the presence of added base. Analysis of a used Pt catalyst by electron microscopy indicated that sintering of the metal was negligible. Moreover, elemental analysis of the reaction medium revealed that only a small amount of Pt leached into solution after 24 h, suggesting dissolution of Pt is not responsible for deactivation. Adsorption of reaction byproducts on the active Pt catalyst was the most likely cause of catalyst deactivation.

(198c) Effect of Bi-Functional Catalysts in Producing High Quality Liquid Fuels From Renewable Feedstocks

Vamshi Krishna Guda[1], Hossein Toghiani[1], Rebecca K. Toghiani[1] and Charles U. Pittman Jr.[2], (1)Dave C. Swalm School of Chemical Engineering, Mississippi State University, Mississippi State, MS, (2)Department of Chemistry, Mississippi State University, Mississippi State, MS

High quality liquid fuels were obtained by the in-situ catalytic treatment of pinewood pyrolysis vapors. An integrated reactor system, comprised of an auger reactor for the pyrolysis of pine wood chips and a packed bed reactor mounted on the top of auger for catalytic upgrading, was employed for the studies carried out at 450 ^{o}C and a WHSV of 10-12 h^{-1}. The catalysts used were HZSM5, 10 wt% Fe-ZSM5, 10 wt% Ni-ZSM5, HY, 10 wt% Fe-Y, Co-Mo/γ-Al_2O_3 and Ni-Mo/γ-Al_2O_3. All experiments were conducted for one hour at a biomass feed rate of approximately 10 g/min. The liquid and gas products were analyzed using GC-MS and GC-TCD-FID, respectively. Physical properties including water content, acid value, pH, viscosity, and elemental analysis of the upgraded bio-oils were measured. While the fresh catalysts were characterized using BET, XRD,TPR, and TEM, coke deposits on the used catalysts were characterized using TGA.

Low liquid yields and high gas yields were observed in all the catalytic experiments. While all the catalysts successfully deoxygenated the pyrolytic vapors, product yields, product selectivity and the production of hydrocarbons varied significantly with the catalyst composition. Metal promoted zeolites were more effective as deoxygenation catalysts compared to the respective proton zeolites, with iron and nickel promoted HZSM5 catalysts producing the highest yields of phenolics and aromatic hydrocarbons, respectively. The oxygen contents and the acid values of the Fe and Ni promoted HZSM5 upgraded bio-oils were substantially lower. TG analysis showed that the metal promoted catalysts coked less than the unpromoted zeolites. Results indicate that the bi-functional catlysts are better deoxygenating catalysts than the unpromoted acidic catalysts.

(198d) Cogeneration of Electricity and Valuable Chemicals: Electrocatalytic Oxidation of Glycerol in Anion-Exchange Membrane Fuel Cell

Zhiyong Zhang, Le Xin, Zhichao Wang and Wenzhen Li, Chemical Engineering, Michigan Technological University, Houghton, MI

Cogeneration of Electricity and Valuable Chemicals: Electrocatalytic Oxidation of Glycerol in Anion-Exchange Membrane Fuel Cell

Zhiyong Zhang, Le Xin, Zhichao Wang, Wenzhen Li

Electrocatalytic oxidation of glycerol for cogeneration electricity and valuable chemicals was investigated in anion-exchange membrane - direct glycerol fuel cells (AEM-DGFCs) with Pt and Au anodes. With the anode Pt loading of 1.0 mg_{Pt} cm^{-2}, the peak power densities of AEM-DGFCs fed with high-purity and crude glycerol can reach 124.5 and 102.8 mW cm^{-2} at 80 °C, respectively. On Au anode catalyst, the peak power densities of AEM-DGFCs can also reach 57.9 and 30.7 mW cm^{-2} under the same conditions. In the meantime, the AEM-DGFC can serve as a catalytic reactor for selective oxidation of glycerol to high-value chemicals (glycerate, tartronate, and mesoxalate) with excellent stability and reusability. The anode overpotentials has been found to be able to regulate the product distributions on both Pt and Au catalysts. This work also found that while Pt favors the generation of glycerate and tartronate, Au offers a unique opportunity for the formation of fully-oxidized C_3 product, mesoxalate.

(198e) Catalytic Depolymerization of a Phenolic β-O-4 Model Compound in Ionic Liquids

Dustin M. Veazey, Chemical Engineering, The Pennsylvania State University, State College, PA and Robert M. Rioux, Chemical Engineering, Pennsylvania State University, University Park, PA

Despite its relatively large abundance (approx. 30 wt. % of biomass), lignin has historically received less attention than cellulose as a renewable source of commodity chemicals. Due to its significant aromatic functionality, lignin is a potential feedstock for the production of aromatic monomers, such as phenol, benzene and toluene. However, the predominant use of lignin recovered from alkaline pulping of wood is as an additive in various adhesives or as a low-value heating fuel. The lack of environmentally benign and selective catalysts for cleavage of major linkages in lignin is one of the challenges inhibiting its utilization for renewable fuels and value-added chemicals. Ionic liquids (ILs) are a potential green reaction media for lignin depolymerization because of their high solubility of milled wood, negligible vapor pressure and tunable solvation properties.

This work investigates commercially available guanidine bases for decomposition of a phenolic β-O-4 lignin model compound in imidazolium ILs from 100 to 150°C. Guanidines successfully cleave the β-O-4 linkage in the aromatic dimer, guaiacylglycerol-β-guaiacyl ether (GG) to varying degrees, though stoichiometric amounts of coniferyl alcohol and guaiacol were not observed during the conversion of GG. Coniferyl alcohol is known to be unstable in aqueous alkaline environment and readily undergoes condensation reactions to produce higher molecular weight products. The stability of coniferyl alcohol in ionic liquids has not been studied extensively, but preliminary results indicate that it is unstable in ionic liquids as well. LC-MS confirms major reaction products are guaiacol (~30% yield) and various higher molecular weight condensation products. This supports formation of a quinone methide intermediate that participates in multiple reaction pathways, such as β-O-4 cleavage and Michael addition, depending on the respective activation energy barrier. This work studies the effect of IL cation/anion pair on the activation energy and selectivity for guaiacol liberation from GG. Future lignin depolymerization strategies should employ optimized base catalyst/ IL systems that favor β-O-4 cleavage, while suppressing the formation of thermodynamically stable condensation products.

(198f) Molecular Reaction Networks of Photocatalytic and Photochemical Reactions of Platform Phenolic Compounds From Lignin

Alexander Samokhvalov, Chemistry Department, Rutgers University, Camden, NJ and Morgan Goldberg, Chemistry Department, Rutgers University, Camden, NJ

Lignin biomass can be converted to the high-value low molecular weight organic compounds. Certain phenolic compounds obtained from lignin find applications as food natural flavors, taste enhancers, antioxidants, fragrances, and fine chemicals for synthesis of sustainable bio-degradable polymers. Photochemical and photocatalytic reactions rely upon the sustainable, abundant and free light of the Sun to convert photon energy to the energy of chemical bonds or to accelerate chemical reactions.

The goal of our research is to investigate molecular reaction networks and mechanisms of heterogeneous photocatalytic and photochemical reactions of few major phenolic compounds from lignin with C=C and C=O bonds in the molecule, such as isoeugenol, eugenol and vanillin. For photooxidation, we use oxygen from air under the "green" ambient conditions, near-UV light, and nanocrystalline TiO_2 as photocatalyst in the non-aqueous solvent. For photolysis and photo-isomerization, we use the UV light vs. near-UV light in the non-aqueous solvent. Molecular products of photoreactions of isoeugenol, eugenol and vanillin were determined qualitatively and quantitatively using thin layer chromatography (TLC), high performance liquid chromatography HPLC with optical detection (HPLC-UV), and with dual UV- and mass-spectrometric detection via electrospray ionization (HPLC-MS-ESI).

Photocatalytic oxidation of isoeugenol with oxygen from air produces dehydrodiisoeugenol at the high (up 80 %) yields, and eugenol and vanillin as minor products. Photooxidation or photoisomerization of eugenol does not proceed under the same conditions. Photooxidation of vanillin leads to vanillic acid and dimeric product. Using photoexcitation with light of the longer wavelengths leads to fewer molecular products of photoreactions; photocatalytic vs. photochemical oxidation (without TiO_2 photocatalyst) leads to the higher yields.

Synthesis of Electroless CuPd Catalyst for Glycerol Hydrogenolysis

Shannon Anderson, Egwu E. Kalu
Department of Chemical & Biomedical Engineering
FAMU-FSU College of Engineering
Tallahassee, FL 32310

Introduction

Glycerol is a byproduct of biodiesel production. It becomes contaminated throughout the biodiesel production process and thus useless in the current state of the industry. Efforts to refine or convert the glycerol are being investigated and it is well known that with the use of catalysts, glycerol can be cracked into many different chemicals – including propylene glycol[123]. Currently 1,2-propanediol (propylene glycol) is prepared from fossil resources and is used in industrial solvents, antifreeze, deicing, food additives, and the pharmaceutical industry. With fossil resources on the decline, the use of glycerol as a renewable feedstock for chemicals such as 1,2-propanediol has been under investigation.

This work seeks to use heterogeneous catalysis to convert glycerol to 1,2-propanediol through hydrogenolysis. Unlike the conventional synthesis (impregnation) approach for catalysts, the present work seeks to explore the use of electroless deposition approach to synthesize the catalyst for the hydrogenolysis of glycerol. The electroless catalyst performance is compared to the performance of catalyst prepared through impregnation method.

The hydrogenolysis of glycerol reaction is thought to occur as shown in Figure 1. Product outcome depends on the catalyst composition used.

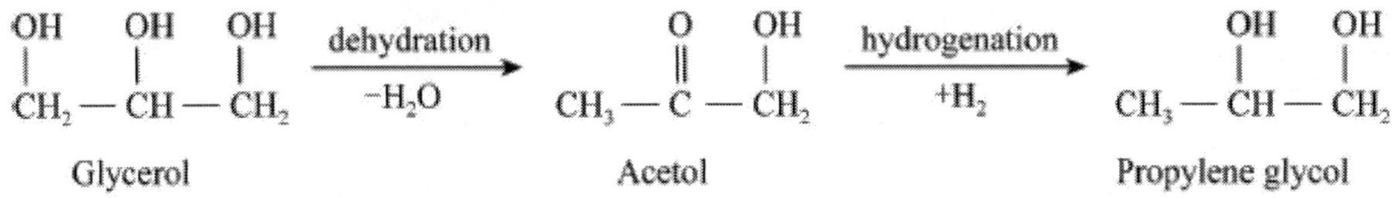

Figure 1. Potential pathway for glycerol hydrogenolysis

Using pure glycerol, recent literature has shown that ruthenium gives good conversion of glycerol while copper gives good selectivity towards propylene glycol[4]. For this work the catalyst formulation was tailored to achieve good selectivity, dehydration and hydrogenolysis. Hence, for the catalyst composition, copper was used to provide good selectivity towards propylene glycol, while alumina (the substrate) was used for the dehydration reaction and palladium was used to generate hydrogen in-situ for the hydrogenation reaction.

---

[1] Li, B, Wang, J, Yuan, Y, Ariga, H, Takakusagi, S, Asakura, K (2011), "Carbon Nanotube-Supported RuFe Bimetallic Nanoparticles as Efficient and Robust Catalysts for Aqueous-Phase Selective Hydrogenolysis of Glycerol to Glycols", *ACS Catal.*, **1**, pp. 1521 – 1528

[2] Roy, D, Subramaniam, B, Chaudhari, R (2010), "Aqueous phase hydrogenolysis of glycerol to 1,2-propanediol without external hydrogen addition", *Catal. Today*, **156**, pp. 31 – 37

[3] Zhou, Z, Li, X, Zeng, T, Hong, W, Cheng, Z, Yuan, W (2010), "Kinetics of Hydrogenolysis of Glycerol to Propylene Glycol over Cu-ZnO-Al_2O_3 Catalysts", *Chin. J. Chem. Eng.*, **18**(3), pp. 384 – 390

[4] Wu, Z, Mao, Y, Wang, X, Zhang, M (2011), "Preparation of a Cu-Ru/carbon nanotube catalyst for hydrogenolysis of glycerol to 1,2-propanediol via hydrogen spillover", *Green Chem.*, **13**, pp. 1311 – 1316

Experimental Method

In preparing electroless CuPd catalyzed alumina ($CuPd/Al_2O_3$), we started with polymer-stabilized Pd nanoparticles for the activation of the substrate for electroless deposition. The polymer-stabilized palladium catalyst solution was prepared by dissolving 0.2 g analytical reagent grade palladium acetate [$CH_3CO_2)_2Pd$] (47.05% Pd) from Aldrich-Sigma Co, in 2.0 mL NH_4OH. The palladium mixture was then added with stirring to a 44 g methanol-poly(vinyl) butyral (Solutia Inc.'s Butvar B-79) mixture formed by dissolving 22 g PVB in 140 mL methanol. This is referred to as the Pd-ink. A known amount of alumina was mixed with known amount of the Pd-ink and sonicated for 24 hours. The Pd-ink/Al_2O_3 mixture was subjected to 275°C temperature in a furnace for 24 hours. The high temperature activation was used to burn-off the polymer and generate zero-valent palladium nanoparticles on the surface of the alumina substrate. The zero-valent palladium nanoparticle initiates the electroless deposition process.

To accomplish co-deposition of Cu/Pd metal catalysts in the electroless process, various reducing and complexing agents and combinations of these agents were used in the electroless bath. For instance, a mixture of formaldehyde and hydrazine was investigated as the reducing agent for the co-deposition of Cu/Pd. Formaldehyde is known to be an excellent reducing agent for electroless copper and similarly hydrazine is known to reduce palladium. Each reducing agent was independently evaluated and their mixtures as well. EDTA and sodium potassium tartrate were used as complexing agents for the copper ions while an ammine complex was used to complex palladium. Several electroless solutions were prepared and investigated. Electroless $CuPd/Al_2O_3$ catalysts were synthesized for different deposition time periods.

A typical CuPd electroless solution using formaldehyde for the deposition of electrocatalyst on the catalyzed alumina substrate was prepared with analytical grade chemicals and deionized water. The solution contained 30 g $CuSO_4$, 1.7 g $PdCl_2$, 50 g Tartrate, 30 mL of formaldehyde (37 wt%), 10 g EDTA in a 1000 mL solution. The electroless plating temperature was fixed between 50 – 60^{0}C and the pH maintained between 10 and 12 pH units with sodium hydroxide used for the adjustment. Electroless plating was carried out for 30 minutes, 45 minutes and 60 minutes with three repeats for each time condition. The plated sample was filtered and washed with DI water. After drying the sample in an oven at 50°C, it was used without annealing at higher temperatures.

Results and Discussion

The prepared catalysts were characterized with SEM and EDAX (Figures 2 and 3) and the SEM results suggest successful co-deposition of copper and palladium on the alumina. SEM images in Fig. 2a – Fig. 2c are electroless deposited catalysts are compared catalyst formed through impregnation in Fig. 2d. The results show that as the electroless metals deposit on the catalyzed Al_2O_3 surface, they form globular clusters. On the other hand, the metals for the impregnated $CuPd/Al_2O_3$ catalyst appear as masses of shredded fluff on the Al_2O_3 surface. As the electroless plating time increased, the metals of the electroless catalysts become less rounded and flakier in appearance. The results in Figure 2 showed that synthesis affects surface morphology and that the electroless process co-deposited both copper and palladium.

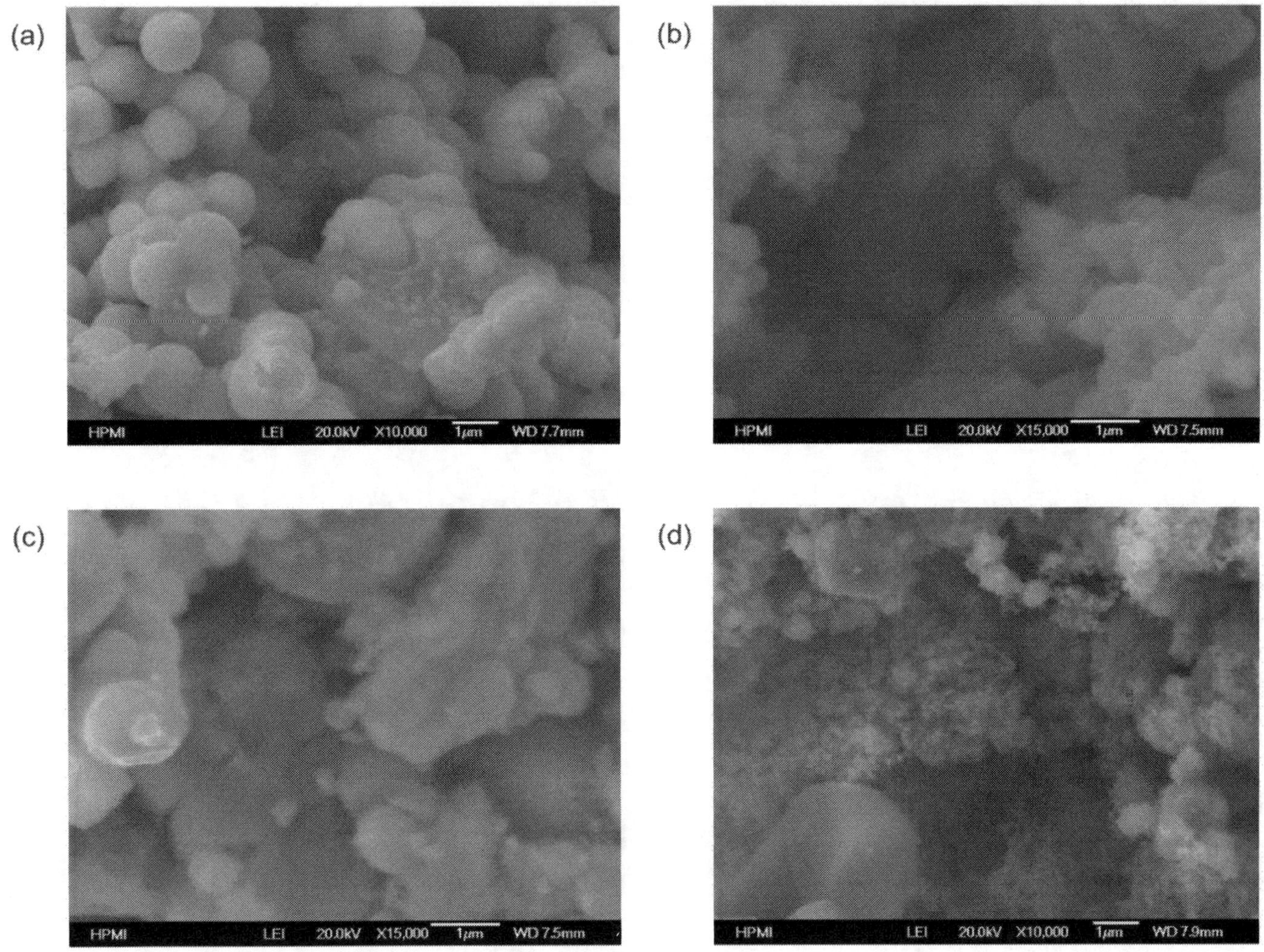

Figure 2. SEM images of Electroless CuPd/Al_2O_3 plated for (a) 30 min (b) 45 min and (c) 60 min (d) Impregnated CuPd/Al_2O_3

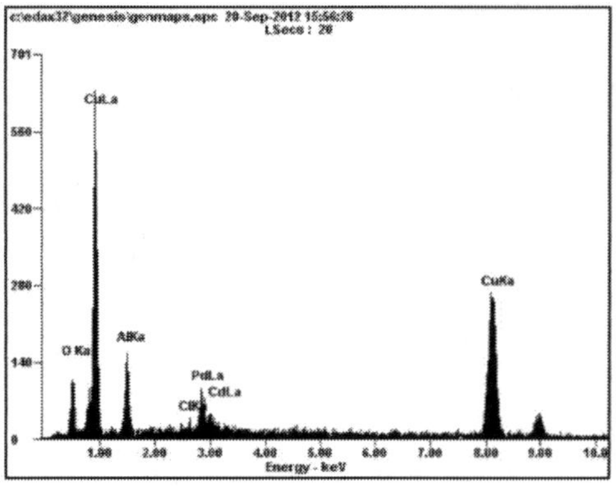

(a)

Element	Wt%	At%
OK	09.72	27.51
AlK	11.16	18.73
ClK	00.66	00.84
PdL	09.18	03.90
CdL	01.13	00.45
CuK	68.16	48.57

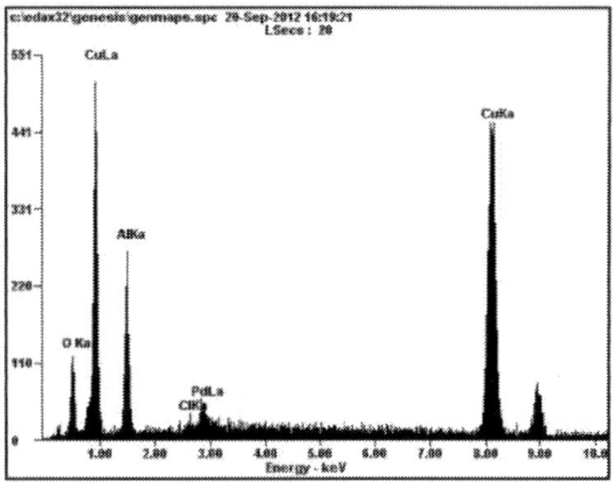

(b)

Element	Wt%	At%
OK	06.44	18.92
AlK	12.42	21.64
ClK	00.56	00.74
PdL	03.09	01.36
CuK	77.49	57.33

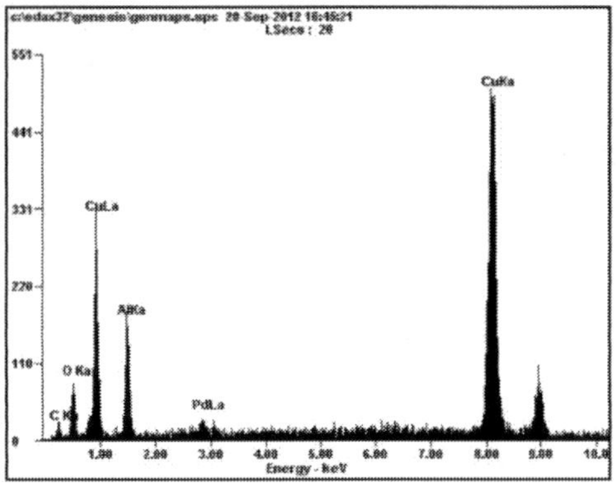

(c)

Element	Wt%	At%
CK	05.89	20.84
OK	04.18	11.11
AlK	09.23	14.54
PdL	01.74	00.69
CuK	78.96	52.82

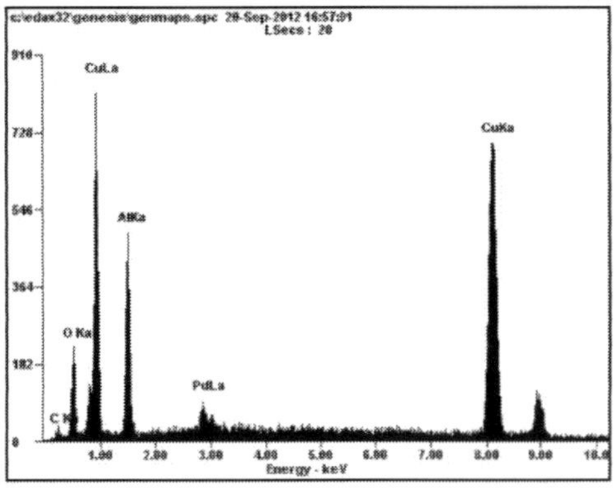

(d)

Element	Wt%	At%
CK	03.29	11.40
OK	06.93	18.00
AlK	14.30	22.03
PdL	03.15	01.23
CuK	72.34	47.34

for

The EDAX plots and compositional tables of Figure 3 confirm the presence of copper and palladium on the alumina surface. For the electroless catalysts copper and palladium had an inverse relationship as plating time increased. Given that there was significantly more copper than palladium in the electroless baths, this decrease in palladium weight percent is attributed to the greater copper concentration of the bath. More copper is likely to be deposited in the long run since formaldehyde is known to be a strong reducing agent for copper but a mild or weak reducing agent for palladium. It is expected that a mixture of formaldehyde and hydrazine will constitute a powerful reducing agent for co-deposition of both metals. The chloride that appears in the EDAX results (Fig. 3a and 3b) may be a result of the palladium-ammine complex used for the synthesis. Palladium chloride is dissolved in hydrochloric acid then ammonium hydroxide is added to form the ammine complex. Some residual chloride may have still been attached to the palladium. It is unclear where the Cd comes from on the CuPd/Al_2O_3 electroless sample plated for 30 min. It should be noted that EDAX plots and compositional charts only present a representative section of a sample, so the composition and area percents given are not completely definitive of the entire catalyst sample.

After the catalysts were synthesized their activity was tested for glycerol hydrogenolysis. Glycerol hydrogenolysis reactions were conducted in 20 mL glass vials with screw tops where 10 mL of 80 wt% glycerol in 1-butanol was charged to the reactor and 0.1 g of catalyst was reacted at 170°C for 12 hours. Figure 4 shows the visual results of this reaction for various catalysts.

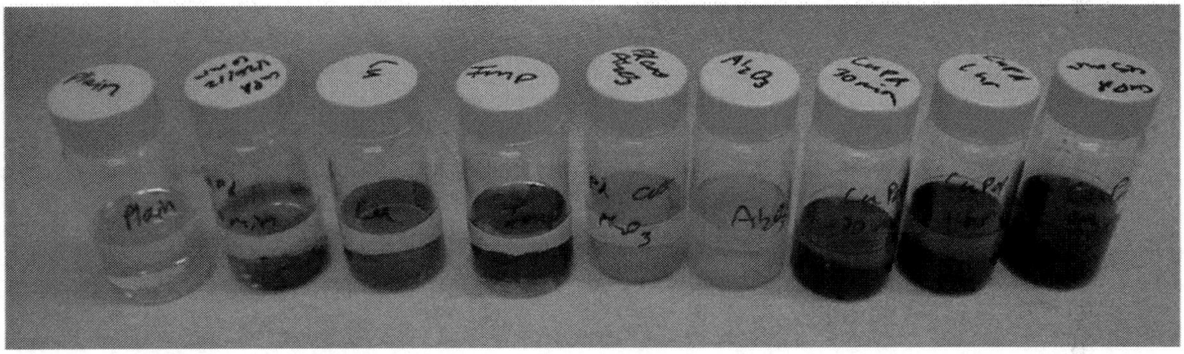

Figure 4. Catalyst screening for glycerol hydrogenolysis. From left to right: None, CuPd/Al_2O_3 plated for 10 min, Cu/Al_2O_3 plated for 10 min, Impregnated CuPd/Al_2O_3, Pd nanoparticles on Al_2O_3, Al_2O_3, CuPd/Al_2O_3 plated for 30 min, CuPd/Al_2O_3 plated for 60 min, CuPd/Al_2O_3 plated for 45 min

Based on reaction solution color, the electroless CuPd/Al_2O_3 catalysts appear to cause a change from the initial reaction composition. The Al_2O_3 as well as the Pd nanoparticle catalyzed and activated Al_2O_3 samples appear to have caused a change in the initial reaction. To insure that glycerol was the reacting species and not the 1-butanol, the reaction was run with the CuPd/Al_2O_3 plated for 60 min at 115°C in 1-butanol. The reaction in 1-butanol did not produce a color change in the reaction solution, thus it was the glycerol that reacted with the catalyst and produced the color change. The difference in volume is a result of failure to tightly secure the reaction vial cap. The impregnated Cu/Pd samples did not show any visual evidence of color change for the preliminary glycerol hydrolysis experiment. Similar results were observed for electroless copper only deposits.

Figure 5. $CuPd/Al_2O_3$ plated for 60 min at 115°C in 1-butanol and 170°C in 80 wt% glycerol in 1-butanol

Conclusion

Electroless deposition was used to synthesize $CuPd/Al_2O_3$ catalysts for glycerol hydrogenolysis. The preliminary results obtained showed that this approach is feasible in synthesizing active catalyst for hydrogenolysis. Further optimization studies are required in order to evaluate the potential of the approach. The use of polymer-stabilized nanoparticles for the catalyzation of substrate for waste conversion to use products is demonstrated.

Acknowledgments

We acknowledge the support of Title III Minority Graduate Engineering at Florida A&M University and ERC Program of the National Science Foundation under Award Number EEC-08212121

(205a) Rare Earth Elements: A US User's Perspective of Issues and Initiatives

Peter Dent, Electron Energy Corporation, Landisville, PA

Strategic and critical mineral supply chains are nationally recognized as being important to the US manufacturing industry, energy supply, and national defense. Currently, China dominates the rare earth market and will continue to do so even after new non-Chinese sources of supply come on line. The17 rare earth elements play a crucial role in international supply chains. The largest US industrial consumers by tonnage are catalyst producers who use rare earths for the cracking of crude petroleum and automotive catalytic converters for treating engine exhaust. Growing demand for alternative energy technologies which utilize rare earths include wind power, hybrid electric vehicles, advanced batteries, fluorescent lighting, and efficient permanent magnet motors and generators. To secure the future of these industries, we must address the challenges associated with supply risks due to their location, vulnerability to disruptions, and lack of suitable substitutes. This presentation will discuss the issues and initiatives in supply and demand in the complete rare earth supply chain and offer a user's perspective on navigating through this uncertain environment.

(205b) The Role of a Transformed National Defense Stockpile in U.S. Strategic Materials Security

Richard A. Lowden, Materials Science and Technology Division, Oak Ridge National Laboratory, Oak Ridge, TN

The United States established a stockpile of strategic materials in 1939 beginning with a handful of ores and minerals which with time was expanded to include additional minerals, metals, chemicals, and other processed materials. The purpose of the stockpile was to preclude a dangerous and costly dependence by the United States upon foreign sources for supplies of scarce but important materials in times of national emergency. Although emphasis has traditionally been on the acquisition and retention of certain strategic and critical materials, the legislation includes provisions that encourage the conservation and development of resources and materials. The authors of the Stockpiling Act realized the value of recycling, improved separation techniques, and alternative methods for refining or processing of materials thus included language to address these issues. Since 1993 the Stockpile has been downsizing, selling off most of its assets, however, recent changes in the global marketplace and increasing risks associated with the supply of important materials such as the rare earths, interest in the stockpile has been renewed. The Stockpile is in the process of being transformed in a more flexible and dynamic strategic materials risk management (security) program with much broader scope and applicability with respect to the entire supply chains for important materials.

(205c) Material Sustainability At General Electric

Anthony Ku and Steven Duclos, GE Global Research, Niskayuna, NY

GE Global Research, working with our GE business partners, has developed a comprehensive strategy to mitigate risks in the availability and sustainability of raw materials that are critical to high technology products. Many such products contain rare elements with limited availability due either to geopolitical constraints or the fact that they are refined as by-products of limited growth commodity metals. We have developed a methodology to assess such risks and prioritize technical solutions. These solutions include the development of advanced manufacturing techniques that minimize waste, recycling technologies to recover materials lost during manufacture or at end of life, and materials engineering that enables substitutions resulting in significant reduction in the use of at-risk elements without sacrificing material properties. Recent examples include our development of superalloys with 50% and 100% less rhenium, and magnet materials with reduced rare earth content. This presentation will discuss GE Global Research's work in this area.

(205d) The Issues and Opportunities Associated with Recycle and Reuse of Strategic and Critical Materials

Eric S. Peterson, Chemical Sciences, Idaho National Laboratory, Idaho Falls, ID

Strategic and critical mineral supply chains are nationally recognized as being important to the US manufacturing industry, energy supply, and national defense. For example, the largest industrial consumers of rare earths in the U.S. are catalyst manufacturers, representing almost 30% of the domestic market by volume in 2008 (40 to 45% in the previous five years). Rare earth elements are also essential to the production of catalysts for the cracking of crude petroleum and of automotive catalytic converters for treating engine exhaust and catalyst that contain rare earth elements influence the production of a diverse range of end user products from aircraft windows to dentures, plastic food packaging, etc. Many facets of the strategic and critical materials supply chain involve "chemistry" or "chemical engineering" from novel digestion and separation processes to reduction and refining to utilization in "old" and "new" products and technologies as well as recycle and reuse. This presentation discusses the issues and opportunities associated with recycle and reuse of critical materials.

(205e) Roundtable Discussion On Critical Materials Supply Chain and Sustainability

Sharon Robinson, Oak Ridge National Laboratory, Oak Ridge, TN

Strategic and critical mineral supply chains are nationally recognized as being important to the US manufacturing industry, energy supply, and national defense. Many facets of the strategic and critical materials supply chain involve chemistry and chemical engineering from novel digestion and separation processes to reduction and refining to utilization in "old" and "new" products and technologies as well as recycle and reuse. After a series of papers from leading experts in the field from both government and industry, the presenters will hold a round-table discussion to identify the issues and opportunities associated with the critical materials supply and sustainability.

(245a) Closed-Loop Decision Making Under Uncertainty for Sustainability Enhancement of Industrial Systems

Zheng Liu and Yinlun Huang, Department of Chemical Engineering and Materials Science, Wayne State University, Detroit, MI

Sustainability performance of manufacturing systems could be significantly improved using carefully selected design, manufacturing, and management technologies. However, due to inherent uncertainty associated with accessible data and information relevant to technology candidates, how to identify most desirable technologies is always a challenge. This becomes more challenging for assuring future sustainability performance of the systems as future market conditions, environmental regulations, etc., could change along the time.

In this paper, we introduce a novel decision-making methodology for sustainability performance improvement under severe of uncertainty, where data uncertainty and scarcity leads to no data distribution information derivable. The introduced methodology is developed by resorting to sustainability principles, control science, and process systems engineering fundamentals. In development, we consider industrial sustainability problems as a general system control problem. In this context, a basic closed-loop feedback control scheme is used to characterize an industrial system under investigation, evaluate and analyze its sustainability performance, and identify most suitable technologies based on the preset sustainability goals, for the improvment of the system's sustainability under budget and other constraints. This sustainability decision making and system performance improvement process will continue as a response to the change of external situations that may be related to market, regulation, policy, etc. Since a large amount of information and data is uncertain, imprecise, and incomplete, an interval-parameter-based approach will be used in system characterization, assessment, and decision making. The effectiveness of the derived decisions and system stability will also be evaluated before and after implementation. The proposed decision making methodology has been successfully used to investigate a number of industrial problems. In this presentation, a comprehensive study on sustainable biodiesel manufacturing in a large geographical region will be illustrated.

(245b) Second Law of Thermodynamics and the Notion of Externalities of Macro-Economics: Two Pillars of Assessing the Sustainability of a Manufacturing System

L. T. Fan, Chemical Engineering, Kansas State University, Manhattan, KS and Tengyan Zhang, Western Research Institute, Laramie, WY

Sustainability engineering, an emerging discipline of engineering, can be regarded as an engineering discipline to craft frameworks, or establish paradigms, for attaining sustainability in any social or human endeavor including sustainable manufacturing. No discipline or branch of engineering can be enduring without theoretical principles underlying it or without a firm theoretical foundation. In this regard, two cornerstones of sustainability assessment of any system or process should be the second law of thermodynamics and the notion of externalities of macro-economics.

The inevitable entropy increase attributable to a process of concern manifests itself in exergy dissipation by the process, which can be evaluated by resorting to the available energy balance; this balance results from integrating the first and second laws of thermodynamics. In reality, however, the available energy balance also implicitly embodies the mass conservation law: It entails the detailed accounting of every material species involved in the process in light of the species' matter-energy.

Economic externalities attributable to a system or process of concern manifest themselves in the process' external costs. These external costs inherent to the system can be evaluated through cost accounting or estimation in conjunction with the process' internal costs, such as the labor and equipment costs.

Various methodologies are available for assessing the sustainability of manufacturing or process systems, which take into account exergy dissipation as well as cost, or can possibly take both into account. One of such methodologies is sustainability potential; this hierarchical approach is proposed to estimate the sustainability-potential of a system based on any of the alternative manufacturing or synthetic routes. This novel notion of sustainability-potential is a generalization of the notion of profit-potential, the estimation of which renders it possible to screen out unsustainable processes derived from the alternative synthetic routes at the earliest stage. The efficacy of sustainability potential is illustrated with the manufacture of vinyl chloride.

(245c) A Shortcut Graphical and Spreadsheet Technique for Identifying Mass Exchange Networks

Russell F. Dunn, Chemical and Biomolecular Engineering , Vanderbilt University, Nashville, TN and Kenneth A. Debelak, Chemical and Biomolecular Engineering, Vanderbilt University, Nashville, TN

Cost-effective waste minimization within chemical processes continues to be a critical component for future sustainability of large chemical and polymer manufacturing companies. Several methods have been developed for the synthesis of mass exchange networks (MENs) for waste minimization. In fact, several of the methods previously developed are based on a two stage targeting approach. The first stage uses the rich (waste) and lean (mass separating agents) composite curves to identify the maximum mass integration possible for an undesirable species to be transferred from a group of waste streams to a group of internal process streams that are used as mass separating agents. Furthermore, the second stage identifies the network of mass exchange technologies that meets the target identified in stage one. Several strategies have been employed for identifying the MEN in stage two including, but not limited to, mathematical optimization and state-space structural approaches.

This paper will present a new graphical/spreadsheet technique and design strategy for identifying the MEN in stage two. This technique allows the simultaneous evaluation of multiple diverse mass separation technologies. Advantages of this strategy will be highlighted and several case studies will be discussed that illustrate the effectiveness of this approach. Finally, experiences using this strategy in teaching undergraduate chemical engineering design courses will also be discussed.

(245d) Synthesis and Performance of Sodium 9,10-Dihydroxy Stearic Acid

Xiao Zhao, Shool of Chemical and material engineering, Jiangnan University, Wuxi, China and Yun Fang, School of Chemical and material engineering, Jiangnan University, Wuxi, China

9, 10-dihydroxy stearic acid (DHSA) and its derivatives are of high value due to their wide range of industrial application, including cosmetic, food, lubricant, plastic additives, personal care, pharmaceutial products and etc. It used to be synthesized by introducing hydroxy groups into carbon-carbon double bond of oleic acid with various catalysts. Almost all those process could not be industrialized for the more excessive co-oxidant and the boring separation process. Here we describe a more convenient method to obtain high purity DHSA by purified oleic acid and hydrogen peroxide with the catalyst of formic acid. The conversion of the oleic acid was over 97%. And the crude DHSA product was obtained by decomposition a small amount of excessive hydrogen peroxide over MnO_2, followed by evaporating water and a residue catalyst. The chemical structure of the recrystallised DHSA was characterized by IR, NMR and MS. Sodium 9,10-dihydroxy stearic acid (DHSA-Na) was prepared by neutralization of DHSA with NaOH. Surface tension, wetting property, lime soap dispersing power (LSDP), Krafft point, detergency and other performances of DHSA-Na were well determined. The results indicated a variety of applications in many fields for the excellent surface and interface chemical performance.

(245e) A Framework for Chemical Risk Management in Product Manufacturing and Its Application to Industrial Cleaning

Masahiko Hirao, Emi Kikuchi and Yasunori Kikuchi, Department of Chemical System Engineering, The University of Tokyo, Tokyo, Japan

For sustainable manufacturing, voluntary efforts are needed in industry to manage risks on human health and the environment associated with the use of chemicals. In this study, a practical framework for chemical risk management in product manufacturing based on simultaneous evaluation of functional, economic, and risk aspects is proposed. The framework is embodied through the case study of industrial cleaning processes, where open-top processes with chemical substances such as chlorinated organic solvents, inorganic compounds, and surfactants, are used.

Process and operation design aimed at risk reductions needs two major activities: evaluation of current and alternative processes, and decision-making on the selection of alternatives. Knowledge and tools, such as technical information, risk evaluation method, and process model, are necessary in order to evaluate risks of processes, as well as their functional and economic aspects. Technical information is systematically organized for cleaning processes, where processes have been designed on the basis of heuristics conventionally. Product defects associated with the careless process modifications aimed at risk reductions can be avoided by evaluating function of alternative processes simultaneously on the basis of the technical information. Risk analysis methods using LCA and risk assessment are applied to the evaluation in order to analyse risk trade-off relationships among alternative processes. A model of open-top cleaning processes is developed to estimate process data needed for evaluating alternative processes.

It is necessary to manage conflicts among evaluated functional, economic, and risk aspects in the decision-making phase. Engineers need to select appropriate alternatives for risk reductions, so that the improved process can achieve not only risk reduction targets but also functional requirements and economic viability. The proposed framework including concrete procedure of such multi-objective decision making is developed through the study on actual cases of process improvement in industrial cleaning. It is revealed that the flexibility assessment of constraints on product quality can be effective in managing conflicts between functional and risk aspects. Furthermore, the engagement of up- and downstream manufacturers in product supply chain can contribute to expanding the range of options of risk reduction measures.

(245f) Scheduling for Multiple Chemical Plant Turnarounds to Minimize Regional Air Quality Impacts

Tianxing Cai and Qiang Xu, Dan F. Smith Department of Chemical Engineering, Lamar University, Beaumont, TX

During the chemical manufacturing process, multiple chemicals are released into the atmosphere including benzene(C_6H_6),carbon monoxide(CO), nitrogen oxides(NO_x), sulfur dioxide(SO_2), and volatile organic compounds(VOC_s).Depending on meteorological conditions, emission quantity, and the fate and transport of chemical species, the environmental impact can affect air quality over a great area. This will cause the regional increased exposure to primary pollutants for neighboring populations with the summation of emission pollutant due to the temporal co-occurrence of start-ups and shutdowns among multiple chemical plants. Furthermore, these emission chemical will react with each other to form other secondary pollutants. The mechanism should follow the principle of reaction kinetics. Therefore, from the emission control point of view, the best practices for improving regional air quality should integrate both efforts on plant-wide emission minimization at every industrial point source, relationship and regional-wide multi-plant emission variability control.

In this paper, a scheduling methodology for coordinating multiple chemical plant start-ups is developed to minimize regional air quality impacts by the combination of reaction kinetics, atmosphere dispersion and scheduling will give the practical and feasible scheme to reduce primary and secondary pollutants in the region where multiple plants are co-located. The whole methodology will include two stages: proactive risk assessment and scheduling optimization. During the risk assessment, the focus is put on the changes in the weather conditions (stability class and wind condition), the pollutant and regional emissions from other sources, especially in the extreme weather or other emergent situations. The simulation result will indicate whether it needs to optimize the temporal sequence of multiple plants' activity. If necessary, the optimization modeling result will give an optimized temporal distribution of start-ups and shut-downs to reduce the chance of emission pairing and the concentration of primary and secondary pollutant. The development provide quantitative decision support for multiple stake holders, including government environment agencies, chemical industry, and local communities.

(250a) Development of a Cold-Quench Bio-Oil Separation System and Its Effect On Cellulose Bio-Oil Yield and Composition

Dustin L. Dalluge, Center for Sustainable Environmental Technologies, Iowa State University, Ames, IA, Yong S. Choi, Iowa State University, Ames, IA, Brent H. Shanks, Chemical and Biological Engineering, Iowa State University, Ames, IA and Robert C. Brown, Mechanical Engineering, Iowa State University, Ames, IA

Renewable energy and sustainable energy production are top priorities for the nation to help provide nation, economic, and environmental security. Among the renewable energy sources, biomass is most promising for production of liquid fuels that can be utilized in the existing infrastructure, so called "drop-in fuels". Biomass offers the advantages that it is renewable in that it grows back every year, it's chemical energy is derived from solar energy, the only truly renewable source of energy for the planet, and it removes CO2 from the atmosphere as it grows thus resulting in a carbon neutral or carbon negative fuel which helps to mitigate concern of atmospheric carbon dioxide concentration.

Two main pathways exist for conversion of biomass to liquid fuels; the biochemical pathway commonly utilized in existing ethanol plants using micro-organisms to ferment starches to ethanol, and the thermochemical pathway which is increasingly gaining attention. Of the many thermochemical pathways, fast pyrolysis, the rapid thermal decomposition of organic compounds in the absence of oxygen to produce liquids fuels and gases, can be used to produce a liquid bio-oil that can be upgraded to fuels. Monosaccharides resulting from the depolymerization of biomass components cellulose and hemicellulose offer advantages in that they can be directly upgraded to liquid fuels by processes such as aqueous phase upgrading, or fermented similar to starch in existing ethanol plants. In either case, cellulose must be depolymerized to monosaccharides before it can be utilized by microorganisms in biological conversion or upgraded to hydrocarbons in aqueous phase upgrading.

Achieving high yields of sugar rich bio-oil is dependent on feedstock, operating conditions, and bio-oil collection systems. Previous research by Kuzhiyil et al.[1] has shown that feedstock can be optimized for sugar production by an acid infusion technique. Operating conditions, such as temperature and sweep gas flow rate, also contribute to the overall sugar yield from biomass. The third piece of the puzzle, bio-oil collection, has also been explored by several other reasearchers, however has not been optimized for sugar production.

Bio-oil compounds are known to be intermediates between the original biomass and the end slow pyrolysis products which are char, essentially carbon, and water. Since these compounds are intermediates they tend to be unstable and continue to react once condensed to form polymers leading to phase separation between the organic and aqueous phases. Therefore, storage of a whole bio-oil can be problematic where separating the reactive species into specific stage fractions during the collection process can prove to be beneficial and has been subject of several researchers[2]. Temperatures of 400-600C and a continuous flow of inert sweep gas must be employed in order to generate the necessary heat and mass transfer characteristics to devolatilize anhydrosugars from biomass. Once the volatile anhydrosugars enter the vapor stream, they may continue to react, where polymerization is commonly an issue as well as continued dehydration leading to formation of char and water. Thus the bio-oil compounds, especially carbohydrates that are easily dehydrated, must be quickly quenched from the reactor temperature to a temperature where they are no longer prone to polymerization or dehydration. To achieve this, a novel bio-oil separation process was co-invented between Iowa State University and the Phillips 66 Company where liquid nitrogen is used to quickly quench the pyrolysis vapors immediately after solids separation to prevent secondary reactions[3]. Quenching the volatile anhydrosugars cools them below their dew point where they form an aerosol in the vapor stream which is then removed in an electrostatic precipitator. It was found that the flow rate of liquid nitrogen could be controlled to selectively quench the pyrolysis stream to temperatures critical to separation of the organic and aqueous phases. In order to quantify changes of bio-oil yields and composition, a 100 gram per hour fluidized bed unit was used to pyrolyze pure cellulose under identical conditions where the bio-oil collection system was then changed between a conventional condenser separation system and this novel cold-quench system. Yield and composition of the resulting bio-oil were then compared to quantify differences in bio-oil due to effect of cooling rate.

1. Kuzhiyil, N., et al., *Pyrolytic Sugars from Cellulosic Biomass.* Submitted.

2. Pollard, A.J.S., *Comparison of bio-oil produced in a fractionated bio-oil collection system*, in *Mechanical Engineering, Biorenewable Resources and Technology*2009, Iowa State University: Ames, Iowa.

3. Daugaard, D.E.S., OK, US), Jones, Samuel T. (Dewey, OK, US), Dalluge, Dustin L. (Ames, IA, US), Brown, Robert C. (Ames, IA, US), *SELECTIVE TEMPERATURE QUENCH AND ELECTROSTATIC RECOVERY OF BIO-OIL FRACTIONS*, 2011, IOWA STATE UNIVERSITY RESEARCH FOUNDATION, INC. (Ames, IA, US),CONOCOPHILLIPS COMPANY - IP SERVICES GROUP (Houston, TX, US): United States.

GENETICALLY MODIFIED BIOMASS: EFFECT OF SINGLE GENE MODIFICATIONS ON THE COMPOSITION OF FAST PYROLYSIS BIO-OILS

Kevin M. Van Geem[1], Yumi Van Wonterghem[1], Eleonora Boren[3,4,5], Ruben Vanholme[3,4], Lorenz Gerber[5], Marko Djokic[1], Frederik Ronsse[2], Wolter Prins[2], Björn Sundberg[5], Wout Boerjan[3,4] and Guy Marin[1]

(1)Laboratory of Chemical Technology, Ghent University, Ghent, Belgium,
(2)Department of Biochemical Engineering, Ghent University, Ghent, Belgium,
(3)Department of Plant Biotechnology and Bioinformatics, Ghent University, Ghent, Belgium,
(4)Department of Plant Systems Biology, VIB, Ghent, Belgium,
(5)Forest Genetics and Plant Physiology, Swedish University of Agricultural Sciences, Umeå, Sweden

Introduction

It has been premised that lignocellulosic biomass is the main renewable energy resource available here on earth and that it can be considered as one of few sources that can provide renewable liquid, gaseous and solid fuels[1]. In contrast to fossil fuels, the use of biomass for energy and chemicals renders significant environmental advantages. Plant growth needed to generate biomass feedstocks removes atmospheric carbon dioxide, which offsets the increase in atmospheric carbon dioxide that results from biomass fuel combustion.

One promising and clean way to acquire bio-oil is by the fast pyrolysis of lignocellulosic biomass, which is considered as a network of hemicellulose and cellulose bound by lignin[2]. Pyrolysis is a thermal decomposition process carried out in the absence of oxygen, or when significantly less oxygen is present than required for complete combustion, yielding gaseous products, liquids (bio-oil and water) and solid charcoal. During the heating in the fast pyrolysis process the biomass constituents are decomposed due to a complicated set of primary and secondary radical reactions. This leads to a complex mixture of water and several hundred organic compounds, with molecular weights ranging from 18 to over 10000 g/mol. These organic compounds present in pyrolysis oil belong to acids, aldehydes, ketones, alcohols, esters, anhydrosugars, furans, phenols, guiacols, syringols, nitrogen compounds as well as large molecular oligomers (holocellulose-derived anhydro-oligosaccharides and lignin-derived oligomers). Pyrolysis oil also contains negligible amounts of ash, and have a volumetric energetic density 5 to 20 times higher than the original biomass.[3] Therefore they offer the potential to be used as a liquid energy carrier or as a renewable raw material for the chemical industry in the production of high-value chemicals and liquid biofuels[4]. Its usage as a substitute for a fuel oil in any stationary heating or electricity generation application and to produce a range of specialty and commodity chemicals has recently gained a lot of interest [5]. However, realizing those potential applications requires having the control over the chemical composition of the bio-oil. Thus the right type of biomass feed and the pyrolysis conditions should be selected. In short, to understand and to be able to predict the pyrolysis behavior, it is essential to understand the kinetics of the thermal reactions that are involved in biomass pyrolysis and relate that to detailed information about the starting material.

Up to now, the main goal of most fast pyrolysis studies was limited to convert as much biomass as possible to liquid bio-oil, neglecting the effect(s) of the biomass composition and/or the process conditions on the bio-oil composition[6]. Therefore in the present contribution the role of small differences in feed composition on the composition of the formed products is investigated. Both pyrolysis GC as well as pilot plant data have been used and advanced characterization techniques such as comprehensive 2D GC (GC×GC) [7] are employed for analysis purpose. Single gene modifications of poplar have been evaluated to assess the effect of these modifications in biomass composition on the produced bio-oil and bio-char.

Experimental

Transgenic lines

The genetically engineered poplar lines are all modified to alter their formation of lignin. The amount of lignin strongly depends on the wood type, ranging from 16% to 33% of the mass of dry wood [8, 9]. Lignin genetic engineering and the lignin biosynthetic pathway are receiving considerable interest. Significant progress in this field has been made by genetic, bioinformatics and biochemical approaches, providing the plant material to research the monolignol biosynthetic pathway [10]. Lignins are complex aromatic heteropolymers derived mainly from three hydroxycinnamyl alcohol monomers, *p*-coumaryl, coniferyl, and sinapyl alcohols, which differ in their degree of methoxylation:. These monolignols result in *p*-hydroxyphenyl (H), guaiacyl (G), and syringyl (S) phenylpropanoid units respectively, when incorporated into the lignin polymer [10]. Hardwood lignins are mainly composed of S and/or G units linked by a series of ether and carbon-carbon bonds, while softwood lignins are composed mostly of G units with low levels of H units [10, 11].

Starting from the metabolic grid, which is represented in Figure 1, the monolignol biosynthesis can be easily understood. It starts with the deamination of phenylalanine and involves successive hydroxylation reactions of the aromatic ring, followed by phenolic *O*-methylation and conversion of the side-chain carboxyl to an alcohol group [10]. Note that the route that is given in Figure 1 is the one that is expected to be the most favored with hardwood types, based on lignin compositional analyses, enzymatic assays and transgenic plants. [10]

Table 1: List of the 15 poplar samples

Sample nr.	Specific type	Branches
1	COMT - ASB10B	1,3,5
2	COMT - ASB10B	2.0,2.1,2.2
3	WT – Biological	1-3
4	WT – Biological	4-5
5	CAD T21	1-3
6	CAD T21	4-6
7	COMT - ASB2B	1-3
8	COMT - ASB2B	4-6
9	CCoAOMT-416	1-3
10	CCoAOMT-416	4-6
11	CCoAOMT-429	1-3
12	CCoAOMT-429	4-6
13	WT – Technical	7-15
14	WT – Technical	7-15
15	WT – Technical	7-15

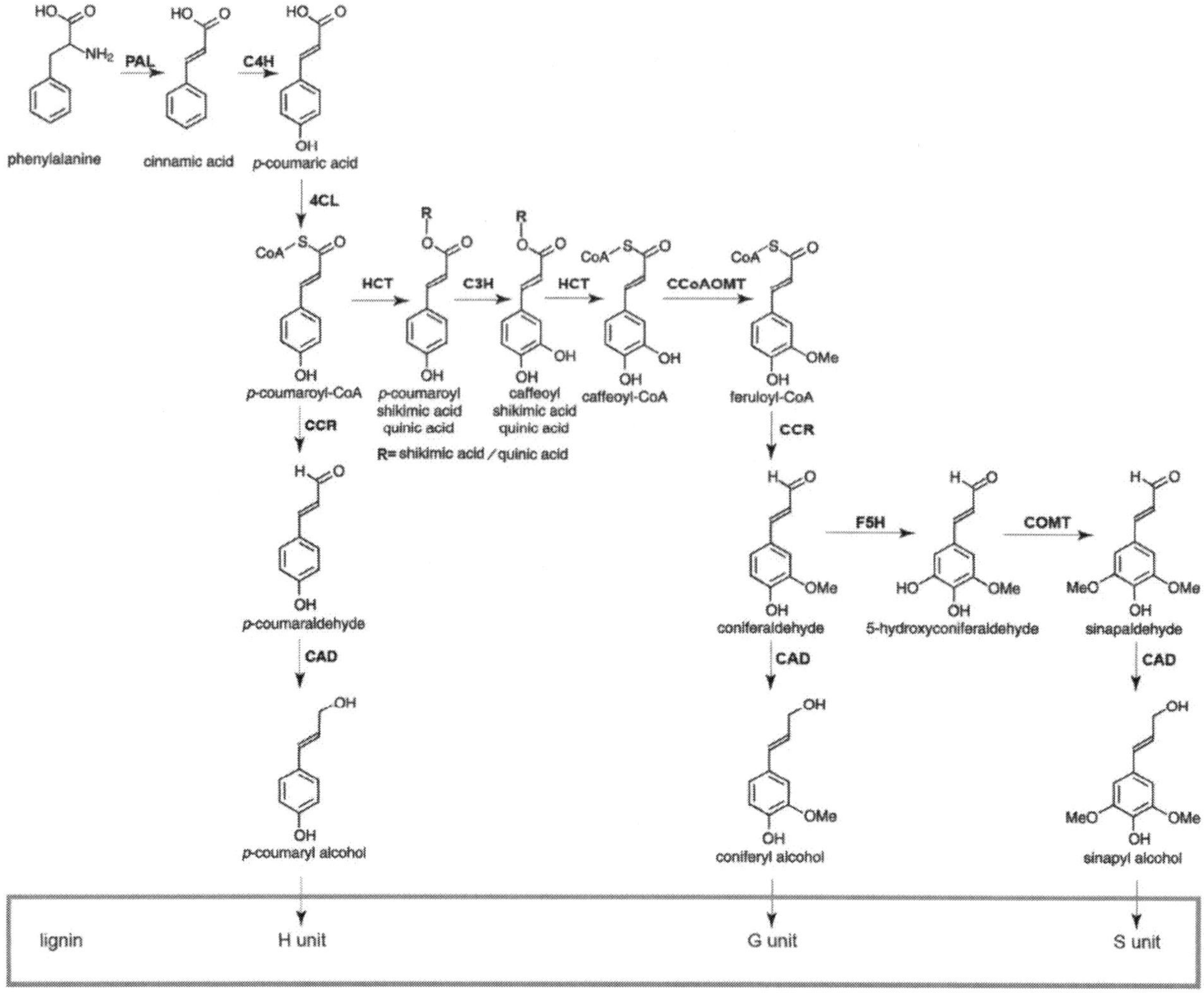

Figure 1: The monolignol biosynthetic pathway.

The transgenic lines were down-regulated in either cinnamyl alcohol dehydrogenase (*CAD*), caffeic acid *O*-methyltransferase (*COMT*) or caffeoyl-CoA *O*-methyltransferase (*CCoAOMT*). Down-regulation of these genes has the consequence that the corresponding steps in the lignin pathway (see Figure 1) are (partly) interrupted, leading to an altered lignin amount and/or composition. CCoAOMT assures the conversion from caffeoyl-CoA to feruloyl-CoA. Down-regulation of this enzyme leads to a decrease of the formation of the G and S units, which is equal to the decrease in the formation of lignin itself. Down-regulation of the multifunctional CAD-enzyme hinders the final reduction of coniferaldehyde and sinapaldehyde to the corresponding alcohols. As a consequence, more aldehydes and fewer alcohols are formed. Previous research [10, 12, 13] indicates that *CAD* down-regulation increases the presence of the free phenolic groups in lignin. *COMT* down-regulation hinders the reaction of 5-

hydroxyconiferaldehyde to sinapaldehyde. As a result the amount of G units increases, while the amount S units remarkably decreases. The used plant materials were one-year-old greenhouse-grown poplars (*Populus tremula* x *Populus alba*). Two lines with reduced COMT activity were used (COMT-ASB10B and COMT-ASB2B-2)[14], two with reduced CCoAOMT activity (CCoAOMT-416 and CCoAOMT-429)[15] and one with reduced CAD activity (CAD T21)[16]. From each genetically engineered line, two biological replicates were made. For the wild type on the other hand, two biological and three technical (i.e. material from the same plants) replicates were made. Note that for each sample three different plants of the same line were used. In total 15 samples were prepared for the experiments, as shown below in Table 1. The material consisted of dried one year old stems that were debarked.

Pilot plant experiments

A schematic overview of the experimental apparatus is shown in Figure 2. The unit can be divided in three main sections, i.e., the feed section, the reactor section and the condensation/collection section. The pyrolysis of a single sample, including collection of the liquid and char fractions, as well as the preheating of the oven, took approximately 3 to 4 hours. For each run, approximately 10 grams of raw material is used. The stem material was chopped into pieces with a length and an internal diameter of max. 10 mm and 5 mm, respectively. The pyrolysis reactions are carried out in a vertical, tubular, stainless steel reactor (d x L = 3,8 cm × 30 cm), which is heated by means of an electric furnace.

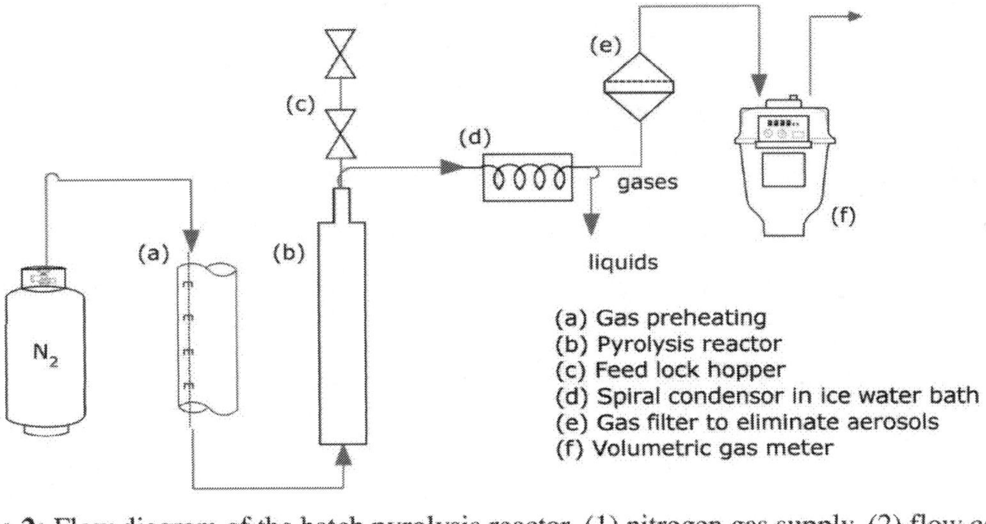

Figure 2: Flow diagram of the batch pyrolysis reactor. (1) nitrogen gas supply, (2) flow control, (3) gas preheater, (4) electric tube furnace, (5) pyrolysis reactor, (6) sintered base plate, (7) fluidised sand/ biomass bed, (8) biomass lock hopper, (9) condenser, (10) condensate/gas separator, (11) cotton filter , (12) diaphragm gas flow meter, (13) gas vent and (14) bio-oil recovery.

In the reactor section itself, the biomass gets into contact with the heat carrier. In this case the heat carrier is pre-heated silica sand with a mean diameter of 250 μm, particle density of 2600 kg/m^3 and a bulk density of 1600 kg/m^3. The pyrolysis reactions take place at 500°C with a residence time of 10 minutes. To acquire the liquid fraction, the spiral condenser is detached after the experiment. Tetrahydrofuran (THF) is injected with a syringe into the spiral to relinquish the bio-oil. After the experiment, the char can be removed from the inner tube of the reactor. A detailed description of the analysis of bio-oils using GC×GC can be found in Djokic et al. [7] A chromatogram of a crude bio-oil is shown in Figure 3. For each sample 100 to 110 components were identified and quantified. The biochar is analyzed using an elemental analyzer of the Thermo Scientific FLASH 2000 Series. Each sample was run at least three times, if the results showed a too large deviation additional runs were carried out.

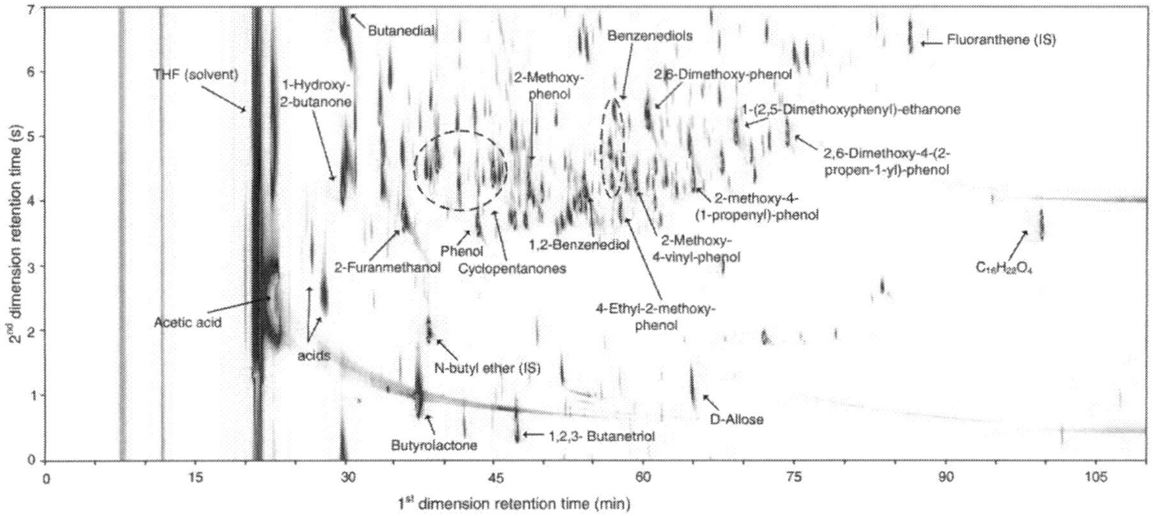

Figure 3: GC×GC-FID colour plot of crude poplar wood bio-oil sample, S13-WT.

Micro-pyrolysis set-up

A schematic overview of the micro-pyrolysis set-up is given in Figure 4. The residence time of the vapours in the micro-pyrolyser is limited to 15-20 ms. The micro-pyrolysis unit (FrontierLab Multi-shot pyrolyzer EGA/PY-3030D) consists of a sampler, a quartz pyrolysis tube that can be preheated to a desired temperature with a furnace, interface and a deactivated needle (inserted into the gas chromatograph (GC) injector).

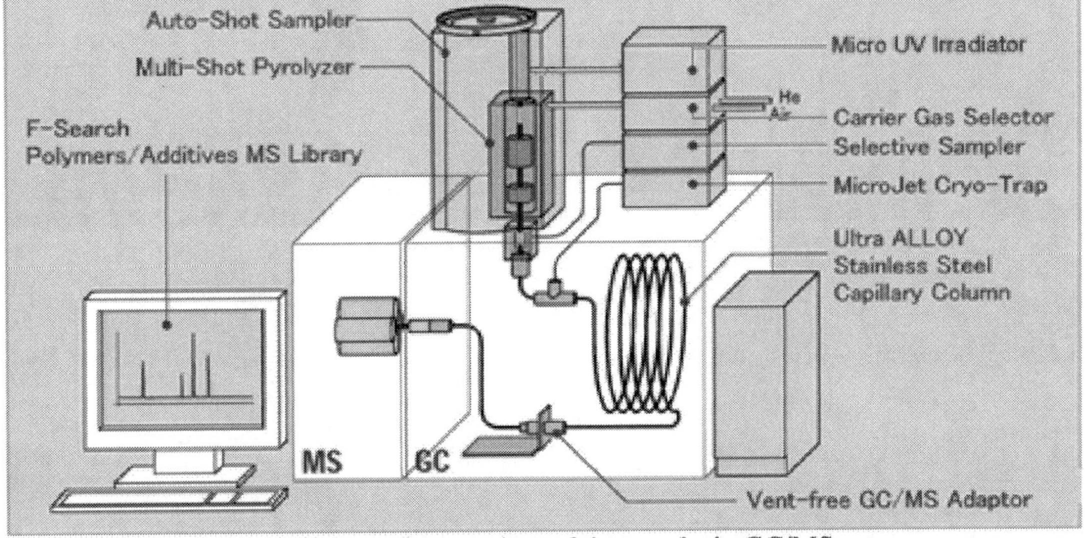

Figure 4: Schematic overview of the pyrolysis-GC/MS set-up

The loaded sample cup (constructed of deactivated stainless steel) is dropped into a furnace, which is preheated to 500°C. The cup contains about 300 micrograms of finely ground biomass. The loaded cup falls freely into the preheated furnace by gravity in a very short time period. During this period the sample is heated to the pyrolysis temperature, ensuring rapid pyrolysis (>2000 °C/s). The pyrolysis vapors are directly swept into the GC using helium as the carrier gas. The constituents of the pyrolysis vapor are separated in the GC column and identified using a mass spectrometer (MS) (Thermo Trace GC Ultra / ISQ MS. The chromatographic separation of pyrolysis products is performed using a Restek capillary column with a stationary phase consisting of 14% cyanopropylphenyl and 86% dimethyl polysiloxane and with a carrier gas flow velocity of 1ml/min.

Results and discussion

Bio-Char and Original Biomass

For the process conditions specified above on average 25.0 ± 0.5 wt% of char was obtained for all samples. The results of the elemental analysis of the biochars and the studied biomass feeds are summarized in Figure 5.

Figure 5 shows that for both the original material and the bio-char the differences in the elemental composition are very subtle. Moreover, the variation in the H and N composition for the COMT lines compared to the others are of the range as the variation in H and N composition for the wild type mutually. Hence, our results indicate that for present study there are no statistically significant differences in elemental composition between the different poplar types. This suggests that even though the feedstock differs in its genetic structure, it is neither traceable in the elemental composition of the original material nor in the produced bio-char.

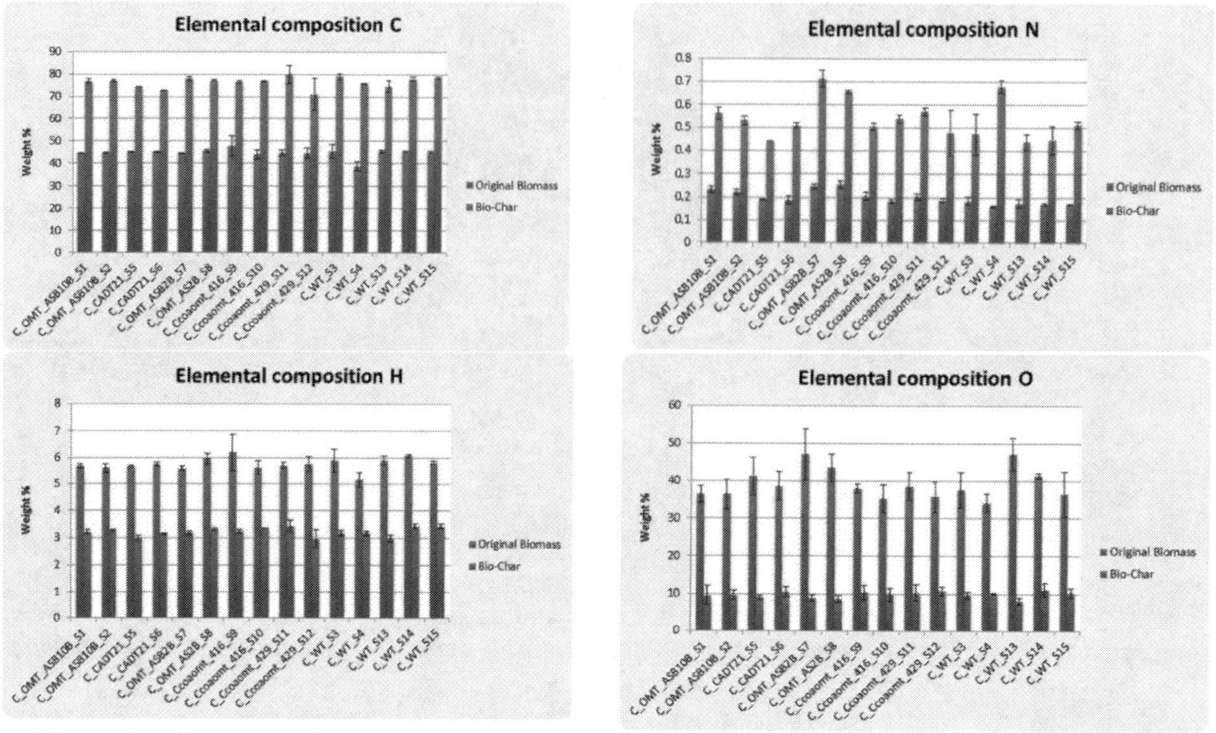

Figure 5: Effect of single gene modifications on the elemental analysis results of original biomass material and formed biochars

Bio-Oil

To assess the differences in composition for the bio-oils (pilot plant set-up) and the pyrolysis vapors (micro-pyrolyser) principal component analysis (PCA) was applied. This is a powerful and facile technique that allows extracting significant information from the detailed compositional information. It also allows to visualize these results and to reveal differences in composition of the bio-oils derived from the different biomass feeds. Both score plots and loading plots were evaluated to interpret the results. Figure 6 shows such a loading plot based on the detailed analysis of the bio-oils analyzed by GC×GC and reveals that indeed the produced differ statistically in composition from each other.

The first three principal components of the PCA explain 65.7% of the variation in the data, with PC1 describing 27.6%, PC2 explaining 25.2% and PC3 explaining 12.9%. The resulting score plot of PC2 versus PC1 (see Figure 6) shows that there are two groups of transgenic lines separated from the wild type, the COMT lines and the CAD lines. The wild type and the CCoAOMT lines are both located around the centre, apart from Sample 12. The separation between COMT lines and wild type is caused by both PC1 and PC2, while the separation between CAD lines and wild type is only caused by PC2. The loading plot reveals that the COMT lines have a significant lower amount of S units compared to the wild type. On the other hand the two S-aldehydes (syringaldehyde and sinapaldehyde) are significantly higher in the CAD samples compared to the wild type and explain why these fractions are statistically different from each other.

The fact that the bio-oils from CCoAOMT lines are not significantly different from the wild type needs to be further studied, considering that down regulation of this enzyme affects mainly the amount of lignin differences between CCoAOMT and the wild type material are expected.

Taking into account the results obtained for the elemental composition of the biomass feeds, it can be concluded that even though the elemental composition of the biomass feedstock is almost identical, the produced bio-oils can have different compositions for a given fixed set of process conditions.

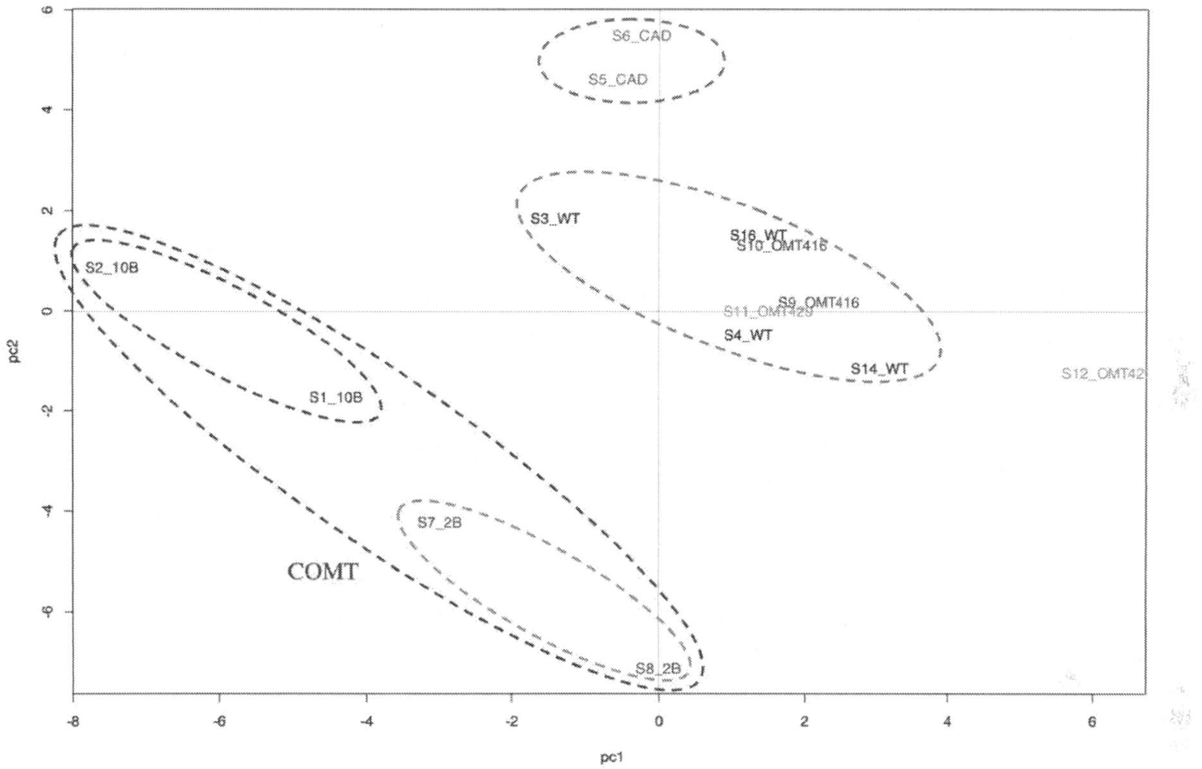

Figure 6: Loading plot for PCA analysis of bio-oils derived from wild and genetically modified poplar

Conclusions

The elemental composition of both the original materials and the obtained bio-chars show that for both the original materials and the bio-chars the differences in the elemental composition are very subtle. There are no noticeable shifts between the different genetic poplar types. This proves that even though the feedstock differs in its genetic structure, it is neither traceable in the elemental composition of the original material nor in the produced bio-char.

The analysis of the bio-oils produced in the pilot plant set-up and the results obtained on the micro-pyrolysis set-up show clear differences between the COMT lines and the wild type, which could be attributed to a lower amount of syringyl (S) units in the lignin of the COMT lines. Subtle shifts between the CAD lines and the wild type were also detected. However, there were no clear shifts traceable for the CCoAOMT lines.

Finally, our results also prove that even though the elemental composition of the biomass feedstock is statistically not different it does not imply that the produced bio-oils have the same composition for identical process conditions. Hence more structural information then elemental analysis is needed to predict bio-oil compositions for a given process condition.

References

1. Bridgwater, A. V., THE TECHNICAL AND ECONOMIC-FEASIBILITY OF BIOMASS GASIFICATION FOR POWER-GENERATION. *Fuel* **1995,** 74, (5), 631-653.
2. Vispute, T. P.; Zhang, H. Y.; Sanna, A.; Xiao, R.; Huber, G. W., Renewable Chemical Commodity Feedstocks from Integrated Catalytic Processing of Pyrolysis Oils. *Science* **2010,** 330, (6008), 1222-1227.
3. Oasmaa, A.; Kuoppala, E.; Ardiyanti, A.; Venderbosch, R. H.; Heeres, H. J., Characterization of Hydrotreated Fast Pyrolysis Liquids. *Energy & Fuels* **2010,** 24, 5264-5272.
4. Wright, M. M.; Daugaard, D. E.; Satrio, J. A.; Brown, R. C., Techno-economic analysis of biomass fast pyrolysis to transportation fuels. *Fuel* **2010,** 89, S11-S19.
5. Nigam, P. S.; Singh, A., Production of liquid biofuels from renewable resources. *Prog Energ Combust* **2011,** 37, (1), 52-68.
6. Venderbosch, R. H.; Prins, W., Fast pyrolysis technology development. *Biofuels Bioprod. Biorefining* **2010,** 4, (2), 178-208.
7. Marko R. Djokic, K. M. V. G., Thomas Dijkmans, Guray Yildiz, Wolter Prins, Guy B. Marin, Quantitative Analysis of Crude and Stabilised Bio-oils by Comprehensive 2D GC. **2012**.
8. Winandy, J. E.; Rowell, R. M., The Chemistry of Wood Strength. *Adv Chem Ser* **1984**, (207), 211-255.
9. Czernik, S.; Bridgwater, A. V., Overview of applications of biomass fast pyrolysis oil. *Energ Fuel* **2004,** 18, (2), 590-598.
10. Boerjan, W.; Ralph, J.; Baucher, M., Lignin biosynthesis. *Annual Review of Plant Biology* **2003,** 54, 519-546.
11. Adler, E., Lignin Chemistry - Past, Present and Future. *Wood Sci Technol* **1977,** 11, (3), 169-218.
12. Leple, J. C.; Dauwe, R.; Morreel, K.; Storme, V.; Lapierre, C.; Pollet, B.; Naumann, A.; Kang, K. Y.; Kim, H.; Ruel, K.; Lefebvre, A.; Joseleau, J. P.; Grima-Pettenati, J.; De Rycke, R.; Andersson-Gunneras, S.; Erban, A.; Fehrle, I.; Petit-Conil, M.; Kopka, J.; Polle, A.; Messens, E.; Sundberg, B.; Mansfield, S. D.; Ralph, J.; Pilate, G.; Boerjan, W., Downregulation of cinnamoyl-coenzyme a reductase in poplar: Multiple-level phenotyping reveals effects on cell wall polymer metabolism and structure. *Plant Cell* **2007,** 19, (11), 3669-3691.
13. Lapierre, C.; Pollet, B.; Petit-Conil, M.; Toval, G.; Romero, J.; Pilate, G.; Leple, J. C.; Boerjan, W.; Ferret, V.; De Nadai, V.; Jouanin, L., Structural alterations of lignins in transgenic poplars with depressed cinnamyl alcohol dehydrogenase or caffeic acid O-methyltransferase activity have an opposite impact on the efficiency of industrial kraft pulping. *Plant Physiol* **1999,** 119, (1), 153-163.
14. Vandoorsselaere, J.; Baucher, M.; Feuillet, C.; Boudet, A. M.; Vanmontagu, M.; Inze, D., ISOLATION OF CINNAMYL ALCOHOL-DEHYDROGENASE CDNAS FROM 2 IMPORTANT ECONOMIC SPECIES - ALFALFA AND POPLAR - DEMONSTRATION OF A HIGH HOMOLOGY OF THE GENE WITHIN ANGIOSPERMS. *Plant Physiol. Biochem.* **1995,** 33, (1), 105-109.
15. Meyermans, H.; Morreel, K.; Lapierre, C.; Pollet, B.; De Bruyn, A.; Busson, R.; Herdewijn, P.; Devreese, B.; Van Beeumen, J.; Marita, J. M.; Ralph, J.; Chen, C. Y.; Burggraeve, B.; Van Montagu, M.; Messens, E.; Boerjan, W., Modifications in lignin and accumulation of phenolic glucosides in poplar xylem upon down-regulation of caffeoyl-coenzyme A O-methyltransferase, an enzyme involved in lignin biosynthesis. *Journal of Biological Chemistry* **2000,** 275, (47), 36899-36909.
16. Baucher, M.; Chabbert, B.; Pilate, G.; VanDoorsselaere, J.; Tollier, M. T.; PetitConil, M.; Cornu, D.; Monties, B.; VanMontagu, M.; Inze, D.; Jouanin, L.; Boerjan, W., Red xylem and higher lignin extractability by down-regulating a cinnamyl alcohol dehydrogenase in poplar. *Plant Physiology* **1996,** 112, (4), 1479-1490.

EXTENDED ABSTRACT

FAST PYROLYSIS OF LARCH, MISCANTHUS SACCHARIFLORUS AND RICE STRAW FOR BIOCRUDE-OIL PRODUCTION BY FLUIDIZED BED REACTOR

Jin Pil BOK[a] and Yeon Seok CHOI[b,*]

[a] Department of Environmental and Energy mechanical Engineering, Korea University of Science and Technology, Daejeon, Republic of Korea

[b] Environment and Energy Systems Research Division, Korea Institute of Machinery and Materials, Daejeon, Republic of Korea

[*] Corresponding author: Tel.: +82 42 868 7344; fax: +82 42 868 7284

E-mail address: yschoi@kimm.re.kr

Woody biomass (larch), energy crop (miscanthus sacchariflorus) and agricultural waste (rice straw) were fast pyrolyzed by a cylindrical fluidized bed reactor to investigate product yields and characteristics of biocrude-oil in this study. For an hour, 300 g of dried each biomass were fed to the fluidized bed reactor with reaction temperature ranging from from 400 ℃ to 550 ℃. The vapor residence time (0.82 s) and the condensing temperature (10 ℃) were not varied during every experiment. The highest biocrude-oil yields were 55.33 wt.% (larch) and 51 wt.% (miscanthus sacchariflorus) at 500 ℃ and 37.27 wt.% (rice straw) at 550 ℃, respectively. According to the experimental results, the fast pyrolysis temperature significantly influenced the product yields and the characteristics of biocrude-oil. The characteristics (higher heating value, viscosity, water content, density, pH, solid residue and ash) of the collected biocrude-oils from each biomass were scrutinized and compared. The composition of the biocrude-oil was also analyzed by GC-MS.

Keywords

Biocrude-oil, Fast pyrolysis, Fluidized bed reactor, Larch, Miscanthus sacchariflorus (MS), Rice straw (RS)

1. INTRODUCTION

To deal with energy supply and environmental problems, a variety of renewable energy have been studied. In particular, conversion techniques from biomass to renewable energy are steadily studied for a long time. Biomass conversion techniques are categorized to biochemical process and thermochemical process [1]. Fast pyrolysis biomass producing biocrude-oil is one of the promising biomass thermochemical conversion processes [2].

Many research groups in the world have been studying fast pyrolysis of biomass and biocrude-oil properties [3]. Main experimental conditions influencing biocrude-oil yield and quality have been studied for two decades. For instance, reaction temperature and vapour residence time significantly influence biocrude-oil yield and quality [2]. Moreover, biocrude-oil has different quality depending on the type of biomass [4]. Hence, we choose larch as woody biomass, miscanthus sacchariflorus as energy crop and rice straw as agricultural waste to compare yield and quality of biocrude-oil from each biomass.

2. RAW MATERIALS

Larch was collected from a woodworking plant in Daejeon area. M. sacchariflorus and rice straw were cultivated and harvested in Muan area and rural area near Daejeon, respectively. As shown in Figs., each biomass were milled and then sieved by using meshes of 1 and 2 mm. These raw materials were dried in an oven at 105 □ for a night to make its water content 1~2 wt.%. Table 1 shows physical properties of raw materials.

Table 1 Physical properties of raw materials

	Proximate analysis (wt.%)[a]			Ultimate analysis (wt.%)[a]				HHV (MJ/kg)[a]
	Volatile	Fixed carbon	Ash	C	H	N	S	
Larch	83.64	15.46	0.9	43.68	5.49	0.03	0	17.63
MS	79.17	15.97	4.86	43.11	5.42	0.25	0	16.54
RS	75.72	13.19	11.09	44.12	5.61	0.49	0	16.06

[a]Dry basis

Fig. 1 shows thermogravimetric analysis (TGA) and differential thermogravimetric (DTG) curves of raw material under a nitrogen atmosphere. Those curves occur by decomposition of hemicelluloses, cellulose and lignin [5, 6].

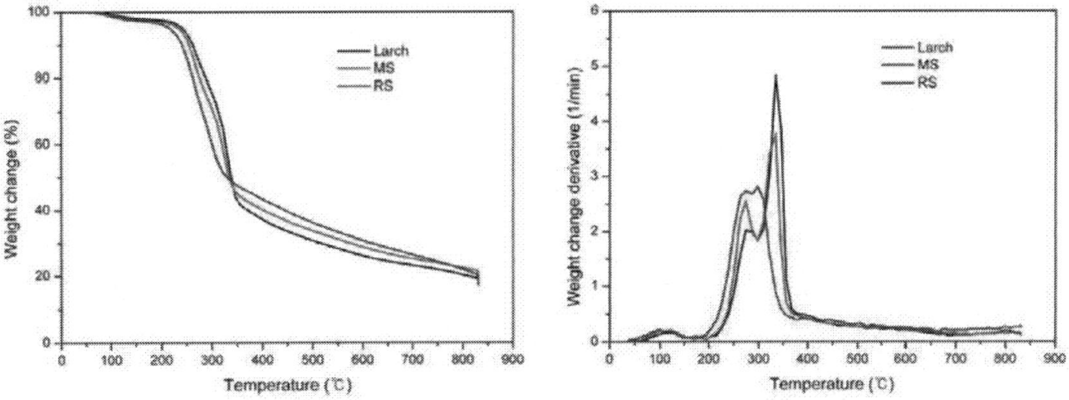

Fig. 1. Thermogravimetric analysis of raw materials

3. EXPERIMENTAL SETUP

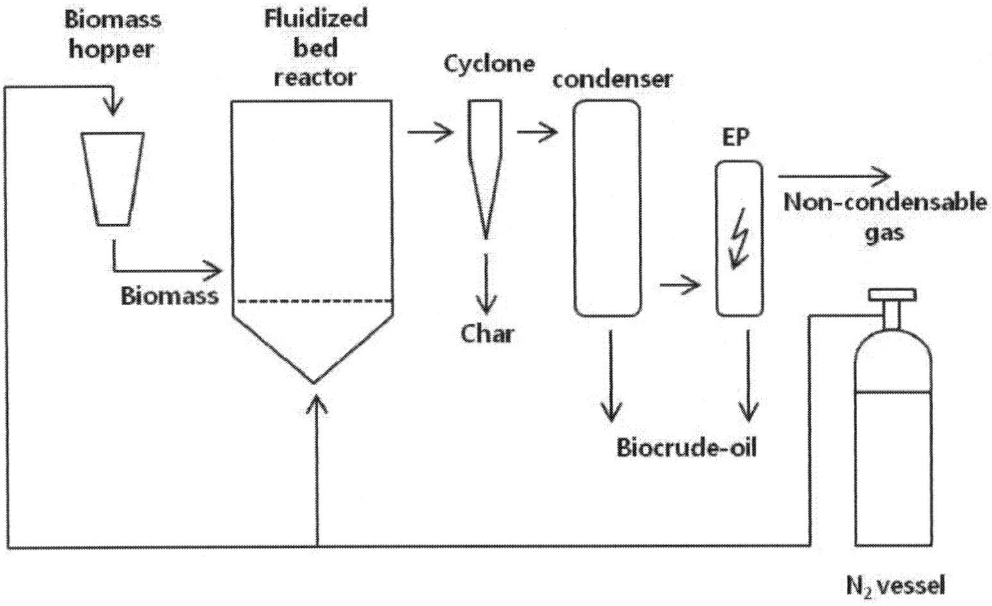

Fig. 2. Schematic diagram of experimental apparatus.

Fig. 2 shows the experimental apparatus for the present fast pyrolysis experiment. A fluidized bed reactor was employed for the fast pyrolysis of raw materials. In the reactor, sand was used for fluidization material. Nitrogen gas was used for the fluidization of the sand. 300 g of each raw material was supplied to the reactor by screw feeder for 1 hour. Raw material was decomposed into volatiles, char and non-condensable gases inside the reactor. A cyclone was employed for char separation from pyrolysis gas. Volatiles condensed to biocrude-oil in the condenser. An electrostatic precipitator was also used to capture biocrude-oil.

4. EXPERIMENTAL RESULTS

4.1 Biocrude-oil yield

As shown in fig. 3, we obtained highest biocrude-oil yield at 500 ℃ in larch and MS cases. Besides, the highest biocrude-oil yield was shown at 550 ℃ in the case of RS.

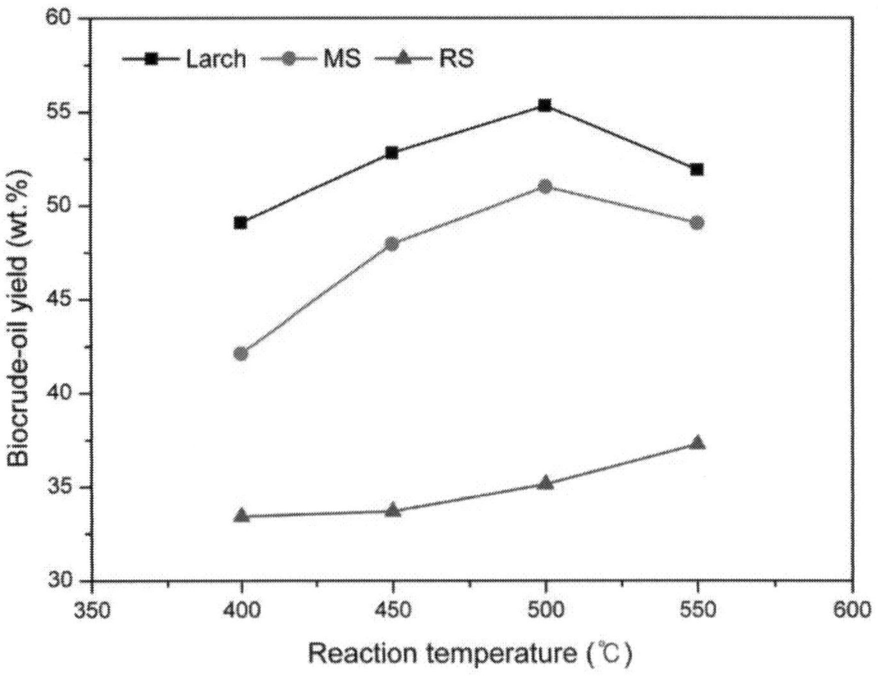

Fig. 3. biocrude-oil yield from each biomass.

4.2 Analysis of biocrude-oil

Figs. 4-6 show higher heating value (HHV), water content and viscosity of biocrude-oil from each biomass. As shown in Fig. 3, the HHV of biocrude-oil from larch was higher than others. Biocrude-oil from rice straw had lowest HHV among them.

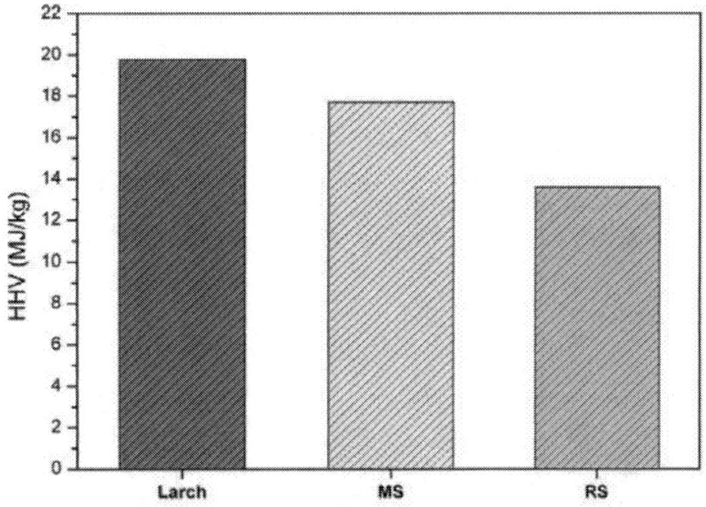

Fig. 4. HHV of biocrude-oils from each biomass.

From the experimental results, biocrude-oil from rice straw contained lots of water. It was over 35 wt.%. However, water content of biocrude-oil from larch was relatively lower than others.

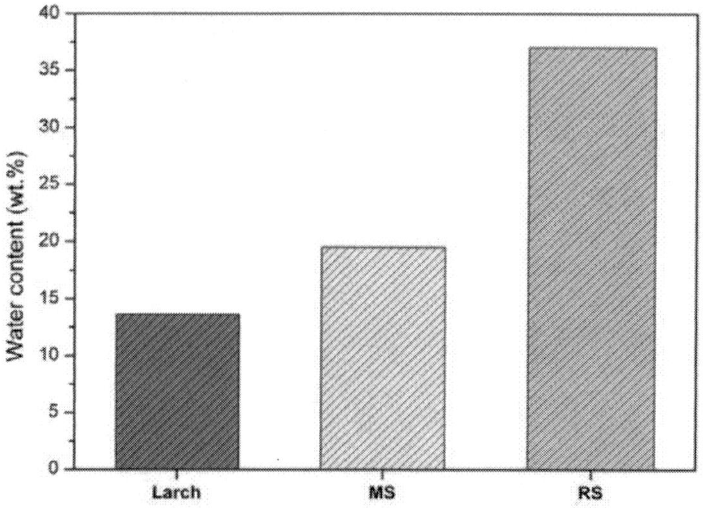

Fig. 5. Water content of biocrude-oils from each biomass.

As shown in Fig. 6, viscosity of biocrude-oil from larch was pretty high.

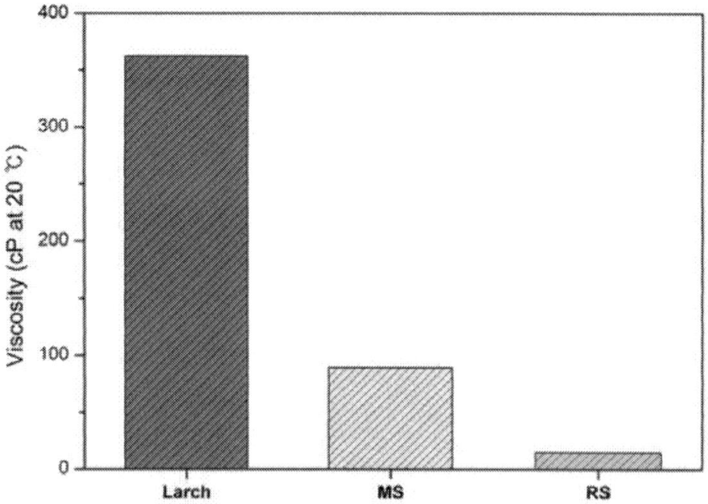

Fig. 6. Viscosity of biocrude-oils from each biomass.

5. CONCLUSIONS

Fast pyrolysis of larch, miscanthus sacchariflorus and rice straw were performed to study product yield and biocrude-oil characteristics. The highest biocrude-oil yields were 55.33 wt.% in larch case, 51 wt.% in the MS case and 37.27 wt.% in the RS case.

Each biocrude-oil had their own characteristics. Biocrude-oil from larch had higher HHV and lower moisture. However, it was much more viscous than others. Biocrude-oil from RS had lower HHV and it viscosity was also low. However, its water content was over 35 wt.%. That would worses the combustion behavior of biocrude-oil.

Reference

[1] A.V. Bridgwater, Principles and practice of biomass fast pyrolysis processes for liquids, Journal of Analytical and Applied pyrolysis, 51 (1999) 3-22

[2] A.V. Bridgwater, D. Meier, D. Radlein, An overview of fast pyrolysis of biomass, Organic Geochemistry, 30 (1999) 1479-1493

[3] A.V. Bridgwater, Review of fast pyrolysis of biomass and product upgrading, Biomass and bioenergy, 38 (2012) 68-94

[4] Akeem M. Azeez, Dietrich Meier, Jurgen Odermatt, Thomas Willner, Fast pyrolysis of african and european lignocellulosic biomasses using Py-GC/MS and fluizied bed reactor, Energy and Fuels, 24 (2010) 2078-2085

[5] Debasish De, Debapriya De, Characterization of an alkali-treated grass fiber by thermogravimetric and X-Ray crystallographic analysis, Journal of Applied Polymer Science, 109 (2007) 2008-2015

[6] Haiping Yang, Rong Yan, Hanping Chen, Chuguang Zheng, Dong Ho Lee, David Tee Liang, In-depth investigation of biomass pyrolysis based on three major components: Hemicellulose, cellulose and lignin, Energy and Fuels, 20 (2006) 388-393

(250d) Production of Biocrude From Biomass by Acidic Subcritical-Water Followed by Alkaline Supercritical-Water Two-Step Liquefaction

Hema Ramsurn and Ram B. Gupta, Chemical Engineering, Auburn University, Auburn, AL

Increased demand in transportation fuels, environmental concerns and depletion of fossil fuel require development of efficient conversion technologies for second-generation biofuels. The main objective of this work is to efficiently liquefy biomass into energy-dense biocrude. A novel two-step process is proposed in which acidic subcritical-water followed by alkaline supercritical-water media are utilized for the liquefaction. The concept is tested with switchgrass. The first step is carried out at 200 °C in acidic subcritical water to liquefy hemicelluloses to biocrude while avoiding the repolymerization reactions which would otherwise produce char. In the second step, the remaining un-liquefied biomass (biomass-H) is subjected to supercritical water at 380 °C with $Ca(OH)_2$ as catalyst for minimizing the formation of char, enhancing lignin solubilization and therefore increasing liquefaction of the remaining polysaccharides toward biocrude. The proposed two-step liquefaction produces significantly higher amount of biocrude as compared to the traditional one-step process. The yield of biocrude from the proposed process is 40% on mass basis and 67% on energy basis of the feedstock biomass.

(250e) A Comparative Study of Different Analytical Methods in Quantifying the Sugar Compositions of Bio-Oil

Xianglan Bai[1], Alex Liu[2], Kwang Ho Kim[3] and Robert Brown[2], (1)Aerospace Engineering, Iowa State University, ames, IA, (2)Center for Sustainable Environmental Technologies, Iowa State University, Ames, IA, (3)Dept. Agricultural and Biosystem Engineering, Iowa State university, Ames, IA

A Comparative Study of Different Analytical Methods in Quantifying the Sugar Compositions of Bio-oil

Xianglan Bai[1, 5], Alex Liu[2, 5], Kwang Ho Kim[3, 5] and Robert C. Brown [4, 5]

1. Department of Aerospace Engineering

2. Department of Chemical Engineering

3. Department of Agricultural and Biosystem Engineering

4. Department of Mechanical Engineering

5. Center for Sustainable Environmental Technologies

Iowa State University, Ames, IA 50010

GC/MS (or FID) is an analytic technique widely used to quantify the compositions of bio-oil that produced from the pyrolysis of biomass. In GC/MS, the temperature of GC column is gradually ramped up from room temperature to an elevated temperature (230 °C -300°C) and the chemical compounds of bio-oil are separated based on the differences of the evaporation temperature for each compound. However, certain compounds in bio-oil are thermally instable and may polymerize or degrade at the high temperature levels, which may cause uncertainties in the quantification results. For example, levoglucosan in liquid phase can polymerize at an elevated temperature. Recently, a new technique was developed in our group to quantify the sugar compositions of bio-oil using HPLC. With this new method, the sample temperature is kept at 75°C during the whole analysis process to prevent levoglucosan from polymerizing. In the present study, the sugar yields in the same bio-oil are quantified by using the two analytic methods in order to evaluate the influence of analytical methods on the quantification of the sugar compositions in bio-oil.

(250f) Hydrogenation of Pyrolysis Oil Derived From Lignin

Wei Mu[1], Haoxi Ben[2], Arthur J. Ragauskas[3] and Yulin Deng[1], (1)School of chemical and biomolecular engineering, Georgia Institute of Technology, Atlanta, GA, (2)School of Chemistry & Biochemistry , Georgia Institute of Technology, Atlanta, GA, (3)School of Chemistry & Biochemistry, Atlanta, GA

Pyrolysis is recognized as one of the most economical ways to convert solid lignin to liquid. However, the oil products from this process cannot be directly used as a fuel. The major problems are instability, high acidity and low heating value. In our study, the heavy oil portion from pyrolysis oil was used for the raw material. Two-step hydrotreatments were applied and three catalysts, Ruthenium, Platinum and Palladium on carbon were evaluated with Ruthenium showing the best result. The other two catalysts could not achieve 100% conversion. After the reaction, all aromatic rings converted to aliphatic structure. Fewer methoxy groups with an increase in C-O aliphatic bonds were observed. The decrease of C=O bonds indicated the carbonyl groups were reduced to hydroxyl groups. During the first-step hydrotreatment, the surface of catalyst was blocked by the tar and coke formed in the reaction. Therefore, a second-step hydrotreatment was appended by replenishing the fresh catalyst. The mechanism of hydrogenation was studied and a reaction path was proposed. How the substituent group on aromatic ring would affect the reaction was also evaluated. NMR was extensively used in our research, including proton NMR, ^{13}C NMR, 2-D HSQC NMR and DEPT-analysis. The NMR technology has non-destructive property, which is superior to the traditional GC-MS analysis. The NMR could also provide statistical distribution of carbon atom and hydrogen atom in different chemical environment, which gives a better insight of the reaction mechanism.

(250g) Techno-Economic Analysis of Biomass Catalytic Pyrolysis

Rajeeva Thilakaratne[1], Yihua Li[2], Tristan Brown[3], Robert C. Brown[1] and Guiping Hu[2], (1)Mechanical Engineering, Iowa State University, Ames, IA, (2)IMSE, Iowa State University, Ames, IA, (3)Bioeconomy Institute, Iowa State University, Ames, IA

Abstract

Due to mounting interest in pyrolysis as a biorenewable substitute for fossil fuels, a study was carried out to quantify the economic feasibility of the catalytic pyrolysis pathway. Catalytic pyrolysis is the only form of pyrolysis to have been commercialized. Catalytic pyrolysis is focused on reducing the oxygen content in the intermediate product (bio-oil) to minimize costly hydrogen consumption during upgrading.

CHEMCAD simulation software was used to model the catalytic pyrolysis process, using publically available data from the companies and national labs involved in pathway research. A discounted cash flow rate of return (DCFROR) spreadsheet was used to evaluate the minimum fuel selling price (MFSP) of the pathway's hydrocarbon-based transportation fuel products. The study revealed that the MFSP obtained is low enough relative to gasoline and diesel fuel to consider catalytic pyrolysis as a prospective alternative to petroleum-based fuels. In addition, the main contributing areas to the capital and operational costs were investigated and a sensitivity analysis was used to identify the factors having the greatest impact on MFSP.

(254a) Integrated Processing of Microalgae for Valuable Food Ingredients, Chemicals and Fuel Production

Kaige Wang[1], Yi Liang[2], Zhiyou Wen[2] and Robert Brown[1], (1)Center for Sustainable Environmental Technologies, Iowa State University, Ames, IA, (2)Food Science and Human Nutrition, Iowa State University, Ames, IA

Lipid-based biofuel production by microalgae has captured interest of researchers due to microalgae's high biomass yield and ability to synthesize and accumulate large quantities of lipid. However, high cost of algal cultivation and processing has limited the commercialization of algal fuel production. Developing a microalgae biorefinery strategy, i.e., using every components of algal biomass for fuels and chemicals production is a feasible approach for commercially utilization of the algal biomass.

This study reports a hybrid processing for treating the microalgae *Schizochytrium limacinum*. Crude glycerol, a major byproduct in biodiesel production plants, was used as substrate for producing high content of algal oil containing DHA (22:6, n-3), a high value omega-3 polyunsaturated fatty acid with beneficial effects on human health. The residues after oil-extraction from *S. limacinum* were then catalytically pyrolyzed on a micro-furnace pyrolyzer with zeolite catalyst being used. Aromatics and ammonium was generated from this catalytic pyrolysis process. Among the aromatics, benzene, toluene and xylene, which are valuable basic petrochemicals,were the dominant ones. Most of the nitrogen in the residues was released in ammonium form, which may be in turn used as nutrient for algal culture. In summary, all the components in the biomass of the *S. limacinum* were utilized for producing chemicals, fuels and value-added products which can be future enhanced the economic for algal industry.

(254b) Pyrolytic Fractionation of Oleaginous Feedstocks

Balakrishna Maddi, Sridhar Viamajala and Sasidhar Varanasi, Chemical & Environmental Engineering, University of Toledo, Toledo, OH

Traditional pyrolysis of oleaginous biomass results in bio-oils that contain a mixture of compounds derived from protein, starch and lipids – constitutive biopolymers of these feedstocks. As a result, such biooils have to undergo energy-intensive denitrogenation and deoxygenation to render them suitable as fuels. We have recently developed a "*pyrolytic fractionation*" technique based on the observation that the protein, starch and lipid portions undergo thermal degradation over distinct non-overlapping temperature intervals. Proteins are most thermally labile followed by starch and lipids. Hence, conducting the pyrolysis in three temperature stages provides an opportunity to recover homogenous and biopolymer-specific biooils. Bio-oils from protein and starch can find other applications, in addition being lower grade fuels. For instance, N-rich biooil from protein can serve as a source of high value N-compounds and levoglucosans from starch pyrolysis could serve as precursors for value-added polymers. Similarly, bio-oil from lipid fraction, predominantly containing free fatty acids, can be converted to fuels (such as biodiesel or hydrocarbons) or other higher value fatty acid derivatives. Finally, the bio-char can be a soil amendment.

In this presentation, results showing component-specific biooil compositions and yields from pyrolytic fractionation of algal biomass and oil seeds will be presented. We will also present a comparison of energy needs for pyrolytic fractionation relative to other themochemical methods for conversion of oleaginous feedstocks.

(254c) Hydrothermal Carbonization and Supercritical Ethanol in Situ Transesterification for the Production of Algal Biodiesel

Robert Levine, Chemical Engineering, University of Michigan, Ann Arbor, MI and Phillip E. Savage, Chemical Engineering, University of Michigan, Ann Arbor, Ann Arbor, MI

The production of transportation fuels from microalgae is complicated by the fact that algae are typically grown suspended in a dilute aqueous media and have a high demand for nitrogen and phosphorous. In this work, we explore a new and environmentally preferable strategy for algal biofuel production. There are three major components of the bio-refinery model we have developed: 1) cultivation of oil-rich algae using recycled water and nutrients; 2) treatment of the harvested algae through hydrothermal carbonization to produce a biochar; and 3) conversion of the carbonized algal biomass into transportation fuels using in situ transesterification with supercritical ethanol.

Lipid-rich *Chlorella* was grown in bioreactors until it contained approximately 50% oil, then carbonized for 15 to 90 min at 180 to 235 °C. Biochar obtained by filtration retained >90% of lipids of the original biomass and was converted into biodiesel by reacting the solids with ethanol at 270-325 °C for 15-180 min. Fatty acid ethyl yields in excess of 90% were obtained while employing 10-100 times less ethanol than previously reported for comparable transesterification reactions. The nutrients liberated during hydrothermal carbonization were used to produce additional algal biomass, thereby recycling N and P. Overall, our work suggests that it is possible to produce net-energy positive algal biofuels with significantly less environmental impact than the petroleum fuels they replace.

(254d) Integrating Photo-Bioreactor and Fermentor to Produce Biofuels and Bioelectricity

Alim Dewan, Marci Kerls and M. Nazmul Karim, Chemical Engineering, Texas Tech University, Lubbock, TX

We have investigated simultaneous production of biofuels and bioelectricity from an integrated bioreactor that combines photo-synthesis and fermentation. In the integrated bioreactor, a fermentor and a photo-bioreactor were combined in a way that the carbon dioxide produced in the fermentation can be used to grow algae in the photo-bioreactor. Also, the reactors were attached to a proton exchange membrane in between so that they form a biological fuel cell. To synchronize the carbon dioxide production in the fermentation and the carbon dioxide utilization in photo-synthesis, the fermentor was run continuously but the photo-bioreactor was run in batch mode. An autotrophic microalgae, *Chlorella vulgaris,* and a bear yeast, *Saccharomyces bayanus,* were grown in the photo-bioreactor and in the fermentor, respectively. The ethanol production by the yeast, oil production by the algae, and electricity generation were monitored over time. The electricity generation was characterized by potentiodynamic polarization. In a two-liter size integrated bioreactor, we observed an ethanol productivity of 0.49±0.37 g l^{-1} hr^{-1}, algae oil productivity 2.5 mg l^{-1} hr^{-1}, power density of 70.61 ± 12.41 mW/m^2 of the electrode, and current density of 155.02 ± 31.23 mA/m^2. Preliminary scale-up studies show a great potential for this integrated system.

(254e) Determination of Lipid Hydrolysis Kinetics in Soybean Oil and Algal Systems

Joshua C. Wissinger[1], Amber L. Bosley[1] and Constance Schall[2], (1)Chemical and Environmental Engineering, University of Toledo, Toledo, OH, (2)Chemical & Environmental Engineering, University of Toledo, Toledo, OH

Determination of lipid hydrolysis kinetics in soybean oil and algal systems

Joshua C. Wissinger, Amber L. Bosley, Constance A. Schall

The third generation of biofuels, algae, is of growing interest as a renewable feedstock due to its high yield of energy per unit area, use of carbon dioxide and other available resources for growth, and minimal competition with the food industry. Biodiesel and jet fuel can be produced from lipid-rich algae. However, extracting lipids from algae is complicated due to the high water content of harvested and mechanically dewatered algae (60 to 90% water). Thus, there is a need for processes that convert wet algal biomass into fuels or fuel precursors directly in the aqueous phase. Such a process alternative is to use the water in the harvested and partially dewatered algae for hydrolysis of lipids to free fatty acids (FFA), forming aqueous and oil rich phases for ease of extraction and separation of the lipid component. Lipid processing techniques include refining FFA by catalytic cracking and hydrotreating for jet fuel production, or the UOP/Eni Ecofining™ process to obtain clean green diesel. In our first step, subcritical water is reacted with lipids producing FFA by lipid hydrolysis eliminating the need for expensive and water sensitive catalysts. In the initial development of an empirical kinetic model, soybean oil, a system much less complex than alga, was used in hydrolysis studies at subcritical water (SCW) conditions. This model was applied to algal lipid hydrolysis. The robustness of our model has been tested by verifying both the autocatalytic behavior of lipid hydrolysis reactions, as well as hydrolysis reactions over a range of temperatures.

(254f) Microwave-Assisted Subcritical Water Extraction of Lipids From Wet Algae

Harvind Kumar Reddy[1], Yingqiang Sun[1], Yin Li[2], Tapaswy Muppaneni[1], Sundaravadivelnathan Ponnusamy[1], Shuguang Deng[1], Tanner Schaub[3], Barry Dungan[4], Francisco Holguin[4], Peter Lammers[5], Wayne Voorhies[6] and Peter Cooke[7], (1)Chemical Engineering, New Mexico State University, Las Cruces, NM, (2)Chemical and Biological Engineering, Zhejiang University, Hangzhou, China, (3)College of Agricultural, Consumer and Environmental Sciences, New Mexico State University, Las Cruces, NM, (4)College of Agricultural, Consumer and Environmental Sciences, New Mexico State university, Las Cruces, NM, (5)Energy Research Laboratory, New Mexico State University, Las Cruces, NM, (6)Molecular Biology, New Mexico State university, Las Cruces, NM, (7)Core University Research Resource Laboratory, New Mexico State University, Las Cruces, NM

Microwave-assisted extraction of algal oil at subcritical water conditions was demonstrated with wet algae samples containing up to 90% of water. The extraction efficiency of this method was compared to those of the conventional solvent extraction methods, Folch and Bligh & Dyer. The extraction parameters including biomass loading, temperature and extraction time are optimized after executing a set of designed experiments. Algal oil was produced after purification of bio crude with activated charcoal. The residual algal biomass was also analyzed for its fatty acid and energy content to validate potential byproduct development as part of algal biorefinery. Energy calculations were performed for microwave assisted subcritical water extraction and compared with those of the conventional methods. Analytical instruments including GC-MS, FT-ICR were used to analyze the biomass and biocrude. Thermogravimetric analysis (TGA), TEM and SEM equipped with EDS were used to characterize the biomass in different stages of the extraction process.

(260a) Characterization of Carbohydrate Accessibility and Enzyme Adsorption Capacity for Diverse Cell Wall Phenotypes Subjected to Alkaline Hydrogen Peroxide Pretreatment

Muyang Li[1], Sivakumar Pattathil[2], Michael G. Hahn[2] and David Hodge[1,3], (1)Biosystems and Agricultural Engineering, Michigan State University, East Lansing, MI, (2)Complex Carbohydrate Research Center, University of Georgia, Athens, GA, (3)Chemical Engineering and Material Science, Michigan State University, East Lansing, MI

This study characterizes the changes to polysaccharides in terms of their abundance, accessibility, and affinity for binding by a range of cellulolytic enzymes and CBMs in the cell walls of grasses and dicots that were alkaline hydrogen peroxide (AHP) pretreated at varied severities or hydrothermally pretreated followed by AHP post-treatment. These bioenergy feedstocks represent relatively diverse cell wall phenotypes, including the monocots switchgrass (*Panicum virgatum* cv. Cave-In-Rock), corn stovers (a commercial hybrid and inbred brown midrib lines bm1 and bm3), and *Miscanthus* spp. and herbaceous and woody dicots including goldenrod (*Solidago* spp.) and hybrid poplar (*Populus* spp.). It was shown that increasing H_2O_2 loading results in increasing digestibility only for monocot grasses, while the digestibility of dicots is saturated at low H_2O_2 loadings. Polysaccharide accessibility in the residual cell wall is characterized through "glycome profile" which involves screening successive cell wall extracts with an array of 156 monoclonal antibodies with affinities for the entire spectrum of non-cellulosic cell wall polysaccharides revealing changes in their accessibility and extractability as a function of pretreatment. Enzyme binding is assessed by characterizing binding isotherms for the proteins contained commercial cellulase enzyme cocktails, purified cellulase components, and a library of GFP-linked CBMs.

(260b) Lignin Structural Changes Associated with Oxidative Pretreatment Catalyzed by Cu-Diimine Complexes

Zhenglun Li, Department of Chemical Engineering and Material Science and DOE Great Lakes Bioenergy Research Center, Michigan State University, East Lansing, MI and David Hodge, Chemical Engineering and Material Science, Michigan State University, East Lansing, MI

A wide range of alkaline-oxidative treatments of lignocellulose are known from the literature and practice of chemical pulping and bleaching and from more recent cellulosic biofuels applications. Alkaline hydrogen peroxide (AHP) has been developed as a pretreatment or post-treatment for increasing biomass digestibility by decreasing the recalcitrance of high-lignin biomass. Traditional AHP pretreatment uses the high pH of 11.5 for deprotonation of hydrogen peroxide which leads to the formation of oxidative radicals ($\cdot OH$ and $\cdot O_2^-$), whereas the low selectivity of the radical reactions and the non-productive decomposition of hydrogen peroxide to oxygen at alkaline pH limit the economic competitiveness of AHP pretreatment. Significantly, oxidative delignification catalyzed by coordinated transition metal ligands within laccases and peroxidases is the route taken for lignin degradation in nature. Catalyzed oxidative pretreatments offer the promise of mild temperatures, high reaction specificities, and high oxidizing efficiency that can simultaneously generate a material that is susceptible to enzymatic carbohydrate depolymerization, be performed at decreased pH with correspondingly less inorganics load, and generate modified low MW aromatics that may be potentially upgraded to petroleum-displacing fuels and chemicals. In this work, the effect of Cu-diimine complexes during catalyzed oxidative pretreatment on model lignins is characterized. Model lignins include a Kraft lignin and dioxane milled-wood and enzymatic lignins from select hardwoods and grasses, while characterization is performed using ^{31}P NMR, ^{1}H-^{13}C HSQC NMR and SEC which show distinct structural changes during pretreatment including, among others, clear evidence of lignin modification and depolymerization and cellulose depolymerization. Additionally, significant improved cellulolytic enzyme digestibility is found for select catalyzed pretreatments in woody dicots.

(260c) Mechanism of Exo-Exo Synergism Between CBH1 and CBH2 During the Hydrolysis of Cellulose

Makoto Ikeo, Shohei Okino, Yoshiki Ueno and Daisuke Taneda, Technology Development Department, JGC Corporation, Higashiibaraki-gun, Ibaraki Prefecture, Japan

In past studies, JGC corporation has reported that the efficiency of enzymatic hydrolysis of cellulose using cellulase with low enzyme loadings can be significantly improved by placing the reaction under static condition (2010 AIChE Annual Meeting, 617c Breaking the Barrier to Obtain High Yields On Enzymatic Hydrolysis Using Low Enzyme Loadings). The hydrolysis of 100ml 10% filter paper slurry at the enzyme loading of 6 mg protein/g-dry substrate was conducted using Cellulase SS (Nagase Chemtex). The agitated condition obtained a hydrolyzate with a final glucose concentration of 60 g/L. In contrast, a hydrolyzate with a final glucose concentration of 90 g/L was obtained under the static condition.

Further investigations found that this improvement occurs by the preservation of CBH2 and the exo-exo synergism between it and CBH1 under the static condition. Cellulase SS solution was agitated without any substrates under 50 °C, and SDS-PAGE analysis was conducted to measure the concentration of individual components. Of all the cellulase components, only the band of CBH2 disappeared within 6 hours of agitation, indicating its deactivation. Further examination using fractionated enzyme solutions, one of which lacked CBH1 and the other lacking CBH2, revealed that both CBH1 and CBH2 are required for the efficient hydrolysis of cellulose.

In this study, the mechanism of the exo-exo synergism between CBH1 and CBH2 was investigated by comparing the characteristics of enzyme adsorption and desorption to the cellulose under agitated and static conditions. The hydrolysis of 10% filter paper slurry with the enzyme loading of 12 mg-protein/ g-dry-substrate under both static and agitated conditions were conducted, and the changes in concentration of each cellulase component in the hydrolyzate was monitored using SDS-PAGE and a densitograph. The glucose concentrations of the hydrolyzates obtained were 88 g/L and 60 g/L under the static and agitated conditions respectively. Under both conditions, the CBH1 and CBH2 concentrations in the hydrolyzate dropped sharply, showing concentration decreases of 100% and 80% respectively, at the initial stage of the reaction, indicating their adsorption to the substrate. Under the static condition in which CBH2 is active, the CBH1 and CBH2 concentrations gradually rose as the reaction progressed, indicating desorption from the substrate. The final concentrations of CBH1 and CBH2 after the reaction were 80% and 90% of the original concentration respectively. Under the agitated condition in which CBH2 is deactivated, no increase in the concentration of CBH1 or CBH2 was observed after desorption. CBH2 is likely to be deactivated by shear stress from agitation. However, CBH1 is stable even under the agitated condition, indicating that, under agitated condition, CBH1 remains adsorbed to the substrate and the hydrolysis remains incomplete.

From this result, the exo-exo synergism was hypothesized as enhancing the hydrolysis by preventing the unproductive binding of CBH1 during the reaction. For further investigation, the adsorption and desorption characteristics of enzymes during the sequential addition of CBH1 and CBH2 fractions was observed. The hydrolysis of 10 wt% filter paper using CBH1 fraction yielded hydrolyzate with glucose concentration of 60 g/L at the end of the reaction, and no desorption of CBH1 from substrate was observed. The addition of CBH2 fraction at the later stage of this reaction reaccelerated the reaction, increasing the final glucose concentration to 90 g/L. Also, CBH1 desorption was observed after the addition of the CBH2 fraction. The results obtained suggested that CBH1 desorbs from the substrate and completes the hydrolysis under the presence of active CBH2, but remains absorbed on the substrate in an inactive form when CBH2 is absent.

(260d) Cellulose Thin Film Hydrolysis by Cellulase Enzymes in the Presence of Inhibitors – a QCM Study

Utshab Chakravorty[1], Stephen E. Rankin[1], Sue E. Nokes[2] and Barbara L. Knutson[1], (1)Chemical and Materials Engineering, University of Kentucky, Lexington, KY, (2)Department of Biosystems and Agricultural Engineering, University of Kentucky, Lexington, KY

Cellulose thin film hydrolysis by cellulase enzymes in the presence of inhibitors – a QCM study

Utshab Chakravorty, Barbara Knutson, Stephen E. Rankin, Sue Nokes

Abstract

Conversion of biomass to chemicals and biofuels using the products of cellulose hydrolysis is an important area of research for the production of the next generation of biofuels. However, the saccharification and biochemical conversion of cellulose from lignocellulosic biomass is potentially limited by the release of inhibitors during the processing steps which are used to make the cellulose accessible. Understanding the role and sensitivity of these inhibitors in altering the activity of cellulase enzymes is necessary to the design of pretreatment and separation approaches that maximize cellulose conversion. Bulk studies of cellulose degradation do not provide sufficiently detailed time-dependent and concentration-dependent information to delineate the effect of inhibitors on the adsorption of cellulases to cellulose surface and the subsequent hydrolytic activity of the cellulases. This study uses the frequency response of model cellulose thin films cast onto Quartz Crystal Microbalance (QCM) sensors to model the binding and hydrolysis of cellulose by cellulases (from *Trichoderma reesei*) in the presence of furfural and acetic acid. The sensitivity of the frequency response during cellulose hydrolysis to the specific inhibitor and inhibitor concentration is demonstrated. The concentration-dependent inhibition frequency response data are modeled using a reaction network that accounts for enzyme adsorption, hydrolysis, and several modes of enzyme inhibition and deactivation. The model results can be used to interpret the mode of inhibition. The rate constants of the binding and hydrolysis reactions determined by regression to sets of QCM data with varying inhibitor concentration indicate a strong binding affinity of the enzyme to the substrate and consistent rate coefficient for hydrolysis. The QCM data are used to understand the possible roles of various modes of active site inhibition and inhibition of binding to cellulose on changes in overall conversion in the presence of the inhibitors.

(260e) Investigation of Hemicellulase Inhibition in the Production of Bioethanol

Erik Kuhn[1], Xiaowen Chen[2], Clare J. Dibble[2], Joseph Shekiro[3], Nicholas J. Nagle[1] and Richard T. Elander[4], (1)Bioprocessing R&D, National Renewable Energy Laboratory, Golden, CO, (2)National Bioenergy Center, National Renewable Energy Laboratory, Golden, CO, (3)National Renewable Energy Laboratory, Golden, CO, (4)National Bioenergy Center, National Renewable Energy Lab, Golden, CO

Many of the lignocellulosic bioconversion routes utilize xylose and/or glucose produced by operations such as pretreatment and enzymatic hydrolysis. The insoluble solids (IS) from pretreatment enter enzymatic hydrolysis and exist either in the presence of the solubilized pretreatment fraction (i.e. whole slurry) or absent of the solubilized fraction (i.e. washed). The difference between washed IS and whole slurry enzymatic hydrolysis is the background concentration of sugars, organic acids, and other compounds prior to enzymatic hydrolysis. Whole slurry enzymatic hydrolysis is preferred as it decreases water and utility use, but higher glucose and xylose yields have been observed utilizing washed solids. Using commercially available hemicellulase and cellulase preparations we have observed little to no xylose production from high solids, whole slurry enzymatic hydrolysis on select dilute acid pretreatment slurries while greater than 80% xylan to xylose yields have been observed from high solids, washed IS enzymatic hydrolysis. Previous work has shown end product inhibition of Aspergillus niger beta-D-xylosidase by D-xylose, and inhibition from phenolic acid compounds on the endoxylanase, XynA, from Clostridium cellulovorans. Organisms that utilize C-5 sugars in fermentation decrease the xylose concentration of the fermentation broth, which may allow for further xylose production during fermentation. In the present work a range of potential xylanase inhibitors observed in dilute-acid pretreatment slurries and fermentation broths are dosed into washed IS and beechwood xylan at varied concentrations prior to enzymatic hydrolysis, and the effect on xylose production is presented.

(260f) Increased Enzyme Binding to Substrates Does Not Always Increase Catalytic Activity

Shishir Chundawat[1], Dahai Gao[1], Anurag Sethi[2], Stephanie Crews[1], Leonardo da costa Sousa[1], Nirmal Uppugundla[1], Venkatesh Balan[1], Sandrasegaram Gnanakaran[2] and Bruce E. Dale[1], (1)Chemical Engineering and Material Science, Great Lakes Bioenergy Center, Michigan State University, Lansing, MI, (2)Theoretical Division, Los Alamos National Laboratory, Los Alamos, NM

Substrate binding affinity is typically thought to be one of the primary rate-limiting steps preceding enzyme catalytic action during most homogeneous reactions. However, interfacial-based enzyme catalysis on insoluble crystalline substrates, like cellulose, has an additional step comprising of individual biopolymer chain decrystallization from the substrate interface. This additional decrystallization step has ramifications on the role of enzyme-substrate binding and its relationship to overall catalytic efficiency. We find that altering the crystalline structure of cellulose from its native allomorphic state called cellulose I to cellulose III results in reduced overall cellulase binding but enhanced hydrolytic activity on the latter substrate. We have constructed a comprehensive kinetic model for processive enzymes acting on insoluble substrates to explain this anomalous finding. Our study indicates that this phenomenon can be explained by the enhanced decrystallization kinetics of individual cellulose chains from the surface of cellulose III that despite reduced overall binding affinity results in enhanced catalytic activity. These findings have implications on our understanding of interfacial enzyme catalysis for biopolymers that incorporate analogous decrystallization events.

(264a) Catalytic Pyrolysis of Microalgae for High-Quality Bio-Oil Production

Zhenyi Du, Bioproducts and Biosystems Engineering, University of Minnesota, St Paul, MN, Michael Mohr, University of Minnesota, St Paul, MN, Roger Ruan, Department of Biosystems & Agricultural Engineering, University of Minnesota, St. Paul, MN and Paul Chen, Dept. of Biosystems & Agricultural Engineering, University of Minnesota, Saint Paul, MN

Pyrolysis is a thermal decomposition process of biomass at high temperatures (ca. 500 °C) in the absence of oxygen with the organic vapors condensed as pyrolysis oil. Our previous studies indicate that microalgae can be converted to an energy-dense bio-oil by pyrolysis. This technique can readily transform the whole algal biomass into biofuels without the need of oil extraction and transesterification typically being used for biodiesel production. However, algal bio-oil (31 to 36 MJ/kg), although with much higher heating values than lignocellulosic bio-oil, still requires significant modification to remove the heteroatoms (N and O) to become an acceptable transportation fuel. Catalytic upgrading using zeolites is a promising process for the direct conversion of microalgae into hydrocarbons. Several commercial and laboratory-prepared catalysts, including Y zeolite, HZSM-5 zeolite, nickel, cobalt, copper and ruthenium-substituted ZSM-5, were investigated on their effects of deoxygenation and denitrogenation. Egg albumin, starch and canola oil were used as the model compounds of protein, polysaccharide and lipid fractions in microalgae. And these model compounds and microalgae Chlorella vulgaris respectively were pyrolyzed in quartz tubes with two catalyst beds at both ends. Pyrolytic vapors passed through the catalysts and were then identified with Gas chromatography-Mass spectrometry. Results indicate that all these catalysts showed the effects of O, N removal to some extent with aromatic hydrocarbons (benzene, toluene and styrene) dominating at the highest yield of approximately 25% based on the starting algal biomass. Based on the results of model compounds, it is proposed that these catalysts eliminated the O, N-containing functional groups and advanced aromatization due to their shape-selectivity and acidic catalytic sites. Details on the reaction pathways and product distribution will be presented. The results implicate that this one-step pyrolysis/upgrading method could be a promising way to make high-quality bio-oil from microalgae.

Conversion of Ethyl Acetate over Acid-Base Catalysts as a Model Reaction for Biooil Refining: Infrared and Flow Reaction Studies

Thanh Khoa Phung, Elisabetta Finocchio, Guido Busca

Dipartimento di Ingegneria Chimica e di Processo, Università di Genova, P.le J.F. Kennedy 1, 16129, Genova, Italy

Introduction

Several processes are under development to convert biomasses into renewable fuels (Huber et al., 2006; Demirbas, 2008). The production processes of bioethanol and "biodiesel" (fatty acid methyl esters) are already commercial, but they need further improvements to reduce their impact to food and agriculture market as well as to reduce their costs. Several other processes, such as pyrolysis and catalytic conversion of different renewable raw materials (wood, vegetable oils, etc.) and also of wastes are under study. The very high oxygen content of the resulting biooils, which prevents their massive use in several countries such as in European union, because it may overcome the limits by law is a common problem of these processes. Limits to oxygen content are due to the need to reduce the amount of toxic oxygenated VOCs in the waste gas and to improve the cold properties. The presence of carboxylic acids in the fuel is particularly noxious because of the high melting point of many acids which reduces cold properties, as well as to the corrosion behaviour associated to their acidity. Carboxylic acids frequently come from incomplete conversion of esters, such as in the case of the conversion of triglyceride-based raw materials.

To improve the qualities of biooils as fuels, further catalytic treatments can be performed (Bulushev and Ross, 2011). Conversion in the presence of acidic catalysts is reported to improve fuels reducing their oxygen content. Solid acids are also reported to be active for the conversion triglyceride-based materials, such as palm oil, into hydrocarbon-rich fuels (Twaiq et al., 2003).

In this work, the conversion of ethyl acetate as a model reaction for biooil refining has been investigated on acid-base catalysts. This ester is much shorter than the typical esters present in vegetable oils, however, it contains the important chemical functionality (R'-O-COR) within a short aliphatic chain (Danuthai et al., 2009). The aim of this work is to have indication on the mechanism involved in the acid- or base- catalyzed conversion of esters.

Experimental

Siralox 1.5/40 and Pural MG70 from Sasol were used as the catalysts. Siralox 1.5/40 (hereinafter indicated as Al100) is a transitional alumina with small amounts of silica as a stabilizing agent and containing also a small amount of α-Al_2O_3 and has the typical Lewis acidic behaviour of transitional alumina (Bevilacqua et al., 2006). Pural MG70 is a hydrotalcite-like Layered Double Hydroxide containing 70% MgO and 30% Al_2O_3. After calcination in air at 800°C within 5 hours, this material is constituted by a mixed oxide MgO-$MgAl_2O_4$, hereinafter indicated as MG70.

The catalytic experiments were carried out in a fixed-bed tubular quartz flow reactor, operating isothermally, loaded with 0.5 g of the catalyst (60-70 mesh sieved), feeding 12.5% vol/vol ethyl acetate (EA) in Nitrogen with total flow rates of either 40 cc/min or 80 cc/min. The carrier gas (Nitrogen) was passed through a bubbler containing high purity EA (99.5%). The temperature in the experiment was varied stepwise from 473 K to 973 K.

EA conversion was defined as usual:

$$X_{EA} = \frac{n_{EA(in)} - n_{EA(out)}}{n_{EA(in)}}$$

While selectivity to product i is defined as follows:

$$S_i = \frac{n_i}{v_i(n_{EA(in)} - n_{EA(out)})}$$

where n_i is the moles number of compound i, and v_i is the ratio of stoichiometric reaction coefficients.

The outlet gases were analyzed by a Gas Chromatograph Agilent 4890 equipped with a Varian capillary column "Molsieve 5A/Porabond A Tandem" and TCD and FID detectors in series. In order to identify the compounds of the outlet gases, a GC/MS (Thermo Scientific) was used.

For FT-IR studies, the adsorption/desorption process has been studied using Nicolet 380 FT-IR Spectrometer. Pressed disks of the pure catalyst powders were activated "in situ" in the IR cell connected with a conventional gas-manipulation apparatus, before any adsorption experiment IR spectra of the surface species as well as of the gas phase were collected upon increasing temperature in static conditions (p_{EA}~ 4 torr). The FT-IR studies of ethyl acetate adsorption/desorption on two catalysts have also been carried out.

Results & discussions

1. Flow reaction

Ethyl acetate conversion over Al100 catalyst starts to be significant (1.2 %) at 573 K and increases further by increasing temperature up to be complete at 773 K and above at 96.5 h^{-1} GHSV (gas hourly space velocity). Following the thermal evolution, the conversion increases with further formation of decomposition products. Ethylene C_2H_4, acetone CH_3COCH_3 and acetic acid CH_3COOH are the major reaction products, while carbon monoxide CO, carbon dioxide CO_2, methane CH_4, acetaldehyde CH_3CHO, ethanol C_2H_5OH, diethyl ether $(C_2H_5)_2O$, acetic anhydride $(CH_3CO)_2O$ are minority products. The conversion also increases with decreasing space velocity, whereas conversion decreases with extending time on stream, due to deactivation phenomena at the catalyst surface (Figure 1).

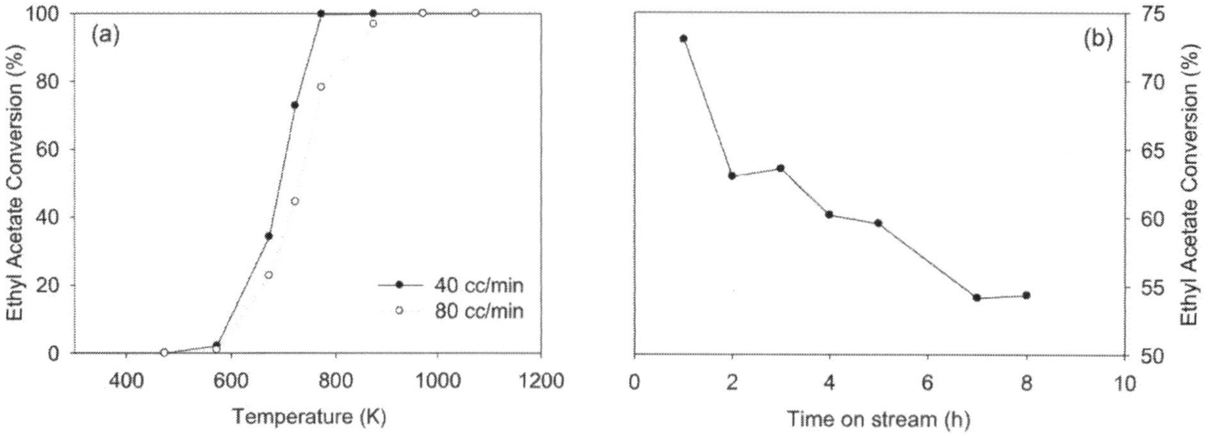

Fig. 1. (a) Conversion of EA over Al100 as a function of flowrate of 40 cc/min and 80 cc/min; (b) Conversion of ethyl acetate over Al100 as a function of time on stream, at 723 K, total flowrate of 40 cc/min.

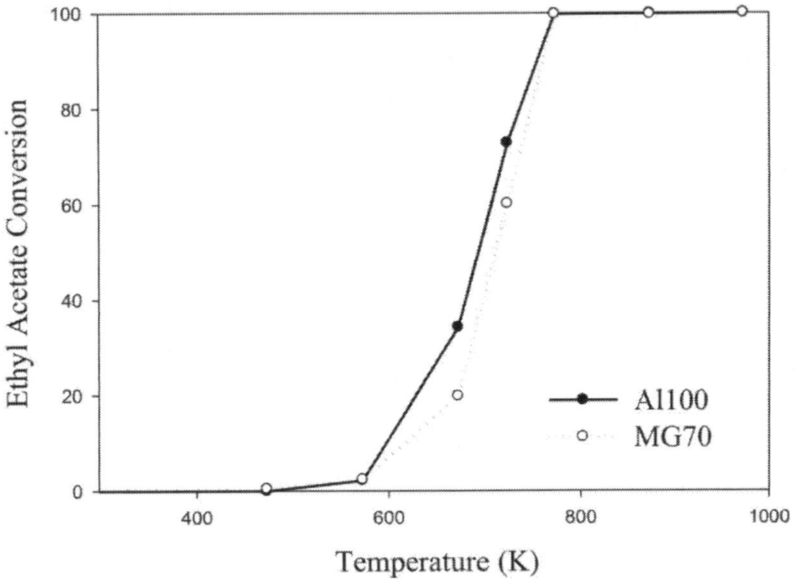

Fig. 2. Conversion of ethyl acetate over Al100 and MG70 at 96.5 h^{-1} GHSV.

Working with 96.5 h^{-1} GHSV, calcinated hydrotalcite MG70 is likely less active than Al100 in the range 550-800 K. Above this temperature both catalyst allow total EA conversion.

Table 1. Product selectivities of ethyl acetate conversion over Al100 and MG70.

Descriptions	Temperature (K)											
	573		673		723		773		873		973	
	Al100	MG70	Al100	MG70	Al100	MG70	Al100	MG70	Al100	MG70	Al100	MG70
Conversion	**2.2**	**2.4**	**34.3**	**19.8**	**72.8**	**60.2**	**99.8**	**99.7**	**99.9**	**99.7**	**100**	**100**
CH_4	0.0	0.0	0.0	0.0	0.1	0.3	0.5	0.5	3.3	0.5	5.5	7.6
CO	0.0	0.0	0.0	0.0	0.2	0.3	1.0	0.5	5.5	0.5	0.2	0.9
CO_2	10.0	19.3	9.3	4.7	9.7	13.0	11.0	13.7	10.3	13.7	11.4	14.6
C_2H_4	45.8	3.3	57.4	4.6	47.9	9.2	40.4	21.0	46.3	21.0	60.8	66.0
CH_3CHO	15.0	0.0	9.1	20.2	9.5	19.8	10.7	8.7	8.1	8.7	7.4	0.1
C_2H_5OH	6.5	29.5	2.8	30.8	1.4	21.1	0.4	12.4	0.0	12.4	0.0	0.0
CH_3COCH_3	8.5	47.9	13.7	29.9	15.7	27.3	21.8	27.0	18.0	27.0	1.4	2.4
$(C_2H_5)_2O$	14.1	0.0	5.5	2.9	3.8	1.2	1.4	0.9	0.0	0.9	0.0	0.0
CH_3COOH	0.0	0.0	1.6	0.4	10.7	0.2	7.9	0.4	4.6	0.4	8.3	0.0
$(CH_3CO)_2O$	0.0	0.0	0.0	0.0	0.0	0.0	1.6	2.3	2.0	2.3	4.3	6.7
$C_5H_{10}O^a$	0.0	0.0	0.0	2.2	0.0	1.3	0.3	2.8	0.1	2.8	0.0	0.0
$C_5H_8O^b$	-	0.0	-	0.0	-	1.3	-	2.8	-	2.8	-	0.0
$C_6H_{12}O_2^c$	-	0.0	-	2.2	-	0.9	-	0.2	-	0.2	-	0.0

Note:
[a]: 2-pentanone
[b]: 3-penten-2-one
[c]: ethyl butanate

In table 1, the selectivities to products obtained over the two catalysts are compared. On Al100 more ethylene and less acetone is obtained than using MG70 catalyst below 723 K. However, increasing temperature makes acetone selectivity of both catalysts are similar, while ethylene selectivity on MG70 catalyst rises until values similar to Al100 at 973 K. Conversion over MG70 also give more ethanol than Al100.

2. IR experiments

The FT-IR spectra of ethyl acetate adsorbed on Al100 is reported in Figure 3. Adsorption at room temperature gives rise to bands due to molecular EA weakly adsorbed at the surface. The band at 1705 cm^{-1} can be attributed to the C=O stretching mode of ester groups, while the sharp band at 1377 cm^{-1} is essentially a CH deformation mode of $-CH_3$ group. The weak and broad CH deformation mode of $-CH_2$ falls at 1450 cm^{-1}, whereas the band at 1280 cm^{-1} is assigned to a COC stretching mode of the ester group.

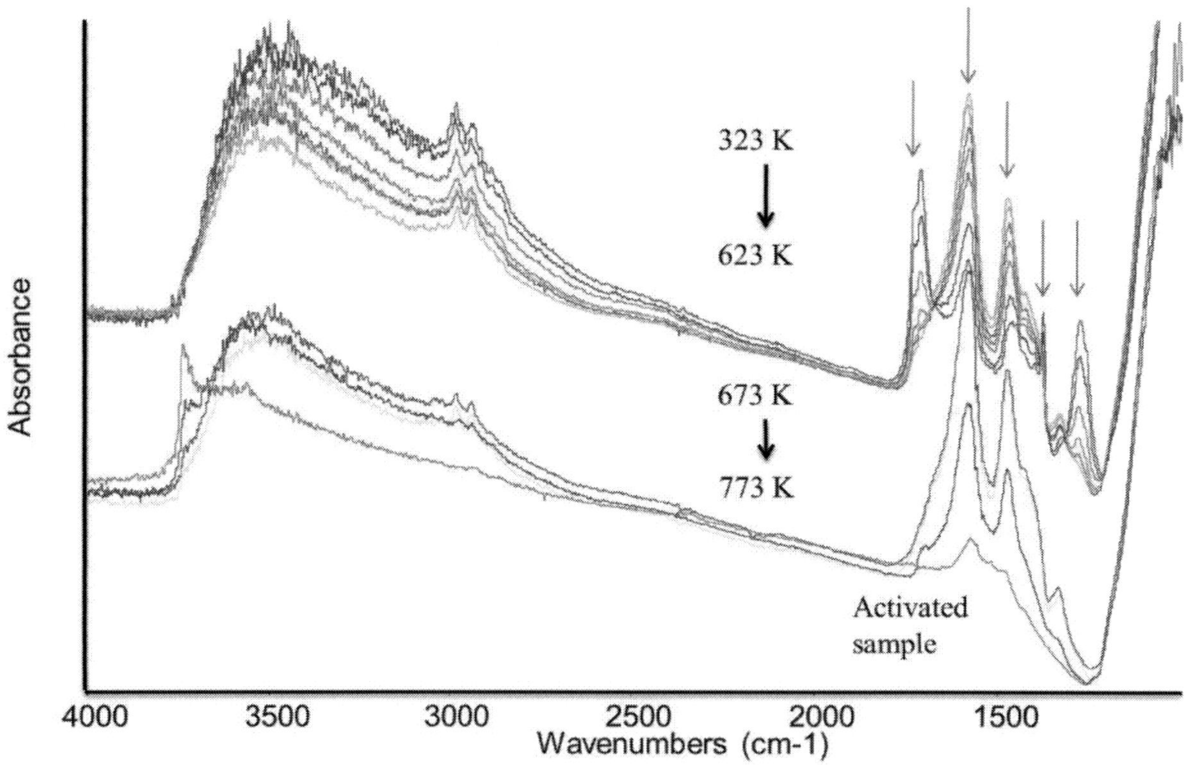

Fig. 3. FT-IR spectra of surface species arising from ethyl acetate adsorbed on Al100 (gas phase spectra subtracted)

In figure 4, the result of the same experiment performed on MG70 is shown. Also in this case we find both molecularly adsorbed EA species and the carboxylate species. The position of the more intense bands of the molecularly adsorbed species on MG70 are observed at 1709 and 1303 cm^{-1} (C=O and COC stretching, respectively) at slightly different position than Al100 at 1705 and 1280 cm^{-1}. However, on MG70 the vibrationally most perturbed species is not the most stable one. In fact, after outgassing 473 K the position of the band due to the molecularly adsorbed species are found at 1711 and 1275 cm^{-1}.

In the gas phase of EA, the corresponding band are found at 1764 and 1238 cm^{-1}. It is evident that adsorption causes shift down of the C=O stretching and shift up of the COC stretching of the EA, this suggests that the main adsorption occurs by interaction of carbonyl oxygen lone pair to Lewis site of the catalyst. However, a more complex interaction can be responsible for the complex behaviour of the low frequency band on MG70, shifting back down at high temperature.

On MG70 we also find the formation of the carboxylate species. The position of these bands at 1592 and 1458 cm^{-1} (asymmetric and symmetric of OCO stretching, respectively) is again different than that observed on Al100 at 1581 and 1473 cm^{-1}. On MG70 the carboxylate bands increase following increasing temperature until 473 K and then are stable, starting to reduce at 673 K until being disappeared at 723 K. In contrast, these species still observed on Al100 surface even adsorbed until 773 K.

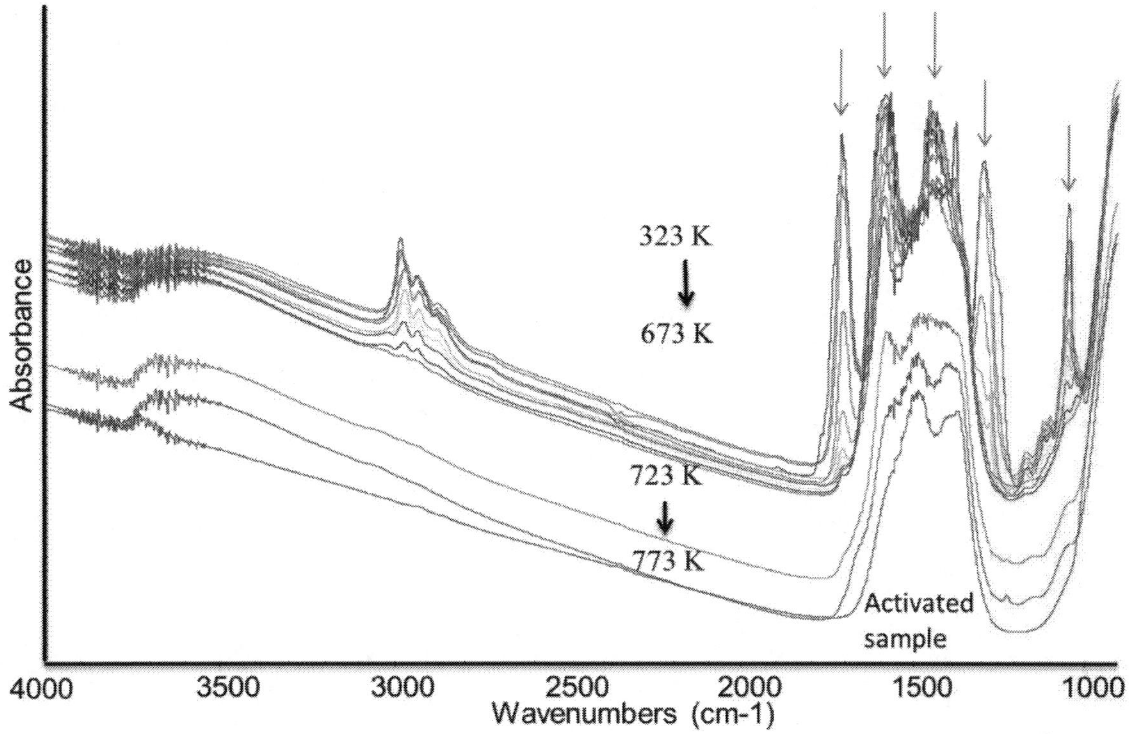

Fig. 4. FT-IR spectra of surface species arising from ethyl acetate adsorbed on MG70 (gas phase spectra subtracted).

The gas phase IR spectra of EA decomposed on acid-base catalysts show ethylene, acetic acid, carbon dioxide, water as the main products. During the IR experiments ketene is also detected among the products of EA decomposition in IR cell. Similar experiments have been carried out also following adsorption and thermal evolution of intermediate reaction products such as acetic acid and ethanol.

Conclusions

The molecularly adsorbed ethyl acetate on alumina undergoes cleavage of ester group, leading to cracking and ketonization. Ethylene, acetic acid, acetone are the major products of ethyl acetate conversion on acid-base catalyst. Acid catalyst Al100 produces mainly ethylene even at low temperature and low conversion, while the basic catalyst MG70 produces more acetone at low temperature by ketonization reaction, those products are similar selectivity at high temperature for both catalysts. This data allow us to propose a detailed discussion of mechanisms of adsorption and conversion of ethyl acetate on acid-base catalysts.

Acknowledgement

This work was supported by the "fund by EMMA in the framework of the EU Erasmus Mundus Action 2". EF acknowledges the University of Genova "Progetto di ricerca di Ateneo 2011" for funding.

References

1. A. Demirbas, 2008. Biofuels sources, biofuel policy, biofuel economy and global biofuel projections. Energy Convers. Manage. 49, 2106-2116.
2. G.W. Huber, S. Iborra, and A. Corma, 2006. Synthesis of Transportation Fuels from Biomass: Chemistry, Catalysts, and Engineering. Chem. Rev. 106 (9), 4044–4098.
3. D.A. Bulushev, J.R.H. Ross, 2011. Catalysis for conversion of biomass to fuels via pyrolysis and gasification: A review.Catal. Today. 171, 1– 13.
4. F.A. Twaiq, N.A.M. Zabidi, A.R. Mohamed, S. Bhatia, 2003. Catalytic conversion of palm oil over mesoporousaluminosilicate MCM-41 for the production of liquid hydrocarbon fuels. Fuel Process. Technol. 84, 105– 120.
5. T. Danuthai, S. Jongpatiwut, T. Rirksomboon, S. Osuwan, D.E. Resasco, 2009. Conversion of methylesters to hydrocarbons over an H-ZSM5 zeolite catalyst. Appl. Catal. A. 361, 99-105.
6. M. Bevilacqua, T. Montanari, E. Finocchio, G. Busca, 2006. Are the active sites of protoniczeolites generated by the cavities? Catal. Today.116, 132-142.

(264c) Kinetics of High Pressure Catalytic Reaction Pathways for Dihydroeugenol Over Pt/ZrO2

Sara L. Yohe, W. Nicholas Delgass, Fabio H. Ribeiro and Rakesh Agrawal, School of Chemical Engineering, Purdue University, West Lafayette, IN

The novel H_2Bioil [1] process involving high pressure biomass fast-hydropyrolysis followed by vapor phase fixed-bed catalytic hydrodeoxygenation (HDO) has been suggested for the conversion of biomass to liquid fuels. The first stage fast-hydropyrolysis generates hundreds of oxygenated compounds, complicating the study and development of new catalysts for the second stage HDO reactor. Therefore, model compound studies provide a means to understand the reactivity of specific oxygen functional groups and to gain fundamental understanding of the link to catalyst functionality.

Kinetic data was obtained for the reaction of the lignin-derived compound dihydroeugenol (2-methoxy-4-propylphenol) in hydrogen over a 2%Pt/ZrO_2 catalyst in a vapor-phase fixed-bed reactor at 24 atm total pressure. Weight hourly space velocities were varied from 100-1000 g hr^{-1} g cat^{-1} to obtain reaction pathways. Reaction orders and activation energies were obtained over a temperature range of 200-400°C, hydrogen partial pressure range of 0-23 atm and dihydroeugenol partial pressure range of 0.4-1.2 atm under differential conditions.

Three main reaction pathways were identified: HDO, transalkylation, and hydrogenation of the aromatic ring. The WHSV studies suggest that the HDO pathway occurs via a series reaction, with methoxy group removal occurring first followed by phenol group removal. Transalkylation occurs via methyl attachment to the aromatic ring. The four isomers of the pure hydrogenation product 2-methoxy-4-propylcyclohexanol have: 1) positive apparent activation energies in the range of 25 to 40 kJ mol^{-1} in the temperature range 200-240°C and negative apparent activation energies of -10 to -40 kJ mol^{-1} for temperatures of 280-320°C 2) dihydroeugenol orders of 0.5-0.7 and 3) hydrogen orders of 1.9-2.5 at 300°C. The similarity of these kinetic results to those for benzene hydrogenation suggest that the mechanism of ring hydrogenation for dihydroeugenol is similar to that of other aromatic compounds [2]. At 300°C the aromatic HDO products 4-propyl phenol and propylbenzene have apparent activation energies of 88 and 128 kJ mol^{-1}, dihydroeugenol orders of 0.6 and 0.6, and hydrogen orders of 0.36 and 0.38, respectively. The analogous hydrogenated HDO products 4-propylcyclohexanol and propylcyclohexane have apparent activation energies of 27 and 67 kJ mol^{-1}, dihydroeugenol orders of 0.33 and 0.43, and hydrogen orders of 3.2 and 2.4 respectively. The reaction sequences and rate expressions will be discussed for the three reaction pathways.

References:

1) R. Agrawal, N. R. Singh, Synergistic Routes to Liquid Fuel for a Petroleum Deprived Future, AIChE Journal,55,7,1898-1905, 2009.

2) S. D. Lin, M. A. Vannice, Hydrogenation of Aromatic Hydrocarbons over Supported Pt Catalysts I. Benzene Hydrogenation, Journal of Catalysis, 143, 539-553, 1993.

(264d) Aqueous-Phase Ketonization of Acetic Acid Over Ru/TiO2/Carbon Catalysts

T. N. Pham, Dachuan Shi and Daniel E. Resasco, School of Chemical, Biological and Materials Engineering, University of Oklahoma, Norman, OK

Ketonization of acetic acid has been accomplished in the liquid phase on Ru/TiO_2 and Ru/TiO_2/C catalysts, at temperatures significantly lower than those typically needed for this reaction. The catalysts were prepared by first impregnating the activated carbon support with titanium (IV) isopropoxide, followed by slow hydrolysis and calcination in the absence of air. The resulting TiO_2/C sample was further impregnated with Ru (III) chloride, dried, and calcined in air. After an in-situ pre-reduction in H_2, the resulting Ru/TiO_2/C exhibited high activity and selectivity to acetone in both, organic and aqueous phases at 180°C. By contrast, when the TiO_2/C sample was precalcined in air before the addition of Ru, the resulting catalyst showed significantly lower activity when it was pre-reduced in-situ, and was essentially inactive without the pre-reduction. While it is usually expected that in water the oxygen vacancies are filled, it is shown here that under the conditions investigated, they remain active for long time. A combination of techniques including XRD, XPS, and TPR were used to characterize the reducibility of the catalysts and the formation of oxygen vacancies. It is proposed that the presence of Ru on TiO_2 facilitates the formation of Ti^{3+} sites, which are responsible for the observed activity. Furthermore, the hydrophobicity of the carbon support is believed to decelerate the typical inhibiting effect of water for reactions catalyzed by oxygen vacancies.

(264e) Catalyst Screening for Upgrading Biomass Fast Pyrolysis Vapors

Mark R. Nimlos[1], Calvin Mukarakate[1], Ryan M Richards[2] and David Robichaud[1], (1)National Bioenergy Center, National Renewable Energy Laboratory, Golden, CO, (2)Department of Chemistry and Geochemistry, Colorado School of Mines, Golden, CO

Oils condensed from fast pyrolysis of biomass are attractive as possible feedstocks or intermediates for existing refinery operations. Unfortunately, these oils often contain too much oxygen to be introduced into existing refineries and they must be upgrade. We have been exploring the strategy of upgrading pyrolysis vapor so as to avoid polymerization reactions that may occur during condensation. We have screened catalysts using an auto sampler apparatus connected to a Molecular Beam Mass Spectrometer (MBMS) and a micro-reactor. In the auto sampler, biomass pyrolysis vapors are passed through a small packed bed of catalysts (~ 1 mg) and into the sampling orifice of the MBMS. This apparatus allows rapid screening of temperature and catalyst formulation and universal sampling to determine loss of oxygen in the products. Using multivariate analysis of the mass spectra from these experiments, we can rank the catalyst effectiveness for producing nonoxygenated products. The micro-reactor uses approximately 1 g of catalyst in a packed bed and enables the study of catalyst deactivation and the determination of carbon balances. Measurement of the mass spectra of the products in real time shows the oxygenated compounds that are formed during deactivation and provides an indication of deactivation mechanisms. Carbon balances allow us to determine coking rates.

(264f) Deoxygenation of a Biofuel Model Compound (2-Methyltetrahydrofuran) On Transition Metal Phosphides

Phuong P. Bui[1], Juan A. Cecilia[2], S. Ted Oyama[1,3], Atsushi Takagaki[3], Antonia Infantes-Molina[4], Haiyan Zhao[1], Dan Li[5], Enrique Rodriguez-Castellon[6] and Antonio J. Lopez[6], (1)Chemical Engineering, Virginia Polytechnic Institute and State University, Blacksburg, VA, (2)Universidad de Málaga, Málaga, Spain, (3)University of Tokyo, Tokyo, Japan, (4)Inorganic Chemistry, Universidad de Malaga, 29071, Malaga, Spain, (5)Virginia Polytechnic Institute and State University, Blacksburg, VA, (6)Quimica Inorganica, Univ Malaga, Malaga, Spain

Second generation bio-fuels are attracting interest as they are produced from non-food sources such as waste products from agriculture, forestry, or paper manufacture. These biomass sources have several advantages over conventional fossil feedstocks for their abundant availability, extremely low sulfur and nitrogen content, localized production, and reduction in greenhouse gas emissions (due to the CO_2 consumption of biomass). The average composition of raw oil from pyrolysis of biomass has a high oxygen content (35-50 wt%) that results in low heating value, immiscibility with hydrocarbon fuels, high acidity, and chemical and thermal instability. The oxygen compounds can be removed through hydrodeoxygenation (HDO) reactions which entail cleavage of C-O bonds in the presence of hydrogen to form hydrocarbons and environmentally benign water. Reactions in HDO are similar to those in hydrodesulfurization (HDS) and hydrodenitrogenation (HDN) processes because they all involve the breaking of carbon-heteroatom bonds. The similarity of HDO with HDS and HDN potentially allows integration of biomass conversion with the existing petroleum refining infrastructure for processing and transportation. Among new catalyst compositions that have been explored in hydrotreating processes, phosphides of transition metals stand out as potential substitutes for conventional sulfides. For example, nickel phosphide, cobalt phosphide, tungsten phosphide and molybdenum phosphide are highly active for HDS and HDN of petroleum feedstocks. This study has two objectives, first, to compare phosphides (Ni_2P/SiO_2 and WP/SiO_2) prepared by the reduction of phosphate and phosphite precursors and second, to evaluate their activity in the hydrodeoxygenation of the bio-oil model compound, 2-methyltetrahydrofuran (2-MTHF). The latter is a five-atom saturated ring compound that has a methyl group which allows distinguishing between the ring-opening products and gives mechanistic information.

Transition metal phosphides, Ni_2P and WP, supported on a high surface area fumed silica (Cab-osil ® EH5) provided by Cabot Corp. were synthesized using the phosphate and phosphite methods. Equal moles of metal per gram of support (1.16 mmol/g support) were used. The phosphate method (method A) produces phosphides via phosphate precursors which contain phosphorus at an oxidation state of +5. Meanwhile, the phosphite method (method I) produces phosphides via phosphite precursors which have phosphorus at a lower oxidation state of +3. This lower oxidation state is expected to make the precursors potentially reducible at a lower temperature.

The catalysts were characterized with temperature programmed reduction, BET surface area measurement, CO chemisorption, X-ray diffraction and X-ray photoelectron spectroscopy. The phosphite precursor of nickel phosphide was reduced at a lower temperature (560°C) than the phosphate precursor (590°C). A similar trend was observed for tungsten phosphide in which the phosphite precursor was reduced at a lower temperature (580°C) than the phosphate precursor (600°C). The BET surface areas range from 149 to 217 m^2/g, significantly lower than the area of the silica support (350 m^2/g). Catalysts from the phosphate method have a lower surface area than those from the phosphite method, probably due to the double sintering from calcination and reduction of the former.

Hydrodeoxygenation of 2-methyltetrahydrofuran (2-MTHF) was carried out in a packed-bed reactor as a function of temperature at atmospheric pressure. Quantities of catalysts loaded into the reactor were equivalent to 30 umol of active surface metal atoms (from CO uptake measurements). The gas-phase feed contained 3.2% of 2MTHF in the H_2

stream. The catalytic activities of all phosphides at different temperatures were presented as turnover frequencies (TOF) based on the same number of CO chemisorption sites. In both cases, nickel phosphide was more active with higher TOF values than tungsten phosphide. Between each method, nickel phosphide by the phosphate method had a TOF value about 1.4 times than that of the phosphite method while tungsten phosphide had practically the same TOF value for both methods.

The main HDO products of supported Ni_2P were pentane and butane whereas the products of supported WP were mostly pentenes and pentadienes, suggesting different reaction pathways on these catalysts. Contact time studies were carried out to identify the intermediate products and determine reaction pathways. Quantities of catalysts loaded into the packed-bed reactor were equivalent to 5 umol of active surface metal atoms. For nickel phosphide from both methods, the phosphate method and the phosphite method, the product selectivities were similar with pentenes as primary products, 2-pentanone as a secondary product, and pentane as a final product; a reaction network involving surface intermediates was proposed in Scheme 1. In contrast, for tungsten phosphide the product selectivities depended greatly on the preparation method. With WP/SiO_2 (phosphite method) pentenes were produced prior to pentadienes indicating sequential hydrogen removal on a single site (Scheme 2). With WP/SiO_2 (phosphate method) pentadienes were produced prior to pentenes, indicating a surface intermediate reacting on two active sites (Scheme 2). This result was consistent with the lower P content of the phosphate-method catalyst which would leave more open metal sites.

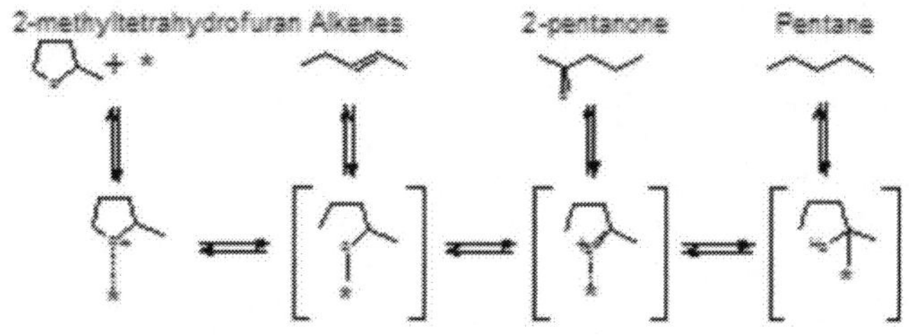

Scheme 1: HDO reaction network of 2-MTHF on Ni_2P/SiO_2

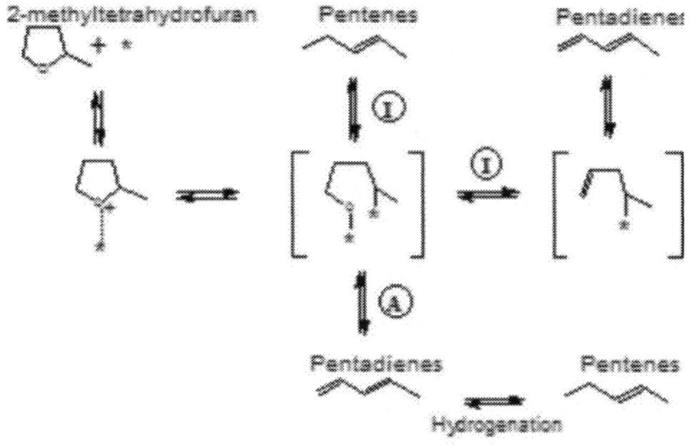

Scheme 2: HDO reaction network of 2-MTHF on WP/SiO_2; (I) – Phosphite method, (A) – Phosphate method

(266a) Selective Electro-Oxidation of Glycerol to Glycolate On Au/CNT Catalysts

Zhiyong Zhang, Le Xin, Ji Qi, Zhichao Wang and Wenzhen Li, Chemical Engineering, Michigan Technological University, Houghton, MI

Selective Electro-Oxidation of Glycerol to Glycolate on Au/CNT Catalysts

Zhiyong Zhang, Le Xin, Ji Qi, Zhichao Wang, Li Wenzhen

Glycolic acid has a broad range of cleanser applications due to its high acidity (pK_a=3.83) and chelation properties with metal ions. However, the current industry glycolic acid production processes involve formaldehyde, carbon monoxide, or hydrogen cyanide, all of them are highly toxic chemicals. In this work, we developed an alternative green approach to produce glycolate with high selectivity through electro-oxidizing bio-renewable glycerol in alkaline solution on carbon nanotube (CNT) supported gold catalyst (average particle size: 2.5 nm; 1.0 - 4.5 nm). CNT was used as electro-catalyst support based on the consideration of its higher electric conductivity, and better durability under high potentials. It has been demonstrated that the product distribution from glycerol electro-oxidation is strongly dependent on the applied anode potential. Under an anode applied potential of 1.6 V (vs. Standard Hydrogen Electrode, SHE) at ambient conditions (room temperature and atmosphere), a selectivity of 85% to glycolate was successfully achieved with a glycerol conversion of 50%, after 3 hour's reaction. The detailed reaction pathways will be presented.

(266b) Cerium-Containing HZSM-5 Catalysts for Biomass Upgrading

Gregory T. Neumann, Chemical and Biomolecular Engineering, University of Notre Dame, Notre Dame, IN and Jason C. Hicks, Department of Chemical and Biomolecular Engineering, University of Notre Dame, Notre Dame, IN

Cerium-Containing HZSM-5 Catalysts for Biomass Upgrading

Gregory T. Neumann, and Jason C. Hicks*

Department of Chemical and Biomolecular Engineering, University of Notre Dame, IN

Lignocellulosic biomass has been widely studied as a renewable source for the sustainable production of chemicals and fuels.[1,2] Catalytic fast pyrolysis, a scalable, thermochemical process, has received particular attention as a method to upgrade lignocellulosic biomass directly to chemicals and fuels. We report the synthesis, characterization, and glucose catalytic fast pyrolysis results of HZSM-5 catalysts modified through the addition of cerium. Cerium was incorporated in the HZSM-5 materials using three different methods: 1) incipient wetness, 2) ion-exchange, and 3) incorporation in the dry gel.[3] Interestingly, of the three materials studied for the catalytic fast pyrolysis of glucose, only the dry gel, cerium incorporated catalyst significantly altered the product distribution.[3] With this catalyst, the production of benzene, toluene and xylenes decreased, while production of furans, aldehydes, and ketones increased.

It was also found that the cerium-containing, dry gel catalyst could stabilize pyrolysis oils through decarbonylation and ketonization reactions. For instance, the cerium incorporated dry gel HZSM-5 catalyst enhanced the decarbonylation reaction pathway of furfural to produce CO and furan.[4] Another component of pyrolysis-oil is acetic acid. We have shown that the cerium incorporated dry gel HZSM-5 catalyst has the ability to couple acetic acid to selectively produce acetone through a ketonization reaction pathway. These materials have been characterized by N_2 physisorption, powder X-ray diffraction (XRD), diffuse reflectance UV-Visible (UV-Vis) spectroscopy, diffuse reflectance FT-IR, scanning electron microscopy (SEM), scanning transmission electron microscopy (STEM), transmission electron microscopy (TEM), ammonia temperature-programmed desorption (NH_3-TPD), and combination focused ion beam (FIB)/ SEM.

1. Hicks, J. C., *J. Phys. Chem. Lett.* **2011,** *2* (18), 2280-2287.

2. Huber, G. W.; Iborra, S.; Corma, A., *Chem. Rev.* **2006,** *106* (9), 4044-4098.

3. Neumann, G. T.; Hicks, J. C., *ACS Catal.* **2012**, 642-646.

4. Neumann, G. T.; Hicks, J. C., *Top. Catal.* **2012,** *55* (3-4), 196-208.

(266c) Cycloaddition of Biomass Derived Furans for the Renewable Production of p-Xylene

Christopher L. Williams[1], Chun-Chih Chang[1], Phuong T. Do[2], Nima Nikbin[2], Stavros Caratzoulas[3], Dionisios G. Vlachos[4], Raul F. Lobo[2], Wei Fan[1] and Paul Dauenhauer[1], (1)Chemical Engineering, University of Massachusetts Amherst, Amherst, MA, (2)Chemical Engineering, University of Delaware, Newark, DE, (3)Catalysis Center for Energy Innovation, University of Delaware, Newark, DE, (4)Catalysis Center for Energy Innovation, Chemical and Biomolecular Engineering, University of Delaware, Newark, DE

Increasing demand for domestic energy and renewable materials has led to an accelerated research effort to develop biomass-derived fuels and chemicals [1]. One sustainable feedstock for these renewable chemicals is sugars produced by the saccharification of biopolymers (e.g., cellulose, hemicellulose) [2]. The catalytic conversion of sugars to high value commodity chemicals, like p-xylene (used in the production of PET plastics [3]), is currently a research area of interest [4]. In this work, we investigate the conversion of dimethylfuran to *p*-xylene by a two-part reaction (Diels-Alder cycloaddition followed by dehydration). By reacting dimethylfuran with high-pressure ethylene, competitive side reactions that reduce *p*-xylene yield are identified and minimized. Multiple catalyst types are investigated providing insight into the relevant catalyst active sites and microstructure. 75% yield of p-xylene is demonstrated using Y-zeolite in combination with an aliphatic solvent at high temperature. Additionally, it is found that Diels-Alder cycloaddition (the first step in conversion of DMF to p-xylene) is promoted by confinement within microstructured zeolites and the second step (dehydration) is catalyzed by Brønsted acid sites, rendering the first step (cycloaddition) rate limiting. Gas phase DFT calculations supporting the necessity of Brønsted acid sites are also presented. This catalytic system introduces a new chemical pathway to convert biomass-derived furans to high-value aromatic feedstocks.

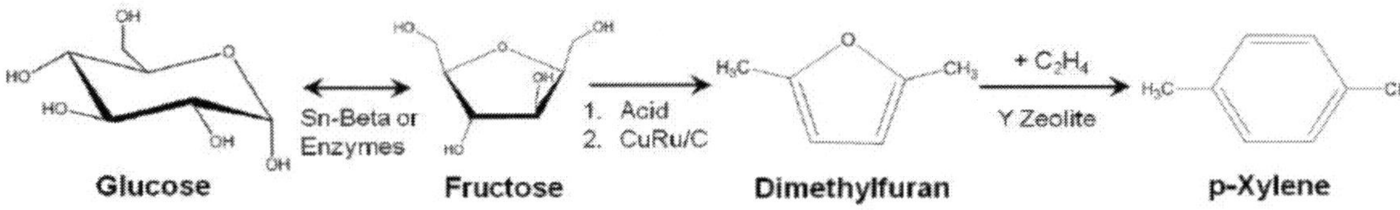

References:

[1] Climent, M. J.; Corma, A.; Iborra, S. *Green Chemistry* **2011**, *13*, 520-540

[2] Vlachos, D. G.; Chen, J. G.; Gorte, R. J.; Huber, G. W.; Tsapatsis, M. *Catalysis Letters* **2010**, *140*, 77-84.

[3] Sheehan, R. J. In *Ulmann's Encyclopedia of Industrial Chemistry*; Wiley-VCH Verlag GmbH & Co. KGaA, Weinheim, 2011; pp. 17-28

[4] Williams, C. L.; Chang, C.-chih; Do, P.; Nikbin, N.; Caratzoulas, S.; Vlachos, D. G.; Lobo, R. F.; Fan, W.; Dauenhauer, P. J. *ACS Catalysis* **2012**, *2*, 935-939

(266d) Understanding the Solvation of Fructose and 5-Hydroxymethylfurfural in Water - Dimethyl Sulfoxide Mixtures Using Molecular Dynamics and Vibrational Spectroscopy

Vladimiros Nikolakis[1], Samir H. Mushrif[2], Stavros Caratzoulas[3], Bryon Herbert[4], Karl S. Booksh[4] and Dionisios G. Vlachos[5], (1)Catalysis Center for Energy Innovation, Chemical Engineering, University of Delaware, Newark, DE, (2)Chemical Engineering, University of Delaware, Newark, DE, (3)Catalysis Center for Energy Innovation, University of Delaware, Newark, DE, (4)Chemistry and Biochemistry, University of Delaware, Newark, DE, (5)Catalysis Center for Energy Innovation, Chemical and Biomolecular Engineering, University of Delaware, Newark, DE

Fructose dehydration to 5-hydroxymethylfurfural (HMF) is a well-studied reaction and it is considered a characteristic example of converting carbohydrates to useful chemical intermediates [1]. The choice of solvent (DMSO, water or DMSO – water mixture) significantly affects its selectivity and yield [1,2]. For example, the acid catalyzed dehydration in an aqueous environment is accompanied by side reactions which lead to the formation of acyclic dehydration and fragmentation products, insoluble humins and HMF rehydration products. On the other hand, HMF yields higher than 90% can be achieved in DMSO [3]. Unfortunately, issues related to the separation of HMF from DMSO render this solvent unattractive for commercial applications. The goal of this work is to understand the solvation of fructose and HMF in DMSO and to provide insights about the effect of solvation on the selectivity of fructose dehydration. To achieve this goal we used a combination of vibrational spectroscopy (ATR/FTIR, Raman) and molecular simulations (Molecular Dynamics).

The analysis of the fructose hydroxyl hydrogen – DMSO oxygen radial distribution function showed that the coordination number of DMSO around fructofuranose is ~3.5. This number is smaller than the number of hydroxyl groups of fructose because one DMSO is shared between two hydroxyl groups and because of the formation of intramolecular hydrogen bonds. In the case of fructose - DMSO mixtures, a red shift of the Raman S=O asymmetric stretching is observed which indicates that fructose breaks the DMSO clusters owing to strong hydrogen bonding between the hydrogen of its hydroxyl groups and the oxygen of DMSO. The Raman scattering cross-sections of the DMSO S=O stretching when a DMSO molecule interacts with another DMSO, fructose, water or HMF were estimated from the spectra of the binary mixtures using the coordination numbers from the MD simulations. It was also possible to use these values together with the MD predicted coordination numbers to satisfactorily predict the effect of water fraction on the Raman scattering intensity of the S=O stretching band in ternary mixtures when the H_2O/DMSO molar ratio was less than ~0.3. MD simulations also showed that with increasing water content the DMSO orientation around fructose changes with the sulfur atom moving away from the carbohydrate. The analysis of the ATR/FTIR spectra reveal that with increasing water content the average hydrogen bond enthalpy of the fructose hydroxyls decreases by ~2.5 kJ/mol. The MD simulations indicate that this decrease is due to the reduction of the strength of DMSO – fructose hydrogen bond, and due to the replacement of a fraction of the stronger DMSO – fructose hydrogen bonds by the weaker fructose – water ones. Finally, MD simulations also show the preferential coordination of DMSO around HMF carbonyl group could shield HMF from further rehydration to levulinic acid and formic acid and from humins formation.

1. Kuster, B. F. M., Starch 1990, 42 (8), 314-321.

2. Chheda, J. N.; Roman-Leshkov, Y.; Dumesic, J. A. Green Chemistry 2007, 9 (4), 342-350

3. Musau, R. M.; Munavu, R. M., J of Chem Tech. and Biotechnology 1987, 32, 920-924

4. Hess, B.; Kutzner, C.; van der Spoel, D.; Lindahl, E., J. Chem Theory and Comp. 2008, 4 (3), 435-447.

(266e) Selective Liquid Phase Oxidation of Glycerol Using (Group I-B)-Pd Bimetallic Catalysts Prepared by Electroless Deposition Method

Abraham A. Rodriguez, Christopher T. Williams and John R. Monnier, Chemical Engineering, University of South Carolina, Columbia, SC

Glycerol (1,2,3-propanetriol) is the byproduct of biodiesel production from the transesterification of vegetable oils or animal fats, produced at a rate of 10 pounds of glycerol for every 100 pounds of biodiesel produced. Increase in production of Biodiesel has led to a surplus of crude glycerol which has destabilized the glycerol market. Thus, new ways of utilizing this excess feedstock are needed in order to cope with the current and future production trends. The heterogeneously catalyzed selective liquid phase oxidation of glycerol has been studied by many research groups using mono and bimetallic Pd, Pt, and Au catalyst on different supports. Literature shows that bimetallic catalysts generally yield higher activities and selectivities when compared to their monometallic counterparts. Using the Electroless Deposition (ED) method, Group I-B metals have been deposited on a commercial Pd/C catalyst. ED permits controlled deposition of a second metal on a primary metal surface, which allows for better correlation of catalyst composition and surface composition with performance. A 5 wt% Pd/C catalyst provided by BASF was used as base catalyst for this study and a series of bimetallic catalysts were prepared, with coverage of Group-IB metal on Pd/C ($\theta_{I\text{-}B}$) ranging from 0 to 100%. Batch reactor kinetic results for glycerol oxidation over these Au-Pd/C bimetallic catalysts under basic conditions showed enhanced activity for bimetallic compositions with θ_{Au}>0.30, exhibiting a maximum activity at θ_{Au}=0.50, with a rate constant ~20 times larger than for palladium and ~6 times larger than for gold. The volcano shaped curve suggests the formation of bimetallic sites composed of Au-Pd in close contact. XPS data shows a shift to lower binding energies in the Au 4f region for all bimetallic compositions when compared to a 1% Au/C catalyst, suggesting electron transfer from Pd→Au. The bimetallic effect could be explained by this electron transfer forming a bimetallic site where the terminal OH group of the glycerol molecule is more easily activated. Selectivity to glyceric acid for the bimetallic catalysts was 80% (±2%) for catalysts below θ_{Au}<0.90, dropping to 60% and lower for catalyst with θ_{Au}>0.98. This can be explained by the ability of Au under this reaction conditions to form H_2O_2, which is an unselective oxidant, combined with the low availability at high coverages of Pd sites which can decompose H_2O_2, causing further oxidation of glyceric acid to 2 and 1 carbon products. Further work with different bimetallic compositions and different reaction conditions is in progress to better understand these effects, and will also be discussed in this presentation.

(266f) Minimizing Humin Formation Via HMF Derivatization

Christian P. Canlas, Nicolas Grosso and Justin M. Notestein, Chemical and Biological Engineering, Northwestern University, Evanston, IL

5-Hydroxymethylfurfural (HMF) is a valuable intermediate towards levulinic acid and 2,5-furandicarboxylic acid, which are both included on the list of DOE's top 12 value-added chemicals. However, like most biomass transformations, diminished yield due to humin formation is a huge challenge. Here, we present a route that utilizes alumina-based catalysts that minimizes humin formation from HMF via acetalization and etherification of the HMF before conversion to methyl levulinate. These catalysts contain vanishingly small amounts of metals (Ti or Si) grafted from the tetrahalide that moderate the reactivity of the Al_2O_3 support. Results have shown that these Al_2O_3 based catalysts are effective in converting HMF into 2-(dimethoxymethyl)-5-(methoxymethyl)furan in nearly quantitative yield, a form we show is resistant toward humin formation and cleanly converts to methyl levulinate. Somewhat surprisingly, the derivatized HMF has very infrequently been reported in the open literature. The catalysts are synthesized from commercially available alumina and are characterized using nitrogen physisorption, TEM, DRUV-vis, TGA, and NMR. Derivatization of HMF and formation of the levulinate is carried out in methanol at 140-160C using sealed, glass batch reactors. Products are analyzed using GC-MS

(266g) Gold Supported On Nanocrystalline Metal Oxides Are Highly Effective Catalysts for the Oxidation of Glucose Under Base-Free Conditions

Yuran Wang, Krishna K. Sharma and Yuriy Román-Leshkov, Department of Chemical Engineering, Massachusetts Institute of Technology, Cambridge, MA

Selective oxidation of glucose to produce mono- and di-carboxylic acids is an attractive catalytic technology for the production of valuable chemical intermediates from biomass. Currently, most heterogeneously-catalyzed processes for the transformation of glucose to gluconic acid require the use of corrosive basic promoters (e.g. NaOH) to regulate the pH of the reaction, given that severe catalyst deactivation occurs once the solution enters the acidic regime. The development of a catalyst that maintains high catalytic activity and product selectivity in the absence of basic promoters is of considerable importance. Here, we demonstrate that gold nanoparticles supported on nano-crystalline metal oxide supports are active and stable catalysts for the oxidation of glucose to gluconic acid under base-free conditions. We show that the size of the metal oxide is a critical factor to achieve catalyst stability. Thus, although Au NPs supported on micro-sized metal oxides are initially as active as Au NPs supported on nano-sized supports, only the latter remain active over several cycles without severe metal aggregation and leaching. Similarly, only Au NPs on nano-crystalline supports are capable of performing the oxidation reaction under acidic conditions (pH = 1.6, H_2SO_4) with rates similar to those obtained under uncontrolled pH conditions, while also maintaining activity over multiple cycles. This unique feature opens an attractive avenue to pursue cascade reactions involving acid/redox sequences (e.g., hydrolysis followed by oxidation).

(277a) Sustainability in Global Green-Economy Development - Challenges for the Energy and Process Industries

Hebab A. Quazi, Strategic Sustainable Development Institute, AngelNet Energy Water & Environmental Systems, Austin, TX and Ana Guevara, AVENTI Associates, Miami Beach, FL

International Development Organizations such as The World Bank have created international "Clean Technology Fund" that are helping developing nations make greater use of Clean Energy Resources. World Bank's mission is to eradicate global proverty and promote sustainable development. Global climate change poses new development challenges, but opens new economic opportunities. Smart mobilization of international resources including Climate Investment are key to addressing mitigation and adaptation to the Climate Change. World Bank's development and climate change "Strategic Framework" includes support for accelerated development and deployment of new technologies amd steping-up the policy research, knowledge, and capacity building. The Clean Technology Fund assists accelerating transformation to low-carbon economies. Special attention is focused on emissions reduction based clean energy technologies. The paper will addres the challenges faced by and the contributions made so far by new energy and industrial technologies in supporting sustainable global green-economy-development. The technologies includes low-carbon fuels, improved energy conversion, solar, wind, and bio-fuels.

(277b) Sustainable Chemical Processing Combining Concentrated Sunlight and Natural Gas

Alan W. Weimer, Chemical and Biological Engineering, University of Colorado at Boulder, Boulder, CO

"The sun has a unique role in sustainable energy production, in that it is the undisputed champion of energy; the resource base presented by terrestrial insolation far exceeds that of all other renewable energy sources combined"[Lewis and Nocera, 2006]. Solarthermal chemical processing is ideally suited for driving high temperature chemical reactions where products don't have to be stored as intermediates for downstream continuous processing. Major technical challenges revolve around the availability of thermal shock and oxidation resistant high temperature materials for fabricating solar receivers/reactors and the design of scalable reactors having high efficiency. Heat recuperation/integration is a key issue for achieving high thermal efficiency. The efficiencies for such processes are typically higher than those of competing solar technologies because the energy from the sun is used to directly drive chemical reactions instead of first being converted to another energy form, e.g. electricity. According to the U.S. Geological Survey, 1 barrel of oil equivalent is roughly equivalent to 6,000 standard cubic feet (SCF) of natural gas (NG) in terms of energy. Therefore, with oil near \$100/barrel and NG near \$3/1000 SCF (\$3/28.3 SCM), \$100 of oil has equivalent energy content as ~\$18 worth of NG. This ~ 5X price differential provides for a solarthermal pathway in which low cost NG can be used as a feedstock or, possibly, as a supplement to solar energy used to heat the chemical process 24/7. Major economic challenges include the cost of installed heliostats and competition from low cost fossil fuels such as NG. Three rather near-term opportunities for commercial application include the decarbonization of NG to produce carbon black and hydrogen, the steam-methane reforming of NG for generating peak power, and the gasification of biomass with additive NG.

(316a) Effects of Carbon Sources, Oxygenation and Ethanol On the Production of Inulinase by K. marxianusYX01

Wenjie Yuan, School of life science and biotechnology, Dalian University of Technology, Dalian, China and Jiaoqi Gao, School of Life Science and Biotechnology, Dalian University of Technology, Dalian, China

Inulinase is one of the most important factors of the consolidated bioprocessing (CBP), which combines enzyme production, inulin saccharification, and ethanol fermentation into a single process. In this article, the inulinase production and cell growth on the *K. marxinaus* YX01 under different conditions were studied. Carbon source was shown to be less significant on the production of inulinase although the activity of inulinase was a little higher using inulin as carbon source than using glucose or fructose. The concentration of the carbon sources had a great repression effect on the activity of inulinase, when the concentration is up to 60g/L, the activity of the inulinase was only 50% of that of 20g/L. The enzyme activity was strongly influenced by aeration rate. It has been shown that the activity of inulinase and cell growth under anaerobic conditions both kept at a low level but aeration at 1.0 vvm (air volume/broth volume·minute) led to higher activity. It is interestingly that the inulinase activity per unit biomass did not show significant differences under different aeration rates. Ethanol had great repression on the cell growth. The cell stopped to grow when the addition ethanol was over 9% (v/v), but ethanol did not restrain the activity of secreted inulinase and the enzyme was stable under 15% ethanol addition.

(316b) Effects of Solvent and Enzyme Source for Transesterification of Waste Oils

Michael Gagnon[1], Jonathan Fogel[2], Subhash Minocha[2] and **Palligarnai Vasudevan**[1], (1)Chemical Engineering, University of New Hampshire, Durham, NH, (2)Biological Sciences, University of New Hampshire, Durham, NH

Enzymatic transesterification with lipase as the catalyst eliminates soap formation. Unlike alkali-based reactions, the products can easily be collected and separated. Moreover, enzymes require much less alcohol to perform the reaction, and can be reused despite some loss in activity at the end of each cycle. Several strains of lipases have been found to have transesterification activity, namely *Candida antarctica*, *Pseudomonas cepacia*, and *Thermomyces lanuginosus*. The two primary obstacles in enzyme-catalyzed reactions are i), the immiscibility of the two substrates, namely hydrophilic methanol and hydrophobic triglyceride, resulting in the formation of an interface leading to mass transfer resistance; and ii), the strong polarity of methanol, which tends to strip the active water from the enzyme's active site leading to enzyme deactivation. The addition of an organic solvent as the medium to the reaction system simultaneously overcomes the two limitations by enhancing the solubility of oil and methanol in the solvent, and by limiting the concentration of methanol surrounding the enzyme.

We are genetically engineering plants to constitutively express a lipase for biodiesel production from spent oils. We have cloned the gene of a lipase with known transesterification activity from *Thermomyces lanuginosus*, a thermophylic fungus. Cloning involved isolation of total RNA, reverse transcription of the mRNA into cDNA and PCR amplification of the lipase gene using specific primers. The gene was first inserted into a cloning vector (pCR8/GW/TOPO) and sequenced to confirm its identity. The gene has been inserted into a plant destination vector (pGWB408 and pMDC83) via LR clonase reaction. *Nicotiana tabacum* (tobacco) leaf was transformed with the lipase gene using *Agrobacterium tumefaciens* (strain GV3101) and transferred onto selection media plates. We have also grown *Arabidopsis thaliana* from seeds that were transformed using the floral dip transformation method. The recombinant enzyme was collected from the genetically engineered plants, purified, and tested for both hydrolytic and synthetic activity. The activity results will be compared with enzyme catalysts from a commercial *Thermomyces lanuginosus,* and the effect of solvent will also be presented.

(316c) Production of Renewable Fuel From Enhanced Activated Sludge Through a Fluidized-Bed Catalytic Cracking (FCC) Process

Emmanuel Revellame, William Holmes, Rafael Hernandez and W. Todd French, Dave C. Swalm School of Chemical Engineering, Mississippi State University, Mississippi State, MS

Annual production of activated sludges in the United States ranges from 7 – 8 million tons. Currently, disposal technologies of these sludges include incineration and landfilling, both of which are heavily regulated and could create adverse environmental impacts. A fraction of the activated sludges can be used as fertilizer once it complies with regulations for Class A biosolids. Activated sludges contain microbial consortia known to produce lipidic materials and many other organic compounds, which could be used as fermentation seed for enhancement of its lipid content. Several studies showed that reasonable yields (2 – 6% weight) of biodiesel could be produced from activated sludge. At these yields however, activated sludge biodiesel is less economical than petroleum diesel. By increasing the fuel yield, which could be accomplished by enhancement through fermentation, in combination with an in situ catalytic conversion to fuel, the economics of this feedstock could be improved. This study describes the production of renewable fuel from enhanced activated sludge using Fluidized Catalytic Cracking (FCC). This process could convert most of the compounds present in sludge into fuel, and thus, increase fuel yields and reduce cost.

(316d) Transformative Catalytic Technology for Advanced Biofuel Production

Manhoe Kim, NextCat Inc, Detroit, MI, Like Yang, NextCat INC, Detroit, MI, Steven O. Salley, Chemical Engineering and Materials Science, Wayne State University, Detroit, MI and K. Y. Simon Ng, Department of Chemical Engineering and Materials Science, Wayne State University, Detroit, MI

Biodiesel producers are facing challenges of high cost feedstocks and lack of enabling technology to process low grade (low cost) lipids having high free fatty acid (FFA) and water levels. In order for biodiesel producers to became commercially viable, they must be able to use less expensive lipid feedstocks, a wide range of crude bio-oil feedstocks, with varying FFA content, including palm oil (FFA ~12%), residual corn oil (FFA ~15%) from the distiller dried grains (DDGs), animal fats (FFA 2~35%), yellow grease (FFA 5 - 15%), and brown grease (50~100%). In order to avoid the problems related to homogeneous catalytic system, a transformative heterogeneous catalytic conversion technology using the heterogeneous metaloxides catalyst supported by ZrO_2 has been studied in continuous flow reactor systems. The catalyst processing factors (aging, drying, and extruding, and drying) and the reactor scale-up factors (reactor diameter and length ratio, residence time, superficial flow velocity) which determine catalytic activity and durability has been studied. Downward feed flow has a plug flow feature and gives higher FAME yield than upward flow mode. Superficial velocity of the feed gives effects on both the degree of mixing of the reactants and dissolution of the active sites on the catalyst surface due to shear force induced by the viscous flow. In order to decrease shear force, the void space of the catalyst bed should be increased and the reactor diameter and length ratio should be optimized. This technology can potentially reduce the feedstock, operating and processing costs and reenergize the latent biodiesel production facilities to enable to be used for biodiesel production with various non-edible waste lipids.

(316e) Transesterification of Triglycerides for Biodiesel Over Selected Zeolite Catalysts

Yu-Yuan Wang, Hsin-Yu Chou and Bing-Hung Chen, Department of Chemical Engineering, National Cheng Kung University, Tainan, Taiwan

Transesterification of triglycerides in excess methanol to biodiesel using selected zeolite catalysts has been carried out. Various zeolites, such as beta-zeolite, zeolite Y and Mordenite, doped with alkali metals either by ion-exchange or impregnation, are employed to catalyze the transesterification at a relatively lower temperature around 60°C. A faster transesterification rate and higher conversion efficiency would be achieved with appropriate pretreatment on catalysts. Under a reaction temperature at 60°C, conversion efficiency of triglycerides to biodiesel as high as 97.3% was obtained. Cyclic durability of zeolite catalyst has been examined by running the same zeolite catalysts for 6 cycles of transesterification of triglyceride in methanol. It is found that catalyzed conversion efficiency of triglycerides did not differ significantly during the first 3 cycles. Additionally, regeneration on spent zeolite catalysts has been attempted as well. The zeolite catalyst can be recycled at least for four times without significant deterioration in terms of conversion efficiency of triglycerides.

(316f) Hydrotreating of Soybean Oil Over NiMo Carbide Catalyst On Five Different Supports

Huali Wang[1], Shuli Yan[2], **Elvan Sari**[1], Steven O. Salley[1] and Simon K. Y. Ng[1], (1)Chemical Engineering and Materials Science, Wayne State University, Detroit, MI, (2)NextCAT inc., Detroit, MI

Abstract

As an alternative to conventional sulfide catalysts, NiMo carbide catalysts were prepared by precipitation method and studied for the hydrotreating of soybean oil to produce hydrocarbons in the gasoline to diesel range. The effect of the catalyst supports on activity and selectivity was investigated by using three different types of materials: mesoporous material (Al-SBA-15), alumina (γ-Al_2O_3) and zeolites (ZSM-5, Zeolite β and USY). The structural characterization of the catalysts was performed using XRD, BET and TEM. Catalytic tests were carried out in a bench scale flow reactor at 400 °C and 650 psi. The results showed that the Al-SBA-15 supported catalyst has the highest surface area of 711.5 m^2 g^{-1} and largest pore volume of 0.71 cm^{-3} g^{-1}. Among the five catalysts, hydrotreating on NiMoC/Al-SBA-15 gave the highest yield of organic liquid product (96%) and highest selectivity (97%) to hydrocarbons in the boiling range of the diesel fraction. For the three zeolite supported catalysts, hydrotreating of soybean oil produced more hydrocarbon products in the boiling range of green gasoline (about 15-40%).

(318a) Effect of Crystallinity and Degree of Polymerization On the Fast Pyrolysis Decomposition Products of Cellulose

Andrew Moore, Jiajia Meng, Richard Venditti, David Tilotta and Sunkyu Park, Forest Biomaterials, North Carolina State University, Raleigh, NC

Fast pyrolysis has received significant attention as a possible pathway for converting biomass to transportation fuels and chemicals. Cellulose is the most prominent component in biomass; however, its degradation pathway is not fully understood. More so, the effects of the fundamental structure of cellulose in fast pyrolysis have not been extensively investigated. This study aims to determine the effect of the crystallinity and degree of polymerization on the primary degradation products of cellulose. Pyroprobe-GC-MS was used to examine the volatile fraction, while a secondary GC measured CO and CO_2, and residual char amount was also recorded. This experimental configuration allowed for a full carbon mass balance. The pyrolysis was performed on a CDS pyroprobe with set-point temperature of 600°C to imitate pilot scale pyrolysis temperatures. The preliminary results showed that levoglucosan and glycolaldehyde are key primary degradation products, and their amounts were quantified on the GC-MS via standard calibrations. The results from these experiments are shedding light on the competing mechanisms of cellulose thermal degradation, as well as offering insights on the optimal biomass that will produce more desirable products.

Experimental and Modeling Study of Woody Biomass Pyrolysis at Low Temperatures

Hayat Bennadji[a], *Krystle Smith*[a], *Shaka Shabangu*[a,b], *Elizabeth M. Fisher*[a,*]

[a] Sibley school of Mechanical and Aerospace Engineering, Cornell University, Ithaca, NY, USA
[b] Department of Soil and Crop Sciences, Cornell University, Ithaca, NY, USA

* Corresponding author. Address: Sibley School of Mechanical and Aerospace Engineering, 289 Grumman Hall, Cornell University, Ithaca, NY 14853, USA.
Tel.: +1 607 255 8309; fax: +1 607 255 1222.
E-mail address:emf4@cornell.edu (E.M. Fisher).

1. Introduction

The energy contained in biomass can be used in several ways including combustion and either thermal or biological conversion to liquid, gaseous or solid fuels. The thermal treatment of biomass produces a complex mixture of products which are generally categorized by their volatility as gases, tars, and char; the proportions of these products vary with thermal conversion conditions, and in general are affected by both chemistry and physical processes such as heat transfer and inter-particle transport (Bridgwater, 2009).

Among thermal treatment products, char has attracted considerable interest as an additive to improve the water retention and productivity of poor agricultural soils, in which context it is known as biochar (Lehmann and Joseph, 2009; Manya 2012). The current study focuses on the slow or conventional pyrolysis conditions favoring char production: peak temperatures < 600 °C, heating rates < 50 $°C.min^{-1}$, particle size > 1 cm. These conditions are in the thermally thick regime where significant temperature gradients exist inside the wood particle.

Schemes for biochar production (Brown, 2010; Manya, 2012) generally use the gases released during pyrolysis either as a combustion fuel or as a feedstock for the production of fuels or chemicals. Information about the composition and release rate of these gases and tars is useful for process design for biochar production, and when coupled with well-defined pyrolysis conditions is valuable for testing pyrolysis heat transfer and kinetics models.

Previous literature includes numerous studies of slow pyrolysis of single wood particles of various simple geometries (sphere, cylinder, and cube) in the thermally thick regime. The majority of these studies report temperature, mass, and/or particle size as functions of time during pyrolysis. Many earlier studies (for example Koufopanos et al., 1989; Bilbao et al., 1993; Liliedahl and Sjostrom, 1998) emphasize the influence of the process conditions on mass loss and temperature histories More recent single particle studies (for example DiBlasi, 2001; Peters and Bruch, 2003; Sadhukhan, 2008; Park et al., 2010) make use of different experimental configurations and sample characteristics, and also include comparison to numerical models.

Although there are some slow pyrolysis studies in which product yields are reported, most of these utilize a complex geometry such as a fixed bed of biomass particles (Williams and Besler, 1996; Wang et al., 2009). In these studies, yields are generally summed over the entire pyrolysis process and lumped into the categories of tars, gases, and chars. There are few experiments reporting time histories of individual species, and these are restricted to a few species. Grønli (1996) carried out pyrolysis experiments on a wood cylinder, heated from one end. He reported the effect of radiant heat flux on the product yields, including time-resolved, speciated permanent gas mole fractions. Gaston et al., (2011) reported the evolution of individual permanent gases upon insertion of wood spheres of different sizes into a fluidized bed reactor.

In the current study, we present time-resolved temperature and species profiles for permanent gases and several volatile species, for dry poplar cylinders undergoing slow pyrolysis under well-defined conditions relevant to biochar production. To our knowledge this constitutes the first experimental data set available in the literature about the slow pyrolysis with both time-resolved species and temperature measurements under a well-defined heat transfer boundary conditions. Results were then compared to predictions of the numerical model proposed by Park et al., 2010.

2. Material and methods

2.1. Materials

The wood samples, obtained from Madison Mill Inc., were 4-cm-long cylindrical sections of a 1.905-cm-diameter tulip poplar dowel with the grain oriented along the axis. The wood cylinders were dried at a temperature of 105 ^{0}C.

2.2. Pyrolysis procedure

The bench-scale reactor and analytical equipment used in this study were used for gas-phase kinetic studies (Zheng et al., 2009), and later modified to carry out wood pyrolysis experiments. The heat for wood pyrolysis was provided by 300 slpm of nitrogen, flowing through a 5.25 cm ID, 1 m long stainless steel tube at approximately atmospheric pressure. Before the start of each experiment, the wood sample was mounted in a holder and stored in a water-cooled side-arm of the reactor under a secondary nitrogen flow. The wood sample was rapidly inserted into the heated main nitrogen flow, with its axis perpendicular to the flow reactor axis. After 30 min inside the reactor, the sample was cooled to 165 ^{0}C or lower in nitrogen, and then cooled in room air.

The temperature profiles inside the sample were measured using a thin (either 1-mm-ID or 0.25-mm-ID) sheathed, K-type thermocouple inserted through holes drilled radially to the centerline of the particle, at one cm axial distance from the end of the particle.

The operating conditions are summarized in Table 1. For each condition, the pyrolysis experiments were repeated 2 to 3 times. Splitting of the wood particle occurred during pyrolysis for some experiments; experiments with particle splitting are omitted from the data set.

Table 1: Experimental pyrolysis conditions and yields of char, gases and tars

Particle Size	Gas T, ^{0}C	Re_D	Thermocouple Sheath OD, mm	Run	m dry wood (g)	Weight % char	Weight % gases	Weight % tars
D = 1.905 cm L= 4 cm	375	5340	1	1	5.583	28.54	11.94	59.52
				2	5.442	28.40	11.65	59.95
				3	5.608	28.64	12.04	59.32
	418	5050	0.25	1	5.623	24.37	13.12	62.51
				2	5.454	24.32	12.56	63.12
				3	5.537	24.21	11.78	64.01

Re_D: Reynolds number based on the reactor internal diameter (ID).
Tar weight % obtained by difference.

2.3. Gas mixing and sampling

The gases and volatiles released during the pyrolysis process were sampled 80 cm downstream of the wood particle, through an open-ended quartz sampling tube of 6.35 mm OD, with a sampling flow rate of 12.5 slpm. On the basis of previous gas-phase experiments (Zheng et al., 2009), this distance was expected to be more than adequate for complete mixing of the pyrolysis gas stream into the bulk nitrogen flow. At the temperatures and residence times reported here, gas-phase reactions outside of the solid particle can be neglected, according to previous experimental observations (Di Blasi, 1996; Morf et al., 2002). This expectation was confirmed through gas-phase plug-flow reactor calculations, with a mechanism of secondary gas-phase pyrolysis developed by Calonaci et al., 2010. Additional confirmation was provided by comparing gas-phase composition obtained at two different sampling locations: the downstream location described above, and a position one mm from the surface of the cylindrical particle.

Sampled gases flowed continuously through Teflon tubing and an ice bath for tar removal, and then into analytical equipment. Although many tar compounds were removed in the ice bath, other condensable species were present at levels well below the saturation conditions in the ice bath, thus passing through the bath without losses. The gas residence time in the entire sampling and analytical system was negligible on the time scale of pyrolysis.

2.4. Analytical methods

Time-resolved profiles of volatile species were obtained by Fourier Transform Infrared (FTIR) analyzer (Nicolet 6700) equipped with a MCT-B detector and a 5-cm-pathlength Teflon gas cell with ZnSe windows. The sample flowed continuously during the experiment, with a cell pressure of 430 torr, producing typical sensitivities of 3 ppm with a data acquisition time of 90 s at 1 cm-1 resolution. Species concentrations were quantified via interactive subtraction of a reference spectrum obtained either from an analyzed calibration gas mixture or from a mixture prepared by injection of a pure condensable species into a nitrogen stream.

At the end of a series of experiments on two or three wood particles, the condensed species were collected from the ice bath by rinsing the sample traps with acetone. Condensed species were analyzed by Gas Chromatograph / Mass Spectrometer (Thermo Trace GC Ultra coupled to a quadrupole mass spectrometer model DSQ II) equipped with a DB5 column, using a method developed for bio-oils (Chen et al., 2010).

2.5 Numerical Simulation

For comparison with experiments, a literature model of single-particle pyrolysis (Park et al., 2010) has been implemented in COMSOL, a commercial Multiphysics software. Park's model is one-dimensional. It models intra-particle convection with velocities following Darcy's law; diffusive fluxes are neglected. Conduction within the particle is represented by an effective conductivity value that represents radiative heat transfer through pores as well as a weighted average of conductivity properties of volatiles, wood, and char. Convection is also modeled within the particle, with values based on relative amounts of gaseous species within the particle and velocities produced through Darcy Flow. Park's kinetic mechanism has parallel endothermic wood decomposition reactions leading to gas, tar, and an intermediate solid, followed by exothermic reactions of the tar and intermediate solid. Kinetic parameters and thermochemistry for these reactions are derived in part from prior literature and in part to match Park's experimental data. We make use of the proposed model with materials properties and chemical parameters unchanged, but with the geometry changed from spherical to axisymmetric.

One important input to the simulation is the imposed heat flux. We performed heating experiments with aluminum cylinders in the flow reactor under conditions similar to those for the wood pyrolysis experiments. Best fits to the resulting interior temperature profiles were used to obtain convective heat transfer coefficients, which agreed within 4 % with a correlation for infinite cylinders (Hilpert, 1933). The same correlation was used to determine heat transfer coefficients for the conditions of the experiment.

3. Results and Discussion

3.1. Temperature results

Figures 1a and 1b show the measured temperature histories at the centerline of the sample, for gas temperatures of 375 °C and 418 °C, respectively. Only the first 20 min of the 30-min pyrolysis run are shown.

The two sets of temperature histories show similar behavior: First, temperature increases nearly linearly. The heating conditions can be characterized by an average rate of temperature rise, W, over the initial, roughly linear part of the curve, as described by Di Blasi et al., 2001, giving values of 1.05 and 1.25 $°C.s^{-1}$ for the two cases shown here. Next, the slope decreases at around 4.5 min for 375 °C gas temperature and 4 min for 418 °C gas temperature, producing a plateau or near-plateau in the temperature profiles at 366°C and at 377 °C respectively. Finally, an increase of the temperature appears around 5.7 min and 5 min leading to an overshoot of 7 °C and 30 °C above the respective final temperatures of 375 °C and 418 °C.

Other researchers have observed similar plateau and maximum behavior, and have interpreted these phenomena in terms of the thermochemistry of production of gases, tars, and char. The plateau is attributed to the endothermic degradation of the low-temperature components (extractives and hemicellulose) leading to vapor-phase tar or other volatiles (Pyle and Zaror, 1984; Bilbao et al., 1993; Morten G.C. and Morten C.M., 2000; Di Blasi, 2001; Park et al., 2010). The maximum (temperature overshooting the final value) must be due to exothermic reactions. Several possible processes have been proposed to explain it: lignin decomposition (Bilbao et al., 1993; Antal and Varhegyi, 1995; Di Blasi et al., 2001), secondary tar cracking reactions (Di Blasi, 2002), or the dehydrocellulose decomposition to gas and char (Kilzer and Broido, 1965; Gullett and Vesilind, 1985). Park et al., (2010) attribute the temperature overshoot to the exothermic conversion of an intermediate solid to char.

Temperature histories are compared to the predictions of the model of Park et al. (2010) in Figures 1a and 1b. At both temperatures, there is qualitative agreement between experiment and prediction: curves are roughly linear at first, followed by a plateau or near-plateau and then a temperature overshoot declining to the final equilibrium temperature. The simulation also agrees with experiment in showing greater temperature overshoot in the higher-temperature case. Quantitatively, agreement is good during the initial, nearly linear, heating process up to 300 °C, suggesting that heat flux and materials properties in the model are reasonably accurate. Above 300 °C, the curves diverge. The predicted temperatures have reduced slope in the 300-375 °C range, while in the experiment the linear portion continues to 365 °C. The slower rise of temperature in the predictions also delays the time of the temperature overshoot beyond that in the experiment. Thus the differences between experiment and prediction appear to be mainly due to the kinetics of the endothermic reactions that produce the reduced slope.

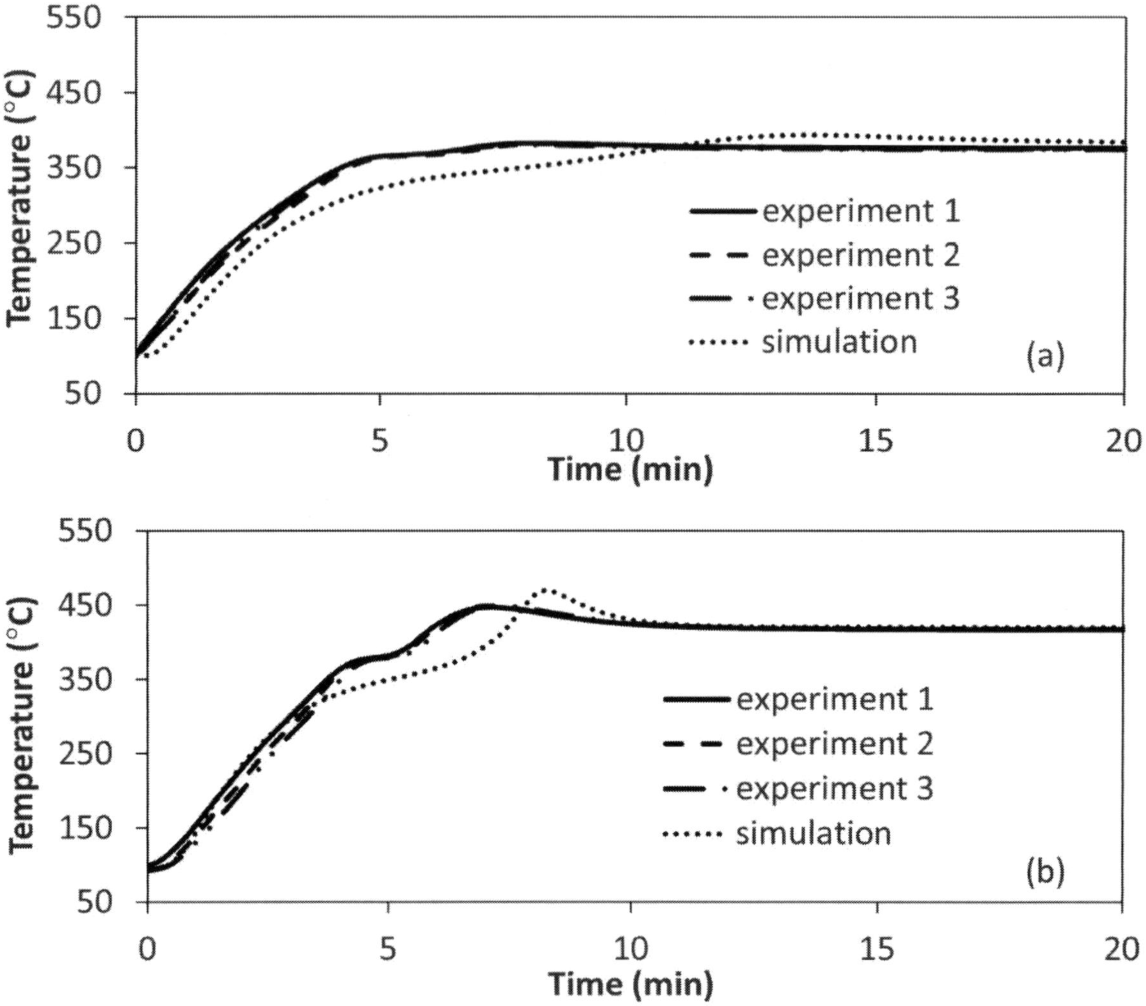

Fig. 1. Measured centerline temperature histories compared to simulation results. 1a: gas T= 375°C; h=57.5 W/(m^2°C)); 1b: gas T= 418°C; h=59.4 W/(m^2°C)).

3.2. Pyrolysis products results

3.2.1. Species concentrations

Figure 2 shows the species time histories for the experiments at 375 ^{0}C. Species mole fractions reflect the dilution of the pyrolysis gases in the large nitrogen flow in the reactor. The time-resolved mole fractions can be integrated to provide relative yields; multiple runs at a given set of operating condition allow a standard deviation to be calculated. On a mass basis, the CO yields relative to CO_2 are 0.38±0.009. This relative yield is in good agreement with previous slow-pyrolysis literature (Di Blasi, 2001).

Each species has a peak occurring in the first ten minutes of the experiment. The acetic acid concentration peaks at 2.45 minutes. The peaks of CO_2, HCHO, and CHOOH, and then those of CO and CH_3OH, appear between 4 and 4.4 minutes. Finally, methane is detected at low concentrations, with a peak appearing at 7 min.

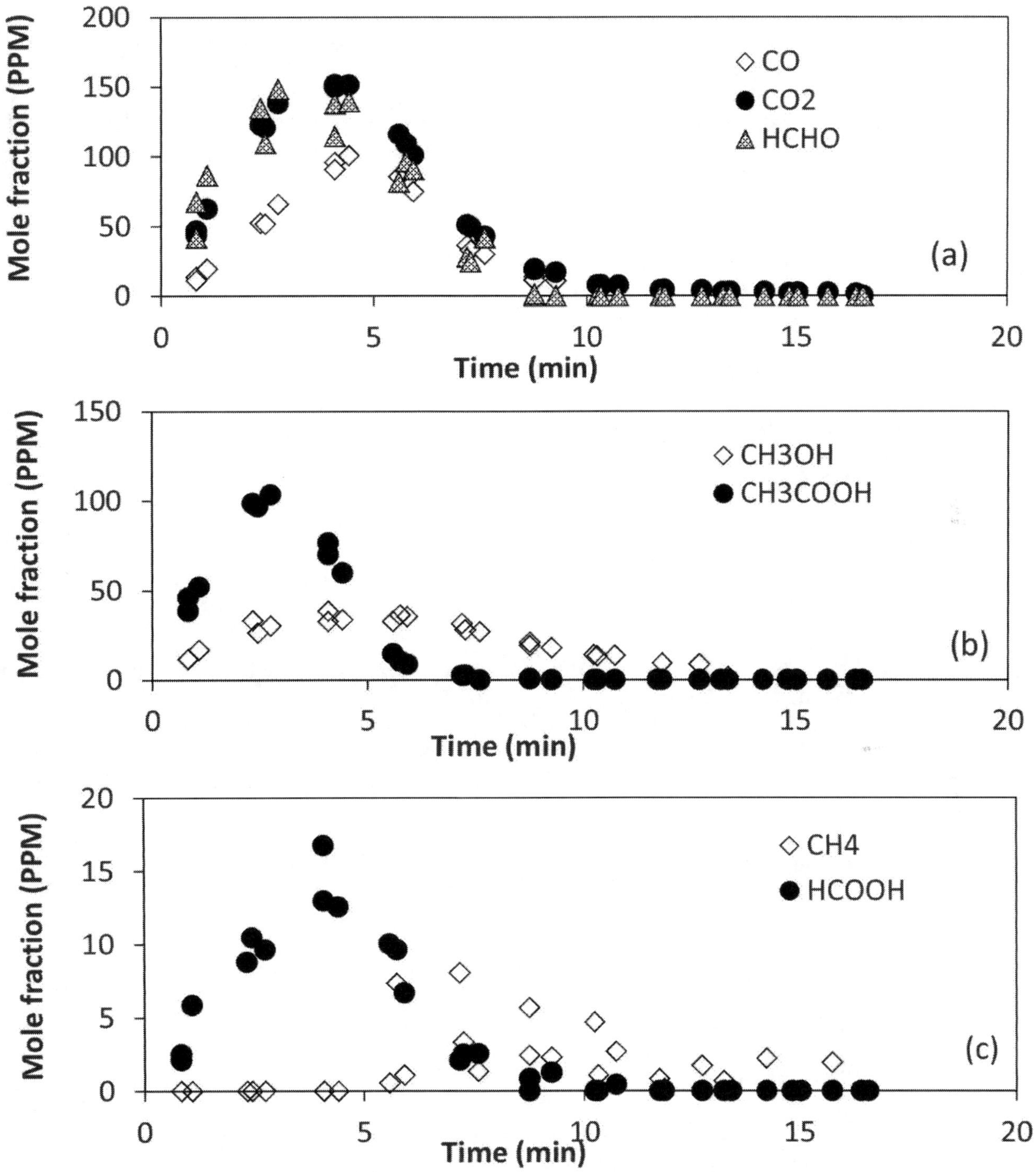

Fig.2. Mole faction history of species released during pyrolysis at gas temperature 375 °C.

3.2.2. Pyrolysis yields

Table 1 shows the char, tar and gases yields obtained for the two different pyrolysis temperatures. Gases are defined as CO, CO_2, and CH_4 only, and tar yields are obtained by difference. Trends agree with literature (Grønli et al., 2000; Di Blasi, 2001; Park et al., 2010). The char yield decreases slightly from 29 % to 24 % as the gas temperature increases from 375 °C to 418 °C; both gas and liquid yields increase with temperature. The liquid fraction accounts for about 60% weight loss of poplar wood, with a gas fraction of about 12%.

4. Conclusions

Pyrolysis of dry poplar cylinders was carried out in a flow reactor under conditions favoring char formation. The intra-particle temperature history together with time-resolved mole fractions of light pyrolysis species CO, CO_2, CH_4, HCHO, CH_3OH, HCOOH, and CH_3COOH were measured. The numerical model proposed by Park et al., 2010 was used to predict the temperature profiles under the present conditions. The model shows qualitative agreement with the present experiments, with quantitative differences due to the temperature dependence of endothermic reactions.

Acknowledgement

The authors are grateful for the financial support to this work from the Fondation des Fondateurs and Yossie Hollander.

References

Antal, M. J., and Varhegyi, G., 1995. Cellulose Pyrolysis Kinetics: the Current State of Knowledge. Ind. Eng. Chem. Res. 34, 703-717.

Bilbao, R., Millera, A., Murillo, M.B., 1993. Temperature profiles and weight loss in the thermal decomposition of large spherical wood particles. Ind. Eng. Chem. Res. 32, 1811-1817.

Bridgwater, A.V., Hofbauer, H., Van Loo, S., 2009. Biomass pyrolysis. In: Thermal biomass conversion. CPL Press. 37-78, 423 - 429.

Brown, T. R., Wright, M.M., Brown, R.C., 2010. Modeling and Analysis: Estimating profitability of two biochar production scenarios: slow pyrolysis vs. fast pyrolysis. Biofuel Bioprod. Bior. 5, 54-68.

Calonaci, M., Grana, R., Hemings, E.B., Bozzano, G., Dente, M., Ranzi, E., 2010. Comprehensive kinetic modeling study of bio-oil formation from fast pyrolysis of biomass. Energy Fuels. 24, 5727-5734.

Chen, T., Ronghou Liu, C.W., Fei, W., Liu, S., 2010. Effect of selective condensation on the characterization of bio-oil from pine sawdust fast pyrolysis using a fluized-bed reactor. Energy Fuels. 24, 6616 - 6623.

Di Blasi, C., 1996. Kinetic and Heat Transfer Control in the Slow and Flash Pyrolysis of Solids. Ind. Eng. Chem. Res. 35, 37-46.

Di Blasi, C., Branca, C., Santoro, A., Hernandez, E.G., 2001. Pyrolytic behavior and products of some wood varieties. Combust. Flame. 124, 165-177.

Di Blasi, C., 2002. Modeling intra- and extra-particle processes of wood fast pyrolysis. AIChE J. 48 (10), 2386-2397.

Gaston, K.R., Jarvis, M.W., Pepiot, P., Smith, K.M., Frederick, W.J., Jr., Nimlos, M.R., 2011. Biomass Pyrolysis and Gasification of Varying Particle Sizes in a Fluidized-Bed Reactor. Energy Fuels. 25, 3747–3757.

Grønli, M.G., Melaaen, M.C., 2000. Mathematical model for wood pyrolysis-Comparison of experimental measurements with model predictions. Energy Fuels. 14, 791-800.

Grønli, M. G., 1996. A theoretical and experimental study of the thermal degradation of biomass. Ph. D. Thesis, NTNU, Trondheim, Norway.

Gullett, B.K. and Vesilind, P.A., 1985. Temperature profiles in thermally decomposing pelletized refuse-derived fuel. Waste Manag. Res. 3, 161-171.

Hilpert, R., 1933. Forsch. Geb. Ingenieurwes. 4, 215.

Kilzer, F. J. and Broido, A., 1965. Speculations on the nature of cellulose pyrolysis. Pyrodynamics. 2, 151-163.

Koufopanos, C.A., Papayannakos, N., Maschio, G., Lucchesi, A., 1991. Modeling of the pyrolysis of biomass particles. Studies on kinetics, thermal and heat transfer effects. Can. J. Chem. Eng. 69, 907-915.

Lehmann, J., Joseph, S., 2009. Biochar for environmental management: science and technology. Earthscan/James & James.

Liliedahl, T.and Sjostrom, K., 1998. Heat transfer controlled pyrolysis kinetics of a biomass slab, rod or sphere. Biomass Bioenergy. 15, 503-509.

Manya, J.J., 2012. Pyrolysis for biochar purposes: A review to establish current research gaps and research needs, Environ. Sci. Technol. 46 (15), 7939-7954.

Morf, P., Hasler, P., Nussbaumer, T., 2002. Mechanisms and kinetics of homogeneous secondary reactions of tar from continuous pyrolysis chips of wood. Fuel. 81. 843-853.

Morten, G.C. and Morten, C.M., 2000. Mathematical model for wood pyrolysis-comparison of experimental measurements with model predictions. Energy Fuels. 14, 791-800.

Park, W.C., Atreya, A., Baum, H.R., 2010. Experimental and theoretical investigation of heat and mass transfer processes during wood pyrolysis. Combust. Flame. 157, 481-494.

Peters, B., Bruch, C., 2003. Drying and pyrolysis of wood particles: Experiments and simulation. J. Anal. Appl. Pyrolysis. 70, 233-250.

Pyle, D. L. and Zaror, C. A., 1984. Heat transfer and kinetics in the low temperature pyrolysis of solids. Chem. Eng. Sci. 39, 147-158.

Sadhukhan, A.K., Gupta, P., Saha, R.K., 2008. Modeling and experimental studies on pyrolysis of biomass particles. Journal of Analytical and Applied Pyrolysis. 81, 183-192.

Wang, Z., Cao, J., Wang, J., 2009. Pyrolytic characteristics of pine wood in a slowly heating and gas sweeping fixed-bed reactor. J. Anal. Appl. Pyrolysis. 84, 179-184.

Williams, P.T., Besler, S., 1996. The influence of temperature and heating rate on the slow pyrolysis of biomass. Renew. Energ. 7, 233-250.

Zheng, X., Fisher, E.M., Gouldin, F.C., Zhu, L., Bozzelli, J.W., 2009. Experimental and computational study of diethyl sulfide pyrolysis and mechanism. Proc. Combus. Inst. 32, 469-476.

Kinetic Modeling of Syngas Production From Bio-Oil Gasification

Fariba Seyedzadeh Khanshan and Richard H. West
Northeastern University, Boston MA

Abstract

Gasification of bio-oil is a desirable process to produce syngas as a renewable resource with no net greenhouse gas emissions. Bio-oil is a mixture of hundreds of chemicals derived from fast pyrolysis of biomass. Production of syngas from bio-oil is usually a high pressure and high temperature process. Optimizing the process conditions (temperature, pressure, residence time, etc.) requires an improved understanding of the chemical kinetics of the thermal cracking reactions involved in bio-oil gasification.

We produced kinetic models for bio-oil gasification, generated using Reaction Mechanism Generator (RMG), an open source software tool that can build detailed kinetic models for hydrocarbon pyrolysis and combustion. Starting with a surrogate bio-oil consisting of ten known species, and reaction conditions (temperature, pressure, reaction time) from the literature, RMG predicts a kinetic model consisting of many thousands of elementary reactions between hundreds of intermediate species.

We describe the challenges involved in using RMG to build a comprehensive model for surrogate fuels containing many large compounds, and how they may be overcome. Finally, we compare the results of the reaction modeling with experimental data on the composition of the syngas produced, and use the model to suggest optimal operating condition for bio-oil gasification.

Introduction

Global warming due to greenhouse gas emissions, increasing energy demands, and decreasing fossil fuel resources have increased interest in renewable fuels. Bio-oil a carbon neutral and renewable fuel, results from the fast pyrolysis of biomass in the absence of oxygen. Biomass can be found as forest residue, animal waste, wood chips and municipal solid waste [1]. The major products of biomass fast pyrolysis under high temperature and pressure are liquid bio-oil, hydrogen, carbon monoxide, carbon dioxide, light hydrocarbons, and solid bio-char [2, 3]. Bio-oil has the potential to be used directly as a fuel, or converted to syngas [4]. Syngas from bio-oil gasification is a valuable chemical mixture and mainly contains carbon monoxide, carbon dioxide, hydrogen and methane. Expensive transportation of biomass as a low density but bulky and polluting material, and problematic direct conversion of biomass to syngas, makes processing via a bio-oil intermediate appealing: bio-oil, a high-density liquid, is easily transportable [5]. The major components of bio-oil are acids, ketones, aldehydes, alcohols, and furans derived from cellulose and hemicellulose, and phenolic and cyclic oxygenates derived from lignin [6, 7]. A typical composition of wood oil includes water (20-30)%, aldehyde (10-20)%, lignin fragments (15-30)%, carboxylic acid, carbohydrates (5-10)% and phenols (2-5)% [5, 8].

There have been several studies of bio-oil gasification, both experimental [9, 10, 11, 12] and modeling [13, 14, 15]. Adjaye *et al.* [16] worked on kinetic modeling of the catalytic conversion of bio-oil in a fixed-bed reactor, calculating yields of products as a function of temperature based on their previous proposed pathways [16], but did not provide a complete kinetic model. To our knowledge there are no complete detailed kinetic models of the bio-oil gasification process.

The efficiency of bio-oil conversion to syngas is highly dependent on operation conditions of the process. Optimization of process conditions requires an improved understanding of the chemical kinetics of thermal cracking reactions involved in bio-oil gasification.

In this work we develop detailed kinetic models for bio-oil gasification using Reaction Mechanism Generator software, RMG. The detailed kinetic models are imported to Cantera, an open source software package for modeling chemical kinetics, thermodynamics, and transport processes, and used to determine the characteristics of bio-oil gasification in batch and CSTR reactors under different operating conditions. The effects of RMG parameters on the model predictions are investigated, as well as the influences of pyrolysis temperature, residence time, and pressure on the syngas yields. The ratios of major gaseous products are compared with experimental works [12] and finally the operation conditions are optimized to maximize syngas yield according to the model.

Reaction Mechanism Generator (RMG)

Usually the number of the species and reactions involved in detailed kinetic models are very large. As manually calculating all the model parameters is impossible, the only reliable way to build large detailed kinetic models is on a computer. In recent years several computational algorithms to build large kinetic models have been developed [17].

RMG, Reaction Mechanism Generator, is an open-source automatic reaction mechanism generator for building large kinetics models [18, 19, 20]. Like other reaction network generators, RMG has to represent chemical species in memory and identify duplicates, create reactions and new species in the network, and estimate the thermochemistry of each species and the rate coefficient of each reaction. In RMG, molecules are represented as graphs, with atoms as nodes and bonds as edges connecting the nodes; standard graph-theory methods are used to identify equivalent graphs and ensure uniqueness. RMG uses "reaction families" to generate all the possible reactions that a species can undergo in the presence of the other species in the chemical mechanism; every reaction family represents a particular type of elementary chemical reaction, such as bond-breaking, or radical addition to a double bond. Each reaction family has a recipe for mutating the graph, and a library of rate expressions for different reacting sites. RMG uses a database of known values wherever possible to find thermochemical data for species, but usually the data are unknown and it estimates parameters using a group contributions method. Thermochemistry estimates are based on Benson's group additivity method [21]. A similar method is used to estimate the rate coefficients for the reactions: functional groups are identified using graph matching and the rates are estimated from a database of rules.

RMG chooses species to include in the model according to reaction flux. It gradually expands a "core" model by adding species from the "edge". The core begins with a trusted seed mechanism of small-molecule chemistry and the initial reactant species (in this case 10 components of bio-oil). All reactions between core species are identified and their rates estimated; any new products are added to the edge. The core is then expanded iteratively, repeatedly adding the edge species with the largest rate of creation, until a user-specified tolerance is reached and the core model is designated complete (for the given tolerance).

There are currently two versions of the RMG software: the original, which is mostly written in Java with some Fortran (RMG-Java) [20], and a more recently developed version written mostly in Python (RMG-Py) [22]. RMG-Java is better established, currently being on version 3.3. RMG-Py is more developmental, but is hoped to offer more flexibility and additional features in the near future. In this work we used both RMG-Java and RMG-Py to build models with similar specifications.

Methods

Kinetic Model Generation in RMG

The kinetic model is generated starting from a 10-component mixture representing bio-oil from wood pyrolysis [23]. The species and their mass fractions are listed in Table 1.

Table 1. Mass fractions of the ten major bio-oil components [23].

Components	Mass fractions (%)
Water	21.10
Hydroxyacetaldehyde	21.77
Acetic acid	9.48
Hydroxypropanone	15.06
Levoglucosan	17.27
Propionic acid	1.25
(5H)-furan-2-one	2.37
Isoeugenol	10.79
Phenol	0.37
Syringol	0.54

The small-molecule chemistry was taken from the GRI seed mechanism, constructed from the GRI-Mech 3.0 methane oxidation model [24]. The inert bath gas (for pressure-dependent rate calculations) was nitrogen. To assess the importance of including pressure-dependent reaction networks in the simulations, RMG jobs were run both without pressure dependence (assuming all reaction rates are in the high-pressure limit) and using the modified strong collision approximation to solve the master equation for unimolecular reaction networks. The termination times of the simulations in RMG were between 0.5 and 20 seconds, and the tolerance was varied from 1 to 0.1.

Bio-oil molecules are large and varied, which usually leads to RMG consuming all the memory in the computer and crashing before the tolerance is reached and the model is complete. Unfortunately, current attempts to circumvent this constraint, e.g. by "pruning" species from the edge, are only partially successful. The only model that ran to completion was from RMG-Py with a very loose tolerance (1.0) that resulted in a small model with only 37 species, that was not an appropriate model for simulation. The largest model we build, on a computer with 4 GB of RAM, contained 307 species and 28197 reactions in the core and 24490 species and 430217 reactions in the edge.

Cantera simulations

The output of RMG is a reaction network consisting of hundreds of species and thousands of reactions, with associated thermodynamic and kinetic data. This is saved in "Chemkin" format, which is also compatible with the open-source alternative Cantera [25]. We used the Python interface to Cantera 2.0 to create simulations of both Plug Flow Reactor (PFR) and Continuous Stirred Tank Reactor (CSTR) conditions for bio-oil pyrolysis, with initial mass fractions taken from Table 1, and with residence times, temperatures, and pressures either corresponding to experimental data [12], or varied as part of an optimization study. The mole fractions of the major gases H_2, CO, CH_4 and CO_2 at the end of the simulation were recorded, as these are the parameters reported by [12]. Many simulations were performed to investigate the effects of varying temperature, residence time, and pressure; simulating CSTR versus PFR; constructing models with or without pressure-dependent reactions; and the influence of model size from a series of incomplete (interrupted) RMG jobs.

Optimization

The optimization of the bio-oil gasification process involves looking for the optimal temperature, pressure, and residence time within given constraints to maximize some objective function. As the pyrolysis of bio-oil is a complex process, there are many possible objective functions. In this work, primarily as proof of concept, we used a very simple objective function to represent syngas yield: the sum of the hydrogen and carbon monoxide mole fractions exiting the reactor. The optimization model is therefore:

$$\begin{aligned} &\textit{Maximize} && F = Y_{H2} + Y_{CO} \\ &\textit{Subject to} && 800\ K < T < 1700\ K \\ & && 0.5\ atm < P < 20\ atm \\ & && 0.5\ sec < t < 30\ sec \end{aligned}$$

The Constrained Optimization by Linear Approximation (COBYLA) method from the SciPy toolkit [26] was used to solve the optimization, with the objective function being evaluated by Cantera [25].

Results

The simulation results from Cantera revealed that choosing a PFR or CSTR reactor doesn't make a significant difference to product yields, with kinetics models from both RMG-Java and RMG-Py. To investigate the influence of model size, a large RMG-Java model was interrupted at three stages of its generation, resulting in incomplete models containing 103 species, 202 species, and 307 species in the core. Again, the PFR and CSTR reactors produced similar results for all three models. Figure 1 reveals how syngas yield, specially H_2 and CO, increases with the model size. The models are quantitatively and qualitatively different, so have not converged with respect to model size. This shows the importance of having a complete and large kinetic model.

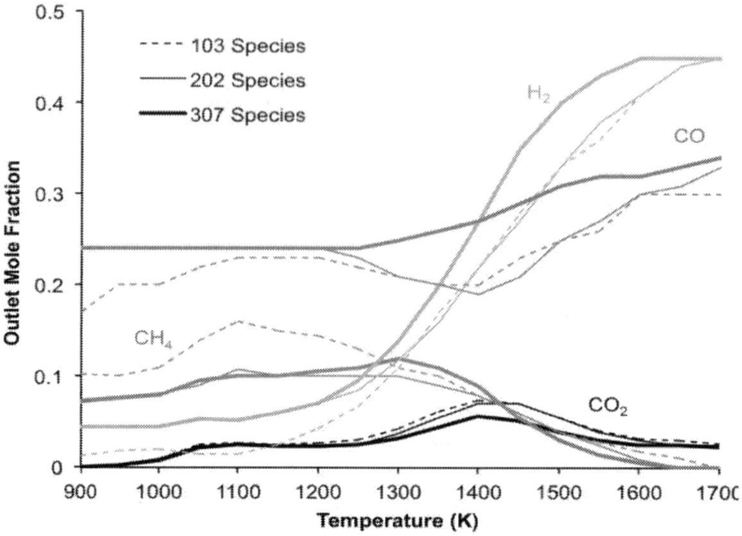

Figure 1. Syngas production varying with model size, from a CSTR with residence time 5 sec.

Figure 2 shows there is an effect of reactor pressure on predicted mole fractions at the reactor exit. However, for this system (unlike small molecule combustion), there doesn't seem to be much difference between results from models *without* pressure dependent calculations (Fig 2a) and with pressure dependent reactions calculated by RMG (Fig 2b). In both models increasing the pressure will increase the syngas yield. The biggest difference is in H_2 and CH_4 yield below 3 atm and from 900 to 1300 K.

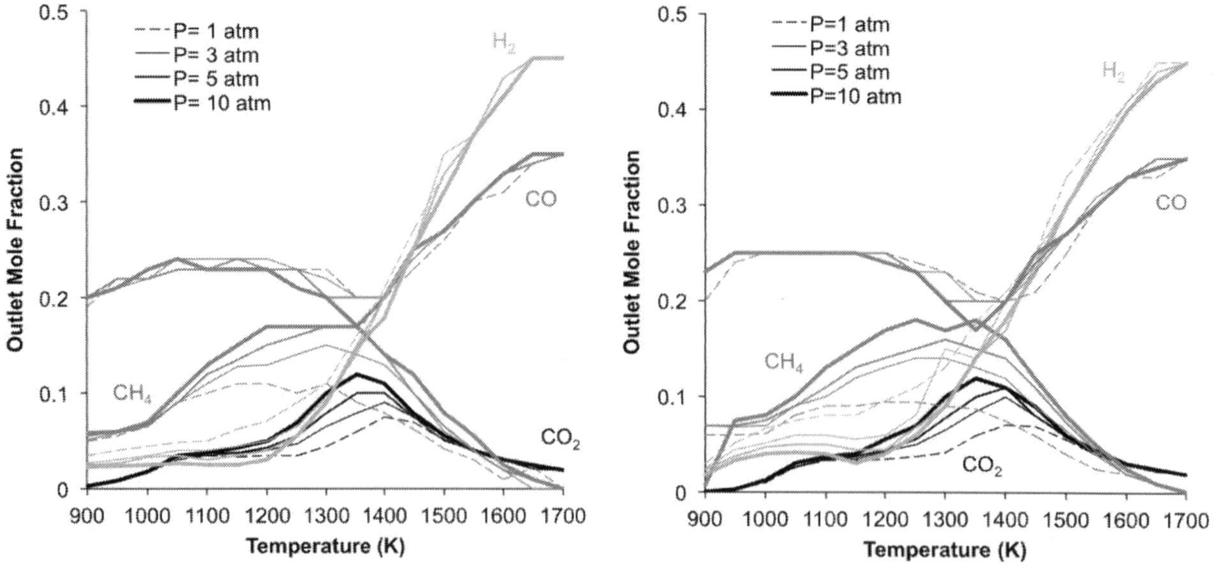

Figure 2. Mole fraction of four major gases at exit of a CSTR with residence time 5 seconds at a range of temperatures and pressures, according to kinetic models built by

RMG- Java (a: left) without pressure-dependence calculations (b: right) with pressure-dependent reaction networks calculated by modified strong collision approximation.

Process optimization predicts the objective function will be maximized at 8 seconds, 1700 K and 19 atm, but there is no available experimental data for bio-oil gasification at high temperature and pressure to compare and validate these results. The available experimental data used in this work is from Zhang et al. [12] for lower temperature bio-oil gasification. They investigated the influence of the temperature and N_2 flow rate on syngas production in a fixed bed reactor at atmospheric pressure between 873 K and 1273 K. They suggested that the optimum temperature for this process in fixed bed reactor is 1073 K and the higher residence time was not effectively helpful in the syngas yield. Figure 3 shows the distribution of the main gas products at in different temperatures from the experimental work and our simulations. There is a difference in CO_2 and CO contents between the experimental and modeling results. Despite this, H_2 and CH_4 are mainly compatible with experimental results.

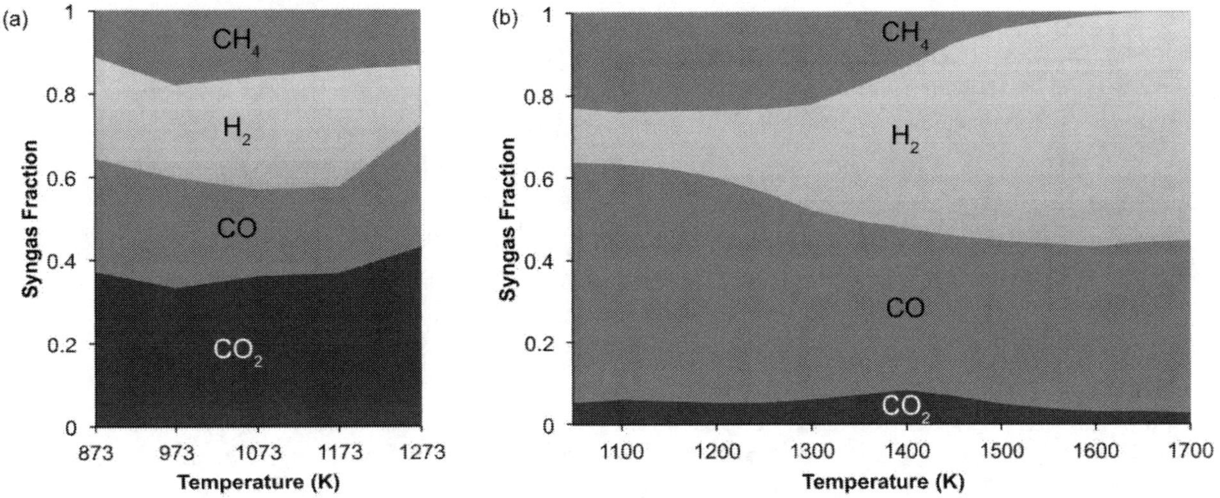

Figure 3. Distribution between four major gas components as a function of temperature, (a) from experimental work by Zhang et al. [12], (b) from modeling results.

Discussion

Simulations suggest that there are not significant differences between kinetic models from RMG-Py and RMG-Java, and that conditions inclusion of pressure dependent reactions doesn't make a remarkable difference in these conditions. As bio-oil contains large molecules unfortunately all the RMG-built models are currently incomplete in both RMG-Java and RMG-Python. The importance of having large and complete models is demonstrated by comparing a series of incomplete models at different sizes: they are significantly different from each other and the larger kinetic models have higher syngas conversion.

Zhang et al. [12] argue the large amount of CO_2 in their experimental work was mainly because the bio-oil contained a lot of organic carboxylic acids, from which carboxyl decomposition was a main source of CO_2. The great underprediction of CO_2 content in the model (Fig 3) could be mainly because the simulation was started with low carboxylic acid content. Despite these differences, there is enough overlap with experimental results in low temperature and pressure gasification to be encouraging. However, optimization results suggest the optimum bio-oil gasification process can be achieved at higher temperature and pressure. As there is no experimental report of bio-oil gasification in high temperature and pressure conditions, further experimental and modeling investigations are needed to improve the understanding.

Conclusion

We have shown that for the incomplete models compared here there are no significant differences between the Python and Java versions of RMG. The largest model we could build in either program had not converged with respect to model size, so a full comparison will require improvements to the memory management of both versions, and increased computing power. A scarcity of kinetic data also causes RMG to make approximate estimates for many of the reaction parameters. Although pressure is predicted to alter product yields inclusion of pressure- dependent kinetics had little effect on these simulation results. At these conditions, none of the simulations were able to predict the fractions of the four major gasses measured by Zhang et al [12] well, but identifying specific deficiencies in the model will require more detailed data. The models, though incomplete, suggest that temperatures higher than available experimental data may be beneficial to the gasification process. Detailed experimental data at higher temperatures and pressures would be very valuable.

References

[1] T. A. Volk, L. P. Abrahamson, and E. H. White, "Developing a willow biomass crop enterprise for bioenergy and bioproducts in the United States," presented at the Bioenergy 2000: Moving technology into the marketplace, Buffalo, NY, 2000.

[2] C. Briens, J. Piskorz, and F. Berruti, "Biomass Valorization for Fuel and Chemicals Production – A Review," *International Journal of Chemical Reactor Engineering* **6 (1)**, 2008.

[3] R. Xu, "*Development of advanced technologies for biomass pyrolysis*," PhD Thesis, University of Western Ontario, 2010.

[4] P. L. Spath and D. C. Dayton, "Preliminary screening-technical and economic assessment of synthesis gas to fuels and chemicals with emphasis on the potential for biomass-derived syngas," National Renewable Energy Laboratory , 2003.

[5] M. Latifi, "Gasification of Bio-oils to Syngas in Fluidized Bed Reactors," PhD Thesis, University of Western Ontario, 2012.

[6] A. C. Basagiannis and X. E. Verykios, "Reforming reactions of acetic acid on nickel catalysts over a wide temperature range," *Applied Catalysis A: General* **308**, 182–193, 2006.

[7] 1. Takanabe, K. *et al.* ",*Steam reforming of acetic acid as a biomass derived oxygenate: Bifunctional pathway for hydrogen formation over Pt/ZrO2 catalysts,"* Journal of Catalysis **243 (2)**, 263–269, 2006.

[8] T. Davidian, N. Guilhaume, E. Iojoiu, and H. Provendier, "Hydrogen production from crude

pyrolysis oil by a sequential catalytic process," *Applied Catalysis B: Environmental* **73 (1-2)**, 116–127, 2007.

[9] S. Panigrahi, S. T. Chaudhari, and N. N. Bakhshi, "Production of synthesis gas/high-Btu gaseous fuel from pyrolysis of biomass-derived oil," *Energy & Fuels* **16 (6)**, 1392–1397, 2002.

[10]S. Panigrahi, A. K. Dalai, and S. T. Chaudhari, "Synthesis gas production from steam gasification of biomass-derived oil," *Energy & Fuels* **17 (3)**, 637–642, 2003.

[11]G. Van Rossum and S. Kersten, "Catalytic and noncatalytic gasification of pyrolysis oil," *Ind. Eng. Chem. Res.* **46 (12)**, 3959–3967, 2007.

[12] M. Zhang, H. P. Chen, Y. Gao, R. X. He, H. P. Yang, and X. H. Wang, "Experimental Study on Bio-oil Pyrolysis/Gasification," *BioResources* **5 (1)**, 135- 46, 2009.

[13] M. R. Hajaligol, J. B. Howard, J. P. Longwell, and W. A. Peters, "Product compositions and kinetics for rapid pyrolysis of cellulose," *Ind. Eng. Chem. Res.* **21 (3)**, 457–465, 1982.

[14]T. R. Nunn, J. B. Howard, J. P. Longwell, and W. A. Peters, "Product compositions and kinetics in the rapid pyrolysis of sweet gum hardwood," *Ind. Eng. Chem. Res.* **24 (3),** 836–844, 1985.

[15] M. L. Boroson, J. B. Howard, J. P. Longwell, and W. A. Peters, "Product yields and kinetics from the vapor phase cracking of wood pyrolysis tars," *AIChE* **35 (1)**, 120–128, 1989.

[16] J. D. Adjaye and N. N. Bakhshi, "Catalytic conversion of a biomass-derived oil to fuels and chemicals I: Model compound studies and reaction pathways," *Biomass and Bioenergy* **8 (4)**, 131-149, 1995.

[17] R. G. Susnow, A. M. Dean, and W. H. Green, "Rate-based construction of kinetic models for complex systems," *The Journal of Physical Chemistry A* **101 (20)**, 3731–3740, 1997.

[18] J. Song, S. Raman, J. Yu, C. D. Wijaya, and G. Stephanopoulos, "RMG: the Next Generation of Automatic Chemical Reaction Mechanism Generator. proceedings," presented at the AIChE Annual Meeting, 2003.

[19] J. Song, "Building Robust Chemical Reaction Mechanisms: Next Generation of Automatic Model Construction Software," PhD Thesis, Massachusetts Institute of Technology, Cambridge MA, 2004.

[20] RMG – Reaction Mechanism Generator. Open-source software package. Version 3.3. http://rmg.sourceforge.net.

[21] S. W. Benson, "Thermochemical kinetics: methods for the estimation of thermochemical data and rate parameters," *Recherche*, 1976.

[22] RMG-Py – Reaction Mechanism Generator in Python. Open-source software package. Development version at *https://github.com/GreenGroup/RMG-Py/*

[23] C. Branca, P. Giudicianni, and C. Di Blasi, "GC/MS characterization of liquids generated from low-temperature pyrolysis of wood," *Ind. Eng. Chem. Res.* **42 (14)**, 3190–3202, 2003.

[24] G. P. Smith, D. M. Golden, M. Frenklach, N. W. Moriarty, B. Eiteneer, M. Goldenberg, C. T. Bowman, R. K. Hanson, S. Song, W. C. Gardiner, Jr., V. V. Lissianski, and Z. Qin, "GRI-Mech 3.0," *http://www.me.berkeley.edu/gri-mech*

[25] Cantera: An object-oriented software toolkit for chemical kinetics, thermodynamics, and transport processes. Version 2.0b. *http://code.google.com/p/cantera*

[26] E. Jones, T. Oliphant, P. Peterson *et al.*, "SciPy: Open Source Scientific Tools for Python", *http://www.scipy.org/*

Insights from Glucose Pyrolysis for Cellulose Pyrolysis

Experimental and predicted kinetics of glucose, the monomer of cellulose, provides quantitative and qualitative insights into the pyrolysis of cellulose. This finding is somewhat surprising, as the monomer does not have the β-1,4-glycosidic linkage between rings that is characteristic of cellulose. More recently, Mettler et al. [1] have reported thin-film pyrolysis data for glucose and cellulose oligomers that quantify how kinetics for glucose (with -OH groups instead of glycosidic linkages) differs from the oligomeric cellulose. However, our earlier work has shown that ab initio modeling of the pericyclic reactions of glucose's C1-O-H group gives a transition state and rate constant that can be related to C1-O-R groups, where R is a hydrocarbon or alcohol group.

Experiments consisted of glucose, cellobiose, cellulose, amylose, and amylopectin pyrolyzed in a thermogravimetric analyzer and a Pyroprobe, where products were analyzed by comprehensive orthogonal gas chromatography/mass spectrometry (GCxGC-TOFMS). Lower reaction temperatures were used to favor the lower-energy reactions predicted by computational techniques, and the GCxGC-TOFMS provided more detailed product speciation. Our work is compared to Patwardhan et al. [2], who pyrolyzed several different glucose-based carbohydrates, and Lin et al. [3], who used analyses by liquid chromatography to suggest that the cellulose chain decomposes by splitting into a cellulose-levoglucosan molecule (cellobiosan, -triosan, etc.) and a shortened cellulose chain.

[1] Mettler, M. S., Paulsen, A. D., Vlachos, D. G., Dauenhauer, P. J. The chain length effect in pyrolysis: bridging the gap between glucose and cellulose, Green Chemistry, 2012, DOI: 10.1039/C2GC35184F.

[2] Patwardhan, Pushkaraj R., Satrio, Justinus A., Brown, Robert C., Shanks, Brent H. Product distribution from fast pyrolysis of glucose-based carbohydrates. J. Anal. Appl. Pyrolysis, 2009, 86 323-330.

[3] Lin, Y. C. Cho, J. Tompsett, G. A. Westmoreland, P. R. Huber, G. W., Kinetics and mechanism of cellulose pyrolysis. J. Phys. Chem. C, 2009, 113, 20097-20107.

(318e) Measuring Trace Components in Syngas Using Gas Chromatography and Ion Chromatography

Patrick Johnston[1], Patrick Woolcock[1], Karl Broer[1] and Robert C. Brown[2], (1)Center for Sustainable Environmental Technologies, Iowa State University, Ames, IA, (2)Mechanical Engineering, Iowa State University, Ames, IA

Measuring Trace Components in Syngas using Gas Chromatography and Ion Chromatography

Patrick A Johnston[1], Patrick Woolcock[1], Karl Broer[1] and Robert C. Brown[1,2]

[1]*Center for Sustainable Environmental Technologies*

[2]*Department of Mechanical Engineering*

Iowa State University

Ames, Iowa, 50011

Specific nitrogen and sulfur containing compounds are unfavorable in upgrading syngas to hydrocarbon based fuels. These compounds are known to poison catalysts at concentrations of sub part-per-million levels in syngas. It is critical to remove and quantify these compounds to provide an efficient and cost effective approach for Fischer-Tropsch synthesis.

The analytical techniques in this study will focus on off-line and on-line gas measurements that were analyzed from a 20kg/hr fluidized bed biomass gasifier with a unique gas cleaning system. Minor constituents of raw and clean biomass gasification syngas were analyzed on-line using a custom designed dual Gas Chromatography (GC) system before and after the cleaning stages. Trace concentrations of hydrogen sulfide, carbonyl sulfide, carbon disulfide, and ammonia were determined with a limit of quantification (LOQ) level of parts-per-billion (ppb) with sulfur and nitrogen chemiluminescence detectors (SCD/NCD).

Hydrogen cyanide concentrations were determined using Ion Chromatography with pulsed amperometric detection (PAD) using a silver working electrode. These samples will be captured in an impinger train using a basic aqueous solution and quantified off-line.

(318f) Rapid Analysis of Feedstock Ash Composition Using Laser-Induced Breakdown Spectroscopy (LIBS)

Tyler Westover, Biofuels and Renewable Energy, Idaho National Laboratory, Idaho Falls, ID, Garold L. Gresham, Interfacial Chemistry, Idaho National Laboratory, Idaho Falls, ID and Richard Boardman, Chemical Engineering, Idaho National Laboratory, Idaho Falls, ID

Inorganic compounds are known to be problematic in the thermochemical conversion of biomass to syngas and ultimately hydrocarbon fuels. The elements Si, K, Ca, Na, S, P, Cl, Mg, Fe and Al are particularly problematic and are known to influence reaction pathways, contribute to fouling and corrosion, poison catalysts, and impact waste streams. Substantial quantities of inorganic species can be entrained in the bark of trees during harvest operations. Herbaceous feedstocks often have even greater quantities inorganic constituents, which can account for as much as one-fifth of the total dry matter.

Current methodologies to measure the concentrations of these elements, such as inductively coupled plasma-optical emission spectrometry/mass spectrometry (ICP-OES/MS) are expensive in time and reagents. This study demonstrates that a new methodology employing laser-induced breakdown spectroscopy (LIBS) can rapidly and accurately analyze the inorganic constituents in a wide range of biomass materials, including both woody and herbaceous examples. This technique requires little or no sample preparation, does not consume any reagents, and the analytical data is available immediately. In addition to comparing LIBS data with results from ICP-OES methods, this work also includes discussions of sample preparation techniques, calibration curves for interpreting LIBS spectra, minimum detection limits, and the use of internal standards and standard reference materials.

(318g) Development and Application of Solid-Phase Microextraction for Pyrolysis and Syngas Analysis

Patrick Woolcock[1], Patrick Johnston[1], Jacek Koziel[2] and Robert Brown[1], (1)Center for Sustainable Environmental Technologies, Iowa State University, Ames, IA, (2)Agricultural and Biosystems Engineering, Iowa State University, Ames, IA

Thermochemical conversion is a well-established approach to converting biomass and waste into valuable products. New developments in technologies such as gasification and pyrolysis are making these techniques increasingly viable as a source of renewable fuels and chemical production. As these processes become more complex and demanding, the ability to characterize product streams becomes more important. Rapid and reliable quantification of key compounds is essential to continuing process development and ensuring optimal performance of commercialized installations.

The primary goal of this work is to develop and demonstrate a new technique for analyzing process gas streams at elevated temperatures. A recently developed technique using solid-phase microextraction (SPME) is extended from the proof-of-concept stage to application in pilot scale gasification and pyrolysis processes. SPME is a sample extraction and preparation technique that is growing in popularity because it eliminates several complicated and costly stages of conventional sample analysis. Using a small needle coated with an extraction phase, sample extraction and preparation are combined into a single step using one device. [i] Directly injecting this coated needle (a.k.a. fiber) directly into the analytical equipment facilitates rapid sample analysis using gas chromatography mass spectrometry (GC-MS) or flame ionization detection (GC-FID).

This work applies the SPME technique for the first time to the quantification of trace tar compounds in a 20kg/h biomass gasification and gas cleaning system. Raw syngas produced from gasification contains numerous contaminants that must be largely removed before it can be used as a fuel or chemical feedstock. Particularly problematic are tars derived from incomplete thermochemical conversion, which condense as syngas cools and cause fouling of pipes and equipment. Conventional tar analysis, which is based on exhaustive extraction and gravimetric measurements, requires many hours to collect and analyze a single sample. This method is also inadequate for analyzing lower tar concentrations that exist downstream of cleaning devices, which may still be problematic to downstream processes like catalysis. After calibrating the SPME technique on a lab scale model-tar/syngas system, its performance was evaluated on a pilot scale for determining trace tar concentrations downstream of an oil-based tar scrubber.

The SPME method is also applied in this work to directly measure vapor compounds during pyrolysis. The liquid bio-oil product created via pyrolysis is composed of hundreds of compounds. Many of these compounds experience rapid changes during the conversion process and when condensed into bio-oil, making them difficult to collect and analyze. The SPME technique makes it possible to capture these compounds at different stages of the pyrolysis reaction and directly analyze them using GC-MS or GC-FID. This work has been focused on the detection of levoglucosan, a prominent sugar product from pyrolysis. Measuring the different levels of levoglucosan and other intermediate vapor compounds provides strong indicators of process reaction chemistry and the ability to optimize the process for desired products.

---

[i] Pawliszyn J. Handbook of solid phase microextraction. 2009, Beijing: Chemical Industry Press. 410 p

(322a) Magnetic Separation of Algal for Biofuel Production

Jeffrey J. Chalmers, William G. Lowrie Department of Chemical and Biomolecular Engineering, The Ohio State University, Columbus, OH, Wei Xue, Chemical and Biomedical Engineering, The Ohio State University, Columbus, Jie Xu, Chemical and Biomolecular Engineering, The Ohio State University, Columbus, OH, Maciej Zborowski, Department of Biomedical Engineering, Cleveland Clinic, Cleveland, OH and Brad Postier, Phycal, St. Louis, MO

Phycal is an algal biofuels company focusing on a hybrid photoautotrophic/heterotrophic open pond system using freshwater algal species. To **potentially reduce costs associated with algal biofuels production and/or improve the quality of intermediate steps through improved efficiencies downstream, Phycal is d**eveloping strains that can transport and store significantly more iron than wild-type strains. The potential reward is that they can out-compete contaminating species (*e.g*., algae, bacteria, protozoa, and *etc*.) for the available iron. Once stored, the iron is in the form of magnetite inside ferritin complexes that are paramagnetic. Thus, in addition to the advantage of iron scavenging, these strains can be manipulated magnetically with the potential of magnetic separation.

In this presentation we will present progress on our Phase II SBIR work to 1) characterize the magnetic susceptibility of genetically engineered and intrinsic magnetic susceptibility as a result of culture conditions, and 2) pilot scale separation systems to "dewater" algae cultures.

(322b) A Novel Taylor-Couette Photobioreactor for Energy Efficient Micro Algae Cultivation

Bo Kong, Department of Chemical and Biological Engineering, Iowa State University, Ames, IA and **R. Dennis Vigil**, Iowa State University, Ames, IA

Numerous studies have demonstrated that algal growth rate and light utilization efficiency is dramatically impacted by the cycling in the intensity of light exposure as cells move between well-illuminated photic regions near a photobioreactor external surface and the darker regions of the reactor. In particular, when the characteristic frequency of the light/dark cycles is greater than approximately 1 Hz, the "flashing light effect" can boost cell growth rates 50~100% compared to growth rates obtained without light/dark cycling. However, high frequency light/dark cycles are not easy to achieve in large-scale reactors and may be one of the reasons for low yields. In commonly used flat panel bubble column and airlift reactors, the random nature of the agitation provided by the gas bubbles produces a wide spectrum of random light/dark frequencies. In the work presented here, we make use of robust and coherent hydrodynamic structures known as Taylor vortices that arise in the annular region of a Taylor-Couette flow cell (concentric cylinders with the inner cylinder rotating) in order to rapidly and periodically shuttle algal cells between photic and non-photic regions of the reactor. Because the fluid agitation in a Taylor vortex algal reactor is not dependent on bubble-induced turbulence, much lower carbon dioxide flow rates can be used and the carbon capture efficiency is significantly higher than in conventional algal reactors. Here, we present experimental results (using Chlorella vulgaris) and analyses based upon computational fluid dynamics simulations to demonstrate that even at relatively low light-dark cycling frequencies, the Taylor vortex algal reactor offers higher biomass yield than can be achieved in corresponding bubble column experiments. Furthermore, we consider important factors for design and scale up of Taylor vortex algal reactors.

(322c) Effect of Growing Conditions On Algal Carbohydrate to Butanol Production

Alice C. Jernigan, Chemical Engineering, University of Arkansas, Fayetteville, AR and Christa N. Hestekin, Ralph E. Martin Department of Chemical Engineering, University of Arkansas, Fayetteville, AR

There are several environmentally sound reasons to use algae as a feedstock for biofuels. The fact that they are readily abundant, can absorb unwanted elements out of waste water, and can use carbon dioxide as a carbon source are just a few of the reasons that algae are being examined as one of the most likely medium to replace petroleum fuels. An issue of some concern is that microalgae have a high rate of growth in the summer and lower rates in the winter, which can cause a depletion of feedstock for a bioreactor. We are using a species of *Clostridium* bacteria to convert algal sugars to butanol in a bioreactor. In order to facilitate localized bioreactors that can operate year-round, we examined dried and stored stocks of algae taken during a six month harvesting period and examined the effects of climate, location, and processing on the amount of carbohydrates extracted from the algae and converted into butanol.

(322d) Photoautotrophic Growth and Lipid Production Kinetics of the Microalgae Scenedesmus Dimorphus

Joanne Belovich, Jacob Schwenk, Christopher Hardulak and John Van Blargan, Cleveland State University, Cleveland, OH

The design and optimization of culture systems for microalgae for biofuel production are complicated by the conflicting requirements for high biomass productivity and high oil content. Reactor models that are needed for this optimization rely on models of the algae growth and lipid productivity kinetics. Unlike heterotrophic growth of most microorganisms, photoautotrophic growth of algae can have multiple limiting nutrients, further complicated by potential light limitation. Factorial experimental designs were conducted to identify the nutrients, and their concentration ranges, that are potentially limiting for growth and oil production of the microalgae *Scendesmus dimorphus*. The individual and interactive effects of six groupings of the components of 3N-BB medium on growth rate, lipid content, and total biomass yield of *S. dimorphus* were determined. Trace metal and vitamin concentrations were reduced to 1/6 the level of 3N-BB medium without adversely affecting growth rates and biomass concentration, while concentrations of $CaCl_2$ and K_2HPO_4/KH_2PO_4 were reduced to 1/10 that of 3N-BB without adversely affecting biomass and lipid concentrations. Lipid productivity was maximized at the lowest $NaNO_3$ concentration (1/10 that in 3N-BB) and independent of $MgSO_4$ concentration, while the interaction of these two chemicals enhanced biomass concentration. Initial substrate experiments were then conducted to determine in greater detail the effects of magnesium sulfate, sodium nitrate, and potassium phosphate on biomass growth. Kinetic models for growth rate and substrate consumption were developed from the data and validated. These fundamental kinetic models can then be used in a variety of reactor styles and configurations in order to identify the system and operating parameters

(322e) Use of Sodium Bicarbonate for Efficient Carbon and Water Management for Autotrophic Microalgae Cultivation in Open Pond System

Jinsoo Kim, School of Energy, Environmental, Biological and Medical Engineering, University of Cincinnati, Cincinnati, OH, Joo-Youp Lee, School of Energy, Environmental, Biological, and Medical Engineering, University of Cincinnati, Cincinnati, OH and Ting Lu, Wastewater Treatment Division, Metropolitan Sewer District of Greater Cincinnati, Cincinnati, OH

Dissolved inorganic carbon including carbon dioxide(aq) and bicarbonate ion (HCO_3^-) is an essential carbon source for the autotrophic growth of microalgae. However, the use of CO_2 gas only cannot provide a sufficient amount of dissolved inorganic carbon concentrations in a narrow pH window suitable for the growth. In this study, sodium bicarbonate was used to increase dissolved inorganic carbon concentrations in conjunction with CO_2 gas. Added CO_2 gas was able to suppress the increased pH values as a result of bicarbonate ion uptake and to supply additional dissolved inorganic carbon. The fates of inorganic carbon and sodium ion were studied by taking mass balances for the two components. The effects of sodium ion on CO_2 gas absorption and algal productivity will be presented for sustainable water reuse and autotrophic microalgal culture in a typical open pond system. A preliminary economic analysis of the use of sodium bicarbonate based on the experimental results will also be presented.

(322f) Transgenic Expression of a Bacterial Exo-Acting Intracellular α-Amylase in the Chlamydomonas Reinhardtii Chloroplast

Xiaoqing Wang[1], Barbara Sears[2], Yan Susie) Liu[3] and Wei Liao[1], (1)Biosystems and Agricultural Engineering, Michigan State University, East Lansing, MI, (2)Plant Biology, Michigan State University, East Lansing, MI, (3)Department of Biosystems and Agricultural Engineering, Michigan State University, East Lansing, MI

Microalgae with their relatively high starch and lipid contents are one of the most promising non-food-crop-based feedstocks for biofuels production. To date, most studies have focused on accumulating algal lipids for biodiesel production and some have focused on hydrogen generation. Very little attention has been paid to simultaneously converting algal carbohydrate and lipids into biofuels. Although some algal species are able to accumulate significant amounts of carbohydrates (mainly starch) and lipid, their rigid cell wall retards the extraction of those carbon-rich compounds. In order to overcome this barrier, this study constructed an amylase-producing strain of the unicellular green alga *Chlamydomonas reinhardtii*, with the goal of converting algal starch into soluble sugars *in vivo*, which would facilitate their recovery. An exo-acting intracellular thermophilic α-amylase AmyB from the hyperthermophilic bacterium *Thermotoga neapolitana* was used as the amylase gene of choice. The AmyB gene was synthesized with codon usage optimized for the *C. reinhardtii* chloroplast; it was cloned into chloroplast expression vector p666 to yield plasmid p666AmyB. The vector p666AmyB was inserted into the site between the *psaB* and *atpA* genes on the chloroplast genome of *C. reinhardtii* by micro-particle bombardment, without interruption of chloroplast gene function. PCR analysis indicated that *AmyB* was integrated into the chloroplast genome of the transformants, and western blotting demonstrated the AmyB protein accumulation. The activity of AmyB can reach 17.4 U/g dry biomass. Thus, α-amylase was successfully expressed within the chloroplast, accomplishing an important step towards improving the ability to harness and harvest the carbon-compounds synthesized by micoalgae.

(327a) Hydrogenation of Fast Pyrolysis Oils From Proteinaceous Biomass

Charles A. Mullen[1], Akwasi A. Boateng[1] and Stephen Reichenbach[2], (1)Eastern Regional Research Center, Agricultural Research Service, USDA, Wyndmoor, PA, (2)University of Nebraska, Lincoln, NE

Fast pyrolysis oils from biomass materials with high protein content have been recently shown to be in most cases more stable and partially deoxygenated compared with those from mostly lignocellulosic biomass (wood, grasses) due to nucleophilic substitution of nitrogen for oxygen. However, in order for them to be used as transportation fuels or petroleum refinery feedstocks these pyrolysis oils still must be upgraded to reduce their heteroatom (N, O, S) content. Because the composition of these proteinaceous pyrolysis oils differs greatly from those from lignocellosic feedstocks, their behavior in various upgrading steps will be different. Therefore, we studied the hydrogenation of fast pyrolysis oils from the presscake of pennycress seeds as a model for the hydrotreating of proteinaceous fast pyrolysis oils. Several heterogeneous precious metal catalysts were screened in a batch mode using a Parr reactor, with standard conditions of 2000 psi H_2 at 300 °C used to hydrogenate the bio-oil. The products analyzed by several methods including elemental analysis, two-dimensional GC/MS and NMR to characterize the chemical transformations that occurred and compare the products with those from hydrogenation of lignocellulosic (wood) pyrolysis oils. Furthermore, to deduce the selectivity of the catalysts for some of the individual reactions taken place in the complex system, the two-dimensional GC-MS data was subjected to advanced data comparison methods for the determination of reliable peaks for discrimination among the products of the different catalysts. This allowed for comparison of corresponding chemicals and chemical families within the products from the various catalysts and identification of the catalytic selectivites for various reactions occurring in the mixture.

(327b) Biorefinery - Liquefaction of Pyrolysis Char

Roland Feiner[1], Hannes Pucher[1], Nikolaus Schwaiger[1], Peter Pucher[2] and Matthaeus Siebenhofer[1], (1)Institute of Chemical Engineering and Environmental Technology, Graz University of Technology, Graz, Austria, (2)R&D, BDI - BioEnergy International AG, Grambach, Austria

The energy policy in Europe in the post-Fokushima-era is faciing a crucial decision whether to stick to conventional energy carriers or to utilize secure renewable energy sources. According to studies of Shell Germany [1] we will anyway need the same amount of energy obtaind from coal today in the field of solar energy and new biomass applications in 25 years to fulfill the future world energy demand.

Additionally the European fuel market is facing the challenge to meet the directives of the European Commission to increase the volume of biofuels within the field of automotive fuels to 10% by 2020 [2]. First generation biofuels like biodiesel will not be able to meet this target, since admixture to standard fuels is limited (EN 590) [3]. Therefore new complementary technologies are required to aim this goal. One of the main suggestions is whole plant usage of wood or straw which is not in competition with the food industry. Gasification of the feedstock and liquefaction by Fischer-Tropsch Synthesis is an established but complex route with an overall efficiency of 50% [4].

An alternative route of producing liquid energy carriers from biomass is Liquid Phase Pyrolysis.Our biorefinery concept (BiomassPyrolysisRefinery) is a two-step approach. In the first step lignocellulose biomass like wood or straw is converted into liquid and solid products. Apart from biocrude condensibles of medium quality combustion properties, biochar is a major product of Liquid Phase Pyrolysis with an amount of up to 80% by mass. In the second step the solid and liquid products are upgraded by biochar hydrogenation and by pyrolysis oil hydrodeoxygenation [5]. Preliminary research work in this field of Liquid Phase Pyrolysis was accomplished at our institute with promising results [6], [7]. Focus of this project is processing of the biochar. Biochar is highly accessible to chemical conversion by hydrogenation including intermediates for fuel synthesis due to its pore structure.

Direct Liquefaction in a slurry reactor is one way for further processing of biochar. Opposite to coal liquefaction, which was applied on an industrial scale already, direct hydrogenation of biochar-slurry in the presence of a catalyst has not been reported. Direct liquefaction is a heterogeneous catalytic process. Several catalysts and coal based feed were investigated [8]. Viability has been proven. A systematic investigation and optimisation of the process has been carried out.

Prior to liquefaction biochar, which has a process dependent particle size distribution is milled in a centrifugal mill to 200 microns and dried in a vacuum drier. The biochar is then slurried with carrier and catalyst. The reactor system is heated and connected to a hydrogen source. Overall kinetics of liquefaction is deduced from monitoring hydrogen consumption at constant pressure or donor solvent consumption by GC-MS.

Experiments with hydrogen donors at 450°C showed biochar conversions of more than 90% by mass.The formation of a wide range of alkanes <C30 as well as cycloalkanes is observed. Latter substances are assumed to be derivatives of the lignin structure. Tar-like liquids, asphaltenes and pre-asphaltenes have been detected. Asphaltenes and pre-asphaltenes are intermediates during conversion of coal to oil by hydrogenation [9]. Yields and composition of the products are determined by multi-stage extraction in a fluidized bed extraction unit and analysed by NDIR continuous gas analysis, ATR-IR, TGA and GC-MS.

[1] **German Shell AG.** *World Energy Consumption to 2060: Scenario sustainable growth*. Presentation. 1999.

[2] **2009/28/EC, EU Directive.** *Directive on the promotion of the use of energy from renewable sources.* Strasbourg : European Union, 2009.

[3] **Standard EN 590.** *Automotive fuels - Diesel - Requirements and test methods.* Brussles : European Committee for Standardization, 2008.

[4] **Coren, Company.** *Information & Press. [Online] 24 01 2011.* http://www.choren.com/informationen/faq/.

[5] **Mercader, F. M. et al.** *Hydrodeoxygenation of pyrolysis oil fractions: process understanding and quality assessment through co-processing in refinery units*. Energy & Environmental Science. 2011, 4, pp. 985 - 997.

[6] **Mertlitz, V.** *Dissertation: Flüssigphasen-Pyrolyse biogener Edukte.* Graz : TU Graz, 2010.

[7] **Nikolaus, Schwaiger.** *Dissertation: Reaktionstechnische Analyse zur Optimierung der Flüssigphasenpyrolyse.* Graz : TU Graz, 2011.

[8] **Sol, Weller and et.al.** *Coal Hydrogenation Catalysts Batch Autoclave Tests*. Industrial & Engineering Chemistry. 42, 1950, 330-334.

[9] **Kaneko, et al.** *Ullmann's Encyclopedia of Industrial Chemistry - Coal Liquefaction.* Weinheim : Wiley-VCH, 2005.

(327c) Process Simulation of a Bi-Phasic Reaction: Hydrogenation of p-Hydroxybenzaldehyde

C. E. Odafin[1], S. B. Gebreyohannes[1], T. N. Pham[2], Jimmy A. Faria[2], B. J. Neely[1] and K. A. M. Gasem[1], (1)School of Chemical Engineering, Oklahoma State University, Stillwater, OK, (2)School of Chemical, Biological and Materials Engineering, University of Oklahoma, Norman, OK

Bi-phasic catalytic reactions provide an efficient method for the conversion of pyrolysis oil to usable biofuel. The simulation of the hydrogenation of p-hydroxybenzaldehyde was used as a prototype to develop a bi-phasic reactor model describing both reaction kinetics and phase separation. The effects of phase behavior modeling on bi-phasic reactor model predictions were then investigated. Specifically, case studies were conducted to examine the prediction accuracy of the Non-Random, Two-Liquid (NRTL) activity coefficient model equipped with parameters from: UNIversal Functional Activity Coefficient (UNIFAC) group contributions, Quantitative Structure–Property Relationships (QSPR) generalized NRTL parameters (NRTL-QSPR), and regressed NRTL model parameters (NRTL_R). Further, sensitivity analyses were conducted to evaluate the influence of the kinetic parameters on reactor model predictions.

Kinetic data on the p-hydroxybenzaldehyde system from the OU CIRE research group was used in the simulation. Aspen Plus and Visual Basic for Application (VBA) were used to model the biphasic reaction process, accounting for both the reaction and phase separation. A user-defined MS Excel (VBA) reaction model was designed, and Aspen Simulation Workbook (ASW) was used to link the reaction model to the phase separation model on Aspen Plus. The NRTL model parameters estimated from the NRTL-QSPR model demonstrate better phase equilibria property predictions than those generated using the UNIFAC–predicted models parameters. The sensitivity analyses indicate that the adsorption and deactivation constants have significant impact on the bi-phasic model predictions of p-hydroxybenzaldehyde hydrogenation.

(327d) Supercritical Upgrading of Algae Paste

Klaus Hellgardt, Christian Richard and Bhavish Patel, Chemical Engineering, Imperial College London, London, United Kingdom

Biodiesel is commonly produced via transesterification of plant derived oils (rape, palm). However, biodiesel produced from algae paste has the potential to be more lucrative due to the intrinsically higher productivity of algae. The current method of only transesterifying extracted algae oil is not particularly economical as it requires water removal, drying and extraction processes prior to transesterification. Thus direct supercritical treatment of wet algal paste is suggested to convert the lipids but also the algal biomass itself into a useful biofuel. Algal biomass might be amenable to this processing technique because algae do not contain lignocellulose and therefore could yield a hydrocarbon fuel with much lower oxygen content. The kinetics of the transformation of algal paste is followed using GC MS, ^{13}C NMR and ^{1}H NMR to observe functional group distribution/concentration within the product mixture. Batch stainless steel and quartz reactors are employed to evaluate the effect of catalyst addition on the rate and selectivity of algae paste conversion. First results of the continuous processing of algae paste in a flow reactor to yield bio-crude will be shown.

(327e) Catalytic Conversion of Pinyon-Juniper Into Upgraded Biooils

Bhuvanesh kumar Yathavan and Foster Agblevor, Biological Engineering, Utah State University, Logan, UT

Pinyon-juniper woodlands occupy over 47 million acres of Southwestern USA including Nevada, Utah, Arizona, New Mexico, Colorado and southwestern California. This consists of various species of pine and juniper. In the State of Utah pinyon-juniper occupies 20% of the land and the dominant species are Utah juniper (*Juniperus utahensis*) and pinyon pine (*Pinus edulis*). The pinyon-juniper woodland is becoming a major problem in Utah because of forest fires. These woody species are being harvested to reduce forest fires but there are no immediate applications for this biomass. A potential application of this biomass is to convert it into thermochemical fuels and chemicals. Samples of pinyon-juniper harvested by the US Bureau of Land Management were obtained for these studies. The biomass mixture which consisted of bark, leaves, and wood were ground to pass 2-mm mesh screen. The samples were pyrolyzed in a bench scale fluidized bed pyrolysis reactor in the presence of sand, red mud, and zeolite catalysts. The yield of biooil produced with sand was highest (63 wt%), but it also had the highest viscosity of almost 500 cP at 40 C. The red mud and zeolite catalysts produced less biooils (40 wt%), but the viscosity of the oil was considerably lower (200 CP) compared to the biooil from the sand. Both the zeolite and the red mud produced more gases than the sand pyrolysis medium. The viscosities of pinyon-juniper biooils were much higher those produced from hybrid poplar wood and this was attributed to the higher nitrogen content of the feed emanating from its higher leaf content.

(327f) Comparison of 1- and 2- Step Biodiesel Production From Activated Sludge Using Supercritical Methanol and Economic Analysis

Adebola T. Coker[1], Rafael Hernandez[1], Emmanuel D. Revellame[1], William Todd French[2], Alexei V. Iretskii[3], William Holmes[1] and Mark G. White[1], (1)Dave C. Swalm School of Chemical Engineering, Mississippi State University, Mississippi State, MS, (2)Chemical Engineering, Mississippi State University, Mississippi State, MS, (3)Chemistry, Lake Superior State University, Sault Sainte Marie, MI

In recent times, demand of renewable fuels like biodiesel has escalated due to spurs from the 2007 Renewable Fuel Standard that mandates the production of 1 billion gallons of biodiesel by 2022, as well as environmental and energy security concerns. One of the reasons why biodiesel is not yet cost-competitive with petroleum fuels is due to the high cost of feedstock. A high-potential feedstock for biodiesel production is activated sludge obtained from wastewater treatment plants, which is relatively cheaper to use. Researchers have shown that the fats present in the sludge can be transformed to biodiesel. However, drying of sludge prior to oil extraction and reaction is a major operating cost of this process since it is so high in water content. Thus, the most efficient process to convert the oils from activated sludge to biodiesel will need to have high water tolerance. In this case, supercritical methanol works suitably as a reactant and catalyst.

A study was conducted to determine the yield of biodiesel produced from reacting wet sludge using only supercritical methanol. Activated sludge was obtained from a Tuscaloosa, AL wastewater treatment plant and an approximate composition of 93% water and 7% solids was reacted with methanol in a 450 ml Parr® reactor using a methanol: solids ratio of 30:1 (mass basis) at 300 °C for 3 - 12 hours. Analysis of the product by Gas Chromatography-Mass Spectrometry (GCMS) showed a number of fatty acid methyl esters (biodiesel) such as methyl linoleate and methyl palmitoleate; and an average biodiesel yield of 7 %. A series of experiments were conducted for the direct transesterification (1-step) optimization based on temperature, reaction time and methanol volume for the production of biodiesel from wet sludge. For this, an oleagineous (oil-storing) yeast - *Rhodotorula glutinis* was used as a model system for sludge (to reduce inconsistencies in sludge batch variations) and gave biodiesel yields as high as 21 %. Furthermore, the 2-step method of hydrolyzing the fats to fatty acids followed by esterification was also examined for changes in yields. This study will show the results of both methods compared and the economic analysis for the prospect of scaling up.

(327g) Upgrading of Pyrolysis Oil by Hydroprocessing in a Packed Bed Flow Reactor

Venkata K. Penmetsa, Dept of Forest Products, Mississippi State University, Statkville, MS, Divya Parapati, Department of forest products, Mississippi state university, Starkville, MS and Philip Steele, Department of Forest Products, Mississippi State University, Mississippi State, MS

Raw bio-oil cannot be combusted as transportation fuel directly because it has a number of deleterious properties such as high acidity, high water content, lower heating value and variable viscosity over time. The negative bio-oil properties largely result from the high proportions of oxygenated compounds present in the bio-oil. For the production of a renewable fuel pyrolysis oil must be upgraded. Hydrodeoxygenation is one upgrading technology applied to produce viable hydrocarbons for transportation fuels. Previous hydrodeoxygenation studies have provided considerable information about upgrading of bio-oil, but further understanding of this route to hydrocarbons required. Information on the effect of bio-oil quality, various process parameters i.e. temperature, flow rate, pressure, effect of catalyst in order to achieve desirable product in terms of quality and quantity, and reduction of catalyst deactivation. In our current study we are hydroprocessing the bio-oil in two stages. For the stage 1 hydrotreating process a batch reactor is utilized. For the hydrocracking stage the bio-oil is treated by a continuous reactor. This route simulates a future method by which we will produce stage 1 hydrotreated bio-oil by a simpler method than is currently available via hydrotreating. This process is being developed concurrently with the hydrodeoxygenation research. The studies for stage 2, hydrocracking was carried out in a continuous reactor using the hydrotreated bio-oil from the batch process to ease the hydrocracking reaction in the continuous reactor by avoiding catalyst coking and continuing the runs for longer time with better yield. The results obtained are analyzed for acid value, oxygen content, heating value. Different process parameters and catalysts were studied to obtain better yield with minimum deactivation of the catalyst.

(350a) Treated Waste Water Source Heat Pump Design and Improvement

Huajun Yuan, Anum Raza, Bo Song and Sohail Murad, Chemical Engineering, University of Illinois at Chicago, Chicago, IL

A water source heat pump is a central heating/cooling system that pumps energy to or from water to provide heating or air conditioning. It is an efficient alternative to conventional heat, ventilation and air conditioning (HVAC) systems. Waste water treatment facilities provide easy access to large bodies of water which can be readily harvested. Such system generally lead to an almost 50% reduction in energy cost, and can therefore play an important role in maintaining and improving a sustainable ecosystem. The high flow rate and almost constant water temperature in such facilities make them especially attractive energy sources for heat pumps. We have collaborated with Metropolitan Water Reclamation Department of Greater Chicago (MWRDGC) to design and test different layouts of water source heat pump including open and close loops systems. The test data obtained will be used to increase the efficiency and lower maintenance costs, which will significantly reduce the cost of energy usage at the site. The test system has been installed at the Kirie Waste Water Treatment Facility in Des Plaines, IL. The system will in the future be further improved by using Phase Change Materials (PCMs) to take advantage of large latent heat of (about 250kJ/kg) PCMs to store energy during off-peak hours and release it during peak hours. This will reduce overdesign of the system and lower the initial capital cost. The entire process has also been simulated using the Aspen Plus simulation package and these simulations will allow us to further increasing the efficiency in the future as well. This project is a modest contribution to the development of more sustainable energy systems.

(350b) Self-Heat Recuperation and Its Applications in Chemical Industries

Yasuki Kansha, Yuping Liu, Yui Kotani, Hiroyuki Mizuno, Masanori Ishizuka and Atsushi Tsutsumi, Institute of Industrial Science, The University of Tokyo, Tokyo, Japan

The combustion of fossil fuels for heating produces a large amount of carbon dioxide (CO2), which is the main contributor to the global greenhouse gas effect. Hence, the reduction of carbon dioxide (CO2) emission and the reduction of energy consumption for heating has nowadays become a very important issue in the efforts to suppress global warming. Heat recovery technology such as pinch technology, which exchanges heat between the hot and cold streams in a process, has been applied to thermal processes to reduce energy consumption. Although the net energy input seems to be reduced by using exhausted heat as the additional heat, the heat is also provided by the combustion of fossil fuels, leading to exergy destruction during energy conversion from chemical energy to heat.

Recently, Kuchonthara and Tsutsumi have proposed an energy and exergy recuperative integrated gasification power generation system and a design method for the system. Authors following on this concept and developed self-heat recuperation technology (SHRT) based on exergy recuperation. The most important characteristic of this technology is that the entire process stream heat can be recirculated into a process without any heat addition, leading to marked reduction of exergy and energy savings for the process.

In this research, the process design methodology based on self-heat recuperation is illustrated with the theoretical minimum energy required. Then, the energy required for processes based on self-heat recuperation and that for conventional processes with heat recovery in chemical industries are compared. From these results, the direction of energy saving process designs with fundamental developments is shown.

(350c) Simulation of Self-Heat Recuperation Using Magnetocaloric Effect

Yui Kotani, Yasuki Kansha and Atsushi Tsutsumi, Institute of Industrial Science, The University of Tokyo, Tokyo, Japan

The reduction of carbon dioxide (CO_2) and the fossil fuel consumption has been one of our major concerns over past few decades. In chemical processes that involve heating, heat is mainly provided by fossil fuel combustion associated with great exergy loss. Recently, self-heat recuperation technology based on the exergy recuperative heat utilization principle has been developed, which can recirculate all heat of the process stream by providing work instead of heat generated by fuel combustion. By compressing gas at its highest temperature in the process for heat elevation, the energy consumption for heating can be reduced drastically compared to conventional heat-recovery technology.

A novel magnetic heat circulator for thermal process based on self-heat recuperation technology is proposed. In magnetic heat circulator, process heat is recirculated by magnetocaloric effect of ferromagnetic materials through cyclic magnetization and demagnetization. The ferromagnetic material is adiabatically magnetized/demagnetized at the highest/lowest temperature of the process to create a temperature difference required for heat exchange. All heat is recirculated inside the system without heat addition. The theoretical energy consumption of magnetic heat circulator was calculated in terms of the temperature-entropy diagram. The simulation result shows large potential for energy saving in thermal processes.

In this research, because the magnetic heat circulator poses a complex problem of heat transfer, magnetic field, and physical properties of ferromagnetic material, a numerical 1D model of magnetic heat circulator has been constructed. The magnetocaloric effect (MCE), varying magnetic field, thermal conduction and heat transfer has been implemented in the governing equation.

(350d) A Novel Seawater Desalination Based On Self-Heat Recuperation

Hiroyuki Mizuno, Yasuki Kansha and Atsushi Tsutsumi, Institute of Industrial Science, The University of Tokyo, Tokyo, Japan

The demand for potable water has been increasing recently due to the rise in population and industrialization of developing country. There are two well-known desalination processes, thermal and membrane processes. Thermal process which separates potable water from seawater by evaporation can be divided into several types such as Multi Stage Flash and Multi Effect Desalination. Thermal desalination consumes large energy due to the heat requirement of water evaporation. But there are also merits; the contents of feed water do not strongly affect the desalination operation, and the salinity concentration of the product water is relatively low.

On the other hand, membrane processes such as Reverse Osmosis (RO) which separates portable water from brine water by semipermeable membrane is the most widely used desalination technique. Its reason is merely because it consumes less energy than conventional thermal process. Thus, most newly desalination plant is constructed based on RO process. However, when feed seawater contains large amount of dissolved solids and impurities, they damage the membrane. For this reason RO requires a greater degree of pretreatment than the thermal desalination process.

Summing up, thermal desalination does not need the greater degree of pretreatment but requires large amount of energy.

To overcome the problem, we proposed an energy efficient thermal desalination process based on self-heat recuperation. In a thermal process based on self-heat recuperation, both latent heat and sensible heat are recirculated into the process without any heat addition. As a result, energy requirement is drastically reduced.

In this study, the energy consumption of thermal process based on self-heat recuperation in the steady state has been calculated using a process simulator. We obtained some results; the energy consumption grows with increase of the minimum temperature difference in the heat exchangers, and increased quantities of production flow rate make the specific energy consumption smaller. The desalination based on self-heat recuperation consumes approximately 1/4 of the energy compared to the conventional thermal desalination process when the minimum internal temperature difference in heat exchangers is assumed to be 5 K and the percentage of the product in the feed stream is assumed as 30%.

By applying the self-heat recuperation technology to the thermal desalination process, we proposed an innovative thermal desalination process.

(350e) Identifying Similar Environmental Objectives in Multi-Objective Optimization Via Clustering Methods

Diego Gabriel Oliva[1,2], **Gonzalo Guillén Gosálbez**[1] and Laureano Jiménez Esteller[1], (1)Department of Chemical Engineering, University Rovira i Virgili, Tarragona, Spain, (2)CONICET UTN, INGAR Instituto de Desarrollo y Diseño, Santa Fe, Argentina

Multi-objective optimization has recently emerged as a useful technique in environmental engineering, as it allows treating environmental aspects as additional objectives to be optimized rather than constraints on the system. One major limitation of these models (as applied to environmental engineering) is that they typically include a large number of objectives, which causes difficulties regarding the computation and visualization of the Pareto solutions.

Previous works by the authors focused on reducing the number of objectives in these problems using rigorous mixed-integer linear programs (MILP). In this piece of research, we have addressed a broader problem: given a multi-objective optimization model, we aim at grouping the environmental objectives into clusters, with the property that the minimization of any objective within a cluster will result in the minimization of the rest of objectives within the same cluster. Our final goal is to aid decision-makers by providing information on the structure of the problem, thereby offering them the opportunity of selecting the objectives to be minimized according to their preferences and the similarities between the environmental indicators of concern.

We present herein several theoretical and algorithmic developments in this area. We first provide a rigorous definition of clustering error in multi-objective optimization to quantify deviations from the Pareto structure of the problem that take place when optimizing in a reduced domain of objectives defined according to some clustering. We then introduce a new algorithm to perform such a clustering, and compare it with other traditional clustering methods available in the literature. Our clustering strategy is based on finding structural dominance relations between objectives, which are quantified according to different metrics

Finally, examples are taken from the literature to perform the comparison with other clustering methods. The main advantages of our method are its low running time and quality (i.e., error) of the clustering generated. The clusters identified provide in turn valuable insight for designers in multi-objective optimization.

(350f) Holistic Multi-Phase Processing of Natural Materials

Aydin K Sunol, Chemical & Biomedical Engineering, University of South Florida, Tampa, FL and **Rosanna Ayala**, University of South Florida, Tampa, FL

A novel and green multi-fluid process for extraction and isolation of value added products from natural plants will be discussed. The process aims to remove components with different functional groups such as nutrients and anti-oxidants as well as utilize the residue beyond its heating value or energy conversion options. The solvents used are high pressure warm water and supercritical carbon dioxide while the plant used is acai berry. The extract fractionation beyond functional groups for isolation of chiral and isomers is part of extract removal component of the approach. The typical plant material residue options traditionally include its use as biomass. However, in this application residue is fermented for fruit wine. The pilot plant studies, product characterization, and process modeling using a flowsheet simulator illustrate the potential and flexibility of this dual green solvent process. The use of dual solvents multi-phase simultaneously where one is supercritical is a first in high pressure natural plant refining.

(370a) Engineering Pellets of Biomass Blends

M. Toufiq Reza[1], David T. Graves[2], Joan G. Lynam[1], Md. Helal Uddin[3] and Charles J Coronella[3], (1)Chemical and Materials Engineering, University of Nevada, Reno, Reno, NV, (2)Chemical & Materials Engineering Department, University of Nevada, Reno, Reno, NV, (3)Chemical & Materials Engineering, University of Nevada, Reno, Reno, NV

Transportation and handling of lignocellulosic biomass are often challenging and expensive due to low bulk density, in the range of 60-80 kg/m^3 for agricultural straws and grasses and 200-600 kg/m^3 for woody biomass. Pelletization results in reduced transportation costs and provides for better handling and feeding. Lignocellulosic biomass contains lignin, a natural binder for pelletizing, with a glass transition temperature of about 140-180°C. However, many agricultural residues have lower lignin content than wood and as a result they are unsuitable for pellet-making without use of an exogenous binder.

The storage properties and hydrophobicity of biomass can be enhanced by thermal pretreatment. Hydrothermal carbonization (HTC, or wet torrefaction) is a pretreatment process for making a homogenized, carbon rich, and energy dense solid fuel, called biochar, from underutilized lignocellulosic biomass. Compared to raw biomass, biochar is both more hydrophobic for better storage and more friable for better processing. In this pretreatment method, the biomass is treated with hot compressed water in an inert environment in the temperature range of 200-260°C. The range of mass yield of biochar is 60-90% and its energy yield is 75-95%, indicating that the process causes an energy densification of 6-25%. Higher reaction temperature decreases mass yield and increases energy densification.

HTC biochar pellets made from woody biomass are very dense, durable and hydrophobic. Because the pellets are so extremely durable, it is reasonable to consider use of the HTC biochar from woody biomass as a possible binder for blending with raw biomass feedstocks.

Switch grass, rice hulls, and corn stover are three examples of lignocellulosic biomass with lignin content too low to allow for production of durable pellets without use of a binder. HTC biochar from miscanthus treated at 260°C was used as a binder and as an energy booster to produce blended biomass pellets. A hydraulic press with a temperature-controllable heated die was used for pelletization. The blended biomass was pressed into a 13 mm die to make cylindrical pellets 5-8 mm long. Pelletizing was done at a temperature corresponding to the glass transition temperature of lignin. The optimal hydraulic pressure was found to be 20 MPa at a temperature of 140 °C. The pressing time to allow the lignin to melt and bind, was 30 seconds.

Engineered pellets made from blends of HTC biochar and raw biomass exhibit an increase in hydrophobicity, abrasion resistance, energy density, and mass density compared to pellets made from raw biomass.

(370b) Torrefied Biomass: Superior Fuel for Co-Firing in Pulverized Coal Fired Power Generation Facilities

Prashanth R. Buchireddy[1], John Guillory[1] and Mark Zappi[2], (1)Mechanical Engineering, University of Louisiana at Lafayette, Lafayette, LA, (2)Chemical Engineering, University of Louisiana at Lafayette, Lafayette, LA

Approximately 54% of the United States power supply is being generated by coal fired electric generators. Due to the abundance of coal reserves in the U. S., the use of coal will be predominant. However, the environmental concerns including greenhouse gas emissions have prompted several agencies to regulate the gaseous emissions. Among the techniques that were proposed to reduce these gaseous emissions and the costs associated with these gaseous emissions, biomass co-firing with coal has drawn huge attention from electric utility producers. Biomass co-firing is considered as one of the most practical strategic options that has a short to medium term implementation potential for renewable power production. It is one of the most attractive options that will help utility companies in complying with the greenhouse gas reductions and more importantly the renewable energy portfolio standards.

Biomass is considered carbon neutral, and has lower sulfur, nitrogen, ash, and heavy metals than coal. However, several disadvantages exist compared with coal, such as lower heating value, lower density and higher moisture content. These properties pose several challenges related to fuel preparation, handling, storage, transportation, milling, and feeding in coal fired power plants. Torrefaction is a thermal pretreatment technique and has a potential to address the above mentioned issues. This process involves heating biomass to temperatures of 230-300°C in the absence of an oxidizing agent and under atmospheric pressure. During this process, 70% of mass is retained in the solid product which contains 90% of the initial energy. The remaining 30% of biomass is converted to volatiles and gases that contain 10% of the biomass energy. Thus, energy densification can be achieved via torrefaction by a factor of 1.3 on mass basis. The product, torrefied biomass, has improved fuel properties in terms of increased mass energy density, possesses hydrophobic nature, and is more uniform, in comparison with unprocessed biomass. Also, the grindability characteristics of the torrefied biomass are greatly improved and are comparable with coal and petroleum coke. This allows for effective utilization of torrefied biomass in coal fired power plants using the existing burners with minor modifications.

Research is being performed on both laboratory scale and pilot scale, using various lignocellulosic materials including pine, willow, arundo, and pecan shells at UL, Lafayette. The effect of various parameters including temperature, residence times, particle size, moisture content on torrefaction process is being evaluated, the results of which will be presented. Also, work is being performed on a pilot scale reactor that was fabricated by an industrial counterpart, which has a capacity of up to 30 lbs/hour. Testing is being carried out on the pilot scale reactor to improve the overall process efficiency. Mass and energy balances on the pilot scale reactor will be performed the results of which will be presented. Also, the effect of parameters tested on the energy density, bulk density, hydrophobicity, and grindability using modified hard grove grindability index (ASTM D409) will be presented and compared with representative samples of coal and petroleum coke.

TORREFIED BIOMASS AS COAL REPLACEMENT FUEL: MARKET ACCEPTANCE CHALLENGES DUE TO THE DISRUPTIVE TECHNOLOGY

William Breneman[1], Hiroshi Morihara, Howard Dawson, David Carter,

HM[3] Energy, Inc., Gresham, OR

Biography

William Breneman, PE has a BS. Chem.E. from Carnegie Mellon U. a M.S. Chem. and MBA from Ohio University. He spent over 30 years in the process development of organosilicon materials for Union Carbide Corp. and is presently an independent consultant specializing in the production technology for ultra-high purity silane and polycrystalline silicon. He recently joined with the other contributors to form HM^3 Energy, Inc. to promote the technology and commercialization of torrefied woody biomass as a replacement for coal in powder coal fired units.

Preface

My topic today is replacing coal as a solid fuel with torrefied biomass. Before I begin I will place one of our toreifed biomass briquettes and an ordinary wood fuel pellet into this bowl of water. It will demonstrate how HM3 Energy's TorrB™ densified torrefied biofuel stands up to an insult which would destroy the form and function of ordinary wood pellets.

The Problem

Coal, the world's first choice for energy production, will become much more expensive as a result of increasingly restrictive environmental requirements.

Some form of "tax" on the use of fossil fuels is becoming more certain.

Combined, these two factors will drive up the cost of electricity by nearly 100%.

But large coal-fired power plants provide the "base-load" power unobtainable from solar or wind derived sources.

[1] Presenter and addressee for communications.

The need is for a fuel which burns clean, is not derived from fossil sources, is economically transportable and stable and, ideally, could be used in any amount in existing coal-fired plants without modification.

The US EPA and similar agencies in many other countries have promulgated regulations that in effect will result in many coal fired power plants to dramatically curtail burning coal (and to either shut down or convert to other fuel – like natural gas).

In addition, local regulators have forced curtailment of coal fired technologies in individual states, including Washington. Even coal producing states like KY, TN, WV and IL are targeted for reduced coal based energy production.

The Japanese federal government will guarantee utilities' power prices for the next 15 to 20 years through Japan's Feed-in Tariff (FiT)[i] regulations when electricity is generated by renewable, non-carbon emitting energy sources.

The Korean government passed RPS regulations to reduce coal use in coal-fired power plants. Coal use must be reduced by 2% in 2012 and by 10% in 2022[ii].

The Chinese government recently announced that all utilities must reduce the use of coal by 20% by 2015[iii].

The Target/Addressable Market

Domestically, for HM^3 Energy the target market is: 25 million tons coal used annually by Portland General Electric, TransAlta Centralia and Pacific Power

Biomass as a Coal Replacement Fuel

Switching a conventional coal fired power plant to traditional wood pellets can be expensive.

100% replacement is not yet practical with traditional wood pellets.

The "soft" hemi-cellulose within the traditional pellets is largely responsible for the low water resistance and the difficulty in grinding the pellets to the fine powder required for rapid and complete combustion.

What is Torrefaction?

Torrefaction has been around for many years. In fact, when coffee beans are "roasted," they are being torrefied.

Torrefaction is the roasting of something at high temperature in the absence of oxygen.

This drastically changes the properties of the material being roasted.

For example, once coffee beans are torrefied, they can easily be ground into a fine powder for efficiently brewing coffee.

When woody biomass is torrefied, its properties are also drastically changed.

It is no longer fibrous, but it becomes brittle.

Note that the material we are torrefying in this sample is juniper – and very small pieces of juniper. We torrefy left over biomass like bark, twigs and branch tops that have no real value.

Torrefaction is an elevated temperature process which converts the hemi-cellulose in biomass to volatile organic compounds (and water).

This transforms the tough, flexible biomass into a friable high energy density, water resistant material.

The high energy density results from the absence of water since coal, as processed by the power plants contains or carries up to 30wt % water.

Torrefied biomass offers a dimensionally stable product form (somewhat "adjustable" in its size) which is easily pulverized to the same consistency as traditional coal.

Fuel Availability

Over 400 million tons of biomass can be economically and sustainably collected in the U.S. each year[iv].

We have made preliminary evaluations of a wide range of feedstocks.

To a large extent the TorrB™ Torrefied Biomass process is a great equalizer to deliver a consistent product in terms of physical and energy properties from a variety of biomass materials.

Most of the "specialized" processing involves feedstock conditioning. We have chosen to focus our primary effort on Woody Biomass and have designed the feed conditioning system in that regard.

Other forms of woody biomass such as agricultural residues, urban wood waste, farmed energy crops and yes, we have actually torrefied barn-yard waste – and rendered it odor-free and friable (although the high moisture content does not make it attractive).

Based on recent survey by Oak Ridge National Laboratory (2011), the amount of available biomass will increase in the U.S. from just over 400 million tons today to over 1100 million tons by 2030. The largest growth would be in the form of "Energy Crops"; plants intentionally grown for fuel. But Forest based resources would also increase by 45% over the same period.

Cost Comparison

TorrB™ torrefied biomass priced at $ 190/ton can produce electrical power at 9.0 ¢/kwh in an existing coal fired power plant. A basic premise is that the coal handling facility would not require any significant modifications to handle the new fuels. Such would be the case only if the new fuel had substantially the same handling characteristics as coal. Specifically, that means the new fuel would be of the same or higher energy density, the same or higher bulk density, water resistant, and have nearly the same resistance to fracture as coal. And yet, while the new fuel would need to be "strong" for transport, it also needs to be grindable to a fine powder without fouling the grinder with essentially under torrefied material. HM^3 Energy's TorrB™ torrefied biomass meets all of these criteria.

An electric energy cost of 8.2 ¢/kwh using $ 72/ton coal in a plant which requires emission controls and an imposed carbon "tax".

The cost of providing equivalent energy in the form of wood pellets is higher than using torrefied wood briquettes. The differential cost for covered storage ($6), Shipment ($ 9), Covered storage at port facilities ($ 12), covered land transportation ($ 2) and Storage and handling facilties at the customer's location ($ 15) together raise the cost of providing 17 Gigajoules of power (1 tonne of pellets, 0.74 tonne of TorrB fuel) by $ 50 per 17 Gigajoule of energy. This is a 216% increase.

The economic impact to the electric consumer of a conversion from coal to TorrB™ Biomass is shown under several scenarios. TorrB™ is priced in this scenario at $ 190/ton relative to coal at $ 72/ton.

The TorrB™ Process

The TorrB™ process includes these six fundamental process operations.

Feed Preparation

To convert woody biomass into high energy density, water resistant briquettes, raw forest waste is shredded in the field and trucked to the processing center. The critical parameter is the collection radius of ~ 60 km to maintain low local transportation costs.

In the Feed Stock Preparation it is further sized to give a consistent maximum cross section and tramp metals and rocks are removed.

This portion of the process utilizes standard commercial hardware.

The emphasis is on relatively small production units (nominally 100,000 ton/yr) located close (within ~ 60-70 km) to the material source to minimize local transportation costs. In the Pacific Northwest, for example there are many shuttered sawmills which have good infrastructure and nearby forest resources – including rail access.

Raw forest waste is shredded in the field and trucked to the processing center. The critical parameter is the collection radius of ~ 60 km to maintain low local transportation costs.

There, in the Feed Stock Preparation it is further sized to give a consistent maximum cross section and tramp metals and rocks are removed.

This portion of the process utilizes standard commercial hardware.

The emphasis is on relatively small production units (nominally 100,000 ton/yr) located close (within ~ 60-70 km) to the material source to minimize transportation costs. In the Pacific Northwest, for example there are many shuttered sawmills which have good infrastructure and nearby forest resources – including rail access.

Feed Stock Drying

The sized and metal and rock free biomass material is dried to less than about 18% moisture in a conventional dryer system. By minimizing the maximum temperature, the volatile organic emissions are kept below regulatory control limits. There are several dryer configurations which are operable. The most favored would be those which use direct hot gas (air or combustion gas) as the source of heat.

Torrefaction

The dry biomass material then enters an oxygen free "retort" or roaster reactor of our proprietary design where residual moisture and volatile organic compounds are removed while converting the hemi-cellulose fraction of the feedstock.

The key to successful torrefaction is to have all material be equally well converted to a specific state which maximizes energy density, and grindability while preserving sufficient hemi-cellulose to facilitate densification.

This area of technology has been extensively studied in several research institutions. The most extensive and practice oriented work is being done in Europe by the Energy Research Center of the Netherlands (ECN).

Thermal Oxidizer

The combustible vapors from the torrefier are burned in a thermal oxidizer which produces the heat energy to drive the torrefier as well as to provide most of the heat required for drying.

As forest biomass contains virtually no sulfur, mercury or other heavy metals the emissions from the thermal oxidizer will meet appropriate standards.

A bag filter would be used to trap particulate emissions.

Natural gas may be used as an auxiliary fuel during start-up and to provide additional heat energy if the incoming biomass is more wet than standard (about 30 – 35% moisture). In a fully integrated commercial facility, a portion of the torrefied mass, mainly the fine fraction which contains the bulk of the non-combustible dirt, rocks, and the like would be burned for on-site energy conservation.

Conditioning

The torrefied biomass is "conditioned" in preparation for densification.

Conditioning, which includes cooling the torrefied material to below its auto-ignition temperature, is a proprietary operation involving specific physical operations, but does not use any synthetic additives or binders.

Densification

Densification to a high density, mechanically stable, water resistant briquette is performed in a customized device under proprietary conditions. The physical form can be altered to some extent, but the current emphasis will be to produce a "briquette" of nominally 0.75" to 1.5" feature size. At HM^3 Energy, we have conducted full commercial scale densification tests which, while not entirely successful have pointed the way toward success. That pathway is not inconsistent with our laboratory produced demonstration samples and does not involve non-commercial hardware.

The samples we have provided are only to show the basic form and physical properties of the TorrB™ Torrefied Biomass and do not necessarily represent the size of the planned commercial product.

Mass & Heat Balance

This is a simplified block flow diagram of the TorrB™ Torrefied Biomass process showing the major energy flows.

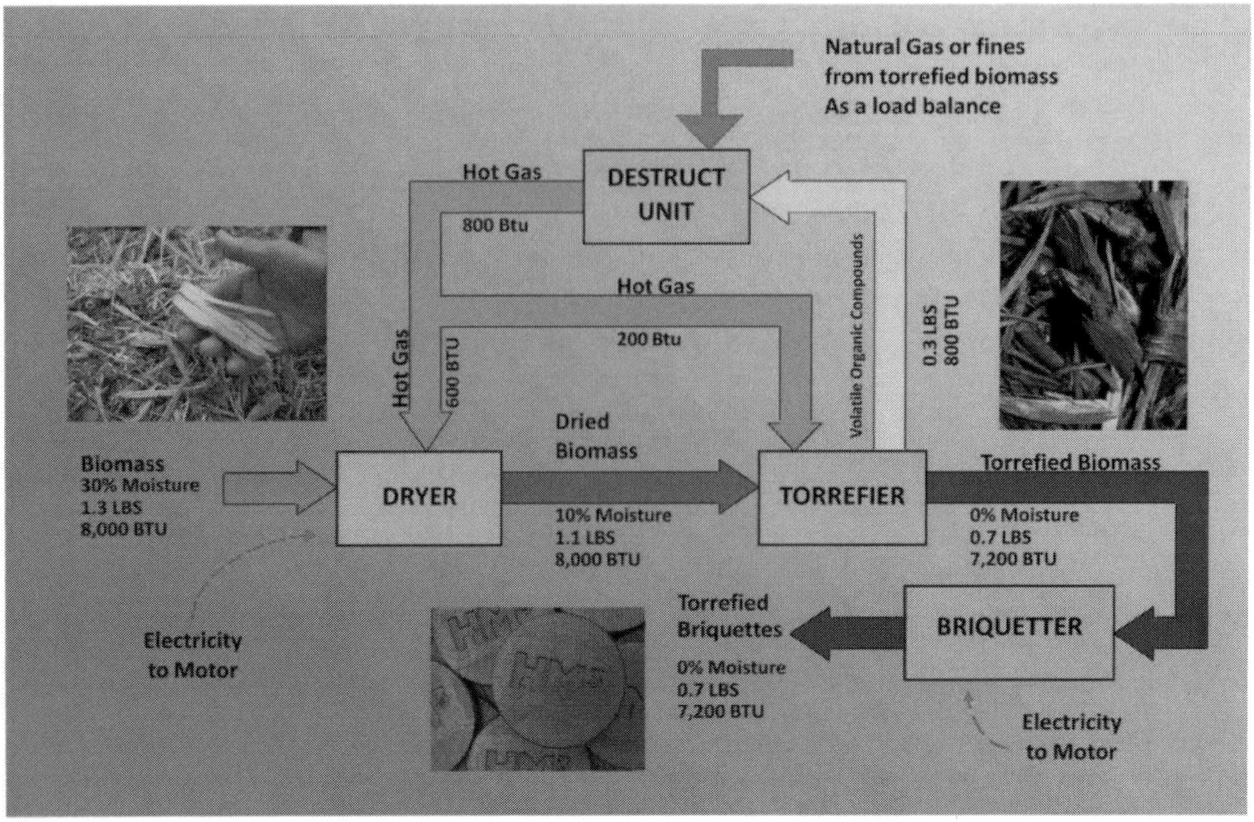

Figure 1. Block Flow Diagram

The ancillary equipment not shown includes any the feedstock pre-treatment, sizing, etc. and the proprietary means we use to remove non-combustibles.

The energy required to drive the process is derived from combustion of the volatile organic compounds evolved during torrefaction. The energy for start-up and to balance the load if the feedstock is overly wet (> ~ 30% moisture) is augmented by natural gas or by use of the dirt laden biomass made available by the waterless dirt removal operation.

Validation

HM³ Energy's TorrB™ torrefied biomass was evaluated by reputable demonstration laboratories.

Both the feed preparation (by Hazen Research, Golden, CO) and combustion testing (by Western Research Institute of Laramie, WY), technicians commented on the ideal properties of the material in this successful endeavor.

As a result of Portland General Electric's (PGE) staff overseeing these successful tests, PGE is in a position to provide critical support.

Job Creation

By utilizing this technology, a whole new infra-structure of employment opportunities will be created:

Engineers for process and equipment design

Procurement specialists for biomass acquisition

Engineers and real estate professionals for site selection

Environmental engineers and scientists for permiting

Mechanics and instrument technicians for operations

Business and logistics managers to deal with the large transportation issues

Accountants, legal staff

And opportunities for internships to provide experiences prior to full employment.

Loggers, truckers and forest managers to provide the feedstock

Other Impacts

The use of TorrB™ Torrefied Biomass can have a large beneficial impact on the environment.

Carbon emissions from fossil fuels are reduced proportionately. Sulfur emissions are nearly eliminated and nitrogen oxides are greatly reduced (less ammonia is needed to further reduce NOx to meet regulatory requirements).

Mercury emissions are eliminated, improving the environmental health of water resources.

A "customer" is provided for biomass which would otherwise be burned in an uncontrolled manner which generates much irritating smoke.

Uncontrolled forest fires destroy valuable animal habitat, wreck havoc on streams and even result in loss of residential property.

Acknowledgements

Acknowledgement is given to:

- Phase I and II grant from Small Business Innovative Research through the U.S. Dept. of Agriculture
- U. S. Endowment for Forest & Communities, Inc.

- A grant from the Nature Conservancy
- A grant Oregon-Built Environmental and Sustainable Technology Center (BEST).

Postface

We'll now examine the torrefied briquette which was earlier placed into the glass of water. As you can see, the briquette is still, firm and has retained its structural integrity. And while it likely has taken up about 3 – 10% moisture, it is still a very useable material and one which can withstand the rigors of open transport means.

---

[i] http://www.renewableenergyworld.com/rea/news/article/2012/06/japan-approves-feed-in tariffs

[ii] http://www.reuters.com/article/2010/03/18/renewable-korea-idUSTOE62H06F20100318

[iii] Emission standard of air pollutants for thermal power plants
GB 13223-2011 replacing GB13223-2003 and putting into effect as of Jan.1, 2012

[iv] http://teeic.anl.gov/er/biomass/restech/dist/index.cfm

(370d) Hydrothermal Carbonization of Lignocellulosic Biomass: Kinetics and Pelletization Study

Wei Yan[1], Charles J. Coronella[2], S. Kent Hoekman[3] and Larry G. Felix[1], (1)Energy Conversion, Gas Technology Institute, Birmingham, AL, (2)Chemical Engineering Dept., University of Nevada, Reno, Reno, NV, (3)Division of Atmospheric Science, Desert Research Institute, Reno, NV

Hydrothermal carbonization (HTC) is a promising technology to convert diverse feedstocks to homogeneous energy-dense solid fuel. The resulting HTC solid fuel has favorable properties such as high energy content, low volatiles content, friability and hydrophobicity. Currently, kinetics of hydrothermal carbonization is not available for performing techno-economic analysis of hydrothermal carbonization process. In this study, a specially-designed two-chamber reactor was used to perform hydrothermal carbonization of lignocellulosic biomass. The hydrothermal carbonization of loblolly pine was performed at various conditions (e.g. HTC temperature, reaction times, and biomass particle size). The decreasing mass yield and increasing fuel density occurred at the first 1- 2 minute of the reaction. The simple reaction mechanism of two parallel first-order reactions was validated, and kinetic parameters were presented. In addition, the excellent properties of HTC pellets were identified like forming robust pellets, stability in water, high volumetric energy density, as well as the ease of pelletization without adding any binder. Hence, this study pointed out hydrothermal carbonization process could help greatly overcome technical and economic barriers in thermochemical conversion of lignocellulosic biomass for fuels, chemicals, and electricity.

(370f) Effects of the Pretreatment Processes of Grassy Biomass Grown for Soil Remediation and Biofuel Production

Maria Nydia Ruiz-Felix[1], Justinus A. Satrio[1], William J. Kelly[1] and Ronald Balsamo[2], (1)Chemical Engineering, Villanova University, Villanova, PA, (2)Biology, Villanova University, Villanova, PA

Biomass is the only renewable source of organic carbon that, when utilized efficiently and optimally, can significantly reduce our present reliance on fossil fuels and, thus reduce the emissions of green house gases from these non-renewable materials. Lignocellulosic biomass is abundant, does not compete with the production of food crops, and considered relatively inexpensive. However, the complex nature of biomass and many ill-defined issues related to biomass utilization pose a substantial challenge to the large-scale biomass utilization. Growing biomass for energy crops, such as switchgrass, means that more land needs to be utilized. It is desired that biomass for energy crops to be grown in land area that is considered marginal.

At Villanova University, we are working on to develop a sustainable processing system for producing biofuels from grassy plants that can be grown on heavy-metal contaminated soils in underutilized areas such as superfund sites. Certain types of perennial grasses, such as switchgrass and timothy grass, may have great potential for improving the quality of heavy metal contaminated soil by taking up the heavy metal contaminant from the soil and subsequent removal by repeatedly harvesting the above-ground biomass. These grasses also have great potential to be used as bio-renewable feedstock for the production of bionergy and biofuels due to their relatively high yields of biomass per acreage of land. On the biomass conversion processes, the focus of the study is on utilizing switchgrass and timothy grasses that is grown in the soil that has been purposely contaminated with lead for their potential as feedstock for biofuel production using a hybrid biochemical and thermochemical approaches, i.e. fermentation and fast pyrolysis. To be reported in this paper is a part of the study in which we are pretreated the lignocellulosic materials in order to optimize the utilization of carbon in the biomass. Lignocellulosic materials consist of cellulose, hemicelluloses and lignin; these three components are strongly bound together which make it difficult to utilize this type of biomass in the production of biofuels on a commercially viable scale. We hypothesize that optimal utilization of carbon is to be done by first converting cellulose into sugars which subsequently fermented to produce ethanol. In the second step, the remaining solid biomass (containing primarily lignin) is pyrolyzed to produce bio-crude oil. Low acid hydrolysis was applied with the intention of disrupting the structure of the biomass by breaking in order to allow enzyme to penetrate the biomass structures more to increase the sugar yields from cellulose during enzymatic treatment.

To be reported in this presentation are the results obtained from the treatment on switchgrass and timothy grass by dilute acid hydrolysis followed by enzymatic hydrolysis. The effects of acid hydrolysis to the sugar yield and the composition of the remaining solids from enzymatic treatment are evaluated. To be reported also is the effect of lead contained in the grasses.

Concentrate Enhanced Reverse Osmosis: A New Process for RO Concentrate and Brackish Water Treatment

Anthony Tarquin, PhD.; Guillermo Delgado, PhD Candidate

University of Texas at El Paso

Advances in reverse osmosis (RO) technology have led to a significant reduction in the production cost of water desalination. Brackish water desalination, however, is often challenging because of the high costs associated with residual RO concentrate management, especially at inland locations. The city of El Paso owns and operates the largest inland desalination plant in the world [1]. The Kay Bailey Hutchison (KBH) desalination plant treats brackish water from the Hueco Bolson and produces, when operated at full capacity, 27.5 million gallons of water every day. The KBH desalination plant uses reverse osmosis (RO) membranes to treat the brackish water. This results in a production of 3 million gallons of RO concentrate every day. The concentrate is disposed via injection wells into the ground [2]. El Paso Water Utilities (EPWU), the Texas Commission of environmental Quality (TCEQ), and the U.S. Bureau of Reclamation have supported several research projects that can reduce the volume of this concentrate, increase the production of fresh water, and minimize the cost of disposal of the concentrate. One project in particular [3] started as pilot Seawater Reverse Osmosis system (SWRO) to evaluate the performance of seawater RO membranes to treat the KBH concentrate. However, the excellent results obtained thanks to the design and not only the membrane performance made this research to evolve to a new treatment process. The Concentrate Enhanced Recovery Reverse Osmosis (CERRO©) process (UTEP Provisional patent application number 61/233,761) is a new batch treatment process to reduce the volume of concentrate produced during RO treatment.

Concentrate Enhanced Recovery Reverse Osmosis Process

The CERRO process consists of one or several seawater reverse osmosis membranes in parallel working in a batch configuration system. It is the batch configuration that allows the process to reach higher recoveries than conventional RO systems. The water is pretreated in the feed tank, treated by the seawater RO membrane, the permeate is recovered, and the concentrate is recirculated to the feed tank. A detailed description of the process is shown in the next sections.

Pretreatment

There is a tank where the feed water is pretreated by the addition of sulfuric acid and antiscalants. The pH of the feed water needs to be between 3.5 and 4.5 in order to eliminate the presence of alkalinity. This has to be done to avoid the precipitation of calcium carbonate which can foul the membranes. Antiscalants are added to the water to maximize the treatment time and to avoid damaging the membranes due precipitation of calcium sulfate and/or polymerization of silica.

Treatment

After pretreatment is done, the feed water is treated by the seawater RO membranes. Several membranes can be used in the process, but they have to be working in parallel. Membranes connected in series are not needed in the CERRO process. The permeate produced during treatment can be recollected. The concentrate is send back to the feed tank. This is done in order to increase the recovery of the system since seawater RO membranes have very low recovery rates. The operating pressure of the system is from 500 psi's to 900 psi's. The system can be operated at a constant pressure or at a constant permeate flow. At a constant pressure, the permeate flow will drop during the process because the osmotic pressure of the water will increase as it gets more concentrated. On the other hand, the operating pressure of the system will rise if the system is operated at a constant flow. Figure 1 shows the permeate flow of a test made using the CERRO process at a constant pressure of 700 psi's.

One important characteristic of the CERRO process is that the membranes can be cleaned after a batch. A fast flush with permeate or feed water between batches increases the life of the membranes and reduces the risk of fouling and scaling the membranes.

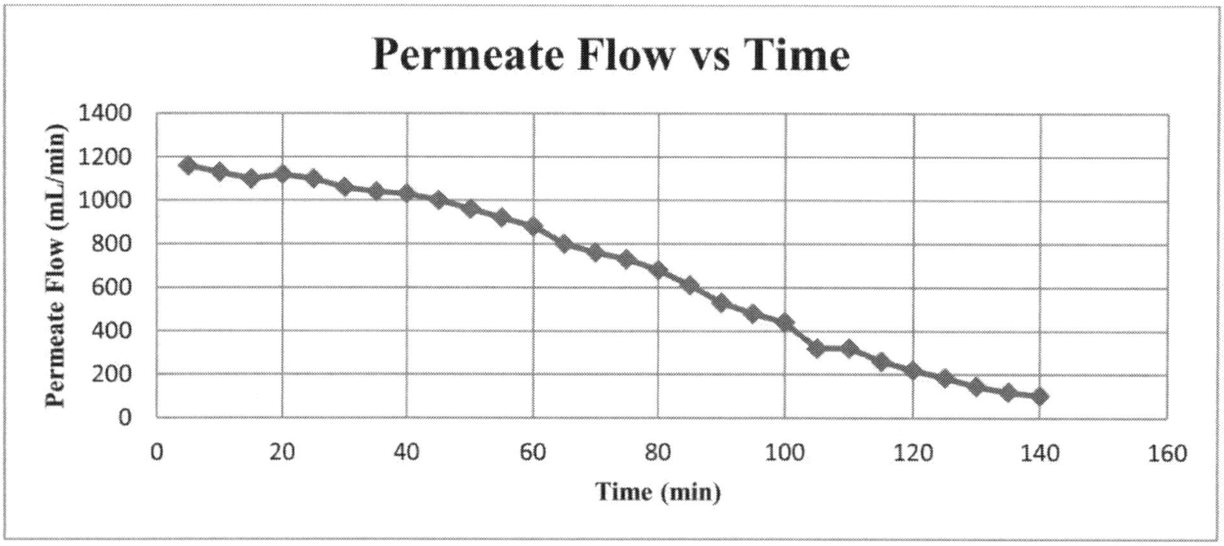

Figure 1: Permeate Flow of the CERRO Process operated at 700 psi's

One limitation for any membrane treatment system is the precipitation of sparingly soluble salts (i.e. $CaSO_4$) [4]. Usually, the presence of this type of salts decreases the recovery of fresh water due precipitation in the system during treatment. Several tests using the CERRO process have shown that the system can work under conditions of supersaturation. This happens because the treatment time of the system can be faster than the induction time of precipitation. This characteristic makes the CERRO process an excellent option for treating water with high concentrations of calcium sulfate and/or silica as will be shown in the case studies. Figure 2 shows a schematic of the design of the CERRO process.

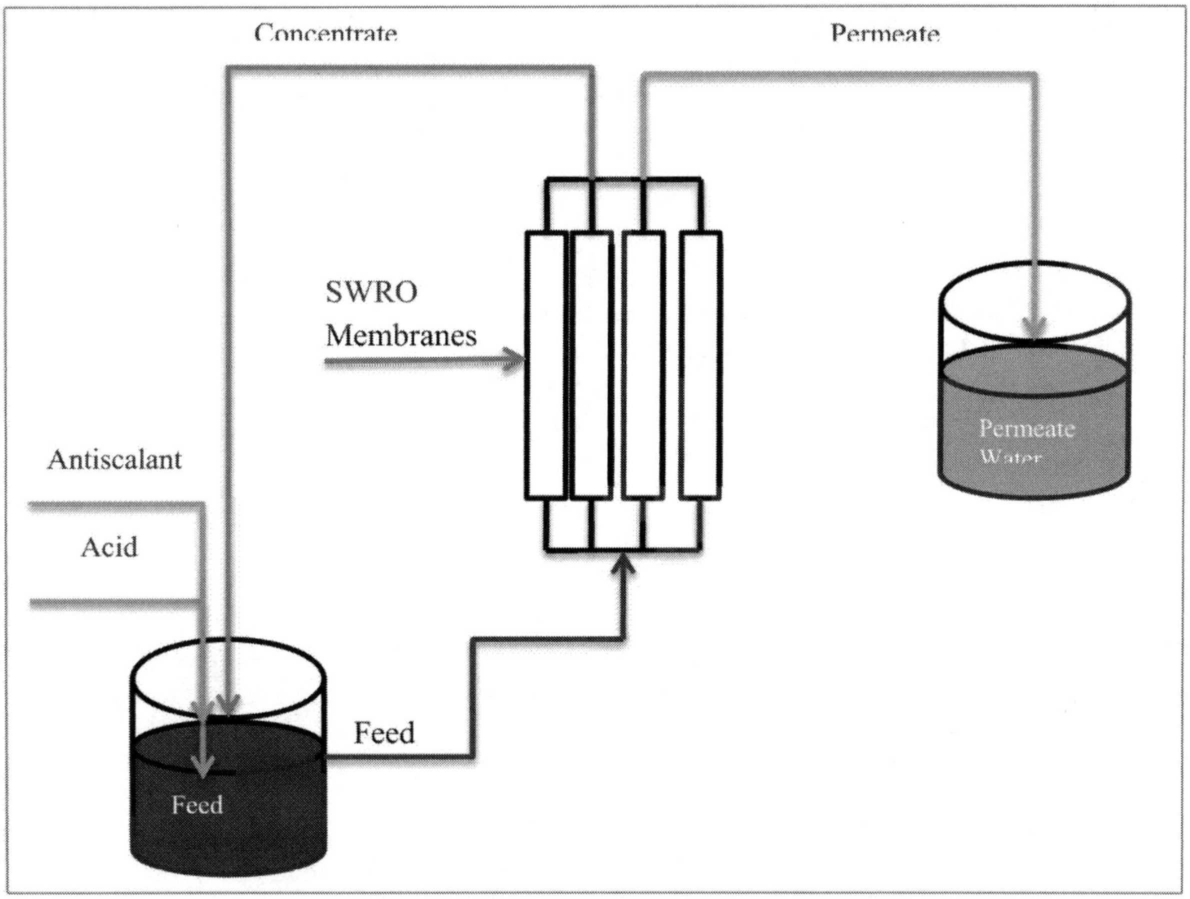

Figure 2: Schematic of the CERRO Process

As mentioned before, the CERRO process was designed as an effort to reduce the volume of RO concentrate generated in the KBH desalination plant. Pilot testing using this system has shown that 70% to 90% of the RO concentrate can be recovered using the CERRO process. However, other research projects have shown that the CERRO process can be used in brackish water containing high concentrations of calcium sulfate. Conventional RO systems will have low recovery rates due calcium sulfate precipitation. However, the CERRO process have shown an exceptional performance recovering water from these type of source waters, working under supersaturation conditions, and, in some cases, even when precipitation occurs during treatment.

Three case studies where the CERRO process was used are presented in the next sections.

Case Study 1: KBH Concentrate Volume Reduction

The KBH Desalination Plant, located in El Paso, Texas, is the largest inland desalination plant in the United States. This plant produces a concentrate volume of two million gallons per day when operated at full capacity. Table 1 shows the characteristics of the KBH concentrate [3].

Table 1: Characteristics of the KBH Concentrate

	Concentration		
Parameter	**Min**	**Average**	**Max**
Cl	4840	5089	5540
SO_4	1050	1111	1200
Alk-T	400	427	445
Ba	0.042	0.31	0.48
EC (uS/cm)	16600	18122	20200
Fe	0.03	0.07	0.12
T Hard	2050	2291	2430
Mn	0.11	0.17	0.22
ortho-P	0.11	0.19	0.49
pH (pH units)	7.6	7.8	8
Ca	281	608	937
K	43	113	759
Mg	85.7	161	183
Na	1730	2810	3260
Sr	8.74	17.1	30.1
SiO_2	26.9	131	173
TDS	10300	10722	11200
Temp (°C)	23.3	25.4	26.3
Turb (NTU)	0.06	0.3	1.94

(Concentrations in mg/L except as noted in table)

Note that the average concentration of silica is 130 mg/L. There is a high risk of membrane fouling due silica polymerization if this concentrate is treated in a conventional RO system. During the testing using the CERRO process, silica concentrations of 800 mg/L and above were reached. This is because the treatment time was faster than the polymerization time of the silica. Figure 3 shows the silica concentration at different times during one experiment.

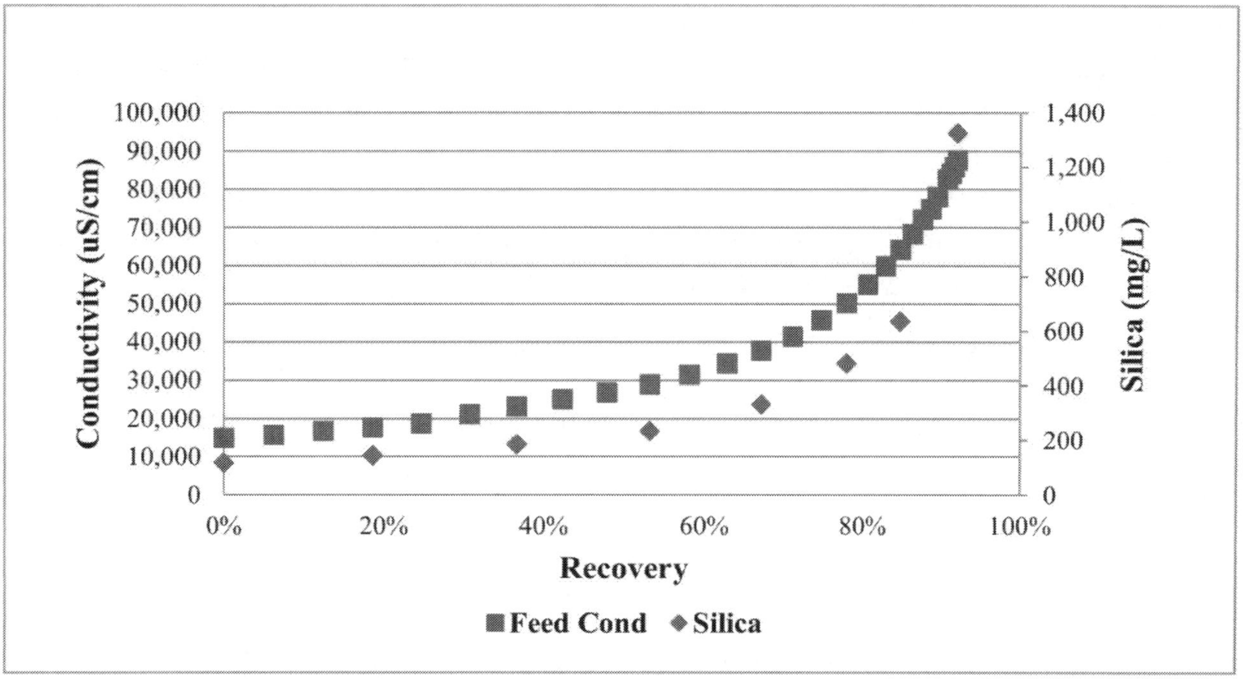

Figure 3: Silica concentration at Different Times during Testing

During this test, the recovery of the system was 92%. The silica concentration was 1,325 mg/L (average of a triplicate). Precipitation occurred after the system was shut down and the final concentrate was removed from the feed tank. After a quick flush with permeate water, the SWRO membrane did not showed any signs of being fouled. Several tests have been made treating KBH concentrate using the CERRO process. Table 2 shows the results of some tests.

Table 2: Results of Several Tests in the KBH Plant

	CERRO Final Concentrate								
Test (date)	**Ca**	**Mg**	**SO4**	**SiO2**	**Cl**	**TDS**	**Cond (uS/(cm)**	**Recovery**	**Notes**
8/17/2010	3,500	851	18,500	597	28,250	71,347	82,600	84%	Spiked with Na_2SO_4
10/1/2010	4,100	1,215	13,540	183	35,900	65,660	93,500	90%	
10/8/2010	3,300	1,337	12,400	178	32,500	67,347	81,800	88%	
11/12/2010	600	851	17,480	368	21,500	60,100	69,800	84%	Spiked with Na_2SO_4
7/9/2010	4,000	972	11,080	247	34,500	69,652	87,500	82%	Spiked with NaCl
6/10/2010	4,080	886	12,900	780	27,750	54,540	79,800	90%	
12/21/2010	3,800	3,645	11,720	382	39,000	76,154	86,500	91%	Spiked with $MgCl_2$

(Concentrations in mg/L except as noted in table)

In some of the experiments, the KBH concentrate was spiked with different salts (i.e. NaCl, Na_2SO_4, $MgCl_2$). This was made in order to change the chemistry of the KBH concentrate to simulate a different water source. As can be seen in table 2, the change in the chemistry of the KBH concentrate did not affect the performance of the CERRO process. More specifically, in the test made in 11/12/2010, it can be seen that the calcium concentration is relatively low. This is because calcium sulfate precipitated during the test. Even though there was precipitation inside the membrane, a fast rinse with permeate after the test avoided any damage to the membrane.

Case Study 2: Treatment of Alamogordo, New Mexico Brackish Water

Alamogordo groundwater is naturally contaminated with gypsum. Analysis made in the groundwater showed that 60% of the total dissolved solids are calcium and sulfate ions. The Brackish Groundwater National Desalination Research Facility (BGNDRF) is a research facility where desalination systems are tested using this specific groundwater. The CERRO process was tested with this brackish water having recoveries from 60% to 90%. Table 3 shows the characteristics of Alamogordo groundwater.

Table 3: Characteristics of Alamogordo Brackish Water

	Cations (mg/L)			**Anions (mg/L)**				
Sample ID	**Ca**	**Mg**	**Na**	**Cl**	**F**	**NO3**	**SO4**	**TDS (mg/L)**
Alamogordo BW	421	254	676	533	2.43	28	3,001	5,942

The presence of high concentrations of calcium and sulfate ions makes the treatment of this groundwater very difficult. The CERRO process was used in order to evaluate the maximum recovery that can be achieved using reverse osmosis. Table 4 shows the results of a test made using the CERRO process. In this experiment, the ground water was treated in a RO pilot unit until a 50% recovery was reached. The concentrate of this unit was treated in the CERRO process until precipitation occurred. The CERRO process achieved a 36% recovery which gave a total recovery of 68% of the ground water. No antiscalants were used in this test.

Table 4: Final Results of a Test Where no Antiscalants Were Used

	Cations (mg/L)			**Anions (mg/L)**				
Sample ID	**Ca**	**Mg**	**Na**	**Cl**	**F**	**NO3**	**SO4**	**TDS (mg/L)**
RO Concentrate	1,020	563	1,320	1,209	2.76	53.55	6,053	10,950
CERRO Final Feed	530	846	1,994	1,817	1.93	77.59	7,599	13,531
CERRO Permeate	5.26	2.89	10.98	35.68	0.33	5.21	26.05	>100

The high content of calcium sulfate in the groundwater is a limiting factor for the treatment process. However, it was found that calcium sulfate is more soluble when sodium chloride is present in the water

[5]. One batch of groundwater was treated using the CERRO process. However, the concentration of sodium chloride was raised to 10,000 mg/L. This was made in order to increase the solubility of calcium sulfate and increase the final recovery. Figure 4 shows two tests made using the CERRO process. One Batch was made with the addition of sodium chloride and the other batch was used as a control with no changes in the characteristics of the groundwater.

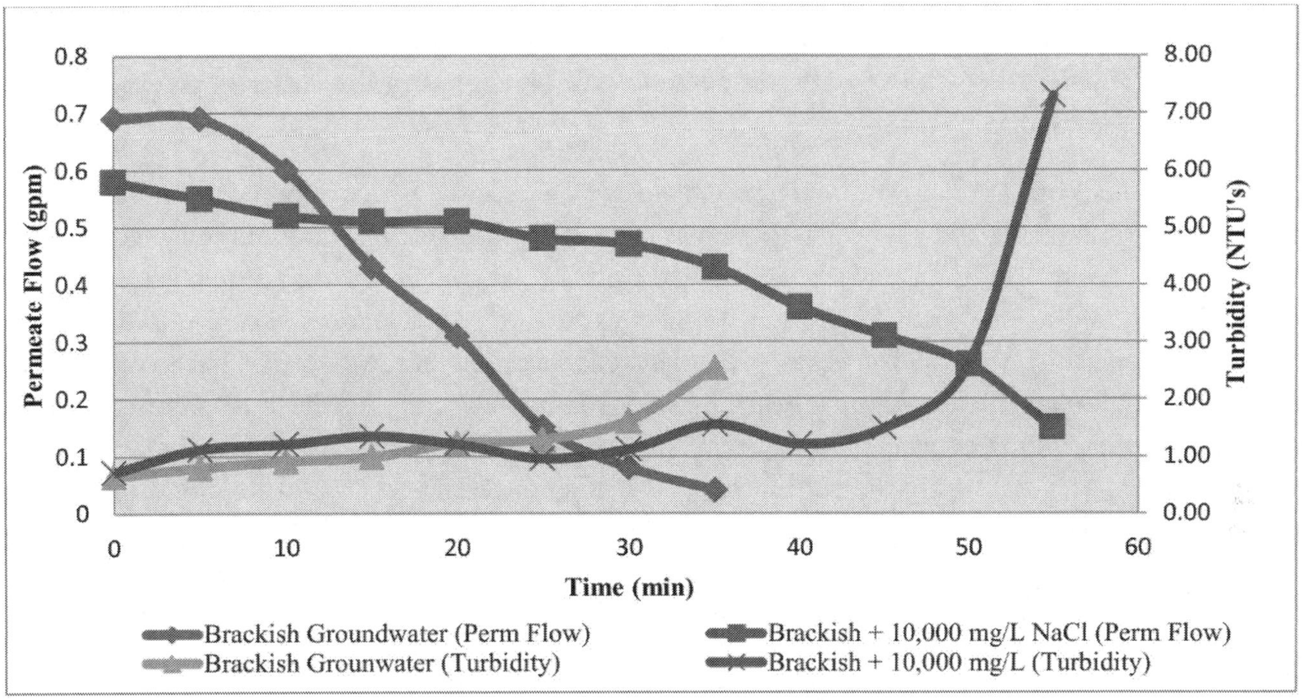

Figure 4: Permeate Flow and Turbidity of Two Batches Using the Cerro Process

The recovery in the experiment used as a control was 68% before precipitation. The recovery in the batch with the addition of sodium chloride was 87%. The addition of sodium chloride in the water clearly increased the recovery of the system. During this research, it was found that the main limitation for the CERRO process was the high concentrations of calcium sulfate. However, the system showed an excellent performance at supersaturation concentrations.

Case Study 3: Treatment of Alto, New Mexico Brackish Water

Brackish groundwater found in Alto, New Mexico was treated using the CERRO process. This water has high concentrations of calcium sulfate. Test done using the CERRO process showed that 96% recovery can be achieved. Table 5 shows the characteristics of the groundwater.

Table 5: Characteristics of Alto Groundwater

	Concentration (mg/L)					
	Ca	Na	Cl	PO4	SO4	TDS
Alto Brackish Water	237	105	165	<1	887	1,779

Several tests were made using the CERRO process with this groundwater as a feed. The system reached a recovery higher that 90% in every test. It was found that precipitation occurred when 94% to 96% recovery was reached. Even though the system reached recoveries above 90%, the production of drinking water from this water source resulted to be cost prohibitive due to the high cost related to the management of the concentrate. In order to be a suitable project, a recovery of 98% or higher is necessary. A research project is being developed where the calcium sulfate is removed from the concentrate by the addition of lime. The excess in calcium forces the precipitation of calcium sulfate. After the calcium sulfate is removed, the concentrate will be treated in the CERRO process a second time increasing the final recovery.

The CERRO process is a novel process where brackish water and RO concentrate can be treated. The system has shown an excellent resistance to membrane fouling because of two main reasons. First, the compact design of the system allows a fast operation time which allows the system to work under supersaturation conditions. Second, the capacity of the system to work in a batch configuration permits that the membranes can be cleaned continuously. This characteristic of the system is particularly helpful to protect the membranes even under supersaturation conditions. Finally, the main limitation of the system is the precipitation of calcium sulfate. Future projects will intent to find a suitable method to remove the calcium sulfate from the concentrate, treat the concentrate a second time and increase the maximum recovery of the system.

Bibliography

1. EPWU. Setting the Stage for the Future. . 2007;2012.

2. Delgado G. Treatment of RO concentrate using VSEP technology. . 2009.

3. Tarquin AJ, Fahy MP, Balliew JE. Concentrate Volume Reduction Research in El Paso, Texas. ASCE Conf Proc. 2010;371:358.

4. Gloede M, Melin T. Physical aspects of membrane scaling. Desalination. 2008;224:71-75.

5. Messinger A, Anna Hahn D. A Dictionary of Chemical Solubilities: Inorganic. In: , 1921:951.

Demonstration of Zero Discharge Desalination Technology for Concentrate Management

Tom Davis[1], Brad Biagini[2], Malynda A. Cappelle[1], Bernie Mack[2], and Emily Gilbert[2]

[1]*Center for Inland Desalination Systems, University of Texas at El Paso*
[2]*Veolia Water Solutions & Technologies*

Zero Discharge Desalination (ZDD) technology utilizes electrodialysis metathesis (EDM) to remove troublesome ions from membrane desalination concentrate so that the solution can be further processed to increase the yield of fresh water from brackish groundwater that is difficult to treat by RO or NF alone. ZDD technology is particularly applicable to treatment of groundwater containing appreciable concentrations of scaling salts such as gypsum ($CaSO_4$), calcium phosphate, barium sulfate, etc. In contrast to conventional electrodialysis, which has two membranes and two solution compartments in a repeating unit, EDM utilizes four ion-exchange membranes, two depleting compartments, and two concentrating compartments. The arrangement of membranes and transport of ions in one EDM repeating unit are illustrated in Figure 1. More than 100 of these repeating units are installed between a single pair of electrodes in a commercial-size EDM stack.

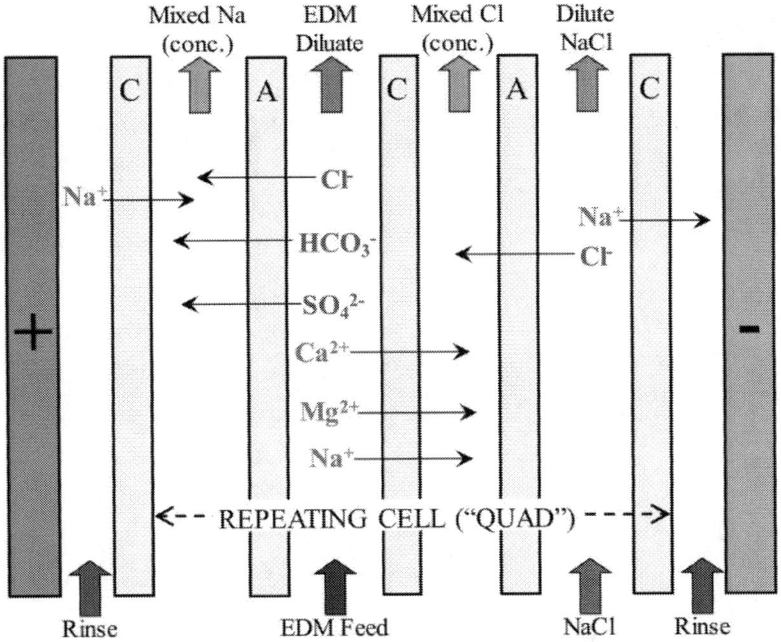

Figure 1. Membrane Arrangement and Ion Transport in EDM.

The RO concentrate flows through one of the depleting compartments, and the diluate can be returned to the system as RO feed or blended with RO permeate depending on project-specific water quality requirements. A solution of NaCl flows through the second depleting compartment. Application of an electric potential to the electrodes of the EDM stack causes the anions from the RO concentrate to migrate through anion-exchange membranes where they meet Na^+ ions from the NaCl compartment to form a concentrated salt stream rich in Na_2SO_4. Similarly, the cations from the RO concentrate meet Cl^- ions from the NaCl stream to form another concentrated salt stream rich in $CaCl_2$. Because Na_2SO_4 and $CaCl_2$ are highly soluble salts, the concentrations of these two streams can achieve very high levels with minimal scale potential. In fact, when $CaSO_4$ is the dominant salt in the groundwater, all of the salts removed from the groundwater can be carried away in the two EDM concentrate streams, each containing only about 1% of the water that was in the RO feed. Thus the ZDD process has the potential to recover 98% of the water from groundwater compared to typical yields of about 75% when the same groundwater is treated by RO alone.

The high recovery obtainable by the ZDD process is particularly beneficial in arid regions where groundwater is scarce and concentrate disposal is problematic. Typical disposal methods – dumping in a sewer, deep-well injection, evaporation ponds, or hauling by truck – are impractical, too expensive, or damaging to the environment in many locations. At the very least, the ZDD technology can reduce the costs and or size of conventional disposal methods leading to lower operational or capital costs. The ZDD technology has the advantage that the salts are in a highly concentrated solution. In this concentrated state, the salts can be separated and recovered as marketable products. In one example of the ZDD process illustrated in Figure 2, the salt-depleted stream from the EDM is returned to the RO feed, and the two EDM concentrate streams are combined to precipitate $CaSO_4$, which has potential markets for soil augmentation or the production of gypsum wallboard. The supernatant from the precipitation contains NaCl that can be recovered and reused in the EDM stack. Continuing laboratory research is directed toward recovery of useful products from the EDM concentrate streams so that even more water can be recovered to further approach Zero Discharge Desalination.

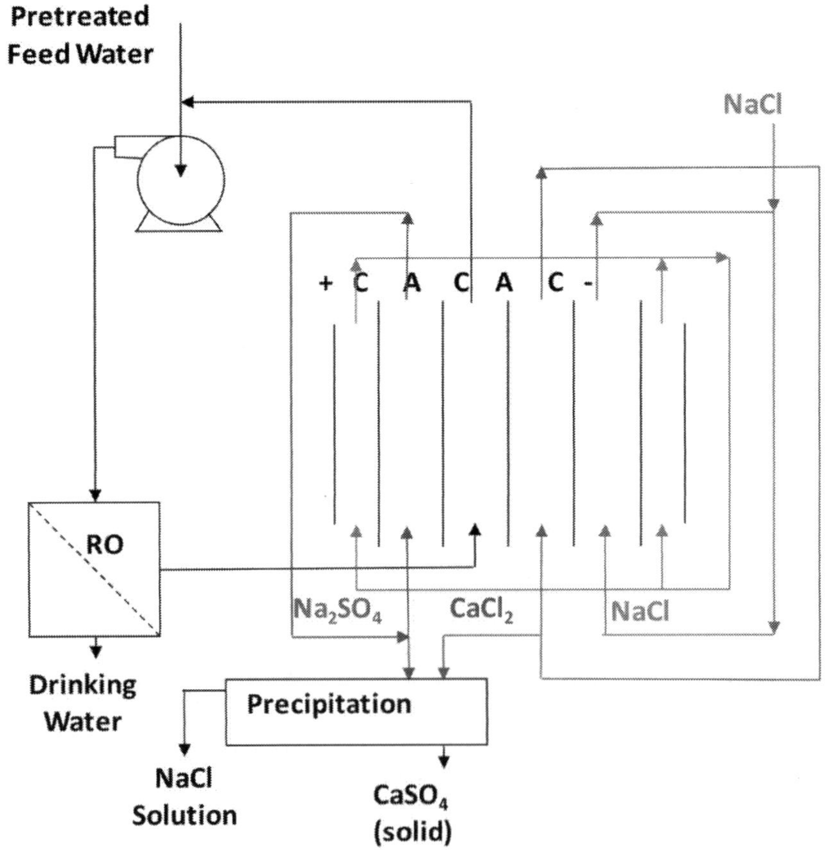

Figure 2. Flow Diagram of ZDD Process

The ZDD technology has been evaluated in progressively larger pilot studies at the Brackish Groundwater National Desalination Research Facility (BGNDRF) in Alamogordo, New Mexico. From the first pilot with 1 gpm capacity to the latest with 20 gpm capacity, refinements to the process have been incorporated. Major efforts have been directed toward dealing with silica that occurs in the groundwater of that region. In the 1-gpm pilot, precipitation of silica was observed in the stream that circulated through the EDM diluate compartments and the RO concentrate. Ions were removed from that stream by the EDM, and water was removed by the RO. However, because silica is not ionized appreciably at neutral pH, the silica accumulated in that circulating stream until it became supersaturated and precipitated. Precipitation of the silica could be prevented by purging some of that stream, but the purge stream represented a loss of water from the process. The water loss was reduced by treating the purge stream to remove silica and then returning the liquid to the process. Although the silica removal was successful, it added complexity and chemical cost to the ZDD process. The ultimate solution was to replace the RO membranes with nanofiltration (NF) membranes. Certain NF membranes are highly permeable to silica, and the presence of silica is not harmful to drinking water. With the selected NF membranes install 20-gpm pilot plant, and the silica problem was completely solved. Elimination of the silica purge allowed 98% of the brackish water to be produced as fresh water.

Table 1. Concentrations (mg/L) of solutes in the ZDD process streams.

		NF Streams			EDM Streams			
CONSTITUENT	**Source**	**Feed**	**Perm**	**Conc**	**Feed**	**Diluate**	**Mixed Cl**	**Mixed Na**
ANIONS								
F	1.7	1.4	0.9	2.5	1.8	0.8	ND	61.3
Cl	290	210	260	160	140	100	89,000	11,766
SO_4	1,300	1,300	9.50	3,600	2,600	1,300	1,600	90,764
CATIONS & METALS								
Ca	250	220	19	NT	380	160	17,000	716
Mg	94	94	7.50	NT	190	86	7,400	310
Na	370	350	150	NT	580	350	21,000	57,078
OTHER CONSTITUENTS								
SiO_2	24.0	27.0	23.0	NT	33.0	33.0	5.7	11.4
pH	7.4	7.0	7.0	7.0	6.9	6.6	1.8	7.9
HCO_3	205	110	70	135	128	43	NT	4,447
Conductivity	2,857	2,724	872	5,021	4,271	2,404	124,800	109,000
TDS (at 180 °C)	2,300	2,180	471	4,980	3,840	1,940	133,000	146,834

The concentrations of ions in the process streams of the 20-gpm pilot plant are shown in Table 1. The concentrations of calcium and sulfate ions in the source water and the EDM concentrate streams are highlighted. The concentrations of both the calcium and the sulfate increased by a factor of almost 70. Also noteworthy is the fact that the NF membrane was so permeable to silica that the silica level in the feed and diluate of the EDM were only slightly elevated.

(378c) Water Purification with Closed Circuit Reverse Osmosis

Richard Stover, Desalitech, Ltd., Newton, MA

The demand for clean water for industrial, agricultural and domestic use is increasing around the world. Reverse osmosis (RO) is an important method for water purification and desalination. Advances in membrane performance have increased permeate recovery, improved reliability and reduced energy consumption, however, broader application of RO has been limited, in part, by its energy requirements, costs, risks of fouling and scaling and operational challenges posed by source water changes. An improved technology that further increases permeate recovery, lowers costs, lowers energy consumption and increases the reliability and flexibility of RO would help advance the beneficial and efficient uses of locally-produced, sustainable water sources.

Closed circuit (CC) RO is being applied for water purification and desalination in commercial installations in several international locations. CC RO uses standard equipment and membranes in a batch-like configuration that operates at a lower average feed pressure than conventional RO, thereby reducing energy consumption. The process exposes membranes to varying feed compositions as the system concentrates and dilutes and provides high cross flow across short membrane arrays. These conditions improve membrane performance by improving flux distribution and enhancing cross flow. The process also has the potential to prevent fouling and scaling and disrupting bio-film formation and scale precipitation which, in turn, may reduce chemical use and membrane clean-in-place frequencies. The permeate recovery percentage can be adjusted at the control panel without modification of system hardware, giving operators the means to adjust system performance as necessary to cope with source water changes. A recovery range of 15 to over 97% has been demonstrated in single-stage units, and energy consumption reduced by 35%.

The objective of this study is to evaluate CC RO as an alternative water treatment process to conventional RO for achieving high recovery and brine minimization. The author presents the technical principles of the CC RO processes and operating data from running plants. Process design and performance models are developed and validated with the purpose of adapting CC RO for new applications. This paper will help stakeholders better understand this emerging technology and the potential general applicability of CC RO processes in desalination and water reuse.

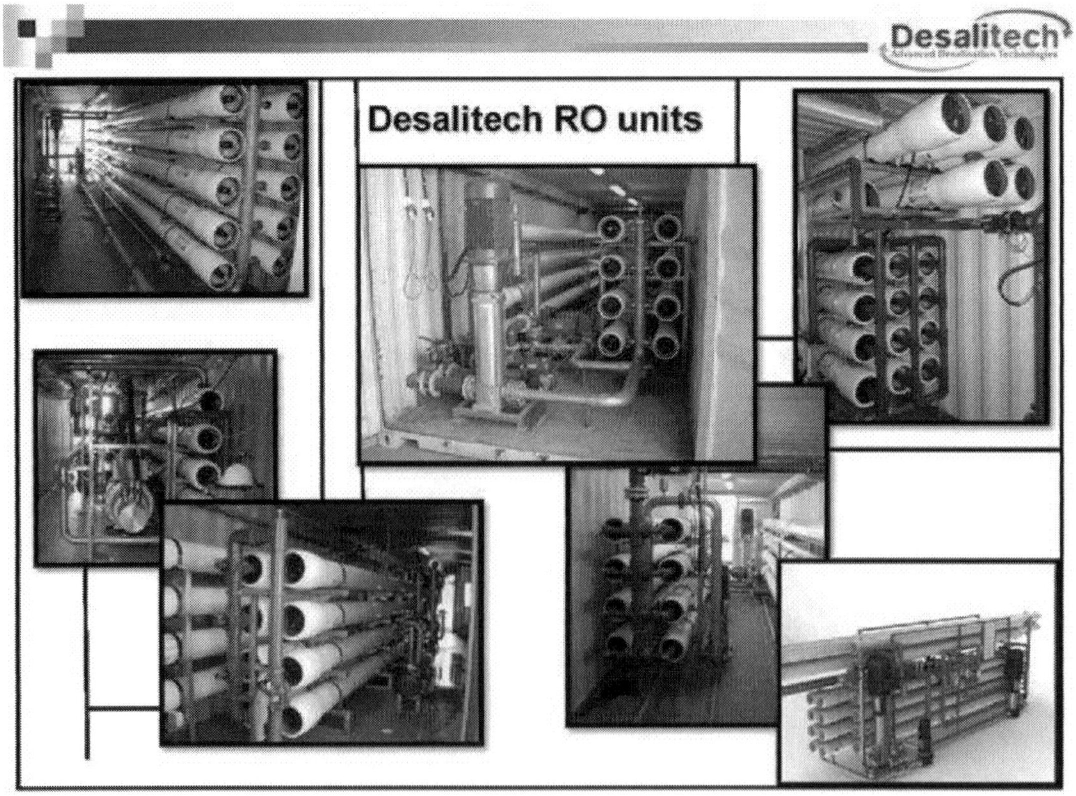

(378d) A Detail Analysis of the Performance of Different Antiscalants in Calcium Sulfate Supersaturated Brackish Water

Anthony Tarquin and **Guillermo Delgado**, Civil Engineering, University of Texas at El Paso, El Paso, TX

A Detail Analysis of the Performance of Different Antiscalants in $CaSO_4$ Supersaturated Brackish Water

Anthony Tarquin, PhD.; Guillermo Delgado, PhD Candidate

University of Texas at El Paso

Reverse osmosis (RO) desalination is an economic method to remove dissolved solids from brackish water. However, the presence of sparingly soluble salts in the water (e.g. $CaSO_4$, $CaCO_3$) reduces the recovery of these systems. If the solubility limit of these salts is reached, precipitation is likely to occur causing fouling and scaling of the membranes. It is at this point where the use of antiscalants greatly improves the recovery of desalination systems.

This research presents the evaluation of different antiscalants in brackish water supersaturated with calcium sulfate. A total of eight different antiscalants from five different manufacturers were tested. Each antiscalant was tested at different calcium sulfate concentrations and different antiscalants dosages. Brackish water from Alamogordo, New Mexico was used for this project. The brackish water was saturated using 1 Molar solutions of sodium sulfate and calcium chloride. The efficiency of the antiscalants was measured by the extent that the antiscalants increased the induction time of calcium sulfate precipitation in the water. The turbidity of the water was monitored in time intervals to identify the time of calcium sulfate precipitation. A saturated solution without antiscalant served as a control. The results showed that the type of antiscalant is an important factor that significantly affects the induction time of calcium sulfate precipitation. Figure one shows the results from one test done using different types of antiscalants.

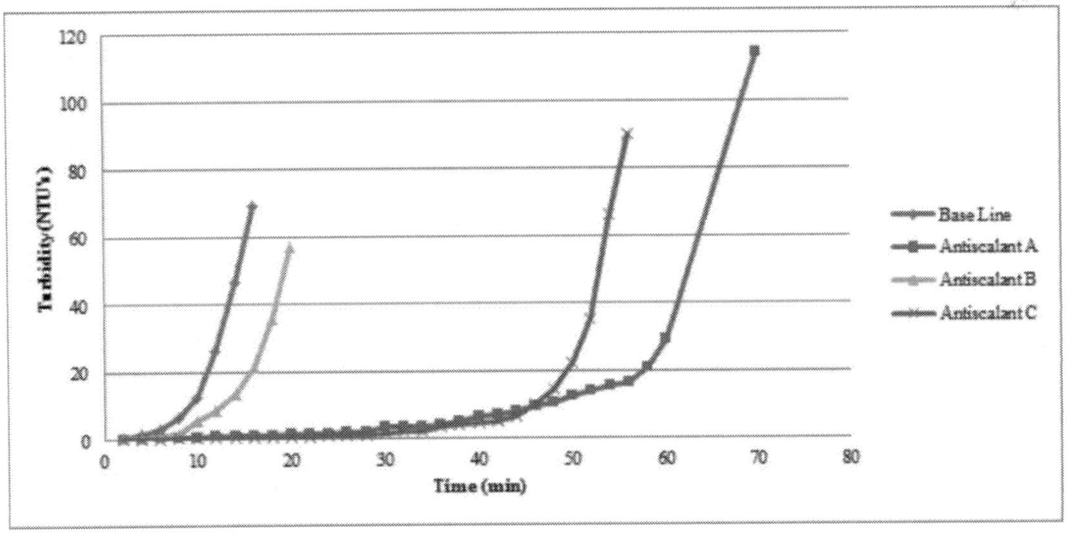

Figure 1: Antiscalant Performance at 0.03 M $CaSO_4$ Concentration and 0.8 ppm of Antiscalant

Additionally, the results showed that the dosage of the antiscalant is a factor for increasing induction time. However, the results showed that antiscalants are not efficient at high supersaturation concentrations (0.1 M) of calcium sulfate, no matter the dose. In the same way, the results showed that antiscalants only increase the induction time of calcium sulfate precipitation, but they do not increase the solubility of calcium sulfate. Figure 2 shows the results of a test done using different concentrations of the same antiscalant.

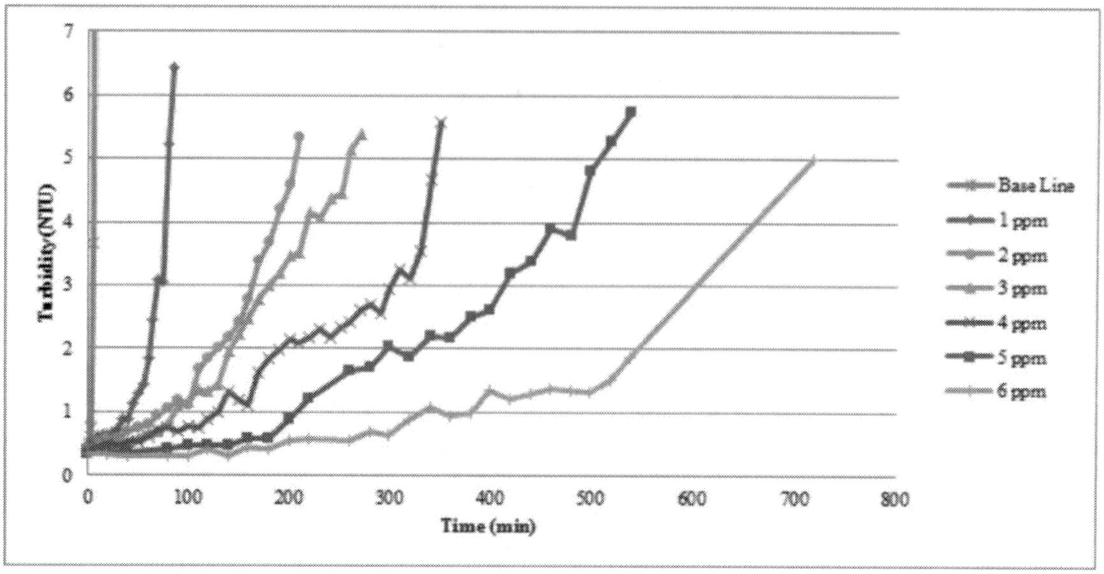

Figure 2: Different Concentrations of the Same Antiscalant at 0.05 M $CaSO_4$ Concentration.

The selection of the best antiscalant is a very important step during the design of a membrane desalination system, especially batch processes such as the CERRO process. This research showed that the selection process should include factors such as type and dosage of antiscalant. Recovery rates can be increased and operation costs can be reduced by optimizing system design.

(378e) Purity of NaCl in the Electrodialysis Metathesis (EDM) Process for Water Desalination

Lucy Mar Camacho[1], Tom Davis[2] and Malynda A. Cappelle[1], (1)Center for Inland Desalination Systems, University of Texas at El Paso, El Paso, TX, (2)Center for Inland Desalination Systems (CIDS), University of Texas at El Paso, El Paso, TX

Electrodialysis metathesis (EDM) is a component in the zero discharge desalination (ZDD) process that maximizes water production and salt recovery from desalination. In the EDM process the concentrated effluent from reverse osmosis (RO) or nanofiltration (NF) is treated to recover minerals, such as gypsum and magnesium chloride, as well as additional pure water. The minerals are separated and collected in two of the four compartments of the EDM stack that are separated by ion exchange membranes. One compartment will contain a concentrated mixture of sodium salts and another compartment will contain a concentrated mixture of chloride salts. The third compartment will contained the minerals-rich feed water to be treated. The fourth compartment is filled with sodium chloride to provide the ions for the metathesis reaction. The concentrated mixtures are generated when a voltage is applied to the repeated four-compartment units of the EDM stack. Once the concentrated streams exit the system they can combine to precipitate gypsum.

The early precipitation of calcium sulfate in one of the concentrate compartments can adversely affect the potential recovery of other useful salts. Furthermore, the EDM stack can be clogged and damaged. Commercially available sodium chloride may contain sodium sulfate and other minerals that will promote the formation of gypsum in the stack. In this study we are adding impurities to the sodium chloride compartment to determine the lowest purity allowed by the system. Preliminary experiments have shown that sodium chloride with concentrations of 1,000 mg/L of sodium sulfate doesn't affect the performance of the salt separation process. This results suggest that the cost of EDM can be further reduced by reducing the need of adding high quality chemicals to the process.

(380a) Multidisciplinary Optimization Model for Region Specific Sustainable Biorefining

Sumesh Sukumara[1], Dr. Jeffrey Seay[1], Dr. Fazleena Badurdeen[2], Joseph Amundson[3] and William Faulkner[3], (1)Chemical and Materials Engineering, University of Kentucky, Lexington, KY, (2)Institute for Sustainable Manufacturing, University of Kentucky, Lexington, KY, (3)University of Kentucky, Lexington, KY

Biofuels are a promising alternative to the fossil based fuels, upon which we have become dependent to supply our energy needs. There are many ways in which biomass can be converted into useful products. Integrated biorefining is a possibility which combines a slate of processes involving various feedstock to produce transportation fuel, electric power and other useful chemical products. Widespread availability, varying feedstock and specific regional conditions makes it challenging to develop an optimization model that can be applied to any region to estimate the overall economics of the process. Based on the literature and our previous experiments it is clear that a multidisciplinary approach is required to deal with the above challenge. This contribution will illustrate a novel approach to develop a region specific optimization tool which links various aspects of the biofuel supply chain like transportation, storage, processing, distribution and environmental effects. This contribution will show how various small-scale optimization models can be combined, resulting in a complete, multivariable economic optimization model. For this model, literature data sources as well as mapping and spatial modeling data are used to locate biomass sources in a given geographic region. Based on the source of biomass and available road network, transportation costs are optimized using ILOG-OPL software. The results can be used to determine a suitable location for potential biorefineries in any given region. Processing and capital costs are other key factors in determining overall economics for biorefining and hence demands a detailed process optimization model.

A multi feedstock process optimization model including both preprocessing and conversion of biomass into biofuels and products will be described in this contribution. The process optimization model to be described consists of two major processes to produce transportation fuel and power as end products. This model has the capability of including various types of biomass such as dedicated energy crops, crop residues and animal wastes as feedstock. Aspen Plus V-7.3 is the software package used for this purpose. Multiple process and transportation optimization models are then interfaced using Visual Basic in order to provide an economic assessment for biorefining in a region. This model will be validated using the Jackson Purchase region of Western Kentucky, which is not only rich in coal, lignocellulosic biomass and crop residue but also has large number of poultry farms, as a case study. Finally an optimum location for the proposed biorefinery, ratio of combined feedstock and profitability can be determined based on the output of this model. The model developed will have an ability to show how multiple parameters in various stages of supply chain effects the overall economic of the process. This innovative linking of various optimization models will help develop a large scale multivariable model for biorefining which will include all possible feedstocks for a given region and provide the option to choose from various available processes in the model based on the cost, demand and environmental regulations in a region. The model can not only be used by investors as a tool to estimate the potential profitability of biorefining in a region but will allow any locale to attain the distant goal of sustainability.

Sustainable Processing of Water Hyacinth through a Distributed Biorefinery System

José Ezequiel Santibañez-Aguilar,[1] J. Betzabe González-Campos,[1] José María Ponce-Ortega,[1] Medardo Serna-González[1], and Mahmoud M. El-Halwagi[2, 3]

[1]*Chemical Engineering Department, Universidad Michocana de San Nicolás de Hidalgo, Morelia, Michoacán, México, 58060*

[2]*Chemical Engineering Department, Texas A&M University, College Station, TX, USA, 77843*

[3]*Adjunct Faculty at the Chemical and Materials Engineering Department, King Abdulaziz University, Jeddah, Saudi Arabia.*

Water hyacinth is a major nuisance of water bodies. In addition to blocking pathways, this plant quickly consumes the available nutrients in the water as well as dissolved oxygen thereby affecting the aquatic life in the water bodies as well as human activities such as fishing. To eliminate this plant from the water bodies, several techniques have been implemented (including chemical, physical and mechanical elimination techniques, many of them with severe environmental consequences). Much more work is needed to address this problem because the very high growth rate of water hyacinth. Recently, several papers have reported the possible use of water hyacinth as feedstock to produce biofuels, compost, chemical products, green manure, and paper, among others. In this regard, this paper presents an economical and environmentally sustainable alternative to address this problem, where water hyacinth is used as a raw material to produce a set of products as well as clean water (see **Figure 1**). Since the water hyacinth is geographically located in a given region, a distributed processing system is required considering the installation of both preprocessing and central facilities to take advantages of the economies of scale. In this work the objective is to present a mathematical programming formulation to address the optimal design of distributed biorefinery systems of water hyacinth according to economic and environmental criteria. The proposed model is based on the superstructure shown in Figure 2 and it takes into account a set of water bodies as sources of water hyacinth for biomass and water production. The biomass can be processed using different technologies to obtain several products, which can be sent to the available markets. On the other hand, the water extracted from water hyacinth can be treated to separate the pollutants (because the roots of water hyacinth absorb several pollutants helping this way to clean the water bodies) and to obtain clean water. The proposed model is applied to a case study in the central part of Mexico, which is a region with several lakes highly congested with water hyacinth. The results show that water hyacinth elimination can be economically attractive when the distributed treatment system proposed in this paper is considered.

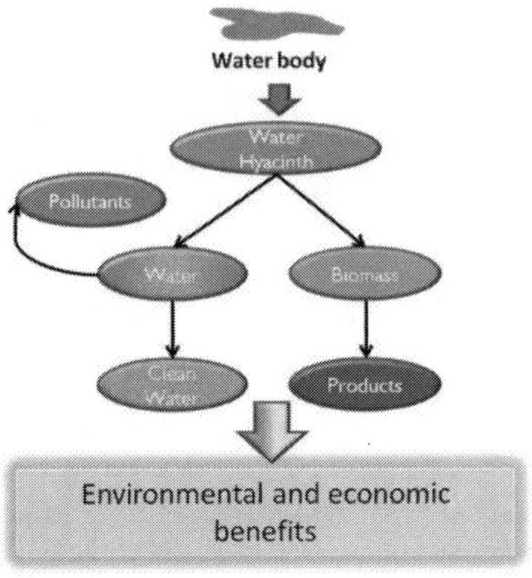

Figure 1. Schematic representation for the addressed problem.

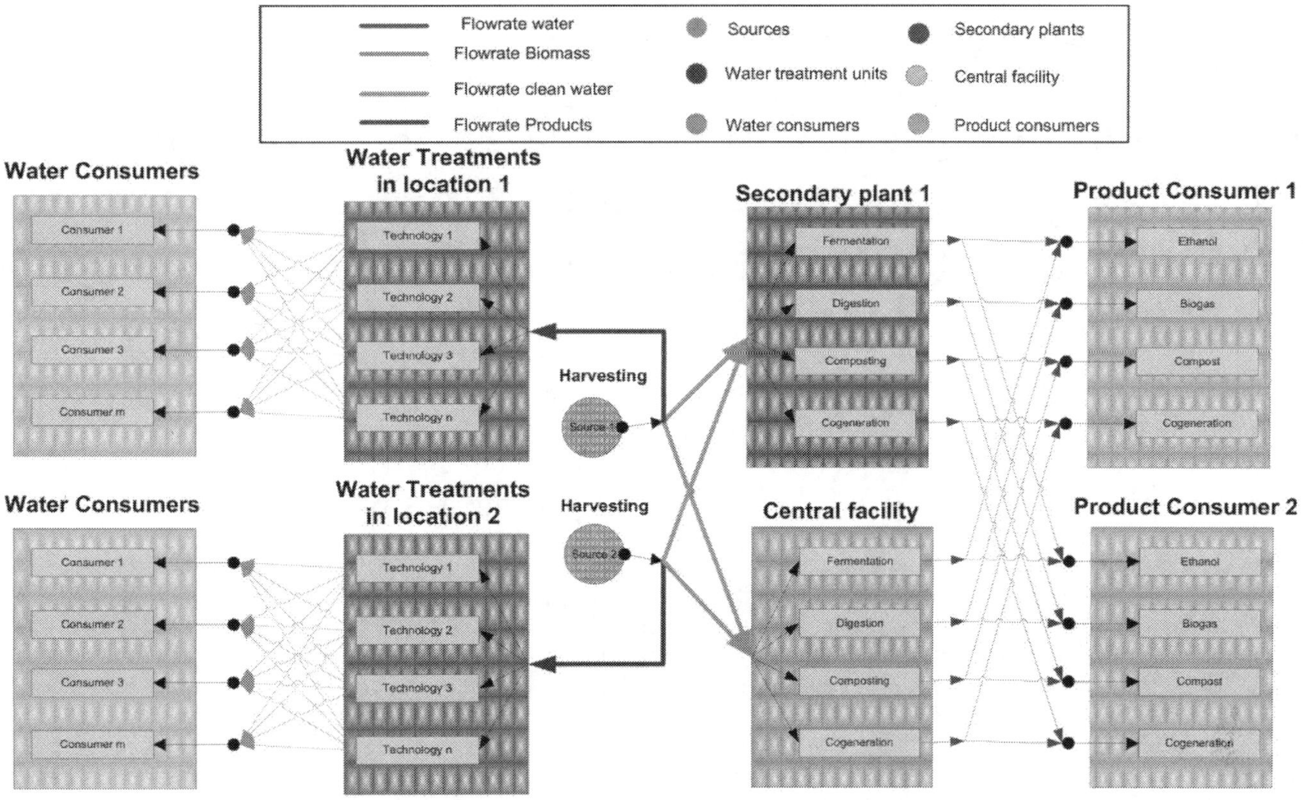

Figure 2. Proposed superstructure for the sustainable use of water hyacinth.

(380c) Optimal Design of County-Level Hydrocarbon Biorefinery Supply Chains Under Uncertainty: A Case Study for the State of Illinois Using Spatially-Explicit Model

Yuan Yao, Berhane H. Gebreslassie and Fengqi You, Department of Chemical and Biological Engineering, Northwestern University, Evanston, IL

Hydrocarbon biorefinery technologies, which convert biomass to hydrocarbon biofuels such as gasoline and diesel, have been considered as promising approaches to replace traditional fossil-fuel-based energy generation technologies. Hydrocarbon biofuels can be potentially used without significant changes of the current fuel distribution and utilization, including pipelines, pumping stations, and vehicles. Compared with their conventional counterparts, hydrocarbon biofuels provide vehicle similar to or better performance.[1] Renewable Fuel Standards (RFS) has put a target of producing 36 billion gallons of biofuels in 2022.[2] In order to achieve this goal in a relatively short time, it is appealing to investigate design of supply chain network for biofuels. Currently, several mathematical models have been developed to address this issue.[3, 4] However, most of them are deterministic approaches that assume all problem parameters can be exactly known in advance. In practice, the hydrocarbon biorefinery supply chain may be affected by a variety of uncertainties, such as seasonal and geographical fluctuation of biomass supplies, population growth and unexpected events.[5] It is necessary to develop a comprehensive model for the optimal design of hydrocarbon biorefinery supply chain under the presence of uncertainties. Another critical issue is the existence of potentially high financial risk when uncertainties are involved in. Risk management should be taken into consideration in the final design of hydrocarbon biorefinery supply chain to reduce the impacts of unexpected outcomes of uncertain parameters.[6]

In this paper, we study the county-level hydrocarbon biofuel supply chain for the State of Illinois. A spatially explicit model is developed to integrate decision making across multiple temporal and spatial scales and simultaneously predict optimal network design, technology selection, capital investment, production operations, and logistics management decisions under hydrocarbon biofuel demand and biomass supply uncertainties. The objective of the model is to minimize the total expected annualized cost and financial risk that is quantified by conditional value-at-risk (CVaR)[7] and downside risk.[8] The State of Illinois consists of 102 counties. Thus, the supply chain network in this case study has 102 harvesting sites, 102 hydrocarbon biorefinery location candidates and 102 demand zones. Three major types of biomass are considered, which are agricultural residues, energy crops, and wood residues. Two types of hydrocarbon biofuels, diesel and gasoline, are produced and shipped to demand zones. We consider two major conversion technologies: gasification followed by FT synthesis and fast pyrolysis followed hydroprocessing. The capacities for both of them have three levels: 1-50 MGY (million gallons per year), 50-100 MGY, and 100-200 MGY.

Case study in the State of Illinois with sample size of 100 scenarios is solved and the minimum total expected annualized cost is \$2,822.565MM. It corresponds to a minimum supply chain cost of biofuels is 4.00 \$/GEG (gasoline-equivalent gallon). In the optimal solution, 11 biorefineries with capacities ranging from 62 MGY to 200 MGY are built in 11 counties that have high population density. Six of them are in the north part of Illinois, another four are located in central Illinois and the one left is built in southern Illinois. For each biorefinery, biomass used is supplied by local harvesting sites or those in adjacent 1 to 8 counties. The number of counties that are involved in supplying biomass to hydrocarbon biorefineries ranges from 23 to 56. When biomass supply decreases or biofuel demand increases, more counties join biomass supply network, leading to more inter-county transportation. According to production capacity and biofuel demand, hydrocarbon biofuels of each biorefinery are shipped by trucks to satisfy local demand or demand in other counties. The biorefinery with capacity of 80 MGY in Cook County can only meet local demands because of the highest demands of biofuels in Chicago area. On the contrary, as a result of the lowest demands in southern Illinois, the biorefinery in Wayne County with capacity of 125 MGY is able to meet biofuel demands in as many as 28 southern counties. When biomass supply reduces or biofuel demand increases too much, the number of counties whose demand can be met by each biorefinery shrinks a lot. Affected by yields of two technologies using different kinds of biomass that are geographically representative, most of biorefineries choose fast pyrolysis except the plant in La Salle County that adopts gasification. The inventory levels of biofuel products and biomass feedstock in a year always reach peaks in November and decrease gradually in other months, this trend is consistent with the seasonality of typical biomass resources corn stovers in the State of Illinois. Increase of either biofuel demand or biomass supply contributes to a high level of biomass inventory. Biofuel inventories are more sensitive to demand, they fall down sharply when demand grows greatly. In the optimal solution after financial risk management, some of biorefineries shift from counties with high population to counties with rich biomass resources to reduce the risk of supply disruption. Based on the optimal solutions of hydrocarbon biorefinery supply chain, counties with both high population density and abundant biomass resources are perfect locations of biorefineries.

References

[1] F. Q. You and B. Wang, "Life Cycle Optimization of Biomass-to-Liquid Supply Chains with Distributed-Centralized Processing Networks," *Industrial & Engineering Chemistry Research,* vol. 50, pp. 10102-10127, 2011.

[2] EIA, "U.S. Energy Information Administration," 2011.

[3] A. Dunnett, C. Adjiman, and N. Shah, "A spatially explicit whole-system model of the lignocellulosic bioethanol supply chain: an assessment of decentralised processing potential," *Biotechnology for Biofuels,* vol. 1, pp. 1-17, 2008.

[4] B. Aksoy, H. Cullinan, D. Webster, K. Gue, S. Sukumaran, M. Eden, and N. Sammons, "Woody biomass and mill waste utilization opportunities in Alabama: Transportation cost minimization, optimum facility location, economic feasibility, and impact," *Environmental Progress & Sustainable Energy,* vol. 30, pp. 720-732, 2011.

[5] B. H. Gebreslassie, Y. Yao, and F. Q. You, "Design under Uncertainty of Hydrocarbon Biorefinery Supply Chains: Multiobjective Stochastic Programming Models, Decomposition Algorithm and A Comparison between CVaR and Downside Risk," *Submited to AIChE Journal* 2012.

[6] F. Q. You, J. M. Wassick, and I. E. Grossmann, "Risk Management for a Global Supply Chain Planning Under Uncertainty: Models and Algorithms," *AIChE Journal,* vol. 55, pp. 931-946, 2009.

[7] S. Uryasev and R. T. Rockafellar, "Conditional Value-at-Risk: Optimization approach," in *Stochastic Optimization: Algorithms and Applications*. vol. 54, S. Uryasev and P. M. Pardalos, Eds., ed, 2001, pp. 411-435.

[8] G. Eppen, and and R. Martin, "A scenario approach to capacity planning," *Oper Res,* vol. 37, pp. 517–527, 1989.

(380d) Life-Cycle Assessments of Forest Biomass Supply in Michigan

Robert M. Handler, Michigan Technological University, Houghton, MI, Dalia Abbas, Tennessee State University, Nashville, TN, TN, Ajit Srivastava, Michigan State University, East Lansing, MI and David R. Shonnard, Chemical Engineering, Michigan Technological University, Houghton, MI

In this study, we developed measures of the environmental footprint for primary forest products supply chain activities (harvesting and hauling wood) within the state of Michigan. A life-cycle assessment procedure is described, relying on a combination of peer-reviewed literature, national databases, and primary data collected from loggers and truckers within the Michigan study area. Several different equipment configurations and operating scenarios for roundwood harvesting have been considered, with greenhouse gas emissions and fossil energy demand per unit of delivered wood utilized as ultimate environmental indicators. Results indicated that a full processor / forwarder is the best choice of harvesting equipment configuration due to relatively low inputs and high reported productivity, although the burdens of harvesting depend strongly on the intensity of harvest being conducted. Multimodal truck + rail transport had roughly 2X lower environmental burdens than typical log truck transport, which was directly related to the increased fuel efficiency of rail transport. Aggregated results for roundwood supply within the state of Michigan are comparable to similar studies in other regions, although the mechanization of the harvesting industry and large size of MI log trucks are factors in the relatively smaller environmental burden.

(380e) Distributed Small Scale Marine Biorefineries: Scale Analyses and Design for Developing Countries

Alexander Golberg[1], Gregory Linshiz[2], Nathan Hillson[2], Jay D. Keasling[3], Alexander Chemodanov[4] and Mark Koudritsky[5], (1)UC Berkeley, Berkeley, CA, (2)Joint BioEnergy Institute, Lawrence Berkeley National Laboratory, Berkeley, CA, (3)Berkeley Center for Synthetic Biology, UC, Berkeley, Berkeley, CA, (4)Independent, Hedera, Israel, (5)Independent, Toronto, Canada

Abstract

Distributed Small Scale Marine Biorefineries: scale analyses and design for developing countries.

Alexander Golberg[1], Gregory Linshiz[2], Nathan J. Hillson[2], J. D. Keasling, M. Koudritsky

Alexander Chemodanov[3].

[1] Department of Mechanical Engineering, *Etcheverry* Hall, 6124, University of California

at Berkeley, Berkeley, CA 94720 USA. and Center for Engineering in Medicine and Department of Surgery, Massachusetts General Hospital, Harvard Medical School, Boston, MA, USA

Email: agolberg@gmail.com.

[2] Joint BioEnergy Institute ,Lawrence Berkeley National Labs, 1 Cyclotron Road, Berkeley, CA 94720.

[3] Independent researcher, 2-236 Duke st. W. Kitchener ON, Canada.

[4] Independent researcher, Milner 16b Hadera P.B.20995, Israel.

The major growth in liquid fuel demand over the next 20 years will be predominantly due to developing countries. While Gross Domestic Product (GDP) and primary energy production in Organization for Economic Co-operation and Development (OECD) countries is not linearly correlated, GDP growth in developing countries requires increases in primary energy production. The World Energy Council predicts that India and China will overtake developed countries in transportation fuel consumption by 2025. Due to climatic, economic, and fossil fuel resource constraints, there is an urgent need for the sustainable, cost-effective production of carbon-neutral transportation fuels.

Biofuels present an alternative to fossil fuels. First generation biofuel technologies utilize established processes and currently produce biofuels on a commercial scale. First generation feedstocks include sugar beet, starch-bearing grains, and conventional vegetable oil crops, and first generation fuel products include ethanol and biodiesel. Second and third generation biofuel technologies, currently in research and development, utilize animal fat, lignocellulosic biomass, and algae feedstocks, and produce hydrotreated vegetable oil, cellulosic-ethanol, biomass-to-liquids (BtL)-diesel, bio-butanol, and advanced drop-in replacement fuels such as fatty-acid ethyl esters, bisabolene, pinene, n-alkanes, n-alkenes, and methyl ketones.

Although biofuels may collectively supply a portion of future transportation fuel demand, competition between “energy crops” and “food crops” for land and water resources is a growing concern. Furthermore, the extents to which land erosion, potable water consumption, fertilizers, pesticides, and climate change impact biofuel sustainability have yet to be evaluated .

While significant efforts are being directed towards developing feedstocks and conversion technologies, biorefinery design remains in its infancy. The optimization of biorefinery size, feedstock, technology, and serviced area will be required to reduce the costs of the resulting biofuel products.

Recent research in constructal design shows that the optimal distribution of flows of products and services across a populated area depends on a balance between the size of the production centers with sizes of distribution networks that connect these centers to end users. Although larger systems are thermodynamically more efficient in production, they serve larger areas; thus, the collection and distribution logistical costs also increase with size. Therefore, a balance exists between efficiencies of scale, and system distribution losses. This balance has been investigated both for thermodynamics of energy sources and economics of agricultural/biofuel systems. Applications of the balance principle have been demonstrated for the thermodynamical optimization of hot water flow and heating, refrigeration, combined solar power and desalination, and agricultural product processing economics. The balance principle has also lead to proposals for "distributed energy systems" to optimize energy production size and service area.

The goal of this work is to show that the balance between thermodynamic efficiency of system size, collection and distribution is valid for biorefineries. This work also aims to demonstrate that population characteristics, such as density and per capita liquid fuel consumption, play a critical role in the design of biorefinery size, technology choice and serviced area.

We report a model macro algae biorefinery design for midsize towns in low to medium income countries with low liquid fuel consumption per capita. We targeted these populations since the majority of new fuel systems built in future years will serve their growing demands. We focus our model on macro algae, a promising biofuel crop feedstock that does not compete with food crops for arable land or potable water. Furthermore, macro algae, which do not contain lignin, are convenient candidates for cost effective processing with current technology. We analyze biorefinery production of a transportation biofuel from the green macro algae *Ulva sp*, and demonstrate the integration of this model biorefinery into the distributed energy systems. For example, we analyze a system that would supply all the liquid fuel needs of a 20,000 person coastal town in a middle income country, where the annual per capita demand is 26 gallons of liquid fuel per year. The designed system consists of an open controllable photobioreactor and a consolidated bioprocessing unit, which yields renewable liquid fuel. We propose to use a green seaweed *Ulva* as a biomass source. The analyses of prior studies shows that *Ulva* yields of 15-22 kg dry weight$\cdot$ m^{-2} $\cdot year^{-1}$, with a heating value of about 19 $MJ\cdot kg^{-1}$, can be achieved with 200 $W\cdot m^{-2}$ solar radiation with a zero pesticide footprint. With current ethanol/seaweed conversion rates, a 30 ha cultivation farm, along with 8 ha of a photo-voltaic (PV) facility to provide operational energy, is required to satisfy the liquid fuel demands of a middle income town in a developing country. The system is easy to operate and maintain, and can be manufactured and installed in multiple locations. Finally, in the context of a single biorefinery, we compare macro algae with corn grain and cassava feedstocks.

IDENTIFICATION OF THE MOST APPROPRIATE METHOD TO USE SEWAGE SLUDGE AS A SOURCE OF ALTERNATIVE ENERGY

Yolanda Pliego-Bravo[1], **Marlene Vergara-Hernández**[1], Galo Rafael Urrea-García[1] and María Elena García-Reyes[1],

[1]Instituto Tecnológico de Orizaba, Veracruz, México.

Key words: **SEWAGE SLUDGE, ALTERNATIVE ENERGY**

Abstract

Currently, Mexican law requires the safe use and the recovery of waste, as in the case of sludge from a wastewater treatment plant that can be used as fertilizer, however, the presence of metals can be a limiting. Therefore another option is energy recovery from sludge. In this paper have been analyzed the sewage sludge from a treatment wastewater plant to meet its potential use as alternative energy source. To determine the physicochemical and microbiological characteristics have been made the analysis of moisture, ash, organic matter, volatile matter, total solids, volatile solids, fixed carbon, density, pH, total nitrogen, organic nitrogen and ammonia nitrogen as well as fecal coliform and Salmonella *spp*. Elemental analysis (C, H, O, N and S). The heating value (higher heating value and lower) for recovery as an energy source. Just as the heavy metal content to estimate the waste generated during incineration.

Introduction

Water is the most useful and important resource of life. When wastewater is treated to return this resource to the environment, a semisolid, nutrient-rich by-product called wastewater sludge is produced. Of the amount of water treated in treatment plants, is generated 1% of sludge. The cost of sludge treatment is higher than the treatment of wastewater due to time spent and equipment needed to perform each treatment. Therefore, the processing of waste water sludge must be done in a cost-effective way , (Turovskiy, 2006).

Wastewater sludge generation has increased considerably in recent years due to the amount of treated water, those sludge represent an important source of pollution. Research has concluded that sludge could be converted to a useful source energy in order to remove all its negative effects and to contribute to the power generation, (Midilli, 2000).

Renewable energy is of growing importance in satisfying environmental concerns over fossil fuel usage. The recovery of biosolids alternatives are intended to recover the chemical value of these materials and find ways direct target for processing or as raw material for second-generation processes, basically focused on the recovery of organic content minimization risks of spontaneous combustion, pests, odors and leachate.

There are many ways to dispose the sludge but it is necessary to choose the correct one, in order to have a benefit and minimize the pollution generated by them. There are mainly four commercial means of sludge disposal; fertilizer, land filling, dumping in the sea, and incineration. However, these means are not designed for waste-to-clean energy conversion. On the other hand, pyrolysis and gasification are considered waste-to-clean energy technologies. Additionally, these means can reduce flue gases volume and fix heavy metal in the solid residues, while producing valuable products of oil and syngas. (Nimit, 2010).

Thermochemical processing of biomass uses heat and catalyst to transform plant polymers into fuels, chemicals, or electric power. This contrasts with biochemical processing of biomass, which uses enzymes and microorganisms for the same purpose.

Gasification is the conversion of carbonaceous solids at elevated temperatures and under oxygen-starved conditions into syngas, a flammable gas mixture of carbon monoxide, hydrogen, methane, nitrogen, carbon dioxide, and smaller quantities of hydrocarbons. Gasification can be used to convert any carbonaceous solid or liquid to low molecular weight gas mixtures. In fact, the high volatile matter content of biomass allows it to be gasified more readily than coal.

Pyrolysis is by definition thermal decomposition of organic compounds occurring in the absence of oxygen to produce liquids, gases and char. It is always also the first step in combustion and gasification processes where it is followed by total or partial oxidation of the primary products.

Methodology

The sewage sludge samples were collected from a waste water treatment plant and analyzed by 2500 and 2540 Standard Methods, NOM-21-RECNAT-2000 and NOM-004-SEMARNAT-2002.

Proximate analysis and the ultimate analysis

Moisture was determined, ash, volatile matter and fixed carbon by the proximate analysis, as well as, organic matter, total and volatile solids and the content of heavy metals. The ultimate analysis provided the content of carbon, hydrogen, nitrogen, sulfur and oxygen.

Microbiological analysis

By microbiological analysis were determined fecal coliform and Salmonella *spp*, based on the technical requirements of NOM-004-SEMARNAT-2002.

Heating Value

The higher and lower heating value were obtained by calorimetery, and the thermochemical characterization was done by thermogravimetry and differential scanning calorimetry.

Results and discussion

The purpose of studies of this nature was to identify the thermochemical properties of sewage sludge and thereby take advantage of these properties in a process of alternative energy generation, which contributes to solve the worldwide energy crisis.

Table 1 shows the proximate analysis, which consist of the content of moisture, ash, volatile matter and fixed carbon. These parameters are important in the characterization, since these will define the quality of the fuel in the combustion to determine the content of combustible material present, and produce smoke when burned and the amount of coke or "char" (intermediate combustion products) that can occur, and the content of unburned carbon as fly ash.

Due to the percentage of moisture in sludge samples, would be required a drying thereof prior to its energetic use by the pyrolysis and gasification technologies, reaching a moisture content of about 10-15%. On other hand, there are technologies as hydrothermal processing in which the high moisture content is favorable for the production of hydrogen-rich fuel gas, (Brown, 2011). The results agree with the values displayed by Mejía, (2011), mentions that the sludge that is treated by filtration and centrifugation, such as sludge studied, has a moisture content of 85 to 50%.

Table 1, presents the values of proximate analysis of sludge with different moisture content, presented by Dominguez, (2006). On a dry basis, the volatile matter content of a fuel varies less than 5% for anthracite over 75% in the case of wood. The reactivity of a solid fuel and its conversion to char within the gasifier depends on its content of volatile matter. Fuels with higher volatile matter content are more reactive and therefore can be converted into gas, more easily producing less char, (Basu, 2006).

Table 1. Proximate analysis*

Parameter	*This worK*	*Domínguez (2006)*	
Moisture	83	71	81
Ash	29.91	31.2	38.1
Volatile matter	67.91	62.3	54.7
Fixed carbon	2.18	6.5	7.2

*Dry basis.

Table 2 shows the results of ultimate analysis of the sludge studied and a comparison with sludge studied by other authors that have characterized sludge, for use as an alternative energy source. It is remarkable the low content of nitrogen and sulfur, which are undesirable elements in thermochemical processes, since they increase the production of compounds that generate environmental problems as NOX and SOX.

Table 2. Ultimate analysis

Element	*This work*	*Lou (2011)*	*Zhang (2011)*
N	5.77	2.97	8.3
C	35.26	28.56	42.3
H	7.23	3.29	3.7
S	0.80	0.11	19.1
O*	50.94	65.07	26.6

*By diference

Table 3 shows the average of some heavy metals analyzed from the sample of sewage sludge, and presents the values set by the standard.

Table 3. Heavy metals content in sewage sludge

Metal	*Content (ppm)*	*Reference NOM-002-SEMARNAT-2002*
Antimonio	22.90	---
Arsenico	< 0.30	41-75
Talio	29.02	---
Cadmio	< 1.50	39-85
Cromo	46.76	1200-3000
Mercurio	< 0.30	17-57
Plomo	< 10.00	300-840
Selenio	< 15.00	---
Vanadio	< 15.00	---
Zinc	54.54	2800-7500
Niquel	< 15.00	420
Cobalto	1.91	---
Manganeso	20.78	---
Bario	15.92	---

Figure 1 shows the results of the count of fecal coliforms and Salmonella *spp*, as expected, the NMP is high, indicating that the sludge has to be stabilized for safe and acomplish with regulations. It has been documented that Salmonella, as well as fecal coliforms die when exposed to temperature of 55 ° C for one hour or between 15 and 20 min at 60 ° C (Cota, 2008), therefore, the use of sludge studied, using a thermochemical process is favorable to the elimination of pathogenic microorganisms.

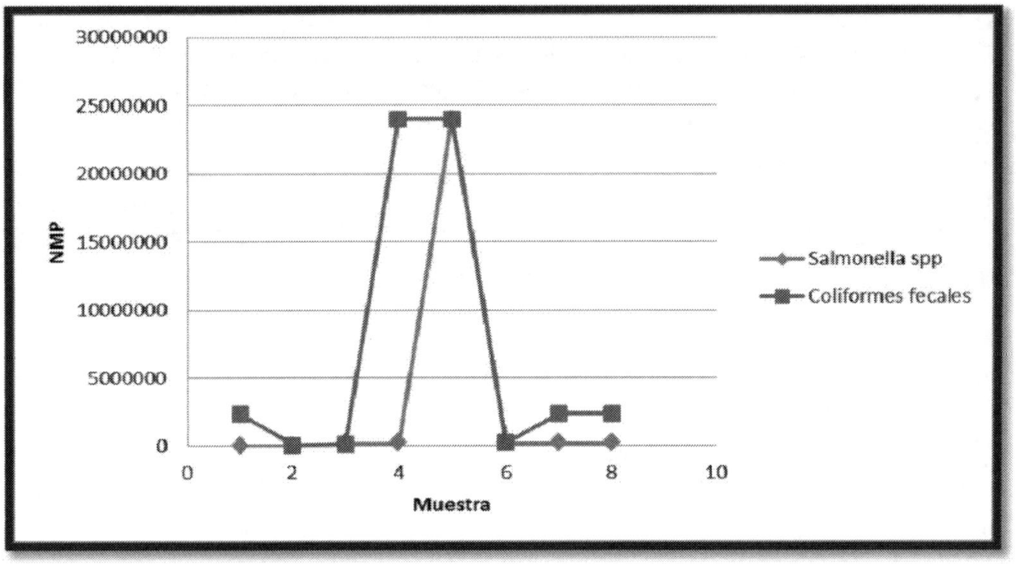

Figure 1. Fecal coliforms and Salmonella *spp*

Table 4 shows higher and lower heating value, obtained by calorimetry

Table 4. Higher and lower heating value

Heating value	*Value (Mj/kg)*
Inferior	1.655
Superior	1.7825

*wet basis

Figure 2 shows the thermogravimetric curve obtained from TGA analysis to which the sludge sample was subjected, it can be appreciated the mass losses when, it was subjected in TGA at heating in a range of 0 to 900 ° C at a rate of 10 ° C / min in a nitrogen atmosphere at a flow of 20 mL / min.

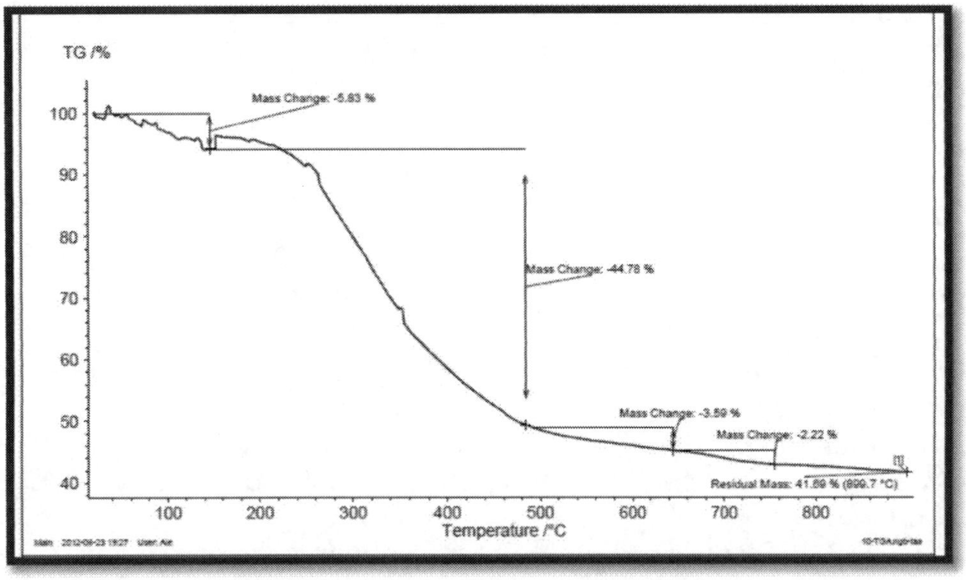

Figure 2. Thermogravimetry curve

Figure 3 shows the results of the curve of DSC performed to the sample of sludge and it was carried out under the same conditions as it was conducted thermogravimetric analysis, in this case was observed endothermic curve because the transformation involves energy absorption, and can observe the ΔH obtained at different temperatures.

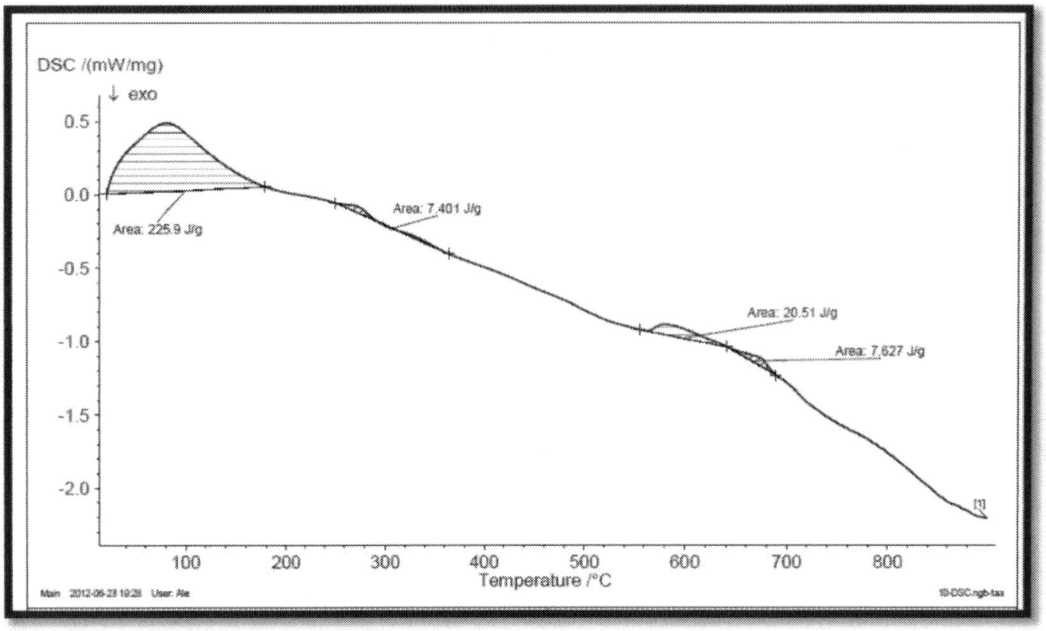

Figure 3. Differential scanning calorimetry (DSC)

Conclusions

There were obtained the general characteristics of the sewage sludge by performing ultimate and proximate analysis, thus fulfilling the objectives set for the characterization thereof. With the characteristics observed in the sludge, it can be concluded that they are feasible for use by a thermochemical process as alternative energy source.

References

1. Beiping, Z., Sijiang, X., Bo, X., Dongke, Y., Xiaoyuan, J. (2010). Mechanism of wet sewage sludge pyrolysis in a tubular furnace. School of Environmental Science and Engineering. International Journal of Hydrogen energy 36(2011)355-363.
2. Brown, R. C. (2011). Thermochemical processing of biomass. Conversion into fuels, chemicals and power. Wiley.
3. Domínguez, A., Menéndez, J. A., Pis, J.J. (2006). Hydrogen rich fuel gas production from the pyrolysis of wet sewage sludge at high temperatura. Instituto Nacional del Carbón, CDIC, apartado 73 Oviedo, España.
4. Girovich, M. J. (1996). Biosolids treatment and management processes for beneficial use. Marcel Dekker, Inc.

5. Grammelis, P., Basinas, P., Malliopoulou, A., Sakellaropoulos, G. (2009). Pyrolysis kinetics and combustion characteristics of waste recovered fuels. Fuel 88 (1), 195–205.

6. Mejía, M. E. (2011). Aspectos técnicos y legales del manejo de lodos en México. academiadeingenieriademexico.mx/...lodos.

7. Nimit, N., Islam, I. A., Somrat, K., Ashwani, K. G. (2010). Hydrogen and syngas production from sewage sludge via steam gasification. The Combustion Laboratory, Dept. of Mechanical Engineering, University of Maryland, College Park, MD, USA. The Waste Incineration Research Center, Dept. of Mechanical and Aerospace Engineering, King Mongkut's University of Technology, North Bangkok, Thailand.

8. Oropeza García Norma. (2006). Lodos residuales: estabilización y manejo. Departamento de ingeniería. Universidad de Quintana Roo.

9. Otero, M., L.F. Calvo, M.V. Gil, A.I. García y A. Morán. (2007). Co-combustion of different sewage sludge and coal: A non-isothermal thermogravimetric kinetic analysis. Bioresource Technology.

10. Principi, P.; Villa, F.; Bernasconi, M. y Zanardini, E. (2006). Metal toxicity in municipal wastewater activated sludge investigated by multivariate analysis and in situ hybridization, Water Research 40, pp 99-106.

11. Ramírez, M. C, Larrubia, M. A, Herrera, M. C., Guerrero, P. y Malpartida, I. (2007). Valorización energética de biosólidos. Algunos aspectos económicos y ambientales en la EDAR Guadalhorce (Málaga). Universidad de Málaga. Facultad de Ciencias. Departamento de Ingeniería Química.

12. Rui, L., Shubin, W., Gaojin, L., Qing, Y. (2011). Energy and resource utilization of deinking sludge pyrolysis. Applied Energy.

13. Turovskiy, I. S., Mathai, P. K. (2006). Wastewater sludge processing. Wiley-Interscience.

14. Vallina, D., Rodrigo, M., Centeno, L. M., Muruais, J., Maílo, A., Ramírez, J., Tolosa, E. (2010). Guía de valorización energética de residuos. Madrid, España.

(381a) Improvements in Bio-Butanol Separation by Adsorption: Adsorbent Screening, Kinetics and Equilibrium

Niloofar Abdehagh[1], F. Handan Tezel[2] and Jules Thibault[2], (1)Chemical and Biological Engineering Department, University of Ottawa, Ottawa, ON, Canada, (2)Department of Chemical Engineering, University of Ottawa, Ottawa, ON, Canada

The rising cost of crude oil combined with depletion of the resources, political instability in oil producing countries and environmental concerns are some of the reasons that have motivated different waves of interest for renewable and sustainable fuels. To achieve a greater use of biofuels, it is paramount to develop innovative techniques to convert biomass to biofuel as efficiently as possible in order to make these production processes more economically viable. Recent developments in biotechnology offer an avenue for bio-production of fuels from renewable and sustainable resources instead of petroleum-based fuels.

Bio-butanol offers several strong attributes and has advantages over other biofuels such as lower volatility, flammability and corrosiveness. The net heat of combustion (NHOC) for butanol is 29.2 MJ/L, which is 83% of gasoline NHOC (which is 32.5 MJ/L) whereas values for ethanol and methanol are 65 and 45% of gasoline NHOC, respectively. Butanol is a very hydrophobic compound with a solubility of 7.7 g/100 mL (at 20°C) in water,whereas lower alcohols (methanol, ethanol and n-propanol) are miscible in water. Also, butanol can be used in car engines without any modification. Butanol was initially produced from fermentation, but in the 1950s, butanol produced from petroleum via the oxo alcohol process became much cheaper and the chemical route remains the only economic source of butanol today. However, the resurgence of interest in the production of butanol from fermentation is being witnessed from research institutions and major players, including large multinational petroleum companies. However, the production of biobutanol via fermentation is facing significant engineering challenges due the very low final concentration, low yield and severe butanol toxicity to microorganisms. It is therefore important to find an efficient separation technique to recover butanol at the end of fermentation or during the fermentation to reduce the level of toxicity and prolong the fermentation in addition to making it an economically feasible process.

Adsorption is an energy-efficient process that can be used to achieve this objective. In selecting a suitable adsorbent for the adsorption process, many factors need to be considered such as : adsorption rate, adsorption capacity, ease of desorption, selectivity for the desired product and cost of the adsorbent. Activated carbon and zeolites are the most common adsorbents used in this process since they are hydrophobic materials having high adsorption capacity for butanol. In this study, two activated carbons (AC F600 and AC F400) with different particles sizes (0.55-0.75 and 1 mm) and two zeolites (ZSM-5 and NaY) with different SiO_2/Al_2O_3 ratios (80 and 1.8) were tested. Adsorbent screening was done based on the kinetic and equilibrium experiment results. In kinetic experiments, AC F400 and AC F600 showed much faster adsorption kinetics in comparison to zeolites., The results obtained from kinetic experiments for both activated carbons showed that the adsorption reaches equilibrium in approximately 100 minutes while for zeolites, the system did not yet reach equilibrium even after 240 min.

Results for equilibrium experiments showed that the highest adsorption capacity for butanol was obtained with AC F400. With an equilibrium butanol concentration of 15 g/L, the butanol adsorption capacity for AC 400 was 283 mg/g while butanol adsorption capacities for AC F600, ZSM-5 and NaY were 150, 155 and 116 mg/g, respectively. Based on these results, AC F400 was selected as the adsorbent to carry out further experiments. Due to the presence of other components such as acetone and ethanol in the fermentation broth in the butanol production process, it is essential to assess the impact of other components o[J1] n butanol adsorption capacity. To investigate the competitive effect of these components on butanol adsorption, ternary solutions of butanol-acetone-water and butanol-ethanol-water with different initial concentrations of acetone and ethanol (1, 5 and 10 g/L) were tested. In this series of experiments, results showed that the presence of ethanol has a negligible effect on butanol adsorption whereas the addition of acetone decreased the butanol adsorption capacity. The negative effect of acetone was more pronounced at higher acetone concentrations. For the 10 g/L butanol equilibrium concentration, in the presence of 1, 5 and 10 g/L initial acetone concentration, butanol adsorption capacity decreased from 258 mg/g to 239, 212 and 206 mg/g, respectively.

In this study, the selectivity of AC F400 for the three main fermentation species (butanol, acetone and ethanol) was investigated using binary solutions of butanol-water, acetone-water and ethanol-water separately. The results showed that AC F400 is more selective to butanol in comparison to the other components. At an equilibrium concentration of 10 g/L, adsorption capacity of AC F-400 for butanol was 258 mg/g, while for acetone and ethanol, the adsorption capacities at the same equilibrium concentration were 115 and 55 mg/g, respectively.

From all the results obtained in this investigation, activated carbon F400 has been chosen to be the best adsorbent for butanol separation due to its high affinity, fast kinetics and higher adsorption capacity.

[J1]Note that acetone is not an alcohol.

(381b) 100% Conversion of Waste Brown Grease to Biodiesel and Syngas As Valuable Products for Heat and Power Cogeneration

Shoucheng Du[1], Fabio Poretti[2], Iman Noshadi[1], Richard Parnas[1] and George M. Bollas[1], (1)Department of Chemical, Materials & Biomolecular Engineering, University of Connecticut, Storrs, CT, (2)Department of Mechanical Engineering, University of Connecticut, Storrs, CT

Hydrocarbon-based biofuels derived from renewable, non-food carbon sources serve as a promising alternative to petroleum-derived fuels and are capable of mitigating greenhouse gas emissions. A sustainable biofuel strategy should utilize widely renewable biomass feedstocks, especially wastes, for which industries carry the high cost of disposal. Fuel production and power cogeneration from waste recovery is a sustainable and potentially profitable option to explore, addressing today's economic challenges of energy independence and waste management. A potentially interesting candidate in this scenario is brown grease. Brown grease is a mixture of high-value hydrocarbons, such as waste vegetable oil, animal fats and grease that is collected in water treatment plants and requires further treatment before disposal. However, brown grease has a higher level of contamination, compared to yellow grease, and cannot be used as animal feedstock or fertilizer. Additionally, brown grease is responsible for about 40% of all sewer overflows, causing back-ups and damage to pipe lines and for about 20,000,000 illnesses each year in the USA [1]. Therefore, special attention must be dedicated to its disposal and the cost associated. The main goal of this project is to explore the possibility of transforming 100% of the raw brown grease into biodiesel and synthesis gas that can be used for power generation. This work comprises a multi-disciplinary study of experimentation and simulation of optimal process options for the conversion of brown grease to energy and energy carriers.

In this study, heat treatment of the raw brown grease results in phase separation of the organic mixture. The light product, mainly characterized by a high free fatty acid (FFA) content (approximately 90%), is extracted and processed to obtain biodiesel. A two-step esterification process, using 5% (v/v) HCL and with a 9 to 1 methanol/oil molar ratio, produces liquid products with an acid value smaller than 0.5% FFA. The triglyceride is then transesterified using 2wt% KOH and a 6 to 1 of methanol/fats molar ratio, reaching higher ester content in the biodiesel. Gas Chromatography (Figure 1) shows that the biodiesel from the extracted brown grease oil not only satisfies the total and free glycerin percentage requirements imposed by the ASTM specifications, but also presents similar value in comparison with the biodiesel derived from Yellow Grease (Table 1), which makes it more competitive as brown grease is more abundant.

Table 1. Comparison of biodiesel produced from brown grease, yellow grease and ASTM standards in total and free glycerin

Sample	Free glycerin	Total glycerin
Biodiesel from brown grease	0.015	0.209
Biodiesel from yellow grease	0.022	0.158
ASTM standard	<0.020%	<0.240%

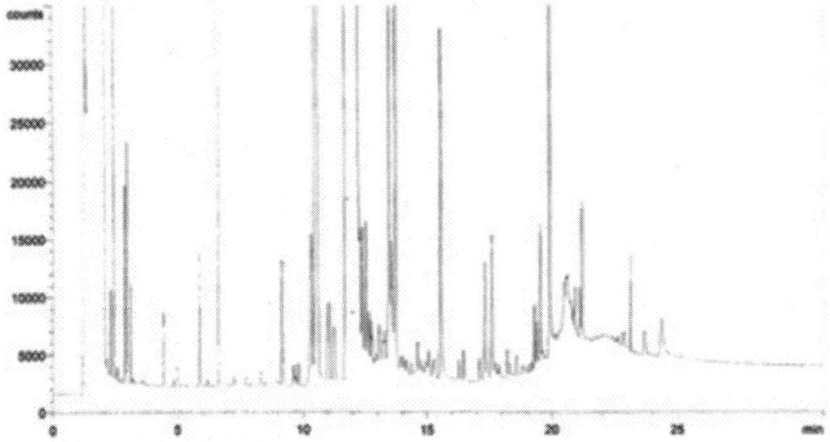

Figure 1. GC analysis of biodiesel produced from brown grease (Signal Count versus time)

The heavy product consists of a solid mixture, which is extracted and gasified to obtain synthesis gas. In Figure 2 thermogravimetric analysis (TGA) of the heavy brown grease product shows that this biosolid is gasifiable in its entirety. However the minor, almost negligible, residue of gasification contain inorganic compounds, of undefined environmental impact. Further analysis of the inorganics and their fate after gasification will be presented and discussed.

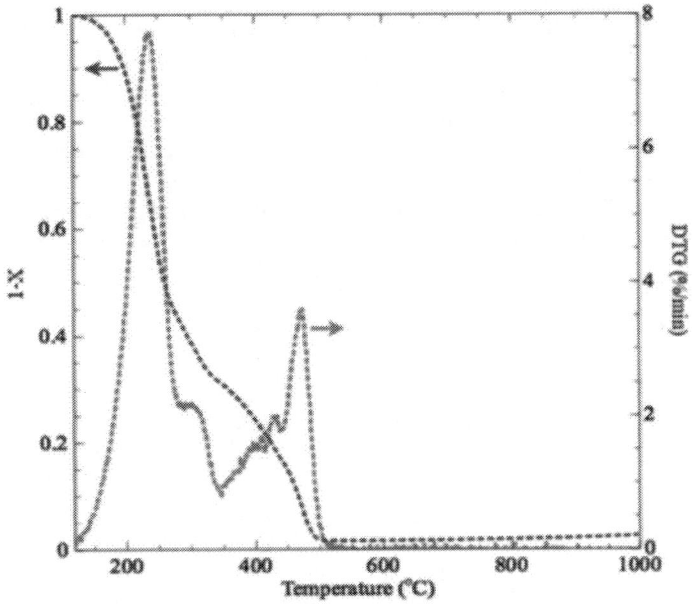

Figure 2. Gasification of biosolid residue from biodiesel production

TGA and gas product analysis (performed in gas chromatography/mass spectroscopy/Fourier transform Infrared, not shown here) show that the heavy product of brown grease is rich in hydrogen; thus, a valuable feedstock. Elemental analysis (Table 2) shows that the carbon to hydrogen ratio of the brown grease heavy fraction biosolid is lower than pine sawdust, which confirms its value as a gasification feedstock.

Table 2. Elemental analyses of biosolid from biodiesel production, biomass and coal (mol%)

Feedstock	N	C	H	S	C:H
Biosolid	0.0263	3.2483	6.4692	0.0157	0.5021
Sawdust	0.0261	4.4880	6.0236	0.0201	0.7451
Coal	0.1007	5.9767	5.0600	0.0881	1.1812

Due to the relatively low available quantity of the brown grease heavy fraction (as opposed to the efficiency increase of large scale gasification), co-gasification of the brown grease residue with fossil fuels, such as coal, is investigated. Co-gasification of coal and brown grease residues is capable of increasing the power generation plant throughput; thus, increasing its overall efficiency [2]. On the other hand, co-gasification reduces the consumption of coal in power generation plants resulting in economic and environmental benefits. Experimental results of co-gasification of biosolids and coal will be compared with the more widely studied co-gasification of woody biomass and coal [3].

On the basis of experimental results, process simulation of an integrated brown grease biorefinery is performed in Aspen Plus. The objective of the simulation study is to identify the cost and profitability of an integrated brown grease refinery with power co-generation, focusing also on the benefits of capturing CO_2 and the overall reduction of greenhouse gas emissions. The energy penalties for drying, O_2 separation, H_2S, SO_x, NO_x, and CO_2 separation are estimated for various co-gasification scenarios. A brief description of the process flow sheet is as follows. Raw brown grease is fed to the system and the first operation consists of heating the mixture in order to evaporate all the containing water and to phase separate the substance in the two layers. The oil at the top layer is extracted and sent to the biodiesel production hierarchy, while the heavier biosolids are taken to the syngas production hierarchy for gasification. In the biodiesel production hierarchy the liquid animal fat enters the pretreatment unit where impurities are filtered and removed. Cleaned animal fat enters an esterification unit where the oil is homogenously mixed at a constant temperature of 60°C with Methanol (MeOH) and Hydrochloric acid (HCl), which serves as a catalyst. The reaction reduces the concentration of the FFA. Then the stream is washed and dried. A second esterification process follows, to convert all the FFA remaining into Methyl ester. The stream is washed and dried again to remove remaining methanol and acid catalyst. The processed flow then enters the Transesterification unit where a base catalyst, Potassium hydroxide (KOH), is used to convert the triglycerides into biodiesel. This reaction also produces glycerol, which can be separated and sold as a chemical. Finally, crude biodiesel enters the post-processing unit where it is refined and tested for quality standards.

In the syngas production hierarchy the dewatered biosolids enter the preparation unit where they are mixed with coal, which has previously passed a process of sizing (crushing and screening) and drying. Air is separated mainly in Nitrogen and Oxygen in the Air Separation Unit (ASU) through cryogenic distillations. The mixed feedstock, together with the stream of almost pure Oxygen, enters the gasifier. The melted ash is collected at the bottom of the reactor and disposed. Then the syngas is cleaned from H_2S and CO_2 in the Acid Gas Cleaning Unit (Rectisol Process). Captured CO_2 is sent to compression and storage, and the H_2S is converted to solid S in the Claus Unit [3]. Finally the syngas is burned in a combined cycle, where heat and electricity are produced. The hot gas, after expansion in the gas turbine, pass through a Heat Recovery Steam Generator (HRSG) where superheated, high pressure steam is produced which can be used for district heating and/or expanded in a steam turbine for additional power production.

The experimental analysis integrated with the ASPEN model allows to examine the feasibility of recovering 100% of the raw brown grease that would otherwise have to be disposed at a high cost, due to its contamination as a water treatment waste. The opportunity of transforming it into valuable energy products opens access to new markets, while the sustainability of the overall process and the improved carbon life cycles of this process, lead to its reassessment from an unwanted waste to a precious resource.

Acknowledgment

This work was supported in part by the Faculty Large Grant of the University of Connecticut and in part by the EAGLES (Engineers as Global Leaders for Energy Sustainability) Atlantis Fellowship.

References:

[1] Tyson, K. S. Brown Grease Feedstocks for Biodiesel, National Renewable Energy Laboratory, 2002.

[2] Saw, W.; McKinnon, H.; Gilmour, I.; Pang, S. Fuel. 2012, 93, 473-478.

[3] Valla, L.; Du, S.; Bollas, G. M. 2012 AICHE Annual Meeting, October 2012, Pittsburgh, PA, USA.

(381c) Techno-Economic Analysis of Mixed Alcohol Production Through Hydrogenation of Volatile Fatty Acids Derived From Seaweed

Peyman Fasahati, Chemical Engineering, Pukyong National University, Busan, South Korea and J. Jay Liu, Department of Chemical Engineering, Pukyong National University, Busan, South Korea

Recently, there has been growing interest regarding use of seaweeds as biomass feedstocks. On bioethanol production from seaweeds, there are two main obstacles that make it hardly applicable in industrial scale. First, high seaweed price greatly effects the economy of bioethanol production from seaweeds. Developing mass cultivation sites and enhancing seaweeds cultivation procedures would be necessary steps to address this problem. Secondly, ethanol production from seaweeds has relatively low yield. Research on finding new enzyme, bacteria, or yeast, which are capable of fermenting a wider range of sugars produced by seaweed saccharification, can be the only solution. An alternative way to produce bioethanol would be finding new substitute processes that produce ethanol with higher conversion and lower final ethanol price.

Volatile fatty acids (VFAs) can be easily produced from seaweeds. They have higher conversion yields in comparison with those of direct ethanol production. VFAs can easily be converted to ethanol by hydrogenation. For these reasons, this study develops a techno-economic model to survey economy of VFAs production from seaweeds followed by hydrogenation to produce bioethanol.

Our previous study on direct bioethanol production from seaweed showed that the maximum dry seaweed price (MDSP) should be lower than 160 US$ per ton to allow ethanol production economically viable. This study also calculates the MDSP for this process and survey possible opportunities to improve the economy of ethanol production.

The whole process was designed and simulated in Aspen plus V7.3 and then sent to aspen process economic analyzer for economical evaluations. Plant scale-up effects on the economy of plant were also examined in detail.

Flex Fuel Polygeneration: Optimizing Cost, Sustainability, and Resiliency

Matthew Kieffer, Tristan Brown, Robert C. Brown

Iowa State University
Center for Sustainable Environmental Technologies

Flex Fuel Polygeneration (FFPG) is a new technology that employs multiple energy sources and produces multiple energy carriers to construct optimal energy plant designs in terms of cost, sustainability, and resiliency. This technology allows for adjustments to be made to market fluctuations and improvements to the energy efficiency and environmental performance of power plants. The diversification of energy pathways also increases resiliency to contingencies such as earth quakes, flooding, reduced winds, or other events that may reduce the availability of energy sources. In this study we develop two FFPG plants that employ combinations of natural gas, biomass, wind, and the use of anaerobic digestion, fuel cells, different types of turbines, and Fischer-Tropsch synthesis. To obtain FFPG plant performance results, we combine techno-economic analyses available in literature of existing energy plants to reflect the FFPG plant designs. By doing high level analysis we can compare costs, products, and emissions of theoretical plants to plants that are currently in operation. What we have found are preliminary results regarding the internal rates of return. If the internal rates of return prove to be positive, we will continue to design and perform high level analysis on additional FFPG plants.

(381e) Feed Flexible Process Optimization Model for Region Specific Sustainable Biorefining

Sumesh Sukumara[1], Jeffrey R. Seay[2], Dr. Fazleena Badurdeen[3], William Faulkner[4] and Joseph Amundson[4], (1)Chemical and Materials Engineering, University of Kentucky, Lexington, KY, (2)Department of Chemical and Materials Engineering, University of Kentucky, Paducah, KY, (3)Institute for Sustainable Manufacturing, University of Kentucky, Lexington, KY, (4)University of Kentucky, Lexington, KY

Biofuels in recent years has emerged as one of the most promising alternatives to replace fossil fuel. Researchers around the world are working on developing new processes and improve existing ones to convert biomass into various marketable products. Integrated biorefining is a possibility which combines a slate of processes involving various feedstock to produce transportation fuel, electric power and other useful chemical products. Also, due to the seasonal and regional variability of feedstock, the overall supply chain has to be looked at carefully. Thus, a multidisciplinary economic model needs to be developed which incorporates various small models and allows each to exchange information. According to the existing literature and preliminary experiments performed its evident that transportation and processing cost share a significant portion of the overall cost of an integrated biorefinery. This novel contribution aims to show how a feed flexible process optimization model can be developed and interact with the overall region specific optimization model.

Processing cost in a biorefinery is mainly due to the utilities and raw materials involved. The model developed applies various concepts of chemical engineering to minimize the utilities and maximize the profitability for any given process in a region. Various modules of the Aspen Engineering Suite are combined to achieve the required objective. The process optimization model to be described consists of two major processes to produce transportation fuel and power as end products. These models are first developed considering all the possible mass and heat streams involved to get various products. A thermal pinch analysis is performed to find the best possible heat integration practically possible for a process. Based on the optimum heat exchanger network the process simulations are changed accordingly and finally the capital costs are estimated for the process. The process optimization model developed transfers its data with a Visual Basic interface to the other models which eventually estimates the overall profitability of the process. This model will be validated using the Jackson Purchase region of Western Kentucky, which is not only rich in coal, lignocellulosic biomass and crop residue but also has large number of poultry farms, as a case study. The robust model developed can be run for various combinations of feedstock possible. These process simulation results coupled with the overall supply chain model can predict some critical conclusions like profitability, optimum feed ratio, utility requirement and also some sensitive process variable that effect the overall supply chain of the biorefining process. This innovative linking of various optimization models will help develop a large scale multivariable model for biorefining which will include all possible feedstocks for a given region and provide the option to choose from various available processes in the model based on the cost, demand and environmental regulations in a region. Hence the multiple feedstock process optimization model developed supports the overall supply chain model based on which investors can check the economics of a process and decide if an investment is worthy in a given region.

(381f) Sustainable Design of Energy Supply Chains From P-Graphs: Cost, Ecological Footprint, and Emergy

Leisha Vance[1], Istvan Heckl[2], Botond Bertok[3], **Heriberto Cabezas**[1] and Ferenc Friedler[4], (1)Office of Research and Development, U.S. Environmental Protection Agency, Cincinnati, OH, (2)Department Computer Science and Systems Technology, University of Pannonia, Veszprem, Hungary, (3)Department of Computer Science, University of Veszprém, Veszprém, Hungary, (4)Department of Computer Science and Systems Technology, University of Pannonia, Veszprém, Hungary

A collaboration consisting of the Office of Research and Development (ORD) of the U.S. EPA and the research group led by the founders of the P-graph framework at the University of Pannoniahas developed a methodology for designing sustainable supply chains based on an optimization using the p-graph framework constrained by integrated sustainability indicators and engineering costs. The result was a powerful methodology for designing cost-effective and environmentally sustainable supply chains. We illustrated the methodology with a prototype supply chain designed to produce heat and electricity for a generic district in Hungary. Possible sources of heat and electricity included electricity from the Hungarian grid, and heat and electricity generated from natural gas, corn, corn silage, grass silage, or wood or some combination of these sources. Twenty-one different supply chains, each capable of producing 18 TJ per year of heat and 7.2 TJ per year of electricity were found. Each supply chain was ranked according to cost, and assessed environmental impacts using the ecological footprint (representing land use burden), and emergy (representing energy resource burden). Decisively feasible supply chains were found with cost variations of +2% to -17% compared to "business as usual" scenarios i.e. using only natural gas and electricity from the Hungarian grid. For these supply chains, the sustainability profile as represented by the ecological footprint varied from +8% to -78%, and the emergy results ranged from -54% to -93%. Both comparisons were done in contrast to the conventional natural gas/electricity from the Hungarian grid. Most importantly, it appeared feasibly possible to design supply chains for heat and electricity generation which were both cheaper and more sustainable than the supply chain currently in use.

(383a) Process Design and Optimization for Hydrocarbon Biofuel Production From Corn Stover Via Gasification

Belinda Wang, Berhane H. Gebreslassie and Fengqi You, Department of Chemical and Biological Engineering, Northwestern University, Evanston, IL

Energy resources are limited but the demand is constantly increasing. In the last decade, researchers have put effort into developing renewable energy sources because of their sustainability and environmental friendly. One of the advantages of biofuel production is it emits less over petroleum-based fuel because it yields less environmental impacts throughout the life cycle compared to their petroleum counterpart. The U.S. Government has proposed the Energy Independence and Security Act (EISA) of 2007 to regulate the development of biofuels in the United States. The main goal is to increase the production of clean renewable fuels that leads to energy independence from imports. EISA requires the total amount of biofuel production to increase to 36 billion gallons per year by 2022, from 4.7 billion gallons per year in 2007 [1]. Among all types of biofuels, biomass-derived hydrocarbon fuels provide the most similar characteristics with fossil fuels. Hydrocarbon biofuels have the advantage of compatibility with the existing transportation, storage, engines, and processing technology infrastructure [2]. Therefore, large-scale production of biomass-derived hydrocarbon fuels can help the nation to achieve the EISA requirement in a sustainable and cost-effective way.

In this paper, we optimize the production process of gasoline and diesel from corn stover. We propose a superstructure which demonstrates the network configuration of process. The purpose is to produce hydrocarbon biofuel through various technologies with at least two options. Sections with technology alternatives are gasification, syngas cooling, hydrogen production, and Fischer-Tropsch synthesis. For gasification, we consider two types of gasifier, high-temperature and low-temperature. The choice for syngas cooling is direct quench or indirect quench. Hydrogen can be produced through two pathways, internal production or steam methane reforming. The alternative in Fischer-Tropsch synthesis is the selection of catalyst among cobalt, iron, and nickel. The maximum moisture content for biomass gasification is in the range of 20% to 30% (wet basis), and less than 15% (wet basis) for normal operation [3]. Biomass with 25% moisture on wet basis enters the plant. Therefore, a rotary dryer is needed to reduce the biomass moisture content to 10% on wet basis. Pressurization biomass is then entering the gasifier with 95% purity of oxygen produced from the air separation unit. An air separation unit is used to separate oxygen from air. The gasifier operates at 28 bar and at two different temperatures which 1300°C for high temperature (HT), and 870 °C for low temperature (LT). According to Henrich, the mass ratio of oxygen to biomass is fixed at 0.35 using entrained-flow gasifier [4]. Steam addition to gasifier is fixed at 0.48 mass ratio of steam to biomass [5]. The low-temperature gasifier requires a different mass ratio of oxygen to biomass at 0.26 and mass ratio of steam to biomass at 0.17 [6]. The high-temperature gasifier does not produce hydrocarbons because of its near equilibrium conditions [7]. On the other hand, LT gasifier produces a significant amount of methane, ethane, and ethylene. The raw syngas exiting gasifier is cooled down to lower temperature. There are two methods for cooling, direct quench and indirect quench. Water is added directly to the syngas for direct quench. While for indirect quench, heat exchangers are required. The next step is to modify syngas to the optimal condition needed for Fischer-Tropsch synthesis. Steam methane reforming (SMR) and water gas shift are used to adjust the amount of carbon monoxide and hydrogen to meet the optimal ratio of H_2 and CO around 2.1 [7]. The purpose of a SMR step is to reduce methane, ethane, and ethylene content. Hydrogen can be produced through two pathways, internal production or steam methane reforming. Internal production uses the hydrogen within the syngas. While steam methane reforming uses additional natural gas to produce hydrogen. A pressure swing adsorption (PSA) process is used to isolate hydrogen. Excess hydrogen from steam methane reforming and water gas shift is split into two, one entering hydrocracking, and the rest to hydrotreating. The Fischer-Tropsch synthesis reactor operated at 200 °C and 25 bar using different catalyst, such as cobalt, iron, and nickel. Per-pass carbon monoxide conversion in the reactor is set at 40% [7]. According to Song et al [8], the production distribution of Fischer-Tropsch synthesis follows the Anderson-Schulz-Flory alpha distribution which the chain growth factor, a, depends of the molar fraction of $H_2$and CO and the temperature of the reactor. The value of alpha is dependent on the type of catalyst. The product composition of C1 to C20 is determined following the Anderson-Schulz-Flory alpha distribution, and the rest is all wax. Methane, ethane, and propane produced from FT synthesis are sent to gas turbine section to combust and generate electricity. The remaining Fischer-Tropsch liquids then distillate and hydrocrack to produce gasoline and diesel. Gasoline includes lighter hydrocarbons (C4~C12) and diesel includes heavier hydrocarbons (C13~C20).

The objective is to maximize the net present value (NPV) and minimize global warming potential (GWP) subject to design and operational constraints that include mass balances, energy balances, economic analysis, and environmental constraints. The equipment costs are based on the capacity of the units which are related to the feed flow rates and sizing factor. Other cost includes annual operational cost and utility cost. The process makes profit from selling gasoline and diesel. The net present value accounts for the investment and discounted annual cost and profit. The environmental impact is measured using the global warming potential metric following the recent development of the life cycle assessment (LCA) procedures. GWP is a relative measure of how much heat a greenhouse gas traps in the atmosphere, and is calculated over a specific time interval that has to be stated, commonly 20, 100 or 500 years. In this work, 100 year is adopted following Kyoto protocol. The process is optimized to determine the optimal decision of alternative selections for the production of hydrocarbon biofuel. The optimal choice is using high temperature gasification, cobalt catalyst in Fischer-Tropsch synthesis, direct quench and internal hydrogen production. To obtain the optimal trade-off between the two contradicting objective functions, we use the epsilon constraint method. The Pareto curve demonstrates the trade-off that exists between the NPV and GWP. The optimal solutions will form a Pareto-optimal curve which represents the set of optimal solutions of the two objectives. The curve separates the region of feasible and infeasible solutions. It has the minimum GWP of 930 ktonne CO_2-equivalent when the NPV is 0. When maximize NPV, GPW has the highest possible value of 3892 ktonne CO_2-equivalent. The results show that the capital investment is 40% of the overall costs over 20 years. Among the capital investment, 39% is equipment and installation cost, 13% is engineering and supervision, 13% is construction expenses, 9% is legal and contractors fee, 14% is project contingency, and 12% is working capital.

Feedstock cost is the main expenditure which is around 43% of overall costs. Utility cost is around 2 %, maintenance cost is 5%, and catalyst cost is 10% of overall costs during the 20-year lifespan. GWP calculated in this process including steam, natural gas, electricity, heat, and CO_2 emissions. Carbon dioxide emission accounts for 78% of overall GWP, and 13% for electricity. While each of the other GWP categories are ranging from 2% to 5%. Therefore, it is important to consider how to reduce CO_2 emissions. One effective method is carbon capture and storage. Carbon capture and storage (CCS) is the technology to avoid releasing CO_2 into the atmosphere by capturing $CO_2$and store it to prevent releasing to atmosphere. Future work could include the alternatives of CCS in the process.

References

[1] *Energy Independence and Security Act of 2007,* U. S. Congress, 2007.

[2] V. Sarisky-Reed, "Advanced Biofuels: Infrastructure Compatible Biofuels," B. P. U.S. Department of Energy, Ed., ed, 2009.

[3] H. Knoef, "Practical Aspects of Biomass Gasification," in *Handbook Biomass Gasification*, H. Knoef, Ed., ed Enschede, Netherlands: BTG Biomass Technology Group, 2005.

[4] E. Henrich and F. Weirich, "Pressurized entrained flow gasifiers for biomass," *Environmental Engineering Science,* vol. 21, pp. 53-64, Jan-Feb 2004.

[5] R. F. Probstein and R. E. Hicks, *Synthetic Fuels*. Mineola, NY: Dover Publications, 2006.

[6] R. L. Bain, "Material and Energy Balances for Methanol from Biomass Using Biomass Gasifiers," National Renewable Energy Laboratory, Golden, CO NREL/TP-510-17098, 1992.

[7] R. M. Swanson, A. Platon, J. A. Satrio, and R. C. Brown, "Techno-economic analysis of biomass-to-liquids production based on gasification," *Fuel,* vol. 89, pp. S2-S10, Nov 1 2010.

[8] H. S. Song, D. Ramkrishna, S. Trinh, and H. Wright, "Operating strategies for Fischer-Tropsch reactors: A model-directed study," *Korean Journal of Chemical Engineering,* vol. 21, pp. 308-317, Mar 2004.

(383b) Characterization of Products From High-Pressure Biomass Pyrolysis in an Entrained-Flow Reactor

Gautami Newalkar[1], Abiola Shitta[1], Scott A. Sinquefield[2], Kristiina Iisa[3], Carsten Sievers[1] and Pradeep K. Agrawal[1], (1)School of Chemical & Biomolecular Engineering, Georgia Institute of Technology, Atlanta, GA, (2)Ipst, Georgia Institute of Technology, Atlanta, GA, (3)National Renewable Energy Laboratory, Boulder, CO

A major advantage of biomass gasification is that any lignocellulosic material can be converted to syngas and subsequently to biofuels. There are techno-economic advantages of using high pressure in gasifying biomass. In a gasifier, pyrolysis and char gasification steps are known to proceed in series. Thus it becomes important to study the characteristics of pyrolysis products, since these have an effect on the gasification kinetics.

Pyrolysis experiments on pine and switchgrass (sieved to 180-250 μm) were performed in entrained flow reactor at high temperatures (600-1000 °C) and high pressures (1-20 bar). Heating rates as high as 10^4 K/s were achieved with solids residence time from 3-10 s. Char, gases and condensable tars were characterized using SEM, N_2 physisorption, GC and MS respectively.

SEM was used to study the influence of pyrolysis temperature and pressure on char morphology. The results show that with increase in pyrolysis temperature and pressure the particles melted and gained more spherical features. At ambient pressures, a significant fraction of particles burst at higher temperatures. At high pressures, pyrolysis gases tend to be trapped in char particles forming distinct gas filled pockets. Char structure and surface area, is strongly impacted by the increased temperature and pressure during pyrolysis, and is likely to affect char gasification activity.

The yields of light gases such as CO, CO_2, H_2, CH_4, C_2H_4 and C_2H_2 were quantified using GC analysis. The condensable organics from pyrolysis were extracted and analyzed using MS. A variety of tar species from phenolics to polynuclear aromatic hydrocarbons (PAHs) were identified and their relative abundance rated as a function of temperature and pressure.

The present study will provide the basis for improved mathematical model for char gasification based on meaningful descriptors.

(383c) High Temperature Gasification of Biochar with Added Methane

Aaron W. Palumbo[1], Erica L. Jorgensen[1] and Alan W. Weimer[2], (1)Chemical & Biological Engineering, University of Colorado at Boulder, Boulder, CO, (2)Chemical Engineering, University of Colorado at Boulder, Boulder, CO

Temperatures above 1300°C may be used to increase conversion and synthesis gas quality and decrease carbon dioxide and hydrocarbon formation in an allothermally heated gasifier. Using methane, or natural gas, as a reactant increases the hydrogen content of the synthesis gas to be compatible with liquid fuel synthesis technologies. Thermodynamic equilibrium results confirm an adjustable H_2:CO given a variable methane-to-biomass ratio as well as superior utilization of hydrogen and carbon monoxide compared to methane reforming. Experimental results show that a complex, kinetic-limited relationship exists between char gasification and methane reforming. Reactions of pyrolysis further complicate experimental controllability of the product gas composition. Char gasification and methane reforming were modeled in a 1-D plug flow reactor configuration assuming a monodisperse particle distribution and an entrained particle flow. Experiments with biochar, methane, and steam were used to obtain global kinetic models of the relevant reactions. Effects of reactant flow rate, particle diameter, and wall temperature were explored.

(383d) Simultaneous Removal of Biomass-Generated Syngas Contaminants Using Biochar-Based Catalysts

Ajay Kumar, Pushpak N. Bhandari and Raymond Huhnke, Biosystems and Agricultural Engineering, Oklahoma State University, Stillwater, OK

In the quest for low-cost and efficient technologies to convert biomass into fuels, chemicals, and power, gasification offers great potential. However, to effectively utilize syngas, the main product of biomass gasification, it must be cleaned of its contaminants. Contaminants such as ammonia, tars, and H_2S cause various problems in downstream applications of the gas. Simultaneous removal of these contaminants using a single catalytic bed and same conditions will significantly reduce costs. The primary objective of this work was to investigate the use of carbon-based catalysts, biochar and activated carbon derived from biochar, for individual and simultaneous removal of tars, ammonia, and H_2S.

Biochar is a low surface area waste product generated during biomass gasification. Biochar has shown good potential for tar removal. To improve upon the low surface area of biochar, activated carbon was synthesized using the biochar derived from switchgrass gasification. Activated carbon was also modified using dilute acid to give an acidic surface to help improve removal of toluene (model tar) and NH_3. The three catalysts (biochar, activated carbon, and modified activated carbon) were evaluated for their contaminant removal efficiency, first separately and individually with each contaminant and then simultaneously with all of the contaminants. Evaluation was carried out at temperature of 700 °C and 800 °C and a gas residence time of 0.001 kg/(m^3/hr) in a fixed-bed catalytic reactor fed with syngas.

Several key findings, to date, are following. First, both the activated carbons and biochar were effective in tar removal (up to 92%). Second, as compared to biochar, activated carbon catalysts resulted in higher toluene conversion due to its higher surface area (~900 m^2/g compared to <10 m^2/g of biochar), larger pore diameter (19 A° compared to 15.5 A ° of biochar) and larger pore volume (0.44 cc/g compared to 0.085 cc/g of biochar). Chemisorption experiments showed that CO_2 was weakly adsorbed (physiosorption) onto activated carbon surface. This indicates that the sorbent can easily be regenerated. The adsorptive capacity was as high as 1.8 mmol/g at 60 kPa indicating a high sorption capacity.

(383e) Effect of Temperature and Oxidizing Medium On Tar Formation in Southern Pine Gasification

Narendra Sadhwani[1], Sushil Adhikari[2] and Mario Richard Eden[1], (1)Department of Chemical Engineering, Auburn University, Auburn, AL, (2)Biosystems Engineering, Auburn University, Auburn, AL

Synthesis gas (a.k.a. syngas) can be used as a feedstock in industrial applications for electricity, chemical and hydrocarbons production. However, each application has stringent syngas cleanliness requirements to avoid operational complications. Consequently, gasification syngas cleanup has emerged as a pivotal research area since the advancements have direct implications on the technical and economic viability of gasification. Tar, a major contaminant in synthesis gas, poses technical and operational issues ranging from equipment corrosion and fouling, environmental pollution and catalyst deactivation. The objective of this study was to analyze the effect of temperature and oxidizing medium on tar formation and its mechanism in southern pine gasification. The effect of three temperature levels (600°C, 700°C, 800°C) and two oxidizing media (steam and carbon dioxide) on the primary gases (CO, CO_2, CH_4 and H_2) and tar were studied using a bench-scale fluidized bed gasifier. A number of gas analysis instruments such as a GC/MS (Agilent 7890/5975), a micro GC (Agilent 490), a portable gas analyzer (Nova 7905AQN4), and an FTIR were used to collect the necessary information. Results from both experimental and mathematical studies will be presented.

(383f) Thermocatalytic Conversion of Pongamia Pinnata Seed Cake Over Ruthenium Impregnated MCM-22 Catalyst

Dinesh Singh, Chemical Engg., I.P Uelhiniversity , Delhi, India and K. K. Pant, Chemical Engineering Department, Indian Institute of Technology, Delhi, New Delhi, India

Thermocatalytic conversion of *pongamia pinnata* seed cake over ruthenium impregnated MCM-22 catalyst

This work studies the pyrolysis of *pongamia pinnata* seed cake for the production of bio oil. India is fast growing economy with over 120 billion people and facing the challenges to meet the need of energy demand. Production of energy crops like jatropha and pongamia pinnata for the production of biodiesel is already a national policy. It is assumed that that seed cake generated after oil extraction will be huge and its disposal thus may be a problem. Direct burning of this cake will not be a beneficial option. Energy generated by burning of this biomass is very low 15-16 Mj/kg. As it is not possible to extract all oil from these seeds so it is expected that these seeds with cellulose, hemicelluloses and lignin will also have some oil components. Pyrolysis is a process wherein we heat biomass in absence of oxygen and the complex molecules of biomass defragment into to smaller carbon number hydrocarbons mostly in the range of C1 to C44. In thermal pyrolysis we mainly get either light gases like CO, CO_2, methane or either very heavy molecules leading to formation of tar both of these hydrocarbon not consider good as transportation grade fuel. Besides this another major factor which hinders the use of thermal pyrolysis for the production of bio oil is the oxygen content which decreases the calorific value of the bio oil. Hence our focus is to reduce the oxygen content as well as crack the molecules in the range of diesel fuel. Research has been done on the cracking of pyro oil using ZSM-5 as catalyst impregnated with nobel metals. In this work we have used Ru supported over MCM-22 as catalyst for the cracking of biooil. MCM22. MCM22 has large pore diameter formed by 12 member rings with a system of 10 member rings. These large pore allows large hydrocarbon molecules to enter in the cage structure and thus provide more catalytic surface area for cracking reaction to occur. Ru-supported catalysts were prepared using conventional wet impregnation technique. An appropriate amount of precursor solution of RuCl3-xH2O, was dropped to the MCM-22 to obtain desired wt% of Ru, followed by drying in an oven at 110 ^{o}C for 3 h. The catalyst was calcined from room temperature to 580 ^{o}C. Prior to catalytic activity testing, all catalysts were reduced by hydrogen at 400 ^{o}C for 3 h. The catalyst was characterized using SEM, XRD, BET and chemisorptions. Oil extracted Pongammia pinnata seed cake has been utilized for the production of bio oil in a fixed bed SS 316 reactor under atmospheric pressure using a semibatch operation. In a typical process two zones were prepared in the reactor, one of biomass with typical mass of around 10 g whereas the weight of catalyst was varied from 1 to 3 grams. Nitrogen gas flow rate was varied from 25 ml/min to 100 ml/min which provide an inert atmosphere as well as swept the volatile products from the reactor to the condenser. The uncondensed gases were analyzed by GC TCD for CO, CO_2, H_2 and methane whereas higher carbon number gases were analyzed by FID Porapak Q packed column. The liquid products were analyzed by capillary column of ZB-1. Prior to liquid analysis bio oil was dissolved in carbon disulphide and was given 30 minutes ultrasonic treatment. The effect of temperature, nitrogen gas flow rate and biomass to catalysis weight has been studied. The pyrolysis products were further passed through condenser so as to separate condensable and non condensable as liquids. Further condensed vapors were separated as aqueous phase and oil phase. Pyrolytic oil obtained were analysed by GC, NMR and FTIR. Gas chromatograph mass spectrometry (GC–MS, PerkinElmer clams 500) equipped with a HP-5 capillary column (30 m x 0.25 m i.d., 0.25 mm film thickness) was used for identifying the compounds in the liquid sample. Helium (99.999%) was used as the carrier gas with a constant flow rate of 1 mL/min. in split less mode. The oven temperature was programmed from 35°C to 100°C in 2 minutes, 100°C to 200 ^{o}C 5 in minutes and 200°C to 250°C up to 30 minutes. MS was conducted with the following operation conditions: transfer line 250°C, ion source 150 ^{o}C and electron energy 70 eV. Bio oil and higher carbon number gases were analyzed by using a combination of packed and capillary column. Thermogravemetric analysis of the seed cake were also performed with different heating rate (2, 5, 10 oC/min) in nitrogen atmosphere prior to the pyrolysis to find out the initial and final temperature for thermal degradation. The solvent extractives, cellulose and lignin content were analyzed using TAPPI standards e.g. T204 cm 97, T203 cm-99 and T236 cm-85 respectively. SEM analysis of biochar was done to see the effect on the change of structure of biomass before and after pyrolysis. The 1H NMR (Proton nuclear magnetic resonance) spectrum of the bio-oil was recorded using a Bruker, 300 MHz High Performance Digital NMR Instruments. The samples were dissolved in deuterated chloroform containing TMS (Tetramethylsilane) as a standard. Bio oil and upgrade bio oil with ruthenium as noble metal shows 1H-NMR spectra with different peaks. These peaks can be divided into three main regions: aromatic, olefinic, and aliphatic, based on the chemical shifts of specific proton types in the regions of 9.0–6.0, 6.0–4.0, and 3.0–0.5 ppm, respectively. FTIR analysis was done using KBr pellets. The bends near the 3350 shows the presence of O-H group whereas those near to 3100 confirm the presence of aromatic rings. Large spectra beyond 3400 indicate presence of water and that of near to 2500 arise due to the presence of paraffinic compounds. Peaks near around 1700 confirms C=O group. GCMS analysis confirms the presence of few of the components octadecanoic acid, pentadecanenitrite, 9 nonadecene, hexatricontane, Pnonyloxybenzaldehyde, 2(4 hydroxybutyl_-2 nitro cycloecanone. TCD data confirms that an increase in the pyrolysis temperature leads to higher production of hydrogen and methane where as lower temperature that is up to 400°C yields more of CO and CO_2. Also it was observed that though there is not much change in the char yield with or without ruthenium but a small increase can be seen in the gas yield in presence of ruthenium. At the same time, all the oil obtained in presence of ruthenium shows a shift towards the increase of lower hydrocarbons (between C6 to C20). Furthermore it was observed that an optimum of 1% ruthenium leads to highest yield (~45.3%) of bio oil with carbon number from C-9 to C18 (boiling point cut 80-320 °C).

(383g) Techno-Economic Analysis of a Biorefinery Utilizing Geothermal Energy for Biomass Drying and Auxilliary Power Generation

Jordan Tiarks[1], Sudhanya Banerjee[1], Yanan Zhang[1], Terry Meyer[1], Song-Charng Kong[1], Robert C. Brown[1] and Brian Anderson[2], (1)Mechanical Engineering, Iowa State University, Ames, IA, (2)Chemical Engineering, West Virginia University, Morgantown, WV

A techno-economic study of a biorefinery based on biomass gasification is conducted using Aspen Plus and corresponding economic analysis tools for three different scenarios. In the first scenario, a 2,000 metric tonne per day biorefinery is modeled utilizing corn stover as the feedstock. The gasification scenario modeled produces a synthesis gas that is upgraded to liquid transportation fuels, with excess hydrogen and electricity produced as co-products. In an effort to reduce the overall greenhouse gas emissions associated with this process, the second scenario replaces steam utilized in the plant originally produced by the combustion of fossil fuels with steam derived from geothermal resources. Finally, an additional power generating step based on the Organic Rankine Cycle is implemented in the third scenario which generates electricity from the low-temperature steam remaining from various processes that is normally exhausted to the atmosphere. Results show that the lowest price of transportation fuels results from scenario one. Despite the modest price increases, scenario two can reduce the overall greenhouse gas emissions resulting from the biorefinery, while the addition of the Organic Rankine Cycle increases the overall plant efficiency.

(406b) Accelerating Hydrolysis of Cellulose and Xylan to Monosugars by Using Hydrolyzing Enzymes

Kun-Lin Yang, Department of Chemical and Biomolecular Engineering, National University of Singapore, Singapore, Singapore and Laura Sutarlie, Chemical and Biomolecular Engineering, National University of Singapore, Singapore, Singapore

Cellulose and xylan are the most abundant renewable biomass in nature. However, they cannot be used directly as carbon sources for bacteria or yeast during fermentation to produce biofuel. To utilize cellulose and xylan, they need to be hydrolyzed to produce glucose and xylose. Nowadays, it is preferable to hydrolyze cellulose or xylan through enzymatic process because of its mild reaction conditions. In the case of cellulose, we hydrolyze filter paper by using cellulase from *Trichoderma reesei* as our model system. By using HPLC, we find two main peaks which correspond to glucose and cellobiose. These results show that the main hydrolytical products are glucose and cellobiose, and no other glucose oligomers are produced. Interestingly, after 12 h of hydrolysis, only 38% of filter paper is converted to glucose while 14 % is accumulated as cellobiose. We also hydrolyze xylan by using xylanase (*Thermomyces lanuginosus*) as our model system. It is found that after 12 h, only 2% of xylan is converted to xylose and 8% of xylan is hydrolyzed to xylobiose.

Accumulation of cellobiose and xylobiose has significant implication since they cannot be used by bacteria or yeast to produce biofuel. In addition, cellobiose and xylobiose can inhibit the enzymatic hydrolysis of cellulose and xylan. Therefore, it is important to avoid the accumulation of cellobiose and xylobiose during enzymatic hydrolysis. To further hydrolyze cellobiose and xylobiose, we add additional β-glucosidase (from *Aspergillus niger*) and β-xylosidase (from *Selenomonas ruminantium*) to cellulase and xylanase, respectively. Our results show that the amount of β-glucosidase and β-xylosidase can be optimized to maximize the concentrations of monosugars and accelerate the hydrolysis of cellulose and xylan.

(406c) Inorganic Analysis of HTC Biochar

M. Toufiq Reza[1], Joan G. Lynam[1] and Charles J Coronella[2], (1)Chemical and Materials Engineering, University of Nevada, Reno, Reno, NV, (2)Chemical & Materials Engineering, University of Nevada, Reno, Reno, NV

The increase in greenhouse gas concentration, primarily due to emissions of CO_2 produced in use of fossil fuel, motivates many efforts to identify renewable energy alternatives. Using fuel from food sources is not optimal, since malnourishment is still prevalent in the developing world. Despite the abundance of lignocellulosic biomass, transportation, handling, low fuel value, and ash content hinder the development of this potential coal replacement. Torrefaction of lignocellulosic biomass can improve the energy value to that of low grade coal, but high inorganic content continues to be a challenge in co-firing processes. Ash is difficult to remove and alkali metals like potassium, sodium, sulfur, and chlorine tend to corrode equipment, as well as cause slugging and fouling.

Hydrothermal carbonization (HTC or wet torrefaction) is a pretreatment process which makes a homogenized, carbon rich, and energy dense solid fuel, called biochar, from underutilized lignocellulosic biomass. Compared to raw biomass, biochar is both more hydrophobic for better storage properties and more friable for better processing. In this pretreatment method, the biomass is treated with hot compressed water in an inert environment in the temperature range of 200-260°C. The ionic constant of water is maximized in this temperature range. HTC at 260°C has an acidic pH of 3.5-3.8, predisposes most of the alkali metals to react. Moreover, the HTC process gives a mass yield of 60-90% and an energy yield 75-95%.

Minerals comprising ash might be dissolved in hot water during HTC. An experimental investigation was undertaken to evaluate which minerals are dissolved by HTC, with an attempt to identify conditions to that maximize leaching. Four biomass feedstocks were studied, including miscanthus, switch grass, corn stover, and rice hulls, and HTC at 200 °C, at 230 °C, and at 260 °C was studied. ICP-OES was used to quantify the inorganic content in each biomass and HTC char, Solid samples underwent acid digestion (HNO_3 and HF) prior to ICP-OES. The chlorine content was estimated from the ash of the solid samples using SEM-EDX.

Every biomass displays different inorganic contents, but HTC pretreatment produces similar trends for each. Potassium, sulfur, magnesium, phosphorus, calcium, and chlorine contents decrease with increasing pretreatment temperature. Sodium content is unaffected by HTC pretreatment, while silicon content apparently increases with increasing reaction temperature.

(406d) Using Non-Cl⁻ Containing Ionic Liquids for Rice Hull Pretreatment

Joan G. Lynam[1], M. Toufiq Reza[1], Victor R. Vasquez[2] and Charles J Coronella[3], (1)Chemical and Materials Engineering, University of Nevada, Reno, Reno, NV, (2)Chemical Engineering, University of Nevada, Reno, Reno, NV, (3)Chemical & Materials Engineering, University of Nevada, Reno, Reno, NV

Rice is the fundamental food for about half the world's population and provides over 20% of calories consumed by humans. Since the rice hull or husk is 20% of the grain, vast quantities of rice hulls are produced worldwide, which are then disposed by either burning, generating pollution and CO_2, or by disposal in landfills. As a secondary agricultural residue, rice hulls are already transported to processing centers, and are an underutilized renewable resource for conversion to fuel. However, their outside silica coating makes use of the cellulose inside challenging.

Ionic liquids are molten salts with melting points below 100 °C. Ionic liquids have been found which can dissolve lignocellulosic biomass, but the more effective biomass-dissolving ionic liquids tend to contain Cl^- ions. Cl^- ions can interfere with downstream processing, such as enzymatic hydrolysis.

In this study, the non- Cl^- containing ionic liquids 1-ethyl-3- methylimidazolium acetate (EMIM Ac), 1-ethyl-3-methylimidazolium formate (EMIM Formate), and 1-ethyl-3- methylimidazolium Lactate (EMIM Lactate) were investigated. A 10% (w/w) loading of rice hulls and a temperature of 110 °C were used, while the effects of mesh size, stirring, and ionic liquid type were investigated. Non- Cl^- containing ionic liquids remove lignin from rice hulls, which would be expected to make cellulose more available for further processing. Reducing mesh size of the rice hulls from between 1.651 and 1.168 mm to 1.168 to 0.589 mm in diameter greatly increased the amount of lignin which could be precipitated from the ionic liquid using ethanol. Faster stirring enhanced removal of lignocellulosic components, indicating that external diffusion may be a significant barrier to fractionation. EMIM Formate was found to remove significant quantities of lignocellulosic components as precipitates with water and ethanol addition.

If non- Cl^- containing ionic liquids can be used to remove lignin from rice hulls, the remaining hulls should be available for enzymatic hydrolysis and fermentation to biofuels. This renewable, nonfood, and accessible agricultural residue could be converted to fuel while reducing its alternative damaging environmental impact.

(406e) Chemical Pretreatment and Fermentation of Water Hyacinth (Eichhornia crassipes) At Its Optimum Age

Alvin R. Caparanga, School of Chemical Engineering and Chemistry, Mapua Institute of Technology, Manila, Philippines and Jan-Ray M. Oxales, School of Chemical Engineering and Chemistry, Manila, Philippines

Water hyacinth (*Eichhornia crassipes*) is an aquatic plant that has been considered an environmental nuisance. But, this lignocellulosic biomass can be a good substrate for ethanol production. The average life-span of this plant is 25 days. In this study, we determined the age at which the plant's holocelulose (cellulose + hemicelluloses) is highest and its lignin lowest. We found out that a 5-day old plant has the highest holocellulose-to-lignin ratio. We pretreated the 5-day-old plant samples with different chemicals (H_2SO_4, NaOH, H_2O_2) at 121 °C and 60 psia for 1 h to convert hollocellulose into fermentable sugars. We also followed the changes in lignin, glucan and xylan contents of the treated plant samples. The NaOH-treated samples had the highest level of delignification (63.32 %) and highest cellulose to glucose conversion (68.00 %). The H_2SO_4-treated samples had the highest xylan reduction (74.65 %) but lowest cellulose to glucose conversion (38.58 %). The H_2O_2-treated samples had the lowest delignification (30.43 %) although its cellulose to glucose conversion was 59.57 %. Using a traditional strain of *Saccharomyces cerevisiae*, we fermented the hydrolyzed samples and produced 13.63 g ethanol per liter (0.16 g ethanol per g dry biomass) for the NaOH-treated sample, 8.13 g ethanol per liter (0.13 g ethanol per g dry biomass) for the H_2SO_4-treated sample, and 9.32 g ethanol per liter (0.14 g ethanol per g dry biomass). These yields are either better or near-equal to those reported for same biomass and other agricultural biomass.

(406f) Alkaline Polyol Pulping - A New Pathway to Chemicals From Biomass

Michael G. Schnitzlein, Process Engineering, MCI Intl. Hochschule, Innsbruck, Austria

Alkaline Polyol Pulping –
Alternative Pathway for Chemicals from Lignocellulose

M. G. Schnitzlein[a], M. Hundt[b], N. Engel[b], D. Rein[a], K. Schnitzlein[b]

[a]MCI Management Center Innsbruck, Maximilianstr. 2, 6020 Innsbruck, AUSTRIA

[b]Brandenburg Technical University Cottbus, Burger Chaussee 2, 03044 Cottbus, GERMANY

The production of quality chemicals from biomass requires high standards on precursors for subsequent specialty products and biochemical processes. Purity and chemical compositions of these precursors, typically classified as cellulose, lignin and sugar fractions, need to match the tight requirements of biochemical pathways or catalytic conversion processes. Essentially all commercially introduced pulping processes and lignocellulose pre-treatment processes do not meet those general criteria as they were designed to focus on specific application requirements. The authors therefore conceptualized and developed a unique pulping process for producing suitable precursors for diverse chemical and biochemical production pathways with a focus on forest-based lignocellulose.

The process conceptualization employed an integrated product and process design approach, which included general quality criteria for possible precursors. Among others, the cellulose fraction needed to allow enzymatic hydrolysis with low enzyme concentrations through a significantly low lignin content and a low amount of undesired modifications of the cellulose fiber. For the lignin product, no hetero-atoms, like sulfur or chlorine, should be introduced to the product, which would compromise subsequent catalytic conversions and possible biochemical pathways. The sugar fraction was to be essentially free of furfural derivatives and to only contain mono- or oligomeric sugars with low molecular weights or sugar salts.

The process itself is perceived as a first stage in a distributed production scheme, being able to allow smaller economic entities to operate the fractionation plants and produce quality precursors as intermediates for larger chemical producers and customers. Subsequently, the process needed to build small, making large pretreatment vessels and high pressure operations prohibitive. Furthermore, bio-hazardous material was not allowed in the process to make operation in forest areas possible and safe. All those requirements were met by the presented process concept, which is using an alkaline polyol solution as a fractionation medium in the thermal conversion of soft- and hardwood lignocellulose. Fractionation is taking place at atmospheric pressure and temperatures of 200 °C and beyond, resulting in an essentially water-free thermochemical decomposition of the soft- and hardwood lignocellulose within minutes. With this the process could be designed using only continuous operating, small sized process equipment, thus allowing it to be implemented in mobile, even trailer based plants. As an important consequence, it is therefore possible to move the fractionation plant to the resources rather than feedstock to a centralized chemical operation. The term AlkaPolP has been coined to describe this fractionation principle using alkaline polyol in an essentially water-free solution.

Results from bench top laboratory work showed unbleached AlkaPolP cellulose fractions with softwood feedstock (pine) typically containing only 2–4 % w/w total lignin, with kappa numbers ranging from 7 to 10 and Klason lignin at about 1.5 %. Tests with softwood (pine, spruce) required fractionation residence times of 15 minutes and below, whereas those for hardwood (birch, common beech) could even be shorter to achieve similar fractionation results. Together with studies on a subsequent enzymatic hydrolysis, glucose recovery rates of up to 83 % of the feed stock glucose content were achieved. Furthermore, significantly lower enzyme concentrations were necessary for hydrolysis as compared to current lignocellulose ethanol processes. The lignin fraction obtained by precipitation from the black liquor showed average molecular weights of 1800-3500 and a PDI of 1.6 to 2.7 for pine and molecular weights of 1000-1500 and a PDI of 1.3 to 1.7 for beech. Due to its low hydrophobic characteristics, AlkaPolP lignin was successfully tested in the production of phenolic resin boards, substituting up to 25 % of phenol.

Based on laboratory data, various AlkaPolP process concepts were modeled using ASPEN Plus simulation software. Thermodynamic data for high alkaline solutions were calculated using an unsymmetric electrolyte model, allowing the evaluation of complex chemical equilibria. Precursors were priced based on current substitution opportunities in Western markets, leaving out possible future economical advantages due to improved product characteristics and new production pathways. Economics of various process schemes were studied which essentially differ in the way complete recycling of all process chemicals is achieved. The results are compared with published economical data for lignocellulose ethanol plants. Results show superior economic performance and substantiate a broad economic viability of the entire process concept. Future research strategies are also derived for the commercialization of AlkaPolP fractionation processes.

(414a) Technological and Economic Analysis of Adsorption Processes for Direct Capture of CO2 From Air

Ambarish R. Kulkarni, Stephanie A. Didas, Christopher W. Jones and David S. Sholl, School of Chemical & Biomolecular Engineering, Georgia Institute of Technology, Atlanta, GA

Direct capture of CO_2 from air is a concept that can be potentially used to account for CO_2 emissions from mobile and disperse sources that are inaccessible to flue gas capture. Initially, we study equilibrium process models to determine the viability of adsorption-based air capture technologies using an amino-modified silica adsorbent, TRI-PE-MCM-41, and a structured monolithic contactor unit. Using the purity of CO_2 and annual product throughput as metrics for comparing process performance; we identify some of the operational parameters, adsorbent characteristics and sources of parasitic energy losses that have a significant effect on the performance of the process. Specifically, we will discuss a process with a daily throughput of ~1.1 tCO_2 at 88.5 $\%CO_2$ purity using standard shipping container sized air capture modules. By accounting for the available sources of energy (low pressure steam and electricity) and geographic location, we obtain a net operating cost for air capture of CO_2 to be ~ $100/$tCO_2$. Recent results from more detailed adsorption models and experiments will also be presented. Finally, potential strategies for reducing the energy and monetary cost of these processes will be discussed. Our analysis supports continued work to establish the technological and economic feasibility of adsorption-based air capture.

(414b) Dynamic Modeling and Control of a Solid-Sorbent CO_2 Capture Process with Two-Stage Bubbling Fluidized Bed Adsorber-Reactor

Srinivasarao Modekurti, Department of Chemical Engineering, West Virginia University, Morgantown, WV, Debangsu Bhattacharyya, Department of Chemical Engg., West Virginia University, Morgantown, WV and Stephen E. Zitney, AVESTAR Center, National Energy Technology Laboratory, Morgantown, WV

Solid-sorbent-based CO_2 capture processes have strong potential for reducing the overall energy penalty for post-combustion capture from the flue gas of a conventional pulverized coal power plant. However, the commercial success of this technology is contingent upon it operating over a wide range of capture rates, transient events, malfunctions, and disturbances, as well as under uncertainties. To study these operational aspects, a dynamic model of a solid-sorbent-based CO_2 capture process has been developed.

In this work, a one-dimensional (1D), non-isothermal, dynamic model of a two-stage bubbling fluidized bed (BFB) adsorber-reactor system with overflow-type weir configuration has been developed in Aspen Custom Modeler (ACM). The physical and chemical properties of the sorbent used in this study are based on a sorbent (32D) developed at National Energy Technology Laboratory (NETL). Each BFB is divided into bubble, emulsion, and cloud-wake regions with the assumptions that the bubble region is free of solids while both gas and solid phases coexist in the emulsion and cloud-wake regions. The BFB dynamic model includes 1D partial differential equations (PDEs) for mass and energy balances, along with comprehensive reaction kinetics. In addition to the two BFB models, the adsorber-reactor system includes 1D PDE-based dynamic models of the downcomer and outlet hopper, as well as models of distributors, control valves, and other pressure-drop devices. Consistent boundary and initial conditions are considered for simulating the dynamic model. Equipment items are sized and appropriate heat transfer options, wherever needed, are provided. Finally, a valid pressure-flow network is developed and a lower-level control system is designed.

Using ACM, the transient responses of various process variables such as flue gas and sorbent temperatures, overall CO_2 capture, level of solids in the downcomer and hopper have been studied by simulating typical disturbances such as change in the temperature, flowrate, and composition of the flue gas. To maintain the overall CO_2 capture at a desired level in face of the typical disturbances, two control strategies were considered–a proportional-integral-derivative (PID)-based feedback control strategy and a feedforward-augmented feedback control strategy. Dynamic simulation results show that both the strategies result in unacceptable overshoot/undershoot and a long settling time. To improve the control system performance, a linear model predictive controller (LMPC) is designed. In summary, the overall results illustrate how optimizing the operation and control of carbon capture systems can have a significant impact on the extent and the rate at which commercial-scale capture processes will be scaled-up, deployed, and used in the years to come.

(414c) CO2 Separation Using Mg(OH)2 Solutions – Dissolution Kinetics and Vapor-Liquid-Solid Equilibrium Studies of CO2- Mg(OH)2 System

Hari Krishna Bharadwaj[1], Joo-Youp Lee[1] and Tim.C Keener[2], (1)Chemical Engineering Program, University of Cincinnati, Cincinnati, OH, (2)Environmental Engineering Program, University of Cincinnati, Cincinnati, OH

Coal-fired power plants are responsible for a considerable amount of total anthropogenic CO_2 emissions throughout the world. CO_2 from flue gases can be absorbed by $Mg(OH)_2$ slurry at 52 °C in a column followed by a regeneration step in a stripper operated at 65 °C.

One of the most important steps in the CO_2 scrubbing process is the rate at which magnesium hydroxide dissolves from the solid part of the slurry and becomes available for reaction with absorbed CO_2. The dissolution rate kinetics of $Mg(OH)_2$ was studied using a pH stat and the dissolution rate was determined based on acid consumption required to maintain a preset pH value. The effects of pH, temperature, mixing and particle size on the dissolution rates of $Mg(OH)_2$ will be presented. A dissolution rate kinetic expression obtained under different operating conditions will also be presented.

Knowledge of Vapor liquid solid equilibrium (VLSE) of CO_2 in $Mg(OH)_2$ is essential for the proper design and operation of absorption and desorption equipment. The VLSE data for the $Mg(OH)_2$ – CO_2 system under various operating conditions of pressure, temperature, and concentration of $Mg(OH)_2$ will be presented. A suitable model for fitting the VLSE data will also presented.

(414d) CO_2 Capture Using Particulate Silica Aerogel Immobilized with Tetraethylenepentamine (TEPA)

Jerry YS Lin, Robert Pfeffer and **Nick Linneen**, Chemical Engineering, Arizona State University, Tempe, AZ

CO_2 capture using particulate silica aerogel immobilized with tetraethylenepentamine (TEPA)

Nick Linneen (presenter), Jerry YS Lin, Robert Pfeffer

School for Engineering of Matter, Transport, and Energy

Arizona State University

Tempe AZ 85289

Abstract:

Post-combustion capture of CO2 using solid amine-based sorbents is attracting significant attention due to their high selectively, reduced corrosive properties, and low energy regeneration requirements. Research has shown that sorbents that have large pore diameter and pore volume generally achieve the highest CO2 adsorption capacities. This is a result of the support's greater potential to retain large amounts of amine, while also allowing effective CO2 transport within the sorbent. We have developed a novel sorbent using particulate aerogel as a support. Aerogel, also known as "solid smoke," is a high porosity, high surface area, silica-based material that has very large pore volumes and pore diameters compared to other supports. Therefore aerogels can retain large amounts of amine while also maintaining some mesoporosity for effective CO2 transport. By impregnating tetraethylenepentamine (TEPA), which is an ultra-low molecular weight linear polyethylenimine (PEI), into commercially available 10 micron size aerogel (Cabot Corp. Nanogel), we have obtained an adsorption capacity as high as 6.1mmol/g which is very competitive relative to other CO2 sorbents. The aerogel-amine based sorbent was also found to achieve fast kinetics reaching 90% of its equilibrium capacity within the first 10min and also maintaining its adsorption capacity after 10 adsorption/desorption cycles.

(414e) Development of High Performance Carbon-Based Adsorbents for CO2 Adsorption

Placidus B. Amama, RXBT/UDRI, Air Force Research Laboratory, Wright-Patterson AFB, OH and Patrick J. Shamberger, RXBT, Air Force Research Laboratory, Wright-Patterson AFB, OH

Thermal management of high-power systems is a pressing challenge for many USAF platforms. These systems require rapid absorption and ultimately rejection of low-quality (low temperature) heat. Activated carbon is capable of adsorbing pressurized CO_2 gas with a high heat of adsorption, offering the possibility of rapidly absorbing and rejecting heat at low temperatures. The activated carbon adsorbent has to meet important performance criteria for high efficiency of the system. The performance criteria include high adsorption capacity, high adsorption kinetics, excellent stability and mechanical properties, and high thermal conductivity. For activated carbon adsorbents to meet all of these criteria, the CO_2 adsorption capacity and the adsorption-desorption kinetics must be improved. The adsorption of CO_2 on activated carbon can be significantly enhanced by the introduction of basic nitrogen functionalities on the surface. In this work, surface modification of a series of activated carbon adsorbents via gas phase amination under various conditions has been performed. The surface area, pore structure and pore volume were characterized using N_2 physisorption while the types of chemical functionalities present on the surface after amination were characterized by X-ray photoelectron spectroscopy (XPS), Fourier transform infrared spectroscopy (FTIR), and temperature programmed desorption (TPD). CO_2 adsorption on the modified activated carbon samples have been measured gravimetrically using a high-pressure TGA.

(417a) Product & Resource Recovery – an Evolution

George Abrahim, VWS, Veolia Water, Moon, PA

The dramatic shift in energy prices, availability of mineral resources, and scarcity of water in certain regions are driving industrial facilities to seek innovative solutions to reduce their risk exposure.

The wastewater plant of the 21st Century is also evolving to become a resource rather than a cost center. Self sufficient in terms of energy, high recovery for water recycle/reuse, produces recoverable substances and generates less waste. Until recently, this was generally regarded as an obscure business, whose growth rate and profitability scarcely deserved attention.

This evolution of the water business will continue to drive the development of new innovative technologies and processes such as desalination of brackish water for recycle/reuse using high recovery ROs; recovery of biogas of high COD wastewater stream using anaerobic MBR; and by-product recovery of metals from mining wastewater using various conventional treatment schemes. This presentation will discuss the drivers behind P&RR and also presents case studies highlighting innovation in new technologies and their value proposition to the market place.

(417b) Development of An Adsorbent Derived From Concrete Sludge for Phosphorus Recovery From Waste Water

Akihiro Yamasaki, Department of Materials and Life Science, Seikei University, Musashino, Japan, Takeshi Sasaki, Department of Materials and Life Science, Seikei University, Musashino, Tokyo, Japan, Atsushi Iizuka, Research Center for Sustainable Materials & Engineering, Institute of Multidisciplinary Research for Advanced Materials, Tohoku University, Sendai, Miyagi, Japan and Yukio Yanagisawa, Department of Environment Systems, The University of Tokyo, Kashiwa, Chiba, Japan

A solid adsorbent for phosphorus recovery was prepared from concrete sludge. Concrete sludge was diluted with water with dilution ratios ranging from 1 to 15. The diluted concrete sludge was then filtered and dried. Two types of drying methods were examined: natural drying under atmospheric conditions and forced drying in an oven at 105°C. The phosphorus recovery performances of prepared adsorbents were examined using a potassium dihydrogen phosphate (KH_2PO_4) solution containing 100 mg-P/L as a model of wastewater, which is close to the water effluent produced in the sewage sludge dewatering process. All the adsorbents prepared showed phosphorus recovery performance. The forced drying treatment at 105°C significantly increased the phosphorus recovery rate from the solution. The adsorbent prepared with dilution ratio 10 and forced drying showed the highest phosphorus recovery rate. The final phosphorus concentration in solution was 0.83 mg/L at 4 h from the addition of the optimized adsorbent. These results indicate that the phosphorus adsorbents prepared from concrete sludge are applicable for practical and inexpensive phosphorus recovery processes.

(417c) Performance of a Continuous Reactor for Electrochemical Treatment of Pulp Mill Wastewater

Indra Mani Mishra[1], S Mahesh[2], Basheshwar Prasad[3] and I. D. Mall[3], (1)Department of Chemical Engineering, Indian Institute of Technology Roorkee, Roorkee, India, (2)Environmental Engineering, S.J. College of Engineering, Mysore, Mysore, India, (3)Chemical Engineering, Indian Institute of Technology Roorkee, Roorkee, India

The present paper reports the performance characteristics of a continuous electrochemical reactor (ECR) for the treatment of black liquor (BL) discharged from an agri-based pulp and paper mill. The ECR had a working volume of 2 dm^3 and used six iron electrode plates with an active electrode area per unit volume of 54 $m^2\ m^{-3}$. The ECR operated at an optimal galvanostatic state at a current density of 55.56 Am^{-2} with a corresponding cell voltage of ~ 3.1V at near neutral pH without any additives. The residence time (t) in the range of 2-4 h favors the degradation of chemical oxygen demand (COD), color and turbidity. A COD removal of 78.2 and 82.15% was achieved at flow rates of 1 and 0.5 $dm^3\ h^{-1}$, respectively, with a corresponding color removal of ~90 and 79%. The ECR showed good performance at flow rate spikes from 0.5®1®2 dm^3h^{-1} and 1®2 dm^3h^{-1} with ~79 and ~73 % COD removal, respectively. The specific energy consumption (SEC) decreased in a narrow range from 16.3 to 14.3 kWh kg^{-1} of COD removed for t varying from 4 to1 h. For a charge of 0.625 $Ahdm^{-3}$, the current efficiency (CE) values of 310 and 274% were obtained for t = 2 and 1 h, respectively. Batch and continuous study results compared well at t = 2 and 1 h. The ECT showed deterioration in the COD removal efficiency on long time runs with the anode plates swarmed with micro flocs which necessitate their removal.

UV-Radiation Through DC-Corona Discharge and Wastewater Treatment

Muhammad SulemanTahir[1,2], MahmoodSaleem[3], Niaz Ahmad Akhtar[4], M. Siebenhofer,[1]

[1]Institute of Chemical Engineering and Environmental technology, University of Technology Graz, Austria
[2]Department of Chemical Engineering, University of Gujrat, Pakistan
[3]Institute of Chemical Engineering and Technology, University of the Punjab, Lahore, Pakistan, [4]National
[4]Textile University Faisalabad, Pakistan

Abstract— DC-Corona discharge is applicable in waste water treatment as well as off gas-treatment to mitigate hazardous gaseous, liquid and particulate pollutants. Target is to investigate the decomposition and degradation of hazardous pollutants like acetone at specific current/voltage characteristics in a DC-Corona discharge reactor equipped with brush type discharge electrode. Depletion of total organic carbon (TOC) and change of pH was determined in a tube type corona discharge reactor under air flow and water circulation. TOC depletion and pH change is quite similar to the results of acetone oxidation under UV-irradiation with a low pressure Hg Lamp (LPML). A DC-Corona discharge reactor equipped with a brush type discharge electrode is capable of oxidizing and disintegrating acetone. Complete degradation of acetone into intermediate products and finally into carbon dioxide is possible.
A DC-Corona discharge reactor was investigated at various applied voltage to determine UV-radiation. Experimental investigation shows that system is capable to generate the UV-radiation over a range of 230-265 nm. DC-Corona reactor with brush type discharge electrodes is capable to produce the oxidizing reagents at ambient operation conditions. It is hard to differentiate either degradation of pollutants is individual effect of intense corona, UV-radiation or oxidizing reagents. An accumulative effect of UV-radiation and oxidizing reagents on degradation of organic pollutants was measured and confirmed.

Keywords; DC-Corona discharge, Current/voltage characteristics, TOC depletion, pH variation,

I. INTRODUCTION

Waste water from industrial sources may contain hazardous and complex pollutants like organic, inorganic and particulate matter. The diversity of pollutants results in a system of water quality criteria based on the interaction between pollutant concentration, environmental impact and human health.

Under UV-irradiation fast decomposition of acetone with ozone is reported in literature [1]. Author also demonstrated the reaction scheme for the formation of ozone, OH radical and decomposition of ozone. The reaction scheme for the formation of oxidizing reagents is given in equation 1-5. Literature [2] also reported similar reaction scheme.

$$H_2O + hv\,(185\ nm) \rightarrow \bullet H + \bullet OH \quad (1)$$

$$O_2 + hv\,(185\ nm) \rightarrow O^{\bullet} + O^{\bullet} \quad (2)$$

$$O_2 + O^{\bullet} \rightarrow O_3 \quad (3)$$

$$O_3 + hv \rightarrow O_2 + O(^1D) \quad (4)$$

$$O(^1D) + H_2O \rightarrow 2.OH^{\bullet} \quad (5)$$

Ying Kang and ZuCheng Wu [2] explained the effect of strong electric field of corona discharge on the OH radical formation due to the collision reaction between electron, air and water. High energy electron (5- 15eV) collisions with water and oxygen molecules generate different reaction products as shown in reaction (1) to (5). Several authors [3, 7] investigated and repotted the process of degradation of organic compounds and formation of various intermediate products. Photochemical purification of water and effect of inorganic and organic chemicals on the human health is demonstrated in literature [4] and [5]. Remarkable decomposition of Toluene with surface discharge was investigated by O. Yujiro et al. [6]. Author of literature [7] conducted the studies and reported the destruction of methane with pulsed corona discharge. Author of Literature [7] reported that methane destruction depends upon the discharge current and time of discharge.

Decay of acetone in aqueous solution and decrease in TOC under UV-irradiation is reported in literature [8]. Dissociation mechanism of acetone and formation of intermediate products (intermediate products further oxidize to carbon dioxide and water) is explained and demonstrated in literature [8]. Won-Tae Shin [9] investigated and compared the oxidation of organic impurities in water by injecting oxygen or ozone at different applied voltage. Literature [10] focused on the reaction mechanism of hydroxyl radicals with acetone at atmospheric conditions. Ranajit K. [10] also determined a higher yield of acetonyl radicals along with other intermediate products and reaction of OH radicals with acetone as given in equation (6).

$$OH + CH_3C(O)CH_3 \rightarrow products \quad (6)$$

Authors of literature [11, 12] reported that corona discharge is a function of the geometry of discharge electrodes and applied voltage. High intensity corona generation is helpful to dissociate water molecules and generate hydroxyl radicals as explained in literature [9], [13] and shown in reaction (1). Effect of applied voltage, water and oxygen concentration on the generation of OH radicals is well explained and investigated by R. Ono [13]. Author of literature [13] reported the measuring and quantification of hydroxyl radicals with pulsed corona discharge. Literature review reflects the importance of corona discharge in effluents treatment.

DC-corona reactor with brush discharge electrode has a potential to generate the enhanced and uniform high intensity corona. High energy (photon) corona can eliminate organic impurities from waste water and off-gas simultaneously. DC-corona reactor is also flexible to operate under low and high load conditions as per process requirement.

II. Materials and Methods

Experimental Setup

Figure 1 shows the experimental set up.

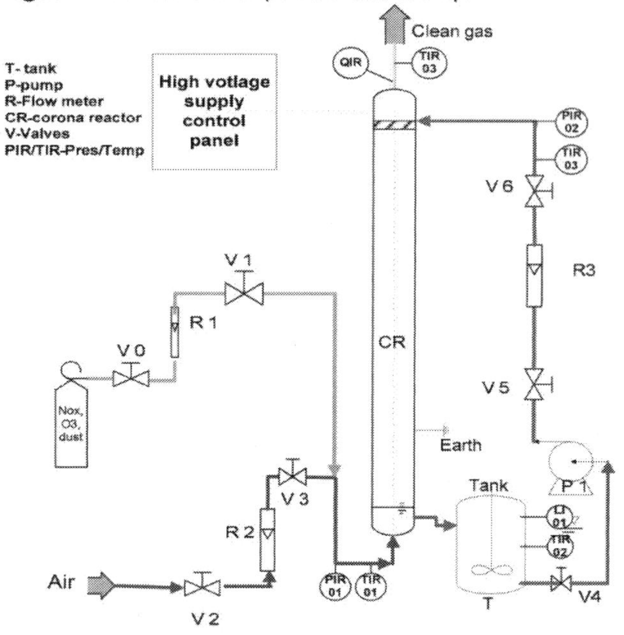

Figure 1: Experimental set up of the test facility; collecting electrode: stainless steel; tube diameter: 66 mm; tube length: 1500 mm; Discharge electrode: stainless steel (brush type electrode length: 700 mm, brush diameter 5 mm and brush wire diameter 0.15 mm).

According to Fig.1 air is fed at the bottom and rinsing water, containing acetone, is continuously fed on top of the reactor. For analytical purposes water samples are taken at the bottom of the reactor or circulation tank (T). To quantify the change in TOC, the samples are analysed with "TOC-5000A SHIMADZU".

0.5g/Liter of acetone was mixed with de-ionised water and water circulation was set at 50 litres per hour with flow meter (R3) and air flow was set at 0.5 m^3/h with flow meter (R2).

Prior to activation of corona discharge the circulation system was operated and after two hours corona discharge was activated and operated at 16 kV/0.6- 0.8mA.

III. Results

The current/voltage characteristics under batch operation conditions of the reactor as well as with air and with air plus water circulation was recorded and shown in Figure 2.

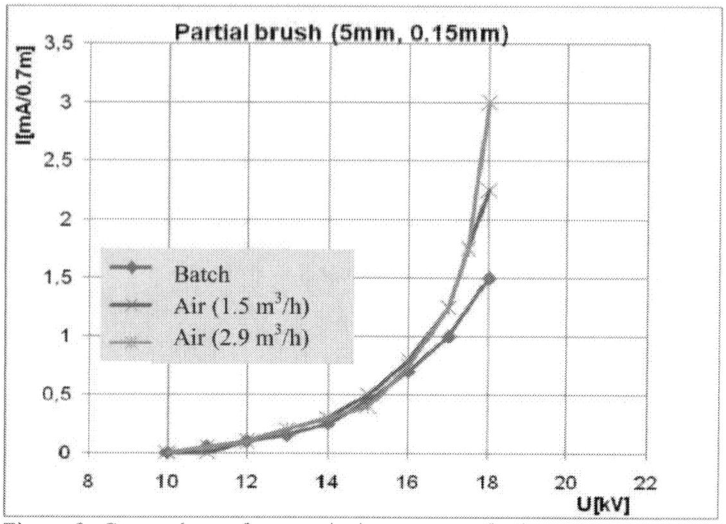

Figure 2: Comparison of current/voltage curves for brush discharge electrode under batch, only air, and air and water flow conditions.

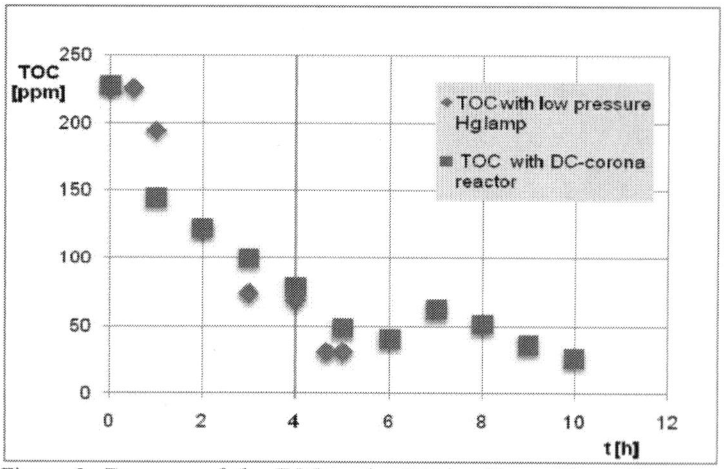

Figure 3: Decrease of the TOC under UV-irradtion with medium pressure Hg lamp and under corona discharge in a DC- corona reactor with partial brush discharge electrode and ambient operation conditions.

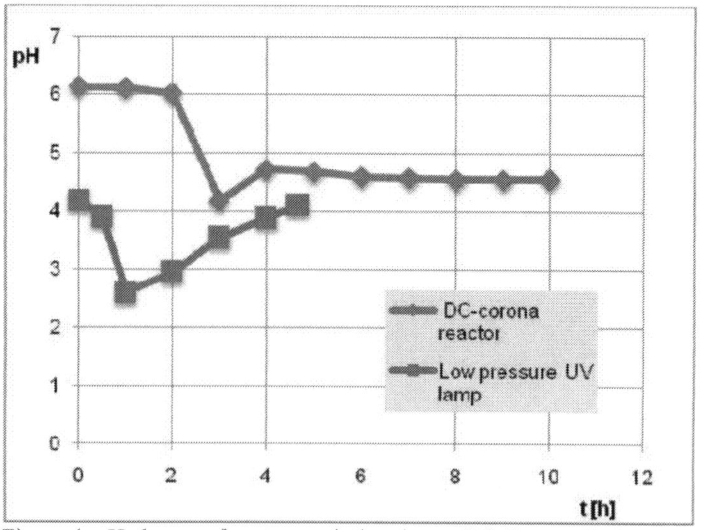

Figure 4: pH change of acetone solution due to DC-Corona discharge with brush discharge electrode and degradation is compared with low pressure Hg lamp irradiation (radiation source).

Power consumption comparison with respect to TOC depletion.

Processes were compared in terms of power consumed to purify the water at similar degradation level of pollutants. Acetone degradation level is nearly same after 5 hours operation with low pressure mercury lamp and DC-Corona reactor. Detail is explained in Table 1 and 2.

Table 1: Comparison of DC-Corona discharge and low pressure mercury lamp process in terms of TOC depletion.

Process used	time [h]	Initial TOC [ppm]	Final TOC [ppm]	Depletion [%]
LPML	5	250	30	88
DC-CD	5	250	49	80.4

Low pressure mercury lamp is capable to degrade representative pollutants in water up to 88 percent on the basis of depletion of total organic carbon as shown in Figure 3 and table 1. Process was continued for 5 hours with continuous monitoring especially of pH changes. Pollutant degradation process was also conducted with DC-Corona reactor for same operation conditions and degradation trend of TOC was recorded calculated 80.4 percent as shown in Figure 3 and table 1.

Comparison of energy consumption

Table 2: Comparison of DC-Corona discharge(CD) and low pressure mercury lamp (LPML) process in terms of energy consumption per ppm reduction.

Process	LPML	DC-CR
Operation Time [h]	5	5
Power [W]	15	11.2
Total energy consumed [Wh]	90	56
Depleted TOC[ppm]	220	201
Energy consumed [Wh] per ppm	0.49	0.28

IV. Discussion

Experiments were conducted to determine the degradation mechanism for acetone in a DC-corona discharge and under irradiation with a low pressure Hg lamp with air purge.

The DC-corona reactor was experimentally investigated under batch and flow conditions to develop the current /voltage relationship as shown in Figure 2. Corona current increases with increase in applied voltage under batch and flow conditions. Higher corona current flow is obtained with air and water circulation than the batch conditions as shown in Fig. 2. For smooth operation and to avoid the sparking, operation voltage was limited to 16kV corresponding with a corona current of 0.6-0.8mA.

Under DC-Corona discharge as well as UV-irradiation, depletion of organic carbon (TOC) was observed as shown in Fig. 3 and also described in table 1. Change in pH indicates formation of acidic intermediates. Variation in pH value during the operation is demonstrated in Fig.4.

The trend of TOC depletion confirms the conversion of acetone into intermediate products like acetic acid, formic acid, acetaldehyde, methanol etc. Further disintegration of intermediate products into carbon dioxide and water is seemingly not affected by acidification because the rate of TOC depletion remains constant. The reaction scheme of acetone degradation is summed up through equations (7-9). Literature [7] also suggests a degradation reaction scheme for acetone to similar primary intermediate

$$CH_3COCH_3 + (hv, OH, O_3) \rightarrow CH_3COOH + CH_3COH + CH_3OH \quad (7)$$

$$CH_3OH + (hv, OH, O_3) \rightarrow HCOOH + HCOH \quad (8)$$

$$HCOOH + (hv, OH, O_3) \rightarrow CO_2 + H_2O \quad (9)$$

According to literature [1] UV-irradiation from low pressure Hg lamp (LPML) is capable of ozone generation and consecutive disintegration of acetone. TOC depletion through corona discharge is comparable with TOC depletion through UV-irradiation. It can not be confirmed with these experiments whether degradation through corona discharge is supported by UV-irradiation or ozone formation only.

Change in pH value through DC-corona discharge and UV-irradiation is different as shown in Fig. 4 indicating different mechanism. Drop of the pH value indicates formation of acetic acid, formic acid, intermediates etc., from acetone. Change in pH has been observed by Taek-Soo Kim et al., [1] too.

Finally an increasing pH value during operation indicates further degradation of acidic intermediates to carbon dioxides according to a representative mechanism shown in equation (9).

Energy consumed for degradation of pollutants with low pressure mercury lamp and DC-Corona reactor equipped with partial brush discharge electrodes was calculated to make a comparative study.

DC-Corona reactor was also operated at various applied voltage to determine the UV-radiation and formation of oxidizing reagents. Experimental recorded data shows that DC-Corona reactor with type discharge electrodes is capable to generate the UV-radiation and oxidizing reagents like ozone (atomic oxygen etc) as in Fig. 5 and 6. Experimentally recorded data for UV radiation with DC-Corona reactor is demonstrated in Fig.5 and UV radiation was in the range of 250-255 nm at peak.

System was also investigated to determine the ozone formation capability and ozone generation was confirmed as in Fig. 6. It is also recorded that ozone generation was increased with the increase in applied voltage like 12 kV, 13 kV and 14 kV. Applied voltage 12 kV, 13 kV and 14 kV corresponds to specific corona current of 0.05 mA, 0.1 mA and 0.35 mA respectively. Ozone generation was recorded with brush type discharge electrodes showed an enhancement of ozone with increase of corona current from 0.05mA to 0.35mA.Ozone generation is optimized at 0.1 mA corona current with respect to energy consumption.

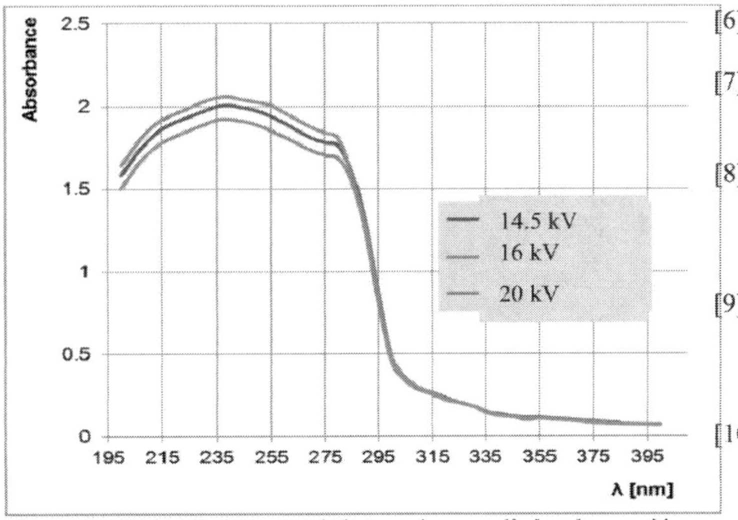

Figure 5: UV-irradiation recorded at various applied voltage with DC-Corona reactor. EDTA solution was circulated to absorb the UV-radiation at various applied voltage and absorbance was measured with photo spectrometer.

V. CONCLUSION

Current voltage behavior of DC-Corona reactor with brush electrode was compared for batch operation and for flow operation. Higher corona current was observed with air flow and water circulation. Corona discharge process was optimized for specific air and water flow to avoid process instabilities and sparking.

DC-Corona discharge reactor is capable of producing UV-radiation and can be utilized as advanced oxidation reactor for effluent treatment [14].

It can be concluded that DC-corona discharge with brush type discharge electrodes is capable of efficiently disintegrating organic compounds in water and waste water. It has highly interesting features for detoxication of waste water. DC-Corona discharge is therefore a highly potential technology for flue gas purification with the major action in particle precipitation as well as wastewater treatment.

VI. REFERENCES

[1] Taek-Soo Kim, Removal of organic impurities in an aqueous solution using uv irradiation produced from high and low pressure hg lamps. Industrial & Engineering Chemistry Research, 46(14):4799–4810, July 2007.

[2] Ying Kang and ZuCheng Wu. Formation of hydroxyl radicals and oxidation of toluene under corona discharge with water vapor as radical source. Chinese Science Bulletin, 53(14):2248–2252, July 2008.

[3] Guy Bablon, Ozone in water treatment, Application and Engineering. *Fundamental aspect*, page 22, 1991

[4] D. Vione, Degradation of phenol and benzoic acid in the presence of a tio-based heterogeneous photocatalyst. *Applied Catalysis B: Environmental*, 58(1-2):79–88, June 2005.

[5] Thomas Oppenlander. Photochemical processes of waste treatment. *Photochemical purification of water and air*, 101–106, November, 2002.

[6] Y. Keping. Electrostatic precipitation, 11Th international conference, Hangzhou, 2008. Page 690-693.

[7] Shibata K, Characteristics of methane destruction using a pulsed corona discharge at atmospheric pressure. *J. Plasma Fusion Res. SERIES,*, 6:760–763, 2004

[8] Mihaela I. Stefan. Reinvestigation of the acetone degradation mechanism in dilute aqueous solution by the uv/H_2O_2 process. Environmental Science & Technology, 33(6):870–873, March 1999

[9] Won-Tae Shin, A pulseless corona-discharge process for the oxidation of organic compounds in water. Industrial & Engineering Chemistry Research, 39(11):4408–4414, November 2000.

[10] Ranajit K. Talukdar, Reaction of hydroxyl radical with acetone. 2. Products and reaction mechanism. The Journal of Physical Chemistry A, 107(25):5021–5032, June 2003.

[11] D. Brocilo, Electrode geometry affects on the collection efficiency of submicron and ultra-fine dust particles in spike-plate electrostatic precipitators. Journal of Physics: Conference Series, 142(1):012032+, 2008.

[12] Kuroda. Effect of electrode shape on discharge current and performance with barrier discharge type electrostatic precipitator Journal of electrostatics Source 2003, vol. 57, no3-4, pp. 407-415 [9 page(s) (article):407–415, 2003.

[13] R. Ono. Measurement of hydroxyl radicals in pulsed corona discharge. Journal of Electrostatics, 55(3-4):333–342, July 2002.

[14] M. S. Tahir, An innovative and advancedoxidationprocess for effluent treatment through wet tube-type electrostatic precipitation, Joural of Chemical Engineering and Processing: Process Intensification, VBol 52 (16-20), Feb 2012

(417e) Production of Novel Adsorbents From Waste Electric and Electronic Equipments

Pejman Hadi[1], John Barford[2] and Gordon McKay[1], (1)Chemical and Biomolecular Engineering, Hong Kong University of Science and Technology, Hong Kong, Hong Kong, (2)Chemical and Biomolecular Engineering, The Hong Kong University of Science and Technology, Clear Water Bay, Hong Kong

The problems regarding ecosystem contamination are increasing with expansion of the technology. Among these problems, toxic heavy metal pollution in air, soil and water is one of the foremost global problems and is a great threat to humanity. Toxic metal compounds coming to the earth's surface not only reach the earth's surface waters, but can also contaminate underground water by leaking from the soil after rain and snow. Therefore, the earth's waters may contain various toxic metals which might result in drinking water contamination and thus poisoning due to bioaccumulation. Bioaccumulation is an increase in the concentration of a chemical in a biological organism over time, compared to the chemical's concentration in the environment. Compounds accumulate in living things any time they are taken up and stored faster than they are excreted or metabolized. In response to the ever-growing problems, federal and state governments have instituted environmental regulations to protect the quality of surface and ground water from heavy metal pollutants, such as Pb and Zn.

A number of methods for the removal of heavy metals from dilute aqueous solutions have been reported, including a reduction process, sulfide treatment, ferrous chloride treatment, magnetic ferrites treatment, ion exchange and ion exchange followed by chelating resin. Among the several treatment technologies developed, adsorption has been found to be superior to other techniques for wastewater treatment in terms of simplicity of design, ease of operation and insensitivity to toxic substances. In general, activated carbons are extensively-applied effective adsorbents for wastewater treatment. Nevertheless, the high cost of commercial activated carbon has led to hunting for more economical sources for the raw material to reduce the capital cost. Among these, agricultural waste based activated carbons are of great interest nowadays. Several researchers have studied the production of activated carbon from Ceiba pentandra hulls, pecan shell, hazelnut shell, peanut hull, almond shell, peach stone and sawdust.

To the best knowledge of the authors, no report has been published about the use of non-metallic fraction of Waste Electric and Electronic Equipments (NM-WEEE) as adsorbents in order to remove the metal contaminants from the dilute aqueous solutions. Since these materials contain a considerable amount of silicon, alumina, and carbon, all of which have a high potential to be good adsorbents when treated well, they were chosen to be our research target. Another reason for opting for these materials is that during the recent years, there is an ever-increasing concern about management of this waste and lack of proper methodology for its disposal. The two prevalent methodologies applied to this waste are either incineration or landfilling both of which have been hindered due to the evolution of potent toxic gases in the case of former technique and lack of enough land space and toxic contamination of the land in the case of later technique. Without the systems and the technology in place to effectively capture and manage these wastes, the accumulation of e-wastes has and will continue to be a significant problem at a global level. The rate of increase in the volume of e-waste now far exceeds the generation of municipal waste.

Preliminary experiments showed that untreated NM-WEEE is a very hydrophobic material with a trivial surface area. Therefore, it was expected that this material could not be used as an adsorbent in wastewater treatment despite its potentiality in terms of its constituent components. Hence, to facilitate the adsorption and development of porosity, chemical modification was carried out by the impregnation of the precursor in KOH for 3hr followed by pyrolysis in an inert atmosphere at 250°C for 3hr. Although the surface area of the material was increased from 2 m^2/g to 240 m^2/g after modification, a high metal removal efficiency could not necessarily be guaranteed. Thus, a full adsorption test was carried out by using modified NM-WEEE as adsorbent and copper ions as adsorbate. The results indicated a significant amount of adsorption which was 3 mmol/g in comparison with 2.2 and 2mmol/g for its famous industrial counterparts, i.e. Suqing D401 and Lewatit TP207, respectively. The study was extended for two more metal ions, that is Zn and Pb. In the case of Zn, our novel adsorbent could adsorb 2.5 mmol/g, whereas this amount was only 1.9 and

1.6 mmol/g of Suqing D401 and Lewatit TP207, respectively. For Pb-bearing effluent, the adsorption is even more encouraging which shows an adsorption capacity of 3.5, 2.3 and 2.2 for our novel adsorbent, Suqing D401 and Lewatit TP207, respectively. All these figures indicate the extraordinary efficiency of the innovatively synthesized material to treat the wastewater effluents containing metal ions.

In terms of financial aspects, it is notable that since the precursor is a waste material and the treatment temperature is very low compared to the ones in the normal modification processes in the industry (600-1000°C), the production cost will be reduced enormously.

To conclude, the production of this novel adsorbent could environmentally be of great interest in two points of views: on one hand, this waste (WEEE) can cause serious environmental contamination problems and this finding can beneficialize this material's usage. On the other hand, this synthesized material can be employed to remove the heavy metal ions from the effluents causing ecological concerns.

Development of a Process for Purification of High TDS Copper Mine Pool Water

Annie Lane[1], *Paul J Usinowicz*[1], *Bruce Monzyk*[2], *Rick Peterson*[1] *and Todd Beers*[3], *(1) Battelle, Columbus, OH, (2) Ascend, Decatur, AL, (3)Winner Global, Sharon, PA*

Abstract

Battelle, with Winner Global Energy and Environmental Services LLC (Winner), has developed a process that uses Battelle's Acid Mine Drainage Value Extraction Process (AMD VEP™) for the purification of copper mine pool water. This Norway-based acid mine drainage (AMD) water represents a surface water contamination source that is damaging to the local fishing industry. The primary contaminants of concern are copper (Cu) and high TDS (especially sulfate), though this water is highly contaminated with other metals, as shown in Table 1, necessitating a process capable of producing a water product of sufficient purity to discharge to surface waters.

Table 1. Composition of AMD water

Parameter	mg/L	lb/1000 gal
Pb	0.02	$1.7x10^{-4}$
Fe	790	6.6
Zn	49.5	0.4
Ni	0.61	$5.1x10^{-3}$
SO_4	4100	34.2
Al	208	1.7
Cu	81.4	0.7
Cd	0.17	$1.4x10^{-3}$

The AMD VEP is based on Flotation-Liquid Liquid Extraction (F-LLX™) water purification separation technology that has been developed to produce discharge quality water from high TDS mine pool water. F-LLX, developed for Winner by Battelle for AMD prevention, features high throughput processing rates, high water recovery, no significant waste generation, removal of metals and sulfate ions to dischargeable levels and, further, produces useful product metal and sulfate concentrates.

In order to develop the process, testing was performed in several stages. The first stage included generation of a surrogate water based on the composition of mine pool water provided by the client (Table 1). The surrogate water was used in bench scale continuous flow testing to demonstrate the feasibility of treating the water with the F-LLX process. Based on the positive results of this feasibility study, a subsequent continuous bench scale test program was conducted with the actual mine pool water. The results of this continuous bench scale test and the requirements for purification of the source water were used to determine the process design and operating parameters for a demonstration plant to be constructed in Norway.

Introduction

Acid mine drainage (AMD) from closed copper mines in the Folldal municipality of Norway is the source of surface water (river) contamination. As a result, the government of the Folldal municipality has set targets for removal of copper, iron, zinc, and sulfate from the Folldal AMD stream, and has contracted with Winner to evaluate the applicability of the Acid Mine Drainage Value

Extraction Process to treatment of this source contamination. The target values for the water purified by the process were identified by the Norwegian Ministry of Mines at the initiation of the test program. The primary water purification targets were for removal of copper (Cu) to <1 mg/L and iron (Fe) to <50 mg/L. Additional desired targets for zinc and sulfate removal were set at 1 mg/L and 100 mg/L, respectively.

The AMD VEP uses the F-LLX separation technology originally developed by Battelle for the purification of coal mine pool water. The F-LLX technology features high-throughput processing rates, no significant waste generation, and removal of metal and sulfate ions to dischargeable levels, all while providing useful product concentrates along with purified water. Figure 1 shows a simplified process flow diagram for the AMD VEP.

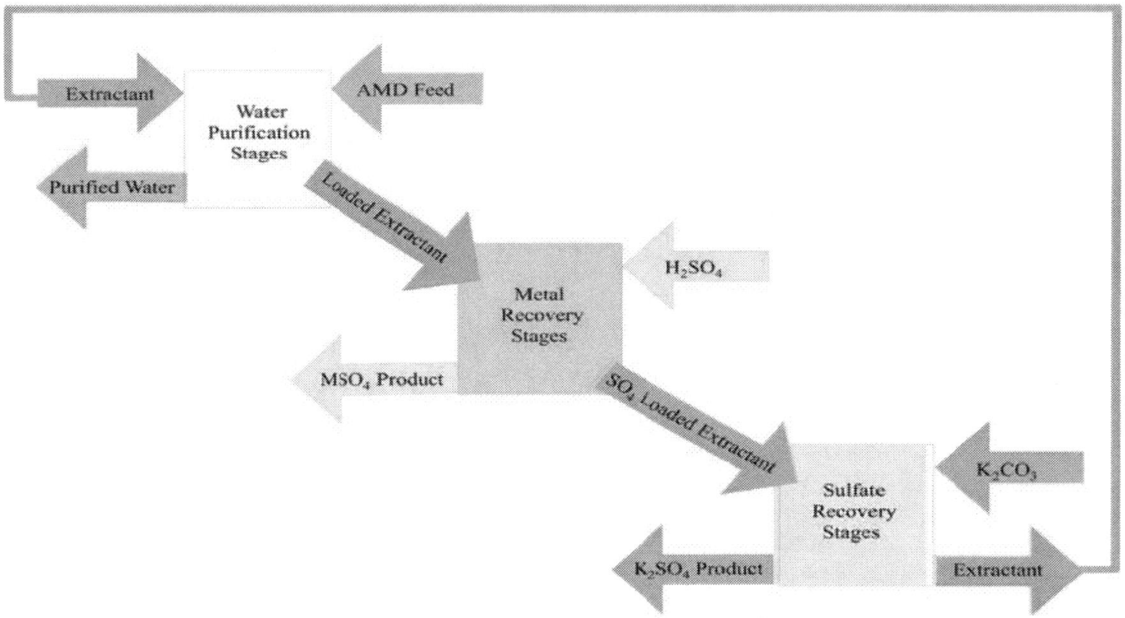

Figure 1. Simplified Process Flow Diagram

In the test program, Battelle first demonstrated the feasibility of using the F-LLX technology for the removal of the metal ions and sulfate constituents in a surrogate AMD feed matching the composition of the Folldal AMD. Due to the promising results of the testing with surrogate water (>99% removal for the majority of the contaminants), and since only the extraction capabilities of the process were demonstrated with the surrogate water, further testing was requested using mine pool water from the Folldal mine.

Methodology

The objective of the continuous bench testing with the actual Foldall mine pool water was to develop the balance of the F-LLX water purification process and to demonstrate its effectiveness on a continuous basis to meet the treatment targets. The balance of the process included full development of the process flow scheme including regeneration of the extractant phase for reuse in the process, and product concentrate generation operations.

Treating the Folldal AMD to meet the water purification targets in a continuous, steady-state, bench-scale system provides the assurance that the process can be successfully scaled up for application at the Folldal site. Data from the bench-scale demonstration allows development of such requirements as suitable equipment types and sizes, estimated raw material consumption needs, process control set points, and control windows. The deliverables include a process flow diagram with mass balances, characterization of process products, measurements of key process parameter set points, minimum residence times for key equipment, and raw material consumption rates (sulfuric acid and potassium or sodium carbonate).

Other critical process operational information relevant to the design of the next-stage field demonstration plant includes:

- The required number of extraction, decanter, and recovery/extractant-phase regeneration stages.
- The needed extractant/aqueous (E/A) flow rate ratio for the extraction stages, with estimates for the decanter and stripping stages.
- An internal recycle flow rate range for the stripping stages.
- Extractor mixing and settling times, and a range of mixer tip speeds for operation.
- The consumption rate of process chemicals (sulfuric acid and potassium or sodium carbonate).
- Product analysis, including the composition of the purified water, metal sulfate concentrate, and potassium sulfate concentrate.

In order to demonstrate the F-LLX water purification process for the Folldal mine pool water, the Tasks described in Table 2 were conducted.

Table 2. Tasks for Development of F-LLX Water Purification Process

Task #	Title	Description	Notes
1	Perform analysis of AMD water received from the Folldal mine.	Characterize "as received" AMD water with respect to sulfate anion and metal cation concentrations, TDS, and pH.	
2	Develop Bench Scale Test Plan.	Based on AMD characteristics, select test configurations and set points for parametric testing to identify best case process parameters.	A base case configuration will be evaluated. An alternate configuration will be included if necessary to achieve water purification targets.
3	Set up bench scale F-LLX test apparatus for testing.	Setup bench scale test apparatus with controls, sampling and analytical support.	
4	Operate the F-LLX process using the Folldal AMD water to demonstrate effectiveness and identify process parameters.	Includes performing "range finding" flow adjustments to determine the set points for which steady state conditions will be attained.	Fix the extractant to aqueous (E/A) flow ratio for the steady state operation. E/A ratios are selected to reduce the residual contaminant levels in the purified water to target levels and to minimize extractant usage.
5	Operate the F-LLX process to achieve steady-state conditions, then collect operational data and product samples.	During approach to steady-state monitor vessel liquid levels, phase disengagements, ensure steady flows, track key control parameters (metal and sulfate ion levels in the raffinate) to ensure process effectiveness.	Operational data collected will include reagent consumption rates and product collection rates. Target will be 3-5 sample sets collected 1-2 hours apart at steady-state. Samples will include feed water, product water, sulfate concentrate, and metal concentrate.

6	Perform IN and OUT sample analyses for all streams of interest and determine mass balance.	Perform sulfate analyses by IC and metals analysis by ICP to identify concentrations of products and impurities.	Main analyses are conductivity/TDS, and concentrations of sulfate and metal contaminants. The water chemistry of the feed and product water will also be determined.
7	Shut down operation.	Shut down bench scale test apparatus and drain and clean up system.	
8	Perform preliminary engineering assessment (PEA).	Based on results of bench testing, define basic F-LLX process for purification of AMD water.	
9	Prepare Report with conclusions from bench scale testing.	The report will summarize the test results, and present the process box flow diagram with experimentally determined flow rates and stream concentrations along with conclusions from the PEA.	Product characterization and production estimates will be included along with expected product concentrations and impurities.

Process Parameter Identification

A base case test configuration and setpoints were determined from the feasibility testing with surrogate water and analysis of the actual mine pool water collected from the Folldal municipality and received at Battelle for this test program. The fundamental approach for developing the F-LLX process parameters uses the characteristics of the water to be processed, the chemistry for removing the target chemicals based on the commercially available extractant, the operating conditions (flow, pH, temperature), and conventional mixer-settler technology. The analytical results for the Folldal mine pool water, including tote-to-tote variability, are presented in Table 3.

Table 3. Properties of Actual AMD Water

Parameter	**Estimated AMD Water Composition (mg/L)**	**Actual AMD Water Composition (mg/L)**	
		Average between totes	Standard deviation
Aluminum	208	250	30
Cadmium	0.17	0.25	0.02
Copper	81.4	93.7	10.4
Iron	790	1260	113
Lead	0.02	0.08	0.06
Nickel	0.61	0.71	0.12
Zinc	49.5	70.6	7.3
Sulfate	4100	6020	269

Range Finding Tests

The base case test configuration was operated in a series of "Range finding" tests, including adjustments to extractant and aqueous feed flows, chemical reagent feed flows, , stirrer speeds, target pH values, interface levels, and internal recycle flow rates. The adjustments were made to define the range of successful operation for the process configuration to select the optimal process parameters at which steady state conditions should be attained. For instance, the extractant to aqueous (E/A) flow ratio for steady state operation is selected to reduce the residual contaminant levels in the purified water to target levels while minimizing extractant usage.

The E/A ratio is based on the capacity of the extractant and the strength of the target chemical in the aqueous phase. For the ion exchange mechanism in the extraction reaction, the requirement is that there be sufficient exchange equivalents for the aqueous phase. The standard approach for setting the E/A ratio is to use the aqueous phase characteristics and the known reaction (exchange capacity in the case of ion exchange) to set the theoretical conditions. Actual E/A requirements are impacted by the reactor configuration, effectiveness of mixing/contact in the extraction reactor, and the effectiveness of the stripping. Minimum extractant usage is desirable since this reduces the costs to construct the field system and the chemical costs for makeup of the extractant feed due to minor losses.

Figure 2: Operation of the Extraction Stages during the Range Finding Tests

Process Validation

For validation of the F-LLX water purification process to meet the targets, the bench system was operated to steady state at the best-case operational parameters as determined from the "Range Finding" tests. The operation of the F-LLX water purification process to steady state consisted of four extraction stages, configured counter-current, with a split extractant-phase feed to E2 and E4, two decanters (following the first and second extraction stages), three metal stripping stages, and four sulfate stripping stages followed by a basic water wash. Steady state is defined as three turnovers of the extractant phase, where one turnover is defined as the total extractant-phase in the system divided by the extractant flow rate. The flow rates were selected during the steady state operation in order to achieve increased residence times (from those in the "range finding" tests), and consequently remove more sulfate from the mine pool water.

Upon reaching steady state, samples were collected from the processed water, end products, and other in-process streams and analyzed for the constituents of interest (sulfate anion and metal cations). The information gathered from the steady state operation was used to generate the F-LLX process flow diagram for development of the F-LLX demonstration plant, including:

- Mass balance
- Characterization of process products
- Measurement of key process parameter set points
- Minimum residence times for key equipment, and
- Reagent feed or consumption rates (sulfuric acid and potassium carbonate).

Summary of Results

The samples collected from the steady state run were sent to an outside laboratory for analysis; the results for the five sets of samples collected during steady state operation are shown in Table 4 below. The AMD Feed refers to the sample of AMD water collected on the last day of the steady state run, the Raffinate (composite) is the composite of the F-LLX effluent or processed water, and E4 is the fourth-stage extractor effluent sampled at the clock time shown in parentheses. Results with "less than" were below the indicated detection level.

Table 4. Steady State Test Results for Removal of Contaminants

Source	Al mg/L	Cd mg/L	Cu mg/L	Fe mg/L	Pb mg/L	Ni mg/L	Zn mg/L	SO_4 mg/L
AMD Feed	271	0.260	101	1340	0.120	0.799	75.7	6210
Raffinate (Composite)	<1	0.0317	0.0628	<1	<0.075	0.508	6.23	1200
E4 (1600)	<1	0.0534	<0.05	<1	<0.075	0.633	9.92	1790
E4 (1700)	<1	0.0494	<0.05	<1	<0.075	0.590	9.49	1440
E4 (1800)	<1	<0.025	<0.05	<1	<0.075	0.353	2.96	1560
E4 (1900)	4.32	0.0907	2.76	6.18	<0.075	0.754	30.4	1710
E4 (2000)	<1	0.0702	0.0602	<1	<0.075	0.685	17.9	1460
E4 Average	**1.66**	**0.0577**	**0.594**	**2.04**	**<0.075**	**0.603**	**14.1**	**1592**

The results for the purified water (raffinate) show that the targets for Fe and Cu were met throughout. While 92% of the zinc was removed, and 82% of the sulfate was removed based on the composite raffinate sample, the removal efficiency during the bench-scale tests did not meet the desired targets for zinc and sulfate. However, the results of the "Range Finding" tests (shown in Table 5) did meet the targets for removal of the Zn and sulfate.

Table 5. Range Finding Test Results

Parameter	AMD Feed (mg/L)	Purified Water Targets (mg/L)	Preliminary Purified Water Results (mg/L)
Cu	93.7	1	<0.1
Zn	70.6	1	<0.1
SO_4 (approach to steady state)	6020	100	60-120
SO_4 (steady state)			~1000 - 1200
Fe	1263	50	<0.25

Due to the success in removing the zinc and sulfate from the Folldal AMD water in the approach to steady state operation of the bench unit and the known limitations of the bench-scale equipment, it is anticipated that the recommended F-LLX process information for the field system will meet the water purification targets for the Folldal AMD water.

Aluminum, cadmium, lead, and nickel had no target levels, but were reduced by 99%, 78%, 38%, and 37%, respectively. Aluminum, in particular, was significant due to its high concentration (271 mg/L) in the AMD feed. Additional significant advantages for removal of these metals are the associated reduction in the total TDS in the purified water and the increase in pH to dischargeable levels. A 74% reduction in the TDS in the purified water is achieved versus the Folldal AMD water.

The F-LLX process also generates concentrated byproducts in the extractant recovery and regeneration stages. For this process a concentrated mixed metal sulfate product is generated in the metal stripping stages, and a concentrated potassium sulfate product is generated in the sulfate stripping stages. Alternatively, sodium carbonate can be used in the sulfate stripping stages, generated a sodium sulfate product. The concentrate product results and generation rates are summarized in Table 6 and Table 7.

Table 6. Metal Sulfate Concentrate Sample Analysis

Sample	Al	Cd	Cu	Fe	Pb	Ni	Zn	SO4
	mg/L	mg/L	mg/L	mg/L	mg/L	mg/L	mg/L	mg/L
MSO_4 Product (1600)	1,230	0.985	460	5040	0.591	1.58	271	29,100
MSO_4 Product (1700)	1,180	0.932	445	4730	0.584	1.57	261	53,900
MSO_4 Product (1800)	1,240	0.991	459	5010	0.591	1.63	275	29,600
MSO_4 Product (1900)	1,250	1.020	428	4920	0.625	1.75	254	57,000
MSO_4 Product (2000)	1,480	1.200	547	5940	0.742	2.03	326	29,300
MSO_4 Product Average	**1,276**	**1.026**	**468**	**5128**	**0.627**	**1.71**	**277**	**39,780**
MSO_4 Generation Rate L/hr (gal/hr)	**0.4-0.5 (0.1-0.13)**							

Table 7. Potassium Sulfate Concentrate Sample Analysis

Sample	SO_4 (mg/L)
K_2SO_4 Product (1600)	46,000
K_2SO_4 Product (1700)	47,200
K_2SO_4 Product (1800)	54,000
K_2SO_4 Product (1900)	49,800
K_2SO_4 Product (2000)	55,000
K_2SO_4 Product Average	**50,400**
K_2SO_4 Generation Rate L/hr (gal/hr)	**0.4-0.5 (0.1-0.13)**

Conclusions

The results of the steady state operation of the bench equipment validated the F-LLX water purification process for treatment of the supplied AMD water to achieve the copper and iron target residuals. Although removal of the zinc and sulfate from this water to their respective targets was not achieved during steady state operation, initial sample results, taken during the approach to steady state operation, indicate that the extraction stages were capable of removing the zinc and sulfate to below their respective target values. The primary cause of the reduced extraction during steady state operation was the insufficient stripping of the sulfate from the extractant, thereby limiting the extraction efficiency when recycled to the extraction stages. Testing of the extractant showed that full regeneration was not achieved in four stripping stages as the final stripped extractant still had sulfate retained, which resulted in the recycled extractant being less effective for sulfate and metals removal in the extraction circuit. A significant reason for this was the limited number of stages in the bench-scale system that were available for the integrated extraction and stripping processes. Based on the test results, it was determined that one additional sulfate stripping stage, for a total of five stages, will achieve the sulfate target of 100 mg/L in the purified water.

Based on the results of the continuous bench-scale testing, it can be confidently stated that the field-scale application of the F-LLX technology on the Folldal AMD water will meet the treatment targets including adequate removal of zinc and sulfate. The sulfate extraction will be achieved in the field-scale system through addition of stripping capacity in the sulfate recovery circuit and improved pH control to achieve total displacement of the sulfate ion on the loaded extractant phase with carbonate ion. As a result, the regenerated extractant will be at its full strength when recycled to the extraction stages, improving the removal of the zinc and sulfate from the Folldal AMD water as demonstrated in initial test data. The zinc extraction efficiency will be further enhanced in the field-scale system because of improved pH control and sufficient residence time in E4, as the zinc removal will follow ferrous iron removal.

(438a) Design Principles of Telluride-Based Nanowire Heterostructures for Thermoelectric Applications

Yue Wu, School of Chemical Engineering, Purdue University, West Lafayette, IN

Thermoelectric (TE) devices, which can perform a direct conversion between thermal and electrical energy, have attracted great attention due to their promising potential in improving the energy efficiency and in solid-state cooling. However, the low efficiency of the TE materials prohibits their wide applications. Certain TE materials, such as Bi_2Te_3/Sb_2Te_3 superlattice film (*ZT*~2.4) and $AgPb_mSbTe_{2+m}$ bulk crystals (*ZT*~2.2), although possessing high performance due to the improved phonon scattering at nanoscale interfaces and grain boundaries, require very complicated material composition or extremely expensive/time-consuming manufacture process such as molecular beam epitaxy. Theoretical predictions and initial experimental results have suggested that one-dimensional (1D) nanostructures, especially the nanowire heterostructures, which take the advantages of both quantum confinement to enhance the power factor and phonon scattering at nanowire surface and compositional interfaces to lower thermal conductivity, could offer a much higher *ZT* value. Meanwhile, the syntheses of various 1D nanowire heterostructures have been demonstrated through the chemical vapor deposition process based on vapor-liquid-solid (VLS) growth mechanism as well as the pulsed electrodeposition, but it is still a great challenge to obtain high-quality thermoelectric nanowire heterostuctures at a simple yet scalable way. Here we present a design principle to develop new categories of telluride-based thermoelectric nanowire heterostructures through rational solution-phase reactions. The synthesis yields "barbell" and "dumbbell" compositional-modulated nanowire heterostructures with narrow diameter and length distribution. Initial characterizations of the hot-pressed nanostructured bulk pellets of these heterostructure show a largely enhanced Seebeck coefficient and greatly reduced thermal conductivity, which lead to an improved thermoelectric figure of merit. This approach opens up new platforms to investigate the phonon scattering and energy filtering.

(438b) Excess Thermopower in Carbon Nanotubes Using Thermopower Waves

Joel T. Abrahamson, Bernat Sempere, Michael P. Walsh, Jared M. Forman, Sayalee G. Mahajan and Michael S. Strano, Department of Chemical Engineering, Massachusetts Institute of Technology, Cambridge, MA

The nonlinear coupling between an exothermic chemical reaction and a nanowire or nanotube with large axial heat conduction results in a self-propagating thermal wave guided along the nano-conduit. The reaction wave induces a concomitant thermopower wave of high power density, resulting in an electrical current in the same direction. Thermopower waves also tend to produce monopolar voltage pulses, although conventional thermoelectric theory clearly predicts bipolar voltage, and they propagate in materials with both high electrical and thermal conductivity. Temperature and voltage measurements of thermopower waves on single-walled carbon nanotubes (SWNTs) show that they can generate voltage and power in excess of the predictions of the Seebeck effect for the temperature gradient, as much as four times greater power. We hypothesize that the excess thermopower stems from a chemical potential gradient across the SWNTs with a new model that describes the electrical properties of the waves. The fuel (prototypically picramide) and ambient oxygen adsorb and dope the SWNTs ahead of the wave, and desorb and react in the hot zone behind the wave front. Raman spectroscopy confirms that picramide dopes SWNTs. Furthermore, the excess thermopower depends on the mass of fuel added (relative to SWNT mass), and the chemical potential difference matches the magnitude of the excess thermopower. This model relies on SWNT property data measured in the literature rather than fit parameters. Thermopower waves thus provide an exciting new class of highly scalable, stable energy storage devices, and offer the possibility of greatly exceeding the power limitations of conventional thermoelectric devices.

(438c) First Principles Model for Predicting the Thermoelectric Properties of Complex Antimonides and Oxides

Maria Stoica, Alireza Faghaninia, Zhou Yu and Cynthia S. Lo, Energy, Environmental and Chemical Engineering, Washington University in St. Louis, St. Louis, MO

Nanostructured inorganic materials have in the past ten years seen a rise in interest for thermoelectric applications, due to the enhanced performance exhibited at low dimensions. In our work, we study the effect of dimensionality on the thermoelectric properties of Co- and Mo- based antimonides and doped oxides. Specifically, we calculate the Seebeck coefficient, "dark" electrical conductivity, and thermal conductivity of these materials. We first perform Molecular Dynamics (MD) simulations and Density Functional Theory (DFT) calculations, where we perform Γ-point calculations as well as integrating over the first Brillouin zone. We then perform post-processing of the momentum/optical matrix elements to calculate the desired properties and predict the behavior of both known (e.g., unfilled bulk) and designed (e.g., filled bulk, nanostructured) materials. For the CoX_3 compounds (X=P, As, Sb), we look at Ag- and Au-filled skutterudites; we also compare their performance to Mo-based antimonides. For the oxides, we have performed transition metal doping of ZnO and $CuAlO_2$ compounds with oxygen vacancies, as well as direct and charge-compensated replacement of dopant atoms. We also perform computations for both bulk and 2-D nanowires for each group of materials, and demonstrate the enhancement of thermoelectric properties as a result of nanostructuring.

(438d) Characterizing Thermal Transport in Self Assembled Monolayer Junctions

Shubhaditya Majumdar[1], Scott N. Schiffres[1], Alan J. H. McGaughey[1] and Jonathan A. Malen[2], (1)Mechanical Engineering, Carnegie Mellon University, Pittsburgh, PA, (2)Department of Mechanical Engineering, Carnegie Mellon University, Pittsburgh, PA

Our investigation involves studies of the thermal transport properties of a single organic-inorganic interface – a self-assembled monolayer (SAM) junction. Molecular junctions have generated excitement as potential forerunners in organic electronics and are fundamental to the development of new hybrid materials. Characterization of their thermal properties is important in understanding how they will dissipate heat, a parasitic byproduct in their application in electronics and energy conversion. Accurate quantification of their thermal conductance, which describes this behavior, is thus essential.

Complementary experimental and computational approaches will address a prototypical Au-SAM-Au interface, where an alkane based SAM is initially considered. Molecular dynamics simulations are being implemented to model and study the effect of molecule length, choice of leads and other factors affecting bonding between the organic and inorganic components. Independent experimental measurements on transfer printed leads on SAMs are being performed using frequency domain thermoreflectance – a laser-based non-contact measurement technique. Initial measurements indicate that the Au-SAM-Au interface has a higher conductance than previously studied SAM interfaces, perhaps because prior measurements considered SAMs sandwiched between dissimilar contacts. Hence, this work is novel because the contacts are identical, and the interface conductance can be directly related to the SAM properties. This study is thus expected to provide a detailed picture on the thermal transport phenomenon across such interfaces.

(438e) Studies of Thermal Conductivity in Hybrid Organic-Inorganic Nanocrystal Superlattices

Wee-Liat Ong[1], Sara Rupich[2], Dmitri Talapin[2], Alan J. H. McGaughey[1] and Jonathan A. Malen[1], (1)Mechanical Engineering, Carnegie Mellon University, Pittsburgh, PA, (2)Chemistry, University of Chicago

Nanocrystal superlattices (NCSLs) are 3-D arrays of inorganic spheres bridged by self-assembled molecular chains. NCSL thin films hold promise for thermal management and energy conversion applications as studies have shown that their electronic and thermal transport properties can be decoupled. This unique feature allows them to be tailored according to the specific needs of the applications (e.g. high figures of merit materials for thermoelectric, high-efficiency photovoltaic materials). An additional advantage of this material comes from their solution-based fabrication technique which is readily scaled up for large volume and cheaper than conventional nanostructuring methods.

Electronic transport in NCSLs has been studied extensively. Little is known, however, about the nature of thermal transport in these complicated materials. Both experimental measurements and preliminary modelling work will be presented to elucidate the thermal transport mechanisms in NCSLs. Several factors affecting the NCSL thermal conductivity including the geometry and chemical compositions of the NCSL, will be shown.

(450a) Molecular Design of Thin Film Composite (TFC) Hollow Fiber Membranes for Isopropanol Dehydration Via Pervaporation

Jian Zuo, Chemical and Biomolecular Engineering, National University of Singapore, Singapore, Singapore, Yan Wang, Department of Chemical & Biomolecular Engineering, National University of Singapore, Singapore, Singapore, Shi Peng Sun, National University of Singapore, Singapore, Singapore and Tai-Shung Chung, Department of Chemical and Biomolecular Engineering, National University of Singapore, Singapore, Singapore

High performance hollow fiber membranes consisting of a thin film composite (TFC) polyamide layer and a porous Torlon® polyamide-imide substrate have been fabricated via interfacial polymerization for pervaporation dehydration of isopropanol. The effects of substrate characteristics on the formation of the thin polyamide layer and the pervaporation performance are investigated. It is found that the resultant TFC membranes show superior separation performance if the substrates have a fine surface pore size and desirable sub-structural morphology. The effects of amine monomers, i.e., m-phenylenediamine (MPD) and hyperbranched polyethyleneimine (HPEI), for interfacial polymerization with trimesoyl chloride (TMC) on separation performance are also compared. The TFC membrane prepared from HPEI exhibits a higher pervaporation flux than that from MPD possibly due to the fact that the former has higher amine content than the latter. As a consequence, the polyamide selective layer of the former has a greater hydrophilicity than the latter. X-ray photoelectron spectroscopy (XPS) and contact angle measurements have confirmed the hypothesis. The molecular weight of HPEI also affects the pervaporation performance of the resultant TFC membranes. With the optimal interfacial polymerization conditions, the TFC membrane fabricated from HPEI with a molecular weight of 2 kg/mol shows the best separation factor of 624 with a flux of 1282 g/m^2-h and a permeate water concentration of 99.1 wt% using a feed composition of 85/15 wt% IPA/water at 50 °C. After post methanol treatment, the flux can be enhanced to 1920 g/m^2-h with a slightly decrease in permeate water concentration to 98.4 wt%. The mechanism of flux enhancement has been examined by slow beam positron annihilation spectroscopy (PAS). It is found that the removal of the excess amine from the TFC layer during the methanol treatment is the main cause.

(450b) Microfiltration of Lignocellulosic Hydrolyzates Using Dead End Filtration to Separate Lignin

Bhavin Bhayani, Empire State Paper Research Institute, Syracuse, NY and Ramarao Bandaru, Paper and Bioprcess Engineering, State University of New York, Syracuse, NY

Microfiltration of lignocellulosic hydrolyzates using dead end filtration to separate lignin and other inhibitors

B. Bhayani, B.V. Ramarao

Department of paper & Bioprocess engineering, SUNY ESF

In the production of biofuels from lignocellulosics, the biomass is first extracted using dilute mineral acids or hot water. This hemicellulose rich oligomers are hydrolyzed under treatments at higher temperature and pressure to yield sugar monomers which are fermented into biofuels such as ethanol, biobutanol or bioplastics. The initial treatment with either dilute acids or hot water also allows some of the lignin to pass in particulate, colloidal or dissolved forms into the sugar solution. Lignin and other phenolic compounds derived from it are strongly inhibitory to microbial fermentation processes. They also adhere to polymeric membranes, carbohydrates and other surfaces, fouling separation membranes. Their removal from the liquor stream not only facilitates downstream separations but improves the fermentation profiles of the resulting sugar hydrolyzates. Therefore, methods to separate lignin from liquors and hydrolyzates form valuable additional operations in the biorefineries.

A new approach has been developed which can separate inhibitory components from the extracts based on the fact that reduction of temperature of the biomass hydrolyzates can result in the native agglomeration and separation of lignin containing solids by simple settling and/or filtration which thus can result in large increases in separation efficiencies for hydrolyzates. By filtering the extracts at lower temperature (10°C) it is possible to obtain higher flux rate than at higher temperatures (20°C). As shown in Figure 1 and Figure 2, the cumulative flow for extracts at 10°C is higher than the extracts at 20°C at two different pressures. The turbidity of permeate for any given combination of pressure and temperature was less than 10 NTU. It has been hypothesized that the natural flocculation tendency of extracts increases with decreasing temperature.

In addition to the temperature, the effect of addition of a flocculent (Polyethylene oxide, PEO) has been investigated. Substantial enhancement in filtrate flux upon addition of 20 ppm of PEO as seen in Figure 3 is observed.

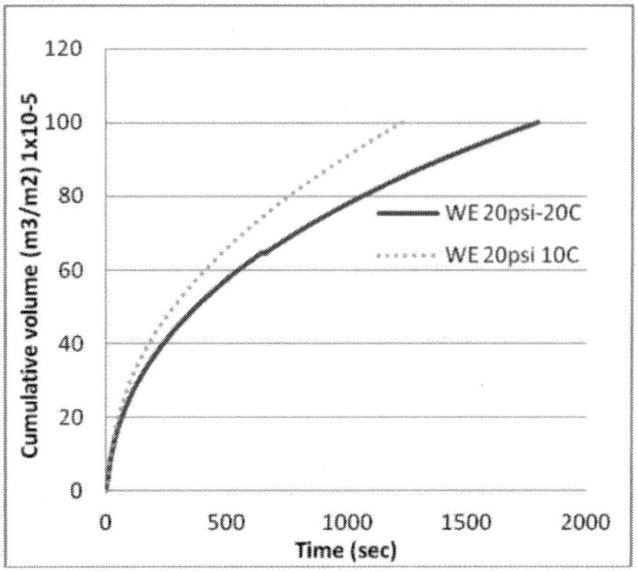

Figure SEQ Figure * ARABIC 1: Cumulative flow at different temperature at 20psi

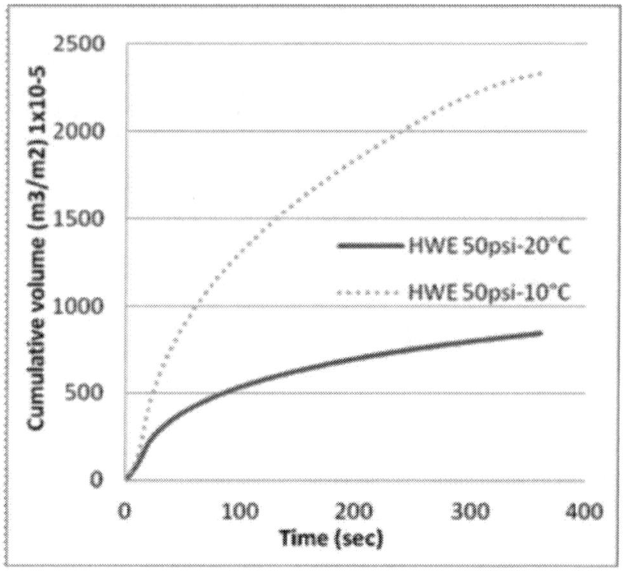

Figure SEQ Figure * ARABIC 2: Cumulative flow at different temperature at 50psi

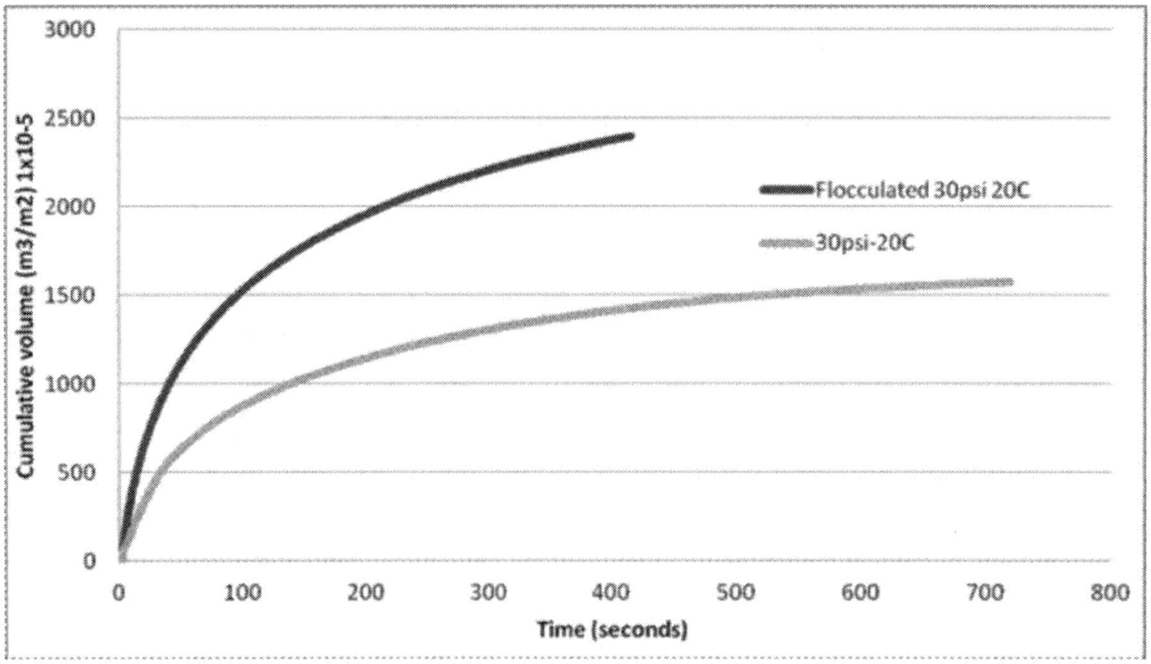

Figure SEQ Figure * ARABIC 3: Effect of addition of PEO

(450c) Mathematical Modeling of the Filtration of Lignocellulosic Hydrolyzates

Bhavin Bhayani, Empire State Paper Research Institute, Syracuse, NY and Ramarao Bandaru, Paper and Bioprcess Engineering, State University of New York, Syracuse, NY

Hot water extracts (HWE) of wood are a complex mixture of various compounds. Depending upon the time of extraction, the size of the colloidal particles decreases from 500 nm to 250 nm. Their surface potential also decreases with increased extraction levels (Duarte et al., 2010). Separation of lignin from lignocellulosic hydrolyzates can be achieved by membrane separation. Dead end filtration of the hydrolyzates has shown the formation of a cake layer consisting of the colloidal particulates in the suspensions. In this work, we present mathematical models of fouling behavior considering the combined actions of membrane pore constriction, pore blockage and cake formation under constant pressure conditions. Parameters representing each of these mechanisms were obtained by fitting to experimental filtrate flux data.

Experimental flux data were obtained for dead end filtrations over ranges of pressure, temperature, concentration, extraction time and molecular weight cutoff of filters. The models provide a clear picture of the transition from the pore blockage to cake filtration regimes. The results provide important insights into the underlying mechanisms of HWE fouling during microfiltration.

(450d) Sugar Concentration for Continuous Enzymatic Saccharification Using Nanofiltration and Reverse Osmosis Membranes

Mohammadmahdi Malmali, University of Arkansas, Fayetteville, AR, Jonathan Stickel, National Bioenergy Center, National Renewable Energy Laboratory, Golden, CO and S. Ranil Wickramasinghe, Department of Chemical Engineering, University of Arkansas

Cost effective conversion of lignocellulosic biomass into bioethanol requires maximizing ethanol yields. Enzymatic hydrolysis of polysaccharides, e.g. cellulose, starch, is usually conducted in a batch reactor. Disadvantages of the classical batch reactor are: product variation from batch to batch, higher overall investment costs due to larger reactor volumes, higher running costs due to frequent startup/shut down, one time use of enzymes as well as catalyst/enzyme separation costs. Development of a continuous saccharification process overcomes the limitations of batch operation. Additional potential advantages include: recovery and reuse of enzymes, improvement of product yield and kinetics, and reduction in inhibition of enzymes. During a continuous saccharification process it is desirable to concentrate the product sugar stream prior to fermentation.

Here we focus on sugar concentration using nanofiltration and reverse osmosis membranes. Nanofiltration, which originated in the 1970s, is one of the newest pressure driven membrane filtration processes. Low pressure reverse osmosis membranes came to be known as nanofiltration membranes with some of the earliest applications being described in the 1980s. Characteristics of nanofiltration membranes include greater than 99% rejection of multivalent ions, 0-70% rejection of monovalent ions and greater than 90% rejection of small organic compounds with molecular weights in the range 150-300.

We have investigated concentration of aqueous sugar streams using a number of different commercially available nanofiltration and low pressure reverse osmosis membranes. Experiments have been conducted using model feed streams as well as corn stover derived biomass hydrolysates. Model feed streams consisted of DI water containing D-Xylose and D-glucose at a ratio of 3:1. In addition compounds that negatively impact ethanol yields during fermentation such as acetic acid, furfural and 5-hydroxymethylfurfural were also added to the feed stream in the ratio 6:2:1. The effect of feed pH was determined. Membrane performance was characterized by determining the level of sugar (xylose and glucose concentration) as well as removal (passage through the membrane) of acetic acid, furfural and 5-hydroxymethylfurfural. The percentage decrease in permeate flux was also determined. Corn stover hydrolysates obtained from the National Renewable Energy Laboratory, Golden Colorado, were also tested.

Membrane surfaces have been characterized using X-ray photoelectron spectroscopy. Differences in membrane performance have been related to interactions between the various species present in the feed and chemical groups present on the membrane surface. The result presented here may be used to identify suitable membranes and operating conditions for sugar concentration and removal of species such as acetic acid, furfural and 5-hydroxymethylfurfural.

(450f) Algae Dewatering Using an Inclined Gravity Settler

Joanne Belovich, Cleveland State University, Cleveland, OH, Zhaowei Wang, Newland Biotech, Cleveland, OH and Jing Hou, Newland Biotech, Inc., Cleveland, OH

One of the main barriers to the cost-effective production of biofuel from microalgae culture is the difficulty of harvesting the algae from dilute cell suspensions. Because of light limitation, the microalgae concentration in large-scale ponds typically does not exceed 1-2 gdw/L. A downward flow inclined gravity settler was evaluated for its effectiveness in algae culture dewatering, using two stains of algae, *Scenedesmus dimorphus* and *Chlorella vulgaris*. Experimental results showed that *S. dimorphus* can be concentrated up to eight-fold using a single settler stage, with a biomass recovery of 80%. Separation efficiency was independent of biomass concentration between 1 and 6 gdw/L, suggesting that a two-stage sequential system of settlers may be effective to maximize biomass recovery and concentration. Efficiency of separation of *C. vulgaris* was slightly lower than that of *S. dimorphus*, due to the fact that *S. dimorphus* cells exist in aggregates of up to eight or more cells and thus settle more easily. The system demonstrated consistent results and steady state operation for up to three weeks of continuous operation, and is low in energy consumption. This separation system may be useful as a first stage for removing 95% of the water from large-scale algae cultivation systems.

(450g) Separation and Identificaiton of Microbial Lipids From Oleaginous Microorganism As Biodiesel Feedstock

Guochang Zhang[1], Todd French[2] and Rafael Hernandez[2], (1)Chemical Engineering, Mississippi State University, Mississippi State, MS, (2)Dave C. Swalm School of Chemical Engineering, Mississippi State University, Mississippi State, MS

Oleaginous yeasts are microorganisms that accumulate high lipids up to over 56% of dry cell mass, and can be used as a feedstock for biodiesel production. However, determination of the oil contents in the fatty cells by lipid extraction places a great role in the research on microbial lipids. The most commonly accepted method of lipid extraction in the laboratory uses chloroform, methanol and water as solvents. In the performance of the method, a lot of solvent is needed and the operation is difficult to handle, and the repetition is not high. The purpose of this experiment is to find a new method that is much easier and simpler for extracting lipid from oleaginous yeast grown on glucose or xylose as the sole carbon and energy source. In the experiments, several lipid extraction methods were investigated to determine the effectiveness of the method in terms of lipid recovery and composition. One method is the gravimetrical Bligh-Dyer method using water, methanol and chloroform as solvent. The second method evaluated used *tert*-butyl methyl ether or hexane as the solvent instead of chloroform. Since the terminal objective is to produce biodiesel, following extraction, the lipids have to be transesterified to fatty acid methyl ester (FAME). Therefore, in-situ transesterification of the fatty cells was also studied, where the lipids were extracted and transesterified simultaneously. The results showed that all the methods are effective to extract the lipids from the *R. glutinis* grown on glucose or xylose. In terms of the FAME yield based on cell mass, difference among the three methods is negligible. While using MTBE as solvent, the process is easier to operate.

(450h) Process Development of Fermentable Cellulosic Sugars from Biomass for Biofuels and Chemical Production

Sarad Parekh, Research and Development, Sweet Water Energy, Rochester, NY and Carl Felice, Sweetwater Energy

The recent volatility of petroleum prices is stimulating the biorefining industry to diversify its raw material feedstock, using a broad-based sugar platform technology. Corn silage and energy crops such as sweet sorghum are widely recognized as the promising high-volume, low-cost lignocellulosic feedstocks from which to derive C5- and C6-rich sugar streams that can be bioconverted into biofuels and chemicals.

This presentation will review Sweetwater Energy's overall approach and concept of integrating a decentralize model, whereby a modular, feedstock-flexible bioprocessing machine pretreats and processes ligocellulosic biomass, and generates adequate quantities of C5 and C6 sugar-rich streams at the biomass-generation site. The sugars are then transferred to a regional refining site for the production of biofuels. This talk will highlight: 1) the importance of innovative biomass harvesting and storage technologies to efficiently and economically transfer ensiled biomass from the field to the bioprocessing site, 2) characterization of the physical and chemical properties that enable optimum bioconversion process, 3) specific pretreatment conditions and various process strategies for deriving streams rich in C6 and C5 sugars for biofuels production, and 4) review recent advances made at Sweetwater Energy in enhancing and improving the ability to process and ferment a wide range of pretreated biomass into ethanol chemicals and advanced fuels, such as bio-jet fuel.

(454a) Leveraged Process Development for Biomass to Ethanol

Brandon Emme, Novozymes North America, Franklinton, NC and M. Torry-Smith, Novozymes North America Inc., Franklinton, NC

Overview: Visionary talks (invited) will outline commercial, governmental, or academic efforts to advance multi-product biorefineries. Topics likely to be addressed include: development or improvement -- including conceptual analysis and sustainability assessments -- of renewable feedstocks-based multiproduct biorefineries, i.e., based on starch or oil crops and/or lignocellulosic biomass; development of enabling catalytic and/or separative processes; discovery or development of enabling enzyme or fermentative/microbial technologies or bio-based materials.

Appropriate integration of process development steps in the biomass to ethanol process is a challenge. Building a fully flexible pilot plant to allow for multiple process configurations is cost prohibitive for most technology developers. For lack of fully integrated data, technoeconomic models can be employed to integrate smaller scale or individual unit operation data into the context of a full process to determine the best operational configuration and production cost. Furthermore, these models can be used to compare competing process concepts, both for state of technology as well as at n-th state conditions.

This talk will present the Novozymes 2G flexible cost model and illustrate through several Novozymes process development examples how it has been used to assimilate results from Pretreatment through Fermentation to establish winning process concepts. The latest small and large scale hydrolysis data using the new commercial cellulase enzyme Cellic CTec3 on several different feedstocks will be presented, as well as some insights into what the models can tell us about the likely short and long term impacts of enzyme development on the Minimum Ethanol Selling Price.

(454b) The Ethics of Biofuels in a World of Nine Billion People

John J. Sheehan, NREL, Golden, CO

Our global agriculture system is failing us:

- Food prices are up.
- A billion people face obesity.
- A billion people face starvation and malnutrition.
- Nitrogen and phosphate pollution from farms continue to threaten the health of our waterways and other natural ecosystems.
- Greenhouse gas emissions from agricultural expansion and land clearing continue to contribute a third to our total emissions.
- Agriculture remains the largest single consumer of land and water on the planet.

And in the midst of all this, we are struggling to realize a vision for a truly sustainable bio-economy that will put more demands on the global agriculture system.

This talks focuses on the ethical as well as technical challenges facing the nascent bio-economy, especially with respect to the challenge of feeding nine billion in the year 2050. Never has there been a time when it was so urgent to untangle these issues. We need a straightforward discussion of the ethics of sustainable development and sustainable biofuels.

Built off a recent paper by Foley et al in *Nature* on how we will feed more people demanding more food, the talk extends this question to include feeding and fueling the world. It focuses on four key strategies that are vital to maintaining a sustainable agricultural system: 1) Elimination of expansion of agricultural lands globally; 2) Reduction of the agricultural yield gap around the world; 3) Improvement of resource utilization (water and fertilizer); and 4) Reduction in meat consumption and food system waste. Successfully pulling off these strategies *may* provide room for agriculture to deliver food, fuel and chemicals to an increasingly wealthy population of nine billion people in the year 2050.

(454c) Biobased Chemical Production Through a Generalized Technology Platform

Brent H. Shanks, Chemical and Biological Engineering, Iowa State University, Ames, IA

Replacement of crude oil-derived chemicals with biorenewables chemicals will necessitate developing capabilities to remove oxygen selectively from carbohydrates. The serial approach of targeting one biorenewable chemical product is both time intensive and expensive. Since biorenewable chemicals must compete against a highly efficient petrochemical production system there is a need to create a generalized oxygen-removal framework in which a range of biorenewable chemicals can be produced from a common technological platform. One such generalized technological framework, being developed by *CBiRC,* depends on the exploitation of a common metabolic pathway leading to a diversity of intermediate chemicals that are subsequently converted to chemical products using chemical catalysts. An overview of the technical strategy being used by *CBiRC* to achieve a generalized chemical production platform will be discussed as well as several specific examples involving varying chain length carboxylic acids with decarboxylation to olefins and biological-produced pyrones with subsequent ring opening or aromatization.

(454d) Progress Toward Sustainable Biofuels — Pilot-Scale Demonstration of Integrated Cellulosic Ethanol Production

Daniel J. Schell, National Renewable Energy Laboratory, Golden, CO

In 2007, the U.S. Department of Energy's (DOE) Office of the Biomass Program, in collaboration with the National Renewable Energy Laboratory (NREL), established the goal of achieving cost competitive cellulosic ethanol production from corn stover by 2012. The baseline conversion process to be improved consisted of dilute sulfuric acid pretreatment followed by enzymatic cellulose hydrolysis and cofermentation of biomass sugars to ethanol using a metabolically-engineered fermentation strain. The technology improvement strategy was based on achieving yearly increases in overall process conversion yields, along with associated reductions in modeled ethanol production costs, from 2007 to 2012. Progress toward this goal was measured using n^{th}-plant technoeconomic models to estimate ethanol production cost based on conversion yields obtained in integrated pilot and bench-scale experiments. The final objective was to demonstrate at pilot scale integrated process conversion yields commensurate with the 2012 cost target.

This presentation will briefly review the history of this technology development effort and highlight results achieved by the DOE Biomass Program at NREL from 2000 to the present. Much of the progress was facilitated by DOE-supported, industry-led research to develop improved hydrolytic enzymes for cellulose conversion and more robust microorganisms for ethanol production from a broader range of biomass sugars. Specifically, improved cellulases increased cellulose-to-glucose yields during enzymatic hydrolysis of pretreated biomass in the presence of high levels of biomass-derived sugars, and improved cofermenting microorganisms were able to convert more of the available biomass-derived sugars to ethanol in the presence of inhibitory compounds. Work at Idaho National Laboratory to improve the efficiency of the corn stover collection and supply chain also lowered the cost of delivered feedstock.

The presentation will highlight work that led to improved process understanding, higher sugar and ethanol titers, and ultimately increased conversion yields and associated cost reductions culminating in pilot scale demonstration runs. Preliminary results from bench-scale experiments are achieving yields above 75 gal ethanol/ton feedstock, ethanol titers above 68 g/L and a modeled cost below $2.20/gal ethanol. Pilot scale results will be presented along with bench-scale testing results performed in parallel characterizing integrated conversion performance over a wider range of operating conditions than could be demonstrated at pilot scale.

Quantitative Structures and Thermal Properties of the Birch Lignins during Ionic Liquid-based Biorefinery

Jia-Long Wen [†], Bai-Liang Xue [†], Feng Xu [†], Run-Cang Sun [*,†,‡]
[†] Institute of Biomass Chemistry and Technology, Beijing Forestry University,+
Beijing, China
[‡] State Key Laboratory of Pulp and Paper Engineering, South China University of Technology,
Guangzhou, China

ABSTRACT. The use of ionic liquid (IL) in biomass pretreatment has received considerable attention recently because of their effectiveness in decreasing biomass recalcitrance to subsequent enzymatic hydrolysis. To understand the structural changes of lignin during pretreatment and enzymatic hydrolysis process, ionic liquid lignin (ILL) and enzymatic mild acidolysis lignin (EMAL) were sequentially isolated from ball-milled birch wood in this study. The quantitative structural features of ILL and EMAL were compared with the corresponding cellulolytic enzyme lignin (CEL) by nondestructive techniques (e.g. FTIR, GPC, Quantitative ^{13}C, 2D and ^{31}P-NMR). The IL pretreatment caused structural modifications of lignin (cleavage of β-*O*-4 ether linkages and formation of conjugated structure and condensed structures). In addition, lignin fragments with lower S/G ratios were initially extracted, while the subsequently extracted lignin is rich in syringyl unit, which was mainly originated from S2 layer of secondary wall. Moreover, the maximum decomposition temperature was increased in the order of ILL<EMAL<CEL, which was related to the corresponding β-*O*-4 ether linkages content and molecular weight. Based on the results observed, a possible dissociation mechanism of lignin during IL pretreatment was proposed.

KEYWORDS: Ionic liquid lignin, nuclear magnetic resonance (NMR), quantitative structures; molecular weight; thermal properties

INTRODUCTION

Recently, ionic liquids (ILs) have attracted a significant attention as the promising green solvents with environmental friendliness in the field of lignocellulosic biomass utilization. Many ILs are able to dissolve wood biomass and its major components, cellulose, hemicelluloses, and lignin (Mäki-Arvela et al., 2010). It has been demonstrated that ILs pretreatment facilitates the "delignification process" and significantly increases the enzymatic hydrolysis efficiency of pretreated biomass (Sun et al., 2009; Lee et al., 2009). However, its commercialization is hindered by the high cost of ILs. The efficient recycling of ILs, as well as the valorization of the lignin byproducts of biomass deconstruction, is essential to the large-scale success of IL pretreatment processes (Klein-Marcuschamer et al., 2011). To achieve the high-value utilization of lignocellulosic biomass, isolation and chemical conversion of the isolated component is extremely needed. Fortunately, some ILs was also reported to extract lignin from biomass (Sun et al., 2009; Tan et al., 2009; Lan et al., 2011; Pinkert et al., 2011). However, the detailed structural information of collected lignin needs to be further confirmed. Among numerous ionic liquids, 1-ethyl-3-methylimidazolium acetate [C_2min][OAc], was demonstrated to effectively extract lignin from wood and to be one of the best solvents for lignocellulosic biomass (Lee et al., 2009; Arora et al., 2010; Li et al., 2010). A recent report has demonstrated that the lignin isolated with ionic liquid from poplar wood have a similar structural features as compared to the corresponding MWL except for small molecular weight (Kim et al., 2011). Torr et al. have investigated the impact of ionic liquid pretreatment on the chemistry and enzymatic digestibility of *Pinus radiata* compression wood and showed that the pretreatment caused some structural modifications to the lignin and polysaccharides. However, to thoroughly understand the [C_2mim][OAc] pretreatment and enzymatic hydrolysis on the structural features of lignin, more representative lignin fractions should be collected and investigated (Torr et al., 2012). It is well known that the structural details of residual lignins are of considerable importance in pulp delignification and bleaching studies (Argyropoulos et al., 2002; Jaaskelainen et al., 2003). Therefore, isolation and detailed structural analysis of the pristine IL lignin and residual lignin should be investigated to understand the effects of IL delignification process on the structures of residual lignin. More importantly, the well-characterized lignin is benefit to develop highly functionalized natural product based on lignin. Furthermore, the thermal properties should also be investigated if the utilization of lignin involves thermo-chemical process, since knowledge of lignin thermal degradation is fundamental for predicting pyrolysis behavior of this material.

In the present study, southwest birch wood, a fast-growing hardwood species after eucalyptus and poplar in China, was comprehensively investigated by ILs pretreatment and subsequent enzymatic hydrolysis processes. During these processes, two lignin fractions, ionic liquid lignin (ILL) and enzymatic mild acidolysis lignin (EMAL) were successively isolated with high yields and purities. In addition, CEL of birch wood was isolated for comparison. To thoroughly investigate the structural features of the isolated birch lignins, the lignins were determined by gel permeation chromatography (GPC). Moreover, lignin structural information, such as S/G ratios and major substructures (inter-coupling bonds, β-*O*-4, β-β, β-5, etc.), were qualitatively and quantitatively obtained according to quantitative ^{13}C, 2D-HSQC NMR and ^{31}P-NMR spectroscopy. Furthermore, to reveal the relationship between structural features and thermal properties, thermo-gravimetric analysis (TGA) was also implemented. The results obtained on the lignins will contribute toward the goal of better understanding the structure of lignin and the IL pretreatment processes for native feedstocks.

MATERIALS AND METHODS

Southwest birch (*Betula alnoide*) was harvested in October 2010 from Yunnan Province (China). Then, raw materials were manually debarked and chipped using a mechanical chipper. After drying at 60°C for 16 h in an oven, the wood chips were screened to obtain wood particles (20-40 mesh, 0.70-0.35 mm). Subsequently, the wood particles were dewaxed with benzene-ethanol (2:1, v/v) in a Soxhlet instrument for 12 h. The extractive-free powder was further oven dried at 60°C for 16 h and stored in a desiccator before use.

Ionic liquid lignin (ILL) and enzymatic mild acidolysis lignin (EMAL) were isolated according to **Figure 1**. ILL was isolated according to previous literatures (Sun et al., 2009; Kim et al., 2011).

The weight average (M_w) and number-average (M_n) molecular weights of the three acetylated lignin preparations were determined by gel permeation chromatography (GPC, Agilent 1200, USA) with a refraction index detector (RID) on a PL-gel 10 µm Mixed-B 7.5 mm ID column, calibrated with PL polystyrene standards. GPC analyses were run at least twice. NMR Spectra of the lignins were performed on a Bruker AVIII 400 MHz spectrometer at 25 °C in DMSO-d_6 according to previous literatures (Wen et al., 2010; Faix et al., 1994).

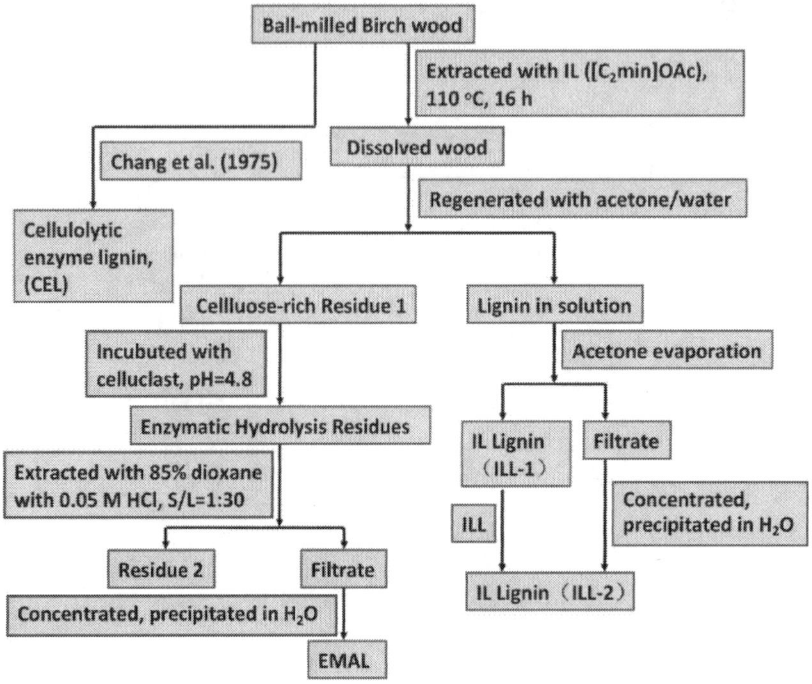

Figure 1. Flow chart of lignin extraction process.

RESULTS AND DISCUSSION

Quantitative ^{13}C-NMR analysis

To obtain quantitative information of different structural units in the lignin fractions, quantitative ^{13}C-NMR calculation was also applied. The integral values for the structural moieties are reported per aryl groups based on a previous literature (Capanema et al., 2005). **Table 1** gives the major signal assignments and the quantitative results of the signals for the lignins. The syringyl/guaiacyl ratio (S/G) was calculated on the basis of the number of carbons per aromatic ring in C2 of guaiacyl and C2/C6 of syringyl units. The integral value between δ 108 and 103 ppm divided by 2 can then estimate the content of syringyl moieties, and the guaiacyl moieties were determined by the integral value between δ 114 and 108 ppm. Interestingly, it was found that the ILL contained less syringyl units as compared to EMAL, which was abundant with syringyl units in the current study. A previous study reported that the lignin isolated during the early part of ball-milling may originate mainly from lignin in middle lamella (Hu et al., 2006), which was reported to contain less methoxyl groups per C9 than that in the secondary wall lignin (Whiting et al., 1982). Therefore, a possible speculation is that the lignins with lower S/G ratios were initially extracted, while the residual lignins with abundant syringyl units mainly in secondary wall were subsequently extracted. The results are in agreement with the data presented in some literatures (Yuan et al., 2011a, 2011b), in which the S/G ratios increased in the lignins isolated from MWL-pre-extracted wood. Similarly, another study also found that ASAM (Alkaline Sulfite Pulping with Anthraquinone and Methanol) residual lignin from beech wood was rich in syringyl units (Choi et al., 2011). The findings confirmed that syringyl-riched lignin fractions were extracted from biomass in the later period of separation process, which are mainly isolated from secondary wall lignin. In consideration of the lower yield and lower S/G ratio of CEL in our study, one supposed tendency is that CEL isolated from birch wood in our study is mainly derived from middle lamella.

Table 1. Main Chemical Shift Assignments in the Quantitative ^{13}C-NMR and 2D-HSQC Spectra of ILL, EMAL and CEL

δ (ppm)	Assignments	ILL	EMAL	CEL
160-140	Aromatic C-O	2.14	2.17	2.21
140-123	Aromatic C-C	1.81	1.72	1.66
123-103	Aromatic C-H	2.12	2.20	2.24
61.2-58	β-*O*-4	0.56 (0.51)	0.59 (0.56)	0.65 (0.58)
58-54	OCH_3	1.36	1.66	1.40
54-53	β-β, β-5	0.08	0.07	0.06
	[a]DC	0.33	0.25	0.20
	[b]S/G ratio	1.14 (1.81)	1.62 (2.72)	1.26 (1.90)

The aromatic region in the ^{13}C NMR spectrum (δ 160.0-103.0 ppm) can be divided into three broad categories: oxygenated aromatic carbons (δ 160.0-140.0 ppm), aromatic carbon-carbon structures (δ 140.0-123.0 ppm), and aromatic methine carbons (δ 123.0-103.0 ppm). It was found that the amount of oxygenated carbons per aryl in ILL, EMAL, and CEL was 2.14, 2.17, and 2.21, respectively. This was in agreement with the FT-IR spectra, in which the content of conjugated C=O is higher in CEL as compared to that of ILL and EMAL. However, the content of carbon-carbon structures was slightly higher in ILL than that of EMAL and CEL. The degree of condensation (DC) was determined by subtracting the observed signal intensity for $C_{Ar\text{-}H}$ (δ 123-103 ppm) from the theoretical value of $C_{Ar\text{-}H}$ (Capanema et al., 2005). It was found that the DC values of ILL, EMAL and CEL were 33, 25 and 20%, respectively. The increased DC in ILL is also supported by the increased number of aromatic C-C and the decreased number of aromatic methine carbons as shown in **Table 1**. In addition, the data clearly showed that the amount of β-*O*-4 in ILL is slightly lower than those of EMAL and CEL. The observation was in agreement with the result of a recent report, in which the β-aryl ether linkage was found to be decreased in abundance as a result of pretreatment (George et al., 2011). However, the amounts of β-5 and β-β increased in ILL than those of EMAL and CEL, suggesting that more condensed lignin presented in ILL sample. Furthermore, the lower S/G ratio of the ILL also added some clues to explain the high content of condensed structure in it because the guaiacyl-type (G) lignin units were found to be reactive toward condensation during the pretreatment (Villaverde et al., 2009).

Quantitative 2D-HSQC spectra

To further investigate the structural features of the lignin fractions, quantitative 2D-HSQC spectra of the lignins were also acquired. **Figure 2** shows the side-chain and aromatic regions of HSQC spectra for the lignins. The important correlations, such as those from substructures of β-ether (β-*O*-4, A), resinol (β-β, B), phenylcoumaran (β-5, C), and some minor structural units (D, spirodienone; F, *p*-hydroxycinnamyl alcohol end groups) can be distinctly assigned according to the recent literatures (Wen et al., 2010; del Río et al., 2009).

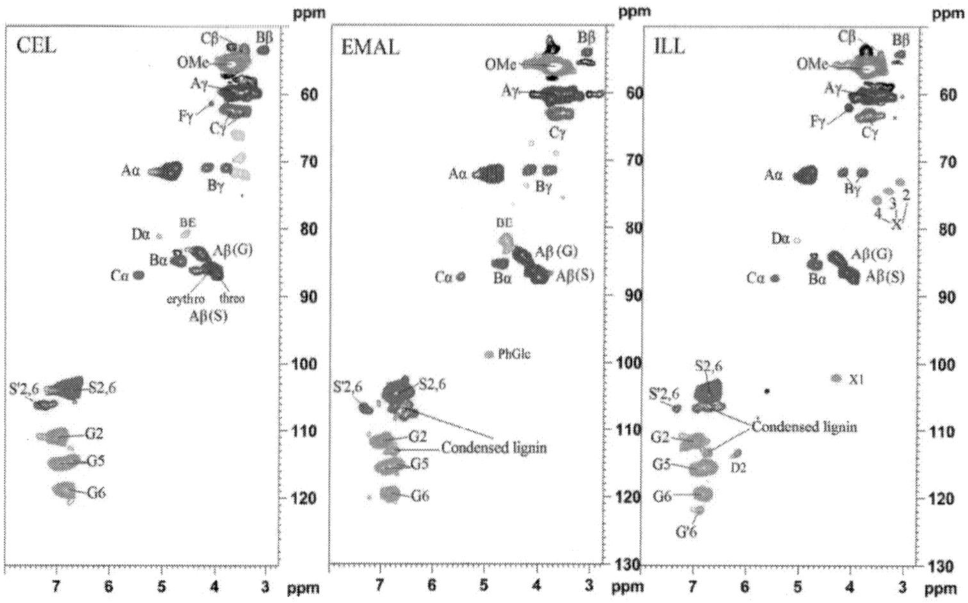

Figure 2. Quantitative 2D-HSQC spectra of ILL, EMAL, and CEL

Comparison of the 2D-HSQC NMR spectra provides more detail information about the fine structural features of the lignin fractions. There have no significant differences in the side-chain region of the spectra. The signals for spirodienone structure (D) presented in ILL and CEL rather than in EMAL suggested that spirodienone unit (D) was not stable in the acidic conditions as the reaction progress. Coincidently, similar results were reported by Hallac et al. when they investigated lignin under acidic condition (Hallac et al., 2010). In addition, the fact that related signals of xylans were observed while the signals for acetylated xylans disappeared, suggesting that the deacetylation of xylans occurred during the IL pretreatment.

In the region of associated carbohydrates, the observed signal at 99.5/4.90 ppm in EMAL was tentatively assigned to phenyl glycoside linkages (PhGlc) according to model compounds data, which probably implied that more lignin-carbohydrate linkages could be found in the S_2 layer of secondary wall (Terashima et al., 1996). Similarly, the fact that benzyl-ether (BE) linkage was detected at δ 81.0/4.65 ppm in CEL and EMAL also revealed that more LCC linkage probably occurred in the secondary wall rather than middle lamella. However, another signal, which appeared at 102.0/4.32 ppm in ILL, is attributed to the β-D-xylopyranoside unit, suggesting that linear xylans were associated to ILL.

In the aromatic regions in the HSQC spectra of the lignins, cross-signals from syringyl (S), and guaiacyl (G) lignin units were distinctly presented. However, *p*-hydroxyphenyl (H) was not observed. The normal S-type lignin units showed a prominent signal for the $C_{2,6}$-$H_{2,6}$ correlation at δ_C/δ_H 104.2/6.72 ppm, whereas the signals corresponding to $C_{2,6}$-$H_{2,6}$ correlations in C_α-oxidized S units (δ_C/δ_H 106.2/7.27 ppm) are also presented in the HSQC spectra of the lignin fractions. By contrast, the G units showed different correlations for C_2-H_2, C_5-H_5, and C_6-H_6 at δ_C/δ_H 111.4/6.96 ppm, 114.8/6.71+6.94 ppm, and 119.8/6.72 ppm, respectively. Two observed signals at C-5 were attributed to different substituents at C-4 position. In addition, some condensed lignin structures likely occurred at mainly G_2 and $S_{2,6}$. The C-H correlations from condensed lignin structures are the 2-position of guaiacyl units linked at the 5-position to other lignin side-chains (Torr et al., 2012). However, the specific condensed lignin formed in S units remains unknown. The possible condensed position in S units probably located at C_α of the side-chain, which can then react with other lignin unit (Li et al., 2007). Similarly, some condensed G_2 was observed in some softwood species in a few just published papers. In a word, the structural features of lignin collected during the IL pre-treatment (ILL and EMAL) slightly condensed as compared to the corresponding CEL.

^{31}P NMR analysis of lignin

The ^{31}P NMR data confirmed that birch lignin belongs to guaiacyl-syringyl type, as evidenced by the presence of syringyl and guaiacyl-type phenolic structures in the ^{31}P NMR spectra of all lignin samples. The decrease of aliphatic hydroxyl groups in ILL can be attributed to the dehydration reaction and acid-catalyzed elimination reactions occurred during ILL pretreatment (Kubo et al., 2008; El Hage et al., 2010). In addition, the reduced content of phenolic hydroxyl groups in EMAL and CEL also add clue to explain the high molecular weights and high content of β-*O*-4 linkages of the lignin fractions. It was also observed that the IL pretreatment resulted in an increase of phenolic hydroxyl and carboxylic groups, which is also corresponding to the low molecular weight and low β-*O*-4 content. The content of carboxyl OH groups was found to be 0.37 mmol/g for ILL and 0.25 mmol/g for EMAL, which was higher than corresponding value of CEL, suggesting that the IL pretreatment resulted in an increase of carboxyl OH. The increased carboxyl OH probably indicated that oxidation of aliphatic OH might have occurred during the IL pre-treatment.

Molecular-weight determination. To investigate the effect of IL pretreatment on the molecular-weights of lignin, GPC analysis of the lignin fractions was performed. As compared with the molecular weights and polydispersity index of CEL, ILL and EMAL exhibited a significant decrease in molecular weight and polydispersity index. The results clearly indicated that IL pretreatment has significantly effect on the molecular weight of lignins due to the different delignification mechanism. A recent published paper also demonstrated that lignin subunits after pretreatment of [C_2mim][OAc] were released via dissociation or depolymerization, thus resulted in the reduced size and shape of lignin (Cheng et al., 2012). Together with the higher content of phenolic OH, the lower M_w value for ILL is a good evidence for lignin fragmentation or depolymerisation by the ionic liquid pretreatment. In addition, the slight increase of phenolic OH in EMAL as compared to CEL also implied that residual lignin extracted as EMAL was depolymerized under the conditions given. Moreover, low polydispersities of the IL lignin fractions indicated that the IL pretreatment resulted in homogeneous lignin fragments with narrow polydispersity (M_w/M_n=1.49-1.62) as compared to that of CEL. Furthermore, it was found that the content of β-*O*-4 is positively related to the molecular weight of the lignins. The amounts of β-*O*-4 units increased in the order of ILL < EMAL < CEL, indicating that ILL is the most degraded preparation whereas CEL undergoes the less degradation from the preparations studied. Specially, more β-*O*-4 content, the higher molecular weight of lignin.

Thermogravimetric analysis (TGA). To study the relationship between quantitative structure and thermal properties, TGA of the lignin fractions were comparatively investigated. It was found that the maximum decomposition temperature (T_M) increased in the order of ILL<EMAL<CEL at the beginning of lignin decomposition (250-350 °C), which was accompanied with the increment of molecular weight, as also revealed by a previous paper (Sun et al, 2000). In other words, higher molecular weights leads to the shift of maximum decomposition temperature (T_M) to a higher temperature region, as also revealed by the data presented here. This is probably that the initial reaction involves the deformation of the weaker C-O bond in the β-*O*-4 structure during the lignin

decomposition. Most of the ether linkages were expected to be cleaved at the stage of 200-350 °C, confirmed by high content of β-*O*-4 in the CEL (Faravelli et al., 2010). The different thermal stabilities are mainly influenced by inherent structure and various functional groups of lignin polymers, degree of branching and condensation of the lignin macromolecule. In addition, typical weight losses in TG curves flatten out at temperatures higher than 500 °C, with a slow release of the final products before final char formation. Cracking of the methoxy (-OCH_3) groups released methane around 400-600 °C. It was found that the char residues at 600 °C were 26% for CEL, 27% for EMAL, and 35% for ILL. This fact suggested that more condensed lignin structures were found in ILL than that of CEL and EMAL, as revealed by the higher DC value obtained from quantitative NMR data. In addition, the "char residue" is inversely proportional to the methoxy groups contained in the lignin samples (**Table 1**), as also reported by other authors (Jakab et al., 1997). From the above observations, it can be concluded that the lignin having more condensed structures and less content of OH_3 result in more char residue. However, lignin with a higher molecular weight and more β-*O*-4 structure has a higher T_M at the major lignin decomposition.

Conclusion

In conclusion, the present study investigated the quantitative structural features of lignins collected during the IL pretreatment and enzymatic hydrolysis. It was found that a significant decrease in the molecular weight of ILL and EMAL as compared to the corresponding CEL; such a decrease was primarily due to the slight homolytic cleavage of β-*O*-4 inter-linkages, as indicated by the quantitative ^{13}C, 2D-HSQC, and ^{31}P-NMR methodologies. Additionally, characteristic dissociation mechanism of ILL could also account for the structural characteristics and decreased molecular weights of ILL. Moreover, there has a possible relationship between quantitative structures and thermal properties, that is, higher molecular weight and higher β-*O*-4 content will lead to a higher T_M at the major lignin decomposition stage, while more condensed lignin structures in ILL than that of EMAL and CEL provides the evidence of higher "carbon residue" remained in ILL rather CEL and EMAL. However, after IL pretreatment and enzymatic hydrolysis, typical lignin structural features in the original lignin, such as β-*O*-4, β-β, and β-5 linkages, are mostly reserved in all the lignins, although some cleavage and condensation occurs during the IL pretreatment and enzymatic hydrolysis. In view of the typical structural units and thermal properties of the lignins, consequently, this would also make it more easily for further utilization of these lignins as starting materials for high-value-added products.

REFERENCES

(1) Argyropoulos, D.S.; Sun, Y.; Paluš E. Isolation of residual kraft lignin in high yield and purity. *J. Pulp Pap. Sci.* **2002**, *28*, 50-54.

(2) Arora, R.; Manisseri, C.; Li, C. L.; Ong, M. D.; Scheller, H. V.; Vogel, K.; Simmons, B. A.; Singh, S. Monitoring and analyzing process streams towards understanding ionic liquid pretreatment of switchgrass (*Panicum virgatum L.*). *BioEnergy Res.* **2010**, *3*, 134-145.

(3) Cheng, G.; Kent, M. S.; He, L.; Varanasi, P.; Dibble, D.; Arora, R.; Deng, K.; Hong, K.; Melnichenko, Y. B.; Simmons, B. A.; Singh, S. Effect of ionic liquid treatment on the structures of lignins in solutions: molecular subunits released from lignin. *Langmuir* **2012**, *28*, 11850-11857.

(4) Choi, J. W.; Faix, O. NMR study on residual lignins isolated from chemical pulps of beech wood by enzymatic hydrolysis. *J. Ind. Eng. Chem.* **2011**, *17*, 25-28.

(5) del Río, J. C.; Rencoret, J.; Marques, G.; Li, J. B.; Gellerstedt, G.; Jiménez-Barbero, J.; Martínez, Á. T.; Gutiérrez, A. Structural characterization of the lignin from jute (Corchorus capsularis) fibers. *J. Agric. Food Chem.* **2009**, *57*, 10271-10281.

(6) El Hage, R.; Brosse, N.; Chrusciel, L.; Sanchez, C.; Sannigrahi, P.; Ragauskas, A. Characterization of milled wood lignin and ethanol organosolv lignin from miscanthus. *Polym. Degrad. Stab.* **2009**, *94*, 1632-1638.

(7) Faix, O.; Argyropoulos, D. S.; Robert, D.; Neirinck, V. Determination of Hydroxyl Groups in Lignins Evaluation of ^{1}H-, ^{13}C-, and ^{31}P-NMR FT-IR and Wet Chemical Methods. *Holzforschung* **1994**, *48*, 387-394.

(8) Faravelli, T.; Frassoldati, A.; Migliavacca, G.; Ranzi, E. Detailed kinetic modeling of the thermal degradation of lignins. *Biomass Bioenerg.* **2010**, *34*, 290-301.

(9) George, A.; Tran, K.; Morgan, T. J.; Benke, P. I.; Berrueco, C.; Lorente, E.; Wu, B. C.; Keasling, J. D.; Simmons, B. A.; Holmes, B. M. The effect of ionic liquid cation and anion combinations on the macromolecular structure of lignins. *Green Chem.* **2011**, *13*, 3375-3385.

(10) Hu, Z. J.; Yeh, T. F.; Chang, H. M.; Matsumoto, Y.; Kadla, J. F. Elucidation of the structure of celluloytic enzyme lignin. *Holzforschung* **2006**, *60*, 389-397.

(11) Jaaskelainen, A. S.; Sun, Y.; Argyropoulos, D. S.; Tamminen, T.; Hortling, B. The effect of isolation method on the chemical structure of residual lignin. *Wood Sci. Technol.* **2003**, *37*, 91-102.

(12) Jakab, E.; Faix, O.; Tilla, F. Thermal decomposition of milled wood lignins studied by thermogravimetry/mass spectrometry. *J. Anal. Appl. Pyrol.* **1997**, *40-41*, 171-186.

(13) Kim J. Y.; Shin E. J.; Eom I. Y.; Won K.; Kim Y. H.; Choi D.; Choi I. G.; Choi J. W. Structural features of lignin macromolecules extracted with ionic liquid from poplar wood. *Bioresour. Technol.* **2011**, *102*, 9020-9025.

(14) Klein-Marcuschamer, D.; Simmons, B. A.; Blanch, H. W. Technoeconomic analysis of a lignocellulosic ethanol biorefinery with ionic liquid pretreatment. *Biofuels Bioprod. Biorefin.* **2011**, *5* (5), 562-569.

(15) Kubo, S.; Hashida, K.; Yamada, T.; Hishiyama, S.; Magara, K.; Kishino, M.; Ohno, H.; Hosoya, S. A characteristic reaction of lignin in ionic liquids: Glycelol type enol-ether as the primary decomposition product of beta-*O*-4 model compound. *J. Wood Chem. Technol.* **2008**, *28*, 84-96.

(16) Lan, W.; Liu, C. F.; Sun, R. C. Fractionation of bagasse into cellulose, hemicelluloses, and lignin with ionic liquid treatment followed by alkaline extraction. *J. Agric. Food Chem.* **2011**, *59*, 8691-8701.

(17) Lee, S. H.; Doherty, T. V.; Linhardt, R. J.; Dordick, J. S. Ionic liquid-mediated selective extraction of lignin from wood leading to enhanced enzymatic cellulose hydrolysis. *Biotechnol. Bioeng.* **2009**, *102*, 1368-1376.

(18) Li, C. L.; Knierim, B.; Manisseri, C.; Arora, R.; Scheller, H. V.; Auer, M.; Vogel, K. P.; Simmons, B. A.; Singh, S. Comparison of dilute acid and ionic liquid pretreatment of switchgrass: Biomass recalcitrance, delignification and enzymatic saccharification. *Bioresour. Technol.* **2010**, *101*, 4900-4906.

(19) Li, J.; Henriksson, G.; Gellerstedt, G. Lignin depolymerization/repolymerization and its critical role for delignification of aspen wood by steam explosion. *Bioresour. Technol.* **2007**, *98*, 3061−3068.

(20) Mäki-Arvela, P.; Anugwom, I.; Virtanen, P.; Sjöholm, R.; Mikkola, J. P. Dissolution of lignocellulosic materials and its constituents using ionic liquids-A review. *Ind. Crops Prod.* **2010**, *32*, 175-201.

(21) Pinkert, A.; Goeke, D. F.; Marsh, K. N.; Pang, S. S. Extracting wood lignin without dissolving or degrading cellulose: investigations on the use of food additive-derived ionic liquids. *Green Chem.* **2011**, *13*, 3124-3136.

(22) Sun, N.; Rahman, M.; Qin, Y.; Maxim, M. L.; Rodriguez, H.; Rogers, R. D. Complete dissolution and partial delignification of wood in the ionic liquid 1-ethyl-3-methylimidazolium acetate. *Green Chem.* **2009**, *11*, 646-655.

(23) Tan, S. S. Y.; MacFarlane, D. R.; Upfal, J.; Edye, L. A.; Doherty, W. O. S.; Patti, A. F.; Pringle, J. M.; Scott. J. L. Extraction of lignin from lignocellulose at atmospheric pressure using alkylbenzenesulfonate ionic liquid. *Green Chem.* **2009**, *11*, 339-345.

(24) Torr, K. M.; Love, K. T.; Cetinkol, O. P.; Donaldson, L. A.; George, A.; Holmes, B. M.; Simmons, B. A. The impact of ionic liquid pretreatment on the chemistry and enzymatic digestibility of *Pinus radiata* compression wood. *Green Chem.* **2012**, *14*, 778-787.

(25) Villaverde, J. J.; Li, J. B.; Ek, M.; Ligero, P.; Vega, A. D. Native lignin structure of miscanthus x giganteus and Its changes during acetic and formic acid fractionation. *J. Agric. Food Chem.* **2009**, *57*, 6262-6270.

(26) Wen, J. L.; Sun, Z. J.; Sun, Y. C.; Sun, S. N.; Xu, F.; Sun, R. C. Structural characterization of alkali-extractable lignin fractions from bamboo. *J. Biobased Mater. Bio.* **2010**, *4*, 408-425.

(27) Whiting, P.; Goring, D. A. I. Chemical characterization of tissue fractions from the middle lamella and secondary wall of black spruce tracheids. *Wood Sci. Technol.* **1982**, *16*, 261-267.

(28) Yuan, T. Q.; Sun, S. N.; Xu, F.; Sun, R. C. Structural Characterization of Lignin from Triploid of *Populus tomentosa Carr. J. Agric. Food Chem.* **2011**, *59* (12), 6605-6615.

Fractionation of lignocellulosic materials for biorefinery: Lignins and Hemicelluloses

Run-Cang Sun [a,b,*]

[a] *Institute of Biomass Chemistry and Technology, Beijing Forestry University, Beijing, 100083, China*

[b] *State Key Laboratory of Pulp and Paper Engineering, South China University of Technology, Guangzhou 510640, China*

ABSTRACT

Lignocellulosic materials represent an abundant carbon-neutral renewable resource for the production of bioenergy and biomaterials. A concept for converting renewable lignocellulosic materials to valuable biofuels, chemicals, and biomaterials generally has been referred to as biorefinery. Lignin, one of the three main constituents of lignocelluloses, is undervalued as a by-product of the biorefinery. Hemicelluloses are heterogeneous polymers of pentoses (xylose and arabinose), hexoses (mannose, glucose and galactose), and sugar acids, which are usually degraded during the biorefinery. The development of value-added products from lignin and hemicelluloses will greatly improve the economics of these processes. However, value added applications from lignins and hemicelluloses need a proper investigation of their inherent properties, which significantly affect their applications. In this paper, the recent research in our group on fractionation and structural characterization of lignin and hemicelluloses from lignocellulosic materials for biorefinery has been reviewed. The structural features of the lignin and hemicelluloses obtained were elucidated by various destructive and non-destructive techniques. It was believed that the in-depth investigation of the structure features of lignin and hemicelluloses will be beneficial to understand the potential of these materials for future value added applications. Moreover, novel functional biopolymers and biomaterials, such as ionic hydrogels and nanocomposite films, from hemicelluloses were developed. The results showed that the ionic hydrogels had high water adsorption capacity and showed rapid and multiple responses to pH, ions, and organic solvents, which may allow their use in several areas such as adsorption, separation, and drug release systems. The effective method developed makes it possible to produce hemicellulose-based biomaterials with improved mechanical properties.

Keywords: Fractionation; lignocelluloses; biorefinery; lignin; hemicelluloses; characterization; biomaterials

* Corresponding author. Tel.: +86 10 62336903; fax: +86 10 62336903.

E-mail address: rcsun3@bjfu.edu.cn (R.C Sun).

(471d) Dissolution of Cornstalk with Acetic-Based Ionic Liquids and Additives

Junli Xu, Institute of Process Engineering, Chinese Academy Of Sciences, Beijing, China

Dissolution of cornstalk with acetic-based Ionic Liquids and additives

J. L. Xu[a], F. Yu[b], X. M. Lu [a,*], S. J. Zhang[a]

[a] *Institute of Process Engineering, Chinese Academy Of Sciences, Beijing, 100190*

[b] *Graduate University Of Chinese Academy Of Sciences, Beijing, 100149*

[*] *Corresponding author: xmlu@home.ipe.ac.cn*

Key words: lignocellulose, pretreatment, ionic liquids, catalyst,additives

Biomass is abundant and renewable, for example wood, cornstalk, grass and so on; as the energy crisis biomass feedstock is widely used for biofuel production, and this kind of energy production is cleaner than fossil fuels. But the production of biofuel from lignocellulose has technical challenges, particularly for the pretreatment of biomass. Cellulose, hemicellulose and lignin are main components of biomass. And these three components are connected by hydrogen bonds, including intramolecular hydrogen bond and intermolecular hydrogen-bonding. It is very hard to break the hydrogen-bonding to separate the components by pretreatment. There are many traditional pretreatment methods, for example acidolysis, alkaline hydrolysis, these methods have some shortcomings like toxicity, volatility. Ionic liquids are different salts, most of them are liquids at room temperature; and some of them have positive attributes like high thermal stability, negligible vapor pressure and non-flammability, so making ionic liquid pretreatment of biomass potentially benign to environment as compared with pretreatment by acids, alkali or other organic solvents. But the ILs used for dissolving biomass also have some shortcomings, for example high viscosity and corrosity; acetic-based ionic liquids are greener to dissolve lignocellusic biomass compared with 1-butyl-3-methylimdazolium chloride([Bmim]Cl), and some diluents or catalysts are added to the ILs in order to dissolve the biomass quickly. In this research, dissolution of cornstalk and regeneration cellulose rich material (CRM) in the acetie-based ionic liquids like 1-allyl-3-methylimdazolium chloride([Amim]Cl) were made, and investigated the impact on the additive TBD, the results showed that the cellulose content of CRM increased from 24.63% to 32.23%, and the characterization of CRM by SEM, XRD, IR indicated that addition of TBD to [Amim][OAc] has definite help for dissolving cornstalk compared with pure [Amim][OAc].

References

1. Wang XJ, Li HQ, Cao Y, Tang Q. Cellulose extraction from wood chip in an ionic liquid 1-allyl-3-methylimidazolium chloride ([Amim]Cl), Bioresource Technology 102(2011) 7959-7965

2. Li WY, Sun N, Stoner B, Rogers RD, Rapid dissolution of lignocellulosic biomass in ionic liquids using temperature above the glass transition of lignin, *Green Chem.*, 2011: 13, 2038- 2047

---

* General Program of National Natural Science Foundation of China(No.20976174),2009-2012

(471e) Characteization of Solubilized Biompolymers Fractionated From Alkali Pulping Liquors

Ryan J. Stoklosa, Chemical Engineering & Materials Science, Michigan State University, East Lansing, MI and David Hodge, Department of Chemical Engineering and Material Science, Michigan State University, East Lansing, MI

Alkaline extraction of lignin and hemicellulose from woody biomass can be integrated with existing chemical pulp mills and can potentially increase revenue by utilizing the non-cellulosic portion of the biomass for fuels and chemicals rather than for combustion for heat and power. The overall purpose of this work is to characterize the extractability, recoverability, and properties of the solubilized hemicellulose (glucuronoxylan) and lignin during the alkali impregnation and heat-up stages of soda pulping for two hardwood species, *Acer saccharum* (sugar maple) and *Betula pendula* (silver birch). Liquors will be sampled throughout the heat-up and during the course of pulping. A material balance will compare the difference between the extracted material in the pulping liquor and the residual wood chips after pre-extraction based on total mass, xylan, and lignin removed. The recoverable yield of the solubilized biopolymers will be determined using a variety of methods including aggregation and precipitation by acidification followed by ultrafiltration or by complete polymer precipitation by ethanol to determine an upper recoverability limit. The properties of the recovered biopolymers as a function of extraction condition will be characterized by a variety of techniques including total composition, total polymer molecular weight distributions by size exclusion chromatography and xylan number average degree of polymerization by a novel reducing end assay. Lignins will be fractionated using sequential solvent extraction and the properties of these lignins will be characterized by SEC and ^{13}C NMR.

Liquid Hot Water Pretreatment of Banana Agricultural Waste for Monomeric Sugars Production

Juan C. Quintana-Marín. Bioprocesos y Flujos Reactivos, Facultad de Minas, Universidad Nacional de Colombia-Sede Medellín. Medellín, Colombia.
Adriana Ruiz. Bioprocesos y Flujos Reactivos, Facultad de Minas, Universidad Nacional de Colombia-Sede Medellín. Medellín, Colombia.
Nathan S. Mosier. Agricultural and Biological Engineering & Laboratory of Renewable Resources Engineering (LORRE), Purdue University, West Lafayette, IN.

Abstract

Banana agricultural waste represents a regionally-specific and abundant cellulosic material that has potential as a feedstock for producing cellulosic biofuels. Efficient conversion of lignocellulosic biomass into monomeric sugars requires particle size reduction, pretreatment, and enzymatic hydrolysis. Liquid hot water (LHW) is an effective hydrothermal pretreatment process with potential for industrial application that must be optimized in terms of the temperature, reaction time, particle size of biomass, and water:solids ratio that can vary depending upon the characteristics of the lignocellulose. We present the optimization of LHW pretreatment of banana agricultural waste for temperatures between 180 and 220 °C, at times between 0 and 60 min, two particle size ranges (<0.42 mm and 2.83 to 6.35 mm), and solid loadings of 5 and 15% dry matter (w/w). The highest saccharification yields of glucose (82%) occurred after pretreatment at 220 °C for 20 min. While these conditions resulted in extraction of nearly all of the hemicellulose, the yield of xylose and arabinose was lowered by the presence of alkali in the raw material which catalyzed the degradation.
Keywords: Banana agricultural waste, Cellulose enzymatic hydrolysis and Pretreatment

Introduction

The imminent exhaustion and the increasing demand of fossil fuels have motivated the search for new substitutes such as ethanol production derived from energy crops and lignocellulosic residues (LCR). These materials are renewable and allow a favorable CO_2 balance when biofuels and fossil fuels are compared. In addition, LCR are cheap and do not compete as raw materials for other industrial processes but are very pollutant when left in the field to decompose after prunning or harvesting [1-3].

Bioconversion of local residues to ethanol may favor rural economic development by offering employment and supplying a local supply of fuel. This plan is applicable in tropical or neartropical regions where banana is produced and millions of tons of waste is generated after fruit production, which corresponds to the end of the plant life cycle [4].

Part of the rejected banana fruit as well as the cut banana plant, composed of pseudostem, rachis and leaves, are generally abandoned on the ground to decompose. This results in spreading diseases and polluting water supplies. Proposed uses of fruit residues has been animal feed, composting, edible flour and vaccines production [2,5] while the lignocellulosic fraction has been processed to produce fibers usable in the textile, paper and fiberboards [4,6,7]. Only in recent years they have been evaluated as energetic materials through using in biodigesters and ethanol production [3,4]. Concerning ethanol production, the starch in the fruit is relatively easy to transform by amylolytic enzymes while plant-derived lignocellulosic wastes need to be pretreated before the enzyme hydrolysis step [3].

Given the differences in composition and structural complexity of every part of the plant, we are optimizing the bioconversion routes of them individually. A study in cooperation between University of

California and National University of Colombia examined steam explosion (SE) and LHW pretreatments on pseudostem, leaves, and rachis at two conditions of time and temperature. One of the results was that LHW was better suited for rachis.

While the preliminary study indicated that LHW was better suited for rachis, further optimization was needed to understand the effect of variables such as time, temperature, pH and solids loading on total sugar yields [8]. In this work we evaluate such variables in LHW on banana rachis and enzymatic hydrolysis in order to find the highest overall sugar production and ultimately to develop a mathematical model to predict the sugar yields from the combination of values that every variable takes during the pretreatment.

Materials and Methods

Banana Rachis

The bunch axis in the plant, known as banana rachis, was collected in the Urabá Province (Antioquia, Colombia) a large banana producer zone in the world, and delivered to our laboratory where it was dried in an oven maintained at 50°C for five days. After drying, the remaining water was measured using an HB43 halogen moisture analyzer (Mettler Toledo Inc., Columbus, OH) at 105 °C which resulted in 5.54±0.18 %. Particle sizes of <0.40 mm and 2.83 to 6.35 mm were obtained by grinding and sieving this dried material.

Chemical Composition Analysis

Unpretreated banana rachis was subjected to water and ethanol extractions consecutively in order to remove non-structural compounds. The resultant solid was dried at 45 °C until the moisture content was less than 5% and then it was submitted to a two-step sulfuric acid hydrolysis as established in NREL-LAP procedures. The hydrolyzate was filtered through preashed fritted crucible and the liquid fraction was analyzed by HPLC while the precipitate was dried to quantified acid insoluble lignin and burnt for ash content correction. The acid soluble lignin was measured using the absorbance of the filtrate solution at $\lambda = 320\ nm$.

Pretreated solids were washed twice with water at 60 °C and dried at 45 °C for two weeks. The same procedure of biomass composition was applied to them taking into account only the carbohydrates content.

Oligosaccharides determination was carried out following the LAP 014 of NREL guidelines. This process was scaled down using glass ampoules of 1.5 ml total volume and sealed with a 0.003-in thick brass circle fitted between the cap and a teflon stopper in order to withstand the high pressure when were placed in the autoclave .

Because glucose, xylose and arabinose are in anhydrous form in the raw material, the computation of sugar conversion must be consider the change in molecular weight and by-products as a result of hydrolytic reactions [9].

LHW pretreatment

The pretreatment was carried out in stainless steel batch tube reactors of 4 ml. This size is sufficient small to reach the maximum target temperature in the center in less than 30 seconds and we filled three fourths (3 ml) of its volume to allow for liquid expansion upon heating. The solid loadings were equivalent to 5 and 15% w/w and also two particle sizes were tested (<0.42 mm and 2.83 to 6.35 mm). Temperature and time were combined as follows: 180 °C with 10, 20, 30 and 60 min reaction times; 200 °C with 5, 10, 15, 30 and 60min reaction times; and 220 °C with 1, 5, 10, 20, 30 and 60 min reaction times. A Tecam sandbath was used for heating-up and maintaining the tubes at the mentioned temperatures. Once the planned time passed, the tubes were moved to an ice-water bath to stop the reactions.

Enzymatic Hydrolysis

Enzymatic hydrolysis conditions were the same for all runs in order to make clear the effect of pretreatment on the overall sugar yield. Without washing the solids, 3ml of Na-citrate buffer pH 4.8 was added bringing the total volume to 6ml. Commercial cellulose (Spezyme CP) and β-glucosidase (Novozyme 188) at 15 FPU/g-glucan and 40 CBU/g-glucan were the dose respectively based on the glucan subjected to pretreatment. The reactions were carried out over 72 hours at 50 °C and 250 rpm in an orbital shaker.

HPLC analysis

Calibration curves were prepared with standards of glucose, xylose, arabinose, acetic acid, furfural and 5-hydroxymethyl furfural (HMF). All of these compounds were quantified after chemical composition analysis, LHW and enzymatic hydrolysis using a Waters 2695 Separation Module with a Waters 2414 refractive index detector and separated in a Bio-Rad Aminex® HPX-87H column at 65 °C and fixed to a 0.6 ml/min flow of 5mM H_2SO4.

Results and Discussion

Chemical composition

The obtained biomass characterization is different to the composition reported by Velásquez-Arredondo et al. [3] who used bromatological procedures. They reported 32.1, 8.8, 1.13 and 19% of cellulose, hemicellulose, starch and lignin respectively. We could characterize more than 97% of the whole rachis (Table 1).

Table 1. Banana rachis composition

Component	Dry weight basis (%)
Glucan	47.32±0.71
Xylan/Glactan	13.12 ±0.25
Arabinan	3.55 ±0.08
Acetyl	0.26 ±0.45
Acid Insoluble Residue	15.83±0.36
Acid Soluble Lignin	1.10±0.02
Total	97.15

Effect of pretreatment

Before the cellulose fraction can be used as a substrate for ethanol production, it must be separated from the confines of its protective hemicellulose-lignin wrapper and the effective surface area available to cellulase enzymes must be increased [3]. LHW helps to achieve this through the release of hemicelluloses which are solubilized into either oligosaccharides or monosaccharides or a mixture of both. Furthermore acetic acid is released because the hemicelluloses are acetylated. Depending upon the conditions of the LHW pretreatment sugar degradation products as furfural may be formed. Short times and no reagent requirements beyond water made LHW a cost-effective pretreatment.

Smaller particle size and 5% solids loading	Bigger particle size and 5% solids loading	Smaller particle size and 15% solids loading

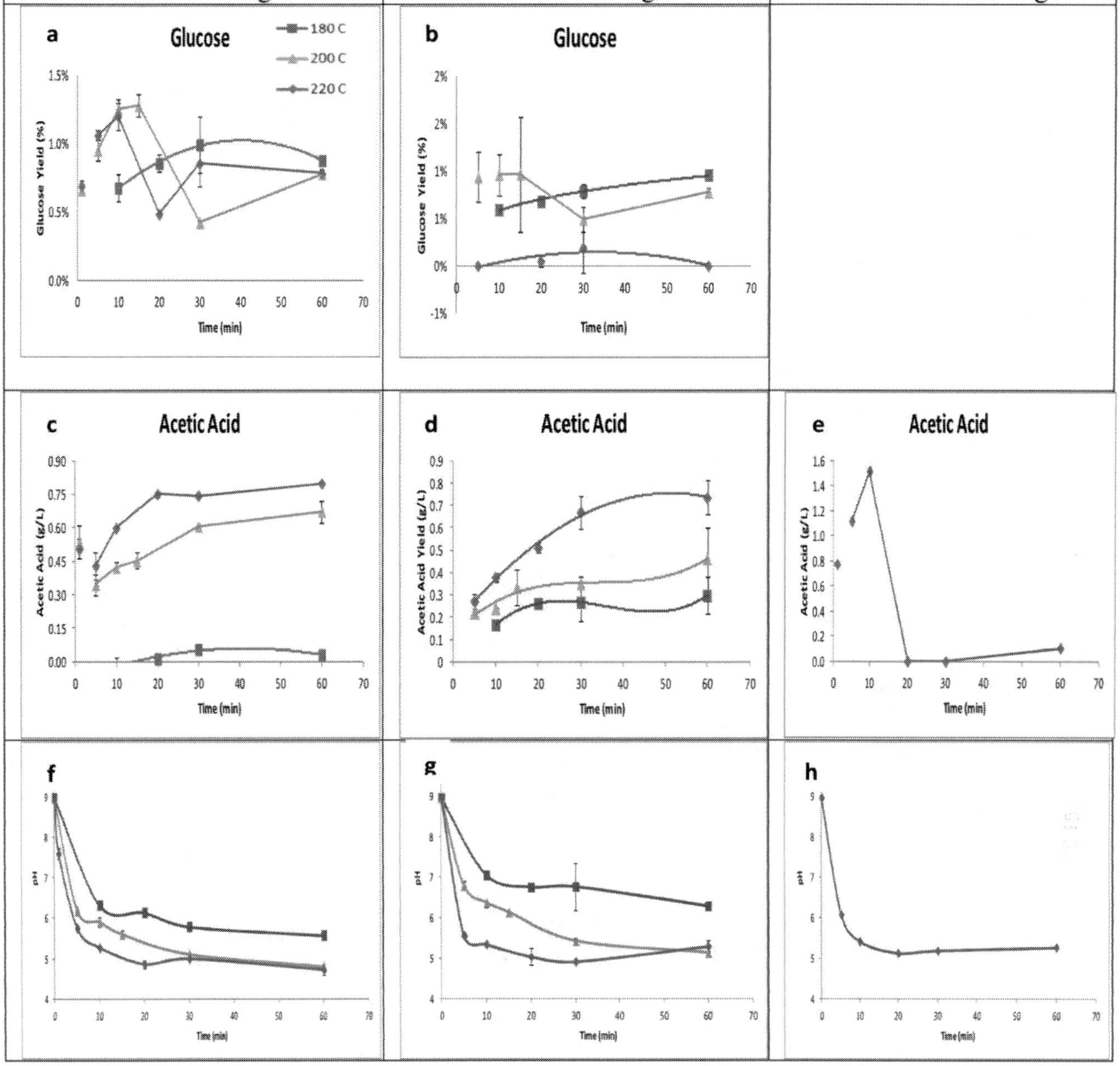

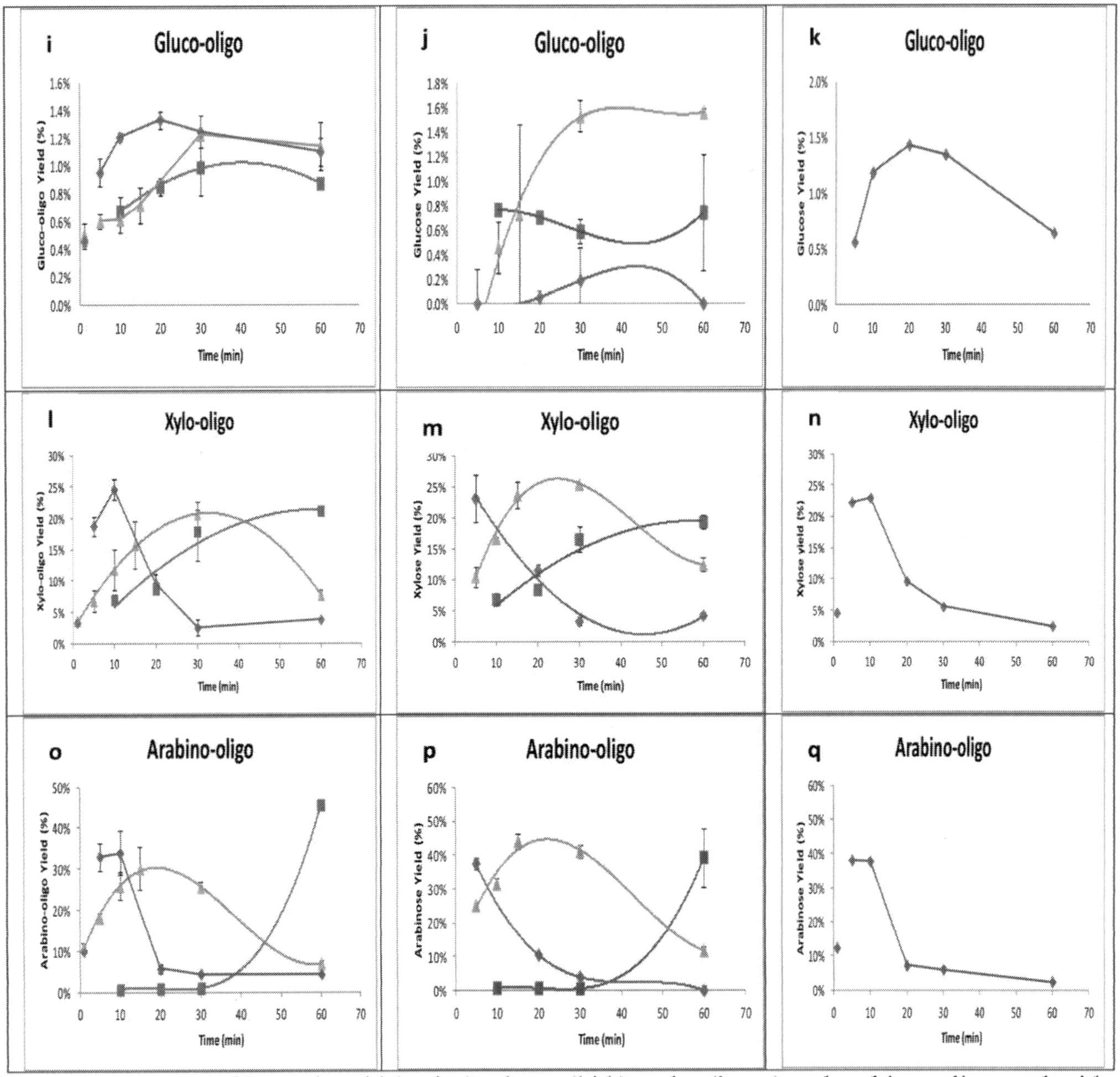

Figure 1. Glucose (a,b), acetic acid (c, d, e), gluco- (i,j,k), xylo- (l,m,n) and arabino- oligosaccharides (o,p,q) produced during LHW pretreatment on banana rachis at 180 °C, 200 °C and 220 °C.

Smaller particle size and 5% solids loading	Bigger particle size and 5% solids loading	Smaller particle size and 15% solids loading

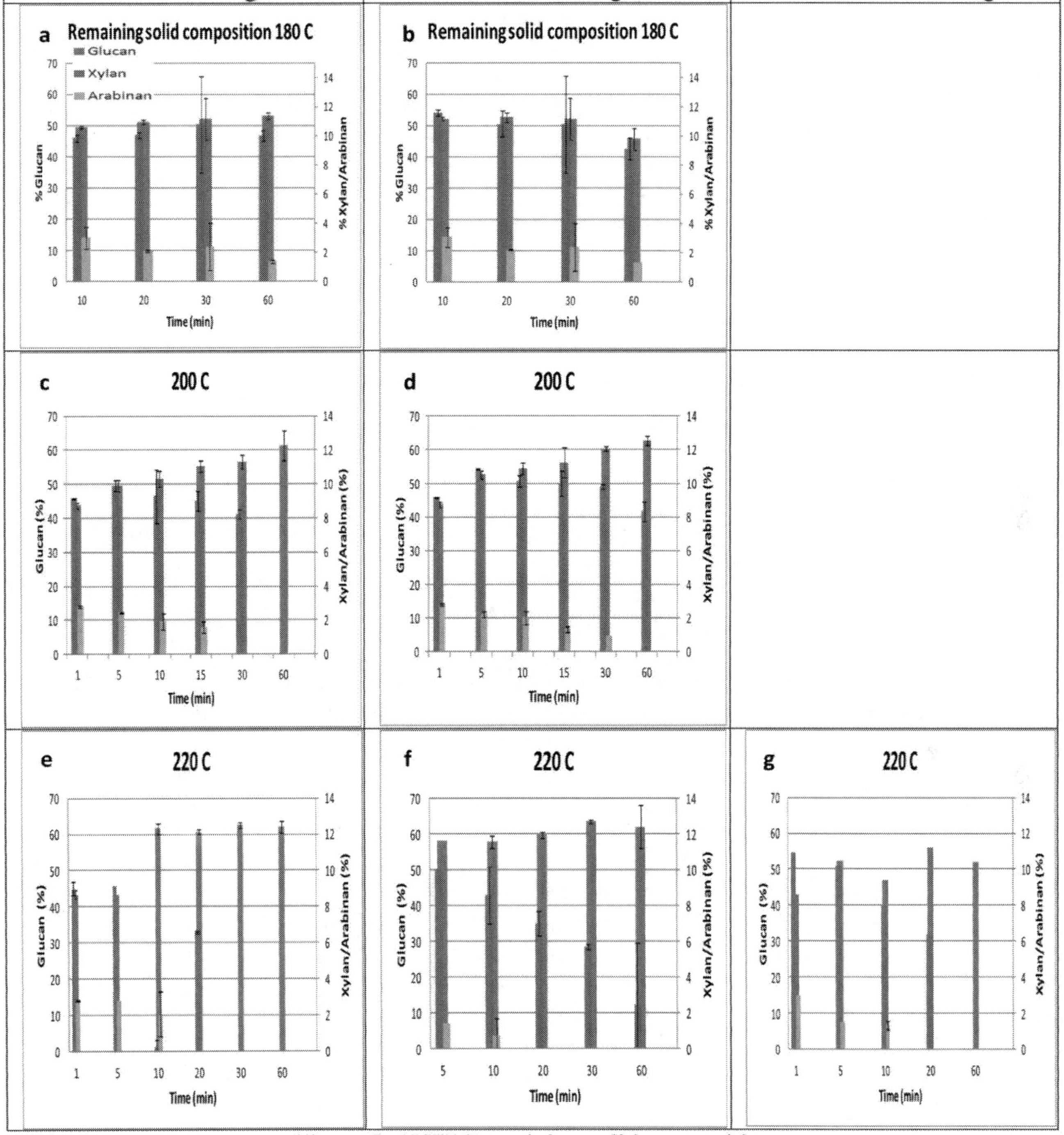

Figure 2. LWH Remaining solid composition

Glucose, xylose and arabinose are fermentable by recombinant yeast strains therefore this work focused on pretreatment, pursues the highest possible C5 and C6 sugar yields [10].

The high initial pH of 9 in Figure 1 (f to h) is due to alkaline compounds present in fresh rachis. Those alkaline compounds combined with the presence of acetyl result in buffering the pretreatment solution to a pH between 4 and 7 for pretreatment times longer than 5 min. It has been found that C5 sugars are degraded quickly into resins at pH 7 [11]. At pHs near 4, the solubilization of hemicellulose into oligomers is favored [9,12]. This along with the low solid loadings are the reason why we didn't observe monomeric xylose, arabinose, or furfural. However oligosaccharides were found in the liquid and were degraded under more severe conditions of higher temperatures and longer reaction times (Figure 1 l to q).

Yield is defined as the product recovered compared to the stoichiometric amount of product that could be produced from the initially available substrate. Xylose and arabinose from xylan and arabinan respectively, while furfural by the contribution of both. When the initial pH was adjusted from 9 to 2.5 using sulfuric acid, the detected furfural yield was equivalent to 1 and 1.5% after pretreatment at 200 °C after 30 and 60 min while at 220 °C the yields were 0.66, 1.63, 1.22 and 1.23 when 10, 20, 30 and 60 min had passed.

In previous work, the maximum yield of xylo-oligomers was 8g/100 g for maple wood in after 60 min at 160 °C with xylose and furfural at 2g and less than 0.4g respectively [13]. Another study shows that the the maximum amount of xylooligosaccharides released from barley husks was obtained after 0.66 h [14]. While these previously reported results are similar to ours (Figure 1 l to q), we cannot compare subsequent enzymatic hydrolysis of the cellulose because the results were not reported.

Enzymatic digestibility

Commercial enzymes appear to have activities besides cellulase and β -glucosidase such as arabinase and xylanase because the amount of these sugars is higher than the obtained after the pretreatment even in oligomeric form.

Although there was not large amounts of glucose nor degradation products of glucose (HMF, levulinic acid, etc.) observed in the liquid fraction after pretreatment, the yield of glucose after the enzymatic hydrolysis of the pretreated glucan is reduced at long time pretreatment times and at high temperatures (Figure 3 a to c). This pattern may be due to other compounds such as humins from pentoses and melted lignin that cover the remainning cellulose and prevent the interaction cellulose-enzyme [15-16]. Another explanation could be inhibition of the cellulase and/or β-glucosidase by phenolic compounds released from lignin during hydrothermal pretreatments [17] because the pretreated slurry was not washed prior to enzymatic hydrolysis.

It seems that temperature is more important than time because the bigger changes in enzymatic hydrolysis are observed at different temperatures. The pattern observed in Fig 3 (d to i) Also shows that xylose is more sensitive at more severe conditions than arabinose. The same was mentioned by Tharhezadeh and Karimi [10] and observed by Koostra et al. [18].No significant differences due to the particle size were observed in glucose yields (Figure 3 a,b). This may be due to the fact mentioned in Mittal's [19] work that a noticeable effect of this variable on pretreatment could occur at diameters higher than 8 mm as a consequence of the mass transfer limitation inside the particle. Higher oligosaccharide yields at bigger particle size (Figure 3 d,e and g,h) can be due to the differences in packing of the reactors and the major facility to moisten fine solids. This is analogous to the lower hemicelluloses solubilization obtained by Cara et al. [20] when they impregnated the raw material with water or sulfuric acid before the pretreatment.

Regarding solids loading, the much amount, the poorest mass transfer and lowest sugar yields (Figura 3 c,f,i). Furthermore we only evaluate this effect at the smaller particle size which is easier to moisten.

Overall sugars after both pretreatment and enzymatic hydrolysis. Smaller particle size and 5% solids loading	Overall sugars after both pretreatment and enzymatic hydrolysis. Bigger particle size and 5% solids loading	Overall sugars after both pretreatment and enzymatic hydrolysis. Smaller particle size and 15% solids loading

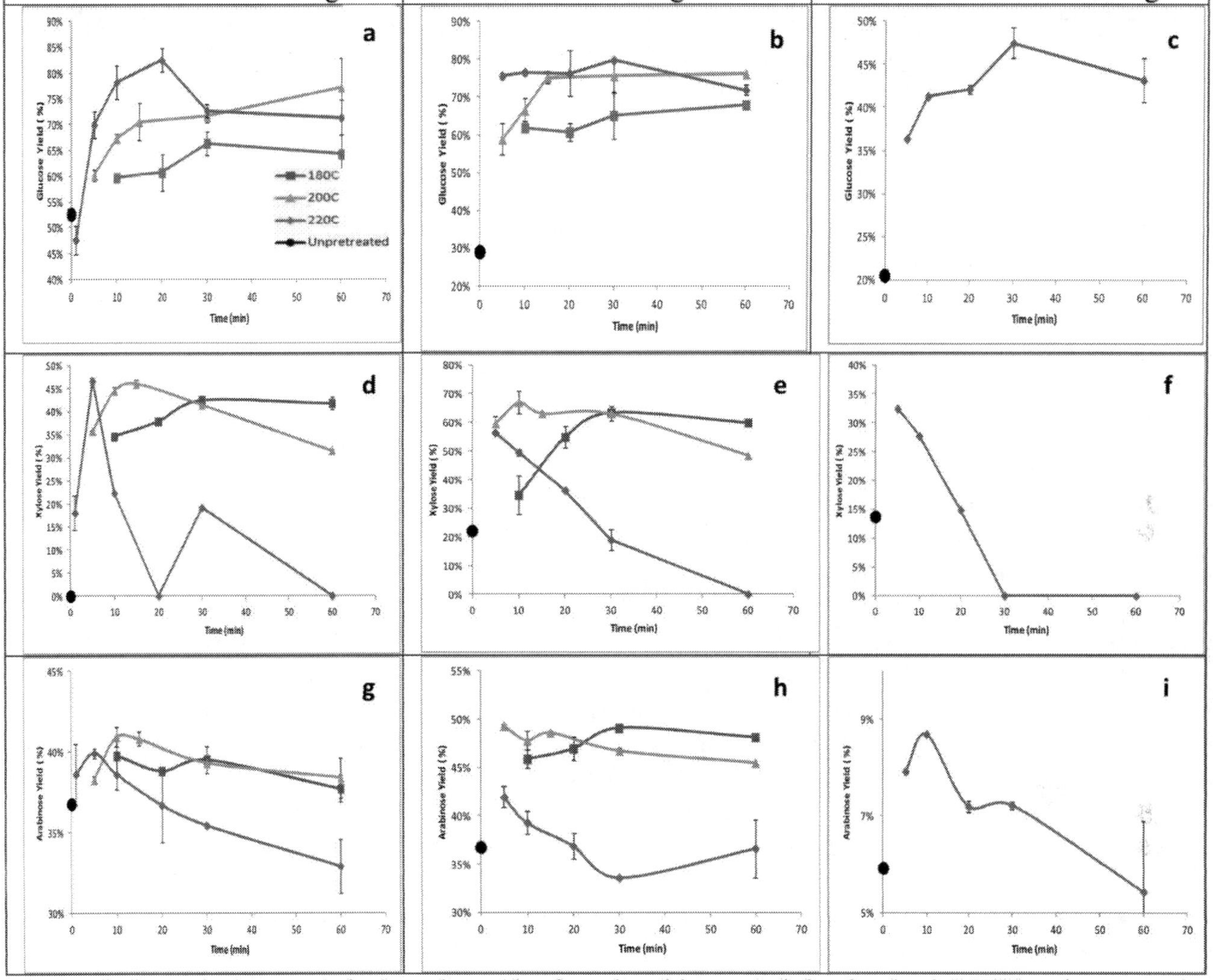

Figure 3. Overal glucose (a,b,c), xylose (d,e,f) and arabinose (g,h,i) after both LHW and enzymatic hydrolysis on banana rachis at —■— 180 °C, —▲— 200 °C and —◆— 220 °C LHW conditions.

In general, higher glucose yields are correlated with higher xylan and arabinan solubilization which occurred much faster at 220 °C and smaller size particles (Figure 2 e,f) except the higher solids loading. At the same temperature the maximum glucose yield was obtained (82 %, Figure 3a) while xylose and arabinose were much degraded. Their higher overall yields were 67% and 49.3% at 200 °C and 10 and 5 min respectively.

Conclusion

On the search for a bioconversion pathway of banana rachis into fermentable sugars, this study has been focused on the pretreatment process. Banana rachis is a current lignocellulosic waste left on the ground after harvesting and has to be modified chemical and structurally for reaching high overall monomeric sugar yields during the enzymatic hydrolysis step. LHW pretreatment was proved under different conditions of temperature, time, pH, particle size and solids loading in order to elucidate their effect on by-products, mono and oligosaccharides production in the overall process. It was founded that banana rachis contains a lot of alkaline compounds that combined with acetic acid released from celluloses favored a buffering effect maintaining the pH between 4 and 7. Also, pentose yields were higher at bigger particle sizes, lower solid loadings, 200 °C and times around 5 to 10 min while the best overall glucose yield was obtained at 220 °C and 20 min.

Acknowledgements

We acknowledge financial support from Colciencias under Jóvenes Investigadores 2010 program as well as Enlazamundos for the financial support for the Juan C. Quintana-Marín internship in the Laboratory of Renewable Resources Engineering (LORRE) – Purdue University. Also, we thank Linda Liu from LORRE for the help with HPLC analysis.

References

[1] Cardona, C.A. y Sánchez, O.J. (2008). Trends in biotechnological porduction of fuel ethanol from different feedstocks. Bioresource Technology. 99: 5270-5295.

[2] Hamaker, B.R., P. Zhang, et al. (2005). Banana starch: production, physicochemical properties, and digestibility – a review. Carbohydrate Polymers. 59: 443-458.

[3]Velásquez-Arredondo, H.I., Ruiz-Colorado, A.A. and De Oliveira junior, S. (2010). Ethanol production process from banana fruit and its lignocellulosic residues. Energy. 35:3081-3087.

[4]Tock JY, Lai CL, Lee KT, Tan KT, Bathia S. (2010). Banana biomass as a potential renewable energy resource: A Malaysian case study. Renewable and Sustainable Energy Reviews. 14: 798-805.

[5]New York Times. Science. (2012). Bananas and genetic engineering: Past, present and future. http://www.latimes.com/news/science/sciencenow/la-sci-sn-banana-genetics-20120712,0,4914046.story

[6]Zuluaga Robin, Putaux Jean-Luc, Restrepo Adriana, Mondragon Iñaki, Gañán Piedad. (2007). Cellulose microfibrils from banana farming residues: isolation and characterization. Cellulose. 14: 585-592.

[7]Quintana M, Velásquez J, Betancourt S, Gañán P. (2009). Binderless fiberboard form steam exploded banana bunch. Industrial Crops and Products. 29:60-66.

[8] Hosseini, S.A. and Shah, N. (2009). Multiscale modelling of hydrothermal biomass pretreatment for chip size optimization. Bioresource Technology. 100:2621-2628.

[9]Zeng, Meijuan, Mosier, Nathan S., Huang, Chia-Ping, Sheman, Debra M., Ladisch, Michael R. (2007). Microscopic Examination of Changes of Plant Cell Structure in Corn Stover Due to Hot Water Pretreatment and Enzymatic Hydrolysis. Biotechnology and Bioengineering. Vol. 97, No. 2: 265-278.

[10]Tarhezadeh MJ and Karimi K. (2007). Acid-based hydrolysis processes for ethanol from lignocellulosic materials: A review. Bioresources. 2(3): 472-479.

[11] Lu Y. (2008). Kinetic and mechanistic studies of a biomimetic catalyst for hemicellulosic biomass hydrolysis. A dissertation submitted to the Faculty of Purdue University. P:87.

[12] Kim Y, Mosier N, Hendrickson R, Ezeji T, Blascheck H, Dien B, Cotta M, Dale B, Ladisch M. (2008). Composition of corn dry-grind ethanol by-products: DDGS, wet cake, and thin stillage. 99: 5165-5176.

[13]Mittal A, Chatterjee SG, Scott GM, Amidon ST. (2009). Modeling xylan solubilization during autohydrolysis of sugar maple wood meal: Reaction kinetics. Holzforschung. Vol. 63: 307-314.
[14]Garrote G, Domíngquez H, Parajó JC. (2004). Production of Substituted Oligosaccharides by Hydrolytic processing of Barley Husks. Ind. Eng. Chem. Res. 43: 1608-1614.
[15]Donohoe BS, Decker SR, Tucker MP, Himmel ME, Vinzant TB. (2008). Visualizing Ligning Coalescense and Migration Through Maize Cell Walls Following Thermochemical Pretreatment. Biotechnology and Bioengineering. Vol. 101, No. 5: 913-925.
[16] Wahlgren M and Arnebrant T. (1991). Protein adsorption to solid surfaces. Tibtech. Vol. 9: 201-208.
[17]Li X, Ximenes E., Kim Y., Slininger M., Meilan R., Ladisch M., Chapple C. (2010). Lignin monomer composition affects Arabidopsis cell-wall degradability after liquid hot water pretreatment. Biotechnology for biofuels. 3:27
[18]Koostra AMJ, Mosier NS, Scott EL, Beeftink HH. (2009). Differential effects of mineral and organic acids on the kinetics of arabinose degradation under lignocellulose pretreatment conditions. Biochemical Engineering Journal. 43:92-97.
[19] Mittal, A., Chatterjee, S.G., et al. (2009). Modeling xylan solubilization during autohydrolysis of sugar and aspen wood chips: Reaction kinetics and mass transfer. Chemical Engineering Science. 64:3031-3041.
[20] Cara C, Ruiz E, Ballesteros M, Manzanares P, Negro M, Castro E. (2008). Production of fuel etanol from steam-explosion pretreated olive tree pruning. Fuel. Vol. 87, No. 6: 692-700.

(481a) Integration of Post-Combustion Capture with Compression and Liquefaction Process of CO2

Yeong Su Jeong[1], Ung Lee[1], Seeyub Yang[1], Youngsub Lim[1], Chi Seob Lee[2] and Chonghun Han[1], (1)School of Chemical and Biological Engineering, Seoul National University, Seoul, South Korea, (2)KEPCO E&C Company, Yongin, South Korea

Along with the worldwide efforts to reduce the greenhouse gases, Carbon Capture and Storage/Sequestration (CCS) has emerged as one of the most promising technologies in South Korea. Capture process is known to occupy 70 to 80 percent of the overall energy demands in CCS technology, and consequently, most of up-to-date research has been dedicated to this aspect by developing innovative CO_2 absorbents or modifying the process itself to reduce the regeneration energy of the absorbent. Each constituent of the technology can then be integrated with one another in order to mitigate the overall energy consumption. However, there is no possible candidate for CO_2 storage site in Korea at this time, so the capture process must be integrated with an intermediate compression process instead which is always required to transport $CO_2$from capture plant to the storage site.

In this study, a process simulation including both post-combustion capture of CO_2 using MEA absorbent and multi-stage compression was performed using Aspen Plus®. Different physical property methods were selected for each process to allow more stable and accurate simulation. [1] It was assumed that the compressed (liquefied) CO_2 will be transported via CO_2 carrier, which requires appropriate specifications of CO_2 for ship loading. [2] In addition, 4-stage compression system with intercooling was adopted where both cold and hot utilities are available from compression as well as isenthalpic expansion. A quantitative analysis of how these utilities can be used was carried out so that the energy required for capture as well as the overall process was reduced.

This research was supported by a grant from the LNG Plant R&D Center funded by the Ministry of Land, Transportation and Maritime Affairs (MLTM) of the Korean government.

References:

[1] Van Wagener, D.H., Rochelle, G.T., 2011. Stripper configurations for CO_2capture by aqueous monoethanolamine, Chem. Eng. Res. Des. 89, 1639-1646.

[2] Aspelund A., Jordal, K., 2007. Gas Conditioning–The interface between CO_2 capture and transport, Int. J. Greenhouse Gas Control, 1(3), 343-354.

(481b) A Hybrid Model-Based Framework for the HAZOP Analysis of Geological Storage of Carbon Dioxide

Greeshma Gadikota[1], Babji Srinivasan[1], Ah-Hyung Alissa Park[2] and Venkat Venkatasubramanian[1], (1)Chemical Engineering, Columbia University, New York, NY, (2)Earth and Environmental Engineering & Chemical Engineering, Columbia University, New York, NY

With ever increasing energy demand, carbon capture, utilization and storage (CCUS) has become one of the greatest challenges facing the global community since carbonaceous fuels are still the most abundant and inexpensive sources of energy. CCUS is a complex multistep process which involves (i) capture of CO_2 from flue gas generated in coal-fired power plants (ii) transportation of CO_2 using long pipelines to the storage site, and (iii) injection of CO_2 into the geological formations. Each of these steps pose a number of potential risks such as impurities associated with CO_2 capture, leakage in pipelines, corrosion in pipelines, leakage and/or migration of CO_2 in the geological formations due to faults and fractures, contamination of freshwater aquifers etc.Therefore, it is essential to perform a systematic HAZOP analysis for these processes before implementation of any of the CCUS technologies.

In this work, a hybrid model-based framework is proposed to perform HAZOP analysis for CCUS. First, detailed signed directed graph-based qualitative models are constructed for various components with specific emphasis on CO_2 storage. Further, dynamic simulations of the various tasks involved in the complex CCUS process are performed using quantitative models. These quantitative dynamic simulations with an emphasis on CO_2 storage, supplement the overall graph theoretical model built using process-knowledge. Together, these models are utilized to understand the behavior of the entire complex process under various conditions and more importantly, to evaluate the risks present in the overall system. Furthermore, these graph theoretic models with dynamic simulations are explored to perform fault detection, diagnosis and monitoring in the overall CCUS process with emphasis on CO_2 storage in geological media.

(481c) Impact of Alkalinity Sources On the Life-Cycle Energy Efficiency of Mineral Carbonation Technologies

Abby Kirchofer, Earth, Energy, and Environmental Sciences, Stanford University, Stanford, CA, Adam Brandt, Energy Resources Engineering, Stanford University, Stanford, CA, Sam Krevor, Petroleum Engineering, Imperial College, London, England, Valentina Prigiobbe, Petroleum and Geosystems Engineering, University of Texas at Austin, Austin, TX and Jennifer Wilcox, Department of Energy Resources Engineering, Stanford University, Stanford, CA

Impact of alkalinity sources on the life-cycle energy efficiency of mineral carbonation technologies

Abby Kirchofer[1], Adam Brandt[2], Sam Krevor[3], Valentina Prigiobbe[4], and Jennifer Wilcox[2]

[1] Earth, Energy, and Environmental Sciences, Stanford University

[2]Energy Resources Engineering, Stanford University

[3]Petroleum Engineering, Imperial College

[4]Petroleum and Geosystems Engineering, University of Texas at Austin

Mineral carbonation is a Carbon Capture and Storage (CSS) technology where gaseous CO_2 is reacted with alkaline materials (such as silicate minerals and alkaline industrial wastes) and converted into stable and environmentally benign carbonate minerals (Metz et al., 2005). Previous studies on mineral carbonation have principally focused on identifying process conditions optimal for enhancing the chemical rates of reactions. No study, however, has yet performed a scheme that optimizes these conditions with respect to engineering and economic consideration for the goal of producing a viable carbonation process.

Here, we present a holistic, transparent life cycle assessment model of aqueous mineral carbonation built using a hybrid process model and economic input-output life cycle assessment approach. We compared the energy efficiency and the net CO_2 storage potential of various mineral carbonation processes based on different feedstock material and process schemes on a consistent basis by determining the energy and material balance of each implementation (Kirchofer et al., 2011). In particular, we evaluated the net CO_2 storage potential of aqueous mineral carbonation for serpentine, olivine, cement kiln dust, fly ash, and steel slag across a range of reaction conditions and process parameters. A preliminary systematic investigation of the tradeoffs inherent in mineral carbonation processes was conducted and guidelines for the optimization of the life-cycle energy efficiency are provided.

The LCA model allows for the evaluation of the tradeoffs between different reaction enhancement processes while considering the larger lifecycle impacts on energy use and material consumption. All main process stages are included in the tool, and comprehensive system boundaries are applied throughout the model. The process-model core of the mineral carbonation LCA tool includes 8 process stages defined generically to be applicable to a variety of mineral carbonation technologies (see Figure 1). Because our tool aims to compare process schemes that vary significantly in input resource and process design, the model is built at a general, first-order level. The methodology accounts for the following three types of energy consumption: on-site energy consumption, energy of material and energy inputs consumed in the sector of interest (embodied direct energy), and energy of material and energy inputs consumed in all other sectors (embodied indirect energy). Including both on-site and embodied energy allows a full accounting of the total greenhouse gas (GHG) reduction benefits of each process scheme.

Figure 1. Life cycle process model schematic for aqueous mineral carbonation based on the Olivine – 155 ûC case; line thickness is scaled to the energy and mass fluxes.

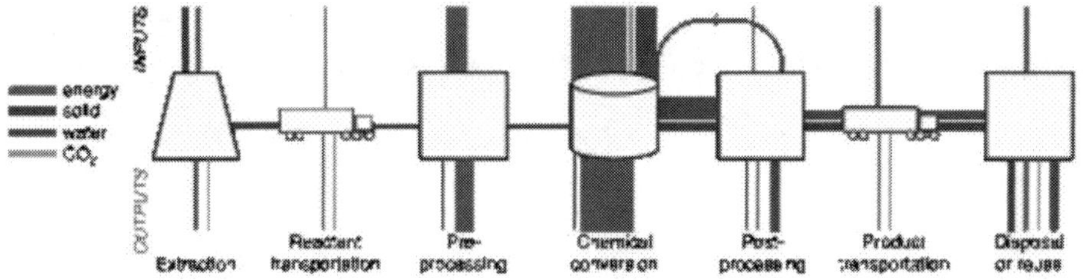

The life-cycle assessment of aqueous mineral carbonation suggests that a variety of alkalinity sources and process configurations are capable of net CO_2 reductions. The maximum carbonation efficiency, defined as mass percent of CO_2 mitigated per CO_2 input, was 83% for CKD at ambient temperature and pressure conditions. In order of decreasing efficiency, the maximum carbonation efficiencies for the other alkalinity sources investigated were: olivine, 66%; SS, 64%; FA, 36%; and serpentine, 13%. Within the range of parameters tested, for all cases maximizing the extent reacted of the alkalinity source unambiguously improves process efficiency. In general, maximizing the extent reacted is critical to optimizing the processes because it minimizes the material handling requirements which contribute negatively to the energy budget. In particular, we observed that:

á the process efficiency is maximized by increasing extent reacted through the most energetically favorable enhancement measure;

á mixing, heating, and grinding are the main energy drivers across all processes;

á reuse of carbonate as aggregate does not necessarily improve life-cycle energy efficiency;

á not all alkalinity sources benefit from high reaction temperatures;

á any steps to increase reaction rates would dramatically improve process efficiency.

Additionally, the CO_2 storage potential of mineral carbonation was estimated using the life-cycle assessment results and alkalinity source availability. The annual storage potential for a given alkalinity source was calculated by multiplying its availability (Mt/yr) by the CO_2 sequestration efficiency of mineral carbonation of that alkalinity source (t-CO_2/t-alkalinity source). For industrial alkalinity sources, availability was based on U.S. production rates (Kelly et al., 2011). For natural alkalinity sources, availability is estimated based on U.S. production rates of a) lime (18 Mt/yr) or b) sand and gravel (760 Mt/yr) (USGS, 2011). The low estimate assumes the maximum sequestration efficiency of the alkalinity source obtained in the current work and the high estimate assumes a sequestration efficiency of 85%. The total CO_2 storage potential for the alkalinity sources considered in the U.S. ranges from 1.3% to 23.7% of U.S. CO_2 emissions, depending on the assumed availability of natural alkalinity sources and efficiency of the mineral carbonation processes.

References

Metz, B., O. Davidson, H. C. de Coninck, M. Loos, and L. A. Meyer (eds.) (2005) IPCC Special Report on Carbon Dioxide Capture and Storage, Prepared by Working Group III of the Intergovernmental Panel on Climate Change. Cambridge University Press, Cambridge, United Kingdom and New York, NY, USA.

Kirchofer, A., A. R. Brandt, S. Krevor, V. Priogiobbe and J. Wilcox (2011) Impact of alkalinity sources on the life-cycle energy efficiency of mineral carbonation technologies. In preparation.

Kelly, K. E., G. D. Silcox, A. F. Sarofim and D. W. Pershing (2011) International Journal of Greenhouse Gas Control, **5**, 1587-1595.

USGS (2011) Mineral Commodity Summaries 2011, United States Geological Survey.

(481d) Uncertainty Quantification in CO2 Sequestration Using Surrogate Models From Polynomial Chaos Expansion

Yan Zhang and Nick Sahinidis, Chemical Engineering, Carnegie Mellon University, Pittsburgh, PA

In this paper, surrogate models are iteratively built using polynomial chaos expansion (PCE) and detailed numerical simulations of a carbon sequestration system. Output variables from the numerical simulator are approximated as polynomial functions of uncertain parameters. Once generated, PCE representations can be used in place of the numerical simulator and often decrease simulation times by several orders of magnitude. However, PCE models are expensive to derive unless the number of terms in the expansion is moderate, which requires a relatively small number of uncertain variables and a low degree of expansion.

To cope with this limitation, instead of using a classical full expansion at each step of an iterative PCE construction method [1], we introduce a mixed-integer programming (MIP) formulation to identify the best subset of basis terms in the expansion. Our approach makes it possible to keep the number of terms small in the expansion. We compare the proposed MIP-based subset selection method against the classical stepwise regression algorithm for building sparse PCE models [2]. The PCE models from our proposed method are simpler than the ones using stepwise regression.

Monte Carlo (MC) simulation is then performed by substituting the values of the uncertain parameters into the closed-form polynomial functions. Based on the results of MC simulation, the uncertainties of injecting CO2 underground are quantified for a saline aquifer. Moreover, based on the PCE model, we formulate an optimization problem to determine the optimal CO2 injection rate so as to maximize the gas saturation (residual trapping) during injection, and thereby minimize the chance of leakage.

References cited:

1. Oladyshkin, S.; Class, H.; Helmig, R.; Nowak, W. An integrative approach to robust design and probabilistic risk assessment for CO2 storage in geological formations. Computational Geosciences 2011, 15, 565–577.

2. Blatman, G.; Sudret, B. An adaptive algorithm to build up sparse polynomial chaos expansions for stochastic finite element analysis. Probabilistic Engineering Mechanics 2010, 25, 183-197.

(481e) Geo-Chemo-Mechanical Studies for Permanent CO2 Storage in Geologic Reservoirs

Greeshma Gadikota[1], Juerg Matter[2], Peter Kelemen[2] and Ah-Hyung Alissa Park[3], (1)Chemical Engineering, Columbia University, New York, NY, (2)Earth and Environmental Sciences, Columbia University, Palisades, NY, (3)Earth and Environmental Engineering & Chemical Engineering, Columbia University, New York, NY

Increasing concentration of CO_2 in the atmosphere is attributed to rising consumption of fossil fuels around the world. The development of solutions to reduce CO_2 emissions to the atmosphere is one of the most urgent needs of today's society. CO_2 injection into geological formations is one of the CO_2 storage options with a large capacity. If non-carbonate minerals such as basalt and olivine exist in the CO_2 injection site, the injected CO_2 can react with those minerals to form thermodynamically stable solid carbonates. In nature, mineral carbonation takes place via dissolution of the mineral to release Ca/Mg ions followed by precipitation of Ca/Mg in the presence of CO_2. The rate limiting step of this carbon mineralization process is considered to be the dissolution of the mineral.A question has been raised regarding the effect of in-situ carbon mineralization on the stability of the geologic storage of CO_2. Both the kinetics and extents of mineral dissolution and carbonation reactions within the geological reservoir are important factors to evaluate for the long term stability of the geologically injected CO_2.Therefore, this work focuses on understanding CO_2-mineral-water interactions at a wide range of reaction conditions (T_{max} = 200°C and $P_{CO2, max}$ = 200 bar). Both deionized water and synthetic brine were used as the aqueous reaction media. Minerals such as olivine (Mg_2SiO_4), labradorite ($(Ca,Na)(Al,Si)_4O_8$) and basalt (mixture of silicate minerals), were selected based on their abundance in nature and both dissolution and single-step carbonation experiments were performed to investigate fast and long-term kinetics of mineral weathering. Two different regimes (i.e., surface reaction vs. mass transfer limited regimes) of mineral weathering were identified based on the analyses of the activation energy and reaction kinetics.

(481f) Comparison of CO2 Storage Resource Methodologies

Angela Goodman, National Energy Technology Laboratory, Pittsburgh, PA and Grant Bromhal, National Energy Technology Laboratory, Morgantown, WV

Estimating storage resources for geologic CO_2 sequestration requires a high-level, first-cut assessment of potential reservoirs that can serve as viable storage sites. CO_2 storage resource estimates are based on physically accessible CO_2 storage volume in formations without consideration of regulatory or economic constraints. These high-level resource assessments may be used for broad energy-related government policy and business decisions. Subsurface storage volume estimates depend on geologic properties (area, thickness, and porosity of formations) and the efficiency of storage (the fraction of the accessible pore volume that is available to injected CO_2). Currently there is no standard method to estimate CO2 storage potential and reported CO2 estimate widely vary. In this work, publically available resource methodologies for predicting how much CO2 can be stored in deep saline formations are compared. Resource methodologies applied are from the United States Department of Energy, United States Geological Survey, Carbon Sequestration Leadership Forum, Massachusetts Institute of Technology, and Lawrence Berkeley National Laboratory. The methodologies are applied to formations throughout the United States.

(482a) Effects of Sodium Sulfide Supplementation On Alkaline Pretreatment of Herbaceous Biomass

Venkata Ramesh Pallapolu, Chemical Engineering, Auburn University, Auburn University, AL and Y. Y. Lee, Department of Chemical Engineering, Auburn University, Auburn University, AL

In alkaline pretreatments of biomass, it is feasible to retain most of carbohydrates while removing a large fraction of lignin. The enhancement of digestibility is caused primarily by lignin removal. Retention of hemicellulose after pretreatment gives a significant economic benefit since it eliminates the need of detoxifying hemicellulose sugars. Sulfur containing reagents are highly reactive with lignin, thus can serve as delignification enhancer. We have investigated the effects of adding small amount of sodium sulfide in alkaline pretreatment of herbaceous biomass. The alkaline reagents tested were either recoverable ones (ammonia, sodium carbonate) or inexpensive one (lime). Pretreatments were applied to switchgrass, corn stover, and palm residue with and without supplementation of sodium sulfide to see the isolated effects of sodium sulfide. Pretreatments were done in batch mode under low severity conditions: 90 - 120 °C, 12 - 24 hr reaction time, 5 - 15 % alkali, 1 - 3% sodium sulfide, and 5~10:1 of L:S ratio. The treated substrates were put through enzymatic digestibility tests using commercial cellulase (Novozyme CTec 2). Addition of sodium sulfide did not change carbohydrates retention, but increased delignification by 20 - 25 %. Addition of sodium sulfide has significantly increased digestibility for all alkali treatments: 15 – 45 % in glucan digestibility and 10 – 40 % in xylan digestibility depending on type of substrate and pretreatment severity. The results were further assessed to determine the optimum set of pretreatment conditions considering carbohydrate retention and digestibility as criteria. Details of the performance data are also presented.

(482b) Impacts of Cell Wall Hydrophilicity On the Enzymatic Deconstruction of Diverse Biomass Feedstocks and Pretreatments

Daniel L. Williams and David Hodge, Chemical Engineering and Material Science, Michigan State University, East Lansing, MI

This work will present research results relating the environment of the water in the cell wall to the potential for enzyme-catalyzed depolymerization of plant cell wall polysaccharides from diverse plant cell walls treated with a range of pretreatments. These biomass feedstocks will include grasses using alkaline hydrogen peroxide (AHP) as a chemical pretreatment or as a post-treatment following liquid hot water pretreatment (LHW) as well as hardwoods delignified by either soda pulping or organosolv pulping. The effect of pretreatment on the carbohydrate and lignin composition after pretreatment and fiber swelling behavior will be shown. We hypothesize that the digestibility improvement resulting from may be attributed to mild oxidation or solubilization of the lignin remaining in the cell wall and, for grasses, destruction of ferulate crosslinks between cell wall polymers which would have the net effect of increasing the hydrophilicity to allow improved water and enzyme penetration into the cell wall. To test this hypothesis, we will correlate the hydrophilicity of the cell wall as measured indirectly by lignin and cell wall carboxylic acid content and total lignin content to water swelling capacity and digestibility. Additionally the effects of shear-induced fibrillation and particle size reduction on water will be quantified as well.

(482c) Demonstration Scale Alkali Passivation for the Pyrolysis of Biomass

Lysle Whitmer, Center for Sustainable Environmental Technologies, Iowa State University, Boone, IA and Robert C. Brown, Mechanical Engineering, Iowa State University, Ames, IA

Demonstration Scale Alkali Passivation for the Pyrolysis of Biomass

Lysle Whitmer and Robert C. Brown

Center for Sustainable Environmental Technologies

Department of Mechanical Engineering

Iowa State University

Ames Iowa – 50010 U.S.A.

Abstract

The objective of this research is to evaluate the ability to passivate alkali laden biomass feedstock for use in thermochemical conversion processing at demonstration scale. Laboratory Pyrolysis-GC/MS results indicate that naturally occurring mineral matter found in all biomass catalyzes the decomposition of cellulose during pyrolysis to yield light oxygenated compounds in bio-oil such as formic acid and acetaldehyde, rather than uncatalyzed pyrolysis which tends to depolymerize the cellulose to produce anhydrosugars[1]. Acid infusion of the biomass has been shown to form stable salts with the naturally occurring alkali and alkaline earth metals (AAEM), rendering the AAEM less catalytically active and producing bio-oil that is rich in anhydrosugars.

This research focuses on the scale up of AAEM passivation technology to produce consistently pretreated biomass feedstock in large quantities suitable for demonstration scale reactors. A number of independent variables have been investigated and a response surface methodology (RSM) developed for the pretreatment technology. Micro-pyrolysis and the subsequent analysis of the products produced from the pretreated biomass samples served as a proxy for effective passivation.

1. Kuzhiyil, N., et al., *Pyrolytic Sugars from Cellulosic Biomass.* Submitted.

(482d) Examination of Biomass Composition, Structure, and Enzymatic Digestibility As a Function of Ionic Liquid Pretreatment Temperature

Christopher J. Barr[1], Kevin Click[2], Paul Langan[3], B. Leif Hanson[4] and Constance Schall[1], (1)Chemical and Environmental Engineering, University of Toledo, Toledo, OH, (2)Chemistry, Kent State University, Kent, OH, (3)Biosciences Division - Los Alamos National Laboratory, Los Alamos, NM, (4)Instrumentation Center, The University of Toledo, Toledo, OH

Deconstruction of lignocellulosic biomass into simple sugars constitutes a core barrier for producing value added products from the sugar platform. Lignocellulosic biomass is primarily comprised of cellulose (a glucose polymer), hemicellulose (a xylose polymer), and lignin. Enzymatic hydrolysis of lignocellulosic biomass is one of the crucial steps in biochemical conversion to monomeric sugars. The native cellulosic portion of biomass has a crystalline and oriented fibrillar structure in the cellulose I crystal form. Cellulose I can be a major impediment for enzymatic hydrolysis. Ionic Liquid (IL) pretreatment can disorder the crystalline cellulosic component to an amorphous material or result in structural changes in the cellulose fibers to other crystal forms, such as cellulose II. In these studies, the effects of IL pretreatment temperature on various sources of lignocellulosic biomass were examined using X-ray powder and fiber diffraction, compositional analysis and comparative enzymatic hydrolysis. Polysaccharide digestibility was related to the observed structural and compositional changes in biomass. The improvement in hydrolysis is hypothesized to be primarily due to structural changes in the cellulose constituent.

(482e) Flash Hydrolysis for Producing Algal Hydrolysate and Its Application in Enzymatic Saccharification of Lignocellulosic Biomass

Jose Garcia, Old Dominion University, Norfolk, VA, Rui Chen, Department of Biosystems & Agricultural Engineering, Michigan State University, East Lansing, MI, Benjamin Thomas, Department of Biosystems & Agricultural Engineering, Michigan State University, East Lansing, VA, Sandeep Kumar, Civil & Environmental Engineering, Old Dominion University, Norfolk, VA and Wei Liao, Biosystems and Agricultural Engineering, Michigan State University, East Lansing, MI

Flash hydrolysis using subcritical water is a novel and versatile process for selectively hydrolyzing polymeric components of algae biomass. In this study, our objective is to use high water content of algal biomass advantageously for extracting proteins and carbohydrates in a short residence time (few seconds). The present study was focused on the extraction of proteins from *Scenedesmus spp.* using subcritical water in a continuous flow reactor. The effect of temperature and residence time on the yield of peptides and free amino acids in liquid hydrolyzate are studied. About 60-80 wt% of proteins in microalgae could be extracted within few seconds in the temperature range of 200-325°C under subcritical water conditions. The algal samples were first characterized for biochemical compositions and then optimal conditions for protein extraction using subcritical water were determined.

Algal hydrolysate obtained via flash hydrolysis at different temperatures was investigated for its use as reaction medium during saccharification of lignocellulosic biomass. The objective was to replace conventionally used buffer solutions and optimize the enzymatic hydrolysis of biomass in the presence of nutrient rich algal hydrolyzate. It is understood that algal hydrolyzate is an additional protein source to the reaction medium that are irreversibly absorbed by lignin present in lignocellulosic biomass. This helps in increasing the hydrolysis rate by reducing the interference in cellulase activity.

(482f) Effects of Dilute Acid Pretreatment On Cellulose DP and Factors Related to Cellulose Digestibility

Wei Wang, Xiaowen Chen, Ashutosh Mittal, Rui Katahira, Melvin P Tucker and David K. Johnson, National Renewable Energy Laboratory, Golden, CO

Several chemical and physical factors have been proposed to be related to biomass recalcitrance. The degree of polymerization (DP) of cellulose has been considered to be one of the most important properties affecting the enzymatic hydrolysis of cellulose. However, there are few studies investigating the effect of pretreatment on cellulose DP in biomass and the relationship between cellulose DP and enzymatic hydrolysis. This study of the effect of dilute acid treatment on cellulose DP was first started with various pure cellulose materials. A substantial reduction in DP was found for all cellulosic materials studied even at conditions that would be considered relatively mild for the pretreatment. Different cellulosic materials decreased to a similar DP upon treatment with acid regardless of the significant differences in the starting DP of the celluloses. The effect of dilute acid pretreatment on cellulose DP in biomass samples was also investigated. The cellulose DP in corn stover was also found to decrease greatly on acid pretreatment, however not to the extent observed with pure cellulosic samples. Acid pretreated corn stover was found to contain cellulose with DPw of 1600 - 3500, significantly higher than the DPw found with acid treated pure celluloses (150 – 300). The digestibilities of various acid-treated and untreated cellulosic and biomass samples with differing cellulose DP were tested with a cellulase complex (Genencor GC 220). Other cellulose properties, such as particle size and crystallinity were also monitored during digestions. There was little difference in the cellulose conversion for the treated and untreated celluloses despite their very large differences in DP. From this study it does not appear that cellulose DP is a main factor affecting cellulose saccharification. Other properties including particle size, xylan removal, pore volume etc., were also measured for the pretreated biomass samples. A statistical analysis was performed on all these factors to determine which were the most significant in determining cellulose digestibility.

(498a) Hydrolysate Fermentation Characterization for Xylose-Fermenting Saccharomyces Cerevisiae Strains Derived From Directed Evolution for Improved Xylose Utilization and Tolerance to Inhibitors

Tongjun Liu, GLBRC, East Lansing, MI; Chemical Engineering and Materials Science, Michigan State University, E. Lansing, MI, Trey Sato, University of Wisconsin, Great Lakes Bioenergy Research Center, Madison, WI and **David Hodge**, Chemical Engineering and Material Science, Michigan State University, East Lansing, MI

The goal of this work is to quantify the spectrum of pretreatment-derived compounds from alkaline hydrogen peroxide (AHP) pretreatment of grasses, to quantify their impact on fermentation of high sugar content hydrolysates on *Saccharomyces cerevisiae* strains engineered to ferment xylose by introduction of xylose fermentation capabilities by introduction of either xylose reductase and xylitol dehydrogenase or the xylose isomerase followed by evolution for improved xylose utilization and AHP inhibitor tolerance, and to demonstrate improved performance and successful co-fermentation of undetoxified hydrolysates. Specifically, we were able to determine that the classes of inhibitors included inorganics (Na^+ from NaOH and sulfate used for pH adjustment) and solubilized organics (aromatics, acetate, and oxidation prodcucts including formate). It was found that inorganics were not significantly inhibitory at concentrations of 0.5 M Na^+ which would be present in hydrolysates from high-solids pretreatment and hydrolysis. We were able to identify the soluble aromatics and especially the hydroxycinnamic acids as the major contributor to fermentation inhibition. We are able to demonstrated complete sugar utilization from undetoxified AHP corn stover hydrolysates from high-solids pretreatment and enzymatic hydrolysis containing more than 120 g/L of glucose and xylose with conversion to ethanol at high yields with minimal byproduct formation.

(498b) Conversion of E-AFEX(TM) Pretreated Corn Stover to Ethanol by Bcrl Fast SHF Process

Ming Jie Jin, Yanping Liu, Nirmal Uppugundla, Cory Sarks, Leonardo da costa Sousa, Bruce E. Dale and **Venkatesh Balan**, Chemical Engineering and Material Science, Great Lakes Bioenergy Center, Michigan State University, Lansing, MI

Extractive Ammonia Fiber Expansion (E-AFEX^{TM}) pretreatment convents cellulose I to cellulose III and removes part of the lignin in the biomass. Hence, E-AFEX^{TM} pretreated biomass typically has a higher enzyme digestibility compared to AFEX^{TM}. So far, E-AFEX^{TM} has not been fully studied in terms of fermentability and ethanol production potential.

Here we present the performance of BCRL fast SHF process (*Jin et.al, 2012, Energy Environ. Sci., 2012, 5(5), 7168-7175.*) on E-AFEX^{TM} pretreated corn stover for ethanol production. We are aiming to achieve 2-3 times higher productivity with an enzyme loading as low as 8 mg/g glucan and similar or higher ethanol yield when compared to current technology. BCRL fast SHF process has been studied on regular AFEX^{TM} pretreated corn stover and showed high ethanol productivity and reduction of more than 1/3 enzyme loading. In the BCRL fast SHF process, the easily hydrolyzed part of biomass is converted to ethanol first in a 24 h enzymatic hydrolysis and 24 h fermentation process. This way the slow rate period of enzymatic hydrolysis is avoided. The liquid hydrolysate after fermentation is subjected to distillation. The residue solids (difficultly hydrolyzed part of biomass (DHPB)) with some enzymes adsorbed are transferred to the next cycle enzymatic hydrolysis for further hydrolysis. By this approach, parts of the enzymes are reused. After five cycles, the accumulated recalcitrant biomass is further hydrolyzed in the last step. With less degradation products and low initial sugar concentration, the hydrolysis of DHPB is easier to be hydrolyzed. For fermentation, we use high inoculation size (OD_{600} =20) and recycle the yeast cells after fermentation. Details about the condition used and process optimization will be discussed during the presentation.

(498c) High Titer Ethanol Production From SPORL Pretreated Lodgepole Pine by Simultaneous Enzymatic Saccharification and Combined Fermentation (SSCombF) of Enzymatic and Pretreatment Hydrolysates

Tian-Qing Lan, School of Light Industry and Food Sciences, South China University of Technology, Guangzhou, China, Bruce S. Dien, USDA-ARS NCAUR, Peoria, IL, Ronald Hector, NCAUR, USDA ARS, Peoria, IL, Rolland Gleisner, US Forest Service, Madison, WI and J.Y. Zhu, USDA Forest Service, Forest Products Laboratory, Madison, WI

We will present high titer ethanol production of 47 g/L from softwood lodgepole pine by SPORL pretreatment using S. cerevisiae YRH400. SPORL pretreatment was conducted at 180C for 20 min with solids loading of 25%. Sulfuric acid and sodium bisulfite charge on oven dry wood were 2.2 and 8 (wt/wt)%, respectively. Simultaneous enzymatic saccharification of SPORL pretretaed solids and combined fermentation with pretretament liquor were conducted at total soldis of 20% to achieve ethanol yield of 285 L/ton. The SPORL pretreated solids were not washed and combined with pretretament liquor directly in enzymatic saccharification. The CTec2 loading was only 10 FPU/g solids. The liquor was conditioned using lime to adjust pH and evaporated to achieve total solids of 20% in processing. This liquor conditioning process reduced acetic acid and furfural concentration, but did not significantly affect HMF, the major inhibitor derived from sofwood. No sugar loss was observed through evaporation. This study demonstrated the fisrt step process integration by eliminating substrate washing, very important to reduce water usage, without using high enzyme loadings.

Acetone-Butanol-Ethanol fermentation in a continuous and closed-circulating fermentation system with PDMS membrane bioreactor

paper number:254905

Chunyan Chen, Zeyi Xiao, Haidi Cui, Junqing Zhang, Weijia Li, Chao Ying

Introduction

The production of n-butanol with solvent-producing strains of *clostridium*, commonly known as ABE (acetone-butanol-ethanol) fermentation is one of the largest biotechnological processes ever developed [1]. Butanol and acetone are excellent organic solvents and important chemical substances, which are widely used in chemical industry, such as pharmaceuticals, agrochemicals and organic syntheses. In recent years, butanol is regarded as genuine biofuel due to its considerably higher combustion value than ethanol and blending in fuels more readily than ethanol [2,3].

Many researches on ABE fermentation have been indexed. Qureshi et al investigated butanol production from wheat straw by simultaneous saccharification and fermentation in five different batch fermentation processes [4]. Cheng et al studied bio-butanol production by solvent-producing bacterial microflora and the maximum butanol concentration of 12.4 gL^{-1} was obtained in batch fermentation [5]. However, as known, the economics of the ABE bio-fermentation are hampered by the high cost of substrates and butanol toxicity leading to cell inhibition, low product accumulation, low space-time-yields and high purification costs [6,7].

To relieve butanol inhibition and improve the fermentation efficiency, several separation technologies have been investigated for extracting products and controlling their concentration in fermentation broth, such as distillation [8], gas stripping [9], extraction [6], adsorption [10], perstraction [11] and pervaporation [12]. Based on energy requirements of the separation steps, adsorption and pervaporation were indentified as most promising unit operations for in situ product recovery of butanol. The use of pervaporation for product recovery from biomass fermentation processes has been comprehensively reviewed by Vane [13].

In the earlier work, a system integrating fermentation with in situ pervaporation, named as continuous and closed-circulating fermentation (CCCF) system was developed for aerobic ethanol production [14,15]. The results showed the feasibility and prospect of the long term CCCF system in ethanol production with improved fermentation performance. However, it was another challenging and interesting subject whether the CCCF system is suitable for the anaerobic butanol fermentation. In this work, two cycles of CCCF experiment with different coupling operation modes were conducted, then both the overall kinetic parameters and time course profiles of fermentation were explored, intending to find the process principles of ABE fermentation in this specific system.

Methods

Culture and inoculum preparation

C.acetobutylicum strain CICC 8012 was obtained from China Center of Industrial Culture Collection (Beijing, China). Initially, sporulated cells were activated by heat shock at 100℃ for 1 min. The activated spore culture of 5 ml was inoculated to 100 ml sterile medium in a glass bottle and cultivated anaerobically for 24 h at 37℃. The medium used for seed culture and fermentation was identical and contained (for 1L) glucose 60 g, yeast extraction 3 g, phosphate buffer (KH_2PO_4 0.5 g, K_2HPO_4 0.5 g, ammonium acetate 2.2g), vitamins (para-amino-benzoic acid 1mg, thiamin 1mg, biotin 0.01mg), mineral salts ($MgSO_4 \cdot 7H_2O$ 0.2 g, $MnSO_4 \cdot H_2O$ 0.01 g, $FeSO_4 \cdot 7H_2O$ 0.01 g, NaCl 0.01 g). Glucose and yeast extraction were sterilized by autoclaving at 121 ℃ for 20 min. Phosphate buffer, vitamins and mineral salts were filter-sterilized through 0.45 μm sterile membrane.

CCCF integrated with pervaporation

The CCCF set-up was adopted from the previous study with some modifications [14]. Experiments were carried out in a 10 L jacketed glass bioreactor with the fermentation volume of 8 L. After sparging with nitrogen, the bioreactor was inoculated with 10% inoculums of highly motile cells of *C.acetobutylicum* CICC 8012. Two cycles of fermentation with different operation modes were conducted in the experiments. For the first cycle, the operation mode was of fermentation intermittent coupling with pervaporation, which means fermentation was running for 24 h every day, while pervaporation was only operated from 8:30 to 20:30 (12 h) each day. For the second cycle, the operation mode for the first 192 h was continuous coupling, which means both fermentation and pervaporation were running for 24 h every day. Then, the operation mode was shifted to intermittent coupling which was the same as the first cycle. During the two cycles of fermentation, the temperature of pervaporation and fermentation was the same 37℃ controlled by a thermostat, and the pH value was adjusted to no lower than 4 with ammonia water of 25 wt%. Glucose and distilled water were added every 2 h in order to maintain the glucose concentration of about 30 gL^{-1} and broth volume of 8 L. The permeation vapor released at the membrane downstream (with an absolute pressure of 2000 Pa) was collected in cold traps immersed in a -30℃ refrigerator.

Pervaporation membrane and module

The pervaporation membrane used here was in-house developed composite polydimethylsiloxane (PDMS) membrane, and two kinds of modules were applied in the two cycles of CCCF respectively. A circular flat-plate membrane module providing an effective membrane area of 0.024 m^2 was used in the first cycle of CCCF, and a square flat-plate membrane module providing an effective membrane area of 0.08 m^2 was used in the second cycle. The preparation procedure of the membrane and the structure of the module had been fully described in the previous work [16,17].

Analysis

Biomass concentration was determined in three terms of dry cell weight, cell number and Optical Density (OD), respectively. For dry cell weight, the broth sample was first centrifuged at 12000 rpm for 5 min, then cells at the bottom was washed with distilled water and finally dried in an oven at 80℃. For cell number, broth sample was moderately diluted and then counted by a microscopic in a counting chamber after staining with lodine solution. For OD, the broth sample was directly determined by a spectrophotometer at 540 nm. Solvents (acetone, butanol, ethanol) were quantified by gas chromatography (GC112A, shanghai, China) equipped with a FFAP capillary column (30m×0.32mm×0.5μm) with flame ionization. Injector temperature 180℃, detector temperature 200℃, column temperature linearly ramping from 60 to 180℃ at 20℃ per min. Nitrogen was used as the carrier gas. Isobutanol was used as an internal standard, and the injection volume was 1 μL. The concentrations of glucose were determined using the 3,5-dinitrosalicyclic acid (DNSA) method [18].

Results and discussion

Cell growth profiles of ABE fermentation

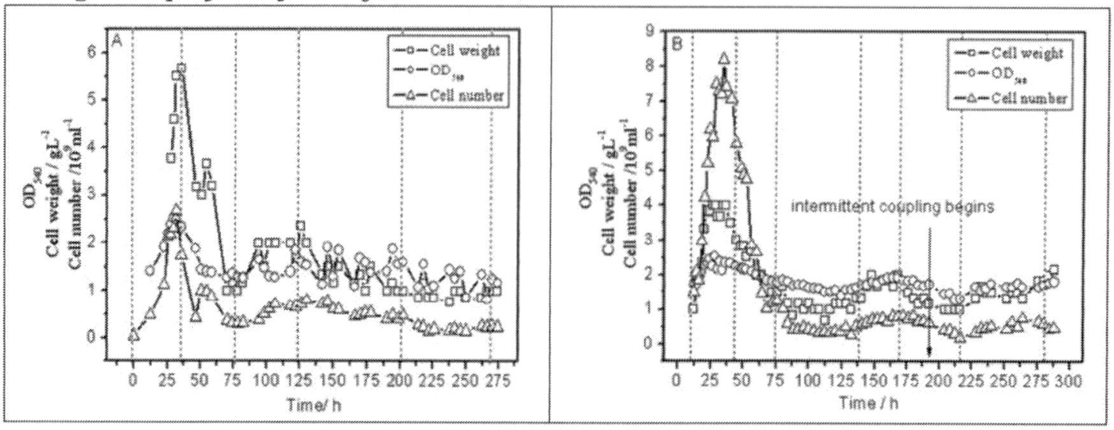

Fig. 1 Time course profiles of dry cell weight, cell number and OD_{540} in two cycles of CCCF. (A) the first cycle; (B) the second cycle.

Three different measurements for microbial concentration/density, dry cell weight, optical density and cell number, were used to describe the cell growth. Time course profiles of those measurements during two cycles of CCCF were shown in Fig. 1. These three measurements were all indicating a similar feature of the cell growth. As described from Fig. 1, the time course profiles of cell growth in the CCCF were very different from those in batch fermentation. For example, the time course profile of cell growth in the first cycle could be roughly divided to 5 phases. From 0 to 30 h (phase 1), cells grew rapidly with dry cell weight increasing to the maximum of 5.6 gL^{-1}, but it dropped sharply to 1.33 gL^{-1} in phase 2 (from 30 to 75 h). Interestingly, the cell concentration increased again from 75 to 125 h (phase 3), and doubled at the maximum of the phase. However, the microorganism decreased again in the phase 4 (from 125 to 225 h). After 225 h (start of phase 5), the cell concentration remained steady, with the average dry cell weight of 1.33 gL^{-1}. For the second cycle, the cell growth profile was similar to the first cycle except for the additionally steady phase 3 between the first decrease phase and the second increase phase (Fig. 1B). However, both cycles had two peaks in the profile of cell growth, which means the microorganism could adapt the CCCF and be surviving in the deteriorated environment in the PDMS membrane bioreactor.

Fermentation kinetic profiles of ABE fermentation

Time course profiles of ABE concentrations in the broth showed two peaks (Fig. 2), at near 40h and at near180h, respectively. This also indicated the adaptable and survival feature of *C.acetobutylicum* CICC 8012 in the long-term CCCF. Besides, compared Fig. 1 to Fig. 2, the cell concentration began to decrease when butanol concentration was over 6.5 gL^{-1} in the first cycle and higher than 8.5 gL^{-1} in the second cycle separately. Obviously, this was the direct evidence that high butanol concentration inhibits the growth of microorganism. Note that the original strain of the second cycle was the descendant of the first cycle, therefore, the higher inhibition concentration of the second cycle may be partly attributed to the adaptive evolution of the microorganism in the long-term fermentation.

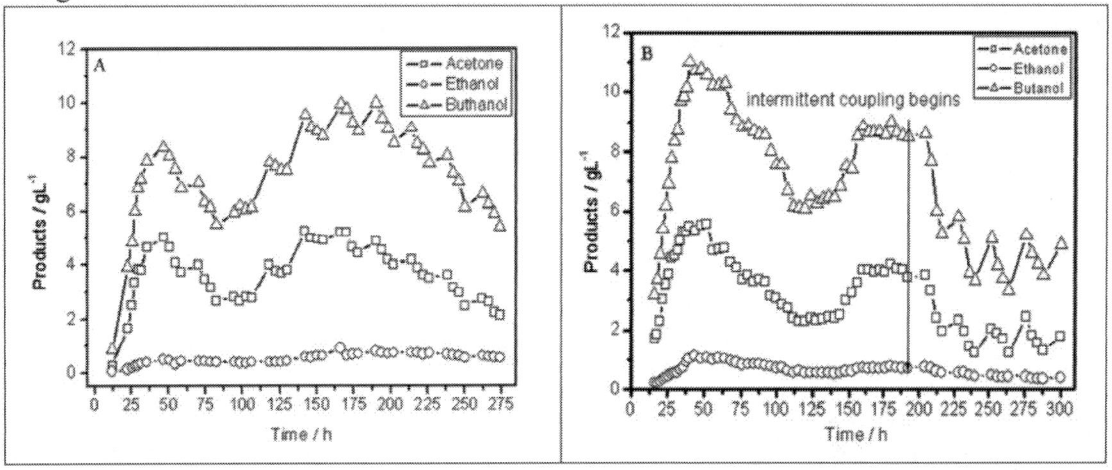

Fig. 2 Time course profiles of ABE concentrations in the broth of two cycles of CCCF. (A) the first cycle; (B) the second cycle.

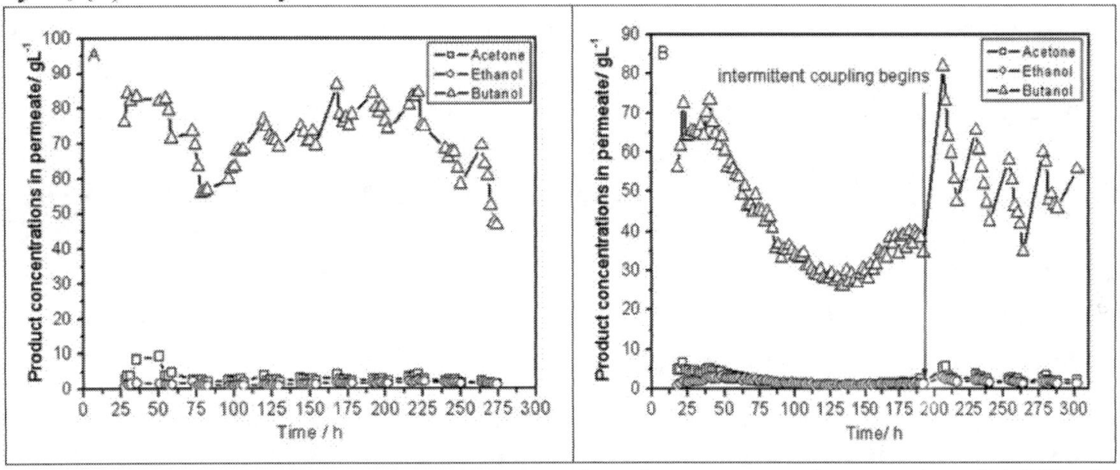

Fig. 3 Time course profiles of ABE concentrations in permeate of two cycles of CCCF. (A) the first cycle; (B) the second cycle.

As mentioned in section 3.1, the butanol in permeate could be totally condensed and collected by the cold trap, but not the acetone and ethanol. It was early proved that the composite PDMS membrane used here had stable inherent performances, saying permeation flux and selectivity, for separation of a given organic substance from its aqueous mixture. In Fig. 2 and Fig. 3, the butanol concentration profiles behaved a similar status, which imply an inherent relationship between the concentrations in broth and in permeate. Nevertheless, it seems that such an inherent relationship for acetone and ethanol could not be found. This supports the observation that the acetone and ethanol in permeate could not be completely condensed and collected. On the other hand, the maximum and average butanol concentration in permeate was 84.41 and 70.96 gL^{-1} in the first cycle, and that in the second cycle was 81.87 and 46.42 gL^{-1} respectively. Compared to the butanol concentration in the broth no higher than 12 gL^{-1}, the highly concentrated production in the permeation could greatly reduce the subsequent cost of recovery. Besides, due to the different membrane modules with different effective membrane areas used in the two cycles, the average butanol concentration in permeate of the first cycle was higher than that in the second cycle as shown in Fig. 3.

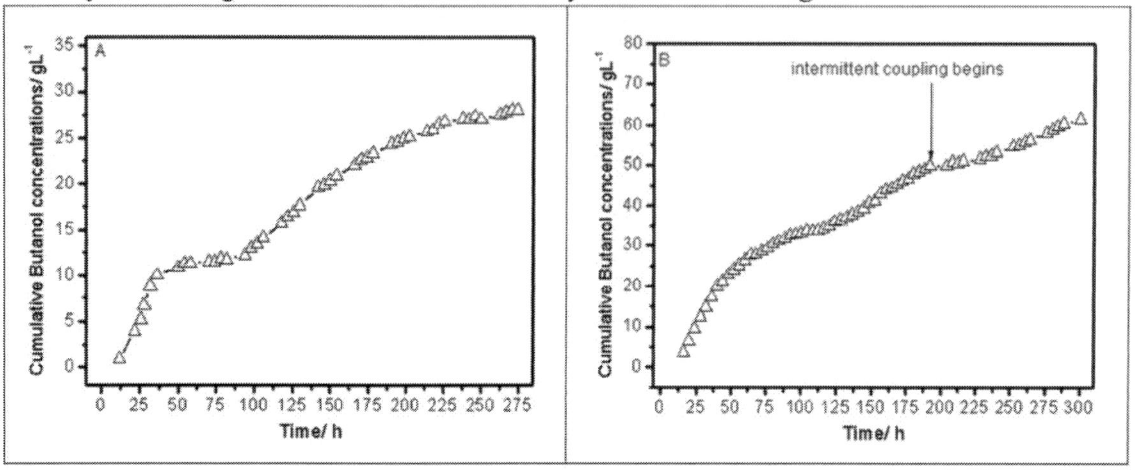

Fig. 4 Time course profiles of cumulative butanol concentrations in two cycles of CCCF. (A) the first cycle; (B) the second cycle.

Figure 4 shows the cumulated butanol amounts in the broth during the fermentation, which was expressed as butanol concentration in the broth. It was reported that butanol concentration of 12.4-21.6 gL^{-1} and butanol productivity of 0.25-0.30 $gL^{-1}h^{-1}$ could be obtained in traditional batch fermentation with duration of about 72 h [4,5]. Correspondingly, during previous 72h, the cumulative butanol concentration of 28.61 gL^{-1} and butanol productivity of 0.40 $gL^{-1}h^{-1}$ had been achieved in the second cycle (Fig. 4B). The higher fermentation efficiency was considered to be attributed to the fermentation-pervaporation coupling in the specific CCCF system. Moreover, the final cumulative butanol concentration of 28.03 and 61.43 gL^{-1} in the two cycles of CCCF far exceeded the butanol inhibition concentration, which showed great superiority over the batch fermentation.

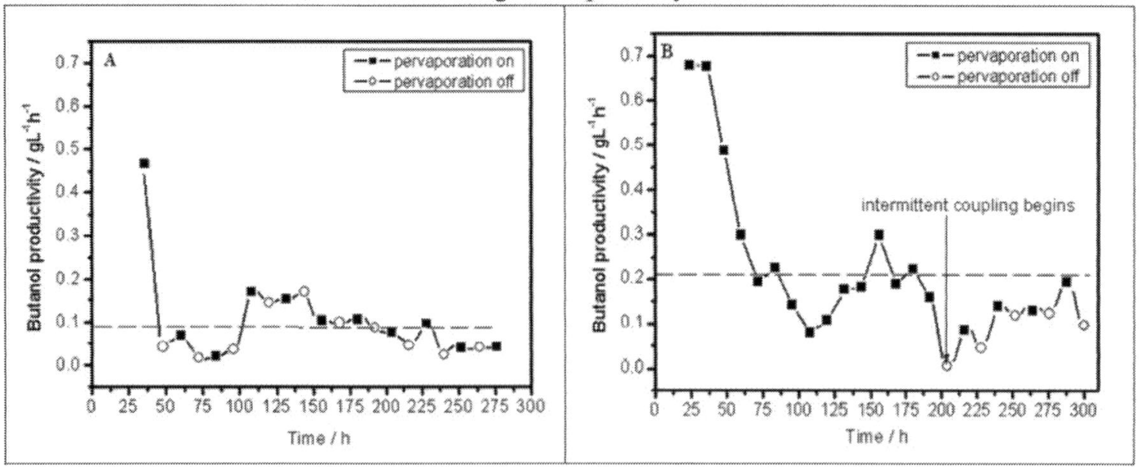

Fig. 5 Time course profiles of butanol productivity in two cycles of CCCF. (A) the first cycle; (B) the second cycle.

Time course profiles of butanol productivity in the two cycles were shown in Fig. 5, the maximum butanol productivity of 0.47 $gL^{-1}h^{-1}$ with an average value of 0.11 $gL^{-1}h^{-1}$ were obtained in the first cycle of CCCF. And the corresponding maximum and average butanol productivity in the second cycle were 0.69 and 0.21 $gL^{-1}h^{-1}$ respectively, which were much higher than those of the first cycle due to the bacterial adaptability and different fermentation-pervaporation coupling modes. The same case was observed for glucose consumption rate in Fig. 6, with the average value of 0.63 and 1.12 $gL^{-1}h^{-1}$ in the two cycles respectively. Similarly, the time course profiles of fermentation kinetic parameters exhibited two peaks, which also indicated the reproducing phenomenon of the clostridium. Besides, the time course profiles of the butanol productivity and glucose consumption rate showed nearly same trend in each cycle, which indicated a constant butanol yield rate and stable fermentation system in the process of CCCF.

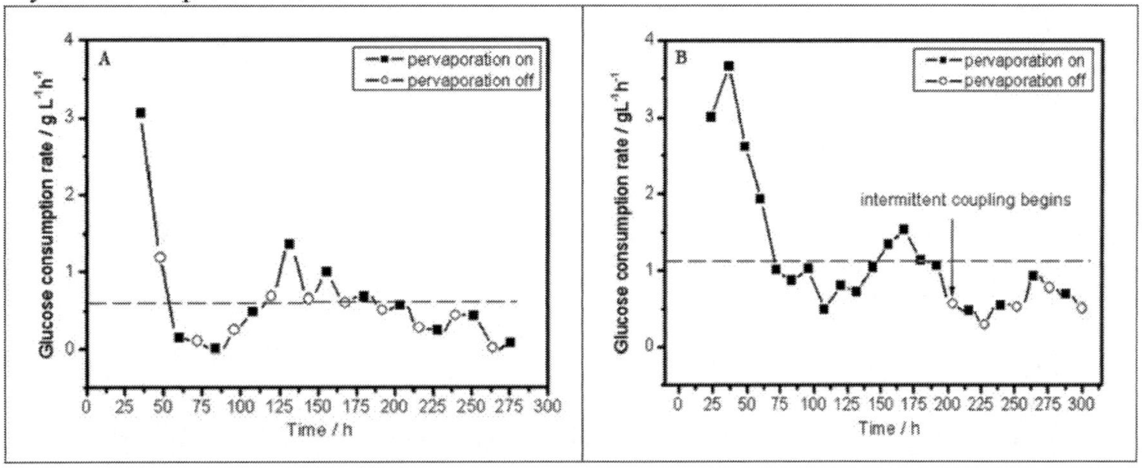

Fig. 6 Time course profiles of glucose consumption rate in two cycles of CCCF. (A) the first cycle; (B) the second cycle.

On the other hand, only glucose was added and ABE removed in the long-term CCCF system. Consequently, in addition to the damaged or dead cells, some metabolites of the microbes which could not permeate the membrane, such as acetic acid, butyric acid, glycerol and lipids, were accumulating in the broth. All these resulted in a gradually worse fermentation condition and caused strong environmental stresses which could be harmful to growth and fermentation of microorganism [5]. Therefore, the fermentation kinetic profiles all demonstrated a generally decreasing trend as fermentation processing (Fig. 5 and Fig. 6).

Conclusions

Continuous and closed-circulating fermentation (CCCF) system integrating fermentation with in situ pervaporation could effectively reduce the butanol inhibition and extend the fermentation duration. The clostridium showed good adaption ability and could be surviving in the deteriorated environment in the PDMS membrane bioreactor. The butanol production in the second cycle was as high as 61.43 gL^{-1}, with the maximum and average butanol productivity of 0.69 and 0.205 $gL^{-1}h^{-1}$. Fermentation performance under continuous coupling mode was superior to that under intermittent coupling mode. The results showed that ABE fermentation in the CCCF system was feasible and promising.

References

1. Van Hecke, W., Vandezande, P., Claes, S., Vangeel, S., Beckers, H., Diels, L., De Wever, H. (2012), "Integrated bioprocess for long-term continuous cultivation of Clostridium acetobutylicum coupled to pervaporation with PDMS composite membranes". Bioresour. Technol. 111, 368-377.
2. Kumar, M., Gayen, K.(2011), "Developments in biobutanol production: New insights". Appl. Energy 88, 1999-2012.
3. Thirmal, C., Dahman, Y.(2012), "Comparisons of existing pretreatment, saccharification, and fermentation processes for butanol production from agricultural residues". Can. J. Chem. Eng. 90, 745-761.
4. Qureshi, N., Sahaa, B. C., Hector, R. E., Hughes, S. R., Cotta, M. A.(2008), "Butanol production from wheat straw by simultaneous saccharification and fermentation using Clostridium beijerinckii: Part I - Batch fermentation". Biomass Bioenerg 32, 168-175.
5. Cheng, C.-L., Che, P.-Y., Chen, B.-Y., Lee, W.-J., Chien, L.-J., Chang, J.-S.(2012), "High yield bio-butanol production by solvent-producing bacterial microflora". Bioresour. Technol. 113, 58-64.
6. Bankar, S. B., Survase, S. A., Singhal, R. S., Granstrom, T.(2012), "Continuous two stage acetone-butanol-ethanol fermentation with integrated solvent removal using Clostridium acetobutylicum B 5313". Bioresour. Technol. 106, 110-116.
7. Garcia, V., Pakkila, J., Ojamo, H., Muurinen, E., Keiski, R. L.(2011), "Challenges in biobutanol production: How to improve the efficiency?" Renew Sust Energ Rev 15, 964-980.
8. Wang, H., Li, C., Li, C.(2012), "RSM optimization of the operating parameters for a butanol distillation column". Asia-Pac. J. Chem. Eng. 7, 117-123.
9. Ezeji, T. C., Qureshi, N., Blaschek, H. P.(2003), "Production of acetone, butanol and ethanol by Clostridium beijerinckii BA101 and in situ recovery by gas stripping". World J Microb Biot 19, 595-603.
10. Qureshi, N., Hughes, S., Maddox, I. S., Cotta, M. A.(2005), "Energy-efficient recovery of butanol from model solutions and fermentation broth by adsorption". Bioprocess. Biosyst. Eng. 27, 215-222.
11. Qureshi, N., Maddox, I. S.(2005), "Reduction in butanol inhibition by perstraction: Utilization of concentrated lactose/whey permeate by Clostridium acetobutylicum to enhance butanol fermentation economics". Food Bioprod. Process. 83, 43-52.
12. Yen, H.-W., Chen, Z.-H., Yang, I. K.(2012), "Use of the composite membrane of poly(ether-block-amide) and carbon nanotubes (CNTs) in a pervaporation system incorporated with fermentation for butanol production by Clostridium acetobutylicum". Bioresour. Technol. 109, 105-109.
13. Vane , L.M.(2005), "a review of pervaporation for product recovery from biomass fermentation processed". J.Chem.Technol.Biotechnol.80,603-629.
14. Chen, C., Tang, X., Xiao, Z., Zhou, Y., Jiang, Y., Fu, S.(2012), "Ethanol fermentation kinetics in a continuous and closed-circulating fermentation system with a pervaporation membrane bioreactor". Bioresour. Technol. 114, 707-710.
15. Ding, W. W., Wu, Y. T., Tang, X. Y., Yuan, L., Xiao, Z. Y.(2011), "Continuous ethanol fermentation in a closed-circulating system using an immobilized cell coupled with PDMS membrane pervaporation". J. Chem. Technol. Biotechnol. 86, 82-87.

16. Shi, E., Huang, W. X., Xiao, Z. Y., Li, D. H., Tang, M.(2007), "Influence of binding interface between active and support layers in composite PDMS membranes on permeation performance". J. Appl. Polym. Sci. 104, 2468-2477.
17. Tang, X. Y., Wang, R., Xiao, Z. Y., Shi, E., Yang, J.(2007), "Preparation and pervaporation performances of fumed-silica-filled polydimethylsiloxane-polyamide (PA) composite membranes". J. Appl. Polym. Sci. 105, 3132-3137.
18. Bergmeyer, H.U., Grassel, M., 1983. Reagents for enzymatic analysis: enzymes—-amylase, third ed. In: Bergmeyer, H.U. (Ed.), Methods of Enzymatic Analysis, vol. 2. Verlag Chemie, Weinhiem, Germany, pp. 151–152.

Adsorption of Galacturonic Acid onto Anion-exchange Resins to enhance the Enzymatic Hydrolysis in a Proposed Citrus Processing Waste Biorefinery

Raul C. Rivas-Cantu[1], Patrick L. Mills[2], Kim D. Jones[1], and Shad Nelson[3]

[1] Department of Environmental Engineering, [2] Department of Chemical and Natural Gas Engineering, [3] Department of Agriculture, Agribusiness & Environmental Sciences. Texas A&M University-Kingsville, Kingsville, Texas.

Extended Abstract

Citrus processing waste (CPW) is considered as a renewable feedstock for biorefining applications. The main structural polysaccharides of CPW as cellulosic biomass are cellulose and pectin. Pectin is one of the main polysaccharides forming the structure of citrus fruits and is composed of monomers of galacturonic acid (GA) forming a polymeric chain. Considering that in USA are generated around 3million metric tons per year of wet CPW (NASS, 2011), and assuming a content of 5-6 wt% of pectin, a considerable amount of 120,000 to 150,000 tons can be converted into galacturonic acid assuming a 80 wt% of the pectin. As part of the development for a CPW-based biorefinery, the enzymatic hydrolysis of grapefruit processing waste (GPW) brake down the structural polysaccharides of the peels and pulp to produce sugars such as glucose (from cellulose and hemicellulose) and galacturonic acid (from pectin) in relatively high concentrations (the total sums approximately 12-14 wt% of the total wet weight of the waste). Currently there is not industrial technology to recover GA from any CPW. One proposed technology for recovering poly-hydroxyl-carboxylic acids is the adsorption-desorption process utilizing anion-exchange resins. Studies in the adsorption of lactic, glutaric and glyoxylic acids utilizing weakly basic anion-exchange resins (WBAER) have been done during the last years by Uslu et al., 2009; Bayazit et al., 2010; and Bayazit, et al., 2011; . Resins such as Amberlite IRA-67 and IRA 96 (Dow Chem. 2008a,b) have been utilized for this purpose. To define the process feasibility and conditions for its application in CPW-Biorefinery is needed to develop a series of tests and experimental procedures.

A proprietary technology for the integrated pretreatment and enzymatic hydrolysis steps is being developed to enhance the hydrolysis reaction rate, and increase the yield of fermentable sugars. In this process, galacturonic acid is produced as monosaccharide during the enzymatic hydrolysis of pectin resulting from the synergistic action of pectinases, cellulases, and β-glucosidase. Issues related to the inhibition of enzymatic hydrolysis due to product and pH take place from the galacturonic acid production. Galacturonic acid cannot be fermented by traditional yeasts and bacteria, and genetically modified microorganisms (i.e. E. Coli K011) have to be utilized for its ethanol fermentation (Stwart et al., 2006; Ingram et al., 1999) making this process more complicated and expensive. Then GA has to be separated from enzymatic hydrolysis liquor to avoid these process inhibitions, and be recovered for its utilization and conversion in value-added products.

The aim of this research is the development of a process for the simultaneous separation of GA during the enzymatic hydrolysis of GPW as model biomass. This process is being performed using anion-exchange resins for the adsorption of GA in batch and semi-continuous systems. Laboratory experiments for the adsorption and recovery of GA from pure solutions and pectin hydrolysis liquor have shown the potential for the application of this technology to enhance the enzymatic hydrolysis rate and the yield of fermentable sugars. Furthermore, this innovative process could alleviate the environmental and economical issues related to wastewater treatment resulting from the biorefining applications of GPW.

Preliminary Results

Initial studies have been completed to investigate the feasibility, and efficiencies of WBAER on the adsorption of model aqueous solutions of GA to simulate its extraction from the enzymatic hydrolysis liquor resulting from GPW conversion. Adsorption experiments were performed in a batch reaction. The aim of these experiments is to determine the effect of important variables on the adsorption of GA onto selected anion-exchange resins, such as quantity of adsorbent (anion-exchange resin), equilibrium time, initial galacturonic acid concentration, and adsorption temperature.

Preliminary results are shown in Figures 1and 2 of the Langmuir adsorption isotherms for galacturonic acid onto a selected WBAER. Figure 3 shows the equilibrium time for the adsorption of GA on WBAER.

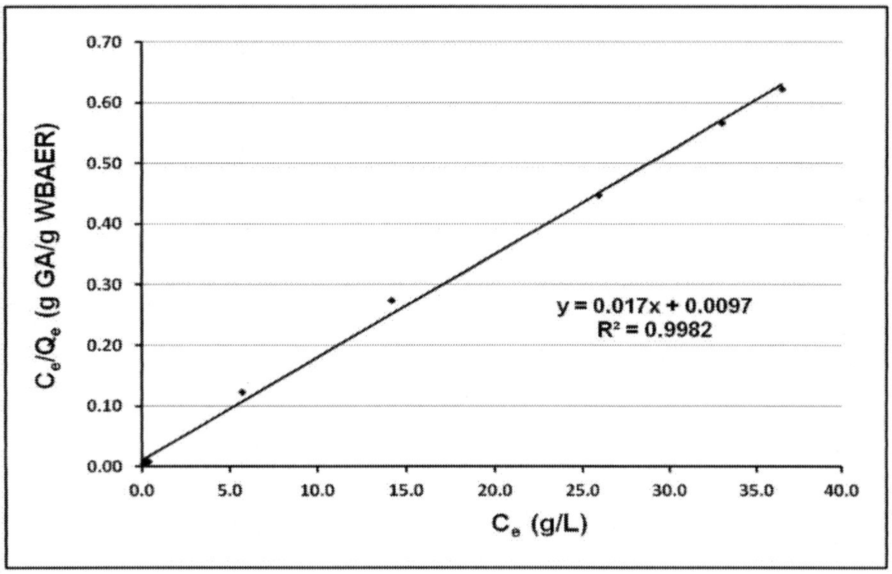

Figure 1. Linear Langmuir adsorption isotherm of galacturonic acid using the weakly basic anion exchange resin Amberlite IRA-67

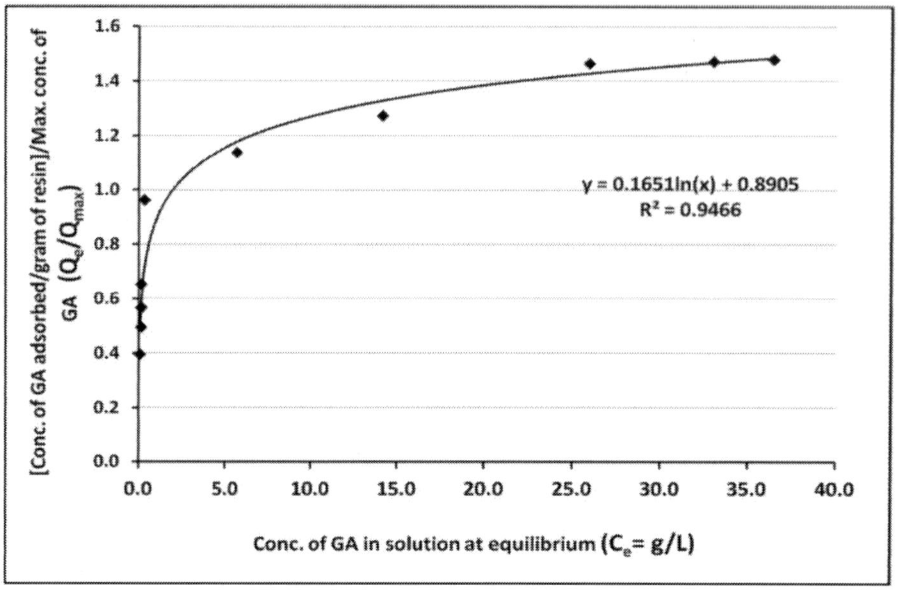

Figure 2. Langmuir adsorption isotherm of galacturonic acid using the weakly basic anion exchange resin Amberlite IRA-67.

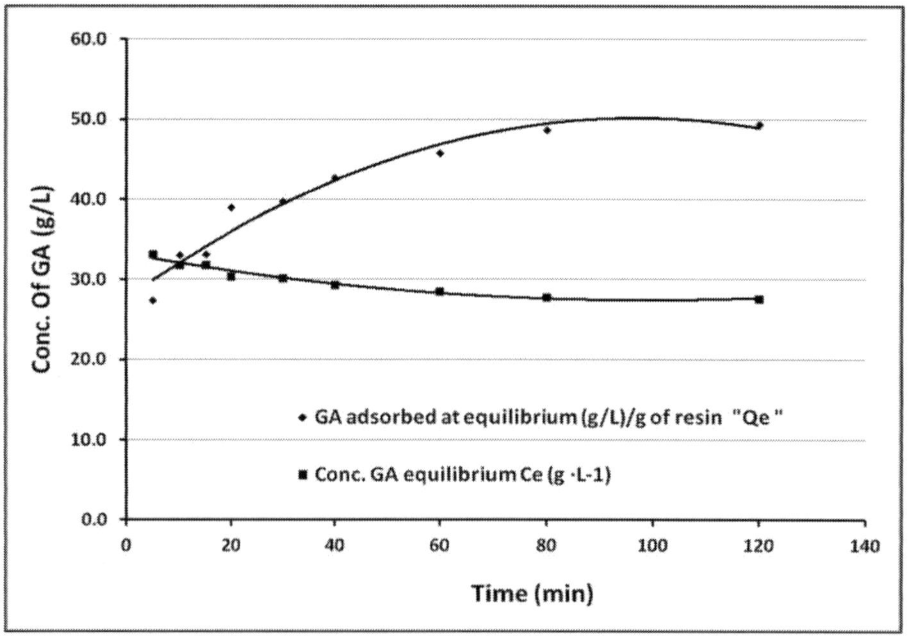

Figure 3. Equilibrium time for the adsorption of galacturonic acid using the weakly basic anion exchange resin Amberlite IRA-67.

Bibliographic References

1. Bayazit, S. S., Inci I., and Uslu, H. (2010). Adsorption of Glutaric Acid and Glyoxylic Acid onto Weakly Basic Ion-Exchange Resin: Equilibrium and Kinetics. J. Chem. Eng. Data 2010, 55, 679–684
2. Bayazit S. S., Inci, I., and Uslu H (2011). Adsorption of Lactic Acid from Model Fermentation Broth onto Activated Carbon and Amberlite IRA-67. J. Chem. Eng. Data 2011, 56, 1751–1754
3. Ingram, L.O., Aldrich, H.C., Borges, A.C., Causey T.B., Martinez, A., Morales, F., Saleh, A., Underwood, S.A., Yomano, L.P., York, S.W., Zaldivar, J., and Zhou, S., (1999), Enteric bacterial catalyst for fuel ethanol production, *Biotechnology Progress,* vol. 15, pp 855-866, 1999.
4. National Agricultural Statistics Service (NASS), (2011), *Citrus Fruit 2011 Summary*. United States Dept. of Agriculture (USDA). Last date accessed on 11.14.2011. http://usda01.library.cornell.edu/usda/ current/CitrFrui/CitrFrui-09-22-2011.pdf.
5. Stewart, D.A., Widmer, W.W., Grohmann, K., and Wilkins, M.R., (2006). Ethanol production from citrus processing wastes. *US Patent Application Publication*, No.US/2006/0177916-A1, 2006.
6. The Dow Chemical Company (2008). AMBERLITE™ IRA-67 Industrial Grade Weak Base Anion Exchanger, Product Data sheet. http://www.dow.com/assets/attachments/business/ier/ier_for_industrial_water_treatment/amberlite_ira67/tds/amberlite_ira67.pdf
7. The Dow Chemical Company (2008). AMBERLITE™ IRA-96 Industrial Grade Weak Base Anion Exchanger, Product Data sheet http://www.dow.com/assets/attachments/business/ier/ier_for_industrial_water_treatment/amberlite_ira96/tds/amberlite_ira96.pdf
8. Uslu H, Inci I, Bayazit S, and Demir G.(2009). Comparison of Solid-Liquid Equilibrium Data for the Adsorption of Propionic Acid and Tartaric Acid from Aqueous Solution onto Amberlite IRA-67. Ind. Eng. Chem. Res., Vol. 48, No. 16, 2009.

(498f) Possibility of Multi-Fuel Production From Trap Grease by Combination of Heat-Driven Upgrading and Methane Fermentation

Hidetoshi Kuramochi[1], Kouji Maeda[2], Takuro Kobayashi[1], Yoshitaka Ebie[1], Keisuke Fukui[2] and Masahiro Osako[1], (1)Center for Material Cycles and Waste Management Research, National Institute for Environmental Studies, Tsukuba, Japan, (2)School of Mechanical and Systems Engineering, University of Hyogo, Himeji, Japan

Trap grease, which is a free-floating aggregate on the surface of the tank between restaurant and sewer, is either an oil or sludge-like grease including food scraps and water. It is a low-quality mixture of fats and oils, and is not recycled as a fuel in Japan. However, oily components such as glycerides and fatty acids in the grease a re considered to be a feedstock for alternatives to fossil fuels such as diesel and heavy oil. Furthermore, we consider that residues discharged during extraction of such oily content from the grease may be converted into biogas via methane fermentation. In the present study, first, we proposed a heating-driven liquefaction and phase separation for recovery of the oily content, and investigated the possibility of the obtained oil to use as an alternative to heavy oil in terms of impurity and property. From the investigation, the oily content was a clean fuel, especially a low-sulfur fuel. Hence, the heat-driven upgrading was useful for recovering a high-quality oil content from trap grease. In addition, we showed this upgrading can be successfully scaled up to a small demonstration phase. Unfortunately, the recovered oil could not meet some of fuel properties for replacement of high-grade heavy oil. Therefore, some improvements were proposed, and then their effectiveness was experimentally evaluated. Second, to evaluate the possibility of the residues to produce biogas, several types of residues from different restaurants were characterized in terms of volatile solid (VS), total nitrogen, etc, and then biogas productivity of the individual residues was examined using a batch-type methane fermentation. The results showed that the residues were applicable for methane fermentation and the biogas productivity ranged from 0.5 to 0.9 L/g-VS. These residues are considered to be good feedstocks for methane fermentation. From a series of investigation, the combination of the heat-driven upgrading of trap grease and methane fermentation of the residue from the grease upgrading can produce not only an alternative to heavy oil but also biogas. We expect both fuels may be used in a dual-fuel engine to generate electricity.

(519a) Advanced Computational Modeling of Photosynthetically Active Radiation (PAR) Distribution in Algae Photobioreactors

Bo Kong, Department of Chemical and Biological Engineering, Iowa State University, Ames, IA and R. Dennis Vigil, Iowa State University, Ames, IA

The development of highly efficient scalable continuous-flow algae photobioreactors (PBRs) could be a major advance towards enabling algae-based biofuel and biorenewable chemicals technology.Conversion of solar energy into chemical energy by photosynthesis is strongly depends on the quantity and quality of light received by microorganisms. Therefore, it is crucial to understand the availability and distribution of light in order to optimize the design and operation of a photobioreactor. In particular, it is important to understand the distribution of photosynthetically active radiation (PAR) for phototrophic microorganisms, which includes wavelengths from 400-700 nm. This radiation is absorbed by various pigments in the organisms, such as chlorophyll, carotenes and xanthophylls, which have different PAR absorption spectra. Despite the fact that only a narrow band of wavelengths are useful for photosynthesis, the most frequently used method for simulating light distribution in PBRs is to employ the Beer-Lambert law to calculate the overall light intensity distribution in one dimension. However, there exist several more advanced CFD techniques to solve the radiation transfer equation, such as the discrete ordinate (DO) and Monte Carlo (MC) methods, which are commonly used in other engineering fields, for example in combustion and HVAC. Although most of these models are used to compute total radiative transfer, in principle they can be extended to include spectral dependencies. Consequently, it is necessary, desirable, and feasible to develop a multidimensional spectral radiation model capable of predicting light distribution in photobioreactors.

In this work we describe a detailed and general numerical method for simulating the spatial distribution of PAR in photobioreactors using the discrete ordinate (DO) method for solving the radiative transfer equation (RTE), which was implemented in a computer program developed using the open source CFD library openFOAM. This non-gray DO implementation divides the radiation spectrum into N wavelength bands, in each of which the treatment of the light is the same as that for the gray DO model.Optical properties of the medium and boundary conditions for the non-gray DO model are applied on a band basis. Thus, the PAR transfer can be modeled by solving RTEs in small wavelength bands in the 400-700 nm wavelength range with a non-gray DO model. In algal cultures, the liquid medium is usually transparent to visible light, and therefore the transport of visible wavelengths are determined mainly by algae cells. The optical properties of the algae cells and the radiative properties of the turbid culture medium were calculated at each wavelength band range using the method developed by Pottier, et.al. 2005, which is based on Mie theory and requires information concerning algal pigmentation, shape, and size distribution. The model was validated in a simple rectangular rector configuration using the experimental data provided in Pottier, et.al. 2005. Subsequently, simulations were carried out in tubular and annular photobioreactor configurations, and implications of the simulation predictions are discussed.

(519b) Mechanistic Modeling of the Enzymatic Saccharification of Cellulose-Iβ and Cellulose-III: Parameter Determination and Model Validation

Andrew J. Griggs[1], Ambarish Nag[2], James J. Lischeske[1], Jonathan J. Stickel[1], Ashutosh Mittal[3], Wei Wang[3] and David K. Johnson[3], (1)National Bioenergy Center, National Renewable Energy Laboratory, Golden, CO, (2)Computational Science Center, National Renewable Energy Laboratory, Golden, CO, (3)Biosciences Center, National Renewable Energy Laboratory, Golden, CO

Cost-effective production of fuels and chemicals from lignocellulosic biomass often involves enzymatic saccharification, which has been the subject of intense research and development. Recently, a mechanistic model for the enzymatic saccharification of cellulose has been developed that accounts for distribution of cellulose-chain lengths, the accessibility of insoluble cellulose to enzymes, and the distinct modes of action of the component cellulases [Griggs *et al.* (2012) *Biotechnol. Bioeng.*, **109(3)**:665–675; Griggs *et al.* (2012) *Biotechnol. Bioeng.*, **109(3)**:676–685]. However, determining the appropriate values for enzyme adsorption, inhibition, and rate parameters required further experimental investigation. In this work, we performed several sets of experiments to aid in parameter estimation and to quantitatively validate the model. Several cellulosic materials having varied degrees of polymerization, crystallinity, and crystalline allomorph (cellulose-I_β and cellulose-III_I) were digested by component enzymes (EGI and CBHI) and by cocktails of these enzymes. Based on information from the literature and the results from these experiments, a single set of model parameters was determined and used to produce model simulation results. The agreement between simulation results and experimental data will be demonstrated. Cases of discrepancy point the way for further model development.

(519c) A Pore-Hindered Diffusion and Reaction Model Can Help Explain the Importance of Pore Size Distribution in Enzymatic Hydrolysis of Biomass

Jeremy S. Luterbacher, Department of Chemical and Biomolecular Engineering, Cornell University, Ithaca, NY and Larry P. Walker, Department of Biological and Environmental Engineering, Cornell University, Ithaca, NY

Until now, most efforts to improve carbohydrate production from biomass through pretreatment and enzymatic hydrolysis have used random optimization, trend extrapolation and/or educated guessing. These efforts could employ a rational design process if they were guided by a modeling framework that captured the key mechanisms governing the relationship between pretreatment and enzymatic hydrolysis. In this study, we propose a pore hindered diffusion and kinetic model for enzymatic hydrolysis of biomass. When compared to data available in the literature, this model accurately predicts the well-known dependence of initial cellulose hydrolysis rates on surface area available to a cellulase-size molecule. Modeling results suggest that, for particles smaller than $5x10^{-3}$ cm, a key rate-limiting step is the exposure of previously unexposed cellulose occurring after cellulose on the surface has hydrolyzed, rather than binding or diffusion. However, for larger particles, according to the model, diffusion plays a more significant role. Therefore, the proposed model can be used to design experiments that produce results that are either affected or unaffected by diffusion. Finally, by using pore size distribution data to predict the biomass fraction that is accessible to degradation, this model can be used to predict cellulose hydrolysis with time using only pore size distribution and initial composition data.

(519d) Anaerobic Digestion of Waste Materials for the Production of Biogas

Ramalingam Subramaniam, Stephen Dufreche, Rakesh Bajpai and Mark Zappi, Chemical Engineering, University of Louisiana at Lafayette, Lafayette, LA

Anaerobic digestion of waste materials for the production of biogas

Ramalingam Subramaniam, Stephen Dufreche, RakeshBajpai, and Mark E. Zappi

Bioprocessing Research Laboratory

Department of Chemical Engineering

University of Louisiana at Lafayette, Lafayette, LA

Biogas energy can be produced from various sources of biomass which include agriculture crops and residues, municipal solid wastes, and animal waste. In this work, anaerobic biogas was produced from waste materials such as shrimp waste, landfill leachate, and crab waste. The biogas produced from these waste materials in an anaerobic digester contains 65-70% methane and 25-30% carbon dioxide. Biogas with a methane composition of 66% has a heating value of 668 Btu/ft^3. Batch experiments were conducted in 500 ml three neck rectors as well as New Brunswick, microfirm MF114 reactor. Kinetic model was developed from the batch experiments. Based on the batch kinetic results, continuous process was investigated in 14 L bioreactor with a working volume of 10L. Batch kinetics, fermentor design and scale up will be discussed in this presentation.

(519e) Mass Transfer in High Solids Biomass Slurries

Samin Rezania[1], Jonathan Michael D. Thomas[2], Amlan Chakraborty[2] and **R. Eric Berson**[2], (1)University of Louisville, Louisville, KY, (2)Chemical Engineering, University of Louisville, Louisville, KY

Enzymatic saccharification rates and yields are hindered in high solids slurries, which is generally attributed to mass transfer limitations resulting from the high viscosity of the slurries. However, other possible mechanisms exist and no evidence has yet been presented directly linking the lower rates and yields to the mass transfer limitations. The objective of this work was to compare saccharification results for identical high solids slurries in a conventional Erlenmeyer flask with another system capable of overcoming mass transfer limitations in high viscosity slurries.

In order to achieve high mass transfer rates in the high viscosity slurries, a resonating acoustic mixer was employed for the saccharification reaction. Dispersion coefficients in the mixing vessel were determined as a function of viscosity in order to quantify the mass transfer improvements and determine if the system is a valid choice for studying reactions in the absence of mass transfer limitations. Since conventional means of measuring dispersion could not be performed due to the slurry characteristics, a computational fluid dynamics model was developed that simulated motion of the slurries in the mixer. The model was validated by comparing experimentally determined mixing times using an electrolytic tracer with mixing times from the simulations. Computational and experimental mixing times differed by 2% or less in all cases. Dispersion values were achieved on the order of values for water flow in a pipe, indicating the system was a valid choice for studying reactions in the absence of mass transfer limitations.

At low initial solids where good mixing could be achieved in the shake flask, the glucose release in the mixer was just 4% higher than in the shake flask, while for 30% initial total solids, the glucose release in the mixer was approximately 40% higher than in the shake flask. These results provide a direct link between low rates and yields of saccharification of viscous biomass slurries and mass transfer limitations.

(519f) Determining Dispersion in High-Solids Biomass Slurries

James J. Lischeske[1], Michael A. Sprague[2] and Jonathan J. Stickel[1], (1)National Bioenergy Center, National Renewable Energy Laboratory, Golden, CO, (2)Computational Science Center, National Renewable Energy Laboratory, Golden, CO

Conversion of lignocellulosic biomass to fuels can be made more cost-effective by operating at high-solids conditions. However, little is known about the transport properties in these systems. A thorough understanding of convection and diffusion in high-solids biomass slurries is necessary in order to properly develop predictive models for the degradation of biomass in reactor systems. In this work, packed-beds of pretreated biomass were prepared in preparative chromatography columns. Step inputs of tracer molecule solutions of varying size and type were passed through the packed beds. Initial residence-time-distribution analyses were performed to determine the extra-particle porosity of the packed beds and to determine apparent pore volume accessibility of the biomass. Secondly, an advection-diffusion-adsorption model was used to distinguish the effects of dispersion and adsorption of tracer molecules, and to measure parameters for the adsorption and diffusion. The relationship between dispersion and flow rate was examined, and diffusion parameters for a given bed were estimated. These diffusion parameters were then related to tortuosity factors, and the relationship between tortuosity and packing density was examined.

Fractionation and characterization of lignin from wood biomass using ionic liquid and alkaline ethanol solvent

*Yong-Chang Sun[a], Ji-Kun Xu[a], Feng Xu[a], Run-Cang Sun[a,b,]**

[a] Institute of Biomass Chemistry and Technology, Beijing Forestry University, Beijing 100083, China

[b] State Key Laboratory of Pulp and Paper Engineering, South China University of Technology, Guangzhou 510640, China

**Corresponding author at: Institute of Biomass Chemistry and Technology, Beijing Forestry University, Qinghua Road No. 35, Beijing 100083, China*

Tel.: +86-10-62336972; fax: +86-10-62336972

E-mail address: rcsun3@bjfu.edu.cn (R.-C. Sun)

Lignocellulosic biomass is a new source of renewable and carbon neutral materials for biofuel and bioenergy production. However, we note the recalcitrance of lignocellulosic for bioconversion. The presence of lignin is a barrier to enzymatic hydrolysis of cellulose by cellulases, as well as inhibiting fermentation to form products such as ethanol and butanediol. Lgnin is a three-dimensional amorphous polymer based on methoxylated phenylpropanoid units, and is crosslinks to different polysaccharides with covalent bond. The use of ionic liquids (ILs) in biomass pretreatment has received considerable attention recently because of their effectiveness at decreasing biomass recalcitrance to subsequent enzymatic hydrolysis. ILs pretreatment has been shown to reduce the lignin and hemicelluloses content, as well as the cellulose crystallinity. Certain ILs completely dissolved wood, agricultural waste, and other biopolymers at moderate temperatures and ambient pressure. In this article, we investigate the lignin extraction uses a purpose designed ionic liquid 1-butyl-3-methylimidazolium acesulfamate [BMIM]Ace at 120 °C and open atmosphere. The IL +organic solvents systems were also studied. The lignin was successfully fractionated from the biomass by precipitation in acetone after IL treatment, allowing the IL to be recycled. The results indicated that exceeding 89.6% of regenerated biomass obtained. After that, the regenerated biomass was subsequently extracted with alkaline ethanol solvent. It was indicated that the IL-organosolv lignin had a higher molecular weight (M_w, 1270-7810 g/mol) than the alkali-ethanol lignin (M_w, 2440-3020g/mol). This comparison revealed that organic solvents considerably affect the M_w. This result probably due to the viscosity and inhomogeneity of IL mixed with different organic solvents. Moreover, the lignin fractions were characterized by high-performance anion exchange chromatography (HPAEC) and Fourier transform infrared (FT-IR). It should be note that a small amount of hemicelluloses loss during the IL extraction procedure. These results indicated a potential approach in further optimization of the choice of ILs/organic solvents for the efficient utilization of the lignocellulosic biomass, and all constituents can be fully recovered and further processed.

Keywords: Lignocellulosic biomass; Fractionation; Lignin; Ionic liquid; Alkali;

Introduction

The depletion of the world's petroleum supply and environmental issues associated with crude oil usage increase a demand for developing alternative, environmental friendly, and non-petroleum dependent energy sources [1]. An alternative and sustainable source for chemicals, materials, and fuels is lignocellulosic biomass. However, the structure of the lignocellulosic biomass is very complex; the rigid structure makes it highly resistant to biological and chemical degradation [2]. Lignocellulosic biomass consists mainly of cellulose (35-50%), hemicelluloses (20-35%), and lignin (5-30%) [3]. Cellulose is a linear polymer of β-1,4-linked D-glucopyranose monomers, and there are large amount of intra- and inter molecular hydrogen bonds in the cell walls [4]. However, cellulose is traditionally material used as a raw material for the production of paper, paperboard, fiberboard, and other similar products. Hemicelluloses are consisted of different five or six sugars, which can be further produced of xylitol and furfural. Hemicelluloses can also be hydrolyzed and used for fermentation, but the usage is more complicated due to the fact that xylose, which is the dominating monosaccharide released, is difficult to ferment [5]. Lignin is a stubborn polymer, which formed covalent bond with hemicelluloses. Lignin is a three dimensional network, and the complex structure is difficult for chemicals and microbial attack. All these factors make lignocellulosic biomass usage limitation. Recently, ionic liquid (IL) has received growing interest as green solvents for lignocellulosic biomass pretreatment [6]. IL is salts that melting point blow 100°C, and can be design for different cation and anion. It has several advantages, such as neligible volatility, wide liquid range, and relatively high thermal stability as well as ability to solubilize organic, inorganic, or polymeric substances [7,8]. In this paper, the ball-milled wood was dissolved in a new IL named 1-butyl-3-methylimidazolium acesulfamate [BMIM]Ace to extract lignin. Organic solvents were firstly added in this IL solution. The lignin obtained was characterized by high-performance anion exchange chromatography (HPAEC), Fourier transform infrared (FT-IR), and gel permeation chromatography (GPC).

Methods

Materials and Reagents

Eucalyptus was received from Guangdong province, China. Leaves and bark were removed, and the stems were chipped into small pieces (1-3 cm). After drying at 60°C for 16 h in an oven, the chips were ground using a mortar and sieved through a 40-60 mm mesh. The content of acid-insoluble lignin was 30.1% according the procedure of the National Renewable Energy Laboratory (NREL/TP-510-42618, TP-510-42622). The ionic liquid [BMIM]Ace was synthesized according to the literature [9].

Wood Dissolution and Regeneration

2.0 g of ground wood was added to 30 g of [Bmim]Ace, and 10mL of DMSO, methanol, acetone, and toluene were added in the mixture, respectively. The mixture was mechanically stirred at 120°C for 3 h in an oil bath in open atmosphere. After that, the wood was regenerated in hot water, and the solution was added into acetone to precipitate lignin after remove water. Lignin was separated by filtration. The acetone was evaporated for the recycle use of IL.

Sugar Analysis

The sugar content of the lignin fractions were detected by high-performance anion exchange chromatography (HPAEC). A 4-6 mg sample was hydrolyzed with 1.475 mL of 10% H_2SO_4 for 2.5 h at 105°C. The mixture was filtered and diluted. The sample was injected into the HPAEC system. L-arebinose, D-glucose, D-xylose, D-rhamnose, D-mannose, D-galactose, uronic acids were detected. All experiments were performed at least in duplicate, and the average values calculated for all of the lignin fractions.

FT-IR Analysis

The Fourier-transform infra-red (FT-IR) spectra of the samples were recorded using a Thermo Scientific Nicolet iN10-MX FT-IR chemical imaging microscope (Thermo Scientific, America) fitted with a narrow-band liquid nitrogen cooled MCT detector. Spectra were recorded with 64 scans at a resolution of 4 cm^{-1} between the wave numbers of 4000 and 700 cm^{-1}.

Molecular Weight

The weight-average and number-average molecular weights of the lignin fractions were determined by GPC according to Sun *et al* [10].

Results and Discussion

Yield of Lignin

The yield of the lignin was showed in Table 1. Interestingly, different organic solvents resulted in different yields. The toluene + IL mixture released a higher content of lignin (15.4%), while the methanol + IL mixture yield 3.0% of lignin. The neat IL extraction resulted in 11.5% of lignin released. We can see that methanol and acetone combined with IL extraction released low amounts of lignin. The results indicated that some organic solvents such as toluene can increase the lignin extraction efficiency, while DMSO, methanol, and acetone will decrease it. This results probably due to the difference of dielectric constants of the organic solvents [11]. Meanwhile, adding of organic solvents is that the anions become more delocalized and the strength of the interaction with lignin decreases with increasing the polarity parameter. The regenerated biomass was subsequently extracted by alkaline ethanol solvent. The results showed that high amount of lignin were extracted ranging from 20.2 to 22.3%. However, the neat IL pretreated residue yielded 8.0% of lignin, indicating that organic solvents with the ability to swell the lignocellulosic framework of the wood particles, is beneficial to enhance the penetration of the IL into the cell walls.

Tbale 1. Yield of lignin (% Klason lignin) released in different solvents systems.

Solvent	DMSO+IL	Methanol+IL	Acetone+IL	Toluene+IL	Neat IL
Yield (%)[a]	8.6	3.0	3.3	15.4	11.5

[a] The yield was based on the Klason lignin

Purity Analysis

To characterize composition of the IL/IL+ organic solvents systems, the five lignin preparations were prepared for determination of their carbohydrate content, and the results are given in Fig. 1. As can be seen, all lignin fractions contained different amount of sugars. The total content of sugars detected in the lignin fractions were in the order of the extraction systems of toluene+IL > neat+IL > methanol+IL > DMSO+IL> acetone+IL, indicating that IL+toluene mixture result in a relatively high content of carbohydrate released. The acetone+IL system has a weak effect on the hemicelluloses moiety in the wood cell. Toluene+IL treatment yielded high amount of xylose. Xylose and uronic acid were the mainly sugars detected in the five lignin fractions. Moreover, we should note that neat IL extraction released more xylose (7.6%) than glucose (2.8%). Therefore, the results suggest that the hemicellulosic polymers in the cell walls might be degraded during the treatments with different organic solvents combined with IL under the conditions given.

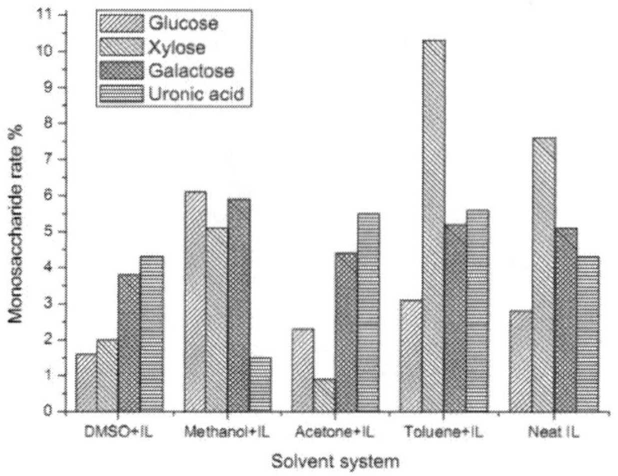

Fig. 1. The monosaccharide rare (wt%) of the different extraction systems.

FT-IR Analysis

To identify the isolated extracts as lignin, the IR spectra of the five extracts obtained were analyzed (Fig. 2). As can be see from the spectra, all the five spectra have typical bands at 1572, 1502, and 1431 cm^{-1}, corresponding to aromatic skeletal vibrations. The band at 2961, 2874, and 1462 cm^{-1} arise from the C-H stretching and asymmertric vibrations of CH_3 and CH_2, respectively [12]. The wide absorption band at 3396 cm^{-1} originated from the O-H stretching vibration in aromatic and aliphatic OH groups. However, the bands at 1648 and1739 cm^{-1} are assigned to the presence of conjugated and nonconjugated carbonyl groups in the lignin structure. The IOL-5 (extracted with neat IL) spectra has a strong absorption band at 1738 cm^{-1}, indicating that more C=O groups present in the lignin fraction, adding organic solvents in IL could decrease the C=O group. Syringyl and condensed guaiacyl absorptions are obviously observed at 1313 cm^{-1}, whereas guaiacyl ring breathing with C=O stretching appears at 1245 cm^{-1}. The absorption at 1056 cm^{-1} is owing to the C-H in-plane deformation. In addition, aromatic C-H out of plane bending appears at 938 and 853 cm^{-1}.

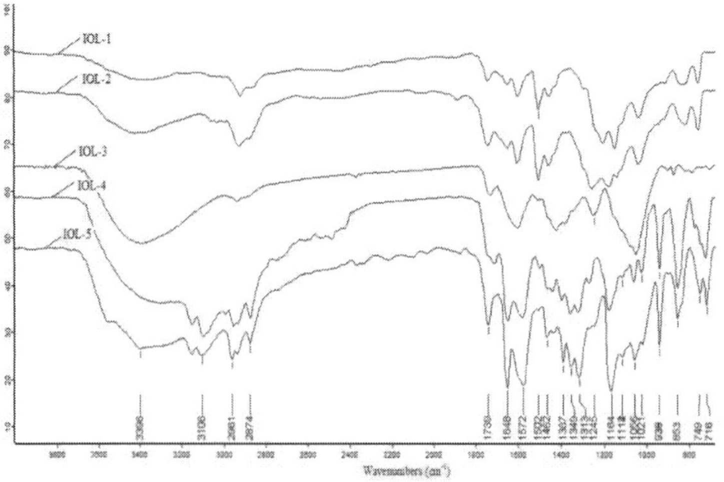

Fig. 2. The FT-IR spectra of the extracted lignin: IOL-1, IL+methanol extracted lignin; IOL-2,

IL+acetone extracted lignin; IOL-3, IL+toluene extracted lignin; IOL-4, IL+DMSO extracted lignin; IL-5, neat ionic liquid extracted lignin.

Molecular Weight

The weight-average and number-average molecular weights of the lignin fractions were determined by GPC. As shown in Table 2, the lignin fractions exhibited an M_w ranging from 7810 to 1270 g mol^{-1}. Obviously, the acetone-IL lignin fraction has a higher M_w (7810 g mol^{-1}) than other lignin fractions, which meant a large fragment of lignin extracted from the cell wall, indicating a weak extaction ability of the mixed solution. However, the fresh IL extracted lignin showed a low M_w (3700 g mol^{-1}), suggestted relatively high extraction ability. The alkaline ethanol lignin has relatively low molecular weights (3020-2440 g mol^{-1}).

Table 3. Molecular weight of the lignin fractions.

	Lignin fraction				
	IOL-1	IOL-2	IOL-3	IOL-4	IOL-5
M_w	1270	7810	6590	4430	3700
M_n	930	1330	1170	1830	720
M_w/M_n	1.37	5.87	5.63	2.42	5.14

Ionic Liquid Recycling

ILs are regarded as a "green solvent". Recycling ionic liquid for reuse was achieved under vacuum. The extraction and stripping process was repeated four cycles. The purity analysis of the IL was checked by the ^{1}H-NMR spectra as shown in Fig. 3. It was found that there was no obvious difference between the fresh IL and the recycled IL. The only difference was the color of the recycled IL, which becoming brown as the recycling number increased. The coloration may be owing to the contaminants present, which arise from the degradation of the wood and sugars.

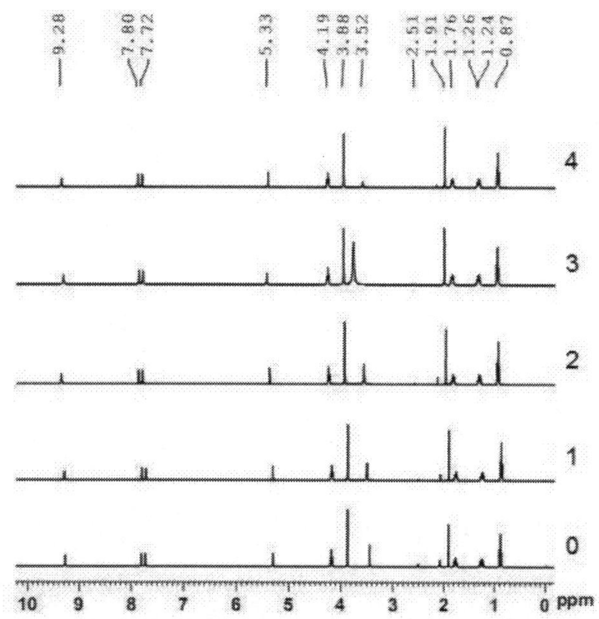

Fig. 3. ^{1}H-NMR spectra of the fresh and recycled [BIMI]Ace in DMOS-d_6. The number of the right side represents the lignin extraction cycle for this ionic liquid.

Conclusion

It was proved that imidazolium acesulfamate ILs has the ability to dissolved wood lignin in an open atmosphere. Different organic solvents+IL systems released different content of lignin from the lignocellulosic biomass. This was probably due to the difference dielectric constants of the organic solvents. The IL+toluene system resulted in a relatively high content of carbohydrate. GPC results showed that the IL-organosolv lignin has a high molecular weight than the alkaline ethanol lignin. The recovered biomass can be used as a valuable source for enzymatic saccharification. These results indicated a potential approach in further optimization of the choice of low-cost ILs and organic solvents for the efficient utilization of the lignocellulosic biomass.

Acknowledgements

The authors are grateful for the financial support of this research from the Specific Programs in Graduate Science and Technology Innovation of Beijing Forestry University (No. BLYJ201213).

References

[1] Farrell, A.E. Plevin, R.J. Turner, B.T. Jones, A.D. O'Hare, M. Kammen, D.M. 2006. Ethanol can contribute to energy and environmental goals. Science 311, 506-508.

[2] Yang, B. Wyman, C.E. 2008. Pretreatment: the key to unlocking low-cost cellulosic ethanol. Biofuels Bioprod. Bioref. 2, 26-40.

[3] Lynd, L.R. Weimer, P.J. van Zyl, W.H. Pretorius, I.S. 2002. Microbial cellulose utilization: fundamentals and biotechnology. Microbiol. Mol. Biol. Rev. 66 (3), 506-577.

[4] Zhbankov, R.G. 1992. Hydrogen-bonds and structure of carbohydrates. J. Mol. Structure 270, 523-539.

[5] Garde, A. Jonsson, G. Schmidt, A.S. Ahring, B.K. 2002. Lactic acid production from wheat straw hemicellulose hydrolysate by Laceobacillus pentosus and Lactobacillus brevis. Bioresour. Technol. 81 (3), 217-223.

[6] Simmons, B.A. Singh, S. Holmes, B.M. Blanch, H.W. 2010. Ionic liquid pretreatment. Chem. Eng. Prog. 106 (3), 50-55.

[7] Macfarlane, D.R. Seddon, K.R. 2007. Ionic liquid-progress on the fundamental issues. Aust. J. Chem. 60, 3-5.

[8] Seddon, K.R. 1997. Ionic liquids for clean technology. J. Chem. Tcehnol. Biotechnol. 68, 351-356.

[9] Pinkert, A. Goeke, D.F. Marsh, K.N. Pang, S.S. 2011. Extracting wood lignin without dissolving or degrading cellulose: investigations on the use of food additive-derived ionic liquids. Green Chem. 13, 3124-3136.

[10] Sun, Y.C. Wen, J.L. Xu, F. Sun, R.C. 2010. Fractional and structural characterization of organosolv and alkaline lignins from Tamarix austromogoliac. Scientific. Res. Essays 5(24): 3850–3864.

[11] Li, W.J. Zhang, Z.F. Zhang, J.L. Han, B.X. Wang, B. Hou, M.Q. Xie, Y. 2006. Fluid Phase Equilibria 248, 211-216.

[12] Tejado, A. Pena, C. Labidi, J. Echeverria, J.M. Mondragon, I. 2007. Physico-chemical characterization of lignins from different sources for use in phenol-formaldehyde ersin synthesis. Bioresour. Technol. 98, 1655-1663.

(535b) Statistical Analysis On Purification of Lignocellulosic Hydrolyzates by Polymer Flocculation

Rakesh Yasarla, Paper & Bioprocess Engineering, State University of New York, Syracuse, NY and Ramarao Bandaru, Paper and Bioprcess Engineering, State University of New York, Syracuse, NY

Lignocellulosic materials such as wood present as renewable and sustainable resources for the production of fuels and plastics. Raw lignocellulosic hydrolyzates produced by acid or water treatment contain significant quantities of colloidal material. Separation of these colloidal particles can be achieved by polymer flocculation. Cationic and anionic polymers were used to separate the lignin fractions. The main objective of the work is to demonstrate the efficiency of the factorial experimental designs to elucidate the factors effecting the separation and purification of hot water hydrolyzates of forest biomass. Two factorial experimental designs of flocculation were studied for the polymer flocculation. The response variable representing flocculation was chosen as the turbidity of the clarified suspension (in NTUs). Although the maximum settling velocity and the maximum height of the sediments were investigated, the turbidity was quite representative of flocculation. The independent variables (factors) were chosen as the polymer dosage, agitation level, and the initial pH of the hydrolyzate. A two level, 3 factor (full factorial) design consisting of 8 experimental runs was analyzed for the application of polymer flocculations. The independent factors were polymer dosage, agitation level, and initial pH of the suspension. Experimental runs were conducted in random order, and the Yates algorithm was applied to display the run data.

(535c) Fractionation of Forest Residues of Douglas-Fir for Fermentable Sugar Production by SPORL Pretreatment

Chao Zhang, Tianjin University , Tianjin, China, Rolland Gleisner, US Forest Service, Madison, WI, J.Y. Zhu, USDA Forest Service, Forest Products Laboratory, Madison, WI and John Sessions, Oregon State University , Corvallis, OR

Douglas-fir (*Pseudotsuga menziesii*) forest residues were physically fractionated through sieving. The bark and wood were separated for large-sized fractions (>12.7 mm) and their contents were determined. The chemical compositions of the large fractions were calculated based on the contents and chemical compositions of the bark and wood. The chemical compositions of the fine fractions were analyzed. The bark and wood content in the fine fractions were calculated based on the measured glucan and lignin contents in each fraction. It was found that fractionation by particle/chip size can effectively fractionate bark and wood and therefore lignin from carbohydrates. The large sized-fractions (>12.7 mm) represent approximately 60% of the collected forest residues but only contain approximately 37% of the total bark and 35% of the total ash, or a selectivity over bark and ash of 1.6 and 1.7, respectively. Pretreatment of forest residues by Sulfite Pretreatment to Overcome Recalcitrance of Lignocelluloses (SPORL) and subsequent enzymatic hydrolysis revealed the presence of 14.3% bark can reduce substrate enzymatic digestibilities (SED) 16% compared with that from a bark free sample. The SED of a bark is 41% compared with 73% for wood when pretreated under the same conditions. Separating pretreatment of bark from wood is beneficial for producing a more enzymatically digestible substrate. The results from the present study could have significant implications for harvesting forest residues.

(535d) Exploration of the Effect of Ionic Liquid Pretreatment Temperature On Biomass Composition, Structure, and Enzymatic Digestibility

Christopher J. Barr[1], Kevin Click[2], Paul Langan[3], B. Leif Hanson[4] and Constance Schall[1], (1)Chemical and Environmental Engineering, University of Toledo, Toledo, OH, (2)Chemistry, Kent State University, Kent, OH, (3)Biosciences Division - Los Alamos National Laboratory, Los Alamos, NM, (4)Instrumentation Center, University of Toledo, Toledo, OH

Deconstruction of lignocellulosic biomass into simple sugars constitutes a core barrier for producing value added products from the sugar platform. Lignocellulosic biomass is primarily comprised of cellulose (a glucose polymer), hemicellulose (a xylose polymer), and lignin. Enzymatic hydrolysis of cellulosic and hemicellulosic portions of biomass is one of the crucial steps in biochemical conversion of terrestrial biomass into monomeric sugars. The native cellulosic portion of biomass has a crystalline and oriented fibrillar structure in the cellulose I crystal form. However, cellulose I can be a major impediment for enzyme-assisted hydrolysis to monomeric sugars which can be used as a feedstock for conversion to fuel or other value added chemicals. Ionic Liquid (IL) pretreatment can disorder the crystalline cellulosic component to an amorphous material or result in structural changes in the cellulose fibers to other crystal forms, such as cellulose II with an improvement in susceptibility to hydrolysis. In these studies, the effects of IL pretreatment temperature on various sources of lignocellulosic biomass were examined using X-ray powder and fiber diffraction, compositional analysis and comparative enzymatic hydrolysis. Polysaccharide digestibility was related to the observed structural and compositional changes in biomass. The improvement in hydrolysis is hypothesized to be primarily due to structural changes in the cellulose constituent.

(535e) pH-Based Fractionation and Characterization of CO2-Precipitated Lignin From Kraft Black Liquor

Mark C. Thies and **Julian Velez**, Chemical and Biomolecular Engineering, Clemson University, Clemson, SC

Selected fractions of lignin were precipitated from a Kraft black liquor using a pH-based fractionation process, with the objective being to produce various lignin cuts for characterization. A semi-batch operating mode was used, with CO_2 being bubbled into approximately 2 L of a black-liquor charge in order to bring about the precipitation process. Experiments were carried out in a stirred, 2-L Parr reactor setup modified for our experiments. The CO_2 feed was carefully controlled such that the pH of the black liquor was decreased in units of 0.5, with the first precipitation experiment taking place at a pH of 13.5 and the last at a pH of 9.5. For each pH decrement, a lignin precipitate was collected for further analysis. It is important to note that each collected fraction of lignin was obtained over a narrow pH change of only 0.5. This was accomplished by removing the lignin precipitate and spent black liquor from the reactor after a given pH decrement (e.g., the 1st run from 13.5 to 13.0 pH), and then returning the spent black liquor back to the reactor for the next CO_2 addition – and for the next pH decrement (e.g., from 13.0 to 12.5 pH for the 2nd run). This process was continued until a pH of 9.5 was reached. A total of seven "narrow-pH" lignin fractions were produced.

Matrix-Assisted Laser Desorption/Ionization Time of Flight (MALDI) mass spectrometry was used to determine the molecular weight distribution of the pH-fractionated lignin cuts described above. The effects of different kinds of matrices, of sample preparation techniques, and of sample/lignin salt content on MALDI analysis was investigated and will also be discussed. For determining the specific metals content of the various lignin cuts, which contribute to their ash content, Inductively Coupled Plasma Atomic Emission Spectroscopy (ICP-AES) was used.

Finally, selected physical properties of the lignin cuts, such as softening points, densities, and viscosities, were also measured.

(535f) Utilization of Lignin Previously Depolymerized From WHITE-Rot FUNGI to Obtain Phenolic Compounds

Natalia Andrea Cano, Universidad Nacional de colombia, Medellín, Colombia; Grupo Bioprocesos y flujos Reactivos, Universidad Nacional de Colombia, Medellín, Colombia, Ángela Adriana Ruiz, Dept. Chemical & Oil engineering, Universidad Nacional de Colombia, Medellín, Colombia, Medellín, Colombia and Camilo alberto Suaréz Méndez, Dept. Chemical & Oil engineering, Universidad Nacional de Colombia, Medellín, Colombia

UTILIZATION OF LIGNIN PREVIOUSLY DEPOLYMERIZED FROM WHITE-ROT FUNGI TO OBTAIN PHENOLIC COMPOUNDS

NATALIACANO[1], ÁNGELA RUÍZ[2], CAMILO SUÁREZ[3]

1Ingeniera Biológica, estudiante maestría Medio ambiente y Desarrollo, Universidad Nacional de Colombia Sede Medellín, nacanol@unal.edu.co, 2,3 Docente Escuela de Procesos y Energía, Universidad Nacional de Colombia Sede Medellín, aaruiz@unal.edu.co , camisua@gmail.com

Abstract: The white-rot fungi play an important role in the depolymerization of lignin because they produce a series of oxidative extracellular enzymes that combining efficiently are capable of completely degrades lignin to CO_2 and H_2O by biological means. The products of enzymatic depolymerization of lignin can be used as intermediates for the production of molecules with aromatic rings in its structure (eg, phenolic compounds).

Given that the lignin is the only renewable resource in the world that contains aromatic compounds and is available in large volumes, has the ability to introduce chemical functionality of these compounds to give a wide industrial application. It is necessary to do research on these processes of degradation from both standpoints: biological and traditional chemical synthesis. Additionally, comparing the lignin with other phenolic compounds, the first exists as a less toxic raw materials and lower price. These phenolics compounds constitute a large number of products and services highly demanded by society.

Keywords: Depolymerized, lignin, laccase, phenolic compounds, white rot fungi

Lignin is the third most abundant natural polymer found in nature, after cellulose and hemicellulose. It is estimated that the amount of lignin in the earth is 300 million metric tons, with an annual biosynthesis production of 20 million metric tons [1]. Its mass content depends on the vegetable species origin. In wood, lignin content varies between 19 and 35% [2].

Lignin is an amorphous polyphenolic material,generated by the copolymerization of three monomers phenyl-propanoic: alcohol coniferyl, alcohol sinaphylco, alcohol and p-coumaryl [3][4]. This three-dimensional polymer is highly branched with a variety of functional groups: phenolic hydroxyl, aliphatic hydroxyl, methoxyl, carbonyl, carboxyl and sulfonates, which provide active sites for chemical and biological interactions [5].

This material comes mainly from industrial processes such as: paper industry (50 million tons of solid per year) [6], and bioethanol production processes. This availability arouses great interest about lignin use, in order to take economic and environmental advantages.

However, the potential use of lignin is far from being exploited and most of the production is burned for energy and chemical recycling process. Only a limited number (1-2%) has been used for different applications [7].

Although lignin is the largest renewable source of aromatic polymers in nature, one of the reasons of its limited use is that lignin is a compound chemically recalcitrant to degradation by most microorganisms, because its complex heterogeneous structure. However, the white-rot fungi produce a range of extracellular oxidative enzymes (being the most common laccases and peroxidases), which are capable of combining efficiently depolymerize lignin completely to CO_2 and H_2O [8]. These types of fungi are involved in the degradation of most chemical structures that compose wood, including cellulose, even when lignin is the main compound degraded [9].

One of the advantages that white rot fungus has is the fact that these microorganisms are extremely efficient at using nitrogen; because in poor condition of it, the strain can use it, especially for the production of extracellular enzymes and essential components of the cell, [10].

Within lignollitic extracellular enzymes, there are two extracellular peroxidases: lignin peroxidase (LiP) and manganese peroxidase (MnP) that generate H_2O_2 in the system, and there are fungi that secrete a combination of peroxidases and oxidases [11]. Within oxidases that produce these fungi are laccases which does not need H_2O_2 to act.

Laccase is a polyphenol (p-diphenol oxidase), it can be inducible or constitutive and intracellular or extracellular depending on growing conditions, which catalyzes the oxidation of a variety of aromatic compounds such as mono, di and polyphenols, aminophenols, and diamines using molecular oxygen as an electron acceptor, reducing it to water [12]. This extracellular glycoprotein has four copper atoms in the oxidation state, giving them a blue color, also has the ability to interact with a redox mediator such as ABTS (2.2 '-azinobis (3 - etilbencenotiazolin 6- sulfonate) for oxidize nonphenolic compounds, which are not substrates of laccases [13]. It also degrades recalcitrant phenols in an oxidative process involving the mediator and the substrate [12]. i.e, the ABTS acts as an activator or co-oxidant of enzyme and is characterized by be a synthetic aromatic compound with nitrogen substitutions and is oxidized by laccase obtaining a stable radical cation $ABTS^+$ by an interaction mechanism enzyme-mediator still unknown [13].

Thus, this enzyme can act as an oxidant-donor from an electron able to penetrate and react within the inner regions of the lignin polymer. The result is the depolymerization of ligning, by biological activity of the laccase enzyme on this material.

The products of enzymatic depolymerization of lignin can be used as intermediates for the production of molecules with aromatic rings in its structure (eg, phenolic compounds).

Given that the lignin is the only renewable resource in the world that contains aromatic compounds and is available in large volumes, has the ability to introduce chemical functionality of these compounds to give a wide industrial application. It is necessary to do research on these processes of degradation from both standpoints: biological and traditional chemical synthesis. Additionally, comparing the lignin with other phenolic compounds, the first exists as a less toxic raw materials and lower price.

This way, this study aimed the isolation of the white-rot fungi from soils of Urabá - Colombia(With potencial lignolytic) with the purpose of determining the ability of these strains to produce the enzyme of interest (laccase) under a variety of culture mediums referenced in the literature for this purpose, by testing coloring cualitative 2,2'-azino-bis(3-etilbenzotiazolin-6- ácido sulfónico) (ABTS). The enzymatic activity was measured and the time of the highest laccase activity under specific conditions of fermentation was determined

After obtaining the enzyme cocktail, it was applied on commercial lignin for identification and quantification of phenolic compounds, formed from the biological pathway of lignin degradation by the enzymatic action of laccases. These kind of phenolic compounds constitute a large number of products and services highly demanded by society (Phenolic resin for industrial abrasives, manufacture of alkyl phenols to additives for lubricating oils, industry Adhesives such as wood and shoe-, Resins for decorative and industrial laminates, insulating varnishes drivers, conservative cosmetic additives, Manufacture of dyes), therefore, this source is considered as an attractive economic and environmental alternative .

The methodology has been organized based on the approach of the overall. The design of the project is done so in 3 stages of development, first tests are performed with enzymes extracted from white-rot fungi (laccases) to determine, by quantitative and qualitative methods, enzymatic activity of these enzymes on lignin (using spectrophotometric technique). Second stage is the depolymerization of lignin and characterization of products formed from the biological degradation of lignin (GC-MS technique) and the third stage faces the industrial application of the products obtained.

REFERENCES

[1] CRESTINI, C., SERMANNI, G.G. AND ARGYROPOULOS D.S. Structural Modifications Induced During Biodegradation of Wheat Lignin by Lentinula edodes: Bioorganic & Medicinal Chemistry. 6, 967- 973,1998.

[2] DENCE, C.W. and LIN, S.Y. General structural features of lignin in: Lin, S.Y., Dence, C.W. (Eds.), Methods in Lignin Chemistry, Springer-Verlag, Berlin Heidelberg, 3–6. 1992.

[3] WANG, M., LEITCH, M. AND XU, C. Synthesis of phenol–formaldehyde resol resins using organosolv pine lignins: European Polymer. 45, 3380–3388, 2009.

[4] JOLIVETA, C., GUILLET, B., KARROUMB, M., ANDREUX, F., BERNOUXD, M. AND ARROUAYSA, D. Les phénols de la lignine et le 13C, traceurs de l'origine des matières organiques du sol: Sciences de la Terre et des planètes / Earth and Planetary Sciences, 651–657, 2001.

[5] VÁZQUEZ, G., ANTORRENA, G., GONZÁLEZ & J. AND MAYOR, J. lignin-phenol-formaldehyde adhesives for exterior grade plywoods: Bioresource Technology . 51,187-192, 1995.

[6] GOSSELINK, R.J.A., DE JONG, E., GURAN, B. AND ABÄCHERLI A. Co-ordination network for lignin-Standardization, production and applications adapted to market requirements (EUROLIGNIN): Industrial Crops and Products. 20,121-129, 2004.

[7] LORA, J.H. AND GLASSER, W.G. Recent application of lignin: A sustainable alternative to nonrenewable materials. J. Polym. Environ. 10, 39-48, 2002.

[8] HAKALAA, T.K., LUNDELLA, T., GALKINA, S., MAIJALAA, P., KALKKINENB, N. AND HATAKKAA, A. Manganese peroxidases, laccases and oxalic acid from the selective white-rot fungus Physisporinus rivulosus grown on spruce wood chips: Enzyme and Microbial Technology. 36, 461–468, 2005.

[9] DOWSON, C.G., RAYNER, A.D.M. AND BODDY, L. Foraging patterns of Phallus impudicus, Phanerochaete laevis and Steccherinum fimbriatum between discontinuous resource units in soil: FEMS Microbiology Letters. 53, 291-298, 1998.

[10] DEACON, J.W. En: Introducción a la Micología Moderna (Editorial Limusa.Primera edición). México, 13-41, 1993.

[11] PERI, F.H. AND GOLD, M.H. Manganese regulation of manganese peroxidase expession and lignin degradation by the white-rot fungus *Dichomitus squalens*: Appl. Environ. Microbiol. 57,2240-2245, 1991.

[12] RODAKIEWICZ-NOWAK, J. Phenols oxidizing enzymes in water-restricted media: Topics in Catalysis. 11, 419-434, 2000.

[13] BOURBONNAIS, R. and PAICE, M. G. Oxidation of non-phenolic substrates An expanded role for laccase in lignin biodegradation: FEBS Letters. 267, 99-106, 1990.

(535g) Quantification of Wheat Straw Lignin by Comprehensive NMR Techniques

Jijiao Zeng[1], Greg Helms[2] and Shulin Chen[1], (1)Department of Biological Systems Engineering, Washington State University, Pullman, WA, (2)Chemistry, WSU

Quantification of wheat straw lignin by comprehensive NMR techniques

Jijiao Zeng[1], Greg Helms[2] and Shulin Chen[1*]

1. Department of Biological Systems Engineering, Washington State University, Pullman, Washington 99164-6120, USA

2. Department of Chemistry, center for NMR spectroscopy, Washington State University, Pullman, Washington 99164-6120, USA

[*]Corresponding author. Tel: +1 509 335 3743; fax: +1 509 335 2722.

E-mail address: chens@wsu.edu

Abstract:

Lignin is secondary natural organic compositions of plant cell wall besides cellulose and hemicelluse. Structurally lignin is linked with hemicelluloses and cellulose to form recalcitrant complex that resist the environmental stress. Aiming to increase efficiency of carbohydrate saccharification, many efforts have been made to remove and/or modify lignin moieties. However, due to limitation of our knowledge on detection and quantification of lignin unit at structural level, the relationship between structural changes and influence on cellulase after these processes is still unclear. In order to provide complete structural information of lignin units, a quantitative method was developed to characterize wheat straw lignin subunits by combination of quantitative ^{13}C NMR and HSQC 2D NMR techniques. Milled wood lignin (MWL) and cellulolytic enzyme-treated lignin (CEL) was isolated from wheat straw and characterized by 500 MHz NMR spectrometer. In addition to common syringyl, guaiacyl and *p*-hydroxyphenol units, various units and side chains including *b*-O-4, dibenzodioxocin, phenylcoumaran, pinoresinol, spirodienone ferrulate, coumarate and *p*-benzonate were identified and quantified. The results also showed that CEL contained more lignin-carbohydrate linkages (LCC) and polysaccharide anomeric compared to MWL. Moreover, the increased amount of ferrulate in CEL illustrated that wheat straw lignin was cross-linked with polysaccharides by ferrulate unit which may limit polysaccharide digestibility in non-woody biomass. *p*-benzonate and *p*-coumarate are both identified as acetylated group in ¦Ã-position of lignin side chains. The approach taken and the information obtained in the study will provide useful information for elucidating mechanisms of pretreatment technologies in terms of changes caused to lignin units.

(542a) Development of a Novel Hot-Carbonate Process for Post-Combustion CO2 Capture: Role of Organic Promoters in Enhanced Absorption and Bicarbonate Crystallization

Manoranjan Sahu, Advanced Energy Technology Initiative, University of Illinois at Urbana-Champaign, Champaign, IL, Qing Ye, University of Illinois at Urbana-Champaign, Champaign, IL and Yongqi Lu, Prairie Research Institute, University of Illinois at Urbana-Champaign, Champaign, IL

Development of cost-effective new technologies are essential for separating and capturing CO_2 from power plant flue gases and retrofitting the CO_2 capture and sequestration (CCS) technology with minimum cost penalty. Currently available monoethanol amine (MEA)-based processes are highly energy-intensive and expensive. A novel hot-carbonate absorption process with bicarbonate crystallization-enabled high pressure stripping (Hot-CAP) is being developed to reduce the energy requirement for CO_2stripping and compression.

In the Hot-CAP, CO_2 is absorbed into a concentrated potassium carbonate solution (about 40wt%) at elevated temperatures (60-80°C). A carbonate–based system tends to have a slower CO_2 absorption rate compared with the MEA. A key technical issue is thus to enhance the absorption rate to a level that is comparable to MEA-based processes. In this study, five organic catalysts and three potassium amino acid salts were investigated as organic promoter to accelerate the absorption rate of CO_2 into the carbonate solution. The rate of CO_2 absorption was measured using a stirred tank reactor (STR) reactor. Results indicated that some selected organic catalysts enhanced the absorption rate significantly and the absorption rates were comparable with that of 5M MEA. The absorption rate depended on the promoter dose and CO_2 loading level in the solution. Additionally, the crystallization of the CO_2 rich solution without and with organic promoters was further investigated under various process conditions. Results from crystallization revealed that potassium bicarbonate prevailed in crystal solids formed.

This presentation will provide a summary of results from the studies of absorption and crystallization in the presence of promoters.

(542b) Process Development for Phase-Changing Aminosilicone-Based CO2-Capture Absorbents

Tiffany Westendorf[1], Benjamin R. Wood[1], Sarah Genovese[1], Irina Sultanova[1], Robert J. Perry[1], Rachel Farnum[1], Michael J. O'Brien[1], Malgorzata Rubinsztajn[1], Ravi-Kumar Vipperla[2] and Robert Enick[3], (1)Global Research, General Electric, Niskayuna, NY, (2)GE Energy, Greenville, SC, (3)University of Pittsburgh, Pittsburgh, PA

Liquid aminosilicone-based absorbents have been developed that have the potential to significantly reduce the cost of CO_2 capture from pulverized coal power plants. Upon exposure to CO_2, these novel absorbents solidify, thereby enabling innovation in CO_2 capture process design. Compared to MEA-based CO_2 capture absorbents, the phase-changing aminosilicones have the potential to reduce CO_2 capture cost through higher CO_2 absorption capacity, greater thermal stability, and reduced solvent losses due to volatility. Furthermore, the phase-changing behavior of these new absorbents and the absence of aqueous diluent enable the use of more energy-efficient separations process units. Finally, by enabling pressurized desorption of CO_2, the energy required to compress the clean CO_2 stream pre-sequestration is significantly reduced.

A process is being developed in which CO_2 is removed from simulated flue gas in a spray absorber, and the solid CO_2-sorbent complex is slurried in liquid absorbent for transport to a pressurized desorber. In the desorber, the solids are heated to release CO_2 and regenerate the liquid absorbent for recycle. Experiments at the lab scale focusing on spray absorption, slurry rheology, and pressurized desorption are in progress to generate data needed to scale up the process. These results and their effect on estimated CO_2 capture economics will be discussed.

(542c) A Techno-Economic Evaluation of Various Stripper Configurations in Amine Absorption Technology for Post-Combustion Carbon Capture

Sumedh S. Warudkar, Kenneth R. Cox, Michael S. Wong and George J. Hirasaki, Chemical and Biomolecular Engineering, Rice University, Houston, TX

As a part of this work, we are comparing the techno-economic performance of the amine absorption process under a variety of stripper operating conditions – ranging from vacuum strippers operating at 30 kPa to high pressure strippers operating at 300 kPa. We are modeling the amine absorption process using the process simulation software ProMax® with TSWEET®. The equipment sizing and capacity data thus generated are the feed parameters for Aspen Process Economic Analyzer which provides the cost estimates for the process equipment.

In our previous studies, we have reported the comparisons of energy consumption and absorber-stripper column sizing for monoethanolamine (MEA), diethanolamine (DEA) and diglycolamine (DGA) at varying concentrations and stripper operating conditions [1]. Of the three amines compared, we've identified 40 wt% DEA and 60 wt% DGA as the more suitable amines for carbon capture application than MEA. We have also shown that an increase in the stripper operating pressure results in a reduced reboiler energy duty and a smaller stripper column [2]. Use of very low pressure steam - waste heat, results in a reduced 'parasitic energy duty' at the power plant but results in larger stripper columns and increased load on the compression train.

It can be seen that there are two contrasting factors involved in deciding the optimal operational configuration – potential energy savings by reduction of parasitic duty and an increase in the capital expenditure with increasing the equipment sizes. Thus, in spite of the detailed technical analysis conducted previously; it hasn't been possible to predict the optimal configuration amongst those compared. The aim of the techno-economic analysis undertaken as a part of this study is to predict with some clarity the better performing absorbent and the corresponding system configuration.

References

1. S. S. Warudkar, K. R. Cox, M. S. Wong and G. J. Hirasaki, "Comparison of alkanolamines for post-combustion carbon capture from a pulverized coal fired power plant", AIChE Fall Meeting, Minneapolis, MN (2011).

2. S. S. Warudkar, K. R. Cox, M. S. Wong and G. J. Hirasaki, "Amine absorption with vacuum stripping for post-combustion carbon capture", AIChE Spring Meeting, Houston, TX (2012).

(542d) Novel Solvent Selection and Solvent Stripping for CO2 Capture From Power Plants

Juan M. Salazar[1], Urmila Diwekar[1], Kevin G. Joback[2] and Abhoyjit Bhown[3], (1)Vishwamitra Research Institute, Center for Uncertain Systems: Tools for Optimization and Management, Clarendon Hills, IL, (2)Molecular Knowledge Systems, Inc., Bedford, NH, (3)Electric Power Research Institute, Palo Alto, CA

Novel Solvent Selection and Solvent Stripping for CO_2 Capture from Power Plants

Juan Salazar and Urmila Diwekar

Center for Uncertain Systems: Tools for Optimization & Management

Vishwamitra Research Institute

Clarendon Hills, IL 60514

E-mail: urmila@vri-custom.org; Tel: 630-886-3047

And

Kevin Joback

Molecular Knowledge Systems, Inc.

Bedford, NH 03110-0755

And

Abhoyjit S. Bhown

Electric Power Research Institute, Inc.

Palo Alto, CA 94304

This paper presents a new approach to solvent selection for post-combustion CO_2 capture from fossil fuel fired power plants. Two main questions arise in a solvent capture process: how to select an effective separating agent and how to design and synthesize this separation process. In previous studies, few solvents were considered for CO_2 capture. Further, researchers arrived at these solvents using experimental methods with limited capacity to screen a large number possible candidates. In this work, we are using computer aided molecular design (CAMD) method to obtain a large number of possible solvent candidates. These candidates are then evaluated by integrating the solvent selection and CO_2 capture process to ensure improved energy, environmental, and economic performance. In this paper we present the first attempt at using group contribution based CAMD to derive new solvents for CO_2 capture.

CO_2 absorption in a solvent-based capture process is essentially based on the reversible, selective nature of the chemical reaction between the liquid solvent and CO_2 in the flue gas. The process flow diagram shown in Figure 1 has been widely used in literature. Flue gas from the boiler is brought into contact with the solvent in the absorber after the removal of impurities such as NOx, SOx and particulate matter using processes that also cool the flue gas. A blower is used to compensate the pressure drop experienced in the absorber. In the absorber, CO_2 selectively absorbs into the solvent, and flue gas leaving the absorber is relatively free of CO_2. The CO_2-rich solvent from the absorber is then pumped to the top of a stripper (or regeneration vessel), via a heat exchanger. Solvent is regenerated in the stripper at elevated temperatures and near atmospheric pressure. The desorption heat required for removing the absorbed CO_2 is provided to the reboiler section of the stripper. The CO_2-lean solvent, containing far less CO_2 is then pumped back to the absorber via a lean-rich heat exchanger to cool it to the operating temperature of the absorber. We used this process diagram to evaluate new solvents.

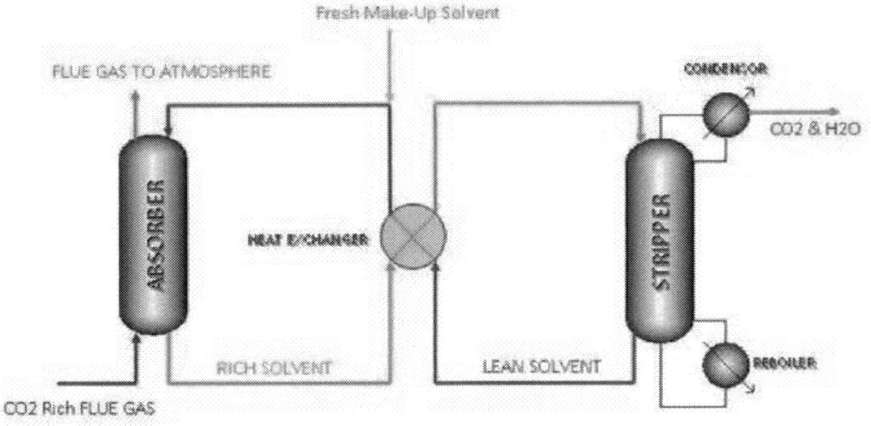

Figure SEQ Figure * ARABIC \s 1 1
Base Absorption Process Flowsheet for CO_2 Capture

In our CAMD approach, we concentrated on amines. In order to use group contribution methods, at first we analyzed the equilibrium performance for these new amines for absorption (CO_2 capture) and for regeneration (solvent stripping). The CAMD resulted in a list that included a group of more than 50 alkyl alcohol amines whose CO_2 solubility properties have not been experimentally determined as far as we know. We evaluated these solvents using ASPEN Plus models and group contribution methods like UNIFAC to evaluate the nonideality. Figure 2 08D0C9EA79F9BACE118C8200AA004BA90B02000000080000000E0000005F005200650066003300320033003000320037003400390030000000 shows the preliminary results of these solvents. The figure shows the energy requirement versus boiling point for 18 solvents simulated both as ideal an nonideal solvents. In general, we observe the nonideality related to these solvents reduces the energy consumption. We will be generating and evaluating additional solvents in order to determine optimum solvents.

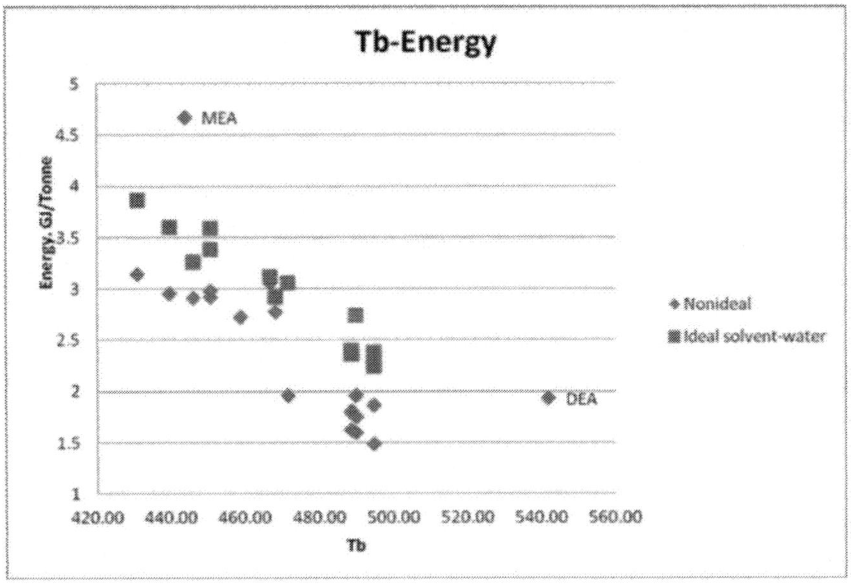

Figure SEQ Figure * ARABIC \s 1 2
Energy Requirements for the New Solvents

(542e) Low Energy Demanding CO2 Capture Process Using Aqueous Amine Solvent

Huiyong Kim, Sogang.ac.kr, Seoul, South Korea and Kwang Soon Lee, Sogang University, Seoul, South Korea

Amine based CO2 capture process has been studied and intensified using the heat integration concept. While many studies on aqueous absorbent-based carbon capture process focus on developing new solvent system, there is much less studies on process intensification to save required energy for CO2 separation.

The purpose of this research has been placed on development of a new process configuration that can achieve a considerable energy saving for CO2 capture. Sensitivity analysis for the absorber and desorber which are two main units of the process provided ideas and possible routes for process modification. With these possibilities, heat flows of the original process were studied for further heat integration to minimize unnecessary heat or coolant usage. With some modifications of the process configuration, heat requirements for solvent regeneration has been reduced considerably. Since minor modifications are proposed, results of this research are belived to be easily applicable to current amine-based carbon capture facilities. The new process has been suggested through a rate-based simulation study using AspenPlus.

(542f) Designing Reversible Ionic Liquids for CO2 Capture

Jackson R. Switzer[1], Amy L. Rohan[1], Kyle M. Flack[2], Emily C. Nixon[2], Amber C. Rumple[2], Elizabeth J. Biddinger[1], Pamela Pollet[2], Charles L. Liotta[2] and Charles A. Eckert[1], (1)School of Chemical and Biomolecular Engineering, Georgia Institute of Technology, Atlanta, GA, (2)School of Chemistry and Biochemistry, Georgia Institute of Technology, Atlanta, GA

Reversible ionic liquids (RevILs) have been developed as alternatives to traditionally cited CO_2 capture solvents such as monoethanolamine (MEA). RevILs capture CO_2 through chemical reaction of CO_2 with an amine functionality to form an ionic liquid. The resulting ionic liquid is then capable of absorbing additional CO_2 through physical absorption. Physical absorption is attractive because it has a relatively low energy penalty for reversal. RevILs also present additional energy advantages over MEA. Whereas MEA is used as a dilute aqueous solution, reversible ionic liquids are designed to be non-aqueous solvents, utilizing only the residual water present in the emission source. This strategy allows for significant energy savings by eliminating the need to heat large amounts of unused water during regeneration, improving the overall efficiency of the CO_2 capture process.

The design of reversible ionic liquids has been an important focal point in our research group. Using structure-property relationships we have shown an ability to tune the relevant CO_2 capture properties by altering the structure of the starting RevIL. Properties targeted include CO_2 capture capacity, ionic liquid viscosity, temperature of CO_2 release, temperature of RevIL evaporation, and the enthalpy of CO_2 absorption. In example, we have observed that by varying the degree and length of alkyl chain branching in the RevIL, we can favorably affect CO_2 capacity, ionic liquid viscosity, as well as the temperatures of CO_2 release and RevIL evaporation. We have also examined several additional modifications to the structure of our RevILs and investigated their effects on CO_2 capture properties. Our progress on developing these structure-property relationships to further the design of optimized solvents for CO_2 capture will be discussed.

(542g) CO2 Post Combustion Capture Using Aminosilicone Solvents

Irina Sultanova[1], Benjamin Wood[2], Sarah E. Genovese[1], Paul Wilson[3], Rachel Farnum[1], Paul Buckley[3] and Anthony Furman[3], (1)Global Research, General Electric, Niskayuna, NY, (2)General Electric, Niskayuna, NY, (3)GE Global Research Center, Niskayuna, NY

CO_2 post combustion capture using aminosilicone solvents

Irina Sultanova, Benjamin R. Wood, Sarah E. Genovese, Paul Wilson, Rachel Farnum, Paul Buckley, Anthony Furman (GE Global Research),

A post-combustion CO_2 capture process using aminosilicone solvents offers advantages over conventional MEA-based aqueous solvent methods. The aminosilicone-based process can potentially decrease the overall size of the system, decrease the amount of required energy for desorption, and therefore reduce the overall cost of CO_2 capture process. A bench-scale system is currently being developed to demonstrate the technical feasibility of continuous CO_2 absorption and desorption with liquid aminosilicones, including recovery and reuse of the solvent. Upon completion of the design and construction, the system will be used to generate the data needed for further scale-up of the process. Some physical properties of the aminosilicones including surface tension, viscosity, heat of CO_2 absorption, and desorption and absorption isotherms were measured and are currently used in design of unit operations. These results and their effect on the cost of CO_2 capture process will be discussed.

(547a) Fractionation and Characterization of Lignin Extractives From E-AFEX(TM) Pretreatment Process

Leonardo da costa Sousa[1], Shishir Chundawat[1], Vijay Bokade[2], Marcus Foston[3], Ali Azarpira[4], Bruce E. Dale[1] and **Venkatesh Balan**[1], (1)Chemical Engineering and Material Science, Great Lakes Bioenergy Center, Michigan State University, Lansing, MI, (2)Catalysis Division, National Chemical Laboratory (NCL), Pune, India, (3)Department of Chemistry and Biochemistry, Georgia Institute of Technology, Atlanta, GA, (4)Great Lakes Bioenergy Center, University of Wisconsin, Madison, WI

Extractive AFEX^{TM} (E-AFEX^{TM}) is a novel pretreatment technology that is able to simultaneously convert cellulose I to cellulose III and partially extract lignin from lignocellulosic biomass. As a result, pretreated feedstocks contain highly digestible carbohydrate content and significantly reduced lignin content. Depending on the pretreatment conditions, it is possible to obtain variations in the selectivity towards lignin solubilization. Temperature, ammonia concentration and co-solvent type are the major factors that contribute to a selective lignin solubilization and subsequent fractionation. Some pretreatment conditions tend to solubilize ash and hemicellulose residues along with lignin. However, since most applications for lignin require low ash and carbohydrate contents, these extractives have been further fractionated to obtain varying purity lignin fractions. For this purpose, solvent precipitation and filtration based methods weredeveloped to fractionate lignin. In this work, multiple fractions generated from the E-AFEX^{TM} extractives generated from pretreated corn stover were characterized by several techniques, which include GPC, TGA, ^{13}C-NMR, ^{1}H-NMR and 2D NMR. A comparison of E-AFEX^{TM} lignin with other commercial lignin sources is also presented here. The level of lignin extraction and cellulose III conversion was correlated with cellulose digestibility for corn stover. Furthermore the benefits of this new pretreatment methodology to downstream processing are presented in comparison with conventional AFEX^{TM}.

(547b) Effects of Mechanical Refining On the Enzymatic Digestibility of Acid Pretreated Corn Stover

Xiaowen Chen[1], Erik Kuhn[1], Wei Wang[1], Todd B. Vinzant[2], Sunkyu Park[3], Keith Flanegan[4] and Melvin Tucker[5], (1)National Renewable Energy Laboratory, Golden, CO, (2)Biosciences Center, National Renewable Energy Laboratory, Golden, CO, (3)Department of Forest Biomaterials, North Carolina state university, Raleigh, NC, (4)KL Energy Corporation, Rapid City, SD, (5)National Biology Center, National Renewable Energy Lab, Golden, CO

Mechanical refining has been applied in the pulp and paper industry for decades. It has been shown that mechanical refining provides better pulp quality through three different mechanisms: 1) shortening the fiber; 2) crushing and fibrillating the fibers internally; and 3) abrasion and fibrillation of the fibers on the exterior surfaces. However, mechanical refining has not been widely studied for the purpose of application to the concept of the bio-refinery. Recent studies by Koo et al. showed that mechanical refining using a laboratory PFI mill could potentially increase glucose yields in enzymatic saccharifications up to 20% using green liquor pretreated hardwood. The refining reduced the enzyme usage by ~50%. However, studies on disk refining and blending reported by Zhu et al and Dibble et al, respectively, showed minor effects on digestibility in terms of increased glucose yields. In this research, we explored the effects of five different refining methods including PFI milling, disk refining, extrusion, Szego milling and blending. The effects of the various refining methods on enzymatic hydrolysis using low severity pretreated corn stover substrates were investigated. The effects of different methods of refining on pretreated biomass are characterized by measurements of particle size distributions, degree of polymerization (DP), and crystallinity of cellulose, ultrastructural changes using SEM imaging, and enzymatic digestibility are reported.

(547c) Isoconversional Methods to Model Oxidative Alkaline Pretreatment of Lignocellulose

Mayra del Pilar Quiroz Galvan[1], Christian Andres Nova Salazar[2], Jonatan Garcia Mape[2], Juan Manuel Rivera Aragon[2] and Rocio Sierra[3], (1)Chemical Engineering Department, Universidad de los Andes, Bogotá, Colombia, (2)Chemical Engineering, Universidad de los Andes, Bogotá, Colombia, (3)Chemical Engineering, Universidad de los Andes, Bogota, Colombia

Oxidative alkaline pretreatment has proven successful to increase lignocellulose digestibility. In order to properly assess its effects, it is necessary to kinetically model the process. Most efforts to model pretreatment obtain kinetic parameters (i.e., activation energy and frequency factors) from model fitting methods using rate laws. The Isoconversional method is an alternative that in addition to being more accurate because does not depend on a so called constant temperature, is also usually simpler. In spite of these facts, isconversional methods have not been applied to kinetically model pretreatment.

The present work applies isoconversional methods to model oxidative alkaline pretreatment of sugar cane bagasse addressing a main difficulty inherent to the method: the need to provide large amount of data. This has been accomplished by implementing a novel procedure to estimate lignin and carbohydrates contents in lignocellulose. It uses the partial least square (PLS) method combined with Near Infrarred (NIR) Spectroscopy Analysis. The method was proven succesful to obtain kinetic parameters that were shown to be temperature dependent.

(547d) Quantitative Predictions of Bioconversion of Aspen by Dilute Acid and SPORL Pretreatments Using a Unified Combined Hydrolysis Factor (CHF)

Wenyuan Zhu[1], Carl J. Houtman[2], J.Y. Zhu[2] and Rolland Gleisner[3], (1)Nanjing Forestry University, Nanjing, China, (2)USDA Forest Service, Forest Products Laboratory, Madison, WI, (3)US Forest Service, Madison, WI

A combined hydrolysis factor (CHF) was developed to predict xylan hydrolysis during pretreatments of native aspen (*populus tremuloides*) wood chips. A natural extension of previously developed kinetic models allowed us to account for the effect of catalysts by dilute acid and two sulfite pretreatments at different pH values. When xylan is modeled as two fractions with different hydrolysis rates as identified as fast and slow xylan, the model closely matches the experimental data. Extent of xylan hydrolysis is strongly correlated with pretreatment solids yield, energy consumption for size reduction, and substrate enzymatic digestibility (SED). Composition of the pretreatment hydrolysate was less correlated with extent of hydrolysis due to carbohydrate decomposition reactions. Substrate cellulose enzymatic conversion and enzymatic hydrolysis glucose yield can be predicted to approximately 10% accuracy using *CHF* alone. The conventional combined severity fcator is a special and simplied case of CHF. The CHF developed has significant practical importance for scale up experiments which cannot afford to numerous experiments for optimization.

(547e) Study of A Novel Biomass Pretreatment Process Utilizing A Modified Taylor-Couette Mixer and the Effect of Cationic Polyelectrolyte On Alkaline Pretreatment and Enzymatic Hydrolysis

Shaowen Ji, Wei Wang and Ilsoon Lee, Department of Chemical Engineering and Materials Science, Michigan State University, East Lansing, MI

Lignocellulosic biomass has been considered as one of the primary candidates as alternative renewable resources for biofuel production. The presence of lignin is widely recognized as the barrier to cost-effective production, which inhibits enzyme accessibility and decreases sugar yields. The costly pretreatments and large consumption of enzymes are needed to reach high cellulose conversion. Therefore, the removal or effective modification of lignin is required for efficient pretreatment process. Here we report an efficient biomass pretreatment conducted via a novel modified Taylor-Couette mixing system, aiming to overcome the natural recalcitrance of lignocellulosic biomass in a short processing time and facilitate better enzyme accessibility to polysaccharides. Both lignin and hemicellulose were significantly removed after 2 minutes processing, leaving rich glucan content in the residue. Studies on the experimental parameters, such as mixing speed, substrate loading, and enzyme digestibility, were systematically performed. We further investigated the effect of cationic polyelectrolyte on the alkaline pretreated biomass. Structural studies revealed the morphology change of the cell wall in the residue, showing the retention of lignin-containing nanoparticles. The cellulose and hemicellulose conversion were enhanced after pretreatment, suggesting the possibly positive effect of polyelectrolyte on significantly modifying biomass morphological properties, such as increased porosity and improved accessibility of cellulose surface.

(547f) Applying Partial Least Square (PLS) to Quantify Lignin and Carbohydrates Content of Lignocellulose

Rocio Sierra, Chemical Engineering, Universidad de los Andes, Bogota, Colombia

Analytical chemistry is often employed to measure lignin and carbohydrates content of lignocellulose. Most traditionally used methods are long, comprise a large number of steps, are intensive in reactants use, require highly sophisticated equipment such as HPLC and most of the time result in a large standar deviation for replicates. If a large set of data is required, for example when determining kinetic parameters of pretreatment, this method is an important limitation.

Partial Least Square (PLS) combined with Near Infrared Spectroscopy (NIR) was used to determine lignin and carbohydrates content of lignocellulose in solid phase. A calibration curve was obtained using about 100 samples of lignocellulosic material with different lignin and carbohydrates content. An ample spectrum of lignin and carbohydrates contents was attained by using different degrees of maturity and plant parts, as well as by some modifications on the lignocellulose compositional structure through the application of different types of pretreatment. The method was shown succesful to quantify lignin and carbohydrates and much more faster than other traditionally used methods. However, an important limitation is the need to obtain a new calibration curve for each type of biomass used.

(571a) Engineering Yeast for Advanced Biofuels

Jose L. Avalos[1,2], Gerald Fink[1] and Gregory Stephanopoulos[2], (1)Whitehead Institute, Cambridge, MA, (2)Department of Chemical Engineering, Massachusetts Institute of Technology, Cambridge, MA

The development of renewable liquid transportation fuels is a key component of the national strategy to reduce CO_2 emissions, slow the progression of global climate change, and achieve energy self-sufficiency. Although bioethanol is the most commercially successful biofuel to date, its physicochemical properties and current methods of production make it an inadequate long-term solution to these problems. We are developing yeast strains that produce advanced biofuels that overcome ethanol's limitations. In order to accelerate strain development, we have developed a set of molecular tools and strategies that facilitate the cloning, assembly, integration, and optimization of engineered pathways. We demonstrate the use of this molecular toolkit to overcome roadblocks in the molecular and cellular biology of yeast that present bottlenecks in metabolic engineering, such as the screening of isoenzymes, vectors, promoters and strains. These molecular tools also facilitate new strategies for the assembly of multigenic pathways, which we used in the combinatorial assembly of multiple "isopathways" for the production of advanced biofuels. Using these tools we have engineered yeast strains that produce three different advanced biofuels with better fuel properties than ethanol, and better compatibility with prevalent infrastructures.

(571b) Phosphate Optimization for Economical Lipid Production From Lipomyces Starkeyi Grown On Starch

Sharif M. Rahman, Ramalingam Subramaniam, Stephen Dufreche, Mark Zappi and Rakesh Bajpai, Chemical Engineering, University of Louisiana at Lafayette, Lafayette, LA

Phosphate optimization for economical lipid production from *Lipomyces starkeyi* grown on starch

Sharif M. Rahman ,Ramalingam Subramaniam, Stephen Dufreche, Mark Zappi and Rakesh Bajpai, Chemical Engineering, University of Louisiana at Lafayette, Lafayette, LA

Microbial lipids can be obtained by the oleaginous yeast *Lipomyces starkeyi* using sweet potato starch as carbon source which is available largely from processors during cutting and canning process. But high cost of production is the major barrier for industrial production of microbial lipids. This cost can be reduced by optimizing nutrient components. Phosphorus is one of the highest cost contributing elements for microbial culture; and it is mostly incorporated with phospholipids and nucleic acids into the cells. It is under investigation that reduction of phosphate concentration affects cell growth as well as lipid accumulation. However, it is predicted that there is an optimum point of phosphate level where lipid accumulation does not change significantly. This work mainly focuses on optimizing phosphate concentration of the medium since it is a vital factor for economical microbial lipid production using waste materials.

(571c) Platform Pathway for the Synthesis of 3-Hydroxyacids As Value Added Products Derived From Biomass

Himanshu H. Dhamankar and Kristala L.J. Prather, Chemical Engineering, Massachusetts Institute of Technology, Cambridge, MA

The dwindling nature of petroleum and other fossil reserves has provided an impetus to microbial synthesis of fuels and value added chemicals from biomass as a renewable resource. The use of biomass as a feedstock requires the design and engineering of novel pathways connecting biomass derived simple sugars to such value added chemicals and fuels. 3-Hydroxyacids (3-HAs) are a versatile class of molecules that can serve as value added products derived from biomass. 3HAs such as 3-hydroxybutyrate and 3-hydroxyvalerate serve as monomers for biodegradable polyhydroxyalkanoates (PHAs) and chiral building blocks for pharmaceuticals and chemicals. We have established a novel platform pathway in *E. coli* that allows stereospecific synthesis of straight and branched chain 3HAs using glucose and small acid molecules as starting materials. Using this pathway we have been successful in establishing the first biosynthetic route towards the valuable pharmaceutical building block 3-hydroxy-γ-butyrolactone (3-HBL) and demonstrated synthesis of two novel PHA monomers: 3-hydroxyhexanoic acid and 4-methyl-3-hydroxyvaleric acid. Exploiting the full potential of this pathway requires exploring alternative pathway enzymes and engineering them to enhance pathway productivity and improve versatility for the synthesis of other valuable 3HAs. In particular, we are interested in enhancing pathway productivity for synthesis of 3-HBL which features as one of the top ten value added chemicals from biomass in a 2004 Department of Energy report and serves as a versatile chiral building block for the synthesis of drugs like Zetia®, cholesterol reducing synthetic statins such as atorvastatin (Lipitor ®) and rosuvastatin (Crestor ®), antibiotics such as Zyvox® and the nutritional supplement L-Carnitine. Ongoing efforts to identify alternative pathway enzymes for extending the 3HA pathway towards additional value added products as also enhancing enzyme activity and specificity through protein engineering towards specific products will be discussed.

(571d) Engineering a Yeast Conversion System for the Novel Production of Alkanes

John Blazeck, Department of Chemical Engineering, University of Texas at Austin, Austin, TX and Hal Alper, Chemical Engineering, The University of Texas at Austin, Austin, TX

Bio-derived alkanes are attractive second generation biofuels that circumvent some of the disadvantages associated with traditional biofuels. Engineering alkane production in yeasts relies upon the metabolic engineering of fatty acid metabolism. Fatty acids consist of a long aliphatic carbon chain decorated with only one carboxylic acid group, and thus serve as excellent substrates for alkane pathway engineering. An attractive host for alkane production is *Yarrowia lipolytica*, an oleaginous yeast adept at producing high levels of fatty acids. Here, we report the novel heterologous production of pentane in a *Y. lipolytica* fungal system by assembling a plant-derived pentane-production pathway. Pentane production was optimized using environmental and genetic techniques including increasing precursor availability through genetic knockouts and media formulation and by employing promoter engineering and codon optimization for high enzyme expression levels. Here, we will present the results of this optimization and discuss strategies for improved pentane production and for long chain alkane production in this yeast.

(571e) Engineering of Fatty Acid Pathway and Hydrocarbon Production in E. Coli

Fengming Lin, chemical engineering, University of michigan, ann arbor, MI, Yu Chen, Department of Chemical Engineering, University of Michigan, Ann Arbor, MI, Robert Levine, Chemical Engineering, University of Michigan, Ann Arbor, MI, Neil Marsh, chemistry, University of michigan, ann arbor, MI and Xiaoxia Nina Lin, Chemical Engineering, University of Michigan - Ann Arbor, Ann Arbor, MI

To improve the efficiency of producing fatty acid derived hydrocarbons, such as alkane, alkene, and alkyl ester, in model microorganism *Escherichia coli,* we are engineering pathways related to fatty acid biosynthesis using several strategies. First, a metabolic network bi-level optimization model was developed and applied to strain design for maximum fatty acid production. The gene manipulations suggested by this model were performed and led to an increase of total fatty acid from 58.5 mg/g dry cell weight in the wild type to 84.9 mg/g dry cell weight in a multiple knockout strain. Second, the triacylglycerol biosynthesis pathway was successfully introduced into *E. coli* by heterogeneous expression of wax ester synthase/acyl-coenzyme A: diacylglycerol acyltransferase (WS/DGAT) enzymes together with the knockout of dgkA, which not only increases the total amount of fatty acids in *E. coli*, but also improves the fuel quality of fatty acids. Third, an alkane biosynthesis pathway of cyanobacterial origin is being explored by utilization of synthetic protein scaffolds which spatially organize the enzymes involved. All combined, these strategies should lead to the improvement of titer and yield of producing fatty acid derived hydrocarbons in microorganisms.

(571f) Metabolic Engineering of Clostridium Tyrobutyricum for Isopropanol Production

Wenyan Jiang, Chemical Engineering, The Ohio State University, Columbus, OH, Mingrui Yu, Department of Chemical and Biomolecular Engineering, The Ohio State University, Columbus, OH and Shang-Tian Yang, The Ohio State University, Columbus, OH

Isopropanol is one of the most widely used solvents in the world. It is also one of the secondary alcohols that can be used as a direct or partial replacement for gasoline. Isopropanol can be produced by anaerobic microorganisms such as *Clostridium beijerinckii*. However, the native metabolic pathways are limited by tight regulation and the isopropanol production is low, less than 30 mM, in these strains. *Clostridium tyrobutyricum* is an anaerobic acidogen which can convert glucose and xylose to acetate and butyrate. Previously, our group had engineered this bacterium to produce n-butanol at a high rate, yield, and titer that are even higher than most of those reported native solventogenic *Clostridia*. In this work, we studied the feasibility of enabling C. *tyrobutyricum* to produce isopropanol. A plasmid was constructed with *C. acetobutylicum ctf* (acetoacetyl-CoA transferase), *adc* (acetoacetate decarboxylase) and *C. beijerinckii adh* (secondary alcohol dehydrogenase) genes under *C. tyrobutyricum* ATCC 25755 *thl* (acetoacetyl-CoA transferase) promoter. The plasmid was transformed into both *C. tyrobutyricum* wild type and *ack* (acetate kinase) knockout mutant strains. Isopropanol production by these mutants were confirmed in batch fermentation with glucose as the carbon source. The effects of isopropanol production on cell growth and fermentation kinetics were studied and will be presented in this paper.

(571g) Renewable Production of 5-Carbon Polyamide Building Blocks Using Engineered E. Coli

Jake Adkins, Chemical Engineering, Arizona State University and **David R. Nielsen**, Chemical Engineering, Arizona State University, Tempe, AZ

Polyamides are a class of homopolymers and copolymers with properties that balance excellent mechanical strength with durability and chemical resistance. Common examples include nylons which are produced from amino acid or diamine and diacid monomer building blocks. Today, such monomers are principally derived from petroleum resources. In recent years, however, the microbial production of 4- through 6-carbon monomers such as succinic acid, adipic acid, putrescine, and cadaverine have been reported. Here we report how, through *de novo* pathway engineering, our lab has been exploring the 'bottom up' development of microbial biocatalysts to produce other remaining polyamide monomer building blocks from renewable feedstocks. Systematic over-expression of the natural lysine degradation pathway enzymes of *Pseudomonas* sp. in a lysine over-producing *Escherichia coli* strain has allowed us to produce, as individual products, both 5-aminovaleric acid and glutaric acid (precursors to nylon-5 and copolymers such as nylon-5,5, for example). To date, genes from the *davBADT* operon of *P. putida* KT2440 has proven to be support the highest productivities in shake flask cultures. Alternative expression strategies been designed to compare the utility of the natural operon versus the individualized expression of pathway enzymes with the latter proving most effective. In shake flask cultures containing glucose in minimal media, our initial strains have already achieved 5-aminovaleric acid and glutaric acid titers as high as at least 350 mg/L. Here we report an overview of our most recent findings in this project. In all cases of strain development, systematic pathway development will be described through the recruitment and screening of candidate isoenzymes from different genetic sources for each step. In addition, to support greater overall productivity, strategies aimed at increasing precursor availability, improving pathway flux, and overcoming product toxicity are also described.

(580a) Performance of a Pilot Scale Membrane Aerated Biofilm Reactor for the Treatment of Landfill Leachate

Eoin Syron, School of Chemical and Bioprocess Eng,, University College Dublin, Belfield, Dublin, Ireland and Eoin Casey, School of Chemical and BioProcess Engineering, University College Dublin, Dublin, Ireland

Performance of a pilot scale membrane aerated biofilm reactor for the treatment of landfill leachate

Eoin SYRON, Eoin CASEY,

School of Chemical and Bioprocess Engineering, University College Dublin, Belfield, Dublin 4, Ireland

Keywords: Oxygen Permeable Membranes, Bubbleless Aeration, Biofilm, Wastewater treatment

Aeration accounts for over 60% of the total energy spend in wastewater treatment plants, and even with recent technical improvements, bubble aeration can only achieve 50% oxygen transfer efficiency. Due to increasing energy costs and increasing regulatory pressure, there is a substantial growing interest in the application of diffusive gas transfer membranes for bubbleless aeration. These membranes avoid the mass transfer limitations associated with bubble-liquid interactions and increase the gas residence time within the reactor allowing more of the supplied oxygen to be transferred. Unlike conventional membranes in wastewater treatment, the concept of biofouling is not discouraged and the resultant biofilm on the membrane plays an active role in the degradation of COD and Nitrogenous constituents in wastewater.

Landfill leachate is water which has percolated through the waste bed in a landfill and leached organics and nutrients from the solid waste. This wastewater must be treated before being discharged to a water course to prevent pollution of the receiving water. Typically leachate has a high ammonia concentration due to the anaerobic degradation which takes place in the landfill. The Membrane Aerated Biofilm Reactor (MABR) is ideally suited for treatment of a high strength ammonia wastewater firstly as to the slow growing nitrifiers are attached to the membranes there is no need of an extra biomass retention operation, secondly the location of the nitrifying organisms on the membrane surface and their high oxygen demand exploit the oxygen transfer advantages of the MABR. While previous studies have reported on the MABR for ammonia nitrification, use of the MABR for the treatment of leachate has not previously been undertaken.

A 60 litre pilot plant (figure 1) was constructed to treat leachate from Arthurstown landfill, Co Kildare, Ireland. The pilot uses hollow fibre PDMS membranes supplied by SilPro (Broadstairs, UK) which have an internal diameter of 300mm and an external diameter of 500 mm giving a wall thickness of 100 mm and a total surface area of 4.2 m^2 . PDMS membranes were chosen for this application due to their durability, ease of supply, and high oxygen permeability. In contrast to porous membranes, PDMS does not suffer from problems such as intra membrane wetting and pore clogging. Although the surface roughness of PDMS is quite low, there have been few reports of poor adhesion or biofilm formation on these membranes. The hollow fibres were potted in bunches of 60 fibres 0.4m long into 10mm PVC piping using polyurethane as the potting agent. Bunches were then arranged in cassettes of 11 bunches and 10 cassettes were placed in the reactor. The pH is controlled via a feedback loop and maintained between 7.5 and 8 with the addition of Sodium Hydroxide or Hydrochloric acid. Initial Oxygen Transfer Rates OTR experiments determined by re-oxygenation test gave results of $50gO_2\ m^{-2}\ bar^{-2}\ day^{-1}$.

Currently the reactor is removing $2.5g\ N\text{-}NH_4\ m^{-2}\ day^{-1}$which is comparable with previously reported systems Syron & Casey 2008, but it is foreseen that this rate of reaction will continue to increase with time and thicker biofilm growth, a theoretical maximum of $15\ g\ N\text{-}NH_4\ m^{-2}\ day^{-1}$ if all of the oxygen supplied is used for oxidation to nitrite. The current pilot is one of the few pilot scale studies of a MABR. Previous attempts have encountered problems with the control of excess biofilm resulting in clogging in the membrane module. In the present study the biofilm control strategy is allowing the pilot to continue with stable performance over a prolonged period.

References

Casey, E.; Glennon, B.; Hamer, G. Oxygen mass transfer characteristics in a membrane-aerated biofilm reactor. *Biotechnol.*

Bioeng. **1999**, *62* (2), 183–192.

Syron, E.; Casey, E. Membrane-aerated biofilms for high ratebiotreatment: Performance appraisal, engineering principles,scale-up, and development requirements. *Environ. Sci. Technol.***2008**, *42*, 1833–1844.

(580b) Symbiotic Hollow Fiber Membrane Photobioreactor for Microalgae Growth and Bacterial Wastewater Treatment

Linh Vu T. K., Department of Chemical and Biomolecular Engineering, National University of Singapore, Singapore 117576, Singapore and Kai-Chee Loh, Department of Chemical and Biomolecular Engineering, National University of Singapore, Singapore, Singapore

For decades, microalgae have gained much interest for production of high value compounds for food, feed, cosmetic and pharmaceutical products. With rapid depletion of fossil fuels and severe effect of greenhouse gas emissions on global climate change, many microalgae strains are potential sources for biodiesel production. Much attention is now focused on the development of a suitable photobioreactor that can assure mass production of high quality microalgal biomass. One of the crucial factors in designing a photobioreactor is the efficient gaseous transfer of CO_2 supply and O_2 removal.

Symbiotic microalgal-bacterial processes which incorporate microalgae and bacteria in one closed system could provide a potential solution for CO_2 supply and O_2 removal with low energy consumption. In this system, microalgae can produce sufficient O_2 for bacteria to biodegrade organic matter, while bacteria concomitantly release CO_2 for microalgae during photosynthesis. This process, so called photosynthetic oxygenation, is a promising approach to reduce the energy cost for intensive mechanical aeration and potentially eliminate CO_2 emission. For years, photosynthetic oxygenation has been applied not only for economical treatment of wastewater but also for degradation of hazardous contaminants. However, when growing microalgae and bacteria together, slow growth rate of microalgae and unexpected interactions between microalgae and bacteria may result in process failure. More importantly, microalgal biomass from the microalgal-bacteria consortia cannot be used flexibly for the production of food, feed or other high value compounds.

In this research program, we proposed to develop a hollow fiber membrane photobioreactor (HFMP) for symbiotic microalgal growth and bacterial wastewater treatment. In the proposed bioreactor, a gas exchange membrane will be used to physically separate microalgae and bacteria cultures as well as to solely facilitate transfer of CO_2 and O_2. The overall benefits of the proposed membrane bioreactor are to reduce the energy cost for mechanical aeration in an activated sludge process, reduce CO_2 emission to the atmosphere and most importantly, harvest microalgal biomass for extracting other value-added products.

Results obtained from the developed hollow fiber membrane photobioreactor (HFMP) demonstrate a proof-of-concept of the symbiotic relationship that existed between microalgae and activated sludge in wastewater treatment. In the absence of aeration in the synthetic waste water, BOD of synthetic wastewater was completely degraded while microalgae grew continuously in the absence of the external CO_2 supply. This concept hence can be applied for existing activated sludge treatment tanks in order to reduce the demand of aeration energy, and reduce CO_2 emission to the atmosphere. Most importantly, additional produced microalgal biomass can be harvested for other purposes such as for food, feed or biodiesel production.

(580c) Increased Operational Flexibility and Robustness of Anaerobic Wastewater Processes Via Bioreactor Coupling

Satish J. Parulekar, Chemical and Biological Engineering, Illinois Institute of Technology, Chicago, IL

Anaerobic wastewater purification processes have been increasingly used in the recent years. Anaerobic digestion is a biological process in which organic matter is transformed by a community of microorganisms into biogas (primarily methane and carbon dioxide) in the absence of oxygen. This process appears to be a promising method to solve some energy and ecological problems in agriculture and industry. Advantages this process has over the other common waste treatment process, the aerobic activated sludge process, include (i) much lower energy requirement; (ii) generation of less sludge, curtailing thereby the cost of energy-intensive post-treatment and disposal; and (iii) generation of biogas, which can be used as an energy source. In the past few years, a number of experimental studies aimed at improving the performance of anerobic digestion and kinetic models for the process have been reported. The process involves a consortium of acid generating bacteria (host population) and methane generating bacteria (commensal population). The initiation of this process requires hydolysis of complex insoluble organics by extracellular enzymes synthesized by the host population to produce simple soluble organics, which can be utilized by the host population. A comprehensive analysis of static and dynamic behavior of a mixed culture of involving acidogens (X) and methanogens (Y) in two coupled bioreactors is presented here. The two-bioreactor system is considered here to make the activated sludge process more robust with respect to wide fluctuations in feed conditions. A single continuous culture may operate at up to seven steady states, including up to four coexistence steady states, with only one coexistence steady state being locally stable. The one-way interaction between X and Y allows for compartmentalization of the system for a stand-alone bioreactor and two coupled bioreactors into two subsystems, which facilitates the analysis of steady state types and stability characteristics of these and classification of dynamic behavior. The bioreactors in the two-reactor system are identical only in terms of feed composition and bioreactor space time. The reactors need not have identical volumes and volumetric feed, effluent, and exchange rates. A two-reactor system may admit up to forty nine steady states, which are comprised of up to forty coexistence steady states, at least at very low interaction rates. The interaction rates between the two bioreactors, R_1 and R_2, are considered to be unequal. The steady states are comprised of symmetric steady states and asymmetric steady states. The symmetric steady states correspond to identical composition in the two bioreactors. The remaining steady states, the asymmetric steady states, correspond to different composition in the two bioreactors and are admissible over a range of interaction rates. The static and dynamic analysis of the two-reactor system is facilitated by appropriate grouping of large number of steady states arising for very low R into nine clusters. Numerical illustrations reveal the rich steady state structure of the bioprocess in coupled bioreactors. While a single bioreactor can operate at only one locally stable coexistence steady state, the coupled bioreactors can operate at up to five locally stable coexistence steady states over certain ranges of R_1 and R_2. The two-reactor system is operationally more flexible and more robust vis-a-vis single reactor as concerns maintenance of mixed culture. Emergence of four additional steady state clusters and additional coexistence and partial washout steady states at intermediate values of interaction rates reveals that the coupled bioreactors are an example of a complex system.

(580d) Two Phase Biodegradation of Phenol in a Hollow Fiber Supported Liquid Membrane Bioreactor

Prashant Praveen, Chemical and Biomolecular Engineering, National University of Singapore, Singapore, Singapore and Kai-Chee Loh, Department of Chemical and Biomolecular Engineering, National University of Singapore, Singapore, Singapore

During biodegradation of toxic aromatic compounds, microorganisms often experience substrate inhibition wherein the cell growth and metabolism are inhibited at higher concentration of the toxic substrate. Several bioreactors have been proposed for the alleviation of substrate inhibition. Most of these are based on the use of immobilized cells but with the advent of two phase partitioning bioreactor (TPPB), this is set to change. TPPBs are based on the equilibrium distribution of pollutants between cell culture medium and an immiscible, biocompatible and non-biodegradable organic solvent with high affinity for the substrate. Although TBBPs attain very high growth and biodegradation rates and would have made an ideal system; but various problems such as foaming and emulsification, arising from phase dispersion limit their applicability.

The operational challenges encountered in conventional TPPBs can be obviated by non-dispersive contacting between the two phases using semi-permeable membranes. In this research, a hollow fiber supported liquid membrane bioreactor (HFSLMB) was developed for simultaneous extraction and biodegradation of phenol from wastewater using *Pseudomonas putida*. In a semi-dispersive approach, extraction was carried out by dispersing the organic solvent into feed solution whereas the transport of substrate from solvent to the cells through the liquid membrane was non-dispersive. The dispersed solvent droplets facilitated liquid membrane renewal and resulted in a stable mass transfer flux while in the absence of any solvent, cell growth environment resembled that in a monophasic system. *P. putida* in suspension was able to biodegrade inhibitory phenol concentrations at 1000-4000 mg/L without experiencing severe substrate inhibition. For example, biodegradation of 4000 mg/L was achieved within 76 hours while the specific growth rate and biomass yield were 0.31 hr^{-1} and 0.26 g/g, respectively. Phenol biodegradation in the HFSLMB occurred in two stages: an exponential removal rate was observed in the beginning which subsequently assumed a linear profile under diffusion limitation. The specific growth rates during the exponential growth phase could be controlled by increasing the volume of the solvent while the removal rate under mass transfer limitation improved by changing the operating conditions and the interfacial area. In evaluating the long-term stability of the HFSLMB, the bioreactor was repeatedly operated in batch mode with 1000 mg/L phenol. The biodegradation performance improved during the first few runs due to the presence of attached cells on the membrane. But the performance deteriorated subsequently at higher biofilm thickness. The biofilms could be removed by alkaline washing lasting less than 5% of the operation time and resulting in the restoration of the biodegradation performance. With two washing cycles, HFSLMB operation could be sustained for 416 hours in which 20 batch runs were conducted.

The HFSLMB offered many advantages that included a better growth environment for the microorganisms, lower energy demand, high separation factor, high removal rates and elimination of secondary waste. The modular design is easier to scale-up and can also be used for continuous operation. Modeling the HFSLMB will provide further insights into substrate mass transfer and metabolism.

The novel anaerobic-circulating fluidized bed bioreactor (A-CFBBR) For High rate biological nutrient removal from high strength wastewater

MEHRAN ANDALIB , GEORGE NAKHLA *, JESSE ZHU

Introduction

High strength wastes produced by fertilizer production, explosive manufacturing, and recovery of nuclear fuels as well as landfill leachate contain in addition to organic matter, nitrogen in the forms of ammonia, nitric and nitrous acids (Chen and Lin, 1993). Similarly, many food processing wastewaters such as rendering wastes also contain high ammonia concentrations as a result of protein digestion. Nitrogen removal from such wastes, which require anaerobic processes to recover energy, can be challenging.

The Circulating Fluidized Bed Bioreactor (CFBBR), introduced and developed by Nakhla and his coworkers (Cui et al., 2004), was tested for biological nutrient removal (BNR) from municipal wastewaters in both lab and pilot scales at OLR of 5 kg COD/m^3·d and NLR of 0.5 kg N/m^3·d (Chowdhury et al., 2010). The CFBBR consists of an anoxic riser and an aerobic downer with fast and conventional fluidization regimes respectively. More than 90% organic, 70%-80% total nitrogen and 50%-70% phosphorous removal were reported, at hydraulic retention times (HRTs) of 2-3 h and an observed biomass yield of 0.12-0.16 g VSS/g COD. The CFBBR was reported to treat high strength wastewater such as landfill leachate and septage as well at the aforementioned OLR and NLR (Eldyasti et al., 2010). However, in order to accomplish high rate biological nutrient removal from high strength waste containing nitrogen, a newly developed bioreactor referred to henceforth as anaerobic-circulating fluidized bed bioreactor (A-CFBBR) was designed and tested to treat synthetic high strength wastewater with TCOD of 10600±270 mg/L, NH_3-N of 250±10 mg/L. The advantages of the A-CFBBR over conventional anaerobic treatment is simultaneous high rate nitrogen and carbon removal as well as energy recovery as opposed to only methanation.

Materials and Methods

System description and operating conditions

The A-CFBBR (**Figure 1**) is comprised of a conventional anaerobic fluidized bed (AF) with an overall height of 3.6 m followed by a CFBBR comprising anoxic and aerobic fluidized bed bioreactors (FBRs) with heights of 2 m and 3.6 m respectively. Zeolite particles (3 kg) with an average diameter (d_m) of 425-610 μm were used as carrier media in the anaerobic column and 3 kg zeolite particles with average diameter of 610-825 μm, were circulated between the aerobic and anoxic fluidized bed columns of CFBBR. The particle weight hold up in the aerobic column of the CFBBR was maintained at 2.4 kg continuously. In the riser, heterotrophic bacteria grow on the media and the biofilm becomes thicker. At a certain biofilm thickness, depending on the superficial liquid velocity, the biofilm-coated particles reach the height where they can be transferred to the downer through the inclined pipe. An intermediate graduated container was placed between the two columns, as shown in **Figure 1,** to monitor the particle transfer rate. After exposure to the high shear force in the gas-liquid-solid phase in the downer, the biofilm detaches and leaves the system along with the effluent. Particles from the bottom dense phase of the downer with a thin biofilm (< 100 μm) were transferred back to the riser manually through a gate valve that was occasionally opened at the time of particle transfer and to make up the particles in the riser. The hydrodynamic pressure difference between the riser and downer facilitated particle transfer.

The system was operated for about six months on synthetic wastewater (SWW). **Table 1** shows the operating conditions of the A-CFBBR system in Mode 3 in three different phases of VI-I to VI-III. Complementary information regarding the operating conditions in other Modes and Phases can be found Elsewhere (Andalib et al., 2012). At the maximum volumetric loading rate (Mode 3), the system was operating at an overall hydraulic retention time (HRT) of 12.8 h, empty bed contact time (EBCT) of 3.2 h and AF and CFBBR estimated sludge retention times (SRTs) of 32 d, and 28 d, respectively. The AF and CFBBR were operated at 37°C and 20°C, respectively.

Acclimatization and start-up

Initially, the clean media was fluidized in the anaerobic, anoxic, and aerobic columns at superficial liquid velocities (u_l) of 1.4 cm/s, 1.2 cm/s and 1.6 cm/s respectively. The anaerobic column was seeded with 16 L of secondary anaerobic digester sludge (ADS) from the St. Mary wastewater treatment plant (Ontario, Canada) with TSS and VSS concentrations of 15900 mg/L and 12180 mg/L respectively. The CFBBR was seeded with enriched nitrifiers, acclimatized in the lab using 10 L of returned activated sludge (RAS) from the Adelaide Pollution Control Plant, London, Canada, with TSS and VSS concentrations of approximately 7000 and 4800 mg/L respectively. Thereafter, the SWW, with the composition shown in **Table 2a**, was fed at a flow rate of 1.8 L/d and 0.1 L/d into the anaerobic and anoxic columns on day eight corresponding to OLR 0.28 kg $COD/m^3_{AF}\cdot d$ and NLR of 0.01 kg $N/m^3_{AF}\cdot d$.

Analytical methods

Samples from the feed tank, the anaerobic column, the anoxic column, and the final effluent were collected for analysis. Total suspended solids (TSS), volatile suspended solids, (VSS) were analyzed in accordance with Standard Methods 2540D, 2540E respectively (APHA 1998). HACH methods and testing kits (HACH Odyssey DR/2800) were used to analyze total and soluble chemical oxygen demand (TCOD and SCOD), total nitrogen (TN), NH_4-N, NO_2-N, NO_3-N, and PO_4. Alkalinity was measured by titration with 0.01 N H_2SO_4 in accordance with the Standard Method no 2320 (APHA 1998). DO and ORP were measured onsite using an Oakton DO 6 meter, and an Oakton ORPTestr 10 (Oakton, Singapore). Based on Standard Method no 2540G (APHA, 1998), the attached biomass on the carrier media was measured and expressed as mg VSS/g clean particles. The rate of biogas produced in the methanogenic column was measured by a wet tip gas meter connected to the top of anaerobic column. Methane, nitrogen, and hydrogen content of the biogas were determined using a GC.

Results and discussion

System biomass inventory

Figure 2a illustrates the buildup of attached biomass in the anaerobic, anoxic, and aerobic columns. The process of biomass buildup comprises two steps: the attachment of biomass during acclimatization and the growth of biomass during the process. The biomass attachment during acclimatization under the same hydrodynamic conditions should result in relatively the same biomass attachment. As shown in **Figure 2a**, the attachment of biomass in anaerobic, anoxic, and aerobic column, labeled as A, B and C, are 12.5 mg VSS/g media, 14 mg VSS/g 10.8 mg VSS/g media respectively. The aforementioned attached biomass densities correlated inversely with u_l of 1.4, 1.2, and 1.6 cm/s respectively. As shown in **Figure 2a**, the concentration of immobilized biomass in the anaerobic reactor increased slowly between days 1 and 120 as a result of the gradual start-up strategy adopted which limited the availability of substrate in the reactor, and the low growth rate associated with the methanogenic bacteria.

Nutrient removal

The anaerobic fluidized bed (AF) was operated as a mesophilic methanogenic reactor at a temperature of 35.7±0.8 °C and alkalinity of 3034±228 as mg $CaCO_3$/L. **Table 2b** illustrates the steady-state characteristics of the anaerobic, anoxic and aerobic effluents in Mode 3 when A-CFBBR was operational. The complementary information regarding the steady-state characteristics in Mode 1 when only the AF was operational and Mode 2 when the AF and only the aerobic fluidized bed were operational were depicted in **Table B** of the SD. As shown in **Table 2b,** although the feed pH was 4.6, pH inside the AF remained constant at 7 throughout the experiment due to the high rate of acetic acid degradation in the column. ORP in the AF was observed in the range of -158 mV to -193 mV.

Figure 2b illustrates the temporal biogas production in AF during phases I to VI. As shown in **Figure 2b**, the percent methane of the produced biogas fluctuated between 45% and 52% resulting in methane production rates of 103±6 L/d, 144±5 L/d, and 188±6 L/d in Phases IV, V and VI respectively. Based on the CH_4 equivalent of COD converted by aceticlastic methanogens at standard conditions (0°C and 1 atm) of 0.35 L CH_4/g COD, the theoretical methane production at 37°C was also calculated and depicted in **Figure 2b**. Discrepancies between the theoretical and measured methane of 3.8%-6.4% were observed. It should be noted that the ratio of biogas produced in the system to the volume of the reactor in Phase VI was 24 L_{gas}/L_{rea}.d which was handled without the need for a dedicated solid/liquid/gas separator at the top of the reactor.

Figure 2c depicts the COD removal rate in AF in different phases. As shown in **Table B** of the **SD**, the influent TCOD concentration was constant at 10600±270 mg/L. The effluent SCOD in phases I to IV, when the anaerobic attached biomass was not fully developed, was observed to be in the range of 218 mg/L to 260 mg/L. In phase IV, SCOD decreased to 124 mg/L at OLR of 26 kg COD/m^3·d. As seen in **Table 2b**, in Phase VI at the maximum OLR of 35 kg COD/m^3·d the system reached 98.3% TCOD removal, with effluent SCOD increasing marginally to 180 mg/L. TSS concentrations in the bulk liquid after seeding in phase I decreased gradually from 927 mg/L to the minimum level of 124 mg/L in Phase V at OLR of 26 kg COD/m^3_{AF}·d. However, increasing the OLR to 35 kg COD/ m^3_{AF}·d in Phase VI resulted in an increase in TSS to 188 mg/L. TCOD decreased from 10600 mg/L to an average of 50±20 mg/L in the A-CFBBR effluent throughout all phases, corresponding to the overall TCOD removal of 99.3%. The influent TCOD to the CFBBR comprised the AF effluent, which were 181 mg/L and 390 mg/L in phases V and VI respectively and 5% feed with a TCOD of 10600 mg/L. The TCOD influent to the CFBBR was 740±20 mg/L in phase VI corresponding to OLR of 3.3-4.8 kg COD/m^3_{CFBBR}·d based on the overall volume of the CFBBR in Phase VI (**Table 1**). The final effluent SCOD of the A-CFBBR was 21±10 mg/L and 35±10 mg/L in phase V and VI respectively. In Phase IV, the suspended VSS in the anaerobic column averaged 45 mg/L, which increased to 65 mg/L and 146 mg/L in Phases V and VI. The increase in the anaerobic effluent VSS as a result of higher VSS detachment rates in Phases V and VI (**Table 1** and **Table A** of the SD) coincided with higher OLRs in those Phases.

In Phase V, the anoxic column was not operational and nitrogen removal was solely based on biomass assimilation in the AF where NH_3-N decreased from 250 mg/L in the feed to 213±10 mg/L in the AF effluent, and nitrification in the aerobic column reduced ammonia from 213±10 mg N/L to < 1 mg NH_3-N/L in the system effluent. Assuming the overall anaerobic yield of 0.016 g VSS/g COD and 9±1.1% nitrogen content of biomass (**Figure 2d**), the nitrogen assimilated into biomass was calculated as 20±3 mg /L. The nitrate and nitrite concentrations in AF remained below 3.9 m Ng/L and 0.2 mg N/L in Phase V respectively. The NH_3-N remaining in the AF effluent was sequentially completely nitrified in the aerobic column to an average concentration of 0.9±0.5 mg N/L at NLR of 0.7 kg N/m^3_{CFBBR}·d. SNR batch tests in Phase V showed nitrification rate of 0.25-0.27 g N/g VSS·d, which confirmed the aerobic online nitrification rate within 10% discrepancy. The influent STN to the aerobic column in Phase V was 217±11 mg N/L in the form of NH_3-N and the effluent STN from the aerobic column was 205±9 mg/L in the of NO_x-N. Nitrogen assimilation in the aerobic biomass was estimated at 1.4 mg N/L in Phase V with OLR of 0.5±0.1 kg COD/m^3_{CFBBR}·d and biomass nitrogen content of 9%, thus, indicating that a partial denitrification, of 10 mg N/L may have occurred simultaneously in the aerobic column. The anoxic column was operated with three aerobic to anoxic recirculation flows to feed ratios of 2.3, 4, and 6.2 in Phases VI-I, VI-II and VI-III respectively (**Table 1**). The influent NH_3-N and TN to CFBBR in Phase VI were 215 mg N/L and 232 mg N/L corresponding to NLR of 1.1 kg N/m^3_{CFBBR}·d. As shown in **Table 2b**, the effluent NH_3-N throughout Phase VI remained 1-3 mg N/L, within 13% of the batch SNR range of 0.25-0.31 g N/g VSS·d. The recirculated nitrate to the anoxic column was denitrified to effluent concentrations of 76±22 mg NO_3-N/L, 39±2 mg NO_3-N/L and 88±5 mg NO_3-N/L in Phases VI-I, VI-II and VI-III respectively. In Phase VI-II when the effluent NO_3-N and NO_2-N were 39 mg N/L and 2 mg N/L, total nitrogen removal of 82.3% was achieved with effluent NH_3-N of 1.3 mg N/L. The batch SDNR test showed a range of 3.4-3.8 g NO_3-N/g VSS·d. In Phase VI-III, when the aerobic to anoxic recirculation flows to feed ratio increased to 6.2, aerobic DO dropped to 5.6 mg O_2/L from the initial value of 6.3 mg O_2/L and anoxic DO increased to 0.7 mg O_2/L in Phase VI-II, resulting in incomplete nitrification in the aerobic column as reflected by the average nitrite concentration of 20 mgN/L (**Table 2b**) as well as insufficient denitrification in the anoxic column. Therefore, effluent nitrite and nitrate increased to 20 mg N/L and 88 mg N/L, respectively (**Table 2b**). Phosphorus removal occurred as a result of biomass assimilation in the three columns. As shown in **Table 2b**, in Phase VI, 2.8±0.2 mg P/L and 3.0±0.3 mg P/L, were removed in AF and CFBBR respectively as compared with 3.3 mg P/L and 0.8 mg P/L estimated based on 1.8% biomass phosphorus content by weight of TSS.

At the optimum performance, Phase VI-II of Mode 3, the CFBBR in this work demonstrated 85% COD removal and 84% nitrogen removal at OLR of 4.8 kg COD/m^3_{CFBBR}·d, NLR of 1.1 kg N/m^3_{CFBBR}·d, EBCT of 1.6 h and HRT of 5 h. In comparison, Chowdhury et al. (2010) reported 90% COD and 80% nitrogen removal from municipal wastewater (MWW) using CFBBR at OLR of 4.2 kg COD/m^3_{CFBBR}·d, NLR of 0.26 kg N/m^3_{CFBBR}·d, EBCT of 1.5 h and HRT of 2.4 h. Elsewhere Eldyasti et al. (2010) observed 87% COD and nitrogen removal from landfill leachate using the same system at OLR of 2.6 kg COD/m^3_{CFBBR}·d, NLR of 0.8 kg N/m^3_{CFBBR}·d, EBCT of 11 h and HRT of 18 h. Other versions of the CFBBR developed by the same research group such as CFBBR with a bigger anoxic column called TCFBBR (Andalib et al., 2011b) and enhanced bioparticle circulation CFBBR (Li et al., 2012) also showed 88% and 95% COD as well as 84 and 88% nitrogen removal from MWW at OLR of 4.3 and 3.4 kg COD/m^3_{CFBBR}·d, NLR of 0.5 and 0.44 kg N/m^3_{CFBBR}·d, EBCT of 1.0 h and HRT of 2.2 h respectively. Thus, the A-CFBBR achieved comparable organics and nitrogen efficiencies, albeit at 2-4 times the NLR of MWW at roughly the same EBCT. Chen et al. (2011) also reported 87% COD and nitrogen removal from MWW using an A2/O system incorporating a biological aerated filter at OLR of 1.23 kg

COD/m^3·d, NLR of 0.23 kg N/m^3·d and HRT of 8 h. A longer HRT in the CFBBR in this study (of 5 h versus 2.2 h) and the study conducted by Eldyasti et al. (2010) seemed to be conducive to higher nitrogen loading rates of 1.1 and 0.8 kg N/m^3_{CFBBR}·d respectively. On the other hand, using a finer aeration system in the aerobic column of this work resulted in a lower biomass detachment rate of 0.035 1/d and a higher attached biomass of 20 mg VSS/ gmedia (**Table 1**) compared to the detachment rate of 0.093 1/d and attached biomass of 7 mg VSS/ gmedia in the work of Chowdhury et al. (2010), and the detachment rate of 0.18 1/d and attached biomass of 4.7 mg VSS/ gmedia in work conducted by Andalib et al. (2011b), which resulted in a higher nitrification rate.

Biomass yield

Figure 2d illustrates the linear regression of cumulative VSS produced, based on the sum of the effluent biomass, the net change in attached biomass and biomass wasted, versus cumulative COD removed in anaerobic fluidized bed, CFBBR and A-CFBBR. The highest observed yield for the AF utilizing acetic acid was 0.026 g VSS/g COD in phase II at OLR of 5.3±0.2 kg COD/m^3_{AF}·d. Along with increasing the OLR to 34.5 kg COD/m^3_{AF}·d in AF, the observed yield decreased to 0.016 g VSS/g COD. The very low observed yield of the anaerobic column was expected, due to the estimated long sludge retention time (SRT) in the column of 27-32 days (**Table 1**) and the true yield for methanogens utilizing acetate of 0.05 g VSS/g COD (Metcalf and Eddy 2003). Very low observed yields of 0.116 g VSS/g COD to 0.138 g VSS/g COD were measured in Phase VI-I to VI-III in the aerobic/anoxic CFBBR. Due to the fact that 95% of the overall COD was consumed in the anaerobic fluidized bed, the overall observed yield of A-CFBBR was calculated as 0.017 g VSS/g COD, close to the yield of AF.

Conclusions

- The newly developed A-CFBBR affected 99.7% COD, 84% nitrogen removal with a very low sludge yield of 0.017 g VSS/g COD while treating a high strength SWW containing 10700 mg COD/L and 250 mg NH_3-N/L. The system performance at a high OLR of 35 kg COD/m^3_{AF}·d and 1.1 kg N/m^3_{CFBBR}·d with an overall HRT of less than 12 h is remarkable.
- Microbial communities analysis using DGGE confirmed the presence of AOB and NOB in both the aerobic and anoxic columns and *Pseudomonas putida* responsible for SND in the aerobic column.

List of Tables and Figures

Table 1- Operating condition of the system in Mode 3 (A-CFBBR under operation)

Table 2- **(a)** Composition of the synthetic wastewater **(b)** Steady-state characteristics of anaerobic, anoxic and aerobic effluents in Mode 3 (A-CFBBR under operation)

Figure 1- Schematic of anaerobic-circulating fluidized bed bioreactor (A-CFBBR)

Figure 2 - **(a)** Biomass inventory **(b)** methane production in AF **(c)** COD removal in AF **(d)** biomass yields in AF, CFBBR and A-CFBBR

References

1. Andalib, M., Nakhla, G., Zhu, J., 2010 b. Dynamic testing of the twin circulating fluidized bed bioreactor (TCFBBR) for nutrient removal from municipal wastewater. Chem. Eng. J. 162, 616-625.
2. Andalib M, Nakhla G, , Zhu J. 2012. High-rate biological nutrient removal from high strength wastewater using novel Anaerobic Circulating Fluidized Bed Bioreactor (A-CFBBR). Bioresource Technology 118, 526-535.
3. APHA; AWWA; WEF. 1998. Standard methods for the examination of water and wastewater. 20th Edition, American Public Health Association, Washington D.C.
4. Chen, K.C., Lin, Y.F., 1993. The relationship between denitrifying bacteria and methanogenic bacteria in a mixed culture system of acclimated sludges. Wat. Res. 27(12), 1749-1759.
5. Chowdhury, N., Nakhla, G., Zhu, J., Islam, M., 2010. Pilot-scale experience with biological nutrient removal and biomass yield reduction in a liquid-solid circulating fluidized bed bioreactor. Wat. Environ. Res. Doi: WER 08-11-1541RR.
6. Cui, Y., Nakhla, G., Zhu, J., Patel, A., 2004. Simultaneous carbon and nitrogen removal in anoxic-aerobic circulating fluidized bed biological reactor (CFBBR). Environ. Technol. 25, 699-712.
7. Eldyasti, A., Chowdhury, N., Nakhla, G, Zhu, J., 2010. Biological nutrient removal from leachate using a pilot liquid–solid circulating fluidized bed bioreactor (LSCFB). J of Haz. Mat. 181, 289–297.
8. Metcalf and Eddy, 2003. Wastewater Engineering: Treatment and reuse, Fourth ed., McGraw-Hill.
9. Chen, Y., Peng, C., Wang, J., Ye, L., Zhang, L., Peng, Y., 2011, Effect of nitrate recycling ratio on simultaneous biological nutrient removal in a novel anaerobic/anoxic/oxic (A2/O)-biological aerated filter (BAF) system. Bioresource Technol. 102(10), 5722-5727.

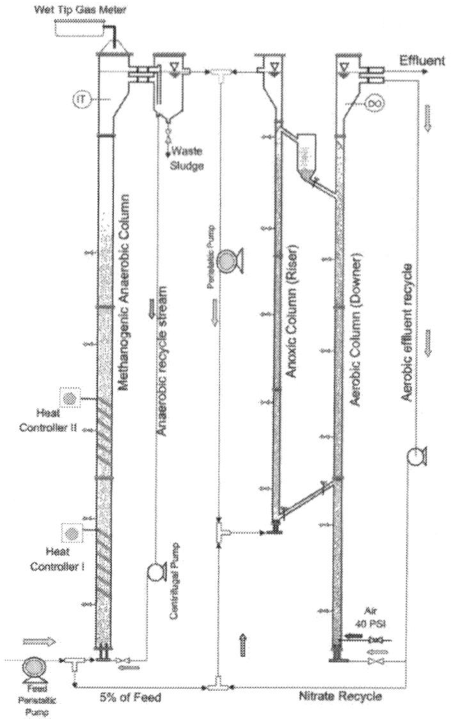

Figure 1

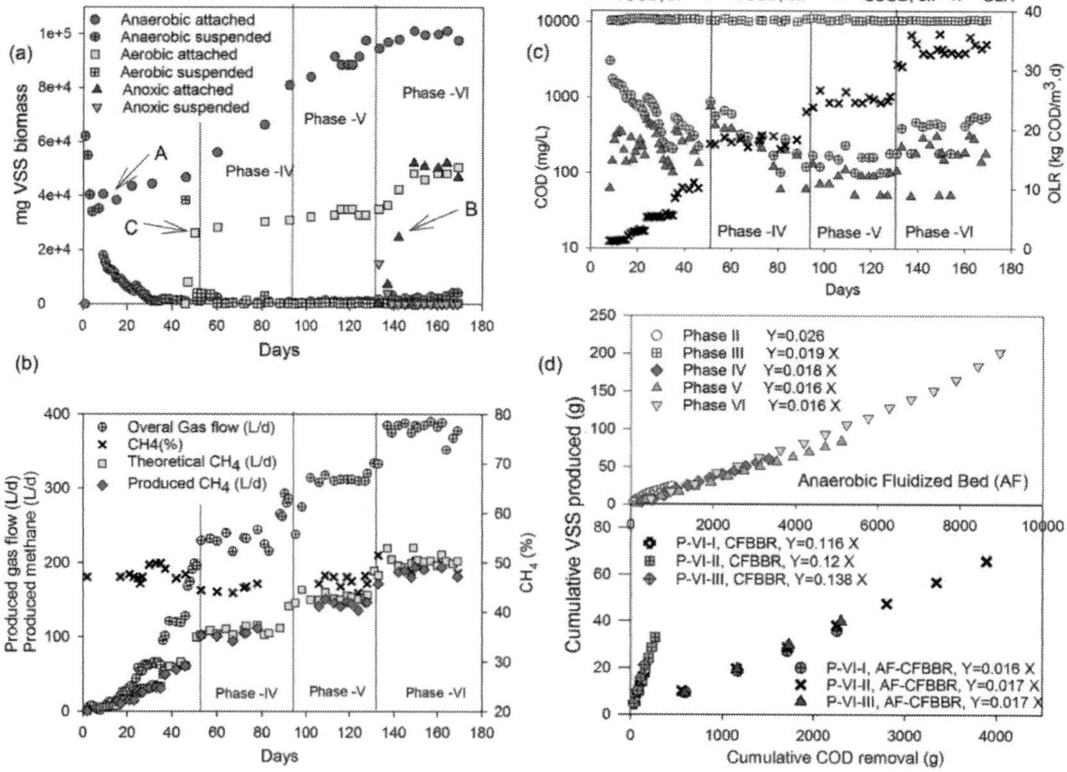

Figure 2

Table 1- System operating conditions in Mode 3 (A-CFBBR under operation)

	A-CFBBR under operation (**Mode 3**)		
	Phase-VI-I	**Phase-VI-II**	**Phase-VI-III**
Day	129-145	145-160	160-172
Feed flow rate (L/d)	47-55[a]	55[a]	55[a]
Anaerobic-Anaerobic circulation flow (Q/Q_{in})	109	107	105
Anoxic-Anoxic Circulation flow (Q/Q_{in})	2	2	2
Aerobic-Aerobic circulation flow (Q/Q_{in})	64	59	64
Aerobic-Anoxic Circulation flow (Q/Q_{in})	2.3	4	6.2
OLR based on anaerobic reactor (kg COD/m^3.d)	33.3±1.5	33.5±1.2	34.5±0.8
NLR based on anaerobic reactor (kg N/m^3.d)	0.8±0.1	0.8±0.1	0.85±0.0
OLR based on CFBBR reactor (kg COD/m^3.d)	3.8±0.7	3.3±0.1	4.8±0.2
NLR based on CFBBR reactor (kg N/m^3.d)	1.11±0.0	1.12±0.0	1.15±0.0
Anaerobic EBCT (h)=$V_{compact}/Q_{in}$	1.6	1.6	1.6
Aerobic EBCT (h)=$V_{compact}/Q_{in}$	1.3	1.3	1.3
Anoxic EBCT (h)=$V_{compact}/Q_{in}$	0.27	0.26	0.26
Anaerobic HRT(d)	0.32	0.32	0.32
Aerobic HRT(d)	0.17	0.17	0.17
Anoxic HRT (d)	0.04	0.04	0.04
Anaerobic attached biomass (mg VSS/g media)	32±0.5	33±0.2	33±0.4
Aerobic attached biomass (mg VSS/g media)	16.5±1.7	20.6±0.5	20.5±0.5
Anoxic attached biomass (mg VSS/g media)	1-41	85±1.5	82.2±2
Total anaerobic attached biomass (g VSS)	96.4	100.2	99.3
Total aerobic attached biomass (g VSS)	37.9	47.5	47.8
Total anoxic attached biomass (g VSS)	0.3-24.6	51.2	49.5
F/M in AF (g COD/g VSS.d)	5.5	5.4	5.5
F/M in CFBBR (g COD/g VSS.d)	0.66	0.35	0.55
Anaerobic col. detachment rate (1/d)		0.035	
Aerobic col. detachment rate (1/d)		0.035	
Anoxic col. detachment rate (1/d)		0.02	
Aerobic DO (mg/L)	6.2±0.1	6.3±0.1	5.6±0.0
Anoxic DO (mg/L)	0.35±0.1	0.4±0.1	0.7±0.2
AF estimated SRT (d)		32	
CFBBR estimated SRT (d)		28	
Superficial liquid velocity in anaerobic col. (cm/s)		1.42±0.04[b]	
Superficial liquid velocity in aerobic col. (cm/s)		1.7±0.1	
Superficial liquid velocity in anoxic col. (cm/s)		1.2±0.1	
Specific nitrification rate (**SNR**) (g NH_3-N/g VSS·d)-Downer		0.25-0.31	
Specific nitrification rate (**SNR**) (g NH_3-N/g VSS·d)-Riser		0.05±0.01	
Specific denitrification rate (**SDNR**) (g NO_3-N/g VSS·d)-Riser		3.4-3.8	
Specific denitrification rate (**SDNR**) (g NO_3-N/g VSS·d)-Downer		0.3±0.1	

[a] 5% of overall feed was directly fed to the bottom of the anoxic column.
[b] The terminal settling velocity of Zeolite with d_m=600 μm and 1000 μm biofilm thickness is 2.8 cm/s, still significantly lower than the operating u_L (Andalib et al., 2010 a).

Table 2a. Composition of the synthetic wastewater

Feed composition [a]	CH_3COOH (mL/L_{Feed})	NH_4Cl (g/L_{Feed})	K_2HPO_4 (g/L_{Feed})	$MgSO_4.7H_2O$ (g/L_{Feed})	$CaCl_2.2H_2O$ (g/L_{Feed})	Yeast extract (g/L_{Feed})	$NaHCO_3$ (g/L_{Feed})	Trace element (mL/L_{Feed})
Concentrations	9.5	1.7 and 0.93	0.1	0.03	0.03	0.03	5.7-6.7	1
Trace element Composition	$FeCl_2.4H_2O$ ($mg/L_{Trace\ E}$)	$MnCl_2.4H_2O$ ($mg/L_{Trace\ E}$)	H_3BO_3 ($mg/L_{Trace\ E}$)	$ZnCl_2$ ($mg/L_{Trace\ E}$)	$CuCl_2$ ($mg/L_{Trace\ E}$)	$AlCl_3$ ($mg/L_{Trace\ E}$)	$CoCl_2.6H_2O$ ($mg/L_{Trace\ E}$)	$NiCl_2$ ($mg/L_{Trace\ E}$)
Concentrations	2000	500	50	50	30	50	50	50

[a] All the above reagent were ACS Grade, 99.5% min.

Table 2b. Steady-state characteristics of anaerobic, anoxic, and aerobic effluents in Mode 3 (A-CFBBR operational)

		A-CFBBR operated (**Mode 3**)						
Parameter (mg/L)		**Phase-VI-I to VI-III**	**Phase-VI-I**		**Phase-VI-II**		**Phase-VI-III**	
Days (d)		*129-172*	*129-145*		*145-160*		*160-172*	
	Influent	Anaerobic Effluent	Anoxic	Effluent	Anoxic	Effluent	Anoxic	Effluent
Produced biogas(L/d)	-	333 to 405	-	-	-	-	-	-
TCOD	10600 ± 270	390± 140	61±14	49±8	80±5	58±7	79±24	67±1
SCOD		181± 86	30±11	31±5	50±16	40±18	46±16	34±12
TSS	-	188±85	28±8	22±10	42±15	32±10	29±8	30±10
VSS	-	146±56	21±5	16±8	27±12	19±7	23±6	23±8
NH_3-N	250	215±8	70±1	3±1	45±12	1.3±0.3	57±19	3.1±0.6
NO_3-N	2	1.8±1	4.7±4	76±22	2±0.5	39±2	16±11	88±5
NO_2-N	1	1.3±1	1.3±0.6	0.9±0.1	2±0.5	2±0.5	38±2	20±2
TN	255(430)[b]	231±12	-	78±21	-	45±10	-	113±10
PO_4-P	26.3	23.5±2.6	19±0.5	18.8±0.8	20.5±2	20±2	21±1	19.3±0.5
Alkalinity[a]	-	3065±115	2115±21	1875±35	2630±180	2017±280	2095±100	1885±120
ORP(mV)	-	-193±8	-	38±13	-	12±2	-	2±1
pH	4.6±0.1	7.09±0.0	8.2±0.1	8.4±0.1	8.2±0.2	8.4±0.1	8.1±0.1	8.3±0.0
Temp (°C)	20	35.7±0.8	22.4±0.1					

[a] as mg $CaCO_3$/L   Nitrogen and phosphorus content of biomass were measured 9±1.2% and 2.1±0.4% respectively

Reduction of VOC Emissions in High Purity Oxygen Activated Sludge Wastewater Treatment Process: Toxchem™ Based Fate & Modeling Case Study

Malcolm Fabiyi, Richard A. Novak, Praxair, Inc., Burr Ridge, IL
Rajeev Goel, Spencer Snowling, Hydromantis, Inc., Hamilton, ON

ABSTRACT

In the last decade, stringent regulations have been imposed on the permissible level of Volatile Organic Compound (VOC) emissions from industrial facilities. The clean air act amendments of 1990 and related regulatory extensions (e.g., National Emissions Standards for Hazardous Air Pollutants, NESHAPs) have significantly reduced the acceptable level of Volatile Organic Compound (VOC) emissions from industrial facilities (Woodward & Curran, 2006). A total of 188 organic compounds have been designated Hazardous Air Pollutants (HAPs); facilities which generate or handle these compounds are subject to permitting, monitoring and reporting requirements. Although extensive emissions controls have been implemented in the production process of many industrial facilities, significant amounts of VOCs may appear in the wastewater. These VOCs can be stripped into the air during conveyance through collection systems or during biological treatment.

Fate and Transport models are used for a variety of tasks including design, emissions analysis and regulatory reporting. The development of fate & transport models in wastewater collection and treatment systems were spurred by regulatory drivers like the Clean Air Act (Melcer, 1994). A variety of computer-based fate & transport models such as BASTE, EPA Water Models (versions 7, 8 and 9), CORAL, PAVE, SIMS, TORONTO, INTERCEPTOR and Toxchem™ have been applied for addressing the fate of volatile contaminants in collection systems, drop structures, weirs, quiescent surfaces, and wastewater treatment processes (Quigley et al, 2006; Melcer, 1994).

EPA Water Models (7, 8 & 9) and Toxchem™ are the most commonly utilized fate and emissions transport modeling platforms in the wastewater industry. In both models, the default aeration options exclusively model the mass transfer behavior as diffused air processes or surface aeration style mechanical mixers. Attempts at modeling the performance of High Purity Oxygen (HPO) based devices have typically required finding the most appropriate approximation to the mass transfer and VOC stripping behavior of HPO aeration systems by modifying default parameters in the diffused air or surface aeration modules (NYSERDA, 2000; Levine et al, 2010; Rodieck et al, 2001). In contrast, an improved modeling approach is employed in the recently released Toxchem™ 4.1 (Hydromantis, 2012), which extends the simulation capabilities of the platform to include the modeling HPO and Sequencing Batch Reactor (SBR) process systems.

This paper discusses results for VOC emissions modeling obtained by applying Toxchem™ 4.1 to VOC reduction projects that utilize SBR and HPO systems. We provide a methodology for modifying the default parameters in diffuser and mechanical surface aeration modules in fate and transport models in order to enable VOC emissions characteristics associated with HPO systems to be modeled, and provide comparative analyses on the effect of the approach adopted on simulation results.

KEY WORDS

VOC emissions, mass transfer, mechanically agitated contactors, Toxchem, Fate & Transport models, High Purity Oxygen

BACKGROUND

The mass transfer models in fate and transport emissions models are based on ASCE standard model parameters (Chern & Yu, 1995). They model the three major mechanisms through which VOC removal from wastewater occurs: (i) biosorption of VOCs onto activated sludge floc and subsequent removal via waste sludge, (ii) volatilization and stripping of VOCs into the atmosphere during aeration, and (iii) biological degradation.

Mechanisms for VOC Stripping

There are two distinct mechanisms for VOC stripping in aerated activated sludge biological wastewater treatment processes: (i) *turbulence induced stripping*, in which turbulence generated by the aeration system induces surface renewal leading to the transfer of VOCs from the bulk liquid across the air-water interface; and (ii) *bubble aided stripping,* in which VOCs in the wastewater diffuse into un-dissolved gas bubbles and are stripped as the bubbles leave the water surface.

Aeration Systems and Their effect on VOC Stripping

Three general classes of aeration devices exist (Melcer, 1994; Mueller et al, 2002): (i) mechanical surface aerators (ii) diffused aeration systems, and (iii) subsurface mechanical aerators with gas sparging.

Mechanical Surface Aerators: Mechanical surface aerators spray water droplets into the headspace above the aerator; gas-liquid exchange occurs as gases diffuse from the surrounding atmosphere into the droplets, as well as from the droplets into the atmosphere. In these devices, VOC emissions occur in three ways: (i) from the droplets to the air, (ii) from the bulk liquid into air bubbles that have been entrained under the surface of the water, as the air bubbles rise and reach the liquid surface, and (iii) across the turbulent free surface of the basin.

Diffused Aeration Systems: Fine Bubble Diffuser systems are commonly utilized for aeration in wastewater treatment processes because they provide relatively high oxygen transfer efficiencies (OTE) (in the range of 20%-30% for most common applications) relative to alternatives like coarse bubble systems and surface aerators (Mueller et al, 2002). Considering the fact that air typically has about 21% oxygen by volume, OTE values for aeration systems in the range of 20%-30% implies that only about 4%-6% of the volumetric air flow supplied to a basin using a diffused aeration system is actually dissolved. The undissolved gases, which make up about 94%-96% of the gas supplied, provide a large volumetric flow of gas which strips VOCs from the basin.

Subsurface mechanical aerators: In subsurface mechanical aeration systems, process gases such as oxygen are injected below the water surface in combination with a mechanical mixer which serves to maintain efficient gas-liquid contacting. Oxygen diffuses into the wastewater, and VOCs transfer from the liquid phase to the gas bubbles. The bubbles eventually rise to the surface and VOC release occurs as the bubbles break the surface. Subsurface mechanical aerators generate some turbulence at the water surface, and also provide subsurface sparging of gases. The mass transfer and VOC stripping behavior of such systems therefore combines elements of both mechanical surface aeration (surface renewal aided VOC diffusion across the air-water interface due to turbulence) as well as diffuser aeration (off-gas VOC stripping.)

High Purity Oxygenation Systems for VOC Emissions Reduction

The VOC stripping potential of the large un-dissolved volume of gases (approximately 94%-96% of gas supplied) associated with the conventional aeration process has long been recognized. Process options that have high oxygen transfer efficiencies, such as high purity oxygenation systems, have been selected for several wastewater treatment applications where reduction in VOC emissions from the wastewater treatment process is a key project goal (e.g., Rodieck et al, 2001; Lubkowitz-Bailey et al, 2002; Levine et al, 2010; NYSERDA, 2000; Melcer, 1994). High purity oxygen (HPO) systems have been used in wastewater treatment applications for over four decades (Mueller et al, 2002). High efficiency oxygenation systems such as Praxair's In-Situ Oxygenation (I-SO™) system (Mueller, 2002) have been demonstrated to provide up to 97% reduction in VOC emissions relative to a comparative diffused air based aeration system (Rodieck et al, 2001).

Praxair's In-Situ Oxygenation (I-SO™) system is a subsurface, down-pumping mixing and oxygenation system used for providing oxygen in open basin HPO operations (Figure 1). The down-pumping action of the helical impeller generates a vortex which entrains pure oxygen gas supplied to a floating hood. The high shear in the draft tube region breaks up the entrained gas into fine bubbles. The intimate gas-liquid contact in the draft tube significantly enhances mass transfer. Dissolved oxygen is dispersed by the mixing action induced by the high velocity liquid jet ejected from the draft tube. Undissolved gas bubbles detach from the liquid flow and rise; some bubbles are recycled through the draft tube, allowing high oxygen transfer efficiencies to be attained. Standard aeration efficiencies of up to 10 Ibs O_2/HP-hr and oxygen transfer efficiencies of up to 90%-95% can be obtained using the I-SO™ mixing system (Praxair, 2003; NYSERDA, 2000). HPO systems have high aeration efficiencies (8-10 Ibs O_2/HP-hr), low mixing intensities (as low as <10 w/m^3) and high standard oxygen transfer efficiencies (+90%), and have been associated with significant reductions in VOC emissions when used in the wastewater treatment process (Mueller et al, 2002; Levine et al, 2010; Lubkowitz-Bailey et al, 2002; Rodieck et al, 2001; and NYSERDA, 2000).

METHODOLOGY

Given that commercially available modeling platforms offer only diffused aeration and mechanical surface aeration options as the default modules, the approach taken in this paper is to adapt aeration models (based on air) for use in modeling the behavior of high purity oxygen systems. Using Toxchem 4.1 as the fate & transport modeling platform, critical model parameters such as comparative gas flow rates, oxygen transfer efficiencies, O_2 mass fraction in gas source, ratio of the gas and liquid film mass transfer coefficients (K_g/K_l), aeration power, standard oxygen transfer rate (lbs O_2/HP-hr), and mass transfer correction factors (alpha) are varied for both diffused air and mechanical surface aeration systems to establish which variables to modify to adequately model the VOC emissions characteristics of pure oxygen systems.

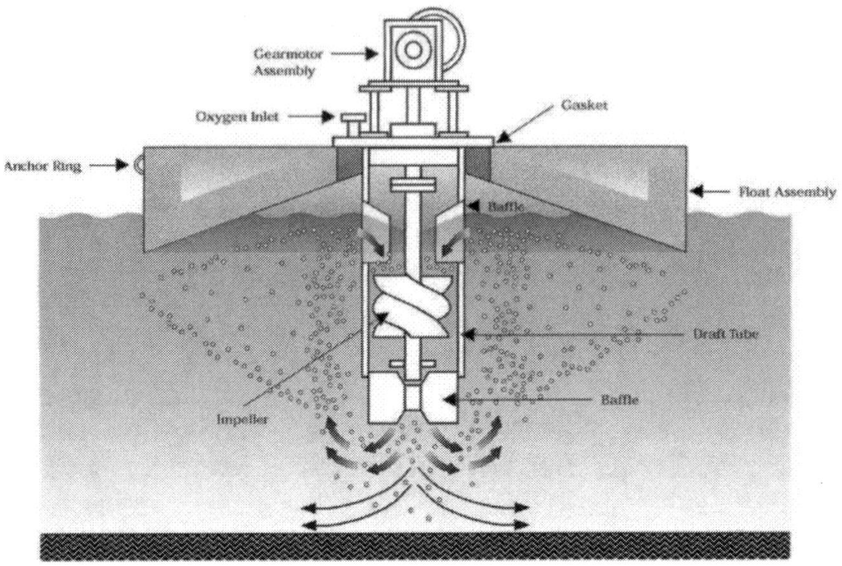

Figure 1: Praxair's In-Situ Oxygenation (I-SO™) system

Published data on comparative VOC emissions reductions from pilot and full scale applications using HPO vs. conventional air-based aeration systems are used as the data source for the analysis (Lubkowitz-Bailey et al, 2002; Rodieck et al, 2001).

In Lubkowitz-Bailey et al, 2002 and Rodieck et al, 2001, pilot and full scale data are presented for a SBR activated sludge process, used for the treatment of Active Pharmaceutical Ingredient (API) wastewater from a pharmaceutical plant. The data evaluated in this study is from the pilot system, which comprised a single SBR system operated in either diffused aeration or HPO mode (see Tables 1 & 2). The pilot scale SBR was operated in 6 hour cycles with 5 phases (Mix/Fill, React/Fill, React, Settle, and Decant). Wastewater flow of 250 gpd was supplied to the process. The reactor diameter was 10 ft, with a height of 15 ft, giving a total SBR volume of 6,000 gals. Fill volume during the tests was 4,650 gals, giving an effective liquid depth of about 11.6 ft. Gas flowrates supplied during the tests ranged from 0.32-0.67 cfm during HPO operations, and 28.5 to 32.5 cfm for diffused aeration operations. Under all aeration modes, enough oxygen was provided to ensure that COD removal targets were met. The VOCs contained in the API wastewater included Acetone, Toluene, MTBE, MEK, Methylene Chloride, 1,1-Dichloroethane, Ethanol, Isopropanol, Methanol, 1,1,1-Trichloroethane, Butyl Acetate and Heptane. During pilot testing, comparative VOC treatment performance was carried out in three baselines tests, and three spike tests (Rodieck et al, 2001). The spike tests were used as the basis for the modeling exercise in this study. Average MLSS was approximately 3,500 mg/l. Detailed monitoring of VOC emissions was undertaken by monitoring VOC in influent, effluent and off-gas samples.

Acetone and toluene were the major HAPs in the influent, with concentrations ranging from 7to 5,2200 and 49 – 320 mg/l respectively at pilot scale. Influent COD ranged from 4,069 mg/l (Rodieck et al, 2001) to 4,800 mg/l (Lubkowitz-Bailey et al, 2002).

The full scale system was designed for 240,000 gpd peak daily flow. Treatment was carried out in three (3) HPO SBR systems, with a design F/M basis of 0. 1 Ib BOD/lb MLSS-day. High Purity Oxygen supply to each SBR was provided using a single 40 HP Praxair I-SO™ system in each reactor, capable of transferring about 300 Ibs O_2/hr.

Model Configuration
VOC fate modeling was undertaken using the new Toxchem™ 4.1 modeling platform. This latest version of Toxchem™ (Hydromantis, 2012) includes High Purity Oxygen and Sequencing Batch Reactor Modules. Given that SBRs were utilized with pure oxygen and conventional aeration in the studies that were used as a basis for this paper (Rodieck et al, 2001; Lubkowitz-Bailey et al, 2002) the updated Toxchem™ platform enables the effective simulation of VOC and odor emission in treatment processes that involve the application of pure oxygen in subsurface mixing and dissolution systems. An earlier version of Toxchem™ which did not have custom SBR and HPO modules was used to support the modeling work in Rodieck et al (2001).

Since an integrated SBR and HPO module does not exist in Toxchem™ 4.1, the HPO SBR system was modeled as a SBR with modified diffused aeration parameters. To capture the dynamics of HPO operations, the effective oxygen transfer efficiencies and gas flowrates that apply to HPO operations were utilized in the diffuser models. Previous work has demonstrated that this approach to modeling HPO systems is effective (Fabiyi, 2012; Rodieck et al, 2001; NYSERDA, 2000). The three Spike tests reported in Rodieck et al (2001) were used as the basis for the simulations. The simulations were set up to evaluate HPO vs. diffused aeration system performance in parallel trains. Toxchem™ also allows visual displays that match emissions values around process units (emissions hotspots) to preset threshold values that are characterized by representative color codes (see Figure 2).

Model Calibration
Test data for the spike test runs were input into the representative model layouts and run using test values for process flowrates, oxygen mass transfer efficiencies and reactor conditions. A systematic analysis of the impact of various process parameters on the predictive accuracy of the model was undertaken using the Sensitivity Analysis tool in Toxchem™. This tool allows the evaluation of the impact of four broad classes of factors on model output, including: contaminant properties (e.g., Van't Hoff coefficients for estimation of Henry's Law Constant, adsorption characteristics, bio-kinetic properties of the compounds); site properties (elevation, wind speed, pH); and biological treatment system properties (e.g. liquid depth, oxygen transfer efficiencies, aeration power input, DO levels, Kg/Kl, aeration, etc.)

The sensitivity analysis indicated that the Van't Hoff estimates were the most important factor for ensuring that the model output was reliable. The default values for the Van't Hoff variables were reviewed and updated where necessary. Given that the reactor was operated at 80-100°F (Rodieck et al, 2001), default Henry's law constant data that were reported outside of this temperature range were dropped and replaced with the Van't Hoff estimates, which allow adjustment of Henry's Law constants for temperature.

A simplified form of the Van't Hoff relationship is given as

$$LnH_{cc,T} = A - \frac{B}{T}$$

Where $H_{cc,T}$ is the Henry's Law Constant, while A and B are the Van't Hoff coefficients. The value for A varies with the exact unit specifications used for Henry's Law Constant, but the numerical value estimated for B does not usually change for the commonly utilized unit expressions for $H_{cc,T}$.

Henry's Law Constant (HLC) values can be affected by a variety of factors such as temperature, pH, compound hydration, compound concentration, co-solutes, co-solvents, dissolved salts, suspended solids, dissolved natural organic materials, surfactants and water composition (Staudinger & Roberts, 1996). Starting Van't Hoff estimates were obtained using average values from a comprehensive compilation of published studies reporting experimental determinations of Henry's Law Constant temperature dependence relationships (Staudinger & Roberts, 2001). Table 3 summarizes the HLCs for the VOCs utilized in the model.

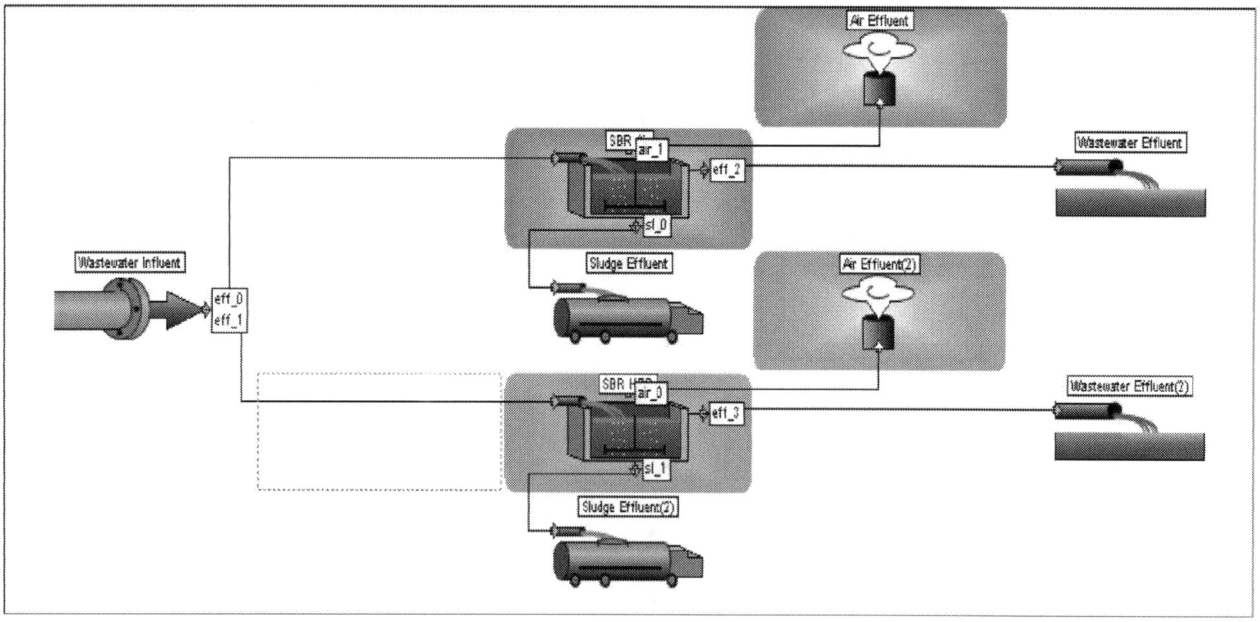

Figure 2: Layout depicting parallel trains used for evaluating VOC emissions behavior of HPO vs conventional diffused aeration SBR processes.

Table 1: Summary of Comparative VOC Emissions from pilot tests used in this study (Source: Rodieck et al, 2001).

FIELD MEASUREMENT RESULTS FOR TOLUENE

Testing Event	Percent of Influent Toluene Load Emitted to the Atmosphere			
	Measured		Model	
	HPO	DA	HPO	DA
HPO Baseline Test #1	0.1	---	0.3	---
HPO Baseline Test #2	0.5	---	0.4	---
DA Baseline Test	---	10.5	---	7.6
Spike Test #1	0.5	18.7	0.2	4.9
Spike Test #2	0.3	9.5	0.2	3.7
Spike Test #3	0.2	11.5	0.2	4.2
Average	0.3	12.6	0.2	5.1

Table 2: Model input variables (Source: Rodieck et al, 2001).

FATE AND TRANSPORT MODEL INPUT PARAMETERS

Category	Parameter	Units	HPO	DA
Ambient Conditions	Elevation	ft	329	
	Temperature	°F	80	
	Wind speed	mph	6.2	
Influent	VOC concentration	mg/L	variable	
	Suspended solids (SS)	mg/L	10	
	Volatile SS Ratio	%	10	
	Temperature	°F	80-100	
Reactor	Reactor Depth	ft	8.33	
	Reactor Surface Area	ft^2	30.6	
	Mixed liquor suspended solids (MLSS)	mg/L	3,000 – 4,400	
	MLSS volatile SS ratio	%	50	
	MLSS dissolved oxygen content	mg/L	2.0	
	Oxygen supply gas flow rate	cfm	0.33-0.80	15-35
	Oxygen transfer efficiency	%	95	10
	Mole fraction of oxygen source	---	1.0	0.2

Table 3: HLC = Henry's Law Constant. S&R (2001) refers to Staudinger, J & Roberts, P.V. A Critical Compilation of Henry's Law Constant Temperature Dependence Relations for Organic Compounds in Dilute Aqueous Solutions. Chemosphere 44 (2001).

		Default			S&R (2001)		Modified	
		HLC	Van't Hoff A	Van't Hoff B	Van't Hoff A	Van't Hoff B	Van't Hoff A	Van't Hoff B
Toluene			5.133	3024	5.271	1745	5.271	2800
	°C/St.Dev				(±1.11)	(±436.3)		
Methyl-Tertiary-Butyl Ether		0.022688			9.07	3178	6	3178
	°C/St.Dev	(@25°C)						
Methylene Chloride		0.122618			4.561	1644	4.561	2900
	°C/St.Dev	(@25°C)			(±0.50)	(±148)		
Butyl Acetate,n-		0.00671			6.4	2486	6.4	3846
	°C/St.Dev	(@25°C)						

RESULTS & OBSERVATIONS

a. Effect of HPO Operations on VOC Emissions

Rodieck et al (2001) carried out three spike tests using numerous VOCs that were intended to stress test the HPO system and evaluate its capacity for VOC reduction versus a comparative air based system. Those spike tests have been modeled in this paper, and good agreement was found between model predictions and test results. The VOCs of particular interest across all the spike tests that were carried out are Toluene, MTBE, $MeCl_2$ and n-Butyl Acetate. The model provided reliable estimations of the effect of aeration mode (conventional vs. HPO) on VOC stripping behavior (see Figures 3 – 6).

Spike Test 1 (Figure 3). 97% and 93% reductions in Toluene and MTBE emissions respectively were observed in Spike Test 1. The model also provides reliable estimates of VOC emissions reduction, with 98% and 97% reductions in Toluene and MTBE emissions respectively. . Comparative gas supply ratio V_{Air}/V_{HPO} (Air volume vs. HPO volume) is 91.2 (i.e., 77 ft3 HPO, 7024 ft3 air).

Spike Test 2 (Figure 4). 97% and 96% reductions in Toluene and Methyl Chloride emissions respectively were observed. Model estimates of VOC emissions reduction tracked test results at 97% for Toluene and 96% for Methyl Chloride. Comparative gas supply ratio V_{Air}/V_{HPO} (Air volume vs. HPO volume) is 54.6 (i.e., 125 ft3 HPO, 6830 ft3 air)

Spike Test 3 (Figure 5). 96% and 98% reductions in Toluene and Butyl Acetate emissions respectively were observed during the test. Model estimates of VOC emissions reduction was 97% for Toluene and 93% for Butyl Acetate. Comparative gas supply ratio V_{Air}/V_{HPO} (Air volume vs. HPO volume) is 48.4 (i.e., 161 ft3 HPO, 7800 ft3 air)

The data and simulation results across all the spike tests consistently demonstrate the significant reduction in VOC emissions obtained by using HPO instead of air as the source of oxygen in the process gas supply. Toluene was common to all the spike tests, and the average reduction in

Toluene related VOC emissions across all tests was 97% for HPO vs. conventional diffused aeration based systems (see Figure 6). Differences in the comparative volumes of gas introduced into the system during air versus HPO operations ranged from 48-92.

b. Comparison of HPO and SBR Modules

The predictive capacity of the new HPO module in Toxchem™ 4.1 was tested by comparing the results from simulations of the Batch 3 spike tests of Rodieck et al (2001) using the SBR and HPO modules. As previously indicated, the modeling of HPO operations in the SBR is done using diffused aeration principles. Although the I-SO™ system is a mechanically agitated mixing and gas dissolution system, it differs fundamentally from surface aerators because it is a subsurface mixer. The stripping behavior of subsurface mixing systems in the wastewater treatment process is modeled most accurately by the off-gas volumes. The mass transfer characteristics of the new HPO module in Toxchem™ is modeled on the hydrodynamics and mass transfer behavior of the I-SO™ system (Praxair, 2003).

Effect of Depth and Surface Area: The default HPO and Activated Sludge Diffusion Aeration modules in Toxchem™ are complete mix systems. In order to model an SBR system using these modules it could be necessary to modify the sequencing batch steps in a SBR as separate process units. We investigated the effect of depth and surface area on the stripping behavior in the HPO and Diffuse Aeration process layouts to determine whether the modification of either of these two parameters could be tolerated when trying to model the SBR using complete mix systems (see Figure 7). For the sensitivity analysis, depth ranged from 2 to 20 ft, and surface area from 30 to 200 sqft. Actual SBR depth and surface area were 11.6ft and 78.5 sqft respectively. The analysis indicated that while depth and surface area had no major effect on the stripping behavior in the HPO system, the diffused aeration system was significantly affected by these parameters. The sensitivity of the VOC stripping behavior in diffused aeration systems to surface area and depth, suggest that the modification of either of these parameters when trying to model the SBR using complete mix systems could lead to issues with the interpretation and accuracy of the model results.

Due to this observation, a simplified layout was selected, which comprised only of the HPO and Diffused Aeration module with no additional process units (Figure 8). The process units had the same depth and surface area values as the actual SBR. A key limitation to the method could stem from high VOC losses during the fill cycle of the SBR. Rodieck et al (2001) minimized VOC losses and volatilization during the fill cycle by feeding the liquid below the liquid surface (Geiling, 2012).

The results obtained for VOC reduction for HPO vs. conventional aeration processes using this approach compared well with results from the SBR modeling approach (see Figure 9).

c. *Effect of Gas Flow rate on VOC Emissions*

There are significant differences in gas flows between HPO and conventional aeration systems. As previously observed, the use of HPO systems can reduce gas flow as much as 50-100 times compared with air based process. Reduced gas supply and off gas rates in HPO systems is a key factor in the enhanced VOC reduction observed in HPO systems (Fabiyi, 2012, Rodieck et al, 2001; Levine et al, 2010).

A sensitivity analysis was run in Toxchem™ to determine the effect of gas flowrates on VOC stripping. A linear correlation was observed between VOC stripping and the volumes of airflow supplied to the wastewater treatment process within the range of flowrates analyzed (see Figure 10). The linear relationship observed between gas flowrates and VOC stripping provides strong support for the identification of bubble-aided stripping as a key mechanism for VOC emissions in diffused aeration and subsurface mechanically aerated systems (Levine et al, 2010).

CONCLUSIONS

VOC emissions are affected by a variety of factors including mixing intensity, gas flowrates, surface renewal rates and surface turbulence (Chern & Yu, 1995). This implies that VOC emissions can be reduced by using aeration devices that (i) have a relatively low power input, hence minimizing excessive turbulence and surface renewal, and (ii) provide high oxygen transfer efficiencies (OTE), thereby minimizing the amount of un-dissolved gas bubbles that will strip VOCs from the liquid.

We find that the VOC emission behavior of the HPO system is predicted accurately by modeling the mass transfer characteristics as that of a fine bubble diffuser system, with adjustments for gas flows (actual O_2 feed to the HPO unit is used), O_2 purity and oxygen transfer efficiency.

Attempts to utilize the default mechanical aeration module indicated that the mass transfer characteristics of surface aeration systems are not well suited to model HPO systems (Fabiyi, 2012). The ready adaptability of diffuse aeration models for capturing the dynamics of VOC emission from mechanical HPO oxygenation systems suggests that the dominant VOC emission mechanism in both diffused air and subsurface aeration systems is similar – VOC transfer to un-dissolved gas bubbles. In contrast, mechanical surface aeration systems introduce a third mechanism into VOC stripping dynamics. In addition to volatilization due to surface renewal from turbulence, and VOC emissions due to volatile transfer to un-dissolved gas bubbles, VOC losses also occur due to exchange between the air and volatiles in the airborne water droplets. Gross air and water conditions such as temperature and wind velocities therefore become more important in accurately estimating VOC emissions in mechanical surface aeration systems (Chern & Chou, 1999; Roberts & Dandliker, 1983).

Despite the potential for VOC emission reduction afforded by high purity oxygen systems for wastewater treatment, commonly available fate & transport models do not have default modules that enable their modeling (Levine et al, 2010; Lubkowitz-Bailey et al, 2002; Rodieck et al, 2001). Toxchem™ 4.1 is the first commercially available fate modeling tool that has a default HPO module. Our work has demonstrated that this new tool is robust and reliable, accurately modeling both pilot and full scale application of HPO aeration for VOC emission reduction.

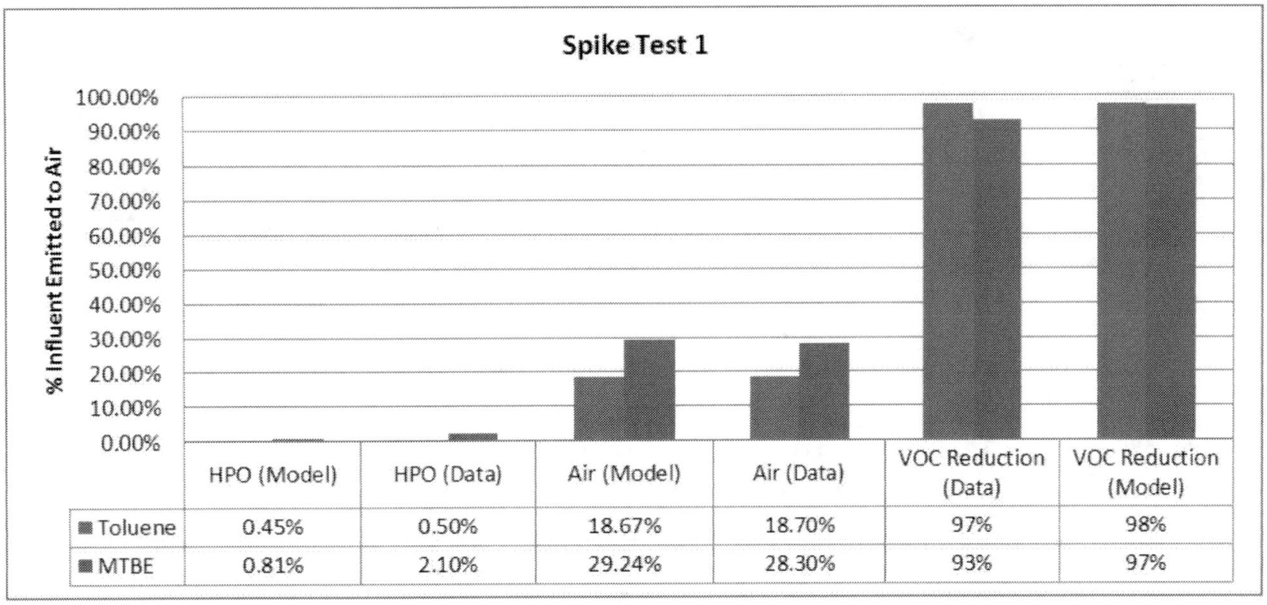

	HPO (Model)	HPO (Data)	Air (Model)	Air (Data)	VOC Reduction (Data)	VOC Reduction (Model)
Toluene	0.45%	0.50%	18.67%	18.70%	97%	98%
MTBE	0.81%	2.10%	29.24%	28.30%	93%	97%

Figure 3: Effect of aeration mode on VOC Stripping in Spike Test 1

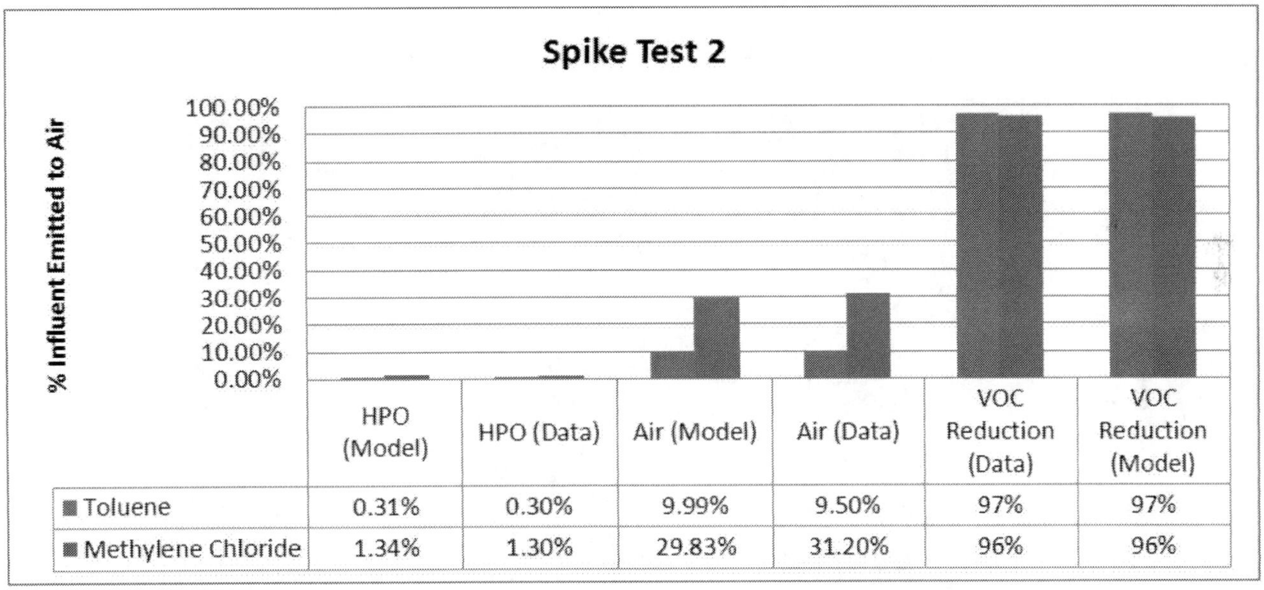

	HPO (Model)	HPO (Data)	Air (Model)	Air (Data)	VOC Reduction (Data)	VOC Reduction (Model)
Toluene	0.31%	0.30%	9.99%	9.50%	97%	97%
Methylene Chloride	1.34%	1.30%	29.83%	31.20%	96%	96%

Figure 4: Effect of aeration mode on VOC Stripping in Spike Test 2

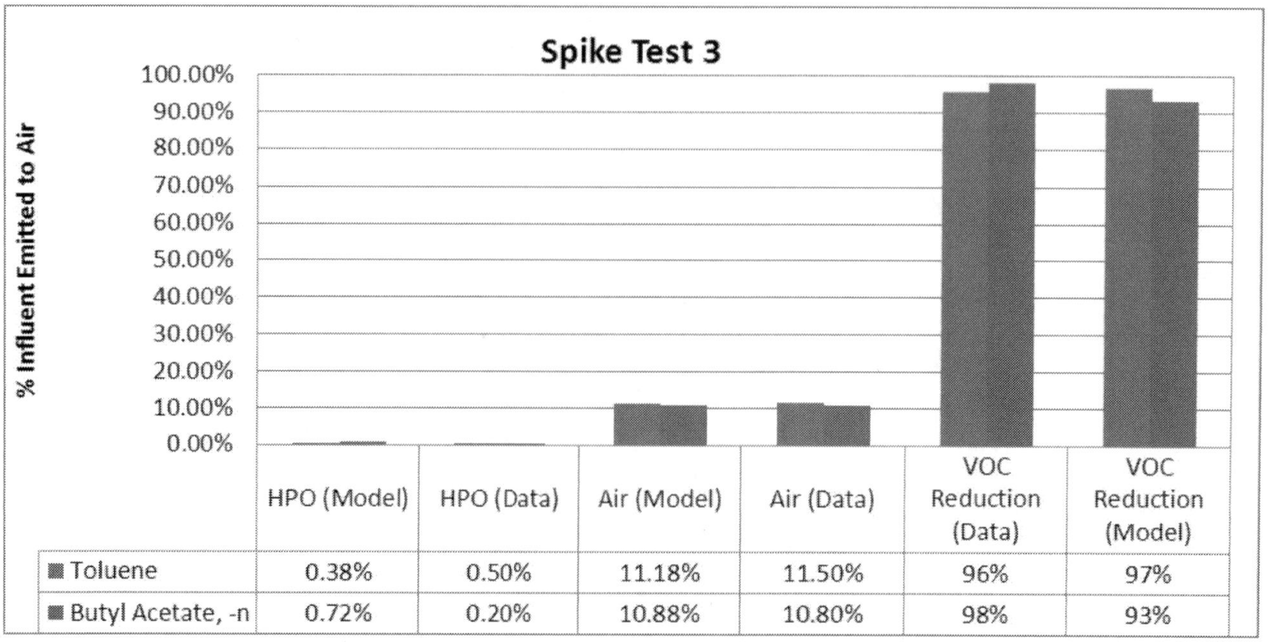

Figure 5: Effect of aeration mode on VOC Stripping in Spike Test 3

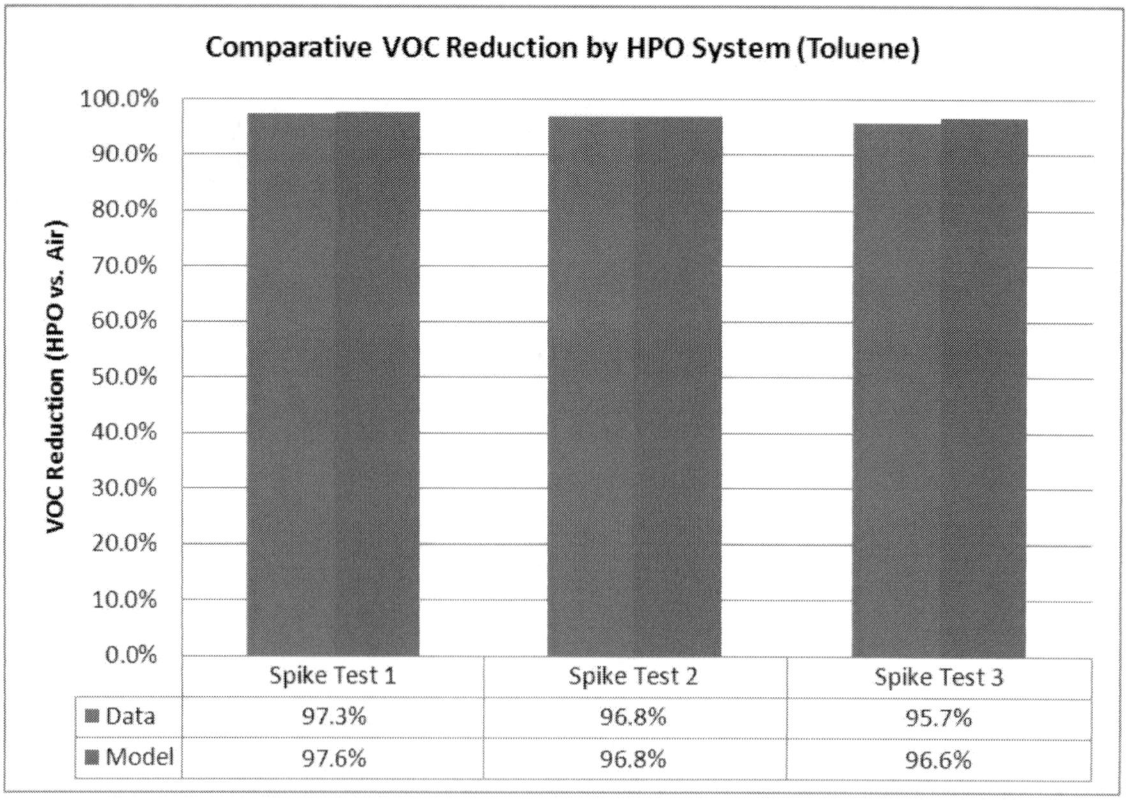

Figure 6: Summary of Modeling and Empirical Values for VOC Emissions reduction obtained using I-SO™ HPO Systems

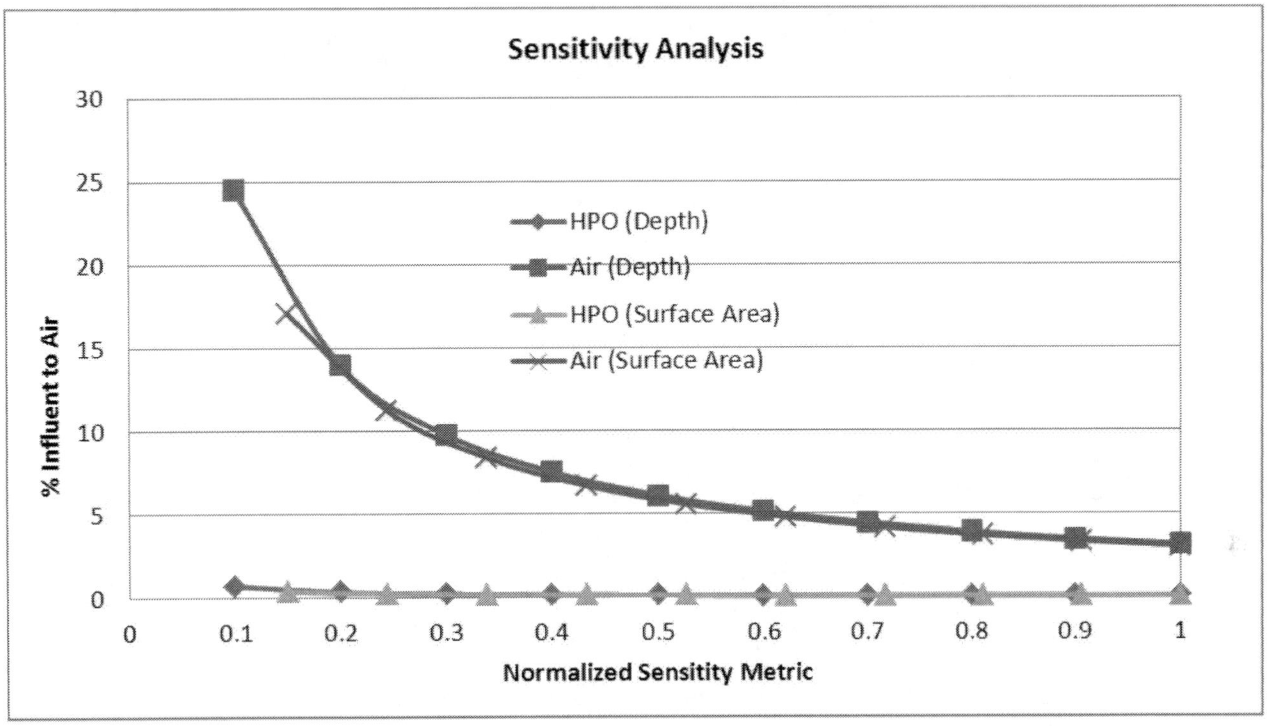

Figure 7: Effect of depth and surface area on VOC emissions for Toluene.

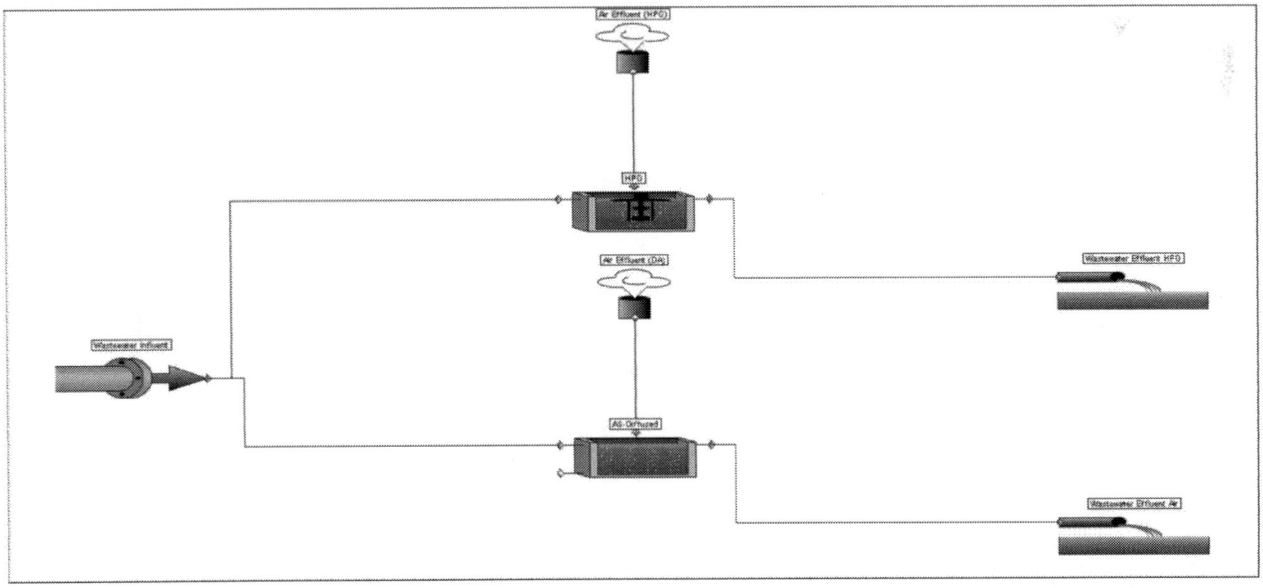

Figure 8: Simplified process layout with HPO and Diffused Aeration modules.

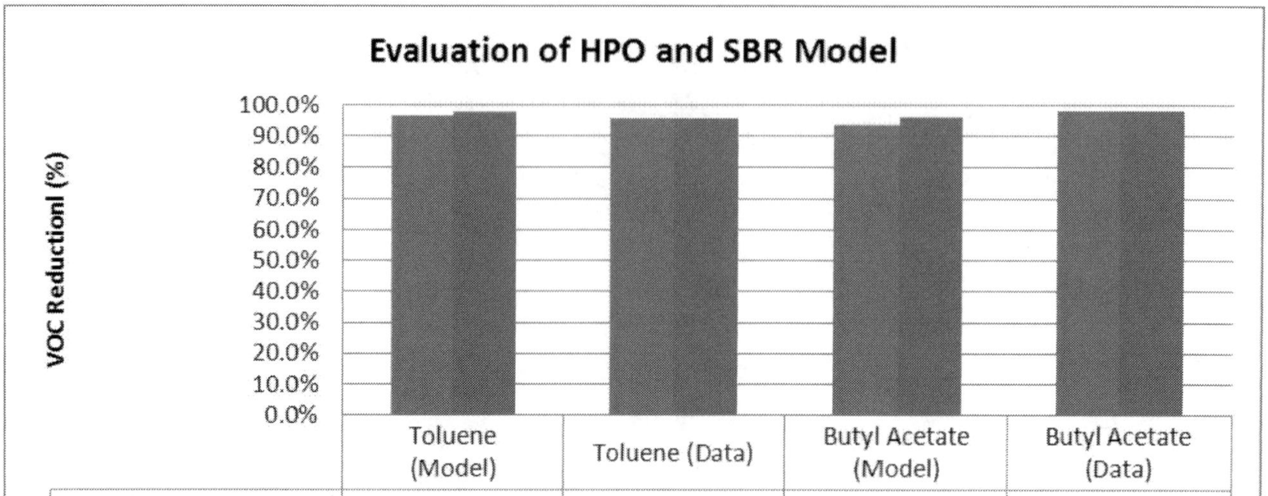

	Toluene (Model)	Toluene (Data)	Butyl Acetate (Model)	Butyl Acetate (Data)
SBR (Modified Diffuser)	96.6%	95.7%	93%	98%
HPO	97.6%	95.7%	96%	98%

Figure 9: VOC reduction for HPO vs. Air based processes using the SBR model with modified diffused aeration parameters and the new HPO process layout.

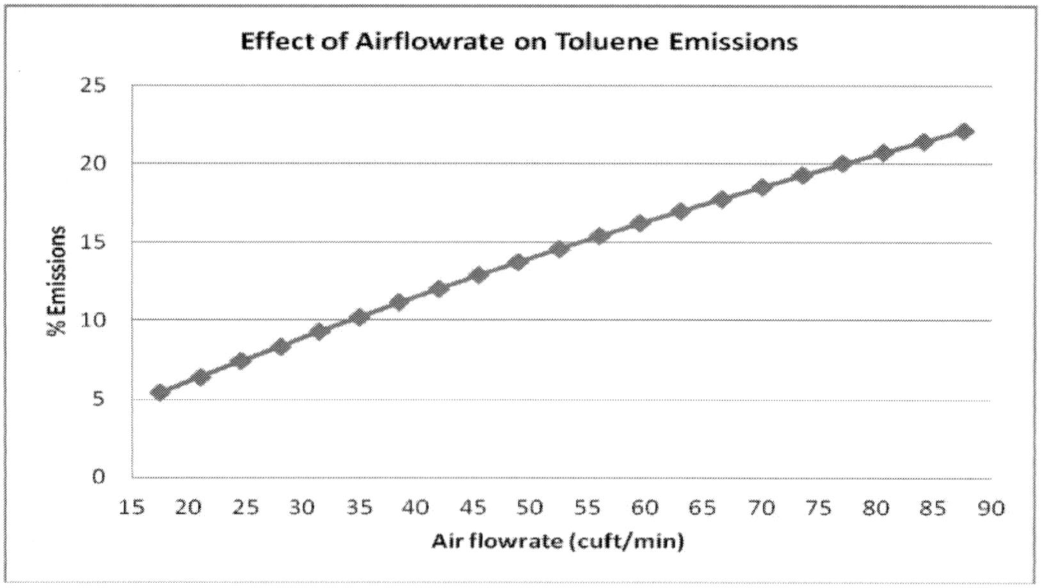

Figure 10: Effect of gas flowrates on Toluene emissions in API pharmaceutical wastewater

REFERENCES

1. Geiling, C. Personal Communications. May 2012.
2. Hydromantis, 2012. Toxchem Version 4.1 Release Notice. http://www.hydromantis.com/news_new.html
3. Jia-Ming Chern and Cheng-Fu Yu. Volatile Organic Compound Emission Rates from Diffused Aeration Systems. 1. Mass Transfer Modeling. Ind. Eng. Chem. Res. 1995, 34, 2634-2643.
4. Jia-Ming Chern and Shun-Ren Chou. Volatile Organic Compound Emission Rates from Mechanical Surface Aerators: Mass Transfer Modeling. Ind. Eng. Chem. Res. 1999, 38, 3176-3185.
5. Levine et al, 2010. Pure Oxygen Activated Sludge Unit Effectively Controls Volatile Organic Compound Emissions from a Mixed Petrochemical Wastewater, WEFTEC 2010
6. Lubkowitz-Bailey et al, 2002. Start-up of a High Purity Oxygen Sequencing Batch Reactor System for Treatment of an Active Pharmaceutical Ingredient Wastewater. WEFTEC 2002
7. Melcer, H. Monitoring and Modeling VOCs in Wastewater Facilities. Environmental Sci. Tech. Vol 28, No. 7, 1994.
8. Mueller et al, 2002. Aeration: Principles & Practice.
9. NYSERDA Report, 2000. In-Situ Oxygenation (I-SO™) for Volatile Compound Emission Control.
10. Praxair Report of Clean Water Tests at Philadelphia Mixing Solutions, 2003.
11. Quigley, C., Card, T., Monteith, H., Witherspoonm, J., Adams, G., Pettit, W., & Torres, E. A Comparison of Three Wastewater Collection System VOC Emissions Models. WEFTEC, 2006.
12. Roberts, P. V., & Dandliker, P. G. Mass Transfer of Volatile Organic Contaminants from Aqueous Solution to the Atmosphere during Surface Aeration. Environmental Sci. Tech. Vol. 17, No. 8, 1983.
13. Rodieck et al, 2001. Alternative Treatment Strategy for High Strength Pharmaceutical Wastewater. WEFTEC 2001.
14. Staudinger, J & Roberts, P.V. 1996. A Critical review of Henry's Constant for Environmental Applications. Critical Rev. Environ. Sci. Technol. 26 (3), 205-297
15. Staudinger, J & Roberts, P.V. A Critical Compilation of Henry's Law Constant Temperature Dependence Relations for Organic Compounds in Dilute Aqueous Solutions. Chemosphere 44 (2001) pp 561 – 576.
16. Woodward & Curran, 2006. Industrial Waste Treatment Handbook

(586a) Comparison of Syngas Fermentation Reactors for Ethanol Production

Hasan K. Atiyeh[1], James J. Orgill[2], Mamatha Devarapalli[1], John R. Phillips[1], Randy S. Lewis[2] and Raymond L. Huhnke[1], (1)Biosystems and Agricultural Engineering, Oklahoma State University, Stillwater, OK, (2)Chemical Engineering, Brigham Young University, Provo, UT

Ethanol can be produced from biomass feedstocks or municipal solid wastes (MSW) using a hybrid thermochemical-biochemical conversion process called gasification-fermentation. In the gasification-fermentation process, biomass is gasified and converted to synthesis gas (syngas; primarily CO, CO_2 and H_2). The advantage of gasification-fermentation over the saccharification-fermentation process is that gasification utilizes cellulose, lignin and hemicellulose components of the biomass, leading to a greater potential for high carbon conversion efficiency and high ethanol yield from the same amount of biomass. Syngas fermentation rates are subject to gas-liquid mass transfer limitations since CO and H_2 have low solubilities in the fermentation broth. Successful syngas fermentation balances the kinetic capacity of the cell inventory with CO and H_2 supply that is set by the mass transfer characteristics of the reactor used. Continuously stirred tank reactor (CSTR), trickle-bed reactor (TBR) and hollow fiber membrane reactor (HFR) were compared for syngas fermentation with regards to gas utilization (including mass transfer assessment), cell growth, and product formation. The volumetric mass transfer coefficient ($K_{tot}A/V_L$) for O_2 in the three reactors was measured. The HFR provided the largest $K_{tot}A/V_L$ followed by the TBR. The mass transfer characteristics in each reactor were affected by agitation speed, and gas and liquid flow rates. Correlations for $K_{tot}A/V_L$ for each reactor were developed to predict mass transfer during syngas fermentation by *Clostridium ragsdalei* to produce ethanol and acetic acid. Although scale-up issues are different for each reactor, the comparison of the three reactors provides guidance as to which reactor(s) should be evaluated for scale-up, especially if one reactor significantly outperforms the others regarding ethanol yield and productivity and syngas utilization efficiency. Further, insight gained from the analysis of the three reactors can be extended to other fermentation processes.

(586b) Electrode Surface Kinetics of Methyl Viologen for Biological Processes

Chang Chen, Chemical Engineering Department, Brigham Young University, Provo, UT and Randy S. Lewis, Chemical Engineering, Brigham Young University, Provo, UT

Electrons are essential for biological processes and are often supplied by reduced species (e.g. glucose, H_2, CO, etc.). However, a recent focus has also been the use of electrodes to supply electrons for biological processes. For electrodes, electrons are either transferred by an extracellular electron mediator or directly transferred to immobilized bacteria on the electrode.

External chemical electron mediators, such as methyl viologen (MV), NADH and neutral red, can support metabolic functions of various bacteria. There is great potential in using electrodes but a more rigorous engineering analysis is needed to assess constraints such as electron efficiency, toxicity, thermodynamics, reaction rates, and availability of species. This work focuses on characterizing the kinetics of methyl viologen (MV) at the surface of an electrode. MV was chosen as a model compound as a result of its reported beneficial effects on various biological processes.

MV has three redox states (MV^{2+}, MV^{+} and MV^{0}). MV^{2+} undergoes a two-step reduction to acquire 2 electrons, forming MV^{0}. MV^{0} is capable of delivering electrons to cells. The reduction by an electrode is an electrochemical reaction and the reaction rate is a function of the applied potential. The relationship can be expressed by the Bulter-Volmer equation. The MV^{2+} reduction is also a surface reaction with three steps of absorption, surface reaction and desorption. The MV^{2+} reduction is demonstrated to be surface reaction limiting. This study combined the surface reaction with the electro-chemical expression which provides a more detailed engineering analysis to improve future design and control for electrode-mediated biological processes. Results of the kinetic study will be presented, along with thermodynamic constraints of utilizing MV. A key component of the presentation will be a discussion of the importance of engineering analysis when applying electrode systems to biological processes.

(586c) Characterization and Kinetic Study of Fast Pyrolysis and Torrefaction Products From Woody Feedstocks Using Direct Py-GC/MS

Jordan Klinger, Chemical Engineering, Michigan Technological University, Houghton, MI, David R. Shonnard, Chemical Engineering, Michigan Technological University & Sustainable Futures Institute, Houghton, MI and Ezra Bar-Ziv, Mechanical Engineering-Engineering Mechanics, Michigan Technological University, Houghton, MI

Pyrolysis and torrefaction are complex thermal degradation pathways that can produce valuable biofuels products – pyrolysis for bio-oils and torrefaction for bio-coals. Due to the complexity of these processes, decomposition mechanisms and product formations can vary based on the starting feedstock. Yet, detailed mechanistic understanding of the native material degradation is essential for process/product formation control. Current kinetic models and speciation as based around global phenomenological and empirical models, with little to no direct relation to actual decomposition mechanisms. This presentation will present chemical speciation and models based on the product identification that provides new insights for model development. Product speciation includes major gaseous and vapor products formed during both torrefaction (200-300˚C) and pyrolysis (500-700˚C). The preliminary models are formed around observed products/product formation rates during torrefaction at fast heating rate (1000˚C/s). The models were developed around temporal histories of two primary observed product groups during torrefaction: (1) CO, CO_2, H_2O and (2) organic acids (acetic acid, formic acid). Group 1 products are likely formed through dehydration and deoxygenation reactions, followed by Group 2 formation through ring scission reactions, depolymerization, and other complex hemicelluloses degradation reactions. The direct observation of these chemical groups leads to the proposed degradation mechanisms, and kinetic parameters. This work was accomplished with the use of a Pyroprobe 5200 micropyrolysis reactor (CDS Analytical, Inc.) in-line with a Trace Gas Chromatograph Ultra (Restek Rxi-5ms capillary column) and Trace DSQ II Mass Spectrometer (Thermo-Fisher Scientific).

(586d) Kinetic Study of Autothermal Reforming of Glycerol Using a Dual Layer Monolith Catalyst

Yujia Liu[1], Bi Chen[2], Bob Farrauto[3] and Adeniyi Lawal[2], (1)Dept of Chemical Engineering and Material Science, Stevens Institute of Technology, Hoboken, NJ, (2)Dept. of Chemical Engineering and Materials Science, New Jersey Center for MicroChemical Systems, Stevens Institute of Technology, Hoboken, NJ, (3)BASF Catalysts LLC, Iselin, NJ

Kinetic study of autothermal reforming of glycerol using dual layer monolith catalyst

Yujia Liu[a], Bi Chen[a], Robert Farrauto[b] Adeniyi Lawal[a]*

[a] New Jersey Center for MicroChemical Systems, Dept. of Chemical Engineering and Materials Science, Stevens Institute of Technology, Castle Point on Hudson, Hoboken, NJ 07030, USA

[b] BASF Catalysts, 25 Middlesex Turnpike, Iselin, NJ 08830, USA

Crude glycerol is a by-product during biodiesel production via transesterification of triglycerides. The conversion of glycerol to synthesis gas and then to methanol represents an important niche for crude glycerol processing that could relieve small biodiesel producers from high glycerol disposal cost while reducing their dependence on methanol from fossil fuels. The autothermal reforming (ATR) of glycerol into synthesis gas (syngas) was studied using the BASF dual layer monolith catalyst.

This paper presents an approach for developing the kinetic model for the ATR of glycerol. The determination of reaction pathways/mechanisms on one hand and chemical kinetics and parameter estimations on the other make up the two vital aspects of kinetic modeling of catalytic processes. A LHHW type rate model is developed and subjected to parameter estimation and model discrimination. The kinetic parameters in conjunction with the kinetic equations can then be used as a design tool for process scale-up.

(586e) Predicting Biomass Fuel Bound Nitrogen (FBN) Conversion in an Oxygen/Steam-Blown Fluidized Bed Gasifier

Karl Broer, Center for Sustainable Environmental Technologies, Iowa State University, Ames, IA and Robert C. Brown, Mechanical Engineering, Iowa State University, Ames, IA

Predicting Biomass Fuel Bound Nitrogen (FBN) Conversion in an Oxygen/Steam-Blown Fluidized Bed Gasifier

Karl M. Broer (kbroer@iastate.edu)

Robert C. Brown (rcbrown@iastate.edu)

When biomass is processed in a fluidized bed (FB) gasifier, its nitrogen content, also known as the fuel bound nitrogen (FBN), is converted primarily into ammonia (NH_3) and nitrogen gas (N_2). Smaller amounts are also present in the raw syngas as hydrogen cyanide (HCN), char-bound nitrogen, and tar-bound nitrogen. NH_3 is of the most interest because it is typically found in the syngas in high enough concentrations to form problematic amounts of NO_X when the syngas is combusted for integrated gasification-combined cycle applications. This NH_3 can also be responsible for poisoning catalysts in biomass-to-liquid applications. Chemical equilibrium calculations indicate that for typical FB biomass gasification temperatures, pressures, and equivalence ratios, nearly all FBN should be converted to harmless N_2. These results contradict experimental results, which show significant NH_3 concentrations. More sophisticated biomass gasification models based on fundamental gasification kinetics have been developed that are capable of predicting NH_3 concentrations. The model developed by de Jong et al. [1] exhibits modest ability to accurately predict NH_3 concentration, but relatively poor ability to predict the concentrations of other important syngas constituents, such as CO and H_2. Additionally, the de Jong model was designed for steam-air gasification, whereas steam-oxygen blown gasification presents a new range of operating conditions that are characterized by much higher initial oxygen concentrations at the bottom of the fluidized bed, and different relative proportions of permanent gases in the resulting syngas.

A revised kinetic model involving fundamental homo- and heterogeneous combustion reactions has been generated to predict the final concentration of the nitrogen compounds in syngas, along with the other major constituents, such as H_2, CO, CO_2, and CH_4, for steam/oxygen-blown gasification. In addition to predicting the syngas composition, the model suggests which reaction pathways of nitrogen evolution are most important under FB gasification conditions. Knowledge of these major mechanisms allows insight into which methods may be worth further investigation for *in situ* reduction of NH_3 concentration in syngas in favor of N_2 and char-bound nitrogen, which are less problematic for downstream syngas applications.

1. de Jong, W., Unal, O., Andries, J., Hein, K., & Spliethoff, H. (2003). Biomass and fossil fuel conversion by pressurized fluidised bed gasification using hot gas ceramic filters as gas cleaning. *Biomass and Bioenergy*, *25*, 59-83.

(586f) An Experimental Investigation of Solid Carbon Conversion to Fuels in the Presence of CO2

Marco J. Castaldi, Department of Earth and Environmental Engineering (HKSM), Columbia University in the City of New York, New York City, NY

To meet the challenges of increased energy demand and waste management we must move to an atom economy where every atom is utilized in the best possible manner. To achieve that goal, a fundamental understanding of the underlying mechanisms and processes of energy and waste generation is necessary. Furthermore, if we consider where the world stands today in terms of energy use and where it will be in 100 years, there is a significant challenge to meet the expected demand in a more CO_2-neutral way. Therefore, a primary objective for new energy technology development must be to produce carbon-free energy through the advancement of technologies that use both conventional and alternative sources.

We have recently reported on the impact of CO_2 on biomass gasification and its ability to more effectively gasify biomass than steam. Continuing this investigation has led to understanding the impact of heating rates and different reaction environments. This paper will present the results from the gasification of various biomass feedstocks. The heating rates were varied from 1°C min^{-1} to 100 °C min^{-1} to ballistic rates (~ 9000°C min^{-1}). The gasification mediums investigated include H_20/N_2, $CO_2/N_2/H_20$, and O_2/N_2. Global activation energies for pyrolysis were found to be significantly higher than for gasification while those for the grasses were significantly lower than the woods, possibly indicating a catalytic effect during pyrolysis of the high mineral content herbaceous feedstocks. CO_2 pyrolysis (110-450°C) activation energy values for lignin, cellulose and biomass averaged 36, 216, and 50 kJ mol^{-1} and CO_2 gasification (500-700°C) values for lignin and biomass were 25 and 33 kJ mol^{-1}, though cellulose did not exhibit significant gasification behavior. Using a least squares fit on the rate of mass loss fraction, the global decomposition reaction during pyrolysis for lignin in either medium was found to be third order while that for cellulose was first order and for the various biomass samples either first or second order. In addition, the impact of pretreatment on biomass will be presented.

In addition, rapid heating rates (greater than or equal to 700°C min^{-1}) were utilized to gasify Clean Wood and two Refuse Derived Fuel (RDF) samples using thermogravimetric analysis coupled to gas chromatography (TGA/GC). Reaction atmospheres included Air, 5% O_2/95% Ar, 10% O_2/90% Ar, 100% Ar and steam and were used to produce gas evolution profiles for hydrocarbons ranging from H_2 to C_4H_{10}. While expected results were obtained using Air reaction atmospheres, some interesting results were observed using steam and Ar. Different concentration profiles and production rates of C_2H_6 compared to C_2H_4 and C_2H_2 enabled some understanding of the reaction sequence occurring during gasification under rapid heating conditions. Kinetic analysis showed pre-exponential factors of 8.00×10^{27}, 2.02×10^{29} and 3.71×10^{23} (sec^{-1} K$^{1/2}$) for samples Clean Wood, RDF C (Industrial Solid Waste basis) and A (Municipal Solid Waste basis), respectively. Furthermore the apparent activation energy was determined to be 22, 71, and 185 (kJ mol^{-1}) for Clean Wood, RDF C and A respectively indicating that the Clean Wood is slightly more reactive than RDF C and more reactive than RDF A.

Finally, it will be shown how CO_2 can adjust the products of gasification and fast pyrolysis to a desired output. This process allows the decoupling of the biomass feedstock from the final product. Data will be presented focusing on the relation between gasification medium, feedstock selection and major chemical species evolution.

(602a) Axial Dispersion in a Horizontally Rotating Scraped Surface Bioreactor

Mahyar Ghorbanian, University of Louisville, Louisville, KY and **R. Eric Berson**, Chemical Engineering, University of Louisville, Louisville, KY

Horizontally rotating bioreactors offer many advantages for high-solids enzymatic saccharification. By operating horizontally: (1) particle settling and dead zone formation are avoided by horizontal rotation of the blades, which scrape against the surface of the reactor. (2) Impellers in conventional stirred tank reactors need to be run above the just suspended speed to maintain an off-bottom/uniform suspension, which demands significant power. The horizontal scraping action of the impellers in the SSBR requires very low rotation rates and, hence, much less power while more effectively preventing particle settling. (3) The scraping action of the blades maintains a clear reactor surface and, thereby, improves heat transfer.

Despite these advantages, all the transport motion occurs only in the angular direction. Multi-directional transport is important for dispersion of enzyme and carrying away product from the reaction sites. The purpose of this study was to investigate the degree of axial dispersion, if any, in a horizontally rotating scraped surface bioreactor (SSBR) during high solids enzymatic saccharification.

Computational fluid dynamic modeling qualitatively showed clear evidence of axial motion within the slurries. Axial dispersion coefficients were then quantified as a function of the solids concentrations and fluid viscosity. The dispersion coefficients were obtained experimentally by measuring the transient concentration of an injected electrolytic tracer and then fitting the concentration/time data to an axial dispersion model. Axial dispersion was strongly dependent on viscosity and there were order of magnitude differences between coefficient values for water and biomass slurries approaching 20% solids content. The degree of axial transport was still significant enough to consider horizontally rotating reactors as a viable option for processing high solids biomass slurries.

Cycle Operation of Laboratory-Scaled Adsorption Heat Pump for Regenerating Steam from Waste Water

Yoshiho Iwama, Kyushu University, Fukuoka, Japan
Yuuki Tanaka, Kyushu University, Fukuoka, Japan
Kazuya Nakashima, Kyushu University, Fukuoka, Japan
Bing Xue, Kyushu University, Fukuoka, Japan
Agung Tri Wijayanta, Kyushu University, Fukuoka, Japan
Koichi Nakaso, Kyushu University, Fukuoka, Japan
Jun Fukai, Kyushu University, Fukuoka, Japan

Introduction

Steam is demanded and generally provided from large quantities of fossil fuel while large amount of waste water with the range of 60-90 °C is released from these industrial processes [1]. The system to generate steam from waste heat is therefore required for the effective utilization of energy. In this study, a novel steam generation process using a direct heat exchange system with adsorbent-water pair is proposed. Contacting water and adsorbent makes excess water evaporate due to the release of adsorption heat from adsorbent. Because this system does not require any heat exchangers, increment in packing density of adsorbent particles in the reactor is expected. The purpose of this study is to investigate the novel steam generation system using water-adsorbent pair. As a basic study, cycle operation consists of steam generation process and regeneration process of adsorbent is experimentally studied under atmospheric pressure. Steam generation at high pressure is also studied.

Experimental

Apparatus

Fig. 1 illustrates schematically the experimental setup for generation in an adsorption heat pump (AHP) cycle. The setup consists of a thermostatic bath, a cylindrical reactor, a heater, a drier, a compressor, a condenser, a reservoir an electronic balance. The connecting metallic tube with diameter of 6.35 mm (0.25 inch) and outer wall of the reactor are insulated with 12.5 mm thickness of ceramic fiber. The reactor is made of stainless steel with height of 100 mm, inner diameter of 71 mm and outer diameter of 80 mm. Cylindrical zeolite 13X particles with diameter of 1.6 mm and lengths 2–6 mm are used. Mass of zeolite is 0.25 kg.

The high-pressure experimental apparatus includes a relief valve, an evaporator and a vacuum pump in addition to the cycle experimental apparatus. A relief valve is installed for high-pressure steam generation and set 0.2 or 0.4 MPa. The relief valve, which is initially closed, opens when the gas (steam plus air) pressure reaches a pre-set value. The evaporator and the vacuum pump are installed for the preheat process. The evaporator temperature is 80°C (saturated water vapour pressure is 0.047 MPa).

Procedure

The cycle operation begins with generation and ends with regeneration. In the steam generation process, water at 80 °C was introduced to the reactor by the feeding pump. Steam was then generated by adsorption heat. The mass of steam was measured by an electric balance as condensed water. The steam generation process was terminated when water interface reached the top of packed bed. After the steam generation process, remained water was drained from the bottom of the reactor. Dried air at 120 °C and 0.62 Nm/s was then introduced to the reactor for a given period of time in the regeneration process. The temperature in the packed bed was measured by thermocouples inserted in the packed bed. Both steam generation process and regeneration process was cycled 5 times. As the basic study, the cycle experiment was carried out under atmospheric pressure.

In the high-pressure experiment, saturated water vapor at 80 °C is introduced as preheating. After that generation process follows the same procedure as steam generation process of cycle experiment.

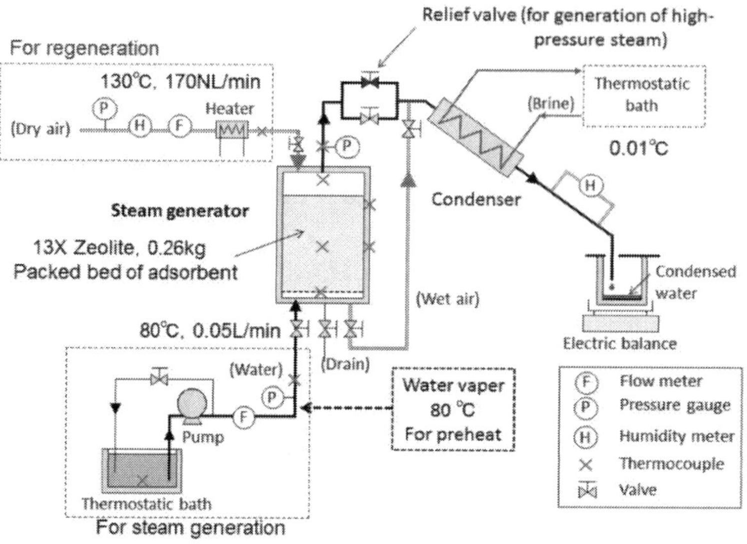

Fig. 1 Schematic illustration of experimental setup

Results

Cycle Operation at 0.1 MPa

Fig. 2 describes transient temperatures for all five cycles based on t_{regen}=3600 s. The thermal response is distinguished in the first cycle from subsequent cycles due to the difference in initial temperatures inside the reactor and mass ratio of water to dry zeolite (x). Results from the second to the fifth cycles depict good agreement in thermal response.

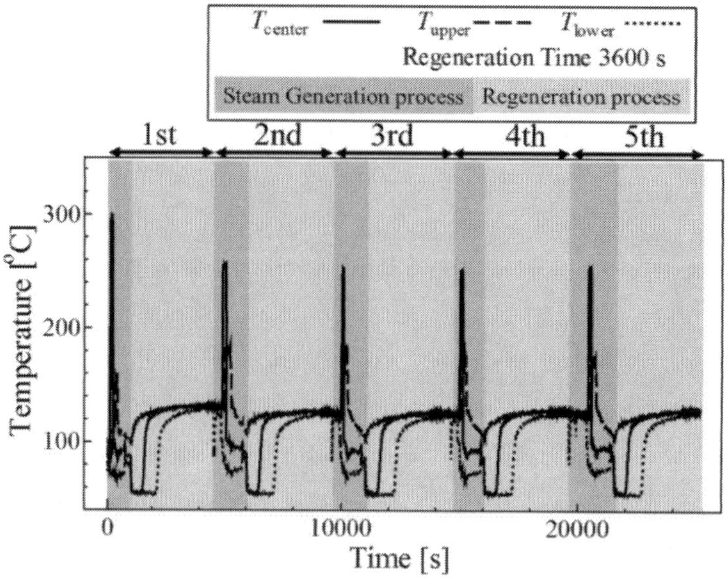

Fig. 2 Transient temperatures over five cycles based on regeneration for 3600 s

Fig. 3 shows the results for the fifth cycle. Fig. 3(a) shows the time variation in temperatures and steam production in the generation process. The feeding water temperature (T_{lower}) initially reaches 100 °C (boiling point under atmospheric condition) due to adsorption heat. Then T_{lower} gradually decreases to 7 °C. T_{center} dramatically increases from 130 °C to 260 °C due to adsorption of steam on zeolite. The maximum is retained for 100 s because adsorption is paused under a local equilibrium condition. As water passes through this part, the temperature rapidly drops to 100 °C. A slow decline to 90 °C follows uniform with water temperature until the end of the process.

Fig. 3(b) shows the transient temperatures during regeneration process. The feeding dry gas temperature (T_{upper}) on the top remains stable at 130 °C during the whole process. The temperature (T_{center}) in the center quickly declines to 50 °C largely from free water evaporating away as a flow of dry gas. T_{center} then quickly rises to 110 °C because of the decreasing desorption rate, followed by a slow increase to 130 °C at the feeding gas temperature (T_{upper}). A similar response to the temperature in the center (T_{center}) is then observed. Finally, regeneration stops as the feeding (T_{upper}) and outlet (T_{lower}) temperatures reach almost the same value.

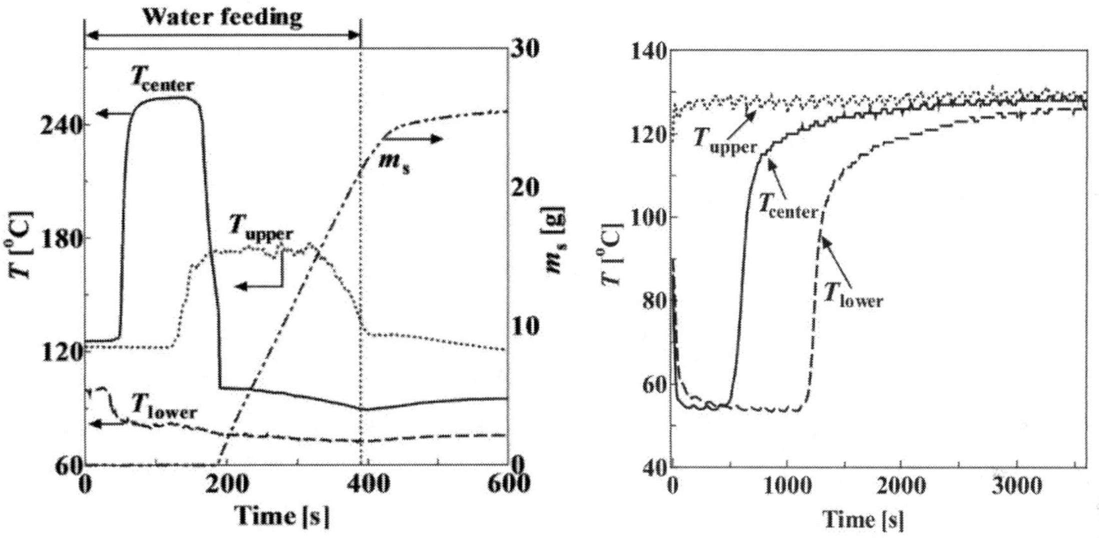

(a) Generation (b)Regeneration

Fig. 3 Transient behavior in the fifth cycle

Table 1 summarizes the cyclic operation conditions and results based on t_{regen}=3600 s. The generation rate per zeolite mass ($\dot{m}_{cycle}/m_z$) from the second to fifth cycle is obviously larger than that at the first cycle, in spite of a decrease in adsorption heat ($\Delta H_{ads,zw}$) at 1126 kJ/kg compared with that at 1350 kJ/kg for the first cycle. $\dot{m}_{cycle}$ is defined as mass of steam generated per the cycle time including generation (t_{gen}) and regeneration (t_{regen}).

$$\dot{m}_{cycle} = m_s /(t_{gen} + t_{regen}) \quad (1)$$

The rise on $\dot{m}_{cycle}/m_z$ is mainly due to a higher initial zeolite ($T_{z,ini}$) and reactor ($T_{rea,ini}$) temperatures. $T_{z,ini}$ increases from 83.3°C at the first cycle to an average at 126°C in the following cycles, while $T_{rea,ini}$ rises from 83 to 121°C. $\dot{m}_{cycle}/m_z$ varies in a range from 2.08 to 2.48×10^{-5} kg-s/kg-z·s at different operation conditions. Table 1 also summarizes cycle operation conditions and results based on t_{regen}=1200 s. Compared with t_{regen}=3600 s, the mass of generated steam (m_s) is expected to decline due to lower $T_{z,ini}$, $T_{rea,ini}$ and a decline in total heat of adsorption ($Q_{ads} = m_z(x_{eq}-x_{ini})\Delta H_{ads,zw}$). Average $T_{z,ini}$ and $T_{rea,ini}$ decrease from 125 to 110 °C and 121 to 94.2°C, respectively. Declines in both Δx (from 0.195 to 0.135) and $\Delta H_{ads,zw}$ (from 1126 to 1080 kJ/kg) with a larger x_{ini} result in a decrease in Q_{ads}. However, average $\dot{m}_{cycle}/m_z$ is improved by 10% from 2.22×10^{-5} 1/s (t_{regen}=3600 s) to 2.44×10^{-5} kg-s/ kg-z s (t_{regen}=1200 s). This is because m_s drop to only a half when the regeneration time is decreased to one-third.

Table 1 Cycle operation conditions and results based on regeneration for 3600 s and 1200 s

		Regeneration for 3600s					Regeneration for 1200s					
		1st	2nd	3rd	4th	5th	1st	2nd	3rd	4th	5th	6th
Generation												
$m_{w,in}$	[g]	327	325	314	306	305	319	216	255	247	245	252
$u_w\times10^4$	[m/s]	1.98	2.02	2.48	2.27	2.10	2.02	3.70	2.19	2.19	2.06	2.23
$T_{w,feed}$	[°C]	69.4	72.4	68.1	68.5	71.5	69.8	71.4	71.1	71.9	71.0	72.0
$T_{rea,ini}$	[°C]	83.0	125	120	119	119	79.7	92.9	94.1	94.7	94.9	94.2
$T_{z,ini}$	[°C]	83.3	130	124	124	124	81	104	110	112	112	113
$x_{ini}\times10^2$	[-]	1.7	6.5	6.5	6.5	6.5	1.7	12.5	12.5	12.5	12.5	12.5
$\Delta H_{ads,zw}$	[kJ/kg-w]	1350	1126	1126	1126	1126	1350	1085	1085	1085	1085	1085
$T_{rea,fin}$	[°C]	85.7	93.8	90.6	89.9	92.5	84.2	86.1	87.0	87.3	86.8	86.9
$T_{z,fin}$	[°C]	87.8	90.2	88.4	88.6	90.3	88.5	88.2	88.2	88.2	87.9	88.0
T_s	[°C]	140	180	180	178	178	184	175	183	184	186	186
m_s	[g]	22.0	31.7	28.8	26.7	26.6	18.9	12.6	13.7	14.3	13.7	13.8
$\dot{m}_{cycle}/m_z \times10^5$	[kg-s/kg-z s]	1.72	2.48	2.25	2.08	2.08	3.39	2.26	2.45	2.56	2.46	2.44
Regeneration												
$T_{g,feed}$	[°C]	131	131	131	131	131	131	131	131	131	131	131
u_g	[Nm/s]	0.64	0.64	0.64	0.64	0.64	0.64	0.64	0.63	0.64	0.64	0.64
$w_{s,feed}\times10^6$	[kg-s/kg-a]	4.77	4.77	4.77	4.77	4.77	4.77	4.77	4.77	4.77	4.77	4.77

Steam Generation at High Pressures

Table 2 shows experimental conditions for high-pressure steam generation without and with preheat. Fig. 4(a) shows the results for $u_w = 2.57\times10^{-4}$ m/s and P_{preset} = 0.4 MPa. There is a difference of 50 degrees in the maximum temperature of T_{z1} and T_{z2} because partial pressure of steam at beginning T_{z2} rising is higher than that at beginning T_{z2} rising. Superheated steam is generated at 220°C and 0.45 MPa. Fig. 4(b) shows the time variations in T_{upper} and m_s for P_{preset} = 0.4 MPa. The effective time for steam generation decreases to 50 s, which is only 14.2% of the water feeding time. At the end of water feeding, m_s has reached 1.4 g and continues to increase to 4.28 g due to the delay of condensed water.

Table 2 Experimental conditions for steam generation without and with preheat

		No preheat	Preheat	No preheat	Preheat
P_{preset}	[MPa]	0.2	0.2	0.4	0.4
$P_{g,ini}$	[kPa]	101.3	≤2.5	101.3	≤2.5
$T_{z,ini}$	[°C]	78.8	78.5	79.2	77.7
x_{ini}	[-]	0.017	0.017	0.017	0.017
$T_{w,feed}$	[°C]	78.8	79.1	79.1	81.3
$u_w\times10^4$	[m/s]	2.44	2.78	2.53	2.61

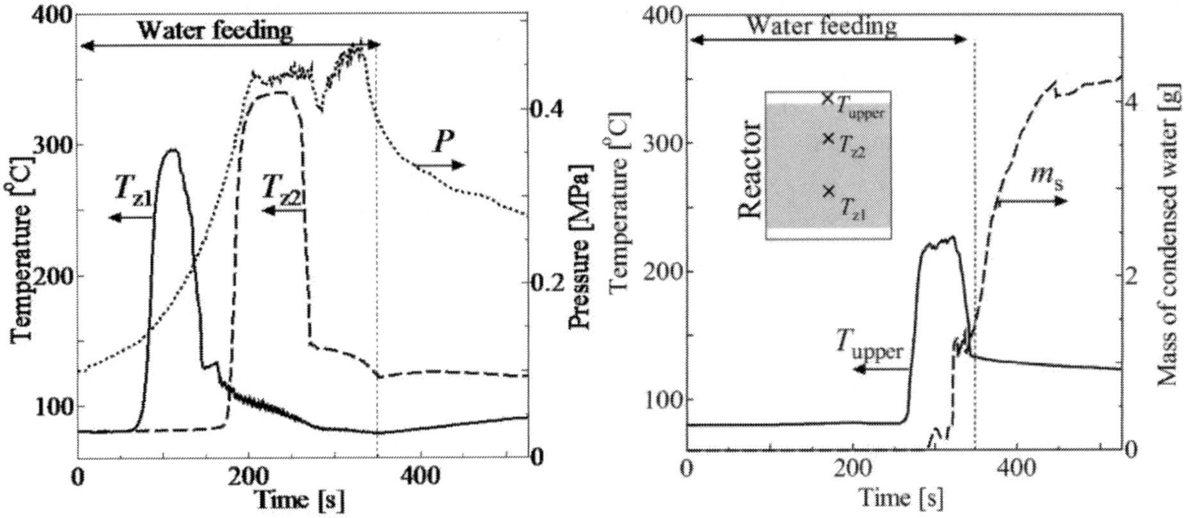

(a)Time variation in temperature and total pressure (b) Time variation in temperature and mass of steam

Fig. 4 Transient behavior in steam generation process at P_{preset} = 0.4MPa

Fig. 5(a) shows the time variations of temperatures (T_{z1}, T_{z2} and T_{upper}) and total pressure (P_g) at P_{preset} = 0.4 MPa. During preheat, both T_{z1} and T_{z2} increase as a result of steam adsorption on zeolite. T_{z1} and T_{z2} rise to their maximum temperatures of 270 and 230°C in 55 s, respectively. During water feeding, T_{z1}, T_{z2} and T_{upper} rise in sequence with increases in position height, because feed water flow originates at the bottom of the reactor.

Fig. 5(b) shows the effect of preheat on normalized time variations in steam temperature. The time ratio of effective steam generation to water feeding ($t_{s,eff}/t_{w,feed}$) is 45%, which is much larger than the 25 % without preheat. The increase is significant because most of the zeolite-gas region achieves local equilibrium at the start of water feeding. A local equilibrium for zeolite and gas is reached in most of the packed bed, and also in the upper part where steam escapes. Therefore, generated steam passes through the packed bed more easily with preheat.

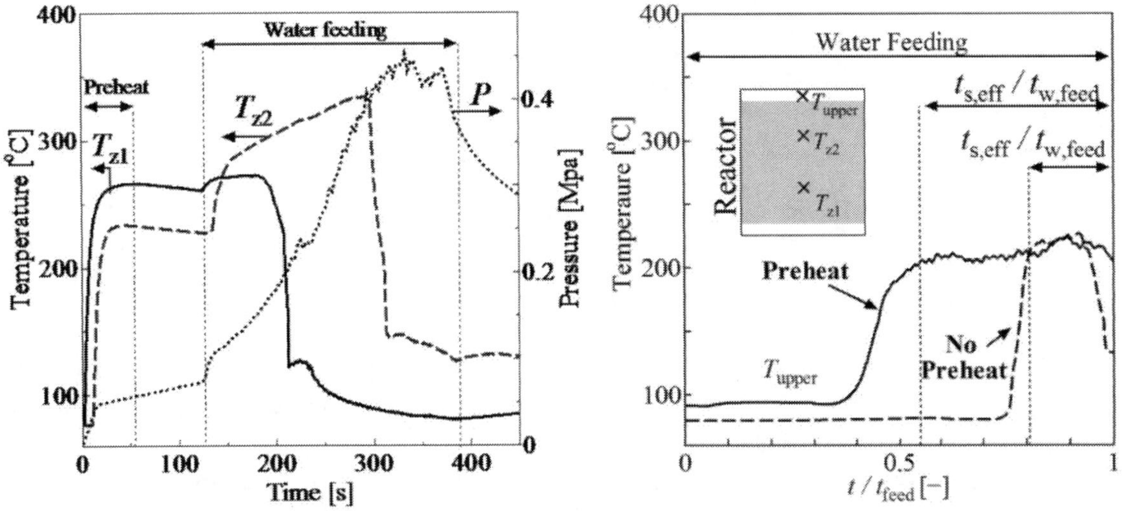

(a)Time variation in temperatures and total pressure with preheat (b) Effect of preheat on normalized time variation of steam temperature

Fig.5 Transient behavior in steam generation process with preheat at P_{preset} = 0.4MPa

Fig. 6 shows the mass of steam (m_s) without and with preheat at two preset pressures (P_{preset}). The increases in m_s of 20.5 and 54.8% are provided with preheat at P_{preset} = 0.2 and 0.4 MPa, respectively. The reason why increase the mass of steam generated with preheat is explained from the following heat balance. During experiment, initial conditions for both with and without preheat are same. According to the heat balance in Eq. (2), Q_{in} (heat flow in) is zero. Q_{acc} (sensible heat change in zeolite, water and reactor) is almost the same for various P_{preset}. Q_{out} (heat of steam flowing out) is determined by Q_{gen} (overall adsorption heat).

$$Q_{in} - Q_{out} + Q_{gen} = Q_{acc} \tag{2}$$

Q_{gen} for the experiment without preheat is given in Eq. (3), and that with preheat is shown in Eq. (4), respectively:

$$Q_{gen} = m_{z,dry}(x_{eq} - x_{ini})\Delta H_{ads,zw} \tag{3}$$

$$\begin{aligned} Q_{gen} &= m_{z,dry}(x_{eq} - x_2)(\Delta H_{ads,zw} + L) + m_{z,dry}(x_2 - x_{ini})\Delta H_{ads,zw} \\ &= m_{z,dry}(x_{eq} - x_{ini})\Delta H_{ads,zw} + m_{z,dry}(x_{eq} - x_2)L \end{aligned} \tag{4}$$

where x_2 is the mass ratio of water to dry zeolite at the end of the preheat process. Although x is measured before the start and after the finish of an experiment, the change in x during the process could not be measured during the experiment. As described above, the reason for the increase in m_s is that Q_{gen} increases by the latent heat of vaporization of steam from preheat.

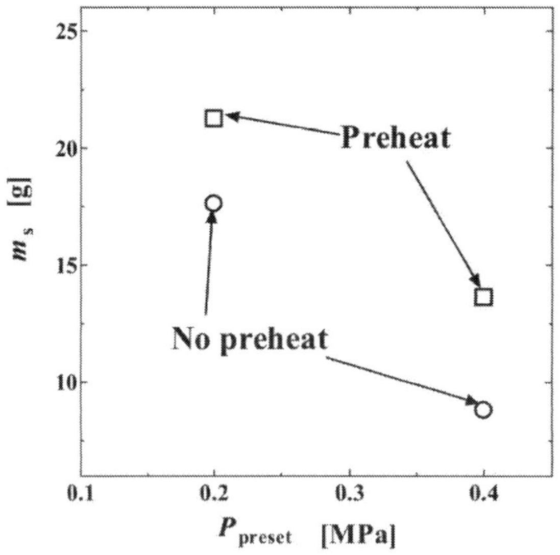

Fig. 6 Steam generation with and without preheat at two preset pressures

Summary

Cycle operation for generation steam using an AHP has been studied experimentally. Experiments have been conducted with and without a preheat process at various pre-set pressures. Concluding remarks in this study are summarized as follows:

1. Steam generation in cycle operation in a zeolite-water AHP using low-grade waste heat has been experimentally demonstrated.
2. Reproducibility in temperature response and mass of generated over cycles proves the durability in practical applications.
3. Experiments demonstrate a significant increase in high-pressure steam generation with a preheat process using low-pressure steam in the zeolite-water AHP system, because the latent heat of vaporization in water is transported to the zeolite before the steam generation process.

Nomenclature

L	= latent heat of vaporization	[kJ/kg]
m	= mass	[kg]
$\dot{m}_{cycle}$	= steam generation rate in cycle operation	[kg-s/s]
P	= total pressure	[Pa]
p	= partial pressure	[Pa]
Q	= heat	[J]
T	= temperature	[K]
t	= time	[s]
u_g	= superficial velocity of feed gas	[m/s]
u_w	= superficial velocity of feed water	[m/s]
V	= volume	[m^3]
w	= mass ratio of steam to dry air	[–]
x	= mass ratio of water to dry zeolite	[kg-w/kg-z]
ΔH_{ads}	= adsorption heat	[kJ/kg]

<Subscripts>

acc = accumulation
center = center of packed bed height
dry = dry
eff = effective
eq = equilibrium
feed = feed
fin = final
g = gas
gen = generation
in = inlet
ini = initial stage
lower = bottom of the packed bed
out = outlet
preset = preset value of relief valve
rea = reactor
regen = regeneration
s = vaporized water (steam)
upper = top of the
w = liquid water
z = zeolite
z1 = one third of the packed bed height
z2 = two thirds of the packed bed height

Acknowledgements

This work was partially supported by the NEDO project under Research and Development Program for Innovative Energy Efficiency Technology, and by a grant from the Global Centre of Excellence in Novel Carbon Resource Sciences, Kyushu University.

Reference

1. Japan Machinery Federation; "Research on new business development through the use of energy technologies", pp. 12-15, Tokyo, Japan (2007) [In Japanese]

(602ab) Investigation of Application for Star Block Copolymer As a Material for Photocatalytic Degradation of Chlorophenols

Seung-Gun Chung[1,2], **Yong-Soo Choi**[1], Jae-Woo Choi[1] and Sang-hyup Lee[1], (1)Water Resource Cycle Research, Korea Institute of Science and Technology, Seoul, South Korea, (2)Earth and Environmental Science, Korea University, Seoul, South Korea

This study investigated the photocatalytic degradation of chlorophenols using two different types of porphyrin core star block copolymer (P-PSD and P-PD) under visible light irradiation. Both P-PSD and P-PD catalysts show similar tendency in degradation of chlorophenols. Pentachlorophenol was degraded faster than other compounds (Pentachlorophenol > 2,4,6-Trichlorophenol > 2,4-Dichlorophenol). The removal rate of chlorophenols by P-PSD was higher than P-PD, which could be attributed by the existence of hydrophobicity on P-PSD. The degradation intermediates and by-products of chlorophenols were also identified by HPLC and LC-MS analysis. The analysis results revealed that the degradation of highly-chlorinated phenols were rapid than the less-chlorinated phenols as confirmed by residual chlorinated compounds and released chloride ions. The toxicity of PCP was greatly reduced after P-PSD catalyzed photocatalytic treatment. The process of photocatalytic degradation using porphyrin core star block copolymer is found to have potentially advantageous on dissolved chlorophenol pollutants in water.

This study investigated the photocatalytic degradation of chlorophenols using two different types of porphyrin core star block copolymer (P-PSD and P-PD) under visible light irradiation. Both P-PSD and P-PD catalysts show similar tendency in degradation of chlorophenols. Pentachlorophenol was degraded faster than other compounds (Pentachlorophenol > 2,4,6-Trichlorophenol > 2,4-Dichlorophenol). The removal rate of chlorophenols by P-PSD was higher than P-PD, which could be attributed by the existence of hydrophobicity on P-PSD. The degradation intermediates and by-products of chlorophenols were also identified by HPLC and LC-MS analysis. The analysis results revealed that the degradation of highly-chlorinated phenols were rapid than the less-chlorinated phenols as confirmed by residual chlorinated compounds and released chloride ions. The toxicity of PCP was greatly reduced after P-PSD catalyzed photocatalytic treatment. The process of photocatalytic degradation using porphyrin core star block copolymer is found to have potentially advantageous on dissolved chlorophenol pollutants in water.

(602ac) Adsorption of Fluoroquinolone Antibiotics Onto Activated Sludge: Implications for Membrane Bioreactor Operation

Ke He, Chemical, Biochemical and Environmental Engineering, University of Maryland, Baltimore County, Halethorpe, MD and Lee Blaney, Chemical, Biochemical and Environmental Engineering, University of Maryland, Baltimore County, Baltimore, MD

Abstract

Given the increasing detection of pharmaceuticals in treated wastewater, the need to optimize wastewater treatment processes for pharmaceutical removal is considerable. In conventional wastewater treatment plants, the activated sludge process contains the greatest potential for pharmaceutical treatment. That potential can be harvested using novel treatment processes, such as membrane bioreactors (MBRs), which allow the operator to uncouple the solids retention time (SRT) and the hydraulic residence time. By increasing the SRT, greater degrees of pharmaceutical removal can be attained through adsorption to biosolids and biological transformation mechanisms. This research focuses on two aspects of such systems: (1) development of analytical methods for detecting trace concentrations of fluoroquinolone antibiotics and (2) description of fluoroquinolone adsorption onto activated sludge solids.

A rapid, accurate, and sensitive method has been developed for the quantitative determination of six fluoroquinolone antibacterial agents (*i.e.*, ciprofloxacin, norfloxacin, levofloxacin, sparfloxacin, sarafloxacin, and orbifloxacin) by online solid-phase-extraction high performance liquid chromatography with fluorescence detection (oSPE-HPLC-FLD). That method was used as a template for development of an ultrahigh performance liquid chromatography tandem mass spectrometry (UHPLC-MS/MS) detection method. The UHPLC-MS/MS method allows quantitation of fluoroquinolone antibiotics to the ng/L level. With this method, the six fluoroquinolone antibiotics of concern were measured in raw wastewater, the activated sludge reactor, and treated wastewater from the Little Patuxent Water Reclamation Plant (LPWRP; Savage, MD). These concentrations of fluoroquinolones were employed in laboratory-based investigations of the adsorption of fluoroquinolones onto activated sludge.

Activated sludge was collected from LPWRP; the specific biomass concentration was determined by weighing the dry solids. Biological degradation of fluoroquinolones was minimized through the addition of 0.1% w/v of sodium azide. Batch adsorption experiments were designed to investigate the impact of pH and ionic strength on fluoroquinolone partitioning between the aqueous and solid phase. Specific masses of biosolids were added to batch reactors containing synthetic wastewater containing environmentally-relevant (from LPWRP) concentrations of fluoroquinolones. At specific times, samples were collected and analyzed using UHPLC-MS/MS to determine the fluoroquinolone concentrations in the aqueous phase; solid phase concentrations were determined via mass balance. The equilibrium data (contact time = 36 hours) was fit to typical isotherms to fully describe adsorption of fluoroquinolones onto activated sludge. These data will further understanding of how SRT affects adsorption of antibiotics in activated sludge / MBR operation. Ultimately, the results of this research will be used to design and operate a laboratory-scale MBR to investigate fluoroquinolone removal through adsorption and biological transformation mechanisms.

(602ad) Ozone Treatment of Tetracycline Antibiotics: Transformation Kinetics and Removal in Water/Wastewater Matrices

Zachary Hopkins, Chemical, Biochemical, and Environmental Engineering, University of Maryland, Baltimore County, Baltimore, MD and Lee Blaney, Chemical, Biochemical and Environmental Engineering, University of Maryland, Baltimore County, Baltimore, MD

Ozone treatment of tetracycline antibiotics: Transformation kinetics and removal in water/wastewater matrices

Abstract by Zachary Hopkins researching under Dr. Lee Blaney

The fate of pharmaceuticals in the environment is of increasing concern. Antibiotics represent unique threats to human and ecological health due to the potential for development of antibiotic resistance, inhibition of select microorganism populations, and various sub-lethal effects stemming from the ability of antibiotics to act as intermicrobial signaling agents. To preclude lasting effects to environmental and human health, advanced water and wastewater treatment processes are required to successfully remove trace concentrations of antibiotics from water supplies.

In this research, three tetracycline antibiotics (*i.e.*, tetracycline, chlortetracycline, and rolitetracycline) were treated using concentrated aqueous ozone solutions (~16 mg/L as O_3), which were generated by bubbling gaseous ozone into DI water. The baseline structure of tetracycline antibiotics contains four potential sites for ozone interaction. All three tetracyclines have pK_a values of approximately 3.3, 7.6, and 9.3; above $pK_{a,2}$, ozone reacts rapidly with the free tertiary amine functionality (~4×10^6 $M^{-1}s^{-1}$). Below pH 7.6, ozone reacts with the tetracycline ring structure. Regardless of pH, ozone was extremely reactive with the three tetracyclines; for example, the apparent second order rate constants ranged from 5×10^4 to 5×10^6 $M^{-1}s^{-1}$ across the pH range of interest (*i.e.*, pH 5–9). Tetracyclines were measured using ultrahigh performance liquid chromatography tandem mass spectrometry (UHPLC-MS/MS).

While pH affects the reaction kinetics of tetracycline transformation by ozone, a variety of water quality parameters, including alkalinity and natural organic matter (NOM), influence the ultimate transformation of tetracyclines in water/wastewater treatment. To draw general conclusions about the impacts of the background water quality on ozonation of the three tetracyclines of concern, three different water matrices, namely tap water, surface water from Herbert Run (Baltimore, MD), and tertiary effluent from the Little Patuxent Water Reclamation Plant (Savage, MD), were employed. These three experimental solutions were chosen to resemble treatment of finished drinking water, raw drinking water, and treated wastewater. Tetracyclines (approximately, 2×10^{-9} to 2×10^{-7} M) were spiked into these background matrices and dosed with ozone (approximately, 3×10^{-4} M) to result in molar ratios of applied ozone to initial (total) tetracycline antibiotics of 0–5 mol/mol. The ultimate transformation of the three tetracycline antibiotics in these matrices will be presented as a function of pH, NOM content (measured as total organic carbon), and alkalinity. This data will be beneficial to the design of ozonation processes aimed at treating trace concentrations of pharmaceuticals in water/wastewater treatment processes.

(602ae) Kinetics Study of Calcium Carbonate Decomposition At High CO2 Environment

Yong-Hua Duan[1], De-Long Xu[2], Hui Li[3], Lele Zhang[4], Xiao Fan[1], Wen-bin Yang[1] and Yong Min[3], (1)Material Institute, Xi'an University of Architecture and Technology,, Xi'an, China, (2)Xi'an University of Architecture and Technology, Xi'an, China, (3)Xi'an University of Architecture &Technology, Institute of Powder Engineering, Xi'an, China, (4)Xi'an University of Architecture &Technology, Institute of Powder Engineering, Xi'an, China, xian, China

Kinetics Study of Calcium Carbonate Decomposition at High CO2 Environment

Yong-Hua Duan[1], Delong Xu[2], Hui Li[3], Le-le Zhang[1], Xiao Fan[1], Wen-bin Yang[1], and Yong Min[3],

(1)Material Institute, Xi'an University of Architecture and Technology, Xi'an, China, (2)Institute for Powder Eng, Univ. of Archi. & Tech., Xi'an, China,

(3)Xi'an University of Architecture &Technology, Institute of Powder Engineering, Xi'an, China

With the concerns of "global warming", CO_2 sequestration from cement industry are becoming significant important for the "green processing" for the environments. This is because the cement processing released about 36% of CO_2 comparing to the 12% from regular coal-power plant. $CaO/CaCO_3$, a major component and raw material from cement, was investigated in order to design suitable processing to sequestrate the CO_2 during the cement production. It was found that the raw material particle size, surface area as well as chemical components were important factors, which will effect on the CO_2 absorption efficiency, cycle times, particle activities, and so on. Various analytical equipments such as SEM, XRD, FTIR, TGA, were used to monitor the processing and analysis materials during the processing.

It was found that there are two competitive mechanisms involved during the processing i.e., chemical reaction and physical diffusion. In order to investigate the dominated effect, we selected the different particles as model compounds for the analysis. As a result, we found the small particle (particle size less than 5 micron) is dominated by chemical reaction mechanism, i.e., CO_2 can completely diffusion into the core of the particle and result in a complete chemical reaction to form $CaCO_3$; and the larger particle is dominated by the physical diffusion mechanism, i.e., only partial CO_2 can diffuse into the CaO particles and formed a "core-shell" particles, where the centre core is CaO and out-shell is $CaCO_3$. In addition, the small particle always accomplished with a high cycle times and more complete chemical reaction, which was recommended to use in the industry.

Based on the cement processing, various chemical looping processes were used to demonstrate the $CaCO_3$ Calcinations / CaO Carbonation Cycle such as, vacuum processing, solids flow processing, fluidization processing, and cocking gas recycling processing. The energy efficiency, raw material activities, and corresponding thermodynamic date will be presented during the presentation.

(602af) Looking for Optimal Conditions to Manufacture Activated Carbon From Barley Husk by Single- or Dual-Optimization

Margarita Loredo Cancino, Eduardo Soto Regalado, Maria Teresa Garza González, Refugio Bernardo García Reyes, Felipe de Jesús Cerino Córdova and Azucena García León, Facultad de Ciencias Químicas, Universidad Autónoma de Nuevo León, San Nicolás de los Garza, NL, Mexico

Activated carbon has a strong affinity for binding organic or inorganic pollutants from aqueous solutions due to its high specific surface area and the possibility of incorporating functional groups on surface. For these reasons, activated carbons are commonly used for wastewater treatment. Despite of the above-mentioned advantages, the use of commercial activated carbon as adsorbent is often limited due to economical reasons, and many researchers have paid attention in the use of alternative precursors, such as agro-waste materials for activated carbon production. In this research, the manufacturing of activated carbon from barley husk (BH) by chemical activation with zinc chloride was optimized using a factorial design with two responses (yield and iodine number) and three factors (activation temperature, activation time and impregnation ratio). Optimal conditions for manufacturing barley husk activated carbon (BHAC) were obtained using a 2^3 factorial design followed by a central composite design by adding axial points to the initial design to explore the space curvature. Single or dual optimization was performed to determine the optimal conditions which improve surface area and yield.

Findings demonstrate that a linear model was not appropriate to represent the design space based on the ANOVA (at a confidence level of 99%) and, therefore, a central composite design was used to fit a quadratic model to the experimental data. According to the ANOVA for the quadratic model of iodine number, the most significant individual factors were temperature and impregnation ratio, double interaction between temperature and impregnation ratio, and the quadratic term of temperature (similar significant factors were determined for yield response except that the quadratic term was the impregnation ratio instead of temperature). It is important to mention that activation time was not significant and, for this reason, it was not considered in the quadratic model but for economical reasons time was set equal to 20 minutes.

Single optimization using the quadratic model of each response gives different optimal conditions depending on the response, for instance: maximal iodine number (1046.26±105.52 mg/g) was predicted at 530°C and impregnation ratio of 1.5 g $ZnCl_2$/g BH, while for maximal yield (61.58±3.32%) were found at 300°C and impregnation ratio of 0.76 g $ZnCl_2$/g BH. Taking into account that maximal responses were found at different conditions and both responses play an important role in manufacturing activated carbon from lignocelullosic precursors, both responses were optimized simultaneously by using the desirability functions approach included in Design expert software.

After simultaneous dual optimization, the maximal desirability function value (0.779) was obtained at 436°C, activation time of 20 minutes, and impregnation ratio of 1.1 g $ZnCl_2$/g BH. At these conditions, predicted responses for iodine number and yield were 829.58±78.30 mg/g and 46.82±2.64%, respectively, whereas experimental tests achieved 901.86 mg/g and 48.48% (values within the predicted interval). Obtained iodine number (901.86 mg/g) is similar or greater than those reported by other researchers dealing with lignocelullosic precursors. Correspondingly, BHAC produced at the optimal conditions (assessed by simultaneous dual optimization) reached a surface area (811.44 m^2/g) similar to commercial activated carbons and lignocelullosic-based activated carbons.

In summary, simultaneous dual optimization seems to be better than single optimization due to the fact that both yield and surface area are very important properties when manufacturing activated carbon from lignocelullosic precursors.

FOULING RATES OF SYNTHETIC THIN STILLAGE

Ravi Challa[1], David Johnston[2], Vijay Singh[1], Mike Tumbleson[1], Nicki Engeseth[3] and Kent Rausch[1,*]

[1]Agricultural and Biological Engineering, University of Illinois at Urbana-Champaign, 1304 West Pennsylvania Avenue, Urbana, IL 61801, USA
[2]Eastern Regional Research Center, Agricultural Research Service, USDA, 600 East Mermaid Lane, Wyndmoor, PA 19038, USA
[3]Food Science and Human Nutrition, University of Illinois at Urbana-Champaign, 1201 West Gregory Drive, Urbana, IL 61801, USA

INTRODUCTION

In the US, fuel ethanol is produced by two processes: dry grind (86%) and wet milling (14%) (Mueller 2010). Ethanol production became more energy efficient with innovative energy saving methods, which reduced the energy requirements per liter ethanol from 33.4 MJ in 1981 to 7.2 MJ in 2008 (Mueller 2010). Following fermentation, whole stillage is separated into thin stillage (5 to 10% soluble solids, w/w) and wet grains (suspended solids) by centrifugation. Thin stillage is concentrated to form condensed distillers solubles (25 to 30% dry solids, w/w) in multiple effect evaporators (Singh et al. 1999).

All dry grind plants use evaporators to concentrate thin stillage to syrup or condensed distillers solubles. For a 40 mmgy ethanol plant, 240 to 280 mmgy thin stillage is produced (Rausch et al 2003). Thin stillage contains 90% water and a portion is concentrated in evaporators and mixed with wet grains to produce distillers dried grains with solubles (DDGS). During evaporation, undesirable deposits form on heat transfer surfaces, increasing resistance to heat transmission and decreasing energy efficiency (Taborek et al. 1972). Fouling deposits increase over time and require periodic evaporator shutdown for cleaning (Wilkins et al. 2006a). Costs associated with fouling include labor and equipment to clean heat transfer surfaces, increased capital, antifoulant chemical costs, production losses and increased environmental impact from cleaning chemical disposal (Epstein 1981).

Process steps involved in converting starch to sucrose are gelatinization, liquefaction and saccharification. Starch is an insoluble granule and resistant to enzymatic breakdown. During ethanol production, first of several steps in starch gelatinization is cooking where the starch granules adsorb water and swell to lose the crystalline structure. Gelatinization is followed by starch enzymatic hydrolysis. Starch hydrolysis is usually expressed in terms of dextrose equivalent (DE) and it is inversely proportional to degree of polymerization. Enzymatic hydrolysis results in starch breakdown into shorter chain oligosaccharides with DE ranging from 0 to 20 (Kenyon 1988). Complete starch hydrolysis to glucose can be represented from the number of reducing ends (DE =100). However, in corn ethanol production by dry grind process, thin stillage composed of unconverted starch, yeast, protein, oil and ash (Singh et al. 1999). Enzymatic hydrolysis results in starch breakdown into shorter oligosaccharides but even under strict reaction conditions, it is difficult to achieve consistent composition at industrial scale and therefore changes occur in the composition of dextrins, maltose and glucose from batch to batch. These compositional changes in process streams could contribute to fouling problems on the evaporator surfaces. Earlier research on thin stillage fouling include the effect of Reynolds number (Wilkins et al. 2006a), pH (Wilkins et al. 2006b), membrane filtration (Arora et al. 2010) and fat content (Singh et al. 1999) but did not identify thin stillage components that accelerate evaporator fouling. In the current study, the effect of sucrose:starch ratio on evaporator fouling was investigated by preparing synthetic thin stillage solutions. Starch is insoluble in water whereas sucrose is soluble at room temperature. The objective is to determine the effect of soluble and insoluble solids on evaporator fouling by varying the starch:sucrose (STA:SUC) in synthetic thin stillage.

MATERIALS AND METHODS

An annular fouling probe was used to perform fouling experiments (Fig. 1). Annular probe design and operation were described previously (Agbisit et al. 2003; Arora et al. 2010; Fischer et al. 1975; Singh et al. 1999; Wilkins et al. 2006a). Fouling experiments were performed using synthetic thin stillage (STS) solutions for 1, 3, 5 and 10% total solids in water. STS components selected were starch and sucrose to demonstrate the individual and interaction effects in evaporator fouling. STS was starch suspended in water, sucrose dissolved in water or sucrose and starch mixed as per the starch to sucrose ratio. Each solution was transferred into a stainless steel tank and tap water was added to make 50L batches. Recycling STS through the fouling system provided for uniform distribution of model components (starch and sucrose) in water.

STS was pumped through the test loop at a flow rate of 15 L/min (Fig. 1). Bulk temperature (T_b) was maintained at 50±2°C for all tests. STS bulk temperature was attained by using water as the heating fluid in a shell and tube heat exchanger (1.5 m, Graver Technologies, DE). The hot water temperature (T_{HW}) was maintained at 55±2°C. A 100 ml sample was collected before start of the experiment to measure test fluid viscosity. Power (410 ± 10W) was supplied to the resistance heater within the probe to an average initial temperature of 100°C after T_b reached 50°C. Probe began to foul immediately, which was reflected in the increased thermocouple temperatures.

The overall heat transfer coefficient (U) (kW/m2.K) was found from the following equation:

$$U = \frac{Q/A}{T_s - T_b}$$

Where T_s is the probe surface temperature (°C) and T_b is the STS bulk temperature (°C).

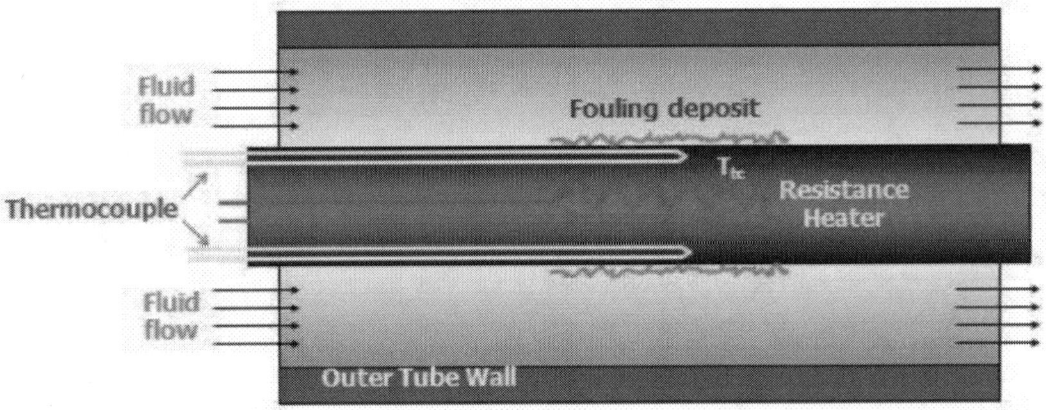

Fig. 1. Annular fouling probe.

Fouling resistance (R_f) was found by the difference of the reciprocals of the overall heat transfer coefficient of the fouled and clean surface.

$$R_f = \frac{1}{U_f} - \frac{1}{U_i}$$

Where U_f is the overall heat transfer coefficients of a fouled probe at time *t* and U_i is the initial overall heat transfer coefficient for a clean probe at time *t*=0. The temperature inside the probe increased as the fouling deposit accumulates on the probe surface.

A schematic of the fouling apparatus is shown in Fig 2. All the tests were terminated when the interior probe temperature reached 170°C or at the end of 10 hr. A 100 ml sample was collected after each test and stored at 4°C for viscosity measurement. Fouling apparatus and annular probe were cleaned following each test. The annular probe along with fouling deposit was soaked overnight in 5% (w/v) NaOH solution. Fouling deposits were removed with a wet sponge; fouling apparatus (batch tank, tubing, heat exchanger, pump and rod housing) was cleaned by recirculating 1% (w/v) hot detergent solution (Alconox, New York, NY), followed by a hot water rinse.

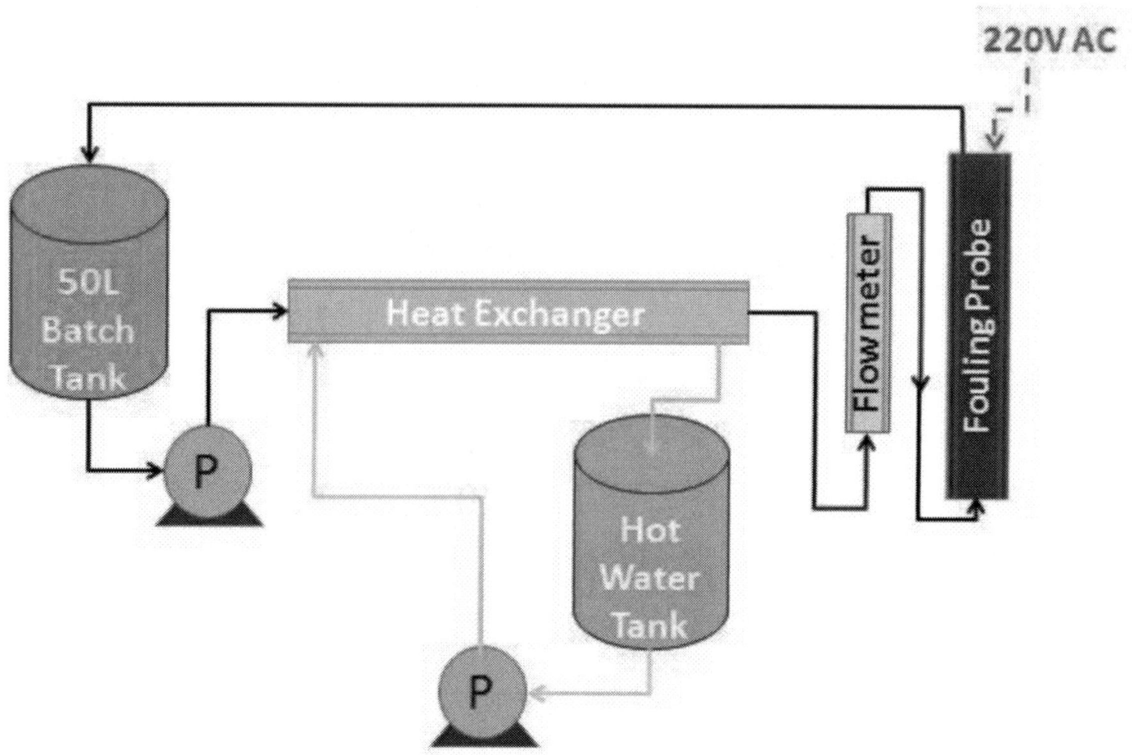

Fig. 2. Fouling apparatus.

RESULTS AND DISCUSSION

1% Total Solids

No induction period was observed for any STA:SUC concentration in synthetic thin stillage (Fig.3). Initial probe temperature (100°C) was too high to get an induction period since the probe did not need any surface conditioning for the onset of fouling. R_f increased rapidly at the and the rate slightly increased with time representing asymptotic fouling. For all STA:SUC concentrations, fouling profiles were similar, increased starch contents increased the fouling resistances except for the STA:SUC of 9:1. Thermocouple readings were recorded for 10 hours and the probe did not reach the maximum temperature, 170°C for STA:SUC treatments at 1% total solids. Fouling profiles followed the trend of increased fouling resistance with increased starch concentration. Fouling profiles had a "saw-tooth" shape which is attributed to signal noise generated from thermocouple recordings in the data logger.

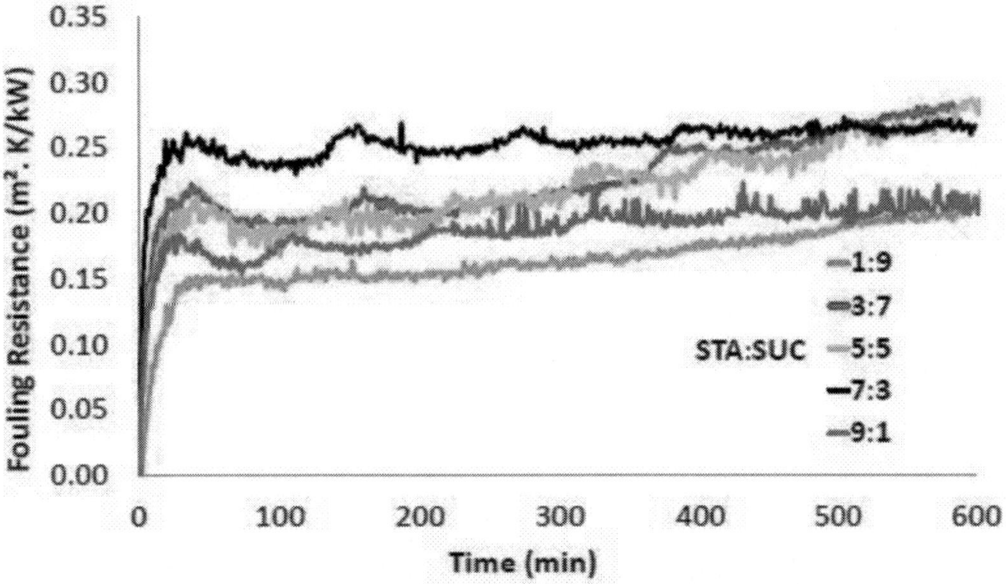

Fig. 3. Effect of STA:SUC on fouling resistance at 1% TS.

10% Total Solids

Fouling profiles for 10% total solids were different from 1% total solids in terms of the slopes (Fig. 4). For STA:SUC of 1:9, $R_f = 0.163$ m^{2}K/kW at the end of 10 hr but did not reach the maximum probe temperature, 170°C. As STA:SUC increased to 3:7, maximum probe temperature was reached in 4.2 hr with final $R_f = 0.264$ m^{2}K/kW. For STA:SUC of 5:5, maximum probe temperature was reached in 0.77 hr with final $R_f = 0.267$ m^{2}K/kW. Treatments having sucrose only (0:10) did not foul measurably by the end of 10 hr. The slopes of the fouling were different for different starch content. Fouling profile for STA:SUC of 5:5 initially had a higher slope and it continued till the probe reached 170°C. The initial slope decreased with decreasing starch content. Sucrose had an effect on fouling in combination with starch but sucrose alone did not foul the probe at the end of 10 hr.

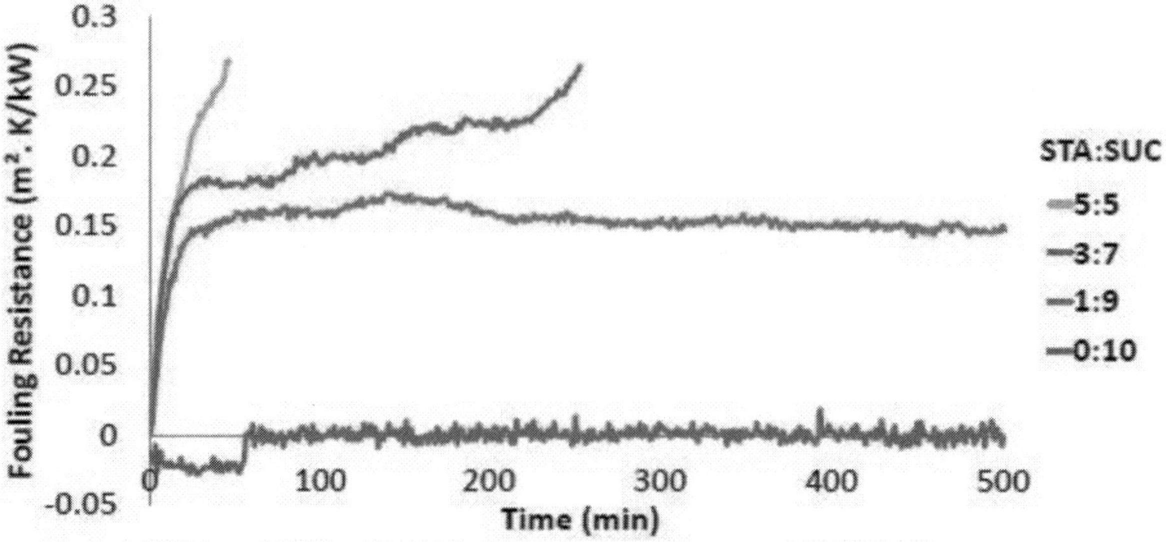

Fig. 4. Effect of STA:SUC on fouling resistance at 10% TS.

Effects of Starch

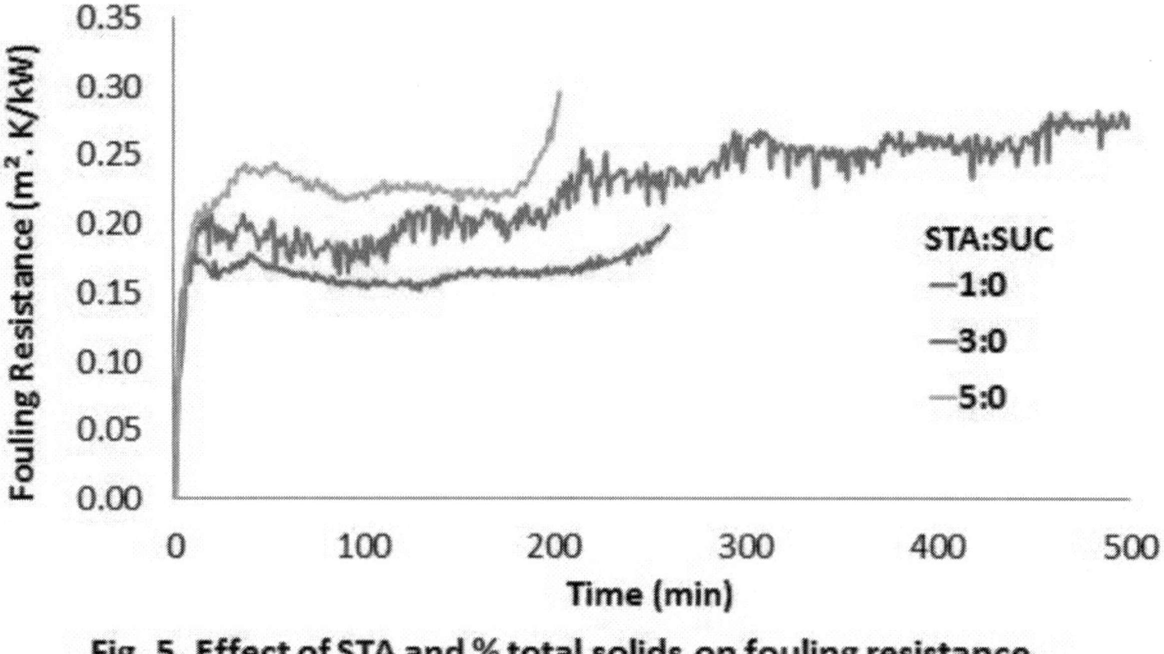

Fig. 5. Effect of STA and % total solids on fouling resistance.

Fouling profiles for 1, 3 and 5% total solids and starch only were obtained (Fig. 5). Synthetic thin stillage with 1% starch had a final R_f of 0.28 m^2K/kW at the end of 10 hr. At 3% starch, probe reached maximum temperature in 4.33 hr with final R_f of 0.19 m^2K/kW and time for 5% starch further decreased to 3.42 hr with final R_f of 0.29 m^2K/kW.

CONCLUSIONS

Higher concentrations of starch in synthetic thin stillage shortened time to reach maximum probe temperature. For 1, 3, 5 and 10% TS, R_f increased mainly because of the starch present in synthetic thin stillage, while glucose had a smaller effect on R_f. Synthetic thin stillage with glucose only did not foul after 10 hr run. Increased starch concentration decreased time to reach maximum probe temperature of 170°C. Glucose had an effect on R_f in combination with starch at 10% TS for STA:GLU of 5:5 compared to STA:GLU of 5:0 in decreasing the time.

REFERENCES

Agbisit, R. M., Singh, V., Valenti, J. J., Kakleas, M., Tumbleson, M. E. and Rausch, K. D. 2003. Technique to measure surface-fouling tendencies of steepwater from corn wet milling. Cereal Chemistry 80:84-86.

Arora, A., Dien, B. S., Belyea, R. L., Singh, V., Tumbleson, M. E. and Rausch, K. D. 2010. Heat transfer fouling characteristics of microfiltered thin stillage from the dry grind process. Bioresource Technology 101:6521-6527.

Epstein, N. 1981. Thinking about heat transfer fouling: a 5 x 5 matrix. Heat Transfer Engineering 4:43-56.

Fischer, P., Suitor, J. W. and Ritter, R. B. 1975. Fouling measurement techniques. Chemical Engineering Progress 71:66-72.

Kenyon. 1988. Maltodextrins and Low-Dextrose-Equivalence Corn Syrup Solids. Pages 7-11 in: Flavor Encapsulation. American Chemical Society.

Mueller, S. 2010. 2008 National dry mill corn ethanol survey. Biotechnology Letters 32:1261-1264.

Singh, V., Panchal, C. B. and Eckhoff, S. R. 1999. Effect of corn oil on thin stillage evaporators. Cereal Chemistry 76:846-849.

Taborek, J., Palen, J. W., Aoki, T., Ritter, R. B. and Knudsen, J. G. 1972. Fouling - major unresolved problem in heat - transfer. Chemical Engineering Progress 68:59-68.

Wilkins, M. R., Belyea, R. L., Singh, V., Buriak, P., Wallig, M. A., Tumbleson, M. E. and Rausch, K. D. 2006a. Analysis of heat transfer fouling by dry-grind maize thin stillage using an annular fouling apparatus. Cereal Chemistry 83:121-126.

Wilkins, M. R., Singh, V., Belyea, R. L., Buriak, P., Wallig, M. A., Tumbleson, M. E. and Rausch, K. D. 2006b. Effect of pH on fouling characteristics and deposit compositions in dry-grind thin stillage. Cereal Chemistry 83:311-314.

(602ai) A Comparative Study of Biodiesel Synthesis From Soybean Oil Via Transesterification Reaction Using Calcium Methoxide in the Absence and Presence of Ultrasound

Vishwanath G Deshmane, Mechanical Engineering, North Carolina Agricultural and Technical State University, Greensboro, NC and Yusuf G. Adewuyi, Chemical and Bioengineering, North Carolina A&T State University, Greensboro, NC

Biodiesel is produced by the reaction of lipid feedstocks with an alcohol such as methanol in the presence of a catalyst (NaOH, KOH or their methoxides) to yield methyl esters (biodiesel) and byproduct glycerol [1,2]. However removal of the base after reaction is a major problem since aqueous quenching results in the formation of stable emulsions and saponification (i.e., soap formation), making separation of the methyl ester difficult and resulting increased production costs. Also, this reaction does not tolerate the presence of water or fatty acids and the use of acids such as H_2SO_4, H_3PO_4, and HCl pose environmental and corrosion problems [3]. On the other hand, use of solid base catalyst offers several process advantages. It can be easily separated from the reaction products and the much more simplified product separation steps result in high yields of methyl esters. Glycerol is also directly produced with high purity (at least 98 %) and is free from any contaminations, thus increasing the profitability of the process. It also gives the opportunity to operate in a continuous process increasing the production capacities and further lowering the production costs. The use of alkaline earth metal oxides as solid base catalysts for transesterification of vegetable oils has been reported in the literature [4]. The basic strength of alkaline earth metal oxides and hydroxides was observed to increase in the order of Mg < Ca < Sr < Ba. From the economical and ecological point of view, calcium derived bases were found to be the most promising as they are inexpensive, exhibit low methanol solubility, and are the least toxic of all. In the present study a commercially available calcium methoxide catalysts has been used in the transesterification of soybean oil to produce biodiesel in a cylindrical jacked glass reactor stirred by mechanical agitation and a 20 kHz ultrasonic, and a novel low-intensity, multi-frequency ultrasonic reactor operated at 611 kHz. In this presentation, the effects of various process parameters such as catalyst concentration, methanol to oil molar ratio and the temperature of the reaction systems on the kinetics of the reaction and the optimization of the process will be discussed. The characterization of the catalytic material (fresh and used) by nitrogen adsorption–desorption (BET) and thermogravimetric and differential scanning calorimetry (TGA-DSC), and the effect of storage time on the activity of the catalyst will also be discussed.

References

1. Mahamuni, N.N.; Adewuyi, Y.G. Optimization of the Synthesis of Biodiesel via Ultrasound-Enhanced Base-Catalyzed Transesterification of Soybean Oil Using a Multifrequency Ultrasonic Reactor". Energy & Fuels. 2009, 23, 2757-2766.

2. Mahamuni, N.N.; Adewuyi, Y.G. Application of Taguchi Method to Investigate the Effects of Process Parameters on the Transesterfication of Soybean Oil Using High Frequency Ultrasound. Energy & Fuels, 2010, 24, 2120-2126.

3. Deshmane, V.G., Gogate, P., Pandit, A. Ultrasound-assisted Synthesis of Biodiesel from Palm Fatty Acid Distillate. Industrial Engineering Chemistry Research. 2008, 48 (17), 7923-7927

4. Watkins, R.S, Lee A.F et al. Li-CaO Catalyzed Triglyceride Transesterification for Biodiesel Application. Green Chemistry. 2004, 6(7), 335-340.

(602aj) Biofuels From Pyrolysis of North-American Grass Species

Shantanu Kelkar, Chemical Engineering, Michigan State University, East Lansing, MI and Christopher M. Saffron, Biosystems & Agricultural Engineering, Department of Forestry, Michigan State University, East Lansing, MI

Native perennial grasses, indigenous to the Midwest, are ideal bioenergy crops due to their potential to be productive on marginal lands. Fast pyrolysis is a simple, flexible process to convert these feedstocks into bio-oil, which is precursor to liquid hydrocarbon fuels. However, organic acids and oxygenates in bio-oil lead to storage instability and limit its early adoption as a transportation fuel. Biomass constituents such as inorganic salts dramatically alter the speciation of pyrolysis products. Therefore, the objective of this study was to prospect amongst several native grasses for cultivars suitable for pyrolysis to biofuels, by investigating the effects of biomass composition on pyrolysis products.

In this work, the composition of native grasses, including big bluestem, coastal panicgrass, deertongue, indiangrass, miscanthus, sandreed, sideoats grama and switchgrass, was determined. Pyrolysis of the grasses was studied using analytical pyrolysis-GC/MS and TGA. The resulting pyrolysis gas contains hundreds of chemical species. Principal component analysis revealed that only six chemical compounds, glycolaldehyde, acetic acid, acetol, methyl glyoxal, 4-vinyl phenol and levoglucosan, were needed to describe more than 90% of the variability associated with pyrolysis products. Correlation analysis revealed that potassium had a significant effect on production of acetic acid, acetol, levoglucosan, char yield and TGA peak reaction rate temperature. Pyrolysis of extractive-free as well as water-washed biomass was studied and compared with untreated biomass. Bio-oil from the native grasses was produced using a screw-conveyor pyrolysis reactor and bio-oil quality was subsequently assessed.

(602ak) Comparing Performance and Re-Acclimation of Two Phase Partitioning Bioreactor and Ordinary Bioreactor Under Starvation Period

Amin Yoosefi Booshehri, Chemical and biomedical engineering, NTU, Singapore, Singapore

Biological treatment plants operated for the removal of pollutants from waste streams are often exposed to periods of no chemical load (starvation periods). The length of such periods can range from a few hours (due to facility operation for a fraction of the day) to a few weeks or months (shut down for factory retooling, equipment maintenance, holiday breaks, process changes) and can negatively affect the elimination capacity of biological processes In the present study, we compare the performance of a two phase partitioning bioreactor(TPPB) and an ordinary bioreactor (OB) after being exposed to hexane starvation period. In OB, after 3 of days starvation period the removal efficiency was reduced by 99%, and about 6 days were needed to recover the initial removal efficiency when the hexane feeding resumed. In contrast, in the TPPB the hexane removal efficiency was not changed significantly and recovered to its initial removal efficiency within 10 hours. The results of this study show that the TPPB system can effectively sustain the biological treatment of waste gas during starvation periods of hydrophobic pollutants.

(602al) Comparison of Biocrude From Micro- and Macroalgae

Griffin W. Roberts[1], Marie-Odile Fortier[2], Belinda S.M. Sturm[2] and Susan M. Stagg-Williams[1], (1)Chemical & Petroleum Engineering, University of Kansas, Lawrence, KS, (2)Civil, Environmental & Architectural Engineering, University of Kansas, Lawrence, KS

Hydrothermal liquefaction (HTL) of algae to a hydrocarbon rich biocrude is growing in popularity primarily because it uses subcritical water as the reaction media eliminating the high energy intensive drying step associated with lipid extraction. HTL efficiently extracts existing lipids in addition to converting other macromolecules such as carbohydrates and proteins to smaller molecules that can be readily upgraded to fuels and other chemicals. Such a process also eliminates the need for a high lipid producing algae and can solely focus on high growth rate algae. Most HTL studies have involved the utilization of microalgae; the few using macroalgae have been marine seaweeds, to our knowledge briocrude has not been produced hydrothermally from fresh water macroalgae.

This study involves pilot scale algae growth tanks utilizing effluent from the second clarifier at the Lawrence, KS wastewater treatment plant. Algae were harvested and hydrothermal treated toward biocrude. Grown under the same conditions both a mixed-culture microalgae and macroalgae identified as *Cladophora sp.* were studied. Hydrothermal treatment was performed with 3g of freeze dried algae and 50mL of water at 350°C for 1 hour. The starting algae and product yields were evaluated on an ash free dry weight basis. High oil yields were obtained for both the micro- and macroalgae (54% and 57%) but had significant chemical and physical differences. Residual solids were determined to have higher heating values of 12.4 and 10.8 MJ/Kg for micro- and macroalgae, respectively. Aqueous co-products had carbon, nitrogen, and phosphorus concentrations of 3100-4000 ppm, 1200-1600 ppm, and 3.5 ppm respectively. The biocrude produced was very similar to that of petroleum crude in terms of energy density (39-42 MJ/Kg) and elemental percentages of carbon, hydrogen, and oxygen.

(602am) Determination of Triacylglyceride Hydrolysis Kinetics in Soybean Oil and Algal Systems

Joshua C. Wissinger[1], Amber L. Bosley[1] and Constance Schall[2], (1)Chemical and Environmental Engineering, University of Toledo, Toledo, OH, (2)Chemical & Environmental Engineering, University of Toledo, Toledo, OH

Determination of triacylglyceride hydrolysis kinetics in soybean oil and algal systems

Joshua C. Wissinger, Amber L. Bosley, Constance A. Schall

The third generation of biofuels, algae, is of growing interest as a renewable feedstock due to its high yield of energy per unit area, use of carbon dioxide and other available resources for growth, and minimal competition with the food industry. Biodiesel and jet fuel can be produced from lipid-rich algae. However, extracting lipids from algae is complicated due to the high water content of harvested and mechanically dewatered algae (60 to 90% water). Thus, there is a need for processes that convert wet algal biomass into fuels or fuel precursors directly in the aqueous phase. Such a process alternative is to use the water in the harvested and partially dewatered algae for hydrolysis of triacylglycerides (TAG) to free fatty acids (FFA), forming aqueous and oil rich phases for ease of extraction and separation of the lipid component. TAG processing techniques include refining FFA by catalytic cracking and hydrotreating for jet fuel production, or the UOP/Eni Ecofining™ process to obtain clean green diesel. In our first step, subcritical water is reacted with TAG producing FFA through hydrolysis eliminating the need for expensive and water sensitive catalysts. In the initial development of an empirical kinetic model, soybean oil, a system much less complex than alga, was used in hydrolysis studies at subcritical water (SCW) conditions. This model was applied to algal TAG hydrolysis. The robustness of our model has been tested by verifying both the autocatalytic behavior of TAG hydrolysis reactions, as well as hydrolysis reactions over a range of temperatures.

(602an) Effect of Particle Size On Enzymatic Hydrolysis of Pretreated Miscanthus

Esha Khullar[1], Bruce S. Dien[2], Kent D. Rausch[1], M. E. Tumbleson[1] and Vijay Singh[1], (1)Agricultural and Biological Engineering, University of Illinois at Urbana-Champaign, Urbana, IL, (2)USDA-ARS NCAUR, Peoria, IL

Effect of Particle Size on Enzymatic Hydrolysis of Pretreated *Miscanthus*

Esha Khullar[1], Bruce S. Dien[2], Kent D. Rausch[1], M. E. Tumbleson[1] and Vijay Singh[1*]

[1]University of Illinois at Urbana-Champaign, Urbana, IL 61801

(217-333-9510) vsingh@uiuc.edu and [2]National Center for Agricultural Utilization Research, ARS, USDA, Peoria, IL 61604

The effect of particle size on enzymatic hydrolysis of pretreated *Miscanthus x giganteus* was determined. *Miscanthus* was ground using a hammer mill equipped with screens having 0.08, 2.0 or 6.0 mm sieve openings. Particle size distribution and geometric mean diameters were determined for all samples. Ground samples were subjected to hot water, dilute acid or dilute ammonium hydroxide pretreatment. Enzyme hydrolysis was conducted on washed pretreated solids; sugar generation was used as a measure for pretreatment efficiency. Glucose and xylose concentrations were monitored using HPLC. Glucose and xylose profiles were generated and hydrolysis rates estimated. Glucan, xylan and total conversion yields were determined by comparing final sugar concentrations obtained to theoretical amounts present in raw biomass.

Geometric mean diameters were the smallest from 0.08 mm sieve screen (56 μm) followed by 2.0 mm (301 μm) and 6.0 mm (695 μm) screens. Across all pretreatments, an increase in total polysaccharide conversion (12 to 26%) was observed when particle size was decreased from 6.0 mm to 0.08 mm. Enzyme hydrolysis of unpretreated biomass samples also resulted in increased total conversions as particle size decreased, although mean conversions (10 to 20%) were much lower than for pretreated biomass samples (40 to 70%), indicating the need for chemical pretreatments in biomass conversion.

(602ao) Fiber Reactor Engineering Design Package for Ultra-High Efficiency Biodiesel Manufacturing

Raul Rivas-Villarreal, Chemical Engineering, Texas A&M University - Kingsville, Kingsville, TX, Patrick L. Mills, Dept of Chemical & Natural Gas Engineering, Texas A&M University-Kingsville, Kingsville, TX and John Massingill Jr., Advanced Materials and Processes, San Marcos, TX

An engineering design package based upon a combination of experimental measurements, data analysis, and reactor modeling tools is described to facilitate development of a new process technology for ultra-high efficiency biodiesel manufacturing. The process design is centered around a new reactor concept called the Fiber Reactor™, which is a novel contactor for conducting reactions involving two partially miscible liquid phases that offers one to two orders-of-magnitude improvement in liquid-liquid mass transfer rates when compared to conventional reactors, such as stirred tanks. This effort is part of a research subproject within the NSF-funded TAMUK CREST on Sustainable Technologies for Environment (STE). The main goal of the STE effort is to develop innovative products by stressing processing of both natural and recycled materials while reducing pollution in a variety of applications through improved process technology. The proposed research on the Fiber Reactor™ is encompassed by the mission and goals of the CREST since the reactor is a key component of sustainable biodiesel processes.

The poster will summarize recent research results whose outcomes enable further understanding of the Fiber Reactor™ by investigating underpinnings of the engineering technology for application to biodiesel transesterification chemistry. Emphasis is placed on developing key reactor scale-up parameters through a combination of experiments, reaction engineering principles, and engineering correlations. These three elements are combined to create a fiber reactor model that is useful for design and scale-up of biodiesel transesterification chemistry from the lab-scale (up to 1-inch OD) to demonstration-scale (ca. 1 m OD) systems. Particular Fiber Reactor™ parameters of interest include fluid pressure drop, fluid flow patterns, fluid flow regimes, liquid-fiber interfacial areas, mass transfer coefficients, heat transfer coefficients, and fiber particulars (material of construction, layout, size, geometry, and surface/wetting properties). Validation of the model-predicted results is also provided using experimental data generated using ½-in OD and 1-inch OD fiber reactor systems.

(602ap) Induction of Microalgal Lipids for Biodiesel Production

Wilbel J. Brewer, Chemical Engineerin, Michigan Technological University, Houghton, MI and Ching-An Peng, Department of Chemical Engineering, Michigan Technological University, Houghton, MI

The production of biofuels such as biodiesel from biomass has gained accelerated interest in recent years as these fuels are becoming increasingly economically-viable, renewable, and carbon-neutral energy sources. One reason for this renewed interest derives from new policy and legislation such as sustainable biofuel targets in the US Energy Policy Act (EPA 2005) and Energy Independence and Security Act (EISA 2007). There have been many attempts in deriving biodiesel fuel from different bioenergy crops such as corn, canola, soybeans, palm, sugar cane and vegetable oil. However there are some significant challenges, these includes: depleting feedstock supplies, land use change impacts, food use competition resulting in high prices and inability to completely displace fossil fuel.

In this study, *Chlorella Vulgaris* microalgae was cultivated under different stressful conditions to produce high-yield biomass with high lipid content which were converted to biodiesel fuel by transesterification. These stressful conditions include the use of high CO_2 concentration, high (25-40 W/m^2) and low (0-10 W/m^2) light intensity. Different operating constraints were investigated to optimize the lipid induction process in *Chlorella Vulgaris*. The approach used in this study was to culture the *Chlorella Vulgaris* microalgae with inoculation $1x10^5$ cells/ml in a 250-ml Erlenmeyer flask, irradiated with fluorescence light bulbs and cultured at room temperature of 25°C. Using these conditions we were able to determine which were suitable for cultivating the algae. During the experiment, Nile red dye was used as a hydrophobic fluorescent probe to detect the intracellular lipid deposits in the cell concentrations.

(602aq) Modeling of High Temperature Biochar Gasification and Methane Reforming

Aaron W. Palumbo[1], Erica L. Jorgensen[1] and Alan W. Weimer[2], (1)Chemical & Biological Engineering, University of Colorado at Boulder, Boulder, CO, (2)Chemical Engineering, University of Colorado at Boulder, Boulder, CO

Temperatures above 1300°C may be used to increase conversion and synthesis gas quality and decrease carbon dioxide and hydrocarbon formation in an allothermally heated gasifier. Using methane, or natural gas, as a reactant increases the hydrogen content of the synthesis gas to be compatible with liquid fuel synthesis technologies. Thermodynamic equilibrium results confirm an adjustable H_2:CO given a variable methane-to-biomass ratio as well as superior utilization of hydrogen and carbon monoxide compared to methane reforming. Experimental results show that a complex, kinetic-limited relationship exists between char gasification and methane reforming. Reactions of pyrolysis further complicate experimental controllability of the product gas composition. Char gasification and methane reforming were modeled in a 1-D plug flow reactor configuration assuming a monodisperse particle distribution and an entrained particle flow. Experiments with biochar, methane, and steam were used to obtain global kinetic models of the relevant reactions. Effects of reactant flow rate, particle diameter, and wall temperature were explored.

(602ar) Process Synthesis for an Organic Rankine Cycle Using Low Enthalpy Geothermal Sources

Jesús D. Arrieta[1], Omar J. Guerra[2,3] and **Ariel Uribe Rodriguez**[2], (1)Chemical Engineering, Industrial University of Santander, Bucaramanga, Colombia, (2)Colombian Petroleum Institute (ICP), ECOPETROL S.A., Piedecuesta, Colombia, (3)Chemical Engineering, Industrial University of Santander, Bucaramanga, Colombia

Energy sustainability is one of the most important issues in the 21th century because of the many problems (global warming, atmospheric pollution, sustainability, etc.) associated with the use of fossil energy sources. In this context, renewable energy sources, like geothermal water, are promising energy sources basically in term of sustainability. In addition, those energy sources can provide energy without negative impacts in our environment. In recent years, many works have been focused on the development of efficient energy conversion processes using renewable sources.

In this work a Mixed Integer Nonlinear Programming (MINLP) model for process synthesis in an Organic Rankine Cycle using a low enthalpy geothermal source is presented. The proposed model was applied to a case study in order to evaluate many design option for the process involved. The results showed that the proposed model is adequate for the optimization of process configuration and operation in Organic Rankine Cycles, resulting in the generation of electricity with a competitive production cost using a low enthalpy geothermal source.

(602as) Rapid Lipid Quantification of Algal Biomass Using Thermo-Gravimetric Analysis

Balakrishna Maddi[1], Patrick Bollin[2], **Agasteswar Vadlamani**[3], Sridhar Viamajala[1] and Sasidhar Varanasi[1], (1)Chemical & Environmental Engineering, University of Toledo, Toledo, OH, (2)Chemical Engineering, Kansas State University, Manhattan, KS, (3)Bioengineering, The University of Toledo, Toledo, OH

Algae, especially green algae and diatoms, have the potential to accumulate high amounts of lipid (upto 50 wt% of dry weight) through chemical and physical stimuli, when grown in laboratory conditions. However, most of these algae cannot withstand the harsh and non-sterile environmental conditions encountered in actual production systems. At present, intense research efforts are devoted to screening for algal species that can survive and accumulate lipids in the open-pond non-sterile production environments. Simple, rapid and accurate lipid quantification techniques are of great value for rapidly screening a large variety of strains. In this regard, traditional laboratory techniques for lipid quantification, such as Bligh and Dyer, Folch, acid/base catalyzed trans-esterification methods are not convenient. They are time-consuming, require volatile chemicals and expensive analytical facilities, and involve several process-steps. In this paper, we will introduce a simple, rapid and accurate lipid quantification technique based on thermo-gravimetric analysis (TGA).

The main components of oleaginous algae are protein, starch and lipids. Our technique is based on the observation that the protein, starch and lipid portions undergo thermal degradation over distinct non-overlapping temperature intervals. Proteins are most thermally labile followed by starch and lipids. Lipids undergo *complete* volatilization over a narrow temperature interval of 370-450 °C, without leaving behind *any* biochar. Therefore, lipid content in the algal sample can easily be correlated to the weight loss of the sample during TGA over this narrow temperature range. Lipid content estimates, for several oleaginous algae and oil seeds, from this approach will be reviewed relative to the results from traditional lipids quantification techniques.

(602at) Scalable Biosynthesis of Quantum Dots for Bioenergy Applications

Victora F. Berard[1], Chris K. Kiely[2], Steven McIntosh[1] and Bryan W. Berger[3], (1)Chemical Engineering, Lehigh University, Bethlehem, PA, (2)Center for Advanced Materials and Nanotechnology, Lehigh University, Bethlehem, PA, (3)Chemical Engineering and Program in Bioengineering, Lehigh University, Bethlehem, PA

The cost-efficient and energy- efficient generation of renewable fuels will be one of the most important factors in the future economic prosperity of the United States. Solar energy is one of the most promising routes to develop a new, domestic, source of energy. While current solar cells generate electrical power, we need a chemical route to store large amounts of energy. The direct conversion of solar energy to a chemical fuel (hydrogen) by splitting water would enable large scale energy storage and create a clean burning fuel for transportation. The possible commercial applications are as large as the current power and fuel markets. The activity of the nanoparticle catalysts utilized for this water-splitting reaction is dictated by their physical size, shape, structure, and their chemical composition. Thus it is imperative that we develop synthesize routes that enable control over all of these parameters. We have created a unique approach to environmentally benign, *in situ* semiconductor nanoparticle biosynthesis from the gram-negative bacteria *Stenotrophomonasmaltophilia*. *S. maltophilia* is able to generate novel nanoparticles of the relevant materials unable to by synthesized by conventional methods, but little is currently known regarding how to control the chemical and physical attributes of the resulting particles from biosynthesis. We will describe our recent efforts to synthesize novel nanostructured materials from *S. maltophilia,* including a detailed description of their physical properties determined using electron microscopy, fluorescence spectroscopy and UV-visible spectroscopy. Applications of the synthesized nanoparticles to photocatalysis will also be discussed.

(602au) Treated Effluent Water Use in Cellulosic Ethanol Production

Divya Ramchandran, Environmental Engineering, University of Illinois, Urbana-Champaign, Urbana, IL, Kishore Rajagopalan, Illinois Sustainability Technology Center, Champaign, IL, Timothy J. Strathmann, Department of Civil and Environmental Engineering, University of Illinois at Urbana-Champaign, Urbana, IL and Vijay Singh, Agricultural and Biological Engineering, University of Illinois at Urbana-Champaign, Urbana, IL

The bioethanol industry exerts a significant demand on water supplies. Current water consumption rate in corn dry grind ethanol plants is 3 to 4 gallons of water per gallon of ethanol produced (gal/gal) and 6 to 10 gal/gal for cellulosic ethanol plants. The main goal of this study is to examine the use of treated wastewater effluent in place of potable freshwater for cellulosic ethanol production. The effects of using two different types of filtered treated effluent; Bloomington- Normal, IL (Residential type) and Decatur, IL (Industrial/Residential Mix type); on the rate of fermentation and final ethanol yield from a pure cellulosic substrate were evaluated. Final ethanol concentration with Bloomington- Normal and Decatur effluent and our control study using de-ionized water were similar, resulting in 4.57 ± 0.22 % v/v (0.36 g/g, db), 4.74 ± 0.13 % v/v (0.37 g/g, db) and 4.55 ± 0.28 % v/v (0.36 g/g, db), respectively. Residual glucose concentrations were <0.04% w/v at 48 hr in all cases, suggesting complete fermentation.

Further study with Decatur effluent using 0.08 mm finely ground *Miscanthus* as the substrate resulted in a final ethanol concentration of 0.46 ± 0.008 % v/v (0.14 g/g db) which was similar to ethanol concentration of 0.52 ± 0.07 % v/v (0.17 g/g db) obtained with control treatment using de-ionized water. These findings suggest that with proper characterization studies and under appropriate conditions, the use of treated effluent water in cellulosic ethanol production is feasible.

(602av) Waste From Beer Fermentation Broth; A Potent Source of Bioethanol and Hydrolyzing Enzymes

Waleed Ahmad Khattak, Mazhar Ul-Islam, Minkyung Kang and Joong Kon Park, Chemical Engineering, Kyungpook National University, Daegu, South Korea

Waste from beer fermentation broth; A potent source of bioethanol and hydrolyzing enzymes

Waleed Ahmad Khattak, Mazhar Ul-Islam, Minkyung Kang, Joong Kon Park*

Department of Chemical Engineering, Kyungpook National University, Daegu 702-701, Korea

(parkjk@knu.ac.kr)

Abstract

A semi solid discharged residue produced at the end of fermentation process, called as waste from beer fermentation broth (WBFB) showed tremendous potential for bioethanol production without the addition of saccharifying enzyme, microbial cells or carbohydrates. Number of hydrolyzing enzymes produced from malt in beer fermentation broth during brewing, were confirmed via sodium dodecyl sulfate polyacrylamide gel electrophoresis (SDS PAGE) analysis. The enzymes extract from WBFB after partial purification was evaluated for their hydrolyzing activities. Various parameters including pH, temperature, thermal stability and buffer system was investigated for hydrolyzing activities. With gradual increase in temperature (25-70°), the presence of hydrolyzing enzymes in WBFB caused the degradation of yeast cell wall and reduced the live cell density to a greater extent. However, elevated temperature enhanced both the hydrolyzing activities and bioethanol production without additional enzyme, microbial cells and carbohydrates along with shortening the simultaneous saccharification and fermentation time. The enhanced bioethanol production at elevated temperature seems to be partially resulted via cell free enzymes system. Comparing to bare supernatant or sediments of the WBFB, the saccharification and bioethanol production efficacy of original WBFB were superior.

Key words; Waste from beer fermentation broth, Bioethanol, Hydrolyzing enzymes, Cell wall degradation, Cell free enzyme system

(602ax) Effect of Engulfing Fire to Thermal Decomposition of Type III Composite Cylinder

Jae Hun Lee, Safety Research Devision, Institute of Gas safety R&D, Gyeonggi-do, South Korea and **Hyo Kim**, University of Seoul, Seoul, South Korea

The characteristics of the thermal decomposition of composite cylinder has been thoroughly investigated when it is exposed to a engulfing fire. In the computations, one-dimensional heat transfer model and the Arrhenius type kinetic equations are incorporated to construct the robust framework for predicting the surface recession profile of the cylinder wall. An experiment was conducted to measure in surface temperatures and internal pressure of a commercial composite cylinder that was exposed to engulfing fire. The cylinder used in the experiment is a Type ¥² NGV composite cylinder which was manufactured with epoxy/carbon fiber by the hoop and helical winding methods. The cylinder liner is made of aluminum alloy, and glass bulb type TPRD is built in the cylinder valve. An engulfing fire test was conducted with kerosene on a NGV composite cylinder. Surface recession depth profiles were calculated using surface temperature data obtained from the experimental results, and they were compared with the time dependent pressure change.

Figure 1. Photograph of engulfing fire test.

Key words: Composite cylinder, Thermal Decomposition, Surface Recession, Engulfing Fire

(602ay) Mixed Metal Oxide Reduction/Oxidation Cycle for Thermochemical Energy Storage in Solid Particles

Brian D. Ehrhart[1,2], Nathan P. Siegel[3], Eric N. Coker[2] and Alan W. Weimer[1], (1)Chemical and Biological Engineering, University of Colorado at Boulder, Boulder, CO, (2)Sandia National Laboratories, Albuquerque, NM, (3)Mechanical Engineering, Bucknell University, Lewisburg, PA

Solid particle heat transfer and storage media have been shown to be able to operate at temperatures >1000°C in concentrated solar power (CSP) applications, much higher than the operational limit of 600°C for currently state-of-the-art molten nitrate salt heat transfer fluid. Solid particles can be endothermically reduced by direct exposure to concentrated solar energy, thus absorbing and storing thermal energy beyond that possible with sensible heating alone. The particles can then be oxidized exothermically at a later time, releasing the stored chemical heat and effectively augmenting the thermal energy storage capacity of the solid particles. A mixed metal oxide spinel material that reduces in the temperature range of interest (1000°C-1200°C) has been examined for applicability to this thermochemical energy storage concept. A description of this application, prospective materials, and details of the thermochemical cycle are presented. The heats of reduction and oxidation for the thermochemical cycle have been determined for various operating conditions to evaluate the amount of thermal energy that may be stored. Additionally, the extent of reduction possible at various operating conditions has been measured.

(602az) Bioinspired Electrets for Solar Energy Conversion

Srigokul Upadhyayula[1], Robert Bonderer[2], Duoduo Bao[1], Brent Millare[3], Khalid Ashraf[4], Amy Ferreira[3], Roger Lake[5] and Valentine I. Vullev[3], (1)Bioengineering, University of California, Riverside, CA, (2)Chemical and Environmental Engineering, University of California, Riverside, Riverside, CA, (3)Department of Bioengineering, University of California, Riverside, Riverside, CA, (4)University of California, Berkeley, CA, (5)Electrical Engineering, University of California, Riverside, CA

Protein-mediated electron-transfer processes sustain a broad range of redox functions in biological systems, such as respiration and photosynthesis. Therefore, such biological systems are inspirational models for materials for energy-conversion applications. This presentation will focus on bioinspired macromolecular electrets with rectifying long-range charge-transfer properties. We have demonstrated molecular designs based on anthranilamide derivatives, which similar to polypeptide α-helices, contain ordered arrangement of amide and hydrogen bonds, but in contrast to polypeptides and proteins, have an extended Π-electronic conjugation for mediating efficient long-range charge transfer. Ab initio calculations, along with dielectric and NMR measurements, revealed that polyanthranilamides possess intrinsic dipole moments in the order of 3 D per residue. Furthermore, the macromolecular dipoles remove the degeneracy between the frontier orbitals of neighboring residues, causing splitting of ~0.2 eV for 10 residues that can readily result in rectification of vectorial charge hopping. Doping the macromolecules with electron withdrawing and electron donating groups did not only alter the bandgap and the energy levels of the frontier orbitals, but also modulated the intrinsic dipole moments, ranging from 0 to 5 D per residue. The design of the presented bioinspired electrets will prove instrumental for the development of materials with rectifying charge-transduction properties essential for solar-energy-conversion applications.

(602az) Integrated Use of Geographic Informational System and MIXED-Integer Linear Programming for Indentifying Optimal Agricultural Areas for Sewage Sludge Amendment: A Case Study of Catalonia

Pavel Vaskan, Chemical engineering, University Rovira i Virgilii, Tarragona, Spain

INTEGRATED USE OF GEOGRAPHIC INFORMATIONAL SYSTEM AND MIXED-INTEGER LINEAR PROGRAMMING FOR INDENTIFYING OPTIMAL AGRICULTURAL AREAS FOR SEWAGE SLUDGE AMENDMENT: A CASE STUDY OF CATALONIA

P. Vaskan[1], A. Passuello[2], G. Guillén-Gosálbez[3], M. Schuhmacher[4], L. Jiménez[5]

Departament d'Enginyeria Química (EQ), Escola Tècnica Superior d'Enginyeria Química (ETSEQ), Universitat Rovira i Virgili (URV), Campus Sescelades, Avinguda Països Catalans, 26, 43007 Tarragona, Spain

Abstract

Deciding the best agricultural soils for sewage sludge amendment is not a trivial task. Several environmental aspects, such as groundwater contamination by nitrates, open waters and soil protection, as well as human exposure to the contaminants present in the sludge matrix must be considered along with economic aspects, making this task quite challenging. This is because the management of this residue concerns different stakeholders, with different views of the problem and conflicting interests. In addition, they may find that their interests are not reached, fact that leads to a low acceptance of the practice.

Reaching a final solution satisfying all the decision-makers involved is indeed very challenging. Systematic tools based on multi-objective mathematical programming are well suited to tackle this type of problems, as they allow generating and assessing a large number of potential solutions. The final goal is to calculate a set of Pareto optimal alternatives, that is, a set of solutions with the property that it is impossible to improve them in one of the objectives without necessarily worsening at least another criterion.

In this work, we study the use of multi-objective optimization for identifying the best geographical areas for sewage sludge (SS) amendment in terms of economic and environmental criteria. Our approach integrates geographic informational systems (GIS) and multi-objective mixed-integer linear programming (MILP) within a unified framework that allows exploring in a rigorous and systematic manner a large number of alternatives for sewage sludge amendment from which the best ones (according to the decision-makers' preferences) are finally identified.

The capabilities of our approach were tested through its application to a case study based on Catalonia. Given a certain amount of sludge to be treated and a set of available agricultural areas. Numerical results show that it is possible to improve the environmental performance of the final solution by compromising the associated cost. Our methodology is intended to assist decision-makers in such a challenging task.

---

[1] pavel.vaskan@urv.cat

[2] ana.passuelo@urv.cat

[3] gonzalo.guillen@urv.cat

[4] laureano.jimenez@urv.cat

[5] laureano.jimenez@urv.cat

(602b) Algal De-Watering Using Temperature Sensitive Hydrogels: A Preliminary Step in the Harvesting Process

Agasteswar Vadlamani, Bioengineering, The University of Toledo, Toledo, OH, Sridhar Viamajala, Chemical & Environmental Engineering, University of Toledo, Toledo, OH and Sasidhar Varanasi, Chemical and Environmental Engineering, University of Toledo, Toledo, OH

Algal cultures from the growth reactors are usually dilute (~0.1-0.5% (w/w)) and direct use of conventional solid-liquid separation methods (e.g. centrifugation, filtration or flocculation) may be cost prohibitive due to large process volumes. In this study, temperature-sensitive gels made of N-isopropylacrylamide (NIPAAm) were investigated as a re-usable absorbent for increasing the concentration of dilute algal suspensions. We observed that when algal cultures were incubated in the presence of NIPAAm gel (at 25°C), cell-free media was absorbed resulting in an increase in cell concentration. The swollen hydrogels were then recovered and heated to 34°C to deswell and release the absorbed media. Since swelling rates were high initially (t<4h) and decreased subsequently (4h < t <15h), cyclic swelling (for 4h) followed by deswelling was applied to more rapidly concentrate the algal cultures. Our experimental results show that the concentration of algal cultures increased from 0.16% to 4.3% (w/w) in 6 cycles. In addition, slurries initially at 3% (w/w) could further be concentrated to 8% (w/w) by incubation with hydrogels without any discernible change in gel swelling/deswelling kinetics. These results suggest that higher solid concentrations could be achieved by applying more swelling/deswelling cycles. In fact, our studies show that removal of nearly all free water is possible to obtain algal pastes with concentrations of nearly 18% (w/w) solids. However, recovery of gels from such concentrated slurries without incurring significant loss of cells is difficult. Overall, our results demonstrate that temperature-sensitive hydrogels can be used to easily dewater dilute algal cultures to concentrations (8%(w/w) or more) where more traditional solid-liquid separation methods can be easily applied. An economic analysis of the hydrogel based dewatering process will be presented.

(602ba) Verification of Biochar Pyrolysis

Lucas J. Rice, Chemical Engineering, University of Missouri, Columbia, MO

Verification of Biochar Pyrolysis

Lucas J. Rice and Tom R. Marrero

Department of Chemical Engineering

University of Missouri

Columbia, MO 65211

Abstract

This study attempts to verify the ability to generate a carbon-negative cycle from the pyrolysis of biomass materials to supplement soil nutrients.

The pyrolysis process is simulated with ChemCAD software with inputs from new of field and bench scale data.

Field data correspond to a kiln approximately 100 feet long with a nominal cross-sectional area of 100 ft^2. The kiln maintains a nitrogen atmosphere to pyrolysis the biomass to biochar. A series of operating characteristics are measured; temperatures in the chamber, moisture of raw materials and biochar, and process residence time. Furthermore, biochar characteristics such as porosity, elements, and carbon content were determined.

Results indicate the maximum and minimum temperatures required for the process. The maximum temperature of the kiln allows for pyrolysis energy balances to determine other factors required such as the residence time. Furthermore, an ideal temperature is calculated in order to create the most efficient.

Biomass materials are used for the biochar include: corn stover, milo stover, and hay.

Recent publication describes the carbon life cycle of normal energy production. There are three potential cycle types mapped out in this process: carbon positive, carbon neutral, and carbon negative.

Carbon positive represents the burning of fossils and release carbon dioxide into the atmosphere. This process results in a net gain of carbon dioxide in the atmosphere, which is why it's called carbon positive.

The burning of biomass in a similar process as fossil fuels also can produce the same carbon dioxide in the atmosphere. During photosynthesis, plant life takes up the produced carbon dioxide facilitating more green growth, which in turn can be used as biofuels. This process is called carbon neutral since the net gain of carbon in the cycle is zero. However, the half-life of carbon dioxide in the atmosphere is significantly longer than 100 years which results in a loss of carbon.

The third type of process discussed is the focus of this presentation, the carbon negative cycle. Biomass is subjected to slow burning pyrolysis which results in two usable products. Similarly to the previous two cycles, carbon dioxide is generated as a bi-product and can be used as energy or recycled back through green life. The biomass is converted to biochar and can be used as soil amendments thereby a portion is sequestered or captured into the soil due to its refractory properties. This sustainable process is called the carbon negative cycle.

To obtain the nutrient rich qualities of biochar for soil enhancement, the biomass is subjected to slow pyrolysis. During pyrolysis, nitrogen and/or nitrogen containing compounds are adsorbed by biochar.

The amount of soil amendment obtained from the biomass is quantified by measurements of the dry weight and the burn-down of the biomass to biochar.

The production of biochar provides fertilizers with a variety of benefits for the environment. The use of biochar instead of traditional fertilizers (ex. ammonia) can potentially provide a more efficient alternative to soil supplements in theory and in practice. By returning the biomass materials to the soil in this manner, the nutrients are recycled and already in a form useable by vegetation. Furthermore, the biochar in the soil helps the ground to maintain its strength preventing the fast erosion due to wind or rain. Water has a higher rate of retention in soil which contains biochar. Biochar is also able to revitalize grounds that have had all the nutrients drained by over cultivation or other methods. This would allow for a larger area base for farmable land. This also leads to the ability forgo such strategies as crop rotations. Lastly, as the carbon negative cycle utilizes more energy from the pyrolysis of the biomass, the efficiency helps to convert energy.

The final results of the study are a new simulation of a sustainable biomass pyrolysis using the ChemCAD software. This simulation is verified and calibrated by field and bench scale data.

(602bb) Microwave-High Temperature Pretreatment and Enzymatic Hydrolysis of Sugarcane Bagasse

Milad R.Esfahani, Chemical Engineering, Tennessee Technological University, Cookeville, TN and Mehrdad Azin, IROST, Tehran, Iran

Sugarcane bagasse is a promising lingocellulosic biomass for production of bioethanol. High-temperature treatment is one of the common prerequisites for almost all of the different kinds of pretreatment methods. Using microwave power, besides of creating high temperature, has physical effects on degradation of cellulosic polymeric structure of lingocellulosic biomass. To study the effect of microwave power for creating high temperature needed for pretreatment of sugarcane bagasse, the degree of hydrolysis was measured by assaying both the released total reducing sugars and the glucose, specifically. Different particle size, temperature and duration of exposures to the heat, were the main studied factors. Afterwards, the enzymatic hydrolysis was studied at different levels of enzyme load and pH. The optimal conditions for microwave pretreatment and enzymatic hydrolysis of sugarcane bagasse were determined by using Taguchi method. The result indicates that while particle size of 35-80 mesh and 1 minute duration time for microwave exposure are common optimal condition for production of both reducing sugars and glucose, other factors like pH, enzyme Load and temperature varies for optimal production of each of them.

(602bc) Life Cycle Assessment of Plasticizers: DEHP Versus "Green" Alternatives

Yuan Li, Department of Chemical and Petroleum Engineering, University of Pittsburgh, Pittsburgh, PA and Eric J. Beckman, University of Pittsburgh, Pittsburgh, PA

Phthalate plasticizers are added to polyvinyl chloride (PVC) during manufacturing to increase resilience, low temperature performance, softness, and ease of processing. Phthalates are widely applied in toys, construction materials, medical devices, food packaging, flooring and other product categories. After several decades of commercial use, recent investigations have prompted concern that phthalate exposure can cause human health issues. At the same time, a number of new plasticizers have been introduced as potentially greener alternatives to the most widely used phthalates. We have employed process life cycle analysis (process LCA) to examine the life cycle impacts of di(2-ethylhexyl)phthalate (DEHP) to those arising from the use of 1,2 cyclohexane dicarboxylic acid diisononyl ester (DINCH) and citrate esters. Initial work has focused on the cradle to gate stage of the life cycle, comparing the raw material acquisition and manufacturing phases. Energy consumption is also an important concern for the environmental impacts. A complete LCA study to get a much "greener" plasticizer substitute is expected.

(602bd) Improved Reproducibility in HPLC Analytical Techniques and Use of MALDI-MS in Biofuel Related Research

Samira Vasheghani Farahani, Chemical and Environmental Engineering, University of Toledo, Toledo, OH and Michael Mayer, University of Toledo, Toledo, OH

Quantitative biomass analyses and molecular weight determination are essential in assessing lignocellulosic feedstocks in biofuel related research. High-performance Liquid Chromatography (HPLC) equipped with a suitable carbohydrate resin column is commonly used for quantitative analysis of polysaccharide constituents in biomass. When sample preparation includes steps such as acid neutralization, introduction of soluble salts can lead to an unstable chromatographic baseline. The contribution of this work is in introducing HPLC sample pretreatment which results in improved baseline stability of resultant chromatograms and therefore reproducible data. Characterizing cellulose, especially its degree of polymerization (DP), is another important step in evaluating its deconstruction to monomeric sugars or in its use for materials. Matrix Assisted Laser Desorption Ionization Time of Flight Mass Spectroscopy (MALDI-TOF) has been used to determine the degree of polymerization for crystalline cellulose with a DP in an excess of 200. A new technique has been developed for MALDI-TOF, utilizing an ionic liquid matrix and solvent system. The technique has potentially wide application in analysis of lignocellulosic substrates.

(602be) Modification of Calcium Based CO2 Sorbents Using Modified Metal Oxides Powders

Wen-bin Yang[1], Hui Li[2], Yong-Hua Duan[1], Xiaodan Fan[3] and Yong Min[2], (1)Material Institute, Xi'an University of Architecture and Technology,, Xi'an, China, (2)Xi'an University of Architecture &Technology, Institute of Powder Engineering, Xi'an, China, (3)Xi'an University of Architecture and Technology, Xi'an 710055, Shaanxi, P.R. China

Modification of Calcium Based CO2 Sorbents Using Modified Metal Oxides Powders

Wenbin Yang, Hui Li, Yonghua Duan, Xiaodan Fan, Yong Min

1Xi'an University of Architecture and Technology, Xi'an 710055, Shaanxi, P.R. China

Abstract:

In order to solve the "sintering phenomena" by using calcium-based sorbents for CO2 sequestration, we has modified calcium based sorbents (MCS) using metal oxides and metal carbonates dry powders, such as MgCO3, Fe2O3, Al2O3, SiO2, ZrO2, etc. The modified calcium based sorbents had significantly improved the cycle times, CO2 sequestration efficiency, particle size distribution, as well as operation temperatures.

Several analytic equipments, such as TGA, DMA, SEM, XRD, particle size analyzer, mercury analyzer, were used to evaluate the particle composite, structure, and material activities. Our preliminary results indicated that the modified calcium based sorbents (MCS) have not only improved the cycle times, but also improved the CO2 sequestration efficiency. SEM morphology studies indicated that the surface of modified calcium based sorbents had formed an stable highly porous surface structure, which has significantly improved the CO2 sequestration and releasing. Comprehensive studies of material composition and scale up test are on the way.

Pretreatment of Wheat Straw Using the H_2O-SO_2 System at Elevated Temperatures

Weina Liu,[a] Yucui Hou,[b] Weize Wu,[a,*] Muge Niu,[a] Wenhua Wang[a]

[a] *State Key Laboratory of Chemical Resource Engineering, Beijing University of Chemical Technology, Beijing 100029, China;* [b] *Department of Chemistry, Taiyuan Normal University, Taiyuan 030031, China*

Extended Abstract

To improve the use value of lignocellulosic materials, which is abundant, renewable and low-cost, intensive research on their conversion is currently carried out[1]. However, Ligmocellulose have an extremely complex components, so a pretreatment process is needed in most cases to get rid of the hemicellulose fraction or the lignin fraction[2, 3].

Many pretreatments, including physical treatments[4, 5]and chemical treatments[6, 7], have been suggested. Among them, sulfuric acid hydrolysis is efficient, but has the disadvantages of pollution, difficulty in recovery, and hard control of products. SO_2 has been validated to be an efficient catalyst for the conversion of lignocellulose[8, 9]. Its greatest advantage lies in its ability to promote the steps of cellulose into glucose rather than further degradation. Moreover, SO_2 can be totally swept out by steam from an aqueous solution, and the recovered SO_2 can be reused after collection. However, no report is available to validate that the pretreatment process of lignocellulose in H_2O-SO_2 system is prior to the formation of glucose rather than furfural.

This work deals with the pretreatment of wheat straw in H_2O-SO_2 system. Kinetics models of the hemicellulose hydrolysis and xylose degradation with time are established, and the reaction rate constants and activation energy are calculated according to Saeman model.

Wheat straw with an hemicellulose content of 30.1 % was dried and milled (less than 1 mm) for use. SO_2 aqueous solution with a mass fraction of 6 % was purchased from Aladdin Reagent Inc. (Shanghai, China). D-glucose anhydrous, xylose, arabinose, furfural and levulinic acid (LA) were all analytical reagent.

---

* Corresponding author. E-mail: wzwu@mail.buct.edu.cn. Tel./Fax: +86 10 6442 7603.

In a typical procedure, 0.3 g of wheat straw, 6 mL of SO_2 aqueous solution and 9 mL of H_2O were loaded into the reactor. The reactor was sealed and purged with N_2. Then, the reactor was submerged into the furnace under an desired temperature for a desired time. After reaction, the reactor was placed into ice water to quench the reaction immediately. At last, the obtained reaction solution was purged by steam produced at 120 °C. The solution was filtrated, washed and analyzed by HPLC and the solid material was dried and weighed to calculate the conversion of reactant.

Treatments were performed in H_2O-SO_2 system with a SO_2 concentration of 0.024 g/mL under the temperature range of 130 °C to 160 °C, and the samples were withdrawn at several reaction times in the range of 0 to 90 min. The yields of products based on hemicellulose are shown in Table 1.

It is clear from Table 1 that temperature influences greatly the distribution of products at a given SO_2 concentration. At 130 °C, the main products are monomers of hemicellulose, such as xylose, arabinose, glucose and acetic acid. As the temperature increased from 130 °C to 160 °C, the monomer sugar is more inclined to further degrade to furfural. The reason is that increasing temperature favors not only the hydrolysis of hemicellulose and also the further degradation of xylose.

At a given temperature, with the prolonging of reaction time, the yield of xylose and arabinose first increase and then decrease for the further degradation. The yield of glucose gradually increased when the temperature was lower than 150 °C. It reached 17.2%, and then changed little at 150 °C from 45 min to 90 min. When the temperature was increased to 160 °C, the glucose yield began to increase again, likely due to the hydrolysis of cellulose.

In this work, the experimental value of hemicellulose hydrolysis in H_2O-SO_2 system is modeled by Saeman model (equation 1).

$$w_X = w_{X,0} e^{-k_2 t} + w_{H,0} \frac{k_1}{k_2 - k_1} (e^{-k_1 t} - e^{-k_2 t}) \quad (1)$$

Where k_1 and k_2 are reaction rate constant of hemicellulose hydrolysis and xylose degradation, respectively, min^{-1}; w_H is the mass fraction of hemicellulose; w_x is the mass fraction of xylose; $w_{X,0}$ and $w_{H,0}$ is the initial value for xylose and hemicellulose, which is designated as 0 and 100 %, respectively in our system. Therefore, the equation becomes as follows in our work:

$$Y = \frac{k_1}{k_2 - k_1} (e^{-k_1 t} - e^{-k_2 t}) \quad (2)$$

Where Y is the yield of xylose at certain reaction time. The fitting curves and the results of reaction rate constants are shown in Table 2.

The kinetics equation was obtained by fitting lnk with 1/T according to the Arrhenius equation. The activations for hemicellulose hydrolysis and xylose dehydration were 7.8 and 9.0 KJ/mol, respectively. It is noted that the activation energy for hemicellulose hydrolysis in our work is lower than that for xylose dehydration. Therefore, the xylose formation proceeded faster than xylose degradation, making it possible to decrease the yield of furfural. The reason is that SO_2 in water not only can react with water to form sulfurous acid to generate H^+ as a Brønsted acid, also can exist in the form of molecular as a Lewis acid, which is benefit for the breakage of glycosidic bond.

In this work, wheat straw is efficiently pretreated in H_2O-SO_2 system with a high xylose yield of 61.1% and a high total products yield of 93.9%. The Kinetic study based on Saeman model and experimental data suggested that SO_2 in water as catalyst favors the conversion of hemicellulose to monosaccharide rather than further degradation.

This work was supported by the National Natural Science Foundation of China (21176020), the Beijing Natural Science Foundation (2082017), and the Program for New Century Excellent Talents in University (NCET-08-0710)

Key Words: Pretreatment; Wheat straw, Hemicellulose, H_2O-SO_2 system, Xylose

Reference

(1) Zhang, Z. H.; Zhao, Z. B. K., Microwave-assisted conversion of lignocellulosic biomass into furans in ionic liquid. *Bioresour. Technol.* **2010,** 101, 1111–1114.

(2) Dadi, A. P.; Varanasi, S.; Schall, C. A., Enhancement of cellulose saccharification kinetics using an ionic liquid pretreatment step. *Biotechnol. Bioeng.* **2006,** 95, 904 – 910.

(3) Kumar, P.; Barrett, D. M.; Delwiche, M. J.; Stroeve, P., Methods for pretreatment of lignocellulosic biomass for efficient hydrolysis and biofuel production. *Ind. Eng. Chem. Res.* **2009,** 48, 3713–3729.

(4) Yang, F.; Li, L. Z.; Li, Q.; Tan, W. Q.; Liu, W.; Xian, M., Enhancement of enzymatic in situ saccharification of cellulose in aqueous-ionic liquid media by ultrasonic intensification. *Carbohydr. Polymers* **2010,** 81, 311–316.

(5) Ha, S. H.; Mai, N. L.; An, G.; Koo, Y. M., Microwave-assisted pretreatment of cellulose in ionic liquid for accelerated enzymatic hydrolysis. *Bioresour. Technol.* **2011,** 102, 1214–1219.

(6) Lu, X. B.; Zhang, Y. M.; Yang, J.; Liang, Y., Enzymatic hydrolysis of corn stover after pretreatment with dilute sulfuric acid. *Chem. Eng. Technol.* **2007,** 30, 938–944.

(7) Guo, G. L.; Chen, W. H.; Chen, W. H.; C., M. L.; Hwang, W. S., Characterization of dilute acid pretreatment of silvergrass for ethanol production. *Bioresour. Technol.* **2008,** 99, 6046–6053.

(8) Shi, J.; Ebrik, M. A.; Wyman, C. E., Sugar yields from dilute sulfuric acid and sulfur dioxide pretreatments and subsequent enzymatic hydrolysis of switchgrass. *Bioresour. Technol.* **2011,** 102, 8930–8938.

(9) Ewanick, S.; Bura, R., The effect of biomass moisture content on bioethanol yields from steam pretreated switchgrass and sugarcane bagasse. *Bioresour. Technol.* **2011,** 102, 2651–2658.

Table 1 The yield of products (Y) on time (t) under different temperatures

	130 °C			140 °C			150 °C			160 °C		
t (min)	Y_x	Y_g	Y_f	Y_x	Y_g	Y_f	Y_x	Y_g	Y_f	Y_x	Y_g	Y_f
15	14.8	1.8	0.0	32.8	2.7	0.0	39.5	7.6	0.0	46.6	8.9	0.0
30	32.8	4.9	0.0	50.7	9.6	0.0	61.0	14.2	0.8	53.5	14.4	2.4
45	45.6	12.7	0.0	54.3	12.6	0.3	59.5	17.2	2.8	48.2	18.7	5.6
60	54.7	13.5	0.0	59.2	14.0	0.7	57.5	17.6	3.9	37.0	21.3	11.9
75	57.5	13.4	0.1	57.2	15.8	1.2	55.8	17.8	4.6	20.6	20.0	15.8
90	54.5	14.9	0.7	55.3	17.5	2.3	49.7	17.7	5.6	13.1	18.3	20.2

Note: Y_x is yield of xylose, %; Y_g is yield of glucose, %; Y_f is yield of furfural, %.

Table 2 The fitting equation of xylose yield (Y) on time (t)

T (°C)	Equation	R^2
130	Y=143.39×[exp(−0.0152t)− exp(−0.0047t)]	0.9493
140	Y=144.08×[exp(−0.0276t)− exp(−0.0084t)]	0.9587
150	Y=136.87×[exp(−0.0392t)− exp(−0.0106t)]	0.9014
160	Y=170.28×[exp(−0.0605t)− exp(−0.0250t)]	0.9405

$$Y = \frac{k_1}{k_2 - k_1}(e^{-k_1 t} - e^{-k_2 t})$$

(602bg) Evaluating the Potentials of Paper Mill Sludge (PMS) As Feedstock for Biofuels Production

Joseph Reckamp[1], Rene Garrido[2], Nydia Ruiz-Felix[2] and Justinus A. Satrio[2], (1)Chemical Engineering, Villanova University, Villanova, PA, (2)Chemical Engineering Department, Villanova University, Villanova, PA

Heightened concerns over escalating greenhouse gas emissions as well as the depletion of petroleum oil and the continuing rise of its prices have resulted in focus on obtaining energy, traditionally produced from the combustion of fossil fuels, from renewable sources. One such renewable source of energy is biomass, in particular lignocellulosic biomass which is rich in lignin, cellulose, and hemicellulose. These carbon-containing components can be converted to liquid fuels by biochemical approaches, thermochemical approaches, or the combination of biochemical and thermochemical approaches.

The ideal biomass materials for energy production are those that are readily available and do not require a lot of energy to produce and collect. More importantly, the biomass materials should not be produced on fertile lands used for growing food crops. Such biomass materials are commonly readily available in the forms of what are considered wastes. These materials are produced from agriculture, forestry, household, or manufacturing activities. Examples of the materials include wheat straws, wood chips, spent mushroom substrate, municipal solid waste (MSW) and paper mill sludge (PMS). This study is focused on PMS materials.

As byproduct of paper making, PMS is produced in the order of 15 to 50 tons per day for a typical paper mill. It is estimated that five million tons of PMS are produced annually in the United States alone. Currently there are not many uses PMS, that most of PMS is disposed to landfills, a financial burden for paper mills. The decline of the landfill space availability has caused the cost of disposing PMS to increase rapidly to the point where the cost of PMS disposal now is about 60% of paper mill's operating costs. Therefore, there is growing interest in the paper industry to research possible uses of PMS, instead of just disposing them.

Thus, the purpose of this study is to show the viability of PMS as a feedstock for the production of liquid biofuels. Exploring the viability first requires characterization of the PMS samples. Characterization by way of Induced Coupled Plasma (ICP), Thermogravimetric Analysis (TGA), Differential Scanning Calorimeter (DSC), Fourier Transform Infrared Spectroscopy (FTIR), and Fiber Analysis will provide information on the physicochemical properties of PMS. The ash content and the mineral composition of the ash material in PMS will be determined. Functional groups and heat capacities will also be determined as well as the amount of lignin, hemicellulose, and cellulose in PMS. The characterization will determine whether PMS is within the optimal specifications as feedstock for biofuel production.

In addition to characterization study, a study on converting PMS to fuels will also be conducted. In order to convert the PMS into a liquid fuel, this study will utilize a method called fast pyrolysis. Fast pyrolysis is a thermochemical conversion technology that rapidly heats organic material to temperatures of 450°C to 600°C in the absence of air which results in rapid decomposition of the material to produce a liquid product typically called bio-crude oil which can be used as a heat source or an intermediate feedstock for producing gasoline or diesel fuels. The fast pyrolysis experiments will be performed by using a micropyrolyzer/GC-MS system. To be reported in this presentation are the results of these characterization and fast pyrolysis studies.

(602bh) Characterization and Fast Pyrolysis Study of Phragmites Australis to Evaluate Its Potential As Feedstock for Biofuels Production

Lindsay Peterson[1], Rene Garrido[2], Nydia Ruiz-Felix[2] and Justinus A. Satrio[2], (1)Chemical Engineering, Villanova University , Villanova, PA, (2)Chemical Engineering Department, Villanova University, Villanova, PA

The use of lignocellulosic biomass as an alternative carbon source for energy production, particularly liquid fuels, is promising due to its widespread availability. A potential biomass source is *Phragmites australis*, commonly known as phragmites or common reeds. It is an invasive wetland weed that is one of the most abundant plant species on the planet, covering approximately ten million hectares, and it can be found on every continent excluding Antarctica. In addition to being abundant, phragmites can produce biomass in high yields. It is reported that it can provide up to 18-28 tons of biomass per acre per year. The phragmites grow easily, producing lignocellulosic biomass in high yield, on land areas that are not used for growing food crops, which potentially makes the plant species an excellent feedstock for biofuel production. It has been reported that *Phragmites australis* is rich in cellulose and lignin, the main components in lignocellulosic biomass that are particularly important for biomass conversions to biofuels.

The purpose of this research is to understand the physical traits of *Phragmites australis* and to understand the feasibility of *Phragmites australis* as a biomass feedstock via a thermochemical conversion process called fast pyrolysis. Fast pyrolysis is a process in which organic materials are rapidly heated to 450 - 600 °C in absence of air. Under these conditions, organic vapors, permanent gases (CO, H2, CO2, CH4), and charcoal are produced. The vapors are condensed to a liquid called bio-crude oil (BCO). The chemical and physical properties of BCO are dependent on the biomass materials used as feedstock. The advantage of pyrolysis over biochemical-based process is that it can transform difficult-to-handle biomass of different natures into clean and uniform liquid (BCO), which then can be used as a heating source or further processed to produce transport fuels, such as gasoline and diesel.

The phragmites samples used for this study were collected from a wetland area located within Villanova University's campus. The material is prepared for physicochemical characterization by drying and grinding. The characterization study includes fiber analysis, proximate analysis (using Thermogravimetric Analyzer (TGA)), mineral content composition (using Induced Coupled Plasma (ICP)), and heating value (by using differential scanning calorimeter (DSC)). DSC is also used to study the thermal behavior of the biomass material when it undergoes fast pyrolysis reaction. Another instrument, Fourier Transform Infrared spectroscopy (FTIR), is used to identify chemical functionality of the biomass material. A limited number of fast pyrolysis experiments by using a micropyrolyzer/GC-MS system were performed to evaluate the yields and chemical product distribution from pyrolyzing phragmites. To be reported in this presentation are the results of these characterization and fast pyrolysis studies.

(602bi) Development of Biodiesel Production Facility for Utilizing Waste Cooking Oil

Brian Delancy[1], Brooks Hopkins[1] and Justinus A. Satrio[2], (1)Chemical Engineering, Villanova University, Villanova, PA, (2)Chemical Engineering Department, Villanova University, Villanova, PA

Development of Biodiesel Production Facility for Utilizing Waste Cooking Oil

Brian Delancy, Brook Hopkins and Justinus Satrio

Department of Chemical Engineering
Villanova University, Villanova, PA 19085

Three years ago, a team of chemical engineering students at Villanova University developed, constructed, and began operating a biodiesel production facility that reuses the waste cooking oil produced from Villanova University dining centers. The production facility is designed to process an annual maximum of 3500 gallons of waste cooking oil. Over the past two years, additional process systems, including a methanol recovery system and soap-production process have been developed as a way to utilize the reaction's glycerol by-product. The focus of our current research is multifaceted, but focuses primarily on the production optimization and output maximization of Villanova's Biodiesel Production Facility. Several quantitative analysis techniques, including Calorimetry and Fourier Transform Infrared Spectroscopy were introduced to measure and maintain quality control of the biodiesel product. Improved methanol recycling techniques and natural soap production minimize any waste, thus making the process more economically efficient and environmentally friendly. Finally, the project increased its relations with Villanova Grounds and Maintenance Department, providing the University with a biodiesel power option in accordance with the University's "Green Energy" initiative. The Villanova Biodiesel Team also worked to expand the program into the local community through a marketing plan, in partnership with the Engineering Advisory Board.

(602bj) Enhanced Lignocellulose Conversion to Biofuels with Thermophilic Bacteria and Thermostable Enzymes

Aditya Bhalla, Chemical and Biological Engineering, South Dakota School of Mines & Technology, Rapid City, SD, Kenneth Bischoff, Bioproducts and Biocatalysis Research Unit, USDA-ARS, NCAUR, Peoria, IL and Rajesh K. Sani, Chemical & Biological Engineering, South Dakota School of Mines & Technology, Rapid City, SD

Enhanced Lignocellulose Conversion to Biofuels with Thermophilic Bacteria and Thermostable Enzymes

Aditya Bhalla[1], Dr. Kenneth Bischoff[2] and Rajesh Sani[1]*

1. Department of Chemical and Biological Engineering, South Dakota School of Mines and Technology, Rapid City, SD.
2. Renewable Product Technology Research Unit, Agricultural Research Service, National Center for Agricultural Utilization Research, U.S. Department of Agriculture, Peoria, IL 61604

Second generation feedstock, especially nonfood lignocellulosic biomass, has been seen as a potential source for biofuel production. Lignocellulose conversion technologies, which are carried out at ≤50°C, have several limitations. Therefore, this study focuses on the importance of thermophiles and thermostable enzymes to overcome the limitations of existing lignocellulosic biomass conversion processes. We have isolated several thermophilic cellulose- and xylan-degrading pure cultures belonging to the genera *Bacillus* and *Geobacillus* using soil samples collected from the Homestake Gold Mine, SD and a local compost facility. Xylanases produced by one of our isolate are found to be highly thermostable as compared to the reported literature. Greater hydrolysis rates of xylan were observed with xylanase from our isolate as compared to commercially available xylanases at high temperatures. In order to produce these enzymes in bulk, we overexpressed thermostable xylanases in a heterologous expression system. Purified xylanase was characterized in terms of pH, temperature, thermostability, and molecular weight. The purified thermostable recombinant enzymes, in the near future, will be used to prepare a thermostable mixed enzyme system for lignocellulose hydrolysis to fermentable sugars.

A Green Carbon Supported Au Catalyst for Acetylene Hydrochlorination Into Vinyl-Chloride-Monomers for Polyvinyl Chloride Production

Kai Zhou, Chmical Engineering, Beijing Key Laboratory of Green Chemical Reaction Engineering and Technology Tsinghua University, Beijing, China

Introduction

Catalysis has been widely used in chemical industries, selecting precious metals or metal oxides as the active components in the majority of the processes. However, the preparation and post-processes of the conventional metal-based catalysts exhaust great resources and environmental cost, which are always toxic and energy-consuming and cannot well meet the demanding of sustainable development for whole society. Therefore, searching promising materials possessing environmental acceptability and low emission is a topic of increased interest and importance for the green chemical processes. Hydrochlorination of C_2H_2 is the dominant method for the synthesis of VCM (vinyl chloride monomer) and further polymerize into poly vinyl chloride. The process takes place in fixed bed reactor in tubes packed with activated carbon loading around 10 ~ 15 wt% $HgCl_2$. The exothermic reaction releases 124.8 $kJ \cdot mol^{-1}$ heat during adding HCl onto C_2H_2. The poor heat transfer of fixed bed reactor resulted in heat spots and hence increasing volatility of $HgCl_2$. The $HgCl_2$ vapor releasing into the environment easily accumulates in the creatures and causes chronic poisoning, which is severely harmful to human beings and environment.

Carbon materials, playing the role as the catalytic substance, have significant advantages over the conventional metal-supported systems. For instance, nanostructured carbon and nitrogen-doped carbon composites have been proved to be the effective alternatives to conventional metal-containing catalysts for heterogeneous reaction processes, such as oxidative dehydrogenation (ODH) of ethyl benzene, Friedel–Crafts Reaction, oxygen reduction reaction (ORR).[1-3]

Therefore, developing new cost-effective and non-mercury catalysts and related process is quite crucial for a sustainable society. In this contribution, a novel carbon based Au catalyst for acetylene hydrochlorination was developed, which showed advanced activity and life time. Active carbon (AC) with hierarchical pore structure was selected as catalyst support. To increase the interaction between the support and active phase, AC was pre-treated by a mixture of H_2SO_4 and HNO_3 to facilitate the anchoring of active metal nanoparticles. The catalyst was prepared by simple co-impregnation in the solution containing $HAuCl_4$ (active phase) and chlorides of metal promoters, following by drying under 130 °C for 8 hrs. The morphology and catalytic performances were characterized by SEM, TEM, XPS, TPR, SEM and XPS results indicated that the existence of promoter metals facilitated the dispersion of Au nanoparticles on AC, which increased the initial reaction activity. Besides, the activity of the main active substance Au^{3+} was also improved. Au-only-based catalyst showed the highest activity with 1wt.% Au content (Au atomic weight to AC weight), while the optimized Au weight content can be reduced to below 0.5 wt.% with the addition of metal promoters. Under the condition of acetylene hydrochlorination of 200 °C and 1200 h^{-1} (Space Velocity, Volume based), the acetylene conversion over 70 % can be achieved and maintained for 10 hrs, which indicated that both the activity and the stability of the catalyst were enhanced. This demonstrated that the carbon based Au catalyst can be a very promising green catalyst for polyvinyl chloride industry.

Table 1. The characterizations of the catalyst based on the different substrates

Catalysts	S_A, $(m^2 \cdot g^{-1})$	X_A*, (%)	S_{VC}*, (%)
Au/MWNT	938	91.25	>99.5
Au/N-CNTs	196	3.64	>98
Au/Celite	178	57.5	>98
Au/ZrO_2	46	-	-

* The Au loading was 1 % in each catalyst and the conversion and selectivity were analyzed under 360 h^{-1}($GHSV(C_2H_2)$) and 180 °C

References

1. M. F. R. Pereira, J. L. Figueiredo, J. J. M. Orfao, P. Serp, P. Kalck, Y. Kihn, *Carbon* 2004, 42, 2807-2813;
2. N. Keller, N. I. Maksimova, V. V. Roddatis, M. Schur, G. Mestl, Y. V. Butenko, V. L. Kuznetsov, R. Schlogl, *Angewandte Chemie-International Edition* 2002, 41, 1885;
3. F. Goettmann, A. Fischer, M. Antonietti, A. Thomas, *Angewandte Chemie-International Edition* 2006, 45, 4467-4471;

(602bl) Preparation of MgO Nanoparticles and Its Utilization for CO2 Chemisorption

Vishwanath G Deshmane, Mechanical Engineering, North Carolina Agricultural and Technical State University, Greensboro, NC and Yusuf G. Adewuyi, Chemical and Bioengineering, North Carolina A&T State University, Greensboro, NC

The increased CO_2 concentration in the atmosphere due to emissions of CO_2 from the utilization of fossil fuels including coal has caused serious concerns for global warming. Of the available CO_2 emission reduction options, CO_2 capture and sequestration has been identified to be the most effective one. In regards to this, development of solid adsorbents to capture the CO_2 at elevated temperature-pressure, as well as atmospheric temperature-pressure, from the natural gas and flue gas effluent has got considerable attention. In the present study, MgO nanoparticles were synthesized and utilized for the adsorption of CO_2 at three different temperatures of 30 °C, 100 °C and 250 °C. The effect of preparation conditions such as surfactant concentration, effect of presence of ultrasound, type of precipitating agent and the precursor concentration on the crystalline sizes and the surface area of the MgO nanoparticles and their subsequent effect on the CO_2 adsorption capacity were studied. Preliminary results showed that the prepared nanoparticles have significant CO_2 adsorption capacities. The prepared material was characterized by nitrogen adsorption–desorption (BET), X-ray diffraction (XRD), scanning electron microscopy (SEM), thermogravimetric and differential scanning calorimetry (TGA-DSC), energy dispersive X-ray spectroscopy (EDX) and Fourier transform infrared spectroscopy (FTIR).

(602bm) Utilization of Process Performance Criteria At the Conceptual Design Stage in Green Production of Organic Chemicals

Sencan Haydaroglu, Chemical Engineering, Northeastern University, Boston, MA and Turker Gurkan, Department of Chemical Engineering, Middle East Technical University, Ankara, Turkey

Biochemical and petrochemical production routes for the derivatives of 12 most promising sugar-based building blocks proposed in the U.S. Department of Energy Top Value Added Chemicals from Biomass Report were evaluated at the conceptual design stage utilizing criteria. Sugar-based building blocks were compared to their petrochemical surrogates consisting of persistence, bioaccumulation, aquatic toxicity, atom and mass efficiency and their economical potential. Raw materials and products do not show a significant difference with respect to bioaccumulation criteria. In contrast, persistence of bio-based raw materials and products in air and their aquatic toxicity results indicate that bio-based routes are more environmentally-benign. The relationship between the economical potential and the standard Gibbs free energy change of major reactions is examined to make a further comparison at this stage. A direct relationship was not observed. However, most of the utilized reactions have a negative standard Gibbs free energy of change indicating that chemical industry exploits the spontaneity of reactions.

(602c) Situ Detection of Fermentative Activity by Redox Potential Measure and Its Applications During Ethanol Fermentation

Chen-Guang Liu, School of Life Science and Biotechnology, Dalian University of Technology, Dalian, China

Redox potential reflects the complex physiological information in cells that can be detected for optimizing the fermentation processes. However, control redox potential by delicate approaches is not easy due to lack of mathematic relationships between redox potential profiles and fermentation parameters. In this work, three important parameters including temperature, pH and dissolved oxygen are chosen to buildup a dynamic model according to Nerst equation. Fermentative activity defined as the specific consummation rate of substrate is connected with the redox potential profiles slope, but not the redox potential value. Base on this model, a feeding scheme was developed to maintain the glucose concentration at a low level in order to eliminate the substrate inhibition, and an aeration scheme was proposed to prolong the duration with rapid ethanol production under high cell viability. Besides, when utilizing the self-flocculating yeast, redox potential controlled repeated batch fermentation realizes the cells recycle and then higher ethanol productivity is achieved because of very high density yeast cells.

Abstract: CFD Modeling of a Gravity Settler for Algae Harvesting

By Scott A. Hug, Karl K. Scott, Jorge E. Gatica, and Joanne M. Belovich

Future world energy needs necessitate the search for alternative and sustainable energy sources. Biodiesel is one such source with many upsides, but current production methods are flawed. Existing large scale biodiesel production facilities rely primarily on soybean and canola oil as the lipid source for the biodiesel. The production of soybean and canola oil is not viable in the long term as a transportation fuel lipid source.

Algae are the future of lipid sources for biodiesel production. While soybean and canola oil can only be grown in lands and climates that support agriculture taking away land that could otherwise be used for food crops, algae can be grown in large open pools on non-arable lands in which they will not compete for space with food crops. Algae can also be grown in seawater, which is more abundant than the fresh water required for other sources of biofuel. Algae cultivation can produce up to 20,000 gallons per acre of algae, whereas canola or soybean based biodiesel cultivation can only produce 38-100 gallons per acre. The cells use carbon dioxide as a carbon source and thus can be used to offset carbon dioxide production of facilities, namely power plants.

The problem with Algae as a lipid source for biodiesel is the cost of dewatering of the cells. Algae concentrations in the best conditions in bioreactors will not surpass 20 g/L due to light limitations. The water must be removed before the lipids can be used in the transesterification process that produces the biodiesel. Typical methods for dewatering the algae involve using settling ponds to concentrate the algae a little and then using centrifugation to do most of the dewatering. The costs of this centrifugation step limit the commercial viability of algae based biodiesel.

A patented novel gravity settler design at Cleveland State University is analyzed in this project. This gravity settler design is displayed in Figure 1 below. Preliminary results obtained with this design show promise to help the commercial viability of the algae biodiesel process by lowering the dewatering costs. A major difference between this and traditional gravity settlers is that the inlet of this gravity settler is at the top, whereas traditional gravity settlers have inlets at the bottom. Two prototypes have been tested and experiments indicate that the technique is effective at dewatering algae.

Figure 1: Novel Algae Settler

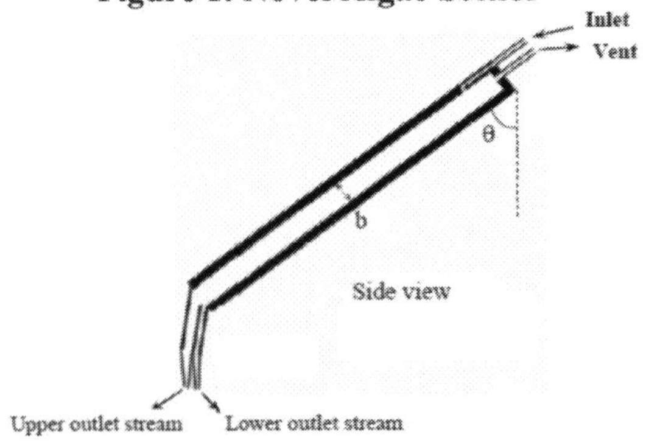

In order to benefit the design process, a computational fluid mechanics model for the system has been developed using COMSOL™ (formerly known as FEMLab), a finite element based multi-physics modeling environment. The model formulation is based on the mixed model that considers a system where there is a dispersed phase of either solids or liquid droplets or bubbles in a continuous liquid phase. This study makes use of COMSOL's particle tracing module, which allows for trajectories of specific particles to be examined over a period of time.

The geometrical model has been developed based on the simplified geometry of the gravity settler shown in Figure 1. After validation experiments are completed and results with simplified relations for settling velocities are evaluated, the simulation environments is to be updated to the more advanced geometry of the prototypes by means of Solidworks® and COMSOL™ LiveLink™ for Solidworks®. This LiveLink™ enables to import complex geometries into COMSOL™ as well as formulation of parametric studies where the CAD software is used to interactively modify geometric parameters that may show relevance in the performance of the settler. Solidworks® parameters can also be used in COMSOL™ to perform parametric sweeps.

Using this LiveLink™ functionality, parametric sweeps will be demonstrated as tools to optimize the settler's design and operation by analyzing geometric (such as length, angle of inclination, and height for dewatering) as well as operating variables (such as inlet and outlet flow rates and velocities). COMSOL™ ant the LiveLink application will be used to optimize these variables in order to create the most efficient model for algae dewatering. The optimized design will be physically prototyped and experimented on to validate the computational fluid dynamics (CFD) modeling.

This study shows how the powerful environments of a detailed CFD modeling interfaced with a commercial computer-aided design (CAD) software can be used to optimize the design of gravity settlers. Comparative studies carried out with alternative designs and results from validation experiments are presented.

The model formulation is based on the mixed model. The mixed model considers a system where there is a dispersed phase of either solids or liquid droplets or bubbles in a continuous liquid phase. It was developed for hard spherical molecules, so some adaption is required to correctly model algae cells which can clump and deform, these modifications will be discussed.

The complex geometry of the gravity settler was mimicked using SolidWorks®. A model of the gravity settler, shown in Figure 2 on the next page, is divided into two parts: The inlet region, and the outlet region. The inlet region is a rectangular prism measuring 215 cm long, 10 cm wide, and 5 cm high. The algae particles enter through the top of the inlet stream. The inlets are placed on the top of the inlet region, and there is a vent on the side of the region. The outlet region is consists of upper and lower outlet streams. The objective of this study to determine the settings in which the water and algae achieve the greatest level of separation in this region based on gravity's influence on the algae particles and water in the inlet region.

The early stages of the project consist of working on each part of the model separately. This allows the simulations to be run on the less complex geometry of each part (i.e, a simple rectangular prism for the inlet region), rather than the complex geometry of the entire gravity settler. These simple geometries allow for the simulation to be run on COMSOL™ much quicker because a more coarse mesh may be used to accurately measure the particle trajectories and other properties in a simulation. For complex geometries, a more fine mesh is needed to accurately measure properties, especially in regions with many edges and corners.

Figure 2: SolidWorks design of the gravity settler. The fluid and particles enter at upper-right corner of the settler, travel through the inlet region, and exit in either the upper or lower outlet stream.

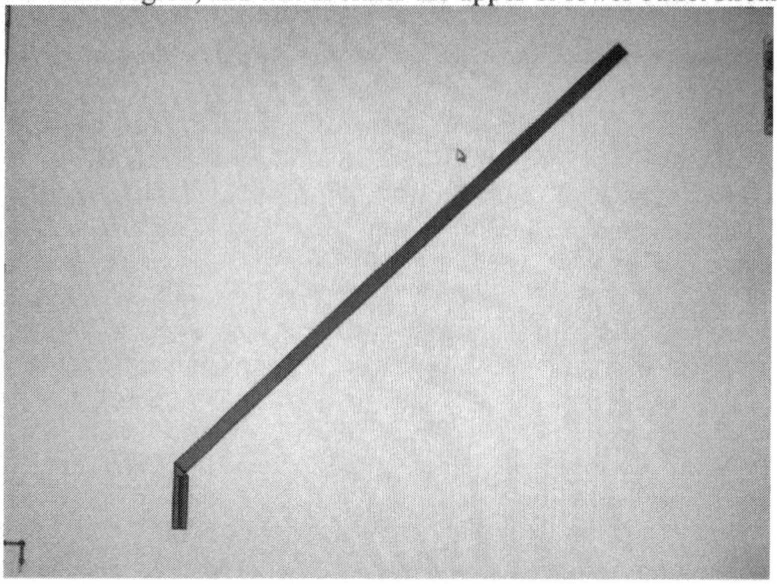

After these runs are made on the simple geometry of the inlet region, the most optimal flow rates, velocities, and length and height of dewatering will be examined throughout the more complex geometry of the entire gravity settler. It is here in which the angle of inclination will be measured

A few modifications must be made from the COMSOL model. The model was developed for hard spherical molecules, so some adaption is required to correctly model algae cells which can clump and deform. These modifications will be discussed.

(602e) Algal Genotyping in an Open System and a Bioreactor

Alice C. Jernigan, Chemical Engineering, University of Arkansas, Fayetteville, AR and Christa N. Hestekin, Ralph E. Martin Department of Chemical Engineering, University of Arkansas, Fayetteville, AR

Algae are a high growth source for biomass with a great potential to be a significantly eco-friendly bio-fuel source. Since different species of algae have different oil and lipid concentrations it is important to genotype the algae in an open environment to determine the best species for biofuel production. Alternatively, in a bioreactor it is important to maintain the dominance of the best-fuel producing species. For genotyping, we examine the 16S and the 18S genes, which are highly conserved among prokaryotes and eukaryotes (respectively), but both have unique small variable regions that can be used to differentiate between species. We examined 2 methods of DNA analysis, capillary electrophoresis single strand conformation polymorphism (CE-SSCP) and single base extension (SBE). For the open system, we were able to use CE-SSCP to identify different algal species, which were confirmed by sequencing. Alternatively, for the bioreactor, we were able to detect and monitor 2 different algal types (*Spirulina* and *Chlorella).*

(602f) Enhancement of H2 and CO2 Uptake for the Production of Biodiesel in Ralstonia Eutropha

Ryan Sullivan, Pin-Ching Maness, Carrie Eckert and Jianping Yu, Biosciences Center, National Renewable Energy Laboratory, Golden, CO

Ralstonia eutropha fixes CO_2 through the Calvin-Benson-Bassham (CBB) cycle using electrons and energy obtained from the oxidation of H_2. Producing biodiesel-equivalent electrofuel from renewable CO_2 and H_2 has immense potential, especially if the fuel is compatible with the existing fuel infrastructure. This research addressed enhanced H_2 and CO_2 substrate utilization by focusing on two strategies: (1) optimizing transcriptional regulations to afford over-expression of Ribulose-1,5-bisphosphate carboxylase/oxygenase (RubisCO), the enzyme responsible for assimilation of CO_2 into the CBB cycle; and (2) hydrogenase over-expression by introduction of additional copies of genes encoding a membrane-bound hydrogenase (MBH), a soluble hydrogenase (SH), and their maturation machinery to enhance oxidation of H_2 to generate NAD(P)H and ATP, both are required for CO_2 fixation. Incorporation of these strategies into a single production *R. eutropha* strain had resulted in 6-fold improvement in CO_2 fixation and 3-fold improvement in H_2 uptake, *in vitro*, with the overarching goal of providing abundant reducing equivalents towards the economic production of biodiesel in *R. eutropha*.

(602g) Biodiesel Production From Cynara Carduncculus Crop: Economic, Environmental and Social Assessment

Carmen M Torres[1], Sergio D. Ríos[1,2], Carles Torras[1], Joan Salvadó[1,2] and **Laureano Jiménez**[1], (1)Departament d'Enginyeria Química, Universitat Rovira i Virgili, Tarragona, Spain, (2)Catalonia Institute for Energy Research (IREC), Tarragona, Spain

The generation, transport and consumption of conventional sources of energy, particularly in the transport sector, have raised many concerns about air pollution, environmental degradation, and petroleum consumption. In this sense, the development of alternative technologies is acquiring importance, and many efforts are being done in the gradual replacement of fossil fuels. Within the renewable sources of energy, the production of liquid biofuels from organic feedstocks sources represents a feasible alternative that avoids important modifications of vehicle engines, maintaining most of the infrastructures for the supply chain and can replace partially the petroleum-based fuels.

Biofuels have many advantages over fossil fuels: as well as being more environmentally friendly and a renewable source of resources, every region in the world has a raw material that can be used, to a greater or lesser extent, to produce some sort of biofuel, thus providing work for the population and improving the economies of the countries that so far have been dependent on petroleum imports.

On the other hand, there is the ethical question of the direct competition for land that is required for producing food. Several groups (Green Peace, Intermon Oxfam, etc) claim that the growing production of biofuels is responsible for the increasing food prices, and the destruction of forests. Second-generation biofuels (those that do not threaten food supplies) were developed to overcome this obstacle.

Of particular interest are the residual parts of agricultural and forest vegetation that contain cellulose for producing bioalcohols, recycled oils, and other plants that contain oils and which do not threaten food supplies (*Jatropha curcas, Cynara cardunculus*, microalgae, etc).

The aim of this work is to analyse the feasibility of biodiesel production obtained by transesterification of vegetable oil extracted from *Cynara Cardunculus* (cardoon), a wild robust perennial plant, native from the Mediterranean basin. The performance assessment of a process with a capacity of 5000 t/year biodiesel production is carried out from the triple bottom line perspective. The characterization of the process includes its rigorous simulation in AspenHysys® from which an automated evaluation tool programmed in Matlab® retrieves the inventory of relevant energy and material inputs, as well as the environmental releases. In the evaluation tool the calculations are grouped into different modules: acquisition of the simulation data, environmental, economic and social assessment, and results analysis.

The simulation performs the alkali-catalyzed transesterification with methanol and a pre-treatment section for the free fatty acids removal by esterification. The operational conditions are based on the cardoon oil characterization and the kinetic data extracted from published experimental studies (Pasqualino, 2006). The simulated process includes the methanol recovery, the catalyst neutralization and the glycerol separation and purification as valuable by-product.

In the environmental module, the Eco-indicator 99 methodology is used for the impact assessment, achieving similar results to other vegetable sources of biodiesel. With respect to the economic criterion, a profitability analysis is performed calculating the net present value, thus considering the initial investment, manufacturing costs and products revenue. As expected, the economic metric points out the necessity of government subsidies to produce positive rates of return. From a social perspective, quantitative indicators (*e. g.*, provision of employment) and qualitative indicators (*e. g.*, nuisance) were selected among the social metrics available in the literature. The three indicators are then aggregated in a single metric using a normalization and weighting procedure based on the geometric mean of the metrics' ratios for pair-wise comparison of alternative processes (Torres et al., 2011).

The performance of the plant is optimized considering the potential alternatives. With this aim, the use of two different catalysts and an additional unit for the recuperation of glycerol in the pre-treatment section are considered, allowing a higher reusing of materials and reducing the waste treatment requirements. Furthermore, a sensitivity analysis is carried out to quantify the robustness of the procedure and identify the critical variables.

References

Pasqualino, J.C., 2006. Cynara cardunculus as an alternative crop for biodiesel production. Ph.D. dissertation. Universitat Rovira i Vigili, Tarragona, Spain.

Torres, C.M., Gadalla, M.A., Mateo-Sanz, J., Jiménez-Esteller, L., 2011. Evaluation Tool for the Environmental Design of Chemical Processes. Ind. Eng. Chem. Res. 50 (23), 13466-13474.

Torres, C.M., Gadalla, M.A, Mateo-Sanz, J.M., Jiménez-Esteller, L., 2012. An automated environmental and economic evaluation methodology for the optimization of a sour water stripping plant. J. Clean. Prod. (in press).

(602h) Reducing the Energy Consumption of Biofuel Production Using a Solar Assisted Steam Generation System with Heat Storage

Robert Brunet[1], Gonzalo Guillén[2] and Laureano Jiménez[1], (1)Departament d'Enginyeria Química, Universitat Rovira i Virgili, Tarragona, Spain, (2)Chem. Enginering, Universitat Politécnica de Catalunya, Barcelona, Spain

In this work, we address the problem of reducing the energy consumption of biofuel plants through their integration with a solar thermal energy system that generates steam. A mathematical model of the solar energy system that includes energy storage is coupled with a rigorous simulation model of the biofuel facility developed in a commercial process simulator. The solar energy system model takes the form of a bi-criteria nonlinear programming (NLP) formulation that accounts for the simultaneous minimization of cost and energy consumption.

We demonstrate the capabilities of this strategy with two case studies in which we address the design of a 12,000 ton/year alkali-catalyzed biodiesel process using vegetable oil modeled in Aspen Plus® and a 120,000 tones/year dry-grind corn-to-ethanol production plant modeled in SuperPro Designer®. In both cases we considered weather data of Tarragona (Spain). A set of Pareto solutions were generated using the epsilon constraint methodology. The results obtained show that is possible to achieve reductions in energy consumption of up to 24.90 % for the biodiesel and 17.24% for the bioethanol with respect to the minimum cost design.

(602i) Biomass-Solvent Catalytic Reactions: Assessing Sugars and Intermediates by HPLC and GCMS

Sipho C. Ndlela, Iowa Energy Center, Iowa State University, Nevada, IA

The first phase of this project evaluated the reaction pathways for biomass feedstock in the presence of various cosolvents including acetone, methanol, CO_2, and water using a 1 liter (7.6cm x 23.1cm, ID x L) high pressure batch reactor. Additionally, the effect of temperature on the extent of the reaction was assessed at 200 to 350°C. In both instances biomass conversion rates compared to the starting material ranged from 70 to 90%, with the remaining portion being the unreacted fraction and lignin. The resulting products consisted of complex chemical fractions: organic acids, organic ester derivatives, sugars, sugar-alcohols, thermal degradation products and more. Although the impact of using various cosolvents was noticeable at lower temperatures, at higher temperatures the extent thermal degradation products increased due to longer heating times to reach the target temperature (2 to 8 h). Despite the challenges with longer residence time when using the batch reactor, results from this study gave an important understanding of the reaction behavior and interaction of the solvent and biomass.

To have better control of the reaction extent at pilot scale, a flow reactor was assembled. The system consisted of a 250 ml/min slurry pump with reactor chamber that is 212 cm longer by 3 cm diameter, a maximum system working pressure up to 69 Mpa, electrical heating jackets and a condenser. Biomass slurry of 5 to 10% was fed into the reactor at varying temperatures from 200 to 350°C and at residence times of 5 to 30 min. Analysis results from the flow reactor study will be main focus for this discussion.

(602j) High Pressure, Oxygen Blown Entrained-Flow Gasification of Bio-Oil

Nicholas Creager, Iowa State University, Ames, IA and Robert C. Brown, Mechanical Engineering, Iowa State University, Ames, IA

The goal of this research is to investigate the gasification of bio-oil in a pressurized entrained flow gasifier. Gasification at high pressure is essential for integrated gasification combined cycle power (IGCC) and efficient syngas upgrading to fuels and chemicals. Feeding solid fuels into high pressure reactors is one of the most difficult challenges of gasification. Dual chamber lock hoppers are an expensive and sometimes unreliable option. Among solid fuels, coal is sufficiently hydrophobic to be mixed with water and pumped at high pressure as high-solids slurry. Biomass is too hydroscopic to be pumped in this fashion. Bio-oil from fast pyrolysis of biomass is essentially "liquid biomass" that can be pumped at high pressures.

This gasifier operates with active electric insulation at temperatures of 700°C to 900°C and under pressures up to 50 bar to simulate the operation of industrial-scale gasifiers. The gasifier consists of two sections: the reactor and the quench vessel. The gasification section is constructed of a 1.0 m long, 3.8 cm dia. silicon carbide tube surrounded by several layers of insulation and encased inside a 15.2 cm dia. stainless steel pressure vessel. Hot syngas entering the quench vessel is sprayed with water to reduce gas temperature to below 100°C followed by pressure let-down and exhaust. Product gases are analyzed by gas chromatography. The system can gasify up to 2.5 kg/hr of bio-oil in oxygen and steam flows of 1.0 kg/hr.

Tests were performed on two bio-oils from the pyrolysis of switchgrass and cornstover using design of experiment protocols. This allowed the construction of statistical models to predict the effect of biomass composition, gasification temperature and pressure, and biomass/oxygen equivalence ratio on gas composition.

OPTIMAL DESIGN OF ALGAE-BASED BIOREFINERIES USING CO_2 FROM DIFFERENT INDUSTRIAL FACILITIES

Rubio-Castro E.[1], *Ponce-Ortega J.M.*[2,*], *Serna-González M.*[2], *Hernández-Calderón O.M.*[1], *Ortiz-del-Castillo J.R.*[1], *El-Halwagi, M.M.*[3]

[1]*Chemical and Biological Sciences Department, Universidad Autónoma de Sinaloa, Culiacán, Sinaloa, México, 80000*

[2]*Chemical Engineering Department, Universidad Michoacana de San Nicolás de Hidalgo, Morelia, Michoacán, México, 58060*

[2]*Chemical Engineering Department, Texas A&M University, College Station, TX, USA, 77843*

* Corresponding author. *E-mail*: jmponce@umich.mx

Abstract

This work proposes an optimization approach for the carbon dioxide sequestration from different industrial facilities to yield an algae-based biorefinery. The proposed approach is based on a distributed system to account for the economies of scale and it involves the site selection for the processing facilities as well as the transportation cost. Additionally, the model considers the optimization for the technologies used in each facility for the CO_2 sequestration, for the algae growth, harvesting, lipid extraction, and different technologies to yield several products like biodiesel, methane, ethanol, between others, depending on the customer demand. The proposed approach considers the mass exchange between different industrial facilities, because the algae from one plant can be used to produce products used in the same or in a different plant. The optimization model considers the maximization for the total net profit, which includes the sales of products plus the tax credit for reduction of CO_2 emissions minus the capital and operational costs for the different processing stages minus the overall transportation costs associated to the system. The constraints for the model include the mass balances for the system, design and operational constraints for the processing units and constraints for the demands for the products. The proposed model is a mixed integer non linear programming model and this was coded in the software GAMS. Finally, two case studies from Mexico were analyzed and the results show several economic and environmental benefits simultaneously.

Keywords: CO_2 sequestration, Optimization, Distributed system, Bio-refineries.

INTRODUCTION

A biorefinery is a system where several processes are integrated to transform the biomass into fuels, energy, chemical products and foods (Ohara, 2003). This concept is analogous to an oil refinery but in a biorefinery the raw material is biomass instead of oil. The biomass can be waste biomass from an industrial or natural process or it can be obtained from the processing of vegetables and animals wastes (for example algae, corn and jatropha (Pokoo-Aikins, et al., 2010)); therefore, the difference between the biomass and the oil is that the former one is a renewable raw material. In the industrial processes there are several CO_2 emissions, and this can be captured for the algae growth; and from this can be obtained oil and biomass that can be used to produce several additional products that can be used in the same industrial facility or sent to a different place where they are required. Therefore, this work proposes an optimization approach for the design of distributed systems for the CO_2 sequestration considering the integration of emissions from different industrial facilities and determining where and what product obtained and the optimal distribution associated through an algae based biorefinery system.

METHODOLOGY

The overall system to obtain alga oil is composed by three process modules (i.e., algae growth, algae harvesting and algae oil separation), like it is shown in **Figure 1a**. In the first module, for the algae growth can be used an open system (raceway pond) or a closed system (photobioreactor). Notice that the growth of algae requires water, sunlight, carbon dioxide and nutrients (phosphorous and nitrogen). Regarding to the carbon dioxide (see **Figure 1b**) the algae growth is an opportunity to sequester this gas by using flue gas emissions from industrial sources. After the algae growth process, a separation process is required. Many separation processes are available for the harvesting method. In the harvesting module there are two outlet streams. One of them is rich in algae oil and low water content and the other one (poor stream) presents a low algae oil content but a high content of water. And the poor stream can be segregated and sent to the waste that is discharged to the environment and/or to the growth module to reduce the fresh water usage while the rich stream is sent to the extraction module for extracting the algae oil (see **Figure 1b**). In this stage of the process, there are three outlet streams: the algae oil, water and the cell paste. Respect to the water and cell paste streams, these ones can be sent to the waste that is discharged to the environment or it can be reused in the algae growth stage (water stream) like it is shown in **Figure 1b**. Finally, like shows the **Figure 1a**, the algae oil can be processed with different technologies to obtain different products (biodiesel, methane, ethanol, etc.) and byproducts (heat, steam, fertilizers, etc.). Additionally, after the oil extraction is important to remove the undesirable components depending of the desirable product.

In this regard, the production (biodiesel, ethanol, methane, etc.) depends on the demand by the specific customers, selling price, process costs and plant capacity. Hence, in this work is proposed an optimization model based on **Figure 1b** to determine the optimal pathway for getting different products from algae oil via the sequestration of carbon dioxide of the flue gas coming from different industries. Then, a novel superstructure is developed to consider multiple technologies on each stage of the algae oil processing. These technologies have specific efficiencies and costs (operating and capital costs). The proposed superstructure is presented in **Figure 2,** which is an extended representation of **Figure 1b**.

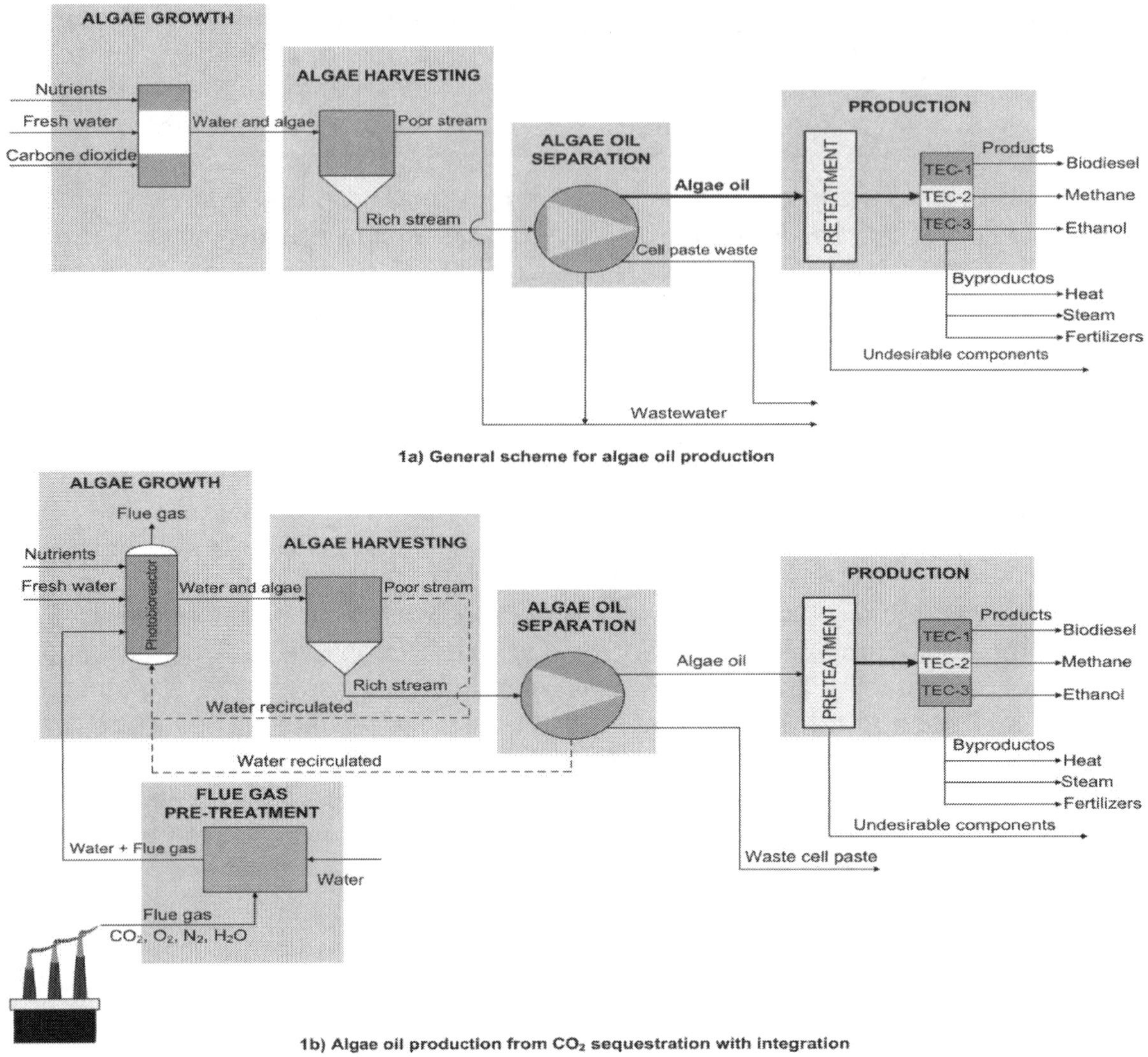

1a) General scheme for algae oil production

1b) Algae oil production from CO_2 sequestration with integration

Figure 1. General production system for the algae oil processing.

Figure 2 is a new scheme to determine the optimal process of the algae oil for obtaining several products to satisfy specific demands of several consumers as well as to determine the feasibility of algae oil processing. Furthermore, there is the opportunity to sequester flue gas to use it like nutrients in the algae growth.

The proposed superstructure considers the optimal allocation of facilities for the algae oil processing through a distributed system. It is because the processing of the algae oil can take place in a *central processing facility* and/or a *hub processing facility*, which could be installed inside the participating plants.

The objective function is the maximization of the overall process profit (*Profit*):

$$Profit = Spc + Gse - Cap - Cop - Cta - Ctp$$

where *Cap* and *Cop* are the capital and operating costs in the system, *Ctp* is the cost to transport the algae oil to the processing facility and to transport the products to the consumers, *Spc* is the profit owing to the product sales and *Gse* is the tax credit for reduction of CO_2 emissions.

RESULTS

The proposed methodology is applied to a case study from Mexico, in this case two industrial facilities located in different geographical places are considered, also transesterification, anaerobic digestion, pyrolisys, gasification and fermentation processing alternatives are considered and the product involved are biodiesel, methane and ethanol which can be sent to two consumers located in different places. Both consumers demand biodiesel and ethanol and just one require methane. The problem then is coded in the GAMS software and the results indicate that the distributed system is able to capture the industrial emissions to obtain economic and environmental benefits. It is because of part of ethanol and all the biodiesel is generated in the central processing facility. And like the methane is demanded just by one consumer this product is generated in the closer processing facility which represents a less transport cost.

CONCLUSIONS

This paper proposes a systematic approach for the optimal design of an algae based distributed biorefinery system to capture the industrial emissions to obtain simultaneously economic and environmental benefits; the proposed approach determines simultaneously the type and location of the processing facilities as well as the distribution of the emissions and products in the distributed system. The application of the proposed methodology to a case study from Mexico shows the potential benefits that can be obtained.

REFERENCES

1. Ohara, H. (2003), "Biorefinery", *Applied Microbiology and Biotechnology*, 62 (5-6) 474-477.
2. Pokoo-Aikins, G., Nadim, A., El-Halwagi, M.M., Mahalec, V. (2010), "Design and analysis of biodiesel production from algae grown through carbon sequestration". *Clean Technologies and Environmental Policy*, 12 (3) 239–254.

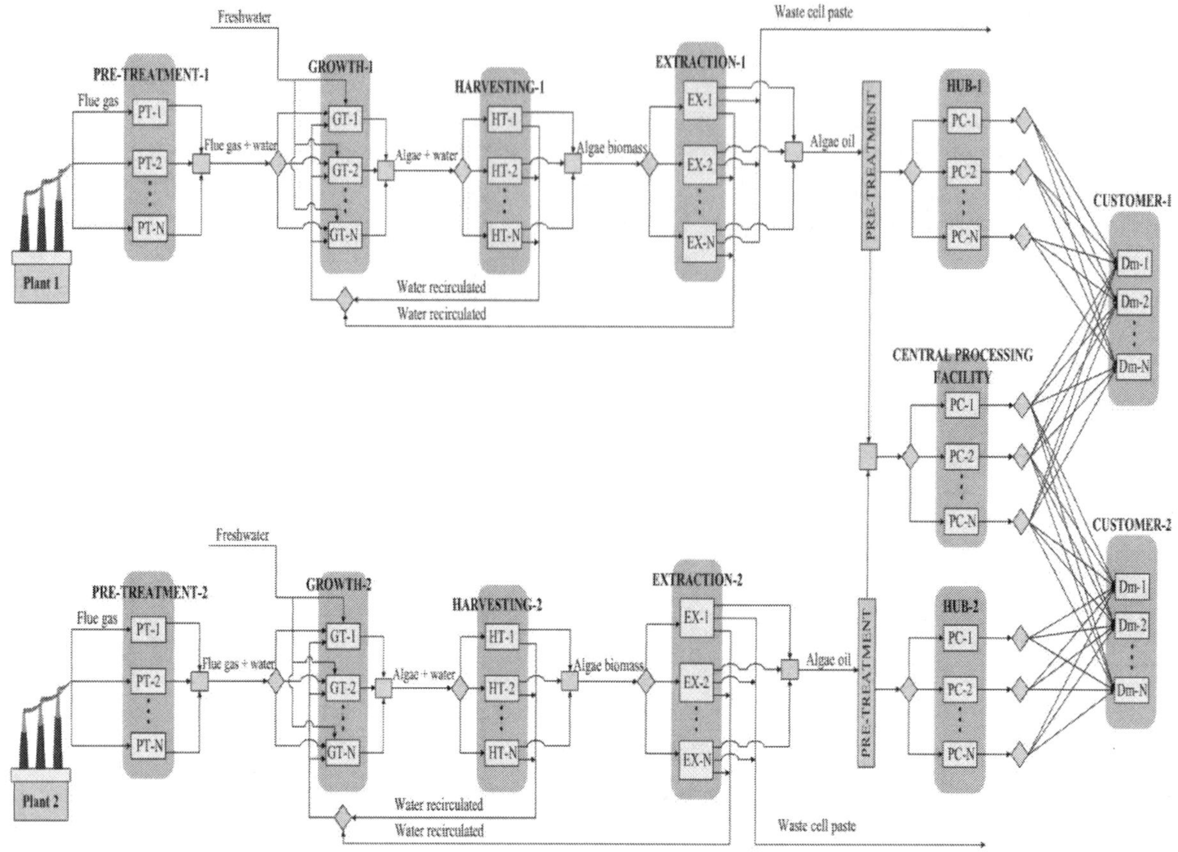

Figure 2. Proposed superstructure for the algae oil processing.

(602l) Production of the Recombinant Trichoderma Reesei Endogluconase Protein Cel7B by Using the Kluyveromyces Lactis Yeast

Zainab Alshoug[1], Michael J. Brodeur-Campbell[2] and David R. Shonnard[2], (1)Chemical Engineering, MTU, Houghton, MI, (2)Chemical Engineering, Michigan Technological University, Houghton, MI

This research is about producing recombinant *Trichoderma reesei* endogluconase protein Cel7B by using *Kluyveromyces lactis* yeast as a host. Cel7B is one of the glycoside hydrolyses family proteins that are produced by *T. reesei*. Cel7B with other endoglucanases, exoglucanases, and β-glucosidases hydrolyze cellulose to glucose, which can then be fermented to fuels or other products.

Examining favorable fermentation conditions for enzyme production and improved activity were the research objectives. Production of protein on different types of media was examined, and measured the activities of the protein by using different tools or procedures. The first method was using different concentration of galactose as a carbon and energy source. The purpose of this method is to determine the relationship between production of enzyme with increasing sugar concentration. The second method was to use different types of media: a complex media-yeast extract, peptone, galactose (YPGal); a minimal media-*yeast nitrogen base* (YNB); and a minimal media with supplement- yeast nitrogen base with casamino acid (YBC). The third method was using different types of reactor or fermenter: a small reactor (shake flask) and a big reactor (BioFlow 3000 fermenter). The purpose of this method is to determine the quantity of the protein produced by using different environments of production.

Different tools to determine the presence and activity of protein were used: for presence of protein sodium dodecyl sulfate-polyacrylamide gel electrophoresis (SDS-PAGE) was used. Secondly, to detect enzyme activity, the carboxymethyl cellulose- 3,5-dinitrosalicylic acid (CMC- DNS) assay was employed.

SDS-PAGE showed that the enzyme band was at 67,000 Da. CMC-DNS activity assay showed that protein activity was greatest with 1% sugar (Galactose) concentration. For the different type of media used in our fermentation, protein was produced from yeast extract peptone galactose (YPGal), and yeast nitrogen base with casamino acid (YBC), but protein was not produced and no activity was detected from *yeast nitrogen base* (YNB). That means the protein production requires the amino acid resources. The quantity of protein which was produced from the rich media (YPGal) over time of production was equal to the quantity of protein that was produced from the minimum media supplemented with casamino acids (YBC) over time with a concentration of 1% galactose.

(602m) Thermal-Catalytic Cracking of Lignin for Production of Value-Added Chemicals

Tae Hyun Kim, Department of Environmental Engineering, Kongju National University, Cheonan, South Korea, Chang Geun Yoo, Agricultural and Biosystems Engineering, Iowa State University, Ames, IA and Kyong-Hwan Lee, (2) Korea Institute of Energy Research, Daejeon, South Korea

Lignin is the second largest components in the lignocellulosic biomass, which can be the valuable byproduct of biorefinery process. If the lignin would be modified or hydrolyzed, high value-added lignin-derived products can be formed.

In the previous reports, ammonia-based pretreatment methods have been introduced as the one of the most effective delignification methods for lignocellulosic biomass. Approximately 60-70% of lignin was hydrolyzed and this chemically modified lignin hydrolysate has a potential for the production of value-added chemicals through thermal-catalytic cracking of lignin because low pH decreases the reaction temperature and increases the reaction rate.

In this paper, various reaction conditions were explored to convert lignin into value-added products. Reaction temperature and pressure were two main factors in this study. From the thermal cracking reaction in the presence of catalysts, several phenolic compounds and carboxylic acids were produced in the form of liquid. Produced gas and solid were also collected and analyzed.

(602n) Potential Carbonyl Condensation Reaction in Detoxification of Phenolic Model Compounds for Bioethanol Fermentation

Maobing Tu[1], Rui Xie[1] and Yonnie Wu[2], (1)Auburn University, Auburn, AL, (2)Chemistry and Biochemistry, Auburn University, Auburn, AL

We investigated the inhibitory effects of ortho-phthalaldehyde (OPA) as a phenolic model compound on fermentation and growth of *Saccharomyces cerevisiae* and its alkaline detoxification. OPA was a potent inhibitor on fermentation and growth of *S. cerevisiae* compared to vanillin, furfural and hydroxymethyl furfural (HMF) at the same concentration level. The inhibition of OPA on fermentation and growth of *S. cerevisiae* was dose dependent. Ethanol production and growth of *S. cerevisiae* were both completely inhibited in the presence of 1.0 mM OPA. The inhibition on fermentation and growth of the yeast decreased with the decrease of the OPA concentration between 0.02 mM-1.0 mM. OPA at 0.02 mM showed no inhibition on both fermentation and growth of yeast, the ethanol final yield was even increased by 4.6% compared to the control. The inhibition of OPA could be overcome by increasing inoculation size of yeast. Most interestingly, we found OPA inhibition could be detoxified under alkaline condition (pH~10) at 60 °C for 2 h in the presence of a reducing sugar (ketose or aldose), but not non-reducing sugar. Mass spectral analysis of OPA reaction products in negative ion mode revealed a high intensity mass at 313.09 m/z. This mass was predicted to be the aldol reaction product of reducing sugar and OPA under alkaline condition. One of the CHO groups on OPA was converted into hydroxyl group by nucleophilic addition of enolate ion of reducing sugar. Loss of one CHO group of OPA could be the key factor for the removal of OPA inhibition.

(602o) Medium and High Through-Put Characterization of Biofuel Feedstocks

Erica Gjersing[1], Angela Ziebell[2], Robert Sykes[1], Geoffrey Turner[3], Renee Happs[1], Melvin Tucker[1], Stephen R. Decker[4], Mark F. Davis[2] and Brian H. Davison[5], (1)National Bioenergy Center, National Renewable Energy Laboratory, Golden, CO, (2)National Renewable Energy Laboratory, Golden, CO, (3)BioSciences Center, National Renewable Energy Laboratory, golden, CO, (4)Biosciences Center, National Renewable Energy Laboratory, Golden, CO, (5)BioEnergy Science Center and Oak Ridge National Laboratory, Oak Ridge, TN

Abstract: The high through-put pipeline has analyzed tens of thousands of samples using NREL's high through-put (HTP) pipeline which characterizes cell wall chemistry and sugar release on pretreatment and enzymatic saccharification (Sykes et al. 2008; Selig et al. 2011). In the last year we have added two new capabilities and worked increasingly closely with plant groups to analyze their data.

The characterization group works together with plant biologist to understand the chemical modifications occurring when plants are genetically modified and why they might cause changes in recalcitrance to pretreatment and saccharification. The characterization group also studies natural populations to understand how natural populations vary and what we might learn from natural diversity.

Plant cell wall sugar content is being determined by a HTP NMR (Nuclear Magnetic Resonance) adaptation to NREL's traditional wet chemistry analysis. This technique is also being used to assess the tendency of a plant to produce inhibitors e.g. furfural and HMF. The method uses proton NMR and Partial Least Squares analysis in conjunction with a set of standards which have the full wet chemistry method. The current NMR facility houses a HTP NMR sample changer which can house >500 samples at a time and sample analysis time take 1-2 minutes. The original full wet chemistry method has also been down scaled to 1/300th by researchers at University of California Riverside and is available for high through-put analysis (DeMartini et al. 2011).

References:

DeMartini, J. D., M. H. Studer, et al. (2011). "Small-scale and automatable high-throughput compositional analysis of biomass." Biotechnology and Bioengineering **108**(2): 306-312.

Selig, M., M. Tucker, et al. (2011). "High throughput determination of glucan and xylan fractions in lignocelluloses." Biotechnology Letters: 1-7.

Sykes, R., B. Kodrzycki, et al. (2008). "Within tree variability of lignin composition in *Populus*." Wood Science and Technology **42**: 649-661.

(602p) Comparison of Enzymatic Reactivity of Corn Stover Solids Prepared by Dilute Acid, AFEX, and Ionic Liquid Pretreatments

Xiadi Gao, Chemical and Environmental Engineering, University of California, Riverside, RIVERSIDE, CA, Rajeev Kumar, Center for Environmental Research and Technology, University of California, Riverside, Riverside, CA and Charles Wyman, Chemical and Environmental Engineering, University of California Riverside, CE-CERT, Riverside, CA

This study is part of a collaboration among the BioEnergy Science Center (BESC), the Great Lakes Biomass Research Center (GLBRC), and the Joint Biological Energy Institute (JBEI), the goal of which is to understand how biomass pretreatments with much different deconstruction patterns impact the chemical and structural features of biomass and its biological conversion to sugars.

Enzymatic hydrolysis was performed on solids resulting from dilute sulfuric acid (DA), ammonia fiber expansion (AFEX), and ionic liquid (IL) pretreatments of corn stover at low to high protein loadings to define how ratios of cellulase, xylanase, and pectinase affected sugar yields. Avicel cellulose, regenerated amorphous cellulose (RAC), and beechwood xylan were also subjected to enzymatic hydrolysis. Initial hydrolysis rates, sugar release patterns over 120 h of hydrolysis, cellulase accessibility to cellulose, and oligomers concentrations were measured over the course of enzymatic hydrolysis to determine enzyme-substrate interactions. These results were then integrated with compositional changes to identify key features controlling enzymatic hydrolysis of solids from these pretreatments and similarities and differences in their impacts on enzyme effectiveness. Ionic liquid pretreated corn stover displayed the highest initial reactivity at all enzyme loadings and the highest final digestibility for a low enzyme loading of 3 mg protein/g glucan in the raw material. However, increasing the enzyme loading to 12 mg/g glucan resulted in dilute acid and AFEX pretreated corn stover attaining the highest overall cellulose digestions. Hydrolyzate from AFEX pretreated corn stover had the highest proportion of xylooligomers, while ionic liquid pretreated produced the most glucooligomers, but amounts of both oligomers dropped with increasing enzyme loadings and hydrolysis times. Ionic liquid pretreated corn stover had the greatest accessibility to enzymes as measured by the maximum cellulase adsorption capacity.

(602q) Understanding How Cell Wall Differences in Agave, Poplar, and Switchgrass Affect Deconstruction in Pretreatment and Enzymatic Hydrolysis

Hongjia Li, Chemical and Environmental Engineering, University of California, Riverside, RIVERSIDE, CA, Sivakumar Pattathil, Complex Carbohydrate Research Center, University of Georgia, Athens, GA, Marcus Foston, Department of Chemistry and Biochemistry, Georgia Institute of Technology, Atlanta, GA, Reichel Samuel, 3Institute of Paper Science and Technology, Georgia Institute of Technology, Atlanta, GA, Rajeev Kumar, Center for Environmental Research and Technology, University of California, Riverside, Riverside, CA, Arthur J. Ragauskas, School of Chemistry & Biochemistry, Georgia Institute of Technology, Atlanta, GA, Michael G. Hahn, University of Georgia, Athens, GA and Charles E. Wyman, BioEnergy Science Center, Oak Ridge, TN; Chemical and Environmental Engineering, University of California Riverside, CE-CERT, Riverside, CA

Agave has emerged as promising biofuels feedstock mainly because it offers high productivity with low water and nutrient demands on semiarid lands. However, our recent results indicated that agave is also significantly less recalcitrant than typical lignocellulosic feedstocks such as poplar and switchgrass. Systematic understanding of structural differences in agave, poplar, and switchgrass and the consequences on cell wall deconstruction can provide clues on paths to reduce recalcitrance that is the key impediment to large-scale commercialization of processes for making fuels from cellulosic biomass. Thus, our high throughput pretreatment and hydrolysis (HTPH) system was used to evaluate the effect of hydrothermal pretreatment conditions and enzyme formulations on sugar yields for four agave materials along with poplar and switchgrass. At the same time, a series of characterization techniques were applied to compare structural differences in cell walls, including lignin composition, non-cellulosic wall components and extractability, cellulose crystalline structure, and surface accessibility. These studies identified important structural features that contribute to lower recalcitrance of agave species and suggested promising directions for more cost effective deconstruction of cell walls of other plants.

(602r) Bacteria Engineered for Bio-Styrene Production

Rebekah McKenna[1], Shawn Pugh[1], Warinsinee Phusitkanchana[2] and **David R. Nielsen**[1], (1)Chemical Engineering, Arizona State University, Tempe, AZ, (2)Chemical Engineering, Arizona State University

Styrene is an important commodity chemical with versatile commercial applications, predominantly as a monomer building block for the production of many useful polymers and co-polymers. Today, all styrene is produced via chemocatalytic routes from petroleum-derived benzene or ethylbenzene according to energy intensive processes. Through *de novo* metabolic pathway engineering, we have pathway engineered a functional pathway to instead produce styrene from renewable feedstocks. By over-expressing both *PAL2* from *Arabidopsis thaliana* and *FDC1* from *Saccharomyces cerevisiae* in a phenylalanine over-producing *Escherichia coli* host we have engineered the first styrene-producing microbe. Initial shake flask cultures produced ~300 mg/L styrene, approaching the determined toxicity threshold. Subsequent works have since sought to overcome several of critical limitations of this preliminary work. Here we present our progress in this regard on several fronts. First, we have increased and balanced pathway flux through the systematic tuning of the expression of *PAL2*, the previous bottleneck enzyme. This was achieved through the systematic tuning of each of gene copy number, promoter strength, and RBS strength. Next, we have sought to overcome the toxicity issue by the 'bottom up' engineering of an aromatic tolerant *Pseudomonas* sp. to produce styrene from glucose and other renewable substrates. At the same time, the ability to circumvent styrene toxicity through its in situ recovery has been explored with the use of solvent extraction, gas stripping, and membrane pervaporation. At last we have sought to improve the sustainability and economics of bio-styrene production by engineering photosynthetic cyanobacteria to produce styrene from CO_2 and light inputs alone.

(602s) Decomposition of Microcrystalline Cellulose by Organic Solvents At Sub- and Supercritical Conditions

Anton Koriakin, Department of Chemical and Biomolecular Engineering, Yonsei University, Seoul, South Korea, Nguyen Van Hai, Chemical and Biomolecular Engineering, Yonsei University, Seoul, South Korea and Chang-Ha Lee, Chemical and Biomolecular Engineering, Yonsei University, Seoul, South Korea

Renewable biomass sources can be converted to fuel, and they are a logical choice to replace oil. By deriving more energy from renewable feedstocks, many countries might be able to decrease their reliance on fossil fuels significantly. In this study, direct thermo-chemical decomposition of microcrystalline cellulose was investigated in various sub and supercritical organic solvents (m-xylene, dodecane, dioxane and methanol). Since the decomposition reaction was performed at 300°C, 350°C and 400°C, the applied pressure was different among the solvents. And the influence of hydrogen and heterogeneous catalysts on decomposition was evaluated for each supercritical solvent. The composition of liquid product was analyzed by Gas Chromatography with a mass selective detector (GC-MS). The char samples were analyzed by X-ray diffraction (XRD).

The addition of hydrogen to the supercritical reaction increased the conversion ratio by about 4-7% in all solvents except non-polar dodecane. At the temperature of 300°C, the conversion ratio was higher in the aprotic solvents than in methanol. At 350°C and 400°C, the conversion in methanol was the highest, but the operating pressure became very high compared to the other solvents. The addition of surface modified silica by acid groups in the methanol solvent led to a significant increase in the conversion ratio. However, the catalyst did not influence the conversion ratio in the aprotic solvents.

(602t) Transgenic Expression of Thermomyces Lanuginosus Lipase in Plants for the Enzymatic Production of Biodiesel

Michael Gagnon[1], Jonathan Fogel[2], Subhash Minocha[2] and Palligarnai Vasudevan[1], (1)Chemical Engineering, University of New Hampshire, Durham, NH, (2)Biological Sciences, University of New Hampshire, Durham, NH

Diminishing petroleum reserves and increasing environmental concerns have led to an urgent need to develop environmentally friendly fuels, such as biodiesel. A relatively new and promising development in the production of biodiesel is via enzymatic transesterification with lipase as the catalyst. The motivation for this research arises from the need to develop inexpensive biocatalysts to replace environmentally harmful chemical catalysts. We are genetically engineering plants to constitutively express a lipase for biodiesel production from spent oils and non-edible plant oils. We have cloned the gene of a lipase with known transesterification activity from *Thermomyces lanuginosus*, a thermophylic fungus. Cloning involved isolation of total RNA, reverse transcription of the mRNA into cDNA and PCR amplification of the lipase gene using specific primers. The gene was first inserted into a cloning vector (pCR8/GW/TOPO) and sequenced to confirm its identity. The gene has been inserted into a plant destination vector (pGWB408 and pMDC83) via LR clonase reaction. Pieces of *Nicotiana tabacum* (tobacco) leaf was transformed with the lipase gene using *Agrobacterium tumefaciens* (strain GV3101) and transferred onto selection media plates. We are currently growing *Arabidopsis thaliana* from seeds that have been transformed using the floral dip transformation method. The recombinant enzyme will be collected from the genetically engineered plants, purified, and tested for its activity. This approach should provide a relatively cheap and environmentally safe source of lipase for use as a biocatalyst.

(602u) Comparison of Catalysts for Applications in Biomass Fast Pyrolysis

Shantanu Kelkar, Chemical Engineering, Michigan State University, East Lansing, MI and Christopher M. Saffron, Biosystems & Agricultural Engineering, Department of Forestry, Michigan State University, East Lansing, MI

Thermochemical methods such as pyrolysis combined with catalysis may be used to produce a hydrocarbon-rich liquid fuel from renewable biomass. The objective of this work was to explore the use of microporous, mesoporous and bi-functional catalysts for catalytic fast pyrolysis of biomass.

ZSM-5, SO_4^{2-} ZrO_2, Al-MSU-S from zeolite beta seeds and sulfated zirconia on mesoporous support 20% SO_4^{2-} ZrO_2 - MCM-41, were synthesized in MSU laboratories. In the first part of this study, microporous catalysts, ZSM-5 and sulfated zirconia, were compared with mesoporous catalysts Al-MSU-S and meso sulfated zirconia for their potential to upgrade oxygenated pyrolysis products. Catalyst properties such as BET surface area, BJH pore volume and acidity using NH3-TPD were measured. Catalyst preparations for upgrading poplar (DN-34) were examined using analytical pyrolysis connected to a GC/MS. Separate experiments were conducted to analyze C1-C9 volatiles using a GC/FID. The product yield, carbon selectivity and yield of hydrocarbon products was evaluated. Although the mesoporous catalysts in this study had many desirable properties such as high pore volume and high acidity, they showed low selectivity for hydrocarbons when compared with conventional ZSM-5 catalyst. The mesoporous catalysts produced a greater amount of non-condensable gases when compared with the microporous catalysts. The char and coke yields of these catalysts were estimated using a TGA. Significant coke production was observed on all catalysts except ZSM-5.

ZSM-5 catalysts may be synthesized with various silica:alumina ratios (SAR) and the change in aluminum content affects the acidity and the hydrophilicity of the catalyst. ZSM-5 catalysts with five different SAR (23, 30, 50, 80, 280) were obtained and tested for the catalytic pyrolysis reaction with the poplar biomass. The catalyst with the lowest SAR was observed to have the highest acidity and production of hydrocarbon molecules. However, the lowest coke yields were observed with SAR=30. The selectivity for various hydrocarbon molecules varied significantly with change in SAR. Metals provide the Lewis acid functionality on zeolite catalysts and may moderate coke production as well as promote olefin production, thus increasing the activity and on-stream lifetime of the catalyst. Bi-functional ZSM-5 (SAR=23) were prepared by loading various dehydrogenating metals. The addition of metals to ZSM-5 enabled a significant increase in hydrocarbon production and decrease in coke production.

(602v) Gas Emissions From the Storage of Western Red Cedar

Xiao He[1], Anthony K. Lau[1], Shahab Sokhansanj[2], C. Jim Lim[1] and Xiaotao T. Bi[1], (1)Chemical & Biological Engineering, University of British Columbia, Vancouver, BC, Canada, (2)Environmental Sciences Division, Oak Ridge National Laboratory, Oak Ridge, TN

The utilization of woody biomass as alternative fuel to generate heat and power has become an important element of sustainability in today's world. Before direct combustion and/or processing in biorefineries, the biomass is usually stored for an extended time period. Gas emission is one of the major problems with biomass storage due to its impact on the surrounding environment and human health, as well as the dry matter losses which would reduce the heating value of the materials. Western Red Cedar (WRC) is a popular wood species in British Columbia, and its harvesting residues can be used as a sustainable biofuel to replace fossil fuels. Aside from the above-mentioned problems associated with biomass storage, WRC has a characteristics pungent smell which would cause concerns about odor nuisance problems. The objective of this paper is to investigate and quantify gaseous emissions from stored WRC woodchips. Fresh WRC chips with an initial moisture content of approximately 50% (wet basis) were used as raw materials in the study. Experiments were conducted in lab-scale reactors under both aerobic and non-aerobic (airtight) conditions at a range of temperatures (5°C, 20°C, 35°C, 45°C and 50°C), which represents cool to hot climate conditions in different geographic locations. Air was pumped into the aerobic reactors after the daily gas sampling event in order to replenish and maintain a high oxygen level in the reactors. The concentrations of gaseous emissions - carbon dioxide (CO_2), carbon monoxide (CO), and methane (CH_4) along with oxygen (O_2) were analyzed by GC, whereas volatile organic compounds (VOCs) were determined by GC/MS.

Results showed that under non-aerobic conditions, the concentrations of CO_2 and CO increased gradually with time for all temperatures. At higher temperatures, higher CO concentrations were measured. In contrast, we observed lower concentrations of CO_2 along with higher O_2 concentrations at higher temperatures. At 20°C, the concentration of CO_2 was the highest; after three weeks, CO_2 concentration reached a plateau of 16% as oxygen content was depleted to 0%, indicating that the storage environment has turned anaerobic. By comparison, oxygen content was lowered to 10-15% for other temperatures. Results further showed that CO_2 and CO emissions from the aerobic reactors exhibit similar trends as the non-aerobic reactors with respect to the effect of temperature. The concentrations of both CO_2 and CO emitted at each temperature decreased slowly with time despite higher oxygen levels of close to 20% being maintained during the entire test period. Biological reaction could be dominant during the storage of wet biomass, leading to the highest level of microbial activity at 20°C rather than at higher temperatures. Microbiological analysis of the WRC woodchips was performed to verify this phenomenon and the extent of microbial activity in the materials. Results from the qualitative GC/MS analysis indicated that the major VOCs emitted include methanol, terpene, aldehydes, acid, alkane, benzene, indoles, acetone, ethers and esters. A TVOC meter was also available for determining the total concentration of VOCs; it was found to be higher at higher temperatures for both aerobic and non-aerobic storage conditions. Under non-aerobic conditions, the initial VOC concentration was approximately 53 ppm and 2 ppm at 50°C and 5°C, respectively; and it gradually decreased with time. Overall, the gas emission results from this study reaffirm the importance of ensuring safe storage conditions for wet woody biomass.

Title: **Wastewater treatment of fluoroquinolone antibiotics using the UV-H_2O_2 advanced oxidation process**

Authors: Sebastian Snowberger; Lee Blaney

Keywords: Advanced oxidation process; pharmaceuticals and personal care products; sustainability; UV-H_2O_2; chemical reactions

---

With recent advancements in medicine, consumption of pharmaceuticals is steadily increasing and, therefore, unprecedented amounts of pharmaceuticals are entering wastewater streams. Currently, wastewater treatment plants do not specifically treat for pharmaceuticals. For that reason, trace concentrations of pharmaceuticals are not only discharged to the environment, but also introduced to drinking water supplies. Health concerns of long-term ingestion of trace concentrations of pharmaceuticals through drinking water are widespread; however, toxicological studies regarding the human health impacts are ongoing. Regardless, numerous studies have identified ecological impacts associated with pharmaceutical contamination. To address these concerns, we studied the transformation of pharmaceuticals under ultraviolet (UV) irradiation and in the UV-hydrogen peroxide (UV-H_2O_2) advanced oxidation process.

This research examined the ability of a lab-scale UV reactor to transform antibiotics into benign compounds. Specifically, a batch-recycle UV reactor was employed to study the transformation of three prominent fluoroquinolone antibiotics: ciprofloxacin, levofloxacin, and norfloxacin. Transformation of fluoroquinolones was studied in DI water and also tertiary effluent from a local wastewater treatment plant. The experimental solutions were prepared by spiking pharmaceuticals (approximate concentrations of 10^{-5} M) into DI and tertiary effluent; hydrogen peroxide (ranging from 10^{-5} to 10^{-3} M) was added during UV-H_2O_2 testing. The pH in the DI solution was controlled using a 0.1M phosphate buffer. Experimental solutions were continuously pumped through the UV cell and samples were generally collected over ten minutes of operation. Fluoroquinolone concentrations were measured using high performance liquid chromatography with fluorescence detection (HPLC-FLD). Transformation of fluoroquinolones occurred through direct photolysis and indirect photolytic mechanisms involving hydroxyl radicals, which were formed via hydrogen peroxide decomposition.

This research has elucidated the quantum yields for transformation of the three fluoroquinolones as a function of solution pH, the impact of the molar ratio of hydrogen peroxide to the fluoroquinolone concentration on transformation kinetics, and the effect of background water quality (*i.e.*, DI water vs. tertiary effluent) on the transformation of fluoroquinolones in the batch-recycle UV-H_2O_2 reactor. Given the widespread adoption of UV technology for disinfection purposes, the results of this research will prove useful to understanding removal of antibiotics in conventional and advanced wastewater treatment processes.

Adsorption Properties of Functionalized PFA-Derived Carbons

Ali Qajar, Maryam Peer, Ramakrishnan Rajagopalan, Henry C. Foley

PFA-derived carbons are synthesized and activated under controlled conditions. The samples are functionalized with concentrated nitric acid under reflux condition to different extends. Surface and structural properties of the samples are characterized by collecting IR spectra, TGA and gas adsorption. Also the samples are compared to commercial activated and functionalized carbons. Our study shows exceptional adsorption properties of PFA-derived carbons for carbon dioxide and ammonia adsorption. Also, functionalization process drastically changes adsorption affinity of the carbons for the adsorption of super-critical gases resulting in higher selectivity for oxygen adsorption over other gases. While, functionalization of the pristine PFA-derived carbon limits pore accessibility resulting in less gas uptake, for the activated PFA-derived carbon the same process promotes adsorption uptakes and dynamics. Ammonia adsorption on the functionalized activated PFA-derived carbon, hits the best reported adsorption uptakes by COF-10 at ~16 mmol/g at 298 K and 1 bar.

(602z) Surfactant Addition to Alkali Solutions for Increased Carbon Dioxide Absorption Rates

Brett P. Spigarelli, Chemical Engineering, Michigan Technological University, Houghton, MI

To meet the growing need for CO_2 capture and storage technology, Michigan Technological University is studying CO_2 capture. Our current approach to CO_2 capture is to bubble CO_2 through alkali solutions. The objective of the present study was to find a way to increase the absorption rate of CO_2 into the alkali solution without reducing the absorption capacity of the solution. One approach to increasing absorption rate is to use a surfactant to chemically alter the gas bubble size. The surfactant chosen for the study was a neutral charge surfactant called DOWFROTH 200 (DF200). Experiments were conducted to study the absorption rate of CO_2 at varying surfactant concentrations of 0, 0.12, 0.24, 0.36, and 0.48 g/L. Results showed that as the concentration of DF200 increased in solution, the absorption rate also increased. The CO
mol/min CO_2 at 0.48 g/L DF200. This amounted to a 14% increase in the CO_2 absorption rate with only a 3% loss in absorption capacity of the solution.

(603a) Cellulosic Ethanol Production From Ionic Liquid Pretreated Biomass with a Recombinant Yeast Expressing Cellulases

Nobuiro Ishida, Biotechnology Laboratory, Toyota Central R&D Labs., Inc., Nagakute, Aichi, Japan, Satoshi Katahira, Toyota Central R&D Labs., Inc., Nagakute, Aichi, Japan, Satoshi Saito, Toyota Motor Co., Toyota, Japan, Noriho Kamiya, Center for Future Chemistry, Kyushu University, Fukuoka, Japan and Chiaki Ogino, Graduate School of Engineering, Kobe University, Kobe, Japan

As the primary polysaccharide of plant cell wall, lignocellulose has received considerable attention as a main feedstock for bio-refinery process such as cellulosic ethanol production. The enzymatic hydrolysis of lignocellulose to soluble sugars is considered to be the most important process in the production of environmental friendly bioethanol. However, the spontaneous crystallization of cellulose due to the chemical uniformity of the glucosyl residues and the high degree of hydrogen bonding can result in densely packed micro fibrils that are inaccessible to cellulolytic enzymes. For the efficient utilization of cellulosic materials in biomass, there is a need to enhance the enzymatic saccharification of this biopolymer.

For deconstruction of biomass with low energy, ionic liquids were employed as pre-treatment media, and the effective degradation and assimilation procedures were investigated [1)]. Subsequently, we developed a transgenic *Saccharomyces cerevisiae* strain that was secretary produced cellulase, and examined the ethanol fermentation from ionic liquid pretreated biomass [2)]. The overall process has been named as 'Ionic liquid-based Consolidated Bio-Processing (i-CBP)'. In this study, we will be presented the effective cellulosic ethanol fermentation from ionic liquid pre-treated biomass using the transgenic *S. cerevisiae* producing the optimal ratio of saccharification enzymes.

1) Uju et al., *Bioresource Technol.*, Vol. 103, p 446 (2012).

2) K. Nakashima et al., *Green Chem.*, Vol. 13, p 2948-2953 (2011).

(603b) A Quantitative Determination of the Effects of Surface/Bulk Lignin and Xylan Contents On Enzymatic Hydrolysis of Lignocellulosic Biomass

Xiaohui Ju, Chemical Engineering and Bioengineering, Washington State University, Richland, WA, Mark Engelhard, Pacific Northwest National Laboratory , Richland, WA and Xiao Zhang, Voiland School of Chemical Engineering and Bioengineering, Bioproducts Science Engineering Laboratory, Washington State University, Richland, WA

Lignocellulosic biomass provides an abundant source of sugars for sustainable biofuels and biochemical production. However, the lack of highly efficient and cost effective cell-wall degrading enzymes is a major obstacle preventing the commercialization of a biologically based lignocellulosic biomass conversion process. The effectiveness of enzymatic hydrolysis is inextricably linked to the structural and chemical characteristics of the biomass material. Due to the diverse nature of lignocellulosic biomass and the heterogeneous structural and chemical characteristics of pretreated biomass substrates, it is difficult to investigate the effect of individual substrate characteristic on enzymatic hydrolysis. We have developed a new methodology based on chemical pulping principles to create a set of biomass "reference substrates" with controlled properties to reflect chemical and structural diversity of biomass materials. The individual effects of surface/bulk lignin and xylan contents on enzymatic hydrolysis of these substrates were determined.

X-ray photoelectron spectroscopy (XPS) has been demonstrated as an effective tool to investigate the surface lignin content and its effect on enzymatic hydrolysis. A Response Surface Methodology (RSM) was employed to attain an aggregated understanding of surface/bulk lignin and xylan contents influence on biomass recalcitrance.

(603c) Use of a Carbon Budget Model to Calculate Land Use Change Greenhouse Gas Emissions Associated with Forest-Based Biofuels Production in Michigan

Jiqing Fan[1], David R. Shonnard[1], Robert E. Froese[2] and Robert Handler[3], (1)Chemical Engineering, Michigan Technological University, Houghton, MI, (2)Michigan Technological University, Houghton, MI, (3)Sustainable Futures Institute, Michigan Technological University, Houghton, MI

Abstract:

Michigan is heavily forested with 19.4 million acres of timberland, and the large volume of timber resources are potential feedstocks for biofuel and bioenergy production. To truly understand the environmental impacts of biofuels production in Michigan, it is essential to evaluate the land use change (LUC) impacts associated with the biomass production practices in this area, namely sustainable harvesting from natural regeneration forest land, and short rotation forestry (SRF) plantation. These two practices are not likely to cause indirect land use change (iLUC) emissions because forest products are not used in the food market. However, the impacts of direct LUC (dLUC) should be examined. The Carbon Budget Model of Canadian Forest Sector (CBM-CFS3) is applied in this study to estimate the dLUC impact associated with the biomass production in Michigan. The CO_2 flux between the atmosphere and soils due to dLUC are quantified and incorporated into the LCA study of the biofuels. We consider three cases for this model; 1. Afforestation of MI forests, 2. Increased harvesting above a business as usual harvest level to satisfy a certain level of biofuel production, and 3. Establishment of SRF plantation on abandoned agricultural lands to satisfy a certain level of biofuel production. A business as usual (BAU) harvesting scenario is first established based on the current harvesting acreage and intensity, a case in which MI forests continue to accrue carbon as forests recover from historic harvest practices (Afforestation case). Because the carbon budget model simulates over a several decade time period, we will also include estimates of changes in fossil greenhouse gas emissions due to changes in extraction practices, for example by relying in inputs from high greenhouse gas intensity crude oils, such as tar sands, using literature values. SRF species, such as poplar, can be planted on abandoned or marginal agricultural land, but tillage at the land preparation stage will disturb the carbon balance and microbial activity in soil, while the growing biomass can sequester carbon from atmosphere into the root system as it grows. Finally, we include these landscape level carbon dynamics into life cycle assessments of biofuels that can be produced from forest feedstocks, for example cellulosic ethanol, to understand the potential impacts on product-based LCA results.

(603d) Gasification Characteristics of Methane Fermentation Residues and Aquatic Weeds

Takeshi Nakajima, Kiyoto Matsumoto, Masanori Ochi, Soshi Watanabe, Junichi Ida, Tatsuki Toda, Tatsushi Matsuyama and Hideo Yamamoto, Department of Environmental Engineering for Symbiosis, Soka University, Hachioji, Tokyo, Japan

In this study, pyrolysis and steam gasification of methane fermentation residues of food waste and aquatic weeds were performed by changing experimental conditions such as reaction temperature and water vapor concentration. In the experiment, the pyrolysis and steam gasification were carried out under two different heating rates (1K/s or instant temperature raise). In the slow heating experiment, time course change of the sample weight, evolved gases and generated tar and char were monitored and analyzed under inert or steam containing atmosphere for the temperature range of 473-973 K. As a result, it was found that under inner gas atmosphere, CO_2 evolution by primary pyrolysis and H_2, CO, CH_4 evolution by secondary pyrolysis were observed for all the samples used. On the other hand, under steam containing atmosphere, rapid H_2 evolution was observed after primary pyrolysis reaction. This may be due to the water-gas shift reaction by steam reforming.

In the rapid heating experiment, samples were inserted into the electric furnace which was pre-heated to desired temperature (873 ~ 1273K) and pyrolysis or steam gasification were performed under inert or steam containing atmosphere. At the exit of the furnace, all the effluent gases were collected to gas bag, and the total yields of the generated gases were examined using GC-MS and GC-TCD. As a result, high H_2 yield was observed from all samples under steam gasification condition. This may also be due to the water-gas shift reaction by steam reforming. In addition, with the increase of the reaction temperature, the yields of the total combustible gases also increased. Futhermore, among all the used samples, Elodea nuttallii (aquatic weed) showed the highest total combustible gases yields.

(603e) Microfiltration of Lignocellulosic Hydrolyzates Using Dead End Filtration to Separate Lignin and Other Inhibitors

Bhavin Bhayani, Empire State Paper Research Institute, Syracuse, NY

Microfiltration of lignocellulosic hydrolyzates using dead end filtration to separate lignin and other inhibitors

B. Bhayani, B.V. Ramarao

Department of paper & Bioprocess engineering, SUNY ESF

In the production of biofuels from lignocellulosics, the biomass is first extracted using dilute mineral acids or hot water. This hemicellulose rich oligomers are hydrolyzed under treatments at higher temperature and pressure to yield sugar monomers which are fermented into biofuels such as ethanol, biobutanol or bioplastics. The initial treatment with either dilute acids or hot water also allows some of the lignin to pass in particulate, colloidal or dissolved forms into the sugar solution. Lignin and other phenolic compounds derived from it are strongly inhibitory to microbial fermentation processes. They also adhere to polymeric membranes, carbohydrates and other surfaces, fouling separation membranes. Their removal from the liquor stream not only facilitates downstream separations but improves the fermentation profiles of the resulting sugar hydrolyzates. Therefore, methods to separate lignin from liquors and hydrolyzates form valuable additional operations in the biorefineries.

A new approach has been developed which can separate inhibitory components from the extracts based on the fact that reduction of temperature of the biomass hydrolyzates can result in the native agglomeration and separation of lignin containing solids by simple settling and/or filtration which thus can result in large increases in separation efficiencies for hydrolyzates. By filtering the extracts at lower temperature (10°C) it is possible to obtain higher flux rate than at higher temperatures (20°C). As shown in Figure 1 and Figure 2, the cumulative flow for extracts at 10°C is higher than the extracts at 20°C at two different pressures. The turbidity of permeate for any given combination of pressure and temperature was less than 10 NTU. It has been hypothesized that the natural flocculation tendency of extracts increases with decreasing temperature.

In addition to the temperature, the effect of addition of a flocculent (Polyethylene oxide, PEO) has been investigated. Substantial enhancement in filtrate flux upon addition of 20 ppm of PEO as seen in Figure 3 is observed.

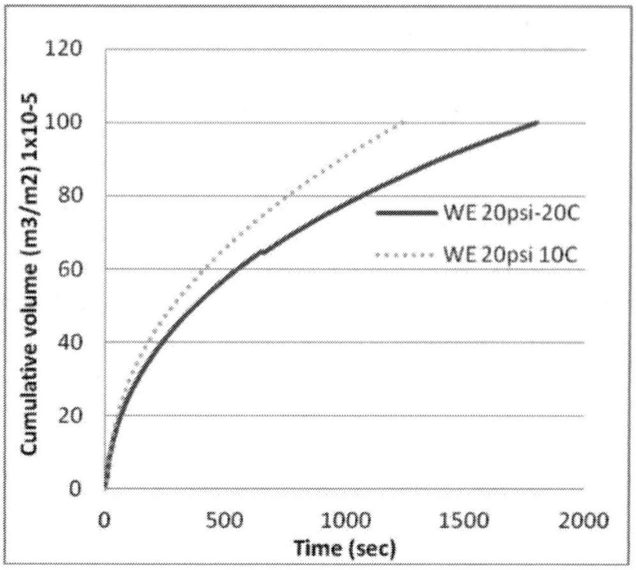

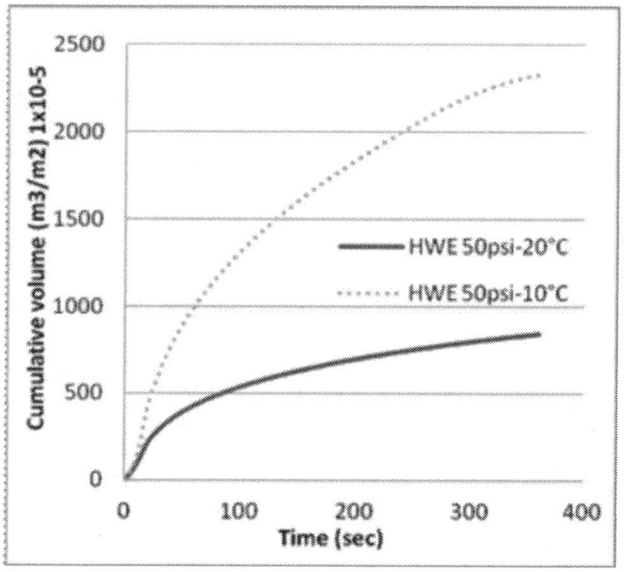

Figure SEQ Figure * ARABIC 1: Cumulative flow at different temperature at 20psi

Figure SEQ Figure * ARABIC 2: Cumulative flow at different temperature at 50psi

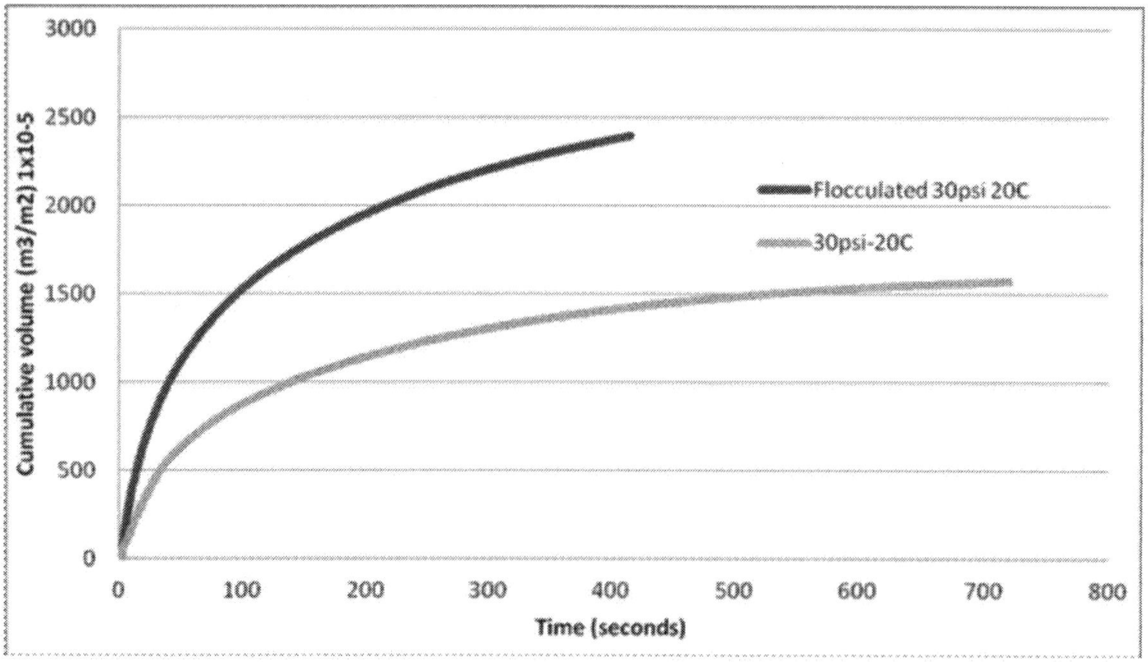

Figure SEQ Figure * ARABIC 3: Effect of addition of PEO

(603f) Lactic Acid Fermentation On Softwood Prehydrolysate by Lactobacillus Delbrueckii

Jing Li, School of Forestry and Wildlife Sciences, Auburn University, Auburn, AL and Maobing Tu, Auburn University, Auburn, AL

Conversion from lignocullulosic biomass into biofuels has been extenstivley investigated recently. However, the whole bioconversion process is still not cost-effective. Developing high-value co-products from biomass prehydrolysate will generate extra revenue to this bioconversion process and become critical to the economic viability of biofuels production. Lactic acid has been classified as valued co-products since it could be used in the food, cosmetic, pharmaceutical chemicals, and biodegradable polymers. The bioconversion process typically includes pretreatment, enzymatic hydrolysis and fermentation. High temperature pretreatment with acid is required to break down the recalcitrant structure of cellulose, hemicellulose and lignin, the harsh condition also generates degraded compounds form carbohydrates and lignin, which are toxic to fermentative microorganisms and inhibit their metabolism, consequently reduces the fermentation rate and yield. The objective of this study is to investigate the fermentability of the softwood prehydrolysate and detoxify the prehydrolysate with different methods to maximize the fermentation yield of lactic acid using *Lactobacillus delbrueckii*. It is has been proved pH control was essential for lactic acid fermentation by *L. delbrueckii* and pH is regulated by adding $CaCO_3$ in this study. Different detoxification methods including alkaline detoxification and H_2O_2 oxidation detoxification has been investigated.

(603g) Production and Properties of Cellulosic Whiskers From Aspen Kraft Pulp

Yang Wang[1], Al Ward[2], Daryl Nichol[2] and Shijie Liu[3], (1)SUNY College of Environmental Science and Forestry, Syracuse, NY, (2)Alberta-Pacific Forest Industries Inc., Boyle, AB, Canada, (3)Paper and Bioprocess Engineering, SUNY College of Environmental Science and Forestry, Syracuse, NY

Cellulose is the most abundant renewable polymer available that can be produced in plants such as woody biomass, cotton, reeds, fruits. A well-known process for preparing cellulosic whiskers from cellulose fibers is acid hydrolysis. In this paper, cellulosic whiskers in the nanometer range were successfully prepared by 16 wt.% sulfuric acid hydrolysis of fully bleached, milled commercial aspen kraft pulp. The effects of preparation conditions, such as hydrolysis time and temperature, on the morphological behavior of cellulosic whiskers were investigated. Higher yields of over 58% were achieved. The chemical composition of the hydrolysis solution was identified and the concentrations of each component were determined by Nuclear Magnetic Resonance (NMR) analysis. The results show the cellulosic whiskers turn the shorter and slimmer when longer hydrolysis time and higher temperature apply. Higher temperature and longer hydrolysis time have negative effects on the yields of cellulosic whiskers, but favor the production of xylose, furfural, glucose, and HMF.

(603h) Identification of Nucleophilic Reactions in Detoxification of Phenolic Model Compounds for Bioethanol Fermentation

Rui Xie and Maobing Tu, Auburn University, Auburn, AL

Biomass pretreatment is typically needed to break down the recalcitrant structure of plant cell wall in bioconversion process. However, during the pretreatment, considerable amount of inhibitors are generated from the degradation of carbohydrates, lignin and extractives; many of these inhibitors significantly reduce the microbial growth and biofuels productivity during the subsequent fermentation process. Overliming detoxification and ammonia conditioning have been widely used for detoxifying biomass hydrolysate. However, gypsum and ammonium salts generated in detoxification will be significant issues for the subsequent processes. Therefore, new detoxification approaches should be investigated for developing cost-effective processes for producing biofuels from lignocellulosic biomass.

In this study, we will use several nucleophiles to detoxify phenolic model compounds and real biomass hydrolysate for bioethanol fermentation with *Saccharomyces cerevisiae*. We have found addition of nucleophiles improved the volumetric ethanol productivity significantly from 0.17 g/L.h to 1.81 g/L.h. This nucleophiles treatment did not change the concentrations of organic acids in hydrolysate, but decreased the furan derivatives by 20%. We will investigate the effects of pH, temperature and reaction time on this nucleophilic reaction. Further research will also be focused on the reaction mechanism between nucleophiles and phenolic model compounds.

(603i) Biomass Characterization Using HPLC and MALDI-MS Techniques

Samira Vasheghani Farahani, Chemical and Environmental Engineering, University of Toledo, Toledo, OH, Michael Mayer, University of Toledo, Toledo, OH, B. Leif Hanson, Instrumentation Center, The University of Toledo, Toledo, OH and Constance Schall, Chemical & Environmental Engineering, University of Toledo, Toledo, OH

Quantitative biomass analyses and molecular weight determination are essential in assessing lignocellulosic feedstocks in biofuel related research. High-performance Liquid Chromatography (HPLC) equipped with a suitable carbohydrate resin column is commonly used for quantitative analysis of polysaccharide constituents in biomass. When sample preparation includes steps such as acid neutralization, introduction of soluble salts can lead to an unstable chromatographic baseline. The contribution of this work is in introducing HPLC sample pretreatment which results in improved baseline stability of resultant chromatograms and therefore reproducible data. Characterizing cellulose, especially its degree of polymerization (DP), is another important step in evaluating its deconstruction to monomeric sugars or in its use for materials. Matrix Assisted Laser Desorption Ionization Time of Flight Mass Spectroscopy (MALDI-TOF) has been used to determine the degree of polymerization for crystalline cellulose with a DP in an excess of 200. A new technique has been developed for MALDI-TOF, utilizing an ionic liquid matrix and solvent system. The technique has potentially wide application in analysis of lignocellulosic substrates.

(603j) Hydrodynamics of a Fluidized Bed Reactor for Fast Pyrolysis of Biomass

Juliana da S. Espindola[1], Fernanda C. Borges[1], Lucas M. Raymundo[1], Luis Antônio R. Muniz[2], Marcelo Farenzena[1], Oscar W. Perez-Lopez[1] and Jorge Otávio Trierweiler[1], (1)Chemical Engineering, Federal University of Rio Grande do Sul, Porto Alegre, Brazil, (2)Universidade de Caxias do Sul, Caxias do Sul, Brazil

Fast pyrolysis is a thermochemical process in which biomass is directly converted into liquid fuels and basic chemicals. Among the technologies applied in fast pyrolysis, the fluidized bed reactors are the most widely used, mainly due to the high mixing performance that this configuration provides, which allows high heat transfer rates and excellent temperature control. In this work the hydrodynamics of a bubbling fluidized bed reactor was investigated, both experimentally and numerically by CFD modeling of fluidizing gas and fluidized particles, in order to allow validation of the model results. The Eulerian approach was used to model the flow of gas and solid phase, both phases treated as interpenetrating continua and governed by a separate set of Navier-Stokes equations. The transport of particles was modeled using drag forces acting on the local volume fraction of solids. The results allow to make a preliminary analysis of the hydrodynamic behavior of a bubbling fluidized bed reactor, helping to understand and develop the process for pyrolysis of biomass.

(603k) Properties and Morphology of Ternary Blends of PLA/PBAT/ Fractionated Lignin

Richard Chen[1], Amar K. Mohanty[2] and Manju Misra[1], (1)School of Engineering and the Department of Plant Agriculture, University of Guelph, Guelph, ON, Canada, (2)Department of Plant Agriculture & School of Engineering, University of Guelph, Guelph, ON, Canada

Lignin is an abundant natural resource that meets every aspect of sustainability since it is highly renewable, low cost, and an industrial by-product of the paper and bioethanol industries. The blending of lignin as a component in a thermoplastic blend poses a challenge in the form of dispersion and compatibility. Polyesters such as poly(lactic acid) and poly(butylene adipate-co-terephthalate) offer the best opportunity of compatibility in melt blending with lignin due to their ability to form hydrogen bonds. The fractionation of lignin into more homogeneous fractions offers better dispersion and more consistent properties. The resulting blend of methanol fractionated lignin, PLA, and PBAT shows highly dispersed lignin particles with particle size of < 5 µm in addition to the presence of lignin between the PLA/PBAT interface inducing improved stress transfer between the two polymer phases. Mechanical properties of the blend showed retention of elongation with slight reduction in strength and modulus for blends containing 30% lignin by weight.

Acknowledgements:

The authors would like to acknowledge the NSERC Biomaterials and Chemicals Research Network for the financial support necessary for this project, along with MeadWestvaco for supplying the lignin used in this project.

(608a) Sustainability of Biofuel Production Systems

Douglas J. Reinemann, Biological Systems Engineering, University of Wisconsin-Madison, Madison, WI

The Great Lakes Bioenergy Research Center is one of three DOE funded Bioenergy Research Centers in the US and the only one that has a major effort devoted to the study of the sustainability of biofuels production systems. We have used a Spatially-Explicit Integrative Modeling Framework (SEIMF) to evaluate the productivity and environmental impacts of biofuel production systems. The modeling framework was implemented to simulate feedstock supply to a cellulosic ethanol bio-refinery located at the center of each of two Regionally-Intensive Modeling Areas (RIMA). Feedstock sources modeled included crop residues and perennial species. Each production system was evaluated in terms of its effect on soil erosion, water-use efficiency, soil carbon stocks, nutrient losses, nitrate leaching, and nitrous oxide emissions. Perennial biomass crops grown on marginal lands can potentially contribute to meet cellulosic feedstock targets while avoiding food-vs.-fuel conflicts and maintaining or even enhancing ecosystem services. The SEIMF approach together with long-term environmental data and analysis was used to model potential cellulosic ethanol production from perennial mixed prairie systems grown on marginal lands across the 10-state U.S. North Central Region. These RIMA modeling results used to develop a spatially-explicit bio-economic model for the study of potential cellulosic biomass supply at regional scale.

We have applied ISO 14000 Life Cycle Assessment (LCA) methods to quantify the environmental impacts of a number of complex biofuels production scenarios including: a) four major ligno-cellulosic agricultural feedstocks (corn stover, switchgrass, poplar and mixed-prairie species) for RIMAs in southern Michigan and Wisconsin, b) biomass feedstocks from native forests in Northern Wisconsin, c) use of the primary AFEX pretreatment process for agricultural residues and SPORL and Dilute Acid pretreatment for woody biomass, d) final fermentation for ethanol production, e) logistics for transportation to regional preprocessing centers and final transport to a centralized ethanol production facility, and f) heat and electric power energy generation scenarios from co-products. Further, we have analyzed three potential breakthrough technologies a) creation of oil-rich grasses as dedicated energy crops, combining both the oil platform and cellulosic platform in one crop, b) feedstock pre-processing to potentially allow the use of high-diversity feedstocks (such as mixed prairie grasses) and other species, and c) use of AFEX-treated lignocellulosic biomass for cellulase production.

A paradigm shift to cellulosic biofuels seems unlikely unless several challenges can be met, including reconciling biofuels with food production, improving system environmental performance, avoiding harmful land use changes, and assembling the biomass supply chains. In particular, local economic and environmental sustainability of the supply system is vital to advance the farmers' interests. Appropriately priced, low supply risk feedstocks are needed to insure the growth of the nascent biofuel industry, as is the ability to inexpensively transport and store the biomass. Local economic and environmental concerns must be resolved in the feedstock supply chain, and diversifying the market for cellulosic feedstocks via multiple co-products would be advantageous as well. Unfortunately, few of these properties are currently demonstrated by cellulosic feedstocks. One promising means to address these and other issues is to process the biomass near where it is produced in so-called local biomass processing depots (LBPDs). RBPDs would increase the stability, uniformity, density, and economic value of a potentially wide variety of cellulosic biomass.

(608b) Neutron Reflectometry, QCM-D , and TIRF Study of the Interaction of Endoglucanases with Films of Amorphous Cellulose

Michael Kent, Joint BioEnergy Institute, Sandia National Laboratory, Emeryville, CA

Cellulase enzyme cocktails include exoglucanases that digest cellulose chain ends and endoglucanases that cleave randomly at interior points along the chains. While it is known that these enzymes work synergistically, the details are not fully understood. In addition, cellulose binding domains (CBDs) are known to play an important role in the digestion of crystalline cellulose but much less is known about the benefit of CBDs in the digestion of amorphous cellulose. Amorphous cellulose is of interest as pretreatment of biomass with ionic liquids, a promising next generation technology, results in a combination of amorphous cellulose and cellulose II. Determining the full effects of endoglucanase activity is challenging because these enzymes can alter the structure of insoluble cellulose in addition to releasing soluble oligomers. To unravel the actions of endoglucanases and the role of cellulose binding domains in enhancing activity on amorphous cellulose, we have combined studies of the profile of water through cellulose films during digestion by neutron reflectivity, measurements of changes in mass and film stiffness using a quartz crystal microbalance (QCM), and visualization of the motion of individual enzymes by total internal reflection fluorescence (TIRF) microscopy. A trend has emerged in that the enzymes possessing a CBM penetrate into and digest within the bulk of the films to a much greater extent than those lacking a CBM. In addition, much different results were observed for a classic endoglucanase and a processive endoglucanase that both possess CBMs, suggesting that penetration is driven by different mechanisms in the two cases.

(608c) Understanding and Redesigning Secondary Cell Wall Deposition in Plants

Dominique Loque, Joint BioEnergy Institute, Berkeley, CA; Physical Biosciences Division, Lawrence Berkeley National Laboratory, Berkeley, CA

The plant cell wall represents a large source of polysaccharides that could be used to substitute for sugar derived from starchy grains, which are currently used to feed and produce biofuels. This lignocellulosic biomass, largely under-utilized, is mainly composed of sugar polymers (cellulose and hemicellulose) embedded very strong aromatic polymer called lignin. Recalcitrant to degradation, lignin inhibits efficient extraction and hydrolysis of the cell wall polysaccharide and prevents cost-effective lignocellulosic-biofuel production. Unfortunately, lignin cannot simply be genetically removed without incurring deleterious consequences on plant productivity. The cost effectiveness of the conversion of the lignocellulosic biomass into sugars is still one of the major components to produce cheap biofuels. Therefore, strategies that can be used to reduce the lignin recalcitrance and that can increase polysaccharide deposition into the cell wall without altering plant growth should be developed.

We used synthetic biology to redesign cell wall biosynthesis and deposition without affecting the plant growth. We generated strategies to reduce lignin recalcitrance either by manipulating its spacio-temporal deposition or by manipulating its composition. We also reengineered the control of cell wall biosynthesis and used it to enhancing cell wall polysaccharides deposition and biomass density. Finally, several of these engineering strategies were combined and translated into an improvement in saccharification efficiency without affecting plant development. Developed approaches and preliminary data will be presented.

¶*This work conducted by the Joint BioEnergy Institute was supported by the Office of Science, Office of Biological and Environmental Research, of the U. S. Department of Energy under Contract No. DE-AC02-05CH11231*

(608e) Development of the Extractive-AFEX^{TM} (E-AFEX^{TM}) Pretreatment Process

Shishir Chundawat[1], Leonardo da costa Sousa[1], Venkatesh Balan[1], Vijay Bokade[2], Mingjie Jin[1], Nirmal Uppugundla[1], James Humpula[1] and Bruce E. Dale[1], (1)Chemical Engineering and Material Science, Great Lakes Bioenergy Center, Michigan State University, Lansing, MI, (2)Catalysis Division, National Chemical Laboratory (NCL), Pune, India

The ammonia fiber expansion or AFEX™ process is a leading thermochemical pretreatment process that facilitates the rapid deconstruction of lignocellulosic biomass to fermentable sugars in an economical fashion. Two major limitations of the conventional AFEX™ process are; (i) it is unable to reduce the inherent recalcitrance of crystalline cellulose I to enzyme biocatalysis, and (ii) it is unable to extract any lignin during the process that are potentially inhibitory to biological catalysts. In order to address these limitations, we have recently developed a novel ammonia based pretreatment called Extractive-AFEX™ (or E-AFEX™). One of the driving forces that led to the development of the E-AFEX™ process was based on our recent discoveries related to unique physicochemical properties of cellulose III (a less recalcitrant allomorph of cellulose; *Chundawat et al, 2011, JACS, 133:11163*). Fortuitously, the conditions that favor cellulose III formation also facilitate selective extraction of lignin from plant cell walls during the pretreatment. In this presentation, we will explore the salient features of the E-AFEX™ technology and its impact on the enzymatic and microbial processability of lignocellulosic biomass. The E-AFEX™ process is able to overcome several of the common bottlenecks to cellulosic biofuel production and can be readily integrated within a non-centralized biomass processing framework. Additionally, the lignin extracted during the process can be converted to valuable chemicals or materials. AFEX™ and E-AFEX™ processing technologies open up exciting new research avenues for cellulosic biorefineries, that allow selective conversion of the aromatic and carbohydrate fractions to a whole suite of fuels, materials and chemicals using desired catalytic or biocatalytic routes analogous to the petrochemical refineries.

(608f) ORNL: Initial Binding of a Cellulose Chain Into the Cellulase Catalytic Tunnel

Pavan GhattyVenkataKrishna, Computational Biology Group, ORNL, Oak Ridge, TN, Edward Uberbacher, Computational Biology Group, ORNL and Xiaolin Cheng, Center for Molecular Biophysics, ORNL

Cel7A proceeds along a single cellulose chain in a processive manner, cleaving one cellobiose unit per catalytic event. We have conducted extensive molecular dynamics (MD) simulations to understand the initial binding of a cellulose chain into the catalytic tunnel of the Trichoderma reesei Family 7 cellobiohydrolase (Cel7A). In unrestrained MD simulations, the cellulose chain diffuses into the tunnel by the length of a cellobiose unit. Among those residues lining the tunnel, the four tryptophan residues exhibit small structural fluctuations, and may thus play a role in guiding the substrate sliding, while the more dynamic polar residues are observed to promote the initial diffusion of the substrate into the tunnel. Further MD simulations show no evident preference for the reducing end or the non-reducing end binding of the cellulose chain, suggesting that the spontaneous initial binding is not driven by the dipole-dipole interaction between Cel7A and the cellulose chain. Rather, the shape complementarity of the cellulose chain end to the tunnel vestibule (involving local hydrophobic stacking and hydrogen bond interactions) plays a more dominant role. The potential of mean force (PMF) calculations, however, show a clear difference at -5 subsite between the reducing end and the non-reducing end binding, implicating that the selective hydrolysis of cellulose from the reducing end can be partially achieved through the preferential initial binding of cellulose to the Cel7A tunnel. The comparison of the PMFs for the wild type and the W40A mutant enzymes highlights the importance of the Trp40 residue in the initial threading of the cellulose chain to the catalytic tunnel.

(613a) Bio-Pulping Assisted Steam Pretreatment of Bagasse by Solid State Cultivation of White Rot Fungi

Jian Liu, Chuannan Long, Hailong Li, Xiaobing Wu and Minnan Long, School of Energy Research, Xiamen University, Xiamen, China

Pretreatment is an important processing step to increase the accessibility of cellulosic material in plant biomass, impacting efficiency of subsequent hydrolysis and fermentation. White rot fungi degrade lignin and have biotechnological applications in conversion of lignocellulose to valuable products. Steam pretreatment is considered to be one of the most practical and versatile pretreatment methods, but to be cost-effective it must produce more accessible fiber for enzymes and prevent the formation of by-products in lower temperature. This study investigated steam pretreatment assisted by bio-pulping of bagasse by solid state fermentation (SSF) using Ceripriopsis sp. to facilitate the conversion into fermentable sugar. The effects of substrate moisture content (68%, 75% and 82%), nutritional addition with metal ions and culture time (0–40 days) on ligninolytic activity, lignin degradation, total weight loss, pulp yield, sugar yield and sugars in soluble fraction were examined. Moisture content significantly affected lignin degradation, with 75% moisture content degrading approximately 2.9-3.7% more lignin than 68% and 82% moisture content after 40 days. Within the same moisture content, fermentations supplemented with metal ions nutrition were statistically better than those without nutrition for lignin degradation and sugar yield from both pulp and soluble fraction, but all 3 ligninolytic activities significantly decreased. Beyond the 21 days pretreatment, additional time resulted in greater lignin degradation, but indicated a decrease in pulp yield and sugar yield. Considering cost, solid state fermentation at 75% moisture content with nutrition was the most preferable pretreatment resulting in 19.1% lignin degradation, 68.2% pulp yield and 25.1% sugar yield over a period of 21 days. It was proved that microbial pretreatment by solid state fermentation has the potential to be a low cost, environmentally friendly alternative to chemical approaches. Moisture relationships will be significant to the design of an effective microbial pretreatment process using SSF technology. Microbial pretreatment employs microorganisms and their enzyme systems to breakdown lignin present in lignocellulosic biomass. This environmentally friendly approach has recently received increased attention and has potential advantages over the prevailing physicochemical pretreatment technologies due to reduced energy and material costs, simplified processes and equipment, and use of biologically based catalysts. Several basidiomycetes such as Phellinus sp., Ceriporiopsis subvermispora, Phlebia subserialis and Pleurotus ostreatus are capable of efficiently metabolizing lignin in a variety of lignocellulosic materials. Ceriporiopsis subvermispora, is a white rot fungus implicated in many diseases of broad-leaved and coniferous trees, causing heart rot, canker and root diseases, and also esca disease of grapevines. It produces oxidative enzymes that degrade lignin. Reducing the lignin content of biomass helps to expose the highly-ordered crystalline structure of cellulose and facilitates substrate access by hydrolytic enzymes. White rot fungi can decompose lignin effectively by excreting unique extra-cellular peroxidases (MnP) as secondary metabolites under nutrient starvation. It has been widely investigated because of its non-specific nature and exceptional oxidation potential for lignin degradation. The degradation efficiency and minimal use of cellulosic polymers relative to other white rot fungi make it a suitable candidate for pretreatment and lignin degradation. However, the efficiency of biodegradation varies with fermentation methods and environmental conditions. Solid state fermentation (SSF) is defined as the growth of microorganisms on solid materials in the absence of free liquid and the technology has been investigated for the development of bioprocesses. Compared to liquid fermentation, where microorganisms are submerged in free liquid, SSF possesses advantages such as simpler operation, lower raw material and input costs and reduced environmental impact from waste water. Moreover, solid state fermentation provides environmental conditions under which white rot fungi grow in nature, thus making it ideal for growth. Considering these merits, applying SSF technology to pretreat lignocellulosic biomass and biopulping may lead to better efficiency and reduce pretreatment costs. However, very limited resources report the application of SSF of Phellinus. sp. for pretreatment of lignocellulosic materials, especially to meet the need for generation of a sugar platform for ethanol production. In the present study, bio-pulping assisted steam pretreatment were applied to determine the efficiency in bagasse. This study examines the microbial pretreatment of wood powder by solid state fermentation. The natural compositions in the wood are not sufficient to meet critical nutrient requirements of fungi. The supplementation of extra nutrients(wheat bran and inorganic salt) is necessary in this study. Moisture content was assumed as a key factor that affected the lignin degrading performance. Moisture content, substrate characteristics, nutrient supplements, aeration and culture duration are important variables for performance of fungi in SSF, impacting growth and metabolite formation, but these processing parameters have not been fully explored. It is reported that moisture content is significant to microbial performance in SSF systems and affects growth as well as product formation. Water in SSF systems functions as a solvent for transport of nutrients, metabolites and other solutes as well as helps maintain stable cellular and molecular structures. Some research have found the quantity and kinetic properties of free water to be more influential to the diffusion of nutrients and metabolites and overall microbial activity. It is reported that the effectiveness of fermentation without nutrition addition on fungal growth and lignin degradation supports the recognized advantages of SSF systems in using wood substrates that provide necessary nutrients in medium preparation. Therefore, the effects of initial substrate moisture content (68%, 75% and 82% wet-basis), nutrition and culture time (0–40 days) on lignin degradation, pulp yield and sugar yield were investigated to evaluate process effectiveness.

(613b) Pretreatment of Corn Stover Using Propyl-Sulfonic Acid Functionalized Nanoparticles

Leidy Peña[1], Feng Xu[1], Keith L. Hohn[2] and Donghai Wang[1], (1)Biological and Agricultural Engineering, Kansas State University, Manhattan, KS, (2)Chemical Engineering, Kansa State University, Manhattan, KS

Propyl-sulfonic acid functionalized (PS) nanoparticles with various levels of functionalization were synthesized, characterized and evaluated as catalysts for pretreatment of corn stover. The degree of functionalization was measured as a function of the hydronium ion concentration, the total organic loading, and the elemental sulfur content. The acid concentration, organic loading, and sulfur concentration of the nanoparticles varied between 0.05 and 0.40 mmol H^+/g, between 5.0 and 30.0%, and between 1.0-18.0%, respectively. FTIR spectra revealed variation on the intensity of the peaks associated with the acid groups on the nanoparticles. The performance of PS nanoparticles on the pretreatment of corn stover was evaluated at 160 °C for 1 h. The biomass and catalyst loads used during the pretreatment were 2.5% and 0.25%, respectively. After pretreatment, the liquid and solid fractions were physically separated. The nanoparticles were magnetically separated from the liquid fraction. The concentration of monomer sugars and oligosaccharides as well as the concentration of degradation products in solution were measured by HPLC. The composition of the solid fraction was analyzed for determination of cellulose and hemicellulose recoveries. The solid fraction was digested with a commercial cocktail of cellulases at 50 °C for 120 h. Biomass digestibility was analyzed against the surface characteristics of the nanoparticles that were used during pretreatment. PS nanoparticles increased the biomass susceptibility to enzyme hydrolysis; therefore, they can be used as an effective catalyst for the pretreatment of corn stover.

Stimulating Cellulase Activity of *Microbulbifer Hydrolyticus* Using Hot-Water Wood Extract Hydrolisate as Substrate.

Karin Arens, Department of Paper & Bioprocess Engineering, State University of New York- College of Environmental Science & Forestry, Syracuse, NY and Shijie Liu, Department of Paper & Bioprocess Engineering, State University of New York- College of Environmental Science & Forestry, Syracuse, NY

ABSTRACT

Among the cellulase producing microorganisms, bacteria often present the advantage of high growth rate, which leads to high enzyme production, and the production of cellulosomes, providing increased function and synergy. In order to improve enzymatic production it is necessary to optimize the fermentation conditions and substrate utilization for a given microorganism. This study uses hydrolisate of sugar maple produced during hot-water extraction (hydrolisate) as substrate to stimulate the production of cellulase from *Microbulbifer Hydrolitycus* IRE-31 (ATCC 700072). Selective strain medium adaptation of the original *M. hydrolyticus* IRE31 was performed by gradual increment of hydrolisate concentration in the marine medium from 2% to 20% (v/v). To improve growth, the pH level was controlled using 5M sodium hydroxide (NaOH) and 0.02M calcium hydroxide ($Ca(OH)_2$) as bases. The effect of the bases on the bacterial growth and activity of the cellulases present in the broth resulted on higher cellulase activity in flask fermentation with 5M NaOH, while in agar the best growth appeared with 0.02M $Ca(OH)_2$ as base. A comparative study to determine the optimum 0.05M citrate buffer pH for enzymatic activity of the *M. hydrolitycus* cellulases present in marine broth supplemented with various hydrolisate concentrations showed best cellulase activity at pH of 4.8. Tests comparing flask fermentations using different substrates such as xylose, glucose, cellobiose, hydrolisate, sodium chloride (NaCl) and yeast extract, exhibited higher cellulase activity in sample broths containing hydrolisate and no discernible activity in broths containing glucose, yeast extract or xylose. The results indicate a positive correlation with the concentration of hydrolisate in the marine broth and the cellulase activity. The addition of 0.25M and 0.20M NaCl in the medium reduced the activity of cellulases in the broth. Scaling up the study to a1L stirred tank fermentation with 10% v/v hydrolysate, set at pH 7.3, 35°C, and 250rpm displayed a maximum enzymatic activity in the broth sample of 0.650 μmol/min-ml after 10 hours. The evidence from this study supports the inference that the use of hydrolysate as a substrate is promising in the stimulation of the production of cellulase by the marine bacteria. However, further study is required regarding optimization of the process.

INTRODUCTION

The current global energy demand and supply of crude oil, combined with projections of fossil fuel depletion, has highlighted the need to find an alternative energy source (EIA Outlook 2011).

The desire to find a renewable and sustainable source of energy has created an increased focus on production of biofuels and bioproducts from lignocellulosic biomass. In an effort to commercialize cellulosic biofuel, new technologies are being explored to increase product yield and reduce the cost in discrete process steps. This new commercial climate and direction for energy supply presents a large market potential for cellulase production (US Dept. of Energy 2011). Currently, the cost of cellulase enzymes accounts for 30 to 50% of the total cost in the production of lignocellulosic ethanol. This cost needs to be reduced in order to make biofuel production more competitive.

Research shows that the enzymatic production can be enhanced by isolating novel strains, improving existing strains, optimizing fermentation conditions and substrate utilization for a given microorganism (Wyman et. al. 2005; Zhang et. al. 2006; Moiser et.al. 2005; Hsu 1980; Eggeman and Elander 2005; Philippidis et.al.1993). Cellulases are mainly produced by fungi, bacteria and protozoans. Production of bacterial cellulases presents some advantages over cellulases produced by fungi. The bacteria growth rate is higher than fungi, leading to higher enzyme production. Moreover, bacterial cellulases are often complex enzyme structures organized in supramolecular complexes, called cellulosomes, which provide increased enzymatic function and synergy (Bayer et al., 1998, Bayer et. al. 2004; Maki et. al. 2009). In this study, hydrolisate of sugar maple produced during hot-water extraction (hydrolisate) is utilized as substrate to stimulate the production of cellulases from the marine bacteria *M. hydrolyticus*. The effect of hydrolisate, xylose, glucose, cellobiose, sodium chloride and yeast extract, on cellulase activity are measured and compared. Initially, a selective strain medium adaptation of the original *M. hydrolyticus* IRE3 was carried out by incremental addition of hydrolisate concentration in the marine medium, (agar and broth). The pH was controlled by using 5M sodium hydroxide (NaOH) and 0.02M calcium hydroxide ($Ca(OH)_2$) as bases, and the effect of the bases on the bacterial growth and the activity of the cellulases present in the broth was measured and compared. A comparative study was performed in order to determine the optimum 0.05M Citrate buffer pH to achieve the best activity for *M. hydrolitycus* cellulases present in broths supplemented with hydrolisate concentrations from 0 to 10% v/v.

METHODOLOGY AND MATERIALS

Microorganism growth, strain adaptation and flask fermentation

Original *Microbulbifer hydrolyticus* IRE31(ATCC700072) was grown in a 250ml Erlenmeyer flask containing 100ml of sterile marine broth 2216 (Difco Laboratories, Detroit, MI). The liquid culture was incubated in the dark at 30°C for 48 hours on a rotary shaker set at a rotation speed of 200rpm. After growth, a loop of liquid culture was transferred to marine agar 2216 (Difco) plates. The inoculated Petri dishes were incubated in the dark at 27°C for 24h. Agar plates were stored in the dark at 8°C for a maximum of 4 weeks.

The strain medium adaptation was performed by exposing the original *M. hydrolyticus* IRE31 to small increments of hydrolisate in the marine medium. The concentrated hydrolysate was filter-sterilized (0.22µm sterile filter, NALGENE, Rochester, NY). Both broth and agar were autoclaved at 121°C for 15 min, cooled (room temperature for broth and approximately to 45°C for agar) and mixed with hydrolysate. The initial selective mediums contained 2% hydrolisate (v/v). The largest colonies were selected, plated on selective mediums with the same concentration as the parent cells, and incubated in the dark at 27°C for 24h. Daughter colonies larger than parent colonies were transfer to a broth containing 3% hydrolisate (v/v) for fermentation tests. After fermentation, the cellulase activity was measured and the broth with the highest cellulase activity was selected for plating on 3% and 4% hydrolisate (v/v) selective medium. The selected strains underwent a gradual increase of hydrolysate concentration at the following increments: 2% to 3%, 4%, 5%, 6%, 10%, 15% and 20%.

The pH level was controlled using 5M NaOH, and 0.02M $Ca(OH)_2$. The effect of each of the bases on bacterial growth and cellulase activity was measured and compared at different concentrations of hydrolisate in flask inoculum and agar plates.

The flask fermentations were carried on 250ml Erlenmeyer glass flasks capped with cotton gauze containing 100ml of sterile fermentation medium. The fermentation was performed at 37°C for 48-72h on a rotary shaker, set at a speed of 300rpm. Samples were periodically drawn to measure the pH and cellulase activity. A comparative fermentation study was performed, maintaining all variables constant, except for the speed of rotation, to determine its impact on the cellulase activity. The fermentation was performed at 35°C for 40-64h on a rotary shaker set at a speed of 300 and 160rpm.

The *M. Hydrolitycus* cellulase activity was measured for microorganisms grown in various concentrations of hydrolisate by volume and compared to those grown in marine medium supplemented with NaCl (0.20M and 0.25M), yeast extract, cellobiose, glucose, and xylose. The marine broth 2216 was used as the blank. During the comparative study between hydrolisate, xylose, glucose and cellobiose, the average concentration of sugars in the hydrolisate was estimated to be $150 g \cdot L^{-1}$. The marine broths were supplemented with sugars at concentrations of 0%, 5% and 10% by volume.

A comparative study to determine the optimum 0.05M Citrate buffer pH for *M. hydrolitycus* cellulase activity, measured the activity of the cellulase present in marine broth with hydrolysate concentrations of 0%, 2%, 4%, 6% and 10% v/v. The cellulase activity was studied at various pH levels of 0.05M Citrate buffer, ranging from 4.8 to 6.0, in order to determine at which pH the enzyme displays a higher activity.

The study was scaled up to a1L stirred tank fermentation with 10% v/v hydrolysate, set at pH 7.3, 35°C, and 250rpm. Samples were periodically drawn to measure the growth and enzymatic activity of the cellulases present in the broth.

The cell growth was determined with a standard curve, in which the dry cell weight at different dilutions was measured and UV absorbance was read at 600nm with results ranging between 0.1 to 0.8A.

Filter paper assay for saccharifying cellulase

The cellulase activity was determined following the International Union of Pure and Applied Chemistry (IUPAC) guidelines, by measuring the enzyme's activity in terms of filter paper units (FPU) per milliliter of original (broth) enzyme solution (Ghose 1987). Dinitrosalicylic acid solution (DNS reagent) was used for color development (Miller 1959). The absorbance was measured against the reagent blank at 540nm with results ranging between 0.1 to 1.0A.

REAGENTS

Citric Buffer

The activity of the *Microbulbifer Hydrolitycus* cellulase was studied at various pH of 0.05M Citrate buffer, ranging from 4.8 to 6.0. The best results were observed at a pH of 4.8. A stock solution of 1M citrate buffer was prepared following the procedures provided by the National Renewable Energy Laboratory Technical Report (NREL/TP-510-42658). The 1M Citrate Buffer stock solution was diluted to 50mM, the pH was checked and adjusted as necessary to 4.8, 5.0, 5.2, 5.6, and 6.0. A glucose standard curve was created for each pH in order to determine the best pH for cellulase activity.

RESULTS AND DISCUSSION

The hydrolisate in the marine medium appears to stimulate cellulase production by *M. hydroliticus*. Initial trials indicate a direct and positive correspondence between the concentration of hydrolisate in the broth and the cellulase activity (see FIG. 1). The production of cellulase is present at specific points of the growth curve. These points of enzymatic activity coincide with a decrease of glucose concentration in the broth. In the various hydrolisate concentrations studied, the highest cellulase activity was shown within the first 20 hours of fermentation (see FIG. 2). Some spikes in cellulase activity are developed after at lower values (see FIG. 3). These results are in agreement to prior studies, which describe cellulase as a growth related enzyme.

FIG. 1. Initial pH 7.5. Results from flask fermentation incubated 36 hours at 30°C and 300rpm.

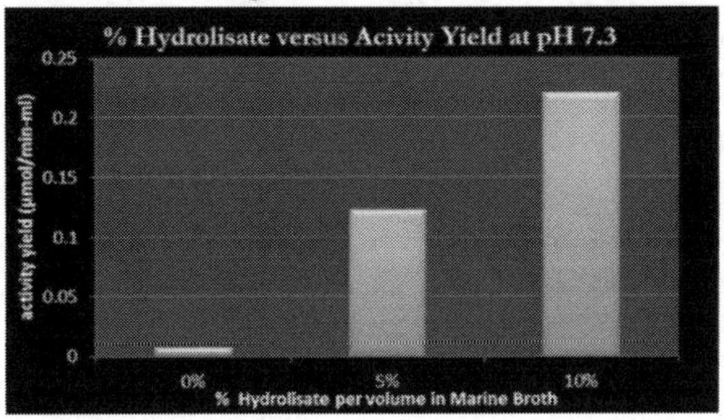

FIG. 2. Results of cell density and cellulase activity in the broth versus time in a 1L stirred tank fermentation with 10% hydrolysate (v/v), set at pH 7.3, 35°C, and 250rpm. Enzymatic activity was measured for broth sample.

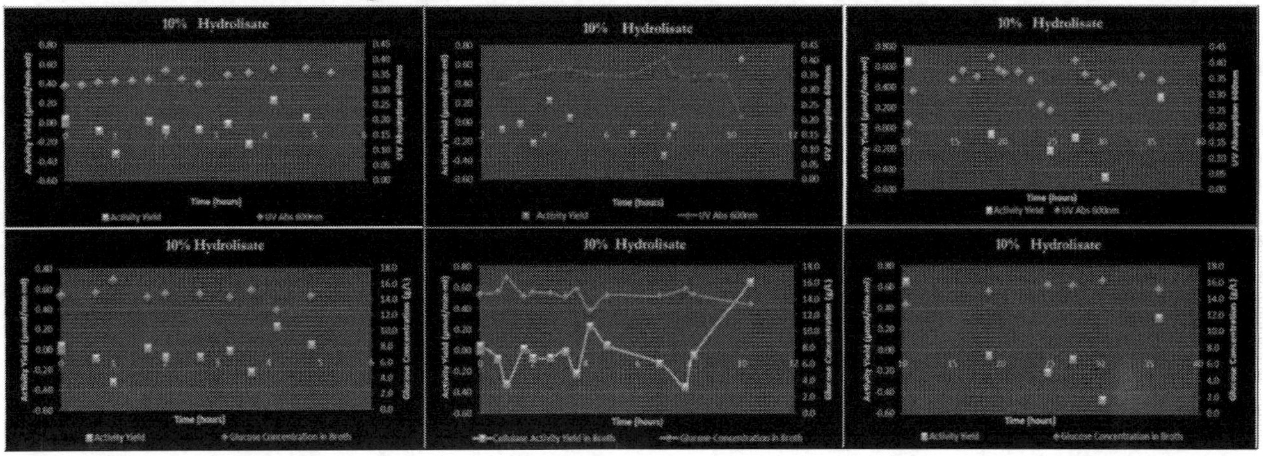

FIG. 3. Cellulase activity over time in a 1L stirred tank fermentation with 10% v/v hydrolysate set at pH 7.3, 35°C, and 250rpm.

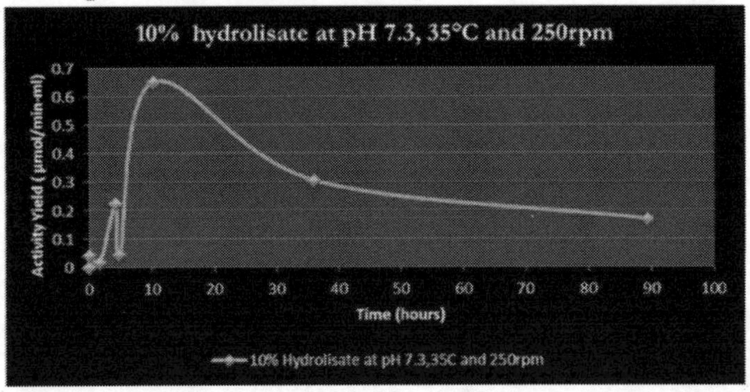

As the concentration of hydrolisate increased in the broth and agar, the importance of controlling the pH was critical. The cells were able to adapt up to 4% v/v of hydrolisate in agar without any addition of base. A further increase of hydrolisate concentration required the introduction of a base. In broth cultures, the cells were able to adapt up to a 20% v/v of hydrolisate. The pH level was controlled using 5M NaOH, and 0.02M $Ca(OH)_2$. The effect of each of the bases on the cellulose activity was measured and compared at different concentrations hydrolisate in flask inoculum. In agar the best results appeared with 0.02M $Ca(OH)_2$. One significant drawback to using 0.02M $Ca(OH)_2$ was identified: it requires greater base volume. For example, to increase the pH of a 100ml medium with 10% v/v hydrolisate to 7.5, a volume of 8ml 0.02M $Ca(OH)_2$ is needed, compared to 50µl of NaOH 5M. In the agar, this volume difference is managed more easily by adjusting the amount of water used during the agar preparation. This volume difference requirement can be problematic during stirred tank fermentation because it dilutes the hydrolisate concentration in the broth as the fermentation progresses with continuous need to add base to the medium, in order to maintain the pH. Another concern is the accumulation of calcium deposits in the feeding lines of the fermentor. Therefore, during the fermentation 5M NaOH is used to control the pH.

M. hydrolyticus optimal growth is at 37°C, pH of 7.5, 300rpm, and NaCl concentrations between 0.1 and 0.5 M (Gonzalez et. al. 1997). The effect of NaCl added to pure marine broth and agar and to marine broth with hydrolizate was also investigated. The addition of 0.25M and 0.20M sodium chloride (NaCl) in the attempt to further the growth of the bacteria in the agar medium resulted in smaller and fewer colonies after 24 hour incubation when compared to agar plates without NaCl. The NaCl addition reduced the pH and the activity of the enzyme in the broth (see FIG. 4). Overall, the pH seemed to be the key variable for an improved enzymatic activity.

FIG. 4. Comparison of *M. hydroliticus* cellulase activity yield and pH in 2% hydrolisate marine broth supplemented with NaCl.

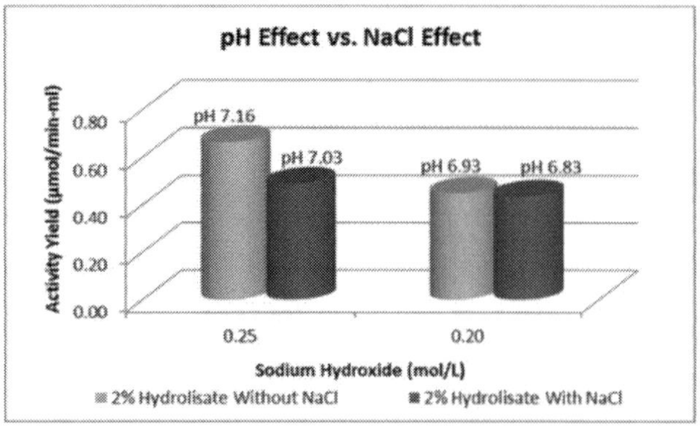

The cellulase activity was measured in citric acid buffer solutions pH of 4.8, 5.0, 5.2, 5.6, and 6.0. The best cellulase activity yield were observed at pH of 4.8 (See FIG. 5a and 5b).

FIG. 5a. Effect of citrate buffer in cellulase activity. Incubation of 72 hours at 30°C and 300rpm.

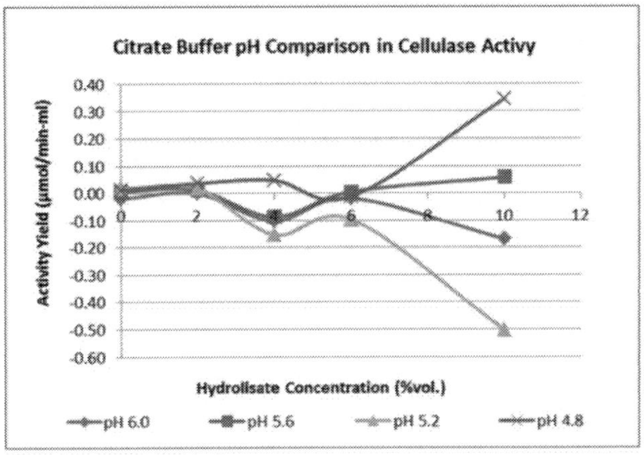

FIG. 5a. Effect of citrate buffer in cellulase activity.

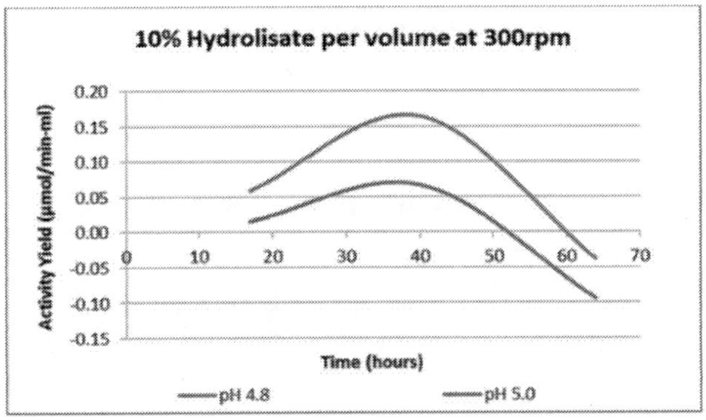

The supplement of the marine broth with glucose and xylose had an inhibitory effect on the production and activity of the enzyme. The addition of yeast extract to the marine broth did not seem to affect the enzyme activity. The presence of cellobiose in the broth stimulated the cellulase activity, but the activity results were less than when supplemented with hydrolisate (see Fig. 6). It was also noted that as the WE concentration increases, so does the activity yield. Better results for cellulase activity are obtained within the first 18hours of fermentation with pH control close to 7.5. At low agitation rate of 160rpm, the cell growth was diminished and did not show cellulase activity for broth supplemented with 10%v/v of hydrolisate. Higher agitation rate of 300rpm during incubation increased the oxygenation of the broth and thus increased the cell production and cellulase activity yield.

FIG.6 Effect of different substrates in the *M. hydroliticus* cellulase activity yield.

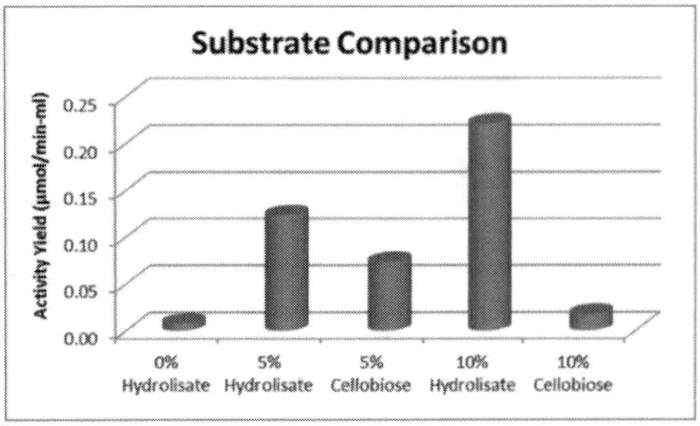

The cellulase activity yield seems to be improved when hydrolisate is used as substrate, and it displays a positive correlation between hydrolisate concentration and activity yield of the enzyme.

CONCLUSION

The marine bacteria *Microbulbifer Hydrolitycus* IRE-31 was adapted to 20% v/v of hydrolisate of sugar maple produced during hot-water extraction using selective strain medium adaptation and pH control. The hydrolisate in the marine medium appears to stimulate cellulase production by *M. hydroliticus*. Initial trials indicate a direct and positive correspondence between the concentration of hydrolisate in the broth and the cellulase activity. Spikes in the production of cellulases are present at specific points of the growth curve. These points of enzymatic activity coincide with a decrease of glucose concentration in the broth. In the various hydrolisate concentrations studied, the highest spikes of cellulase activity were uniformly shown within the first 20 hours of fermentation. The evidence in this study supports the inference that the use of hydrolysate as a substrate is promising in the stimulation of the production of cellulase by the marine bacteria. Such a result may be of significant import relative to decreasing the cost of cellulase enzymes during production of biological fuels. However, further study is required regarding optimization of the process.

REFERENCES

1. Bayer, E.A., H. Chanzy, R. Lamed and Y. Shoham. 1998. Cellulose, Cellulases and Cellulosomes. *Current Opinion Structural Biology* 8:548-557
2. Bayer, E.A., J.P. Belaich, Y. Shoham, and R. Lamed. 2004. The cellulosomes: multienzyme machines for degradation of plant cell wall polysaccharides. *Annu Rev Microbiol.* 58:521-554

3. Wyman, C.E., B.E. Dale, R.T. Elander, M. Hotzapple, M.R. Ladisch, and Y.Y. Lee. 2005. Coordinated development of leading biomass pretreatment technologies. *Bioresour Technol.* 96:1959-1966
4. Philippidis, G.P., T.K. Smith, and C.E. Wyman. 1993. Study of the enzymatic hydrolysis of cellulose for production of fuel ethanol by the simultaneous saccharification and fermentation process. *Biotechnol Bioeng.* 41(9):846-853
5. Lynd, L.R., W.H. van Zyl, J.E. McBride, and M. Laser. 2005. Consolidated bioprocessing of cellulosic biomass: an update. *Curr Opin Biotechnol.* 16(5):577-583
6. Eggeman, T., and R.T. Elander. 2005. Process and economic analysis of pretreatment technologies. *Bioresource Technology* 96: 2019–2025
7. Ghose, T.K. 1987. Measurement of Cellulase Activities. *Pure and Appl. Chem.* 59(2):257-268.
8. Gonzalez, J.M., F. Mayer, M.A. Moran, R.E. Hodson, and W.B. Whitman. 1997. Microbulbifer hydrolyticus gen. nov., sp. nov., and Marinobacterium georgiense gen. nov., sp. nov., Two Marine Bacteria from a Lignin-Rich Pulp Mill Waste Enrichment Community. *International Journal of Systemetic Bacteriology*. 47(2), 369-376
9. Hsu, T.A., 1996. Pretreatment of Biomass. In: Wyman, C.E. (Ed.), Handbook on Bioethanol, Production and Utilization. Taylor & Francis, Washington, DC.
10. Hsu, T.A., M.R. Ladisch, and G.T. Tsao. 1980. Alcohol from cellulose. *Chem. Technol.*, 10 (5), 315–319.
11. Kumar, P., D.M. Barrett, M.J Delwiche, and P. Stroeve. 2009. Methods for Pretreatment of Lignocellulosic Biomass for Efficient Hydrolysis and Biofuel Production. *Ind. Eng. Chem. Res.* 48: 3713–3729
12. Maki, M., K.T. Leung, and W. Qin. 2009. The prospects of cellulase-producing bacteria for the bioconversion of lignocellulosic biomass. *Int J Biol Sci.* 5(5):500-516.
13. Miller, G.C. 1959. Use of the Dinitrosalicylic Acid Reagent for the determination of reducing sugar. *Anal. Chem.* 31: 420–428.
14. Mosier, N., C. Wyman, B. Dale, R. Elander, Y.Y. Lee, M. Holtzapple, and M. Ladisch. 2005. Features of promising technologies for pre-treatment of lignocellulosic biomass. *Bioresource Technology.* 96: 673-686.
15. Zhang, Y.H.P., M.E. Himmel, and J.R. Mielenz. 2006. Outlook for cellulase improvement: screening and selection strategies. *Biotechnology Advances*. 24: 452-481
16. Corn Production: USDA National States Data - Crops query; corn grain, 1986-2006, United States www.nass.usda.gov/QuickStats/Create_FedeBiomass Program: US Department of Energy http://www1.eere.energy.gov/biomass/about.html
17. U.S. Billion Ton Update: Biomass Supply For A Bioenergy and Bioproducts Industry. http://www1.eere.energy.gov/biomass/pdfs/billion_ton_update.pdf
18. International Energy Outlook 2011. Figure 15. http://www.eia.gov/forecasts/ieo/pdf/0484(2011).pdf

(613d) Catalytic Conversion of Glucose Into Fuels and Chemicals Via 5-Hydroxymethylfurfural

Lu Lin, School of Energy Research, Xiamen University, Xiamen, China

With the depletion of fossil resources and growing concerns about global warming and environmental pollution, the efficient conversion of renewable biomass into fuels and chemicals has recently attracted considerable attention. Among the many possible biomass-derived chemicals, 5-hydroxymethylfurfural (HMF) has been recognized as a versatile intermediate that can be used for the production of a wide variety of fine chemicals, polymeric materials and biofuels. In the past few years, many researchers have selected fructose as feedstock and excellent HMF yields have been readily achieved using many methods. However, it should be pointed out that fructose is not abundant in nature and its cost is very high, which limit large-scale and sustainable production of HMF. In contrast, glucose is more abundant, available and cheaper than fructose, which is considered as the preferred feedstock for the production of HMF. Disappointingly, many catalytic systems in which fructose is readily converted into HMF are normally ineffective for glucose, which is attributed to stable pyranoside ring structure of glucose. Therefore, there is a strong incentive to develop an efficient process for the production of HMF using glucose as starting material. In this review, the state-of-the-art of catalytic conversion approaches is systematically summarized and discussed. Moreover, some emphasis is also put on the synthesis of HMF-based derivatives including levulinic acid (LA), 2,5-dimethylfuran (DMF) and 2,5-furandicarboxylic acid (FDCA). Furthermore, we also point out some promising research trends in further studies.

(613e) Comparison of Oleaginous Fungal Lipid Accumulation On Different Lignocellulosic Feedstocks

Zhenhua Ruan[1], Mike Zanotti[1], Yuan Zhong[2], Chad Ducey[3], Wei Liao[1] and Yan Liu[1], (1)Biosystems and Agricultural Engineering, Michigan State University, East Lansing, MI, (2)Biosystems and Agricultural Engineering, Michigan State Univeristy, East Lansing, MI, (3)Werks Management, LLC, Fishers, IN

The perennial grasses miscanthus, giant reed, and switchgrass, along with the agricultural residue corn stover, were evaluated for their bioconversion potential for fungal lipid accumulation. A co-hydrolysis process was adopted to convert lignocellulose into fermentable sugars. The process utilized dilute sulfuric acid pretreatment without detoxification and liquid-solid separation, directly followed by enzymatic hydrolysis. A factorial experiment in a randomized block design was employed to optimize the co-hydrolysis process. The influence of combined severity on the susceptibility of pretreated substrates by enzymatic hydrolysis was clearly discernible, showing that co-hydrolysis can be applied to lignocellulosic biomass to enhance the efficiency of sugar conversion, and reduce the amount of processing water.. The experimental results demonstrated that under the optimal conditions, corn stover had the highest sugar yield followed by switchgrass, giant reed and miscanthus. Then, an oleaginous fungus *Mortierella isabellina* ATCC42613 was selected and applied on hydrolysates to accumulate fungal lipids due to its capability of utilizing both C5 and C6 sugars. The cultivation using hydrolysates as the carbon source exhibited greater cell mass, and equivalent lipid production compared to synthetic medium with pure glucose and xylose. These results elucidated that combining fungal fermentation and co-hydrolysis to accumulate lipids could have potential to enhance utilization efficiency of lignocellulosic feedstocks for advanced fuels production.

(613f) Investigation of Accessory Hemicellulases and Pectinases for Polysaccharide Hydrolysis of Ionic Liquid Pretreated Biomass

Christopher J. Barr[1], Jeffery Mertens[2] and Constance Schall[1], (1)Chemical and Environmental Engineering, University of Toledo, Toledo, OH, (2)NCAUR, USDA-ARS, Peoria, IL

The polysaccharides, cellulose, hemicellulose, and other additional carbohydrate polymers of terrestrial biomass, comprise renewable feedstocks for carbon-based chemicals and fuels. Biomass pretreatment is required to overcome its recalcitrance to biochemical deconstruction to monomeric sugars for sugar platform fuels and chemicals. Ionic liquid (IL) pretreatment followed by enzymatic hydrolysis can be used to saccharify the biomass into its monomeric sugars at a rate much higher than many pretreatment methods. IL pretreatment creates unique biomass substrates consisting of primarily amorphous cellulose with largely intact hemicellulose and lignin. These substrates differ from many other pretreatments which remove hemicellulose and retain cellulose crystallinity. The unique substrate resulting from IL-pretreatment poses an interesting challenge for enzymatic deconstruction using traditional commercial mixtures.

Enzyme mixtures using *A. nidulans* mono-component enzymes with accessory hemicellulases and pectinases were formulated and used in hydrolysis experiments. Exploration of the accessory enzyme requirements yields insight into differences in hemicellulose structure and composition between biomass substrates. Accessory hemicellulases examined include acetyl xylan esterase (AXE), ferulic acid esterase, and arabinofuranosidase. In addition to mono-component enzyme mixtures, commercially-available enzyme mixtures were supplemented with accessory enzymes to optimize hydrolysis for lower enzyme loadings. As part of these studies, analysis of variance (ANOVA) was used to statistically assess and model the main and interaction effects of key enzyme activities.

(613g) Kinetics of Adsorption, Desorption, and Re-Adsorption of a Commercial Endoglucanase in Lignocellulosic Suspensions

Qianqian Wang, South China University of Technology, Guangzhou, China, J.Y. Zhu, USDA Forest Service, Forest Products Laboratory, Madison, WI and Chris Hunt, US Forest Service, Madison, WI

This study conducted quantitative kinetic modeling and in-situ and temporally resolved measurements of adsorption, desorption, and re-adsorption of a commercial endoglucanase in lignocellulosic suspensions. The study defined a cellulase adsorption and desorption competition parameter, a pseudo rate of binding and desorption, binding and desorption capacity, as well as cellulase binding reversibility (a thermodynamic property) and recyclability (a engineering parameter). The results indicate that both substrate chemical and physical structures play important roles in cellulase binding and desorption. Binding of a commercial cellulase onto a cellulosic substrate was reversible. Bindings to two different lignocellulosic substrates were almost irreversible. While lignin and its structure positively affect binding capacity to substrate, they negatively affect cellulase recyclability. Collapsing of substrate pores reduced cellulose accessibility and cellulase binding capacity and increased reversibility and recyclability. Increasing temperature and pH increase cellulase desorption and increased binding reversibility and capacity. This study lays the foundation for developing effective cellulase recycling strategies.

(626a) Production of Drop-in Biofuels Using a Hybrid Biological/Chemical Biomass Conversion Process

Cesar B. Granda, Terrabon, Inc., College Station, TX

Conversion technologies to biofuels can be biochemical in nature, such as fermentations, or thermochemical conversions such as gasification and pyrolysis. Both fermentation and, in particular, pyrolysis show great promise, however, they also have several disadvantages. Fermentation produces well-defined products, which are typically easily recoverable, but often requires the pretreatment and hydrolysis of polymeric sugars prior to the conversion of the resulting soluble sugars into the desired product. Fermentation is also often performed using pure cultures that are prone to contamination and thus require aseptic conditions, which are difficult and expensive to maintain. In addition, the pure cultures are commonly able to uptake only one type of substrate, such as sugars, which limits the types of feedstocks that can be converted. On the other hand, pyrolysis often does away with the need to priorly break down the feedstock into simpler molecules and is able to convert different types of substrates without the need for aseptic conditions. But, by the same token, the product generated by pyrolysis is much undefined and highly corrosive, as a result, it is very difficult to upgrade to an acceptable biofuel quality. In addition, high-moisture feedstocks cannot be efficiently processed through pyrolysis. In an effort to capture the best of both worlds without falling victim of their limitations, Terrabon, Inc. is commercializing a process that can convert any anaerobically biodegradable material into liquid drop-in biofuels by firstly fermenting the feedstock with natural mixed cultures of microorganisms, which are able to degrade many different types of substrates (e.g., proteins, cellulose, hemicellulose, fats, pectin, chitin, glycerol) with little differentiation and that, as a result, do not need aseptic conditions, making the fermentation very robust. The product from the fermentation, which is a mixture of carboxylate salts, is a much more defined feedstock for pyrolysis than raw biomass; therefore, the product from the pyrolysis of these salts is more defined, with its composition being mostly ketones such acetone. The ketones are very easy to upgrade and convert into liquid biofuels. The feedstock flexibility that this hybrid biochemical/thermochemical provides, has allowed Terrabon, Inc. to initially focus on converting the low-hanging fruit in biomass resources, municipal and agricultural solids wastes, into drop-in biofuels, such as bio-gasoline and bio-jet fuel. Terrabon has successfully demonstrated the science of this process and is making great strides towards its commercialization. Terrabon operates demonstration facilities with the capacity to process the equivalent to 5 dry ton/day of biomass and is able to generate enough fermentation products to produce 300 gal/day of bio-gasoline. The demonstration plant monitors different parameters in the fermentation such as acid yield and biogas production. In the downstream chemical processing into hydrocarbons, many parameters are monitored, such as yields and selectivity, as well as the quality (e.g., octane number, IBP, FBP) of the final product. Results from these studies and their implications in the economics will be presented.

Development of a C5 and C6 to Ethanol Consolidated Bioprocess with An "Old New" Bug; Use of the Deinococci Platform

Dr. Michael KREL, Deinove,

Introduction

Deinove is a green technology company specialized in development and commercial exploitation of innovative, environmentally friendly, high-performance processes for production of biofuels and other compounds of industrial or pharmaceutical value. The processes developed by Deinove use bacteria with exceptional natural properties - **the deinococci.**

Deinove was incorporated in late 2006 under the joint impetus of Philippe Pouletty, MD (Managing Partner of the Paris-based private equity firm Truffle Capital), and Miroslav Radman (Professor at Descartes University of Paris, a member of the French Academy of Science and winner of the 2003 Grand Prix *INSERM* science prize). Miroslav Radman discovered the genetic basis of the extraordinary biodiversity and robustness of the deinococci (the bacteria on which Deinove's innovation strategy is founded) and, in particular, the molecular mechanism which enables *Deinococcus radiodurans* to repair its genome after the latter has been shattered by ionizing radiation.

Deinove is the only company in the world with an activity based on systematic exploration of the deinococci - a bacterial genus with exceptional biodiversity but which has rarely been studied and never exploited commercially.

By combining the selection of wild-type strains (with natural properties that are already of industrial value) with bacterial engineering, Deinove is building a unique portfolio of intellectual property and innovative biomanufacturing processes. Deinove is aiming to create breakthrough technologies in second-generation biofuel production and green chemistry.

Technological Approach

Deinove has decided to develop a process based on the potentiality of the Deinococci micro-organisms as we detected that these microorganisms are very well suited for industrial purpose even though there was no existing metabolic engineering tools. In order to do that Deinove had to develop the whole genetic toolbox needed for strain optimization (including insertion, knock out, plasmid, promoters….).

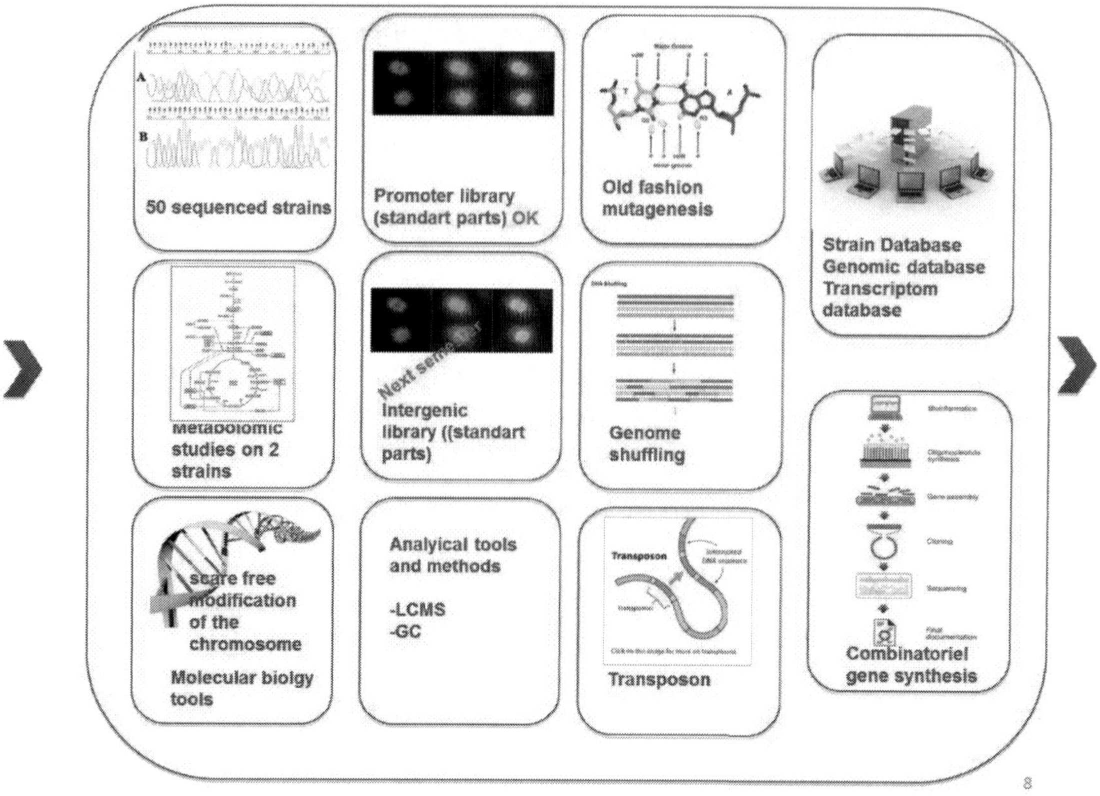

Now that this had been achieved, we are currently working on developing a bacterium that is capable of degrading and digesting both hemicellulose and cellulose (i.e. C5 and C6 sugars) and transforming it into ethanol, in a process known as Consolidated Bio Process (CBP) in order to implement it on industrial scale in the three coming years.

Deinove's research work has confirmed the extreme robustness of the deinococci vis-à-vis various types of stress (ionizing radiation, desiccation, solvents, ethanol, butanol, acid and alkaline milieus, high temperature, etc.) and their ability to integrate large genomic fragments from other micro-organisms. These properties are key success factors for the company's target industrial applications.

The deinococci's properties (robustness, biodiversity and metabolic properties) and Deinove's research results explain these bacteria's huge potential as specialized "cell factories" for the production of ethanol and other components of industrial value in a cost-effective, environmentally friendly way.

Deinove project is a breakthrough scientific challenge as it opened the path to production of cellulosic ethanol from cellulose, hemicellulose and their derivatives (starch, glucose and pentoses) in a simultaneous way. There are no such process running in the world today and this would be of great interest as it is very difficult to get a higher valuation of the C5 stream of the biomass for a mass market (as biofuel). Being able to develop a process that give a higher value to C5 stream (by transforming it into ethanol for instance) and using it in a parallel manner to C6 fermentation is a technical and scientific challenge that has not yet been achieved

Results

- Development of metabolic engineering tools efficient of the Deinococci species that allow insertion on the chromosome or knock out of genes in a non GMO approach (no antibiotic markers, no plasmid and no scars left)
- First ever proof of insertion and/or knock out of selected functionality in Deinococci species
- Development of a bi-functional microorganism able to digest cellulose and hemicellulose material and produce ethanol.

The improvement of the industrial strain regarding the ethanol production can be seen on the following chart:

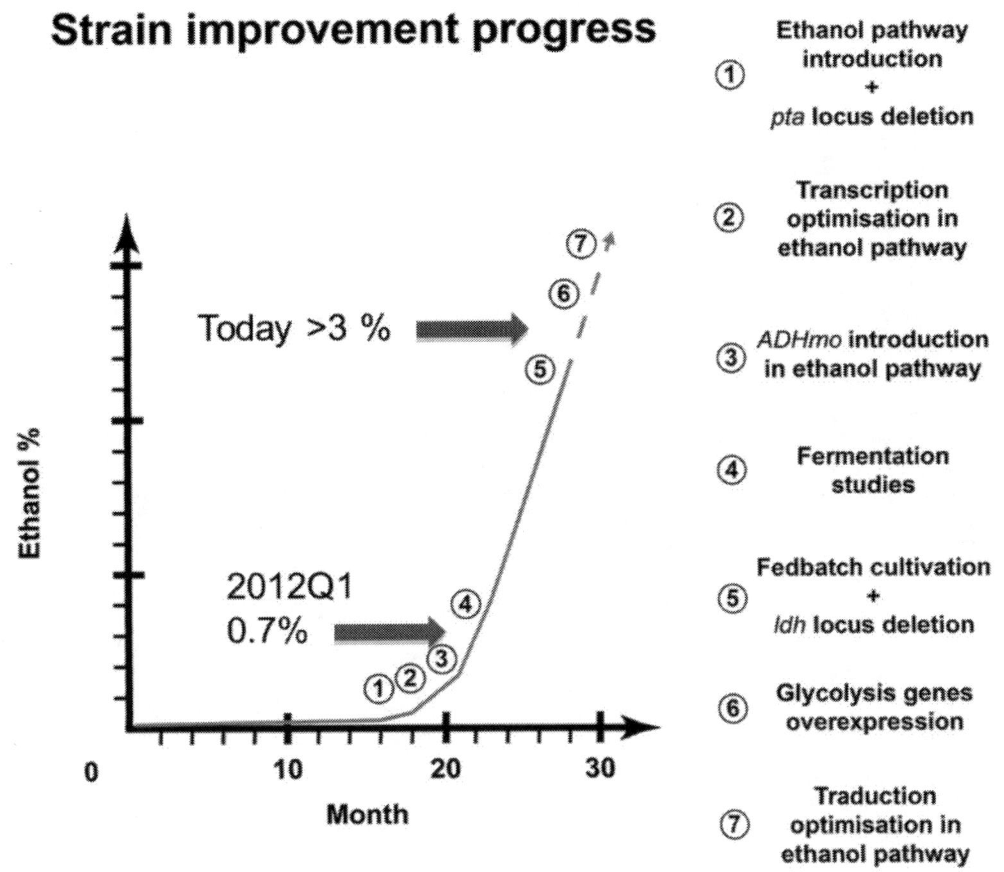

Deinove has already a partnership going on with one of the leading European sugar company, Tereos, on the use of wheat and wheat derivatives for ethanol production as Tereos runs the world biggest wheat-to-ethanol production facility. In the coming months, Deinove will be able to transfer its process to higher scale facilities where the development will be perform with its industrial partners Tereos

(626c) Biomass Particle Size Influences Sugar Conversion

Venkatesh Balan[1], Shuhaida Harun[2], Leonardo da costa Sousa[1], Seema Singh[3], Ming Jie Jin[1] and Bruce Dale[1], (1)Chemical Engineering and Material Science, Great Lakes Bioenergy Center, Michigan State University, Lansing, MI, (2)Department of Chemical Engineering, Universiti Kebangsaan Malaysia, Selangor, LA, Malaysia, (3)Joint BioEnergy Institute, Emeryville, CA

The lignocellulosic biomass (LCB) from crop residues contains about 65-75% of cellulose and hemicellulose that could be converted into fermentable sugars, a renewable source for producing biofuel. It is widely believed that when we reduce the particle size, the surface area increases which results in higher digestibility. We designed experiments to test the effect of particle size (2mm, 5mm, 2cm and 5cm) on rice straw conversion. We pretreated rice straw using two different AFEX™ pretreatment conditions C1 (100°C; 2:1 ammonia to biomass ratio; 80% moisture and 30 minutes residence time) and C2 (140°C; 1:1 ammonia to biomass ratio; 130% moisture and 50 minutes residence time). Subsequently, we hydrolysed the pretreated biomass at different glucan loading (1%, 3% and 6%) using commercial enzymes (Spezyme CP and Novozyme 188) and determined the sugar yield. AFEX™ C2 conditions made the treated rice straw substrate more susceptible to enzymatic hydrolysis than AFEX™ C1 conditions. Interestingly, the largest particle size of rice straw (5 cm) was found to be more digestible when compared to smaller particle sizes (2mm, 5mm and 2cm) under the given enzyme hydrolysis conditions. Viscosity measurements and imaging has been done to explain the phenomenon. To the best of our knowledge, this is the first time such a trend has been observed.

(626d) Efficient Bioconversion of Glucose/Xylose Mixtures for Ethanol Production Using a Novel Co-Culture System

Min Hea Kim[1], Meng Liang[1], Q. Peter He[2] and Jin Wang[1], (1)Chemical Engineering, Auburn University, Auburn, AL, (2)Chemical Engineering, Tuskegee University, Tuskegee, AL

Efficient bioconversion of glucose/xylose mixtures for ethanol production using a novel co-culture system

With the inevitable depletion of the world's conventional oil and gas supply, there has been an increasing worldwide interest in alternative sources of energy. In recent years, growing attention has been devoted to the conversion of lignocellulosic biomass into fuel ethanol, which offers great environmental and potential economic benefits over fossil fuel. The cellulosic biomass is cost-effective compared to low-cost petroleum; the cellulosic biomass is widely available at $40 per dry ton, and this price is equivalent to petroleum at $17.5 per barrel on both a mass and energy basis (Lynd et al., 1999). This potentially abundant lignocellulosic biomass contains fermentable sugars, mainly composed of glucose and xylose. It is worth noting that ethanol production from lignocellulosic hydrolysates in an economically feasible process requires complete utilization of both glucose and xylose. Specifically, the efficient utilization of xylose offers an opportunity to reduce the cost of producing ethanol by 25% (Aristidou and Penttila, 2000). The conventional yeast strain, *Saccharomyces cerevisiae*, is the most frequently used microorganism for ethanol production and it has both high ethanol tolerance and high fermentation yield and rate, but its inability to ferment xylose limits its use in ethanol production with lignocellulosic biomass. Although successful cycles of metabolic engineering have improved xylose utilization in recombinant *S. cerevisiae*, the ethanol production from xylose is still low (Madhavan et al., 2009; Bengtsson et al., 2009; Matsushika et al., 2008). The xylose-fermenting yeast, *Pichia stipitis* shows the highest native capacity for xylose fermentation of any known microbes and can convert both glucose and xylose into ethanol. However, the ethanol productivity of this strain is at least five to ten-folds lower than that obtained with *S. cerevisiae* during glucose fermentation (Alfenore et al., 2004; Agbogbo et al., 2006). Furthermore, *P. stipitis* shows a diauxic growth on the mixed substrates and has low ethanol tolerance, which inhibits its metabolism when ethanol concentration reaches above 30 g/L.

To achieve efficient utilization of glucose/xylose mixtures for ethanol production, most existing research has focused on developing a single recombinant strain which can ferment the mixed sugars derived from lignocelluloses to produce ethanol. In parallel with the ongoing development of the genetic recombinant strategy, co-culture strategy has been recognized as a promising and cost-effective way to co-ferment a mixture of glucose and xylose for ethanol production. However, efficient ethanol production by conventional co-culture systems faces several difficulties and the major ones are the competition for dissolved oxygen between the two microbes, the diauxic behavior of xylose-fermenting strain (i.e., they cannot use xylose in the presence of glucose) and their low ethanol tolerance (Gutierrez-River et al., 2012; Chandel et al., 2011; Li et al., 2011). In addition, very limited research has been done to investigate the dynamic properties of co-culture systems. Since the interactions between two microorganisms (glucose-fermenting and xylose-fermenting strains) are complex due to different assimilation pathways and regulatory mechanisms, it is important to develop new approaches to help understand and explore the dynamic behaviors of the co-culture system.

In this study, we designed and customized a novel bioreactor with an innovative fermentation scheme to offer a promising alternative of overcoming the difficulties of the existing co-culture systems. With this innovative fermentation scheme, the simultaneous complete utilization of both glucose and xylose was achieved. In addition, both diauxic phenomenon and oxygen deficiency of *P. stipitis*, were circumvented by providing the optimal dissolved oxygen required for efficient xylose fermentation. Also, the pseudo-continuous fermentation, i.e. continuous fermentation with cell retention, enhanced ethanol tolerance of both strains significantly by allowing the cells to adapt to a high ethanol concentration and environmental pressure. Specifically, the ethanol tolerance of *P. stipitis* (strain from pseudo-continuous fermentation) had increased 3.5-fold compared with unadapted strain of *P. stipitis* in the presence of 60g/L of ethanol. Furthermore, the dynamic properties of co-culture system were studied under pseudo-continuous fermentation to identify the optimal co-culture conditions by manipulating oxygen transfer rate ($mmolg^{-1}hr^{-1}$), substrate feed rate (ghr^{-1}), and substrate feed ratio (glucose/xylose ratio).

Reference

Agbogbo FK, Coward-Kelly G, Torry-Smith M, Wenger KS. (2006). Fermentation of glucose/xylose mixtures using *Pichia stipitis*. Process Biochem, 41, 2333-2336.

Alfenore S, Cameleyre X, Benbadis L, Bideaux, C, Uribelarrea JL, Goma G, Molina-Jouve C, Guillouet SE. (2004). Aeration strategy: a need for very high ethanol performance in *Saccharomyces cerevisiae* fed-batch process. Appl Microbiol Biotechnol, 63, 537-542.

Aristidou A, Penttila M. (2000). Metabolic engineering applications to renewable resource utilization, Curr Opin Biotechnol, 11, 187-198.

Bengttson O, Hahn-Hagerdal B, Gorwa-Grauslund ME. (2009). Xylose reductase from *Pichia stipitis* with altered coenzyme preference improves ethanolic xylose fermentation by recombinant *Saccharomyces cerevisiae*. Biotechnol Biofuels, 2, 9.

Chandel AK, Singh OV, Narasy ML, and Rao V. (2011). Bioconversion of *Saccharum spontaneum* (wild sugarcane) hemicellulosic hydrolysate into ethanol by mono and co-cultures of *Pichia stipitis* NCIM3498 and thermotolerant Saccharomyces cerevisiae-VS_3, New Biotechnol, 28, 593-599

Gutierrez-Rivera B, Waliszewski-Kubiak K, Carvajal-Zarrabal O, and Aguilar-Uscanba MG. (2012). Conversion efficiency of glucose/xylose mixtures for ethanol production using *Saccharomyces cerevisiae* ITV01 and *Pichia stipitis* NRRL Y-7124. J Chem Technol Biotechnol, 87, 263-270.

Li Y, Park J, Shiroma R, and Tokuyasu K. (2011). Bioethanol production from rice straw by a sequential use of *Saccharomyces cerevisiae* and *Pichia stipitis* with heat inactivation of *Saccharomyces cerevisiae* prior to xylose fermentation, J Biosci Bioeng, 111, 682-686.

Lynd LR, Wyman CE, and Gerngross TU. (1999). Biocommodity Engineering. Biotechnol Prog, 15, 777-793.

Madhavan A, Tamalampudi S, Srivastava A, Fukuda H, Bisaria VS, and Kondo A. (2009). Alcholic fermentation of xylose and mixed sugars using recombinant Saccharomyces cerevisiae engineered for xylose utilization. Appl Microbiol Biotechnol, 82, 1037-1047

Matasushika A and Sawayama S. (2008). Efficient bioethanol production from xylose by recombinant *Saccharomyces cerevisiae* requires high activity of xylose reductase and moderate xylulokinase activity. J Biosci Bioeng, 106, 306-309.

(626e) Upgrading Biomass Derived Lignin for Hydrocarbon Production

Dhrubojyoti D. Laskar, Center for Bioproducts & Bioenergy and Biosystems Engimeering, Washington State University, Richland, WA, Melvin Tucker, National Biology Center, National Renewable Energy Lab, Golden, CO and **Bin Yang**, Center for Bioproducts & Bioenergy and Biosystems Engimeering, Washington State University , Richland, WA

The current vision of biofuels production undervalues lignin's potential to address the nation's high quality liquid fuel requirements. The utilization of waste lignin as a feedstock for conversions to hydrocarbons offers a significant opportunity for enhancing the overall operational efficiency, carbon conversion rate, economic viability, and sustainability of biofuels production. Despite the potential, the conversion of lignin to biofuels has proven to be challenging. Several solubilized and depolymerized lignin fractions with a range of structural features were separated from cellulosic biomass as intermediates for further conversion to hydrocarbons. The detailed release and fate of lignin and lignin mono/oligomers during hydrothermal pretreatment with or without additives (e.g. catalysts) was characterized under various conditions (e.g., temperature, time, flow rate etc.). Various advanced analytical techniques, such as 1D and 2D NMR, FTIR, GC/MS, and GPC, were applied to derive the associated chemical changes in lignin, and lignin-carbohydrate complexes during the various pretreatment and depolymerization processes. The strategies for developing catalysts that can both selectively depolymerize the lignin framework and remove oxygen by hydrodeoxygenation will be addressed in the presentation.

(626f) A New Energy Densification Approach to Convert Lipids to Hydrocarbons

Jonathan Dubinsky, Department of Civil Engineering, University of Coloraod Denver, Denver, CO and **Arunprakash T. Karunanithi**, Department of Civil Engineering, University of Colorado Denver, Denver, CO

We present a thermo chemical process to convert lipids to renewable diesel fuel. Several thermo-chemical processes, such as hydrodeoxygenation (HDO), have been previously proposed for energy densification of fuels from bio feedstocks. However, HDO is an energy intensive process and it requires high pressure hydrogen stream. The use of hydrogen gas in the conversion process has two major drawbacks 1) typical hydrogen production processes are based on fossil fuel and hence are non-renewable 2) significant safety issues related to storage and transportation of hydrogen gas. We present a decarboxylation based approach that does not require the use of gaseous hydrogen and hence is more sustainable, safe and cheap as compared to the current-state of the art thermo-chemical-catalytic processes. We show that this process produces hydrocarbons which can be further refined to produce renewable diesel fuel. A comparative life-cycle assessment of renewable diesel produced from a commercial renewable diesel process and our proposed approach will also be presented.

(629a) IDEAS Minimization of Entropy Generation for Isothermal/Isobaric/Constant Density Reactor Networks with Bounded Utility Consumption

Paul Ghougassian, Chemical Engineering, UCLA, Los Angeles, CA and Vasilios Manousiouthakis, Chemical & Biomolecular Engineering Department,, University of California Los Angeles, Los Angeles, Los Angeles, CA

In this paper, a methodology for the globally optimal synthesis of isothermal/isobaric/constant density reactor networks featuring minimum entropy generation and bounded utility consumption is presented. The proposed methodology employs the Infinite DimEnsionAl State-space (IDEAS) conceptual framework, which is shown to be applicable to the problem under consideration. The IDEAS framework considers all possible reactor units, and all possible mixing and splitting interconnections among them. The resulting mathematical formulation is an infinite-dimensional program with linear constraints and a possibly non-linear nonconvex objective function to be minimized. It will be shown mathematically that, under certain conditions, entropy generation and utility consumption are functions of only the inlet and outlet stream compositions and flowrates and do not depend on the reactor network structure, as long as there exists a network able to deliver the considered outlets from the known inlets. This theoretical result provides the foundation for a graphical method that can identify the global optimum of the considered optimization, by first quantifying the reactor network's Attainable Region (AR) and then depicting the behavior of the entropy generation and utility consumption functions within the AR. The proposed method is demonstrated on a case study. Finally, conclusions are drawn.

(629b) A New Approach to the Use of Exergy Analysis in Synthesis and Design - Focus On Low Temperature Processes

Truls Gundersen and Danahe Marmolejo-Correa, Energy and Process Engineering, Norwegian University of Science and Technology, Trondheim, Norway

A new Approach to the Use of Exergy Analysis in Synthesis and Design – Focus on Low Temperature Processes

Truls Gundersen[1], Danahe Marmolejo Correa[2]

Department of Energy and Process Engineering, Norwegian University of Science and Technology, Kolbjoern Hejes v 1B; NO-7491, Trondheim, Norway.

[1]corresponding author: truls.gundersen@ntnu.no;

[2]danahe.marmolejo@gmail.com.

Abstract

This paper presents a summary of some recent discoveries and developments related to the use of Exergy Analysis for the synthesis and design of industrial production plants with special emphasis on sub-ambient processes. The key elements of the paper are (1) the importance of decomposing certain exergy forms into their exergy components, (2) the importance of decomposing thermo-mechanical exergy at ambient temperature, (3) the systematic definition of exergy sources and exergy sinks for use in exergy efficiencies, (4) the establishment of linear Exergy diagrams using basic thermodynamic equations and simplifying assumptions, (5) the concept of targeting for Exergy similar to Energy targeting in Pinch Analysis, and (6) the use of Exergy as a design tool in conceptual process development. Since Exergy is defined as the maximum work that can be produced when a system reaches equilibrium with its environment through reversible processes, the concept of Exergy is particularly useful for sub-ambient processes. Here, external cooling is provided by refrigeration cycles, where the cooling duty is produced by a sequence of expansion and compression. This requires work or mechanical energy, which is pure exergy. The paper will use a very small example to illustrate the basic concepts, and the Reverse Brayton process for liquefaction of natural gas as an industrial case study.

Background

Despite a rich literature on Exergy (a term that was "coined" by Rant in 1956) in the last decades, and the fact that Exergy is closely related to the 2nd Law of Thermodynamics (often formulated by the use of the logical counterpart Entropy), the use of Exergy in the process industries has been rather limited. One contributing factor to this situation can have been the lack of standardization in the field of Exergy. Different symbols and different names have been used across disciplines (such as chemical versus mechanical engineering) and across continents (such as Europe versus the US). This lack of standardization also includes issues of classification, reference conditions and exergy efficiency definitions. Another contributing factor can have been the fact that Exergy considerations in quite a few cases are in conflict with economic considerations. In the process industries, the bottom line is always economy, expressed either as maximum annual profit or minimum total annualized cost. Since, however, there are considerable uncertainties in both operating cost and investment cost, using measures such as Exergy at least has the advantage of developing a "sound" design from a thermodynamic point of view. For sub-ambient process, exergy losses translate into compression work, which again translates directly into economy.

When using the Exergy concept for sub-ambient processes such as liquefaction of natural gas (LNG) and separation of air by cryogenic separation (ASU), more limitations have been encountered (Marmolejo-Correa and Gundersen, 2011) in the available literature on Exergy Analysis that limit the ability to properly evaluate the quality of process designs. Some of the proposed exergy efficiencies are unable to account for the fact that sub-ambient processes utilize the option to switch between two components of thermo-mechanical exergy, i.e. the temperature based part and the pressure based part (Marmolejo-Correa and Gundersen, 2012a). By expansion in a turbine below ambient temperature, pressure based exergy can be transformed into temperature based exergy (refrigeration or cold exergy) and work. Similarly, compression can be used to "store" work (i.e. exergy) in the form of pressure based exergy. A very important part of this picture is the fact that temperature based exergy for a material stream as well as the exergy of heat exhibits a remarkable behavior below ambient temperature. The exergy of heat is always positive, and when passing ambient temperature from above to below, there is a discontinuity at ambient temperature. Temperature based exergy for a material stream is also always positive, and even though there is no discontinuity at ambient temperature, the exergy value of the stream increases more rapidly per degree Celsius or Kelvin when the temperature is reduced below ambient compared to increasing the temperature above ambient.

This is why decomposition of exergy into different components and to decompose exergy evaluations at ambient temperature are important for understanding the behavior of certain process equipment as well as for the development of exergy efficiencies that properly measures the quality of process designs.

Contributions and Conclusions

The paper describes some modest contributions to the field of Exergy Analysis with emphasis on sub-ambient processes. The issues of standardization, classification and decomposition were discussed in the Background section. Based on insight developed by studying the behavior of exergy components and certain pieces of equipment above and below ambient temperature, a new general exergy efficiency definition called Exergy Transfer Effectiveness (ETE) has been developed (Marmolejo-Correa and Gundersen, 2012c) that focus on exergy change and exergy transformation in processes. Changes in exergy, supply or removal of exergy and transformation of exergy are carefully divided into Exergy Sources and Exergy Sinks. A set of rules has been developed to make a proper classification into sources and sinks.

Existing Exergy diagrams (Linnhoff and Dhole, 1992) for overall processes have been developed from temperature vs. enthalpy diagrams in Pinch Analysis (i.e. from Composite and Grand Composite Curves) by replacing temperature with Carnot factor as the y-axis (i.e. Exergy Composite Curves and the Exergy Grand Composite Curve). These diagrams have a number of shortcomings. First, in their construction, a considerable number of points have to be calculated due to the nonlinear shape of the curves (the Carnot factor is a non-linear function of temperature). Then, in the interpretation of the curves, targets for maximum exergy recovery, minimum exergy losses, etc. are not readily available from the curves. Exergy losses can be measured in a cumbersome way geometrically by measuring the area between the Exergy Composite Curves.

Using basic thermodynamic equations and simplifying assumptions, new Exergetic Temperatures have been proposed (Marmolejo-Correa and Gundersen, 2012b) that can be used to develop (piece-wise) linear Exergy Diagrams that are (1) easy to construct from the process stream data, and (2) provide easily and explicitly various targets for Exergy such as maximum exergy recovery, minimum exergy losses, minimum exergy requirement, and minimum exergy rejection. These diagrams enable the use of Exergy Analysis in early stages of process design, while currently it is used as a post-design measure of quality. More specifically, the paper will discuss its use in combination with the ExPAnD methodology (Aspelund et al., 2007, and Wechsung et al., 2011).

There is considerable scope for increased use of Exergy in analysis, synthesis and design of processing plants; however, there is a lack of expertise in industry and a lack of suitable software for application in industrial environments. The fact that Exergy Analysis must be used with care, since reducing exergy losses caused by irreversibilities often is in conflict with investment cost minimization, may also have contributed to its rather limited industrial use in the past.

References

ADDIN EN.REFLIST

Aspelund, A., Berstad, D.O., Gundersen, T. (2007). An extended pinch analysis and design procedure utilizing pressure based exergy for subambient cooling. *Applied Thermal Engineering*, **27**(16), 2633-2649.

Linnhoff, B., Dhole, V.R. (1992). Shaftwork targets for low-temperature process design. *Chem. Engng. Sci.*, **47**(8), 2081-2091.

Marmolejo-Correa, D., Gundersen, T. (2011). Low temperature process design: Challenges and approaches for using exergy efficiencies, in Pistikopoulos, E.N. et al. (editors). Proceedings of the 21st European Symposium on Computer Aided Process Engineering - ESCAPE 21, Thessaloniki, Greece, vol. 2, pp. 1909-13.

Marmolejo-Correa, D., Gundersen, T. (2012a). A comparison of exergy efficiency definitions with focus on low temperature processes. In review for *Energy*.

Marmolejo-Correa, D., Gundersen, T. (2012b). A new graphical representation of exergy applied to low temperature process design, in Karimi I.A., Srinivasan, R. (editors). Proceedings of the 11th International Symposium on Process Systems Engineering - PSE 2012, Singapore.

Marmolejo-Correa, D., Gundersen, T. (2012c). Exergy transfer effectiveness for low temperature processes. To be submitted to *International Journal of Thermodynamics*.

Rant, Z. (1956). Exergy, a new word for technical work potential. *Forsch. Ing. Wes.*, **22**, 36-37 (in German).

Wechsung, A., Aspelund, A., Gundersen, T. (2011). Synthesis of Heat Exchanger Networks at Sub-ambient Conditions with Compression and Expansion of Process Streams. *AIChE Journal*, **57**(8), 2090-2108.

(629c) Retrofit of Heat Exchanger Networks Using Superstructure Modeling and Global Optimization

Sung Young Kim, Debora Faria and Miguel Bagajewicz, Chemical Engineering, University of Oklahoma, Norman, OK

Retrofit of Heat Exchanger Networks using superstructure modeling and Global Optimization

Sung Young Kim, Debora Faria and *Miguel Bagajewicz

School of Chemical Engineering and Material Science, University of Oklahoma

100 East Boyd Street, T-335 – Norman, OK 73019-0628 USA

** Corresponding Author*

Abstract

In this paper, we present a robust global optimization method to design heat exchanger networks using superstructures and a recently developed method of domain and imgae partitioning of nonconvex functions. In previous work, we applied this method to the Synheat model. When using superstructures bilinear terms are also used, so we extend the bound contraction model to include these. The procedure we propose uses an MILP lower bound constructed using domain/image partitioning (Faria, Kim and Bagajewicz, 2012) and bilinear decomposition (Faria and Bagajewicz, 2011a,b). Then a newly developed bound contraction procedure is applied and compared to branch and bound as well as branch and bound with bound contraction at each node. We finally solve the retrofit problem, by proposing that certain heat exchangers in the superstructure exist, and new exchangers can be added.

1. D. Faria and M. Bagajewicz. A New Approach for Global Optimization of a Class of MINLP Problems with Applications to Water Management and Pooling Problems. AIChE Journal (2011a).
2. D. Faria and M. Bagajewicz. Global Optimization of Water Management Problems Using Linear Relaxations and Bound Contraction Methods. Industrial and Engineering Chemistry Research. 50, 7, 3738–3753 (2011b).
3. D. Faria, S.Y. Kim and M. Bagajewicz. Global Optimization of Nonconvex MINLP Problems by Domain and Image Partitioning : Applications to Heat Exchanger Network. Computers and Chemical Engineering, Submitted (2012)

A NEW APPROACH TO HEAT EXCHANGER NETWORK SYNTHESIS PROBLEM

Abstract ID# 279571

Işıl KİRKİZOĞLU[1], *Sinan GÜREL, Haldun SÜRAL*

Industrial Engineering Department, Middle East Technical University, 06800, Ankara, TURKEY

Introduction

Heat exchanger networks (HENs) bring several fluid streams into their desired temperatures by using available heat in the process. Since water and energy that are wasted to produce utility streams cause environmental and economic costs, processing a fluid to be heated with the one to be cooled is better than producing utility streams for efficient usage of energy.

In this study, we present a new solution approach for the HEN synthesis problem. Our aim is to provide a complete network design coupled with a detailed equipment design for heat exchangers (HEs). The approach can generate all possible HE alternatives instead of structuring the HEN at the very beginning as usually done in the literature (see Yee and Grossmann, 1990). Alternatives are generated systematically by discretizing the inlet and outlet temperatures for hot and cold stream pairs due to heat load equalities for a given solution quality. A mixed-integer nonlinear programming (MINLP) model is solved to design each HE alternative in detail. The model decides on the number of shell-and-tube passes and tubes, diameter of shell, inner/outer tube diameters, and tube pattern while minimizing HE area or the total construction and operation cost.

The Yee-Grossmann approach (Yee and Grossmann, 1990), based on developing a MINLP formulation for a proposed superstructure of the design of network, dominates the HEN design literature since 1990 (see Furman and Sahinidis, 2002; Morar and Agachi, 2010; and references therein). Related studies that keep the superstructure mostly focus on reviewing viability of the formulation assumptions, improving the methodology for solving the MINLP, and/or extending the framework. Mizutani *et al.* (2003), an example of those who extend the Yee-Grossmann approach, combine detailed HE design with the HEN superstructure. Our study differentiates from previous studies in the literature because we produce a detailed design for every heat exchange network alternative and evaluate possible alternatives on a shortest-path network problem whose each path represents a HEN design alternative. Considering all thermo-physical and transport properties, the heat transfer coefficients are calculated for every stream at the environment temperature and pressure. One can also

[1] Correspondence address: isil.kirkizoglu@metu.edu.tr

control the solution quality by deciding on the minimum temperature difference and heat load between inlet and outlet streams as the main parameters.

Problem Formulation and Solution Approach

The HEN synthesis problem is to achieve maximum heat transfer between streams and to use utility streams minimally. The decisions are:

- Which cold and hot streams meet in each HE?
- How much heat will transfer each HE?
- What should be the design properties for each HE?
- How many HEs should be in the network?
- How should HEs be arranged in the network?

The main idea behind the solution approach is to generate all possible HE alternatives, each of which can be represented by a node on a network. We call a stream 'hot (cold) stream' if that would be cooled (heated) during the process. 'System inlet (target) temperature' refers to the temperature of a stream at the beginning (after being processed) in the system. Start and end nodes of the network represent the initial and target status of the streams, respectively. Arcs between node pairs are generated by considering the inlet and outlet temperatures and energy requirements of the streams and have associated costs. The minimum-cost path on the network gives the final HEN design (Figure-1).

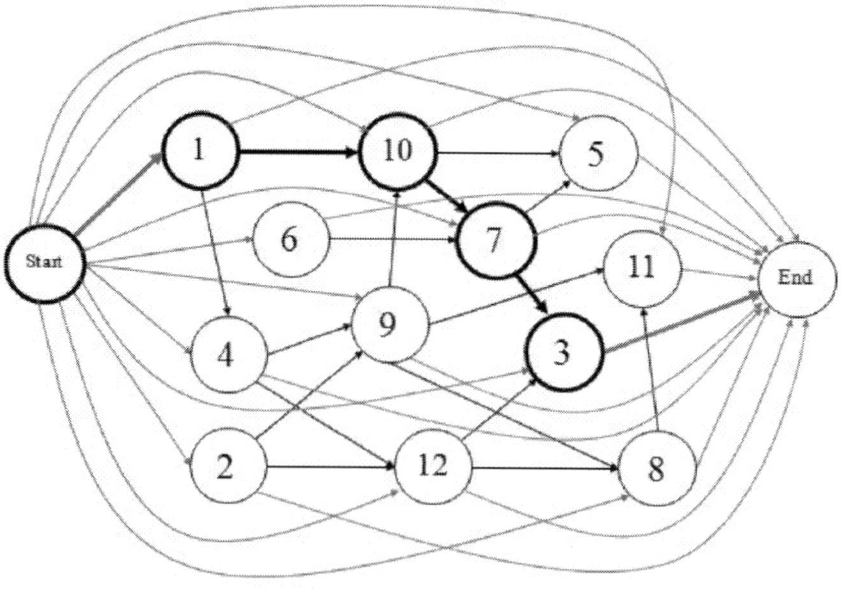

Figure-1. Representation of the network and the selected HEN shown in bold.

Note that nodes (i.e. alternatives HE designs) are generated for all hot and cold stream pairs and for every step change of their temperatures. For example, let us consider the example given in Yee and Grossmann (1990). There are two hot (H1, H2) and two cold (C1, C2) streams and their system inlet and target temperatures are given. If the minimum heat load difference ΔQ is selected as 300 kW, by discretization, the number of steps (β) between system inlet and target temperatures is set as represented in Table-1. After that, from the step heat loads, the step temperature differences (α) are calculated accordingly. Table-2 shows some possible alternatives of HEs for this example.

For every pair of hot and cold streams (i.e., alternative HE), we introduce a design error, γ, that defines the solution quality. Let us consider the example given in Table-1. Every step of C1 corresponds to 287.5 kW where each step of the other streams is 300 kW. For a HE alternative of H1 and C1 with one step change, the design error will be γ= (300 – 287.5)/300 = 4.167%, which implies that the energy balance in the HE is established with this amount of an error. Hence, γ reflects the quality of the solution; where lower γ values imply better solution quality.

At each node on the graph, we keep the temperature information for all streams so that we can keep track of them along the network. Figure-2 shows what data is stored at a node if there are m hot streams (i=1,…,m) and n cold streams (j=1,…,n).

Table-1. Data for a HEN synthesis problem instance

	System inlet (°C)	Target (°C)	Fcp (kW/°C)	Total T diff. (°C)	Total Q diff. (kW)	ΔQ (kW)	A (°C)	B	ΔQ - ΔQ^{max} (kW)
H1	170	60	30	110	3300	300	10	11	0
H2	150	30	15	120	1800	300	20	6	0
C1	20	135	20	115	2300	287.5	14.375	8	12.5
C2	80	140	40	60	2400	300	7.5	8	0

Table-2. Several HE alternatives generated for the example at Table-1.

Hot stream	Cold stream	Hot inlet	Cold inlet	Hot outlet	Cold outlet
H1	C1	11	6	5	0
H1	C1	9	4	7	2
H1	C2	8	5	3	0
H2	C1	6	8	0	2
H2	C2	4	7	1	4

The solution approach consists of 5 main steps summarized below:

a) ***Node Generation***: Generate all necessary nodes by pairing all possible discretized inlet and outlet temperatures for every stream in the system. The discretization of temperatures would be performed as explained above.

	Inlet temperatures of hot streams					Outlet temperatures of hot streams				
Hot	$T_1^{h(in)}$	...	$T_i^{h(in)}$	...	$T_I^{h(in)}$	$T_1^{h(out)}$	...	$T_i^{h(out)}$	...	$T_I^{h(out)}$
i										
	Inlet temperatures of cold streams					Outlet temperatures of cold streams				
Cold	$T_1^{c(in)}$	...	$T_j^{c(in)}$	...	$T_J^{c(in)}$	$T_1^{c(out)}$	...	$T_j^{c(out)}$	...	$T_J^{c(out)}$
j										

Figure-2. Node content of the network.

b) ***Eliminating Infeasible Nodes:*** Calculate the thermo-physical properties for every processed stream for each HE. Note that at a node (HE), the inlet temperature of a processed hot stream should be higher than its outlet temperature, and inverse is also true for a processed cold stream. For each unprocessed stream, the outlet temperature should be equal to its inlet temperature. Accordingly, eliminate the node if the temperature feasibility and energy balance is not satisfied. The heat load required by the cold fluid should be equal to the heat load generated by the hot fluid.

c) ***Individual HE Design:*** For every node in the network, solve a cost-minimizing MINLP model for the detailed design of shell-and-tube heat exchangers with TEMA standards. The MINLP model consists of necessary design (Geankoplis, 2003) and cost equations for each HE (i.e. node). If the model returns infeasible, eliminate that HE alternative.

d) ***Connecting Nodes***

i. *Dummy nodes:* There are two dummy nodes (Start, End) in the network. Every node v should be directly connected to the start and end dummy nodes. If the inlet (outlet) temperatures of streams are not at their system inlet (target) temperatures, they are connected to the start (end) node by using hot/cold utilities (such as cold water for cooling and superheated steam for heating). The utility usage and HE costs are associated to the arc between nodes and dummy nodes. If the streams are at the system inlet temperature, the cost of an arc from the starting node to that node should be 0. Similarly, when the streams reach their target temperatures, the cost of an arc from that node to the ending node is zero.

ii. *Connecting HE Nodes:* For any two nodes v and w, if outlet temperatures of all streams in node v is equal to the inlet temperatures of all streams in node w, arc (v,w) is generated. The cost associated to that arc is the capital cost of HE in node w.

e) ***Shortest Path:*** Solve the shortest-path problem on the whole network to find the best HEN with detailed design.

Computational Results and Discussion

For implementation of our solution approach we use MATLAB. For solving the MINLP models, MATLAB calls GAMS/BARON solver. We solve the shortest-path problem by using the Dijkstra's algorithm. Tests are done with two different PC settings: Intel® Core™ 2 Duo, 1.30 GHz, 3.50 GB RAM and Intel® Core™ i5-2400, 3.10 GHz, 6.0 GB RAM.

Some preliminary experiments are performed for determining the solution parameters and their behavior. As expected, it is seen that better results are obtained when the step heat load difference is decreased (i.e., the number of steps is increased) for each stream. But generally, increasing number of steps results in computational difficulties due to technical limitations related with increasing CPU time and memory requirements. Consequently, the number of steps for each stream should be arranged by taking this trade-off into account.

To test the solution approach, 12 test instances from the literature are solved. As it is explained earlier, our solution approach includes detailed design for every HE in the network. However, the test instances available in literature that solve for detailed HE designs in a network setting are very limited. Therefore, we also tested our approach on problems without detailed design, as well.

The approach we suggest is designed to handle all the physical properties of all streams, namely specific heats, thermal conductivities, viscosities and densities, at their environment temperature. In order to be able to identify and use all these properties at the substance's related temperature level, the composition of each stream (i.e., molar or mass fractions of substances within the stream) should be known. However, the composition of streams is usually not provided for the test instances in the literature. Therefore, we use the average values given for the physical properties. In addition, for the case that no detailed design is included, the overall heat transfer coefficients are also provided for the streams, which are used as an input to the model, rather than a decision variable.

Unfortunatelly, every instance has a different cost function (both capital investment and utility costs), overall heat transfer coefficient calculations (for instances that do not consider detailed

design of every HE), and utility environments. These changes are implemented for every instance. More experiments are done for some instances in order to find better solutions. Such experiments are developed by decreasing heat load difference ΔQ.
In Table-3, the comparison of our results that are obtained with the current best solution setup (discretization configuration) with the results in literature is given. The properties of the instances are also shown. The number of hot and cold streams explains the problem size. The stated notes show which features are allowed in the literature for the given instance.

The HENS problem has a multiobjective aspect. The heat exchange area, hence the capital cost of HEs, and the utility usage are tried to be minimized. This is why different measures can be compared to show the performance of the solution approach. For the comparison of results, the performance measures are the number of process and utility HEs and their areas, total heat exchange area, operating cost (cost of utility usage), total capital cost and the total cost. Percentage improvements are given for these measures. Positive values show the enhancement in results.

According to the results given in Table-3, our solution approach performs better for the cases where a detailed design is done. Heat exchange areas are mostly improved. Increase in usage of utilities are due to the discretization effect. Of course these results can be improved by increasing the number of steps for each stream, thus a greater portion of the solution space is explored. Note that stream splitting and placing utility HEs between process HEs as done by Ponce-Ortega et al. (2010) are not allowed in our approach. Even so, it is also seen from the results that the performance of the solution approach is comparable for the instances where detailed design is not considered.

Conclusion

In this study, we suggest a new solution approach to the heat exchanger network problem with detailed equipment design. By its nature, it is a complex and hard-to solve problem. Our approach solves the problem using discretization. It provides us to easily change and add features to the equipment design part without a need to change the entire HEN solution procedure.

Our approach is flexible and successfully finds the required number of heat exchangers and their connections. Note that the decision maker can change the parameter ΔQ to manage the HEN synthesis achieved at the end. As many solutions for HEN are created, decision makers can select any other design among these solutions according to his/her multiple preferences.

References

Furman, K.C. and Sahinidis, N.V. (2002). A critical review and annotated bibliography for heat exchanger network synthesis in the 20th century. *Ind. Eng. Chem. Res.*, 41(10), 2335-2370.

Geankoplis, C.J. (2003). Transport Processes and Separation Process Principles (4th ed.). New Jersey: Prentice Hall.

Gupta, A. and Ghosh, P. (2010). "A randomized algorithm for the efficient synthesis of heat exchanger networks". *Comput. Chem. Eng.*, 34(10), 1632-1639.

Huang, S.Y. and Chang, C.T. (2012). "An improved mathematical programming model for generating optimal heat-exchanger network (HEN) designs". *Ind. Eng. Chem. Res.*, 51(8), 3508-3515.

Mizutani, F.T., Pessoa, F.L.P., Queiroz, E.M., Hauan, S. and Grossmann, I.E. (2003), "Mathematical programming model for heat exchanger network synthesis including detailed heat-exchanger designs. 2. Network synthesis", *Ind. Eng. Chem. Res.*, 42(17), 4019-4027.

Morar, M. and Agachi, P.S. (2010). Review: Important contributions in development and improvement of the heat integration techniques. *Comput. Chem. Eng.*, 34(8), 1171-1179.

Ponce-Ortega, J.M., Serna-Gonzalez, M. and Jimenez-Gutierrez, A. (2007). "Heat exchanger network synthesis including detailed heat exchanger design using genetic algorithms". *Ind. Eng. Chem. Res.*, 46(25), 8767-8780.

Ponce-Ortega, J.M., Serna-Gonzalez, M. and Jimenez-Gutierrez, A. (2010). "Synthesis of heat exchanger networks with optimal placement of multiple utilities". *Ind. Eng. Chem. Res.*, 49(6), 2849-2856.

Turton, R., Bailie, R.C., Whiting, W.B. and Shaeiwitz, J.A. (2003). Analysis, Synthesis, and Design of Chemical Processes (2nd ed.). New Jersey: Prentice Hall.

Yee, T.F. and Grossmann, I.E. (1990). Simultaneous optimization models for heat integration – II. Heat exchanger network synthesis. *Comput. Chem. Eng.*, 14(10), 1165-1184.

Table-3. Computational results – % Improvement

Instance	Size	Notes	Current best solution setting			# of Process HEs	# of Utility HEs	Total Process HE Area (%)	Total Utility HE Area (%)	Total Area (%)	Total Utility (kW)	Total Capital Cost (%)	Utility Operating Cost (%)	Total Annual Cost (%)
			Hot streams	Cold streams	ΔQ (kW)									
HC-E1[a](R1)	2H2C	SS	14-22	18-10	200	same	same	-	-	8.90	-4.24	4.58	-9.47	-2.25
HC-E2[a] (R1)	1H2C	SS	70	26-40	30	same	1	-	-	13.46	-300.0	-3.62	-750.00	-40.70
HC-E3[a] (R1)	1H2C	SS	22	18-4	90	same	2	-	-	17.11	-	-8.89	-	-17.49
HC-E1[a] (R2)*	2H2C	SS	14-22	18-10	200	same	-1	-	-	8.90	-3.63	5.85	-8.38	-1.05
HC-E2[a] (R2)*	1H2C	SS	70	26-40	30	same	same	-	-	2.18	-3.45	10.03	-3.91	6.26%
HC-E3[a] (R2)*	1H2C	SS	22	18-4	90	same	same	-	-	-2.48	13.46	0.81	14.10	1.92%
YG-E1[b]	2H2C	-	33-18	23-24	100	same	1	-165.3	-6.26	-	62.33	-71.81	39.33	-13.98
YG-E2[b]	2H2C	-	12-16	18-15	150	-1	same	46.01	-107.24	-	-110.7	19.37	-87.95	6.92%
YG-E3[b]	5H1C	SS	9-3-5-3-6	55	120	same	1	41.01	-16.61	-	-21.17	14.35	-12.68	-9.88
PO-E2[c]	2H1C	MU - SS	50-47	75	16	-1	-1	-5.80	22.86	-	8.19	43.18	-124.80	-3.77
PO-E3[c]	2H3C	MU - SS	9-14	12-6-6	1150	2	-1	6.50	2.31	-	-6.45	-0.50	-21.86	-11.56
MG-E1[d]	2H2C	DD	4-10	24-5	100	same	same	67.24	0.00	-	0.00	10.12	0.00	0.59
4S1[e]	2H2C	DD	4-8	9-4	300	-3	same	65.08	-	-	-32.74	87.83	-46.24	10.51
Chp13-E2[f]	3H3C	-	12-2-5	5-8-6	100	-2	same	11.41	-4.96	-	-26.67	19.33	-3.81	13.14
5SP[g]	3H2C	SS	10-2-7	5-15	19	-2	2	-	-	-	-39.70	7.74	-38.63	-6.55

[a]: Huang & Chang, 2012
[b]: Yee & Grossmann, 1990
[c]: Ponce-Ortega et al., 2010
[d]: Mizutani et al., 2003
[e]: Ponce-Ortega et al., 2007
[f]: Turton et al., 2003
[g]: Gupta & Ghosh, 2010

*Result by using Yee and Grossmann (1990) methodology

SS: Stream splitting allowed
DD: Detailed design
MU: Multiutility + intermediate utility HE

(629e) Analysis of Different Sampling Methods for the Simulation-Based Optimization Approach to Model Feedstock Development for Chemical Process Industry

Ismail Fahmi, Chemical Engineering, The University of Tulsa, Tulsa, OK and Selen Cremaschi, Department of Chemical Engineering, University of Tulsa, Tulsa, OK

Analysis of Different Sampling Methods for the Simulation-based Optimization Approach to Model Feedstock Development for Chemical Process Industry

Ismail Fahmi, Selen Cremaschi

Department of Chemical Engineering, The University of Tulsa, 800 South Tucker Drive, Tulsa, Oklahoma 74104, USA

10C04: Energy and Sustainability in Operations

Abstract

Chemical Process Industry (CPI) has been highly dependent on the fossil-based materials as the feedstock. As the reserve for this non-renewable resource depletes, alternative resources need to be discovered. Among many options, biomass is considered very promising because it is renewable, locally available, and abundant. However, incorporating biomass as a feedstock for CPI requires significant amounts of investments for both research and development (R&D) and production capacity expansions. How these investments will shape the evolution of the biomass to commodity chemicals (BTCC) system should be investigated. Moreover, the uncertain decision-dependent endogenous technology evolutions of the BTCC system complicate the analysis. Fahmi and Cremaschi [1] have presented a prototype simulation-based optimization (SIMOPT) approach to study the feedstock development for CPI. The SIMOPT framework explained in [1] uses the concept of generating multiple unique timelines to represent various controlled evolution of the system under uncertainties. The SIMOPT framework [1] uses deterministic optimization with stochastic simulation in an integrated fashion to study the impact of uncertainties on the system performance. Deterministic optimization is used to produce the optimum decision set for the system at its current state. The optimization formulation for the BTCC investment planning used in the SIMOPT framework is explained in detail in [2]. Stochastic simulation is used to predict the system performance under uncertainties. In a single SIMOPT run, the simulation proceeds through the time steps based on the decision set determined by the deterministic optimization. Because of the accumulated differences between the expected and the realized behavior of the system, the system performance under the original decisions may no longer be applicable or feasible at one point in time. This condition is called a trigger event. In case of a trigger event, the simulation is halted and deterministic optimization is recalled to determine a new set of optimal decisions for the remainder of the timeline. The control is passed back to simulation along with the new set of decisions. The iteration between the deterministic optimization and stochastic simulation continues until the final time step. After the completion of the final step, a unique controlled evolution of the system performance under uncertainties (i.e., one unique timeline) is generated. Multiple unique timelines are produced because of different realizations of the uncertain parameters. To have a thorough analysis of the BTCC system performance accounting for different uncertainties, a statistically significant number of timelines are required, i.e. the quality of the results depends on the quality of the uncertain variables' space coverage. This can simply be done by gathering a virtually infinite number of timelines. However, such an approach would be prohibitively expensive in computational costs. Therefore, the minimum number of timelines that will achieve preset statistical significance in the results should be determined. In this work, a systematic analysis using different sampling methods to cover the uncertain parameters space of the BTCC investment planning problem is presented. The sampling methods considered in this work are Monte Carlo Sampling, Latin Hypercube Sampling, Univariate Dimension Reduction, Hammersley Sampling, and Halton Sampling. The performance metrics used for the comparison of these methods are the stabilization of four statistical moments (i.e., mean, variance, skewness, and kurtosis) of the objective and the overall objective distribution shape using the Smirnov test. A BTCC system of ethylene production is presented as the case study.

Keywords: BTCC investment planning, simulation-based optimization, sampling methods comparison

References:

[1] I. Fahmi and S. Cremaschi, "A Prototype Simulation-based Optimization Approach to Model Feedstock Development for Chemical Process Industry," *22nd European Symposium on Computer Aided Process Engineering,* 2012.

[2] I. Fahmi and S. Cremaschi, "Stage-gate Representation of Feedstock Development for Chemical Process Industry," *Foundations of Computer-Aided Process Operations 2012,* vol. , 2011.

(629f) Strategic Model with Transmission Network Design for the Energy Supply Chain Optimization in Petroleum Industries

Ariel Uribe[1], Ivan Ordoñez Sr.[1], Raul Rodriguez Peña Sr.[2] and Omar J. Guerra[3], (1)Colombian Petroleum Institute, ECOPETROL S.A., Piedecuesta, Colombia, (2)Colombian Petroleum Institute (ICP), ECOPETROL S.A, Piedecuesta, Colombia, (3)Colombian Petroleum Institute (ICP), ECOPETROL S.A., Piedecuesta, Colombia

Strategic model with transmission network design for the energy supply chain optimization in petroleum industries

Nowadays, energy management has become an increasingly important issue in each consumption sector (commercial, industrial, residential and transportation sectors). It is basically because energy plays an important role in every human activity and the need for more environment friendly energy resources (less greenhouse gas (GHG) emissions) like biomass, sun and wind power, etc has increased. In the particular case of oil and gas supply chain (production, transformation and transportation), the consumption of energy is estimated to approximately 10 % of gross oil and gas production (IPIECA 2007). About 90 % (48 % natural gas and 42 % oil) of the primary energy used by this industry takes the form of oil and gas, as supplies are available on site and are typically the cheapest source of energy. Some of the oil and gas used directly by the industry is transformed into electricity and heat, especially in the case of oil refineries and other large facilities. In total, about 10 % of the electricity and heat consumed by the industry is supplied from the grid (IPIECA 2007).

In this work, a strategic model for the energy supply chain design and optimization in oil industries is developed. The model considers GHG emissions, renewable energy sources energy efficiency, generation based on Gas, electricity purchase in a highly dependable market of the hydropower (64%), reliability and security constraints in the electric energy supply. The model was applied for the energy planning in an oil field production facility (GEC) at ECOPETROL. This facility include: Apiay field (4 active production fields and 48 oil wells), Castilla field (3 active fields and 126 oil wells), Chichimene field (2 active fields and 56 oil wells), Libertad field (3 active fields and 5 oil wells), Suria field (9 active field and 44 oil wells).

Due to an increased in oil production at GEC, an expansion plan in the energy supply system at GEC should be consider, this expansion plan include energy generation, transmission, distribution and final users. In the current situation at GEC, most of the energy requirements are provided by self-generation with two thermal plants (TermoOcoa (47 Mw) and TermoSuria (47 Mw)). The remaining of the energy requirements is purchased from an external energy supply system SIN (Interconnected National System) at 115 kV. The energy transmission take place at 115 kV with two distribution systems 115/34,5 kV – CDC and CDS. The distribution system for oil fields, facilities and oil Wells is at 34,5 kV.

The analysis is performed for a planning horizon of 5 -10 years, in the analyzed scenarios has an estimated growth of the demand from 120 Mw in 2012 to almost 220Mw in 2015, this huge growth is driven by an intensive campaign of development of all areas at GEC, which is estimated to grow from a production of 147 BOPD in 2012 to more than 300 kbopd in the next years, with an associated production of water slightly above from 400 kbwpd. For the energy supply the combining of different sources was analyzed: SIN and Self-generation (gas, oil, deasphalted, and other renewable alternative sources), in order to ensure energy supply required with an adequate reliability of the system so timely and cost-competitive, thus ensuring the expected volumes.

[3] IPIECA. 2007. "The global oil and gas industry association for environmental and social issues." (http://www.ipieca.org/sites/default/files/publications/Saving_Energy.pdf).

(630a) Design, Analysis, Operation and Control of a Photovoltaic/PEM Fuel Cell/Nicd Battery Hybrid Renewable Energy System (HRES) for Urban Transit Applications

Zachary S. Whiteman[1], Piyush Bubna[2], Ajay K. Prasad[3] and Babatunde A. Ogunnaike[1], (1)Department of Chemical and Biomolecular Engineering, University of Delaware, Newark, DE, (2)Department of Mechanical Engineering, University of Delaware, Newark, DE, (3)Department of Mechanical Engineering, Univeristy of Delaware, Newark, DE

Design, Analysis, Operation and Control of a Photovoltaic/PEM Fuel Cell/NiCd Battery Hybrid Renewable Energy System (HRES) for Urban Transit Applications

Authors: Zachary S. Whiteman, Piyush Bubna, Ajay K. Prasad and Babatunde A. Ogunnaike

A hybrid renewable energy system (HRES), a system consisting of two or more individual renewable energy systems utilized simultaneously to meet an energy demand, potentially allows one to take advantage of complementary strengths of renewable resources in meeting energy demands in both stationary and mobile applications. Because they take advantage of multiple renewable resources, HRESs may represent a cheaper and more dependable alternative to stand-alone renewable energy systems. This presentation is concerned with a hybrid system for mobile applications, specifically, the combination of a roof-installed photovoltaic array with a PEM fuel cell/NiCd Battery bus currently operating shuttle routes on the University of Delaware campus. The primary objectives are: (i) to design the hybrid system (and an appropriate control system) to satisfy required energy demands effectively but with reduced hydrogen consumption; and (ii) to analyze the economic impact of the installation of a photovoltaic array on the PEM fuel cell/NiCd battery bus.

The study is based on actual solar irradiance data collected from a measurement device installed on the roof of the existing HRES bus which was incorporated into a Simulink model developed for the Photovoltaic/PEM fuel cell/NiCd Battery HRES. Control strategies for maintaining the desired state of charge of the NiCd battery while meeting the total energy demand of the bus were designed and implemented on the complete HRES model simulating real shuttle runs under various operating conditions. An economic analysis of the photovoltaic investment required for realizing the HRES process design was also carried out to determine the break-even period for photovoltaic array efficiencies and photovoltaic array surface areas ranging between 5% - 100% and $5m^2$ - $100m^2$ respectively. The presentation will highlight the energy management and operational performance of the control strategies as well as the economic implications of modifying the existing PEM fuel cell/NiCd HRES bus.

(630b) Issues On Energy Sustainability: Consumption, Efficiency, and Environmental Impact

Leisha Vance, Tarsha Eason and **Heriberto Cabezas**, Office of Research and Development, U.S. Environmental Protection Agency, Cincinnati, OH

One of the more important challenges in achieving sustainability is how to meet the energy consumption needs of a growing human population in the face of increasing economic prosperity. In this talk, we present a very simple but rigorous procedure for exploring the relationship between energy consumption, environmental impact, human population, gross domestic product, and energy efficiency. The underlying theory basically assumes that energy use increases with time in proportion to two factors: gross domestic product and human population. Other assumptions are that energy use has a non-zero impact on the environment and decreases only with energy efficiency. This approach allows us to pose and evaluate: (1) the limitations given the known behavior of societies and economies, practical technology, and the Laws of Thermodynamics and (2) possible pathways to sustainability. The goal is essentially to look for a way to meet expected energy consumption requirements without increasing the environmental impact beyond that seen in an appropriate reference year. We have chosen the year 1977 as our reference year because it is the year when the combined ecological footprint of the human population world-wide was equal to the biocapacity of the Earth. That is, the capacity of the Earth was able to sustain the global demand of human beings as measured by the ecological footprint. Although we use this metric as a benchmark, we are not suggesting that the ecological footprint is necessarily an ideal measure of environmental impact. Rather, we use it simply as a convenient means of establishing a reasonable reference point. Based on the aforementioned assumptions, basic calculus, and published values for energy use, gross domestic product, human population, and energy efficiency over time are used to explore energy scenarios. Because many of the sustainability issues associated with energy use, such as climate change, are global in nature, we begin with an analysis at the global scale. Given the expected growth in World gross domestic product as well as human population growth and energy use over the next 20+ years, our results indicate that the kinds of energy-use efficiencies needed to keep the environmental impact at or below the 1977 level seem unrealistic. From our calculations, energy-use efficiency would need to rise to approximately 78% by 2030 to maintain the 1977 impact level. Not only is this level nearly impossible to achieve technologically and thermodynamically, but the amount of energy consumed per unit of gross domestic product (and per capita) would need to cumulatively decrease by 1.5% to 2% per year beginning in 2012 for environmental impact of energy use not to exceed that of 1977. Again, this will not not an easy task. The situation, although quite grim, is not hopeless, and the feasibility of an integrated approach to managing the environmental impact of energy use using all of the aforementioned options will be further discussed.

(630c) Surrogate Model Based Optimal Synthesis of Solid Sorbent Carbon Capture Process

Zhihong Yuan[1], Alison Cozad[1], Nick Sahinidis[1] and David C. Miller[2], (1)Chemical Engineering, Carnegie Mellon University, Pittsburgh, PA, (2)U.S. Department Of Energy, National Energy Technology Laboratory, Morgantown, WV

Concerns about climate change have led countries to set ambitious targets for the reduction of anthropogenic greenhouse gas emissions and a number of initiatives have been set up to reduce CO2 emissions[1]. Carbon capture, utilization, and storage technology is an essential route to achieving a meaningful reduction of global CO2 emissions in the context of continued fossil fuel use in the power sector. For example, the U.S. Department of Energy (DOE) has initiated the Carbon Capture Simulation Initiative (CCSI) to accelerate the development of carbon capture technology. Solid sorbent based technologies are being used to demonstrate the capabilities of the computational tools being developed under this initiative.

Obviously, process synthesis, which has had a conspicuous impact on the development, design and operation of (petro) chemical processes[2], will play a vital role for the development of a cost-effective carbon capture process. In this study, a superstructure optimization formulation for the optimal synthesis of a carbon capture system has been established, various reactors including fluidized bed reactors and moving bed reactors together with different potential topologies have been included in the optimization formulation. Since detailed first principle models of the reactors are computationally intractable for large scale superstructure optimization, we have developed the Automated Learning of Algebraic Models for Optimization (ALAMO) approach to generate a set of low complexity algebraic surrogate models of different reactors to reduce the computational burden associated with reactor models developed in commercial process simulation packages.

Through solving the above formulated mixed integer nonlinear optimization problem which aims at minimizing the increase in cost of electricity due to capture while achieving 90% capture target the potential equipment configurations, interconnections and design/operation levels of the selected equipment can be identified. This presentation will demonstrate the application of this superstructure-based process synthesis approach for a solid sorbent-based carbon capture system.

References

1. MacDowell N, Adjiman CS, Shah N. An overview of CO2 capture technologies. *Energy Environ Sci.* 2010; 3; 1645-1669.
2. Westerberg AW. A retrospective on design and process synthesis. *Comput Chem Eng. 2004; 28: 447-458.*

(630d) Modeling of Process Steam Consumption in Chemical Batch Plants Based On Standard Production Documentation

Cecilia Pereira[1], Stavros Papadokonstantakis[2] and Konrad Hungerbühler[2], (1)Chemistry and Applied Biosciences, Institute for Chemical- and Bioengineering, Swiss Federal Institute of Technology, Zurich (ETHZ), Zürich, Switzerland, (2)Chemistry and Applied Biosciences, Institute for Chemical- and Bioengineering, Swiss Federal Institute of Technology, Zurich (ETHZ), Zurich, Switzerland

Minimization of energy consumption in the chemical industry is one of the key principles of green chemistry. Although traditionally most efforts on energy saving studies have been done for continuous production processes, batch processes have recently become the focus of many research projects [1-3]. However, a shortcut modeling approach for screening of energy consumption in multipurpose batch plants, based on a simple and general data extraction procedure from process documentation is missing. In this work a new methodology for modeling of steam consumption in multipurpose chemical batch plants based on standard process documentation, rules of thumb, expert opinion, classical thermodynamics, and some model parameterization according to Bieler et al [3] is introduced. Two case studies using production data from multipurpose batch plants of two different chemical companies were carried out for the development and validation of this new shortcut modeling approach. Several reaction and work-up processes performed in different equipments were considered for the modeling of steam consumption, and the model results were validated against reference data in each case study. Additionally, the validation results were used to derive uncertainty intervals for the model outputs based on fuzzy set theory. It has been shown that the documentation based models proposed in this work provide acceptable estimations of steam consumption in multipurpose batch plants, and that the batch-to-batch variability of the steam consumption is well captured by the uncertainty intervals resulting from the validation procedure. The possibility of allocating steam consumption to individual processes or products in multipurpose batch plants by means of this new short cut modeling approach enables the identification of energy optimization potentials.

1. Andrej Szïjjarto, S.P., * Ulrich Fischer, and Konrad Hungerbühler, *Bottom-up Modeling of the Steam Consumption in Multipurpose Chemical Batch Plants Focusing on Identification of the Optimization Potential.* Industrial & Engineering Chemistry Research, 2008. **47**: p. 7323-7334.

2. Bieler, P.S., U. Fischer, and K. Hungerbuhler, *Modeling the energy consumption of chemical batch plants - Top-down approach.* Industrial & Engineering Chemistry Research, 2003. **42**(24): p. 6135-6144.

3. Bieler, P.S., U. Fischer, and K. Hungerbuhler, *Modeling the energy consumption of chemical batch plants: Bottom-up approach.* Industrial & Engineering Chemistry Research, 2004. **43**(24): p. 7785-7795.

(630e) A Shortcut Graphical and Spreadsheet Technique for Identifying Heat Exchange Networks

Russell F. Dunn, Chemical and Biomolecular Engineering , Vanderbilt University, Nashville, TN and Kenneth A. Debelak, Chemical and Biomolecular Engineering, Vanderbilt University, Nashville, TN

Incorporating heat integration within chemical processes continues to be a critical component for future sustainability of large chemical and polymer manufacturing companies. Numerous design methodologies have been developed for the synthesis of heat exchange networks (HENs) over the past three decades. Several of the methods previously developed are based on a two stage targeting approach. The first stage uses the hot and cold composite curves to identify the maximum heat integration possible for a selected group of streams to be cooled and streams to be heated. This approach is incorporated in several commercially available software packages such as Aspen Energy Analyzer. The second stage identifies the network of heat exchangers that meets the target identified in stage one. Several strategies have been employed for identifying the HEN in stage two including, but not limited to, heuristics, mathematical optimization, and superstructure approaches.

This paper will present a new graphical/spreadsheet technique and design strategy for identifying the HEN in stage two for large industrial problems. Advantages of this strategy will be provided along with several case studies that illustrate the effectiveness of this approach. Finally, experiences using this strategy in teaching undergraduate chemical engineering design courses will also be discussed.

Sustainabilty Assessment and Design of Centralized and Decentralized Energy Solutions

Healthy economic growth in developing countries combined with increasing population and affluence are resulting in significant increase in total energy consumption. At the same time, growing concern about environmental problems such as air pollution, global warming and depletion of fresh water resources is encouraging efforts for supplying this increasing energy demand in an environmentally friendly and sustainable manner. Depletion and price fluctuation of fossil fuels also mandates finding alternative technologies to satisfy this energy demand.

Especially in rural areas of developing countries, the burden of energy poverty reduces opportunities for enhancing their quality of life. This poverty is due to factors such as absence of electrification or inconsistent availability. Utilization of inefficient energy sources like cow dung, kerosene or biomass residues causes local environmental problems such as indoor smoke further triggering health problems and related healthcare costs. Consequently, many efforts are focused on introducing decentralized and cleaner energy solutions. Such solutions are also being adopted by the developed world due to the desire for enhanced sustainability.

Given these reasons, it is essential to design energy systems consisting of centralized and localized options which create the optimum energy mix to meet energy demand in a manner that is environmentally, socially and economically feasible.

In this study, we evaluate the environmental and energetic performances of different energy technologies utilizing thermodynamic methods such as ecological cumulative exergy consumption or emergy analysis and value chain scale life cycle assessment (LCA). These technologies are two centralized coal gasification technologies (conventional and calcium looping technologies) implemented with CO_2 capture, decentralized multi-crystalline solar cells, biomass gasification, and floating drum biogas power plants. Furthermore, using a multi-objective mixed integer programming (MIP) framework and setting environmental, economic, social and thermodynamic objectives, we plan to find optimum energy mix for the selected society. This framework is applied to Rampura, a village in Central India, which is representative of a developing society.

Based on our preliminary results based on data from Rampura, biogas power plant is most attractive from a life cycle thermodynamic point of view, followed by solar power plant, biomass gasifier and with least environmental performance of coal technologies.

For rural biogas plant, we calculate its renewability index to be around 75 % for only biogas production and slightly lower for electricity generation from biogas because of machinery and maintenance requirements. A quality corrected thermodynamic return on investment is calculated to be around 5 for both biogas and electricity. A sensitivity analysis identifies manure and brick (for construction) to be the most important resources. The photovoltaic solar power plant is found to be much less attractive than the biogas plant, with the most dominant inputs being resources for open ground mounting, inverter, maintenance inputs. The renewability index of solar panel is around 2.5 % and thermodynamic return on investment is slightly over 1. Emergy analysis results of biomass gasifier unit are similar to that of solar PV, just renewability index being slightly higher. Biomass, transportation of biomass to the plant site and preparation of biomass as feedstock are the dominant inputs to the system. Not surprisingly, coal gasification is least attractive with least percentage of renewability and similar energetic performance to biomass and solar systems from an environmental point of view, but may be essential for satisfying energy needs. For coal gasification processes; mining of coal, or coal input constitutes more than half of the emergy input to the systems.

LCA enables us to capture downstream effects of processes investigated which emergy analysis cannot evaluate. According to our results, solar PV utilizes least amount of land followed by centralized coal gasification technologies and biogas plant. Likewise biogas plant is most water utilizing system followed by solar PV and coal gasification systems. While conventional coal gasification technology has higher level of GWP compared to solar PV and biogas digester technologies, CLP has lowest GWP. This situation points out that improved coal gasification technologies implemented with CO_2 capture can significantly ease environmental burden of generating electricity from coal.

Life cycle assessment of biomass gasifier plant is the next steps in this work followed by the formulation of the optimization problem to identify the "optimum" energy mix while accounting for multiple objectives. This framework will be relevant to choosing combinations of centralized and decentralized options in developing and developed economies.

"Challenges and Opportunities for Enhancing Exergy Efficiency of Industrial Sectors in Developing Economies – Insight From India's Growth"

Sachin B. Jadhao1, Aniruddha B. Pandit1 and Bhavik R. Bakshi2,

1. Institute of Chemical Technology, Mumbai, India

2. William G. Lowrie Department of Chemical and Biomolecular Engineering The Ohio State University, Columbus, OH.

Abstract

Given that the development of emerging economies is invariably accompanied by an increase in energy and resource consumption, goals of sustainable development require a parallel increase in resource efficiency. This talk will focus on the trends and opportunities for enhancing the energy and material efficiency of industrial sectors in India. It uses the approach of exergy analysis since it has been found to be an appropriate measure for resource accounting. Studies on industrial sectors have been carried out for several economies and the resulting insight can be used for identifying opportunities for enhancing resource use and reducing environmental impact. The aim of this work is to understand the metabolism of Industry sector of India as it has evolved from a controlled to a more open free market economy, and its implications for sustainable development. We analyzed various industries in India on the basis of, resource consumption and environmental pollution. We estimate second law efficiencies for various industries in India, including Chemical, Iron and steel, Fertilizer, cement, sugar and others. We also estimate the efficiency of the India's industry sector as a whole. Trends of resource consumption and its conversion into useful work are analyzed using exergy analysis and the results are compared with those of other developing and developed economies. In addition, using techniques of network analysis, opportunities for enhancing exergy efficiency are identified.

Introduction

Industry sector in India is one of the major resource consuming sector. Figure 1 and figure 2 shows the fuel exergy and electric exergy consumed in Industry sector of India since 1970. The physical resources (fuel and non fuel minerals) consumption in Industry sector in India has been observed to increasing rapidly from the period of start if industrialization and economic liberalization in India. Among the various industries Inron and steel, fertilizer, Chemical and Cement are the major resource consuming industries. In general the industries are evaluated on the basis of First Law efficiency (calculated by comparing the energy content of the useful output with the energy content of the inputs, including materials), however the first law does not take in to account the quality which leds to the misleading conclusions [1].

Exergy efficiency is the ratio of potentially useful (exergy) output to potentially useful (exergy) input. [1] Second-law efficiency is usually less than first law efficiency.[1] Exergy efficiency captures the degree to which energy inputs are converted into useful work [3]. Exergy analysis provides insights not available via other methods, especially for identifying unit operations with the greatest potential for improvement [2].

A number of efforts have been undertaken to apply exergy analysis to various sectrors in the economy that includes the industrial sector. Since Industry sector accounts for major share in total resource consumption of many economies (particularly with respect to developing economies like India and China), it is important to examine the efficiency of resource utilization in this sector. Many studies in past have analysed the Industrial sector of various economies on the basis of exergy efficiency. These studies presented exergy accounts for individual process stages, in order to identify major losses and evaluate the potential for further technical improvements in the processes. Industrial sector in India have been compared and analyzed on the basis of efficiency or resource conversion in different industries. Most of the existing research on thermodynamic analysis of industry mainly accounts for fossil fuel consumption and its conversion to useful work. Few studies have been conducted using exergy assessment of industries that takes in to account fuel, non-fuel minerals resources. In this study we have analyzed the industries on the basis of fuel and non-fuel resource consumption and emissions.

Research method

In this work, conversion and utilization of all natural resources including energy and materials are described in terms of exergy flow in various Industries in India. Exergy utilization trends and efficiencies for various industries are analyzed and compared with those of other studies. Exergy losses in transformation of material resources and in conversion of various forms of energy into useful work are described in detail.

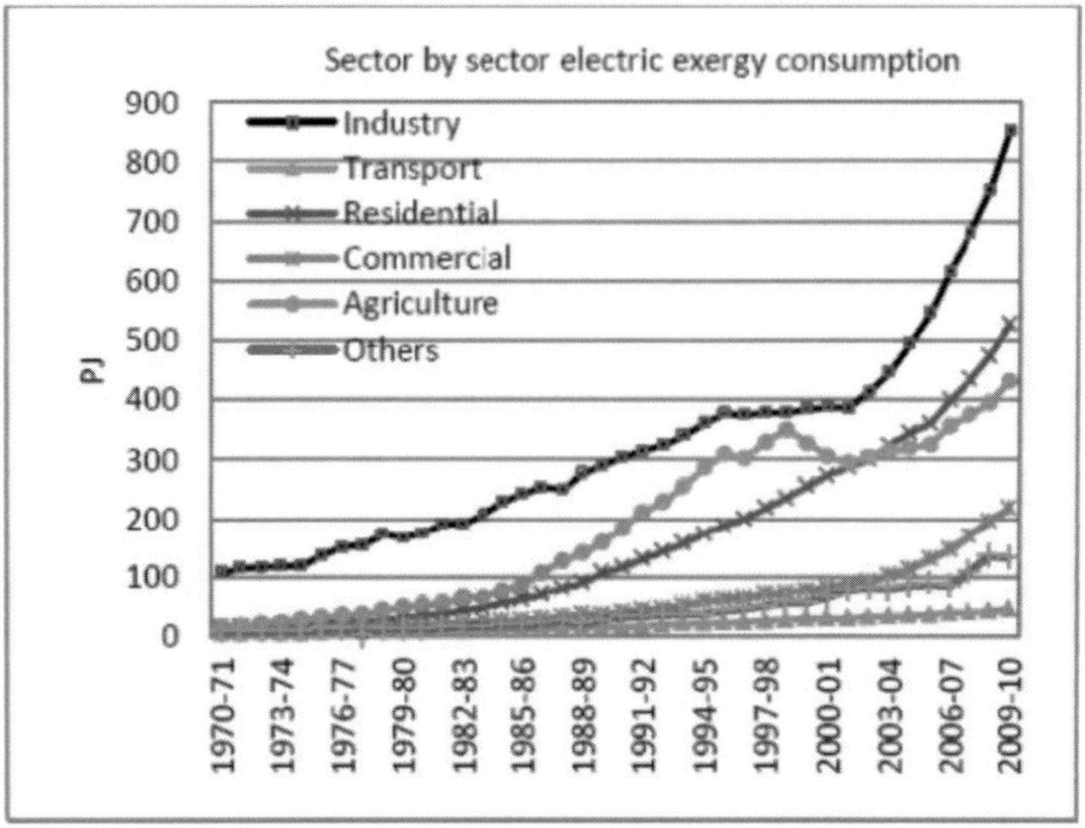

Figure 1 Electric exergy consumed in Industry sector of India

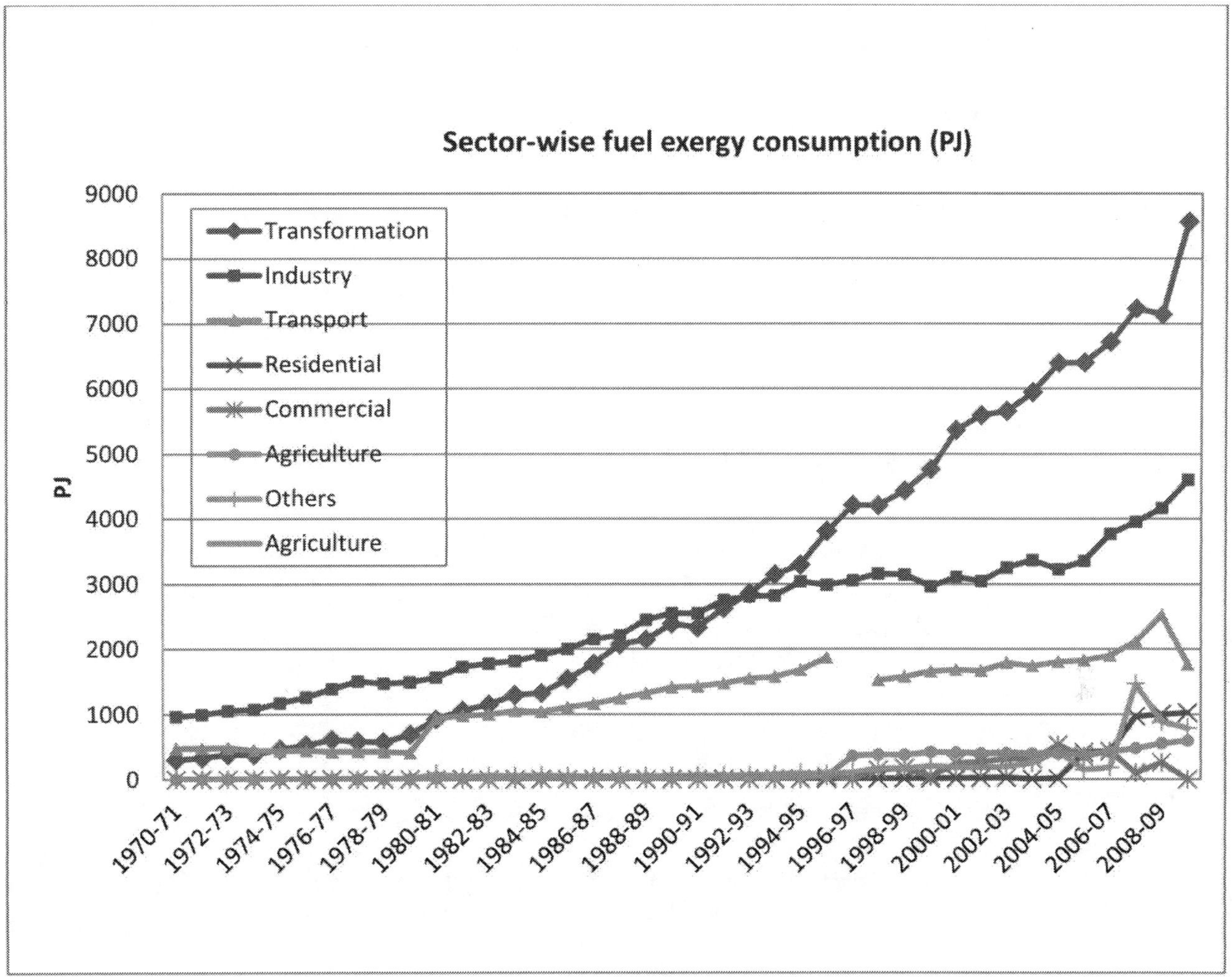

Figure 1 Fuel exergy consumed in Industry sector of India

Sample calculation of exergy efficiency for Iron & Steel Industry in India (2007-08)

Energy utilization in the industry sector is categorized according to the energy use as follows

Energy use in the industrial sector in India by type of Use (%)		
Use Category	Electrical	All Fuels
Mechanical Drive	65	0
Heating	28	100
Others (Assumed Lighting)	7	0
TOTAL	100	100

INPUTS

Total Electric energy consumed in the Iron & Steel Industry= 75.78 **PJ**

Total Electric exergy consumed in the Iron & Steel Industry= 75.78 **PJ**

Total Fuel energy consumed in the Iron & Steel Industry= 369.21 **PJ**

Total Fuel exergy consumed in the Iron & Steel Industry= 391.37 **PJ**

[I] Mechanical Drive

Since 65 % of the electric energy is used for mechanical drive application

$$Energy\ \&\ exergy\ input\ for\ machenical\ drive = 75.78 \times 0.65 = 49.26\ PJ$$

Mechanical drives are assumed to be 90 % efficient

$$Enegy\ \&\ exergy\ output\ for\ machenical\ drive = 49.26 \times 0.90 = 44.33\ PJ$$

[II] Heating

Process heating in the industry is subdivided in to three categories based on the temperature ranges as follows. The temperature range and the efficiency values are taken from the study of Dincer et al.

Process heating temperatures and efficiencies for the industrial sector			
Category	Temperature	Electrical	Fuel

	(° C)	(%)	(%)
Low	<121	100	65.5
Medium	121–399	90	60
High	>399	70	50

Breakdown of the energy used for each temperature range in the industry is referred from the Brown et al. study for 108 industrial processes

From this study the breakdown for Iron & steel Industry is given as follows

Breakdown of energy used for each temperature (%)					
Industry	T range	TO	Mean T (°C)	Electricity	Fuel
Iron and steel	Low	25	45	4.2	0
	High	25	983	95.8	100

Exergy efficiency of low temperature electric heating = 45+273 = 318°K and **η= 100 %**

$$\Psi_{e\,h\,Low} = 1 \times \left[1 - \left(\frac{298}{318}\right)\right] = 6.28\ \%$$

Exergy efficiency of high temperature electric heating: T = 983+273 = 1256 °K and **η= 70 %**

$$\Psi_{e\,h\,high} = 0.7 \times \left[1 - \left(\frac{298}{1256}\right)\right] = 53.39\%$$

Exergy efficiency of low temperature fuel heating: T = 45+273 = 318 °K and **η = 65 %**

$$\Psi_{f\,h\,Low} = 0.65 \times \left[1 - \left(\frac{298}{318}\right)\right] = 4.08\ \%$$

Exergy efficiency of high temperature fuel heating: T = 983+273 = 1256 °K and **η = 50 %**

$$\Psi_{f\,h\,high} = 0.5 \times \left[1 - \left(\frac{298}{1256}\right)\right] = 38.13\ \%$$

Energy efficiency of electric heating= efficiency×percentage of electric energy used

η_e = 100%×4.2%+90%×0%+70%×95.8%=71.26%

Energy efficiency of fuel heating= efficiency× percentage of fuel energy used in iron & steel industry

η_f= 65%× 0%+ 60%×0%+50%×100% = 50%

Exergy efficiency of electric heating= efficiency×% electric energy used in iron & steel industry

Ψ_e= 6.28%× 4.2%+ 53.39%×95.8% = 51.0%

Exergy efficiency of fuel heating= efficiency×% fuel energy used in iron & steel industry

Ψ_f= 7.2%× 0%+ 38.7%×100% = 38.7%

In Industry 28 % of the electricity is used for heating applications.

Energy Efficiency of heating

$$\eta_q = \frac{(Electric\ energy\ used \times efficiency + Fuel\ energy\ used\ \times efficienfy)}{(Electric\ energy\ + Fuel\ Energy\ input)}$$

$$\eta_q = \frac{(75.78 \times 0.28 \times 0.713 + 369.21\ \times 0.50)}{(75.78 \times 0.28\ + 369.21)}$$

$$\eta_q = 51.15\ \%$$

Exergy Efficiency of heating

$$\Psi_q = \frac{(Electric\ exergy\ input \times efficiency + Fuel\ exergy\ input\ \times efficienfy)}{(Electric\ exergy\ + Fuel\ Exergy\ input)}$$

$$\Psi_q = \frac{(75.78 \times 0.28 \times 0.514 + 391.37\ \times 0.387)}{(75.78 \times 0.28\ + 391.37)}$$

$$\Psi_q = 39.35\ \%$$

$$Energy\ Output\ of\ heating = Energy\ Input \times Energy\ Efficiency\ for\ heating$$

$$Energy\ Output\ of\ heating = 75.78 \times 0.28 \times 0.72 + 369.21 \times 0.50 = 199.88\ PJ$$

$$Exergy\ Output\ of\ heating = Exergy\ Input \times Exergy\ Efficiency\ for\ heating$$

$$Exergy\ Output\ of\ heating = 75.78 \times 0.28 \times 0.52 + 391.37 \times 0.387 = 162.35\ PJ$$

[III] Lighting

7 % of electricity is used for lighting application

$$Energy\ \&\ exergy\ input\ for\ lighting = 75.78 \times 0.07 = 5.30\ PJ$$

Lighting energy and exergy efficiencies are assumed as **18.5% and 17.1%**

$$Enegy\ output\ for\ lighting = 5.30 \times 0.185 = 0.98\ PJ$$

$$Exegy\ output\ for\ lighting = 5.30 \times 0.17 = 0.90\ PJ$$

$$Total\ Energy\ Input$$
$$= Aggrigate\ energy\ input\ for\ heating, lightening\ and\ mechanical\ drive$$

$$Total\ Energy\ Input = 369.21 + 75.78 = 444.99\ PJ$$

$$Total\ Exergy\ Input = 391.37 + 75.78 = 466.78\ PJ$$

$$Total\ Energy\ output = 199.88 + 44.33 + 0.98 = 245.19\ PJ$$

$$Total\ exergy\ output = 162.35 + 44.33 + 0.90 = 207.58\ PJ$$

$$\boldsymbol{Overall\ energy\ efficiency} = \frac{Total\ energy\ output}{Total\ energy\ input} \times 100$$

$$= \frac{245.19}{444.99} \times 100 = \mathbf{55.10}\ \%$$

$$\boldsymbol{Overall\ exergy\ efficiency} = \frac{Total\ exergy\ output}{Total\ exergy\ input} \times 100$$

$$= \frac{207.58}{466.78} \times 100 = \mathbf{44.34}\ \%$$

Results and discussion

This work provides insight into areas where major improvements are required to enhance resource efficiency in the various industries. Effects of factors such as increasing consumption and depletion of resource quality on overall efficiency are demonstrated.

REFERENCES

[1] R. U. Ayres, L. T. Peir, and G. V. Mendez. Exergy e-fficiency in industry where do we stand? Environ. Sci. Technol, 2011.

[2] B. R. Bakshi and G. F. Grubb. Appreciating the role of thermodynamics in lca improvement analysis via an application to titanium dioxide nanoparticles. *Environmental Science and Technology,* 45, 7, 3054–3061, 2011.

[3] E. Williams, B. Warr, and R. U. Ayres. Efficiency dilution: Long-term exergy conversion trends in japan. Environ. Sci. Technol., 42:49644970, 2008.

(644a) Life Cycle Assessment of Bio-Jet Fuel From Hydrothermal Liquefaction of Microalgae

Marie-Odile Fortier[1], Griffin W. Roberts[2], Belinda S.M. Sturm[1] and Susan M. Stagg-Williams[3], (1)Civil, Environmental & Architectural Engineering, University of Kansas, Lawrence, KS, (2)Department of Chemical & Petroleum Engineering, University of Kansas, Lawrence, KS, (3)Chemical and Petroleum Engineering, University of Kansas, Lawrence, KS

In 2011, ASTM standard D7566 was released for aviation fuel containing synthesized hydrocarbons such as those from algal feedstocks. Several flights on blends of conventional jet fuel and bio-jet fuel have been conducted on commercial airlines, including a Continental commercial flight on 40% algal bio-jet fuel. Government policy is also influencing the adoption of biofuels; the Energy Independence and Security Act of 2007 calls for the annual production of 36 billion gallons of renewable fuels by 2022. Twenty-one billion gallons of this target must be "advanced biofuel," which is defined as a renewable fuel that produces 50% less greenhouse gas emissions on a life-cycle basis than petroleum-derived fuels. At these early stages in the production of algal bio-jet fuel, a life cycle assessment (LCA) is necessary to ensure that bio-jet fuel produced from algal feedstocks does not result in higher life-cycle greenhouse gas emissions than conventional jet fuel, and to determine whether algal bio-jet fuel would qualify as an "advanced biofuel" in energy policy.

A "well-to-wake" LCA was performed using SimaPro software and the ISO 14040 and 14044 guidelines to compare algal bio-jet fuel to conventional jet fuel. The functional unit is 1 gigajoule of jet fuel. The main algal bio-jet fuel production system analysed consists of algal production in open ponds of wastewater effluent, harvesting through flocculation and sedimentation, dewatering by solar drying, transport to a refinery, hydrothermal liquefaction, upgrading, distillation, transport to an airport, and combustion in a PT6A jet engine. Pilot-scale data collected from experiments at the Lawrence Wastewater Treatment Plant were utilized for the algal production process step and data from lab-scale hydrothermal liquefaction of microalgae was used in constructing the life cycle inventory. Average distances from United States wastewater treatment plants to petroleum refineries, and from petroleum refineries to public use airports, were determined using ArcGIS for the transportation steps. The life cycle impact assessment (LCIA) model was the EPA Tool for the Reduction and Assessment of Chemical and other environmental Impacts 2 (TRACI 2). Two additional algal bio-jet fuel production systems were modeled to compare to hydrothermal liquefaction of algal biomass: hydrotreatment of extracted algal lipids, and gasification and Fischer-Tropsch synthesis using the entire algal biomass. To the author's knowledge, this is the first LCA which involves hydrothermal liquefaction of algae to produce biofuels. The results of this LCA were also compared to other published LCAs for bio-jet fuel from camelina and *Jatropha* feedstocks.

(644b) Evaluating Environmental Sustainability of Microalgal Biofuels: A Life Cycle Thermodynamic View

Matthew G. Borkowski, Civil & Environmental Engineering, Sustainability & Green Design, University of Pittsburgh, Pittsburgh, PA and Vikas Khanna, Department of Civil and Environmental Engineering, University of Pittsburgh, Pittsburgh, PA

Renewable fuels represent an important area for research due to increasing global energy demand, oil price and supply volatility, energy import reliance and security, as well as concerns over resource depletion and environmental impacts. Transportation fuels with reduced resource utilization and environmental impacts are of particular interest as the transportation sector represents 28% of overall U.S. primary energy consumption. Further motivation is provided by the Energy Independence and Security Act of 2007 (EISA), which mandates the domestic production and use of 21 billion gallons of cellulosic and advanced biofuels by 2022. Microalgae represent a promising feedstock for next-generation biofuels due to their high growth rates, high lipid content and rapid harvest cycle. In addition, algae production does not exert direct market pressure on food crop prices and can utilize marginal lands, resulting in reduced land-use change impacts. Much of microalgal biofuels research focuses on either open-pond or photobioreactor cultivation systems, coupled with a variety of harvesting and oil-extraction schemes, and culminating in the production of biodiesel via transesterification along with glycerin and large quantity of residual biomass. Less attention has been placed on alternative conversion technologies and fuel products, or on scalable uses for the residual biomass and other co-products. Renewable diesel (RD) is attractive as a fuel product because of its high energy density and Cetane number, as well as its good cold-flow properties and storage stability. Also referred to as "green diesel" or "hydrotreated vegetable oil," RD is produced by hydrotreating bio-oils to remove oxygen molecules and to fully saturate double bonds. Holistic, comparative evaluation of emerging algae-to-fuel systems considering resource consumption, emissions and their impact across the entire life cycle is critical for ensuring the long-term sustainability of emerging algae-based energy systems.

Much of the existing work on life cycle assessment (LCA) of microalgae has focused on quantifying the life-cycle greenhouse gas (GHG) emissions and the net energy balance for cultivation and biofuel production. Net energy analysis suffers from several limitations such as ignoring the differences in energy quality of different resources and accounting only for non-renewable energy sources. Such aggregation without attention to the quality of resources has led some researchers to question the utility of resulting metrics. Thermodynamic-based methods have been suggested to address the limitations of traditional LCA and to better quantify resource utilization in biofuel production systems. This work proposes to integrate traditional LCA with exergy-based thermodynamic analysis to better compare the sustainability of biofuel systems. Exergy, or available work, represents a common metric for evaluating the quality of energy carriers based not on gross energy content, but on *ability to do work*. Defined as the maximum work a system can produce if it is brought to thermal, chemical, and mechanical equilibrium with its reference environment, exergy represents the maximum *useful work potential* of a system. Exergy analysis also serves to aggregate and compare material flows with energy flows of differing qualities using one common unit (Joules). Recent studies into the exergetic life cycle assessment (ELCA) of biofuels consider the transesterification of waste oils to produce biodiesel, however no complete exergy-based assessment of the microalgal-derived biodiesel and renewable diesel production chains is available.

This study undertakes a well-to-pump traditional LCA and exergetic analysis of *renewable diesel* and *biodiesel* fuels derived from algae cultivated in open ponds. Traditional LCA methodology is first used to examine resource consumption and environmental impacts, focusing in particular on net-energy balance and GHG emissions of the two systems. Preliminary LCA results for both fuels indicate a net negative energy balance with fossil energy ratios ranging from 0.26-1.26 depending on cultivation variables, processing technologies and biomass composition. Fuel upgrade technology choice is found to have a comparatively small contribution to overall life-cycle energy use and emissions.

GHG emissions are comparable for both fuel conversion pathways, but vary strongly with the chosen co-product allocation methodologies.

For the exergetic LCA, various techniques and metrics are used at multiple scales, providing irreversibility-based measures for quantifying the consumption, use efficiency and depletion of material and energy resources. Quantifying exergy losses at the process scale indicates thermodynamic irreversibilities (entropy generation). This results in a simple exergetic efficiency factor for each process, providing insights for where to minimize exergy losses to improve efficiency across the entire production chain. A variety of hierarchical sustainability metrics spanning the detailed data level to aggregate metrics based on thermodynamics are developed in this work. This includes measures such as exergetic return on investment (ExROI), renewability index, exergy breeding factor, and life cycle exergetic efficiency for microalgal biofuels. Uncertainty and sensitivity to changes in key parameters including the biomass composition, co-product options, and alternative processing technologies will be discussed. Such insights are essential in determining which biofuel products and processes are most sustainable and best suited for development at industrial scale. The research will result in a more holistic life-cycle evaluation of the considered biofuels and will address trade-offs between process technologies, fuel products and co-products, and environmental impacts at both the life-cycle and process scales.

(644c) Well-to-Wheel Life Cycle Assessment for the Production of Biodiesel From Trap Grease

Megan Hums[1], Colin J. Stacy[1], Richard A. Cairncross[1] and Sabrina Spatari[2], (1)Department of Chemical and Biological Engineering, Drexel University, Philadelphia, PA, (2)Department of Civil, Architectural and Environmental Engineering, Drexel University, Philadelphia, PA

Biodiesel is a renewable fuel that can be produced from a variety of vegetable oils, animal fats, and waste greases. A bubble column reactor has been developed at Drexel University that efficiently converts brown grease (high acid number lipids extracted from trap grease) into biodiesel. The use of waste grease as a feedstock has lower cost and potentially lower environmental impacts than agricultural oils but requires additional pre-treatment and product purification. This paper uses Life Cycle Assessment techniques to compare the energy requirements and environmental impacts of biodiesel produced from trap grease to low-sulfur petroleum diesel. Well-to-wheel impacts are calculated using GREET, SimaPro7, and GaBi in conjunction with an ASPEN model of the biodiesel production process. Predicted environmental impacts are compared through greenhouse gas, nitrous oxide, and particulate emissions. The life cycle impacts of waste grease biodiesel are compared to traditional processes producing biodiesel from refined vegetable oils and waste cooking oils.

(644d) Life-Cycle Assessment of Lanzatech Waste Gas Conversion to Ethanol and Co-Products

Robert M. Handler, Michigan Technological University, Houghton, MI, David R. Shonnard, Chemical Engineering, Michigan Technological University, Houghton, MI, Evan M. Griffing, LanzaTech, Roselle,, IL and Ignasi Palou-Rivera, LanzaTech, Roselle, IL

A novel method of ethanol production has been developed by LanzaTech, relying on readily abundant carbon monoxide-containing gas feedstocks. Feedstock gases may be industrial waste gases (e.g., steel mill exhaust gas) or may be produced through gasification of biomass feedstocks. In this presentation we will give an overview of the LanzaTech process and summarize initial findings related to the environmental footprint of ethanol production in this manner. Greenhouse gas emissions and fossil energy demand are the key impacts discussed, but water use is also compared to other ethanol production routes. The importance of allocation method used to attribute environmental burdens between ethanol and chemical co-products will be highlighted, along with key assumptions concerning potential alternative use of feedstock gases. Bio-based feedstock results will be summarized and comapred to waste gas feedstocks, using common bio-based feedstocks and current information on gasification of biomass. Important considerations from the initial life cycle assessment of potential LanzaTech operating scenarios will be used to provide guidance about the proper role of this technology in the future alternative fuels marketplace.

Life Cycle System Analysis of Bio ethanol from Agro-Industrial Lignocellulosic biomass in India

Prasad V. Mandade[1], Bhavik R. Bakshi[2], and G.D.Yadav[1]
1. Institute of Chemical Technology, Mumbai, India
2. William G. Lowrie Department of Chemical and Biomolecular Engineering The Ohio State University, Columbus, OH.

Introduction

With recent volatility in oil prices and the mounting evidence of upcoming peaks in global oil and gas production, much attention has been given to the search for alternative energy sources and technologies.[1] Concerns about global warming, pollution, and sustainability are becoming increasingly widespread, many people are looking for ways to decrease the effect of human activities place on the environment.[2] Increasing demand for energy and the depleting petroleum resources stimulates interest in alternative fuels, especially liquid transportation fuels. As the energy strategy of a country aims at efficiency and security and to provide access which being environment friendly and achievement of an optimum mix of primary resources for energy generation.

The motivations for governments to aggressively pursue biofuel development are complex and multidimensional because it has potentials in addressing issues related to energy and food security, climate change, and rural development. Presently it represent minor portion of total energy consumption in the world but it is expected to grow sharply. One of the principal reasons for switching to biofuels is that they should have a negative GHG balance, i.e. offer emissions reductions compared to fossil fuels. If this criterion is met, biofuels will be a useful tool in the fight to avoid dangerous climate change.[3] Amid the soaring energy demand, price volatility and concerns on climate change caused by greenhouse gas emission, biofuel as an alternative energy source has received much attention in recent years among national and international policy makers and the business world, as well as the scientific community and the general public.[4]

Methodology

Data Sources and Challenges

Farming data has wide variation and sometimes complete data set is not available (partial availability of data). Large fluctuation in data is observed due to spatial and temporal variation hence it becomes difficult to analyse data. Following methodology is adapted to deal with the variation in data.

Boundary of analysis

For getting comparable data and results it is essential to have clear system boundaries. The focus in this study is on sustainability indicators of biofuels such as Energy return on investment (EROI), Greenhouse gas emissions, and resources consumed like land and water. Biofuels are assessed considering various phases of life cycle which includes feedstock production, transportation of feedstock to process site, process of conversion of feedstock to ethanol. Feedstock production input includes energy or energy-related material inputs such as fertilizers, herbicides, diesel fuel, electricity for irrigation, machinery and labour used. Process stage includes various inputs which are needed for conversion of biofeedstock to fuel such as Lime, Water, NH_3, Diesel, Cellulose, H_2SO_4 consumed in the process which was developed by NREL.[5]

Allocation

Allocation consists of assigning a fraction of inputs and by-products to main product and co-product based on some rule. Allocation has been one of the most discussed difficulties in LCA. We have considered three allocation methods based on energy, mass, and economic allocation in this case study. Each allocation method (mass, economic or energy) has its advantages and disadvantages and the choice of allocation procedure depends on the goal of study. Allocations were allocated between desired agricultural product and lignocellulosic biomass residue.

Metrics Analyzed

Numerous technologies have been proposed as partial solutions to our declining fossil energy stocks. There is a significant need for consistent metrics to compare the desirability of different technologies.[6] As production of biofuel continues to expand, especially given that this expansion is largely driven by policy, the environmental and social implications of these fuels have come under scrutiny, especially their life cycle greenhouse gas (GHG) emissions.[7] In assessing possible replacements for oil and natural gas, each alternative will present unique trade-offs between energy quantity, energy quality, and other inputs and impacts such as land, water, labour, and environmental health. Several criteria were used in the past to assess energy production technologies based on their absolute and relative yields and assorted costs.[1] Low greenhouse gas emissions in particular are a frequent measure of the desirability of an alternative technology.[8] For sustainable production of biofuel different objectives should go hand in hand with energy production.

Energy return on investment (EROI)

Energy return on investment (EROI) is the ratio of the energy delivered by a process to the energy used directly and indirectly in that process (4). Energy used in the supply process (E_p) refers to the energy consumed in processing the feedstock to fuel, and does not become a part of the product energy. Energy return on energy investment has been the most used metric to compare energetic competitiveness of alternative fuels and accounts for only fuel value. EROI is defined as:

$$\text{EROI}=\frac{\text{Quantity of energy supplied}}{\text{Quantity of energy used in the supply process}}$$

Life cycle GHG Emission per lit

The great attention that has been given to ethanol in the last few years as an important fuel for greenhouse emissions mitigation which is leading to many studies about Life cycle GHG emissions from different resources worldwide.[9] The analysis includes GHG emission balances of lignocellulosic bio-energy systems can thus be compared. Our strong dependence on fossil fuels results on the intensive use and consumption of petroleum derivatives which, combined with diminishing petroleum resources, causes environmental and political concerns. There is clear scientific evidence that emissions of greenhouse gases (GHGs), such as carbon dioxide (CO_2), methane (CH_4) and nitrous oxide (N_2O), arising from fossil fuel combustion and land-use change as a result of human activities, are perturbing the Earth's climate. The variety of processes leading to Bioenergy and the numerous assumptions on which bioenergy systems are based, outcomes of GHG balances can differ even for apparently similar systems. The main reasons for diverging results are: type of biomass sources, conversion technologies, input data, end-use technologies, allocation method, system boundaries, and reference energy system.[10] It has been proposed to calculate a so-called life-cycle-based greenhouse gas (GHG) indicator, which expresses the reduction of GHG emissions of a bio-based fuel chain in comparison with a fossil-based fuel chain.[11] The impact of the replacement of conventional transportation fuels by biofuels on greenhouse gas (GHG) emissions and fossil energy use is subject of fierce debate. The goal of this study is to investigate the GHG and energy balance of biofuels using a consistent set of system boundaries in time and space so that the results of each chain are comparable.[12]

Life cycle water Consumption per lit

Water use data often ignored while push on energy aspect of biofuel but it needs to assessed along with various metrics of the assessment that will help in decision making. Biofuels are the most popular renewable resource option for the transport sector. However, biofuel production at such a scale requires a considerable amount of water resources, and this water–energy nexus is rarely taken into account. Despite the increase of biofuel production and the amount of research about this issue, the potential effects of associated water consumption have not been rigorously analysed particularly in the Indian context.[13] The major uses of water are in both the agricultural and industrial phases of biofuel production. The overall water footprint associated with biofuels must recognize the impact of increased agricultural activity on water quality as well as water consumption.

Land use per lit

Many countries and regions in the world already experience pressure on land needed for converting existing cropland to or developing new land for biofuel production raises immediate concerns. The pressing question in this debate is whether or not the world has sufficient land available for biofuel crop cultivation that could be utilized without compromising the concerns of food vs fuel debate. Land resource is likely to become focus of increasing competition due to its use for different purposes .[1] As the land utilised for farming phase is comparatively larger than other phases of life cycle. It is often ignored land used for production of lignocellulosic crop residue as it has less economic value as compared to desired product. Though it may be undesired product due to increasing interest in lignocellulosic biofuel it needs to assess land utilised for production of crop residue. Present study attempts to allocate the land used for growth of crop residue on the basis of mass, energy and economic value. As the land used per lit of ethanol produced for the transportation of feedstock and process phase is comparatively negligible than land used for farming phase, only land used in farming phase is considered. Land used per lit of ethanol production for potential agro-industrial feedstocks are computed and effect of allocation method is analysed.

Result and discussion

The results focus on various aspects of sustainability of biofuels such as energy return on investment, life cycle greenhouse gas emissions, land use and water use per litre of ethanol produced. Our aim is to highlight comparative Sustainability of lignocellulosic biofuels in Indian context which will help in policy making of nation. This analysis considers Farming, Transportation and Processing phase for ethanol production.

Life cycle EROI

Life cycle energy return on investment of ethanol production from different lignocellulosic biomass feedstocks are compared among themselves. Error bar shows sensitivity of EROI which is due to variation in biomass yield in farming stage of life cycle. When total process energy is considered, energy return on investment for lingo-cellulosic ethanol ranges from 1.59 to 12.45 under different allocation approaches. Economic value allocation shows higher EROI than Mass and Energy Allocation due to comparatively lesser economic value allocation for biomass residue. Sorghum stalk shows highest life cycle energy return on investment amongst all lignocellulosic biomass feedstocks considered. Farming phase consumes major portion of life cycle energy consumption.

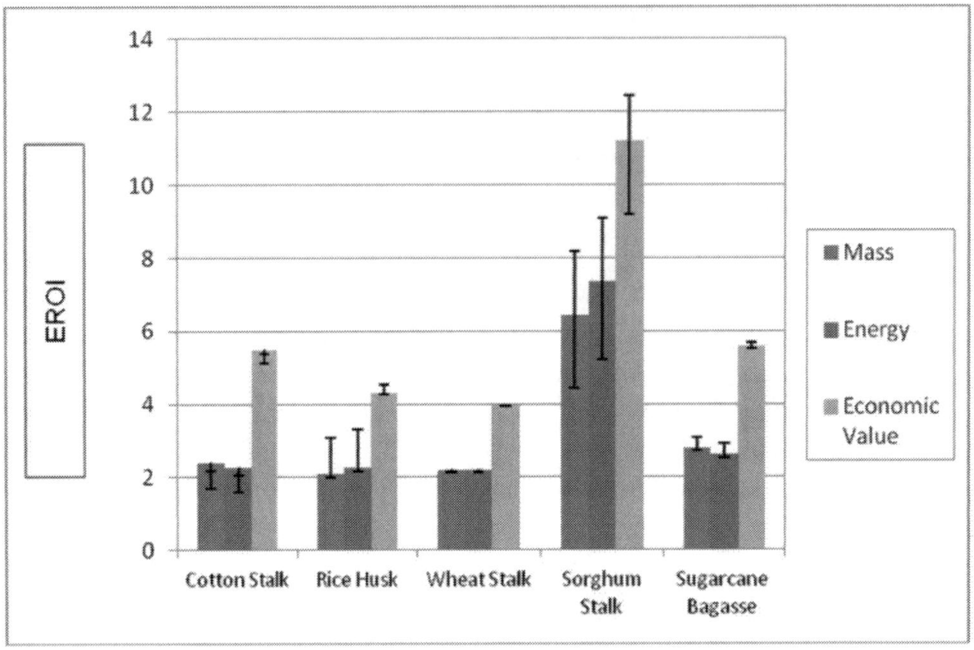

Figure 1: Energy return on investment (EROI) for different lignocellulosic feedstocks

GHG Emission per lit of EtOH

Life cycle GHG emission for lignocellulosic biofuel are computed which are shown in fig.2. Rice husk emits comparatively larger life cycle emission varies from 0.6 – 1.6 kg/lit of ethanol. Variation is due to difference in yield of rice husk per ha. Mass allocation shows higher GHG emission than energy and economic value allocation for most of the feedstocks. Sorghum stalk feedstock shows lowest GHG emission among all other lignocellulosic feedstocks. Results of our GHG emission balances are expressed in kg of CO_2 emitted per lit of ethanol energy unit (kg CO_2 eq/Lit of ethanol produced).

Macedo et al., shows that Life cycle GHG emission for sugarcane to ethanol 0.401-0.436 kg of CO_2 eq per lit of ethanol for Brazilian context.[9] MacLean and Spatari shows GHG emissions for corn to ethanol process 1.8 while for switchgrass 0.31-0.63 kg of CO_2 eq/lit of ethanol production.[14] For municipal solid waste Kalago et al., estimates life cycle GHG emissions are 0.18kg of CO_2 eq /lit of ethanol.[15] Marcelo et al., shows Life cycle GHG emission of sugarcane based ethanol 0.461 0.572 kg of CO_2 /lit of ethanol in Brazil while corn based ethanol ranges 1.39-1.45 for US.[16]

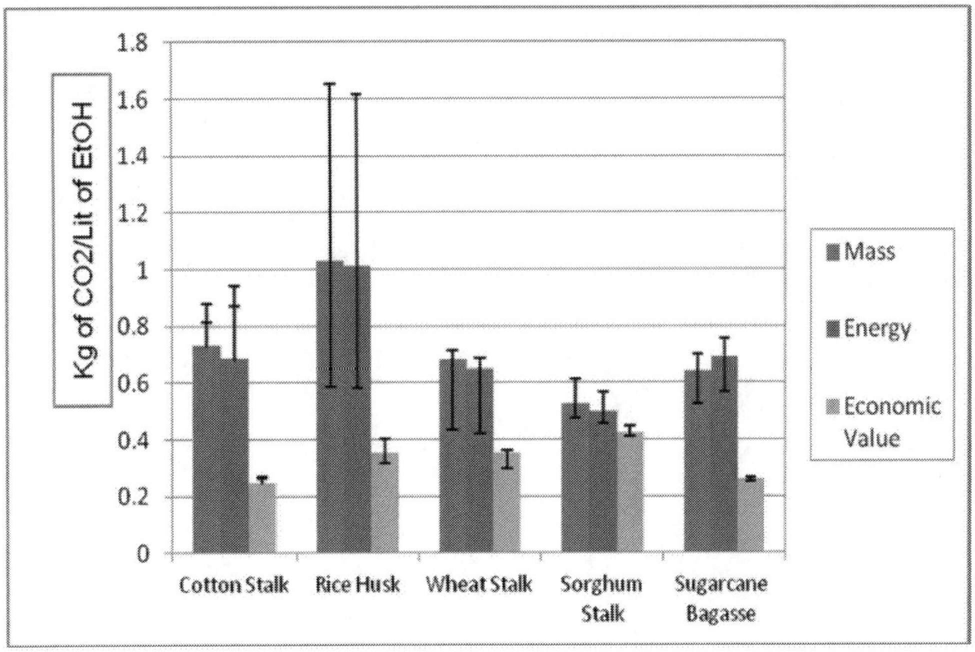

Figure 2: Life cycle GHG emissions per lit of ethanol for different lignocellulosic feedstocks

Life cycle Water Use per lit

Life cycle Water consumed per lit of ethanol is calculated for various lignocellulosic biomass feedstocks. Error bar shows variability in biomass yield due to water used in farming. Farming phase consumes major portion of water of Life cycle of ethanol production. As water data for each dataset is not available it is assumed that water used is directly proportional to crop yield. Accordingly water use is allocated to low, mean and high values of crop yield. From the graph sorghum stalk seems to be most efficient lignocellulosic biomass feedstock than others. As rice crop requires lot of water i.e. 900-2500 mm in agricultural phase rice husk is a less efficient feedstock than other lignocellulosic feedstocks which ranges 230-7150 lit of water for production of one lit of ethanol under different allocation approaches. Agricultural phase consumes major portion of water of life cycle phases for all feedstocks. Economic allocation shows comparatively less water use per lit of ethanol than mass and energy allocation methods. This is due to less economic value of biomass residue than desired product.

Yang et al., estimated average water use per lit of ethanol ranges from 1830-2640 lit for starchy feedstocks like maize, cassava and sweet potato while for sugarcane crop it is 1470 lit in China.[4] Gerbens-Leenes et al., computed global water footprint for ethanol from starchy crop which are in the range 2516-9812 lit of water use while for sugar crop 1388-2516 lit of water for production of one lit of ethanol.[17] D O Minguez-Fau et al., shows water footprint for switchgrass feedstock 1401lit/lit of ethanol for United States which is comparable for wheat stalk and sorghum stalk in the present study but former considers only evapotranspiration in the water footprint calculation.[18] Average water footprint of ethanol (liters water/liter of ethanol) for corn ethanol studies in U. S. ranges from 1174-1492 by Mubako and Lant.[19]

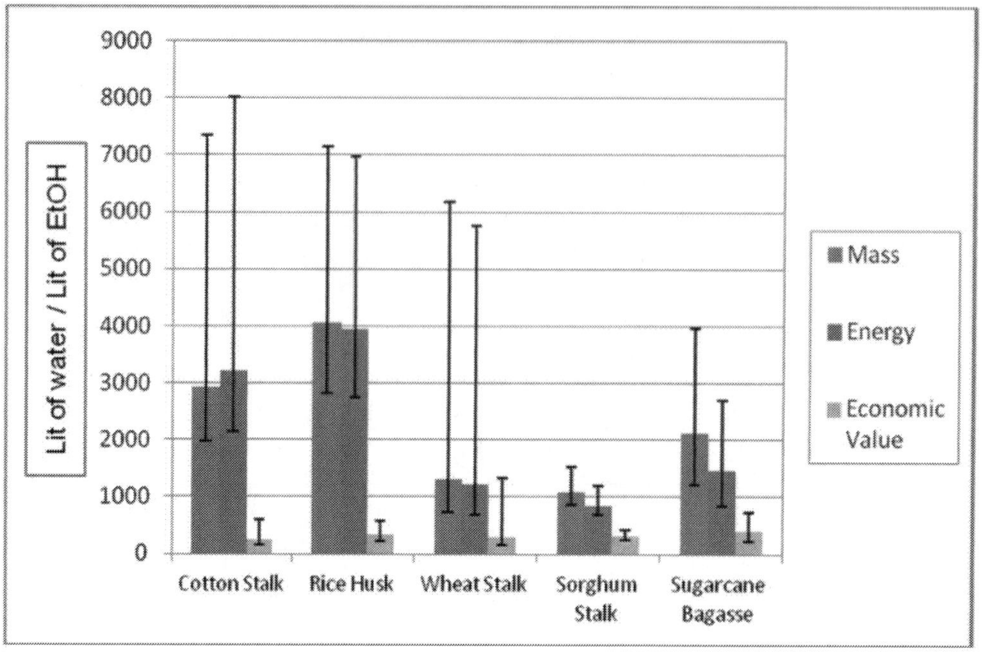

Figure 3: Life cycle water use per lit of ethanol for different lignocellulosic feedstocks

Land use per lit

The land used for feedstock production is a key factor in determining biofuel sustainability. One of the key constraint to expand biofuel production to reduce our dependence on fossil fuels is likely the limited amount of land available for producing energy crops As the land required per lit of ethanol for transportation and processing phase is comparatively less than agricultural phase of the life cycle. Though lignocellulosic biomass residue is by-product of agricultural activity it needs to be assessed without considering it as free source of energy.

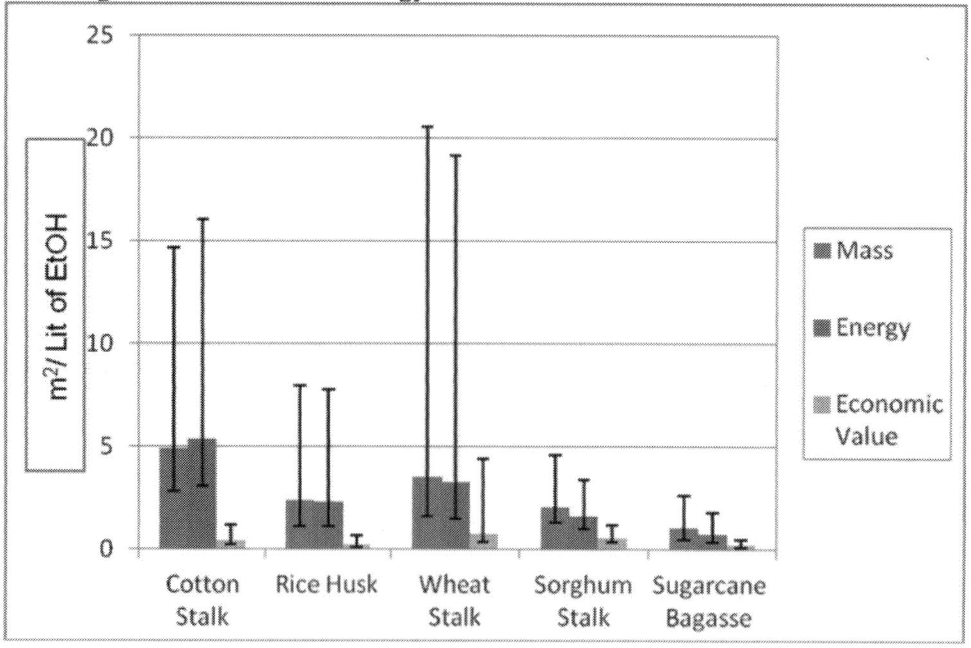

Figure 4: Land use per lit of ethanol for different lignocellulosic feedstocks

In the present study land use per lit of ethanol ranges from 0.33 to 20.54 m^2 for considered lignocellulosic feedstocks under different allocation approaches. Land use for production of a lit of ethanol is comparatively smaller than other feedstocks considered in the current study. Wheat stalk shows wider range for land use per lit because of yield variation. Cotton stalk is the most land intensive amongst all other feedstock considered in this study. Yang et al., estimated land use per lit of ethanol production ranges from 2.93- 4.75 m^2 for starchy feedstocks and 1.9 m^2 for sugarcane feedstocks for China.[4] Only land utilised for farming is considered because for other stages of life cycle such as transportation, processing comparatively negligible. Indirect effects such as land use change due to increase in bioethanol use are not considered in this work.

Discussion

This paper offers comprehensive analysis of lignocellulosic biofuel by considering life cycle EROI and GHG emissions along with resource consumption such as water and land. The driving force to consider agro-industrial lignocellulosic feedstock is the abundant availability which can be exploited for fuel production. This study provides extensive assessment of biofuel scenario which could be useful to researchers, policy makers, government etc.

Sorghum stalk shows more suitability than other cellulosic feedstocks considered in the present study as it has highest life cycle EROI, lowest life cycle GHG emissions. But it has been feed value so the potential availability has been questionable. Sugarcane Bagasse seems to be most land efficient cellulosic feedstock due to high yield of sugarcane crop. Present analysis does not address the quality issue in the Life cycle EROI indicator.

References

1. Cai, X., Zhang, X. & Wang, D. Land availability for biofuel production. *Environmental science & technology* **45**, 334–9 (2011).
2. Felix, E. R. Integrated energy, environmental and financial analysis of biofuel production from switchgrass, hybrid poplar, soybean and casterbean.
3. Phalan, B. The social and environmental impacts of biofuels in Asia: An overview. *Applied Energy* **86**, S21–S29 (2009).
4. Yang, H., Zhou, Y. & Liu, J. Land and water requirements of biofuel and implications for food supply and the environment in China. *Energy Policy* **37**, 1876–1885 (2009).
5. Kadam, K. L. Environmental benefits on a life cycle basis of using bagasse-derived ethanol as a gasoline oxygenate in India. *Energy Policy* **30**, 371–384 (2002).
6. Hammerschlag, R. Ethanol's Energy Return on Investment: A Survey of the Literature 1990–Present. *Environmental Science & Technology* **40**, 1744–1750 (2006).
7. Fingerman, K. R., Torn, M. S., O'Hare, M. H. & Kammen, D. M. Accounting for the water impacts of ethanol production. *Environmental Research Letters* **5**, 014020 (2010).
8. Mulder, K. & Hagens, N. J. Energy return on investment: toward a consistent framework. *Ambio* **37**, 74–9 (2008).
9. Macedo, I. C., Seabra, J. E. a. & Silva, J. E. a. R. Green house gases emissions in the production and use of ethanol from sugarcane in Brazil: The 2005/2006 averages and a prediction for 2020. *Biomass and Bioenergy* **32**, 582–595 (2008).
10. Cherubini, F. GHG balances of bioenergy systems – Overview of key steps in the production chain and methodological concerns. *Renewable Energy* **35**, 1565–1573 (2010).
11. Guinée, J. B., Heijungs, R. & Voet, E. A greenhouse gas indicator for bioenergy: some theoretical issues with practical implications. *The International Journal of Life Cycle Assessment* **14**, 328–339 (2009).
12. Hoefnagels, R., Smeets, E. & Faaij, A. Greenhouse gas footprints of different biofuel production systems. *Renewable and Sustainable Energy Reviews* **14**, 1661–1694 (2010).

13. Elena, G.-C. & Esther, V. From water to energy: The virtual water content and water footprint of biofuel consumption in Spain. *Energy Policy* **38**, 1345–1352 (2010).
14. MacLean, H. L. & Spatari, S. The contribution of enzymes and process chemicals to the life cycle of ethanol. *Environmental Research Letters* **4**, 014001 (2009).
15. Kalogo, Y. & Habibi, S. Policy Analysis Environmental Implications of Municipal Solid Waste-Derived Ethanol. **41**, 35–41 (2007).
16. Oliveira, M. E. D. D. E., Vaughan, B. E. & Jr, E. J. R. Ethanol as Fuel : Energy , Carbon Dioxide Balances , and Ecological Footprint. **55**, (2005).
17. Gerbens-Leenes, W., Hoekstra, A. Y. & van der Meer, T. H. The water footprint of bioenergy. *Proceedings of the National Academy of Sciences of the United States of America* **106**, 10219–23 (2009).
18. Powers, S. E., Burken, J. G. & Alvarez, P. J. The Water Footprint of Biofuels : A Drink or Drive Issue ? **43**, 3005–3010 (2009).
19. Mubako, S. & Lant, C. Water resource requirements of corn-based ethanol. *Water Resources Research* **44**, 1–5 (2008).

(644f) Life CYCLE Assessment of Biofuels Produced by the NEW Integrated Hydropyrolysis-Hydroconversion (IH2) Process

Edwin.A Maleche[1], David R. Shonnard[1] and Terry Marker[2], (1)Chemical Engineering, Michigan Technological University, Houghton, MI, (2)Gas Technology Institute, Des Plaines, IL

LIFE CYCLE ASSESSMENT OF BIOFUELS PRODUCED BY THE NEW INTEGRATED HYDROPYROLYSIS-HYDROCONVERSION (IH^2) PROCESS

Edwin. A. Maleche[1], David Shonnard[1] and Terry Marker[2], (1) Chemical Engineering, Michigan Technological University, Houghton, MI, (2)Gas Technology Institute, Des Plaines, IL

David Shonnard[1], PhD: 906-487-3468, fax: 906-487-3213 email: drshonna@mtu.edu

Edwin Maleche[2], Graduate Student: email: eamalech@mtu.edu

Terry Marker Gas Technology Institute, Des Plaines, IL

1 Department of Chemical Engineering and Sustainable Futures Institute

2 Department of Chemical Engineering

Michigan Technological University

1400 Townsend drive

Houghton, MI 49931 USA

Abstract

Biofuels are alternative fuels that have the promise of reducing reliance on imported fossil fuels and decreasing emission of greenhouse gases from energy consumption. The presentation analyses the environmental impacts focusing on the greenhouse gas (GHG) emissions associated with the production, delivery, and use of biofuel using the new Integrated Hydropyrolysis and Hydroconversion (IH^2) process. The IH^{2}process is an innovative process for the conversion of woody biomass into hydrocarbon liquid transportation fuels in the range of gasoline and diesel.

A cradle-to-grave life cycle assessment (LCA) was used to calculate the greenhouse gas emissions associated with diverse feedstocks production systems and delivery to the IH^{2}facility plus producing and using these new renewable liquid fuels. The biomass feedstocks analyzed include algae (microalgae), bagasse from a sugar cane-producing locations such as Brazil or extreme southern US, corn stover from Midwest US locations, and forest feedstocks from a northern Wisconsin location.

The life cycle greenhouse gas (GHG) emissions savings of 58%–98% were calculated for IH^2 gasoline and diesel production and combustion use in vehicles compared to fossil fuels. The range of savings is due to different biomass feedstocks and transportation modes and distances. Different scenarios were conducted to understand the uncertainties in input data to the LCA model, particularly in the feedstock production section, the IH^2 biofuel production section, and transportation sections.

(659a) Simplified Life Cycle Analysis As a Route to Fast Decision Making in Product and Process Development for Sustainability

Santiago Faucher, Technologies, Hatch, Mississauga, ON, Canada

The accounting of cradle-to-grave life cycle inventories for specific products often requires a substantial investment in time and resources. For this reason, few products are sold with an accompanying inventory of their life cycle impact on the environment. This lack of a baseline and the onerous effort required to obtain it can often lead to confusion in the development of new processes or products designed to be more sustainable. The new product or process is often designed based on the inventor's baseline knowledge and perception of what is more sustainable. After the new product or process has been shown to achieve the desired performance attributes, a full life cycle inventory of the new product and the baseline product is then commissioned as it is now seen as more justified. Ideally, the life cycle impact of the new product developed is shown to be lower than the original product it is intended to replace.

This approach can be successful if the inventor understands intimately the effect of both material and process selection on the life cycle inventory and can predict the results of a life cycle assessment study. However, in most cases the products and processes are complicated and comprised of many parts making such estimations difficult. Thus the risk of designing a new product that is less sustainable than the baseline product it is intended to replace can be high. To address this, simplified life cycle analyses can be applied in the product/process design phase. While this approach does not map out the usage phase of the product, it can capture the substantial differences that process and material selection may have on the life cycle impact of the new product. With the use of simplified life cycle analysis prior to experimental product or process development, the life cycle impacts of several new hypothetical design options can be compared without investing considerable time in experimental development and later full life cycle analyses. This approach increases the speed and efficiency of product and process development for sustainability.

(659b) Process Development for a Higher Quality and Cost Optimized Monomer

Ahmed A. Youssef, SABIC, Innovative Plastics, Mt. Vernon, IN and Mohan Khadilkar, SABIC Innovative Plastics, Mt. Vernon, IN

This study addresses a process improvement that involves the development of a new treatment system for a purge (tar) stream resulting as a by-product in the condensation reaction of phenol with a ketone to produce a bisphenol monomer at one of SABIC's sites. The conventional tar recovery section is cumbersome and includes a multiplicity of vacuum distillation units.

This work discusses the installation of an acid catalyzed cracking process where the monomer is mainly converted to its original constituent (phenol) and the impurities are separated out.

Such improvement yields higher final product purity as a result of a decrease in the total impurities recycled into the main system. It also presents a cost effective process driven by the amelioration in the raw material usage.

The transformation of the recovery section empowers the manufacturing unit at SABIC (Innovative Plastics business) to operate with a reduction in the purge stream by ~ 30%.

(659c) Carbon Footprint Analysis of Dairy Feed From a Mill in Michigan, USA

Felix Adom[1], Greg Thoma[2] and David R. Shonnard[1], (1)Chemical Engineering, Michigan Technological University & Sustainable Futures Institute, Houghton, MI, (2)Department of Chemical Engineering, University of Arkansas, Fayetteville, AR

A carbon footprint analysis was conducted for a single dairy feed mill located in Michigan USA with the aim of developing a preliminary assessment of dairy feed mill operations. The goal was to develop life cycle assessment methodology applicable to the animal feed mill industry and also to gain understanding of the relative importance of milled dairy feed inputs on the greenhouse gas (GHG) emissions of the outputs of the mill. In the final analysis, GHGs emitted from feed mill operations on the basis of one kilogram of dairy feed output from the mill (kg CO_2 eq kg^{-1} of milled dairy feed) was estimated. Scenarios were investigated to consider the prevalence of different feed ingredient inputs likely to represent different U.S. mill locations.

The functional unit used for this study was 1 kg of milled dairy feed at its exit moisture content (an average feed formulation for dairy animal nutrition). Inputs and activities identified in this cradle-to-mill-to-dairy farm analysis included; feed ingredients, onsite energy, and transportation of feed inputs to the milling site and transportation mill output to dairy farms. Sources of Life cycle inventory (LCI) input data include; peer reviewed journal articles, purchase history documents of the milling facility, internet searches (ISI, Google scholar, ProQuest, etc), and a site visit. Using Ecoinvent™ unit processes in SimaPro, an open IO model and other peer reviewed articles, the cradle-to-mill-to-dairy farm (receiving) gate analysis was estimated using two allocation methods (economic and mass).

Feed mill GHG emissions were estimated to be 0.62 and 0.93 kg CO_2 eq (equivalent) kg^{-1} of milled dairy feed for economic and mass allocation, respectively. The highest emissions were due to the feed ingredient inputs which contributed 73-82% towards the carbon footprint depending on the allocation method. Energy and transportation impacts together contributed between 8-12%. Scenario analyses were include to simulate different mill locations where dominant feed inputs change compared to the MI location, for example to a prevalence of dry distiller grains and solids for mills located in Iowa and a prevalence of oats for a mill located in North Dakota.

(659d) Urban Mining: A Process and Technology Review

Daniel Brosig, Hatch Ltd., Mississauga, ON, Canada

The recovery of non-ferrous metals (e.g., copper, aluminum, lead, nickel and precious metals) from urban waste can be economically attractive and represents a more sustainable alternative to the extraction of metals from naturally-occurring deposits. A brief overview of the economic and environmental performance behind the metal recovery processes from urban waste is provided to shed some light on the nature of the challenge and opportunity. Although in principle it could appear to be a simple concept, in practice the mining of "urban ore" is accompanied by a complex host of process and technology problems that span from logistical issues around the separation of the target waste to metallurgical recovery inefficiencies. Currently practiced urban mining processes and the associated state-of-the-art technological solutions are reviewed in this presentation. A conceptual roadmap is presented to explore the development of the processes and technologies that may one day aid in making urban mining ubiquitous around the world.

(659e) Sizing of a Packed Reactive Distillation Column for Isoamyl Butyrate Production Using an Equilibrium Model

Jairo A. Duran[1], Karen A. Cañon-Rubio[1], Alejandro Lopez Gomez[2], Alvaro Orjuela[1] and Gerardo Rodriguez[3], (1)Department of Chemical and Environmental Engineering, Universidad Nacional de Colombia, Bogota, Colombia, (2)Departament of Chemical and Environmental Engineering, Universidad Nacional de Colombia, Bogota, Colombia, (3)Departamento de Ingeniería Química y Ambiental, Universidad Nacional de Colombia, Bogotá, Colombia

Sizing of a Packed Reactive Distillation Column for Isoamyl Butyrate Production Using an Equilibrium Model

Jairo Duran, Karen Cañon, Alejandro López, Álvaro Orjuela

Grupo de Procesos Químicos y Bioquímicos. Department of Chemical and Environmental Engineering. Universidad Nacional de Colombia. Bogotá. Colombia

Reactive distillation has been considered as an alternative for sustainable process in the production of green chemicals because their capabilities of reaching high conversion in equilibrium reactions, flexibility and relatively low costs. Several studies have pointed the importance of taking into account high-quality thermodynamic and kinetic parameters in order to obtain reliable estimations on the performance of the integrated reaction-separation operation. Among different biobased esters, isoamyl butyrate that is used as green solvent and food additive can be obtained from renewable raw materials. For instance, fermentation broths from fuel ethanol production contain up to 2.5% (volume basis) of amyl alcohols which are currently blended with fuel ethanol or used as fuel for steam boilers. As a potential upgrading alternative, these alcohols can be esterified with different acids using a reactive distillation scheme to improve reaction conversions and energy savings with respect to the traditional esterification processes.

In this work, kinetic and phase equilibrium parameters from experimental data were used to simulate and size a reactive distillation column. NRTL and Hayden O'Connell models from regressed experimental data were used in phase equilibrium calculations. Kinetics was measured in a batch reactor using Amberlyst® 35 as catalyst. A full factorial design was established to carry out the kinetic tests. Experimental data were fitted adequately with a pseudo-homogeneous model. In addition, autocatalytic kinetics was measured using an identical methodology. The module RADFRAC of Aspen Plus® was used in modeling of reactive distillation. A basis of 200.000 kg per year of isoamyl butyrate (food grade, >99% weight) was considered in sizing and sensibility analyses. Amberlyst® 35 contained in KATAPAK SP-11® was considered in the simulation. MELLAPAK-250Y® was used in non catalytic packing sections, where autocatalytic kinetics was considered. Column height was calculated assuming an HETP of 0.5 meters, as reported by several authors.

According with simulation results a conversion of butyric acid higher than 96% was possible. Nevertheless, including a pre-reactor in the scheme, a conversion higher than 99.1% was observed. Best performance of operation was found when using a mole excess around 1.5 of isoamyl alcohol to butyric acid in the system.

Keywords: reactive distillation, esterification, catalytic packing, ion exchange resin, flooding factor.

(659f) Analysis of Alternative Processes for Anhydrous Ethanol Production, an Economic and Environmental Approach

Michel Kahwaji Janho[1], Jorge E. Gatica[2], María Rosa Hernández[3], Fernando Daniel Mele[4] and Mauricio Colombo[4], (1)Chemical And Biomedical Engineering, Cleveland State University, Cleveland, OH, (2)Chemical and Biomedical Engineering, Cleveland State University, Cleveland, OH, (3)Chemistry and Chemical Engineering, Universidad Nacional de Tucuman, Tucuman, Argentina, (4)Ingenieria de Procesos y Gestion Industrial, Universidad Nacional de Tucuman, Tucuman, Argentina

The production of sugar is one of the oldest agro-industrial activities. It includes the production of sugar cane, sugar and dairy products: paper, alcohol, energy, etc. In addition to the production of sugar for domestic and international markets, one of the activities that is growing in importance is the production of alcohol from molasses. Energetic integration has also resulted in bringing power generation from bagasse as a secondary activity in modern sugar-cane processing industries.

A new chapter for bio-energy products associated with the manufacture of sugar has gained relevance as blending gasoline with ethanol is becoming part of energetic policies in most industrialized countries. Indeed, energy products from renewable sources are receiving increasing attention, and this has generated a renewed stimulus to examine alternative processes for bio-ethanol production. It has become common for sugar mills to have complementary distilleries with production capacities ranging from 50 to 400 liters per day of non-anhydrous alcohol.

At sugar plants, steam is generated in boilers using as fuel bagasse-cellulosic fiber waste obtained crushing the sugar cane- and it is used to give useful heat and shaft work to the plant, but it can also be used to generate electricity with export opportunities to the electrical network. The great number of process alternatives outlines a serious decision making problem in order to take advantage of the resources.

At steady state conditions the amount of bagasse obtained is more than enough to supply the demands of useful heat and energy. Thus these industries can advantage of the emergence of a new potential market, as they have an increased potential for generating a surplus of electricity for sale to the grid.

In this paper, a classic bio-ethanol distillery is adopted as case-study and simulated using AspenPlus® and UniSim ® commercial simulators. Alternative process flow sheets of varying complexity are analyzed aiming to reduce power consumption and effluent generation.

The process simulation allows a systematic analysis that leads to a good combination of all the values of process variables. In particular, since they demand significant energy influx, an important subsystem for analyzing is the separation of alcohol-water mixtures, There are also different technologies for this dehydration. In this paper we consider azeotropic distillation and hybrid schemes with membrane separation. The development of a simulation module for calculating the membranes performance, typically not included in commercial process simulators,

The consumed electric energy and steam depends on the adopted scheme and they have, depending to their source or generation policy, an associated cost and environmental impact. Among the energy systems that deserve special attention, cogeneration in the sugar industry must be pointed out. This part of the process makes efficient use of a common fuel for the generation of heat and power. As a result, the environmental impact and costs associated with the use of electricity and useful heat change appreciably, such as anhydrous alcohol production schemes can be achieved with a lesser environmental at the expense of decreasing the economic benefit which would result from selling electricity to the grid.

Cogeneration is not new in the sugar-cane processing industry. Indeed, typically co-generation refers to the generation of heat as the main objective and the generation of electrical and mechanical power as secondary objectives. To date, however, no versatile tool that would allow analyzing economical feasible alternatives of anhydrous alcohol production, bearing in mind environmental issues, is available.

In addition, the explicit consideration of environmental issues offers an alternative route to explore trade-offs. Life Cycle Analysis (LCA) is used in this work. Unlike most environmental approaches to design that focus on reducing the generation of residuals, but without taking into account the impact associated to related processes, LCA allows considering environmental impact as an integral part of the problem. The software used to evaluate the environmental impact is SimaPro®.

Alternative schemes that meet the same specifications of purity and recovery in effluent streams, are examined in this paper. Of particular relevance is the assessment of costs and environmental impact of these processes. A baseline case is defined in our case-study, and structural and parametric changes are made. Different scenarios with respect to economic and environmental issues taking into account the specifications of the final product. Near-optimal scenarios are formulated by defining an operating window for some of environmentally friendly technologies available. The impact of different operating policies for the production of anhydrous alcohol on the environment is examined in detail.

The membrane area, the number of plates in the distillation units, the heat exchange area, and the pressure manipulators are considered as structural variables. Among the most relevant operational variables considered one can mention reflux ratio, pressure ratio, as well as compositions and flow rates of streams.

(665a) Sustainable Design and Operations At the David L. Lawrence Convention Center

Angelica Ciranni, Sports & Exhibition Authority of Pittsburgh and Allegheny County, Pittsburgh, PA

In 2003, the David L. Lawrence Convention Center (DLCC) received a Gold rating under the LEED for New Construction rating system, making it the first convention center to receive LEED certification. At that time, the DLCC was also the world's largest green building at 1.5 million square feet. From the early stages of design through current operations, the DLCC has remained committed to sustainability. In order to demonstrate the sustainable operations of the facility, the DLCC registered under the LEED for Existing Buildings: Operations & Maintenance (O&M) rating system and was certified Platinum in April 2012. The DLCC is the first convention center to receive two LEED ratings.

Concurrent with the LEED for Existing Buildings: O&M documentation process, the DLCC underwent a "building in operation" study that examined building performance, facility management, occupant satisfaction, and organizational sustainability. A thorough set of analyses contributed to the findings, including quantitative and qualitative assessments of energy, natural ventilation, water, site, transportation, waste, purchasing, commissioning, occupant comfort, and indoor air quality.

This session will provide an overview of the case study findings and the LEED for Existing Buildings: O&M process for the host facility of the AIChE Annual Meeting, including specific strategies and lessons learned.

Variations in Temperature in Mini-Roof Structures Employing Different Roofing Materials

Jason T. Kirby[1], Ph.D., Ronald D. Sherrod[2], Matt Winslett[3], and Robert W. Peters[4*], Ph.D., P.E.

[1]Associate Professor of Environmental Engineering
Department of Civil, Construction and Environmental Engineering
University of Alabama at Birmingham
1075 13th Street South
Birmingham, Alabama 35294-4440
Phone: (205)-934-8434; e-mail: rwpeters@uab.edu

[2]Undergraduate Student
Department of Chemical Engineering
Worcester Polytechnic Institute
100 Institute Road
Worcester, Massachusetts 01609-2280

[3]Engineering Manager, Energy Management
Facilities Management Department
University of Alabama at Birmingham
822 Facilities Support Building
Birmingham, Alabama 35294-4564

[4*]Professor of Environmental Engineering
Department of Civil, Construction and Environmental Engineering
University of Alabama at Birmingham
1075 13th Street South
Birmingham, Alabama 35294-4440
Phone: (205)-934-8434; e-mail: rwpeters@uab.edu

ABSTRACT

The overall goal of green roof research at the University of Alabama at Birmingham (UAB) campus is to identify a suitable roofing material combination that reduces energy cooling costs. Since 2008, ambient temperature data has been recorded to better understand how the roof heats and cools on a diurnal, seasonal, and yearly basis. During this study, 15 mini-roof combinations were observed for trends in internal temperatures. Conventional roofing systems included river rock, crushed marble chips, black roofs, reflective (white) roofs, concrete pavers coated with epoxy, and an alternative roofing system involving vegetated roofs (planted with sedum plants). To investigate the thermal properties of the roofing structures, an ambient temperature probe was placed inside of each roof to record data every 10 minutes.

The University of Alabama at Birmingham (UAB) is located in the Southeastern United States. This region is characterized by its hot summers and relatively mild winters. Due to these hot summers during which the temperatures can approach 37.8°C (100°F), along with conditions of high humidity, it is crucial to have a roof which minimizes cooling costs. These hot summers also create an urban heat island effect in the Birmingham area due to the current roofing structures and facility usage. An urban heat island is defined on the basis of a temperature difference between urban and rural area, caused by human modifications to the natural geographic landscape. The human modifications observed here are the buildings and roads added to the natural habitat. These structures retain the heat transferred from the sunlight increasing the temperature. This property results in an increased temperature in the highly modified areas (urban downtown areas) versus a less modified area (rural areas).

Various roofing materials are being studied for their thermal efficiency properties; roofing materials studied include black roofs, white reflective roofs, river rock, crushed marble chips, cement blocks, and vegetated roofs [having sedum plants (*Sedum hispanicum*)]. The temperature behavior both on a daily basis and on an annual basis vary in a sinusoidal fashion. Additionally, a pilot vegetated roof was installed on top of Hulsey Center in July 2008. This pilot green roof occupies ~1388.0 m^2 (14,940 ft^2) and contains ~20,000 sedum type plants. Utility bill information has been collected on natural gas usage, water usage, and electricity usage for ~3 years prior to the installation of the vegetative roof, and for more than a year since installation of the green roof. Reductions in the usage of natural gas, water, and electricity has been observed and documented since installation of the green roof on top of Hulsey Center.

Historically, studies on green roofs have explored their energy performance compared with traditional roofs. Thermal performance indicated a significant reduction (~40%) of a building cooling load during the summer period [Spala *et al.*, 2008]. Similar results were achieved for a nursery school, with reductions ranging from 6% to 49%, and reduction ranging from 12% to 87% on the last floor of the nursery school [Santamouris *et al.*, 2007]. Wong *et al.* [2007] note that green roofs tend to experience lower surface temperatures than the original exposed roof, especially in areas well covered by vegetation. When green roofs are well covered by vegetation, the resulting substrate moisture will tend to keep substrate temperature lower than the original exposed bare roof. These studies determined that over 60% of the heat gain was mitigated by vegetated roof systems. Summertime data have indicated significant lower peak roof surface temperature and higher nighttime surface temperature for green roofs as compared to conventional roofs [Sonne, 2006].

After acquiring the necessary raw data from the mini-roof sensors, we noticed that the temperatures cycled in a sinusoidal fashion on yearly and daily time frames, so using the general form sine function was an appropriate model to describe the data. We determined the frequency and amplitude for all fifteen mini-roof systems. The determined functions supported that most roofs are statistically different from one another from an amplitude aspect but the phase angles are statistically the same. We also have discovered that most roofs are significantly different from the standpoint of the average mean roof temperature, but the significance is most prevalent in the summer months and during the other times of the year the roofs behave in a more similar fashion to one another. The model does well in describing the average daily temperature during the year.

Key Words: Roofing materials, temperatures of roofing structures, thermal properties
*Corresponding author

Introduction/Background

Green building design, as defined by the United States Environmental Protection Agency (USEPA), is "the practice of creating structures and using processes that are environmentally responsible and resource-efficient throughout a building's life-cycle from design, to construction, operation, maintenance, renovation and ultimately deconstruction" [EPA, 2010]. However, according to the United States Green Building Council (USGBC), commercial buildings in the US alone account for:

- 65% of electricity consumption,
- 30% of greenhouse gas emissions,
- 30% of raw materials use,
- 30% of waste output (136 million tons annually), and
- 12% of potable water consumption [USGBC, 2010].

The building infrastructure (Residential, Commercial, Institutional, and Industrial) in the United States (US) consumes over two-thirds of the nation's electricity demand and accounts for one-third of all domestic energy consumption [EPA, 2009]. Regrettably, conventional forms of energy production have an adverse impact on natural ecosystems. Collectively, our buildings contribute to 38.9% of the nation's total greenhouse gas emissions. Faced with rising energy costs, diminishing fuel resources and emerging environmental concerns, scientific research has begun to address these challenges by adopting sustainable or green building alternatives.

Common Roofing Systems

The choice and type of urban roofing system is dependent on environmental concerns such as sun exposure and meteorological factors such as temperature, wind and rain. Some roofs may have covenants to determine their usage. Different systems have been developed and designed to perform at its most effective given the exposure conditions of the building location. As a result, many roofing systems exist and are commonly used by the construction community.

There are many roofing systems used today that depend on the building type, whether is for residential or for commercial applications. The most common systems are listed as follows [Reid, 2000].

- Built-up,
- Preformed metal, and
- Shingles.

A brief description of the three roofing system is provided involving advantages, disadvantages and durability characteristics of each.

Built-Up

Built-up roofing systems are most common for flat roof applications. They consist of several layers of material built-up from the internal structural frame of the building. They are most commonly made of alternating layers of plywood, felt, asphalt, tar and gravel [Reid, 2000]. The major advantage of this system is its waterproofing characteristics, and out of all roofing systems, built-up roofs are considered to have the highest degree of waterproofing. However, the system has poor fire resistance, generally requires a professional to install, and is difficult to locate roof leaks. The system typically lasts for 22.5 years [Cash, 1999; Editors of Time Life Books, 1979].

Preformed Metal

Another common roofing system is made of preformed metal panels. The panels are generally made from aluminum, steel and copper [Reid, 2000]. They are most commonly found with contemporary designs. They come in flat, ridged, ribbed or corrugated forms. These systems are easy to install and repair, and can be painted any color. They are considered fire resistant. The panels are easily damaged by wind, falling trees and tree limbs, or any other type of contact. They generally last for 25 years [Cash, 1999; Editors of Time Life Books, 1979].

Shingles

Shingle roofs come in many forms. Shingles are commonly made of asphalt, slate, wood and clay tiles [Reid, 2000]. Asphalt shingles are the most popular, especially for residential construction. They are available in a variety of sizes, weights, and colors. They require little maintenance and are easy to install. However, they are considered to have poor fire resistive qualities. Asphalt shingles generally last for 16–17 years with proper maintenance [Cash, 1999]. Slate and ceramic tiles are highly expensive due to their aesthetic and durability properties. They are fire resistant and have attractive appearances. Yet, they are extremely heavy and require strong structural support. They are also very brittle, and require increased amount of time to install which often entails specialized tools. Repairs are in most cases difficult. Slate and ceramic tiles are considered the most durable, lasting for 20–100 years with proper maintenance. Wood shingles and shakes are also attractive options. They are easy to install and are a natural insulator. However, they are highly flammable and require treatments for weather and insect protections. They generally last for 10 to 20 years with proper maintenance [Cash, 1999; Editors of Time Life Books, 1979].

Urban Heat–Island Reduction and Building Energy Conservation

For millions of Americans living in and around cities, heat islands are of growing concern. This phenomenon describes urban and suburban temperatures that are 2 to 10°F (1 to 6°C) hotter than nearby rural areas. Elevated temperatures can impact communities by increasing peak energy demand, air conditioning costs, air pollution levels, and heat-related illness and mortality. The Environmental Protection Agency (EPA) recommends installing cool or vegetative green roofs, planting trees and vegetation and switching to cool paving materials as a way of reducing the negative effects of urban heat islands. The EPA says green roofs, if installed widely in a city, can contribute to heat island reduction by replacing heat-absorbing surfaces with plants, shrubs, and small trees. The vegetation cools the air through evapotranspiration (or evaporation of water from leaves). Planted rooftops remain significantly cooler than a rooftop constructed from traditional heat-absorbing materials. Further, green roofs reduce summertime air conditioning demand by lowering heat gain to the building.

Energy Modeling

Energy modeling (i.e., energy simulation) is a method for predicting the energy consumption of an occupied structure. Building energy analysis must consider numerous thermal characteristics including: wall and roof materials, the size and orientation of the building, how the building is occupied and operated, as well as influences from the local climate.

The surface temperature of a roof exposed to solar radiation, the resulting heat flow into the building, along with associated indoor temperatures and cooling needs depend on the effect of solar radiation, surface absorptivity, ambient air temperature and wind speeds adjacent to the surface [Givoni, 1994]. When vegetative roofs are considered, because of added thermal mass, it is also important to take into account hourly heat transfer when determining energy consumption; as the heat flux through a vegetated roof can be quite different from conventional roofing.

R-values and *U*-values have been used for many years as a measurement of a building envelope's thermal performance. However, these attributes do not fully take into account the effects of thermal mass, and by themselves, are inadequate in describing the heat transfer properties of construction assemblies with significant amounts of thermal mass [Kosny *et al.*, 2007]. Vegetated roofs are more dependent on the interaction between the roofing systems' unit weight, density, thermal conductivity, moisture content, vegetal coverage and specific heat. Therefore, it is often necessary to utilize computer software, which incorporates these elements into the analysis of high thermal mass roof structures and associated energy consumption. The steady-state *R*-values traditionally used to measure energy performance will not accurately capture the complex, dynamic thermal behavior of vegetated roof systems.

Vegetated (Green) Roof Systems

Green roofs are engineered ecosystems that rely on vegetation to provide benefits such as reduction of roof temperatures and stormwater retention [Wolf and Lundholm, 2008]. Green roofs offer benefits of reducing stormwater runoff, improving air and water quality, and providing habitat and biodiversity for urban centers [Wong *et al.*, 2003][0]. Hydrologic modeling has demonstrated that widespread green roof implementation can significantly reduce peak runoff rates, particularly for small storm events [Carter and Jackson, 2007]. By combining the Green-Ampt method with evapotranspiration of green roofs, Roehr and Kong [2010] estimated the potential runoff reduction achieved by green roofs is 20%. Green roofs provide an excellent option to improve stormwater runoff [Hathaway *et al.*, 2008]. Green roofs are primarily valued based on their increased roof longevity, reduced stormwater runoff, and decreased building energy consumption [Clark *et al.*, 2008]. Carter and Jackson [2007] noted that research studies have primarily been focused on roof-scale processes such as individual roof stormwater retention, plant growth, or growing medium composition. Few studies have examined the impact that widespread green roof application could have on the hydrology of a real-world watershed [Carter and Jackson, 2007]. A major barrier to increasing the prevalence of green roofs is the lack of scientific data available to evaluate their applicability to local conditions [Williams *et al.*, 2010].

Green roofs are typically classified as being either an intensive or extensive roof [Green Roof Plan, undated]. Intensive green roofs are often used on commercial buildings in order to have large green areas that incorporate all sizes and types of plants. These roofs use grasses, ground covers, flowers, shrubs and even trees. They often include paths and walkways that travel between different architectural features to provide space where people can interact with the natural surroundings. Intensive green roofs, sometimes termed "rooftop gardens", utilize planting mediums that have greater depth than extensive green roofs; the deeper soil allows intensive roofs to accommodate large plants and various plant groupings. Intensive green roofs require more maintenance than extensive green roofs because of the plant varieties they will support.

Extensive green roofs have a planting medium that ranges from 1.6 to 6 inches deep. Typically, drought-tolerant sedums (succulent plants) and grasses are used since they are shallow-rooted and use little water. Plant diversity on these roofs is kept low to simplify care and to be sure all plants have similar moisture requirements.

Extensive green roofs can significantly reduce both the timing and magnitude of stormwater runoff relative to a typical impervious roof [Schroll *et al.*, 2011]. They note, however, that regional climatic conditions such as seasonality in rainfall and potential evapotranspiration can strongly alter the stormwater performance of vegetated roofs. Factors such as type of green roof and its geometrical properties (slope), soil moisture characteristics, season, weather and rainfall characteristics, age of the vegetated roof, and vegetation affect the runoff dynamics from green roofs [Berndtsson, 2010]. Fioretti *et al.* [2010] noted that green roofs significantly mitigate storm water runoff generation, as well reducing the daily energy demand. Aitkenhead-Peterson *et al.* [2011] note that most studies on runoff quality from green roofs have been conducted in cooler northern climates. Villarreal and Bengtsson [2004] recommended the use of a combination of best management practices; additionally, they observed that green roofs are effective at lowering the total runoff from Augestenborg (Sweden) and that detention ponds should successfully attenuate storm peal flows. Niu *et al.* [2010] noted that over the lifetime of a green roof (~40 years), the net present value is ~30% to 40% less for a green roofs as compared with conventional roofs (not including green roof maintenance costs). Kirby *et al.* [2011] note that extensive vegetated roof systems offer at least 1 16% enhancement in reducing stormwater runoff as compared to conventional roofs. Clark *et al.* [2008] further note that the additional upfront investment of a green roof is recovered at the time when a conventional roof would be replaced. Rosatto *et al.* [2010] concluded that green roofs contribute positively in reducing runoff, with greater retention with vegetated plots and thicker substrate.

Vegetated roof systems have a number of advantages over that of conventional roof systems. Benefits associated with green roof systems include [Green Roofs for Healthy Cities, undated]:

- Urban greening has long been promoted as an easy and effective strategy for beautifying the built environment and increasing investment opportunity.
- With green roofs, water is stored by the substrate and then taken up by the plants from where it is returned to the atmosphere through transpiration and evaporation.
- Depending on the plants and depth of growing medium, during the summer, green roofs retain 70% to 90% of the precipitation that falls on them; in winter they retain between 25% to 40%.

- Green roofs not only retain stormwater, but also moderate the temperature of the water and act as natural filters for any of the water that runs off.
- Green roofs reduce the amount of stormwater runoff and delay the time at which runoff occurs, resulting in decreased stress on sewer systems at peak flow periods.
- Through the daily dew and evaporation cycle, plants on vertical and horizontal surfaces can cool cities during hot summer months and reduce the Urban Heat Island (UHI) effect. The UHI is also mitigated by the covering some of the hottest surfaces in the urban environment, such as black rooftops.
- Green roofs can also help reduce the amount of dust and particulate matter throughout the city, as well as the production of smog. This plays a role in reducing greenhouse gas emissions and adapting urban areas to a future climate with warmer summers.
- Green roofs help to achieve the principles of smart growth and positively affect the urban environment by increasing amenity and green space.
- The greater insulation offered by green roofs can reduce the amount of energy needed to moderate the temperature of a building, as roofs provide the greatest heat loss in the winter and the hottest temperatures in the summer.
- The presence of a green roof decreases the exposure of waterproofing membranes to large temperature fluctuations, which can cause micro-tearing, and ultraviolet radiation.
- Green roofs have excellent noise attenuation, especially for low frequency sounds. An extensive green roof can reduce sound from outside by 40 decibels, while an intensive one can reduce sound by 46-50 decibels.
- Green roofs can sustain a variety of plants and invertebrates, and provide a habitat for various bird species.

Historically, studies on green roofs have explored their energy performance compared with traditional roofs. Thermal performance indicated a significant reduction (~40%) of a building cooling load during the summer period [Spala *et al.*, 2008]. Similar results were achieved for a nursery school, with reductions ranging from 6% to 49%, and reduction ranging from 12% to 87% on the last floor of the nursery school [Santamouris *et al.*, 2007]. Wong *et al.* [2007] note that green roofs tend to experience lower surface temperatures than the original exposed roof, especially in areas well covered by vegetation. When green roofs are well covered by vegetation, the resulting substrate moisture will tend to keep substrate temperature lower than the original exposed bare roof. These studies determined that over 60% of the heat gain was mitigated by vegetated roof systems. Summertime data have indicated significant lower peak roof surface temperature and higher nighttime surface temperature for green roofs as compared to conventional roofs [Sonne, 2006]. The maximum average daily temperature seen for the conventional roof surface was 54.4°C (129.9°F) in his study, while the maximum average day green roof surface temperature was 32.8°C (~21.7°C lower than the conventional roof). Green roofs offer cooling potential (~3.02 kWh/day) to maintain an average room air temperature of 25.7°C (78.3°F) [Kumar, 2005]. Green roofs help minimize environmental burdens, conserve energy, and extend the life span of the roofing system in overall sustainability [Cavanaugh, 2008]. Up to 30% of total rooftop cooling is due to plant transpiration [Takakura *et al.*, 2000]. Bell and Spolek [2009] compared different types of plants for use in increasing the thermal resistance (*R*-value) of green roofs, and found that ryegrass delivered the highest effective *R*-value compared with bare soil, *Vinca major, Trifolium repens,* and *Sedum hispanicum.* Also, though increasing the depth of bare soil from 5 to 14 cm (2.0 to 5.5 inches) increased the *R*-value, no difference was found for different depths of planted soil. This implies that the bulk of benefit toward *R*-value is from evapotranspiration and leaf shading, rather than the moist soil [Bell and Spolek, 2009].

There are several detailed building simulation programs (BSPs) that take into consideration the complete interaction between all thermal-based elements. The most popular BSPs are A Simplified Energy Analysis Method (ASEAM), Building Design Advisor (BDA), Building Load Analysis and

Systems Thermodynamics (BLAST), Builder Guide, Bus^{++}, Dynamic Energy Response of Buildings (DEROB), DOE-2, Energy-10, Energy Plus, ENERPASS, ENER-Win, ESP, FEDs, Home Energy Saver, Hot 2000, TRNSYS, and VisualDOE [Selim, 2003; Selim, 2008; Mansour, 2001].

UAB has utilized Visual DOE in the past with great success in the analysis of innovative structures designed for energy efficiency. VisualDOE uses the DOE 2 calculating core and provides output in both numerical and graphical forms. This software is a preferred calculation method due to its cost, previous verification/validation success, ease of use, database support and reasonable input/output requirements. We envision that this computer simulation tool will be able to effective capture the differences in roof types being explored in the purposed research.

Results and Discussion

<u>Thermal Performance of Mini-Roof Structures</u>

<u>Mini-Roofs</u>

During this study, 15 mini-roof combinations were observed for trends in internal temperatures. The various 15 mini-roof combinations are summarized in Table 1. Several of the mini-roof structures are depicted in Figure 1. This photo shows the layout of the 15 mini-roofs, and a vegetated roof from which surface temperatures of the mini-roofs were measured periodically using an infrared thermometer (see Figure 1).

(a) (b)

Figure 1. (a) Layout of the 15 mini-roods;
(b) vegetated mini-roof (surface temperatures were measured using an IR thermometer.

The roofing materials used are all standard commercial flat roof materials. Flat roof materials were only looked at during the study, since the primary application for the roofing combinations will be on a commercial flat roof top, and not a slanted roof structure. Each mini-roof is 2.4-m (8.0-ft) long x 1.2-m (4.0-ft) wide x 1.2-m (4.0-ft) deep (see Figure 1). A number of different roofing systems are being examined for their energy performance. All roofs are insulated with 5.1 cm (2.0-in) of extruded polystyrene. Then the particular roofing combination being investigated is applied over the insulation and sealed. The roofs also include a proper drainage spout, to ensure correct water evacuation, such as on a real roof.

The roofing systems being studied using the 15 mini-roofs are listed in Table 1.

Table 1. Mini-roof descriptions.

Mini-Roof No.	Mini-Roof Description
1 (Sensor B)	White TPO/PVC/ Elvaloy fully adhered, FiberTite Membrane.
2 (Sensor C)	Black 60-mil EPDM fully adhered/coated/white urethane, Mule Hide Membrane.
3 (Sensor D)	Black 60-mil EPDM fully adhered, Mule Hide Membrane.
4 (Sensor E)	Beige TPO/PVC/Elvaloy fully adhered, FiberTite Membrane.
5 (Sensor F)	White granular modified, Firestone SBS Modified Membrane.
6 (Sensor G)	Black granular modified, Firestone SBS Modified Membrane.
7 (Sensor H)	Black granular modified coated/white urethane, Firestone SBS Modified Membrane.
8 (Sensor I)	Bituthene IRMA with lightweight "T Clear" pavers.
9 (Sensor J)	Bituthene IRMA with river rock ballast.
10 (Sensor K)	Bituthene IRMA with vegetative green roof, ½-in. drain mat, 350-lbs dry soil.
11 (Sensor L)	Bituthene IRMA with vegetative green roof, 1-in. drain mat, 350-lbs dry soil.
12 (Sensor M)	Black 60-mil EPDM loose, ballasted with river rock, Mule Hide Membrane.
13 (Sensor N)	Black 60-mil EPDM loose, ballasted with #300 marble chips, Mule Hide Membrane.
14 (Sensor O)	White TPO/PVC/Elvaroy loose laid, ballasted with river rock, FiberTite Membrane.
15 (Sensor P)	Bituthene IRMA with vegetative green roof, ½-in. drain mat, 350-lbs dry soil.
Sensor S	Sensor inside mini-roof No.10 (inside the soil of the green roof).
Sensor T	Sensor under the white TPO/PVC/Elvaloy loose laid, ballasted with river rock.

To investigate the thermal properties of the roofing structures an ambient temperature probe was placed inside of each roof (see Figures 2 and 3) recording temperature data every 10 minutes of each day, for more than 3 years. This data was then automatically sent to a data logger and placed into an Excel file for review later. The temperature probe reports the temperature to the nearest hundredth of a degree Centigrade.

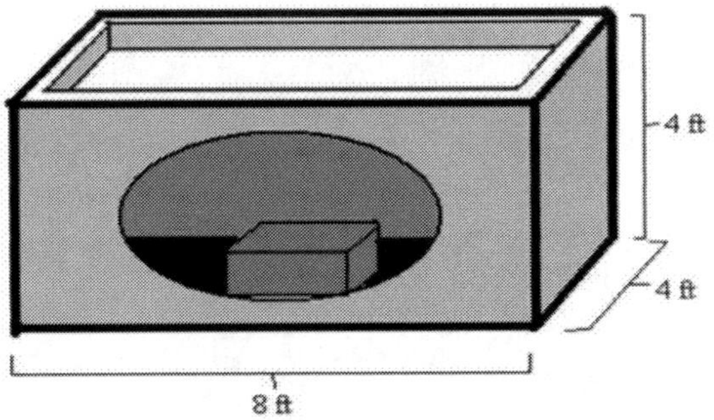

Figure 2. Depiction of a Typical Mini-Roof System.

Figure 3. Mini-roof system showing temperature sensor installed inside a mini-roof.

Figure 4 presents some typical temperature profiles on several different days.

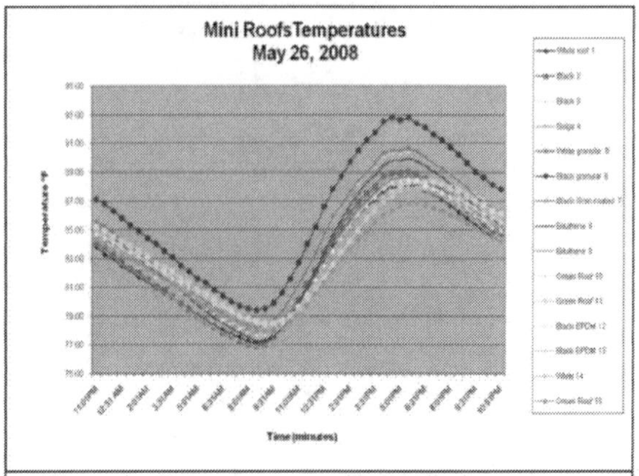

Figure 4a. Temperature profile inside the various mini-roofs on May 26, 2008.

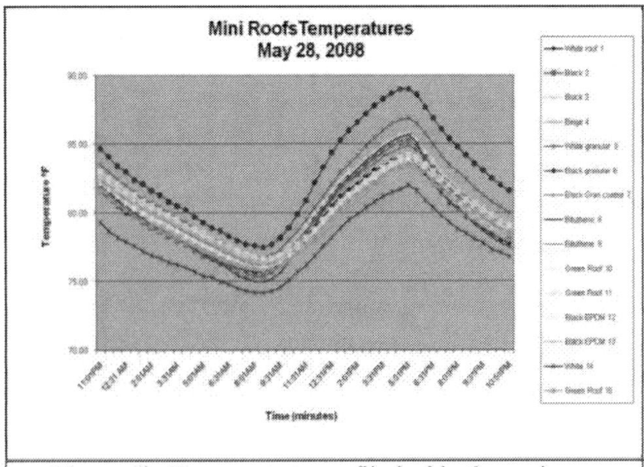

Figure 4b. Temperature profile inside the various mini-roofs on May 28, 2008[Nunez and Peters, 2008].

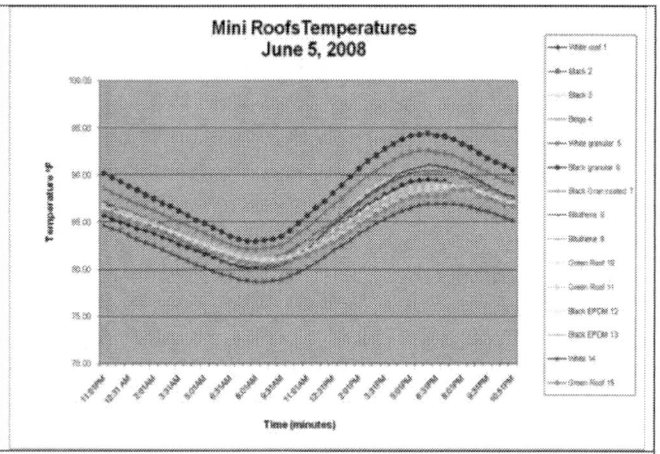

Figure 4c. Temperature profile inside the various mini-roofs on June 5, 2008.

The results from these mini-roof structures have shown the following trends [Nunez and Peters, 2008]:

- Clean white roofs resulted in consistently lower temperatures inside the mini roof than the other roofing materials.
- Black roofs resulted in the highest temperature readings.
- Green roofs resulted in temperatures typically ~1.1-1.7°C (2-3°F) higher than the white roofs; however, they will dampen the drainage of rainfall during a rain storm through the retention of water onto the soil.
- Bituthane (river rock) performs only slightly better than black roofing materials.
- White granular roofing behaved similarly to black granular roofing materials.
- Over time, the reflective (white) roofs become dirty, losing some of their reflectivity, resulting in the roof being less energy efficient.
- White marble chips behaved slightly worse than green roofing materials, but considerably better than black roofing and granular roofing materials.
- While it is too early to come to a definitive conclusion, preliminary evaluations indicate that the "white" and "green" roofs both significantly reduce the roofs surface temperature and therefore, the air temperatures above and around the roof.

Surface temperature measurements during the months of June and July on the various roofing materials used with our mini-roof systems were collected. During this time period, roofing surface temperatures ranging from 20.6°F to 82.2°C were observed. The lower surface temperature values, 20.6°C to 48.3°C, are found in a loose rock or stone roof combination. The higher temperature values occurred on the roofing combinations primarily made of a coating or membrane; these roofs exhibited temperatures roughly between 25.6°C to 82.2°C. The vegetative mini-roofs exhibited surface temperatures ranging between 21.7°C to 52.2°C, while their large counterpart, the pilot roof on top of Hulsey Center, exhibited higher temperatures ranging from 31.7°C to 61.1°C. When looking at the specific roofs contained in each subset of Table 1, it was observed that generally, the lighter colorization of the roof resulted in cooler temperatures. For example, the white Firestone SBS is cooler than the black Firestone SBS on any given day. This thermal property is observed due to the reflectivity of the roof. The darker colored roofs absorb more incoming light radiation than the light colored roofs causing the dark roofs to become hotter. (Since the roofs temperatures observed are taken during the late spring and early summer, it can be inferred that overall roof temperature will increase during late summer).

Photographs of several of these mini-roofs are presented in Figures 5 through 9. Figure 10 presents a typical temperature profile of the 15 mini-roofs during the course of a typical summer week. Series 1 through 5 denoted in the figure refer to the fifteen mini-roofs listed in Table 1. This figure shows a cyclical nature of the temperature readings over each day, generally showing a sinusoidal behavior of temperature; the temperature is cool in the morning, warms up, and is at its hottest during mid-afternoon, and then cools down during evening hours.

Trends seen in statistical comparisons of internal temperatures of similar roofs (using the null hypotheses: $\mu_1 \geq \mu_2$) are summarized below [Peters *et al.*, 2011]:

- Between the two river rock roofs, Roof 9 will most likely always be hotter or equal to roof 12.
- Between the 3 vegetative roofs, all the roofs are statistically equal to each other in thermal properties.
- Both SBS Firestone roofs are statistically the same, but roof 6 is usually hotter.
- The TPO/PVC/Elvaloy Roofs are statistically the same.
- The 60-mil EPDM roofs are also statistically the same.

Figure 5. Black mini-roof.

Figure 6. White (reflective) mini-roof.

Figure 7. Mini-roof equipped with river rocks.

Figure 8. Mini-roof equipped with crushed marble.

Figure 9. Vegetative mini-roof equipped with *Sedum* plants.

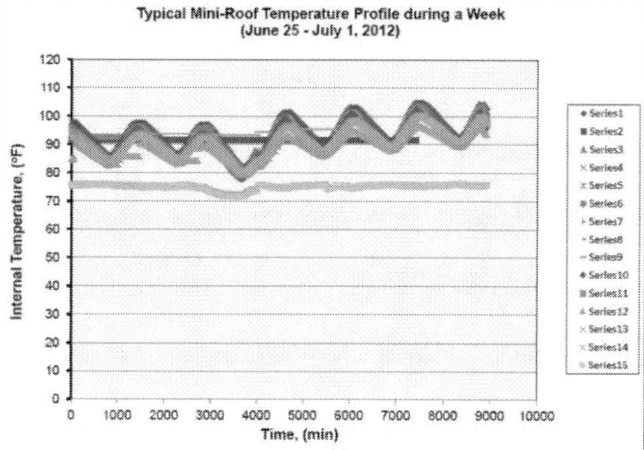

Figure 10. Typical internal mini-roof temperatures during the course of a week.

After acquiring the necessary raw data from the mini-roof sensors, Microsoft Excel (version 2010) was used to plot the data to discover trends in the temperature readings. Subsequent to discovering the trend, a mathematical model was fit to the data. It was noticed that the temperatures cycled in a sinusoidal fashion on yearly and daily time frames, and therefore a general form sine function was utilized as a potential model. Fourier transforms were utilized in order to determine the oscillation frequency of the temperature (ω_0 from the general form $x = A \sin(\omega_0 t + \phi) + C$) by transforming the time domain of the collected raw data into a frequency domain. Two major peaks were discovered from the spectrum: one representing the yearly frequency and the other representing the daily oscillations. After calculating the frequencies, the phase angle (ϕ) and the amplitude (A) of the general form were determined through a regression analysis.

The modeling procedure was applied for all 15 mini-roofs in the study. The developed sine-wave functions indicated that most roofs were statistically different from one another from an amplitude aspect but the phase angles were statistically the same. It was also discovered that almost all roofs had significantly different average mean roof temperatures, but the significance was mostly prevalent in the summer months. During other times of the year, the roofs behaved in a similar fashion (see Tables 2 and 3 for statistics and means of roofs). The fitted sine wave functionalities for the 15 mini-roofs are listed in Table 4.

Table 2.	**Summary of statistics for the phase angles from curve fitting.**		**Table 3.**	**Summary of statistics for the means from curve fitting.**
Statistics of Phase Angles			Statistics of Means	
Mean	4.0303		Mean	66.67
Standard Error	0.00343		Stand Error	0.18
Median	4.0276		Median	66.62
Mode	4.0242		Mode	N/A
Standard Deviation	0.01329		Standard Deviation	0.68
Sample Variance	0.000177		Sample Variance	0.46
Kurtosis	12.0174		Kurtosis	3.11
Skewness	3.3228		Skewness	1.11
Range	0.05512		Range	3.06
Minimum	4.0211		Minimum	65.42
Maximum	4.0762		Maximum	68.47
Sum	60.4541		Sum	1000.12
Number (count)	15		Number (count)	15

Table 4. Fitted sine-wave functionalities describing the internal temperatures in the mini-roofs.

Roof	**Final Equation**	**Amplitude from sinefind.exe**	**Phase Angle**	r^2	**Mean**
1	T = 20.1632*sin ((2*π*3.17894e(-8)*t) +4.021975)+66.68283	20.163	4.022	0.696	66.683
2	T = 20.5499*sin ((2*π*3.18e(-8)*t) +4.02272)+66.63564	20.550	4.023	0.701	66.636
3	T = 20.392*sin ((2*π*3.18e(-8)*t) +4.07624)+66.47947	20.392	4.076	0.712	66.479
4	T = 20.482*sin ((2*π*3.18e(-8)*t) +4.021125)+66.35505	20.482	4.021	0.698	66.355
5	T = 20.4891*sin ((2*π*3.18e(-8)*t) +4.03214)+67.54093	20.349	4.032	0.701	67.541
6	T = 20.7812*sin ((2*π*3.18e(-8)*t) +4.03142)+68.46883	20.781	4.031	0.699	68.469
7	T = 20.2886*sin ((2*π*3.18e(-8)*t) +4.02415)+67.08382	20.289	4.024	0.701	67.084
8	T = 20.7812*sin ((2*π*3.18e(-8)*t) +4.02415)+66.18969	20.781	4.024	0.695	66.190
9	T = 20.3285*sin ((2*π*3.18e(-8)*t) +4.024975)+66.91944	20.329	4.025	0.701	66.919
10	T = 19.8608*sin ((2*π*3.18e(-8)*t) +4.02758)+66.6981	19.861	4.028	0.697	66.698
11	T = 20.0128*sin ((2*π*3.18e(-8)*t) +4.026215)+66.62328	20.013	4.026	0.697	66.623
12	T = 20.0479*sin ((2*π*3.18e(-8)*t) +4.027853)+66.27383	20.048	4.028	0.694	66.274

13	T = 19.9914*sin ((2*π*3.18e(-8)*t) +4.0325)+66.47272	19.991	4.033	0.696	66.473
14	T = 20.0661*sin ((2*π*3.18e(-8)*t) +4.032494)+65.42343	20.066	4.032	0.689	65.423
15	T = 19.4628*sin ((2*π*3.18e(-8)*t) +4.028672)+66.27177	19.463	4.029	0.692	66.272

These results of this study were in agreement with research conducted by Watson [2011] addressing the urban heat island effect for the City of Birmingham by studying the effect from different building facades, building materials and seasonal traits. The data indicates amplitudes (*A*) ranging from 19.5 to 20.7 degrees and phases angles roughly 4.2 to 4.3 radians. The r^2 values ranged from 0.69 to 0.71showing the data modeling was applicable to understanding the trends. The best model developed was for roof 3 (see Figure 11), having the highest correlation coefficient of all 15 mini-roofs.

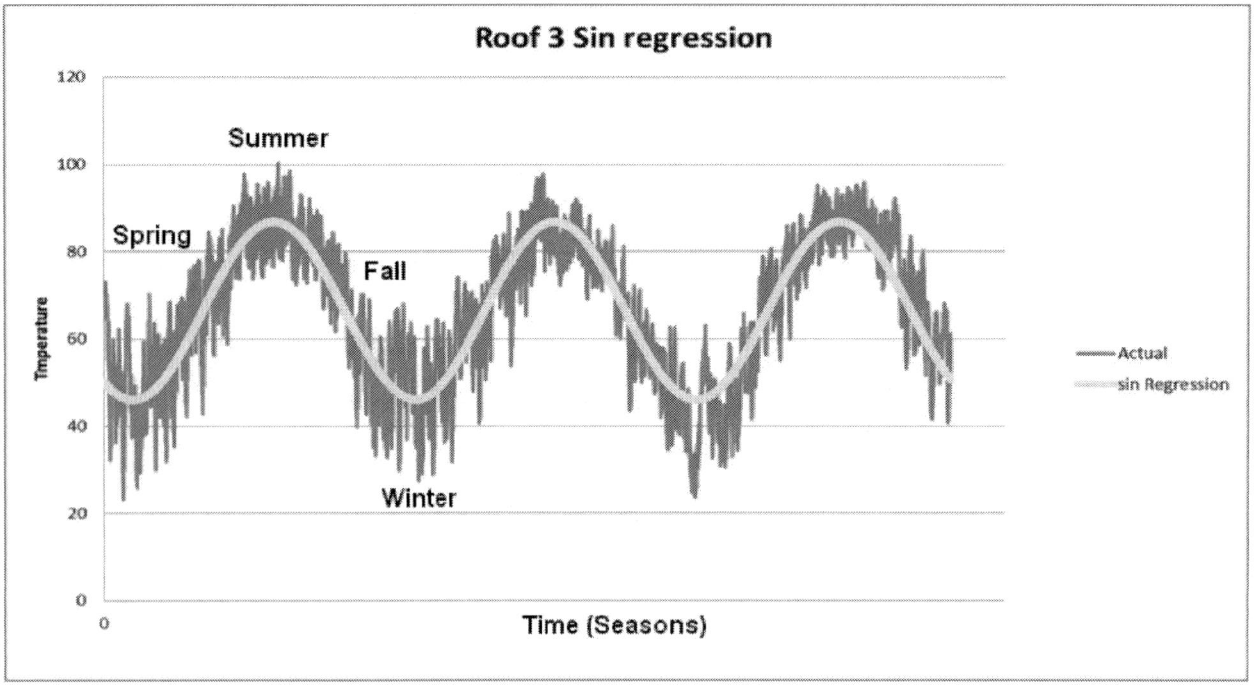

Figure 11. Fitted sine wave regression for a mini-roof system.

Based on the information collected on these mini-roof systems, the UAB Facilities Management Department decided to install an extensive vegetated roof on top of Hulsey Center as a roofing retrofit. UAB wants to obtain more fundamental information and knowledge for establishing green roofs in the southeastern U.S. In the Birmingham, Alabama area, fairly high rainfalls [approximately 132.1 cm/year (52 inches/year) on average] are obtained. However, during the summer months, it is common to have periods of drought with minimal rainfall and very hot and humid days [with temperatures in the 32.2^{+o}C (90+^{o}F range]. Such climatic conditions require plants that can withstand both significant rainfall events and drought conditions. The current pilot green roof on top of Hulsey Center was installed in July 2008. Research conducted at Penn State University shows that green roofs planted with sedum plants reduce the building energy costs [Penn State Center for Green Roof Research, undated (a, b)]. The Facilities Management Department at UAB maintains records of the utilities bills (e.g., natural gas, water, and electricity) for each building on campus. The quantities and costs of utilities prior to (5 years) and after implementation (~3 years) of the pilot green roof have resulted in building energy reductions of 20% to 25%.

Summary and Conclusions

The behavior of different roofing materials affects the heating loads placed upon the buildings and their heating, ventilation, and air conditioning (HVAC) systems. Black roofs resulted in the highest temperature readings. Black roofing materials and bituthane (river rock) perform the poorest of the roofing materials tested, resulting in the highest heating load being placed on the building infrastructure. White granular roofing behaved similarly to black granular roofing materials. Clean white roofs resulted in consistently lower temperatures inside the mini-roof than the other roofing materials; however, over time, the reflective (white) roofs become dirty, losing some of their reflectivity, resulting in the roof being less energy efficient. Green (vegetated) roofs are fairly efficient in terms of energy performance due to evapotranspiration effects. Green roofs resulted in temperatures typically ~1.1-1.7°C (2-3°F) higher than the white roofs; however, they will dampen the drainage of rainfall during a rain storm through the retention of water onto the soil, and thereby lessen the discharge into stormdrains. Green roofs can significantly reduce stormwater runoff, reduce peak flow quantities, and lengthen the time of concentration from roofing structure [Kirby *et al.*, 2011]. "White" and "green" roofs both significantly reduce the roofs surface temperature and therefore, the air temperatures above and around the roof.

Hypothesis testing indicated that, between the 3 vegetative mini-roofs, all the min-roofs are statistically equal to each other in thermal properties. Both SBS Firestone mini-roofs are statistically the same, but mini-roof 6 is usually hotter. The TPO/PVC/Elvaloy mini-roofs are statistically the same. The 60-mil EPDM mini-roofs are also statistically the same.

The temperature varies in a sinusoidal fashion both during the course of the day and on an annual basis. For the various mini-roof structures, the phase angle (ϕ) and the amplitude (A) of the general form were determined through a regression analysis. The developed sine-wave functions indicated that most roofs were statistically different from one another from an amplitude aspect but the phase angles were statistically the same. It was also observed that almost all roofs had significantly different average mean roof temperatures, but the significance was mostly prevalent in the summer months. During other times of the year, the roofs behaved in a similar fashion.

Based on the information collected on these mini-roof systems, the UAB Facilities Management Department installed a vegetated roof on top of Hulsey Center as a roofing retrofit. The vegetated roof is ~1388.0 m^2 (14,940 ft^2) in area, and contains approximately 20,000 sedum plants. Utility bill information both prior to and after implementation of the green roof were gathered for electricity, natural gas, and chilled water were collected. The costs of utilities prior to and after implementation of the pilot green roof indicated utility bill (energy) savings of ~20% to 25% (compared to the case prior to implementation of the vegetated roof system).

Acknowledgments

The financial support and technical support of the UAB's Facilities Management Department is appreciated, including the fabrication and installation of the mini-roofs (on top of the Chemistry Building) and pilot roof on top of Hulsey Center. The input and technical assistance provided by Matt Winslett and Kelly Winnett of the Facilities Management Department in supplying the utility bill information is gratefully acknowledged.

References Cited:

Aitkenhead-Peterson, J.A., B.D. Dvorak, A. Voider, and N.C. Stanley, 2011. "Chemistry of Growth Medium and Leachate from Green Roof Systems in South-Central Texas", ***Urban Ecosystems***, ***14***(1): 17–33.

Bell, H., and G. Spolek, 2009. "Measured Energy Performance of Greenroofs", in ***Proc. Seventh Annual Greening Rooftops for Sustainable Communities Conference***, The Cardinal Group, Toronto, Canada.

Berndtsson, J.C., 2010. "Green Roof Performance towards Management of Runoff Water Quantity and Quality: A Review", ***Ecol. Engrg.***, ***36***(4): 351–360.

Carter, T., and C.R. Jackson, 2007. "Vegetated Roofs for Stormwater Management at Multiple Spatial Scales", ***Landscape and Urban Planning***, ***80***(1–2): 84–94.

Cash, C.G., 1999. "The Relative Durability of Low-Slope Roofing", pp. 119–124 in ***Proceedings of the Fourth International Symposium on Roofing Technology***, Gaithersburg, MD, (September 17–19).

Cavanaugh, L.M., 2008. "Redefining the Green Roof", ***J. Archit. Eng.***, ***14***(1): 4–6.

Clark, C., P. Adriaens, and F.B. Talbot, 2008. "Green Roof Valuation: A Probabilistic Economic Analysis of Environmental Benefits", ***Environ. Sci. Technol.***, ***42***(6): 2155–2161.

Editors of Time Life Books, 1979. ***Roofs and Siding***, Time-Life Books, Alexandria, VA.

Environmental Protection Agency (EPA), 2010. ***Environmental Protection Agency Green Building Design***. www.epa.gov/greenbuilding/.

Environmental Protection Agency (EPA), 2009. ***Buildings and their Impact on the Environment: A Statistical Summary***. www.epa.gov/greenbuilding/pubs/gbstats.pdf.

Fioretti, R., A. Palla, L.G. Lanza, and P. Principi, 2010. "Green Roof Energy and Water Related Performance in the Mediterranean Climate", ***Build. and Environ.***, ***45***(8): 1890–1904.

Givoni, B., 1994. ***Passive & Low Energy Cooling of Buildings***, John Wiley & Sons, Inc., New York, NY.

Green Roofs for Healthy Cities, undated. Website: http://www.greenroofs.org/index.php/about-green-roofs/2577-aboutgrnroofs/, accessed August 3, 2012.

Green Roof Plan, undated. Website: http://www.greenroofplan.com/intensive-vs-extensive-green-roofs/, accessed August 3, 2012.

Hathaway, A.M., W.F. Hunt, and G.D. Jennings, 2008. "A Field Study of Green Roof Hydrologic and Water Quality Performance", ***Transactions Am. Soc. Agric. & Biolog. Engineers***, ***51***(1): 37–44.

Kirby, J.T., R. Peters, S. Watts, and M. Winslett, 2011. "Experimental Investigation of Runoff Mitigation from Vegetated Roofs", Final report submitted to the Alabama Water Resources Research Institute, Auburn, AL.

Kosny, J., D. Yarbrough, P. Childs, and S. Mohiuddin, 2007. "How the Same Wall Can Have Several Different *R*-Values: Relations between Amount of Framing and Overall Thermal Performance in Wood And Steel-Framed Walls", *Oak Ridge National Laboratory*.

Kumar, R., 2005. "Performance Evaluation of Green Roof and Shading for Thermal Protection of Buildings", ***Build. Environ.***, ***40***(11): 1505–1511.

Mansour, H.M.A., 2001. "Sustainable Housing Development with Special Reference to Environmental Aspects in Low Cost Housing in Egypt". M.S. Thesis, Cairo University, Cairo, Egypt. 117 pages.

Niu, H., C. Clark, J. Zhou, and P. Adriaens, 2010. "Scaling of Economic Benefits from Green Roof Implementation in Washington, D.C.", ***Environ. Sci. Technol.***, ***44***(11): 4302–4308.

Nunez, S., and R.W. Peters, 2008. "Effect of Roofing Materials on Building Energy Efficiency", Paper included in the conference proceedings of the 2008 Annual AIChE Meeting, Philadelphia, PA, (November 16–21).

Penn State Center for Green Roof Research, undated (a). Pennsylvania State University website: http://horticulture.psu.edu/cms/greenroofcenter/about_ctr.html; accessed July 5, 2009.

Penn State Center for Green Roof Research, undated (b). "Utility Bills Getting You Down? Plant Stuff Up on the Roof", http://aginfo.psu.edu/news/2000/9/roof.html; accessed July 5, 2009.

Peters, R.W., J.G. Price, R.D. Sherrod, S.A. Watts, J.M. Gohlke, J.T. Kirby, and M. Winslett, 2011. "Roofing Types and Their Effects on the Thermal Properties of Runoff", ***Proc. Alabama Water Resources Conference***, Orange Beach, AL, (September 8–9).

Reid, R.N., 2000. ***Roofing and Cladding Systems: A Guide for Facility Managers***, Fairmont Press, Lilburn, GA.

Roehr, D., and Y. Kong, 2010. "Stormwater Runoff Reduction Achieved by Green Roofs: Comparing SWMM Method to TR-55 Method", pages 1012–1021 in ***Low Impact Development 2010: Redefining Water in the City, Proceedings of the 2010 International Low Impact Development Conference***, San Francisco, CA, (April 11–14).

Rosatto, H.G., D. Laureda, D. Perez, D. Nerrera, M. Meyer, P. Gamboa, G. Villalba, M. Friedrich, M. Bargieta, L.R. Plaza, G. Calvo, M. Miranda, M. Inigo, and E. Quaintenne, 2010. "Water Retention Efficiency of Green Roof Systems", ***Revista de la Facultad de Ciencias Agarias***, ***42*** (1): 213–219.

Santamouris, M., C. Pavlou, P. Doukas, G. Mihalakakou, A. Synneta, A. Hatzibiros, and P. Patargias, 2007. "Investigating and Analysing the Energy and Environmental Performance of an Experimental Green Roof System Installed in a Nursery School Building in Athens, Greece", ***Energy***, ***32*** (9): 178–1788.

Schroll, E., J. Lambrinos, T. Righetti, and D. Sandrock, 2011. "The Role of Vegetation in Regulating Stormwater Runoff from Green Roofs in a Winter Rainfall Climate", ***Ecol. Engrg.***, ***37*** (4): 595–600.

Selim, M.A., 2003. "The Effect of Building Envelope Technology on Energy Conservation in Buildings", M.S. Thesis. Cairo University. Cairo, Egypt.

Selim, R., 2008. "Investigation of Thermal Performance, Indoor Air Quality and Energy Consumption within Tuskegee Healthy House Demonstrative Model", Ph.D. Dissertation,. University of Alabama at Birmingham, Birmingham, AL.

Sonne, J., 2006. "Evaluating Green Roof Energy Performance", ***ASHRAE J.***, ***48*** (2): 59–61.

Spala, A., H.S. Bagiorgas, M.N. Assimakopoulos, J. Kalavrouziotis, D. Matthopoulos, and G. Mihalakakou, 2008. "On the Green Roof System. Selection, State of the Art and Energy Potential Investigation of a System Installed in an Office Building in Athens, Greece", ***Renew. Energy***, ***33*** (1): 173–177.

Takakura, T., S. Kitade, and E. Goto, 2000. "Cooling Effect of Greenery Cover over a Building", ***Energy Build.***, ***31***: 1–6.

United States Green Building Council (USGBC), 2010. United States Green Building Council. www.usgbc.org.

Villarreal, E.L., and A.S.D.L. Bengtsson, 2004. "Inner City Stormwater Control Using a Combination of Best Management Practices", ***Ecol. Engrg.***, ***22*** (4–5): 279–298.

Watson, C.J., 2011. "Sustainable Urban Design Practices that Mitigate Urban Heat Islands and Reduce Energy Consumption", Ph.D. Dissertation, University of Alabama at Birmingham, Birmingham, AL.

Williams N.S.G., J.P. Nicholas, and K.J. Raynor, 2010. "Green Roofs for a Wide Brown Land: Opportunities and barriers for Rooftop Greening in Australia", ***Urban Forestry & Urban Greening***, ***9***(3): 245–251.

Wolf, D., and J.T. Lundholm, 2008. "Water Uptake in Green Roof Microcosms: Effects of Plant Species and water Availability", ***Ecol. Engrg.***, ***33*** (2): 179–186.

Wong, N.H., P.Y. Tan, and C. Yu, 2007. "Study of Thermal Performance of Extensive Rooftop Greenery Systems in the Tropical Climate", ***Build. Environ.***, ***42*** (1): 25–54.

Wong, N.H., S.F. Tay, R. Wong, C.L. Ong, and A. Sia, 2003. "Life Cycle Cost Analysis of Rooftop Gardens in Singapore", ***Building and Environment***, ***38*** (3): 499–509.

(665c) Low-Cost Phase Change Material for Building Envelopes

Ramin Abhari, Syntroleum, Tulsa, OK and Laura Ford, University of Tulsa, Tulsa, OK

Studies have shown that incorporating phase change material (PCM) in building envelopes—attic insulation, wall boards, etc.—can reduce energy consumption for home heating and cooling by more than 30%. However, high manufacturing costs have so far limited PCM use to novelty products. This work has developed a low-cost PCM through a selective process for n-paraffin production and continuous high-throughput equipment for its encapsulation. The n-paraffins are made sustainably via hydrodeoxygenation of bio-oils. The heat of fusion and melting point of our n-paraffins made from bio-oils compare well with those of traditional n-paraffins. Heat cycling tests at 40°C indicate that the n-paraffins encapsulated in high density polyethylene will retain at least 90% of the n-paraffins. The encapsulated n-paraffins have been incorporated into cellulose insulation and exposed to outside conditions for over a month. Results include a more than 10% reduction in heat flux through insulation with our PCM and a 1°C reduction in peak temperature compared with plain insulation. An economic analysis will be presented as well.

Sustainability through Durable Building Materials and Assemblies

Dr. Theresa Weston
DuPont Building Innovations

An opportunity for chemical engineers in transforming the sustainable or green building industry is in developing understanding of building durability. Building durability is a key sustainability attribute. A building cannot be truly energy efficient or sustainable unless it lasts. One of the most important components of building durability is the prevention of moisture damage. Chemical engineering knowledge of combined heat and mass transfer can provide key insights in to moisture damage risk analysis, and can therefore aid in the design on energy efficient and durable buildings.

Sustainable Building Codes & Standards

Despite the overall slow-down of the construction industry, especially the home building industry in North America, one growth area has been green and energy efficient construction. This trend has coincided with a large number of changes in the standards, codes and regulations governing the building construction. More specifically, the industry has introduced new energy efficiency and sustainability codes and standards. While these codes in many ways are an extension of existing building codes, the combination of more stringent, and in some cases new, requirements, lead to the use of new building materials, assemblies and systems. Although a significant amount of research has been conducted moisture durability of traditional wall systems, this research needs to be extended to apply to these new energy efficient and sustainable building systems

Modeling Moisture Damage Risk

Modeling combined heat and mass (water) transport or hygrothermal modeling is a useful tool, which is becoming more prevalent in the construction industry. This type of modeling can highlight the risk of moisture accumulation in building systems. An example (shown below) is the higher risk of accumulation of water from building assemblies when additional layers of low vapor permeability materials such as exterior insulation are added to a wall system.

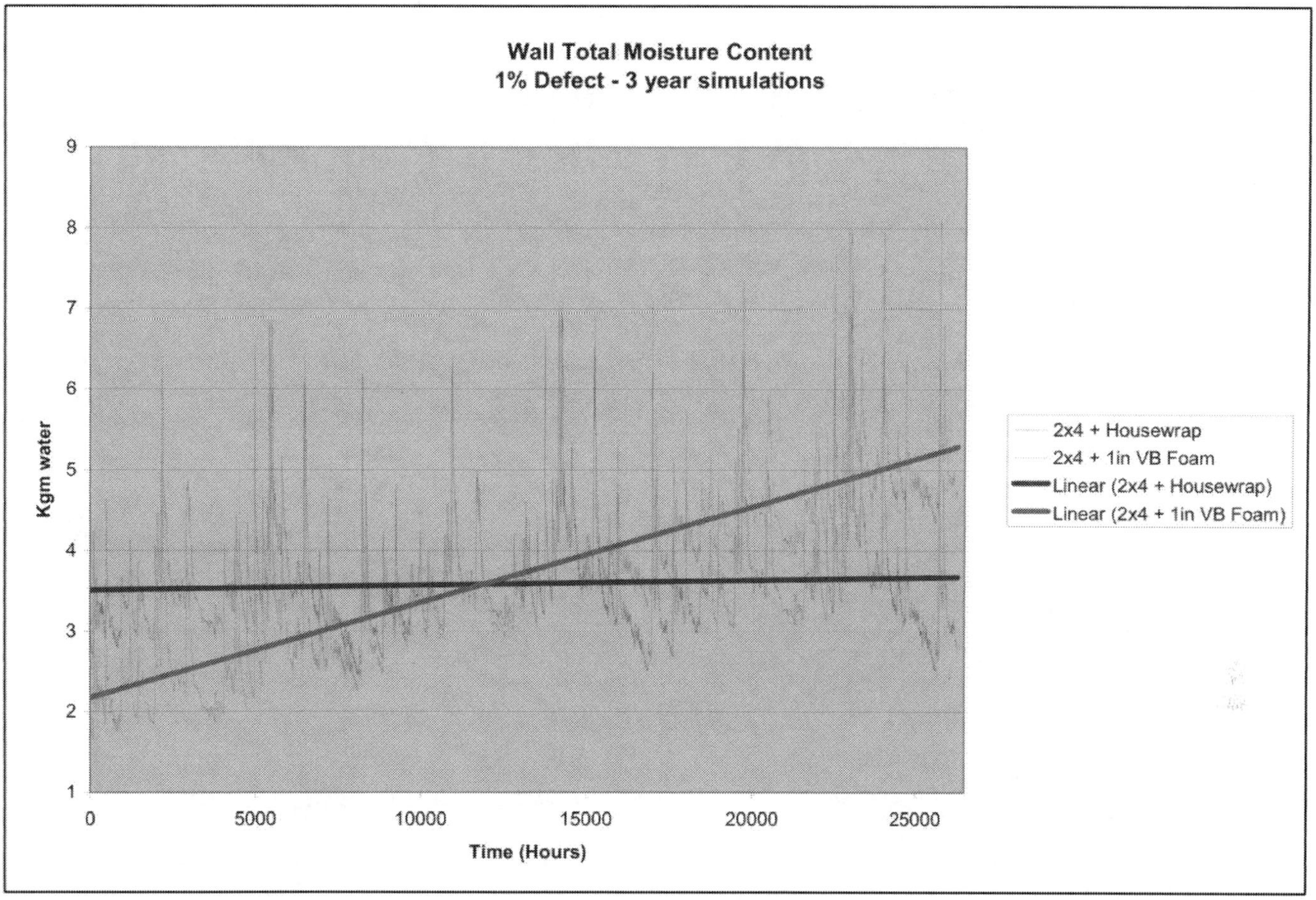

This presentation will provide information on the status of moisture modeling in the construction industry including:

- Standardization of protocols for the use of moisture risk analysis by the development of ASHRAE Standard 160 – *Criteria for Moisture-Control Design Analysis in Buildings*

- Sponsorship of hygrothermal modeling software development by the US Department of Energy, and

- Development of new methodologies to extend the use of hygrothermal simulation to include the management of drying of incidental water leaks.

Examples of specific materials used to help provide durable, energy efficiency to buildings will be discussed.

(667a) A Two-Scale 13C-Based Method for Metabolic Flux Measurement and Prediction in Genome-Scale Models

Hector Garcia-Martin, Fuel Synthesis Division, Joint Bioenergy Institute, Emeryville, CA

Systems biology aims to provide a predictive and quantitative understanding of cell behaviour as the outcome of the interaction of its comprising parts. Metabolic flux profiles (i.e. the number of molecules traversing each biochemical reaction encoded in its genome per unit time) are not only a key phenotypic characteristic but also embody the essence of this complexity since they represent the final functional output of the interactions of all the molecular machinery studied by all the other "omics" fields. Two of the most popular methods for studying metabolic fluxes are Flux Balance Analysis (FBA) and 13C Metabolic Flux Analysis (13C MFA), each of them displaying its own advantages and disadvantages. In this talk I will present a new method: Two scale 13C Metabolic Flux Analysis (2S-13CMFA), which combines the advantages of FBA and 13C MFA. I will showcase its applications and possibilities with data from the KEIO knockout collection.

(667b) Phenotypic and Multi-Omic Approaches to Address Molecular Bottlenecks in the Fermentation of Lignocellulose Into Ethanol by *Saccharomyces Cerevisiae*

Trey Sato[1], Dana Wohlbach[2], Jeffrey Lewis[2], Yaoping Zhang[1], Mingjie Jin[3], Yury Bukhman[1], Wendy Schackwitz[4], Christa Pennacchio[4], David Hodge[3], Venkatesh Balan[3], Bruce E. Dale[3] and Audrey Gasch[2], (1)University of Wisconsin, Great Lakes Bioenergy Research Center, Madison, WI, (2)University of Wisconsin, Madison, WI, (3)Chemical Engineering and Material Science, Michigan State University, East Lansing, MI, (4)The Joint Genome Institute, Walnut Creek, CA

While cellulosic ethanol is being looked to for relief of the global energy demand, a number of molecular bottlenecks currently exist that prevent the efficient bioconversion of lignocellulose into ethanol. For example, it is well known that: 1) native *Saccharomyces cerevisiae* yeast strains cannot sufficiently ferment xylose, and 2) side products generated from pretreatment, including Ammonia Fiber Expansion (AFEX™) and alkaline hydrogen peroxide (AHP), of plant biomass illicit a cellular stress response, which further limits fermentation productivity. At the Great Lakes Bioenergy Research Center, we have taken a multi-comparative approach to facilitate the discovery and understanding of molecular bottlenecks in the fermentation of lignocellulosic hydrolysates by yeast. Through multi-phenotypic and bioinformatic analysis of 111 natural and domesticated isolates, we have identified a wild *S. cerevisiae* strain that maintains rapid growth and cell viability in a variety of distinctly prepared hydrolysates. Following directed engineering of a xylose metabolism pathway, we performed directed evolution that yielded mutants able to ferment 2 to 3-fold more xylose from AFEX™ corn stover hydrolysate (ACSH) than unevolved parents. Furthermore, we employed temporal profiling of gene expression levels during ACSH fermentations, which identified differences in cell physiology between evolved and unevolved strains. Analysis of extracellular metabolite, amino acid and metal concentrations additionally identified limiting nutrients during fermentation. Coupled with comparative genome resequencing of parental and evolved strains, this suite of Omic data is being integrated in metabolic network models to identify and understand genetic differences that impact xylose fermentation in stress-inducing lignocellulosic hydrolysates.

(667c) Towards Industrial Robustness: Understanding of Wild-Type and Populus Hydrolysate- Tolerant Mutant Strains of Clostridium Thermocellum

Jessica Linville[1], Miguel Rodriguez[2], Jonathan Mielenz[3] and Chris D. Cox[1,4], (1)Department of Civil and Environmental Engineering , University of Tennessee, Knoxville, TN, (2)Bioscieces Division, Oak Ridge National Laboratory, Oak Ridge, TN, (3)Biosciences Division, Oak Ridge National Laboratory, Oak Ridge, TN, (4)Center for Environmental Biotechnology, University of Tennessee, Knoxville, TN

An industrially robust microorganism capable of degrading lignocellulosic biomass in the presence of inhibitory compounds produced from the pretreatment process in a CBP scheme is needed to reduce the cost of cellulosic biofuels. *Clostridium thermocellum* has the ability to convert biomasses complex carbohydrates into ethanol by fermentation. However, *C. thermocellum* is inhibited by by-products from the biomass pretreatment process. To increase *C. thermocellum*'s industrial robustness, a 17.5% v/v *Populus* hydrolysate-tolerant mutant (PM) was produced by 117 successive transfers of *C. thermocellum* 27405 strain into increasing concentrations of the hydrolysate. A single colony culture of the PM was then produced and the genome has been sequenced. A study comparing the wild-type (WT) and PM strains of *C. thermocellum* was completed with growth on cellobiose examining sugar utilization and end product formation, and growth on Avicel examining protein concentration. The growth of the PM in 17.5% v/v *Populus* hydrolysate had a 20% higher optical density, a 33% faster sugar utilization rate (g glucan/L/hr), and produced 34% more ethanol (g ethanol/ g glucan) than the WT grown without hydrolysate. The PM retained these characteristics when grown without hydrolysate with a 37% higher optical density, a 40% faster sugar utilization rate and produced 20% more ethanol than the WT grown under the same conditions. The WT was severely inhibited when grown in the presence of 17.5% v/v Populus hydrolysate with a barely detectable optical density, 7% of sugar utilization rate, and produced 7.5% of the expected ethanol versus when it was grown without hydrolysate. A combination of the genomic mutations of the PM and gene expression profiles from triplicate fermentations in various concentrations of the *Populus* hydrolysate will be used to determine which mutations in the genome might be the cause of the increased tolerance to the hydrolysate.

(667d) Development of Advanced Technologies for Feedstock Chracterization and HTP Pretreatment Process Monitoring

Seema Singh, Joint BioEnergy Institute, Sandia National Laboratory, Emeryville, CA

Today, carbon-rich fossil fuels, primarily oil, coal and natural gas, provide 85 percent of the energy consumed in the United States. Fossil fuel use increases CO2 emissions, increasing the concentration of greenhouse gases and raising the risk of global warming. The high energy content of liquid hydrocarbon fuels makes them the preferred energy source for all modes of transportation. In the US alone, transportation consumes around 13.8 million barrels of oil per day and generates over 0.5 gigatons of carbon per year. This has spurred research into alternative, non-fossil energy sources. Among the options (nuclear, concentrated solar thermal, geothermal, hydroelectric, wind, solar and biomass), only biomass has the potential to provide a high-energy-content transportation fuel. Biomass is renewable resource that is carbon-neutral.

Currently, biofuels such as ethanol are produced largely from grains, but there is a large, untapped resource (estimated at more than a billion tons per year) of plant biomass that could be utilized as a renewable, domestic source of liquid fuels. Well-established processes convert the starch content of the grain into sugars that can be fermented to ethanol. Plant-derived biomass contains cellulose, which is more difficult to convert to sugars. The development of cost-effective and energy-efficient processes to transform cellulose in biomass into fuels is hampered by significant roadblocks, including the lack of specifically developed energy crops, the difficulty in separating biomass components, low activity of enzymes used to deconstruct biomass, and the inhibitory effect of fuels and processing byproducts on organisms responsible for producing fuels from biomass monomers.

The Joint BioEnergy Institute (JBEI) is one of three US Department of Energy Bioenergy Research Centers that will address these roadblocks in biofuels production. JBEI draws on the expertise and capabilities of three national laboratories (Lawrence Berkeley National Laboratory (LBNL), Sandia National Laboratories (SNL), and Lawrence Livermore National Laboratory (LLNL)), two leading US universities (University of California campuses at Berkeley (UCB) and Davis (UCD)), and a foundation (Carnegie Institute of Washington at Stanford University) to develop the scientific and technological base needed to convert the energy stored in cellulose into transportation fuels and commodity chemicals. Established scientists from the participating organizations are leading teams of researchers to solve the key scientific problems and develop the tools and infrastructure that will enable other researchers and companies to rapidly develop new biofuels and scale production to meet US transportation needs, and to develop and rapidly transition new technologies to the commercial sector. This talk will present a brief summary of the organization and efforts at JBEI, highlight the efforts on the development and deployment of novel technologies for biomass analysis (compositional and mechanical) and HTP methods for monitoring pretreatment process efficacies.

(667e) Discovery and Functional Determinations of Biomass-Degrading Enzymes by Robotic Cell-Free Translation

Brian G. Fox[1], Taichi E. Takasuka[1], Lai Bergeman[2], Walker Johnnie A.[3], Kai Deng[4] and Trent Northen[4], (1)Great Lakes Bioenergy Center, University of Wisconsin, Madison, WI, (2)Department of Biochemistry, University of Wisconsin-Madison, Madison, WI, (3)Department of Biochemistry, University of Wisconsin-Madison, mad, (4)Joint BioEnergy Institute, Emeryville, CA

Our use of a platform for functional determination of enzymes will be presented. Robotic wheat germ cell-free protein translation system was used for protein synthesis and enzymatic assays were performed using the translation reactions obtained directly from the robot. Fluorescent sugar analogs and purified polysaccharides were used for initial characterizations. Nanostructure initiator mass spectrometry (NIMS)-based assay provided additional functional assignments and also revealed diagnostic differences in time-dependent cascades of products given by different enzymes. The approach was used to evaluate a majority of secreted cellulosomal proteins from *C. thermocellum* in different enzyme formats, their comparative single reactivities, and their synergistic behavior for hydrolysis of various polysaccharide-containing materials. Combinatorial enzyme activity assays revealed synergistic activities between enzymes with different assigned functions

(667f) Characterization of the Interactions of Cellulose, Hemicellulose, and Lignin During Pretreatment Through the Use of Flowthrough Pretreatment

Heather L. McKenzie[1,2], Jaclyn D. DeMartini[1,2], Nancy L. Engle[1,3], Marcus Foston[1,4], Sivakumar Pattathil[1,5], Bruce A. Tomkins[1,6], Michael G. Hahn[1,5], Arthur J. Ragauskas[1,4], Timothy J. Tschaplinski[1,3], Gary A. Van Berkel[1,6] and Charles Wyman[2,7], (1)BioEnergy Science Center, Oak Ridge, TN, (2)Chemical and Environmental Engineering, University of California Riverside, CE-CERT, Riverside, CA, (3)Biosciences Division, Oak Ridge National Laboratory, Oak Ridge, TN, (4)Department of Chemistry and Biochemistry, Georgia Institute of Technology, Atlanta, GA, (5)Complex Carbohydrate Research Center, University of Georgia, Athens, GA, (6)Chemical Sciences Division, Oak Ridge National Laboratory, Oak Ridge, TN, (7)BioEnergy Science Center, Oak Ridge National Laboratory, Oak Ridge, TN

The conversion of lignocellulosic biomass to ethanol requires recovery of the sugars contained in hemicellulose and cellulose at high yields. Hydrothermal pretreatment can be used to hydrolyze and solubilize hemicellulose prior to enzymatic hydrolysis of the cellulose left in the residual solids to glucose. An understanding of the deconstruction of cell wall during pretreatment would support the development of improved energy crops and pretreatment strategies. The results of flowthrough pretreatment of native poplar and model substrates were examined for evidence of the interaction of cellulose, hemicellulose, and lignin during pretreatment. The liquid effluent was analyzed for oligomers and phenols, while the composition of the residual solids was tested using HPLC and NMR. Differences in the release of xylooligomers from native poplar, holocellulose, and isolated xylan demonstrated the importance of the interactions of cellulose-hemicellulose and lignin-hemicellulose on hemicellulose hydrolysis. Pretreatment of the isolated lignin indicates that depolymerization is the primary mechanism for lignin extraction while differences in the release of lignin from native poplar and the isolated lignin indicate that lignin-hemicellulose interactions influence relative reactivity. Recent work has identified the removal of specific cell wall structures, such as methyl glucuronoxylan, as key to improvements in enzymatic digestibility. The removal of these components during pretreatment will be examined closely.

(674a) Soluble Inert Template for Preparing Nanostructured Carbon Materials From Forest and Plant Bioproducts

Zhengrong Gu, Agricultural and Biosystems Engineering, South Dakota State University, Brookings, SD and Xiaomin Wang, South Dakota State University, Brookings, SD

Nanostructured carbon materials is critical for multiple purification/separation processes such as water purification, natural gas purification, as well as used as catalyst supports for catalytic chemical processes, and advanced energy storage materials. Soluble inert templates, which reduce preparation cost by simplifying post template removal steps, are most hopeful substitutes for silica or other insoluble porous templates in preparation of nanostructured carbon materials. However, until now, the role of inert salts, such as NaCl, in preparation of nanostructured carbon materials is not clear. In references for ultrasound spray pyrolysis, soluble salts, generated in degradation of precursors or added into precursor solutions, have been recognized as temporary template for synthesis mesoporous or nanostructured carbon materials, while NaCl was identified as chemical activation catalyst for dehydration and irritation in preparing activated carbon with high surface from nuts shell.

In this project, porous structure of carbon materials was also modified by changing salt composition and preparation parameters such as feedstock, temperature procedure, gas composition of atmosphere, and gas flow rate. The pore structure of generated carbon materials were characterized with N_2 isothermal adsorption using a Micromeritics ASAP2010 automated gas adsorption analyzer. Specific surface areas were calculated using the Brunauer–Emmett–Teller (BET) model for the nitrogen isotherm.

Inert soluble salts such as NaCl and KCl were used as templates, while carbon dioxide (CO_2), carbon monoxide (CO), and steam (H_2O) were used as the activating agents to simulate syngas and flue gas process streams in heat treatment of feedstock. The amount of CO_2 in the gas mixture was diluted with inert N_2 while the total flow was held constant to verify catalytic effect of salts. The influence of syngas composition with varying ratios of CO_2:CO is studied to determine the optimum and tolerable amount of CO in the activating agent.

Therefore, nanostructure of carbon materials has been rationally manipulated by changing salt composition and preparation parameters such as feedstock, temperature procedure, gas composition of atmosphere, and gas flow rate in this project. Furthermore, the nanostructured carbon material, with hierarchical porous structure, was prepared from different herbaceous feedstock in this project. The results of this project shown that inert soluble salts, such as NaCl and KCl, not only functioned as template for forming mesoporous structure but also as catalysts for speeding up generation of microporous structure through gasification of biomass in gas containing steam, CO_2 or O_2.

(674b) Alternative Production of Calcium Carbide

Chinedu Umeozor, Chemical Engineering, University of Missouri, Columbia, MO and Tom R. Marrero, chemical Engineering, University of Missouri, Columbia, MO

ALTERNATIVE METHOD OF PRODUCING CALCIUM CARBIDE

Chinedu J. Umeozor and Thomas R. Marrero

Department of Chemical Engineering, University of Missouri, Columbia, MO 65211

ABSTRACT

The primary purpose of this research is to determine the effect of ultrafine reactant particle size on the preparation of calcium carbide. Reactants are lime and biochar, a relatively pure source of carbon. The carbon has a particle size distribution of 10 to 100 micrometers and reacts with lime in a bench-scale furnace. Reaction temperatures are above 1000°C. The anticipated results of these bench-scale studies are (1) the effect of ultrafine particle size on calcium carbide yields as a function of temperature, and (2) determination of the costs/benefits of biochar in the preparation of calcium carbide. These results are significant with respect to calcium carbide quality and the resulting environmental impact from CO_2 emissions for calcium carbide manufacture.

Current CO_2 emissions associated with calcium carbide production are about 1/2 metric ton per ton of calcium carbide. With the global production of calcium carbide exceeding 15 million tons per year, CO_2 emissions to the atmosphere exceed 8 million tons per year. In addition to this issue, for each metric ton of calcium carbide produced, the actual energy consumed is from 3000 to 3800 kWh which differs significantly from the theoretical energy requirement of 2000 kWh. The difference between the actual energy consumed and theoretical energy required to produce a ton of calcium carbide is therefore large and provides an opportunity to reduce the required energy for calcium carbide production and also reduce the associated CO_2 emissions to the atmosphere. This opportunity is addressed in this study by using purer reactants and ultrafine particles. The ultrafine particle size reactants form a eutectic of calcium oxide and carbon which allows calcium carbide manufacture at 500°C less than normal practice.

(674c) Novel Bacterial Isolate for the Conversion of Lignocellulosic Mimics to Bio-Succinate

Jason W. Soares[1], Sanchao Liu[2], Christian Sund[2], Matthew Servinsky[2], Arthur Ragauskas[3], Laurel A. Doherty[1], Cherie Ziemer[4] and Steven Arcidiacono[1], (1)U.S. Army Natick Soldier Research Development and Engineering Center, Natick, MA, (2)U.S. Army Research Laboratory, Adelphi, MD, (3)Georgia Institute of Technology, Atlanta, GA, (4)ARS-USDA, Ames, IA

Bioconversion of lignocellulosic biomass to useful byproducts offers an opportunity to introduce alternative bio-based products into both the commercial and military sectors. Our focus has been on the discovery of microorganisms that exhibit unique lignocellulose hydrolysis behavior while generating useful co-products. A novel isolate, SD CC 1b, has demonstrated hydrolysis of model substrates that mimic the constituents of lignocellulosic biomass, namely cellulose, of varying degrees of crystallinity, and xylan, the major component of hemicellulose. Anaerobic fermentation of SD CC 1b in a mixed gas (85% nitrogen, 10% carbon dioxide and 5% hydrogen) atmosphere, with the respective substrates as the sole carbon source, revealed preferential hydrolysis of the crystalline domains of cellulose relative to amorphous regions after 72 hours. Growth studies on SD CC 1b hydrolysis of xylan, specifically Oat Spelt Xylan (OSX), revealed 6x more growth of SD CC 1b on OSX and 2x more growth on washed OSX (washed to remove monosaccharides present in commercial powders) relative to substrate-deficient cultures after 96 hours. Metabolic analyses predominantly indicated the production of succinic acid, particularly for OSX-containing cultures. During the hydrolysis of OSX, SD CC 1b produced 3 moles of succinate for each mole of acetate while consuming carbon dioxide, which is present in excess in the gaseous environment. Efforts to increase the succinate:acetate ratio by altering the anaerobic atmosphere composition will be discussed. Our novel isolate could contribute to the emerging market of sustainable biomaterials as a renewable, continual source for bio-succinate production through direct bioconversion of lignocellulosic biomass. Bio-succinate could potentially replace petroleum-based succinic acid, which is used in pigments, detergents and as biodegradable polymers in packaging and film applications.

(674d) Polyols and Polyurethane Foams From Crude Glycerol Based Liquefaction of Lignocellulosic Biomass: Effects of Crude Glycerol Impurities On Product Properties

Shengjun Hu, Food, Agricultural, and Biological Engineering, The Ohio State University-OARDC, Wooster, OH, Xiaolan Luo, Department of Food, Agricultural and Biological Engineering, The Ohio State University-OARDC, Wooster, OH and Yebo Li, Department of Food, Agricultural and Biological Engineering, , The Ohio State University-OARDC, Wooster, OH

Crude glycerol, a major byproduct from the biodiesel production process, can be used as an effective and cost-effective biomass liquefaction solvent for the production of polyols and PU foams. However, crude glycerol consists of not only glycerol but also some impurities compounds such as soap and methanol and its exact composition varies widely with different biodiesel production processes. To achieve polyol and PU production of consistent and controllable properties, it is vital to evaluate the effects of crude glycerol impurities on the properties of polyols and PU foams. In this study, the chemical compositions of five crude glycerol samples from different biodiesel plants were first determined, and then the effects of crude glycerol impurities on polyol and PU properties were studied. Our results showed that the composition of different crude glycerol samples varied widely from each other, but all samples were shown to contain glycerol, soap, methanol, fatty acid methyl esters (FAMEs), water, glycerides, free fatty acids (FFAs), and ash as their main components. To investigate the effects of crude glycerol impurities on polyol and PU properties, crude glycerol impurities were mixed with pure glycerol at different formulations and then the mixtures were used as biomass liquefaction solvents for polyol and PU production. Our results showed that the properties of polyols and PU foams were significantly affected by soap, FFAs, and FAMEs: the hydroxyl number of polyols and the density and compressive strength of PU foams decreased significantly with increasing levels of these compounds in liquefaction solvents. Other crude glycerol impurities such as methanol and water were shown to have insignificant effects on the properties of polyols and PU foams.

Sugar Cane refinery -production of sugarcane wax from sulphitation press mud

A.K.Ray , R.B.Nigam* S.K.Sharma*and P.Sanyal

Indian Institute of Technology Roorkee, Saharanpur Campus, Saharanpur , India

***Former Director, National Sugar Institute, Kanpur, Govt. of India**

**** Former Advisor (By-products),National federation of Cooperation Sugar Factories Ltd.**

ABSTRACT

India is the largest cane sugar producing country in the World. Apart from sugar it discharges a large number of byproducts such as bagasse, molasses, press mud , cane tops, leaves and straw. Economy of Indian sugar industry is not as expected due to variety of reasons. Therefore, there are all round attempts all over the cane sugar producing countries to establish a biorefinery to make this industry profitable. One of the waste products of sugar industry is filter mud which is approximately 3.4 % (at 77% moisture) on the basis of 100 tonne cane crushed. The quantity of filter mud varies from country to country, from mill to mill and even within the same mill and from field to field. The parameters influencing the variation include systems of clarification, extraction, cane variety, degree of mechanization etc. Filter mud is not only a source of biological fertilizer, but also helps in the disposal of distillery spent wash after recovering useful fuel in the form of Methane gas. The press mud of sulphitation sugar factories contains a very valuable chemical ,wax which is a suitable substitute of carnauba wax. A number of modified waxes e.g. mountain wax , can also be derived from it. Till today most of the sugarcane industries leave this waste to the field which makes the soil impervious to water penetration if repeated use is made. No attempt has been made to recover wax from the press mud in industrial scale.With ever increasing shortage of vegetable wax, e.g. Carnauba for its multiple uses, these fine chemicals are mostly imported.

If the economy of recovering and refining wax from press mud is established in a sugar refinery , it can be a finest chemical the sugar refinery can produce comparable to the commercialized wax from petroleum refinery on one hand and solving solid waste disposal problem in the sugar industry on the other and considerable foreign exchange can be saved from this exploitation.

In this present investigation ,an attempt has been made to review the potential of various bye products to produce value added chemicals in a sugar refinery with special emphasis on recovery, refining and modification of sugarcane wax . Details of extraction process in laboratory experiments as well as in pilot plant are reviewed. Further investigation is made on economic feasibility studies in order to establish a full sized plant based on data of a pilot plant study.

1.0 INTRODUCTION:

Sugar industry, worth Rs 60,0000 millions is also one of the largest agro-based industries in India-a country that happens to be also the largest consumer of white plantation sugar.[1-2] It has manufactured 26 million MT in 2007-08 which has declined over the following years due to varieties of reasons, 14.7 million tonnes 2008-2009 and again increased to 16.0 million tonnes 2009-2010. With expected 25 million tonnes of sugar production in 2010-12 and 24 million tonnes of consumption, both sugar production and consumption are bound to increase lineally. Indian sugar consumption is growing at a CAGR of 3.75%.Sugar production in India has been on occupies pride of place in the Indian economy being one of the five largest industries comprising of more than 571 units spread all over the country. Sugar industry has many byproducts like bagasse, molasses, press mud, and leaves (trash). Generally for every 100 tons of Cane the products in tons are as follows [1-2]:

Sugar of 98.5o Pol, tons	11.0- 12.5
Cane top	18.4
Leaves and straw	42.9
Final molasses 88^0 Brix	3.0- 3.6
Filter mud 77% -80% moisture	3.0- 3.4
Bagasse 50% moisture	27.5
Flue gases (180^oC)	120
Electricity, kWh	1500

Surplus bagasse of 50% moisture content may be of the order of 4 tons.

1.1 Constituents of Sugar Cane:

Sugar cane contains varying amounts of various constituents, approximately 0.2 Per Cent fats and wax, 0.2 PerCent pectins(gums),0.588 percent total nitrogeneous bodies,0.08 Per Cent free acids and 0.12 PerCent combined acids , mostly malic and succinic acid though organic acids present in the cane juice consists of glycollic acid, malic acid, oxalic acids,succinic acid and a small quantity offanoic acid, butyric acid and aconitic acid. These vary from 0.5 to 1.0 per cent of cane by weight. The amount of constituents however depends to a great extent on the variety of cane, its maturity, the condition of the soil, the climate and agricultural practices followed.

Small amount of pentosans along with those constituents find their way into juice during milling process; these together with some coloring matters such as Chlorophyll.,Xanthophyll.,Anthocynin.,Saccharatin,Tannins, polyphenols, nitrogenous compounds such as amino acids(as aspartic.) , albuminoids , asparagin, amides,amino-acids, nuleins,Peptones and xanthin bodies make up the organic solids , not sugar of the juice. Most of the fat and waxes , the greater amount of albuminoids, and part of the gums of the juice are removed during clarification by precipitation. These precipitated impurities ,after removal by filtration, form a cake of varying moisture content called filter muds. This cake contains much of the colloidal organic matter, anions that precipitate during clarification, as well as certain non-sugars occluded in these precipitates. The weight of wet filter muds (80 percent water) averages about 3.4 percent cane. Filter mud contains, on a dry basis, about 1 percent by weight of phosphate (p_2O_5) and about 1 percent of nitrogen. As a result it has been used, especially since the turn of the century, as a fertilizer. The filter mud also contains a mixture of waxy and fatty lipids in a ratio of 5:2 and refined wax can be extracted by appropriate treatment by use of solvents. Sugarcane is also a source of sterols, particularly β-sitosterol, the active ingredient in saw palmetto berries, a nutritional supplement for prostate problems.The active ingredients generally are 80%-98% Octacosanol alone or with Policosanol. Sugarcane is also a source of sterols, particularly β-sitosterol, the active ingredient in saw palmetto berries, a nutritional supplement for prostate problems. The various chemicals can be produced from from sugar industry bye-products. These are depicted in Table 1.

Table.1: Various chemicals from a sugar refinery[1-36]

Raw Sugar /juics	Molasses	Bagasse	Filter mud	Sucrose
Ethanol Refined sugar Sugar syrup Dextran Dextranase Epoxy polymer Glycollic acid, Malic acid, Oxalic acids, succinic acid offanoic acid, butyric acid aconitic acid .	Animal feed Ethanol –(Aldehydes &Fusel oil)Rum Vinegar Glycerin Inositol Lactic acid Succinic acid Glutamic acid Aconitic acid Kojic acid Acetic acid, Single cell protein	Xylitol Paper and Paper Board Ethanol Furfural Diacetyl 2,3-pentanedione Composite board(epoxy and bagasse) Compost Animal feed Poultry litter	Fertilizer Animal Feed Compost Waxes and Fats Molasses L-Lysine, Gluconic Acid, Glutamic Acid/Monosodium Glutamate and Itaconic Acid	Methyl *tert*-butyl ether (MTBE) Ethyl *tert*-butyl ether Sucrose syrup Epoxy polymers

1.2 Use of Molasses

(a) Direct utilization: Fertilizer,animal feed
(b) Distilling industry: Rum,ethyl alcohol, rectified spirit,power alcohol,alcohol derivatives
(c) Other fermentation industries: Vinegar and acetic acids, acetone and butanol,citric acid, lactic acid,glycerine, yeast
(d) Other chemical industries: Oxalic acid, and tartaric acid
(e) Miscellaneous industries:Aconitic acid,monosodium glucamate, dextran,L-Lysine, Xanthenes gum,2-keto gluconic,5-keto gluconic , a-ketoglutaric, fumaric acids , glycolic acid,kojic acid and glucaric acid, aldaric acids, Fumaric acid

Molasses contains sucrose. Therefore , a large number of chemicals can be obtained from molasses also from its both sucrose and reducing sugar contents. These are discussed elsewhere[1]

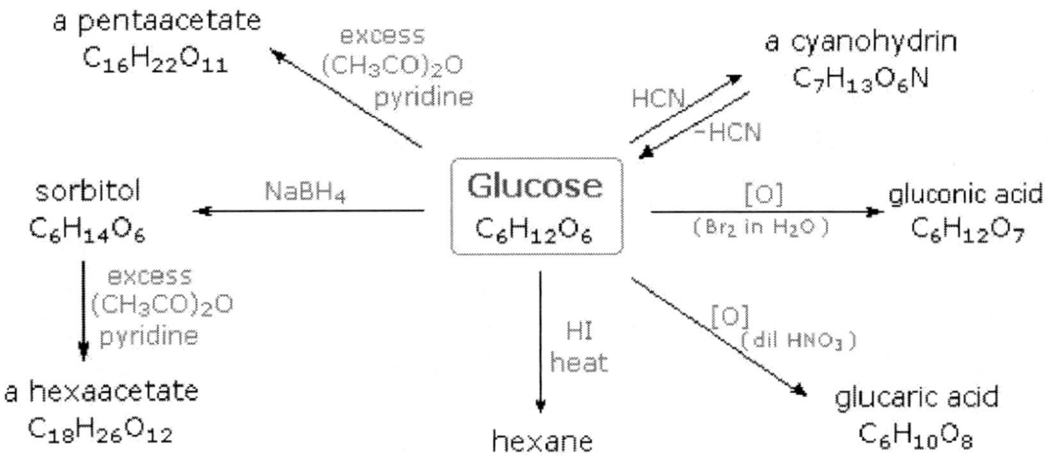

Fig.1 Type of Sugar Industries in India and composition of filter mud:

In sugar Industry in India, generally three types of chemical treatments are accomplished-sulphitation, carbonation and simultaneous sulphitation and carbonation to remove the impurities in cane juice. These impurities are precipitated by the above treatments by using optimum conditions of temperature, and pH, the later after removal by filtration, form a cake of varying moisture content called filter muds. On the above basis , the air dried filter mud will be produced of the order of 5 million tonnes depending upon the moisture content. The weight of wet filter muds (80 percent water) averages about 3.4 percent cane. This mud/ cake contains much of the colloidal organic matter anions that precipitate during clarification, as well as certain non-sugars occluded in these precipitates. Filter mud contains, on a dry basis, about 1 percent by weight of phosphate (P_2O_5) and about 1 percent of nitrogen. As a result it has been used, especially since the turn of the century, as a fertilizer.

The use of filter muds as animal feed has been tried by a number of sugarcane producer territories but so far has not proved economically rewarding, the main constraints being the magnitude of the drying process involved and the low digestibility of the dried scums. The raw material, press mud is a rejected waste material of sugarcane industries, that causes the unwanted problem of pollution to the surrounding suburbs of the sugar mills on accumulation. Also, the availability of the press mud is no problem at all throughout the year. While carnauba plant is a seasonal plant that produces extractable waxes only in marked period of a year.

Press mud of sulphitation sugar factories[16]:

The press mud of sulphitation sugar factories contains a very valuable component in the form of wax. With the increasing shortage of vegetable waxes in the world and new uses coming up for such waxes, the sugarcane wax presents a good source for meeting the future demand for wax not only in India but in other parts of the world.

Carnauba and other vegetable waxes imported into India per annum are valued at a very high price. This shows that a considerable amount of foreign exchange can be saved if sugarcane wax is manufactured in India.

Filter cake or mud press is a chocolate, sticky waste during juice extraction of cane stalks containing 5-15 % protein, calcium and phosphorus; 5-15 % sugar; 10-30% fiber; 10-20% ash and other micronutrients. It is a by-product of sugar industry with high potential as important component of organic fertilizer and animal feed and as source of sugarcane wax and methane (gas for fuel) when process. Sugar cane filter cake mud can also be used in animal feeds. Not only filter mud, but also other supposedly "useless" parts of sugar cane, like molasses and cane tops. Molasses and filter mud in animal feed gives the animals nutrients they otherwise might not be getting, including a carbohydrate (sugar) and B-complex vitamins.

The amount of filter mud varies from country to country, from mill to mill and even within the same mill and from field to field. The multiple factors involved include systems of clarification, extraction, cane variety, degree of mechanization etc. The following table shows the composition of filter mud.

Table 2. Composition of Filter mud[16]

Constituents	A	B	C	D	E
Moisture	72.14	67.73	69.26	69.72	60.331
Sucrose	13.20	10.50	9.96	10.20	0.647
Red. Sugar	0.60	-	-	0.71	0.011
Ash	-	-	7.59	8.10	
Albuminoids	1.93	0.87	1.94	1.80	5.775
Wax	-	3.04	5.13	4.12	7.760

Fibre	-	11.00	-	5.35	10.860
Lime	0.66	0.67	-	1.53	-
Magnesia	-	0.08	-	0.22	0.027
Iron Oxide and Alumina	-	3.06	-	1.40	2.666
Phosphoric Acid	0.57	1.49	1.42	1.39	1.073*
Silica	-	0.21	-	0.37	
Sand & Clay	1.38	1.70	-	2.82	6.506**

*Cal. Phosphate,** Sand and Silica

Sugar cane Wax:

Sugar cane wax is a material that occurs as a thin layer on the outside of the stem of the sugarcane can be obtained as a by-product in the manufacture of cane sugar as a hard green to brown or tan solid.

The filter mud also contains a mixture of waxy and fatty lipids in a ratio of 5:2 and refined wax can be extracted by appropriate treatment by solvents. This wax melts at 82°C, and boils at 146°C, is insoluble in water and in cold alcohol , slightly soluble in cold ether or chloroform, but dissolves easily in hot alcohol, ether, and chloroform, and in benzene [**16**]. This waxy material is not a single body but a mixture of varying proportions of a primary alcohol of the formula $C_{30}H_{62}O$ a with a second body , having the formula $C32H_{68}O$, while possibly yet other alcohols or esters of a high melting point are present in it. It should be noted, however, that only 386 kg of refined wax, which could be roughly equated to carnauba wax, can be obtained from 1000 tonnes of cane. The process is not commercially of interest under existing conditions and, as far as we know, only one plant is operating presently in India and on a small scale.

.How Cane wax goes to the mud?

Cane wax present on the stalk of sugar cane gets partially extracted in to the juice during crushing and gets accumulated in all waste materials , namely juice, bagasse, molasses and the filter mud. In the mud the wax is predominantly higher amount accumulated after defecation and clarification of juice.It is also obtained from vinasses from distilleries and rum factories These solid waste are generally disposed of inlandfills, causing nuisance and pollution due to their high organic and mineral content. The wax from bagasse is of the low order . The yield of wax from sugar cane peels was found to be 0.95% (w/w) of dried sugarcane peels. This yield was less compared to wax extracted from press mud (3.50 to 4.10%) reported by Phukan and Boruah [14].

The percentage of wax in the sulphitation mud is much higher than that in the carbonation mud and therefore makes its recovery from the mud can be partially feasible. The quality of wax present in the filter mud depends on various factors and varies within a wide limit (5 to 21% on dry mud basis).The scale in evaporator typically contains large amount of wax, 40.2 % of the scale in the first evaporator.

Approximate composition of cane wax (carnauba wax) can be as under:

Chemical Composition,%	Values	Fatty Acid,%	Values
Hydrocarbon	51.8	Saturated solid fatty acids	24
Mynicyl alcohol	35.8	Unsaturated liquid acid	30
Saturated fatty acid	12.0	Sterol	42
Sterol	Traces		

Uses of Sugarcane Wax:

Numerous attempts have been made by many researchers [3-35] to use sugarcane wax, after extraction, purification, refining and modification for many applications. .The refined wax has thus to be modified in order to increase its solvent retentivity and gellying properties. Modified cane wax is in every way comparable to carnauba wax. It gives as high a gloss value as carnauba wax and at the same time shows fairly high solvent retentivity.The wax obtained from sugarcane , after refining and modification can be well substituted for the imported materials of the carnauba wax type. Some are illustrated below:

This wax portion finds applications in cosmetics, paper coating, textiles, fruit and vegetable coating, leather sizing, lubricants, adhesives, polishes, and pharmaceutical industry.Thus, it can be considered as an alternative to costly carnauba wax, candelilla wax, and chitosan for edible coating applications. Sreenivas et al.[13] found ash gourd peel a good source of wax which was used as edible coat for strawberries to enhance their shelf life. As the presence of hard wax and vegetable wax is high ,with the industrial expansion of the country; the demand of hard wax is bound to go up. These may be summarized as under[3-15]:

(i) the hard variety sugarcane wax is a replacement of the commonly used carnauba wax. By nature sugarcane wax (the hard variety) is a type of microcrystalline wax having a high potential used in preparation of several important industrial products. Its potential use is wide in industry. The refined variety of the hard sugarcane wax having its potential use in: (a) Medicines and pharmaceuticals (b) foods,in sweet meats and pastries, (c) in tinned food products, (d) in castings and recordings, (e) in carbon papers, (f) in preparation of carbonless copy papers, and carbon papers (g) in electrical insulation of cables and wires, (h) in preparation of various types of paints, varnishes and polishes including boot /shoe polishes, car polishes and floor polishes, (i) in preparation of coloured pencils, crayons, water and oil paints, (j) in cosmetics (k) in preparation of emulsion paints for spraying, ,stencil papers and textile aids, photographic preparations , lubricants and flame proofing agents, corrosion protection, food preservation, tyre plasticizers, printing inks ,toners for copiers and printers, as growth stimulant for seedlings, a source of long chain primary aliphatic alcohols, which find applications as cholesterol-lowering products and other uses. such as protective coatings on fruits for their preservation and for the preparation of waxed papers. Crude sugarcane wax can be used directly for the manufacture of carbon paper

In fact, it has its most likely entries into all the branches of industries where carnauba wax is mostly used. In bulk also it can be produced many times more than any other plant waxes combined together.

(ii) Soft wax portion: It is a mixture of straight chained lipids as established by NMR spectras. It is a source of useful steroids that have their potential and vital uses in medicines and in birth control drugs.

(iii) Pitch: A dark coloured compound of calcium and magnesium salts of wax acids and resin acids. It bears a sticky adhesive character. It has its potential uses in adhesive industries. (iv) The fibrous residues left over: It forms the biggest part of materials in the constitution of the press mud. The de-waxed residues are enriched with useful organic and inorganic salts like organic matters, nitrogenous salts, phosphates, phosphatides and phosphites, potash bearing elements and salts. Its possible use as a agricultural manure is promising.

With supercritical fluid extraction, the wax is a source of long-chain aliphatic alcohols. Octacosanol (C_{28}) is reported to "increase physical stamina" and "remedy damaged nerve cells" and even to stimulate sex hormones.

Composition of sugarcane wax:

There two varieties of wax available in cane wax, hatd wax and soft wax.

Inarkar and Lele [9June 2012] determined the physical properties of both the varieties. The melting point of crude wax was observed to be 62.1°C. Molecular weight of wax was estimated to be 1706 Dalton. Composition of crude wax found using gas chromatography-mass spectroscopy (GC-MS) was alkanes (28.83%), ester (66.26%), fatty acids (4.58%), aldehyde (0.11%), and alcohol (0.22%). Soft wax portion consists of straight chained lipids as established by NMR spectras. It is a source of useful steroids that have their potential and vital uses in medicines and in birth control drugs

Cane wax from Indian sugar factories:

Press- mud is generally used as a fertilizer in India . Dewaxed mud relations fertilizers properties in a superior manner to the unextracted mud. A survey of wax content in sulphitation press- mud carried out at the National sugar Institute, Kanpur[17,19-36] showed that wax percent dry mud varied from about 20 percent in the various regions of India as given below:

Table 1 Wax content and ratio of Hard wax and soft wax in sulphitation press mud

S. No.	Regions	Average wax content in dry press-mud present,%	Hard wax,%	Soft wax,%
1.	North Bihar	12.3	77.5	22.5
2.	East U.P.	16.1	46.1	53.9
3.	Central U.P.	15.2	69.6	30.4
4.	North U.P.	15.4	66.1	33.9
5.	West U.P.	11.2	70.2	29.8
6.	Bombay	8.7	54.0	46.0
7.	Madhya Bharat	10.0	36.0	64.0
8.	Andhra	20.9	21.8	78.2
9.	Madras	17.3	59.2	40.8

Assuming therefore, that the wax recovery is 10.0 percent on dry mud, there is a potential source in India for the production of **20,000 tons** of sugarcane wax. Sulphitation press mud is at present largely used in India as manure. As the extraction wax from it would improve its manurial value it would be advantageous to extract the wax first before using it as manure.

Extraction of Wax from press mud

The sugar cane wax extraction from press mud and subsequently its modification are very important aspects of by-product exploitation for purposeful economical gain.

For recovering wax, the press mud has to be dried in the sun or by other suitable means and the moisture content is to be reduced to about 3.0 percent or less. The dried press mud is then subjected to a solvent extraction process either of the batch type or of the continuous type similar to the solvent extraction of oil from oil cakes.

Solvents :

Many researches have been carried out on the solvent for wax extraction .Phukan and Boruah [9] used soxhlet extraction of peel powder with carbon tetrachloride (CCl_4) for 8–10 hours. The residue was dissolved in isopropanol and was refluxed with charcoal for 1-2 hours to remove any undesirable color or pigments present. The extract was then filtered to remove charcoal and evaporated at 60°C to obtain crude wax. Inarkar and Lele[9] also used carbon tetrachloride,isopropyl alcohol,toluene. Some other investigators used benzene, cyclohexane at room temperature. For further pextractionof raw wax solvents of reduced polarity such as MeOH, acetone and isooctane were used.

However, for industrial exploitation solvents generally used are (i) n-Hexane (ii) solvent oil B.P. 90—120°C or (iii) mineral turpentine, B.P. 135—190°C. The former two namely (i) hexane and (ii) solvent oil come under the category of 'Dangerous petroleum' and are therefore, subject to strict control by the Inspectorate of Explosives under the Indian Explosives Act. It would be necessary to take all required precautions to avoid explosion, such as the use of explosion- proof motors, gas-proof electrical fittings and switches, storage of solvent in underground tanks, earthing of all steel vessels etc.

Mineral Terpenting

Mineral Terpentine is a safe solvent but requires high pressure steam for solvent recovery inspite of which sometimes the solvent losses are high. The low boiling solvents, however, require apart from safety precautions and special equipment and fittings, suitable solvent recovery equipment involving refrigeration and absorption. The solvent which is to be preferred for the recovery of cane wax should have a boiling range higher than the melting point of wax and in view of the fact that some of the wax fractions have a higher melting point than 80 °C, solvent oil with boiling point range 90°C —120°C is considered the most suitable.

Equipment and Indian Entrepreneur

The simplest equipment would be according to the batch process. The equipment for the batch process would mainly consist of the main extraction vessel with suitable fittings and connections for feeding the mud, feeding the solvent, a perforated false bottom, steam coils for de-solventising the mud, vapour connections to condenser etc.,storage tanks for extracts of different concentrations, a boiling pot for recovering the solvent from the extract with steam connections and vapour connections to condensers, solvent separators, refrigeration condensers, carbon absorption systems etc.

Thus equipment would also consists of solvent pumps with explosion proof motors, wiring through gas - proof conduits and other gas proof electrical fittings. A solvent unit by the batch process has been working at the Ravalgaon Sugar Farm Ltd., Ravalgaon (Distt. Nasik, India) for the last several years and about 400 lbs. of wax is recovered daily by the batch process using mineral turpentine as the solvent. Arrangements have been made to change over to solvent oil in the 90°C-120 °C range but the actual changeover has not taken place.

Various Processes and Operations

The Merz Procss:

While the Ravalgaon plant has been designed and fabricated locally, the Merz plant which also operates on the batch process is as imported plant using n-heptane or n—hexane as the solvent. The usual plant consists of batch type extractors, each extractor having an upper extraction chamber and a lower collecting and evaporating chamber separated by an intermediate head provided with an external tubular heater used to evaporate the solvent, heating and stripping coils, solvent distributor coils, solvent condensers, solvent coolers, water separators etc. The Merz plant is designed to get the fatty acid and hard wax portion from the filter mud separately using the same solvent. The principle on which this plant is based is that the fatty portion is dissolved in the cold petroleum spirit which is separately recovered, after which the same solvent when heated recovers the hard portion of the wax.
Each plant consists of an extractor, one fat and one wax still, condensers, cooler, chilled cooler, two automatic water separators, solvent tank, complete refrigeration equipment with chilled water pump etc. The equipment is complete with interconnecting pipes, valves and instruments. The plant can be supplied in various sizes for dealing with 1,2,3 and 5 tons of dry press-mud in 24 hours. The system of extraction has the following advantages:

a. The fatty matter and hard wax can be recovered in the same operation.
b. The same solvent is used for both constituents and solvent losses are negligible.
c. Latent heat of solvent vapour emanating from the **fact** still is utilized for pre- heating of the material in the extractor, thus reducing steam consumption.
d. The working procedure is simple.

One Merz extraction unit for the recovery of cane wax has been installed at Vuyyuru Sugar Factory of M/s K.C.P. Ltd., in Andhra Pradesh (India) several years ago.In this industry ,

Continuous extraction of cane wax.

There are several plants which have boon designed for the continuous solvent extraction of oil cake which with minor modifications can be used for the solvent extraction of cane wax. Manufacturesrs of these units would be in a position to offer suitable plants for these continuous processes. Mention may be made of:

(i) De Smet (ii) Lurgi and (iii) Anderson, Solvent extraction units. The De Smet plant appears to be very suitable for wax recovery from press mud.

The De Smet process:

The De Smet process which is already well known for the extraction of oil from oilseeds may be conveniently applied for the extraction of wax from the press mud. The basic principle of extraction is that of a moving screen upon which the material to be extracted is fed in an adjustable layer. A series of nozzles placed about the belt, spray the solvent on this moving layer of material. The action is countercurrent i.e. the material moves in one direction and the solvent moves in the opposite direction. In this manner the material entering the extractor is first washed by wax containing solvent. Subsequent washes are by solvent which contains less and less wax

until the final wash which is fresh solvent from the still. The complete De Smet extraction plant may be divided by function, into three sections.
1. The extractor and its accessory equipment.
2. The miscella distillation unit with stripping column.
3. The meal drying and solvent recovery system.

The De Smet System has the following advantages: Low steam consumption, low solvent consumption, and ease of maintenance, all parts being easily accessible and utmost safety in all operations. De Smet extraction units for extraction of oil are in operation at Kanpur, Calcutta and a few other places in India.

Availability of plant[17]:

The equipment required for the extraction of wax by the batch process can be designed and fabricated locally. The plant would be simple if mineral turpentine is used. If, however, any of the solvents classified as dangerous petroleum are used it would be necessary to have explosive proof motors and gas proof fittings, switches, wiring etc. which may have to be imported. Particulars relating to imported equipment and their suppliers are given below:

Solvent	**Plant**	**Manufacturers**	**Agent in India**
n- heptane	Merz (batch)	Posedown and Thomson Ltd., Old foundary, cannon street, Hull England	M/S Larson and Toubro Ltd. J. K. Building, Dougall Road, Ballard Estate, Bombay, Maharashtra State, India
n- hexane	De Smet (Continuous)	De Smrt S. A. 38, Avenue De France Antworp, Belgium	,,
n- hexane	Anderson (do)	M/S Lurgi West Voltas Germany	,,
n- heptane	Lurgi	,,	IAEC 43, Forbes St. Fort, Bombay- 1

Production of wax from Press mud and the process of refining

The cost of production of crude wax from press mud would depend on. (i) the wax content of mud (ii) the solvent used *(iii)* the type of plant (ii) the capacity of the unit (v) other local factors such as cost and availability of steam, labour etc.
The refining of crude wax, would consist of first de-ashing with concentrated hydrochloric acids,de-fatting it by using suitable solvents such as alcohol, acetone etc., and subsequently bleaching it by chromic or chloric acids according to requirements, cost of chemicals and market for by- products. The bleached and refined wax will have to be modified to suit different end

uses. For polish compositions requiring high solvent retention, and strengths, the wax will have to be modified into the suitable ester and amido derivatives and properly blended.

Scope for developing the wax industry:

As already stated earlier, there is quite a potential for the development of the cane wax industry, for replacing imported waxes, to meet the increasing demand of vegetable waxes in the country and for export, but considerable initial work is necessary before this industry could be established. In the beginning the installed capacity will have to be necessarily small, and there may not be any scope for more than one or two units to be established in the whole country. From the cane wax, a large number of different types of refined and modified waxes for different end uses as has been done by the German manufacrurers by modifying the naturally occurring Montain wax. All the different types of waxes will have to be made and their uses and markets need to be established. In order to establish this industry, the Government might be approached to give the necessary assistance and support for putting up one or two for plants.

Research

A fair amount of research work .has been done in India on various aspects of cane wax extraction refining and modification.

Sugar Cane Wax from Sugarcane Peels:

Wax was extracted from sugarcane peels as reported by Phukan and Boruah [14]. The process involves soxhlet extraction of peel powder with carbon tetrachloride (CCl_4) for 8–10 hours. The solvent was later evaporated using Rota evaporator (Buchi, Switzerland), and the residue primarily contained wax mixtures with some impurities. The residue was dissolved in isopropanol and was refluxed with charcoal for 1-2 hours to remove any undesirable color or pigments present. The extract was then filtered to remove charcoal and evaporated at 60°C to obtain crude wax.
The amounts of Hardwax, softwax and pitch depend on the type of solvents, their ratios and the reactant ratio as well as the type of adsorbents used.

1. Hard wax = 5.20% Softwax= 4.10% Pitch obtained = 0.12%
2. Carbon Tetrachloride+ IPA, Activated Charcoal:Hard wax obtained = 5.35% Soft wax obtained = 4.00% Pitch obtained = 0.12%
3. Toluene+IPA,ActivatedCharcoal:
   Hardwax=5.50%Softwax=4.20%,Pitch=0.15%
   Further,redissolved in carbon tetrachloride,Hard-cream-coloured-wax=5.05%,softwax=3.85%,pitch=0.15%
4. Toluene +IPA, Activated Carbon,: Hard wax = 5.40% (ii) Soft wax = 3.80% (iv) Pitch = 0.12%
5. Benzene+IPA,ActivatedCharcoal : Hard wax = 5.50% Soft wax = 4.20% Pitch = 0.12%

Benzene+IPA,ActivatedCharcoal:Hardwax=5.25%,Softwax=3.85%,Pitch=0.11%

Pilot Plant study:

Das and Rao**[18]** designed and fabricated a continuous fluidized extractor for the exxtractionof sugar cane wax for a capacity of 5 tons of press-mud per day corresponding to a sugar factory of 500 tons cane crushing capacity. The design includes the accessories of pilot plant.

The wax content of press-mud vaies significantly in India. The input material, i.e. crude wax , even if of the order of 1% press- mud on cane it is easy to estimate that with 7.15% of wax in the press- mud a minimum of 96 tons of crude wax for 1000 tons C.T.D. for a working hrs of 120-150 days could be obtained. In India out of 550 sugar factories nearly 460 sulphitation sugar mill will produce minimum45000 tonnes of crude wax in a season.Dewaxed mud retains fertilizing properties superior to the unextracted mud.

The design calculations were based on the following data:

- 20 hrs. per day
- Wax content in press mud=10%
- Ratio of solvent / solid=5:1
- No vapour phase
- Solvent: Petroleum ether-top 80°C-120°C
- Temperature of extraction=75°C

Pilot plant for Modification of Wax:

Nigam[26-35]proposed the design of the pilot plant scale glasslined reaction vessel of 50 gallons capacity to handle 1 ton/day of press mud (dry) for deashing and defatting accompanied by bleaching of wax. The process involving modification of wax involves two steps:

Firstly, deashing is done in which crude wax, water and HCl are mixed by agitation at 90-95 °C for about 2 to 3 hours, it is cooled at room temperature and water separated. Finally a hot water wash is given several times to remove all the acid. Then defatting of the deashed wax is done by thinning the wax with alcohol. Alcohol takes away the fatty portion and granular hardwax is taken out by centrifugation. The alcohol phase is sent to the alcohol recovery unit to recover alcohol and soft wax is used in isolation of sterol. Finally the wax is bleached by chromic acid followed by washings with water to remove all free acid . A part of bleached wax is converted to form ester wax by glycerol and the other to amide wax by ethyl diamine . Then two waxes so obtained can be mixed in proper proportion to get final modified wax.

The vessel is steam jacketed batch type fitted with paddle type agitator which is suitable for deashing and defatting process treatments. As HCl acid is very corrosive acid and that too at an elevated temperature of 90°C,it is suggested to use a glasslined vessel for the purpose.Using experimental heat transfer coefficients available in literature and the natural convective heat transfer coefficients available the steam consumption and heat transfer rates are estimated.

A paddle type of agitator as suggested is a most frequently used type of agitator used for thick paste like this one of wax plus reactants, the scraping action of the agitator dislodges pasty materials from the walls of the vessel . Anchor paddle sweeps the tank bottom also and the entire peripherial area of the tank in contact with the fluid. The standard design variables are selected from the literature for paddle type agitator design and the power required for for it has also been estimated. The schemes for filter-mud bye-products , flow diagram of the solvent extraction of sugarcane wax, refining of wax and modification of wax are shown under:

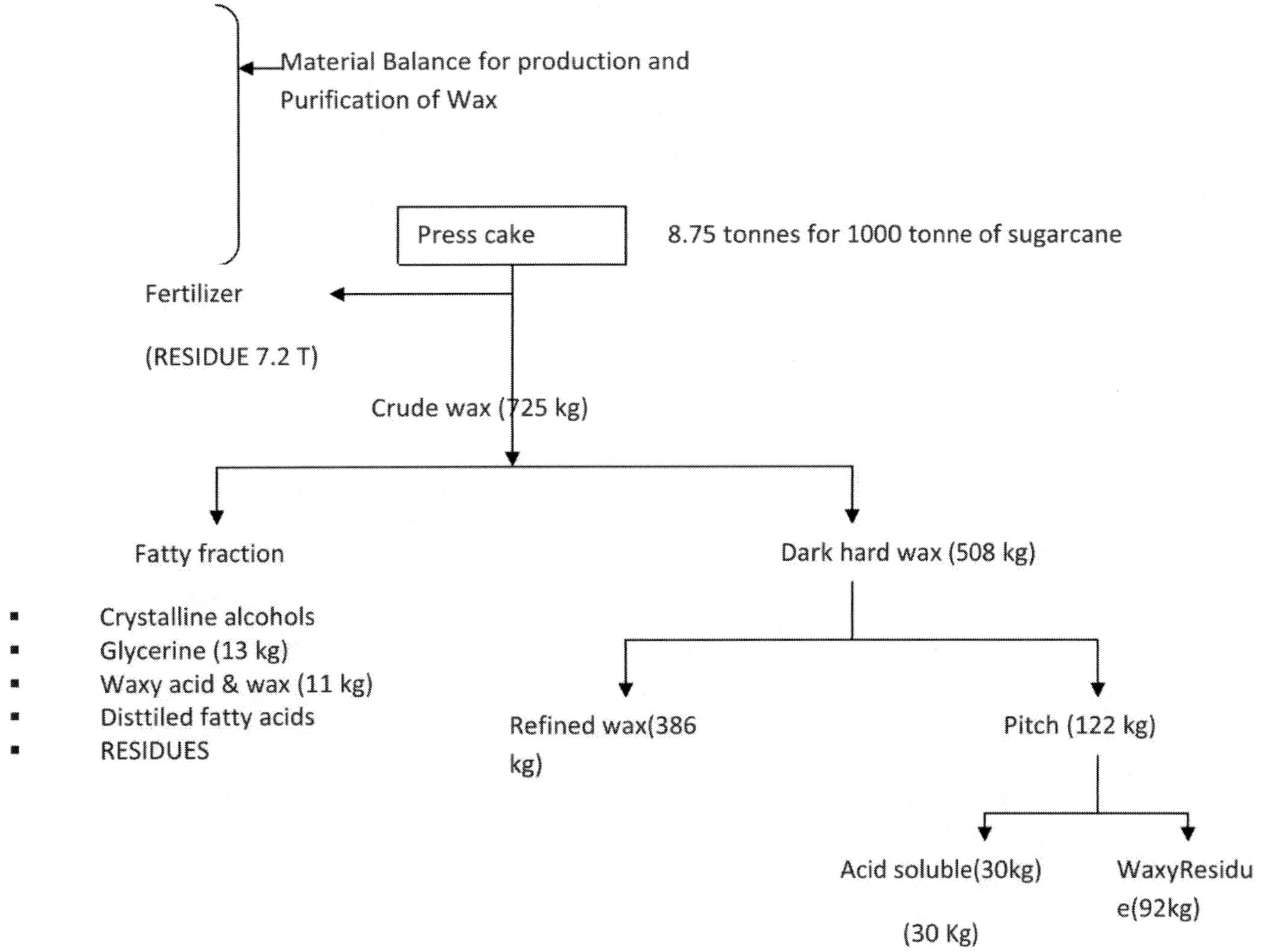

Fig.2 Filter Mud By Products

Flow diagram of the solvent extraction of sugarcane wax

A. Crude Wax Sulphitation factory

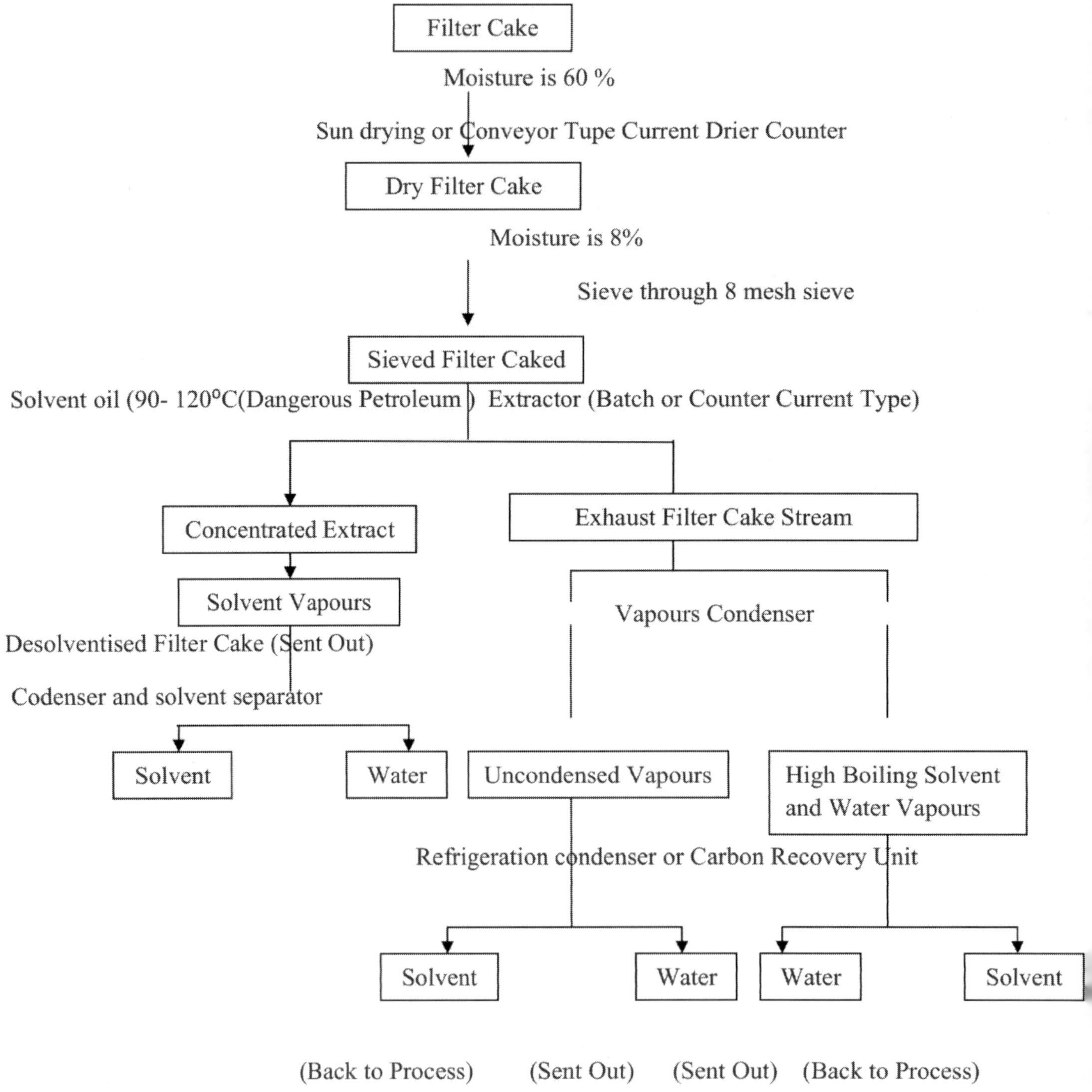

B. Refining of wax

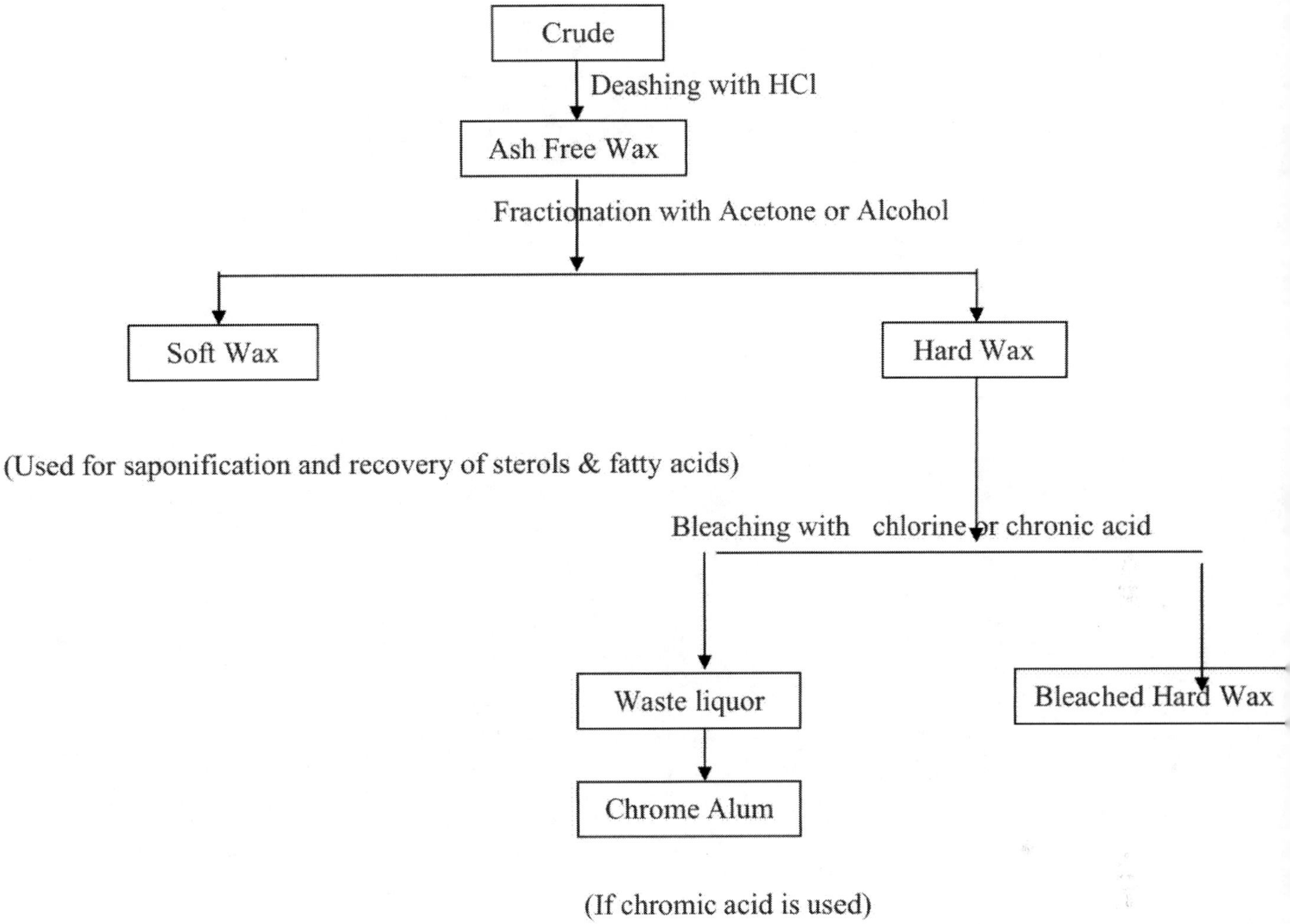

C. Modification of wax

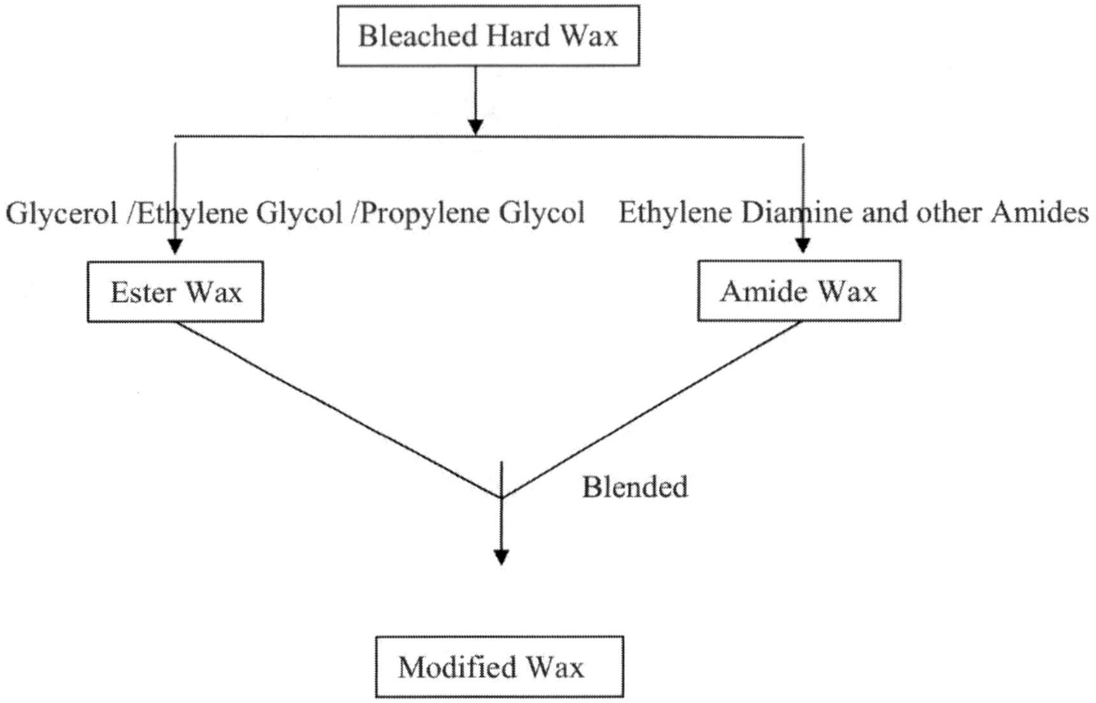

CONCLUSIONS

Extraction of Sugarcane wax has always been a matter of interest to the engineers, due to its enormous industrial applications. But economy of wax recovery has to be studied in detail. Because of these limitations to date, filter mud has not been put to industrial use in a wide scale , although its disposal as a solid waste of the sugar mills is necessary. In most countries, it is spread on the fields. Wax should be recovered from the filter mud which could otherwise make the soil impervious to water if used repeatedly. If properly refined, the wax from filter mud can be similar to the finest vegetable waxed commercialized throughout the world.

Although the chemical possibilities for many bye products from sugarcane seem limitless, few of these products are currently in production. The tradition-bound sugar industry is slow to adopt the new technologies required for implementation. The potential processes require urgent attention. Extraction of Sugar cane wax is one of the processes which has attracted recent attraction.

REFERENCES

1. **A.K.Ray**, Pradosh Sanyal,Alternative Uses of Sugarcane Molasses and Jagerry for production of value added fatty acids ,especially Oxalic acid- Sustainability and Biobased Materials, 17002, Proceedings of 11 AIChE Annual Meeting, held between Oct.16-Oct.21, at Minneapolis 2011,USA,Paper 338e,pp.238

2. **Amiya Kumar Ray**,Pradosh Sanyal,Biorefinery based on Indian Distillery- Innovation of Forest Products, 17000, Proceedings of 11 AIChE Annual Meeting, held between Oct.16-Oct.21, at Minneapolis 2011,USA,Paper 228d,pp.206
3. Sugar Ddyog.com
4. Shrivastava, A. K.Sugarcane wax: physiological significance - extraction and industrial uses, Sugarcane: agro-industrial alternatives. 1995 pp. 333-347, **ISBN** 81-204-0948-5

5. Phukan, Amal C.,Patent 6486335 Issued on November 26,2002
6. Albin H. Warth ,The Chemistry and Technology of Waxes, Reinhold Publishing Corporation,New York,Chapman & Hall Ltd., London
7. http://www.patentstorm.us/patents/6486335/description.html,
8. http://www.ajol.info/index.php/tjpr/article/viewFile/75872/66377
9. Mangesh B. Inarkar and S. S. Lele "Extraction and Characterization of Sugarcane Peel Wax" ISRN Agronomy,Volume 2012 (2012), Article ID 340158, 6 pages doi:10.5402/2012/340158
10. TIFAC,http://www.tifac.org.in/index.php?option=com_content&view=article&id=685&Itemid=205

11. http://bicol.da.gov.ph/Opportunities/sugarcane%20profile/topic11.html
12. A. K. Taylor, "From raw sugar to raw materials," *Chemical Innovation*, vol. 30, no. 11, pp. 45–48, 2000. View at Scopus
13. K. M. Sreenivas, K. Chaudhari, and S. S. Lele, "Ash gourd peel wax: extraction, characterization, and application as an edible coat for fruits," *Food Science and Biotechnology*, vol. 20, no. 2, pp. 383–387, 2011. View at Publisher · View at Google Scholar · View at Scopus

14. A. C. Phukan and R. K. Boruah, "Extraction and evaluation of microcrystalline wax from press mud waste of the sugar industry," *Separation and Purification Technology*, vol. 17, no. 3, pp. 189–194, 1999. View at Publisher · View at Google Scholar · View at Scopus
15. N. Partha and V. Sivasubramanian, "Recovery of chemicals from pressmud—a sugar industry waste," *Indian Chemical Engineering Section A*, vol. 48, no. 3, pp. 160–163, 2006.
16. H.C.Prinsen Geerligs,Cane Sugar and its manufacture,Second Edition,Norman Rodger, 1924
17. Nigam R.B., Unpublished manuscript, National Sugar Institute, 2008
18. Das C.N. and M.N.Rao,Continuous extraction of sugarcane wax(Design of Fluidized Extractor),Journal of Science and Engineering Research, Vol.VI,Part 1, Jan. 1962,pp.127-132,
19. Rao, K.A.N. and Gupta, G. N. Sugar Cane wax from filter press cake, Proc. S.T.A. (ii),79-82,1941,Journal. Sci. and Tech. India VI. 13—15 (1942)
20. Chaturvedi, H.S. ,Gupta G.N. and Sen Gupta, M.N.,Sugar Cane wax from press mud, Indian Sugar, 429-431, 1943.
21. Shrikhande, J. G., A Pilot plant for the extraction of cane wax from press mud, Indian Sugar 73, 1945.
22. Dass, K.S.G. and Kirpa Shankar, Wax Content elimination and subsidar, Scums, J.Sci. and Ind. Res. India,3,462(1945).
23. Ram Dev and Shrikhande, J.G., Chemical Examination of Sugarcane wax from press mud, Proc. STA XV (i), 32—38,1946.
24. Vidyrathi, N.L. and Nansinha Rao, Chemical Investigation of sugarcane Wax ,J. Ind. Chem. Sec. 16, 1939.
25. Ibid, Sugarcane Wax, Indian Sugar, Oct. 1941.
26. Kulkarni, A. B. Indian Patent, 47179 (18th Nov. 1952).
27. Gundu Rao, S.N. Kulkarni, H.G. Kulkarni, D.P., Sugarcane wax, Proc. Pt. I, 62—67 (1949).
28. Kapil, Y.P. and Mukherjee, S., Sugarcane wax and its purification and properties, Part I Proc. STA 1I7 (1952).
29. Kapil Y.P. and Mukherjee, S. Constitution of sugarcane wax composition of Hard Wax, Proc. STA (Part II), 220,228 (1954).
30. Kapil Y.P. and Mukherjee, S., Sugarcane wax— Part II. Purification and properties, Proc. STA (Part II) 229—231 (1954).
31. Mukherjee, S. A survey of cane wax contents of press cake from different sugar factories in India, I.S.S.C.T. Vol.II, 505—512 (1956).
32. Bose, S. and Gupta, K.D., Sugarcane wax — Part III -An examination of fatty lipids for sterols, Proc. STA (Part II), 70—76 (1961).
33. Bose, S. and Shrivastava, A.N. A note on the solvent retentivity characteristics of' sugarcane wax and modified derivatives. Proc. STA. 129—131(1961).
34. Bose, S. and Shrivastava, A.N, Solvent retentivity experiments with sugar Cane Wax and its modified derivatives, Sharkara, 4, 67—72 (1961).
35. Bose,S. and Shrivastava A.N. ,Solvent retentivity experiments with sugarcane wax and its modified derivatives,Part II, Effect of blending different modified derivatives and addition of metallic stearates to the polish composition, Sharkara, 5, 126 (1962).
36. Bose, S., Kapil, Y.P., Mukherjee, S., and Shrivastava,A..N., Purification armd modification of sugarcane wax and study of the characteristics of modified wax derivatives, Proc. I.S.S.C. T., 12th Congress, 1885 (1965).

() 674f: Novel Biodegradable Composites From Poly (Butylene Succinate) Bioplastic and Co-Products From Biofuel Industries: Processing and Properties Evaluation

Andrew Anstey[1], Sudhakar Muniyasamy[1], Murali M. Reddy[2], Amar K. Mohanty[3] and Manju Misra[4], (1)Bioproducts Discovery & Development Centre (BDDC), Department of Plant Agriculture, Crop Science Building,, University of Guelph, Guelph, ON, Canada, (2)Bioproducts Discovery & Development Centre (BDDC), Department of Plant Agriculture, University of Guelph, Guelph, ON, Canada, (3)Department of Plant Agriculture & School of Engineering, University of Guelph, Guelph, ON, Canada, (4)School of Engineering and the Department of Plant Agriculture, University of Guelph, Guelph, ON, Canada

Biocomposites can provide a sustainable platform for biofuel co-products which are available in large quantities due to rapid expansion of biodiesel industries. Co-products function as both a shock absorber and a price adjuster for the biofuel industry by providing an additional income stream. These co-products can be suitable candidates for the development of new biocomposites due to their inherent biodegradability and renewable carbon content.

The present study involves developing biocomposites of poly(butylene succinate) (PBS) using biofuel co-products (canola meal, soy meal, corn gluten meal) as fillers and comparing them with natural fiber-based biocomposites (switchgrass). This work aims to establish the role of these co-products in the design and engineering of new biocomposites, and their influence on material properties and biodegradability under controlled composting conditions. Conventional melt extrusion and injection molding techniques were implemented in fabricating these biocomposites. The mechanical properties (tensile, flexural and impact), thermal properties (TGA, DSC, DMA), structural properties (FT-IR) and microscopic observations (i.e. state of dispersion) (SEM) of these biocomposites were analyzed. A biodegradation study was carried out in order to determine the relative biodegradation rate of the composites and neat bioplastic and to examine the biodegradation process of the materials.

The results demonstrated that the PBS-switchgrass biocomposites had better mechanical behaviour than the meal-based PBS biocomposites. However, the PBS-meal biocomposites displayed better toughness and ductile behavior than the PBS-switchgrass biocomposites. The compostability study revealed that biodegradation rates of the meals were nearly double to that of switchgrass. The same trend was observed in the biocomposites; the PBS-meal biocomposites biodegraded much faster than the PBS matrix and PBS-switchgrass biocomposites. The presence of meals helped the initial growth of microorganisms, leading to hydrolysis of the PBS matrix. This study opens up new knowledge on the comparative compostability of two distinct types of PBS-based biocomposites; one with meal co-products, and another with perennial grass.

Acknowledgements: *This research is financially supported by the OMAFRA – 2009, 2010 New Directions & Alternative Renewable Fuels 'Plus' Research Program-SR9223 and the Ontario Research Fund (ORF) Research Excellence (RE) Round-4 from the Ontario Ministry of Economic Development and Innovations (MEDI) and Hannam Soybean Utilization Fund (HSUF).*

INTEGRATION OF ACID-ENZYME HYDROLYSIS OF LIGNOCELLULOSIC BIOMASS TO FERMENTABLE SUGARS

Suping Zhang , Yan Li , Baozhen Zhang, Yongjie Yan

(East China University of Science and Technology, Shanghai, People's Republic of China)

Abstract

Although many previous studies determined dilute acid pretreatment conditions that maximize xylose yields from pretreatment or glucose yields from subsequent enzymatic hydrolysis of the pretreated cellulose, this paper emphasizes the use of different conditions to realize the highest total fermentable sugar yields from both stages. The acid pretreatment tests were performed under the conditions of acid concentration at 0.05%, 0.25%, 0.50%, 1.0% and the temperature in the range of 160□ to 180□. The maximum yield for just xylose in acid pretreatment occurred at 170□ with the acid concentration of 0.25%. However, the total fermentable sugar yields from the two stages occurred at the pretreatment conditions of temperature 180□ and acid concentration 0.25%. The results show that the conditions for the highest xylose yield in pretreatment step did not give the highest glucose yield in enzymatic hydrolysis step nor total fermentable sugars yield from both stages. Therefore, the effects of dilute acid hydrolysis conditions on the following enzymatic hydrolysis yields were investigated.

Keywords: Lignocellulosic Biomass; Total Fermentable Sugars; Dilute Acid Pretreatment

Introduction

With depletion of oil reserves and the green house effect caused by the consumption of fossil fuels, bio-ethanol production through conversion of biomass has attracted much attention in the whole world [1, 2]. The commercialized production of biofuel up-to-date is mostly based on sugar or grain as raw materials [3, 4]. However, the limited production capacity can not satisfy the large demands from the energy market. The abundant lignocellulosic biomass is a good substitute raw material, including agricultural residue, forest residue, bagasse from sugar industry and even municipal garbage.

Corn stalk is an abundant agricultural residue in China. It is estimated that annual harvest of crop stalk amounts to more than 700 million tons (dried base) [5]. On one hand, energy security can be realized by one more way; on the other hand, the problem of environment pollution caused by direct burning the crop stalks can find a solution.

Production of fermentable sugars from corn stalk is an important step for ethanol production [6, 7]. Cellulose is the main component in corn stalk which can be turned into glucose through hydrolysis. Glucose is the important monosaccharide in production of ethanol through fermentation. With the development of bacterial strains specific for fermentation of xylose and other pentose [8, 9] yielded from hydrolysis of hemicellulose, Xylose can thus be used to produce ethanol through fermentation. Utilization of xylose can greatly improve the economics of hydrolysis or saccharification process [10]. Therefore, research on improvement of the total fermentable sugar yield becomes especially important.

Sacharification can be realized through hydrolysis either acid or enzymatic; either one has advantages and disadvantages [11]. Dilute acid pretreatment is regard as one of the most promising and cost-effective pretreatment process [12].A number of previous studies determined dilute acid pretreatment conditions that maximize xylose yields from pretreatment or glucose yields from subsequent enzymatic hydrolysis [13]. The effect of acid pretreatment on subsequent enzymatic hydrolysis performance has not been studied in great detail, especially in regards to total fermentable sugars yield as the target [14]. This study was focused on the coupling acid-enzyme hydrolysis to maximize the total fermentable sugar yield in a percolation reactor and corn stalk as material.

Experimental

Materials

Corn stalk used in experiments was first washed clean, cut short and got naturally dried in air. After further crushing, the stalk dust through the sieve of 20-40 mesh was selected as materials and stored in dried place for standby. In Table 1, the composition of the material used in experiment is listed.

Table 1 The Composition of Corn Stover on Dry Basis.

Soluble Carbohydrates, Proteins and Tannins (%)	Hemicellulose (%)	Cellulose (%)	Lignin (%)	Ash (%)
10.34	25.78	43.65	17.21	3.02

Hydrolysis Process

Acid pretreatment was carried in a percolation reactor made of stainless steel (316L). The bulk of the reactor is of cylinder-shape with the inner diameter 10cm, total capacity of 1.5 L. The reactor is connected with a steam generator as shown in Fig 1

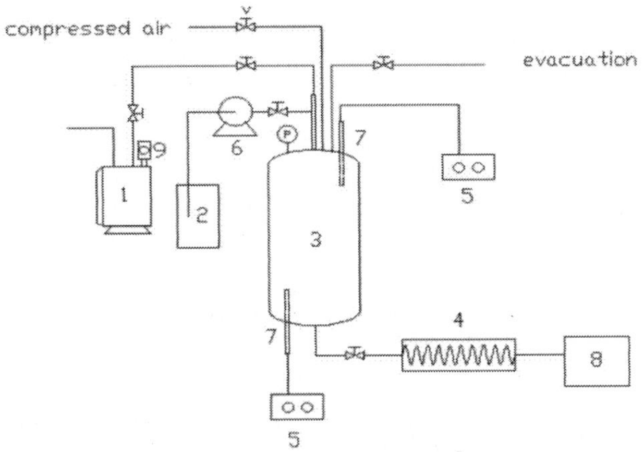

1.steam generator; 2.dilute acid tank; 3. hydrolyzer; 4.condenser; 5. temperature indicator; 6. metering pump; 7、thermal couple ; 8. container for hydrolysis products; 9. pressure controller; V. valve; P.pressure gauge

Fig. 1. Schematic diagram of acid pretreatment experimental apparatus

The residue from the acid pretreatment was subject to the sequent enzymatic hydrolysis by adding enzyme-containing solution and PH-adjusted hydrolysate from pretreatment acid hydrolysate to the predetermined solid-to-liquid ratio.

Analysis method

Monosaccharides were analyzed through external standard method on HPLC (SHIMADZU, LC-20AT) and the detector was Polymer Laboratories PL-ELS 2100 (move phase: acetonitrile: water=23:77(φ), total flow 0.8 mL/min, nebulization temperature 40□, gasification temperature 70□, column: intersil-NH_2, 4.6 mm×250 mm, 5-Micron).

Results and Discussions

Effects of Acid Concentration

The effects of acid concentration on xylose yield were investigated with the acid concentrations varied from 0.05% through 0.1%, 0.25% to 0.50% under the determined temperature 160℃ and the fixed liquid to solid ratio 10:1.

Data in Fig.2 indicates that while the acid concentration reduces to 0.05%, the xylose concentration is very lower due to a low level of hemicellulose hydrolysis rate under the lower acid concentration. It should be noted that the acid concentration in the reactor practically not equal to the acid concentration that is fed because of the buffer action from some natural cellulosic substances [15]. Maloney [16] and coworkers have concluded from their research that the buffer action by the white birch can be 3.5 g of sulfuric acid /kg.

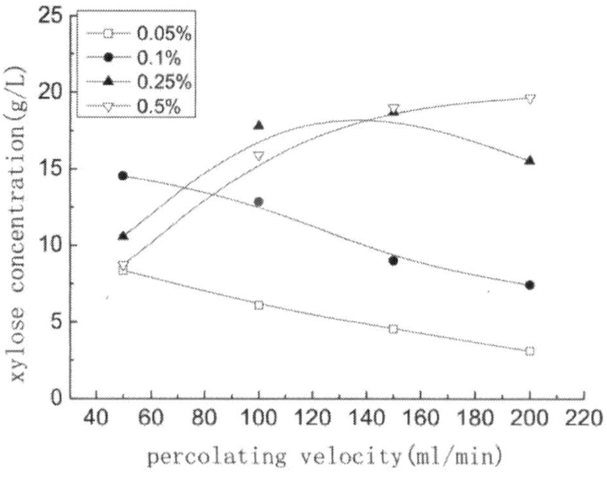

Fig.2. The effect of sulfuric acid concentration on the xylose concentration

When the acid concentration rise to 0.25%, it is evident from observing Fig 2 that decreasing the liquid percolation rate will increase the xylose concentration initially and then decline. The results can be explained as percolation rate decreases, hemicellulose is hydrolyzed more completely, and resulting in xylose in the hydrolysate is accumulated and then reaches the peak. But, a prolonged reaction time also results in a decrease in xylose concentration for its degradation.

The increasing acid concentration dramatically increases the hemicellulose hydrolysis rate and also speeds up the xylose degradation rate. Hydrolysis under the acid concentration of 0.5% did not realize much increase in the xylose concentration. Furthermore, high acid concentration means that operation cost increases and equipment corrosion speeds up. So, the acid concentration of 0.25% was determined as the condition for the highest xylose yield.

Effect of Hydrolysis Temperature

The experiments was proceed within the temperature range of 160-180℃ in the presence of acid with concentration of 0.25%. The results are listed in Fig 3.

At 160□ the rate of xylose generation is slower to reach its peak than those at other temperatures; however sugar preservation is sustained for a longer period of time due to a low level of sugar decomposition. The maximum xylose yields occurred at the temperature of 170℃ with acid concentration of 0.25%.

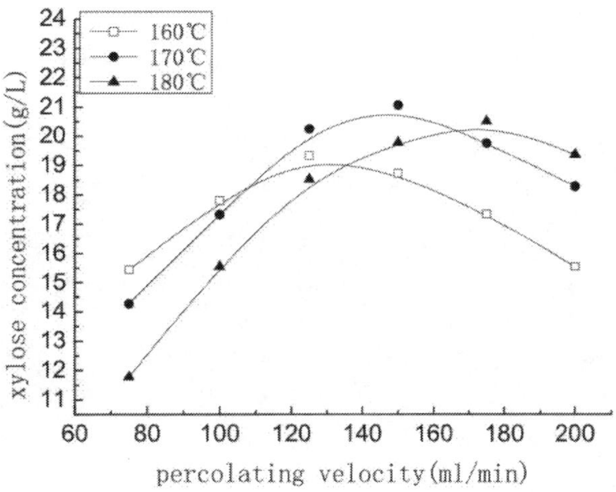

Fig.3. The xylose concentration varys with the temperature and the percolating velocity

Effects of Acidic Pretreatment on Enzymatic Hydrolysis

Fig 4 gives the cellulose contents in the residues after the acid pretreatment. Only a small portion of cellulose is converted when hydrolysis proceeds at 160℃ or 170℃, both with acid concentration of 0.25%. But when temperature is raised to 180℃, the cellulose lost in the residue is noticeable for the cellulose hydrolysis at the pretreatment step. With the hemicellulose and cellulose hydrolysis, the crystalline structure of the cellulose is destructed which favors for the enzymatic hydrolysis (see Fig 5).

Residue of corn stalk after pretreatment operated under various conditions was subsequently hydrolyzed enzymatically. Here, the degree of enzymatic hydrolysis is defined as follows: the amount of glucose so formed is divided by the amount of glucose that is theoretically calculated from the cellulose content in the residue.

Fig 6 indicates that the more severe conditions for the acid hydrolysis can lead to the higher degree of enzymatic hydrolysis. In the runs of tests, the highest degree of enzymatic hydrolysis was over 90%, which implies that a small fraction of cellulose is still unconverted, for the cellulose contained in the residue is not identical.

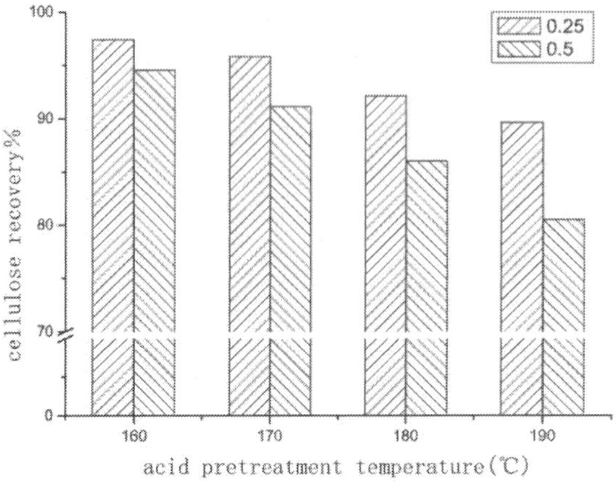

Fig. 4. Cellulose recovery in the water-insoluble solid at different pretreatment conditions, expressed as a percentage of the initial content in the raw material.

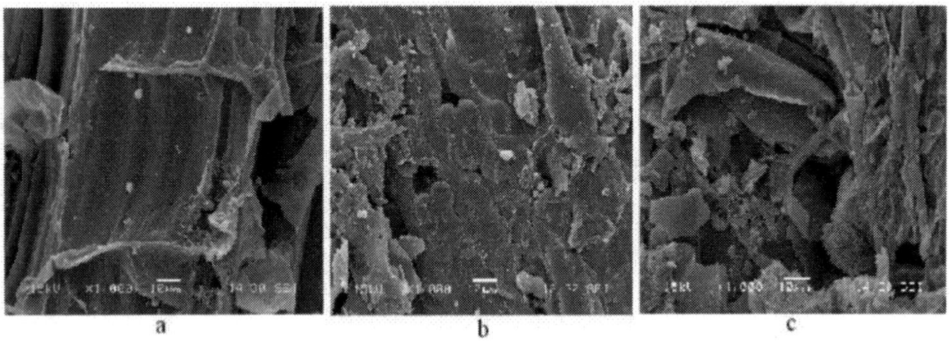

a: corn stalk; b: the residue treated at the pretreatment condition of 170℃; c: the residue treated at the pretreatment condition of 180℃

Fig. 5 SEM images of corn stalk (a) and acid pretreatment residues (b and c)

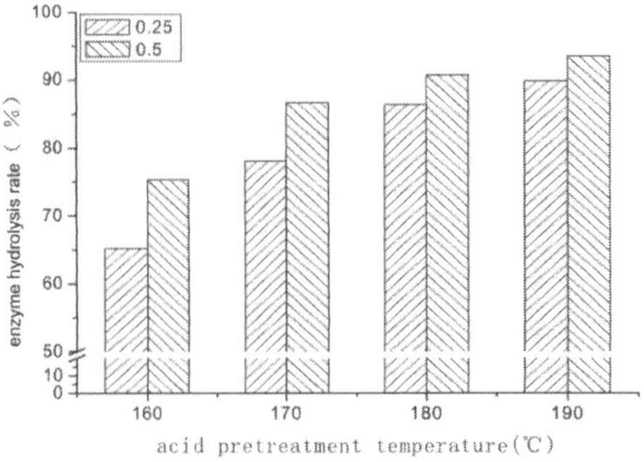

Fig. 6. Enzymatic hydrolysis yield at different pretreatment

Combination of Acid-enzyme Hydrolysis

Concentrations of the total fermentable sugar obtained from both steps of the acid pretreatment and enzymatic hydrolysis are illustrated in Fig 7. Two columns consist of one group with the temperature of pretreatment as parameter; the column in front indicates the concentration of the total fermentable sugar obtained at acid concentration 0.25% as well as the shares of xylose and glucose while the column behind indicates the data obtained at acid concentration of 0.5%.

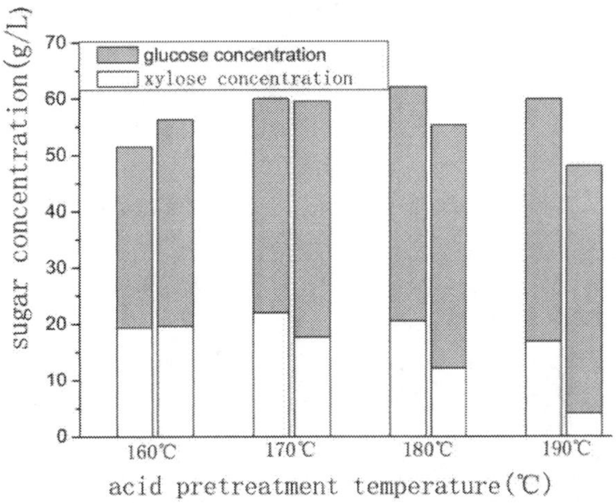

Fig. 7. Overall sugar concentration at different pretreatment temperature

It is known from the Fig 7 that the concentration of total fermentable sugar is the highest when the pretreatment proceeds at 180℃ and acid concentration 0.25%. In case that the xylose concentration is taken as the desired product, the preferential pretreatment conditions are at 170℃ with a sulfuric acid of 0.25%. Compared with the lower temperature (170℃) experiments, there was a 3.1g/L of glucose increased. The results show that the pretreatment conditions for the highest xylose yield did not coincide with that of for the highest total fermentable sugar yield. Therefore, it is essential to control the pretreatment severity when a process of acid-enzyme coupling hydrolysis shall be employed for the lignocellulosic biomass.

Conclusions

In this paper, conclusions are made for the acid-enzyme coupling hydrolysis of corn stalk with the concentration of total fermentable sugar as the target.

(1) The preferential conditions for the pretreatment by the dilute acid are as follows: reaction temperature, 170℃; concentration of sulfuric acid 0.25%; liquid to solid ratio 10:1.

(2) The preferential conditions for acid-enzyme coupling hydrolysis of corn stalk are as follows: pretreatment temperature 180℃; acid concentration 0.25% when enzymatic hydrolysis proceeds.

(3) The highest xylose yield did not coincide with the highest total fermentable sugar yield. The severity of acid pretreatment affects the enzymatic hydrolysis results.

Acknowledgement

The study was supported by National Key Technology Support Program (2011BAD22B01，2012BAD32B07) and Chinese National High-tech R&D Program (863 Program) (2012AA101806) are greatly appreciated.

References

[1] Brethauer S., Wyman C. E., **(2010)** *Bioresour. Technol.*, 101, 4862–4874.
[2] Qureshi N., Ezeji T. C., Ebener J., Dien B. S., **(2008)** *Bioresour. Technol.*, 99, 5915–5922.
[3] Bai F.W., Anderson W. A., Young M. M., **(2008)** *Biotechnol. Adv.*, 26, 89– 105.
[4] Talebnia F., Karakashev D., Angelidaki I., **(2010)** *Bioresour. Technol.*, 101, 4744–4753.
[5] Zhang S. P., Marechal F., Garssner M., **(2009)** *Energy Fuels*, 23(3), 1759-1765.
[6] Ewanick S. M., Bura R., Saddler J. N., **(2007)** *Biotechnol. Bioeng.*, 98 (4), 736-746.
[7] Kumar P.; Barrett D. M., Delwiche M. J., Stroeve P. **(2009)**, *Ind. Eng. Chem. Res.*, 48 (8), 3713–3729.
[8] Saitoh S., Hasunuma T., Tanaka T., **(2010)** *Appl. Microbiol. Biotechnol.*, 87, 1975-1982.
[9] Rubin E. M., Nat. **2008,** 454, 841-845.
[10] Ruiz C. E., Oliva J. M., Saez F., Castro E., **(2008)** *Bioresour. Technol.*, 99(6), 1869-1876.
[11] Gírio F. M., Fonseca C., Carvalheiro F., **(2010)** Bioresour. Technol., 101, 4775–4800.
[12] Juan E. Morinelly, Jill R. Jensen, Matthew Browne, et al. **(2009)** *Ind. Eng. Chem. Res.*, 28, 9877-9884.
[13] Cara C., Ruiz E., Oliva J. M., Saez F., Castro E., **(2008)** *Bioresour. Technol.*, 99, 1869-1876.
[14] Jensen J. R., Morinelly J. E., Gossen K. R., Brodeur-Campbell M. J., Shonnard D. R., **(2010)** *Bioresour. Technol.*, 101, 2317–2325.
[15] Hsu D. D., **(1995)** Biomass Chemical Analysis and Testing Laboratory Analytical Procedures. NREL,.
[16] Maloney M. T., Chapman T. W., Baker A. J. **(1985)** *Biotechnol. Bioeng.*, XXVII, 355-361.

(675b) Hybrid Thermochemical Processing: Fermentation of Pyrolytic Substrates

Tao Jin[1], Yi Liang[2], Donovan S. Layton[1], Mark M. Deaton[3], Zhanyou Chi[4], Robert C. Brown[5], Zhiyou Wen[2] and **Laura R. Jarboe**[6], (1)Chemical and Biological Engineering, Iowa State University, Ames, IA, (2)Food Science and Human Nutrition, Iowa State University, Ames, IA, (3)Chemical & Biological Engineering, Iowa State University, Ames, IA, (4)Department of Biological Systems Engineering, Washington State University, Pullman, WA, (5)Mechanical Engineering, Iowa State University, Ames, IA, (6)Department of Chemical and Biological Engineering, Iowa State University, Ames, IA

As a temporary storage unit of sunlight-derived energy and atmospheric carbon, biomass is an excellent source of carbon and energy for the production of biorenewable fuels and chemicals. However, the cost- and energy-efficient release of fermentable sugars from this biomass is challenging, largely due to the complex structure of lignocellulose. Thermochemical processing of biomass by fast pyrolysis provides a non-enzymatic route for depolymerization of biomass into fermentable substrates that can be used for the biological production of fuels and chemicals. Fermentative utilization of these pyrolytic substrates faces two formidable challenges. First is the fact that most bio-oil-associated sugars are present in the anhydrous form. Metabolic engineering has enabled utilization of the main anhydrosugar, levoglucosan, in workhorse biocatalysts. The second challenge is the fact that bio-oil is rich in microbial inhibitors. Biomass processing, collection of the pyrolytic substrates, detoxification of the substrates prior to fermentation and increased robustness of the biocatalyst are effective methods for addressing this inhibition. Here we describe metabolic engineering efforts to enable utilization of these pyrolytic substrates.

(675c) Influence of Impeller Design On Power Input and Performance During High Solid Enzymatic Hydrolysis of Pretreated Lignocellulose

Benny Palmqvist and Gunnar Lidén, Department of Chemical Engineering, Lund University, Lund, Sweden

Reaching high ethanol titers after the fermentation step is crucial in cellulosic ethanol production in order to make the process cost competitive and commercially available. To achieve this, a high amount of water insoluble solids (WIS), i.e. lignocellulose fibers, is needed in the hydrolysis step. However, a high WIS content in the hydrolysis step is likely to give a high viscosity. The power requirements needed to operate the impeller at high viscosities can be quite substantial, and high WIS processes therefore may need to be operated with a poor mass- and/or heat transfer. We have previously shown that stirrer speed (or more accurately – mixing-power) strongly affects the hydrolysis of certain lignocellulosic materials, in particular pretreated spruce. An almost 100 % yield increase was achieved by increasing the stirrer speed from 75 to 500 rpm. In contrast, the power input seems to have only a very marginal impact on the hydrolysis performance when using for example the energy crop Arundo Donax (Giant Reed). Another difference between these particular materials is their relative change in rheology due to enzymatic hydrolysis. For example, Arundo Donax quickly (within 2-4 hours) losses its network strength, leaving no significant difference in needed power input between different WIS contents (within the range of 10-20 % WIS). Spruce on the other hand retains its strength, i.e. high viscosity, during a much longer time which results in significantly higher power inputs during a longer time. For the same stirrer rate, approximately, four times as much energy is needed for mixing of spruce compared to Arundo Donax at 20 % WIS.

In the present study, enzymatic hydrolysis at high solid loadings (<20% WIS) was investigated using different impeller types. In addition to sugar yields, impeller torques (and hence power input) were monitored throughout the hydrolysis to compare the suitability of the impeller type for this process. Furthermore, it was possible to control the power input at a fixed set-point value throughout the hydrolysis period by adjusting the stirrer speed. This enables a proper comparison between the different impellers based on an energy input basis. The factors affecting high solid hydrolysis will be discussed.

Enzymatic recovery of polyhydroxybutyrate (PHB) from *Bukerholderia cepacia* by pancreatin

Yuanzhen Wang, Shijie Liu*

Department of Paper and Bioprocess Engineering, SUNY College of Environmental Science and Forestry, 1 Forestry Drive, Syracuse, NY. 13210

*Author for correspondence (E-mail: sliu@esf.edu)

1 Abstract

Enzymatic digestion by pancreatin has been developed as an effective and cost efficient method to recover polydroxybutyrate (PHB) from fermentation culture of *Bukerholdria cepacia.* Pretreatment methods including heat shock, surfactant digestion, and oxidizing agent digestion were screened, and heat pretreatment at 95 °C for 45 min was chosen for high efficiency and no introduction of chemicals. Transmission electron microscopy (TEM) indicated the efficient cell lysis and PHB granules release by pancreatin. Optimized enzymatic digestion condition was found to be 50 °C in a pH 8.0 Tris-HCl buffer, with the pancreatin/biomass ratio fixed as 0.02. PHB recovery and purity reached up to 93.5% and 90.6%, respectively. Successful protein removal was achieved. Polymer degradation was negligible because no significant molecular weight reduction was observed.

Keywords: Polyhydroxybutyrate; Enzymatic recovery; Pancreatin; *Bukerholdria cepacia*

(675e) Screening Wood-Decay Fungi for Demethylation of Kraft Lignin Using Two Novel Assays with Applications in Producing Bio-Methanol and Formaldehyde-Based Polymers

Andrew Gibson, Biotechnology, Lakehead University, Thunder Bay, ON, Canada

Screening Wood-decay Fungi for Demethylation of Kraft Lignin using Two Novel Assays with Applications in Producing Bio-Methanol and Formaldehyde-based Polymers

Andrew Gibson [1,2], Balaji Venkatesagowda [1], Iryna Kandybovich [1],

Aneli M. Barbosa [1], Brian Ross [2], Lada Malek [2], Robert F.H. Dekker [1]

[1] Biorefining Research Institute, [2] Department of Biology, [3] Northern Ontario School of Medicine, Lakehead University, Thunder Bay, Ontario, Canada P7B 5E1

Lignin, the most abundant aromatic polymer present in the biosphere, is found in the cell walls of plants rendering the plant resistant to microbial attack and degradation. This biopolymer has potential applications, yet is currently considered a waste product (black liquor) in Kraft pulp and paper mills. By enzymatically modifying Kraft lignin, two immediate products can be obtained: a modified lignin with a higher phenolic content, and bio-methanol. The first can be used in formaldehyde-based polymer applications, while the second can be used in industrial processes and as biofuel. We have developed two specific assay procedures to screen for enzymatic activity that would produce these products; selected ion flow tube-mass spectrometry (SIFT-MS) and Ti(III)-nitriloacetic acid colorimetry. SIFT-MS is a highly sensitive technique capable of measuring trace amounts of gases including methanol (ppb levels) in the headspace of flask-cultivated fungi, while the Ti(III)-NTA assay measures the changes in vicinal hydroxyl groups during the modification of lignin by various fungal species. The Mycological Herbarium at Lakehead University in conjunction with the Biorefining Research Institute has a collection of >450 fungal isolates from the Boreal Forest in northern Canada; 20 of which were screened using these techniques. Fungal cultures were grown under stationary conditions on a Kraft lignin-minimum salts (Vogel) nutrient medium for 3 weeks at room temperature. The fungal isolates showed a reduction of up to 45% of *O*-methyl content using the Ti(III)-NTA assay, and a 10-fold increase in methanol concentration as measured by SIFT-MS compared to controls. The research presents evidence of *new* enzyme activities that specifically cleave *O*-methyl groups from the lignin molecule. Further work will continue and confirm the existence of these enzyme activities through their isolation and purification.

Supported by funds from NSERC-CRD (CRDPJ 38079-09: Dekker) Canada

(675f) Biological and Enzymatic Demethylation of Kraft Lignin and Lignin-Like Model Compounds by Boreal Forest Fungi Using a Novel Technique: Selected Ion Flow Tube-Mass Spectrometry (SIFT-MS)

Balaji Venkatesagowda[1], Andrew Gibson[2], Aneli M. Barbosa[1], Brian Ross[3], Lada Malek[4] and Robert F.H. Dekker[1], (1)Biorefining Research Institute, Lakehead University, Thunder Bay, ON, Canada, (2)Biorefining Research Institute, Department of Biology, Lakehead University, Thunder Bay, ON, Canada, (3)Northern Ontario School of Medicine, Lakehead University, Thunder Bay, ON, Canada, (4)Department of Biology, Lakehead University, Thunder Bay, ON, Canada

Selected Ion Flow Tube-Mass Spectrometry (SIFT-MS) is a rapid and highly sensitive analytical method for detecting volatile organic compounds by chemical ionization using selected ions. The objective of this work was to detect methanol released from Kraft lignin (KL) and lignin-like model compounds (LMCs) by the action of enzymes produced by ligninolytic fungi that specifically cleave-off *O*-methyl groups. Lignin demethylation is considered a *new* enzyme activity, and such enzymes can be described specifically as lignin demethylases. The culture collection at Lakehead University harboring some 450 wood-decay fungi collected from the Boreal Forest in northern Canada were grown on minimum salts (Vogel) nutrient medium containing Kraft lignin and different low MW LMCs under stationary conditions at room temperature for 4 weeks. Growth on these substrates induced the formation of enzymes that specifically liberated methanol as determined by SIFT-MS. The most promising fungi examined producing methanol from KL were isolates LU-86 (84 ppb), LU-122 (125 ppb), BRI-269 (112 ppb) and BRI-270 (180 ppb), and with LMCs as substrates: BRI-122 (30 ppb), BRI-269 (140 ppb) and BRI-270 (171 ppb). The low levels of methanol produced are attributable to these isolates also producing an enzyme (methanol oxidase) that metabolizes the methanol released. Confirmation of demethylation of lignin was supported by an increase in phenolic (vicinal hydroxyl) content in the treated lignin. In another series of experiments, the same fungal isolates (6) were cultivated on media containing KL for 4 weeks. Mycelia were removed by centrifugation, and the supernatants recovered (ECF) and used as the source of enzyme. Enzymatic demethylation of KL and different LMCs was determined by adding ECF and measuring the amount of methanol liberated by SIFT-MS. Different lignin substrates, such as lignosulfonate (Sigma), periodate-oxidized lignin, Kraft lignin, and lignins extracted from softwood and hardwood black liquors, were also examined for the release of methanol by enzymatic demethylation. The release of methanol from these ligninolytic substrates as measured by SIFT-MS demonstrated unequivocally that *O*-methyl groups were cleaved from these compounds by the fungal and enzyme systems studied. The results obtained will be discussed in detail.

Supported by funds from NSERC-CRD (CRDPJ 38079-09: Dekker) Canada

(689a) Development of Intensified Catalytic Reactors for DME and Fischer-Tropsch Biofuels Synthesis in Future Integrated Biorefineries

Martin O'Connell[1], Costa Komodromos[2], Gunther Kolb[3], Ralf Zapf[3], Irene Resa[4], Manickam Jayamurthy[4] and Ashok Bhattacharya[5], (1)1Institut für Mikrotechnik Mainz GmbH, D-55129 Mainz, Germany, (2)Wlfson Centre for Materials processing, Brunel University, Middlesex, United Kingdom, (3)Energy Technology and Catalysis Department, Institut für Mikrotechnik Mainz GmbH , D-55129 Mainz , Germany, (4)Wolfson Centre for Materials Processing, Brunel University, Uxbridge, Middlesex, United Kingdom, (5)Wolfson Centre for Materials Processing, Brunel University, Uxbridge Middlesex , United Kingdom

Intensified catalytic reactors have been developed within a European Union funded project (www.suprabio.eu) to carry out conversion of syngas, derived from thermochemical conversion of biomass, to dimethyl ether (DME) via a direct process (without an intermediate methanol stage) and to middle distillate biofuels via Fischer-Tropsch synthesis. The aim is to provide efficient processes that are scalable to the distributed biomass treatment and conversion processes within an integrated biorefinery system. The approach is to achieve process and engineering integration and to utilise the enhanced heat and mass transfer properties of the reactor designs to overcome the cost penalties associated with downscaling plants.

The catalysts are prepared in a form that allows coating onto arrays of metal plates that are stacked and laser welded to form a mini-reactor. The challenges are to optimise catalyst properties, to achieve stable, adherent and uniform coatings and to design and optimise the reactor geometry to achieve fine control of reaction temperatures.

A range of highly selective, novel nano-catalysts have been prepared and tested for these reactions.

The experiments have been carried out over a wide range of reaction temperatures (210–300 °C), pressure (10–30 bar), $H_2/CO = 2$. The results indicate very promising levels of DME and hydrocarbon yields.

(689b) Drop-in Gasoline From Carbona Wood Gasification, Morphysorb CO2 Removal, and the Tigas Process

Rick Knight, Gas Technology Institute, Des Plaines, IL, Niels R. Udengaard, Syngas Catalyst & Technology, Haldor Topsoe, Inc., Houston, TX, Finn Joensen, Haldor Topsøe A/S, DK-2800 Lyngby, Denmark, Jesper Jensen, Technology Division, Haldor Topsoe A/S, Lyngby, Denmark and Jim Patel, Andritz/Carbona, Vallejo, CA

Drop-In Gasoline from Wood Gasification, CO2 Removal, and the TIGAS Process

ABSTRACT

Biomass conversion is the only near-term option to replace petroleum-based motor fuels with renewable alternatives. One of the pathways under development is thermochemical gasification followed by conversion of the resulting synthesis gas to gasoline. Under U.S. Department of Energy sponsorship, Haldor Topsoe Inc. (HTI), the Gas Technology Institute (GTI), Andritz-Carbona, UPM-Kymmene, and ConocoPhillips have teamed up to build and operate a pilot biorefinery based on this approach. Woody biomass will be converted to gasoline blendstock, offering product compatibility with existing infrastructure. Startup of the 21 bbl/day pilot plant is scheduled for November 2012.

The biorefinery integrates Andritz-Carbona's pressurized, oxygen-blown bubbling-bed gasifier, which features high thermal efficiency and low syngas tar contamination. A cleanup system featuring hot gas filter and catalytic reformer eliminates particulates and heavy organics while also converting residual methane to carbon monoxide and hydrogen. GTI's acid-gas removal process uses Morphysorb® solvent to remove carbon dioxide and hydrogen sulfide with high selectivity that minimizes loss of desired syngas components. The clean, conditioned syngas is then converted to gasoline in Haldor Topsoe's TIGAS process, which employs a unique high-conversion once-through methanol/dimethylether synthesis step. Recycling a portion of the tail gas from the synthesis back to the gasifier results in an essentially power-neutral process, resulting in an overall syngas conversion of well over 90 per cent.

This integrated process is capable of converting any type of biomass, but the commercial case under consideration is to supply the plant with woody biomass such as mill wastes, non-merchantable roundwood, whole tree chips, or other available forestry byproducts. This strategy will leverage the resources, facilities, and expertise of the pulp and paper industry to provide value-added revenue streams while helping meet U.S. transportation fuel needs with home-grown sustainability.

In this presentation, we will discuss the proposed process layout of the wood-to-gasoline process, summarize the engineering and construction status, and present an updated prediction of operating cost for the early commercial plants.

(689c) The Effects of Heating Rate and Reaction Time On Hydrothermal Liquefaction of Microalgae

Julia L. Faeth and Phillip E. Savage, Chemical Engineering, University of Michigan, Ann Arbor, MI

Hydrothermal liquefaction (HTL) is one of many methods for biomass conversion to biofuels or biofuel precursors. HTL avoids energy-intensive drying steps, and is therefore more energy efficient for biomass with very high moisture content, like microalgae. The effects of many different parameters (including biomass concentration, nominal reaction temperature, catalysis, and reaction time) on product yields for HTL of microalgae have been reported in the literature. However, a majority of the reaction times reported in HTL literature have been defined to ignore the time necessary to reach the nominal reaction temperature. Furthermore, the heating times reported vary greatly, from several minutes to a few hours.

The time necessary to heat an HTL reaction mixture to the nominal reaction temperature is dependent on the heating rate. The heating rate of a reaction mixture has been observed to influence the product yields for HTL of woody biomass. Product yields for supercritical water gasification of glucose and cabbage are also influenced by heating rate. This presentation explores the influence of heating rate on the product yields for HTL of microalga *Nannochloropsis* sp. The influence of reaction time, including time necessary for the reaction mixture to reach the nominal reaction temperature, will also be discussed. These effects need to be considered during the development of an efficient HTL flow reaction process.

(689d) Liquefaction of Low-Lipid Macroalgae for Production of Liquid Fuel

Shicheng Zhang, Department of Environmental Science and Engineering, Fudan University, Shanghai, China

Liquefaction of Low-Lipid Macroalgae for Production of Liquid Fuel

Shicheng Zhang*, Dong Zhou, Changjun Li, Xiao Yang, Jianmin Chen

Department of Environmental Science and Engineering, Fudan University, Shanghai 200433, China

*Corresponding author phone: +86-21-65642297; fax: +86-21-65642080; e-mail:

zhangsc@fudan.edu.cn

Low-lipid marine macroalgae Enteromorpha prolifera, one of the main algae genera for green tide, was converted to bio-oil and ester compounds in sub-/supercritical water or alcohols in a batch reactor. Effects of the temperature, reaction time, Na_2CO_3 catalyst, and algae/solvent ratio on the yields of liquefaction products were investigated. The raw algae and liquefaction products were analyzed using elemental analysis, Fourier transform infrared (FTIR) spectroscopy, gas chromatography-mass spectrometry (GC-MS), and 1Hnuclear magnetic resonance (NMR). The bio-oil obtained in water was a complex mixture of ketones, aldehydes, phenols, alkenes, fatty acids, esters, aromatics, and nitrogen containing heterocyclic compounds. And acetic acid was the main component of the water-soluble products. Otherwise, the bio-oils obtained in alcohols were mainly composed of ester compounds. A variety of fatty acid (C3—C22) esters (methyl or ethyl) in the bio-oils respectively obtained in methanol and ethanol were qualified by GC-MS, and their relative contents are above 60% of the total area for each bio-oil. In addition, some N-containing compounds, sugars, fatty alcohols/ketones and very few hydrocarbons were also qualified. Overall, bio-oils obtained in two alcohols are much similar to biodiesel on the composition. Moreover, the bio-oils are found to contain a considerable fraction of light components using thermogravimetric analysis (TGA), and the contents of low-boiling (bp<350 ^{o}C) compounds are up to 70% of the weight for both bio-oils, therefore, it might help to the further separation and refining of bio-oils to produce fuels and chemicals.

References

1. Dong Zhou, Liang Zhang, Shicheng Zhang, Hongbo Fu, Jianmin Chen. Hydrothermal Liquefaction of Macroalgae Enteromorpha prolifera to Bio-oil. Energy & Fuels. 2010, 24, 4054–4061.
2. Dong Zhou, Shicheng Zhang, Hongbo Fu, Jianmin Chen. Liquefaction of Macroalgae Enteromorpha prolifera in Sub-/Supercritical Alcohols: Direct Production of Ester Compounds. Energy & Fuels. 2012. 26 (4), 2342-2351.

(689e) Renewable Liquid Alkanes Production in Continuous Flow Microchannel Reactor

Lalita Attanatho, Department of Chemical, Biological and Environmental Engineering, Oregon State University, Corvallis, OR

The study focused on the development of microtechnology-based chemical reaction process for liquid biofuel production from oil-based biofuel feedstock. A hydrotreating reaction of 5 wt% of oleic acid in dodecane was performed in continuous flow microchannel reactor. The microchannel reactor was fabricated from SS-316 and the wall of the microreactor was coated with a thin layer of NiMoP/Al_2O_3 catalyst. The reactions were carried out in the temperature range of 275-325 °C, residence time in the range of 36-40 s and at constant system pressure of 500 psi.

The results showed that microchannel reactor was well suitable for the hydrotreating process. The almost complete conversion of fatty acid hydrotreating reaction was achieved at a reaction temperature of 300 °C and above. The conversion of fatty acid increased with increasing temperatures and decreasing initial feed concentrations. Hydrotreating of fatty acid occurred primarily via hydrogenation and dehydration and the main liquid products were octadecane and heptadecane. Fatty alcohol, fatty acid and long chain ester were formed as intermediates.

(689f) Novel Reactors with Intensified Catalytic Properties for Renewable Diesel Production

Martin O'Connell[1], Costa Komodromos[2], **Jens Ole Wiborg**[3], Manickam Jayamurthy[4], Hans-Joachim , Kost[1] and Ashok Bhattacharya[5], (1)1Institut für Mikrotechnik Mainz GmbH, D-55129 Mainz, Germany, (2)Wlfson Centre for Materials processing, Brunel University, Middlesex, United Kingdom, (3)Energy Technology and Catalysis Department, Institut für Mikrotechnik Mainz GmbH , D-55129 Mainz , Germany, (4)Wolfson Centre for Materials Processing, Brunel University, Uxbridge, Middlesex, United Kingdom, (5)Wolfson Centre for Materials Processing, Brunel University, Uxbridge Middlesex , United Kingdom

Before feedstocks derived from renewable organic material can be used in conventional automobile engines and distributed using existing fuel infrastructure, it is desirable to convert the material into hydrocarbons whose composition resembles that of petroleum-derived fuels. A well-established method is the conversion of vegetable seed oils into green diesel by hydrotreating, an exothermic process involving catalytic hydrogenation of the double bonds present in fatty acid chains of triglycerides in the seed oils.

This paper describes the development and use of improved catalysts and reactor design concepts to alleviate the mass transfer limitations in conventional fixed bed reactors, enable very careful control of reaction temperature, and enhanced yields of green diesel. The focus is on minimising side reactions, and improving selectivity towards C_{15}-C_{18} alkanes.

Two reactor concepts have been developed and fabricated: a tubular foam reactor designed to allow staged hydrogen addition, and a membrane reactor.

Experimental work was carried out in the temperature range of 280-325°C, pressures of 30-50 bar using rapeseed and Jatropha oils. The results show that the reactor designs enable better control of temperature, with improved mixing, mass and heat transfer, leading to an increased selectivity of the hydrogenation process compared to published data.

(704a) Key Ingredients and Co-Products in Biofuel Life Cycle Analysis

Jennifer B. Dunn[1], Steffen Mueller[2], Zhichao Wang[1], Jeongwoo Han[1] and Michael Q. Wang[1], (1)Energy Systems Division, Argonne National Laboratory, Argonne, IL, (2)Universityof Illinois at Chicago, Chicago, IL

In many cases, life cycle analyses (LCA) of biofuels have treated key ingredients and co-products less rigorously than feedstocks and fuel products. Key ingredients and co-products, however, can significantly impact LCA results and merit attention. For example, we have recently examined the contribution of enzymes to the life cycle energy consumption and greenhouse gas (GHG) emissions of corn and cellulosic ethanol with Argonne National Laboratory's Greenhouse gases, Regulated Emissions, and Energy use in Transportation (GREET) model. We estimate that the contribution to farm-to-pump GHG emissions in the case of corn ethanol is approximately 1%. In the case of cellulosic ethanol, this contribution is much higher (up to 20%) because enzymes are consumed at a much higher rate. We have also added new data to GREET for the production of yeast and determined that it can contribute up to 11% of farm-to-pump emissions for cellulosic ethanol.

Additionally, we examine key co-products from biofuel production with a focus on biochar, which has not received rigorous treatment in previous biofuel LCAs. We discuss the properties of biochar including GHG mitigation, yield increase, and fertilizer offsets and how these properties can be incorporated into LCAs of biofuels that produce biochar as a co-product.

(704b) Life Cycle Assessments of Pyrolysis-Based Biofuels From Diverse Biomass Feedstocks

Matt Mihalek[1], Jiqing Fan[2], Gaurav Bhardwaj[1], Robert Handler[3], Tom N. Kalnes[4], David R. Shonnard[2] and Edwin.A Maleche[2], (1)Michigan Technological University, Houghton, MI, (2)Chemical Engineering, Michigan Technological University, Houghton, MI, (3)Sustainable Futures Institute, Michigan Technological University, Houghton, MI, (4)Research and Development, UOP LLC, Des Plaines, IL

Life Cycle Assessments of Pyrolysis-Based Biofuels from Diverse Biomass Feedstocks

Matt Mihalek[1], Jiqing Fan[1], Gaurav Bhardwaj[1], Kyle Andrews[1], Robert Handler[2], Tom Kalnes[3], David Shonnard[1,2]

1 Department of Chemical Engineering, Michigan Technological University, Houghton, MI

2 Sustainable Futures Institute, Michigan Technological University, Houghton, MI

3 UOP LLC (Honeywell), Des Plaines, IL

Abstract

Renewable hydrocarbon biofuels are being investigated as possible alternatives to conventional fossil fuels like gasoline, kerosene and diesel. Feedstocks such as corn stover, sugarcane bagasse, switchgrass, waste wood, and algae, are being evaluated as candidates for pyrolysis and upgrading to drop-in hydrocarbon fuels. This research is developing preliminary life cycle assessments (LCA) for each feedstock and comparing the greenhouse gas (GHG) emissions of the hydrocarbon biofuels to current fossil fuels. As a comprehensive study, this analysis attempts to account for all of the emissions associated with each feedstock through the entire life cycle. Emissions from all stages including feedstock production, land use change, pyrolysis, stabilizing the pyrolysis oil for transport and storage, and upgrading the pyrolysis oil to a hydrocarbon fuel are included. In addition to GHG emissions, the energy requirements and water use have been evaluated over the entire life cycle. The goal of this research is to help rank the feedstocks and the resultant hydrocarbon biofuels based on three environmental indicators; GHG emissions, energy demand, and water utilization. Preliminary results indicate that liquid hydrocarbon biofuels produced through this pyrolysis-based pathway can achieve greenhouse gas emission savings of greater than 50% compared to petroleum fuels, thus qualifying these biofuels under the US EPA RFS2 program.

(704c) Sustainability Analyses of Production Scale Processes for Forming Biofuels From Algal Biomass Based On Experimental Data

Paul Blowers, Chemical and Environmental Engineering, The University of Arizona, Tucson, AZ, Christina Canter, Dept. of Chem. and Env.Engr., The University of Arizona, Tucson, AZ, Robert Handler, Sustainable Futures Institute, Michigan Technological University, Houghton, MI, David R. Shonnard, Chemical Engineering, Michigan Technological University, Houghton, MI, Micaela Taborga, Engr. Management, U of Arizona, Tucson, AZ and Michael Galka, U of Arizona, Tucson

Algal biomass is being investigated widely as a route to generating liquid transportation fuels using a variety of technologies. The sustainability analyses and life cycle assessments that have been completed in the literature are often based on a range of assumptions regarding the processes and technologies that cultivate, harvest, separate, and transform the cells into fuels and high value products. In this work, we use both bench and pilot scale data of several algal species to rigorously analyze novel harvesting, dewatering, oil extraction, and conversion technologies regarding the material and energy balance requirements. This life cycle inventory for bench scale and pilot plant data is transformed into an environmental footprint using life cycle assessment principles so opportunities and challenges regarding sustainability can be characterized.

Results show that energy requirements for harvesting range over three orders of magnitude on a kWh/gram of algae basis for some emerging technologies, including membrane technologies, electrochemical flocculation, and flocculation with chitosan, which has lower energies than some existing technologies. However, material needs for these processes incur other embodied energy, water, and greenhouse gas emissions that increase the environmental footprint by several factors for some of the technologies. Environmental impacts for extraction methods are generally larger than for algae harvesting, due to lower efficiencies and more energy intensive stages.

Scenarios are investigated that combine different commercially available technologies that have been previously modeled environmental footprints to the experimentally based emerging technologies. Modified processing trains have the potential of enhancing performance and increasing environmental sustainability so sustainable production of algal based biofuels is possible.

(704d) Evaluating the Sustainability and Resource Consumption of Biomaterials Using Eco-LCA and Spatial Analysis

Erin F. Landers and Bhavik R. Bakshi, William G. Lowrie Department of Chemical and Biomolecular Engineering, The Ohio State University, Columbus, OH

Ecosystem goods and services are essential to all human activity. In spite of this importance, the Millennium Ecosystem Assessment has found that many ecosystem services are highly degraded. In many engineering analyses, these necessary goods and services are undervalued, or even ignored completely. It is therefore crucial to develop and modify existing methods to allow these services to be valued and accounted for in decision making. Often, conventional life cycle assessment (LCA) tends to only look at energy consumption and emissions. However, Ecologically Based LCA (Eco-LCA) is a thermodynamically based LCA that allows for many ecosystem goods and services used throughout our economy to be quantified, in addition to the traditional LCA results of energy consumption and emissions. Eco-LCA looks at resource use throughout the entire economy to determine how much various industries and commodities consume. The commodities that we directly get from ecosystems are called provisioning services and include valuables such as food, water, energy, and other materials. There have been recent additions to the Eco-LCA model and it now accounts for even more ecosystem services (particularly provisioning services) than it used to, including many valuable intermediate flows throughout the economy, such as grains, livestock, and other food sources. This enables us to present our results on sustainability from an engineering perspective to others in the sustainability field with different backgrounds, such as ecology, business, or public policy.

In addition to the insight we can get from the economy-scale Eco-LCA, we have also begun incorporating spatial analysis into LCA. By using specific, spatial data, we can achieve levels of precision that would otherwise be difficult to realize. We do not need to make assumptions about the general conditions of different land uses, we can choose a specific plot of land and know the exact conditions on the piece of land and how that affects the sustainability of the resources coming from that land. We can also introduce time through multiple sets of spatial data, allowing us to see the effects of our decisions on ecosystems. The ecological models that we are using come from the InVEST (Integrated Valuation of Ecosystem Services and Tradeoffs) software from the Natural Capital group. InVEST uses spatial data such as Digital Elevation Models (DEM) and Land Use/Land Change (LULC) maps to feed into the ecological models, which include marine, freshwater, and terrestrial models. From InVEST, we can get new maps of different possible future scenarios as well as aggregated data on total ecosystem service use for the parcel of land.

A case study will be presented using both of these techniques to achieve a very comprehensive LCA. First, we will use spatial analysis to look at the farmland that will be growing our feedstock. It will allow us to better quantify the ecosystem services required for our product, such as pollination, water use, carbon sequestration, and sediment runoff. We will then use the results from that to feed into Eco-LCA, which will allow us to see what resources and valuable intermediates will be necessary for both the inputs to the farming as well as the industrial processing of our feedstock to various biomaterials. From these two methods, we can get information to calculate metrics such as Carbon, Nitrogen, and Water Footprints; Return on Investment; and Renewability. In using both of these methods, we will get a more complete picture of ecosystem service requirements for biomaterials and their relative sustainability.

(704e) Economic and Environmental Assessment of Bio-Based Chemicals Production in Singapore's Petrochemical Cluster

Ang Li[1], Arief Adhitya[2], Iskandar Halim[2], Balaji Balagurunathan[2] and Rajagopalan Srinivasan[1], (1)Department of Chemical and Biomolecular Engineering, National University of Singapore, Singapore, Singapore, (2)Process Science and Modeling, Institute of Chemical and Engineering Sciences, Singapore, Singapore

Singapore is one of the major centers for oil, gas, and petrochemicals in the world. The centerpiece of Singapore's petrochemical industry is the Jurong Island complex, which houses many of the world's leading petroleum and petrochemical companies. Currently, the total refining capacity of Singapore is 1.3 million bpd [1] and the ethylene capacity is 2.1 million tons per year, projected to reach 4 million tons per year by 2014 [2]. The petrochemical industry in Singapore is heavily dependent on fossil fuel-based feedstock, mostly crude oil and natural gas. Use of fossil fuels has negative environmental impact due to the greenhouse gas emission. Another crisis arising from overdependence on fossil fuels is their rapid depletion. Fossil fuels are non-renewable, which means they are finite and might well run out in the near future – a study found that fossil fuels reserve depletion times for oil, coal, and gas are approximately 35, 107, and 37 years, respectively [3]. This rapid depletion might render petrochemical production increasingly unsustainable as fossil fuels prices continue to rise. Thus, alternative sources have to be found for continued production and growth of the petrochemical industry.

Biomass is one possible source of renewable feedstock. Biomass is abundantly available on earth in the form of plants and organisms and considered to be carbon neutral [4]. Due to the complex carbon backbone and existing chemical functionality, the processing of biomass along the chemical synthesis chain can significantly reduce the energy requirement as compared with fossil fuels-based feedstock [5]. In this paper, we investigate the feasibility of incorporating bio-based raw materials into the slate of feedstocks available for chemicals production in Jurong Island. We survey the chemicals currently produced in the cluster and identify potential bio-based chemicals that can be produced. A techno-economic assessment based on current technology is then conducted taking into account production quantities, costs, and conversion yields. After the necessary supply chain is established, cost and carbon footprint assessment of the supply chain using various bio-renewable feedstocks sourced from the region is presented.

References:

1. http://www.edb.gov.sg/edb/sg/en_uk/index/industry_sectors/energy/industry_background.html, accessed on 1 May 2012.
2. http://www.siew.sg/energy-perspectives/energy-singapore/singapores-role-global-ethylene-production, accessed on 1 May 2012.
3. Shafiee, S. and Topal, E. (2009). Energ Pol, 37(1), 181-189.
4. Boom, R., Brehmer, B., and Sanders, J. (2009). Maximum fossil fuel feedstock replacement potential of petrochemicals via biorefineries. The Institution of Chemical Engineers. Elsevier B. V.
5. Sanders, J., Scott, E., Weusthuis, R., and Mooibroek, H. (2007). Macromol Biosci, 7(2), 105-117.

(704f) Analysis of the Life Cycle of 2-Methyltetrahydrofuran Produced From Agricultural Waste

Mariano J. Savelski[1], C. Stewart Slater[1], Dave Aycock[2], Steve Prescott[2] and Jeff Shifflete[2], (1)Chemical Engineering, Rowan University, Glassboro, NJ, (2)PENNAKEM, Memphis, TN

The life cycle emissions to manufacture 2-methyltetrahydrofuran from agricultural by-products have been evaluated. This solvent has the potential as a substitute for other ether and chlorinated solvents commonly used in pharmaceutical and fine chemical syntheses. The starting material, 2-furfuraldehyde, is produced from corn cob waste by acid hydrolysis. The process to convert 2-furfuraldehyde to 2-methyltetrahydrofuran occurs in two hydrogenation steps. Cradle to gate life cycle emissions for 2-methyltetrahydrofuran are not available in commercially used environmental databases, so a new inventory was created based on manufacturing information. The raw materials, energy and wastes have been analyzed for each production stage and an evaluation of the life cycle emissions performed. Our study shows that 2-methyltetrahydrofuran exhibits lower impact scores in the emissions categories when compared to solvents such as tetrahydrofuran, made through a conventional chemical route. A further assessment has been performed on the methods to manufacture tetrahydrofuran. A production pathway starting with the bio-based 2-furfuraldehyde process is greener than a 1,4-butanediol route.

Optimal Design of Integrated Electric-Power an Algal Biofuel Generation Systems

C. G. Gutiérrez-Arriaga[1], M. Serna-González[1], J.M. Ponce-Ortega[1], and Mahmoud M. El-Halwagi[2,3]*
[1]*Facultad de Ingeniería Química, Universidad Michoacana de San Nicolás de Hidalgo, Morelia, Michoacán, 58060, México.*
[2]*Chemical Engineering Department, Texas A&M University, College Station Texas, 77843, USA.*
[3]*Adjunct Faculty at the Chemical and Materials Engineering Department, Faculty of Engineering, King Abdulaziz University, Jeddah, 21589. P. O. Box 80204, Saudi Arabia.*

Abstract

This paper presents a multi-objective optimization methodology to determine simultaneously, the optimal design of steam power plants, considering simultaneously economic and environmental criteria, in addition to capturing the CO_2 from the combustion of different fossil fuels (coal, oil and natural gas) and biofuels (biogas, biomass, softwood and hardwood) selected in the boiler for the production of microalgae and the subsequent biomass conversion to biofuel (biodiesel), also part of this algal biomass produced is used as biofuel in the boiler. This system allows the sustainable reduction of the huge CO_2 emissions in the power plants. The methodology considers the optimal selection of the type of fuel to be used because this affects drastically and simultaneously the economic and environmental functions. The methodology is based on a search procedure through genetic algorithms due to the large number of non-convex terms present and to avoid being trapped in local solutions. The environmental objective function is based on the eco-indicator 99 constituted by harmful materials for the health and the ecosystem to measure the global environmental impact of these systems, following the approach of life cycle analysis (LCA), while the economic objective function considers the maximization of gross annual profit, which includes the profit from the electricity produced, tax credit obtained by the reduction of GHG emissions, the revenue generated by the production of biodiesel, and consider the total annual cost including the fuel costs, turbine, boiler, condenser, pumps, feed-water heaters and deaerator, as well as the costs to operate such equipment (fuel, electricity, cooling water, etc.). To show the applicability of the proposed methodology is considered a case study in Mexico. The results are shown through Pareto curves that help to identify a set of optimal solutions that show the relationship between economic and environmental objectives. The problem can be easily solved using the proposed methodology in a relatively small computation time.

Keywords: Life Cycle Analysis; Sustainable Energy; Solar Energy; Biofuels; Greenhouse Gas Emissions; Multi-Objective Optimization; Power Plant.

Introduction.

At present, the increase in the concentration of CO_2 in the atmosphere, as well as other greenhouse gases (i.e. CO, CH_4, N_2O and NO_X) have caused the so-called global climate change. The overall energy consumption and the associated CO_2 emissions is increasing. Due to this problem, the Kyoto Protocol (1998) was signed to reduce the greenhouse gases (GHG) emissions. Furthermore, according to the Intergovernmental Panel on Climate Change (IPCC 2007), the higher CO_2 emission sources are power plants based on fossil fuels, where mainly coal and oil are used. Therefore, there is needed to develop alternative methods to obtain the required energy at the minimum costs and without the environmental deterioration.

One of the technologies available to combat the problem associated to the GHG emissions is the CO_2 microalgae biofixation. A microalgae biofixation system is suitable for the reduction of carbon dioxide emissions in flue gases, and this is presented as a natural and environmentally friendly alternative, where the biomass produced can be used as source for biofuels production (Chisti, 1980; Nagle and Lemke, 1990; Sawayama et al., 1995). Therefore, this paper presents a new methodology that simultaneously considers environmental and economic aspects, and provides the optimal design of power plant systems. Furthermore, this methodology also allows the optimal selection of the different energy sources such as fossil fuels (coal, oil and natural gas) and biofuels (biogas, biomass, softwood and hardwood and algal biomass) in the boiler, where a microalgae systems is used to sequestrate the CO_2 emissions and integrated into a biofuels production system obtaining biodiesel and biomass used as heat source. The environmental impact is measured through the eco-indicator-99 accounting for the life cycle, which considers the indicators for emissions, materials harmful to health and the ecosystem (see Azapagic and Clift 1999). To void get trapped prematurely, the proposed approach is based on a combination of genetic algorithms and a nonlinear programming model, where the simultaneous minimization for the economic and environmental impact is considered.

Problem Statement.

The problem is to determine the operational parameters of a steam power plant as shown in Figure 1, where the model must determine the optimal operational conditions for the steam power plant such that the maximum profit is obtained accounting for the simultaneous integration of the CO2 emissions captured in the microalgae system which is optimized simultaneously. Next section presents the main components of the considered model.

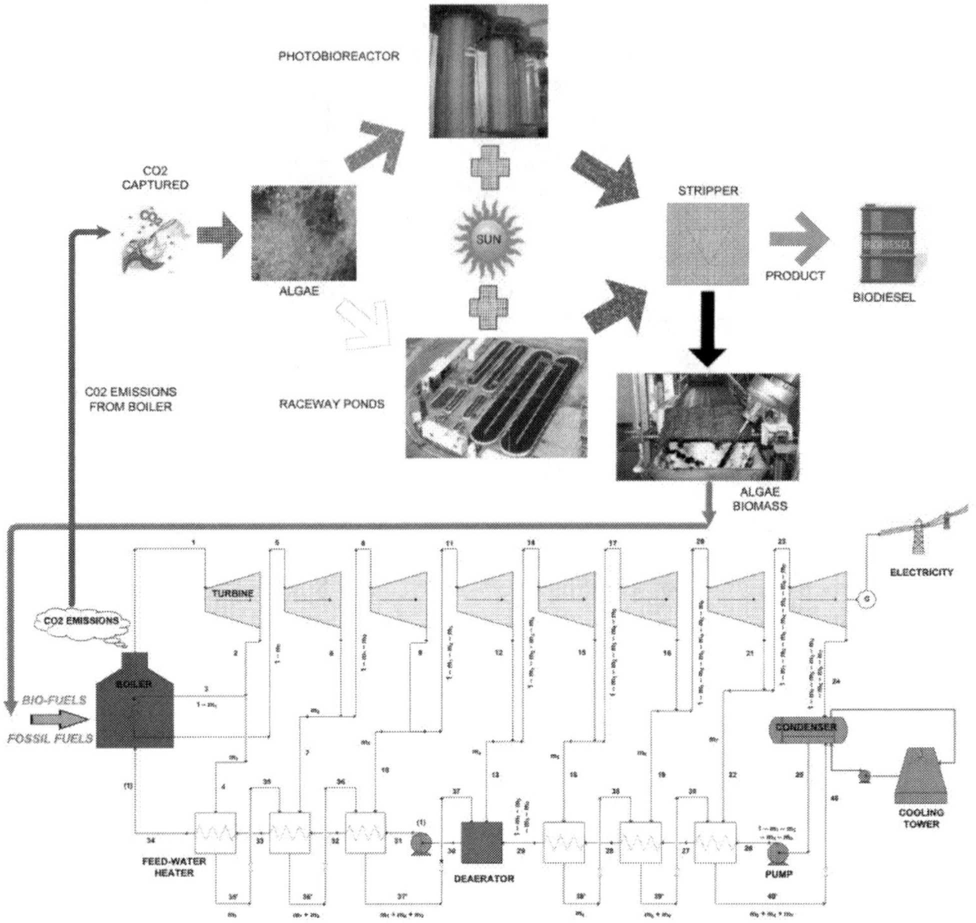

Fig. 1. Proposed scheme for the CO_2 capture in steam power plants.

Methodology

Objective Function

The problem consists of a multi-objective problem, in which the objective function considers two contradict objectives, the first f_1 is to maximize the annual gross profit (*GROSS PROFIT*) and the second f_2 is the minimization of the environmental impact (*EI*), as follows:

$$Objective\,Function = \{\max f_1;\ \min Ef_2\}$$

Economic Objective Function

The economic objective function is the maximization of gross annual profit for the design of the steam power system integrated to the microalgae system as follows:

$$Max\, f_1 = \left(PROFIT + TAX\ CREDIT + ALGAL_{Profit}\right) - \left(TAC + ALGAL_{Cost}\right) \tag{1}$$

This objective function considers the profit from the electricity produced in the generator Rankine cycle which results in a economic benefit (*PROFIT*), and tax subsidies (*TAX CREDIT*) obtained by the reduction of GHG emissions by using biofuels and algal biomass as an additional fuel in the boiler, and the revenue generated by the production of biodiesel from algae is also considered ($ALGAL_{Profit}$); minus the total annual cost (*TAC*), which considers the cost of capital for equipment (boiler, turbine, condenser, pump, boiler feedwater and deaerator), the cost of fossil fuel or biofuel used in the boiler and the cost energy used in the Rankine cycle in the pump, moreover, the cost of production of algae ($ALGAL_{Cost}$) accounting for the cost of nutrients, the cost of the system used for the growth of algae (Photobioreactor or Raceway Ponds) and water used in these units.

Environmental Objective Function

The environmental objective function is to minimize the overall environmental impact generated by the production of electricity; this is done through the eco-indicator-99, which is based on the methodology of the life cycle analysis. This environmental objective functions accounts for the eco-indicator for the fuels used in the boiler, the construction of the boiler, the construction and operation of the condenser, the cooling tower cooling tower, pumps, and other units, as gollows:

$$\begin{aligned} Min\, f_2 = & \left(Ec_1 P_{pump}\right) + \left(Ec_2 P_{fan}\right) + \left(Ec_3 w_P\right) + \left(Ec_4 we_{boiler}\right) + \left(Ec_5 we_{condenser}\right) \\ & + \sum_{t \in T} Ec_t we_t + \sum_{t \in T} Ec_h we_h \\ & + \left[\sum_{f \in F} Ec_f^{Fossil} F_f^{Fossil} + \sum_{bf \in BF} Ec_{bf}^{Biofuel} F_{bf}^{Biofuel} + \sum_{bm \in BM} Ec_{bm}^{Biofuel} F_{bm}^{Biomass} \right] \end{aligned} \quad (2)$$

where Ec_1 is the eco-indicator of the energy used in the pump, Ec_2 is the eco-indicator of the energy used in the fan of the cooling tower, Ec_3 is the eco-indicator of the energy used in the pump of the Rankine cycle, Ec_4 and Ec_5 are the eco-indicators for the production of stainless steel required by the boiler and the condenser, respectively, Ec_t and Ec_h are the eco-indicators for the production of stainless steel required by the turbine t, and heater h, respectively, $we_{condenser}$ is the amount of steel in the condenser, we_{boiler} is the steel used for the boiler, w_P is the power needed to operate the feedwater pumps, P_{fan} is the power of the cooling tower fan, P_{pump} is the pumping power of the cooling tower, $Ec_f^{Fossil}, Ec_{bf}^{Biofuel}$ and $Ec_{bm}^{Biomass}$ are the eco-indicators for fossil fuel f, biofuel bf and biomass bm, respectively. Also, $F_f^{Fossil}, F_{bf}^{Biofuel}$ and $F_{bm}^{Biomass}$ are the flowrates of fossil fuel f, biofuel bf, and biomass bm, respectively.

Variables to be Optimized

To generate an efficient optimization method, genetic algorithms (GA) together with a nonlinear programming mode as it is shown in Figure 2 are used as a search method.

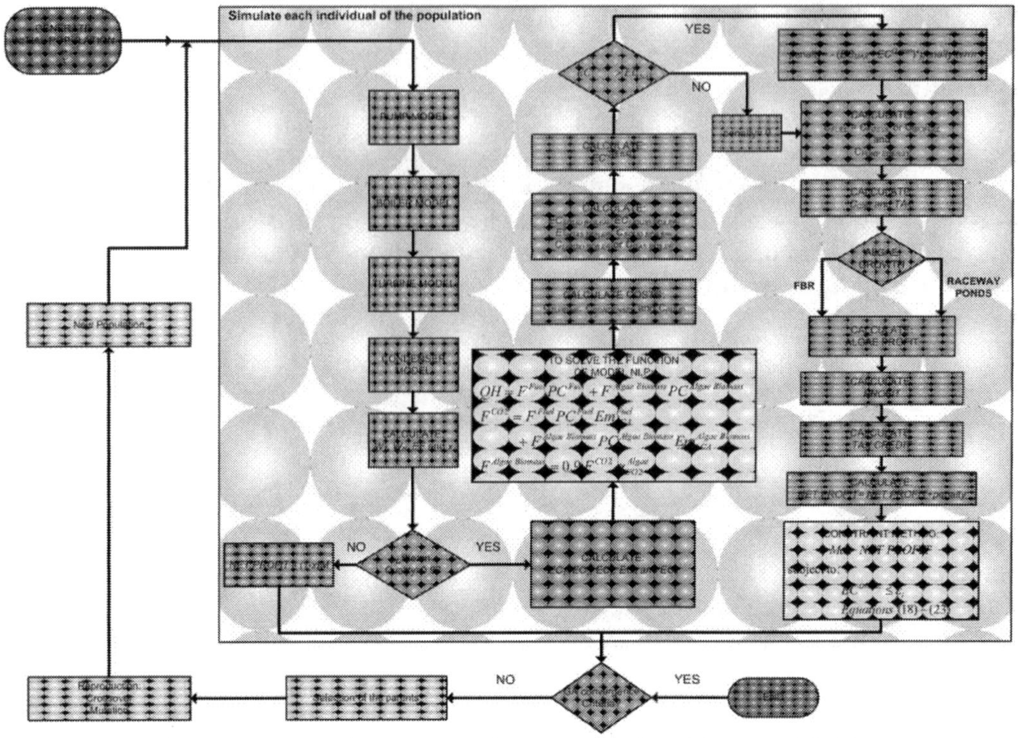

Fig. 2. Solution algorithm, using genetic algorithms and NLP.

The modules used in the simulation of steam power plant are solved sequentially for each component of the system (pump, boiler, turbine and condenser). To solve these modules, the proposed method uses several subroutines to obtain the thermodynamic properties of water and steam (i.e. enthalpy, entropy, volume) through the IAPWS formulation (Wagner et al. 2000), this formulation is the most accepted worldwide for calculations thermodynamic properties in the industry for steam power plants.

Results and Discussion.

A case study is used to show the applicability of the proposed methodology. Figure 3 shows the results obtained for the case study. Fist Figure 3a shows the Pareto curve for the profit versus eco-indicator for a photo-bioreactor, whereas Figure 3b shows the Pareto curve for Raceway ponds integrated to the steam power plant. These figures are implemented for several values of tax credits for the CO_2 reductions. It is noteworthy that in significant economic benefits can be observed in the integrated solutions and at the same time the associated environmental impact is not considerable.

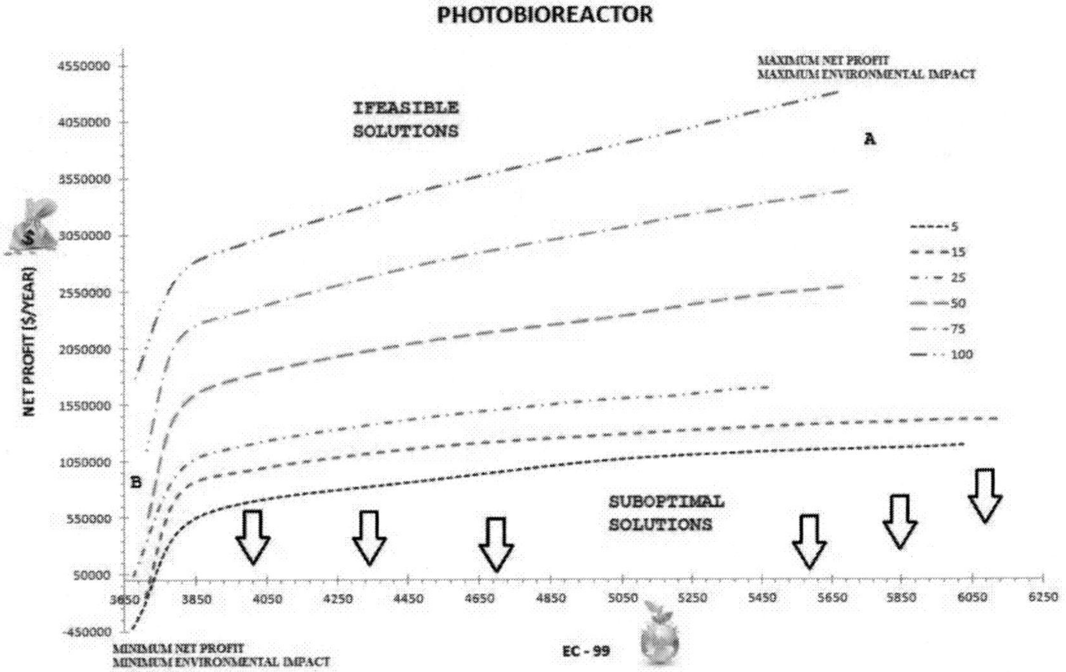

Figure 3a. Pareto curve for a *TAX* credit of \$5/ton CO_2 eq to \$100/ton CO_2 eq for the Photobioreactor.

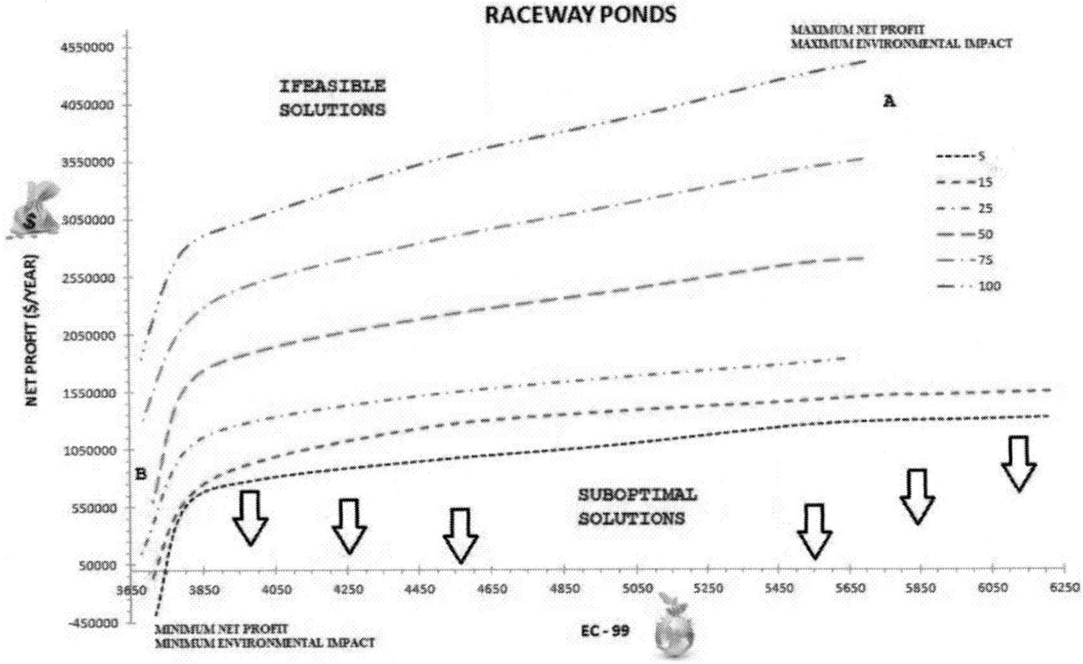

Figure 3b. Pareto curve for a *TAX* credit of \$5/ton CO_2 eq to \$100/ton CO_2 eq for the Raceway ponds.

Conclusions.

This paper proposes integrated approach for the design of a steam power system integrated with a microalgae system, where both systems are optimized simultaneously to maximize the net profit and at the same time to reduce the associated environmental impact. To avoid get trapped prematurely in a suboptimal solution, a solution approach based on GA and NLP is implemented. The thermodynamic correlations as well as the design equations are based on rigorous models. Results show that it is possible to get good economic solution and at the same time to reduce the associated environmental problems through the proposed methodology. Also, the results can be represented in Pareto curves where the decision maker can select the solution that best compromise between the contradict objectives shows.

References

Azapagic A, Clift R (1999) Application of life cycle assessment to process optimization. Comput Chem Eng 23:1509-1526.

Benemann JR (1997) CO2 mitigation with microalgae systems. Energy Convers Manage 38:S475-S479.

Bruno JC, Fernandez F, Castells F, Grossmann IE (1998) A rigorous MINLP model for the optimal synthesis and operation of utility plants. Chem Eng Res Des 76:246-258.

Chisti Y (1980) An unusual hydrocarbon. J Ramsay Soc 27-28:24-26.

Chisti Y (2008) Biodiesel from microalgae beats bioethanol. Trends Biotechnol 26(3):126-131.

Goldberg DE (1989) Genetic algorithms in search, optimization, and machine learning. Addison-Wesley, Reading, MA.

IPCC (2007) Intergovernmental Panel on Climate Change. IPCC fourth assessment report: Climate Change.

Liska AJ, Yang HS, Bremer V, Walters DT, Erickson G, Klopfenstein T, Kenney D, Tracy P, Koelsch R, Cassman KG (2009) BESS: Biofuel energy systems simulator; Life cycle energy and emissions analysis model for corn-ethanol biofuel. Vers. 2008.3. 1, www.bess.unl.edu.

Molina Grima E, Acién Fernández FG, García Camacho F, Chisti Y (1999) Photobioreactors: light regime, mass transfer, and scale up. Progress Ind Microbiology 35:231-247.

Nagle N, Lemke P (1990) Production of methyl-ester fuel from microalgae. Appl Biochem Biotechnol 24-25:355-361.

Protocolo de Kyoto de la Convención Marco de las Naciones Unidas sobre el Cambio Climático. Naciones Unidas, 1998. Disponible en: www.unfcc.int.

Sawayama S, Inoue S, Dote Y, Yokoyama S-Y (1995) CO2 fixation and oil production through microalgae. Energy Convers Manag 36:729-731.

Wagner W, Cooper JR, Dittmann A, Kijima J, Krestzschmar HJ, Kruse A, Mares R, Oguchi K, Sato H, Stocker I (2000) The IAPWS industrial formulation 1997 for the thermodynamic properties of water and steam. J Eng Gas Turb Power 122:150.

Wang M, Wu Y, Elgowainy A (2007) Operating manual for GREET: Version 1.7. Center for Transportation Research, Energy Systems Division, Argonne National Laboratory, Iowa.

(716b) Degradation of of Rhodamine B Dye Over Mn-Doped Titania Nanotubes

Manohar Palla[1], Rakesh Sadu[2], R. Asapu[2], George W. Spencer[2], Jewel Andrew Gomes[3], Zhanhu Guo[3] and **Daniel H. Chen**[3], (1)Dan F. Smith Dept. of Chemical Engineering, Lamar University, Beaumont, TX, (2)Chemical Engineering, Lamar University, Beaumont, TX, (3)Dan F. Smith Department of Chemical Engineering, Lamar University, Beaumont, TX

Mn-doped titania nanotubes were fabricated by solution combustion synthesis followed by the hydrothermal treatment in the removal of Rhodamine B, a dye. Combustion synthesis has emerged as a facile and economically viable technique for the preparation of advanced ceramics, catalysts and nanomaterials. Stoichiometric amounts of transition metal nitrates and fuel (Glycine) are mixed in a solution for the preparation of nanomaterials by using the single step solution combustion method at 350° C. The $Mn_xTi_{(1-x)}\,O_{(2-\delta)}$ (x ranges from 0.05-0.25) nanomaterials are made in the anatase phase by the solution combustion synthesis. The nanotubes are then fabricated using the hydrothermal method using 10 N NaOH at temperature of 130°C for 24 hours.

The prepared Mn-doped titania nanotubes were characterized using JEOL 2100 field emission Gun Transmission Electron Microscope (JEM 2100F TEM) and HITACHI S-3400 N Scanning Electron Microscope (SEM) attached with Energy Dispersive X-ray Spectrometry (EDX). X-ray diffraction (XRD) analysis was performed on a Bruker D8 Focus diffractometer equipped with a Sol-X detector using a copper radiation source, and specific surface area analysis was done using the Brunauer, Emmett, and Teller (BET) method based on N_2 adsorption at 77 K with a NOVA 1000 series analyzer. The thus prepared Mn-doped titania nanotubes contain both monoclinic and anatase phases as evidenced by the XRD and EDX results. The surface area for the nanotubes decreases with increased doping of Mn. Further, TEM analysis shows the formation of nanotubes, which are open ended and are approximately 100-120 nm length with inside and outside diameter estimated to be 6 nm and 12 nm.

The UV-Vis spectrum for the degradation of 4 ppmw Rhodamine B (RhB) over the 0.1 g (0.067 wt %) $Mn_{0.15}Ti_{0.85}O_{(2-\delta)}$ nanotubes shows the successive blue shift from the absorption peak at 554 nm (Rhodamine B) to 498 nm (Rhodamine), indicating the stepwise N'de-ethylation from dark pink to light green. All the experiments were conducted in dark without external energy supply. Degradation of Rhodamine B follows a second order kinetics as shown by the linear plots between $1/C_a$ vs time. 15% Mn is the optimum doping with the hioghest degradation rate constant. The activity of $Mn_{0.15}Ti_{0.85}O_{(2-\delta)}$ increases with temperature with an activation energy of 40777 kJ/kmol for the degradation of RhB.

(716c) Bio-Inspired Green Approach to Manufacture Fibers

Kadhiravan Shanmuganathan, Robert K. Sankhagowit, Prashanth Iyer and Christopher J. Ellison, Department of Chemical Engineering, The University of Texas at Austin, Austin, TX

Recent advancements in fiber production techniques have facilitated new applications for fibrous materials in diverse fields such as optoelectronics, regenerative medicine, piezoelectrics, ceramic materials, etc. Current polymer fiber manufacturing methods almost always use solvents or heat to lower polymer viscosity for processing. A greener approach to making fibers remains a challenge. However, spiders and silkworms have developed benign ways of making silk fibers with high strength and toughness. Their approach of chemically linking small functional units into long chain molecules and solid fibrillar structures while simultaneously extruding the fibers is fundamentally different from current synthetic fiber manufacturing methods, where extrusion of pre-formed long chain polymers is facilitated with organic solvents or heat. Drawing inspiration from nature, a method will be described which uses light to trigger a thiol-ene chemical interaction to rapidly transform small reactive liquid mixtures into solid thread-like structures as they are forced out of a capillary at high speeds. Besides being manufactured without using solvents/volatile components or heat, these fibers are mechanically robust and have excellent chemical and thermal stability.

(716d) Reducing Water Usage in Bioprocessing

Demetri Petrides, INTELLIGEN, INC., Scotch Plains, NJ and Charles Siletti, INTELLIGEN, INC., Mount Laurel, NJ

Water is the primary material used in bioprocessing. While water is not the most expensive raw material the availability of large amounts of purified water and subsequent waste treatment capacity may limit site selection opportunities. This presentation uses bioprocess simulation to explore several strategies. Single-use systems, for example, provide significant reductions in water usage, but there is a practical scale limit. Platform selection can have a significant impact on water usage. For example microbial product expression may be less water-intensive than cell culture. For the majority of products, water reduction can be accomplished through process improvements. Titer optimization, can reduce upstream water usage by 40%. Single-use technologies allow for considerable water savings by eliminating the need for equipment and line cleaning. Single use systems have practical size limitations. Buffer and media preparation strategies like in-line dilution reduce vessel size and resulting water consumption. Alternative downstream technologies can also provide significant reductions in water usage.

(716e) Green Approach to Produce Bio-Jet Fuel From Microalgae

Marian Elmoraghy and Ihab Farag, Chemical Engineering, University of New Hampshire, Durham, NH

Green Approach to Produce Bio-jet Fuel from Microalgae

Marian Elmoraghy, Ihab H. Farag*

Chemical Engineering Department, University of New Hampshire, Durham, NH

* Correspondence, ihab.farag@unh.edu

Introduction

Renewable jet fuel for the aviation industry, also termed bio-jet fuels could reduce flight-related greenhouse-gas emissions by 60 to 80 percent compared to fossil fuel based jet fuel. The green bio-jet fuel is made by blending microalgae bio-fuels with conventional petroleum-derived jet fuel to provide the necessary specification properties. The major advantage of using microalgae oil for biodiesel is the high oil production capacity by microalgae, as they could produce up to 58,700 L oil per hectare.

Sustainable production of bio-jet fuel requires minimizing the energy requirements and reducing fresh water usage while simultaneously lowering the production cost of bio-jet fuel. These are challenges to the biodiesel production from microalgae. Considerable efforts have been made to develop efficient and cost-effective photobioreactors for microalgae growth. Yet, the high cost of installing and operating artificial light sources in conventional photobioreactors with artificial illumination systems remains a major problem.

Objectives

The goal of this research is to develop an economical process to produce bio-jet fuel while minimizing the energy requirements, reducing water usage and meeting the jet fuel specifications. To accomplish this goal the following objectives were defined. 1- Investigate minimizing the energy requirements by replacing fluorescent lights with light emitting diodes (LEDs); 2- Investigate the use of municipal waste water in growing microalgae to reduce fresh water usage; 3- Study the one-step production of biodiesel using in situ algal biomass transesterification process to reduce production time and cost, and 4-investigate biodiesel blending with jet-fuel in order to obtain bio-jet fuel that has specific properties.

Microalgae Growth Using LEDs:

LEDs are lighter than the fluorescent lights and small enough to fit into virtually any photobioreactor (PBR). Other advantages of LEDs include a longer life-expectancy, lower heat generation, and a greater tolerance for switching on and off. LEDs use less electric energy than fluorescent light to produce the same light intensity hence their use lowers the energy requirement for algae growth, and make the process greener by eliminating the CO2 emitted in generating the excess electric power needed for the fluorescent light.

Production of bio-jet fuel:

There are three routes to produce bio-jet fuel from microalgae. Figure 1 shows these three routes. The first route involves using microalgae oil to produce Bio-SPK (Bio derived synthetic paraffinic Kerosene) by cracking and hydro-Processing. This can be used for kerosene-type fuels include jet A, jet A-1, JP-5 and JP-8. FT-SPK (Fischer–Tropsch Synthetic Paraffinic Kerosene) is the second route which involves pyrolysis of solid biomass to produce pyrolysis oil or gasification to produce a syngas which is then possessed into FT SPK. The third route involves algae growth, harvesting, oil extraction and transesterification or (in situ process) to produce biodiesel. The biodiesel will be blended with Jet fuel to produce bio-jet fuel.

Approach:

One of the purposes of this research is to improve energy efficiency of the microalgae cultivation in photobioreactor by comparing LEDSs with fluorescent light sources, and the use of wastewater versus reverse osmosis (RO) water. Chlorella Vulgaris microalgae were grown in wastewater and in RO water. Two LED color panels were used for this comparison: a red panel and a red-blue Panel. The measured light intensities were maintained at a value of 2000 Lux. The absorptivity and the cell counts measurements were recorded while the algae were growing using the spectrophotometer and the microscope. Then, the algae were harvested after the maximum growth was reached. The biomass freeze dried algae were determined after centrifugation and freeze drying processes. Research is still progressing towards production of biodiesel through the in situ process and production of bio-jet fuel.

Discussion:

The highest algae growth (3.2 g freeze dried algae/L over a growth period of 18 days = 178 mg. freeze dried algae/L-day) was obtained using red-blue LEDs and RO water. The lowest algae growth (1.21 g of freeze dried algae/L over a growth period of 18 days = 67 mg freeze dried algae/L-day) was obtained using fluorescent light and waste water. Algae growth (3.2 and 1.63 gm. freeze dried algae/L over a growth period of 18 days = 178 and 91 mg. freeze dried algae/L-day) were obtained using RO water and waste water respectively with red-blue LEDs. Also, Algae growth (1.95 and 1.25 gm. freeze dried algae/L over a growth period of 18 days = 108 and 69 mg. freeze dried algae/L-day) were obtained using RO water and waste water respectively with red LEDs. The results indicate that red-blue LEDs are more effective than red LEDs using either RO water or waste water. Since the power requirements of the fluorescent light and the LEDs are 68 and 45 Watts respectively, the total saving energy by LEDs is about 34%. Light capture efficiencies of LEDs and fluorescent light, based on the heating value of the algae produced compared to the incident light energy over the growth period, are 0.72% and 0.48% respectively.

Figure 1 Production of Bio-jet fuel through three different routes.

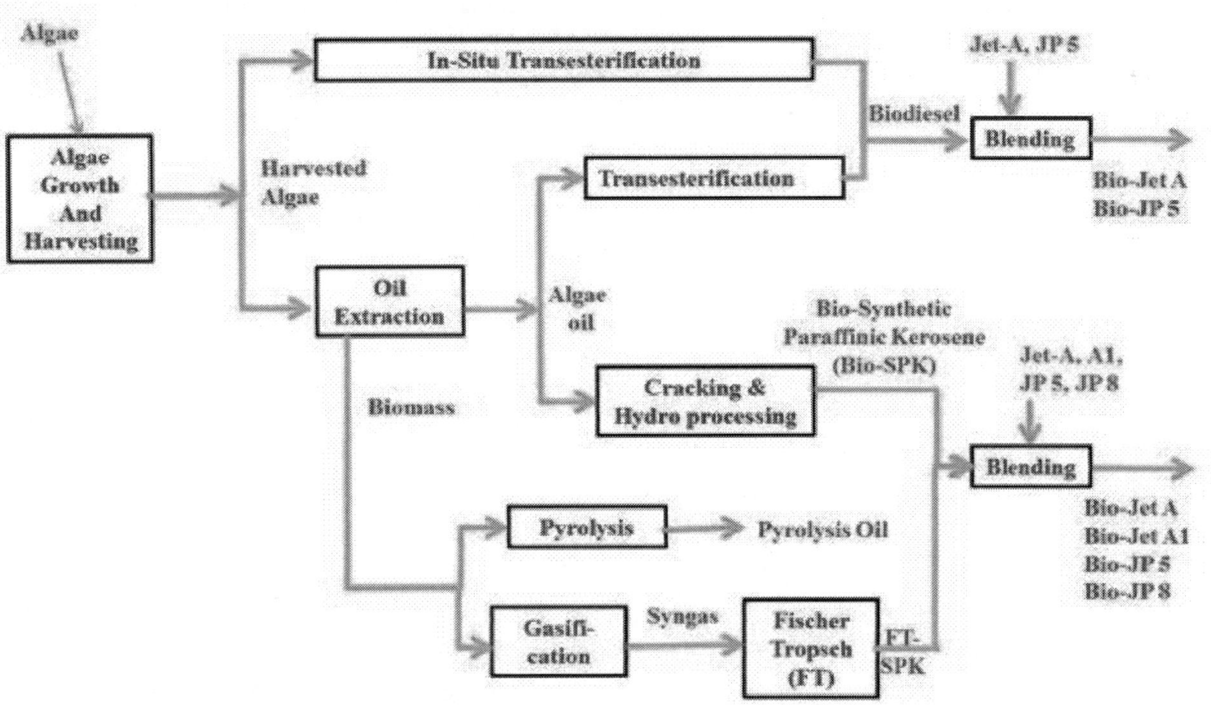

(716f) Impact of Floc Size On Ethanol Tolerance of the Self-Flocculating Yeast SPSC01 and Underlying Mechanism

Chuang Xue and Feng-wu Bai, School of Bioscience and Biotechnology, Dalian University of Technology, Dalian, China

High cost is the biggest challenge in fuel ethanol production, in which energy consumption comprises about 30%, the second largest only after feedstock consumption, and thus saving energy consumption is one of the most effective strategies to make fuel ethanol more economically competitive. Very high gravity (VHG) fermentation can significantly increase ethanol titer in the fermentation broth, which not only saves energy consumption for the downstream distillation but also for the waste distillage treatment due to its significant reduction. However, high ethanol concentration severely inhibits yeast cells, and stuck fermentation frequently occurs, with more sugars unfermented and ethanol yield compromised correspondingly. Therefore, exploration of the mechanism underlying the ethanol tolerance of yeast cells is one of the prerequisites for developing effective VHG ethanol fermentation process.

The impact of the size of the yeast flocs on the ethanol fermentation was studied by controlling their size at 100, 200 and 300 mm, which was characterized by their average chord monitored by the focused beam reflectance measurement. The experimental results illustrated that the yeast flocs with the size of 300 mm exhibited maximum ethanol production, ethanol tolerance and glucose consumption, with 110.0 g l^{-1} ethanol produced, 58.0% viability remained for the yeast cells after 20% (v/v) ethanol shock for 5 h and glucose uptake rate of 286.69 C-mmol l^{-1} h^{-1}. Moreover, the retention of the self-flocculating yeast was improved for the yeast flocs with relatively large size and higher biomass concentration was achieved within the bioreactor, contributing to more glucose consumed and carbon flux to ethanol production. The decrease in the fluxes to the growth of the yeast cells such as DNA, RNA, protein and lipid biosynthesis as well as the increase of ATP consumption for the maintenance indicated that improved ethanol production resulted from the increased yeast floc size raised more energy burden on the yeast cells, and glucose uptake of the yeast flcos was compromised under VHG conditions, making the improvement of glucose uptake rate the key to improve VHG ethanol fermentation.

(724a) Simultaneous Production of Liquid Fuels and Hydrogen From Hybrid Feedstock: Switchgrass and Shale Gas

Mariano Martín, Chemical Engineering, University of Salamanca, Salamanca, Spain and Ignacio E. Grossmann, Chemical Engineering Department, Carnegie Mellon University, Pittsburgh, PA

Over the last decades there has been an enormous effort to develop alternative fuels for transportation. Bioethanol and biodiesel are the most promising ones due to their compatibility with the automobiles and the crude supply chain. However, there are a number of challenges that biofuels must overcome mainly due to the large demand of liquid fuels and the limited production capacity of biofuels from biomass resulting in the need for large harvesting areas and for raw materials that do not compete with the food market. In the mean time hybrid first and second generation biofuels, using corn grain and stover for the production of bioethanol (Cucek et al 2011) or hybrid biomass – fossil fuel feedstock (Baliban et al 2012) can be used to reduce the dependency on fossil fuels and to serve as a bridge between technologies. Recently a new and plentiful source has been found, shale gas. Shale gas is natural gas that is trapped within shale formations. Of the natural gas consumed in the United States in 2009, 87% was produced domestically; thus, the supply of natural gas is not as dependent on foreign producers as is the supply of crude oil, and the delivery system is less subject to interruption and can be used as raw material for liquid fuels.

In this paper, we focus on the conceptual optimal design of multiproduct facilities for the production of green diesel from lignocellulosic biomass and shale gas feedstocks in an attempt to increase the yield to hybrid fuels while reducing the dependency on foreign supply in order to reduce its production cost to make it competitive with bioethanol from switchgrass (Martin & Grossmann 2011). We consider swichgrass as lignocellulosic feedstock since it is the raw material of choice by the DOE due to the harvesting possibilities in the US and its high yield to biofuels. We propose a limited superstructure optimization approach where we first construct a flowsheet embedding alternative routes for the production of liquid fuels. The process is based on FT technology in which the shale gas is reformed with steam to generate raw syngas. In parallel the switchgrass is gasified, reformed (we consider steam reforming or partial oxidation) to produce raw syngas. The total raw syngas is cleaned up from solids and sour gases. Next its composition may need to be adjusted for the optimal operation at the FT- reactor (using either WGSR, or PSA). The syngas is transformed into liquid fuels in an FT - reactor. The heavy liquids are upgraded using hydrocraking to increase the yield towards FT-diesel. The goal is to optimize the flowsheet, the product distribution (hydrogen, FT –diesel and gasoline production) and the operating conditions to maximize FT-diesel and hydrogen production determining the optimal consumption of raw materials (biomass and shale gas) as function of their price while minimizing the energy input for a number of scenarios where we fix the production level of liquid fuels and/or the maximum availability of biomass and shale gas devoted to fuels production to evaluate the profitability and competitiveness of this kind of hybrid facilities.

It turns out that the selected technology for reforming the raw syngas from the gasifier is steam reforming due to the higher yield to hydrogen. The process operates in such a way that the shale gas that is not required to meet the demand established for liquid fuels is used for the production of hydrogen that is sold to help in the economy of the plant as long as its price is low, otherwise it is better not to produce it. Thus, for this process to produce liquid fuels at around $1/gal, the biomass price must be kept below $100/t and the shale gas should not go beyond $11.5/MMBTU.

References:

Baliban, R.C.; Elia, J.A., Floudas, C. A. (2012) Simultaneous process synthesis, heat, power, and water integration of thermochemical hybrid biomass, coal, and natural gas facilities. Comp. Chem. Eng. 37, 10, 297–327

Cucek, L., Martín, M., Kravanja, Z., Grossmann, I.E., (2011) Integration of Process Technologies for the Simultaneous Production of fuel Ethanol and food from Corn grain and stover. Computers and chemical engineering 35,8, 1547-1557

Martín, M., Grossmann, I.E. (2011) Energy optimization of lignocellulosic bioethanol production via gasification. AIChE J. 57, 12, 3408, 3428

(724b) Streamlining Lifecycle Greenhouse Gas Emissions of Coal, Conventional and Unconventional Natural Gas for Electricity Generation

Ozge Kaplan, Office of Research and Development, U.S. Environmental Protection Agency, Research Triangle Park, NC

An analysis of the lifecycle greenhouse gas (GHG) emissions associated with natural gas use recently published by Howarth et al. (2011) stated that use of natural gas produced from shale formations via hydraulic fracturing would generate greater lifecycle GHG emissions than petroleum per unit of energy input. The analysis also suggested that the lifecycle GHG emissions from natural gas (at the point of input into a power plant) could exceed (or be "quite similar to") those from coal. We re-evaluated the estimated lifecycle GHG emissions from Howarth's study, reviewed other published literature on lifecycle analysis of natural gas and coal, and calculated normalized full lifecycle emission factors per MWh of base load electricity generated. The CO_2equivalents were presented on both 20- and 100-yr time horizons to evaluate both near- and long-term impacts. Howarth et al. (2011) raise valid points regarding the need for considering emissions over the full lifecycle, including fugitive CH_4 emissions from hydraulically-fractured natural gas wells and distribution systems, as well as the need to recognize the time basis (e.g., 20-yr, 100-yr, 500-yr) for comparing the climate impacts of different energy sources. However, because their analysis did not account for the efficient use of natural gas in advanced combustion technologies, their estimates appear to be much more representative of worst-case conditions than average industry emissions.

(724c) Biomass and Natural Gas to Liquid Transportation Fuels (BGTL): Process Synthesis, Global Optimization, and Topology Analysis

Richard C. Baliban, Josephine A. Elia and Christodoulos A. Floudas, Chemical and Biological Engineering, Princeton University, Princeton, NJ

Heavy dependence on petroleum and high greenhouse gas (GHG) emissions from the production, distribution, and consumption of hydrocarbon fuels pose serious challenges for the United States transportation sector [1]. Depletion of domestic petroleum sources combined with a volatile global oil market prompt the need to discover alternative fuel-producing technologies that utilize domestically abundant sources. A recent review highlights the single and hybrid feedstock design alternatives that can produce gasoline, diesel, and kerosene from coal, biomass, or natural gas [2]. Biomass and natural gas are two domestic energy sources that have received considerable attention as feedstock candidates for processes that can generate synthetic liquid fuels. The combination of these two feedstocks offers an important advantage over single feedstock systems because the benefits of both natural gas (e.g., cheaper, large domestic reserves) and biomass (e.g., sustainable, reduces GHG emissions) can be harnessed in a hybrid refinery.

We present an optimization based framework to perform a comprehensive technoeconomic and environmental assessment of an alternative energy refinery that will convert biomass and natural gas to liquid fuels (BGTL). The framework will incorporate process design, global optimization, and process synthesis strategies [3-10] to determine the optimal plant design of a BGTL refinery under different scenarios (e.g., feedstock composition, plant capacity, liquid fuel type/composition, etc.). Using this framework, thousands of distinct process designs will be analyzed at once to determine the design that produces the liquid fuels in the most optimal way (i.e., lowest cost or highest profit). This superstructure of postulated topologies will include multiple components for biomass conversion, natural gas conversion, hydrocarbon production, and liquid fuels upgrading and will identify the economical and environmental tradeoffs between each component. A simultaneous heat, power, and water integration will be included in the optimization model to ensure that the cost of utility generation and wastewater treatment is adequately compared to the cost of producing the liquid fuels. Nine case studies are presented to investigate the effect of biomass type and plant capacity on the optimal process topology, total plant cost, and break-even oil price. The key topological decisions from the mathematical model along with the major components of the plant capital and operating costs will be discussed.

[1] National Academy of Sciences, National Academy of Engineering, and National Research

Council. Liquid Transportation Fuels from Coal and Biomass: Technological Status, Costs, and Environmental Issues. Washington, D. C., EPA, 2009.

[2] C.A. Floudas, J.A. Elia, R.C Baliban. Hybrid and Single Feedstock Energy Processes for Liquid Transportation Fuels: A Critical Review. *Comp. Chem. Eng.*, 2012: 41(11), 24-51.

[3] R.C. Baliban, J.A. Elia, C.A. Floudas. Toward Novel Hybrid Biomass, Coal, and Natural Gas Processes for Satisfying Current Transportation Fuel Demands, 1: Process Alternatives, Gasification Modeling, Process Simulation, and Economic Analysis. *Ind. Eng. Chem. Res.*, 2010:49(16), 7343-7370.

[4] J.A. Elia, R.C. Baliban, C.A. Floudas. Toward Novel Hybrid Biomass, Coal, and Natural Gas Processes for Satisfying Current Transportation Fuel Demands, 2: Simultaneous Heat and Power Integration. *Ind. Eng. Chem. Res.*, 2010:49(16), 7371-7388.

[5] R.C. Baliban, J.A. Elia, C.A. Floudas. Optimization Framework for the Simultaneous Process Synthesis, Heat and Power Integration of a Thermochemical Hybrid Biomass, Coal, and Natural Gas Facility. *Comp. Chem. Eng.*, 2011:35(9), 1647-1690.

[6] J. A. Elia, R.C. Baliban, X. Xiao, C.A. Floudas. Optimal Energy Supply Network Determination and Life Cycle Analysis for Hybrid Coal, Biomass, and Natural Gas to Liquid (CBGTL) Plants Using Carbon-based Hydrogen Production. *Comp. Chem. Eng.*, 2011:35(8), 1399-1430.

[7] R.C. Baliban, J.A. Elia, C.A. Floudas. Simultaneous Process Synthesis, Heat, Power, and Water Integration of Thermochemical Hybrid Biomass, Coal, and Natural Gas Facilities. *Comp. Chem. Eng.*, 2012:37(10), 297.

[8] R.C. Baliban, J.A. Elia, R. Misener, C.A. Floudas. Global Optimization of a Thermochemical Based Hybrid Coal, Biomass, and Natural Gas to Liquids Facility. *Comp. Chem. Eng.*, 2012, http://dx.doi.org/10.1016/j.compchemeng.2012.03.008.

[9] R.C. Baliban, J.A. Elia, V.W. Weekman, C.A. Floudas. Process synthesis of hybrid coal, biomass, and natural gas to liquids via Fischer-Tropsch synthesis, ZSM-5 catalytic conversion, methanol synthesis, methanol-to-gasoline, and methanol-to-olefins/distillate technologies. *Comp. Chem. Eng.*, submitted.

[10] J.A. Elia, R.C. Baliban, C.A. Floudas. Nationwide Energy Supply Chain Analysis for Hybrid Feedstock Processes with Significant CO_2 Emissions Reduction. *AIChE J.*, submitted.

Investment Opportunities in the Liquefied Natural Gas (LNG) Projects and Development: The Emerging Cleaner Fossil Fuel of Higher Energy Efficiency

M. N. Idris

Process Refining Technology, Department of Chemical Engineering, University of Maiduguri, Nigeria

ABSTRACT

The LNG remains one of the cleanest, cheapest (except electricity) and higher energy efficiency products toward the actualisation of global energy and utilisations. In these studies, concise effort was used to review the LNG process description and most importantly, the liquefaction processes in a modern scope. The focus on this paper is to highlight the various LNG investment opportunities globally. Apart from the usefulness of LNG production for economic gains, it help in reducing the flaring of global natural hydrocarbon which are further transform into economy values and environmental friendliness. In term of financial gain, Nigeria alone generates an approximate sum of revenue above $200 million US in a month, while the global world through LNG production has an approximate benefit to the tune of $5 billion US monthly. The interest on this study is to highlights the investment opportunities and possible gains to the investors in the oil and gas projects, because of its viability and high economic advantages in various ways when LNG is compared to other fossil fuel.

Keywords: LNG, NLG, MCHE, PMR, Investment Opportunities

Correspondence Author: *Dr. (Engr.) M. N. Idris; MAIChE, AMIChemE, MNSE (muhinu@hotmail.co.uk or idrismn@hotmail.com), Department of Chemical Engineering, University of Maiduguri, Nigeria and/or IPSE-ERRI, SPEME, University of Leeds, United Kingdom, LS2 9JT, Tel: +234(0)705 8771 096

1.0 INTRODUCTION

Natural gases (NG) can be classified as associated or non-associated. The associated natural gases are those that are present with liquid fossil fuel, while non-associated gases are those that occur naturally without any impurities. The various importance uses of natural gas is in lighting cooker/stove and heating houses domestically and industrially use in the high powered engine etc. Interestingly, the exportation of natural gas is increasingly growing, as domestic production has lagged behind this growing demand for this clean fuel. Liquefied natural gas (LNG) processing is typically created using three-step processes Idris et al (2007). Firstly, the subterranean gaseous-form natural gas is 'frozen' into a liquid state through a complex cryogenic process system (called '*liquefaction*'), this involve temperatures as low as -161 ^{0}C. The essence of the liquefaction process of LNG is to make transportation of natural gas easier, cheaper and convenient (Idris and Abowei 2004). Secondly, after the gas is liquefied, it can be stored in cryogenic holding tanks or pumped directly from the cooling vestibule into special insulated transportation vessels, such as railcars, trucks, or ships. Finally, upon delivery of the vessel to its final destination, LNG is pumped from the vessel into either another cryogenic storage tank for later delivery or directly into a re-gasification unit that uses sea water or air to reheat the LNG in order to convert it back into gaseous natural gas (Idris 2003).

The LNG processing sounds fluid, but usually is not quite as simple. Raw natural gas can be 'heavy' (also called 'wet' or 'hot' gas) with various higher carbon molecules. These heavier particles create a higher heat content (ranging from 11.28 – 13.66 kWh/m^3) than permissible by many pipeline distribution standards (PDSs). Although this heavier form of gas can be frozen, stored and shipped, the extra heat content may have to be 'stripped off' prior to entering another distribution system to conform to the gas heat standards of the pipeline network Idris (2003).

In the United States, state public utility commissions (PUCs) oversee the gas pipeline heat content standards. It is estimated that the average heat content of current distributable gas in the United States is approximately 10.66 – 10.97 kWh/m^3, HEA (2011). Any gas disseminated into a system that does not conform to these independent standards will not be permitted into the inherent system without the heavier particles being stripped (through a process called '*fractionation*') or diluted with inert gas. This fractionation process is an additional cost to the overall LNG project. Fractionation can occur either upstream or downstream, but is usually contingent upon which location is most cost effective, politically sound, environmental-friendly or where there may be access to a viable purchasing market for the fractionated heavier gas, although guaranteed the economies and long-plant life. Fractionated gas products can either be marketed or injected back into the ground. Any marketing and sales of this heavier gas can serve as a revenue enhancement of or a contra-expense item to the total LNG projects.

In a similar vein, such adopted standards on gas pipeline heat content measurements can be initiated in the Nigeria LNG project through the appropriate Federal Ministries in order to establish resource data for future research and development on the LNG project and all other energy efficiency projects. The primary aim of this study is to discuss the global investment opportunities in LNG projects and further buttress the cleaner and the environmental friendliness of natural gas over the conventional hydrocarbon products. Once this LNG turns into gaseous gas, it is pumped through a pipeline system that leads to an ultimate market. Natural gases have various important uses which are described as follows:

a) **Domestically:** They are use in heating building, heating water, cooking, drying clothes, lightning etc. Household appliances that use natural gas include: heater (furnace), pool and spa heaters, clothes dryers, outdoor lights, barbecues, water heaters, stoves etc.
b) **Industrially:** They are use in high powered engine, HPE; transport & locomotive engines etc.

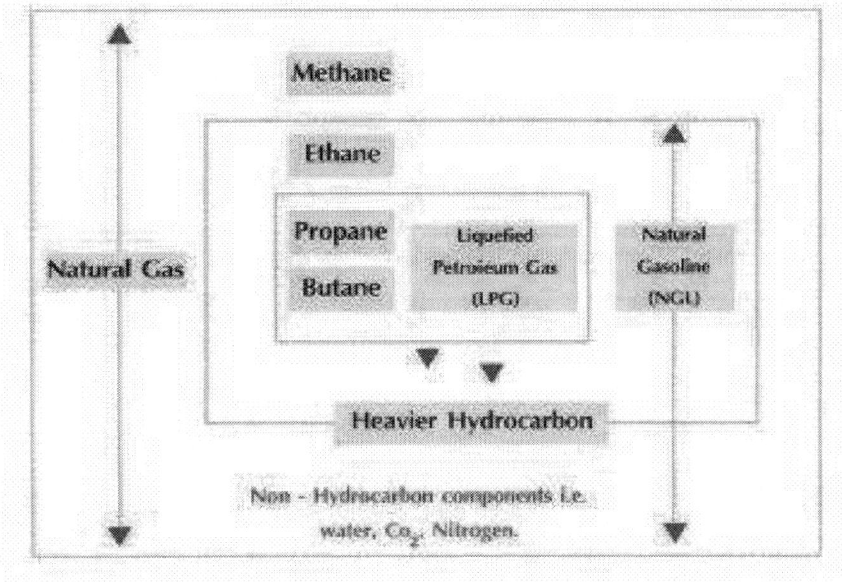

Figure 1: Composition of Natural Gas (Natural Gas Information, 2010)

Table 1 represent a detailed comparison of NG with other conventional fossil fuel. From this table it is clearly observed that NG is easy on engine with non-toxic explosive limit. NG gives a longer service life with lower maintenance costs involved. In addition, NG is the least expensive alternative fuel (except electricity) and it has a higher octane rating (Idris et al., 2006).

Table 1: Comparison of natural gas with gasoline, diesel oil and liquefied petroleum gas, LPG

Properties	Physical state	Boiling Range (°C @ 1 atm)	Density kg/m^3	Net Content Energy $*10^6 kJ/m^3$	Auto Ignition Temperature (°C)	Flash Point (°C)	Octane Number Range N_R=(R+M)2	Flammability Limits	Volume % in Air	Explosive limit for Fuel (PPM)
Gasoline	Liquid	26.67 - 215.56	688.79 - 784.90	5.212 - 5.323	232.2 - 482.2	-42.7	87 - 93	L = 1.4	H = 7.6	500
Diesel No. 2	Liquid	160 - 382.2	784.90 - 881.01	5.268	204.4 - 260	51.67	n/a	L = 0.7	H = 5.0	n/a
LPG (HD-5)	Gas	-42 - 0.56	496.57[b]	5.519	494.3 - 548.89	73.3 - -65.5	104[c]	L = 2.4	H = 9.6	n/a
NG	Gas	-161.67	128.15	5.937	732.2	-148.89	120[c]	L = 5.3	H = 14	nontoxic

Source: NRIS (2005)

Table 2 expresses the thermo-physical properties of methane, ethane and ethylene. This comparison is to further represent qualitatively the high grade and environmental acceptability of methane with other conventional fuel products.

Table 2: Thermophysical and chemical properties of methane, ethane and ethylene

Property	Methane	Ethane	Ethylene
Chemical Symbol	CH_6	C_2H_6	C_2H_4
Molecular weight	16	30	28
Normal boiling point (K)	111.7	184.6	169.3
Freezing temperature (K)	89.9	90.6	104.2
Critical temperature (K)	191	305	283
Critical pressure (bar absolute)	47	49.7	52.5
Expansion ratio-increase in volume as liquid as 1 bar abs boil to gas as 1 bar, 15°C	626	437	489
Density of standard liquid at 1 bar abs (kg/m^3)	424	546	565
Latent heat of vaporisation (cooling potential of phase change) (kJ/kg)	512.4	488.3	483.4
Relative gas density (referred to dry air at 1 bar abs,) 15°C at density 1.21kg/m^3	0.56	1.05	0.97
Air liquefaction hazard	No	No	No
Spontaneous ignition temperature in air (K)	813	783	753
Flammability limits in air (K)	5 - 15	3 - 12.4	2.7 - 36
Minimum ignition energy (MJ)	0.28	0.24	0.085
Flame temperature (K)	2153	2168	2248
Limiting oxygen index (vol %)	11.5	11.0	11.5

Source: Tiratsoo (1979) and BCC (2004)

2.0 THE LNG PROCESS AND OPERATIONS

2.1 The Main Cryogenic Heat Exchanger (MCHE) and LNG Production

The MCHE is the liquefaction unit in an LNG process that cryogenise the treated natural gas from -32 ^{0}C to the desirable temperature -161 ^{0}C by a physical process using the principles of appropriate simulation model. The MCHE is the most important and delicate unit of LNG process plant, because of it significant values in the LNG processing. The treated and dried natural gas (NG) from the mercury removal unit is further processed and liquefied in the liquefaction unit. The processing includes removal of heavy ends in the natural gas section, to help preserve the life of the process equipment and the quality of the LNG. The essence of the MCHE unit helps in liquefaction of the LNG by one-six hundredth of the ideal volume of treated natural gas for economic and easy transportation overseas (Idris et al., 2006).

However, the LNG production from every process train is strongly influenced by the ambient air and the cooling water temperature. At high site temperatures, the power required to produce a quantity of LNG is greater than that required to produce the same quantity at lower site temperatures. Also, at a high site temperature, the power available from the gas turbines to drive the refrigerant compressor is less than that available at lower site temperatures. These factors result in variations in the maximum LNG production rate from day to night and from season to season Idris and Abowei (2004). Figure 1 represents the flow sheet of the Main Cryogenic Heat Exchanger MCHE) or liquefaction unit of LNG Train, while Figure 2 depicts the existing train of the Nigeria LNG Production unit.

2.2 The LNG Energy Efficiency and Opportunity

The LNG energy efficiency can be easily being compared with the conventional gasoline and diesel oil as represented in Table 3. The natural gas which is being liquefied in the LNG plant is an emergent fuel of choice, and considered environmentally friendly, higher economic value and reduced hazard effects.

Table 3: Fossil fuel consumption per 1.05587×10^{-9} kg of gas emission

Gases	Natural Gas	Gasoline Oil	Diesel Oil
Carbon Dioxide	52.153	74.828	95.235
Carbon Monoxide	7.710	14.966	9.070
Hydrocarbons	0.635	2.2675	10.431
Nitrogen Oxide	45.350	149.655	378.219
Sulphur Dioxide	0.272	453.500	770.950
Particulates	2.268	37.641	1405.850

Sources: Modified approach to GEH (1965) and HEA (2011)

From Table 3, it is clearly observed that the natural gas production has various advantages over the conventional gasoline and diesel oil. Therefore, for economics, environmental and health advantages, increase in global production of natural gas could give the energy consumption markets greater opportunities.

3.0 OPPORTUNITIES AND CAPITAL INVESTMENT IN LNG PRODUCTION

3.1 The LNG and the Global Investment

In the year 2012 and beyond, it has been reported that the new LNG-related capital investment would be above \$62 billion US as LNG becomes a significant component of the energy industry's marketing strategy HEA (2011). Historically, the price of LNG has been prohibitive when compared to United States (US) gas prices reference. The costs of delivery ranged are \$0.726 to \$0.871/mkWh range (not including the netback price to the owner of the stranded gas reserves from which the gas was initially purged). Assuming a \$0.363mkWh netback to such owner, a total deliverable gas price of around \$0.726 to \$0.871/mkWh could possibly be attained on a cost basis HEA (2011).

In the 1970's and 1980's, physical gas prices had spiked in the US and Europe, feeding fears of gas shortage. This phenomenon inspired the construction of four major LNG import re-gasification terminals, all on the United States East Coast (Everett, Massachusetts MA, Elba Island, Georgia GA, Cove Point, Maryland MD, and Lake Charles, Los Angeles LA, in total constituting approximately 1.05×10^2 billion m^3/day of installed capacity).

However, deregulation led the producers of natural gas to become motivated to explore and develop newer, easy access gas reserves in the mountain states and the Gulf Coast, flooding the market with excess gas inventory. This was exacerbated by Canadian policy changes that allowed increased gas exports, in effect emigrating additional low-cost gas from the Western Canadian Sedimentary Basin (WCSB) to compete for demand in the United States. The result was a gas supply 'bubble' which kept prices very low for quite some time. Imagine how times have changed. Based on the Department of Energy's Energy Information Administration (EIA), it was anticipated in 2011 medium-term Henry Hub gas prices to range from \$0.769/mkWh to \$0.726/mkWh, and long-term gas prices to be close to \$0.871/mkWh.

In analytical forecasting of natural gas prices, one can label the expected gas price a 'dependent' variable and the underlying fundamentals as 'independent' variables that influence the dependent variable. Using quantitative techniques, Hybrid Energy Advisors, Inc. (HEAI) predicted in the spring of 2002 that Henry Hub financial prices, the dependent variable, would range between \$3.35 and \$4.10 through the winter months of 2002/2003. They also calculated ten-year price forecasts of \$0.5445 - \$0.6776/mkWh during the summer months and \$0.8954 - \$1.0285/mkWh during the winter months. HEAI believes that ten-year city gate prices of gas on the West Coast and Midwest United States will trade approximately equivalent ('flat') to Henry Hub prices, and upper East Coast prices to trade at approximately 13.06 – 18.87 basis points (cents/mkWh) above Henry Hub prices HEA (2011). These forecasts include a greater gas import as a percentage of total domestic gas. HEAI forecasted that by the year 2011, approximately 6 - 10% of the total US gas supply will come from LNG produced from foreign stranded gas reserves. In a similar vein, a forecast in the European and Asian gas consumption of similar trend has been predicted.

Today the total cost of LNG production has been quite streamlined and reduced to approximately $0.484/mkWh due to competition and improvements in technology. Much of the technology pertaining to lower-cost land-based terminals and offshore and ship-board re-gasification units are still relatively unproven. However, major energy industry players and engineering firms do not foresee problems with the design and implementation of these methods HEA (2011).

Therefore, assuming a netback of $3.44/mkWh, the total delivered price of approximately $10.45/mkWh can now be theoretically achieved. This LNG price is almost $3.44/mkWh less than a decade ago. When comparing this new price to industry average price forecasts for Henry Hub and East and West coast city gates, LNG can certainly be considered an economical source of natural gas supply.

3.2 The Nigeria LNG Gas Reserve and Reduce Flaring

Nigeria is highly blessed with massive reserves of associated and non-associated gas, estimated in excess of 160 trillion cubic feet (tcf). It is ranked amongst the 10th largest in terms of proven natural gas reserves in the world, and its natural gas reserves / production is estimated at 109 years. The expert geologists has forecast that there is a lot more gas still to be found, if companies deliberately explore for gas, as opposed to finding it by chance whilst in search of oil NLNG (2012). Unfortunately, Nigeria is one of the biggest gas flares in the world, in spite of the associated environmental hazards of gas flaring. The government is, however, working to achieve a target of zero flares in the near future. It is expected that all the NLNG trains, including the proposed Train 7, operating at full capacity will play a significant part to help put out these flares NLNG (2012).

3.3 The Nigeria LNG and the Progress So Far

The Nigeria LNG Ltd remains the largest consumer of the associated gas in the Nigeria gas project through the stakeholders companies (NNPC, SPDC Shell Gas B.V., Total LNG, and ENI International). A brief of the stakeholders are described as follows:

The Nigerian National Petroleum Corporation (NNPC) was established in 1977 under the constitution of the Federal Republic of Nigeria. It is the corporate entity through which the Nigerian government participates in the oil and gas industry. NNPC and its subsidiaries dominate all sectors of the industry - exploration, production, refining, pipelines, marketing, crude /product exports, and petrochemicals. NNPC owns 49% of the shares in Nigeria LNG Limited.

The Shell Gas B.V (SGBV) is a company incorporated under the laws of the Netherlands. Besides its interest in Nigeria LNG, Shell Gas and Power has interests in many of the world's major LNG production companies including those in Brunei, Malaysia, Australia and Oman and is one of the Royal Dutch Shell Group of Companies which operate throughout the world in all aspects of the petroleum industry. It sells over 80 billion cubic metres of gas yearly. SGBV owns 25.6% of the shares in Nigeria LNG Limited.

Total is a major integrated oil and gas company active in all aspects of the petroleum industry. Total explores in 25 countries, and has reported proven reserves of 3,639 million barrels of oil equivalent (492 million tons). Of these, crude oil accounts for 71% and natural gas for 29%. Total owns 15% of the shares in Nigeria LNG Limited.

Eni is a fully integrated oil and gas company engaged in all aspects of the petroleum business. Eni is involved in exploration, development and production of oil and natural gas. Eni owns 10.4% of the shares in Nigeria LNG Limited.

Currently, the Nigeria LNG Ltd is made up of 6-functional cryogenic plant in Bonny Island, Rivers State. Other proposed LNG projects include Olokola LNG and Brass LNG among others. The Nigerian Gas Company currently caters for Independent Power Producers, and local industries in Lagos and other parts of the country. The proposed West African Gas Pipeline Project (WAGPP) will make gas available to other West African countries for domestic and industrial uses.

4.0 MAJOR SUPPLY AND DEMAND ZONES OF LNG

New technologies in LNG and the upward gas price trends are creating the opportunities for re-gasification terminals to spring up in a variety of areas. These areas include the United States East and West Coasts, Northwest Mexico, Italy, Germany, England, Spain and even emerging markets such as China and India. Asian counterparties such as Japan, Korea and Taiwan, who have historically constituted approximately 70% of demand in the LNG market, are also building new terminals. The gas supply to meet expected growth in LNG demand is from stranded gas reserves associated with crude oil production that historically did not have much hope of coming to market. These reserves are typically owned and operated by many major worldwide energy companies. Longstanding production zones include Algeria, Libya, Nigeria, Qatar, Malaysia, Australia, Brunei, Indonesia and the United States (Alaska, Cook Inlet). Many proposed new areas of LNG liquefactions include Norway, Trinidad & Tobago, Venezuela, Bolivia and various countries of the Middle East.

The ideal economic scenario when evaluating the LNG terminal projects make sense with contingent upon studies related to where and how much gas demand is going to be, where the supply will come from, and how much gas to produce.

5.0 THE FUTURE OF GLOBAL LNG PRODUCTION AND CONSUMPTION

5.1 The LNG Production and Consumption

While natural gas consumption growth rates are expected to increase over the next twenty years by 50% as compared to a current level of 1.84 billion m^3/day, the solution to providing the supply to satiate such growth is not quite as transparent as just evaluating the face-value economics of alternative sources.

In any energy infrastructure project, due diligence on regulatory, permitting, and environmental factors is quite important. A hiccup in procuring the necessary political approvals, permits and rights-of-way can be devastating to the economics and viability of a project. Projects have been known to come to a dead halt, much to the chagrin and expense of the lead project developer and production consortium, over matters such as residential noise or particulate pollution, wildlife endangerment, political vacillation, or just plain general nuisance.

LNG re-gasification terminals are viewed as large, obstructive and generally displeasing to the eye for local residents and businesses. Although designs to bury such large facilities underground have been forthcoming, general apprehension related to the "don't build in my backyard" philosophy still emphatically exists. One mitigant to that problem is sighting the terminal off-shore or aboard the LNG transport ship itself, far from residential or commercial areas. Although such new technologies regarding offshore and ship-board re-gasification is being studied and developed by a variety of industry players, the educating process required with political parties is an on-going and sometimes frustrating process.

The guidelines regarding the oversight, regulatory standards and environmental compliance of on-shore or off-shore terminals are still in debate. In the United States, three of the four import terminals are overseen by the federal energy regulatory commission, FERC. Regulatory oversight can be an issue to contend with as most merchant energy firms and pipeline companies vie for ownership, control, and equity interest. For such market participants, FERC jurisdiction impedes profitability and ease of management control of the owners and its merchant gas off-take counterparties. Many of the industry participants building and sighting these terminals contend to the FERC that if they have the burden of finding, constructing, funding, and permitting a terminal, they should not be subject to losing the economic potential to sell such re-gasified gas into open market by subjecting the capacity to a FERC-controlled 'open-season'. After all, if one firm does all of the work, why would that firm surrender the market opportunity to profit from the off-take gas sales to another firm who has not incurred the same 'finders' costs? For many of these project leaders, equity ownership of a FERC-jurisdiction terminal at a regulated rate of return is not enough. They want the most facile way to attain the real 'juice', that is, they prefer the sole ability to sell their upstream LNG into the interstate domestic markets as gaseous gas and take merchant positions in key locations.

5.2 The LNG non-Operational Risks

The two LNG-related non-operational risks that are worth mentioning at this time are terrorism and/or accident risk and sovereign government risk. Terrorism has been shown to be a more significant influence on energy endeavours than ever before. With the threat of oil tanker, nuclear reactor and LNG ship sabotage, political risk to sighting an on- or off-shore re-gasification unit is high. Man-made or accidental explosion of such a tanker can be catastrophic in the mind of the common person, and although studies have shown that LNG volatility is much lower in its liquid cooled form, the image of a giant vapour cloud can have devastating effects on the human psyche and at the very least create an apocalyptic image that few are willing to internalize. However, the Nigeria LNG Ltd has been having smooth operations of the production trains since the official commissioning in October 2000. This success is due to a trusted and fulfil memorandum of understanding, MOU with the hosting communities.

Secondly, since much of the gas that is being liquefied in the western countries (US and Europe) comes from foreign and third-world countries, sovereign risk can be quite disenchanting as compared to reliable, high-credit, highly-liquid United States gas. Changing governments, political coups, local customs, and draconian regimes can portend inconsistent supply. Due to this factor, price concessions are usually made in relation to United States-priced gas supply, however, the sovereign risk inherent in such supply must to be evaluated concomitantly with the discounted pricing.

6.0 CONCLUSIONS

The global world is in need of additional natural gas sources based on forward fundamental supply/demand and price expectations. More so LNG appears to be one of the viable and constantly improving choices of energy source. While the economic viability of LNG as a necessary gas supply-stack contributor as compared other hydrocarbon in many ways. However, the issues on politics, sovereignty and regulatory risk can impede the progress of such beneficial projects and turn what was once a supply saviour into a bureaucratic roundabout of the hosting communities. To the global world, the LNG and the reduced gas flaring has an approximate benefit to the tune of $5 billion US in a month. Nigeria alone, generate a sum of $200 - $300 million US monthly. This study is to serve as an eye-opener to investors in the oil and gas projects because of its viability and high economic advantages in various ways over the conventional hydrocarbon products (gasoline and diesel oil etc.).

Acknowledgement

The authors wish to acknowledge the support of the Hybrid Energy Advisors HEA, Inc. based in Houston, Texas USA for providing some of the information used in this report. The support of the petroleum technology development fund (PTDF) toward the research work carried at the University of Leeds is highly appreciated.

Abbreviations

EIA:	-	Energy Information Administration
FERC:	-	Federal Energy Regulatory Commission
HEAI	-	Hybrid Energy Advisors Inc.
LNG:	-	Liquefied Natural Gas
NLG:	-	Natural Gas Liquid
MCHE:	-	Main Cryogenic Heat Exchanger
PMR:	-	Propane Mixed-Refrigerant
PUCs:	-	Public Utility Commissions

REFERENCES

1. BCC British Cryogenic Council (2004), 'Cryogenic Safety Manual: A guide to good practise', 3rd Ed., pp52-76
2. Gas Engineers Handbook (1965), American Gas Association, New York, USA
3. M. N. Idris, Z. Abdu and A. S. Ahmed. (2007), 'Compressed Natural Gas (CNG) development in Nigeria's growing economy: A review', International Petroleum Technology Training Journal, Jan 2007, vol. 4, No.1
4. M. N. Idris and M. F. N. Abowei (2004), 'Performance optimization of a liquefied natural gas (LNG) process train', Journal series of Engineering, University of Maiduguri, Oct 2004, vol. 1, No. 3
5. M. N. Idris, A. S. Ahmed and S. S. Adefila (2006), 'Advances in Compressed Natural Gas (CNG) development in Nigeria's growing economy', Journal of Engineering series of university of Maiduguri, Oct 2006, vol. 4 No.1
6. Natural Gas Information
7. NRIS (2005), Natural Resources Information Services 2005, P.C. McKenzie Company, Pittsburgh, USA PA 15241
8. NLNG (2012), the copyright reserve of the Nigeria LNG Limited.
9. Idris M. N. (2003), 'Modelling and Simulation of an Main Cryogenic Heat Exchanger (MCHE) of an existing Liquefied Natural Gas (LNG) Process Unit', An M.Tech Thesis submitted to the Rivers State University of Science and Technology (RSUST) Port Harcourt, Rivers State, Nigeria.

10. HEA (2011) Hybrid Energy Advisors, Inc. based in Houston, Texas USA (2011) www.hybrid-advisors.com and directly by e-mail at information@hybrid-advisors.com

11. Tiratsoo E. N. Natural Gas (1979), Beaconsfield Science Press Int. London 3rd Ed pp15-89

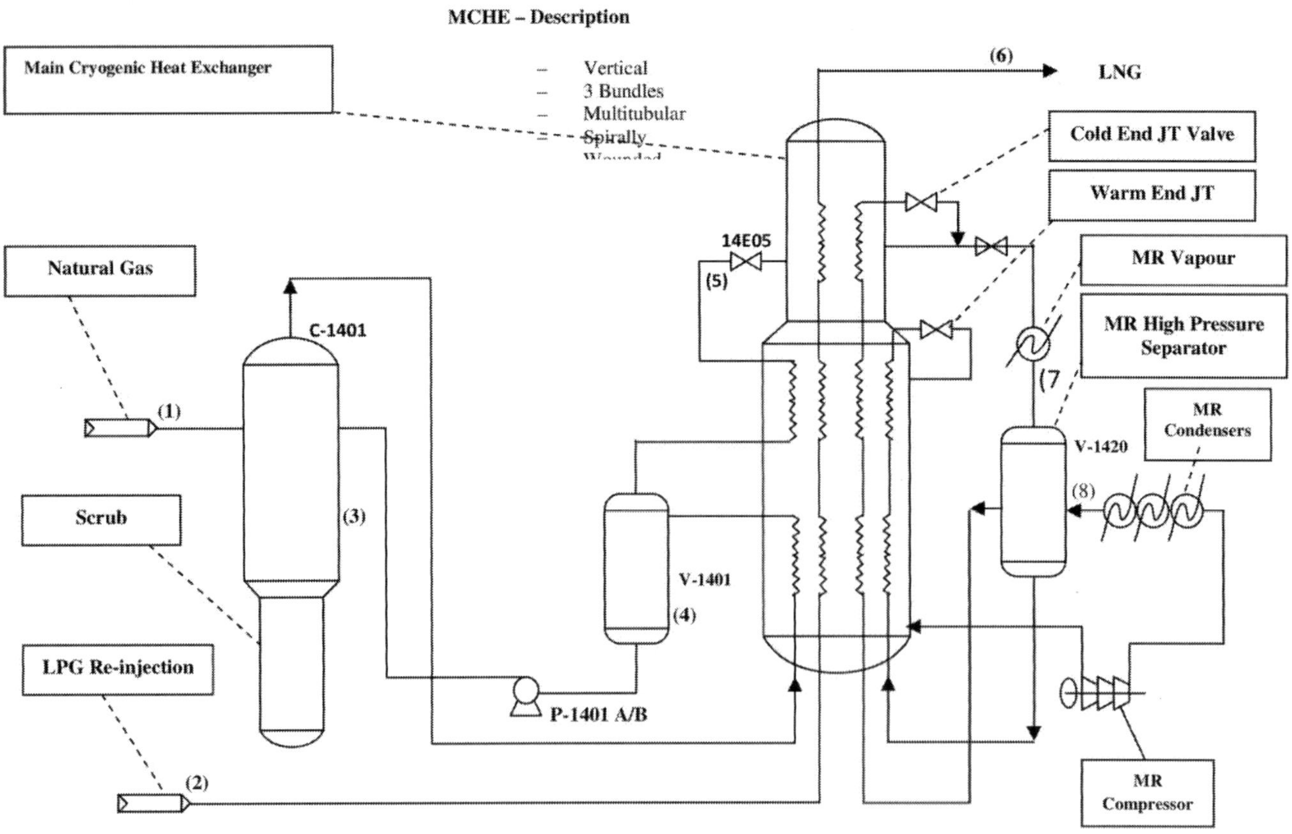

Figure 2: Flow sheet of the Main Cryogenic Heat Exchanger MCHE) or liquefaction unit of LNG Train

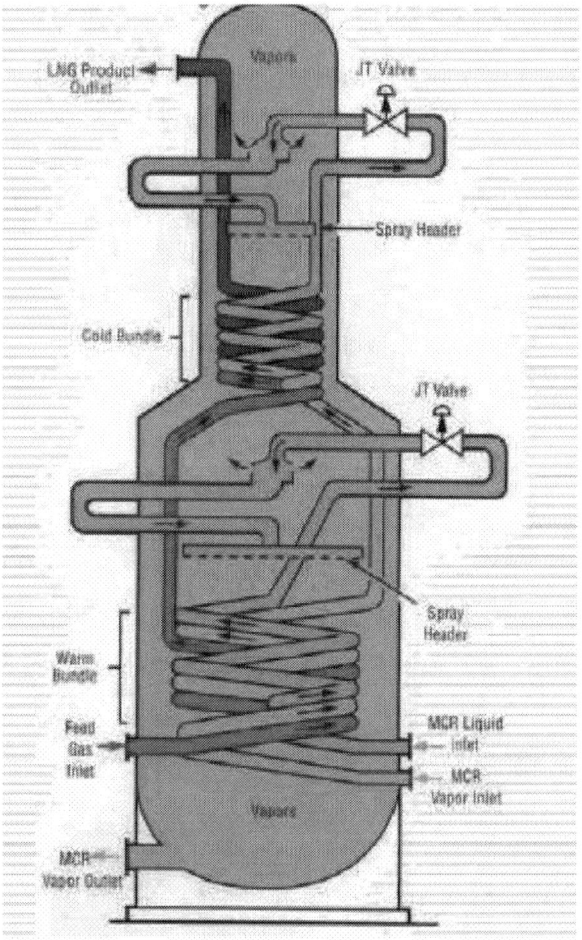

Figure 3: Design flow of MCHE (LNG and World of Energy, 2000)

Figure 4: LNG Production train, the Nigeria LNG Ltd. Source: http://www.nlng.com/

(724e) Co-Fire Pellets Consisting of Coal and Pyrolysis-Derived Binder

Andrew Friend[1], Robert Brown[1] and R. Christoper Williams[2], (1)Center for Sustainable Environmental Technologies, Iowa State University, Ames, IA, (2)Department of Civil, Construction, and Environmental Engineering , Iowa State University, Ames, IA

The goal of this research is to allow utilities with coal-fired plants to meet renewable energy mandates and reduce greenhouse gas emissions through the utilization of a co-fire pellet consisting of coal and a fast pyrolysis-derived binder. In the past decade, carbon dioxide and other greenhouse gases (GHGs) have been subjected to intense study for their contribution to climate change. Electrical utilities are facing federal, state, and local mandates for renewable power production. At the same time, utilities are also facing existing and upcoming regulations that demand emissions reduction from their current sources of power generation. In order to meet these mandates and regulations in a timely and inexpensive manner, electrical utilities are adding more "green" options to their power portfolios. Biomass is a near term option due to its low cost and easy integration into the existing coal-fired infrastructure.

Biomass co-firing is heralded as a simple and inexpensive way to reduce CO_2 and other greenhouse gas emissions from power plants. However, there are many practical difficulties in storing and co-firing two fuels with large differences in heating value, bulk density, moisture content, and ash composition as often occurs for coal and biomass. This can lead to reductions in boiler capacity and combustion efficiency, increases in emissions of regulated pollutants, and loss of markets for coal ash.

Our approach to solving several of the problems of conventional co-firing is to create a pellet from coal and a co-product of fast pyrolysis. These pellets have many advantages over both traditional coal combustion and co-firing with biomass. Co-fire pellets have approximately the same or greater heating value than the coal used in the pellets. They can be handled with existing coal infrastructure, injected into stoker-grate boilers or pulverized for use in PC boilers with little loss in boiler capacity or ash properties. They also contain substantially lower alkali levels than biomass, allowing for a wide utilization of annual or perennial biomasses without fear of boiler corrosion or fouling. Greenhouse gas emissions are much lower for the co-fire pellets than for coal because the biomass contributes relatively little net carbon dioxide emissions. Co-fire pellets allow electrical utilities to harness all the advantages of biomass co-firing without experiencing the downsides. They have the potential for utilities to meet both their renewable energy mandates and their emission reduction regulations without expensive facility retrofits.

(724f) Hybrid Energy Systems for Photosynthetic Capture and Re-Utilization of Fossil-Fuel Derived CO2

Erik Shuster, National Energy Technology Laboratory, U.S. Department of Energy, Pittsburgh, PA and John J. Marano, JM Energy Consulting, Inc., Gibsonia, PA

A broad examination of fossil energy production and use, applying established principles of Industrial Ecology would indicate that society is currently inefficiently using fossil carbon resources. Carbon efficiency could be dramatically improved, if it were possible to re-use the carbon in CO_2 versus the current practice of disposing of it directly to the atmosphere. Large-scale re-use could significantly reduce anthropogenic CO_2 emissions, as one aspect of any comprehensive strategy to reduce impacts of climate change over the next fifty to one hundred years. In a separate paper at this conference, arguments were presented for utilizing CO_2 versus disposing of it via sequestration. There are three major requirements for CO_2 re-use: 1) a source of non-fossil energy, identified as direct solar or nuclear energy, or electricity generated from these primary energy resources; 2) a source of fossil fuel-derived CO_2, identified primarily, but not exclusively as large, stationary, coal or natural gas-fired power plants; and 3) a useful product from the conversion of CO_2, identified as a substitute for petroleum, a "synthetic" oil that may be refined and further processed into a wide variety of finished products.

This paper examines the hybridization of primary energy systems to capture CO_2 from existing power plants and convert it into supplemental feedstocks for existing petroleum refineries and petrochemical facilities. It is argued that policy choices along with successful, targeted R&D could make this approach cost effective versus other options. Both biological (algae-based) and non-biological (syngas-based) conversion schemes employing current and near term technologies are described, analyzed, and compared.

The potential of Synthetic Biology in the longer term to go beyond algae to produce what we refer to as synthetic biofuels produced from sunlight and waste CO_2 is then explored. Desirable attributes of this merging of the biological and the chemical into a truly hybrid technology are described. Systems envisioned would employ process intensification to convert sunlight and CO_2 directly to liquid products and to separate these products, within a single solar-driven reaction chamber. Recent advances within the new field of Synthetic Biology will be reviewed that provide credence to these possibilities. Finally, results of a sensitivity analysis are presented, in order to quantify critical process performance targets, such as process yield, energy requirements, and production costs, which will need to be met for these fuels to become a commercial reality.

(724g) Polygeneration of Fischer-Tropsch Fuels and Electricity by Hybridised Solar Gasification of Coal – a Pseudo-Dynamic Process Model

Ashok A Kaniyal[1], Philip van Eyk[2], Graham J Nathan[2], Peter J Ashman[2] and Jonathan J Pincus[2], (1)School of Mechanical Engineering, University of Adelaide, Adelaide, Australia, (2)University of Adelaide, Adelaide, Australia

Polygeneration of Fischer-Tropsch fuels and electricity by hybridised solar gasification of coal – a pseudo-dynamic process model.

Ashok A Kaniyal[1,2*], Philip J van Eyk[1,3], Graham J NATHAN[1,2], Peter J Ashman[1,3], Jonathan J Pincus[1,4]

*Email: ashok.kaniyal@adelaide.edu.au; Phone: +61403691321.

[1] Centre for Energy Technology, The University of Adelaide, South Australia, AUSTRALIA, 5005

[2] School of Mechanical Engineering, The University of Adelaide, South Australia, AUSTRALIA 5005

[3] School of Chemical Engineering, The University of Adelaide, South Australia, AUSTRALIA, 5005

[4] School of Economics, The University of Adelaide, South Australia, Australia 5005

<>Abstract

Presented is a comparative energetic and environmental performance analysis of a solar hybridised gasification, coal to liquids polygeneration system and a non-solar reference. Using AspenPlus and HYSYS software, the reference system was configured, assuming the integration of a pressurised, Shell entrained flow gasifier with a Fischer Tropsch liquids (FTL) polygeneration facility. The hybrid plant assumes the feasibility of integrating an atmospheric, continuously operational, directly irradiated, oxygen blown hybrid solar reactor with this polygeneration facility. To mitigate the diurnal and stochastic impacts of the solar-boosted production of syngas, the hybrid polygeneration model was also assumed to be configured with a pressurised syngas storage plant. Here, the dynamic operation of the polygeneration system was modeled using a pseudo-steady state approximation for two, six day time-series of validated solar insolation model data. The two time-series were selected to represent a period in 'summer' characterised by a steady, low cloud period of solar insolation, and a period in 'winter' marked by intermittent solar insolation and high cloud. Using these data, a MATLAB model was used to predict the maximum steady rate of liquids that could be produced for the given solar insolation scenario.

For the summer time series, the hybrid solar gasification system was shown to improve the steady rate of liquids production by 32% and decrease the source-to-wheel (STW) GHG emissions per GJ FTL by 26%, relative to the reference. For the winter time-series, only a marginal improvement in liquids production was predicted .This result followed from the assumed difference in gasification temperatures between the hybrid system and the reference. On a total energetic output basis, GHG emissions were shown to decrease relative to the reference by 17-21 percentage points for the summer-time series, but increase by 18 percentage points for the winter time-series. This increase in GHG emissions follows the large parasitic impact of SG compression on the hybrid system's net electrical output. Additional sensitivity analyses identified the potentially significant energetic and environmental value of either operating the hybrid gasifier at just 2 bar-g instead of 1 bar-g or incorporating an O_2 storage system to reduce the parasitic load of the hybrid system's ASU.

Importantly, the present analysis found the difference in the steady output from the FT reactor and electricity generating plant between the summer and winter time-series to be within the feasible operational limits of the respective plant components. This provides assurance as to the feasibility of developing a solar hybridised C2L polygeneration facility using commercially available plant components.

Keywords

Coal gasification, concentrated solar energy, hybrid solar gasification, Fischer-Tropsch liquids production, pseudo-dynamic process modeling, lifecycle GHG assessment.

(730a) Investigate Soy Protein As Plastic Component in Polymer Blends

Feng Chen, Washington State University, Richland, WA and Jinwen Zhang, Composite Materials and Engineering Center, college of Engineering and Architecture, Washington State University, Pullman, WA

Investigate Soy Protein as Plastic Component in Polymer Blends

Feng Chen, Jinwen Zhang

School of Mechanical and Materials Engineering & Composite Materials and Engineering Center

Washington State University, Pullman, WA 99164

Abstract

Soy protein (SP) has been recognized as a promising alternative to petroleum-based plastic materials. However, owing to some obstacles such as water/moisture sensitivity, high melt viscosity, narrow process window and/or brittleness, SP plastics based-on neat SP have little practical values. To overcome these disadvantages, blending SP with other biodegradable polymers has received increasing attention. However, SP has only been investigated as organic filler in these blends. Although these blends display higher water resistance, processibility and/or toughness than neat SP plastics, the *stiffness* of the resulting blends are not satisfactory and the *strength* is often lower than that of both the neat SP plastics and the neat polymer.

In this work, soy protein concentrate (SPC) was blended with poly(butylene adipate-*co*-terephthalate) (PBAT) in a single-step compounding process. The mixing of the blends was carried out by extrusion compounding and test specimens were prepared by injection molding. Different influence of SPC as a plastic component and filler on phase structure and mechanical properties of the resulting blends were investigated. The intrinsic cause-deformability of soy protein and extrinsic cause-melt shear stress and interfacial tension in phase morphology development of PBAT/SPC blends were discussed. SP percolation network structure was noted in the blends with SP processed as plastic component. Owing to the extensive crosslinking of the SP molecules, the resulting blends actually became in situ formed polymer/SP composites. The composition-processing-structure-property relationship in soy protein blends were studied in detail. Resulting blends demonstrate an overall superior performance when SPC is used as plastic.

(730b) Urethane Foam Modeling with Soy-Based Polyols

Galen Suppes, Chemical Engineering, University of Missouri – Columbia, Columbia, MO and Yusheng Zhao, University of Missouri, Columbia, MO

Polyurethanes are very important polymers and have many commercial applications. In recent years, polyurethane (PU) foams remain the largest demand sector, accounting for two-thirds of the total PU demand. In addition, this demand for PU foams, especially in construction and transportation, continues to strengthen across the globe.

The polyurethane polymer-forming reaction occurs between an isocyanate and an alcohol. To make foams, the polyurethane polymer must be expanded or blown by the introduction of bubbles and a gas. A convenient source of gas was the carbon dioxide produced from the reaction of an isocyanate group with water. The main purpose of this paper was to initiate Matlab programming to simultaneously model the reacting, volume, and even cell formation during the foaming process.

Modeling is used to advance product development; specifically to expedite the identification of foam formulations where soy-based polyols work well. In these studies, a soy-based polyol prepared by the oligomerization of epoxidized soybean oil with ethylene glycol and a toluene sulfonic acid catalyst.

(730c) Co-Injection Molded Biodegradable Green Composites From Poly (3-hydroxybutyrate-co-3-hydroxyvalerate), Poly (butylene succinate) and Natural Fiber

Vidhya Nagarajan[1,2], Kunyu Zhang[3], **Nima Zarrinbakhsh**[4], Manju Misra[4] and Amar K. Mohanty[5], (1)School of Engineering, University of Guelph, Guelph, ON, Canada, (2)Bioproducts Discovery and Development Centre, Department of Plant Agriculture, University of Guelph, Guelph, ON, Canada, (3)Bioproducts Discovery and Development centre (BDDC), Department of Plant Agriculture, Crop Science Building,, University of Guelph, Guelph, ON, Canada, (4)School of Engineering and the Department of Plant Agriculture, University of Guelph, Guelph, ON, Canada, (5)Department of Plant Agriculture & School of Engineering, University of Guelph, Guelph, ON, Canada

To the best of our knowledge, this work reports the co-injection molding of bioplastics and natural fibers for the first time in the open literature. This approach opens up a novel method of manufacturing green composites for multitude of uses. To strike the stiffness-toughness balance of co-injection molded part, poly (3-hydroxybutyrate-co-3-hydroxyvalerate) (PHBV) and its composites with miscanthus were chosen as the core material while poly (butylene succinate) (PBS) was selected as the suitable skin material. The process of co-injection molding of PBS and PHBV, and also co-injection of PBS and PHBV with miscanthus has been established. The mechanical properties of these co-injection molded parts were compared with those of neat polymers. As expected, the co-injection molded parts' properties were found to be located between the properties of single injection molded skin and core materials. The unnotched impact strength of coinjection molded green composite has drastically increased compared to the PHBV miscanthus injection molded composite. Thickness of the skin and core in the co-injection molded part was measured and flexural modulus was predicted using composite beam theory. SEM micrographs revealed good adhesion between the skin and the core layers. The deformed core confirmed the stress transfer from the core to the skin. These results suggest that co-injection molding technique has been very effective in improving the strength of the composites and the targeted stiffness-toughness balance was achieved.

This research is financially supported by Ontario Ministry of Agriculture, Food and Rural Affairs (OMAFRA)-2008 New Directions research Program Project number SR9211, the OMAFRA, Highly Qualified Personnel Program, Natural Sciences and Engineering Research Council (NSERC) NCE AUTO21 research program and the Ministry of Economic Development and Innovation (MEDI), Ontario Research Fund - Research Excellence Round 4 program.

SYNTHESIS OF VEGETABLE OIL DERIVED POLYURETHANES

Amber R. Tupper, Barbara A. Wheelden, Sunggyu Lee
Ohio University, Athens, OH

Abstract

Bio-based polyurethanes are desired to replace traditional polyurethanes made from petrochemicals in order to be more environmentally friendly and decrease dependence on nonrenewable fossil energy sources. Vegetable oils are a suitable renewable starting material for bio-based polyurethanes due to their availability and low cost. Vegetable oils are triglycerides composed of different fatty acid chains which depend on the plant and the growing conditions. The chemical and physical properties of the oil depend on the fatty acid composition and degree of unsaturation. In this work, polyols are derived from castor oil through epoxidation of the double bonds followed by ring opening of the epoxide group. The thermal properties of the resulting polyurethanes are studied using differential scanning calorimetry (DSC) and thermogravimetric analysis (TGA).

Introduction

Traditionally, polyurethanes are made from petroleum based polyols. The replacement of conventional, petroleum based polymer plastics with bio-based plastics has been a topic of much research and interest over the past decade. The EPA estimates that 31.04 million tons of plastic materials were generated as waste in America in 2010.[1] Roughly 7% of the world's oil and gas consumption goes to the manufacturing of polymers.[2] Polyols from vegetable oils are suitable bio-based alternatives to petroleum polyols.

The process of making polyols and polyurethanes from vegetable oils, including soybean, canola, sunflower, corn and linseed oil, has been the focus of many studies.[3,4] Many studies create polyurethane from unmodified castor oil polyols or castor oil modified by either a transesterfication reaction with another polyol or conversion to an amide.[2,5] One study determined the thermal stability of several polyurethanes from different vegetable oils.[6]

Castor oil was used as it is readily available and has a fairly uniform structure due to the fatty acids composition of up to 90% ricinoleic acid.[5] Due to the hydroxyl group in these ricinoleic acid chains, castor oil is naturally a polyol and polyurethanes can be formed directly from unmodified castor oil. There are about 2.7 evenly distributed hydroxyl groups per triglyceride in castor oil resulting in a uniform crosslinked structure leading to good mechanical properties.[2] Two disadvantages to using unmodified castor oil polyols are that it has a low hydroxyl number and it has a slow cure rate due to the secondary hydroxyls.[7] However, there are other double bonds that can be converted to hydroxyl groups which would result in a higher crosslink density.

Experimental

Materials

Castor oil (Eur. Pharmaceutical Grade), acetic acid (99.8% for analysis), hydrogen peroxide (ACS 30%), methanol (99+%), and fluoroboric acid (50 wt. % in water) were purchased from Acros Organics. Toluene (Histological Grade) was purchased from Fisher Scientific. Amberlite IR-120 (H) (Analytical) was purchased from Supelco. Lewatit MP-64 (Chloride Form) was purchased from Fluka. 4,4'-Methylenebis(phenyl isocyanate) (MDI) (98%) was purchased from Aldrich.

Methods

The procedure to form a polyurethane by the reaction of the polyol, created though the ring opening of epoxidized castor oil, with MDI is similar to that done for other oils and triglycerides.[3,4,8]

Epoxidation of Castor Oil

The epoxidation of castor oil occurs when the oil reacts with peracetic acid which is formed by acetic acid and hydrogen peroxide. Toluene (50ml), castor oil (100g), acetic acid (10g) and Amberlite IR-120 (17g) were added to a 1000 ml 3-neck round bottom flask with a thermometer, drop funnel, condenser and magnetic stirrer. The mixture was heated to 55°C. Hydrogen peroxide was added drop by drop so as to not exceed 60°C and reacted for 7 hours. The Amberlite was filtered off.

Oxirane Ring Opening

The epoxidized castor oil went through oxirane ring opening by reacting with methanol using fluoroboric acid as a catalyst. Methanol (123ml) and fluoroboric acid (260mg) were boiled in a 3-neck round bottom flask. The epoxidized castor oil was added drop by drop into the flask. The mixture was kept boiling for 35 minutes. Lewatit MP-64 was added to neutralize the catalyst and then removed by filtration. The solvent was removed by vacuum distillation followed by high vacuum at 60°C.

Polyurethane Formation

The polyurethane was synthesized by reacting the resulting castor oil based polyols with MDI. The polyols and MDI (158g) were heated to 60°C separately then mixed and poured into molds preheated to 100°C. The polyurethane was cured overnight at 100°C.

DSC analysis

Differential scanning calorimetry was conducted using TA Instruments DSC Q2000 controlled by Qseries Advantage software. The polyurethanes were heated, cooled, and heated to determine the glass transition temperature.

TGA

Thermogravimetric Analysis was done with a TA Instruments SDT Q600 controlled by Qseries Advantage software. Experiments were carried out in an alumina crucible under N_2 atmosphere. Both standard and stepwise isothermal TGA was done.

References

1. Municipal Solid Waste Generation, Recycling, and Disposal in the United States: Facts and Figures for 2010. *United States Environmental Protection Agency*. 2011.

2. Pfister DP, Xia Y, Larock RC. Recent Advances in Vegetable Oil-Based Polyurethanes. *ChemSusChem*. 2011;4:703-717.

3. Zlatanić A, Lava C, Zhang W, Petrović ZS. Effect of Structure on Properties of Polyols and Polyurethanes Based on Different Vegetable Oils. *Journal of Polymer Science Part B: Polymer Physics*. 2004;42:809-819.

4. Zlatanić A, Petrović ZS, Dušek K. Structure and Properties of Triolein-Based Polyurethane Networks. *Biomacromolecules*. 2002;3:1048-1056.

5. Mutlu H, Meier MAR. Castor Oil as a Renewable Resource for the Chemical Industry. *European Journal of Lipid Science and Technology*. 2010;112:10-30.

6. Javni I, Petrović ZS, Guo A, Fuller R. Thermal Stability of Polyurethanes Based on Vegetable Oils. *J Appl Polym Sci*. 2000;77:1723-1734.

7. Ferrer MCC, Babb D, Ryan AJ. Characterisation of Polyurethane Networks Based on Vegetable Derived Polyol. *Polymer*. 2008;49:3279-3287.

8. Petrovic Z, Guo A, Javni I. Process for the Preparation of Vegetable Oil-Based Polyols and Electroninsulating Casting Compounds Created from Vegetable Oil-Based Polyols. . 2000;09/187,992.

Mechanical and Biological Characterization of Corn-Derived Poly-L-Lactic Acid

Leslie Rea[1], Nishath Khan[2], Manju Misra[2], Amar K. Mohanty[3] and Sujata K. Bhatia[4]
(1)Harvard University, Cambridge, MA, (2)School of Engineering and the Department of Plant Agriculture, University of Guelph, Guelph, ON, Canada, (3)Department of Plant Agriculture & School of Engineering, University of Guelph, Guelph, ON, Canada, (4)School of Engineering and Applied Sciences, Harvard University, Cambridge, MA

The ability to create biocompatible materials, and use such materials for medical purposes in the human body, will help treat diseases that result from loss of mechanical function in tissues. Biopolymers are an intuitive choice for biomedical applications such as tissue repair and regeneration, given that bio-based materials are constructed from naturally-derived materials, and may be expected to be friendly to biological tissues. Moreover, bio-based materials possess tunable chemical, physical, and mechanical properties, so that these materials can be readily constructed to match the native properties of a variety of target tissues, and ultimately be implanted in the human body to enable re-growth of cells and tissues.

Congestive heart failure and emphysema are examples of two diseases that result from a loss of elasticity in the muscles and tissue. Both of these diseases are potentially treatable with implantable, naturally-derived biocompatible materials. The purpose of this study is to characterize the mechanical and biological properties of electrospun corn-derived poly-L-lactic acid. We find that corn-derived poly-L-lactic acid has modulus of elasticity of 0.2-0.4 MPa. These properties are compatible with human myocardium, which has a modulus of elasticity of 0.2-0.5 MPa. Further, we find that corn-derived poly-L-lactic acid is non-cytotoxic to clinically relevant cell lines, including cardiac fibroblasts and cardiac myocytes. These results are promising for the use of bio-derived materials in biomedical applications.

Table 1. Processing parameters for electrospinning corn-derived PLLA nanofiber mats.

Parameter	Sample 1	Sample 2
Wt%	6	7
Weight (g)	6.034	7.2929
Volume (ml)	5	6
Feed Rate (ml/hr)	1.0	1.2
Voltage (kV)	13	15
Distance (cm)	12	12.5
Temperature (^{0}C)	24-25	24
Humidity (%)	63	40-43

Figure 1. Schematic diagram of electrospinning apparatus.

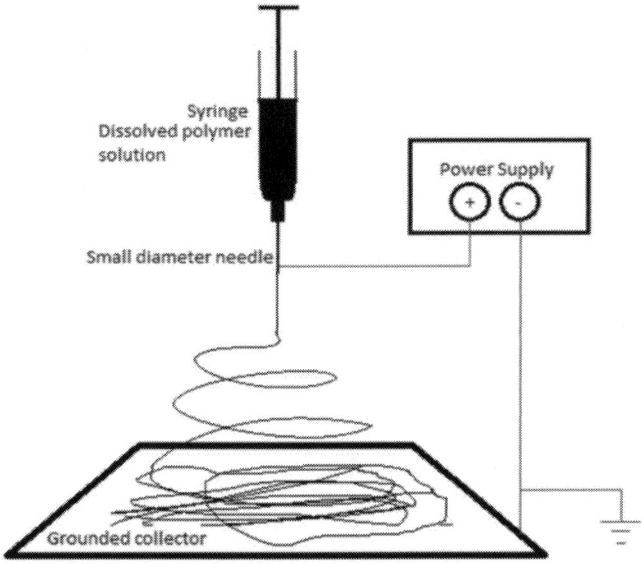

Figure 2. SEM image of an electrospun nanofiber mat constructed from corn-derived PLLA.

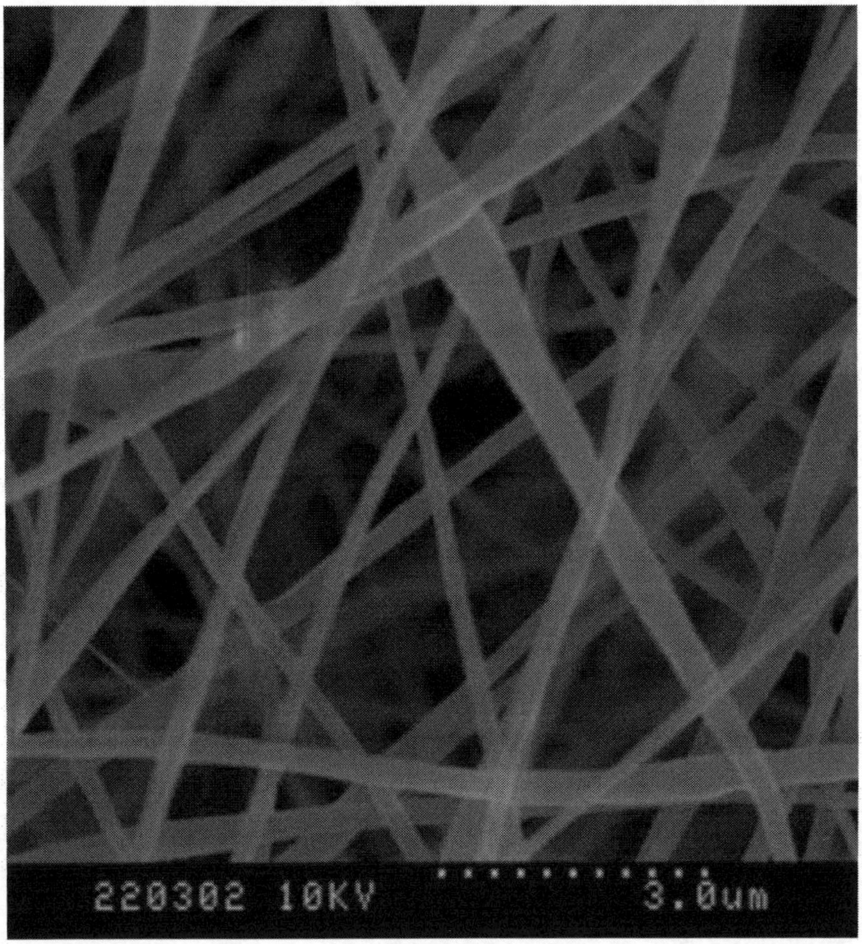

Figure 3. Representative stress–strain plots generated from tensile testing of electrospun nanofiber mats constructed from (A) 6 wt% corn-derived PLLA and (B) 7 wt% corn-derived PLLA.

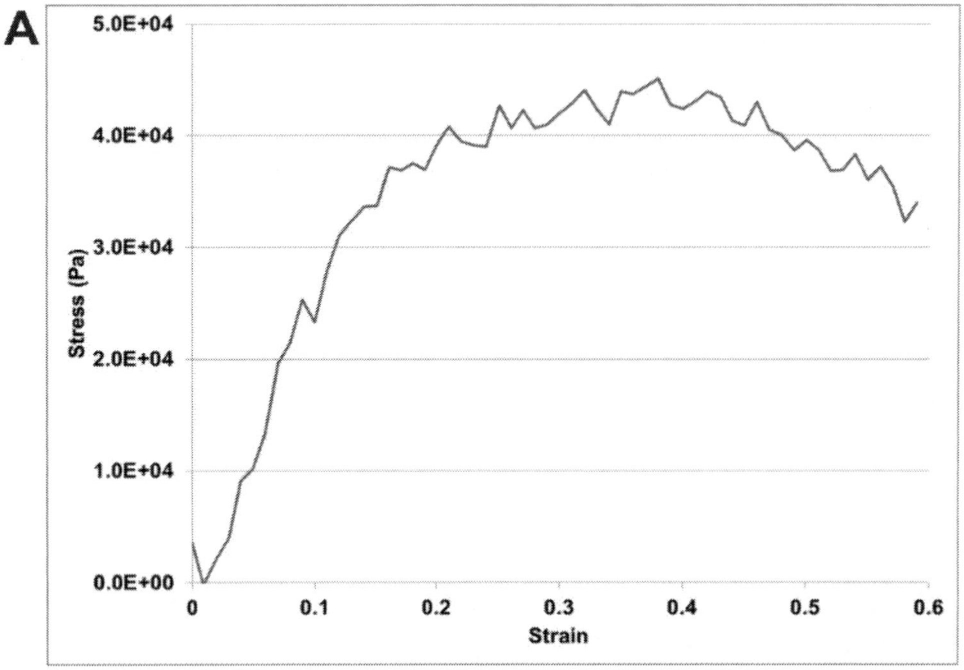

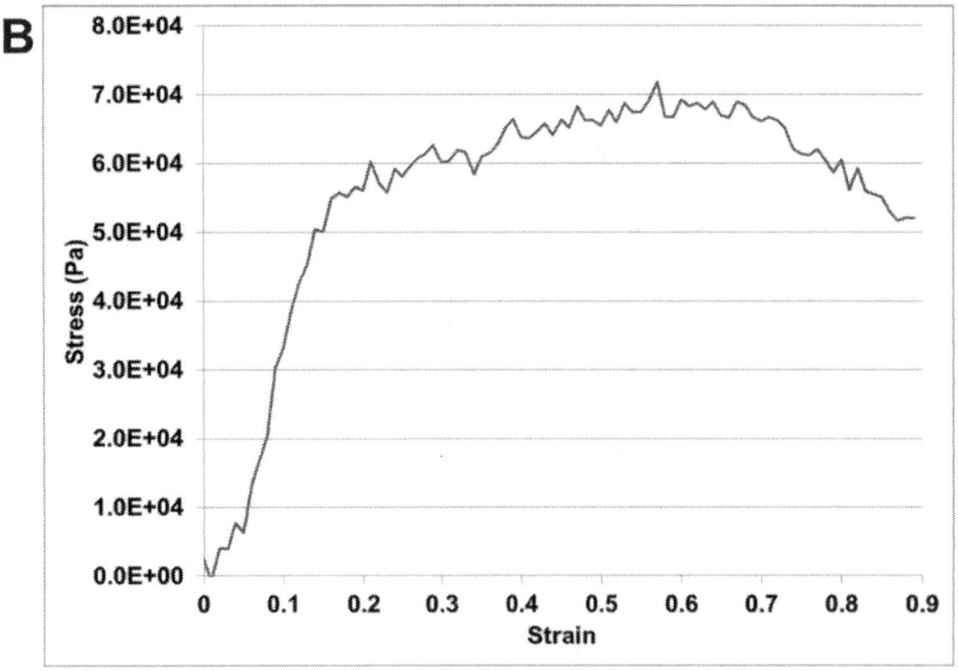

Figure 4. Responses of cardiac myocytes to electrospun nanofiber mats. Myocytes were incubated at 37°C with (A) untreated (control) at 24 h; (B) nanofiber mat constructed from 6 wt% corn-derived PLLA at 24 h; (C) nanofiber mat constructed from 7 wt% corn-derived PLLA at 24 h; (D) untreated (control) at 48 h; (E) nanofiber mat constructed from 6 wt% corn-derived PLLA at 48 h; (F) nanofiber mat constructed from 7 wt% corn-derived PLLA at 48 h.

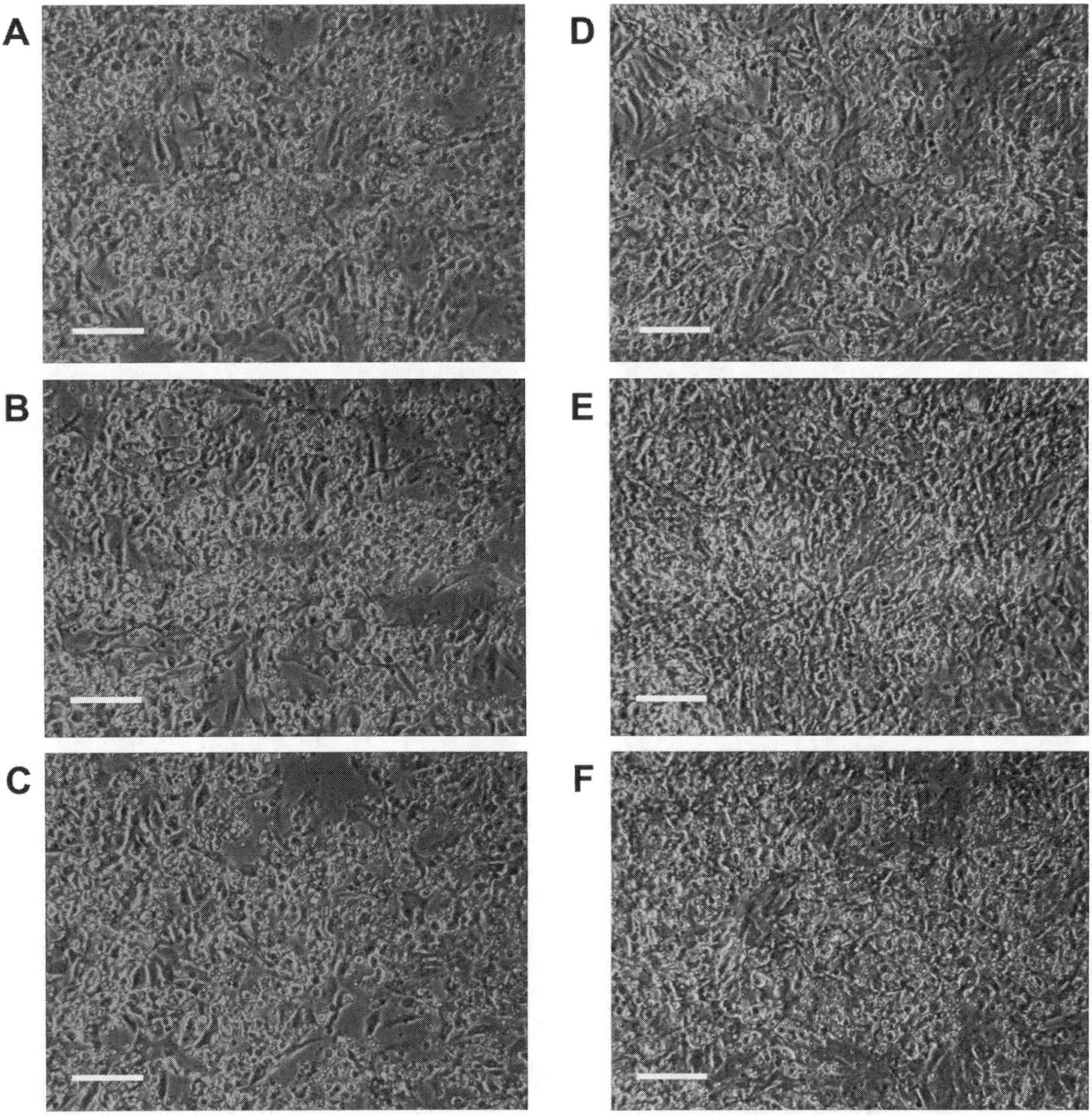

(730f) The Interactions of the Corn Protein Zein and Ionic Liquids

Sean Tomlinson, Jennifer L. Anthony and John R. Schlup, Chemical Engineering, Kansas State University, Manhattan, KS

The corn protein zein is a commercially interesting biopolymer because it is insoluble in water but biodegradable. Ionic liquids are able to dissolve zein and can be used as an extraction solvent for zein recovery. However there has been no research into if or how the zein is affected by the ionic liquid. Using infrared spectroscopy to examine the structure of zein in both protic and aprotic imidazolium-based ionic liquids, we demonstrate how the solvent affects the protein's secondary structure and identify whether the protein is denatured by the ionic liquid. We will also show how the properties of the ionic liquids affect the protein's solubility. It has been found that the ionic liquid's properties are correlated with the structure and solubility of zein in the ionic liquid, particularly the ionic liquid's hydrogen bond accepting and donating abilities. The hydrogen bond accepting and donating abilities of ionic liquids are then used to predict the solubility and the secondary structure of the protein in the ionic liquid.

POLYMER APPLICATIONS IN PULP AND PAPERMAKING- A REVIEW

A.K. Ray*,Sujay Chattopadhyaya and A.K. Mukherjee**
***Department of Paper Technology, Indian Institute of Technology Roorkee**
**** Indian Institute of Technology Delhi**

ABSTRACT

Polymer has numerous applications in the paper manufacturing process either as chemical aid (wet strength and dry strength resins etc. or as a machine components.
In this paper mainly the polymer as wet strength or speciality chemicals are emphasized.
For many practical uses the water and moisture sensitivity of ordinary paper proves to be an insurmountable obstacle. Water leaf paper show excellent dry strength properties, but almost all of this strength is lost when the paper is wetted by water because dry strength is due primarily to hydrogen bonding (and probably some ionic bonding) and these bonds are destroyed by water, which is very effective at forming hydrogen bonds itself. Therefore, development of wet strength requires the formation of different types of bonds which are not adversely affected by exposure to water .The earliest method for developing wet strength was to fuse the cellulose fibers to one another, either by heating the paper to very high temperature or by partially solubilizing the paper with 75% sulfuric acid. The later is the method by which parchment paper is produced. The first chemical method for improving wet strength involved the use of formaldehyde, which can form methylene bridges between hydroxyl groups on the surface of cellulose fibers.

Wet strength is usually reported in terms of the ratio of wet tensile to dry tensile strength, expressed as a percentage. Paper possessing a wet tensile strength greater than 15% of dry tensile is considered to be wet strength papers.

The most important wet-strength agents are applied by beater or wet end addition; i.e., they are added to the pulp before the sheet is formed at the "wet-end" of the machine. Aluminium sulphate is often added to cure resin or to retain anionic Urea-formaldehyde resins. Additives that are not substantive to (absorbed by) paper fibers must be added to the paper after sheet formation. This procedure is usually less cumbersome than addition to the paper stock.

Saturation of paper with water reduces its strength to about 3-10% of its dry strength. Animal glue decreases the rate of absorption of water: if the paper is treated with both formaldehyde and animal glue it will have true wet strength. Treatment with 75% sulfuric acid produces "vegetable parchment," which has high wet strength, but the treatment seriously changes other sheet properties. By means of modern commercial resins, wet-strength of 20- 50% of dry strength can be obtained. Addition of 1% resin can cause a three to six fold increase in wet strength.

INTRODUCTION

Polymers are extensively used in making and manufacturing paper. These are employed either in as chemical aid to increase strengths(wet or dry), as retention aids, drainage aids, de-scaling and flocculating agents , as surface sizing chemicals, coating binders and co-binders,as surfactants and dispersing agents or as a machine parts, paper achine clothing(Nylon,polyester), doctor's blade,roll cover, vacuum boxes,hydrafoil etc. and even in ink for printing[1-10]. In this paper mainly the polymer employed as wet strength or as speciality chemicals are discussed her under:.

WET STRENGTH PAPER:

For many practical uses the water and moisture sensitivity of ordinary paper proves to be an insurmountable obstacle. Water leaf paper show excellent dry strength properties, but almost all of this strength is lost when the paper is wetted by water because dry strength is due primarily to hydrogen bonding (and probably some ionic bonding) and these bonds are destroyed by water, which is very effective at forming hydrogen bonds itself. Therefore, development of wet strength requires the formation of different types of bonds which are not adversely affected by exposure to water .The earliest method for developing wet strength was to fuse the cellulose fibers to one another, either by heating the paper to very high temperature or by partially solubilizing the paper with 75% sulfuric acid. The later is the method by which parchment paper is produced. The first chemical method for improving wet strength involved the use of formaldehyde, which can form methylene bridges between hydroxyl groups on the surface of cellulose fibers.

Wet strength is usually reported in terms of the ratio of wet tensile to dry tensile strength, expressed as a percentage. Paper possessing a wet tensile strength greater than 15% of dry tensile is considered to be wet strength papers.

The most important wet-strength agents are applied by beater or wet end addition; i.e., they are added to the pulp before the sheet is formed at the "wet-end" of the machine. Aluminium sulphate is often added to cure resin or to retain anionic Urea-formaldehyde resins. Additives that are not substantive to (absorbed by) paper fibers must be added to the paper after sheet formation. This procedure is usually less cumbersome than addition to the paper stock.

Saturation of paper with water reduces its strength to about 3-10% of its dry strength. Animal glue decreases the rate of absorption of water: if the paper is treated with both formaldehyde and animal glue it will have true wet strength. Treatment with 75% sulfuric acid produces "vegetable parchment," which has high wet strength, but the treatment seriously changes other sheet properties. By means of modern commercial resins, wet-strength of 20- 50% of dry strength can be obtained. Addition of 1% resin can cause a three to six fold increase in wet strength.

A. USES OF WET STRENGTH PAPERS

Wet strength papers are widely used for packing purpose, for making multi-wall paper sacks, paper bags and wrapping papers and containers employed for protection of flowers and moist or frozen food, such as, fish, vegetable and ice-creams, towels, paper handkerchiefs and toilet tissues. Wet strength paper is now frequently used for horticultural labels, currency notes, laundry tags, maps, blue prints, posters and other articles subject to out-of-doors exposure.

Various Wet strength agents and their Curing Conditions.

Type of Resin	Type of Curing
Urea-Formaldehyde (UF)	Acid
Melamine -Formaldehyde (MF)	Acid
Polyamine and Polyamide : Polyamide Epichlorohydrin(PAE)/Polyamine epichloro alkaline hydrazine	Acid
Glyoxal	Acid
Dialdehyde starch (DAS)	Acid
Chitosan	Alkaline
Miscellaneous(neoprene,Styrene–Butadiene,Acrylic butadiene arynitrile latex	Alkaline
Newer Resin (Polyethylenimine, Polyacrolein,	Alkaline
Quarternary Ammonium Epoxide	
Polyacrylamide-Glyoxal (PAMG)	Acid
Quarternary Ammonium Epoxide	Alkaline

B. 1. **UREA -FORMALDEHYDE RESINS (U.F.)**

Nonionic, anionic and cationic urea formaldehyde resins are used as wet strength resins for the manufacture of paper.

a)Nonionic resins are unsuitable for beater addition because they are not retained by pulp. These resins, therefore, must be applied by surface treatment of the paper after formation on the machine. Surface application of urea-formaldehyde resins yields a stiff paper with reduced absorbency of water surface application and is used where properties, such as, wet and dry abrasion resistance, wet-rub resistance, and improved creep are desired.

The bulk of wet strength paper is produced by wet-end addition because of these advantages; it is not necessary to re-dry the paper after treatment of the web with wet-strength agent, no special application equipment is required; uniform treatment is more easily achieved; folding endurance is increased (surface application of U-F resin reduces folding endurance); tearing strength is reduced by surface treatment. Furthermore, wet-strength resins together with aluminium sulphate promote retention of 'fines' (short cellulose fibers or fragments) improve the 'formation' (uniformity of distribution of fibers in the sheet) and decrease fiber wastes.

b)Anionic Resin: Reaction of urea-formaldehyde polymer with sodium bisulfite produces an anionic bisulfite resin.

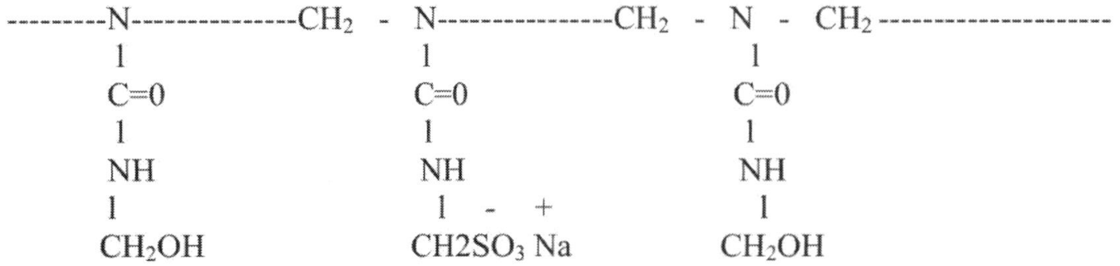

The product is water-soluble, but addition of aluminium sulfate provides good retention of material by unbleached kraft-pulp. Paper pulp itself is anionic, and the aluminium sulphate functions as a cationic mordant in approximately the same way that it anchors resin size and anionic dye to paper pulp. Although the anionic U-F resin provides a low-cost means of imparting wet -strength to unbleached kraft paper, they are not especially effective with other pulps and are absorbed somewhat slowly. Its retention with the pulp when added to aqueous pulp suspension is about 40-50%.

Anionic U-F resin is made by reacting urea and formaldehyde in the molar ratio 1:1.8 - 2.0. The pH is made neutral and to it added 11 gm sodium Meta bi-sulphite per mole of urea. The mixture is heated to boiling point, under reflux. After boiling for 30 minutes and boiling must be continued and then the solution is acidified to pH 2.5 with dil. H_2SO_4 for further one hour. The solution is then neutralized with aqueous sodium hydroxide and cooled.

c) Cationic resin is preferred for other than bleached Kraft pulps. They owe their cationic nature to amine groups, which are introduced by reacting ethylene diamine, diethylene-triamine, or other amino compounds with urea-formaldehyde resins. At low pH, the amino group picks up protons, and the resin becomes cationic. In addition to making the resin cationic, the amine increases the solubility of the polymer permitting the polymerization to be carried to higher molecular weight.

The resin is prepared by reacting urea and formaldehyde under mild alkaline condition to give methylol urea, adding the base, or a suitable salt of the base and excess acid and then condensing at elevated temperature under acid conditions until the solution has attained an appropriate viscosity. The syrup is then neutralized and cooled.

Cationic U-F resins available commercially are to be diluted with water before use to ensure a better distribution of resin in fiber. Sufficient alum should be added to the stock to ensure maximum retention of resin in the paper. The acidity of the stock should also be adjusted to pH 4-5 by adding alum and hydrochloric acid. The quantity of resin should not be more than 3-4% of the dry weight of fiber. 60-70% of the resin added to the aqueous pulp suspension is normally retained on the fiber.

Urea-formaldehyde resins generally react less rapidly than melamine-formaldehyde resins, which are, in general, lower than epichlorhydrine-amine resins. The use of a rapidly curing resin permits the mill to sell the paper as soon as it is manufactured, without storage and warehousing. However, repulping of "broke" (trimmings and rejected papers produced in the mill) is more difficult if wet-strength resin cures too rapidly. A cationic U-F resin requires ten to fourteen days to develop full cure (at 20^0C) whereas the anionic resin requires about twice as long. Paper containing U-F resin can be repulped.

Urea formaldehyde resins do not show great resistance to hydrolysis. At a pH of 4-6, these resins can be hydrolyzed fairly readily, but they resist hydrolysis at pH 8-10. It has been reported, however, that under favorable conditions, paper treated with U-F resins will retain its full wet-strength for six to eight years.

Characteristics

- Low viscosity liquid Or Powder
- Solid content 68 ± 2(%)
- Shelf life (20^0C):

Liquid: 3-6 months
Powder: 1 year
Pot life (20^0C) :0-48 hrs.
At usual room temperature:Setting time 3-10 hrs.
Hardener 5-10%Starch to give wet resistance.
Disadvantages

1. Slow cure
2. Cationic resin coated - 10 - 14 days -full cure
   Anionic resin coated - 21 - 28 days -full cure
3. Hydrolyses readily at pH 4 - 6

Advantages

1. Cheap
2. Repulping possible
3. Retains property (wet strength) 6-8 years.

Use: Urea resin adhesive mixed with starch is used in making water resistant and high humidity resistant paper board boxes and corrugated paper board used in packing industries.
Generally starch and dextrin pastes are used but these are not water resistant.Urea resin and hardener to the extent of 5 to 10% (dry wet) on the weight / starch / depending on the degree of water resistance desired and nature of the starch is used.

B. 2. **MELAMINE FORMALDEHYDE RESIN**

The melamine resins colloid are the first materials to be successfully applied to paper by introduction at the wet end of the paper making process.

The characteristics of a typical melamine colloid is as under:

Form-white, Free flowing Powder

Reactive resin content	: 100%
Solid Content	: 93%
Nitrogen Content	: 38.9%
Shelf life	: 12 months (See fig-1-3)

Melamine acid colloid solution is made as follows:
8.4 ml. of concentrated Hydrochloric Acid is added to 184-ml. of water heated to 35-38^{0}C. The acid solution is vigorously agitated; and 25 g of a spray-dried melamine formaldehyde powder resin (Molar ratio 1:3 MF) as above is added. When dissolution is complete the solution is allowed standing for 3 hours for the colloid to develop. The solid content of this colloidal solution is about 12%. It remains stable at least 3 days, but then rapidly increases in viscosity and gels. After dilution of the solution further to 6% solid, where the life will be indefinitely higher is to be added to a suitable point between the beaters and the head box. The quantity of resin will usually not exceed 3% solids on dry fibre.

Cationic Acid colloid solution has a direct affinity for cellulose pulp, which normally carries a negative charge, and the use of alum to assist retention is, therefore, unnecessary. The acidity of the stock may require adjustment however, and alum, hydrochloric acid or sodium carbonate may be used for this purpose according to circumstances. Optimum wet strength is obtained if the acidity of the stock is adjusted so that the pH of the paper as it leaves the machine is within the range of 4.5 - 5.0.
Between 70 and 85% of the resin added is retained in the paper web after formation and fairly rapidly on passing over the drying cylinders. A further improvement in wet strength usually occurs during storage if the paper is reeled hot.

Table 1 Comparison of U-F & M-F

Parameter	U-F	M-F
Rate of cure	Moderate	Fast
pH tolerance	Low	High
Nature of pulp	Selective	Any kind
Resistance of hydrolysis	Poor	Good
Beating period	More	Less
Effect of humidity	Dimensional change	No change
Sticking to Yankee drier	-	Low
Gluing problem	None	None
Resistance to delamination	Good	Excellent
Folding endurance	Good	Excellent
Dry strength	Increase	More increase
Stain resistance	Poor	Good
Abrasion resistance	-Poor	Good
Heat resistance	Poor	Good
Curing conditions	147 – 163°C in 3 min	130- 190° C in 1.5 min

Table 2 Application of Amino Resins By Beater Addition

	Bleached sulphite pulp				Un-bleached kraft pulp			
	Without alum		With alum		Without alum		With alum	
	Dry	Wet	Dry	Wet	Dry	Wet	Dry	Wet
Resin 3% weight	Tensile strength (1b/in)		Tensile strength (1b/in)		Tensile strength (1b/in)		Tensile strength (1b/in)	
None	5.4	1.2	5.3	1.2	11.0	1.2	10.2	1.2
U-F	6.8	1.7	9.0	3.3	15.8	3.6	20.5	8.2
Anionic U-F	10.0	5.3	10.0	5.3	21.4	7.5	21.0	9.1
Cationic M-F acid colloid	9.4	4.1	10.0	4.4	18.0	5.5	19.4	6.6

Applications of M.F. resins

M-F resins are used for making superior grade wet-strength paper wherever U-F coating is not suitable. One of the major applications of M.-F is to make decorative laminates. The various steps in making decorative laminates are given below:

1. Assemble 6-8 layers (core) of p-f bonded Kraft paper.
2. Put a M-F impregnated alpha-cellulose paper.
3. Put the decorative layer (0.008-0.010"): Highly refined alpha-cellulose Alpha-cellulose paper + 40-50% U-F or M-F resin.
4. Put protective overly (0.001-0.003"): pulp + 65-75% M-F resin.
5. Cure under heat and pressure.

Disadvantages of aminos

1. Acidic pH (around-4.5) during curing
2. Deterioration of paper:
   Embrittlement
   Reduction in Absorbency
   Ha-rsh Hand
3. Resins hydrolyze at low pH
4. Gradual wet-strength reduction during storage.

B.3 POLYMINE AND POLYAMIDE WET- STRENGTH RESINS

The amino-formaldehyde resins have the disadvantage of requiring acid conditions for curing. The recommended pH of around 4.5 causes embittlement and deterioration of paper, reduces absorbency and develops a harsh hand. These characteristics are especially undesirable in tissues and toweling. In addition, the wet-strength resins are hydrolyzed at low pH resulting in a gradual reduction of wet strength on storage, especially when the paper is wet. It was apparent that a wet-strength agent, which would cure at pH 7 or higher, was desired.

In 1952, Daniel and Landes patented a polymer of poly-alkylene Polyamine and epichlorohydrin as a wet-strength resin, which could be cured at pH levels as high as 10. Several years later, a polyamine-epichlorohydrin product was introduced to paper mills, and this type of resin has found general acceptance as a wet-strength agent for absorbent papers and other grades made at pH around 7. The resin is produced from an aliphatic dicarboxylic acid, such as, adipic acid, an alkylenepolyamins, such as, diethylene triamine, and epichlorohydrin. The product is an efficient wet-strength resin, which cures at pH higher than that required for the formal dehyde resins.

A typical polyamide resin produced from adipic acid, diethylene-triamine, and epichlorohydrin would have the structure:

CH_2-Cl
1
CH-OH
1
CH_2
1
O O
(+) 11 11
------HN - H_2C - H_2C - N - CH_2 - CH_2 - NH - C - $(CH_2)_4$ - C ------
1
CH-OH
1
CH_2
1
O O
11 11 (+)
-- C - $(CH_2)_4$ C - HN - H_2C - H_2C - N - CH_2 - CH_2 - NH-------------
1
CH_2
1
CH - OH
1
CH_2 - Cl

Loss of hydrogen chloride is accelerated under basic conditions, and the resulting epoxide group is free to react with cellulose or other amino groups.
The polyamide resins, like other wet-strength resins, may be introduced to paper stock at any convenient location after the major refining work is done and before the sheet is formed, as for example, at the fan pump. The resins are retained efficiently, and absorption occurs rapidly. The resin may be added at any pH from 4 to 10, although the range 6-8 is usually preferred.

There are several significant differences between polyamide resins and the amino-formaldehyde resins. The polyamide resins cure over a broad pH range, and the rate of cure is affected by a lesser extent by change in pH. At the recommended pH conditions amino-formaldehyde resins reduce the absorbancy of paper, but the polyamide resins have little effect. At low levels of add-on, the polyamide resins are more efficient than the amino-formaldehyde resins, but higher ultimate wet strength can usually be obtained using urea-or melamine formaldehyde adducts. The polyamide resins are well retained without the use of aluminium sulfate although addition of 25-ppm sulfate ion is reported to help in achieving maximum efficiency at high levels of addition.

Paper containing U-F and M-F wet strength agents is usually repulped with acid, whereas that containing polyamide resins require hydrochlorite or a high pH. The amino resin increases the dry strength more than do the polyamide resin.

Amino polyamide-epichlorohydrin resins are significantly more expensive than formaldehyde based resins, but the higher cost is offset by other advantages, mostly the advantage of making paper at neutral to alkaline pH. These include properties, such as, improved softness and absorbency, no tendency towards embrittlement on aging, and reduced paper machine corrosion. Another important advantage is the elimination of formaldehyde usage.

Characteristics

1.	Gel time (200 g)	:	2-4 hr.
	Viscosity	:	120 Ns /m2 at 210°C
	Shelf life	:	6 months, 200°C
	Curing pH	:	Higher than Aminos

Table 3 Comparison With Aminos

Parameters	Amino	PA
ph range	Acidic	4 - 10
Rate of cure (effect of pH)	Great	No significant
Paper Absorbency	Reduction	No effect
Wet- strength (at comparable add on)	High	Higher
Alum	Necessary	Not necessary
Repulping conditions	Normal	Special
Dry strength	Increase	More increase
Softness	Less	More
Aging	Brittle	No change
Machine corrosion	High	Low
Cost	Low	High

B-4 GLYOXAL RESINS

Another class of wet strength resin is based on the reaction product of polyacrylamide and glyoxal. This resin can react with hydroxyl groups on cellulose fibers through the glyoxal functionality to form hemiacetal bonds, resulting in improvements in paper wet strength, although, not to the same extent as amino-polyamide-epichlorohydrin resins. The base polyacrylamide usually contain a small amount of cationic co-monomer, such as, dimethyldiallyl ammonium chloride to make the resin substantive to pulp fibers.

```
------- CH2  -   CH   -  CH2  -  CH   -  CH2  -   CH ------------------
                 1                1                1
                 C=0              C=0              C=0
                 1                1                1
                 NH2              NH               NH2
                                  1
                                  CH2
                                  1
                                  OH
```

B - 5 DIALDEHYDE STARCH

Dialdehyde starch, made by periodic oxidation of cornstarch, is the newest type of resin to achieve commercial attention as a wet-strength resin. Dialdehyde starch is anionic in nature and can be used as wet-end additive only if the paper fibers have been treated with cationic starch or synthetic cationic retention acids. High amounts of aluminum sulphate are required for retention of dialdehyde starch. Dialdehyde starch produces only a temporary wet strength, but cationic dialdehyde starch confers more permanent wet strength. Temporary wet strength is useful in towels and sanitary tissues where wet strength is required for only a few seconds and where Permanent strength can cause clogging of plumbing and septic tanks. In addition to temporary wet strength, dialdehyde starch offers extremely easy broke recovery, high wet strength without loss in absorptivity, rapid development of wet -strength, and excellent improvement in dry strength properties. It is the only paper additive that imparts significant improvement in wet and dry strength without causing broke recovery problem.

B - 6 **CHITOSAN**

Chitosan (poly 2- aminodeoxy - 1,4 glucoside) is not a single substance, but a group of partially deacetylated products of various degrees of polymerization. Chitosan is insoluble in water, concentrated alkalies, alcohol, and acetone, but dissolves readily in dilute acid. It has also been shown to impart wet strength to paper. The polymer, although potentially available in large quantities from shell-fish wastes has never become a commercial significant product.

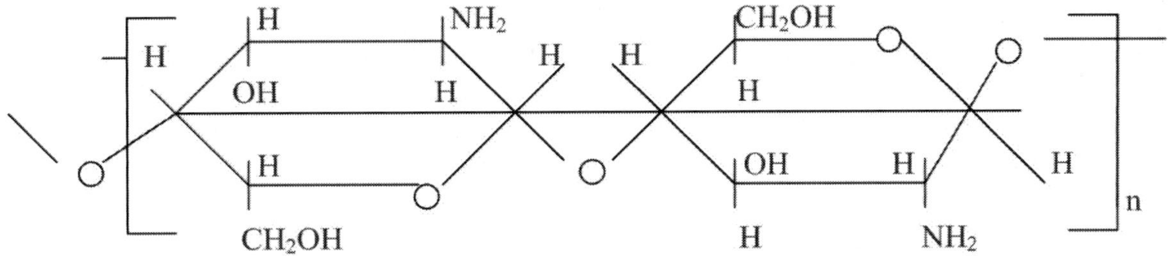

B -7 **MISCELLANEOUS**

In addition to the urea-formaldehyde, melamine- formaldehyde, polyamide, and dialdehyde-starch resins, other resins may impart some wet strength, but are used primarily for purposes other than wet strength.

Small amount of neoprene latex added at the wet end of the papermaking machine gives high wet strength and resistance to a variety of chemicals. Saturation of paper with styrene- butadiene, acrylic or butadiene arylonitrile latex for use in manufacturing book covers, marking tape and gaskets confers wet strength and other properties to paper. Latexes also change many other properties of paper, e.g., hand (the feel of the paper), elongation, stiffness, dimensional stability, edge and internal tearing resistance, folding endurance and tensile strength. Therefore, they are not used when wet strength improvement is the only change desired.

Because latex particles are not readily retained by pulp; their use is frequently more troublesome, but aluminium sulphate aids their retention. Saturation with phenolic resin, like saturation with latex, produces a resin-filled sheet that has high wet strength. However, the treatment of paper with latex or phenolic resin changes its properties in such a fundamental way that it is not usually called " wet- strength treatment".

B- **8 NEWER RESINS**

Polyethylenimine was used in Germany during World War-II as a wet strength agent. It has never become commercially significant, however, it confers several unique properties. It retards the rate of beating of pulp and should be added after the pulp is beaten. Wet strength develops as the pulp is dried, and water absorbency is not affected.

Cereal xanthates, produced from cereal starches or flours, produce permanent wet strength and increase dry tensile strength and folding resistance.

Polyacrolein gives paper with high, permanent wet strength and good dimensional stability. The polycryolein must be treated with sulfur dioxide or a sulphite to make it water-soluble.

C. MECHANISM OF WET STRENGTH FORMATION

With the exception of polymeric amines, all effective wet- strength resins undergo cross-linking reactions and some react with functionality present on the surface of cellulose pulp fibers. Two mechanisms for wet strength development have been proposed.

The protection theory suggests that the resin forms a cross-linked network around existing fiber-fiber bonds and prevents swelling and disruption of these bonds.

The new bond theory proposes that the wet strength resin creates new fiber-fiber bonds that are covalent and thus are not broken on exposure to water. The protection theory is supported by the fact that a given concentration of amino polyamide- epichlorohydrin resin imparts wet strength that is a relatively constant percentage of dry strength in various pulp furnishers, even though the absolute values of wet and dry strength might vary substantially.

The new bond- theory is supported by the fact that wet and dry strength are increased by approximately the same increment with either formaldehyde- based or amino polyamide-epichlorohydrin wet- strength resins.

D. REPULPING

Repulping of paper is necessary for two reasons. During paper manufacture broke (waste) is produced either during breaks on the paper machine or as trim produced during the rewinding process. For the paper making process to be economical, this must be recycled. Also, the use of waste paper furnishes is increasing as the cost of virgin-pulp furnishes increase.

Paper containing wet- strength resin is more difficult to repulp than non-wet-strength paper. Therefore, the repulpability properties of wet strength resins must be considered when choosing a resin. U- F and M-F resins can be readily repulped under acid conditions. Alum is used in the repulping operation when these resins are used. Paper containing polyamide- epichlorohydrin resins are not readily repulped under either acidic or basic condition. Oxidizing agents, such as, sodium hypochlorite is usually required for repulping in such cases. Presumably, the oxidant attacks the polyamide backbone and breaks down the resin. Paper containing poly (methyl diallyamine) epichlorohydrin wet-strength resin is very difficult to repulp. The hydrocarbon backbone of this resin is not subject to attack by reagents commonly used in aqueous systems.

E .APPLICATION OF RESIN COATING IN COLOUR ELECTROPHOTOGRAPHY

A. Basic idea

A recent suggestion has been made that emphasis in the dye and pigment industry is now shifting from the replacement of natural products, quality control and process improvements. Some newer ideas, such as, colour- formers for carbonless copy paper are already in the market place and dyes are becoming of increasing use in information technology.

Recording and reproduction of information using photography and photocopying are already well known. Laser- writing and reading, liquid- crystal displays and the use of compact discs, in all these processes, at some stage either in manufacture or use, electromagnetic radiation is absorbed and hence, can be influenced by the presence of dye or pigment. Other applications, which may prove more beneficial to society in the long run, may be found in converting solar to electrical energy and in dye- related drugs.

B. **Principle**

The principle involves in carbonless copy processes and colour formers is the rupture of microcapsules by impact and release of a virtually colorless pre-cursor which can interest with an underlying copy paper layer to form a colour.

Microcapsules are very small (1-500 μm dia) containers in which a fluid or solid is enclosed in a polymer shell. Spherical microcapsules containing liquid are more easily ruptured than those containing solid. They have many useful functions and are to be found in many different fields. The reasons for micro encapsulation include the following:

a) Protection of the enclosed material and improved storage life.
b) Conversion of a liquid component to a dry solid system.
c) Ensuring separation of incompatible components.
d) Odour masking, dust control and pH control, and,
e) Controlled diffusion of active component through the shell, as for example, in delayed drug release. The general principle of micro-encapsulation is shown in Fig.-4

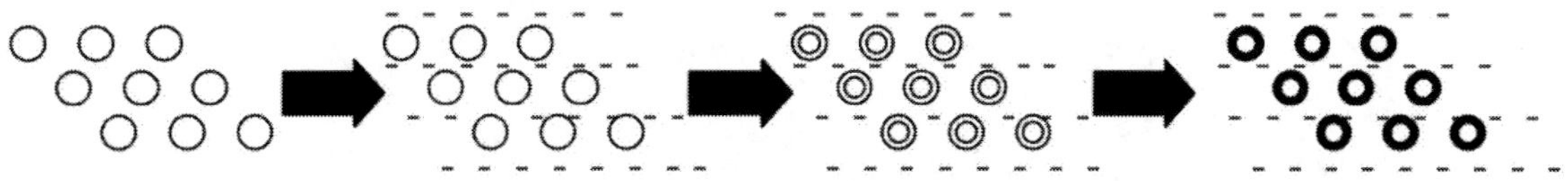

a) Dispersion or emulsion | b) Wall material dissolved in continuous phase | c) Wall material deposited on core | d) Deposited wall solidified & hardened

Formation of Microcapsules.

Phase separation results from the interaction of two oppositely charged, water-soluble polymers, such as, gelatin and gum Arabic. Both polymers exist as negatively charged species in neutral aqueous solution. Where this mixed solution forms the continuous phase in an oil - in - water emulsion and when the pH is reduced to 4.8 only the gelatin becomes positively charged and mutual electrostatic attraction between the polymers causes precipitation around the organic droplets.

For microcapsule use in carbonless copy paper the organic phase, which may be chlorinated paraffin contains a colorless dry pre-cursor known as a colour former. Other means, such as, interfacial polymerization to form the polymer shell, in which the reacting monomers are separately contained in the oil and water phases of the emulsion, is a later development.

Among organic coating are to be found phenol-formaldehyde resins, bisphenols, such as, 2, bis (p-hydroxyphenyl) propane and 4-alkoxy - or 4-phenoxy-1- naphthol. Phenolic resins in which some of the phenolic positions have been replaced by zinc give improved shelf life and inhibit yellowing.

C. APPLICATION OF POLYMER COATED PAPERS IN ELECTROPHOTOGRAPHY.

1. Colour photocopying and photography:

A recent development is colour photocopying which does not employ electrostatic attraction of toners. The method makes use of microcapsules containing, in addition to a colour former, a monomer and photosensitive initiator. By the suitable choice of initiator the capsules can be made sensitive to blue (420 mm), green (550 mm) or red (650 mm) radiation. After exposure, the cores of those capsules, which have been polymerized, do not release their colour formers when subject to pressure. By coating a paper with all three types of capsule and exposing under a transparent document, or under a transmissive liquid crystal display, a latent colour image is produced. This image is then placed in face-to face contact with an acidified clay- coated layer and squeezed through the nip of pressure rollers to produce a final colour print on the clay coated-layer.

REFERENCES

1. Mark, H.F., Goylord, N.G., Bikales, N.M., "Encyclopedia of Polymer Science and Technology." Interscience Publishers, NY, Vol. 9, 748-93 (1965).

2. Mark, H.F., Mcketta, J.J., Othemer, D.F., "Encyclopedia of Chemical Technology." Interscience Publishers, NY, Vol. 14, 500-31 (1963).

3. Mark, H.F., Gaylord, N.G., Bikales, N.M., "Encyclopedia of Polymer Science and Technology." Interscience Publishers, NY, Vol. 12, 494- 531 (1987).

4. Shields, J., " Adhesives Handbook", Published by Butter Worths, London, 74-5 (1970).

5. Vale, G.P., Taylor, W.G.K., "Amino plastics," Published by ILIFFE Book Ltd., London, 113-20 (1964).

6. Jones, F., Rev. Prog. Coloration, Vol. 19, 20 - 32 (1989)

7. Kokurek M.J., Pulp and paper Manufacture, Vol.7 , Joint text book Committee of the Paper Industry, TAPPI Press, 1989

8. Biermann, C.J., 1996, "Handbook of Pulping and Papermaking", Academic Press, CA.

9. Scott W.E., Principles of Wet End Chemistry, TAPPI Press, 1996

10. Personal Communication with Prof. A.K.Mukherjee, IIT Delhi,1995

(734a) Accellerase® Trio: Advances in Cellulosic Enzymes for Commercial Biorefineries

Vinod Srinivasan, Biomass Applications Group, DuPont Industrial Biosciences, Palo Alto, CA

Accellerase® TRIO: Advances in Cellulosic Enzymes for Commercial Biorefineries

Vinod Srinivasan

Biomass Applications Group

DuPont Industrial Biosciences

925 Page Mill Rd

Palo Alto, CA 94304

USA

vinnie.srinivasan@danisco.com

Providing nearly 400 enzyme products ranging from food and beverages to household cleaning and biofuels, Genencor has been a world leader in industrial biotechnology and a pioneer in enzyme innovation. Now a part of DuPont, DuPont Industrial Biosciences provides enzyme products for biomass hydrolysis and offers complete solutions along the entire biorefinery value chain. DuPont's latest generation of biomass enzyme complex, Accellerase® TRIO, is a breakthrough in enzymatic hydrolysis that coverts biomass into fermentable C5 and C6 sugars, often up to 2-fold lower dose. The performance results using Accellerase® TRIO in enzymatic saccharification of various biomass substrates with different types of pretreatment will be presented. Additionally, key learnings and feedback from customers in the emerging biomass industry will be discussed, in relation to the commercialization of 2nd generation ethanol technologies.

With substantial investments planned for the emerging biomass industry, DuPont is building its first commercial cellulosic ethanol plant in Nevada, Iowa. This demonstrates a strategy for success by touching every point along the value chain from farm to factory, to the development of a bio-based economy.

(734b) A Kinetic Study of Bioethanol Fermentation by E. Coli FBWHR Using Wood Hemicellulosic Hydrolysate As Substrate

Yang Wang, SUNY College of Environmental Science and Forestry, Syracuse, NY and Shijie Liu, Paper and Bioprocess Engineering, SUNY College of Environmental Science and Forestry, Syracuse, NY

Hemicellulose is currently a by-product currently under-utilized in the papermaking industry and represents the largest polysaccharides fraction wasted in most cellulosic ethanol pilot and demonstration plants around the world. It is a hetero-carbohydrate polymer, with β-xylose as the major component upon depolymerization. At the State University of New York College of Environmental Science and Forestry, wood extracts were obtained by extracting sugar maple wood chips with hot water at 160 °C for 120 min. The wood extracts were then concentrated and acid hydrolyzed. Ethanologenic bacteria, E. Coli FBWHR, a new improved strain which was obtained from repeatedly adaptation of E. coli FBR5, had a good performance in wood hemicellulosic hydrolysate. This paper discusses the kinetic of bioethanol batch fermentation by E. Coli FBWHR using wood hemicellulosic hydrolysate as substrate. Due to the toxicity of hydrolysate in the fermentation media of concentrated hydrolysate, the ethanol production decreases with the increase concentration of wood hemicellulosic hydrolysate.

(734c) Enhancement in Enzymatic Hydrolysis by Mechanical Refining of Pretreated Biomass

Brandon W. Jones, Sunkyu Park, Richard Venditti, Hasan Jameel and Bonwook Koo, Forest Biomaterials, North Carolina State University, Raleigh, NC

Cellulose is the most abundant natural polymer. Therefore, cellulosic bioethanol has been identified as a fuel source that has tremendous potential to supplement the United States consumption of non-renewable transportation fuel sources. Although cellulosic bioethanol is poised for market entry, several processing problems and high processing costs have has been identified as barriers to commercialization. In order to make cellulosic bioethanol economically viable, these barriers (including high enzyme and raw material costs, and the natural resistance of biological degradation of a plant; "biomass recalcitrance") need to be overcome. This study looks at the effectiveness of refining (a mechanical post-treatment) in addition to the traditional pretreatment step to assist in overcoming the biomass recalcitrance barrier. Laboratory scale refining was conducted using a disk refiner, a PFI mill, and a valley beater at different refining severities to determine which refining mechanisms allowed for the most substantial gains in enzymatic hydrolysis conversion. Studies were conducted using green liquor pretreated hardwood, and also various lignin content kraft (NaOH + Na_2S) pretreated hardwood. Initial results show that refining allows for significant improvements in enzymatic hydrolysis yield. Enzymatic sugar recovery of alkaline pretreated hardwood increased from 55.9% to 81.9% with 5 FPU/g after mechanical refining. Enzyme reduction of more than 50% was achieved at the same level of sugar recovery. To investigate the reason for increased digestibility, pore volume and surface area were evaluated in a wet state using three independent analyses; Water retention value, Simons' staining, and DSC thermoporometry methods. A strong correlation was found between sugar recovery and measured properties.

(734d) Kinetics of Microbial Mass and Lipid Production by Rhodotorula Glutinis From Hydrolyzate of Switchgrass

Guochang Zhang[1], Mary Hetrick[2], Todd French[2] and Rafael Hernandez[2], (1)Chemical Engineering, Mississippi State University, Mississippi State, MS, (2)Dave C. Swalm School of Chemical Engineering, Mississippi State University, Mississippi State, MS

Microbial oils derived from lignocellulosic biomass could potentially be a cheap source of biodiesel. Previous experiments showed that *Rhodotorula glutinis* (ATCC 15125) could convert the sugars in acid hydrolysate of switchgrass into microbial lipids for use as a biodiesel feedstock. The objective of this research is to investigate kinetic behaviors of the microbial mass and lipid production by *R. glutinis* using hydrolyzate of switchgrss as sole carbon and energy source. In the experiments, the C to N ratio and C to P ratio as well as the temperature were evaluated on the microbial growth and lipid production of *R. glutinis*. The lipid content in the cells reached 32% of the cell dry mass. Effects of temperature, nitrogen and phosphate concentration on the net specific growth rate were also studied. In the kinetic models, temperature, substrate concentrations of carbon, nitrogen and phosphorus will be considered in the models for both biomass and lipid production. Also in the study of the models, the overall inhibition of the byproducts from hydrolysis was also be investigated in terms of the microbial growth and lipid production rate.

(734e) Immobilized-Cells Hollow Fiber Membrane Reactor for Fermentation of Inhibitors in Lignocellulosic Hydrolysates

Duong TT Nguyen and Kai Chee Loh, Department of Chemical & Biomolecular Engineering, National University of Singapore, Singapore, Singapore

Production of bioethanol from agricultural and forest residues have been attracted considerable efforts as this lignocellulosic biomass is the most abundant renewable material and has the potential to serve as a sustainable feedstock for biofuel. Conversion process of lignocelluloses to bioethanol typically involves three mains steps: (i) pretreatment to break down recalcitrant lignocellulosic structure; (ii) enzymatic hydrolysis of lignocellulose to produce sugars; and (iii) fermentation of sugars to bioethanol. In the enzymatic hydrolysis of lignocellulosic biomass, sugars mainly hexose and pentose are produced that can be fermented into ethanol. However, at the same time a range of toxic compounds such as furans, carboxylic acids, and phenolic compound are also generated. Furfuran and hydroxymethyl furfuran formed as by-products of the hydrolysis process have been known to inhibit growth rate and ethanol productivity. Carboxylic acids have been identified to diminish the cell membrane function. Phenolic compound released from lignin structures could decrease ethanol yield. Various detoxification strategies have been studied to remove the inhibitors including enzyme treatment, neutralization, overliming, use of molecular sieves, vapour extraction, or liquid-liquid extraction.

In this project, a hollow fiber membrane reactor immobilized with *Zymomonas mobilis* ZM4 is proposed to overcome the inhibitory effects from toxic compounds. Actively growing *Z. mobilis* ZM4 immobilized within the porous matrix of asymmetric-wall hollow fiber membrane is protected by the membrane barrier between the toxic and the cells. Toxic compounds are added in varying concentrations and subsequent inhibitions on growth and ethanol production of the system are evaluated. The advantage of this system is that it could remove the pre-fermentation treatment, potentially reducing investment and operating cost.

(737a) Catalytic Performance of Char for Reduction of Tars in Biomass Gasification

Naomi Klinghoffer, Earth and Environmental Engineering, Columbia University, New York, NY, **Marco J. Castaldi**, Department of Earth and Environmental Engineering (HKSM), Columbia University in the City of New York, New York City, NY and Ange Nzihou, LGPSD UMR CNRS 2392, Ecole des Mines d'Albi-Carmaux, ALBI, France

Biomass has the potential for meeting a significant portion of the demand for energy and transportation fuels in the near future. Gasification is a thermochemical conversion method that creates synthesis gas from biomass, which enables the production of a wide variety of products such as fuels, electricity, or chemicals. Two of the main by-products of gasification are tar anc ash or char. Tar can be converted to synthesis gas via catalytic steam reforming, but the tars cause rapid catalyst deactivation. Char has been shown to have catalytic activity for tar decomposition. However, char properties vary based on gasification conditions, which will impact its catalytic performance. In our experiments, poplar wood was gasified in a fluidized bed reactor under 10% CO_2/90% N_2 and 90% H_2O/10% N_2 at temperatures ranging from 550-920°C. Data is presented that demonstrates the char's catalytic performance for decomposition of hydrocarbons such as methane and propane. The surface area of the char ranged from 400-680 m^2g^{-1}, which is significantly higher than typical catalyst carriers. For example, the surface area of gamma alumina is around 150m^2g^{-1}. The catalytic performance of the char was related to the surface area, with higher surface area giving higher catalytic performance. The presence of iron was identified in the char and carbon deposition was observed on the iron, which suggests it may be a catalytic site. However, the performance of ash is lower than that of char and therefore it is likely that the metal-carbon interactions are important to the catalytic performance of the char. This presentation will discuss the catalytic performance of the char as well as the relevant properties of the char and how they impact its performance.

(737b) Coupling Biomass Conversion with CO2 Reduction in Hot Temperature Water

Hongfei Lin, Chemical and Materials Engineering, University of Nevada, Reno, Reno, NV

The global carbon balance is broken by anthropogenic activities as humans consume fossil fuels and release a significant amount of CO_2 within several hundred years, while the fossil fuel formation takes hundreds of millions of years. To mitigate the imbalance of global carbon, it is appealing to develop technologies that can efficiently and rapidly convert CO_2 into fuels and chemicals without a high excess energy input. Herein our approach is to couple biomass conversion with CO_2 reduction over heterogeneous catalysts under hydrothermal environment. A variety of biomass feedstocks and metal and metal oxide catalysts were evaluated for the simultaneous biomass partial oxidation and CO_2 reduction. In presence of CO_2, increase of yields of the carboxylic acids was observed and a high formic acid yield was attributed to CO_2 reduction. The results suggest that CO_2 reduction coupling with biomass partial oxidation is an attractive alternative to CO_2 capture and sequestration processes.

(737c) Alkene Metathesis of Camelina FAME with Supercritical Ethylene

Randy Maglinao[1], Akash Narani[2] and Nestor Soriano Jr.[1], (1)Bio-Energy Center, Montana State University-Northern, Havre, MT, (2)State University of New York at Buffalo and Bio-Energy Center, Montana State University - Northern, Amherst/Havre, NY/MT

Akene metathesis of unsaturated FAME with ethylene is a powerful synthesis reactor to produce terminal alkenes which can be used for synthesis of polymers and other valuable compounds. Although commercial metathesis catalysts have been proven to be efficient in converting pure FAME, this process has not been used effectively in a mixture of methyle esters like biodiesel. Moreover, there has been no study in the literature that has used supercritical ehtylene in alkene metathesis. We anticipate to minimize mass transfer limitations with the use of supercritical ethylene during metathesis with FAME.

In this study FAME derived from camelina oil, which contains as much as 90S% unsaturated components, was allowed to react with supercritical ethylene at room temperature in the presence of Grubb's catalyst. A high pressure syringe pump is used to deliver 750 psi of ethylene and the reaction is performed in a 500 mL high pressure and high temperature reactor fro 4 hours. Different commercial first and second generation Grubbs catalysts were tested. Depending on the nature of the catalyst, self metathesis of FAME, isomerization and cyclization were observed as well at varying degrees as evidenced by GC/MS analyses. Percent conversion of unsaturated FAME as high as 87% was observed.

(737d) Kinetics of Mixed Alcohol Ethanol/Butanol Esterification of Butyric Acid

Arati Santhanakrishnan[1], Dennis Miller[1], Lars Peereboom[1], Carl T. Lira[1] and Abigail Shannon[2], (1)Chemical Engineering and Materials Science, Michigan State University, East Lansing, MI, (2)MichiganState University, East Lansing, MI

The growing need to reduce our dependence on fossil sources for fuels and chemicals has led to the exploration of alternate pathways for the use of greener chemicals and their manufacture from bio based sources. Esters of higher alcohols (alcohols with more than two carbons) are an industrially important class of compounds and economically viable processes for making them need to be designed. Blends of esters are being studied as fuel additives to biofuels for their high energy density and favorable fuel properties(1). One route for producing esters that is currently being extensively studied is the esterification of carboxylic acids with alcohols using acid catalysts(2). The catalyst used may be homogeneous such as sulfuric acid, or para-toluene sulfonic acid (pTSA). Heterogeneous catalysts used include cationic exchange resins.

Mixed alcohols are obtained from the condensation of alcohols to higher alcohols(3), the Fischer-Tropsch synthesis of alcohols from syn gas(4) and fusel alcohols produced from ethanol fermentation. The simultaneous esterification of these alcohol mixtures would be a good use of these mixed alcohol streams and improve the profitability of these processes. Since these reactions are equilibrium limited, one option for driving them forward is reactive distillation where the products are separated from the reaction mixture as and when formed. The design of a reactive distillation process requires a good understanding of the kinetics of the reaction system.

It is not known whether the presence of one alcohol would accelerate or inhibit the rate of reaction of another, or if the formation of these mixed esters over complicates the recovery process. In this study, the kinetic behavior of butyric acid undergoing esterification by ethanol and butanol has been investigated under homogeneous (pTSA) and heterogeneous catalysis (Amberlyst 70). The fermentation of butyric acid from biomass has sparked interest in the production of many chemicals using butyric acid as a building block. Apart from their potential as biofuel components, ethyl butyrate and butyl butyrate serve as food flavoring agents and green solvents(5).

The kinetics were first studied separately, and then in alcohol mixtures of varying relative compositions of ethanol and n-butanol. Reaction equilibrium constants were first determined experimentally from samples taken after reactions had reached equilibrium. UNIFAC (UNIversal Functional Activity Coefficient) was used to calculate the activity coefficient values. Concentration profiles of activity-based fits and experimental data of mixed alcohol esterification runs show that the models fit experimental data reasonably well. Activity and mole fraction based fits were found to be indistinguishable. An additive combination of individual models was found to predict the mixed alcohol esterification system reasonably well. The kinetic model fitted to experimental data is useful for simulating reactive distillation columns to drive the equilibrium limited simultaneous esterification reactions forward.

References

1. I. K. Mbaraka, D. R. Radu, V. S. Y. Lin, B. H. Shanks, Journal of Catalysis 219, 329 (2003).

2. A. Chakrabarti, M. M. Sharma, Reactive Polymers 20, 1 (1993).

3. W. Ueda, T. Ohshida, T. Kuwabara, Y. Morikawa, Catalysis Letters 12, 97 (1992).

4. T. J. Mazanec, Journal of Catalysis 98, 115 (1986).

5. D. W. Armstrong, H. Yamazaki, Trends in Biotechnology 4, 264 (1986).

(737e) In-Situ Catalytic Upgrading of Pinewood Pyrolysis Vapors: Effect of Process Variables

Vamshi Krishna Guda[1], Hossein Toghiani[1], Rebecca K. Toghiani[1] and Charles U. Pittman Jr.[2], (1)Dave C. Swalm School of Chemical Engineering, Mississippi State University, Mississippi State, MS, (2)Department of Chemistry, Mississippi State University, Mississippi State, MS

The main objective of this research effort was to produce high quality (i.e., deoxygenated, less acidic) bio-oil through in-situ catalytic modification (upgrading) of pine wood pyrolysis vapor. In-situ catalytic upgrading of pine wood pyrolysis vapors was performed using an integrated reactor set-up where fast pyrolysis of pine wood chips was carried out in an auger reactor followed by catalytic upgrading in a packed bed reactor mounted on the top of auger reactor. The pyrolysis process was carried out at 450 °C under a nitrogen atmosphere. The effects of process variables such as catalyst composition (γ-Al_2O_3, Si/Al, HY, HZSM-5), temperature and Weight Hourly Space Velocity (WHSV) were studied. The WHSV was varied by changing the biomass feed rate while holding the amount of catalyst constant. Compared to non-catalytic experiments, deoxygenated products increased over all the tested acid catalysts. While Si/Al and γ-Al_2O_3 yielded higher amounts of liquid product, the zeolites (HZSM-5, HY) favored the formation of aromatic hydrocarbons and higher gaseous yield. The GC-MS results, elemental analysis, acid value and Karl-Fisher titration analysis showed that the liquid fuel was deoxygenated to the largest degree. The largest concentrations of gasoline-range hydrocarbons were obtained using the HZSM5 catalyst at low WHSV (6 h^{-1}) and moderate reaction temperatures (370-400 °C). The occurrence of deoxygenation could be noticed through the reduced oxygen content in the form of H_2O (significant increase in water yield), CO, CO_2.

(737f) Catalytic Hydrodeoxygenation of Guaiacol

Danni Gao, Hyun Tae Hwang and Arvind Varma, School of Chemical Engineering, Purdue University, West Lafayette, IN

Pyrolysis of biomass is a promising technology to convert solid biomass into liquid bio-oil. However, bio-oils have high water and oxygen content which subsequently lowers their energy density relative to conventional hydrocarbons. For this reason, an upgrading process is required. Catalytic hydrodeoxygenation (HDO) is a rapidly developing technology for achieving oxygen removal from bio-oils. While various catalysts have been studied for this purpose, due to the presence of more than 400 oxygenated organic compounds, the catalytic activity and reaction pathways in the HDO process are not well understood. In this context, it is important to select model compounds which represent the raw bio-oil. In this study, Guaiacol, a phenol derived compound produced by the thermal degradation of lignin, is selected as a model compound, because it is one of the major components of bio-oil and, in addition, is thermally unstable and leads to coke formation. Conducted in a fixed-bed reactor, this work identifies the optimum noble metal catalyst(s) along with the optimum operating conditions which minimize hydrogen consumption while maximizing oxygen removal. Further, the HDO reaction kinetics are measured and, based on analysis of both the liquid and gas the gas phase products, the reaction pathways are proposed.

(745a) Plug-Flow Reactor Design for Hydrothermal Liquefaction of Biomass

Mitchell James Minarick, Agricultural and Biological Engineering, University of Illinois at Urbana-Champaign, Urbana, IL

Direct liquefaction of biomass is a technology that has shown good potential for producing oil from biomass, especially for wet biomass sources like algae and livestock waste. Previous research at the University of Illinois at Urbana-Champaign proved that up to 70% of swine manure volatile solids could be converted into oil while reducing the manure's chemical oxygen demand by up to 75%. A continuous stirred-tank reactor design was implemented and produced similar conversion rates to the batch process. Algae conversion rates are similarly high for batch processes, and integrated systems that utilize a manure source to provide nutrients for enhanced algae growth could simultaneously clean the waste stream while producing algae for oil production.

However, the high pressures required for the direct liquefaction process preclude the scale-up of continuous stirred tank reactor systems. A plug-flow reactor was designed and tested to explore the feasibility of a scaled-up continuous reactor for the liquefaction of the biomass from this integrated bioenergy system.

(745b) Conceptual Design of a Tubular Coil Reactor for Biodiesel Production

Rogelio Sotelo-Boyás, Alberto Castillo-Bertran, Rubi Morales-Avila and Lilia Arteaga-Espinoza, Instituto Politécnico Nacional (National Polytechnic Institute of Mexico), Mexico City, Mexico

The continuous production of biodiesel is aimed to be performed in a tubular reactor. The transesterification reaction of vegetable oil was carried out with methanol and potassium hydroxide as catalyst. Tubular reactors are commonly used for relatively fast reactions, however the alkaline transesterification of vegetable oils requires at least 30 minutes to reach a high oil conversion, and it is rather performed in tank reactors, either in batch or continuous modes. To reduce the space needed for the reactor, a coil tubular reactor was conceptualized. Simulation tools used in this work are essential for chemical engineers to perform innovative designs for the chemical industry. In this context, a biodiesel reactor was simulated and optimized. The first stage in this work was an experimental design to determine the best reaction conditions at which the vegetable oil could be effectively transformed into methyl esters. Optimization was performed by Taguchi method and using the response surface methodology. The results taken from here were the methanol to oil molar ratio and the amount of catalyst that maximized the yield of methyl esters. In a second stage the kinetic parameters of the transesterification reactions were estimated with Marquardt algorithm, i.e. the activation energy and the pre-exponential factors for each reaction considered. Thirdly, a simulation of the tubular reactor was performed in ASPEN Plus as a plug flow reactor. From here, the size of the reactor was calculated. The transport phenomena were then evaluated by a multiphysics simulation. A prototype of the reactor was finally build. The reactor shows a very good performance for a production of 60 L of biodiesel in one hour. More innovative solutions could be given at industrial scale by following this methodology.

PRODUCTION OF BIODIESEL FUEL FROM MICROALGAE OIL VIA TRANSESTERIFICATION REACTIONS IN SUPERCRITICAL METHANOL

Jiuxu Liu, Ronghong Lin and Lawrence L. Tavlarides, Syracuse University, Syracuse, NY, USA

Introduction

Biodiesel production is mostly produced from refined vegetable oils which are expensive, and which makes the cost of commercial biodiesel prohibitively high without government subsidies. Furthermore, biodiesel plants for feedstock oil must also compete for food, cosmetics, industrial, and livestock feed demands (McNeff et al., 2008). Although countries that use biodiesel are not expected to take full responsibility, international concerns about the upward pressure on global food prices and intensified competition for cropland currently make the use of crop-based biodiesel a politically stressful situation (Xu & Mi, 2011). To decrease competition with food stock and production costs to make biodiesel profitable in comparison with petroleum fuels, it is inevitable for production processes to use inexpensive triglyceride sources, for example microalgae oil (Chisti, 2007) which potentially offers greatest opportunities in the longer term (Schenk et al., 2008). Also biodiesel synthesis under supercritical alcohols conditions provides both energy and economic benefits over the conventional base catalyzed biodiesel production process (Deshpande, Anitescu, Rice, & Tavlarides, 2010).

Experimental

Materials

Microalgae oil was provide by Soley Institute in Istanbul Turkey. N-Methyl-N-(trimethylsilyl) trifluoroacetamide (MSTFA), reference FAME standards (methyl linoleate 20 wt%, methyl linolenate 20 wt%, methyl oleate 20 wt%, methyl palmitate 20 wt%, and methyl stearate 20wt%) for gas chromatography, a kit with gas chromatography (GC) standard solutions and internal standards to analyze total and free glycerol (according to the ASTM method D-6584) were purchased from Sigma Aldrich. *N*-heptane and HPLC-grade methanol were purchased from Thermo Fisher Scientific and used without further purification.

Experimental setup

A schematic diagram of the experimental setup is illustrated in **Figure 1**. The experimental setup consists of two high pressure syringe pumps (Teledyne ISCO) for triolein and methanol delivery, two preheaters (red dot rectangle), a coiled tubular reactor (316 SS, ID1.753mm×6m) placed in a furnace (Briskheat)(red rectangle), a micrometering valve (Autoclave Engineers) for pressure control, a water bath to cool the products, several thermocouples (T1-T5), a pressure transducer, and a data acquisition system (National Instruments) for recording and monitoring temperature and pressure.

Experimental Conditions

In this study, supercritical transesterification of microalgae oil with methanol was investigated at various temperatures (350, 385, and 400 °C), pressures (150, 200, and 300 bar), methanol-to-microalgae oil molar ratios (from 6:1 to 12:1), and residence times (3-12 min) conditions.

Analytical Methods

Components of microalgae oil, free and total glycerol were analyzed by GC-FID (HP 5890 series II) equipped with a Restek Rtx-Biodiesel triglyceride column (10 m × 0.32 mm I.D. × 0.1 μm film thickness). The ASTM standard method D6584 was followed to prepare calibration curves and perform sample analysis.

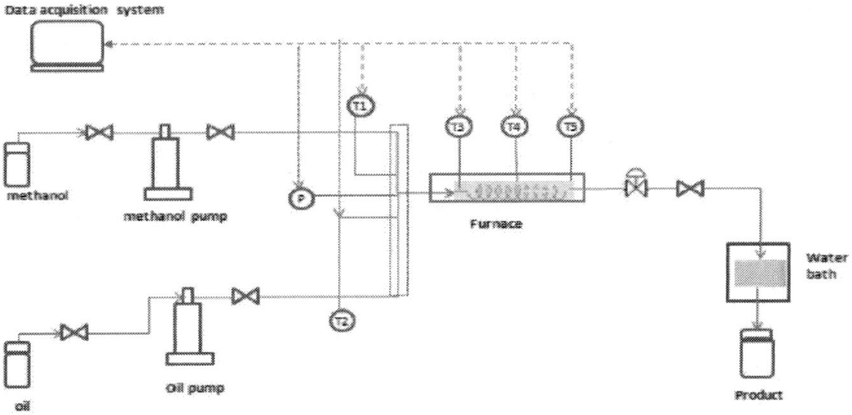

Figure 1. Schematic diagram of experimental setup for supercritical transesterification reactions.

Results and Discussion

Component profiles

Table 1 summarizes experimental reaction conditions and content profiles of different components in biodiesel product samples. Runs 1-9 were performed at 385°C, 200 bar, molar ratio of 9:1, and residence time from 3 to 12 min. Runs 10-18 were performed at 400 °C, 200 bar, molar ratio of 9:1, and residence time from 3 to 12 min. Runs 19-26 were performed at 400°C, 200 bar, molar ratio of 12:1, and residence time from 3 to 12 min. The transesterification conversion was defined as the ratio of mass fraction of the reacted triglyceride to the mass fraction of initial triglyceride (assumed to be 99.9% purity):

$$conversion(\%) = 100\left(1 - \frac{C_{TO}+C_{DO}+C_{MO}}{C_0}\right) \quad (1)$$

Here C_{TO}, C_{DO}, C_{MO} are concentrations of triglyceride, diglyceride and monoglyceride after reaction, respectively, in weight %. C_0 is the initial concentration of triglyceride before the reaction (99.9 wt%).

As expected, reaction conversion increases with temperature and residence time, while glycerol content decreases with temperature and residence time. The reaction conversion was almost complete at 400 °C and residence time longer than five minutes. For 400 °C, 200 bar, molar ratio of 9, residence time longer than 6 min, both free glycerol concentration and total glycerol concentration satisfies the limitation of ASTM standards which is 0.024% and 0.24%, respectively.

Table 1. Content of each compound in final biodiesel product.

Run #	T (°C)	P(bar)	Molar ratio	τ[a] (min)	GL[b] (wt%)	MG[c] (wt%)	DG[d] (wt%)	TG[e] (wt%)	conversion[f] (%)	BG[g] (wt%)	TG[h] (wt%)
1	385	200	9:1	3	0.96	4.26	5.76	3.09	86.87	2.28	3.24
2	385	200	9:1	4	0.79	3.08	1.61	0.23	95.07	1.06	1.85
3	385	200	9:1	5	0.56	2.89	1.16	0.13	95.82	0.93	1.5
4	385	200	9:1	6	0.37	1.8	0.42	0.06	97.72	0.54	0.91
5	385	200	9:1	7	0.25	1.54	0.25	0.03	98.18	0.44	0.69
6	385	200	9:1	8	0.23	1.3	0.19	0.02	98.49	0.37	0.6
7	385	200	9:1	9	0.22	1.23	0.16	0.01	98.59	0.35	0.56
8	385	200	9:1	10	0.22	1.32	0.16	0.01	98.5	0.37	0.58
9	385	200	9:1	12	0.1	0.66	0.03	0.01	99.29	0.18	0.28
10	400	200	9:1	3	0.24	2.06	0.58	0.06	97.3	0.63	0.86
11	400	200	9:1	4	0.12	1.41	0.32	0.04	98.23	0.42	0.54
12	400	200	9:1	5	0.05	0.65	0.05	0.03	99.27	0.18	0.23
13	400	200	9:1	6	0.04	0.59	0.05	0.02	99.34	0.16	0.21
14	400	200	9:1	7	0.02	0.51	0.06	0.02	99.41	0.14	0.17
15	400	200	9:1	8	0.02	0.37	0.04	0.01	99.58	0.1	0.12
16	400	200	9:1	9	0.02	0.31	0.01	0.01	99.67	0.08	0.1
17	400	200	9:1	10	0.02	0.32	0.02	0.01	99.65	0.09	0.1
18	400	200	9:1	12	0.01	0.2	0.02	0.01	99.77	0.06	0.07
19	400	200	12:1	3	0.57	2.48	0.76	0.51	96.25	0.81	1.38
20	400	200	12:1	4	0.3	1.54	0.51	0.36	97.59	0.51	0.82
21	400	200	12:1	5	0.18	1.07	0.12	0.06	98.74	0.3	0.48
22	400	200	12:1	6	0.12	0.79	0.11	0.06	99.04	0.23	0.35
23	400	200	12:1	7	0.11	0.69	0.12	0.04	99.15	0.2	0.31
24	400	200	12:1	9	0.09	0.57	0.08	0.04	99.31	0.16	0.26
25	400	200	12:1	10	0.12	0.52	0.09	0.04	99.35	0.15	0.27
26	400	200	12:1	12	0.06	0.38	0.04	0.02	99.56	0.11	0.16

a- the residence time; b- free glycerol; c- monoglyceride; d- diglyceride; e- triglyceride; f- reaction conversion; g- bound glycerol; and h- total glycerol (GL+BG).

Temperature effect

Temperature is an important factor which affects the biodiesel synthesis process. The reaction conversion and content variation of triglyceride and the consequent reaction products (diglyceride, monoglyceride and glycerol) with temperature at 200 bar, molar ratio of 9, residence time 8 min are shown in **Figure 2** indicating, as expected, higher conversion with temperature.

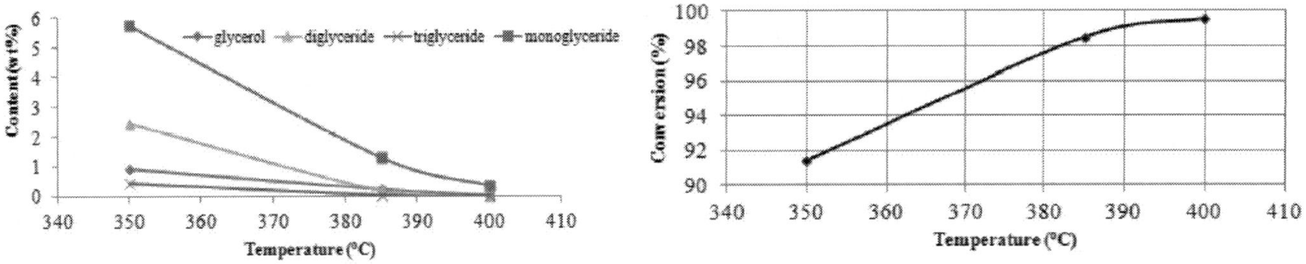

Figure 2. Temperature effect (350, 385, 400 °C) at 200 bar, molar ratio of 9:1, residence time of 8 min on conversion and content of each component.

Figure 3 clearly shows that with increasing temperature, the content of glycerol and unreacted glycerides decreased, and FAME partially decomposed into smaller molecules, which is thought to improve biodiesel fuel quality (Marulanda, Anitescu, & Tavlarides, 2010).

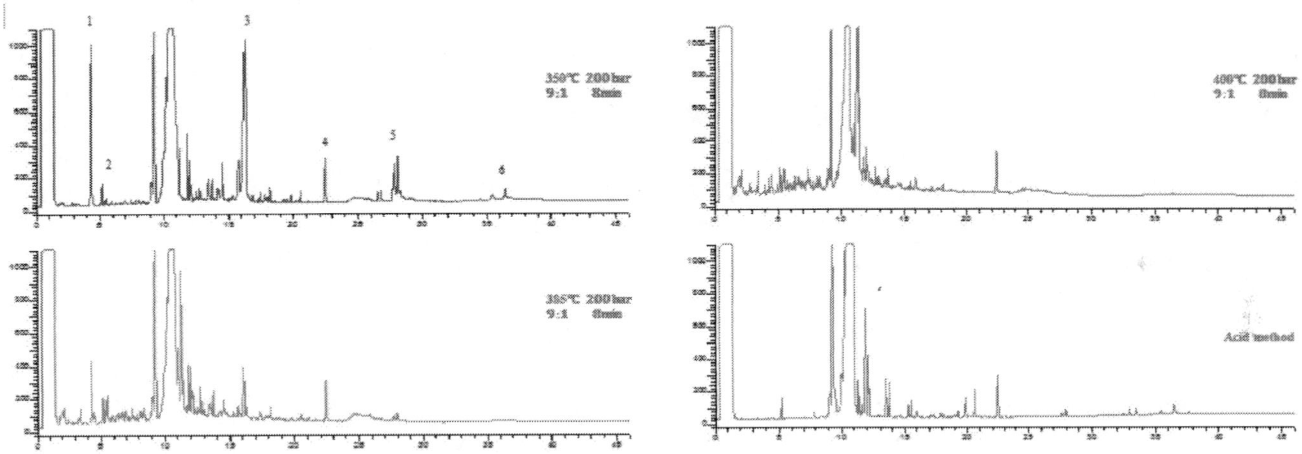

Figure 3. GC-FID chromatography of biodiesel product made at different temperatures. Peak identification : (1) Glycerol, (2) Internal Standard, Buanetriol, (3) Monoglyceride, (4) Internal Standard, Tricaprin, (5) Diglyceride, (6) Triglyceride.

Residence time effect

Figure 4 illustrates how residence time influences conversion and content of components in biodiesel samples made at 400 °C, 200bar, and molar ratio of 9:1. Free glycerol content decreased with residence time, and when residence time is longer than 6 min, free glycerol concentration is lower than 0.024 wt%, which satisfies ASTM standards limit. The reaction conversion increased with residence time, and was near complete (> 99%) after 6 min.

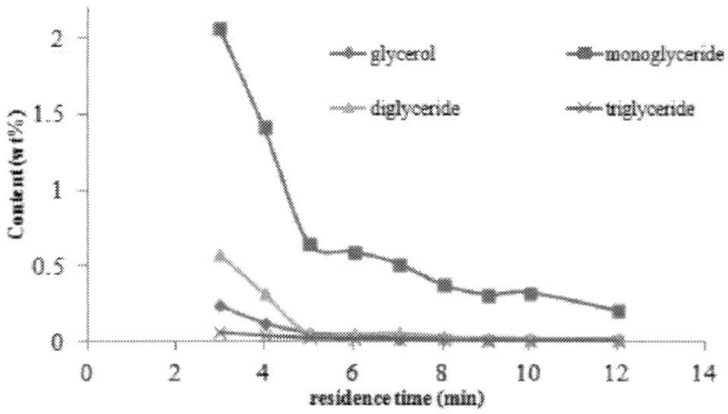

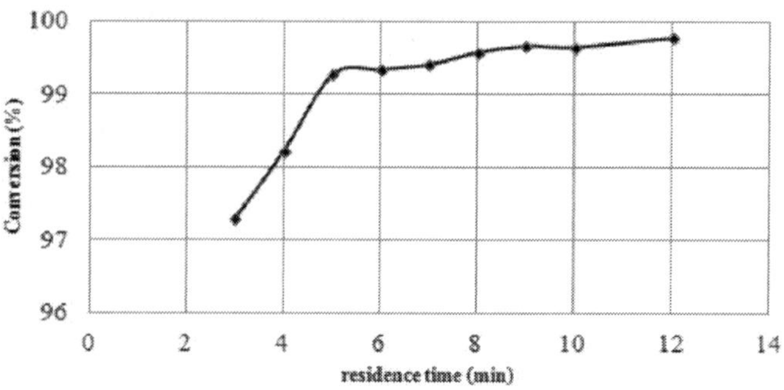

Figure 4. Residence time effect (3-12min) at 400 °C, 200bar, molar ratio of 9:1.

Molar ratio effect

From the observation of product samples, transesterification runs performed at a molar ratio of 9:1 produced an upper biodiesel phase, and a lower second film phase which was considered to be water, glycerol, and some unreacted methanol. With a higher molar ratio of 12:1, the biodiesel phase is present as a lower phase. Without analysis, the upper phase was thought to be an aqueous solution of some of the glycerol decomposition products and unreacted methanol. Sample products made at molar ratio 9:1 have less bound glycerol. There is no obvious difference in the free glycerol content of the sample products with molar ratios of 9:1 and 12:1.

Conclusions

The transesterification of microalgae oil in supercritical methanol was conducted at temperature of 350, 385, and 400 °C, pressure of 150, 200, 300 bar, methanol to oil molar ratio of 6:1, 9:1, and 12:1, and residence times of 3 – 12 min. According to experimental data, a good quality biodiesel was produced at temperature of 400 °C, pressure of 200 bar, a molar ratio of 9:1, and residence time of 7 – 12 min. Unreacted glycerides content decrease with temperature and residence time. Free glycerol decomposed into smaller molecules with temperature and residence time. Free glycerol content satisfies ASTM standards limitation for reactions with residence time of 7-12 min, temperature of 400 °C, and 200 bar. Because of the high temperature, biodiesel molecules will be decomposed into smaller molecules, which are thought to improve viscosity and cold flow properties of the biodiesel product.

References

1. McNeff, C. V., McNeff, L. C., Yan, B., Nowlan, D. T., Rasmussen, M., Gyberg, A. E., et al. (2008). A continuous catalytic system for biodiesel production. *Applied Catalysis A-General, 343*(1-2), 39-48.

2. Xu, R., & Mi, Y. (2011). Simplifying the process of microalgal biodiesel production through in situ transesterification technology. *Journal of the American Oil Chemists Society, 88*(1), 91-99.

3. Chisti, Y. (2007). Biodiesel from microalgae. *Biotechnology Advances, 25*(3), 294-306.

4. Schenk, P. M., Thomas-Hall, S. R., Stephens, E., Marx, U. C., Mussgnug, J. H., Posten, C., et al. (2008). Second generation biofuels: High-efficiency microalgae for biodiesel production. *Bioenergy Research, 1*(1), 20-43.
5. Deshpande, A., Anitescu, G., Rice, P. A., & Tavlarides, L. L. (2010). Supercritical biodiesel production and power cogeneration: Technical and economic feasibilities. *Bioresource Technology, 101*(6), 1834-1843.
6. Marulanda, V. F., Anitescu, G., & Tavlarides, L. L. (2010). Investigations on supercritical transesterification of chicken fat for biodiesel production from low-cost lipid feedstocks. *Journal of Supercritical Fluids, 54*(1), 53-60.

(745d) Simulation and Heat-Integration of Glycerol-Free Biodiesel Plant From Canola Oil with Dimethyl Carbonate

Mohammad Rafiqul Islam, Yogesh Kurle and Tracy J. Benson, Dan F. Smith Department of Chemical Engineering, Lamar University, Beaumont, TX

Biodiesel, a mixture of fatty acid methyl esters (FAMEs), has grown in the marketplace over the last couple of decades. However, the production process produces 1 lb of glycerol byproduct for every 10 lb of biofuel produced. Hence, the glycerol market has become saturated, with little market value for this byproduct. Replacement of methanol with dimethyl carbonate (DMC) eliminates glycerol production, along with process wastewater during manufacturing. The DMC reaction produces 2 FAMEs and 1 fatty acid glycerol carbonate (FAGC) for every triglyceride molecule converted. In early attempts, sodium methoxide, the usual catalyst for base-catalyzed biodiesel production, was found to have poor conversions. However, triazabicyclodecene (TBD) was found to easily ionize the dimethyl carbonate and ease of separation post reaction.

This paper reports on the development of specific processing steps to convert canola oil to FAMEs and FAGCs, meeting ASTM D6751 standard. A process flow diagram was developed using Aspen Plus simulation software and verified through laboratory testing of reactions and downstream unit operations. Heat integration techniques have been applied via pinch analysis to identify the optimal heat load to be removed and added from the hot and cold streams, respectively. It was found that about 31.6% energy savings was possible through heat integration techniques. A comparison with similar process plants indicates that this process is simple, environmently friendly, and energy efficient. Complete reaction details and Aspen modeling results will be reported, illustrating a commercially viable biofuel process.

(745e) Znla Catalyst Durability and Regeneration Method in the Process of Converting DDGS Oil and Yellow Grease Into Biodiesel

Shuli Yan[1], Like Yang[2], Huali Wang[3], Manhoe Kim[4], Steven O. Salley[3] and Simon K. Y. Ng[3], (1)NextCAT inc., Detroit, MI, (2)NextCat INC, Detroit, MI, (3)Chemical Engineering and Materials Science, Wayne State University, Detroit, MI, (4)NextCat Inc, Detroit, MI

This research focuses on using heterogeneous catalyst for biodiesel production. Currently biodiesel producer suffer from high oil feedstock cost. Development of catalytic technologies using some crude and inexpensive oil as feedstocks becomes attractive to all biodiesel producers. However, these cheap oil feedstocks generally contains a high content of impurities, such as free fatty acids, moisture, phosphor lipid, steroidal, some lipid soluble protein and polysaccharide. These impurities not only poison the catalysts but also complicate the whole biodiesel production process. Therefore, investigate the catalyst tolerance to poisons, the regenerate method for the foiled catalysts and the structure of the regenerated catalyst become important. In last decades, there are multiple studies on the effect of FFA and water to homogeneous biodiesel catalysts. However, few focus on heterogeneous catalysts. In our previous studies, a series of Zn based catalysts were developed with high activity and long life in refined soybean oil system. In this study, crude extracted corn oil from DDGS and yellow grease from several restaurants were used as raw materials. The compositions of oil feedstocks were tested by HPLC, GC-MS, ICP, UV-VIS etc. Surface coverage of catalyst active site by long carbon chain species are considered as the reason for catalyst deactivation. A deactivation mechanism is proposed here. The foiled catalyst was regenerated by simple calcination method and hydrothermal calcination method. The regenerated catalysts were characterized by XRD, XPS, EDS, HRTEM etc. and evaluated in batch and continuous reactors. It was found that the regenerated catalyst particles have the same structure and activity as fresh catalyst.

(745f) Theoretical Yields and Properties of Hydrotreated Renewable Diesel From Common Lipid Feedstocks

Andrew M. Duncan, Chemical and Petroleum Engineering, University of Kansas, Lawrence, KS, David A. Slade, Renewable Energy Group, Ames, IA and Susan M. Stagg-Williams, Chemical & Petroleum Engineering, University of Kansas, Lawrence, KS

Rising petroleum prices and growing concerns about greenhouse gas emissions have led to increased biofuel usage globally. In the United States, the Environmental Protection Agency's Renewable Fuel Standard specifies annual target quantities of various bio-derived fuels for inclusion in the U.S. fuel supply, including two types of biomass-based diesel: biodiesel and hydrotreated renewable diesel (HRD). Although biodiesel has been studied extensively and has been in commercial production and use for a relatively long time, HRD fuels have only recently been produced in the United States in commercial quantities and little is known about the physical properties, feedstock requirements, and theoretical yields of HRD.

This work presents a model that uses pure triglycerides and hydrogen as its feedstocks for HRD production from three common lipid feedstocks. The HRD model is derived from mass balances and produces theoretical feedstock requirements, fundamental production constraints, and yield estimates for HRD and its co-products. The corresponding theoretical limitations for biodiesel are also included.

Key fuel properties, including density, viscosity, flash point, and cloud point, are predicted using literature methods for the theoretical HRD fuels determined by the model. The analysis also includes estimates of the net energy content and renewable energy content of each of the fuels considered.

Energy Conservation in Higher Education Facilities

Robert W. Peters, Ph.D., P.E.
Department of Civil, Construction, and Environmental Engineering
University of Alabama at Birmingham
1075 13th Street South
Birmingham, Alabama 35294-4440

Atul Kajale and Zhuo Li
Department of Civil, Construction, and Environmental Engineering
University of Alabama at Birmingham
1075 13th Street South
Birmingham, Alabama 35294-4440

Matthew Winslett
Engineering Manager, Energy Management
Facilities Management Department
University of Alabama at Birmingham
1530 3rd Avenue South
Birmingham, Alabama 35294-4564

Abstract

Sustainability has become an important aspect in planning and design of many new construction projects. Initially, commercial and industrial buildings became the main focus of the sustainability and green house reduction programs as these buildings had great potential for improvements. Leadership in Energy and Environment Design (LEED) certification helps in incorporating energy and water conservation techniques in the new and already existing structures. Higher education buildings such as universities and research institutes that are built prior to 10 years ago have tremendous potential for energy and water conservation. Higher education facilities operate buildings which can be categorized as administration buildings, education buildings (classrooms), research buildings (laboratories both medical/clinical and engineering laboratories), libraries, dormitories, and cafeterias. Resource consumption and occupancy of these buildings vary from each other as each building type serves different purposes.

Introduction

The purpose of this paper is to highlight the importance of sustainability in higher education institution buildings and also demonstrate green initiatives of the University of Alabama at Birmingham (UAB). UAB is one of the leading universities in the medical research field throughout the United States and it operates more than 155 buildings in its Birmingham, Alabama campus. The energy Management Division (EMD) of the Facilities Management Department at UAB is very proactive and is constantly incorporating new technologies in operating the campus buildings efficiently to create more sustainable environment for education, and research. Facilities Management and the various departments spent billions of dollars on utilities every year to ensure the smooth operation of the campus buildings. EMD realized that in order to minimize the money spent on the utilities, it is necessary to reduce the utility demand of the buildings without compromising the normal operations of the buildings. EMD started to analyze the performance of the campus buildings using historical utility bills and utility consumption data for every season.

Many universities are currently employing the conservation practices in operating their campus buildings. However, the green initiatives of the University of Wyoming are discussed briefly as background studies. The University of Wyoming (UW) has performed similar projects in its campus buildings to reduce greenhouse gas emissions in the energy, water, transportation, purchasing and procurement, recycling, and waste management, and behavioral change fields [Sharma and Peters, 2011]. In a three phase energy efficiency program, the UW has completed first phase with installation of condensate, water, and electric meters for core campus buildings. It involves replacement of T-12 bulbs with T-8 bulbs, lighting control upgrades, installation of vending misers, and steam trap upgrade [Peters *et al.*, 2011]. All new buildings will be constructed with the LEED standards which will further reduce the UW's greenhouse gas emissions. Two buildings are already gold and silver certified. Other projects include campus wide tree plantation, use of environmentally friendly fertilizers and pesticides, community awareness, upgrading plumbing fixtures, purchase of new electronic fleet, campus shuttle system, and installation of solar shed [Peters *et al.*, 2011].

Significance of the study: UAB expanded dramatically since the Fiscal Year 2001–2002, which can be explained using the increase in the gross square footage of the campus buildings. Figure 1 shows the total gross square footage of the campus buildings from fiscal year 2001–2002 through 2010–2011. There has been a 36.6% increase in the gross square footage from the fiscal year 2001–2002 through 2010–2011.

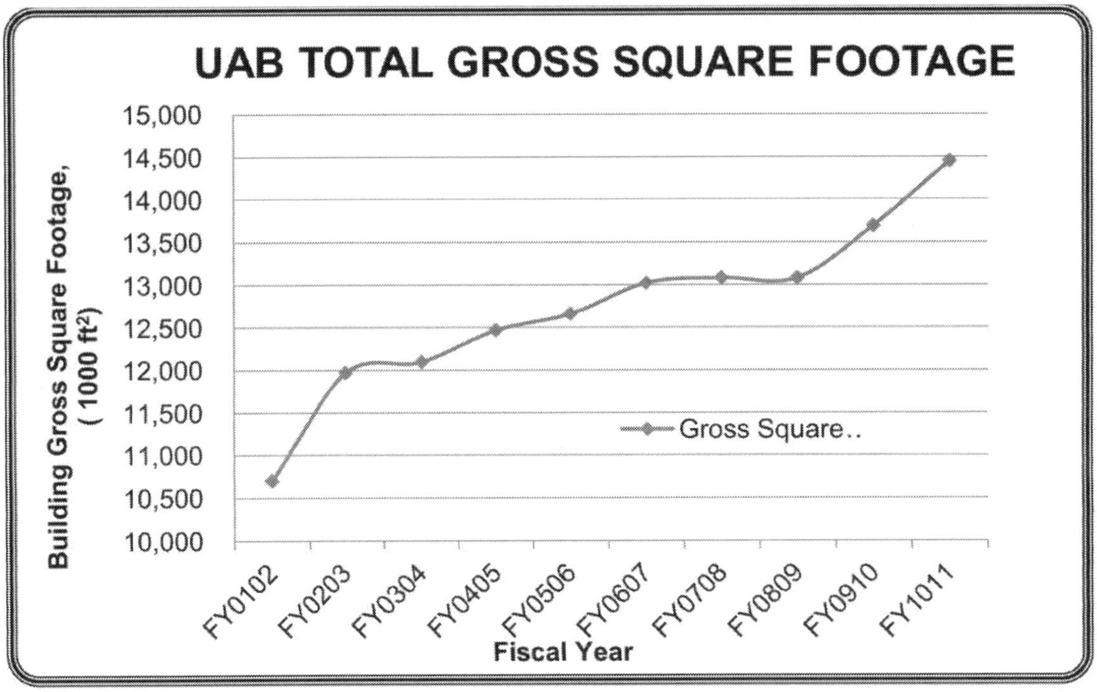

Figure 1. UAB total gross square footage of campus buildings.

As a result, the energy consumption and utility cost have also significantly increased. These utilities include electricity, water and sewer, natural gas, chilled water. Detailed consumption and utility costs of each utility are shown in Figures 2 to 6. In addition, the combined total utilities cost including each component is also shown in Figure 7. It can be seen that with the increase in the gross square footage of the campus buildings, the utility consumption and the utility costs have also increased from fiscal year 2001–2002 to fiscal year 2010–2011, except for steam. Steam consumption decreased because of a change in boiler operation at the hospital, while natural gas consumption increased accordingly. In addition, these trend lines generally are in accord with the total gross square footage as well. The consumption of steam has a different trend with other utilities that declined in the FY 2007–2008 and kept decreasing since then. If the utility cost and consumption value of FY 2001–2002 is assumed as the baseline utility cost and consumption for the study period, an average 80% increase in the utility cost and consumption has been observed at the end of fiscal year 2010–2011. Among all the utilities, consumption and cost of natural gas is the largest and that of steam is lowest. Electricity costs decline after the fiscal year 2009–2010, since EMD implemented energy conservation projects during this period. It is interesting to note that the difference between the electricity cost during fiscal year 2008–2009 and 2009–2010 equals the cost of electricity saved during this period. Thus, if the consumption of all the utilities is reduced using energy efficient coupled with control equipment, the utility cost will reduce significantly. Therefore, incorporation and implementation of energy and water conservation projects is very important in higher educational institutional buildings.

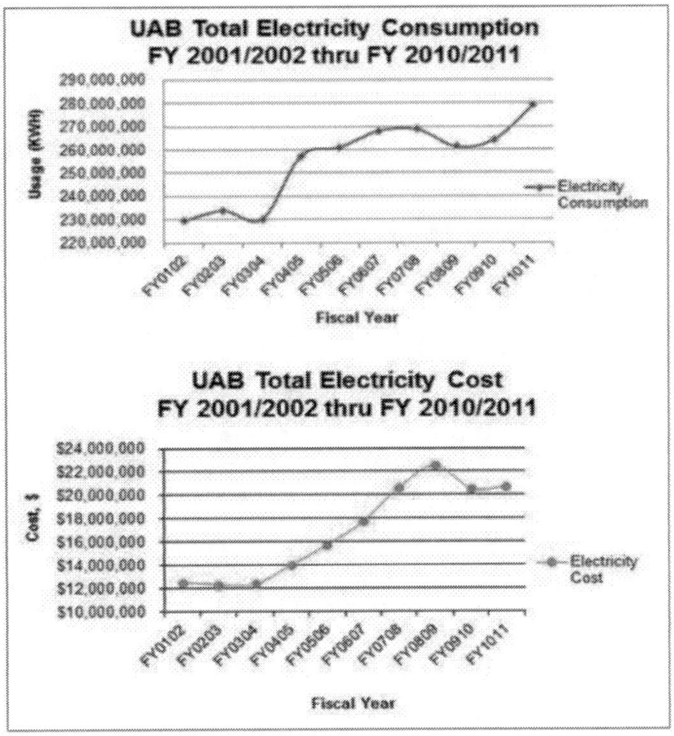

Figure 2. UAB total electricity consumption and cost through FY 2001-2002 and FY 2010-2011.

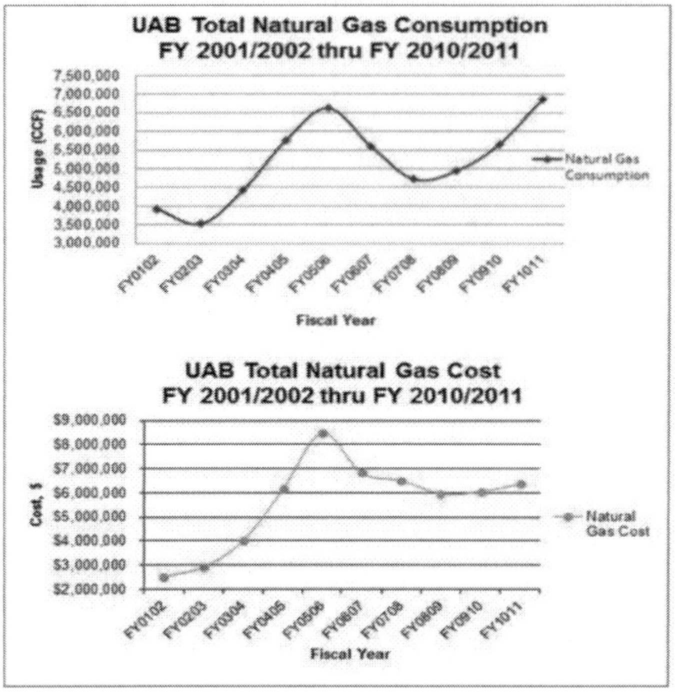

Figure 3. UAB total natural gas consumption and cost through FY 2001/2002 and FY 2010/2011.

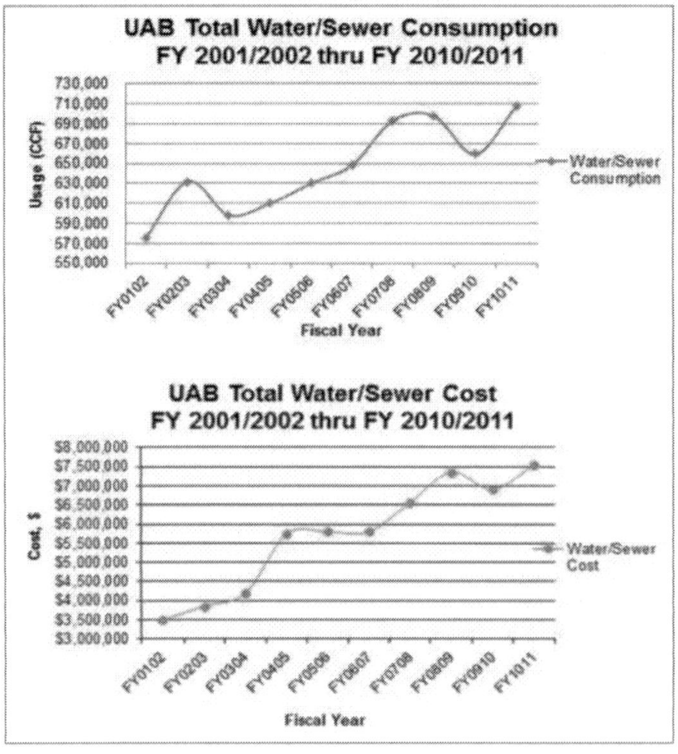

Figure 4. UAB total water/sewer consumption and cost through FY 2001–2002 and FY 2010–2011.

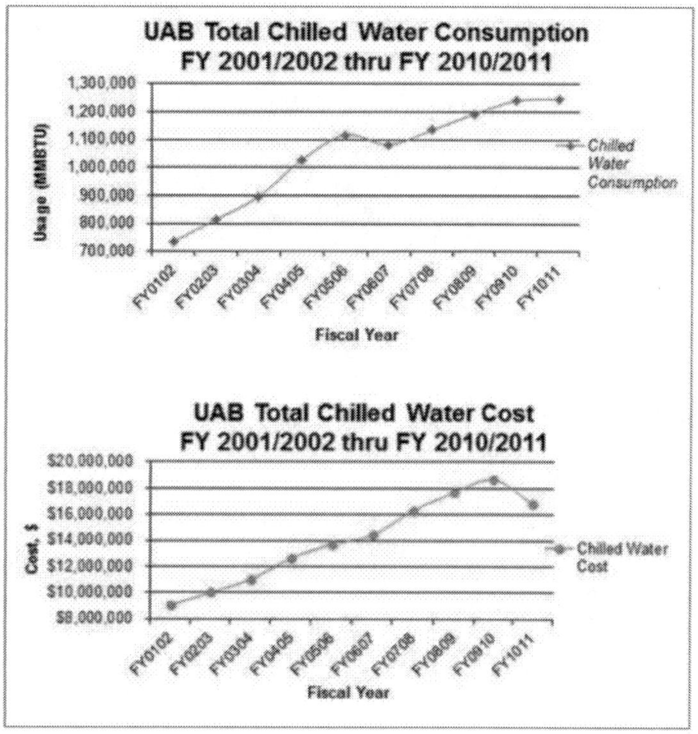

Figure 5. UAB total chilled water consumption and cost through FY 2001/2002 and FY 2010/2011.

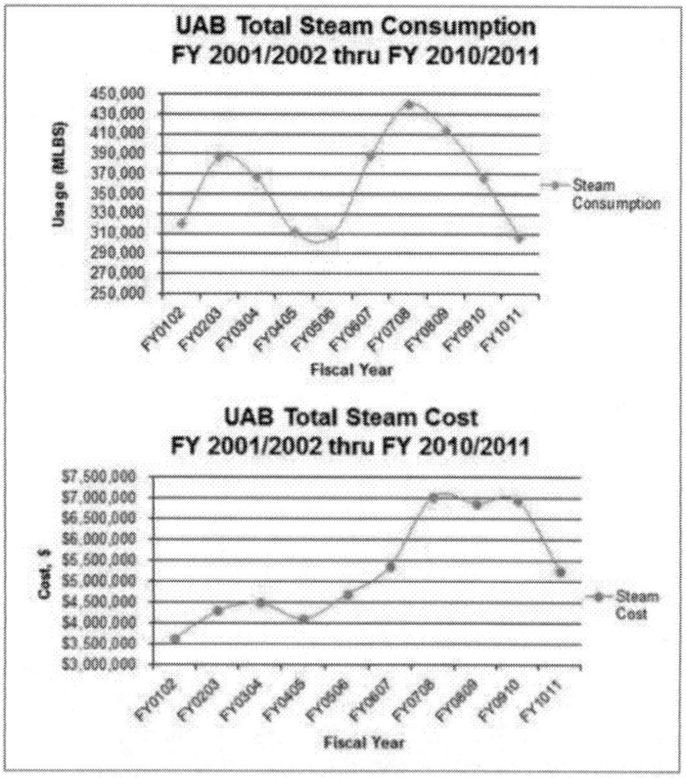

Figure 6. UAB total steam consumption and cost through FY 2001/2002 and FY 2010/2011.

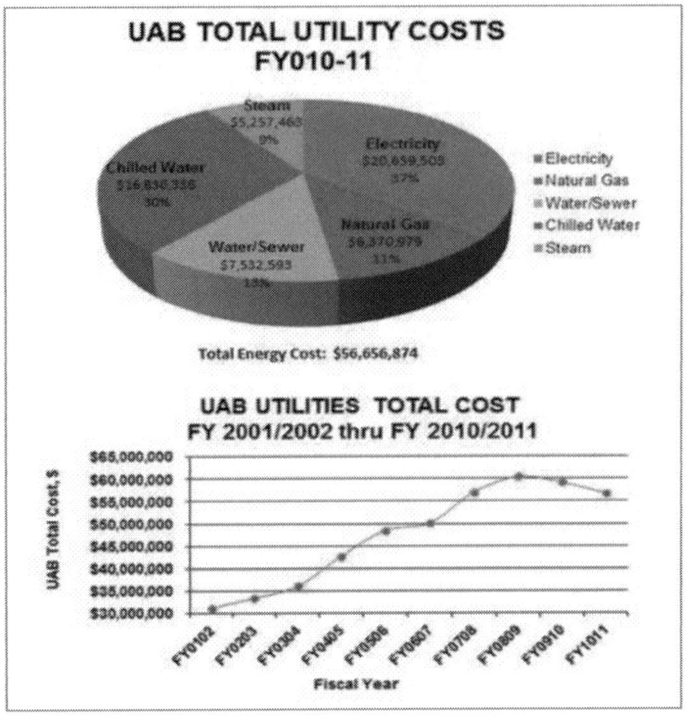

Figure 7. UAB total utilities cost including each utility.

Based on the preliminary analysis of the utility consumption, new conservation approaches have been researched and energy and water conservation strategies were evaluated which are summarized below and are explained briefly in the next paragraph.

- Programmed equipment setbacks during unoccupied times;
- Installation of efficient lighting upgrades with occupancy sensed lighting controls;
- Installation of variable frequency drives (VFD's precisely control electric motor speeds based on needed load);
- Installation of more accurate and efficient control valves for hot and chilled water distribution;
- Installation of more efficient vacuum and compressed air pumping systems that drastically reduce water usage;
- Recovery of condensate water (from conditioned air) and ground water for use in evaporative cooling towers;
- Grid-free solar-charged electric cars [Winslett, 2011]; and
- Pilot vegetated roof on Hulsey Center.

1. **Programmed Equipment Setback during Unoccupied Time:** Air conditioning equipment setbacks during unoccupied times and reduction of air exchanges during unoccupied times have achieved a great payback after being carried out. Since the year of 2007, numerous buildings at UAB were surveyed by graduate students from the Civil, Construction, and Environmental Engineering Department at UAB at intervals around the clock to determine occupancy frequency habits for both students and employees. After gathering the occupancy data, the graduate students analyzed and determined the buildings' least-occupied times are from 10 p.m. to 6 a.m., and holidays and weekends. Every surveyed building was setup with a programmed automation system to shutoff air handling motors and decentralized chillers to conserve unnecessary energy consumption. The installed sensors automatically detect the motion, and turn off lightings if the room or corridor is not occupied for 30 minutes. In addition, a timer is also used to control lighting fixtures. There is individual system to

insure safe operation when internal temperature exceeds 85°F in the summer or below 55°F in the winter. Currently, there are 43 buildings at UAB employed energy during the unoccupied time leading to a cost saving over $1,500,000 per year.

2. **Installation of efficient lighting upgrades with Sensor Lighting Control:** LED exit signs (Light-Emitting Diode) use 1/10 the energy for the same lumen output, utilize 2 to 5 watts of power, and have an expected service life of 20 to 25 years, which greatly decreases required maintenance. In the last two years, over 60,000 fluorescent lamps has been upgraded from T-12* lamps to a more energy efficient T-8* lamps saving over 420,000 kWh's of electricity and $368,000 annually. With energy savings of $20/fixture, the labor/time savings for maintenance can be greater. The UAB Sterne Library pedestrian ramp LED Lighting has been upgraded by installing 13, 46/50 watt 4,000/4,500 lumen LED lights to replace the 250 2' x 2' dual tube fluorescents overhead. The UAB Campus Recreation Center (REC) Pool Lighting has been upgraded for replacing 36 - 1000 watt metal-halide lamps with 200 watt 'compact' fluorescents reducing the wattage by 80% resulting in savings of almost $20,000 per year in electricity costs. At $65 per lamp, the lighting upgrade of the REC center will result in less than a 2 month payback period.

3. **Installation of Variable Frequency Drives (VDF) to Precisely Control Electric Motors:** Installation of VDFs aims to achieve a precise control for electrical motor speeds based on the needed load. VFDs are used to vary the speed of an electric motor by changing the frequency of the electric power going to the motor. A typical application includes air handling units (AHUs) and water pumping systems. UAB has replaced more than 40 motors with variable frequency drives in the last four years. The drive provides a three phase voltage with output varying between 3 and 66 Hz. The voltage/hertz pattern is selectable to that best suited to the motor. The drives automatically adjust this relationship during operation for the maximum operating efficiency according to speed and load [UAB DFP, 1998]. Nowadays, VDFs are very economical and save energy by allowing the motor horsepower to be used only as needed.

4. **Installation of Accurate and Efficient Valves for Chilled and Hot Water Distribution:** UAB has conducted another project to utilize accurate and efficient valving for hot and chilled water distribution. In the chilled water distribution system, butterfly valves used above manhole or other limited access space were high-performance type (Section 231400, UAB Department of Facility Management). UAB has upgraded to ball valves from butterfly valves to provide water. Butterfly and globe valves can precisely control water flow. Depending on different weather conditions, UAB Central Utilities Plants (CUP) varies chilled water temperature set point. The chilled water supply temperatures at the CUP generally are 43°F in summer and 45°F in winter (Section 231400, UAB Department of Facility Management). The Direct Digital Controls have been installed to automatically adjust the heating hot water temperature based on outside air temperature. The heating hot water temperature is adjusted to a lower temperature during the warmer months thereby reducing heat loss from the piping and lowering the cooling load that would be incurred to that heat loss.

5. **Installation of more efficient vacuum and compressed air pumping systems that drastically reduce water usage:** Domestic water conservation and recovery projects have also been very beneficial in reducing costs at UAB. Compressed air and vacuum pumping systems are found throughout the UAB campus laboratories and hospital settings. These traditionally inefficient pumps require water for compression as well as a cooling source. With escalating water and sewer costs, this method of cooling was deemed unfeasible. These projects replace liquid-ring pumps with pumps that do not require water for cooling including claw-type vacuum systems and scroll-type compressed air systems. In most applications less electricity is required to do the same amount of work due to increased pump efficiencies. Depending on building type which determines electrical rates ($0.08–$0.10kW-hr), for every horsepower that is saved by a more efficient pump, electrical costs are reduced approximately $400 - 500 per year for continuous duty operations. A constant single-pass through water flow pump for cooling is especially expensive if sewer costs are also required. The

local cost for each constant year-round one gallon per minute cooling water source is approximately $7,500 per year including sewer charges. Therefore, system paybacks are very short – usually within one to two years. Table 1 shows the annual savings of water and energy due to this upgrade.

Table 1. Savings associated with water and energy using efficient vacuum, and compressed air pumping.

Building Location	Pump System Type	Required Cooling Water (gpm)	Annual Savings		Horse Power Reduction	Annual Electrical Savings ($)
			Volume (gallons)	Water-Sewer, ($)		
Tinsley-Harrison Research	Vacuum	4.5	2,366,820	33,750	0	0
School of Dentistry	Medical Air	4.5	2,366,820	33,750	50	20,000
Bevill Biomedical Sciences	Lab Air	3.5	1,840,860	26,250	5	2,000
McCallum Health Science	Vacuum	2.5	1,314,900	18,750	0	0
Spain-Wallace Hospital	Medical Air	3.5	1,840,860	26,250	15	6,000
Kaul Human Genetics	Vacuum	2.5	1,314,900	18,750	15	6,000
Biomedical Research II	Vacuum	2.5	1,314,900	18,750	15	6,000
TOTAL		23.5	12,360,060	176,250	100	40,000

6. **Recovery of Fin water condensate (from conditioned air) and ground water for use in evaporative cooling towers:** EMD has successfully installed pilot condensate recovery systems in five buildings on the UAB campus: McCallum Building (MCM), Kaul Building (KB), Zeigler Building (ZB), Tinsley Harrison Tower (THT), and Bevill Building (BB). Condensate is formed on the outside surfaces of the pipes and other surfaces in the close proximity of the cooling coils (heat exchanger) in the Air Handling Units (AHU). The AHUs at the medical campus buildings use 100% outside air with no recirculation. Condensate from each AHU is then taken to small tanks from which it is pumped to a 500-gallon storage tank. The recovered condensate water is supplied to the chilled water plant to reduce the make-up water requirement. The temperature of the condensate water is nearly same as that of chilled water and hence less energy is required to bring this condensate water to the desired temperature. Table 2 summarizes building information, the condensate collection system, and initial cost of installation. The installation of condensate collection system (CCS) in the MCM, KB, ZB, and THT was completed before the cooling season of 2011. Therefore, there was an opportunity to collect data during this period except for the BB where system installation work was finished in August of 2011.

Table 2. Summary of the condensate recovery system at UAB.

Building Name	System	Installation Cost, ($)	Remarks
McCallum Building (MCM)	2 AHUs drains into 2 storage tanks	25,000	McCallum and Kaul share one 500-gallon tank; started in May 2011
Kaul Building (KB)	5 AHUs drains into one 500-gallons tank	16,220	
Zeigler Building (ZB)	3 AHUs drains into two storage tanks	5,500	Zeigler and Tinsley Harrison Tower share a one 500-gallon tank; started in May 2011
Tinsley Harrison Tower (THT)	10 AHUs drain into a 500-gallon tank	2,475	
Bevill Building (BB)	6 AHUs drains into two 500-gallon tanks	29,965	Started in August 2011
Total		79,160	

[Kajale *et al.*, 2012]

During the pilot study of the project, monthly totals of the condensate water were collected from the water meter installed across each storage tank. The monthly totals of the water recovered in each building are summarized in Table 3.

Table 3. Condensate recovery at the pilot CRS plant during the cooling season of 2011.

Month	Buildings					
	MCM & KB		ZB & THT		BB	
	Vol. of water, (gallons)	Savings, (dollars)	Vol. of water, (gallons)	Savings, (dollars)	Vol. of water, (gallons)	Savings, (dollars)
May	166,200	$ 688.70	183,700	$ 773.50		
June	430,300	$ 1,811.85	285,100	$ 1,200.46		
July	615,604	$ 2,592.10	410,497	$ 1,728.47		
August	456,040	$ 1,920.23	311,379	$ 1,311.12	125,434	$ 528.16
September	221,457	$ 932.50	122,730	$ 516.77	129,175	$ 543.91
Total	1,889,601	$ 7,945.38	1,313,406	$ 5,530.32	254,609	$ 1,072.07

[Kajale *et al.*, 2012]

Table 4 shows the amount of condensate recovered during 2012. The first three columns 2,3, and 4 represents the condensate data from one air handling unit whereas the column 5 and 6 represent condensate data for the group of air handling units. It can be seen that more than 1.5 million gallons of condensate water has been recovered from June 14th, 2012 through August 30th, 2012 which is equivalent to a saving of more than $5,900.

Table 4. Condensate recovery at CRS plant during cooling season 2012.

Month	Kaul #4	MCM #1	MCM #2	MCM Total		THT Total	
	Total, (gallons)	Total, (gallons)	Total, (gallons)	Total. (gallons)	Savings, ($)	Total, (gallons)	Savings ($)
June	17,776	27,540	33,422	101,888	432	93,703	397
July	53,309	82,795	85,982	343,226	1,455	232,297	984
August	46,817	78,276	88,270	385,282	1,633	235,157	997
Total	117,902	188,611	207,674	830,396	3,520	561,157	2,378

[Kajale *et al.*, 2012]

7. **UAB's Solar Powered Vehicles:** The University of Alabama at Birmingham Facilities Management Department wanted to incorporate a solar collector project as a part of its sustainability efforts but the local cost of electricity, combined with the initial cost of solar panels and ancillary equipment, made project payback times undesirable when only considering energy costs. One way achieve an acceptable return on investment was to utilize solar electricity to recharge batteries in electrical vehicles. Most electric vehicles only transfer their carbon footprint from petroleum fuel to the fossil fuel required to generate the electricity required during a battery recharge. By using solar derived electricity, the carbon is totally eliminated, yielding a mode of grid-free transportation around campus. The solar charging system was installed in November 2008. It consists of 12 Kyocera KDKD205GX-LPPhotovotaic Modules where each module can produce 205 watts totaling 2,460 watts. The panels are cover 192,075 square feet of roof area. WattPlot™ software is used for performance monitoring of renewable energy components from Outback Power Systems. Cars are driven during the day while the solar panels charge the 8-battery bank located in the Facilities Administration Building (FAB) Parking Garage. Car batteries are recharged during the night from the solar charged battery bank.

8. **Pilot vegetated roof on Hulsey Center:** A pilot green roof system has been installed on the top of the Hulsey Center in July 2008. The roof occupies approximately 14,940 ft^2 (i.e., occupying ~81.2% of the roofing area, or approximately 0.34 acres. Prior to installing the pilot green roof, the civil engineering senior design class investigated the loading of a wet vegetative roof on top of Hulsey Center, and found that the current building infrastructure could withstand the loading associated with the vegetative roof. Hulsey Center was originally designed to have two additional floors in the building. This pilot green roof contains more than 20,000 sedum plants. In the construction of this vegetated roof, the existing roof was removed down to the structural concrete deck. A waterproof roofing membrane was installed directly to the concrete deck. A layer of 3-inches of extruded polystyrene roofing insulation was then applied, to which an 8-oz non-woven geo-textile scrim sheet was applied. Figure 8 shows the green roof during its construction phase, and Figure 9 shows the pilot vegetated roof system on top of Hulsey Center, while Figures 10 and 11 show close-up photos of the green roof.

Figure 8. Photograph of roof system on top of Hulsey Center on the UAB campus prior to installation of the green roof.

Figure 9. Close-up of general layout of pilot vegetated roof system on top of Hulsey Center on the UAB campus.

Figure 10. Pilot vegetated roof system on top of Hulsey Center.

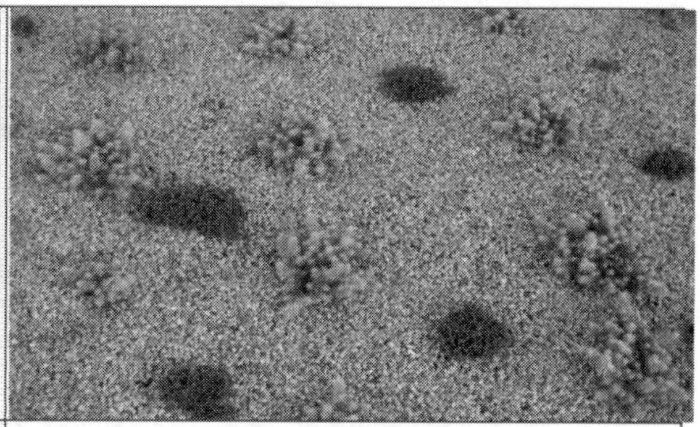

Figure 11. Close-up of pilot vegetated roof system on top of Hulsey Center.

Then, after application of the roofing membrane, 2-ft x 2-ft x 2-in prestressed pavers or brick pavers were installed for design, decoration, and access to the vegetative roof. Then, 3½-inches of light weight engineered soil were applied. Sedum plants were planted at the rate of 150 plants per 100-ft^2. The cost of retrofitting the roof and installing a pilot green roof on top of Hulsey Center was ~\$150,000. The usage of natural gas, water, and electricity are significantly lower than that compared to the utility usage prior to installation of the green roof on top of Hulsey Center. Implementation of the vegetated roof system has reduced the utility bills of Hulsey Center by 20–25% compared with the utility bills prior to implementation of the vegetated roof system [Peters *et al.*, 2012; Kirby *et al.*, 2012; Nunez *et al.*, 2009].

Discussion

The University of Alabama at Birmingham has been actively incorporating the new sustainability and conservation techniques in the campus buildings to improve the building performance and reduce its carbon footprints. Since 2008, several projects were successfully implemented to reduce the utility consumption even when new buildings were added to campus. Starting with the energy conservation, EMD has broadened their conservation scope and implemented projects related to water conservation and reduction in utility consumption. New lighting bulbs and fixtures along with the occupancy sensors have reduced the electricity consumption significantly after year fiscal year 2008-2009. Air handler condensate recovery and recycling is another example of reduction in water consumption which also helps in saving energy cost and sewer bills. All air conditioning equipment and pumps has been upgraded and repaired which will further help in lowering the utility consumption in campus buildings.

As a result, higher education buildings have great potential to reduce their energy and water demand as these buildings are occupied for certain hours. Also, given the nature of university campuses these savings are multiplied by the number of buildings.

This paper has shown that the University of Alabama at Birmingham (UAB) is very enthusiastic in implementing new ideas to lower the carbon footprints and utility consumption to create environment friendly campus for education and research.

Acknowledgement: The authors would like to acknowledge the Facilities Management Department of UAB and the Energy Management Division for providing the technical information in preparing this paper.

References Cited

Kajale, A., R.W. Peters, and M. Winslett, 2012. "Estimation of Air Handling Condensate Recovery in the Medical Campus Buildings at University of Alabama at Birmingham", ***Proc. 2012 Annual AIChE Meeting***, Pittsburgh, PA, (October 28–November 2).

Kirby, J.T., R.D. Sherrod, M. Winslett, and R.W. Peters, 2012. "Variations in Temperature in Mini-Roof Structures Employing Different Roofing Materials", ***Proc. 2012 Annual AIChE Meeting***, Pittsburgh, PA, (October 28–November 2).

Nunez, S., R.W. Peters, and M. Winslett, 2009. "Effect of Green Roofs on Building Energy Costs", ***Proc. 2009 Annual AIChE Meeting***, Nashville, TN, (November 9–14).

Peters, R.W., R.D. Sherrod, and M. Winslett, 2012. "Thermal Performance and Energy Conservation Associated with Implementation of Vegetated Roofs in the Soutgheastern United States", Book chapter in ***New Developments in Renewable Energy***, InTech Publishers, Rejeka, Croatia, [*in press*].

Peters, R.W., M. Winslett, and M.P. Sharma, 2011. "Sustainability Initiatives at Several U.S. Institutions of Higher Education", ***Proc. 2011 AIChE Annual Meeting***, Minneapolis, MN, (October 16–21).

UAB Deparment of Facilities Management, undated. "*Design and Construction Standards*: *Section 231400 Chilled Water*", University of Alabama at Birmingham, Birmingham, AL.

UAB Department of Facilities Planning, 1998. *Facility standard: Variable Frequence Drives.* University of Alabama at Birmingham, Birmingham, AL.

Winslett, M., 2011. Personal communication with Mr. Matt Winslett, Energy Management Division of UAB Facilities Management. Birmingham, AL, (September 30).

ENVIRONMENTAL PERFORMANCE FOR THE ELABORATION PROCESSES FOR CATALYSTS OF Á - AL2O3 AND SILICA

Johana Garcia[1], Gabriel Camargo[2], Luisa Martin[1], Lady Gaitan[1], Daniel Portela[3], Nikolay Agudelo Valencia[1] and Juan Carlos Moreno[4], (1)Environmental Engineering, Universidad Libre, Bogotá, Colombia, (2)Mechanical Engineering, Universidad Libre, Bogotá, Colombia, (3)Mechanical Engineering, Universidad Libre, Bogotá , Colombia, (4)Chemistry, Universidad de los Andes, Bogota

ABSTRACT

The processes for catalyst's preparation can be used as a tool for green engineering concepts in several courses in the engineering curriculum. In this job two catalysts were prepared and the raw materials, utilities (water, electrical power) and waste were measured. The water used in entire processes was registered for evaluated the process impact in the waste generation. The materials balance was realized and the Green Chemistry Expert System was used for the environmental impact.

The catalysts manufactured were α - Al_2O_3 and Silica. In the case of alumina the raw material was aluminum shaving (recycled materials). The quantities of catalyst prepared were 1 g. The aluminum was treated with nitric acid and ammonia. For the silica, the starting material was sodium silicate. The raw materials were characterized by atomic absorption. The final catalysts were analyzed by area BET, TGA, XRD, and infrared.

The processes waste waters was treated for minimizing its environmental impact

INTRODUCTION

The chemical processes impacts on the environment and the different ways to evaluate and avoid can be used to teach and learn about green engineering concepts, and use every tools for measure the different variables and formulate new process benign with the environment. The idea for the present work was originated when one of the authors was teaching a catalysts heterogeneous course for chemical engineering students and simultaneously material balance course for environmental engineering program. The project is an interdisciplinary work between different universities in Bogota Colombia.

The elaboration of catalytic supports can be used as a tool to measure the water footprint, and at the same time to evaluate important items such as waste disposal strategies, material balance, metals recycling in environmental engineering program. On the other hand for chemical engineering courses other aspects can be illustrated with these processes: the reactions efficiency, atoms economy reaction, materials characterization, and the effect of the variables process in the properties of the products.

The main purpose in this project is evaluating the environmental impact of the preparation catalytic supports from raw materials to end products. To reach this goal, two different catalytic supports were chosen α-Al_2O_3 and silica gel. These materials can be obtained from easily raw materials, aluminum shaving (from solid wastes) and sodium silicate. The experimental procedure is available from different sources.

EXPERIMENTAL PROCEDURE

For the α- Al_2O_3, there are several sources for raw material, soft drinks can, mechanical and industrial wastes, etc. the first step was a cleaning process with toluene for degreasing process after that several acid treatments steps are followed for obtain de catalytic support. The block diagram for this process is shown in fig No 1.

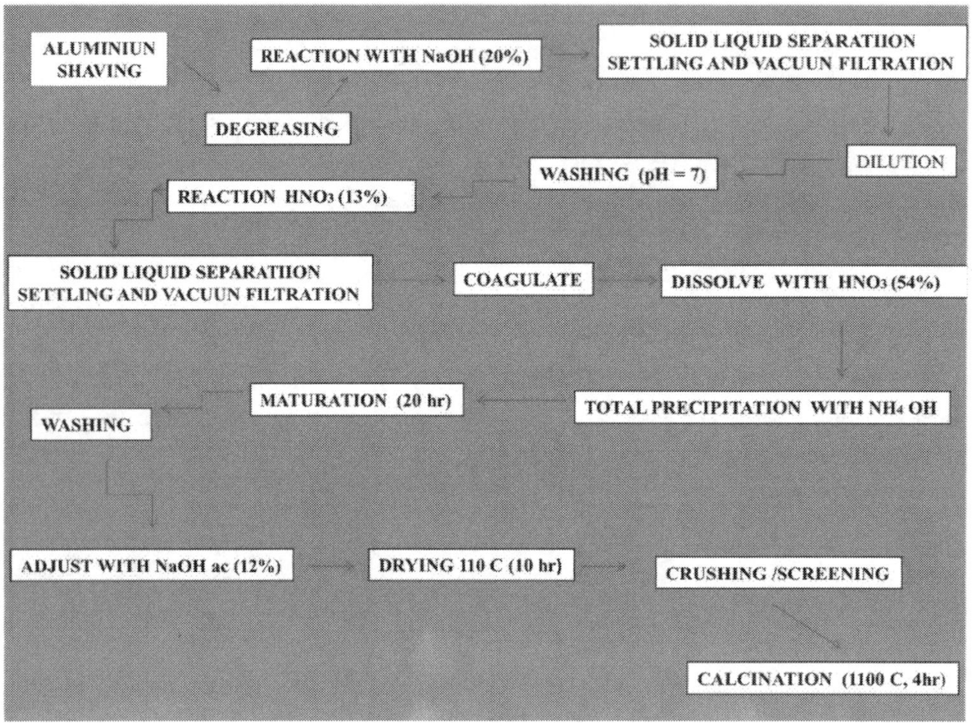

Block Flow Diagram for α- Al_2O_3 Process

For the silica gel support, the raw material was sodium silicate, and the first step in the process was a bleach using diatomaceous earth as adsorbent with constant agitation by 30 minutes. After that, the sodium silicate reacted with HCl (6N) adding the silicate to the acid until the pH was 8, at temperature of 20 c with permanent agitation. The block flow diagram for the entire process is shown in the figure No 2.

RESULTS AND DISCUTION

In the all processes the water employed in every step was measure and both catalytic support properties were determinate by XRD, area BET and TGA analysis. The water quantities were 3.5 Liters for 0.22 grams of α- Al_2O_3 from 3 grams of aluminum shaving. And for the silica gel the final support obtained was 0.6 gram and the water employed was 4.8 L.

The XRD result for the α- Al_2O_3 s material is shown in the figure 3. In this figure the characteristic peaks for the alumina can be observed. The area bet was evaluated and the TGA analysis was obtained

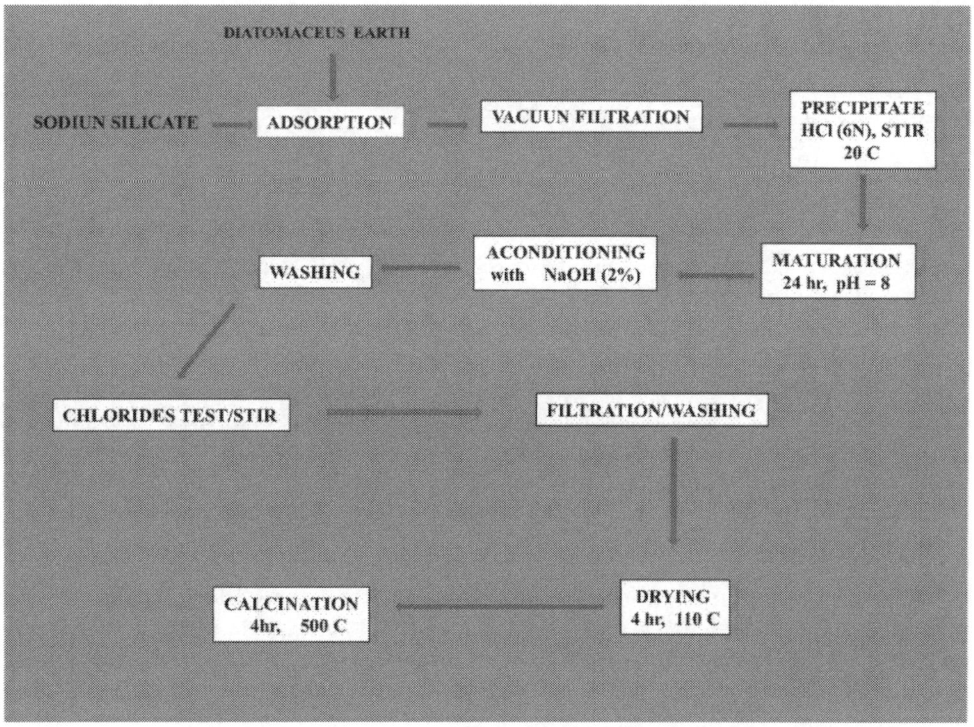

Figure No 2. Block Flow Diagram for Silica Gel process.

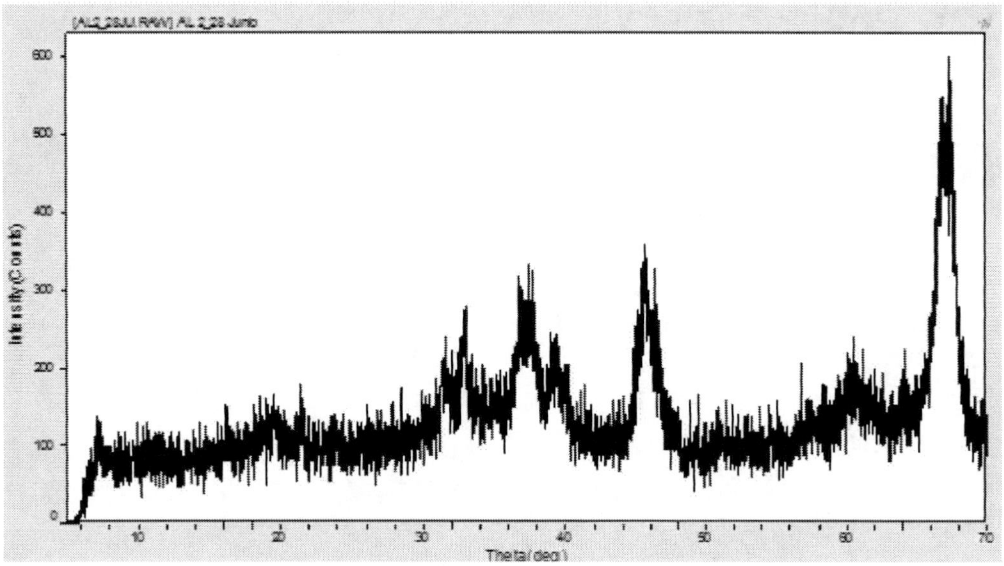

Figure No 3 XRD for α-Al_2O_3

The higher water quantities, in both processes were in the washing steps.

(758a) Thermochemical Conversion to Biocrude From Pilot Scale Grown Wastewater Fed Algae

Griffin W. Roberts[1], Marie-Odile Fortier[2], Belinda S.M. Sturm[2] and Susan M. Stagg-Williams[1], (1)Chemical & Petroleum Engineering, University of Kansas, Lawrence, KS, (2)Civil, Environmental & Architectural Engineering, University of Kansas, Lawrence, KS

Advantages of algae as a biofuel feedstock are well known; high growth rate, high lipid content, and CO_2 neutral. Thermochemical conversion or hydrothermal liquefaction (HTL) of algae to a hydrocarbon rich biocrude is growing in popularity primarily because it uses subcritical water as the reaction media eliminating the high energy intensive drying step associated with lipid extraction. HTL efficiently extracts existing lipids in addition to converting other macromolecules such as carbohydrates and proteins to smaller molecules that can be readily upgraded to fuels and other chemicals. Such a process also eliminates the need for a high lipid producing algae and can solely focus on high growth rate algae. Recent studies have shown that biofuels from algal feedstocks are most economically favorable if the algae is grown in conjunction with wastewater treatment plants; using free nutrients such as nitrogen and phosphorous found in the wastewater effluent rather than purchasing fertilizers for algal growth. There is an additional environmental benefit of removing such nutrients before they enter rivers, lakes and oceans by minimizing unnatural algal blooms which can cause anoxic zones.

This study involves pilot scale algae growth tanks utilizing effluent from the second clarifier at the Lawrence, KS wastewater treatment plant. Algae was harvested and hydrothermal treated toward biocrude. Both a mixed-culture microalgae and macroalgae identified as *Cladophora sp.* were studied. Hydrothermal treatment was performed with 3g of freeze dried algae and 50mL of water at 350°C for 1 hour. The starting algae and product yields were evaluated on an ash free dry weight basis. High oil yields were obtained for both the micro- and macroalgae (54% and 57%). Residual solids were determined to have higher heating values of 12.4 and 10.8 MJ/Kg for micro- and macroalgae, respectively. Aqueous co-products had carbon, nitrogen, and phosphorus concentrations of 3100-4000 ppm, 1200-1600 ppm, and 3.5 ppm respectively. The biocrude produced was very similar to that of petroleum crude in terms of energy density (39-42 MJ/Kg) and elemental percentages of carbon, hydrogen, and oxygen.

(758b) Hybrid Thermochemical/Biochemical Gas Fermentation Process for Production of Fuels and Chemicals

Derek W. Griffin and Michael A. Schultz, LanzaTech Limited, Roselle, IL

The ever increasing demand for renewable feedstock-based fuels and chemicals is driving the interest and rapid development of 2nd generation biofuel processes worldwide. There is significant research into improving the overall carbon and energy efficiency of biorefineries in order to compete with petroleum derived products in terms of economics and sustainability. A range of 1st generation renewable thermochemical or biochemical processes face various challenges including high energy demands, high capital costs, poor carbon utilization, or a combination of these factors. Integrated thermochemical/biochemical processes have the potential advantage to overcome these challenges to produce sustainable fuels and chemicals from renewable feedstocks.

The hybrid thermochemical/biochemical LanzaTech process captures various gas streams with renewable resources and converts them into low-carbon fuels and chemicals. The gas feedstock can come from a variety of sources including gaseous waste carbon from industrial processes or from thermochemical processes such as gasification of biomass/MSW or reformed natural gas. The CO, H_2, and/or CO_2 contained in the feed gas are fermented via a biocatalyst to chemicals/fuels such as ethanol, acetic acid, or 2,3-butanediol which also serve as building blocks to longer chained hydrocarbon drop-in fuels. The LanzaTech process offers superior carbon conversion, energy efficiency, and greenhouse gas emission performance compared to conventional and emerging routes to the same products.

The LanzaTech's integrated thermochemical/biochemical gas fermentation process is described including a comparison to current stand-alone biochemical and thermochemical technologies for the case of producing ethanol from renewable feedstocks, such as biomass. The different routes are compared in terms of product yields, overall carbon capture, thermal efficiencies, and green house gas (GHG) emissions based on overall mass and energy balances of the various processes. A previous analysis comparing gas fermentation to the thermochemical route to ethanol[1] is expanded here to include an assessment of ethanol production via biochemical routes. It is demonstrated that engineering combined thermochemical & biochemical processes can result in benefits that neither process can achieve independently.

[1] Griffin, D.W. and Schultz, M.A. (2012). Fuel and Chemical Products from Biomass Syngas: A Comparison of Gas Fermentation to Thermochemical Conversion Routes. Environmental Progress & Sustainable Energy. *31*, 219-224.

(758c) Effects of Hydrolysis and Torrefaction On Pyrolysis Product Distribution of Spent Mushroom Compost (SMC)

Rene Garrido, Chemical Engineering Department, Villanova University, Villanova, PA, Maria Nydia Ruiz-Felix, Chemical Engineering, Villanova University, Villanova, PA and Justinus A. Satrio, Department of Chemical Engineering, Villanova University, Villanova, PA

Energy based in biomass is a source of renewable energy from organic and inorganic matter formed in a biological or mechanical process. Although fossil fuels, such as petroleum, oil and coal, have their foundation in ancient biomass, they have been "out" of the carbon cycle for a long time. Their combustion consequently disturbs the carbon dioxide content in the atmosphere.

Lignocellulosic biomass can be classified as natural (without human intervention), residual (by-product of waste generated by agricultural activities, forestry, and industry wood processing), and energy-based crops (for the production of biofuels); it is the only carbon-containing renewable energy source. The complex structure of lignocellulosic biomass has been extensively studied and poses highly valuable carbohydrates (cellulose and hemicelluloses) and lignin, whereas there is a substantial challenge to the large-scale biomass utilization. The carbohydrate polymers are strongly bound to the lignin, which makes the separation process technically difficult.

While it is scientifically viable to exploit lignocellulosic materials and organic wastes into energy, chemicals and fuels, the commercial production is far from an economical process and the cost needs to be lowered. It also must be demonstrated that a commercial scale biomass utilization process is environmentally sound, and there is a need to find an inexpensive and widely available lignocellulosic source of biomass

This study investigates the potential use of a unique biomass material, produced from mushroom farms, known as spent mushroom compost (SMC) as a biomass-based energy feedstock. It was found that, even though SMC has a highly probability for bio-oil conversion its high moisture content and high ash content in feedstock results in low yield and quality of bio-oil. The removal of these recalcitrants is a key element to make SMC a desirable feedstock for biomass conversion. This investigation has revealed that using torrefaction and hydrolysis for SMC treatment after pyrolysis shows a higher yield with a wider product distribution after the combined pre-treatments proposed.

Analytical pyrolysis combined with gas chromatography and mass spectrometry (Py-GCMS) is used to study the chemical composition of volatile products from previously hydrolyzed and torrefied biomass. Effects of the severity level of hydrolysis pre-treatment and torrefaction temperatures on the chemical composition and yield distribution of the fast pyrolysis products will be evaluated. The chemical composition study of pyrolyzed biomass would then aid in the understanding of the chemistry behind the different compounds to further enhance the process of upgrading bio-crude oil. By this, ground work is laid as a preliminary attempt to understand SMC as a potential energy feedstock.

(758d) Mixed Catalysts for the Conversion of Hemicellulose and Cellulose

Nathan S. Mosier[1], Eurick Kim[2], Shuo Liu[2] and Mahdi Abu-Omar[3], (1)Agricultural and Biological Engineering & Laboratory of Renewable Resources Engineering (LORRE), Purdue University, West Lafayette, IN, (2)Chemistry, Purdue University, West Lafayette, IN, (3)Chemistry Department, Purdue University, West Lafayette, IN

Direct catalytic conversion of lignocellulosic biomass to biofuels could improve the carbon efficiency of biofuel production. We report the use of maleic acid, a dicarboxylic acid, mixed with Lewis acids such as lithium chloride and aluminum chloride to catalyze the fractionation of biomass into an aqueous solution of pentose (primarily xylose) and insoluble cellulose and lignin, followed by the conversion of the xylose to furfural under higher temperature and pressure. This method achieved 80-90% yield of xylose through hydrolysis of the hemicellulose from various biomass sources (switchgrass, poplar, pine) and achieved 54-61% yield of furfural (based on original biomass). The residual solids contain cellulose that has increased reactivity for hydrolysis by cellulolytic enzymes. SEM imaging suggests that maleic acid pretreatment is effective in fibrillating biomass, thus significantly increasing the accessible surface area of cellulose for enzymatic or catalytic conversion. This hybrid approach combines pretreatment for biochemical conversion with thermochemcial processing. We present a kinetic analysis of biomass hydrolysis and formation of aldehydes and discuss application of results from pure sugars to results from biomass conversion.

(758e) Aspen PLUS Modeling of the Utilization of Equine Waste for Localized Heating Via Fast Pyrolysis

Nicole L. Hammer[1], Akwasi A. Boateng[2], Charles A. Mullen[1], M. Clayton Wheeler[3] and Jose L. Carrasco[4], (1)Eastern Regional Research Center, Agricultural Research Service, USDA, Wyndmoor, PA, (2)U.S. Department of Agriculture, Crop Conversion Science & Engineering, Agricultural Research Service, Wyndmoor, PA, (3)Chemical and Biological Engineering and Forest Bioproducts Institute, University of maine, Orono, ME, (4)Chemical and Biological Engineering, University of Maine, Orono, ME

ARS/USDA is investigating the local utilization of equine waste as an alternative and replacement fuel source for heat generation using bio-oil produced via fast pyrolysis at the equine rehabilitation facility at Morrisville State College (MSC) with potential expansion to local farms. The Aspen Plus based simulation models the production of pyrolysis oil and its combustion in an existing boiler for the generation of hot water. Ultimate and proximate analysis on the manure and its pyrolysis products enabled the defining of the manure as a non-conventional compound and the pyrolysis products as conventional compounds based on stoichiometric formula. This results in quantifying the production of pyrolysis oil and bio-char which are valuable and sustainable energy sources that can be utilized at the local facility. Pyrolysis oil can be used as fuel in an existing hot water boiler system at MSC that produces hot water for the facility via diesel. The results show potential for displacing the diesel fuel by the pyrolysis oil produced from the locally generated equine waste, while also alleviating a waste disposal problem. Based on mass and energy balances of the system, heat integration is possible. Using a portion of the co-product bio-char for combustion provides all the heat required to run the pyrolyser. The fluidized bed pyrolyser will circulate sand from the pyrolysis unit through a combustion unit for reheating and then return it hot to the pyrolysis unit. Excess bio-char can be utilized in surrounding soils acting as a soil amendment. By incorporating the heat sources and heat integration in the model, energy requirements to run the system are significantly reduced. The mass and energy balances generated by Aspen Plus will be the basis for a Techno Economic Analysis for locating such a system at MSC; this will be presented and discussed further.

Critical and Adequate pretreatment for Biomass Sugar Release – Selective Lignin Unlocking Process Inspired from Wood-Feeding Termite

Jing Ke, Shulin Chen

Department of Biological Systems Engineering, Washington State University

Currently, the limiting factor for lignocellulosic biofuels is the lack of efficient pretreatment technology within an industrially acceptable economics. The heterogeneous and irregular arrangement of lignin construction results in resistance to saccharification, thus severely limiting biochemical conversion of lignocellulosic biomass to fermentable sugars. Lignin framework also inhibits cellulase function by competitive adsorption, chemical inhibition, and steric hindrance. Since lignocellulosic cellulose can only be digested after the lignocellulose is exposed out of lignin framework. It is therefore important to understand to which extent of lignin content and composition can be modified in useful ways without deleterious consequences to plant growth and development. Existing biomass conversion schemes typically rely on a combination of thermochemical treatments that are not sustainable and environmentally friendly as the processes still consume a great deal of energy and require input of chemicals because of the needs for high temperature and pressure. Nature, on the other hand, has developed through evolution efficient biological systems for extracting energy from plant materials to power the lives by effectively deconstructing lignocellulose under natural conditions. Biological systems are the proof-of-concept of what can be physically achieved as they are built of the same fundamental elements of matter and follow the same laws of physics as inorganic and human-engineered systems. With adequate understanding, these biological systems can be harnessed for the production of bioenergy sustainably from renewable lignocellulosic biomass. The high efficiency of cellulose utilization by wood-feeding termite in terms of rate and extent (95% of cellulose in wood within 24 hrs) makes it a promising model system for biofuel production by understanding and mimicking its unique pretreatment system. This work explored a unique lignin unlocking process in wood-feeding lower termite workers *Coptotermes formosanus* (Shiraki) for efficient woody carbohydrate utilization and how the termite workers manage the selective lignin modification. The rationale is based on its superior performance on cellulose utilization and its stepwise introduction of specific reactions on lignin, as well as the varied micro-parameters for lignin modification in each gut section. Obtained results provide that termite-induced lignin pretreatment is not achieved by savage degradation, but selectively modification of functional group, cleavage of ether linkage within lignin and ester/ether linkage between lignin-hemicellulose and lignin-cellulose. The demonstrated structural modification results in alleviation of both physical and chemical inhibition effect of lignin on cellulases, together with alteration of the three dimensional structure of lignin, which unlocks cellulose fiber from protection. The results also demonstrate that during the lignin structural modification by termite, a non-enzymatic system of $\cdot O_2^-$ and catalysts, generating from the initial enzymatic lignin aliphatic dehydroxylation and carbonylation, penetrate inside the plant cell wall to induce reactions profiting from their small molecular sizes. This non-enzymatic system is proved to take a lead for the selectivity of lignin modification as what termites do, including modification on the specific functional group and ether/ester linkages within lignin framework or between lignin-hemicellulose and lignin-cellulose, which is partially reverse reaction of lignin synthesis. The *in-vitro* mimicking of this non-enzymatic system with umol level concentrations of the chemicals results in alleviation of the lignin inhibition effect on enzymatic cellulose hydrolysis.

Notably, the lignin-modifying system mimicking from termite is advanced in its low chemical concentration, rate, selectivity, and neutral working condition. It leaves lignin backbone alone, thus consuming less energy and producing negligible lignin-derived inhibitors for subsequent cellulose hydrolysis and sugar fermentation. Meanwhile, the residual lignin backbone could be applied for other purposes such as value-added lignolytic co-products. On the basis of this, this new pretreatment process inspired by termites is from but beyond the nature, which performs lignin pretreatment as efficiently, robust and environmentally-friendly as termites. This effort also made major contributions to advance the fundamental science by supporting the new concept of critical and selective lignin modification for farthest alleviation of the inhibition effect, which is critical for filling the existing knowledge and technology gaps in sustainable and economical biofuel and value-added chemicals production from lignocellulosic biomass.

(774a) Development of Proton Selective Membranes for Efficient and Economical Vanadium Redox Flow Batteries

Justin Walls[1], Pedram Jahanian[2], Guangzhao Mao[2], Dr. Steven Salley[3] and Dr.Simon Ng[3], (1)Chemical Engineering, Wayne State University, Detroit, MI, (2)Chemical Engineering and Materials Science, Wayne State University, Detroit, MI, (3)Chemical Engg., Wayne State University, Detroit, MI

Development of Proton Selective Membranes for Efficient and Economical Vanadium Redox Flow Batteries

Justin Walls, Padram Jahanian, Guangzhao Mao, Steven Salley, K. Y. Simon Ng

Department of chemical Engineering and Materials Science, College of Engineering,

Wayne State University, Detroit MI 48202

Flow Redox Batteries (FRBs), especially the Vanadium Redox Flow Battery (VRFB), provide a promising chemistry for storing intermittent renewable energy on a very large scale. While these batteries have already been developed on the megawatt hour level; none of the FRB technologies has seen significant market penetration. This is mainly because the technology is still very expensive, due to high material costs, and less than satisfactory performance parameters and reliability.

The key component of the VRFB, which is one of the leading causes of the elevated cost and unreliability, is the ion exchange membrane (IEM). The membranes need to possess good ion conductivity while maintaining high mechanical and chemical stability. Perfluorosulfonic polymers such as Nafion are the most common membranes for proton exchange in the VRFB due to their good chemical stability and high proton conductivity. However, due to the tendency of vanadium to pass through the membrane, Nafion membranes suffer from decreased energy efficiency. This project is focused on decreasing the vanadium ion diffusion through the membrane by modifying the nanostructure of Nafion. There has been extensive research activity in the modification of Nafion membranes, for example, modification by layer-by-layer polyelectrolyte multilayer films and in situ sol-gel reaction to close the pores by silica nano particles. Our project combines these two methods using hyper-branched polyelectrolyte multilayer films to further lower the vanadium ion diffusion rate. Nafion modification by the sol-gel chemistry was conducted by treating the Nafion in the tetraethylorthosilicate (TEOS) methanol solution. Poly(diallyldimethylammonium chloride) (PDDA) and poly(sodium styrene sulfonate) was used in the layer-by-layer deposition. Diffusivity and ion crossover of these membranes was tested and compared to that of the unmodified membranes as well as other modification techniques that have been reported. A small scale VRFB system was constructed to test the performance of these modified membranes, and efficiency and reliability of the system as a whole was measured.

(774b) Microporous Separators On Fe/V Redox Flow Battery: A Valuable Opportunity for Cost Reduction

Xiaoliang Wei[1], Qingtao Luo[1], Bin Li[1], Zimin Nie[1], Yuyan Shao[1], Feng Chen[1], Baowei Chen[1], Guanguang Xia[1], Liyu Li[2], Zhenguo Yang[2] and Wei Wang[1], (1)Pacific Northwest National Laboratory, Richland, WA, (2)UniEnergy Technologies, LLC, Mukilteo, WA

Microporous Separators on Fe/V Redox Flow Battery: A Valuable Opportunity for Cost Reduction

Wei, Xiaoliang; Luo, Qingtao; Li, Bin; Nie, Zimin; Shao, Yuyan; Chen, Feng; Chen, Baowei; Xia, Gordon; Li, Liyu; Yang, Z Gary; Wang, Wei

Pacific Northwest National Laboratory, 902 Battelle Boulevard, PO Box 999, Richland, WA 99354

Redox flow batteries are considered as one of the most promising medium-to-large scale energy storage technologies and have attracted much attention both academically and industrially.[1],[2] A number of redox chemistries have been proposed, and significant progress has been achieved by today.[3],[4] However, broad market penetration of redox flow batteries is still hindered by their intrinsic limitations such as high cost and low durability of components, narrow operational temperature range, limited solubility of active species, and/or low electrochemical activity.[5] The recently invented iron-vanadium (Fe/V) redox flow battery employing $Fe^{2+/3+}$ – $V^{2+/3+}$ redox couples have shown to be a promising option for stationary energy storage.[6],[7] The Fe/V flow battery demonstrated stable cycling performance with a nearly 100% utilization ratio over a broad temperature range of 0-50°C.

The positive charged species, Fe^{3+}, is a relatively weak oxidant. Therefore, hydrocarbon-based ion exchange membranes and/or separators are possible options for use in Fe/V flow battery system. This contribution investigated the cycling performance of a variety of polyethylene microporous separators on Fe/V flow cell. Among them, some separators exhibited energy efficiency of around 70% at temperatures ranging from 5-50°C and at current densities up to 80mA/cm^2. Because these separators are very inexpensive, their use significantly reduces the capital cost of Fe/V flow battery, delivering great potential for developing a low-cost energy storage system.

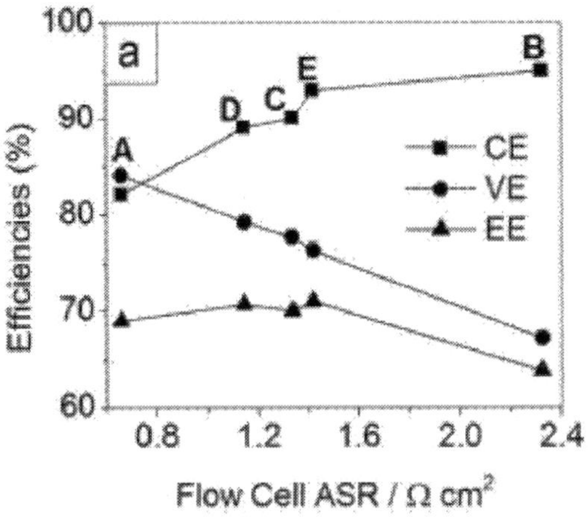

Figure 1 Flow cell cycling efficiencies (CE, VE, and EE) of microporous separators A, B, C, D, and E.

[1] Z. Yang, J. Zhang, M. C. W. Kintner-Meyer, X. Lu, D. Choi, J. P. Lemmon, and J. Liu, *Chem. Rev.* 2011, *111*, 3577-3613.

[2] J. Rugolo, and M. J. Aziz *Energy Environ. Sci.* 2012, **DOI:** 10.1039/C2EE02542F.

[3] M. Skyllas-Kazacos, M. H. Chakrabarti, S. A. Hajimolana, F. S. Mjalli, and M. Saleem *J. Electrochem. Soc.* 2011, *158*, R55-R79.

[4] L. Li, S. Kim, W. Wang, M. Vijayakumar, Z. Nie, B. Chen, J. Zhang, G. Xia, J. Hu, G. Graff, J. Liu, and Z. Yang *Adv. Energy Mater.* 2011, *1*, 394–400.

[5] S. Eckroad, *Vanadium Redox Flow Batteries: An In-Depth Analysis.* EPRI, Palo Alto, CA: 2007. 1014836.

[6] W. Wang, S. Kim, B. Chen, Z. Nie, J. Zhang, G. Xia, L. Li, and Z. Yang *Energy Environ. Sci.* 2011, *4*, 4068-4073.

[7] W. Wang, Z. Nie, B. Chen, F. Chen, Q. Luo, X. Wei, G. Xia, M. Skyllas-Kazacos, L. Li, and Z. Yang *Adv. Energy Mater*. 2012, *2*, 487–493.

(774c) In Situ Investigation of Vanadium Ions Transport in Redox Flow Battery

Qingtao Luo[1], Liyu Li[2], Zimin Nie[3], Wei Wang[1], Xiaoliang Wei[1], Feng Chen[1], Bin Li[4], Guanguang Xia[1], Baowei Chen[5], Gary Yang[1] and Sprenkle Vincent[1], (1)Pacific Northwest National Laboratory, Richland, WA, (2)Hydrocarbon Processing Group, Pacific Northwest National Laboratory, Richland, WA, (3)Pacific Northwest National Labratory, Richland, WA, (4)Pacific Northwest National Laboratory, Richland , WA, (5)Biological Sciences, Pacific Northwest National Laboratory, Richland, WA

The all-vanadium redox flow battery (VRB) is probably the most promising and extensively researched flow battery system. In order to shed light on the VRB capacity fading mechanism, it is necessary to investigate the transport behaviors of vanadium ions across the membrane. In this work, flow batteries with vanadium and iron redox couples as the electroactive species were employed to investigate the transport behavior of vanadium ions in the presence of an electric field. It was shown that the transport of vanadium ions was affected by a combination of different factors, such as the charge/discharge process, current density, and SOC. The electric field accelerated the positive-to-negative and reduced the negative-to-positive transport of vanadium ions in the charging process and affected the vanadium ion transport in the opposite way during discharge. With the increase of current density, the vanadium ion transport rate increases during the charge process but decreases during discharge. At each given current density, the vanadium ion transport rate, in general, demonstrated a trend of decreasing with the increase of SOC.

(774d) Electrolyte Optimization for Fe/V Redox Flow Battery System

Bin Li[1], Wei Wang[1], Zimin Nie[2], Vincent Sprenkle[1], Baowei Chen[3], Xiaoliang Wei[1], Qingtao Luo[1], Guanguang Xia[1], Liyu Li[4] and Zhenguo Yang[4], (1)Pacific Northwest National Laboratory, Richland, WA, (2)Pacific Northwest National Labratory, Richland, WA, (3)Biological Sciences, Pacific Northwest National Laboratory, Richland, WA, (4)UniEnergy Technologies, LLC, Mukilteo, WA

Electrolyte optimization for Fe/V redox flow battery system

Bin Li, Wei Wang, Zimin Nie, Vince Sprenkle, Baowei Chen, Xiaoliang Wei, Qingtao Luo, Guanguang Xia, Liyu Li and Zhenguo Yang

Pacific Northwest National Laboratory

902 Battelle Boulevard, PO Box 999, Richland, WA 99354

Abstract

Redox flow batteries (RFBs) research and development have recently enjoyed a renaissance primarily due to their ability to store large amounts of power and energy, up to multi-MW and multi-MWh, respectively, for renewable energy integration or smart-grid deployment.[1] Broad market penetration, however, has not been achieved due to several limitations such as high cost, instability in long-term operation, and narrow operational temperature range, etc. Recently invented Fe/V redox flow battery (IVRFB) has attracted more and more attentions due to its long-term cycling stability and the freedom to utilize membranes other than expensive Nafion.[2] In this study, extensive matrix study was performed on the Fe/V electrolyte in order to determine the relationships of the composition, state of charge (SOC), and temperature stability. As we know, the temperature stability of the RFB systems is of paramount importance for practical use. A RFB system with wider temperature stability will not only enable the RFB to be deployed with less geographical limitation, but also possibly eliminate the active temperature management system reducing parasitic loss. As previously reported, the current Fe/V electrolyte cannot be operated or stored at lower temperatures ($\leq 0^oC$) due to the precipitations ($FeCl_2$). In this research, factors influencing the stability of electrolytes in both positive and negative half-cells with different SOCs at different temperatures from -10^oC to 40^oC were systematically studied. Through the optimization of the electrolyte composition, an electrolyte stable in the temperature range from -5^oC to 40^oC was identified. Electrolyte stability data and IVRFB flow cell cycling performances using low-cost separators will be presented (Fig.1).

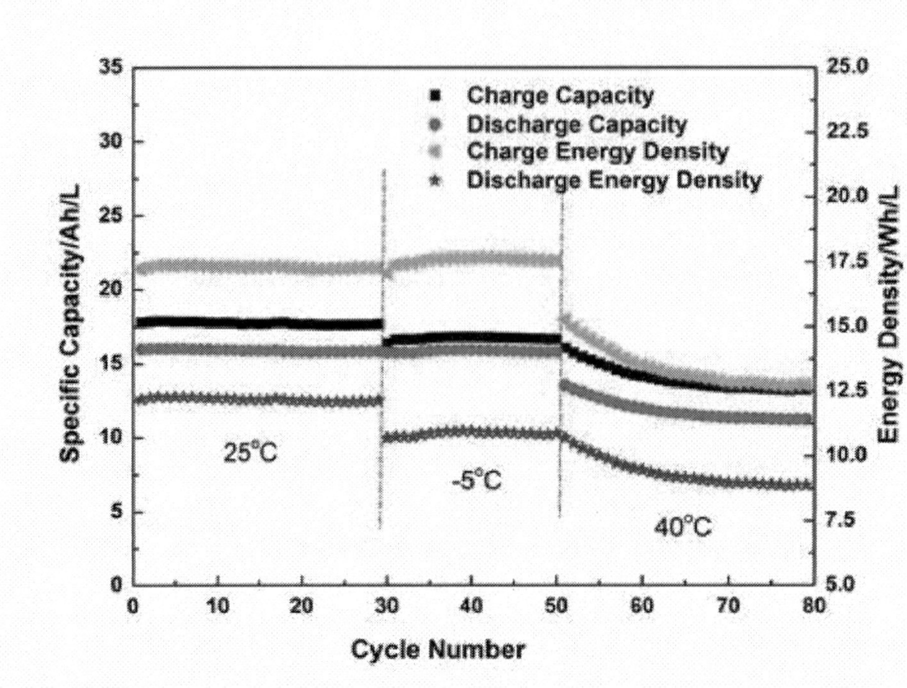

Fig.1 Cycling performances of optimized Fe/V electrolytes

References

(1) Yang, Z.; Zhang, J.; Kintner-Meyer, M. C. W.; Lu, X.; Choi, D.; Lemmon, J. P.; Liu, J. *Chemical Reviews* **2011**, *111*, 3577.

(2) Wang, W.; Kim, S.; Chen, B.; Nie, Z.; Zhang, J.; Xia, G.-G.; Li, L.; Yang, Z. *Energy & Environmental Science* **2011**, *4*, 4068.

(774e) Electrostatically Assembled Carbon Nanoparticle Electrodes for Vanadium Redox Flow Batteries

Abhinandh Sankar and **Anastasios Angelopoulos**, School of Energy, Environmental, Biological, and Medical Engineering, University of Cincinnati, Cincinnati, OH

Efficient charge and mass transfer at the surface of the electrode is critical to overall redox flow battery (RFB) performance. Fundamental structure-property correlations for RFB electrodes cannot presently be obtained using electrodes produced from carbon felts or carbon nanoparticle slurries and pastes due to the ill-defined distribution of active sites in these materials. For example, a recent study using high surface area multi-wall carbon nanotubes as electrode materials yielded only a two to three-fold increase in peak current density relative to glassy carbon. We here report on the first use of electrostatic layer by layer (LbL) assembly to fabricate RFB electrodes. Since its introduction by Iler over four decades ago, LbL assembly has been repeatedly shown capable of producing well-defined arrangements of nanoparticles and the method is now ubiquitous in scientific research. However, the approach has yet to be applied to produce well-defined electrode structures for RFB electrodes and has only rarely been explored in the general field of electrocatalysis. Key areas of concern include: 1) the impact of the polymer counter-ion layer that is typically required for assembly on ohmic resistance, 2) accessibility of active sites in tightly packed layered structures, and 3) durability. In this investigation, we demonstrate how LbL assembly of graphite nanoplatelets may be used to achieve unprecedented high current densities for the all-vanadium system through fine control over the thickness and morphology of the electrode. In particular, we show how nanoparticle packing density, size, and size distribution in addition to local polymer binder concentration and crystalline order all impact active site concentration and promote reagent accessibility to these sites in a manner that has not been previously recognized. Such basic understanding permits rational optimization of electrode chemistry and structure using carbon-based electrocatalysts as well as detailed interpretation of the promotional effects of inorganic electrode additives. Finally, we demonstrate through electrochemical cycling how these coatings are exceptionally stable relative to electrodes fabricated without LbL assembly.

(774f) Fractional Degradation Cost for Cycling a Zinc/Bromine Flow Cell Battery

Alex Bistrika[1], Alexandre F.T. Yokochi[1], Michael Antonishen[2], James Davidson[2], Douglas Halamay[2], David Naviaux[3], Jiajia Song[2], Ted Brekken[3] and Annette von Jouanne[3], (1)School of Chemical, Biological and Environmental Engineering, Oregon State University, Corvallis, OR, (2)Electrical Engineering and Computer Science, Oregon State University, Corvallis, OR, (3)School of Electrical Engineering and Computer Science, Oregon State University, Corvallis, OR

It is widely understood that intermittent resources such as solar and wind are dependent on environmental factors outside our ability to control. This allows for the possibility of either power output exceeding or falling short of forecast levels which leads to more frequent cycling of regional hydro generators and greater overall grid instability under the current infrastructure. Quantifying the cost associated with added maintenance attributed to an increased penetration of wind and the physical limitations of a growing dependence on Pumped Hydro Energy Storage (PHES), becomes essential in painting a more complete picture. However, alternative means of Large Scale Energy Storage (LSES), for storing excess power when the outputs exceeds demand and to supplement output power when it falls short of demand, is a widely accepted route for mitigating intermittent power generation.[1] Thus, it is crucial to establish an accurate and relative cost function for the commercially available and most economically viable storage technologies. The most promising electrical energy storage technology, to replace PHES, is flow cell batteries.[2] This begs the question what are the degradation mechanisms in these building size chemical systems and what do we do when they stop working. In this study we report on the key degradation mechanisms that could be expected for standard operation of a zinc/bromine flow cell battery using graphite electrodes and assign a cost function dependent of system operating parameters. In a previous study we reported the oxidation (or charging) reaction as the most damaging to the electrode material, see **Figure 1**. We discuss aging trends and electrode performance by considering surface chemistry as observed by Raman spectroscopy and scanning electron microscopy. Theoretical system performance based on steady state current density measurements is used to establish a normalized operating cost function for a zinc/bromine battery, as can be seen in **Figure 2**.

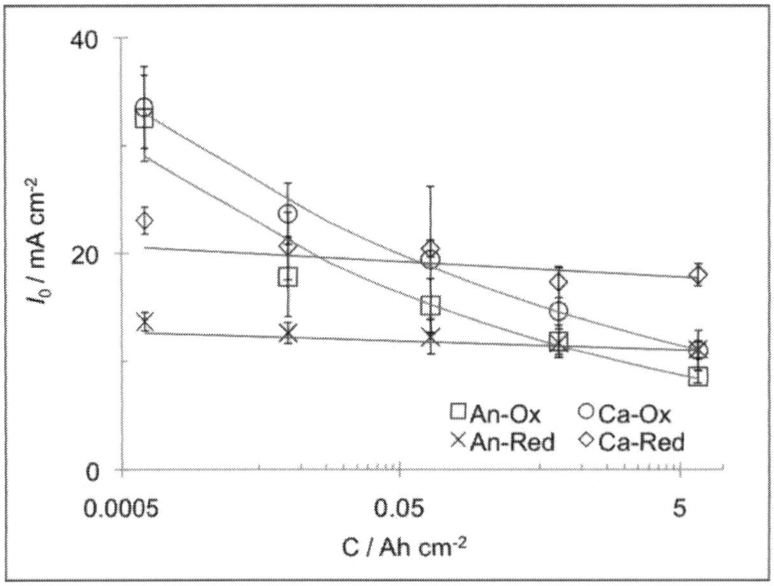

Figure 1 Oxidation and reduction exchange current; measured for both the cathode and anode in 6 [M] NH4Br. The x-axis represents the electrode aging, depicted by total coulombs passed through the electrode per unit surface.

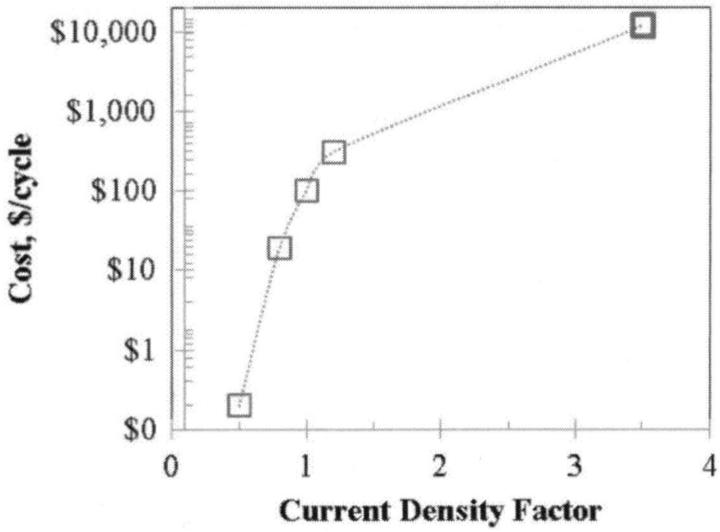

Figure 2 Cost of a 10 % state of charge cycle with respect to a representative operating parameter (current density factor).

1. Yang Z, Zhang J, Kintner-Meyer MCW, et al. Electrochemical Energy Storage for Green Grid. *Chemical Reviews.* 2012/04/07 **2011**;111(5):3577-3613.

2. Soloveichik GL. Battery Technologies for Large-Scale Stationary Energy Storage. *Annual Review of Chemical and Biomolecular Engineering.* 2012/04/24 **2011**;2(1):503-527.

(774g) Design Mesostructured LiMn2O4 As Low-Cost Cathode Materials for Lithium Ion Batteries

Jian Zhu, Chemical Engineering, Wayne State University, Detroit, MI, K. Y. Simon Ng, Department of Chemical Engineering and Materials Science, Wayne State University, Detroit, MI and Da Deng, Chemical Engineering and Materials Science, Wayne State University, Detroit, MI

Lithium ion batteries are the dominant power sources for mobile electronic devices and will play increasingly important role in expanding market of electric vehicles and green energy storage. Spinel $LiMn_2O_4$, alternative to expensive and toxic $LiCoO_2$, as cathode materials in lithium ion batteries are attracting much attention for advantages of low cost, environmental benignness, and abundance. However, the low lithium-ion diffusion coefficient and capacity fading caused by dismutation reaction of Mn^{3+} restrict the applications of $LiMn_2O_4$ at large/industrial scale where bulky batteries are required to have good cycling performance at high current rate with no significant capacity fading. Our strategy is to design and tailor the structure and composition of $LiMn_2O_4$ to improve its conductivity and cycling performance. In this presentation, our latest efforts on the design and facile fabrication of $LiMn_2O_4$ as high-power cathode materials will be reported. The unique structure is tailored and controlled to allow fast lithium ion diffusion while maintaining structure stability. The unique mesostructure is fully characterized by electron microscopy and x-ray diffraction and their electrochemical performance in lithium ion batteries is thoroughly evaluated.

(778a) Immobilization of Vesicles Incorporated with Aquaporins Onto Polymeric Membranes for Low Energy Water Purification

Wenyuan Xie and Yen Wah Tong, Department of Chemical and Biomolecular Engineering, National University of Singapore, Singapore, Singapore

AQPs are transmembrane water channel proteins. Their unique selectivity, high water transport capability, and low activation energy have garnered strong scientific interests with many studies concentrating on the fabrication of biomimetic membranes based on the reconstitution of water channel proteins into self-assembled amphiphilic lipid or polymer bilayers. It has been reported that the osmotic permeability of a single aquaporin is in the range from 6×10^{-14} to 24×10^{-14} m^3 s^{-1}, and the permeability of an AQPs reconstituted biomimetic membrane is 167 um/s/bar, which is by two orders of magnitude greater than the commercial polymeric membranes. It is clear that aquaporin-incorporated biomimetic membranes possess an extremely high potential for water purification and desalination. However, the chief limitation of the biomimetic membrane is that the ultrathin biomimetic bilayers are too fragile to withstand high hydraulic or osmotic pressure in the water purification application. Here we show an innovative yet simple and easy-to-implement method to overcome the membrane fragility by immobilizing water-channel proteins incorporated vesicles into pre-fabricated membranes to develop advanced membrane materials for purifying water at low pressures and low energy. We show that this novel membrane exhibits high mechanical strength and stability during nanofiltration process and AQPs made contribution to the water performance.

(778b) Membrane Bioreactor Process and Fouling Control for Water Reclamation and Groundwater Recharge Applications

Woonhoe Kim[1], Varadarajan Ravindran[2], Ryan Thacher[1] and Massoud Pirbazari[1], (1)Sonny Astani Department of Civil and Environmental Engineering, University of Southern California, Los Angeles, CA, (2)University of Southern California, Los Angeles, CA

The use of membrane bioreactor systems (MBR) for water treatment, wastewater treatment and water reclamation and reuse has drawn great interest from researchers and engineers. Despite its many advantages such as smaller footprint and better product quality, their wider application require better control of membrane fouling and permeate flux decline as foulant removal depends on chemical or physical cleaning. In spite of their excellent retention characteristics, there are still problems that slow down use of membranes in these appliactions. Over time, membrane fouling and subsequent cleaning cause deterioration of membrane materials, resulting in compromised permeate quality and reduced membrane life spans. Chemical and biological fouling are major problems in membrane filtration due to reduced permeate flux, increased energy costs, and system downtime for maintenance. Biofouling in MBR systems is caused by the buildup of organic chemicals, microorganisms, and microbial communities at the membrane surface. Biofilms attached to a surface begin with cell adhesion, and progress to thick layers of extracellular polymeric substances (EPS), other organic chemicals, and a complex community of microbial cells that are often difficult to remove. Organic fouling due to the presence of natural organic matter (NOM) often leads to surface and internal pore fouling. More importantly, organic fouling progressively leads to biological fouling in so far as providing the organic nutrients for biofilm growth and sustenance.

The emerging use of advanced oxidation processes such as ozone, hydrogen peroxide, and UV radiation as well as catalysts (such as titanium dioxide) in water treatment offers new opportunities, because the AOPs are able to decompose a number of membrane foulants very efficiently, particularly the NOM. The effect of these processes on membrane fouling is difficult to predict due to the complex nature of NOM, and the major influence of the conformation of NOM and the decomposition products on organic and biological fouling.

In the context of water reclamation and reuse, the MBR process with ultrafiltration can be made more energy efficient, economical and versatile than reverse osmosis or nanofiltration by using exhibiting better transport properties, higher permeate flux, and greater fouling resistance. The MBR process is specially designed for the control of membrane fouling and permeate flux decline by a combination of powder activated carbon (PAC) sorption and fluid management. The present research is intended to overcome organic and biological fouling problems, and to improve aqueous transport properties. This aspect of the study required the investigation of fouling mechanisms in the operation of MBR systems.

The flat-sheet membrane evaluation tests in batch-mode represent an economic means for evaluating membrane materials for fouling potential and rejection characteristics. Preliminary flat-sheet membrane tests were performed in plate-and-frame cells using polyether sulfone (PES) membranes to obtain *a priori* information on the nature of organic and biological fouling. These tests also included subjecting the feed solution to AOPs including ozonation, and combinations of ozone and hydrogen peroxide. The breakdown of NOM constituents into organics of lower molecular weights and different structural characteristics shall influence the extent of surface fouling and internal pore fouling. Another aspect was the evaluation of microbial fouling using Escherichia coli (E. coli) as model bacteria to obtain prior estimates of microbial fouling and permeate flux decline at different bacterial concentrations in the feed. In these tests, the permeate flux and the effluent total organic carbon (TOC) and UV absorbance at 254 nm were determined to evaluate membrane rejection and flux decline patterns. The reclaimed water used in these tests was obtained from a water reclamation facility in Southern California.

The MBR continuous flow tests that followed the flat-sheet membrane tests employed polyether sulfone hollow-fiber membranes fabricated from the same material as the flat-sheet membranes. These experiments simulated the conditions in an MBR system using PAC and acclimated microorganisms. In these studies, the permeate flux and the effluent TOC and UV absorbance (at 254 nm) were determined to evaluate flux decline patterns in the hollow-fiber membranes. The information has been useful from several standpoints including the nature and mechanisms of membrane fouling, and its impact on permeate flux decline, which will eventually lead to the design of superior hollow-fiber membranes for different MBR applications. Furthermore, the MBR tests using reclaimed water subjected to AOPs would provide a good evaluation on the impact of NOM oxidation on membrane fouling and rejection. It is anticipated that the insights gained from these studies will be useful in the development of novel membranes with superior fouling resistant and aqueous transport characteristics through material modification techniques.

(778c) Optimal Design and Operation of Real-Time Response Systems for Water Security

Angelica Mann[1], Gabriel Hackebeil[1], Jose Rodriguez[1], Katherine A. Klise[2], Carl Laird[1] and Terranna Haxton[3], (1)Chemical Engineering, Texas A&M University, College Station, TX, (2)Sandia National Laboratories, Albuquerque, NM, (3)National Homeland Security Research Center, U.S. Environmental Protection Agency, Cincinnati, OH

Drinking water distribution systems are vulnerable to accidental or intentional contamination. Contaminants can ingress into the system through pipe breakages, water sources, or outlet (e.g. fire hydrant, household faucets) backflows. One proposed approach is the installation of an contaminant warning detection system – a sparse layout of sensors to detect the presence of contaminant. Effective control and cleanup of the system is also essential to reduce population exposure to the contaminant and to return the system to an operational state. Modern mixed-integer programming strategies play a role in the design of protection systems and optimal execution of a response strategy. In conjunction with researchers at Sandia National Laboratories and the Environmental Protection Agency, we have developed several advanced problem formulations to protect the public against potential contamination events. These tools have been integrated into the EPA's Water Security Toolkit, an open-source software package being developed for use by water utilities. In this presentation, we will show three components of the Water Security Toolkit.

First, we developed a novel water quality simulation package called Merlion that increases the simulation speed of TEVASIM by about an order of magnitude for large numbers of contamination. This package is also used to efficiently generate an explicit mathematical model of the contaminant spread for use in mixed-integer programming formulations.

Second, we present a pair of mixed-integer programming (MILP) formulations for source inversion and optimal sampling. The source inversion problem involves locating the contamination source given a small set of measurements. Advancing existing techniques, we have developed an MILP formulation that makes use of discrete (yes/no) measurements from both fixed sensors and manual grab samples to successfully invert for the potential contamination sources. Using an integrated strategy, we then solve another MILP formulation to find the optimal manual grab sample locations for the next sampling cycle. This integrated strategy is effective in finding the contamination source using a small number of sampling cycles.

While identifying the contamination source, population exposure is continuing. One measure to mitigate this exposure is to increase chlorination of the system from a set of disinfectant booster stations. Finally, we propose a MILP formulation for optimally placing booster stations to help neutralize possible contamination events. Due to the large uncertainty in the contaminant injection location and time, this placement problem is a stochastic problem that would normally be solved using decomposition techniques. However, we propose a formulation and solution approach that is both fast and memory efficient. We use Merlion to rapidly solve the large number of possible contamination scenarios and formulate the stochastic programming problem. Using a set of reduction techniques that combine scenarios, but yield an exact mathematical transformation of the original stochastic problem, we can dramatically reduce the size of the complete stochastic programming problem. This approach has been used to place booster stations in a network with more than 3000 nodes and a stochastic programming formulation with more than 400,000 scenarios.

Sandia National Laboratories is a multi-program laboratory managed and operated by Sandia Corporation, a wholly owned subsidiary of Lockheed Martin Corporation, for the U.S. Department of Energy's National Nuclear Security Administration under contract DE-AC04-94AL85000.

(778d) Effects of the Different Stages of Superstructure Development On the Efficiencies and Designs of Heat-Integrated Process-Water Networks

Elvis Ahmetović[1,2] and Zdravko Kravanja[2], (1)Faculty of Technology, University of Tuzla, Tuzla, Bosnia, (2)Faculty of Chemistry and Chemical Engineering, University of Maribor, Maribor, Slovenia

In a recent paper by Ahmetović and Kravanja (2012), a novel superstructure and mixed-integer non-linear programming (MINLP) model was proposed for the simultaneous synthesis of process-water and heat exchanger networks. This superstructure combines the water network (WN) given by Ahmetović and Grossmann (2011) and the heat exchanger network (HEN) introduced by Yee et al. (1990). It includes both direct and indirect heat exchanges (isothermal and non-isothermal heat transfers) between hot and cold streams, as well as some additional opportunities for the splitting and mixing of freshwater and wastewater streams. This model's objective is to minimize the total annual costs of the overall network.

This contribution describes and discusses the usage of the simultaneous optimization model proposed by Ahmetović and Kravanja (2012) for studying its effects, during different stages of superstructure development, on the efficiencies and designs of heat-integrated process-water networks. We first present a base case-study of a water network synthesis problem and reports its freshwater and utilities' consumption. Then, different designs of heat-integrated process-water networks are synthesized and compared from the viewpoint of freshwater and utility consumption, total amount of indirect heat exchanged, and the total eat exchanger area, as well as the total annual cost. In addition, the complexities of the resulting different network designs are also discussed. Using the above-mentioned synthesis methodology, some novel water-network designs have been reproduced for certain literature case-studies with minimum total annual cost significantly better than that reported in the literature. We used different GAMS (2012) MINLP solvers (i.e. BARON, SBB, DICOPT) for model-solving during various stages of the superstructure's development. It is worth pointing out that only small-size MINLP problems regarding heat-integrated water-networks could be solved with global optimality. In other cases we used local solvers SBB and DICOPT to solve problems over reasonable computational times, and compared the obtained results. The results clearly indicated that, after allowing for additional degrees of freedom for heat transfer, and water-splitting and mixing within the network, better network designs with reduced total annual cost resulted, in comparison with those reported results in the literature.

Keywords:water and heat-exchanger networks, MINLP model, simultaneous superstructure optimization.

Acknowledgement: The authors are grateful to the Scholarship scheme for academic exchange between the EU and Western Balkan countries for providing a JoinEU-SEE postdoctoral fellowship. Financial support is also gratefully acknowledged from the Slovenian Research Agency (Program No. P2-0032), and Federal Ministry of Education and Science Bosnia and Herzegovina.

References:

Ahmetović, E. & Grossmann, I. E. (2011). Global superstructure optimization for the design of integrated process water networks. AIChE Journal 57(2): 434-457.

Ahmetović, E. & Kravanja, Z. (2012). Simultaneous synthesis of process water and heat exchanger networks. Manuscript in the preparation.

GAMS (2012). GAMS solvers. http://www.gams.com/solvers/index.htm.

Yee, T. F., Grossmann, I. E. & Kravanja, Z. (1990). Simultaneous optimization models for heat integration—III. Process and heat exchanger network optimization. Computers and Chemical Engineering 14(11): 1185-1200.

(778e) Global Optimization of Water Network Synthesis Under Uncertainty with Risk Management

Cheng Seong Khor[1,2] and Nilay Shah[2], (1)Chemical Engineering Department, Universiti Teknologi PETRONAS, Tronoh, Malaysia, (2)Centre for Process Systems Engineering, Department of Chemical Engineering, Imperial College London, London, United Kingdom

This work addresses the problem of integrated water network synthesis under uncertainty with risk management. We develop a superstructure composed of water sources and sinks with membrane and non-membrane separation-based water regenerators, which leads to a mixed-integer nonlinear program (MINLP). The MINLP involves continuous variables on water flowrates and contaminant concentrations, and 0–1 variables on existence of interconnections of the network elements. Detailed nonlinear models are adopted for membrane regenerators such as reverse osmosis. To address uncertainty in the problem, we employ the classical recourse-based two-stage stochastic programming formulation by using discrete scenarios to approximate the underlying probability distribution of the uncertain parameters. The representation gives rise to a multiscenario two-stage stochastic MINLP, or more specifically, a mixed-integer quadratically-constrained quadratic program (MIQCQP). The framework comprises network design in the first stage and its operation in the second stage acting as recourse. In addition, we extend the model to address risk management using the Conditional Value-at-Risk (CVaR) metric. However, a large number of scenarios are often required to capture the underlying uncertainty of the problem, causing the model to suffer from the curse of dimensionality. To circumvent this issue, we propose a stepwise solution strategy mainly driven by solving the risk-neutral model using a small number of scenarios to reduce the computational load in the scenario-ordering step of simulating the Value-at Risk (VaR) parameter, as well as to provide initial values for solving the risk-averse mean–CvaR model. The presence of nonconvex bilinear terms necessitates the implementation of global optimization techniques to ensure reliable solutions. In this regard, we conduct computational experiments on both the risk-neutral and risk-averse formulations to study the performance of a new global MIQCQP solver, GAMS 23.8.1/GloMIQO 1.0.0, in comparison with the latest releases of two state-of-the-art commercial global solvers, namely BARON (version 10.1.3) and LindoGLOBAL (version 6.1.1.488). By varying the relative optimality tolerance, the solver performances are assessed mainly in terms of CPU time and solution quality of the resulting water network generated. Where possible, we analyze the performances by inferring on the strengths of the algorithmic components in the solvers.

(778f) Cooling Tower Wastewater Treatment and Reuse

Anditya Rahardianto, Han Gu, Larry Gao, John Thompson, Panagiotis D. Christofides and Yoram Cohen, Chemical & Biomolecular Engineering, University of California, Los Angeles, Los Angeles, CA

Cooling towers are ubiquitous across many industries, from large-scale power generation to commercial buildings. Cooling towers consume large volumes of water due to evaporative and windage/drift losses. To limit the accumulation of dissolved and suspended solids (and thus to mitigate mineral scaling, corrosion, and biological growth) in cooling water systems, periodic discharge of cooling water (i.e., blowdown) is needed, contributing to additional loss of cooling water that must also be replenished. Applications of advanced membrane treatment technologies (ultrafiltration (UF) and reverse osmosis (RO)) for cooling tower wastewater (CTW, i.e. blowdown water) treatment and reuse, while environmentally and economically desirable, remain limited. CTW treatment and reuse reduce blowdown water loss, while at the same maintain cooling water quality at levels necessary for effective mitigation of cooling water system corrosion/scaling. In the present work, a fundamental computational framework was developed to quantify and assess the operational costs and benefits of CTW treatment via integrated UF-RO processes. Potentially effective UF-RO operational strategies for CTW treatment were assessed and their optimal process conditions were identified. The potential benefits of advanced adaptive UF-RO operation that optimally handles fluctuations in cooling tower system conditions and UF-RO operational costs (e.g., cooling water quality, peak/non-peak electricity costs) were examined, including both steady-state and intermittent operational modes. Examination of UF-RO process operational requirements for CTW treatment will be presented focusing on the results of a recent field study using a novel UF-RO pilot system operation in a co-generation plant at the University of California, Los Angeles. Results from the field study demonstrated both a significant level of water savings as well as monetary savings for the facility. The overall analysis approach and field study results demonstrate that there is a range of scenario for beneficial CTW treatment and reuse, which would be best served by a self-adaptive treatment process.

Improving the quality of Pulp & Paper Mill waste water employing Electroflocculation Technology

Rita Tandon, Satya Dev Negi, R.M.Mathur

Abstract

Pulp and Paper Industry is one of the highest polluting industries. The wastewater from this industry has very high proportion of organic pollutants and significant color loads. Although there is no stringent legislation to the tolerance limits for discharged color in effluents but some state pollution control authorities have imposed tolerance limits for the color in discharged effluents. The present practice of color removal involves precipitation and coagulation of colored substances using aluminium salts. Although such chemical treatment methods are expensive but the industry has perforce adapted these methods to contain the tolerance limits for color discharge.

Lot of research has been carried out worldwide on various EOP options for removal of color from pulp & paper mill effluent. One such option is electroflocculation (EF) technology which has been extensively studied by Central Pulp & Paper Research Institute (CPPRI). The present paper highlights the findings of pilot plant trials conducted by CPPRI on electroflocculation technology at a mill site to assess the techno-economic viability of the technology for treating different colored streams for improving the quality of pulp & paper mill waste water.

Introduction

Public demand for cleaner environment is forcing the industry to take adequate steps to reduce pollution loads in discharge effluents. In the recent times, the presence of color in pulp & paper mill effluents has emerged as an major environmental issues due to public perception and this has prompted the Regulatory Bodies to include color as a pollutant in the charter on **Corporate Responsibility for Environmental Protection (CREP)** for compliance by Indian Pulp & Paper Industry.

The Indian paper industry uses on an average 100 to 200 m^3 of fresh water per ton of paper and nearly 75% of water used is discharged as effluent. In addition to high proportion of inorganic and organic pollutant the effluent from paper industry contains significant color loads which varies from mill to mill depending upon the raw material used, process employed, type of end products and extent of system closure in the mill.

Broadly the color reduction strategies include at source control through process modification or through end-of-pipe (EOP) treatment methods. Since the process modification demands proper planning and timely execution with substantial investments, the focus is more on EOP options as an immediate measure to control effluent color. Lot of research was done worldwide on various EOP options, which included physico-chemical treatment method, physical separation method, UV irradiation method, biological method, etc., however, among various options tried, chemical precipitation using coagulants and flocculants is the most accepted option for color removal as it is easy to implement the process in existing ETP's without requiring any major investment, but the process has limitation particularly in sludge handling.

In an endeavor to identify a suitable technology for color removal from pulp & paper mill effluents, CPPRI focused on EF technology and extensively studied the technology on lab and pilot scale to establish the techno-economic viability of the process for removal of color from pulp & paper mill effluents.

The present paper highlights the findings of up scaling of the electro-flocculation technology to pilot scale and the viability of the technology for treating pulp & paper mill effluents for color removal.

Results & Discussions

Sources and Magnitude of Color in Paper Industry

Pulp and paper mill effluents have a characteristic brownish black color, which is mainly due to lignins, tannins and other extractives bearing, chromophoric groups. The magnitude of color depends on the raw material used process employed and type of end products produced. Lignin and extractives are highly polymerized substances and are difficult to biodegrade.. The color in small mills is mainly due to lignin going through spent liquors. It is estimated that the quantity of lignin going through spent liquors varies from 300-400 kg/ton of pulp, generating color load of about 1400-1500 kg PCU / ton of pulp. It is estimated that 90% of the color is due to lignin.

In alkali extraction stage effluent only about 50-60 kg of lignin / ton of paper is going into effluents and the combine effluents has color load of about 1500 PCU. In newsprint mills where eucalyptus constitutes the raw material for production of mechanical pulp component, very high color load are noticed in the effluents. Eucalyptus contains about 3-6 % extractives, mostly tannins, which are leached out during pre steaming and refining stages. The washing of CMP pulp is highly colored and the color intensity is several times higher when compared to color due to lignin compounds. **Table-1** summarizes the magnitude of color load in different effluent streams of pulp & paper industry. **Table-2** summarizes the molecular size of chromophores present in the effluent which indicates that except for black liquor, the molecular size of color bodies varies between 5,000-13,000 Dalton units.

Table- 1
Magnitude of Color Load and respective Chromophores in different Effluent Streams of Indian Pulp & Paper Industry

Streams	Total Solids (g/l)	Color (PCU)	Chromophores
Agro-based Mills (Chemical Pulping)			
Black liquor (Bleached Grade)	**40**	**82000**	**Lignin & its degraded products**
Black liquor (Unbleached Grade)	**13**	**6333**	**Lignin & its degraded products**
Combined Mill Effluent (Bleached Grade)	**3.3**	**3100**	**Lignin & its degraded products**
Combined Mill Effluent (Unbleached Grade)	**3.5**	**3030**	**Lignin & its degraded products**
Wood Based Mills (Chemical Pulping)			
Ep Stage Effluent (Eucalyptus, Poplar)	**3.8**	**1600**	**Chlorolignins**
Combined Mill Effluent (Eucalyptus, Poplar)	**3.2**	**2000**	**Degraded lignin products**
Rayon Grade Washing filtrate (Mixed hardwood)	**0.9**	**1200**	**Degraded lignin products**
Wood Based Mills (Mechanical Pulping)			
Pulping Effluent (Eucalyptus)	**1.3**	**8816**	**Extractives (Poly phenols & tannins)**

Source- CPPRI

Table- 2

Molecular Size of Chromophores in different Effluent Streams of Indian Pulp & Paper Industry

Streams	Total Solids (g/l)	Color (PCU)	Molecular Weight* (Dalton)
Agro-based Mills (Chemical Pulping)			
Black liquor Bleached	**40**	**82000**	**5,000-50,000**
E/Ep Stage Effluent	**2.6**	**3800**	**12,000-13,000**
Combined Mill Effluent	**3.3**	**3100**	**7,000**
Wood Based Mills (Chemical Pulping)			
Ep Stage Effluent	**3.8**	**1600**	**7,000-13,000**
Combined Mill Effluent	**3.2**	**2000**	**5,000-7,500**

Source- CPPRI

Application of Electroflocculation (EF) Technology for Color Removal from Pulp & Paper Mill Effluents

Electro-coagulation and electro-flocculation are two techniques involving the electrolytic addition of coagulating metal ions directly from sacrificial electrodes. These ions coagulate with pollutants in water, in a similar manner to that of adding coagulating chemicals such as alum and ferric chloride and allow the easier removal of the pollutants. The system produces about half to one third volume of the sludge.

The Electroflocculation (EF) process involves the application of an electric current to sacrificed metal electrodes having coagulating properties, like Al, Fe, or Mg, generating metal ions and gas bubbles simultaneously. The metal ion released combines with pollutant and gas bubbles generated capture the coagulant agglomerates resulting in most of the pollutant being floated to surface.

The total process is accomplished in three steps as shown below.

1st Step

At Anode $$M \xrightarrow{\text{Electrolytic Dissociation}} M^{2+} + 2e^-$$

At Cathode $$2e^- + 2H_2O \xrightarrow{\text{Electrolytic Dissociation}} 2OH^- + H_2 \uparrow$$

2nd Step

$$M2+ \ + OH^- \longrightarrow M(OH)_2 . MOH^+$$

Insoluble metal hydroxide

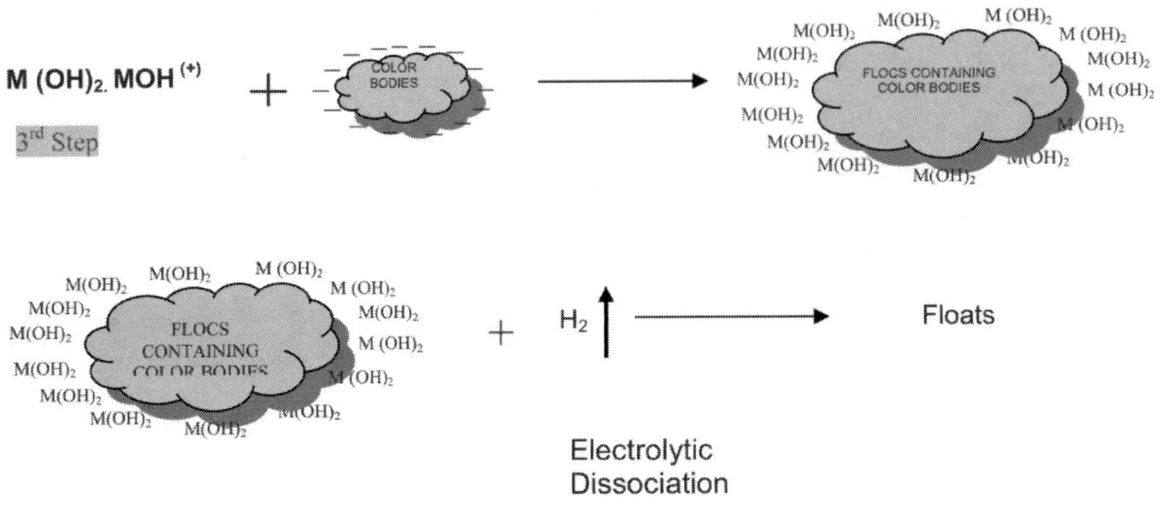

Optimization of Process Variables

To optimize the process conditions extensive research was conducted at lab scale and following variables were studied extensively-

- Type of electrode
- pH
- Temperature
- Total Solids Concentration
- Reaction time
- Optimum Voltage
- Optimum Current
- Electrode Spacing
- Current Density
- Conductivity

The effect of pH, temperature, retention time is shown in **fig. 1, 2 & 3**.

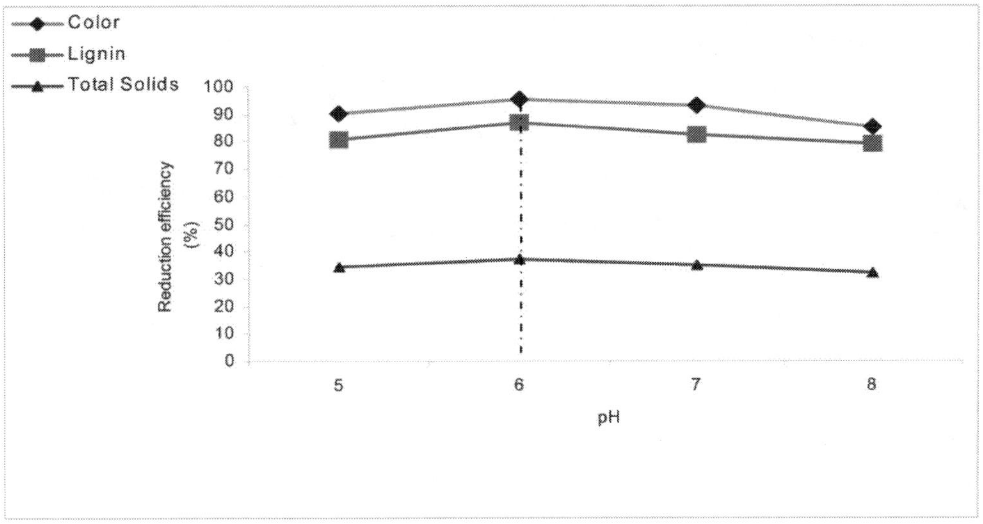

Fig-1 pH VS Process Efficiency

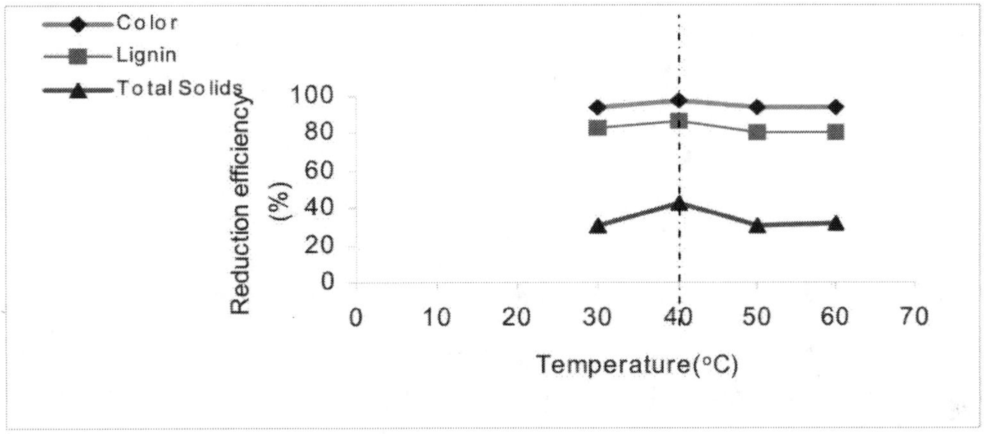

Fig-2 Temperature Vs Process Efficiency

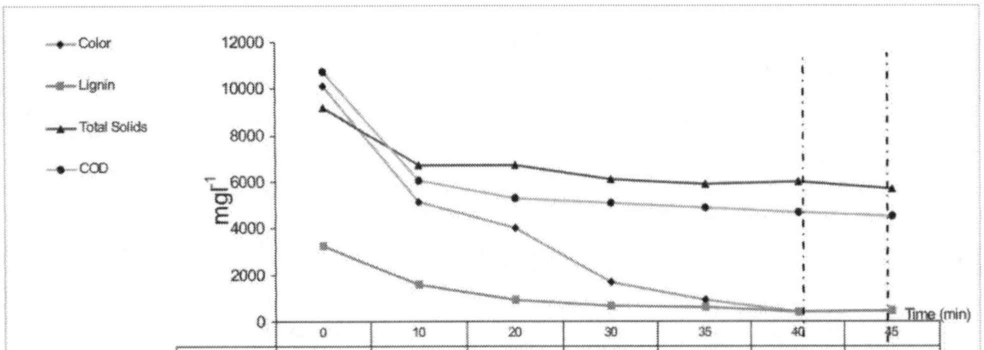

	0	10	20	30	35	40	45
Color	10124	5137	4000	1686	932	405	436
Lignin	3240	1600	890	640	630	410	480
Total Solids	9200	6700	6700	6100	5900	6000	5700
COD	10712	6034	5299	5106	4874	4680	4526

Fig-3 Reaction Time Vs Process Efficiency

Optimization of electrode spacing

Experiments were conducted using different electrode spacing viz 0.5 cm, 1cm and 2 cm. It is observed that the efficiency with respect to removal of color and lignin has increased from 59%, 71% at 0.5 cm spacing to 95%, 84% to 1 cm spacing and then leveled off. As shown in **fig. 4** the optimum electrode spacing was found to be 1 cm.

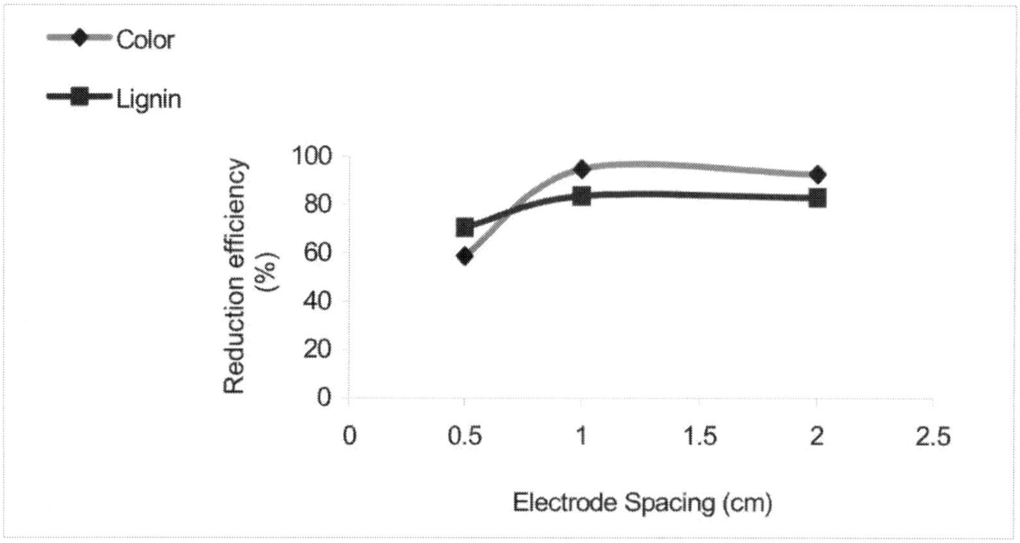

Fig-4 Electrode Spacing Vs Process Efficiency

Pilot Plant Trials on Electroflocculation Technology Conducted in wood based Paper Mill, India

Having established the technical feasibility of the process in lab scale, it was felt necessary to assess the economic viability of the technology on pilot scale. A pilot plant of $1m^3/hr$ capacity was fabricated, installed and commissioned at Wood based paper mill, India and pilot plant trials were conducted on different colored effluent streams.

A schematic of the pilot plant is depicted in **fig.5**

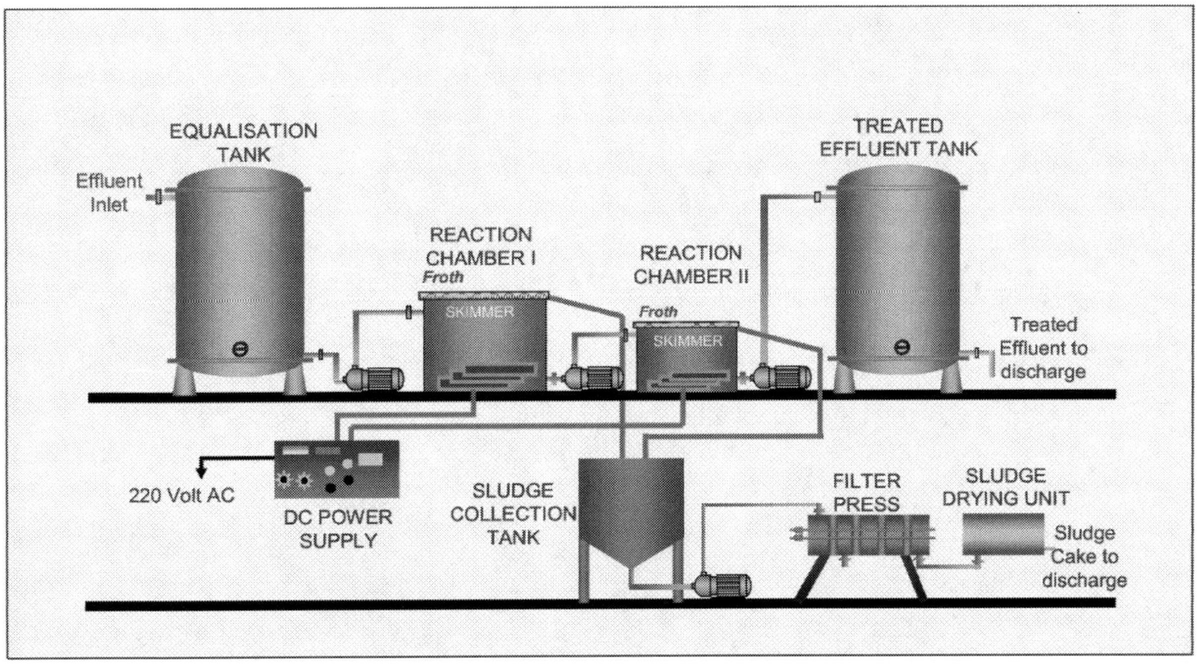

Fig-5 Schematic Of Pilot Electroflocculation Plant of 1m3/hr Capacity Installed in a Wood based Paper Mill, India

Pilot Trials with Bleach Plant Effluent

Trials conducted with bleach plant effluent having initial color load of 518 PCU had resulted into reduction on color by 86%, lignin by 86%, COD by 77% and AOX by 73%. The final achievable color load was 70 PCU. Finding of the trials are depicted in **table- 3**.

Table-3

Trials on Bleach Plant Effluent

	Untreated	**Treated**	**% Redn.**
Color, PCU	518	70	86
Lignin, mg/l	137	19	86
COD, mg/l	768	176	77
AOX, mg/l	15	4	73
Flow rate- $1m^3/hr$, Voltage- 6.0 volts, Current Density- $11\text{-}44 mA/cm^2$, Power consumed- $1.4 kWh/m^3$			

Pilot Trials with RGP Washing Effluent

Trials conducted with RGP washing effluent having initial color load of 1143 PCU had resulted into reduction on color by 88%, lignin by 67% and COD by 38%. The final achievable color load was 140 PCU. Finding of the trials are depicted in **table- 4**.

Table-4

Trials on RGP Washing Effluent

	Untreated	Treated	% Redn.
Color, PCU	1143	140	88
Lignin, mg/l	153	51	67
COD, mg/l	600	372	38
Flow rate- 1m³/hr, Voltage- 6.0 volts, Current Density- 11-44mA/cm², Power consumed- 1.16 kWh/m³			

Pilot Trials with Bagasse Black Liquor

Trials conducted with bagasse black liquor having initial color load of 1178 PCU had resulted into reduction on color by 89%, lignin by 71% and COD by 47%. The final achievable color load was 135 PCU. Finding of the trials are depicted in **table- 5**.

Table-5

Trials on Bagasse Black Liquor

	Untreated	Treated	% Redn.
Color, PCU	1178	135	89
Lignin, mg/l	350	100	71
COD, mg/l	1151	611	47
Flow rate- 1m³/hr, Voltage- 6.0 volts, Current Density- 11-44mA/cm², Power consumed- 1.55 kWh/m³			

Techno-Economic Viability of EF Technology

Based on the findings of the pilot plant trials the techno-economic viability of the process was assessed for treatment of different colored effluents. The results are summarized in **table -6**.

It is estimated that-

- **The cost of treatment for treating bleach plant effluent with input color of 518 PCU and discharge color of 70 PCU works out to be Rs.22/ m^3.**

- **The treatment cost for treating combined mill effluent with input color of 1800 PCU and discharge color of 250 PCU works out to be Rs. 18/m^3.**

- **The treatment cost for treating agro-based black liquor with input color of 3000 PCU and discharge color of 450 PCU works out to be Rs. 53/m^3.**

Table-6
Economic viability of EF Technology for treating different colored effluents

Effluent type	Bleach Plant Effluent	Combined Mill Effluent	Agro-based black liquor
Initial Color, PCU	518	1800	3000
Final Achievable Color, PCU	70	250	450
Color Redn. Efficiency, %	85	85	85
Total Treatment Cost Rs./m^3	22	18	53

Conclusion

1. The electro-flocculation process is technically feasible to remove color from all types of colored effluents generated in Pulp & Paper Industry. Color removal efficiency achieved is in the tune of 85-90%. The voltage requirement is 6.0 volts.
2. Besides color, the process also removes lignin to the tune of 65-80%, COD 45-65%, and Toxicity in terms of AOX 70-75%.
3. Based on pilot scale studies, economics for treatment of pulp & paper mill effluent has been worked out to be between Rs. 18 to Rs.53 per m^3 to achieve a discharge color in the range of 70-450 PCU, depending upon the input color load.
4. The technology needs to be demonstrated on commercial scale, however, it is recommended that individual pulp & paper mills shall make their own evaluation and conduct pre-feasibility studies before adoption in each case.
5. EF technology may also be tried for other industries and small scale industries as well, which generates colored effluents viz. textile, dyes manufacturing units, tanneries, ink manufacturing units, etc.

References

1. Subrahmanyam, P. V. R., Parekh, R.C. and Mohan Rao, j. G., Ippta, Vol. 9, No. 1, 1972, p. 14.
2. Sharma, G. D., Singh, M.N. and Agarwal, N.R., Ippta , Convention: 124(78).
3. 'Pollution Engineering practice Handbook' Ann Arbor Science Publishers. p. 419-426 (1975).
4. Hillis, W. E., Wood extractives and their significance to pulp and paper industry. p.59, Academic Press, New York-1962.
5. Kulkarni A.G., Mathur R. M., Gupta Abha and Pant, Rajesh. Ippta, Vol. 24, No. 3, !987.

6. Metijevic, E. and Janauer, G. E., J. Colloidal Interface Science, 21:68 (1966).

7. Matijevic, E. and Stryker, L. J., J. Colloidal Interface Science, 22:68 (1966).

8. Fernandes, J. D., Thampi, K. P., Prahaladhan, V. M. Singh, Sarju, Ippta, Vol. 22, No. 2, June 1985.

9. Kulkarni, A. G., Mathur, R. M., Tandon Rita, Naithani, S. and Pant, Rajesh, Ippta, Vol.25, No. 1, 1988.

10. Kulkarni, A. G., Mathur, R. M., Abha Gupta, Naithani, S. and Pant, Rajesh, Ippta, Seminar, 1989.

(778h) A Combined Cost Model for Analysis of Degraded Water Utilization in Thermoelectric Power Plants

Iman Safari[1], Michael E. Walker[1], Javad Abbasian[2], Hamid Arastoopour[3], Ming-Kai Hsieh[4], Ranjani B. Theregowda[4], David A. Dzombak[4], Liu Wenshi[5], Radisav Vidic[6] and David C. Miller[7], (1)Chemical and Biological Engineering, Illinois Institute of Technology, Chicago, IL, (2)Chemical and Biological Engineering Department, Illinois Institute of Technology, Chicago, IL, (3)IIT, Chicago, IL, (4)Department of Civil and Environmental Engineering, Carnegie Mellon University, Pittsburgh, PA, (5)Civil and Environmental Engineering, Carnegie Mellon University, Pittsburgh, PA, (6)Department of Civil and Environmental Engineering, University of Pittsburgh, Pittsburgh, PA, (7)U.S. Department Of Energy, National Energy Technology Laboratory, Morgantown, WV

The thermoelectric power industry in the U.S. uses a large amount of freshwater. The large water demand is increasingly a problem, especially for new power plant development, as availability of freshwater for new uses diminishes in the United States. Reusing degraded water sources, such as treated municipal wastewater, provides one option to mitigate freshwater usage in the thermoelectric power industry. The amount of freshwater withdrawal that can be displaced with degraded water sources at a particular location requires evaluation of the water management and treatment requirements, considering the quality and abundance of the non-traditional water sources.

This paper presents a modular framework for evaluation of the economic impacts of utilization of degraded waters for cooling in thermoelectric power systems. Various models are presented for estimation of water quality in a recirculating cooling system, fouling rate in the condenser, and the costs associated with fouling given the make-up water quality, the treatment strategy, and the operating conditions. The results of the overall economic analysis performed on several case studies are discussed.

(779a) Understanding the Pretreatment Chemistry of Softwoods

Carlos Alvarez-Vasco and Xiao Zhang, Voiland School of Chemical Engineering and Bioengineering, Bioproducts Science Engineering Laboratory, Washington State University, Richland, WA

Recalcitrance of lignocellulose is a major barrier to the economical development of biobased fuels and products. To overcome the recalcitrance of lignocellulosic biomass, many chemical, thermal and biological pretreatments have been proposed. The effectiveness of the pretreatments has been done mainly based on the sugar yield and the hydrolysability of resulting substrate. There is a little understanding of the reaction chemistry occurred during these pretreatment methods.

In this study, we evaluated the pretreatment of softwood species in the Pacific Northwest region by using sulfite, wet oxidation and diluted acid pretreatment methods. The effects of different pretreatment chemistries on the structure of biomass, production of chemical intermediates, carbohydrate recovery and enzymatic hydrolysability of pretreated substrate biomass were determined. The structural changes occurred on the biomass lignin and carbohydrate during various pretreatment conditions were investigated by HPLC, GC/MS, and NMR. The results obtained from this study can help us to optimize softwood pretreatment process to maximum the production of biofuels and biobased products.

Upgrading of Biomass-Derived Liquid to Clean Fuels

Siauw H. Ng[1], *Cecile Lay*[1], *Sooraj Bhatt*[1], *Barry Freel*[2], *and Robert Graham*[2], *(1) CanmetENERGY, Natural Resources Canada, Devon, AB, Canada, (2) ENSYN Technologies Inc., Ottawa, ON, Canada*

Abstract

A conventional heavy gas oil (HGO, used as a base oil) and a blend containing 5 wt% Renewable Fuel Oil (RFO, homogeneous, extremely low in sulfur but high in water) obtained from fast thermal conversion of wood residue, were catalytically cracked at CanmetENERGY, Devon. Tests were conducted in a fluid-bed microactivity test (MAT) reactor loaded with a commercial equilibrium catalyst at 510 °C with varying weight hourly space velocity (WHSV, 10–30 gh^{-1}/g) with constant oil injection time at 30 s. The water-free total liquid products (TLPs) were characterized for hydrocarbon type, oxygen, sulfur, and nitrogen. The key observation on catalytic cracking of the blend was the formation/retention of water and carbon dioxide (no carbon monoxide) plus possible oxygenates as products at the expense of gasoline (IBP–221 °C) and liquefied petroleum gas (LPG, i.e., C_3+C_4) production. At constant test severity, the blend gave higher conversion than the base oil. At a given conversion, yields of dry gas (H_2–C_2), diesel (221–343 °C), heavy fuel oil (343 °C–FBP), and coke were approximately equal for the two feeds while yields of LPG and gasoline were slightly lower for the blend. Given the characteristics of the RFO (i.e., lower carbon and energy content than HGO), yields of gasoline and diesel in particular were higher than expected. In fact, on an energy equivalency basis, yields of gasoline and diesel were significantly higher than their corresponding counterparts obtained from HGO alone. Characterization results demonstrated that concentrations of all crackable components in gasoline, i.e., saturated and unsaturated iso-paraffins, n-paraffins, and naphthenes decreased with increasing conversion. In contrast, the concentration of uncrackable aromatics increased. As a result, the sulfur content, octane numbers, and specific gravity of the gasoline also increased. After cracking, most of the oxygen in the blend appeared as H_2O, with the rest in the form of CO_2. The liquid products were analyzed for oxygen content and found to be below the detection limit of 0.25 wt%. Cracking phenomena and the potential of processing conventional FCC feed mixed with RFO were discussed.

Key Words: Fast thermal conversion, fast pyrolysis, biomass, renewable fuel oil, forest waste, residual wood, agriculture residue.

1. Introduction

Global warming stemming from the increase in carbon footprint or greenhouse gas (GHG) emissions has become a worldwide concern in recent years. Oil production and processing, especially when involving nontraditional origins such as oil sands and oil shales are considered major contributors to atmospheric CO_2 buildup. To reduce the negative impact along this line and consider the declining of fossil resources, biofuels are gaining industrial and scientific interests. Biofuels, a source of renewable energy, refer to a wide range of fuels from biological origins, especially biomass. In 2008, biofuels provided 1.8% of the world's transportation fuels.[1] The European Union set targets for biofuel usage of 2% in 2005 and 5.75% in 2010.[2] In Canada, regulations also require an average renewable fuel content of five per cent in gasoline effective December 15, 2010.[3]

In 2010, we reported a study[4] in which the oil from Canola, a cultivated variety of rapeseed, was catalytically cracked in the presence or absence of a conventional fluid catalytic cracking (FCC) feed, a heavy gas oil (HGO). In the present study, the focus is on the production of "fungible" transportation fuels from cellulosic biomass considered to be an ideal feedstock since it is abundant, relatively inexpensive, and does not compete with the global food supply. In summary, the objective of this study is to develop and demonstrate a viable economical pathway to produce renewable transportation fuels from cellulosic biomass such as wood chips, forest wastes, and agriculture residues. The ultimate goal is to mitigate the contribution of petroleum fuels to global warming and to offset the declining fossil resources thereby increasing future national energy security.

2. Experimental

A biomass-derived liquid was obtained from Ensyn's Rapid Thermal Processing (RTP™) plant where wood residue was thermally cracked at mild temperature in a short duration (typically less than 5 seconds) with about 80 wt% liquid yield. The homogeneous liquid, for use as a fuel in boilers or in diesel and stationary turbines, is called "Renewable Fuel Oil" (RFO). RFO was blended with a conventional HGO from a US refinery in 5:95 weight ratio. Properties of the two parent feeds and their blend are given in Table 1. It is important to note that the RFO contains approximately half of the energy and carbon content of HGO, and the total energy and carbon content of a blended feedstock is therefore less than that of a pure HGO on an equivalent weight basis. This must be considered when product yields are reported and compared. The HGO (base oil) and the 5 wt% RFO blend were cracked in a fluid-bed microactivity test (MAT) unit at 510 °C and 10–30 gh^{-1}/g WHSV, with a constant oil injection time of 30 s using the same equilibrium catalyst in the previous study.[4] Details of the experiment have been reported elsewhere.[5]

3. Results and Discussion

3.1. Conversion

In FCC operation, conversion is defined as the portion of the feed converted to 221 °C– products, including gas and coke. Figure 1 shows that conversion increased with catalyst/oil (C/O) ratio. At a given ratio, the 5 wt% RFO blend gave 1–3 wt% higher conversion than the base oil. The gap between the two curves diminished at high C/O ratios.

---

[1] "Towards Sustainable Production and Use of Resources: Assessing Biofuels". United Nations Environment Programme 2009-10-16.

[2] Selçuk, S., Kaygusuz, K., Sari, A., Energ. Sour. 2004, *26* 1119–1129.

[3] The "Renewable Fuels Regulations", published on September 1, 2010 in the "Canada Gazette, Part II".

[4] Ng, S.H., Shi, Y., Ding, L., Chen, S., Gieleciak, R., Hager, D. "Catalytic Cracking of a Rapeseed Oil for Production of Transportation Fuels and Chemicals: Yield Structure", Proceedings of the 2010 AIChE Annual Meeting, Salt Lake City, UT, 2010.

[5] Ng, S.H., Zhu, Y., Humphries, A., Zheng, L., Ding, F., Gentzis, T., Charland, J. P., Yui, S. "FCC Study of Canadian Oil-Sands Derived Vacuum Gas Oils. 1. Feed and Catalyst Effects on Yield Structure", *Energy Fuels* 2002, *16*, 1196–1208.

Table 1. Feed Properties

Feed Name	RFO		Base Feed	5 wt% RFO
	As is	MAF	As is	As is
Water Content, wt%	26.58	0	–	1.896
H in H2O, wt%	2.95	–	–	–
O in H2O, wt%	23.63	–	–	–
Viscosity @ 40°C, cSt	58.9	–	–	–
Ash Content, wt%	0.02	0	–	–
Solids Content, wt%	0.04	–	–	–
Density @ 20°C, kg/dm^3	1.2043	–	0.9200	0.9321
P^h	2.44	–	–	–
Carbon Content, wt%	47.22	64.33	88.02	–
Hydrogen Content, wt%	7.37	–	12.17	–
H in H2O, wt%	2.95	–	–	–
H in Organic, wt%	4.42	6.02	–	–
Nitrogen Content, wt%	0.21	0.29	0.09556	–
Sulfur Content, wt%	0.018	0.025	0.179	–
Oxygen Content, wt%	39.85	–	–	–
O in H2O, wt%	23.63	–	–	–
O in Organic, wt%	16.22	22.10	–	–
HHV, BTU/lb	7369	–	17986	17456
Chlorine (Cl), ppm	29.8	–	–	–
MCRT, wt%	5.33	7.26	0.00	0.27
Distillation (D1160), wt%	–	–	–	–
Water Phase (up to 105°C)	41.23	–	–	–
Oil Phase (Residue)	58.34	–	–	–
Loss	0.43	–	–	–
SARA	–	–	–	–
Saturates	–	–	57.7	–
Aromatics	–	–	45.2	–
Polars	–	–	3.1	–

Note – MAF: Moisture and ash free basis

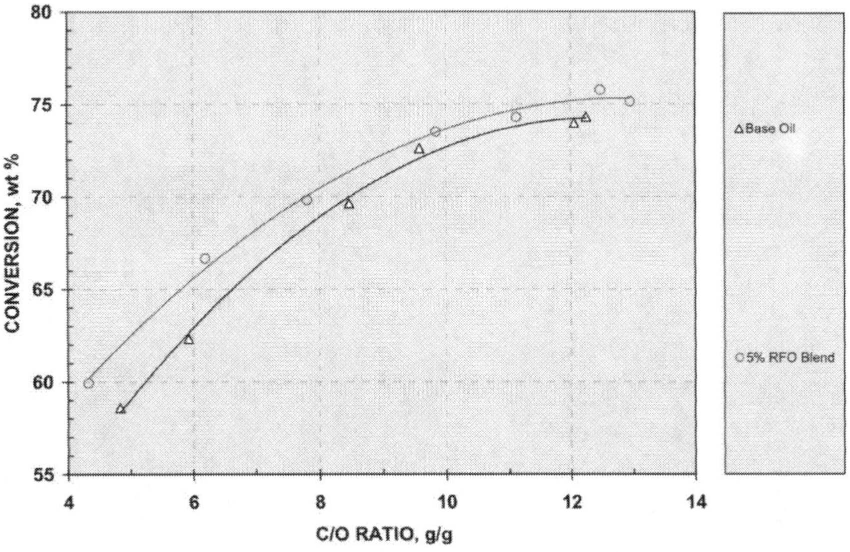

Figure 1. Relationship between Conversion and C/O Ratio

3.2. Dry Gas

Dry gas is composed of H_2, H_2S, CO, CO_2, and C_1–C_2 hydrocarbons. Figure 2 shows that dry gas yield increased exponentially with conversion. At a given conversion, the two feeds gave almost identical dry gas yields. Only CO_2 but not CO was detected during cracking of the two feeds with 0.02–0.08 wt% CO_2 yield higher for the blend at 65–75 wt% conversion indicating the decomposition or combustion of the oxygenates in the blend. However, the blend produced less H_2 by 0.06 wt% throughout the entire conversion in this study possibly due to water formation.

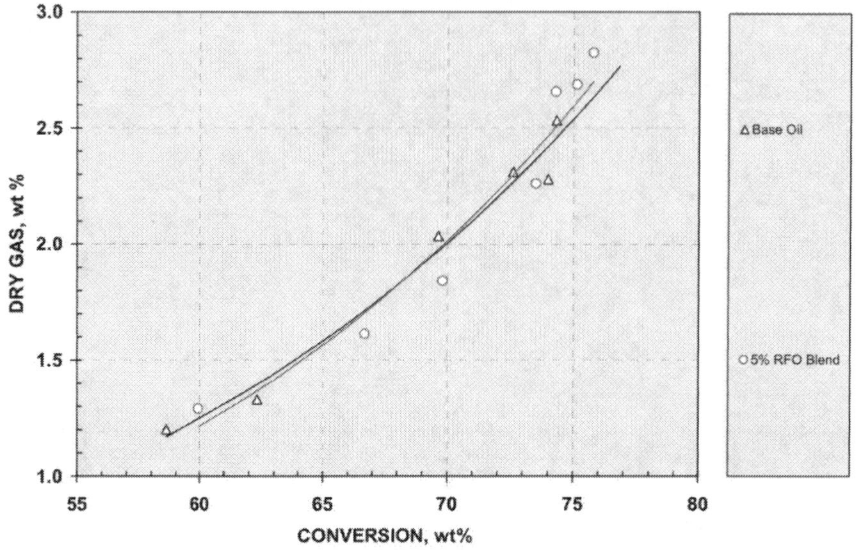

Figure 2. Relationship between Dry Gas Yield and Conversion

3.3. Liquefied Petroleum Gas

Liquefied petroleum gas (LPG) consists of C_3 and C_4 gaseous hydrocarbons. LPG is considered a valuable product since its components can be used as alkylation and petrochemical feedstocks. Figure 3 shows that, in contrast to the dry gas yield, the blend exhibited lower LPG yield by 0.3–0.8 wt% at a given conversion than the HGO feed. This was probably caused by the reduced gasoline yield (to be discussed later) of the former as catalytic cracking is a stepwise process.

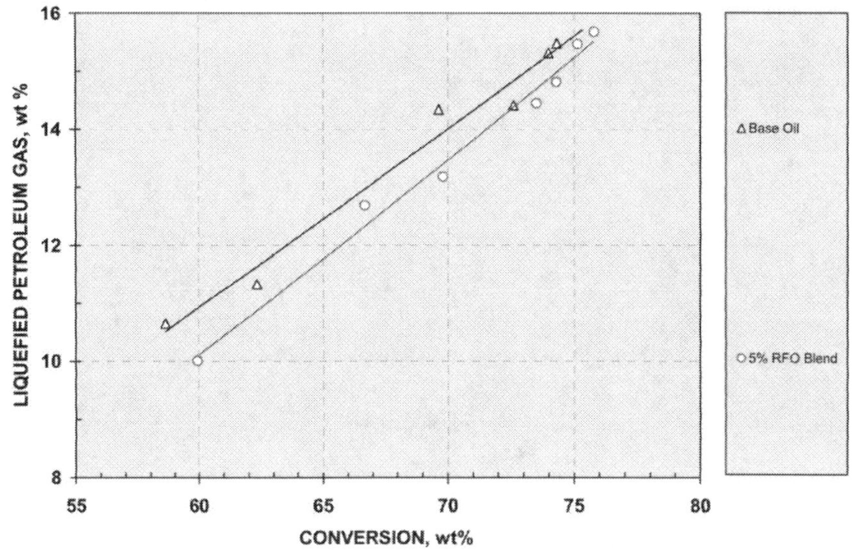

Figure 3. Variation of Liquefied Petroleum Gas Yield with Conversion

3.4 Gasoline

Gasoline (IBP–221 °C boiling point) is the major and the most desirable product in FCC operation. Figure 4 demonstrates that at a given conversion, the blend lowered the gasoline yield by less than 1 wt% until the conversion was higher than 70 wt%. Note that the blend itself contained 1.33 (calculated from RFO analysis) to 1.90 wt% (Table 1) H_2O which could partially explain the drop in gasoline. Overcracking was observed for the blend at 75–80 wt% conversion.

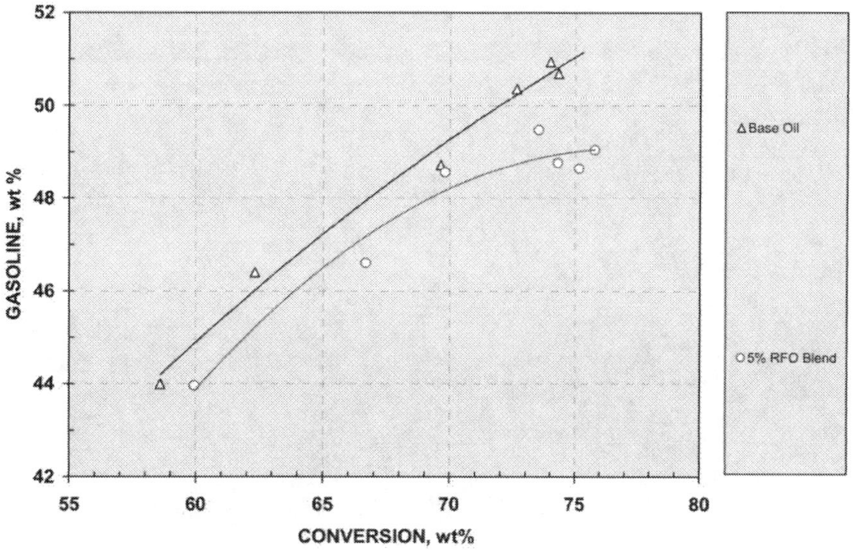

Figure 4. Relationship between Gasoline Yield and Conversion

3.5 Coke

In FCC operation, coke is used to supply heat for feed preheating and cracking. However, too much coke can seriously poison the catalyst and overload the air blower during catalyst regeneration, causing excessively high temperatures in the regenerator. Figure 5 shows that, similar to the dry gas, both feeds gave almost identical coke yield at a given conversion although the blend had 0.27 wt% higher Conradson Carbon Residue (CCR or MCRT in Table 1).

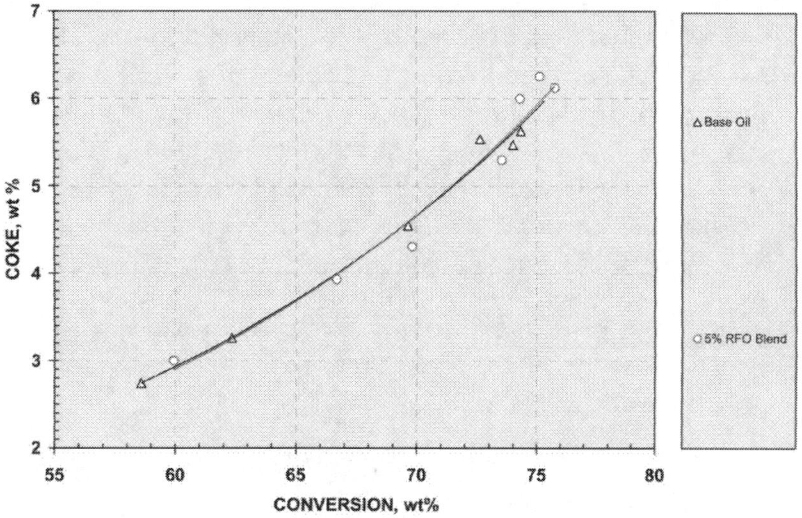

Figure 5. Relationship between Coke Yield and Conversion

3.6 Oxygen Balance

Table 2 gives the oxygen distribution in the gaseous and liquid products. After cracking, most of the oxygen in the blend appeared as H_2O (74.6–94.1 wt%), with the rest forming CO_2 (0.7–5.3 wt%). The liquid products were analyzed for oxygen content and found to be below the detection limit (0.25 wt%). More study is necessary to clarify the existence of oxygen compounds (oxygenates) in TLP.

Table 2. Oxygen Balance in Cracking 5 wt% RFO Blend

Run Number		3222D	3223D	3225DS	3230S	3232S	3233S	3227S
Feed		5 wt% RFO Blend						
Catalyst		EC-2007						
Coke Determination		In situ						
Catalyst Contact Time	**sec**	**30**	**30**	**30**	**30**	**30**	**30**	**30**
Feed Charge	g	2.0647	1.4407	1.1440	0.9075	0.8035	0.7163	0.6899
Catalyst/Oil Ratio	**g/g**	**4.33**	**6.20**	**7.81**	**9.84**	**11.12**	**12.47**	**12.95**
WHSV	g/h/g	27.74	19.36	15.37	12.19	10.79	9.62	9.27
Liquid yield (incl. H2O)	wt %	73.49	67.17	66.36	60.77	59.56	59.33	60.43
Normalized Mass Balance	wt % of Feed							
CO		0.00	0.00	0.00	0.00	0.00	0.00	0.00
CO2		0.29	0.24	0.29	0.41	0.46	0.42	0.45
H2O		**1.67**	**1.84**	**1.92**	**2.03**	**2.07**	**2.11**	**2.09**
Conversion		**59.94**	**66.67**	**69.80**	**73.51**	**74.30**	**75.78**	**75.14**
O2 conc. in TLP	wt%	< 0.25	< 0.25	< 0.25	< 0.25	< 0.25	< 0.25	< 0.25
O2 as CO (rel to O2 in feed)	**wt%**	**0.00**	**0.00**	**0.00**	**0.00**	**0.00**	**0.00**	**0.00**
O2 as CO2 (rel to O2 in feed)	wt%	5.25	0.70	0.69	3.02	3.43	2.57	2.85
O2 as H2O (rel to O2 in feed)	**wt%**	**74.61**	**81.83**	**85.70**	**90.44**	**92.17**	**94.10**	**93.25**
O2 in TLP (rel to O2 in feed)	wt%	< 9.54	< 8.71	< 8.59	< 7.78	< 7.66	< 7.60	< 7.75
O2 balance (approximate)	wt%	89.41	91.24	94.97	101.24	103.26	104.28	103.85

4. Conclusions

- Catalytic cracking of the blend containing 5 wt% Renewable Fuel Oil resulted in the formation of water and carbon dioxide, plus the possible retention or conversion of some oxygenates in the total liquid product, at the expense of gasoline and liquefied petroleum gas production.
- At a given severity and compared with the base oil, the blend gave 1–3 wt% higher conversion which increased with catalyst/oil ratio.
- At a given conversion, the blend gave lower yields of liquefied petroleum gas and gasoline than the base oil. Other yields including those of dry gas, light cycle oil (diesel), heavy cycle oil (heavy fuel oil), and coke were almost the same for the two feeds. However, among the dry gas components, higher CO_2 but lower H_2 yields were observed for the blend.
- After cracking, most of the oxygen in the blend appeared as H_2O with the rest in the form of CO_2. The liquid products were analyzed for oxygen content and found to be below the detection limit.
- Primary indication from this study suggested a high potential of coprocessing the conventional fluid catalytic cracking feed and Renewable Fuel Oil in an adequate ratio on account of the significant contribution of the organic ingredient in RFO to valuable products upon cracking.
- When yields of a Renewable Fuel Oil blend and heavy gas oil are compared on the basis of equivalent energy input to the microactivity test (MAT) system, gasoline and light cycle oil yields from the Renewable Fuel Oil blend are higher than corresponding yields from 100% heavy gas oil.

Acknowledgments

The authors wish to thank the analytical laboratory of CanmetENERGY.

(779c) The Effects of Solvent On Glucose Conversion to 5-Hydroxymethylfurfural

Xianghong Qian, Ralph E Martin Department fo Chemical Engineering, University of Arkansas, Fayetteville, AR and Hongbo Du, Chemical Engineering, University of Arkansas, Fayetteville, AR

Cellulosic biomass represents an abundant renewable resource for the production of bio-based products and biofuels. 5-hydroxymethylfurfural (HMF) is a critical and versatile intermediate for converting biomass to liquid alkanes and many other value-added products. However, cost effective conversion of glucose, the dominant biomass monomer sugar, to HMF remains a critical issue. Experimentally, it was found that glucose conversion and selectivity are especially sensitive to the type of catalysts present and the solvent media. During brønsted acid-catalyzed glucose reactions including mutarotation, condensation, dehyration and isomerization reactions , our simulation results show that protonation of one of the hydroxyl groups on the glucose ring or the ring O and the subsequent breakage of the C-O bond is the critical step. Our results demonstrate that the barriers for acid-catalyzed glucose reactions are largely solvent induced due to the competition for proton between the hydroxyl groups on glucose and the solvent molecules. The barriers and reaction free energies were calculated for glucose dehydration to HMF in various DMSO-H_2O mixtures using *ab initio* molecular dynamics simulations (CPMD) coupled with metadynamics simulations. The effects of solvent on the reaction barrier for glucose to HMF conversion in various DMSO-H_2O mixtures were determined and presented here.

(779d) Catalyst Design and Optimization for Selective Hydrodeoxygenation

Wei Mu and Yulin Deng, School of chemical and biomolecular engineering, Georgia Institute of Technology, Atlanta, GA

Upgrading pyrolysis oil by using hydrodeoxygenation(HDO) process is intensively studied recently. However, as an expensive resource, the large consumption of hydrogen is not desired. Therefore, HDO with selectivity is the target for such type of reaction. Varies catalysts and processes were developed. However, most of these studies either dealt with one certain type of model compounds with some catalyst, or one type of catalyst with some model compounds, which could not provide a systematic comparison between catalysts. In this study, the four most commonly used noble metals (Ru, Rh, Pt, Pd) plus nickel and three types of support materials (acid/netural/alkaline) were tested with three model compounds which has oxygen in different structure. The hydrogenation (HYD) and selective HDO were two competitive reaction in the system so that temperature was critical to the selectivity. The analysis of variance (ANOVA) was used to study the interaction between experiment factors. The optimal reaction condition and proper catalyst was obtained through the data analysis. The reaction mechanism of the deoxygenation with different model compounds was proposed and how the compound structure would affect the kinetics of HYD/HDO reaction was also evaluated.

(779e) Depolymerization of Lignin Extracted From Kraft Black Liquor

Bethany Carter, Chemical and Biomolecular Engineering, Clemson University, Central, SC

Depolymerization of Lignin Extracted from Kraft Black Liquor

Bethany D. Carter and David A. Bruce*

Department of Chemical and Biomolecular Engineering, Center for Advanced Engineering Fibers and Films, Clemson University, Clemson, SC, 29634-0909, USA.

* *Corresponding author. E-mail: dbruce@clemson.edu*

Lignin was extracted from Kraft black liquor via the continuous Sequential Liquid Lignin Recovery and Purification (SLRP) process whereby the pH of the incoming black liquor (pH ~ 13.5) was lowered to a pH ~ 9.5 by acidification with CO_2 resulting in lignin precipitation. The precipitated, liquid lignin phase is then acidified with sulfuric acid until a final pH ~ 2.5 is achieved, whereby the lignin is fully acidified. The acidified lignin is then leached counter currently with water to lower the lignin's ash content.

Lignin extracted from Kraft black liquor via the SLRP process was then depolymerized into smaller more monomeric units that may serve as potential precursors for higher valued chemical. The depolymerization occurs through cleave of the many ether linkage that connect guaiacyl (G), *p*-hydroxyphenyl (H), and syringyl (S) units. The depolymerization of lignin was done in small batch reactors in the presence of varying heterogeneous catalysts, which include Co/Mo on alumina as well as some noble metal supported catalysts such as Pt/Al_2O_3. The $Co/Mo/Al_2O_3$ catalyst are typical for use in hydrodesulfurization processes which has good sulfur tolerance, as the Kraft lignin may have residual sulfur from the Kraft pulping process.

Characterization of depolymerized lignin fractions was done using gas chromatography-Mass spectrometry to analyze lighter molecular weight fractions. Higher molecular weight fractions, which are not easily volatilized, were analyzed using NMR spectroscopy.

SUBSTANTIAL INVOLVEMENT BY DOE TO ADDRESS SCALE-UP CHALLENGES FOR BIOREFINERIES

Gene Petersen and Glenn Doyle, US Department of Energy, Golden, CO; Art Wiselogel, CNJV, Golden, CO; Carol Babb, SAIC, Denver, CO; and Andras Marton, Independent Project Analysis, Inc., Ashburn, VA.

Extended Abstract

Since 2003, the Department of Energy (Department) has been substantially involved in cost-sharing the development of over thirty integrated biorefinery projects to process biomass feedstocks into transportation fuels and biobased products. These projects range in scale from developing engineering designs for pilot operations, to constructing integrated pilot- and demonstration-scale biorefineries, as well as commercial biorefinery demonstrations. Under direction of public law (Energy Policy Act of 2005), the Department was authorized to "carry out a program to demonstrate the commercial application of integrated biorefineries[1]." Under that mandate and subsequent law (e.g., the American Recovery and Reinvestment Act of 2009), the Department's Biomass Program conducted three separate competitive Funding Opportunity Announcements (FOAs) from 2006-2009 to solicit applications from the community to develop integrated biorefineries at these various scales. This presentation provides a brief overview of the careful selection process employed to meet the mandates of public law and describes how the Department utilizes project management best practices to address the challenges of scale-up and completion of these projects. An oft voiced criticism is that the federal government tries to pick technology winners. In fact, the Department's overall objective is to foster development of a robust, commercially-competitive biorefinery industry by selecting a broad suite of technology types that appear technically mature enough to be successful, not to advance any particular technology. A further objective is to reduce the risk associated with designing, constructing and operating first-of-a-kind biorefinery demonstrations such that the return on the investment can become attractive to private investment. In this vein, the overall benefit of the industry-federal partnerships will be discussed along with the practices and processes used to help both recipients of federal funds and the Department achieve successful development of Integrated Biorefineries.

Among the key best practices the Department has utilized to assist in the scale up of these new technologies and to manage the competitive process and oversight of these projects are the following:

Competitive Process and Selection

- Utilized competitive FOAs that identified the Department's minimum feedstock throughput and biofuels/bioproduct production threshold objectives and outlined the evaluation criteria for selecting integrated biorefinery projects. This allowed the industry the flexibility to propose projects intended to meet the criteria. In all three FOAs, the Department did not specify any particular technology or approach to meeting the objectives especially since the FOAs acknowledged the different levels

[1] Energy Policy Act of 2005, §932

of technology maturity by accepting applications for facilities ranging from pilot to demonstration scale.

- Conducted comprehensive merit reviews that compared responsive proposals against the published evaluation criteria using non-conflicted experts to determine the competitive range (i.e., technically meritorious applications) to be recommended to the Selection Official.
- Independent Project Analysis, Inc. prepared risk assessments on responsive proposals to compare them to its industry project management best practice benchmark database to determine each proposal's risk profile and state of readiness to begin construction.
- Program policy factors as outlined in each FOA were then used to assess the balance of the Biomass Program's integrated biorefinery investment portfolio and, along with the merit review results, recommendations for selection were submitted to the Selection Official.
- Based on the above factors and availability of Federal funds, the Selection Official made selections and notified the Contracting Officer to negotiate financial assistance awards.

Post-selection Award and Project Management Practices of DOE

- Upon selection, DOE chose to execute Federal financial assistance agreements instead of contracts to allow for greater flexibility in adjusting to technological, financial and project management challenges.
- The Biomass Program adapted and implemented relevant sections of Department of Energy Order 413.3B, *Program and Project Management for the Acquisition of Capital Assets* (DOE Order 413.3B) to the management of integrated biorefineries, which are process and capital intensive projects. Figure 1 compares the guidelines in DOE Order 413.3B with the development of a typical capital construction project and DOE's substantial involvement.
- DOE employed the industrial consulting firm, Independent Project Analysis, Inc. to comprehensively compare each selected proposal against its project management best practices database to assess the project cost and schedule risks and to provide the Department with an estimate of the amount of contingency needed to account for the identified risks.
- DOE's Golden Field Office contracted with RW Beck (now SAIC) to be the Department's Independent Engineer (IE) to assist in review of initial applications, and then evaluate and monitor progress on the selected projects. Also, the IE performs independent performance test analysis to assess actual biorefinery operations as compared to the target performance criteria. The IE's support to the Department is comparable to the role it would fulfill for a private financial institution involved in the project.
- Best project management practices were invoked, where applicable, to strengthen the Department's and the recipient's controls over the execution of the project. This includes the use of:
    - Separate budget periods to control the flow of federal cost share funds
        - Budget period 1: Engineering, Front-end loading design 2 and 3, National Environmental Policy Act analysis, permitting, secure intellectual property, secure site location and infrastructure needs,

    - secure feedstock agreements, identify and secure off-take agreements, and secure non-Federal cost share commitments
    - Budget period 2: Final design, construction, commissioning, startup
    - Budget period 3: Operations
  - DOE Critical Decision(CD) points
    - CD-2: Approve the performance baseline
    - CD-3: Approve start of construction
    - CD-4: Approve start of operations
  - Project management plans that invoke a Work Breakdown Structure framework and identify risks and barriers to be overcome
  - Resource-loaded Gantt charts and critical path schedules
  - Milestones
  - Go/No go decision points
  - Stage-gates to address technological maturity challenges
  - Earned value management
  - Responsible, Accountable, Consulted, Informed (RACI) charts
  - Risk Mitigation Plans and Risk registers
  - Contingency
  - Continual review of lessons learned to facilitate improving the process of federal and industry co-development of these biorefineries.

Examples of Good Practices and Lessons Learned

- The foremost challenge experienced by integrated biorefineries is successfully and continuously feeding solid biomass to the throat of the reactor. Adaptation of "commercial" technology to these first-of-a-kind technologies has been plagued by numerous challenges, resulting in changes in design, materials of construction, operating parameters, etc.
- The value of integrated pilot plant tests for scale up purposes cannot be overemphasized: For these first-of-a-kind demonstration- and commercial-scale IBRs, integrated pilot plant tests have proven to be invaluable in validating the design and establishing a credible performance baseline that controls scope creep, re-engineering, and minimizes change control actions during construction, all of which increase schedule delays and capital costs. Without the benefit of fully integrated pilot plant data to inform the demonstration or commercial designs, the risks to IBR cost and schedule are magnified. This reinforces key guidance provided to DOE by IPA, Inc.
- A major lesson learned is not to "assume" that a commercially available feed system will successfully integrate with a conversion technology. Similarly, unless "commercial" technology has been successfully integrated with new process technology at an appropriate scale before being commercialized, it cannot be assumed that it will work at the larger scales.
- The more new technology process steps, the higher the risk to successfully executing the project. A corollary to this would be, the more new technology process steps, the greater the need for fully integrated pilot plant tests of durations sufficient to validate and "lock" the demonstration- or commercial-scale design.
- Convening the appropriate team with the right skill and resource mix to successfully execute major capital projects depends on the stage of technical and organizational maturity: Oftentimes, due to the perceived need to protect intellectual property rights

and/or to keep costs at a minimum, R&D/entrepreneurial entities try to take on the challenges of scaling-up bench- or pilot-scale projects without having the necessary skills mix and resources to succeed. Delays in project schedules and cost increases often reflect the inability to quickly ramp up on the learning curve to execute major capital projects. R&D and entrepreneurial organizations need to educate themselves on the risks associated with scaling and plan to secure the appropriate skills and experience mix, either internally or through contracts, to improve their chances of success.

- Key R&D and/or pilot plant testing not complete: The urgency to be "first to market" sometimes results in R&D and/or integrated pilot plant testing not being performed or completed in order to inform a comprehensive ±10% design upon which to base EPC bids that are competitive and less likely to experience significant change actions.
- Adopting formal project management best practices improves the chances of success: Once the decision is made to build a demonstration or commercial biorefinery, the early adoption of formal project management practices and best practices will significantly strengthen the chances for a successful outcome. Assuring alignment with business objectives, establishing a well-defined performance baseline (scope, schedule, cost and risk), managing to this baseline, and controlling scope creep will greatly enhance the execution of the project.
- Pre-award meetings to clarify expectations for negotiating awards and meeting federal requirements were very valuable and provided companies with better roadmaps for proceeding with their project.
- Regular teleconferences (weekly or fortnightly) and face-to-face meetings were valuable tools to facilitate negotiations and monitor progress.
- Utilizing separate budget periods aligned with Critical Decisions (e.g., Budget Period 2 aligned with Critical Decision 3 – Approve Start of Construction) approach provided the requisite time needed to work out issues prior to actual start of construction. In one case, where a Budget Period 1 approach (see above for description of the activities typically involved in BP1) was not used, the project did not achieve successful operation.
- Annual comprehensive project reviews that involved DOE and the recipients allowed DOE to be better informed about actual progress and allowed DOE to provide some useful lessons from the collective experiences of other biorefinery efforts.

Benefits of Industry-Federal Partnerships

- Due to the reluctance of the private financing industry to invest in the "first" plant, the Department's overall objective in cost sharing in pilot-, demonstration- and commercial-scale integrated biorefineries is to "buy down" some of that inherent risk, advance the industry closer to being competitive with fossil-based fuels, and help the industry attract private investment for the 2nd through Nth plants.
- Through these financial assistance agreements, the Department and industry have demonstrated that the Federal government can play an effective role in seeding development of technologies that can help the nation move closer to energy independence while creating domestic employment opportunities.

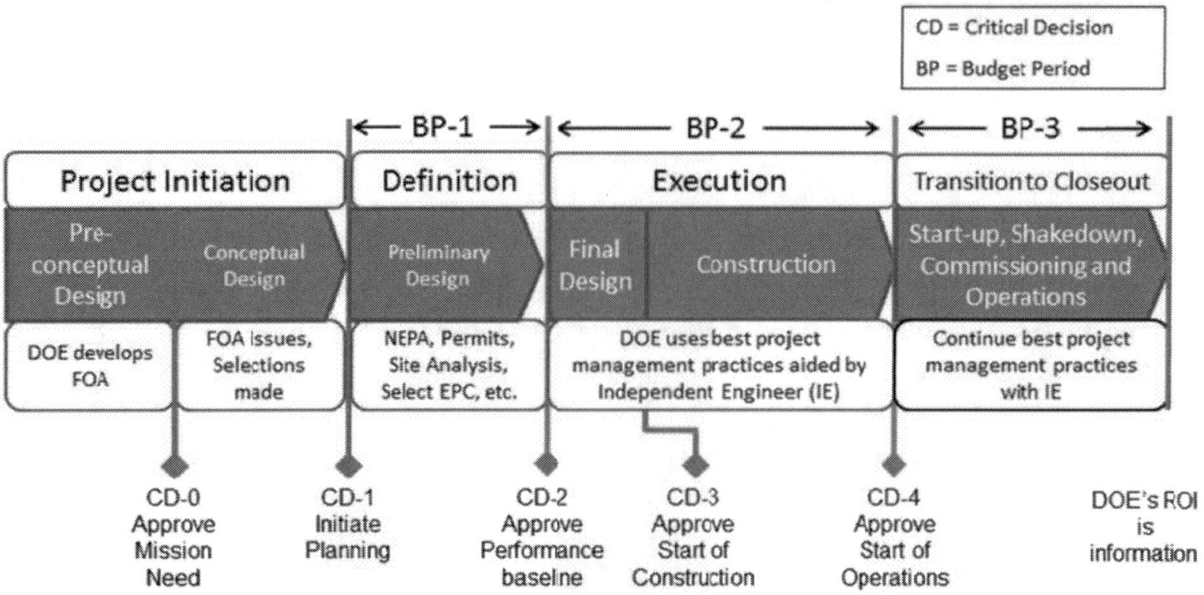

Figure 1 – Integration of Project Management for Biorefinery Development

(780b) Scale up of IH2 Process for Converting Biomass to Gasoline and Diesel

Terry Marker, Martin Linck, Michael Roberts, Pedro Ortiz-Toral, Jim Wangerow and Larry Felix, Gas Technology Institute, Des Plaines, IL

The IH^2 process is a new technology which directly converts biomass such as wood, cornstover, and lemna, into high quality liquid hydrocarbon transportation fuels using hydropyrolysis. Initial work was done on a small bench scale semi-continuous unit, which processes 0.45kg/hr of biomass. Technoeconomic and LCA studies of a commercial process based on these bench tests is very favorable, showing that gasoline and diesel can be produced from biomass at less than $2.00/gallon and have greater than 90% greenhouse gas reduction compared to fossel fuels.

Now the IH^2 process has been scaled up to a continuous unit which processes 50 kg/day of biomass. The new pilot plant unit helped us define key design parameters and also determine catalyst life and stability. GTI will discuss controlling process variables and show continuous data from the 50kg/day pilot plant. The 50kg/day tests lay the ground work for our design of a 50 t/day demo IH^2 unit which should lead directly to 2000 t/day commercial scale process in 2015-2016.

(780c) Piloting a Multiproduct Biorefinery

James J. Foster, Archer Daniels Midland Company, Decatur, IL and Michael Erpelding, Research, ADM, Decatur, IL

For some biorefineries, a single product is the goal of the project. Sometimes it is ethanol or an oil or perhaps a chemical compound, but it is focused on one product that will make or break the project. While it is good to be focused, it can lead to tunnel vision that can prematurely crash a project.

We believe a better method is run the biorefinery as having a number of co-products which allow a number of different outlets for the conversion of biomass. In our current biomass project, partially funded by the Department of Energy, we can produce ethanol, butyl acrylate, fiber and lignin. Butyl acrylate can be used as a drop in replacement for current petroleum based acrylates, fibers can be used for further conversions, and the lignin can be used as fuel source for power or steam generation while the fibers can be used as a feedstock for further conversation or sold as animal feed.

As part of the presentation we will list some of the challenges with the start-up of the pilot plant and the concerns of building a full scale plant including the technical issues and regulatory issues.

(780d) Scale-up of Ethanol Production From Sweet Sorghum Stems Using Advanced Solid-State Fermentation Technology

Shi-zhong Li, Institute of Nuclear and New Energy Technology, Tsinghua University, Beijing, China and Lei Zhang, Institute of New Energy Technology, Tsinghua University, Beijing, China

Compared with conventional submerged liquid fermentation (SLF), solid-state fermentation (SSF) can considerably reduce the process costs of substrate pretreatment and water usage, avoid the pollution problems resulted from the disposal of liquid residues. However, due to the inherent barriers of heat and mass transfer, SSF has not yet proved to be feasible in large-scale ethanol production. Accordingly, Tsinghua University had developed and operated a process of advanced solid-state fermentation (ASSF) at 127 m^3 scale using fermentable sugars from sweet sorghum stems in Inner Mongolia, China, in 2010. The field trial achieved high ethanol productivity that were comparable to conventional SLF, indicating that mass ethanol production in SSF with lower costs and less pollution problems has the great potential for its industrial application. In this study, to further optimize the process of ASSF, an improved rotary-drum fermenter at 555 m^3, the largest solid-state fermenter in the world so far, was constructed and deployed recently in Inner Mongolia, China, which was operated as continuous way for high process efficiency. During the process, the solid substrate and seed culture continue to be fed into the reactor and the fermented residues was conveyed immediately into distiller, making the whole process keep running without interval that improve the process efficiency to a great extent. Consequently, the continuous process achieved ~91% of theoretical ethanol yield that was parallel to 127 m^3 scale in the batch fermentation. The retention time of solid substrate (fermentation time) was only less than 30 h. In the meantime, tech-economic analysis (TEA) from National Renewable Energy Laboratory (NREL) indicated that 16 tons of sweet sorghum stems (with 13.5 % sugar content and 70% moisture) can be used to produce 1 ton ethanol; the fermented residues can be used to feed 1 cattle; the fuel ethanol production cost is $615.4/ton (49cent/litter) and the sweet sorghum stem price is at $30/t. In this study, the results of continuous fermentation and distillation at 555 m^3 scale are agree with the data from bench scale and pilot plant well. The TEA further demonstrates that ASSF is a cost-effective process that is highly feasible to be applied at industrial and commercial scale for fuel ethanol production, and eventually is able to improve the biofuel industry.

(780e) Evaluation of Bio-Oil Corrosion Characteristics

Patrick Johnston, Center for Sustainable Environmental Technologies, Iowa State University, Ames, IA and Robert C. Brown, Mechanical Engineering, Iowa State University, Ames, IA

Evaluation of Bio-oil Corrosion Characteristics

Patrick A. Johnston[1]and Robert C. Brown[1,2]

[1]*Center for Sustainable Environmental Technologies;* [2]*Department of Mechanical Engineering*

Iowa State University

Ames, IA 50011

Petroleum crude oils contain naphthenic acids that are one of the main causes of corrosion in the petroleum industry during storage, transport, and refining. "Naphthenic Acid Corrosion" NAC has been related to three parameters: Total Acid number (TAN), temperature, and flow rate [1, 2, 3, 4]. Bio-oils do not contain naphthenic acids but do however contain other organic acids such as acetic, formic, glycolic and propanoic acid that may have similar consequences to metal surfaces. A comprehensive corrosion study needs to be completed to determine the actual effects of the organic acids in bio-oil on select metal surfaces. As a result of this study, acid number and corrosion potential of pyrolysis bio-oils can be correlated and ranked to produce a fraction that is non-problematic with typical petroleum refining equipment.

The proposed project will evaluate and compare corrosion and acid number (AN) on fast pyrolysis bio-oil stage fractions produced at Iowa State University (ISU). The corrosion characteristics will be evaluated using a vacuum distillation apparatus that will simulate a distillation tower in a petroleum refinery. Metal specimens will be subjected to both the liquid and vapor phase of the bio-oil. Method D664 "Standard Test Method for Acid Number of Petroleum Products by Potentiometric Titration" will be used to determine acidy of the bio-oil samples. Ion Chromatography (IC) and High Performance Liquid Chromatography (HPLC) will be used to quantify organic acid concentration. The corrosiveness will be ranked gravimetrically according to mass loss during the study. Selected metal specimens will also be examined by Scanning Electron Microscopy (SEM) equipped with energy-dispersive spectroscopy (EDS) to determine morphological and composition changes on the metal surfaces.

References:

1. A. Jayaraman, H. Singh, and Y. Lefebvre, Rev. Inst. Fr.Pet.**41**, 265 (1986).

2. C. M. Cooper, Hydrocarbon Process.**51**, 75 (1972).

3. E. Babaian-Kibala, H. L. Craig, G. L. Rusk, et al., Mater.Perform. **32**, 50 (1993).

4. E.B. Zeinalov, V.M Abbasov, and L.I. Alieva, Petro. Chem.**49**, 185 (2009).

(782a) Lactic Acid Production From the Mixture of Kraft Paper Mill Sludge and Hemicellulose Prehydrolysate

Suan Shi[1], V. S. K. Sajja[1] and Y. Y. Lee[2], (1)Chemical Engineering, Auburn University, Auburn, AL, (2)Department of Chemical Engineering, Auburn University, Auburn University, AL

Paper mill sludge is a solid waste composed of pulp residues and ash generated from pulping and paper making processes. The carbohydrate portion of the sludge has chemical and physical characteristics similar to pulp. It can be biologically converted to value-added products without pretreatment. Most of the hemicellulose fraction of pulp mill feedstock is released into black liquor during the pulping process and combusted to recover energy. It is feasible to recover this fraction of carbohydrate for bioconversion. The prehydrolysate can be produced by steam or hot-water treatment. The sugar concentration, however, is far below the level usable as fermentation feed. In this study, we have investigated a bioconversion scheme in which the mixture of these two feedstocks is converted to lactic acid. For this purpose, simultaneous saccharificiaton and fermentation (SSF) was applied using *Lactobacillus delbrueckii* (ATCC 7830) and cellulose enzyme (Novozyme CTec-2). The bioreactor feed was prepared mixing Kraft pulp mill sludge and the prehydrolysate generated from pine wood by hot-water extraction. The prehydrolysate was detoxified by overliming and charcoal treatment. The SSF was carried out without pH control since calcium carbonate in the sludge, the main inorganic ingredient in the ash, acted as a neutralizing reagent. The main carbohydrate in the prehydrolysate, mannose, and the Kraft sludge were converted efficiently giving above 80% of theoretical yield with enzyme loading less than 10 FPU/g-hexose. The performance data and the details of the bioprocess, prehydrolysate preparation, and detoxification procedures are presented in this paper.

(782b) Hydrothermal Processing of Un-Hydrolyzed Biomass for Biocrude and Biochar Production

Sergiy Popov[1], Sandeep Kumar[1], Venkatesh Balan[2], Nirmal Uppugundla[2], Leonardo da costa Sousa[2] and Bruce E. Dale[2], (1)Civil & Environmental Engineering, Old Dominion University, Norfolk, VA, (2)Chemical Engineering and Material Science, Great Lakes Bioenergy Center, Michigan State University, Lansing, MI

Un-hydrolyzed biomass (UHS) resulting from enzymatic hydrolysis during cellulosic ethanol production is a wet and carbon-rich residue. It mainly contains lignin and unconverted sugars depending upon the types of biomass pretreatment used in the earlier stages. Energy Independence and Security Act, 2007 of the USA mandates 21 billion gallons of advanced biofuels production by 2022 from cellulosic biomass e.g. energy crops, agricultural residues, and forest residues. Majority of the biofuels productions are focused on cellulosic ethanol. Several tons of UHS will be generated during cellulosic ethanol/butanol production. The efficient utilization of this residue will have a positive impact in making the biofuels cost-competitive.

Hydrothermal processing of biomass under subcritical water conditions is considered as an environmentally benign method for producing biocrude and biochar together. The process is particularly suitable for a feedstock with very high water/moisture content. Hydrothermal processes capitalize on the extraordinary solvent properties of water at elevated temperature for converting biomass to high energy density liquid and solid products. Here, biocrude is defined as a mixture of oxygenated hydrocarbons produced from the liquefaction of biomass which can be upgraded to liquid fuels and chemicals. Biochar is carbon rich, high energy density solid product resulting from the carbonization of biomass components during hydrothermal processing. Corn stovers pretreated by Packed Bed Ammonia Fiber Expansion pretreatment (PB-AFEX^{TM}) and Extractive AFEX^{TM} (E-AFEX^{TM}) methods were hydrolyzed using commercial enzymes. The UHS generated after enzymatic saccharification was used as a feedstock for producing biocrude and biochar in a 500 ml batch reactor under subcritical water conditions. The effect of temperature in the range of 250-350°C, homogeneous catalyst, and reaction time on biocrude and biochar yields and their heating values were studied. The complete mass and energy balance in the conversion of UHS to products were developed.

Utilization of Citrus Processing Waste as Renewable Feedstock for Biorefinery Applications: Technical Advances, and Analysis of Engineering Challenges for the pretreatment and Enzymatic Hydrolysis Steps

Raul C. Rivas[1], Kim D. Jones[1], Patrick L. Mills[2], and Shad D. Nelson[3]
(1)Department of Environmental Engineering, (2) Department of Chemical & Natural Gas Engineering, and (3) Department of Agriculture, Agribusiness & Environmental Sciences, Texas A&M University-Kingsville, Kingsville, TX

Abstract

An assessment of recent technical advances on pretreatment processes and its effects on enzymatic hydrolysis as the main steps of a proposed citrus processing waste (CPW) biorefinery is presented. Engineering challenges and relevant gaps in scientific and technical information for reliable design, modeling, and scale up of a CPW based biorefinery are also discussed. Some integrated physico-chemical pretreatments are proposed for testing for CPW biomass, including high speed knife-grinding and simultaneous alkali addition. These new proposed processes and the effects of parameters such as particle size, surface area and morphology, pore volume, and chemical composition of the diverse fractions resulting from pretreatment and enzymatic hydrolysis need to be evaluated and compared for pre-treated and untreated samples of grapefruit processing waste. This assessment suggests the potential for filling the data gaps, and preliminary results demonstrate that the reduction of particle size and the increased surface area for the CPW will result in higher reaction rates and monosaccharide yields for the pre-treated waste material.

Keywords

Citrus processing waste; pectin-rich biomass; cellulosic biorefinery; bioethanol; physico-chemical pretreatments; d-limonene; grapefruit.

Introduction

The utilization of cellulosic biomass is currently receiving great interest as a renewable source of raw material for biorefinery processes. According to a National Renewable Energy Laboratory definition (NREL, 2012), biomass is defined as any plant-derived organic matter. Biomass available for energy on a sustainable basis includes herbaceous and woody energy crops, agricultural food and feed crops, agricultural crop wastes and residues, wood wastes and residues, aquatic plants, and other waste materials including some municipal wastes. Thus, biomass represents a very heterogeneous and chemically complex renewable resource.

Some of the best sources of lignocellulosic biomass are agricultural residues produced from cereal crops, and the agro-industrial wastes generated during the processing of fruits and vegetables. These residues and by-products usually represent an economic burden as an environmental disposal problem. Economic and environmental concerns arise from the processing of fruits and vegetables into juices, including the treatment and disposition of solid wastes, and the recovery of byproducts. Low value citrus processing waste (CPW) is the major waste generated by the fruit processing industry and has been suggested as a sustainable source of cellulosic biomass for biorefinery applications.

The citrus processing industry (mainly orange and grapefruit juice) produces millions of tons of solid wastes as by-products composed of peel, pulps, seeds, essential oils, and with a water content of 75-80 wt% (ARS, 1962). Over the three season period 2008-2011, the U.S. production of orange and grapefruit, processed primarily as juice, was approximately 23.3 million tons (21.16 metric tons) (NASS, 2011) with an associated solid waste generation of between 10.0 to 10.6 millions of tons.

Rio Red Grapefruit (*Citrus paradisi* Macf.) is the primary citrus fruit processed in Texas. This is processed to produce juice and canned fruit at the Texas Citrus Exchange facility located in Mission, TX. This plant processed an estimated of 299,720 tons of grapefruit (271,900 metric tons) (NASS, 2011), and subsequently generated approximately an estimated 150,000 tons of solid wastes during the same seasonal periods. Currently, grapefruit wastes are partially processed by means of a high-cost drying process to produce a low-value cattle feedstock which contains a low percentage of protein, and has a bitter flavor (Antwerpen, 1941). Due to these characteristics, its use as a direct feed is limited and must be mixed with other feedstocks, resulting in low utilization rates (Wing, 2003). A consequence of this limited use results in unprocessed byproducts which must be disposed of in municipal landfills. This practice creates additional environmental and economic costs due to the space limitations of landfills, and to the additional burdens of transportation and disposal (Braddock, 1999).

The objective of this paper is to provide an assessment of the state-of-the-science of the existing technologies for the utilization of CPW as a renewable source of cellulosic biomass, and an updated comparison based on preliminary laboratory scale data generated from the pretreatment and enzymatic hydrolysis of Rio Red grapefruit processing waste (GPW) as a potential source for biorefinery applications. This assessment reports on the development of more reliable methods for the production of sugars and value-added products from CPW at bench-top scale, and on improved yields of the most important sugars produced from several types of pretreatments including particle size reduction and caustic addition for enhanced hydrolysis.

Citrus Processing Waste (CPW) as renewable source of pectin-rich and low lignin content cellulosic biomass.

A potential sustainable utilization of citrus processing wastes is their conversion to bioethanol to be used as biofuel, and other value-added products by means of biorefinery processes which target specific chemical composition Grohmann and Baldwin, 1992; Grohmann and Bothast, 1994; Grohmann et al., 1995; Wilkins et al., 2007). The chemical composition of citrus solid wastes consists of soluble sugars (mainly fructose, glucose, and sucrose), and structural polysaccharides such as cellulose, hemicelluloses, and pectin, including lignin-like compounds such as flavonoids, and essential oils (ARS,1962; Ting and Deszyck, 1961; Goodrich and Braddock, 2004; Fernandez-Lopez et al., 2004; Marin et al., 2007; and Ververis et al., 2007)

The relative amount of these components for CPW are shown in Figure 1, and are of interest in the production of bioethanol, value-added derived products, and energy; while flavonoids and pectin are used to produce pharmaceutical, nutraceutical, and industrial products. For these reasons, CPW can be considered a potentially important source of biomass for biorefinery processes, in which the concentration of lignin is very low compared to the lignocellulosic biomass from agricultural wastes such as wood chips, corn stover, sugarcane bagasse, and diverse cereal straws. This specific characteristic results in citrus structural polysaccharides that may be more readily hydrolyzed and fermented to biofuels (Marin et al., 2007; Ingram et al., 1999; Gong et al., 1999) or processed in a biorefinery to value-added products. Approximate composition of orange processing waste, corn stover, and wheat straw are compared and shown in Figure 2.

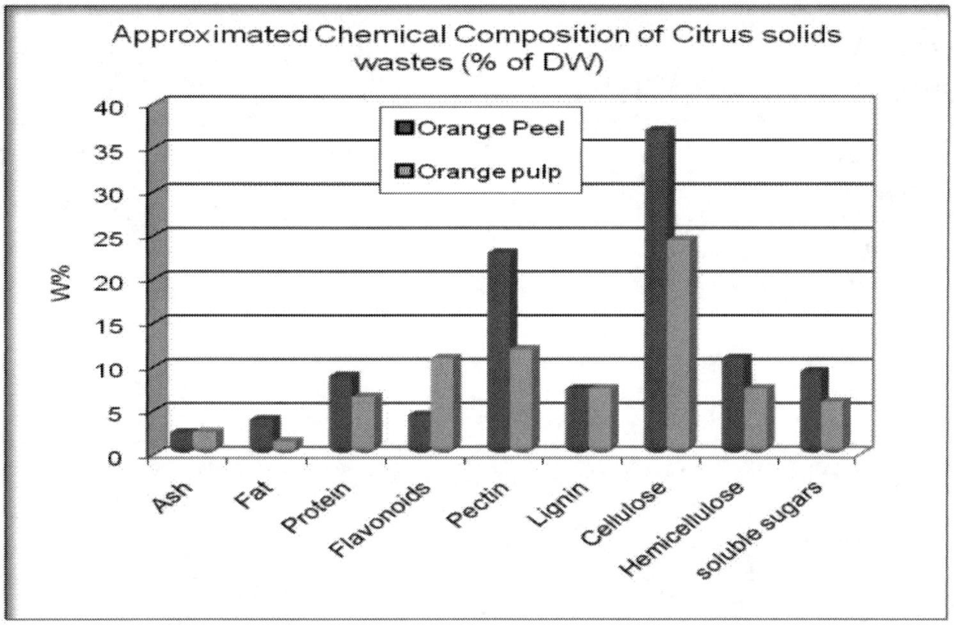

Figure 1. Approximate composition of orange processing waste as % of dry weight (Values extracted from Goodrich and Braddock, 2004; Fernandez-Lopez et al., 2004; Marin et al., 2007; and Ververis et al., 2007)

Current state-of-the-science for a proposed Citrus Processing Waste Biorefinery

The last few decades have seen significant research interest targeting the conversion of citrus processing wastes into fermentable sugars by means of acid and enzymatic hydrolysis (Grohmann and Baldwin, 1992; Grohmann and Bothast, 1994; Grohmann et al., 1994; and Grohmann et al., 1995). In most of these investigations low concentrations of sulfuric acid were used, along with diverse commercial enzymes complexes of pectinases, cellulases, and β-glucosidase to achieve hydrolysis of orange peel with acceptable success for sugar yield, but with long reaction times, on the order of days in most cases. During the last several years new combinations of enzymes have been developed to accelerate the reaction and increase product yield. In 2006, advances in the efficiency of enzymatic hydrolysis of grapefruit peel were demonstrated under similar conditions at bench-top scale (Wilkins et al., 2007). A US patent application publication describing the production of ethanol from citrus processing waste has been awarded to Stewart *et al.* in 2006, based on the research achieved by Grohmann *et al.* (Grohmann and Baldwin, 1992; Grohmann and Bothast, 1994; Grohmann et al., 1994; and Grohmann et al., 1995) at ARS-USDA research laboratories and has been under evaluation for several years. This patent application is based on the following summarized general process steps which describe the current state of the science with an effective but as yet uncommercialized process:

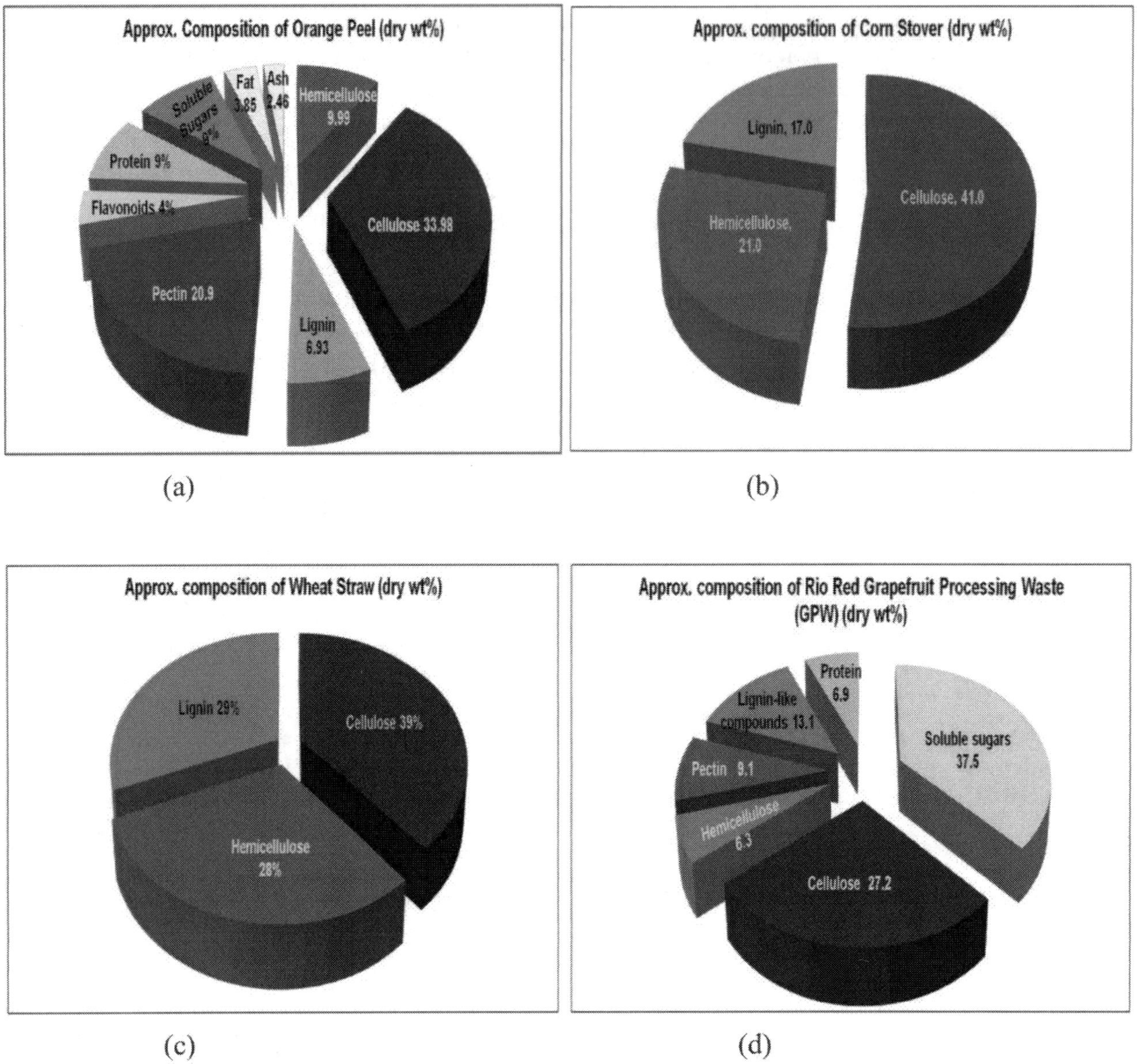

Figure 2. Approximate composition of (a) citrus (orange) peel waste as pectin-rich cellulosic biomass compared versus (b) corn stover, (c) wheat straw, and (d) Rio Red grapefruit processing waste (GPW), in which the lignin-like compounds concentration includes the flavonoids concentration. (first 3 values were extracted from references (Marin et al.,2007; Ingram et al., 1999; Gong et al., 1999)], and GPW was analyzed by the authors)

1) Pretreatment of the citrus wastes by means of shredding and grinding processes to a particle size less than one-half inch, and suspension of the wastes in water to obtain a slurry which is partially hydrolyzed by heating with steam; after this treatment, the slurry is subjected to a flash distillation with steam to remove limonene (citrus oil) with the idea to avoid the growth inhibition of yeasts and other microorganisms during subsequent fermentation processes. An alternative method of limonene removal is applied after the enzymatic hydrolysis by means of a centrifugation process.
2) Enzymatic hydrolysis of the pretreated citrus waste slurry using an enzyme mixture (not described). The enzymatic hydrolysis process is achieved using high solids pumps or high solid mixers under

controlled temperature and pH range. If the limonene has not been removed, a centrifugation process is applied to separate the suspended solids.
3) An alternative simultaneous saccharification and fermentation process (SSF) is described which uses the pretreated citrus wastes to be converted into ethanol in one hydrolysis-fermentation tank.
4) Fermentation of the sugars produced during the enzymatic hydrolysis is accomplished in a fermentation tank using yeasts, and/or Escherichia coli KO11 (a genetically modified microorganism GMO) in a simultaneous or separated fermentation process. The yeast ferments only the six-carbon sugars, and Escherichia coli KO11 ferment five- and six-carbon sugars, and galacturonic acid to produce ethanol and ethyl acetate.
5) Distillation process to recover the ethanol from the fermentation broth.

More recent studies related to the conversion of citrus processing waste (CPW) into bioethanol have been published by Pourbafrani (2010), Pourbafrani et al. (2010), and Lohrasbi et al. (2010). The studies are based on the pretreatment of CPW utilizing a steam explosion process (energy intensive high pressure steam to physically break down the material) and the diluted-acid hydrolysis process using sulfuric acid similar to the method developed by Grohmann et al. (1995), and by Stewart et al. (2006). Their investigations have focused on the application of a flocculating strain of yeast, *Saccharomyces cerevisiae*, for the continuous fermentation of produced sugars into bioethanol. The processes, for the recovery of d-limonene after the steam explosion of CPW and for methane production through anaerobic digestion of the wastes, have also been proposed to improve commercial potential. Thus far, all the developed processes for the conversion of CPW into ethanol and other value-added products have only been evaluated at bench-top and pilot scales. There is no industrial scale CPW biorefinery facility currently in operation.

Biorefinery Concept for Citrus Processing Waste Biomass

According to the National Renewable Energy Laboratory (NREL, 2011), a biorefinery is a facility that integrates biomass conversion processes and equipment to produce fuels, power and chemicals from biomass; such a concept may fit well with the citrus waste biomass model previously described.

A biorefining process concept applied to citrus waste based on current technology would consist of the following four principal stages:

I) Pretreatment processes for citrus processing waste biomass to separate the material into components such as structural polysaccharides and compounds (cellulose, hemicelluloses, pectin, and lignin), proteins, amino acids, essential oils, fine chemicals and pharmaceutical compounds (flavonoids) in a primary fractionation process. Typical technologies applied in this stage are traditional separation processes like shredding and milling, filtration, centrifugation, solvent extraction and distillation. However, novel concepts like compressed hot water (Ueno et al., 2008) and supercritical CO_2 extractions are also being explored.

II) Chemical and/or enzymatic hydrolysis of the polysaccharide matrix may be performed in a secondary process to obtain a liquid blend containing monosaccharides, organic acids, and alcohols which can be utilized as a raw material to subsequent chemical and biochemical conversions.

III) A tertiary refinery process step to convert the monosaccharide blend into valuable end products and chemical intermediates utilizing biochemical processes such as fermentation to produce bioethanol and lactic acid; and thermo-chemical catalytic technologies to produce intermediates like levulinic acid (Bozell et al., 2000). These technologies may be combined with additiional catalytic technologies to produce high added value end-products (Serrano-Ruiz et al., 2010). The catalytic direction of future research for processing this material is very promising. The wastes and byproducts of each step may be used for the production of power and heat or used as raw material for subsequent processes.

IV) A final step which includes the treatment and reuse of wastewater, solid wastes and control of air emissions.

Physical and Chemical Changes in Cellulosic Biomass due to Pretreatment Processes

All biorefinery processes applied for the conversion of any cellulosic biomass into monosaccharides will involve a physico-chemical pretreatment step to improve the acid adsorption capacity and enzymatic accessibility of the cellulosic fraction. The pretreatment allows the optimal decomposition of the structural polysaccharides during the subsequent hydrolysis step to produce a blend composed of five- and six–carbon sugars (neutral, reducing, and acid sugars). Improved pretreatment methods are being actively sought to optimize these steps. For lignocellulosic biomass, several investigations concerning the effect of particle size on the hydrolysis of cellulose have been published over the last 20 years using high purity cellulose, and in some cases lignocellulosic materials such as cotton residues, soft wood, and newspaper (Neilson et al., 1982; Furcht and Silla, 1990). These results have demonstrated that the particle size reduction caused the time for maximum hydrolysis conversion to be reduced on the order of hours.

Recent work on sawdust slurries developed by Dasari and Berson (2007) has demonstrated that particle size reduction may provide a means for reducing the long residence time required for the enzymatic hydrolysis step in the conversion of lignocellulosic biomass to ethanol. Furthermore, the corresponding reduction in viscosity may allow for higher solid loadings and reduced reactor sizes during large-scale processing. There is no reported research involving the utilization of diverse particle size fractions of citrus processing waste biomass, emphasizing the engineering data gap to evaluate this effect of the changing physical-chemical properties due to particle size reduction.

Environmental Issues Surrounding the Generation and Biorefinery of Citrus Processing Waste

Currently, citrus processing waste (CPW) is most often transformed into a low value cattle feed by means of energy intensive drying processes. Due to the low nutritional values of the feedstock and high energy consumption, this operation becomes merely a remediation process for waste disposal with marginal or null economic revenue. Furthermore, during the drying process, large quantities of volatile organic compounds (VOCs) and carbon dioxide can be emitted due to the decomposition of the material, and the burning of natural gas utilized for the drying of CPW respectively. These factors have negative impacts on the environmental burdens and process economics.

Environmental issues arising from the drying process include air emissions estimated at 10-20,000 tons/year of d-limonene as part of volatile organic compounds (VOCs) (Grohmann, 2007., The Federal Clean Air Act Amendments of 1990, and the Environmental Protection Agency (USEPA, 1990 and 2004) require citrus juice processors to file a Title V permit if limonene emissions exceed 100 tons/year (Buff, 1997; Odio, 1996; Braddock and Miller, 2001).

Additional issues arise due to regulated emissions of carbon monoxide (CO), particulate matter (PM), sulfur dioxide (SO_2), nitrogen oxides (NO_x) and hazardous air pollutants (HAP) (primarily methanol and formaldehyde), and carbon dioxide (CO_2) (USEPA,2004; Braddock and Miller, 2001) from the burning of natural gas and other fuels used in the drying process combined with the lost revenue from un-recovered limonene oil, estimated at 20-40 million USD/year (Grohmann, 2007), and the high treatment costs from the consumption of gas and electricity, which has left many citrus juice processors searching for more sustainable alternative methods for treatment and disposal.

The development of new products and technologies such as biofuels, environment-friendly chemicals, and sustainable production processes based on renewable resources is of critical importance to reduce modern society's dependence on non-renewable resources such as oil and coal. Furthermore, the application of bioethanol in fuel formulations may help to ameliorate greenhouse gases emissions, and the production of bio-based compounds and polymers may help reduce additional environmental contamination through product substitution in polluting industries. From an environmental engineering point of view, the utilization of citrus processing wastes as a model biomass could become a local source of renewable feedstock to produce bioethanol and specialty chemicals, and reduce the emissions of greenhouse gases from its transportation to other locations, and can substitute for non-renewable fuel consumption when is utilized as biomass to produce bioethanol. The implementation of a commercial process could eliminate the emissions generated by the wet waste drying process utilized to convert it into cattle feed, and conserve available capacities for other solid wastes in regional municipal landfills.

Critical Engineering Challenges and Technical Limitations for the Implementation of a Citrus Waste Biorefinery

Current pretreatment and hydrolysis technologies applied for biorefining of lignocellulosic biomass cannot be directly applied to pectin-rich CPW biomass. Pretreatment applications of the processing of lignocellulosic biomass have been shown to function variably for waste materials with different physical characteristics. Lignocellulosic biomass is a relatively dry material (approx. 10-15 wt% of water), while CPW biomass is a wet material with 75-85 wt% of water. In addition, when compared to lignocellulosic biomass, CPW has a low concentration of lignin-like compounds, and a high concentration of pectin. These factors indicate that a process established for lignocellulosic biomass may not be totally suitable for the biorefining of CPW. Process conditions, kinetic parameters, reactor design, and hydrolysis reactions for CPW will need to be adapted to the material's specific physical-chemical characteristics in the overall engineering process. Clearly new technology and advances are needed.

Published information is lacking on the relevant scientific and engineering parameters to design for the scaled up pretreatment and enzymatic hydrolysis steps of a proposed CPW-based biorefinery. Important data gaps have been identified and targeted to complete the development of a feasible CPW biorefinery technology.

These can be summed up as follows:

1) Effect of particle size and surface area on enzymatic hydrolysis rates and sugar yields.

2) Development of chemical pretreatments to improve the accessibility of pectinases and cellulases on structural polysaccharides.

3) Evaluation of the effect of initial soluble free-sugars in the enzymatic hydrolysis rates.

4) Utilization of improved enzyme complexes and accessory enzymes to improve enzymatic hydrolysis efficiencies.

5) Fundamental engineering factors for scaling up and optimization of the processes such as reaction kinetics, reactor design, and mass and energy balances of each refinery step.

To begin the process to fill the data gaps related to the effects of particle size, laboratory experiments were developed focused on the grinding of wet GPW and its classification in various particle size fraction distributions. These fractions were obtained by means of a wet sieve shaking procedure. Slurries of each fraction containing 2% wt of insoluble solids were hydrolyzed using a cocktail of cellulases, pectinases and β-glucosidase following a procedure similar to that described by Grohmann *et al.* (Grohmann and Baldwin, 1992; Grohmann and Bothast, 1994; Grohmann et al., 1994; and Grohmann et al., 1995). In other experiments, chemical treatments with diverse alkali concentrations were simultaneously applied during the grinding process. Some of the results from these experiments are respectively shown in Figures 3 and 4.

The graph in Figure 3 illustrates the relative benefit of the particle size reduction on the yield of fermentable total sugars from the GPW, and the graph in Figure 4 describes the effect of alkali treatment of GPW on the enzymatic hydrolysis reaction. The reaction behavior in this graph shows that an optimum alkali concentration could be determined and subsequent separation of treated GPW-fractions could be developed to achieve a maximum glucose yield during enzymatic hydrolysis. Additional research with more innovative pretreatment combinations and process sequencing has the potential to further improve the yield and economics of a proposed biorefinery.

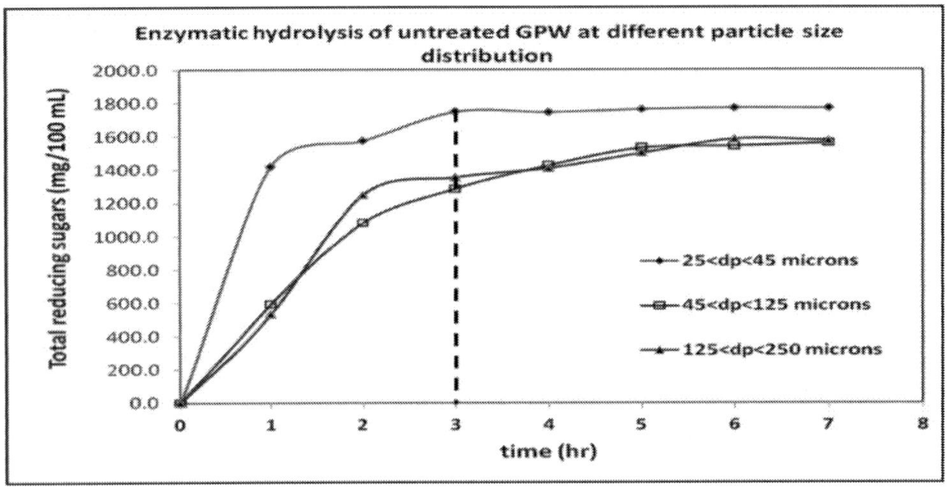

Figure 3. Preliminary results on the effect of the particle size distribution of ground grapefruit processing waste (GPW) in the reaction rate and sugars yield of their enzymatic hydrolysis; the smaller the particle size the shorter reaction time to achieve the maximum sugar yield.

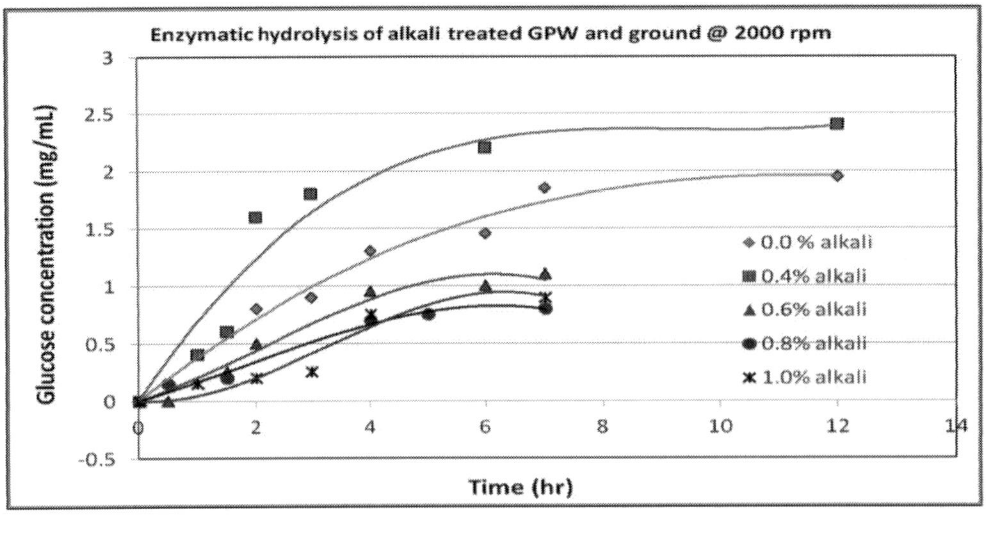

Figure 4. Graph shows the effect on enzymatic hydrolysis of GPW after its grinding and simultaneous treatment with different alkali concentrations at the same distribution of particle size (in this case the samples contained an approximated particle size distribution of 60 wt% for a 500μm >d_p>1000μm, and 40 wt% of 250μm >d_p< 25μm on dry basis).

Economic Constraints for Citrus Processing Waste Biorefinery Scale Up

Almost all of the reported studies and patents for the conversion of CPW into bioethanol have based their process economic viability on the recovery of co-products. Grohmann (1992, 2007), Stewart et al. (2007), and Zhou et al. (2007) reported the co-production of d-limonene, and Pourbafrani (2010), and Pourbafrani et al. (2010) reported the co-production of d-limonene, methane, and pectin to improve the economics for bioethanol production utilizing enzymatic and acid hydrolysis, respectively. This fact implies an additional economic constraint and potential benefit in the scale up of such technologies based on the co-dependence of bioethanol production costs along with the recovery of some co-products. Based on these assumptions, and assuming a plant capacity of 40 million gallons per year and 1.80 US dollars/gal for ethanol price, Grohmann (2007) reported gross margin revenues of 0.34 USD per gallon without d-limonene and 0.64 USD per gallon when d-limonene is included as co-product; similar data were reported by Zhou et al. (2007). In addition to the co-product cost dependence, the calculations indicate that there would be a need for 3.64 million tons of fresh wet CPW to make the process economic. This situation creates a dependence on regional production and recovery of stranded biomass from diverse processing facilities increasing the cost of the feedstock due to transportation.

In another case, Pourbafrani (2010), and Pourbafrani et al. (2010) estimated an approximate necessary sales price of 3.44 USD/ gal for ethanol and 2.81 USD/gal for production costs including the production of methane and d-limonene for a plant capacity of 100,000 tons of wet CPW per year; even so, with the current price of ethanol (approx. 2.79 USD/gal in 2012) this process is not economically viable under current conditions. None of the above mentioned studies consider the added production of additional unique value-added compounds such as flavonoids, carotenoids, and galacturonic acid contained in the CPW, and did not specify what they did with these products, which may add increased costs for high strength wastewater treatment and disposal.

According to Jorgensen et al. (2007) in the general concept of a lignocellulosic biorefinery, there are a number of technical and scientific issues within pretreatment and hydrolysis processes which remain unresolved. However, significant improvements in yield and cost reductions are expected, thus making large-scale biorefining of lignocellulosic substrates possible in the future. Based on such predictions, a similar technology development can be projected for a biorefinery based on citrus processing waste biomass with the additional advantage that CPW contains diverse value-added compounds such as flavonoids, d-limonene and galacturonic acid that may be utilized as platform materials to produce nutraceutics, pharmaceutics, water softeners, and industrial compounds.

To design and implement projects to acquire the essential engineering data for the scale up of the technology, several steps will be needed. The detailed classification and more precise chemical characterization of citrus processing waste before and after pretreatment, as a renewable biomass will require additional investigative studies. The Rio Red grapefruit from the Rio Grande Valley growing region in Texas, USA, can provide a feedstock waste from the Texas Citrus Mutual juice plant in Mission, Texas. The research team from Texas A&M University-Kingsville has begun the necessary characterization of the material before and after processing, and has also developed new methods for measurements of the distribution of sugars released for fermentation into ethanol. Based on the measured properties and composition, specific procedures for the pretreatment and hydrolysis (acid or enzymatic) of the structural polysaccharides are being developed and tested.

Once the Rio Red GPW is well characterized as cellulosic biomass, the engineering process data gaps are being identified and studies implemented to obtain the relevant information needed for scale up from bench-top experiments to pilot plant and industrial levels. Figure 5 illustrates the main biorefinery steps which are in need of research, development, and improvement to resolve the engineering challenges, and fill the technical data gaps required for the scale up and commercialization of a (CPW)-based biorefinery.

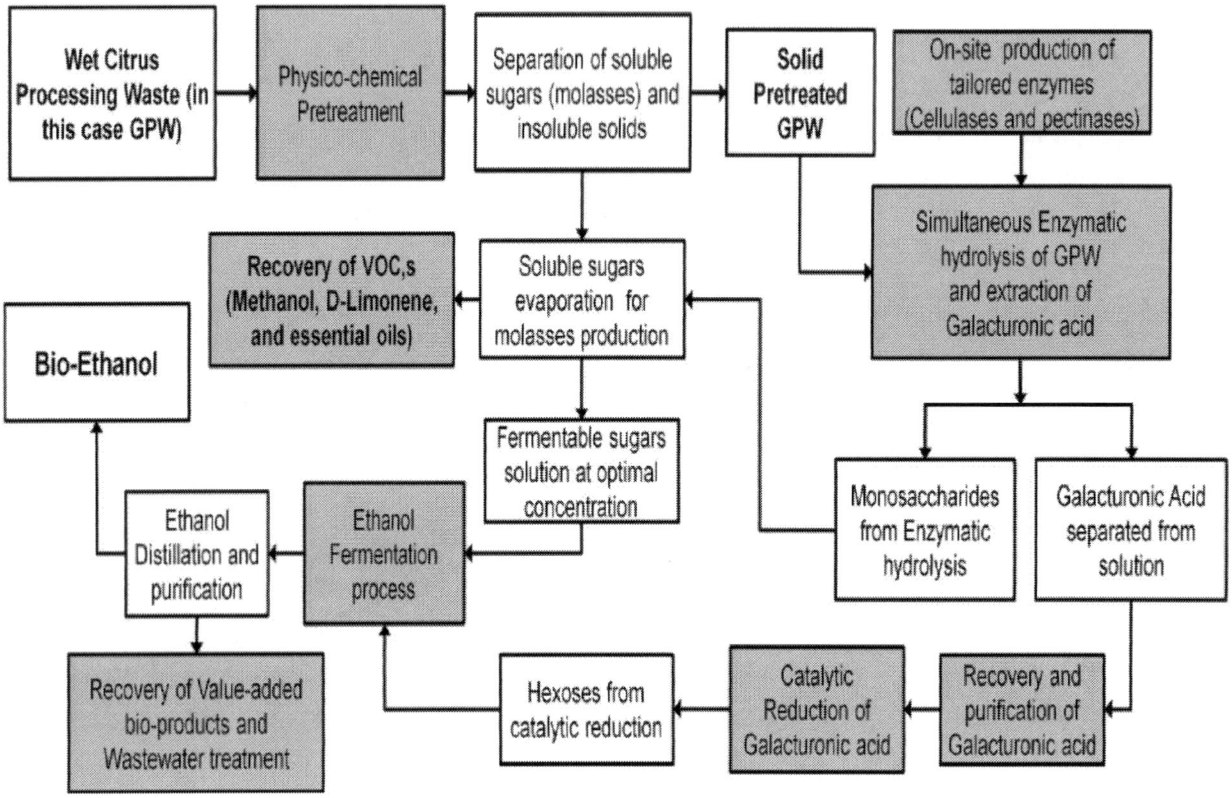

Figure 5. A schematic description of the most important steps of a proposed (CPW)-based biorefinery. Those which need additional research and development for filling the required engineering and scientific data gaps are highlighted.

Conclusions

Preliminary research studies have indicated promising new results from the pretreatment of CPW and provided some initial estimates of the potential for increased fermentable sugar yields (up to 25% in some cases).

Some of the most important findings in this study are the identification of the need for improved research and development in the following critical areas for a commercial scale citrus waste biorefinery:

- Kinetics of pretreatment and hydrolysis processes (yields, reaction rates, enzyme loadings, mathematical models, and others),
- Thermodynamic properties of raw and pretreated GPW (i.e. temperature of degradation),

- The effect of physico-chemical properties produced due to pretreatment (particle size, surface area, and surface morphology).
- Development of advanced processes for recovering value-added products such as flavonoids, pentoses, d-limonene, methanol, and galacturonic acid.
- Critical process data for small to medium size biorefinery facilities linked to the actual waste generation capacity of citrus processing plants (namely 40,000 to 400,000 tons /year).

Acknowledgments

This work thus far has been primarily supported by the South Texas Institute for Energy and the Environment, the Office of Graduate Studies at Texas A&M University-Kingsville, the Texas Citrus Producers Board, and by Mexican National Council of Science and Technology (CONACYT). Portions of this research have been also supported by the US National Science Foundation under Grant No. HRD-020659 (CREST-RESSACA) as part of Sustainable Technologies for the Environment section, and funded partially in part by USDA/NIFA Hispanic Serving Institution grant awards: 2011-30826 and 2011-38422-30803. Any opinions, findings, and conclusions or recommendations expressed in this material are those of the authors do not necessarily reflect the views of the National Science Foundation and USDA/NIFA.

Bibliographic References

1. ARS (Agricultural Research Service) (1962). Chemistry and Technology of Citrus, Citrus products and Byproducts. *Agriculture Handbook 98*. United States Department of Agriculture (USDA).
2. Antwerpen, F. J. (1941) Utilization of citrus wastes. *Industrial and Engineering Chemistry.* November 1941, pp 1422–1427.
3. Bozell, J.J., Moens L., Elliott, D.C., Wang, Y. (2000). Neuenscwander, G.G., Production of levulinic acid and use as a platform chemical for derived products. *Resour. Conserv. Recycl.* Vol. 28, pp 227–39.
4. Braddock, R.J. (1999). *Handbook of Citrus By-Products and Processing Technology*. John Wiley & Sons. Inc. New York, NY.
5. Braddock, R.J., and Miller, W.M. (2001). Volatile Organic Compounds from Citrus Feed Mill Emissions. *Journal of Food Process Engineering,* vol. 24, pp 1-15.
6. Buff, D.A. (1997). VOC Emissions from Citrus Processing Plants. *Transactions of the 1997 Citrus Engineering Conference; ASME*, pp 65.
7. Dasari R.K., and Berson, R.E. (2007). The effect of particle size on hydrolysis reaction rates and rheological properties in cellulosic slurries. *Applied Biochemistry and Biotechnology*, Vol. 136–140, pp.289-299.
8. Fernandez-López, J., Fernandez-Gines, J.M., Aleson-Carbonell, L., Sendra, E., Sayas-Barbera, E., Perez-Alvarez, J. A., Application of functional citrus by-products to meat products. *Trends in Food Science & Technology* Vol. 15, pp 176–185, 2004.
9. Furcht, P.W., Silla, H., Comparison of Simultaneous wet milling and enzymatic hydrolysis of cellulose in ball mill and attrition mill reactors. *Biotechnology and Bioengineering*, Vol. 35, pp 630-645, 1990.
10. Goodrich, R.M., and Braddock, R.J., Major by-products of the Florida citrus processing industry. Document FSHN05-22, Series of the Food Science and Human Nutrition Department, Florida

Cooperative Extension Service, Institute of Food and Agricultural Sciences, University of Florida, Original publication date October 2004. Revised February 2006 http://edis.ifas.ufl.edu.
11. Gong, C.S., Cao, N.J., Du, J., Tsao, G.T., Ethanol production from renewable resources. *Advances in Biochemical Engineering Biotechnology*, 65, 234-235, 1999.
12. Grohmann, K., Baldwin, E.A., Buslig, B.S., Production of ethanol from enzymatically hydrolyzed orange peel by the yeast Saccharomyces cerevisiae, *Applied Biochemistry and biotechnology* (45/46), pp 315-327, 1994.
13. Grohmann, K., and Baldwin, E. A., Fractionation and pretreatment of orange peel by dilute acid hydrolysis. *Biotechnology Letters* Vol. 14, No.12, pp 1169-1174, 1992.
14. Grohmann, K. and Bothast, R.J., Pectin rich residues generated by processing of citrus fruits, apples and sugar beets. In: *Enzymatic Conversion of Biomass for Fuel Production*, Himmel, M.E., Baker, J.O., and Overend, R.P., eds, pp 372–390, *ACS Symp. Ser. No. 566*, Washington, DC: American Chemical Society, 1994.
15. Grohmann, K., Cameron, R.G., and Buslig, B.S., Fractionation and pretreatment of orange peel by dilute acid hydrolysis, *Bioresource Technology 54*, pp 129-141, 1995.
16. Grohmann, K., Final Technical Report, Project Title: *Citrus Waste Biomass Program*. USDA Agricultural Research Service CRADA 58-3K95-4-1053, 2007. www.osti.gov/bridge/servlets/purl/898345-9P4Slw/898345.pdf, Last date accessed on 01.12.2012.
17. Ingram, L.O., Aldrich, H.C., Borges, A.C., Causey T.B., Martinez, A., Morales, F., Saleh, A., Underwood, S.A., Yomano, L.P., York, S.W., Zaldivar, J., and Zhou, S., Enteric bacterial catalyst for fuel ethanol production. *Biotechnology Proress,* vol. 15, pp 855-866, 1999.
18. Jørgensen, H, Kristensen, J.B., Felby, C., Enzymatic conversion of lignocellulose into fermentable sugars: challenges and opportunities, *Biofuels, Bioprod. Bioref.* Vol. 1, pp 119–134, 2007.
19. Lohrasbi M., Pourbafrani M., Niklasson C., Taherzadeh M.J., Process design and economic analysis of a citrus waste biorefinery with biofuels and limonene as products, *Bioresource Technology*, vol.101, pp 7382–7388, 2010.
20. Marın, F.R., Soler-Rivas, C., Benavente-Garcıa, O., Castillo, J., and Perez-Alvarez, J. A., By-products from different citrus processes as a source of customized functional fibers. *Food Chemistry* vol. 100, pp. 736–741, 2007.
21. National Agricultural Statistics Service (NASS), *Citrus Fruit 2011 Summary*. United States Dept. of Agriculture (USDA) http://usda01.library.cornell.edu/usda/current/CitrFrui/CitrFrui-09-22-2011.pdf. Last date accessed on 11.14.201.
22. National Renewable Energy Laboratory (NREL), *Glossary of Biomass Terms*, http://www.nrel.gov/biomass/glossary.html. Last date accessed on 01.12.2012.
23. National Renewable Energy Laboratory (NREL). *What is a Biorefinery?* http://www.nrel.gov/biomass/biorefinery.html. Last date accessed: 11.14.2011.
24. Neilson, M.J., Kelsey, R. G., Shafizadhe, F., Enhancement of enzymatic hydrolysis by simultaneous attrition of cellulosic substrates, *Biotechnology and Bioengineering Symp.*, Vol. 24, No.2, pp 293-304, 1982.
25. Odio, C., D-limonene recovery. *Citrus Eng. Conf. 42*, pp 45–54, 1996.
26. Pourbafrani, M., Citrus Waste Biorefinery: Process Development, Simulation and Economic Analysis, *PhD Thesis*, Department of Chemical and Biological Engineering, Chalmers University of Technology, Göteborg, Sweden, 2010.
27. Pourbafrani, M., Forgács, G., Horváth, I.S., Niklasson, C., Taherzadeh, M.J., Production of biofuels, limonene and pectin from citrus wastes. *Bioresource Technology,* vol. 101, pp 4246–4250, 2010.

28. Serrano-Ruiz, J.C., West, R.M., Dumesic, J.A., Catalytic Conversion of Renewable Biomass Resources to Fuels and Chemicals. *Annu. Rev. Chem. Biomol. Eng.* Vol. 1, pp 79–100, 2010.
29. Stewart, D.A., Widmer, W.W., Grohmann, K., Wilkins, M.R., Ethanol production from citrus processing wastes, *US Patent Application Publication*, No.US/2006/0177916-A1, 2006.
30. Ueno, H., Tanaka, M., Hosino, M., Sasaki, M., Gota, M., Extraction of valuable compounds from the flavedo of Citrus junos using subcritical water. *Separation and Purification Technology*, vol. 62, pp 513–516, 2008.
31. USEPA (United States Environmental Protection Agency), (1990). Clean Air Act Amendment Provisions Related to Specific Stationary Source Categories, 12pp. http://www.epa.gov/ttncaaa1/gen/stasor.pdf. (Last date accessed on 02.23.2012).
32. USEPA, (2004), 40 CFR Part 52, Approval and Promulgation of Implementation Plans: Florida: Citrus Juice Processing, *Federal Register, Vol. 69, No. 20*, pp 4459-4463, Friday, January 30, 2004 / Proposed Rules.
33. Ververis, C., Georghiou, K., Danielidis, D., Hatzinikolaou, D.G., Santas, P., Santas, R., Corleti, V., Cellulose, hemicelluloses, lignin and ash content of some organic materials and their suitability for use as paper pulp supplements. *Bioresource Technology*, vol. 98, pp 296–301, 2007.
34. Wilkins, M.R., Widmer, W.W., Grohmann, K., Cameron, R. G., Hydrolysis of grapefruit peel waste with cellulase and pectinase enzymes, *Bioresource Technology 98*, pp 1596–1601, 2007.
35. Ting, S.V., Deszyck, E.J., The Carbohydrates in the Peel of Oranges and Grapefruit, *Journal of Food Science*, Vol. 26, Issue 2, pp 146–152, March, 1961.
36. Wing, J.M., Citrus Feedstuffs for Dairy Cattle, Document BUL 829, *Series of the Animal Science Department*, Florida Cooperative Extension Service, Institute of Food and Agricultural Sciences, University of Florida. Original publication date November 1982. Reviewed June 2003. http://edis.ifas.ufl.edu. And http://ufdc.ufl.edu/UF00027465/00001 Last date accessed on 01.12.2012.
37. Zhou, W., Widmer, W.W., Grohmann, K., Economic Analysis of Ethanol Production from Citrus Peel Waste. *Proc. Fla. State Hort. Soc.* Vol. 120, pp 310-315, 2007.

(782d) Life Cycle Assessment of Ethanol Produced in an Integrated Biorefinery From a Forest Product Industry Wastewater Stream

Jifei Liu and David R. Shonnard, Chemical Engineering, Michigan Technological University, Houghton, MI

Integrated technology systems are designed based on the concept of "zero waste" to fully utilize natural resources by building industries next to each other when the waste of one is capable to fulfill the input requirement of another. The forest products industry produces wastewater streams of different characteristics, and this study considers the conversion of a wastewater stream from one manufacturing facility. This facility generates a wastewater stream with wood extractives suspended in it, including lignincellulosic components. Integrated with this facility, a biorefinery produces cellulosic ethanol as well as co-products; an organic acid deicer compound and steam. At the same time, this biorefinery strategy partially displaces the need for a wastewater treatment plant (WWTP) at the forest products facility.

For cellulosic ethanol produced at this integrated biorefinery, an original life cycle environmental impact assessment has been conducted to determine greenhouse gas emissions for a functional unit of 1 MJ fuel. In this analysis, there is a reduction of WWTP impacts and these "credits" have been allocated to the main product - ethanol, and also to a co-product acetate de-icer compound based on mass production ratios. However, in this analysis other credit allocations have been investigated. For example, if a carbon market is anticipated in the future, both the original forest product facility and the new biorefinery would benefit monetarily from the reductions of GHG emissions from the avoidance of WWTP emissions. Three scenarios for crediting are analyzed and discussed: **i** all credit allocated to the original forest products manufacturing facility, **ii** all to the biorefinery facility, and **iii** 50%-50% to each. In addition, sensitivity analysis applied to key LCA inputs is also investigated.

(782e) Evaluation of Industrial Wastewaters As Fermentation Substrate for Biocrude Production by Enhanced Activated Sludge Cultures

Andro Mondala, Rafael Hernandez, Todd French, Kamal Lamichhane, Magan Green, Linda McFarland and William Holmes, Dave C. Swalm School of Chemical Engineering, Mississippi State University, Mississippi State, MS

Current studies on evaluating a wastewater treatment plant-based biorefinery concept producing biofuel feedstock lipids and other potentially high-value bioproducts (biocrude) have shown that the carbon source represents a majority of the associated production costs. The process also uses large amounts of wastewater as the fermentation media and nutrient source. This paper will present results of our investigation on the use of high-strength industrial wastewaters as a combined carbon and nutrient source and fermentation medium for activated sludge cultures producing bio-crude. Wastewaters from a pulp and paper mill, landfill leachate, and a dissolved air floatation (DAF) – treated poultry processing wastewater were characterized for biochemical- (BOD) and chemical oxygen demand (COD), nutrient levels, and concentration of major contaminants. These were then tested as stand-alone media or amended with glucose in flask and 3-L bioreactor batch fermentation experiments. Samples were collected and analyzed for biomass and lipid yields, residual sugars and nutrient ions, and COD. Fermentation kinetics modeling will be conducted to determine the biomass growth and lipid accumulation performance of the enhanced activated sludge culture and tolerances towards potential microbial growth inhibitors in the wastewaters. The results will help determine the applicability of industrial wastewaters in the fermentation process and develop pretreatment strategies to remove inhibitory and recalctirant compounds.

(782f) Volatile Fatty Acids As a Sole Carbon Source for Biocrude Production by Enhanced Activated Sludge Cultures

Andro Mondala, Rafael Hernandez, Todd French, Magan Green, Linda McFarland, William Holmes and Gita Ramakrishnan, Dave C. Swalm School of Chemical Engineering, Mississippi State University, Mississippi State, MS

A study on the use of short-chain volatile fatty acids (VFA) as substrates for biofuel feedstock lipids (biocrude) production in enhanced activated sludge cultures will be presented. As part of the overall goal of establishing a sustainable wastewater treatment plant – based biorefinery, it is suggested to treat waste lignocellulose biomass by anaerobic digestion to yield volatile fatty acids such as acetate, butyrate, and propionate; after which the said VFAs will be fed into activated sludge cultures to produce biocrude via fermentation. In this study, batch fermentation experiments using individual VFAs and synthetic ternary mixtures of acetate, propionate, and butyrate with varying ratios as substrates will be conducted in flask cultures and 3-L bench-scale bioreactors. Samples will be analyzed for biomass and lipid yields, residual VFAs in the fermentation broth, and fatty acid profile of the biocrude product. Kinetic analysis of the fermentation data will be conducted to evaluate the growth and lipid production performance of the activated sludge microorganisms using VFAs as the carbon source. This approach could potentially be a more cost-effective alternative to using sugars obtained by acid hydrolysis of biomass since anaerobic digestion is commonly a built-in component within the wastewater treatment process. VFAs fermentation could also provide a relatively higher carbon yield than direct sugar fermentation due to their relatively easier assimilation into the biochemical pathways leading to lipid biosynthesis in microorganisms than sugars.

Session 17006 Recalcitrance of Woody Biomass
AIChE 2012
October 28 ~ November 2
Pittsburgh, PA

A study in lignin inhibition for enzymatic hydrolysis of woody biomass

Zhiying Yu*, Hasan Jameel, Hou-min Chang, Sunkyu Park

Department of Forest Biomaterials, North Carolina State University
Campus Box 8005, Raleigh, NC 27695-8005, USA
Presenting author: zyu2@ncsu.edu

The presence of lignin in woody biomass is a great obstacle that limits the bioconversion efficiency through enzymatic hydrolysis, especially for softwood. It has been speculated that the differences in lignin content, structure, and distribution between hardwood and softwood, will influence lignin-enzyme interactions such as non-specific adsorption of enzyme onto lignin, physical blockage of lignin that restrict the enzyme penetration into biomass structure, and chemical blockage of binding sites through lignin-carbohydrate linkages. In this research, the milled wood lignins (MWL) isolated from eucalyptus, maple and loblolly pine as well as kraft lignin (indulin AT) was investigated for their impact on enzymatic hydrolysis. The woody biomass structure was mimicked by 1) physically mixing the lignins and bleached pulps and, 2) reconstructing the woody biomass via impregnating the lignins into bleached pulps.

The adsorption kinetics of cellulase on individual bleached pulps and lignins showed that the bleached pulps adsorbed 2 to 3 folds of cellulase than lignins. MWL adsorbed 17.5-22.5% cellulase, while kraft lignin only had less than 3.5 % cellulase adsorption. When 20% lignins were physically mixed with bleached pulps for enzymatic hydrolysis, it was found that the presence of 20% MWL had a minor impact on enzymatic hydrolysis, while 20% kraft lignin showed 8.3-11.6% limitation on the hydrolysis. In contrast, when 20% lignins were impregnated into the bleached pulps, 6.2%, 12.0%, 17.6% and 59.0% decrease in 96 h hydrolysis yield was observed for reconstructed eucalyptus, maple, pine and kraft pulps, respectively. Therefore, MWL isolated from untreated hardwood and softwood was found to limit enzymatic hydrolysis due to the effect of physical blockage. A more significant impact of physical blockage of lignin was found with the increasing of guaiacyl unit of lignins in the reconstructed woody biomass with the same lignin content. Kraft lignin inhibited the enzymatic hydrolysis through slightly deactivating the enzymes and significantly blocking the accessible area.

It was speculated that the smaller molecules of lignin had a greater physical blockage effect on enzymatic hydrolysis. Thus, different molecular weight of kraft lignins was fractionated by selective pH and ultrafiltration, respectively. The lignins of bigger molecular weight were found to have less physical blockage than the lignin with smaller molecular weight. In order to investigate the effect of lignin from pretreated lignocelluloses on enzymatic hydrolysis, the cellulolytic enzyme lignins were isolated from acid pretreated and alkali pretreated hardwood and softwood pulps. The effect of acid and alkali pretreatment on lignin structure as well as the effect of hardwood and softwood lignins on enzymatic digestibility will be discussed in terms of three inhibitory mechanisms suggested above.

(787b) Distinct Effects of Lignin and Xylan On the Enzymatic Hydrolysis of Organosolv Pretreated Biomass

Mi Li, School of forestry and wildlife science, Auburn University, Auburn, AL and Maobing Tu, Auburn University, Auburn, AL

Forestry lignocellulosic biomass is a great potential source of sustainable materials for biofuels and chemicals. Technically, bioconversion process of cellulosic ethanol from forestry feedstock requires pretreatment, enzymatic hydrolysis and fermentation. In order to improve accessibility of lignocellulose to enzymes for hydrolysis, high temperature pretreatment with acid, base or solvent is typically needed to break down the recalcitrant structure of cellulose. Besides the degree of cellulose crystallinity, the interaction between xylan/lignin and cellulase enzymes played a key role in the effective hydrolysis of lignocellulosic biomass. Elucidation of the distinct roles of residual xylan and lignin has been investigated in this study. Loblolly pine (*Pinus taeda*) and sweetgum (*Liquidambar*) were pretreated by organosolv and have been quantitatively characterized by the enzymatic hydrolysis based on initial hydrolysis rate and final hydrolysis yield. By correlating the residual lignin and xylan to the initial hydrolysis rate and final yield in organosolv pretreated softwood and hardwood, a more accurate fundamental understanding the roles of xylan and lignin in limiting the enzymatic hydrolysis has been developed. Our results show that Novozyme 22C showed very good performance of hydrolysis on organosolv pretreated biomass. Moreover, the residual xylan in pretreated substrates actually would affect the initial hydrolysis rate much more than that from residual lignin. It suggested that xylan firstly inhibited the cellulase because it's closer to the cellulose micro-fibril. The lignin residual interacted with cellulase after xylan and decreased the overall hydrolysis yield. The addition of xylanases could increase the initial hydrolysis rate by removing xylan. We also found that in the simultaneous saccharification and fermentation the ethyl xylopyranoside was produced by the enzymatic catalysis of xylose and ethanol.

(787c) Ionic Liquid Pretreatment of Lignocellulosic Biomass – Recovery and Reuse of IL

Anantharam P. Dadi, SuGanit Systems Inc, Toledo, OH, Marcus Stacy, SuGanit Systems Inc., Toledo, OH, Guneet Kumar, SuGanit Systems Inc., Reston, VA, Noureen Faizee, Synterra, Toledo, OH and Sasidhar Varanasi, Chemical & Environmental Engineering, University of Toledo, Toledo, OH

University of Toledo and SuGanit Systems Inc. have jointly developed a novel Ionic Liquid pretreatment method for deconstructing lignocellulosic biomass feedstocks to promote rapid enzyme-catalyzed saccharification of the carbohydrates . The IL pretreatment is a facile method that works at moderate temperatures and atmospheric pressure conditions, and was successfully demonstrated on various lignocellulosic biomass feedstocks. SuGanit Systems Inc. is scaling up this technology of sugar production from lignocellulosic biomass by ionic liquid pretreatment. A critical economic driver of the process is efficient recovery of IL from wash streams following biomass pretreatment. We have explored the viability of electrodialysis (ED) for recovering and concentrating IL following biomass pretreatment. It was shown that dissolved nonionic extractibles entering IL wash solutions from biomass could successfully be removed via ED. This presentation will focus on the effectiveness and economics aspects of ED for this purpose, and the important differences between ED-based separation of IL versus salt solutions.

(787d) Solvent Based Pretreatment (N-methyl morpholine oxide) for Enhanced Enzymatic Hydrolysis of Dissolving Pulp and Sugarcane Bagasse

Subramanian Ramakrishnan, Department of Chemical and Biomedical Engineering, Florida A&M University - Florida State University, Tallahassee, FL, Gary Brodeur, Chemical & Biomedical Engineering, Florida State University, Tallahassee, FL, John Telotte, Department of Chemical and Biomedical Engineering, Florida State University, Tallahassee, FL and John Collier, Chemical and Biomedical Engineering, Florida State University, Tallahassee, FL

The first step in the conversion process of biomass to biofuels or hydrogen lies in picking an environmentally friendly solvent which is capable of breaking up the crystalline microstructure of cellulose. Once the cellulose is in suspension, enzymes (cellulases) are added to the reactor to hydrolyze the cellulose to simple sugars that are then fermented by bacteria to biofuels. Till date, a number of solvents have been used for pretreatment of biomass and for reprecipitating it by the addition of an anti-solvent thus making it a multi-stage process. In this work, it is shown that N-methyl morpholine oxide is a good solvent for pretreatment of crystalling pulp and sugarcane bagasse. Sugarcane bagasse and dissolving pulp of degree of polymerization 1160 are used as the substrate and Accellerase 1000 and 1500 purchased from Genentech are the cellulases used. The rates of enzymatic hydrolysis and total sugars released are studied as a function of solution conditions – solution pH, temperature and enzyme loading. These studies reveal that the enzymes are active in NMMO and that the rates of hydrolysis of cellulose pretreated with NMMO are comparable to the rates of hydrolysis of regenerated cellulose suspended in aqueous buffer solutions. A comparison of enzymatic hydrolysis after NMMO treatment of sugarcane bagasse reveals that the rates and yields are comparable and in certain instances higher than acid pretreatment followed by steam explosion. A mathematical model to predict enzymatic hydrolysis based on lumped kinetic modeling is also proposed.

(787e) Role of Tween-80 in Reduction of Nonproductive Cellulase Binding to Lignin

Ravinder Garlapalli[1], Barbara, L Knutson[2], Sue E. Nokes[3] and Stephen E. Rankin[1], (1)Chemical and Materials Engineering, University of Kentucky, Lexington, KY, (2)Chemical & Materials Engineering Department, University of Kentucky, Lexington, KY, (3)Department of Biosystems and Agricultural Engineering, University of Kentucky, Lexington, KY

A key limitation in the enzymatic hydrolysis of lignocellulose to produce soluble sugars is the nonproductive binding of the enzymes to biomass components such as lignin. Bulk studies on hydrolysis of cellulosic substrates have suggested that additives enhance both conversion of sugars and enzyme recovery when lignin is present, where nonionic surfactants are the most widely studied additives. However, the specific mechanism underlying the use of surfactants is not well understood. To study the effect of nonionic surfactant Tween-80 and the interactions with cellulase enzyme on lignin, model lignin thin films are prepared and investigated as a function of the adsorbing solution using a Quartz Crystal Microbalance with Dissipation monitoring (QCM-D). Model lignin films were synthesized by dissolution of dry lignin powder in ammonium hydroxide solution followed by coating onto a quartz crystal resonator. The cellulase for this study is a commercially available mixture of enzymes derived from *Trichoderma reesei*. In-situ studies of single-component binding show that cellulase is bound to the lignin surface irreversibly, while Tween-80 is partially (but not completely) reversible upon rinsing with buffer. Co-adsorption of Tween-80 and cellulase (where both are introduced together) resulted in an apparent net reduction in the amount of cellulase adsorbed on lignin. Sequential adsorption experiments were carried out by varying the order and concentration of Tween-80 and cellulase introduction to lignin surfaces to explore the possibility for displacement of cellulase or pretreatment of biomass using Tween-80. The results of the sequential adsorption experiments suggest that Tween-80 was able to displace adsorbed cellulases, but quantifying the amount of cellulase displaced was difficult with QCM-D as it measures the net change in piezoelectric properties due to mass adsorbed and desorbed on the sensor. Further studies are underway to use analytical methods to quantify the amounts of cellulase desorbed under various conditions and to better understand the mechanism by which the surfactants reduce the nonproductive binding of cellulases through their interaction with lignin.

(787f) Green and High Efficient Fuel Ethanol Production From Sugarcane Bagasse

Xinshu Zhuang, Bio-chem conversion lab, Guangzhou institute of energy conversion,Chinese Academy of Sciences, Guangzhou, WI, China

Sugarcane bagasse (SCB), leftover stalks after juice extraction, were pretreated using liquid hot water (LHW) to remove most hemicellulose and part lignin of lignocellulose. In order to improve the enzymatic hydrolysis efficiency of LHW-pretreated SCB, Tween80, a non-ionic surfactant proved to enhance enzymatic hydrolysis of pretreated biomass in high lignin content, was applied in this research. It was found that the optimal concentration of Tween80 can increase glycan conversion of LHW-pretreated SCB from 44.9% (control) to 78.1%. A laboratory facility with plate-and-frame impeller (PFI) was applied to conduct batch and fed-batch hydrolysis and the process of fed-batch hydrolysis prior to simultaneous saccharification and fermentation (SSF). After hydrolyzing LHW-pretreated SCB for 120 h, the total sugar concentration and glycan conversion obtained from fed-batch hydrolysis are 91.6 g/L and 68.3%, respectively, which are 9.7 g/L and 7.3% more than those from batch hydrolysis. The ethanol production and theoretical yield obtained from the process of fed-batch hydrolysis prior to SSF are 55.4 g/L and 88.3%, respectively, which are 15.5 g/L and 24.7% higher than those from separate fed-batch hydrolys and fermentation.

These research work are financially supported by the National Key Technology R&D Program (2011BAD22B01).

(787g) Crystallinity – A Rate-Determining Parameter for Pretreated Lignocellulosic Biomass

Yuzhi Kang[1], Prabuddha Bansal[1], Bryan Vowell[2], Matthew Realff[1], Jay H. Lee[3] and Andreas S. Bommarius[4], (1)School of Chemical and Biomolecular Engineering, Georgia Institute of Technology, Atlanta, GA, (2)Georgia Institute of Technology, Atlanta, GA, (3)Chemical and Biomolecular Engineering, Korea Advanced Institute of Science and Technology, Daejeon, South Korea, (4)Chemical & Biomolecular Engineering, Georgia Institute of Technology, Atlanta, GA

Crystallinity – A Rate-determining Parameter for Pretreated Lignocellulosic Biomass

Yuzhi Kang, Prabuddha Bansal, Bryan Vowell, Matthew Realff, Jay Lee, Andreas Bommarius

Abstract

Steam explosion is one of the most effective pretreatment methods in disrupting lignocellulosics structure and enhancing its accessibility. The effects of steam explosion on woody biomass and agricultural residues (i.e. bagasse, wheat straw and loblolly pine) are evaluated. The key operational parameters including temperature, residence time and acid concentration have pronounced effects on crystallinity index (CrI) of the feedstock. The CrI of steam-pretreated lignocellulosic biomass was calculated from X-ray diffractometry with the method of Least-squares and was found to decrease as pretreatment severity increases. A series of steam-exploded lignocellulosic samples with varying CrI was generated and the significance of CrI is confirmed by a linear relationship between initial hydrolysis rates and CrI. The present study provides evidence that the CrI is one of key parameters determining the enzymatic hydrolysis rate of pretreated lignocellulosic biomass.

To more fully explore the connection between accessibility, hydrolyzability and reactivity, we have undertaken experiments with phosphoric acid swollen Avicel® at different levels of conversion. We will present the quantitative trends between the rate of conversion and these underlying substrate properties.

(790a) A Comparison of the Recoverability of Organic Solvent Mixtures in Waste Streams Using Process Simulation and Life Cycle Assessment Software

Eduardo J. Cavanagh[1], Mariano J. Savelski[2] and C. Stewart Slater[2], (1)Rowan University, Glassboro, NJ, (2)Chemical Engineering, Rowan University, Glassboro, NJ

Through the support of the U.S. Environmental Protection Agency and in partnership with Pfizer Inc., a flexible modeling tool that calculates avoided Life Cycle Assessment (LCA) emissions and Operative Cost Savings of solvent recovery systems for low volume pharmaceutical synthesis waste streams was developed. This new tool allows for a rapid assessment of the recoverability of solvent mixtures. ASPEN® Simulation Workbook was used for process modeling. Preliminary environmental impact was determined using SimaPro® and EcoSolvent® by comparing the proposed solvent recovery to the incineration of solvent waste and the manufacture of virgin solvent. Different homogeneous binary (azeotropic and non-azeotropic) solvent-solvent and water-solvent systems were analyzed. The proposed solvent mixtures are representative of those found in pharmaceutical manufacturing and include acetone, acetonitrile, toluene, tetrahydrofuran (THF), isopropyl alcohol, n-hexane, and methyl ethyl ketone (MEK).

The solvent recovery process model includes distillation, pervaporation (PV) and a combination of both. Sensitivity analysis for different system parameters such as reflux ratio and feed location was performed in several case studies, and points of minimum emissions and operating costs were also determined.

(790b) Sustainability of Battery Electric Vehicles Propelled by Lithium Ion Batteries : A Life Cycle Perspective

Qiang Dai and Christian M. Lastoskie, Civil & Environmental Engineering, University of Michigan, Ann Arbor, MI

With the foreseeable increasing penetration of battery electric vehicles (BEVs) into the U.S. market, the environmental impacts associated with the delivery of driven vehicle miles are worth investigating for this alternative mode of transportation as compared with conventional Internal Combustion Engine Vehicles (ICEVs). In this study, life cycle assessment has been conducted on Lithium Ion Battery (LIB)-powered BEVs representing the status-quo of current technology. The total life cycle impacts of BEVs, evaluated by Eco-indicator 99, Cumulative Energy Demand and Global Warming Potential, were compared with that of ICEVs burning petroleum to quantitatively demonstrate the environmental friendliness of BEVs. The importance of battery material chemistries, synthesis pathways and electricity mix was also investigated respectively. The results confirmed that BEVs equipped with LIBs offer substantial environmental benefits over ICEVs, including curtailing energy consumption, mitigating global warming, and reducing emissions capable of damaging human health and ecosystem quality. As use phase dominates the environmental burdens of BEVs, the switch to a greener electricity mix would be the key to a more sustainable transportation mode by BEV. Manufacturing and assembly phase of LIB, contributing to up to 20% of the total environmental footprints of BEVs, also provide windows of opportunity to alleviate environmental impacts through future BEV improvements, especially by extending cycle life and energy density of a battery pack.

(790c) Decision Making for Sustainable Processes

Raymond L. Smith, Gerardo J. Ruiz-Mercado and Michael A. Gonzalez, U.S. Environmental Protection Agency, Cincinnati, OH

Multi-objective decision support tools aim to represent the thought process of decision makers in the development and design of chemical processes. These tools often use calculations that mathematically describe preferences to combine various utility scores with weighting factors. Once utility scores are combined into a single objective, the alternative solutions for process development and design can be ranked. The creation of a single objective for a multi-attribute decision necessarily generates tradeoffs (except in the simplest win-win scenarios). While the intention in choosing (equal) weighting factors may be to make them value free, this contribution will show how combining utility scores is always a statement of a relationship between the weighted scores. The relationships between utility scores can be rewritten as a series of equations in terms of the weighting factors. Calculations can be used to determine the weighting factors for different sets of relationships, allowing for explicit representation of the implied tradeoffs for decision makers. In addition, a method will be described for using marginal rates of substitution between utility scores to calculate total utility. This method will simplify the relationships between the utility scores and present a powerful and intuitive way to calculate total utility.

The methods presented will be demonstrated in an example utilizing indicators derived from EPA's sustainability evaluation and design tool, GREENSCOPE (Gauging Reaction Effectiveness for the Environmental Sustainability of Chemistries with a multi-Objective Process Evaluator). GREENSCOPE has 140+ indicators in four basis areas: environment, economics, energy, and efficiency. Previous work has developed absolute sustainability limits for these indicators and provided data needs and solutions for calculations. The combination of GREENSCOPE with weighting / marginal rate of substitution methods and explicit representations of implied tradeoffs will provide users with tools to support multi-objective decisions.

(790d) Cradle-to-Gate Life-Cycle Assessment of Ionic Liquids

Amirhossein Mehrkesh and Arunprakash T. Karunanithi, Department of Civil Engineering, University of Colorado Denver, Denver, CO

Ionic liquids (ILs) are new class of "green" chemicals that are being developed as a replacement for volatile organic solvents for a range of industrial applications. The inherent greenness of ILs is often associated with negligible vapor pressures that they possess. Although, ILs lessens the risk of air pollution due to their insignificant vapor pressure, many are highly soluble in water and can become persistent pollutants. Existing studies show that certain ionic liquids have negative environmental and ecosystem impacts while others do not. Current strategies for designing novel ionic liquids tend to focus on tailoring their structures with an aim to impart specific functionalities (e.g. high CO2 solubility for carbon capture applications). However -in line with the assumption that all ionic liquids are green- they either do not consider the environmental and health impacts during the design or consider them after the fact. In this work, we present systems analysis view of certain ionic liquids and compare them with common organic solvents from a cradle-to gate life-cycle assessment (LCA) perspective. We will focus on the environment, ecosystem and health impacts associated with the production of different ionic liquids.

(790e) Life Cycle Assessment of Thin-Film CdTe Photovoltaics

Yuan Yao and Fengqi You, Department of Chemical and Biological Engineering, Northwestern University, Evanston, IL

Growing concern to adverse health effects and severe atmospheric pollution caused by the operation of fossil-fuel-burning power plants has fostered the research on more sustainable energy resources. Solar energy is one of the most promising alternatives for large-scale electricity production because of its abundance.[1] In the past 12 years, photovoltaics (PV) experienced an average growth of approximately 45% per year that was affected only slightly by the recent global financial crisis.[2] Among a variety of photovoltaic technologies, thin film is relatively new but developing rapidly for low cost, large flexibility[3] and small amount of materials used in modules. CdTe is one of the thin film solar cells that have the great potential to be produced in large scale with low cost.[4] However, any energy production, no matter how "green" they look like, will lead to environmental impacts in their entire life cycle.[5] Thus, it is necessary to apply Life Cycle Assessment (LCA) tools to CdTe photovoltaics to assess their environmental performance, including GHG and heavy metal emissions. Currently, most assessments are conducted based on the data from OECD countries (e.g., United States, Canada and Germany).[6, 7] However, most of solar cell companies prefer to move their manufacturing plants to non-OECD countries (e.g., China, India and Malaysia) because of low cost and abundant raw materials. Current LCA results of CdTe PV fail to provide decision makers with appropriate references on whether it is environmentally sustainable to have CdTe PV manufactured abroad, especially in none-OECD countries. It is of great importance to conduct comparative LCAs in two cases, where CdTe solar cells are manufactured domestically and oversea respectively.

There are several challenges to conduct this comparative study. The first one is to figure out all possible differences between two cases in each stage of CdTe PV life cycle. Since cadmium and tellurium are byproducts of zinc and copper production, LCAs should start with zinc and copper ore mining. Different grades of ores, efficiencies of metallurgical operation, electricity mixes, and some other factors may contribute to the difference of emission factors of life cycle stages between CdTe PV made in OECD and none-OECD countries. Another challenge is that the data for none-OECD countries are not as accessible as those for OECD members. A comprehensive understanding of material flows and emissions, and a deep data digging for non-OECD countries in all stages are required.

In this paper, a "Cradle to Gate" LCA, which includes raw material exploitation, material processing and PV manufacturing, of CdTe PV in two case studies is carried out. In the first case study, China is chosen as a representative of none-OECD countries. A CdTe PV manufacture plant is built in Sichuan Province where the first Chinese CdTe PV manufacturing line was located, [8] and CdTe solar panels are shipped to venders in the United States. In the second case, a local plant is built in Chicago, United States. The geographically representative data mainly come from industrial reports, literatures and Gabi.[9] GHG and heavy metal emissions in all life cycle stages are estimated in two case studies. Total life-cycle GHG emissions of CdTe PV made in China are 113 $kgCO_2$-eq/m^2 and heavy metal emissions are 1.17 mg Cd/m^2. While GHG emissions and heavy metal emissions of CdTe PV made in the United States are 74 $kgCO_2$-eq/m^2 and 0.37 mg Cd/m^2 respectively, which are much smaller than that in China. Emissions of other heavy metals, such as arsenic, lead, mercury and nickel are also estimated in two case studies. In order to integrate heavy metal and GHG emissions to the corresponding environmental impacts, Eco-indicator 99 [10] is applied to two case studies. The total indicator of CdTe PV in the United States is 4.32 points/m^2; while the indicator in China is 8.67 points/m^2, which indicates that the environmental damages caused by investing CdTe PV manufacturing in China is twice as severe as that in the United States. Thus CdTe PV made in the United States are more sustainable. Energy payback times (EPBT) are calculated based on 10.7% module efficiency, 80% performance ratio, 30 years life time and 1825 kWh/m^2/year that is forecast of radiation for Chicago area from National Solar Radiation Data Base by using time-series methods with Oracle Crystal ball.[11] EPBT for two case studies are very close, 0.7 years for CdTe PV in China and 0.62 years for that in the United States. This suggests that CdTe PV made in the United States spend relatively shorter time to generate the same amount of energy used in total life-cycle of CdTe PV. Based on LCAs results, CdTe PV made in the United States obtain a relatively smaller EPBT with much less environmental damages. Investing CdTe PV plants in non-OECD is not able to get a faster energy payback, while generates more emissions and causes severer environmental damages. Therefore, domestic manufacturing in OECD countries is a more environmentally sustainable approach suitable for CdTe PV.

References

[1] S. B. Darling, F. You, T. Veselka, and A. Velosa, "Assumptions and the levelized cost of energy for photovoltaics," *Energy & Environmental Science,* vol. 4, pp. 3133-3139, 2011.

[2] V. M. Fthenakis, "Sustainability metrics for extending thin-film photovoltaics to terawatt levels," presented at the MRS Bulletin, 2012.

[3] K. L. Chopra, P. D. Paulson, and V. Dutta, "Thin-film solar cells: An overview," *Progress in Photovoltaics,* vol. 12, pp. 69-92, 2004.

[4] K. Kato, T. Hibino, K. Komoto, S. Ihara, S. Yamamoto, and H. Fujihara, "A life-cycle analysis on thin-film CdS/CdTe PV modules," *Solar Energy Materials and Solar Cells,* vol. 67, pp. 279-287, 2001.

[5] V. M. Fthenakis and E. Alsema, "Photovoltaics energy payback times, greenhouse gas emissions and external costs: 2004–early 2005 status," *Progress in Photovoltaics: Research and Applications,* vol. 14, pp. 275-280, 2006.

[6] V. M. Fthenakis, W. M. Wang, and H. C. Kim, "Life cycle inventory analysis of the production of metals used in photovoltaics," *Renewable & Sustainable Energy Reviews,* vol. 13, pp. 493-517, 2009.

[7] V. M. Fthenakis and H. C. Kim, "Photovoltaics: Life-cycle analyses," *Solar Energy,* vol. 85, pp. 1609-1628, 2011.

[8] (Apr. 10th). *Chengdu COE Apollo Solar Co ., Ltd.* Available: http://www.coeapollo.com/en/about.asp

[9] Gabi, "Software and database contents for Life Cycle Engineering.," 2011 ed. Stuttgart: PE INTERNATIONAL AG.

[10] M. Goedkoop, R. Spriensma *et al*, "The Eco-indicator 99 a damage oriented method for life cycle impact assessment. Methodology report.," PRé-Consultants, Amersfoort2001.

[11] (24th, March). *Oracal Crystal Ball*. Available: http://www.oracle.com/us/products/applications/crystalball/index.html

(790f) Accounting for Ecosystem Services in Life Cycle Assessment by Eco-LCA: Advances in Methodology and Software

Bhavik R. Bakshi[1], Erin F. Landers[1], Shweta Singh[1], Laura A. Merugula[1], Oleg Mishchenko[2] and Joseph Fiksel[3], (1)William G. Lowrie Department of Chemical and Biomolecular Engineering, The Ohio State University, Columbus, OH, (2)Computer Science and Engineering, Ohio State University, Columbus, OH, (3)Center for Resilience, The Ohio State University, Columbus, OH

Ecologically-Based Life Cycle Assessment (Eco-LCA) is a framework to account for the role of ecosystem goods and services in supporting economic activities (Zhang et al., 2010). Ecosystem services are contributions from nature without which sustainability of any activity on earth would not be possible. They include provisioning services like food, biomass and water; regulating services like regulation of floods, climate and pests; supporting services like biogeochemical cycles and photosynthesis; and cultural services like tourism, aesthetics and spiritual development. Despite their importance, most sustainability methods have not accounted for their role.

In our previous research, we have developed ways of accounting for many ecosystem services, and have developed an environmentally extended input-output (EEIO) model of the United States economy to permit a quick Eco-LCA of economic sectors (Zhang et al., 2010). Hybrid Eco-LCA models have also been developed by combining detailed process level data with the IO model. Thermodynamic methods based on exergy and emergy analysis are used to aggregate various ecosystem services (Baral et al., 2012).

This presentation will describe some significant recent developments in the Eco-LCA data, method and software. A large number of provisioning services are now quantified in Eco-LCA as intermediate flows from relevant sectors by using the algebra of EEIO models. This relies on physical data about products such as grains, legumes, fish, biomass and medicinal plants. The Phosphorous cycle is also included in Eco-LCA by incorporating its flows into relevant sectors. This adds to the C and N cycles that are already included in Eco-LCA. Many ecosystem services, especially regulating services, are difficult to quantify. This challenge is overcome in this work by means of a novel approach of qualitative EEIO modeling and LCA. This approach relies on gathering information about the extent to which each sector of the economy depends on the selected ecosystem service. This dependence is represented in qualitative terms as "high," "medium," and "low," and this information is propagated through the IO model to get qualitative information about the extent to which the selected economic product depends on the ecosystem service.

These new developments have been implemented in the Eco-LCA software available at http://resilience.osu.edu/ecolca/. The software permits three types of analyses: thermodynamic, ecosystem services, and footprints. Thermodynamic analysis includes cumulative exergy consumption analysis with and without ecosystems. This analysis with ecosystems becomes analogous to emergy analysis. The tool enables calculation of metrics such as renewability index and thermodynamic return on investment. Analysis of ecosystem services provides information about these services by following classification systems like the one described in the first paragraph. These data are represented in multiple units, but may be aggregated by a hierarchical scheme that relies on thermodynamic methods. Footprint methods include univariate indicators like carbon, nitrogen, land use and water footprints. Use of this software for performing tiered hybrid LCA by combining process level data with the economy scale EEIO model will also be described and demonstrated, if time permits.

References

Baral, A., B. R. Bakshi, R. Smith, Assessing Resource Intensity and Renewability of Cellulosic Ethanol Technologies using Eco-LCA, *Environmental Science and Technology*, 46, 4, 2436–2444, 2012

Zhang, Y., A. Baral, B. R. Bakshi, Accounting for Ecosystem Services in Life Cycle Assessment, Part II: Toward an Ecologically-Based LCA, *Environmental Science and Technology*, 44, 7, 2624-2631, 2010

(790g) Assessing the Critical Role of Natural Capital in Microalgal Biofuel Production

George Zaimes, Civil and Environmental Engineering, University of Pittsburgh, Pittsburgh, PA and Vikas Khanna, Department of Civil and Environmental Engineering, University of Pittsburgh, Pittsburgh, PA

Current global trends in resource consumption and fossil fuel depletion have driven research in biomass-to-energy systems. In recent years, microalgal biofuel have gained increased attention as a sustainable liquid-fuel alternative. Microalgal derived fuels do not suffer from many of the major drawbacks associated with first or second generation biofuels, and have the potential to act as "drop in" replacements for traditional petroleum fuels. As an emerging field, holistic evaluation of algae-to-fuel systems that considers the resource consumption, emissions, and their impact across the entire life cycle is critical to ensure the long-term sustainability of emerging algae-based energy systems. Current interest in microalgal biofuels has led to several microalgae based life cycle assessments (LCA). Most of these studies have focused on quantifying the greenhouse gas emissions and the net energy balance for microalgal biofuel production without sufficient attention to the spatial variability in algal biomass growth rates, climatological conditions, and stresses on ecological goods and services. As ecosystem goods and services (natural capital) provide the foundation for manmade capital, incorporating the broader role of ecosystem goods and services is critical to any accurate measure of sustainability. Proper quantification of ecosystem goods and services is also critical to assess the vulnerability of emerging microalgal biofuels to the depletion of natural capital and the role that nature plays in making ecological resources available for biofuel supply chains.

This work focuses on a comprehensive life cycle evaluation of 48 microalgal biomass production pathways. We build upon our prior research and examine multiple cultivation locations, culture conditions, cultivation and harvesting options, and industrial symbiosis within microalgal biomass production for the continental United States. Process flows and life cycle inventories are constructed based on first principles, peer reviewed literature, LCA databases, and best available engineering knowledge. The role of natural capital for emerging algae-to-energy systems is accounted for via the use of EcoLCA model. EcoLCA is a recent environmentally extended input-output life cycle oriented approach capable of accounting for the role of natural capital such as ecosystem goods and services in LCA. EcoLCA model considers a wide array of goods and services derived from nature and a hierarchical, thermodynamic aggregation scheme to permit meaningful interpretation. We quantify and aggregate the contribution of ecosystem goods and services in terms of mass, energy, and exergy (available energy) for the algae-to-energy systems. A variety of hierarchical metrics such as quality corrected thermodynamic return on investment and renewability index for the algae-to-energy systems are also developed. Tradeoffs between life cycle environmental impacts are analyzed via a series of two-dimensional pareto optimal surfaces. Synthesis of these results provides insights regarding the vulnerability of the emerging algae-to-energy systems to the depletion of specific resources.

(790h) Multi-Objective Synthesis of Biofuels' Processes by a Novel Concept of Eco-Profit, Extended to Eco-Net Present Value

Lidija Čuček and Zdravko Kravanja, University of Maribor, Faculty of Chemistry and Chemical Engineering, Maribor, Slovenia

Abstract

This paper's contribution introduces and discusses a novel concept of eco-net present value (Eco-NPV), also called eco-net present worth (Eco-NPW). Our recently proposed sustainability metric, based on the concept of eco-profit (Čuček et al., 2012; Kravanja and Čuček, 2012), which is on a yearly basis, has now been extended to Eco-NPV, which is based on the full lifetime of a particular project. Eco-profit is defined as the difference between unburdening (eco-benefit) and burdening (eco-cost) of the environment, where eco-benefit and eco-cost calculations are based on life cycle assessment (LCA). Since eco-profit is expressed in monetary units similarly to economic profit, they can be merged together into so-called 'total profit'. When the synthesis of a process, the complex of processes, or even a regional supply network is performed using e.g. the mixed-integer programming approach, the optimal solution is the one with the maximal total profit. The intention is not only to find economically the more attractive and environmentally least-burdening solutions, but also those that unburden the environment the most.

This concept has now been extended to the net present value (NPV). NPV is an economic measure which properly takes into account the complete economics of the project throughout the project's life-cycle; the initial investment expenditure, all the expected future annual cash flows, the time value of money, and the salvage value. Maximization of the NPV with a discount rate equal to the minimum acceptable rate of return (MARR), is probably the more appropriate method for making capital budgeting decisions, and for the optimization of process flow-sheets or their subsystems (Novak Pintarič and Kravanja, 2006). The Eco-NPV is an analogy of the economic NPV, where income in the NPV's yearly cash flow is now represented by unburdening (eco-benefit) and outcome by burdening (eco-cost) of the environment. Similarly as economic and eco-profit are merged into total profit, the economic NPV and Eco-NPV form total NPV, and the preferred solutions are those with maximal total NPV.

In this paper, this novel concept of eco-NPV is compared with the concept of eco-profit. This approach is illustrated through a couple of case studies of the supply chain synthesis for producing biofuels.

Keywords: Net present value, Eco-net present value, Eco-profit, LCA, Optimization, Biofuels

References

Čuček L., Drobež R., Pahor B., Kravanja Z., 2012, Sustainable synthesis of biogas processes using a novel concept of eco-profit, Computers & Chemical Engineering, doi: 10.1016/j.compchemeng.2012.01.010

Kravanja Z., Čuček L., 2012, Multi-objective optimization for generating sustainable solutions considering total effects on the environment, Applied Energy, doi: 10.1016/j.apenergy.2012.04.025

Novak Pintarič Z., Kravanja Z., 2006, Selection of the objective function for the optimization of process flow sheets, Industrial & Engineering Chemistry Research, 45, 4222-4232

(791a) Nanomaterials Enabled Lithium Chemistry for Advanced Rechargeable Batteries

Chao Wang, Materials Science Division, Argonne National Laboratory, Argonne, IL

Employment of nanomaterials in electrode assembly is promising for the development of advanced lithium-ion batteries. Nanostructured electrodes can potentially achiever higher power and better cyclability than their bulk counterparts, due to the facilitated charge transfer at electrode/electrolyte interface, electronic/ionic transport in the electrodes, and accommodation of strain during lithium insertion/extraction by reduction of material dimensions. Moreover, design of electrodes based on nanomaterials can enable new lithium insertion/extraction reactions that are not possible with bulk materials, and thereby expand our scope of material choices for energy storage. These advantages have lead to substantial literature effort on the synthesis of electrochemically active nanomaterials for battery applications.

Here we report the development of novel manganese oxide nanoparticles (10 - 100 nm) by organic solution synthesis and their solid chemical conversion into lithium manganese oxide ($Li_xMn_2O_4$) spinels for cathode application in lithium ion batteries. We managed to demonstrate that by reducing the material dimensions to nanoscale, charge/discharge up to $0<x<2$ can be achieved which however is not possible for conventional bulk or microscale materials (with only $0<x<1$). Our work shows the promise of expanding the scope of material choices for energy storage by size scaling and the great of potential of improving battery performnace by employing advanced nanomaterials.

(791b) Dynamics of Pumped Heat Energy Storage Systems

Hugo S. Caram and Fan Ni, Chemical Engineering, Lehigh University, Bethlehem, PA

The eventual success and widespread utilization of photovoltaic and wind energy generation will depend on the availability of efficient and economical storage capability. The current technology for long term (diurnal), high efficiency large scale energy storage of electrical energy appears to have been limited to Pumped Hydroelectric Storage (PHS) and Compressed Air Storage (CAES). Both of them are dependent of the availability of geological features such as elevated lakes or empty deep mines for PHS or caverns for the CAES. This features are not available everywhere and other technologies such flow batteries while capable of large scale storage are limited in their power output.

Pumped heat energy storage systems can provide a valuable alternative PHS or CAES. The proposed process uses a Brayton cycle where during the loading cycle heat is pumped to heat up a high pressure thermal regenerator while a low pressure thermal regenerator is cooled down to a low temperature. During the unloading period the gas circulation is reversed. Cold gas from the low pressure regenerator is compressed and heated in the high pressure regenerator to be then expanded to the low pressure vessel. Electrical energy storage capacities in the 30-50kWh/m^3 range appear possible and are comparable to lead acid batteries (25-50 kWh/m^3).

Analysis and optimization of these systems are complicated by the difficult dynamics of thermal regenerators. A novel matrix algorithm is proposed that allows us to directly calculate cyclic steady state as well as transients. The effects of different design parameters such as pressure ratios, compressor and expander efficiencies, type of gases, solid heat capacities and gas solid heat transfer coefficients are explored. Possible synergistic integration with natural gas combustion combined cycles will be discussed.

The effective utilization System for Renewable Energy Making use of Regional Characteristic

Masahiro Yamamoto, Tsuguhiko Nakagawa

Graduate school of *Okayama Prefectural University,*

Kuboki 111, Soja-shi, 719-1197, Okayama, Japan

ABSTRACT

The SMART factory project has been developing a process to reduce CO_2 emissions through researching and developing a process to make plastic from biomass at the biomass accumulated factory in Maniwa, Okayama.

At the biomass accumulated factory, wood chips have been made from shavings and unused wood parts from sawmills. The biomass factory consists of a building chips plant which makes the wood chips, a building silo which stores chips, and a barkyard which crushes bark. The energy used at the biomass accumulated factory consists of electric power and liquid fossil fuels. Liquid fossil fuels have been used by forklift trucks and others vehicles for wood chips. The CO_2 emissions by liquid fossil fuels are accounted for about 43% of the total from the biomass accumulated factory. In order to reduce 25% CO_2 emissions, the SMART factory is planned to be supplied by renewable energy from the wood chips plant and photovoltaic power generation.

In this study, in order to achieve this aim, the mathematical simulation model which calculates the energy demand and supply per hour in the SMART factory has been developed. Based on the mathematical simulation model, a system which is able to reduce consumption of liquid fossil fuels and electric power has been designed.

As a result, it has been proved that it is possible to reduce 25% or more CO_2 emissions using the system that combines photovoltaic power generation with electric forklift trucks.

1. INTRODUCTION

The SMART factory project in Maniwa has been developing a process to reduce CO_2 emissions using renewable energy through researching and developing a process to make plastic from biomass at the biomass accumulated factory.

In this study, we have examined the system which substituted fossil fuels to be used as fuel for mobile equipment such as forklift trucks for renewable energy such as photovoltaic power generation (hereafter refered to as "PV") which will be constructed at the biomass accumulated factory.

2. OUTLINE OF THE NEW ENERGY SYSTEM

Outline of the new energy system is shown in Fig.1. At the biomass accumulated factory, mobile equipment such as two forklift trucks, shovel loaders and two grapples have been used. These vehicles have previously used liquid fossil fuels. Among this mobile equipments we considered that forklift trucks be substituted for electric forklift trucks and the system that charges the PV electricity to the battery of electric forklift trucks.

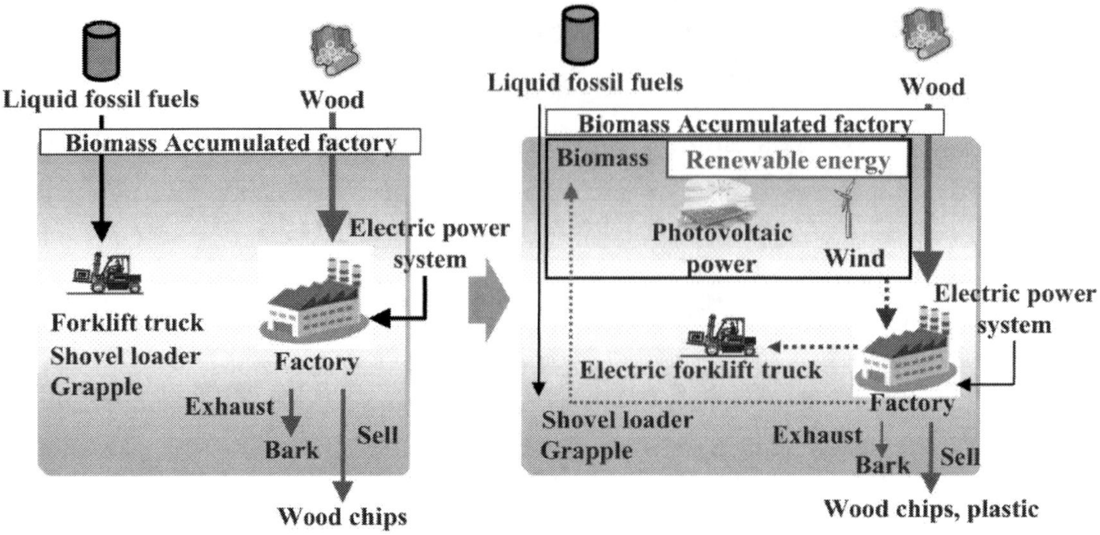

Fig.1. Images of the new energy system

3. THE EXAMINATION METHOD

3.1 Model Equation

For each production of wood chips, using equation (1), the mathematical simulation model which is able to calculate the energy demand and supply per hour in a year (8760hr) has been designed.*[1]

$$\sum_{i=1}^{n}\alpha_i = \frac{X_i}{\eta} + E_i - \beta_i \quad \cdots\cdots (1)$$

where α_i : electric power from electric power system [kWh]

E_i : electric power demand of factory [kWh]

X_i : electric power demand of electric forklift truck [kWh]

β_i : electric power generation of PV [kWh]

η : Charging&discharging efficiency (=0.85)

3.2 Production of wood chips and energy consumption

The relationship of monthly production of wood chips and electric power demand, liquid fossil fuels consumption of forklift truck at the biomass accumulated factory is shown in Fig.2 and Fig.3.

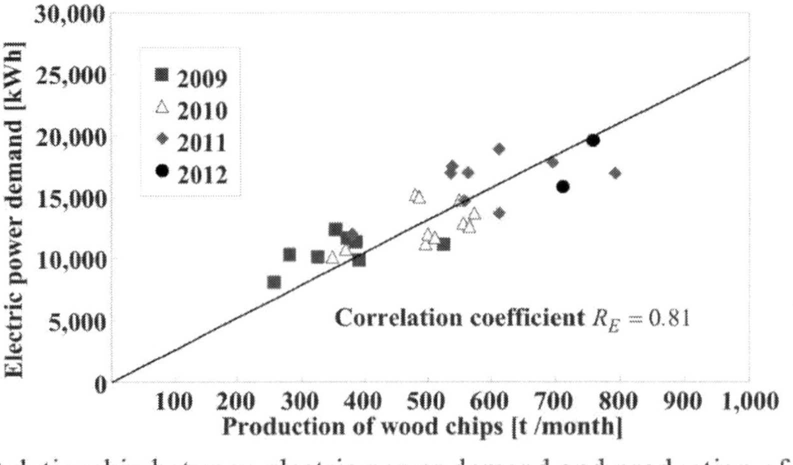

Fig.2. Relationship between electric power demand and production of wood chips

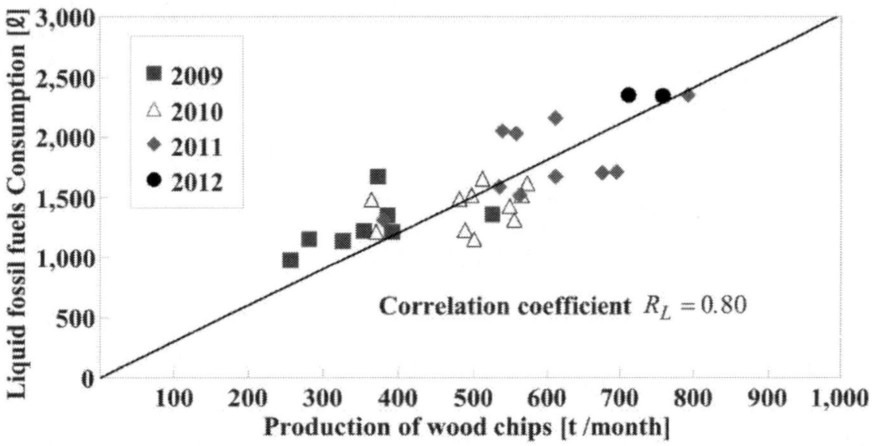

Fig.3. Relationship between liquid fossil fuels consumption of forklift trucks and production of wood chips

In Fig.2, it can be seen that there is a correlation between electric power demand and production of wood chips. For least-squares method,

$$y = 26.3x \quad \cdots \quad (2)$$

From equation (2), it can be estimated electric power demand from production of wood chips. On the other hand, in Fig.3, similarly, it can be obtained by equation (3).

$$Z = 0.9x \quad \cdots \quad (3)$$

where x : Production of wood chips [t/month]
y : electric power demand [kWh]
Z : liquid fossil fuels consumption of forklift trucks [ℓ]

Using equations (2) and (3) in this new energy system the optimal operation performance can be calculated depending on the production level of wood chips.

3.3 Result of examination

Regarding production of wood chips at 400[t/month] and 700[t/month], the relationship of CO_2 emissions and area of PV at the biomass accumulated factory is shown in Fig.4.

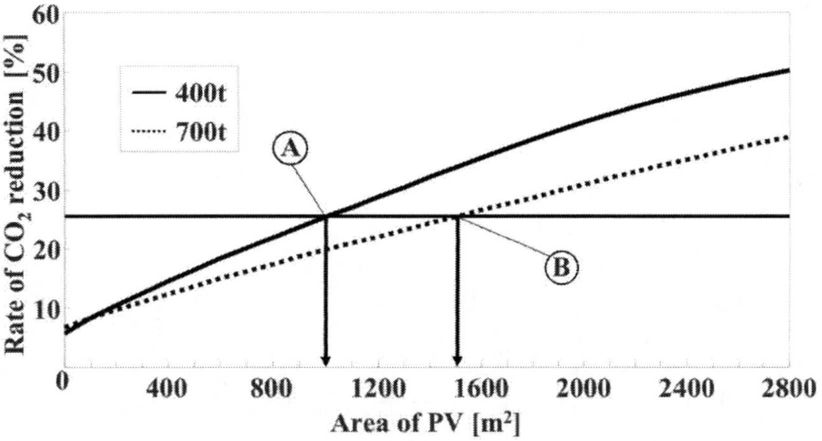

Fig.4.Relationship area of PV and rate of CO_2 reduction per production of wood chips

In Fig.4Ⓐ,Ⓑ the breakdown in the case of reducing 25% CO_2 emissions is shown Fig.5 Ⓐ,Ⓑ. In Fig.4, PV is of low electric power generation efficiency. But, it is able to be installed on a sloping site e.g. the factory roofs. Also, it is amorphous that installation costs are low. Maximum output per unit area is 69.1[W/m^2].

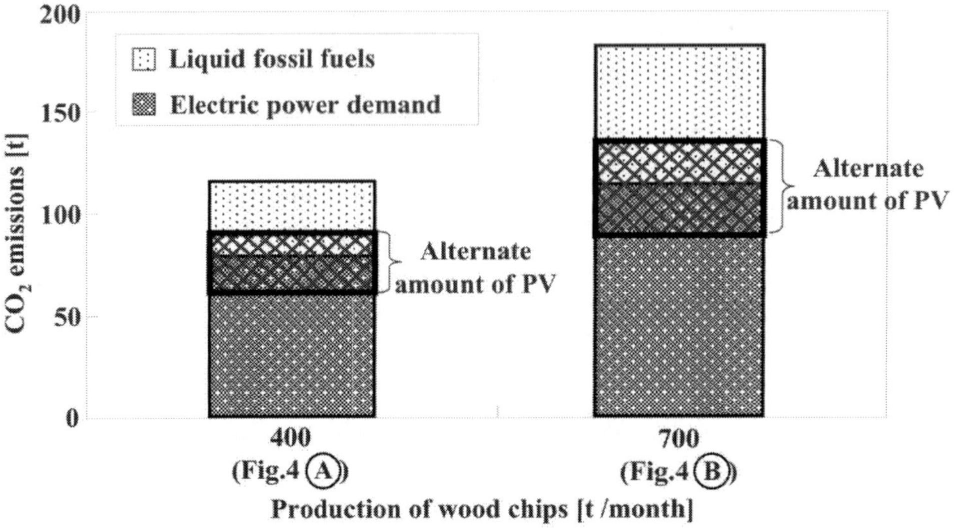

Fig.5. The breakdown in the case of reducing 25% CO_2 emissions

As a result, when production of wood chips is 700[t/month], it has been proved that the PV installation area should be greater than 1500m^2 to reduce 25% or more CO_2 emissions at the biomass accumulated factory.

4. CONCLUSION

Depending on the rate of Production of wood chips, it has been proved that it is possible to estimate liquid fossil fuels and electric power demand and calculate the design area of PV depending on the target of CO_2 emissions reduction.

In this study, we are subsidized by the funds form the Overall Promotion of Ministry of Education, Culture, Sports, Science and Technology. I am deeply grateful to them for their supports.

Reference

1. Shimohara, K., Y. Notoji, T. Nakagawa “Distributed Energy System Combination with Energy Conservation, PV and EV”, *Kagaku kogaku Ronbunshu* Vol.38 No.4, pp.255－262, Japan (2012)

(791d) Thermochemical Cycle of a Mixed Metal Oxide for Augmentation of Thermal Energy Storage in Solid Particles

Brian D. Ehrhart[1,2], Nathan P. Siegel[3], Eric N. Coker[2] and Alan W. Weimer[1], (1)Chemical and Biological Engineering, University of Colorado at Boulder, Boulder, CO, (2)Sandia National Laboratories, Albuquerque, NM, (3)Mechanical Engineering, Bucknell University, Lewisburg, PA

Solid particle heat transfer and storage media have been shown to be able to operate at temperatures >1000°C in concentrated solar power (CSP) applications, much higher than the operational limit of 600°C for currently state-of-the-art molten nitrate salt heat transfer fluid. Solid particles can be endothermically reduced by direct exposure to concentrated solar energy, thus absorbing and storing thermal energy beyond that possible with sensible heating alone. The particles can then be oxidized exothermically at a later time, releasing the stored chemical heat and effectively augmenting the thermal energy storage capacity of the solid particles. A mixed metal oxide spinel material that reduces in the temperature range of interest (1000°C-1200°C) has been examined for applicability to this thermochemical energy storage concept. A description of this application, prospective materials, and details of the thermochemical cycle are presented. The heats of reduction and oxidation for the thermochemical cycle have been determined for various operating conditions to evaluate the amount of thermal energy that may be stored. Additionally, the extent of reduction possible at various operating conditions has been measured.

(791e) Carbohydrate Could Be the Best Electricity Storage Compound and a High-Density Hydrogen Carrier

Y.-H. Percival Zhang, Biological Systems Engineering, Virginia Polytechnic Institute and State University, Blacksburg, VA and Chun You, Biological Systems Engineering, Virginia Polytechnic Institute and State University

Electricity is a high-quality secondary energy, which can be generated from renewable primary energies and fossil fuels. Carbon-neutral electricity generation is becoming more and more important, but most of them, such as wind and solar electricity, suffer from large output variations. Therefore, it is vital to store electricity in the form of chemical compounds. Among numerous potential chemical compounds, such as hydrogen, isobutanol, methanol and hydrocarbons, we suggest that carbohydrates (e.g., starch, cellulose, glucose) could be the best electricity storage compound with a combustion energy density of 17 MJ/kg and an electricity output density of >3,000 Wh/kg [1]. We designed a novel biological CO2 fixation pathway that can fix CO2 by using hydrogen or electricity for producing carbohydrate mediated by cell-free cascade enzymes [1]. In this talk, we will present our efforts in developing building blocks and building modules for this ambitious project and conduct proof-of-concept experiments. Through the above bioconversion plus our previous achievements such as theoretical conversion of carbohydrate to hydrogen or electricity [2] and low-cost enzymatic conversion of cellulose to starch (unpublished), the electricity-carbohydrate-hydrogen cycle would not only supplement current and future primary energy utilization systems for facilitating electricity and hydrogen storage and enhancing secondary energy conversion efficiencies, but also address such sustainability challenges as transportation fuel production, CO_2 utilization, fresh water conservation, food and feed production, and maintenance of a small closed ecosystem in emergency situations [1]. In a word, carbohydrate would become a new oil in the future.

1. Zhang Y-HP, Huang W-D: **Constructing the electricity-carbohydrate-hydrogen cycle for a sustainability revolution**. *Trends Biotechnol.* 2012, **Epub. DOI: 10.1016/j.tibtech.2012.02.006**.

2. Zhang Y-HP, Evans BR, Mielenz JR, Hopkins RC, Adams MWW: **High-yield hydrogen production from starch and water by a synthetic enzymatic pathway**. *PLoS One* 2007, **2**:e456.

(791f) Energy Storage Design for Electric Transmission Systems

Donald J. Chmielewski, Illinois Institute of Technology, Chicago, IL

It is widely recognized that a major concern with the installation of renewable energy systems is the fact that wind and solar sources are non-dispatchable. That is, the power produced from renewable systems is dependent on environmental conditions and is likely uncorrelated with the power demand from load centers. In contrast, fossil based sources typically have the ability to track demand, but depending on the technology possess limits with respect to response time. Thus, many have advocated the use of massive energy storage systems to capture energy produced from renewable sources and then dispatch as load centers demand. An additional concern is the fact that renewable sources tend to be located far from the load centers (again in contrast to most fossil sources which are typically placed near demand). This combination of spatial and temporal mismatch with load center demand has caused many to worry about the stability of the electric power grid. Clearly, wide swings in power throughput over great distances will cause the weakest parts of the grid to fail.

To address both the spatial and temporal issues surrounding renewable power generation we have developed a scheme aimed at analyzing the capabilities of a regional power generation and distribution system. In particular, the scheme proposes a supervisory control law aimed at utilizing a mix of renewable and fossil sources along with energy storage systems to meet power load demands. This proposed control law takes into consideration not only the power output capacity of each source but also the rate at which power output can change as well as the storage capacity and efficiency losses associated with the storage units. In addition, the proposed scheme incorporates the maximum and minimum throughput limits on the distribution lines. If time allows, we will present an extension of the scheme aimed at sizing and placing energy storage systems as well as locating new power distribution lines.

AUTHOR INDEX

AUTHOR INDEX

AUTHOR INDEX

AUTHOR INDEX

AUTHOR INDEX

AUTHOR INDEX

AUTHOR INDEX

AUTHOR INDEX

AUTHOR INDEX

AUTHOR INDEX

AUTHOR INDEX

AUTHOR INDEX

AUTHOR INDEX

AUTHOR INDEX